Available in MyMathLab® for Your Course

Achieve Your Potential

Success in math can make a difference in your life. MyMathLab is a learning experience with resources to help you achieve your potential in this course and beyond. MyMathLab will help you learn the new skills required, and also help you learn the concepts and make connections for future courses and careers.

Visualization and Conceptual Understanding

These MyMathLab resources will help you think visually and connect the concepts.

Updated Videos

Updated videos are available to support students outside of the classroom and cover all topics in the text. Quick Review videos cover key definitions and procedures. Example Solution videos offer a detailed solution process for every example in the textbook.

Find the sum of $6-2i$ and $-4-3i$. Graph both complex numbers and their resultant.

Sum = $2-5i$

00:30/01:37 CC ESP

Video Assessment Exercises

Assignable in MyMathLab, Example Solution Videos present the detailed solution process for every example in the text. Additional Quick Reviews cover definitions and procedures for each section. Assessment exercises check conceptual understanding of the mathematics.

www.mymathlab.com

Trigonometry

ELEVENTH EDITION

Annotated Instructor's Edition

Margaret L. Lial
American River College

John Hornsby
University of New Orleans

David I. Schneider
University of Maryland

Callie J. Daniels
St. Charles Community College

PEARSON

Boston Columbus Indianapolis New York San Francisco
Amsterdam Cape Town Dubai London Madrid Milan Munich Paris Montréal Toronto
Delhi Mexico City São Paulo Sydney Hong Kong Seoul Singapore Taipei Tokyo

Editorial Director: Chris Hoag
Editor in Chief: Anne Kelly
Editorial Assistant: Ashley Gordon
Program Manager: Danielle Simbajon
Project Manager: Christine O'Brien
Program Management Team Lead: Karen Wernholm
Project Management Team Lead: Peter Silvia
Media Producer: Jonathan Wooding
TestGen Content Manager: John Flanagan
MathXL Content Manager: Kristina Evans
Marketing Manager: Claire Kozar
Marketing Assistant: Fiona Murray
Senior Author Support/Technology Specialist: Joe Vetere
Rights and Permissions Project Manager: Gina Cheselka
Procurement Specialist: Carol Melville
Associate Director of Design: Andrea Nix
Program Design Lead: Beth Paquin
Text Design: Cenveo® Publisher Services
Composition: Cenveo® Publisher Services
Illustrations: Cenveo® Publisher Services
Cover Design: Cenveo® Publisher Services
Cover Image: Moaan/Getty Images

Library of Congress Cataloging-in-Publication Data

Trigonometry / Margaret L. Lial, American River College [and three others].—Eleventh edition. pages cm

Includes indexes

ISBN 978-0-13-421743-7 (hardcover)—ISBN 0-13-421743-8 (hardcover)—
ISBN 978-0-13-421764-2 (hardcover)—ISBN 0-13-421764-0 (hardcover)

1. Trigonometry. I. Lial, Margaret L.

QA531.L5 2017

516.24--dc23 2015036391

Annotated Instructor's Edition
ISBN 10: 0-13-421764-0
ISBN 13: 978-0-13-421764-2

www.pearsonhighered.com

ISBN 13: 978-0-13-421743-7
ISBN 10: 0-13-421743-8

To Butch, Peggy, Natalie, and Alexis—and in memory of Mark

E.J.H.

To Coach Lonnie Myers—thank you for your leadership on and off the court.

C.J.D.

Contents

5 Trigonometric Identities 195

6 Inverse Circular Functions and Trigonometric Equations 251

Preface

WELCOME TO THE 11TH EDITION

In the eleventh edition of *Trigonometry,* we continue our ongoing commitment to providing the best possible text to help instructors teach and students succeed. In this edition, we have remained true to the pedagogical style of the past while staying focused on the needs of today's students. Support for all classroom types (traditional, hybrid, and online) may be found in this classic text and its supplements backed by the power of Pearson's MyMathLab.

In this edition, we have drawn upon the extensive teaching experience of the Lial team, with special consideration given to reviewer suggestions. General updates include enhanced readability with improved layout of examples, better use of color in displays, and language written with students in mind. All calculator screenshots have been updated and now provide color displays to enhance students' conceptual understanding. Each homework section now begins with a group of *Concept Preview* exercises, assignable in MyMathLab, which may be used to ensure students' understanding of vocabulary and basic concepts prior to beginning the regular homework exercises.

Further enhancements include numerous current data examples and exercises that have been updated to reflect current information. Additional real-life exercises have been included to pique student interest; answers to writing exercises have been provided; better consistency has been achieved between the directions that introduce examples and those that introduce the corresponding exercises; and better guidance for rounding of answers has been provided in the exercise sets.

The Lial team believes this to be our best *Trigonometry* edition yet, and we sincerely hope that you enjoy using it as much as we have enjoyed writing it. Additional textbooks in this series are as follows:

College Algebra, Twelfth Edition
College Algebra & Trigonometry, Sixth Edition
Precalculus, Sixth Edition

HIGHLIGHTS OF NEW CONTENT

- Discussion of the Pythagorean theorem and the distance formula has been moved from an appendix to **Chapter 1.**
- In **Chapter 2,** the two sections devoted to applications of right triangles now begin with short historical vignettes, to provide motivation and illustrate how trigonometry developed as a tool for astronomers.
- The example solutions of applications of angular speed in **Chapter 3** have been rewritten to illustrate the use of unit fractions.
- In **Chapter 4,** we have included new applications of periodic functions. They involve modeling monthly temperatures of regions in the southern hemisphere and fractional part of the moon illuminated for each day of a particular month. The example of addition of ordinates in **Section 4.4** has been rewritten, and a new example of analysis of damped oscillatory motion has been included in **Section 4.5.**
- **Chapter 5** now presents a derivation of the product-to-sum identity for the product $\sin A \cos B$.
- In **Chapter 6,** we include several new screens of periodic function graphs that differ in appearance from typical ones. They pertain to the music phenomena of pressure of a plucked spring, beats, and upper harmonics.

- The two sections in **Chapter 7** on vectors have been reorganized but still cover the same material as in the previous edition. **Section 7.4** now introduces geometrically defined vectors and applications, and **Section 7.5** follows with algebraically defined vectors and the dot product.
- In **Chapter 8,** the examples in **Section 8.1** have been reordered for a better flow with respect to solving quadratic equations with complex solutions.
- For visual learners, numbered **Figure** and **Example** references within the text are set using the same typeface as the figure number itself and bold print for the example. This makes it easier for the students to identify and connect them. We also have increased our use of a "drop down" style, when appropriate, to distinguish between simplifying expressions and solving equations, and we have added many more explanatory side comments. Guided Visualizations, with accompanying exercises and explorations, are now available and assignable in MyMathLab.
- *Trigonometry* is widely recognized for the quality of its exercises. In the eleventh edition, nearly 500 are new or modified, and many present updated real-life data. Furthermore, the MyMathLab course has expanded coverage of all exercise types appearing in the exercise sets, as well as the mid-chapter Quizzes and Summary Exercises.

FEATURES OF THIS TEXT

SUPPORT FOR LEARNING CONCEPTS

We provide a variety of features to support students' learning of the essential topics of trigonometry. Explanations that are written in understandable terms, figures and graphs that illustrate examples and concepts, graphing technology that supports and enhances algebraic manipulations, and real-life applications that enrich the topics with meaning all provide opportunities for students to deepen their understanding of mathematics. These features help students make mathematical connections and expand their own knowledge base.

- **Examples** Numbered examples that illustrate the techniques for working exercises are found in every section. We use traditional explanations, side comments, and pointers to describe the steps taken—and to warn students about common pitfalls. Some examples provide additional graphing calculator solutions, although these can be omitted if desired.
- **Now Try Exercises** Following each numbered example, the student is directed to try a corresponding odd-numbered exercise (or exercises). This feature allows for quick feedback to determine whether the student has understood the principles illustrated in the example.
- **Real-Life Applications** We have included hundreds of real-life applications, many with data updated from the previous edition. They come from fields such as sports, biology, astronomy, geology, music, and environmental studies.
- **Function Boxes** Special function boxes offer a comprehensive, visual introduction to each type of trigonometric function and also serve as an excellent resource for reference and review. Each function box includes a table of values, traditional and calculator-generated graphs, the domain, the range, and other special information about the function. These boxes are assignable in MyMathLab.
- **Figures and Photos** Today's students are more visually oriented than ever before, and we have updated the figures and photos in this edition to

promote visual appeal. Guided Visualizations with accompanying exercises and explorations are now available and assignable in MyMathLab.

- **Use of Graphing Technology** We have integrated the use of graphing calculators where appropriate, although ***this technology is completely optional and can be omitted without loss of continuity.*** We continue to stress that graphing calculators support understanding but that students must first master the underlying mathematical concepts. Exercises that require the use of a graphing calculator are marked with the icon .

- **Cautions and Notes** Text that is marked **CAUTION** warns students of common errors, and NOTE comments point out explanations that should receive particular attention.

- **Looking Ahead to Calculus** These margin notes offer glimpses of how the topics currently being studied are used in calculus.

SUPPORT FOR PRACTICING CONCEPTS

This text offers a wide variety of exercises to help students master trigonometry. The extensive exercise sets provide ample opportunity for practice, and the exercise problems generally increase in difficulty so that students at every level of understanding are challenged. The variety of exercise types promotes understanding of the concepts and reduces the need for rote memorization.

- **NEW Concept Preview** Each exercise set now begins with a group of **CONCEPT PREVIEW** exercises designed to promote understanding of vocabulary and basic concepts of each section. These new exercises are assignable in MyMathLab and will provide support especially for hybrid, online, and flipped courses.

- **Exercise Sets** In addition to traditional drill exercises, this text includes writing exercises, optional graphing calculator problems , and multiple-choice, matching, true/false, and completion exercises. *Concept Check* exercises focus on conceptual thinking. *Connecting Graphs with Equations* exercises challenge students to write equations that correspond to given graphs.

- **Relating Concepts Exercises** Appearing at the end of selected exercise sets, these groups of exercises are designed so that students who work them in numerical order will follow a line of reasoning that leads to an understanding of how various topics and concepts are related. All answers to these exercises appear in the student answer section, and these exercises are assignable in MyMathLab.

- **Complete Solutions to Selected Exercises** Complete solutions to all exercises marked are available in the eText. These are often exercises that extend the skills and concepts presented in the numbered examples.

SUPPORT FOR REVIEW AND TEST PREP

Ample opportunities for review are found within the chapters and at the ends of chapters. Quizzes that are interspersed within chapters provide a quick assessment of students' understanding of the material presented up to that point in the chapter. Chapter "Test Preps" provide comprehensive study aids to help students prepare for tests.

- **Quizzes** Students can periodically check their progress with in-chapter quizzes that appear in all chapters. All answers, with corresponding section references, appear in the student answer section. These quizzes are assignable in MyMathLab.

- **Summary Exercises** These sets of in-chapter exercises give students the all-important opportunity to work *mixed* review exercises, requiring them to synthesize concepts and select appropriate solution methods.
- **End-of-Chapter Test Prep** Following the final numbered section in each chapter, the Test Prep provides a list of **Key Terms,** a list of **New Symbols** (if applicable), and a two-column **Quick Review** that includes a section-by-section summary of concepts and examples. This feature concludes with a comprehensive set of **Review Exercises** and a **Chapter Test.** The Test Prep, Review Exercises, and Chapter Test are assignable in MyMathLab. Additional Cumulative Review homework assignments are available in MyMathLab, following every chapter.

Get the most out of MyMathLab®

MyMathLab is the world's leading online resource for teaching and learning mathematics. MyMathLab helps students and instructors improve results, and it provides engaging experiences and personalized learning for each student so learning can happen in any environment. Plus, it offers flexible and time-saving course management features to allow instructors to easily manage their classes while remaining in complete control, regardless of course format.

Personalized Support for Students

- MyMathLab comes with many learning resources–eText, animations, videos, and more–all designed to support your students as they progress through their course.
- The Adaptive Study Plan acts as a personal tutor, updating in real time based on student performance to provide personalized recommendations on what to work on next. With the new Companion Study Plan assignments, instructors can now assign the Study Plan as a prerequisite to a test or quiz, helping to guide students through concepts they need to master.
- Personalized Homework enables instructors to create homework assignments tailored to each student's specific needs and focused on the topics they have not yet mastered.

Used by nearly 4 million students each year, the MyMathLab and MyStatLab family of products delivers consistent, measurable gains in student learning outcomes, retention, and subsequent course success.

BREAKTHROUGH
To improving results

Resources for Success

MyMathLab® Online Course for *Trigonometry* by Lial, Hornsby, Schneider, and Daniels

MyMathLab delivers proven results in helping individual students succeed. The authors Lial, Hornsby, Schneider, and Daniels have developed specific content in MyMathLab to give students the practice they need to develop a conceptual understanding of Trigonometry and the analytical skills necessary for success in mathematics. The MyMathLab features described here support Trigonometry students in a variety of classroom formats (traditional, hybrid, and online).

Concept Preview Exercises

Exercise sets now begin with a group of Concept Preview Exercises, assignable in MyMathLab and also available in Learning Catalytics. These may be used to ensure that students understand the related vocabulary and basic concepts before beginning the regular homework problems. Learning Catalytics is a "bring your own device" system of prebuilt questions designed to enhance student engagement and facilitate assessment.

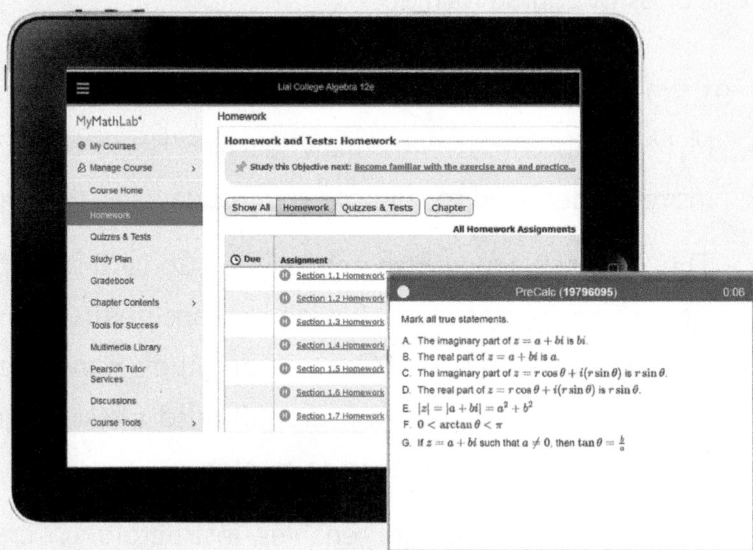

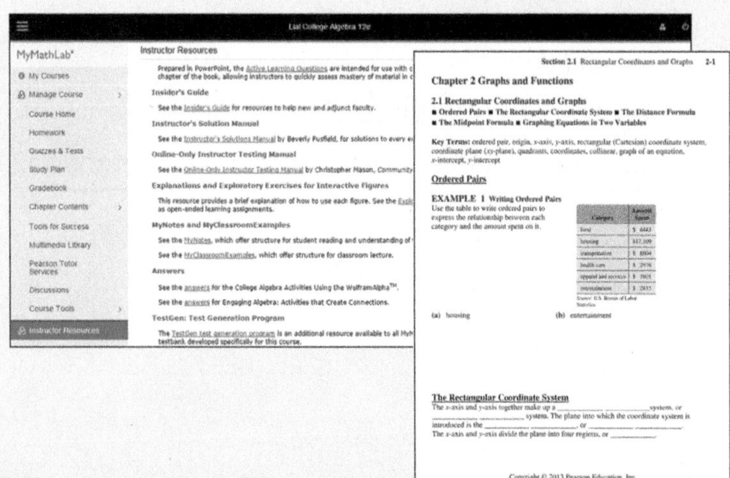

MyNotes and MyClassroomExamples

MyNotes provide a note-taking structure for students to use while they read the text or watch the MyMathLab videos. MyClassroom Examples offer structure for notes taken during lecture and are for use with the Classroom Examples found in the Annotated Instructor Edition.

Both sets of notes are available in MyMathLab and can be customized by the instructor.

BREAK THROUGH
To improving results

Resources for Success

Student Supplements

Student's Solutions Manual

By Beverly Fusfield

- Provides detailed solutions to all odd-numbered text exercises

ISBN: 0-13-431021-7 & 978-0-13-431021-3

Video Lectures with Optional Captioning

- Feature Quick Reviews and Example Solutions:
  Quick Reviews cover key definitions and procedures from each section.
  Example Solutions walk students through the detailed solution process for every example in the textbook.
- Ideal for distance learning or supplemental instruction at home or on campus
- Include optional text captioning
- Available in MyMathLab®

MyNotes

- Available in MyMathLab and offer structure for students as they watch videos or read the text
- Include textbook examples along with ample space for students to write solutions and notes
- Include key concepts along with prompts for students to read, write, and reflect on what they have just learned
- **Customizable** so that instructors can add their own examples or remove examples that are not covered in their courses

MyClassroomExamples

- Available in MyMathLab and offer structure for classroom lecture
- Include Classroom Examples along with ample space for students to write solutions and notes
- Include key concepts along with fill in the blank opportunities to keep students engaged
- **Customizable** so that instructors can add their own examples or remove Classroom Examples that are not covered in their courses

Instructor Supplements

Annotated Instructor's Edition

- Provides answers in the margins to almost all text exercises, as well as helpful Teaching Tips and Classroom Examples
- Includes sample homework assignments indicated by exercise numbers underlined in blue within each end-of-section exercise set
- Sample homework exercises assignable in MyMathLab

ISBN: 0-13-421764-0 & 978-0-13-421764-2

Online Instructor's Solutions Manual

By Beverly Fusfield

- Provides complete solutions to all text exercises
- Available in MyMathLab or downloadable from Pearson Education's online catalog

Online Instructor's Testing Manual

By David Atwood

- Includes diagnostic pretests, chapter tests, final exams, and additional test items, grouped by section, with answers provided
- Available in MyMathLab or downloadable from Pearson Education's online catalog

TestGen®

- Enables instructors to build, edit, print, and administer tests
- Features a computerized bank of questions developed to cover all text objectives
- Available in MyMathLab or downloadable from Pearson Education's online catalog

Online PowerPoint Presentation and Classroom Example PowerPoints

- Written and designed specifically for this text
- Include figures and examples from the text
- Provide Classroom Example PowerPoints that include full worked-out solutions to all Classroom Examples
- Available in MyMathLab or downloadable from Pearson Education's online catalog

ACKNOWLEDGMENTS

We wish to thank the following individuals who provided valuable input into this edition of the text.

John E. Daniels – Central Michigan University
Mary Hill – College of DuPage
Rene Lumampao – Austin Community College
Randy Nichols – Delta College
Patty Schovanec – Texas Tech University
Deanna M. Welsch – Illinois Central College

Our sincere thanks to those individuals at Pearson Education who have supported us throughout this revision: Anne Kelly, Christine O'Brien, Joe Vetere, and Danielle Simbajon. Terry McGinnis continues to provide behind-the-scenes guidance for both content and production. We have come to rely on her expertise during all phases of the revision process. Marilyn Dwyer of Cenveo® Publishing Services, with the assistance of Carol Merrigan, provided excellent production work. Special thanks go out to Paul Lorczak and Perian Herring for their excellent accuracy-checking. We thank Lucie Haskins, who provided an accurate index, and Jack Hornsby, who provided assistance in creating calculator screens, researching data updates, and proofreading.

As an author team, we are committed to providing the best possible college algebra course to help instructors teach and students succeed. As we continue to work toward this goal, we welcome any comments or suggestions you might send, via e-mail, to math@pearson.com.

Margaret L. Lial
John Hornsby
David I. Schneider
Callie J. Daniels

1 Trigonometric Functions

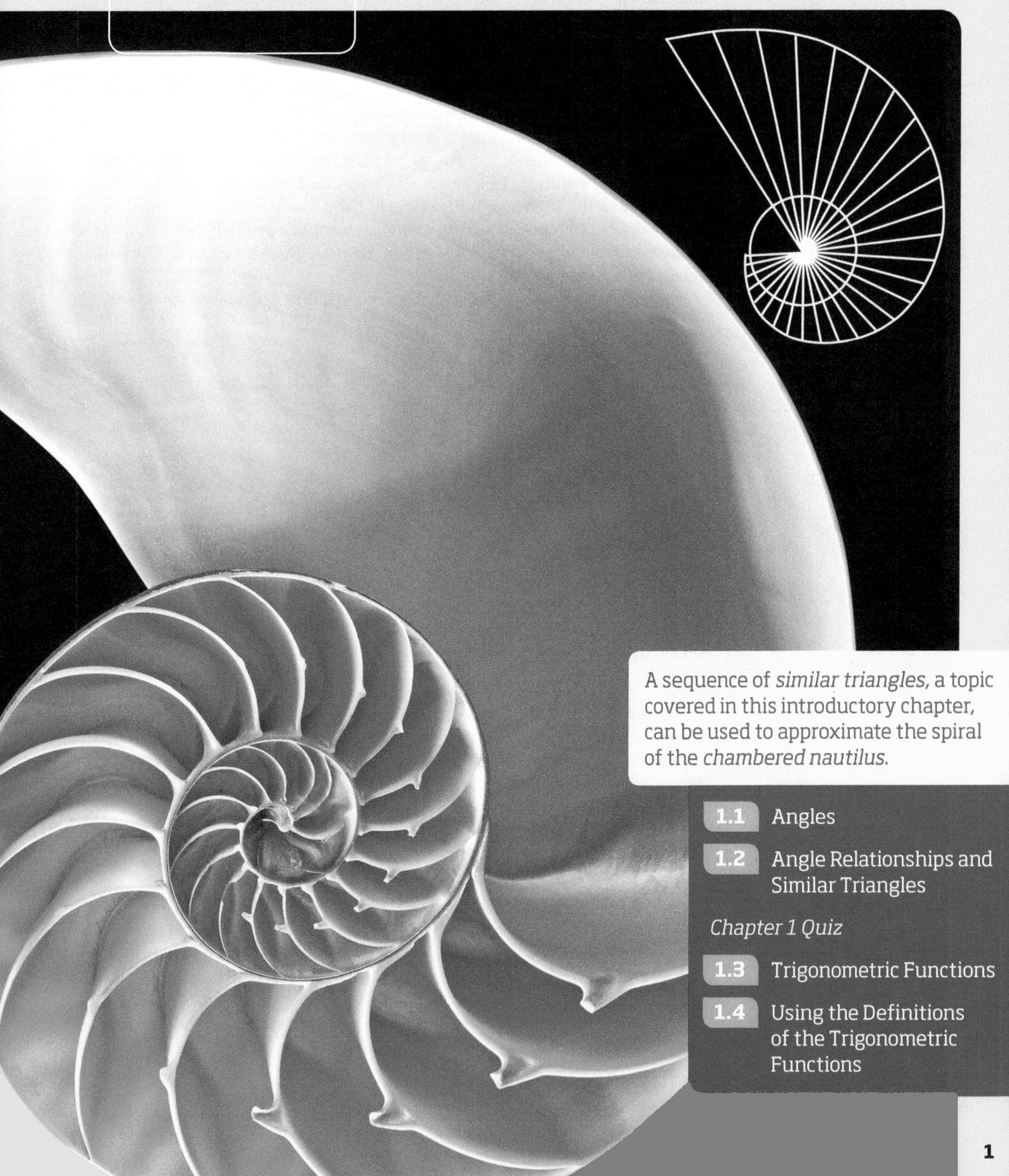

A sequence of *similar triangles,* a topic covered in this introductory chapter, can be used to approximate the spiral of the *chambered nautilus.*

1.1 Angles

- Basic Terminology
- Degree Measure
- Standard Position
- Coterminal Angles

Basic Terminology Two distinct points A and B determine a line called **line AB.** The portion of the line between A and B, including points A and B themselves, is **line segment AB,** or simply **segment AB.** The portion of line AB that starts at A and continues through B, and on past B, is the **ray AB.** Point A is the **endpoint of the ray.** See **Figure 1.**

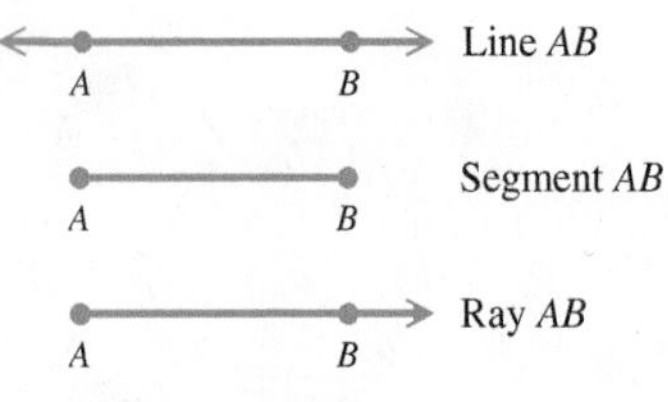

Figure 1

In trigonometry, an **angle** consists of two rays in a plane with a common endpoint, or two line segments with a common endpoint. These two rays (or segments) are the **sides** of the angle, and the common endpoint is the **vertex** of the angle. Associated with an angle is its measure, generated by a rotation about the vertex. See **Figure 2.** This measure is determined by rotating a ray starting at one side of the angle, the **initial side,** to the position of the other side, the **terminal side.** ***A counterclockwise rotation generates a positive measure, and a clockwise rotation generates a negative measure.*** The rotation can consist of more than one complete revolution.

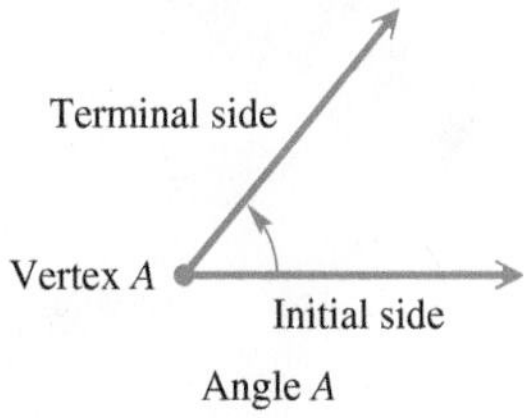

Terminal side

Vertex A

Initial side

Angle A

Figure 2

Figure 3 shows two angles, one **positive** and one **negative.**

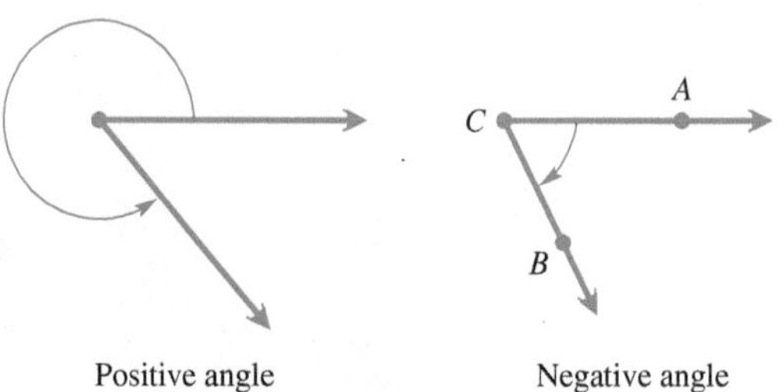

Figure 3

An angle can be named by using the name of its vertex. For example, the angle on the right in **Figure 3** can be named angle C. Alternatively, an angle can be named using three letters, with the vertex letter in the middle. Thus, the angle on the right also could be named angle ACB or angle BCA.

Teaching Tip Explain to students that we often say "angle" when we are actually referring to "angle measure."

Degree Measure The most common unit for measuring angles is the **degree.** Degree measure was developed by the Babylonians 4000 yr ago. To use degree measure, we assign 360 degrees to a complete rotation of a ray.* In **Figure 4,** notice that the terminal side of the angle corresponds to its initial side when it makes a complete rotation.

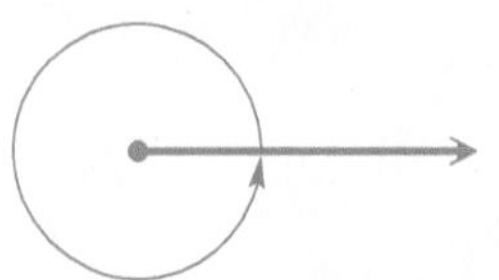

A complete rotation of a ray gives an angle whose measure is 360°. $\frac{1}{360}$ of a complete rotation gives an angle whose measure is 1°.

Figure 4

$$\text{One degree, written } 1^\circ\text{, represents } \frac{1}{360} \text{ of a complete rotation.}$$

Therefore, 90° represents $\frac{90}{360} = \frac{1}{4}$ of a complete rotation, and 180° represents $\frac{180}{360} = \frac{1}{2}$ of a complete rotation.

An angle measuring between 0° and 90° is an **acute angle.** An angle measuring exactly 90° is a **right angle.** The symbol ⅂ is often used at the vertex of a right angle to denote the 90° measure. An angle measuring more than 90° but less than 180° is an **obtuse angle,** and an angle of exactly 180° is a **straight angle.**

*The Babylonians were the first to subdivide the circumference of a circle into 360 parts. There are various theories about why the number 360 was chosen. One is that it is approximately the number of days in a year, and it has many divisors, which makes it convenient to work with in computations.

The Greek Letters		
Α	α	alpha
Β	β	beta
Γ	γ	gamma
Δ	δ	delta
Ε	ε	epsilon
Ζ	ζ	zeta
Η	η	eta
Θ	θ	theta
Ι	ι	iota
Κ	κ	kappa
Λ	λ	lambda
Μ	μ	mu
Ν	ν	nu
Ξ	ξ	xi
Ο	o	omicron
Π	π	pi
Ρ	ρ	rho
Σ	σ	sigma
Τ	τ	tau
Υ	υ	upsilon
Φ	ϕ	phi
Χ	χ	chi
Ψ	ψ	psi
Ω	ω	omega

In **Figure 5**, we use the **Greek letter θ (theta)*** to name each angle. The table in the margin lists the upper- and lowercase Greek letters, which are often used in trigonometry.

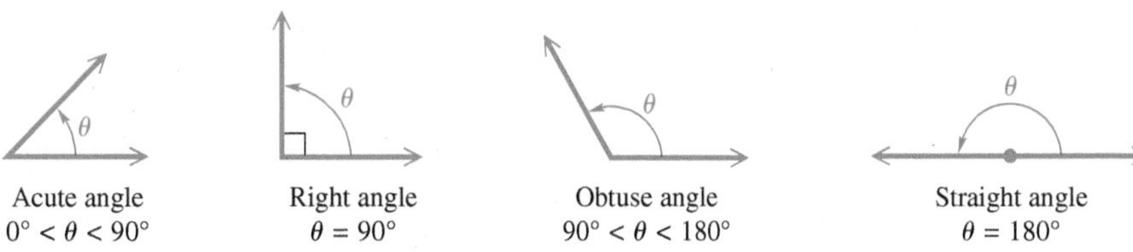

Figure 5

If the sum of the measures of two positive angles is 90°, the angles are **complementary** and the angles are **complements** of each other. Two positive angles with measures whose sum is 180° are **supplementary,** and the angles are **supplements.**

EXAMPLE 1 Finding the Complement and the Supplement of an Angle

Find the measure of **(a)** the complement and **(b)** the supplement of an angle measuring 40°.

SOLUTION

(a) To find the measure of its complement, subtract the measure of the angle from 90°.

$$90^\circ - 40^\circ = 50^\circ \quad \text{Complement of } 40^\circ$$

(b) To find the measure of its supplement, subtract the measure of the angle from 180°.

$$180^\circ - 40^\circ = 140^\circ \quad \text{Supplement of } 40^\circ$$

✔ **Now Try Exercise 11.**

Classroom Example 1
Find the measure of **(a)** the complement and **(b)** the supplement of an angle measuring 55°.

Answers:
(a) 35° (b) 125°

EXAMPLE 2 Finding Measures of Complementary and Supplementary Angles

Find the measure of each marked angle in **Figure 6.**

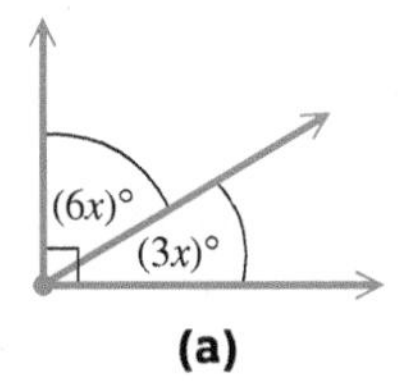

(a)

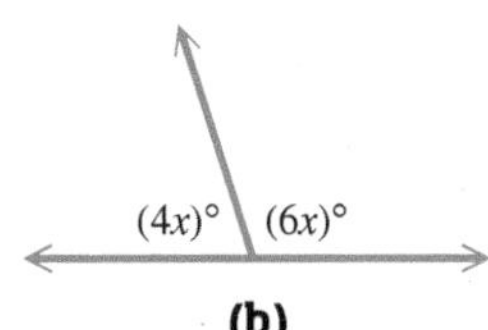

(b)

Figure 6

SOLUTION

(a) Because the two angles in **Figure 6(a)** form a right angle, they are complementary angles.

$$6x + 3x = 90 \quad \text{Complementary angles sum to } 90^\circ.$$
$$9x = 90 \quad \text{Combine like terms.}$$
$$x = 10 \quad \text{Divide by 9.}$$

Don't stop here.

Be sure to determine the measure of each angle by substituting 10 for x in $6x$ and $3x$. The two angles have measures of $6(10) = 60^\circ$ and $3(10) = 30^\circ$.

(b) The angles in **Figure 6(b)** are supplementary, so their sum must be 180°.

$$4x + 6x = 180 \quad \text{Supplementary angles sum to } 180^\circ.$$
$$10x = 180 \quad \text{Combine like terms.}$$
$$x = 18 \quad \text{Divide by 10.}$$

The angle measures are $4x = 4(18) = 72^\circ$ and $6x = 6(18) = 108^\circ$.

✔ **Now Try Exercises 23 and 25.**

Classroom Example 2
Find the measure of each angle.
(a) (b)

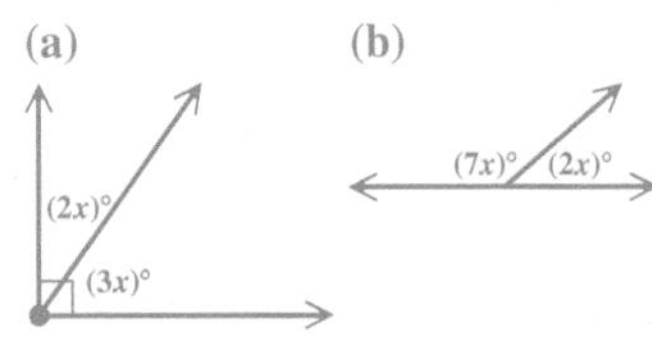

Answers:
(a) 36°; 54° (b) 140°; 40°

* In addition to θ (theta), other Greek letters such as α (alpha) and β (beta) are used to name angles.

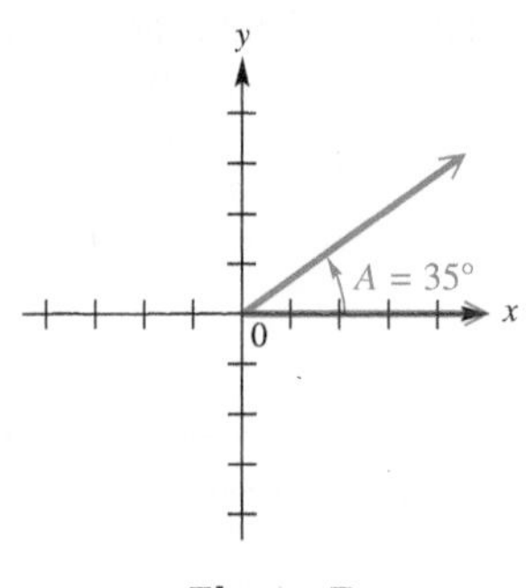

Figure 7

The measure of angle A in **Figure 7** is 35°. This measure is often expressed by saying that **m(angle A)** is 35°, where m(angle A) is read **"the measure of angle A."** The symbolism $m(\text{angle } A) = 35°$ is abbreviated as $A = 35°$.

Traditionally, portions of a degree have been measured with minutes and seconds. One **minute,** written **1′**, is $\frac{1}{60}$ of a degree.

$$\mathbf{1' = \left(\frac{1}{60}\right)^{\circ}} \quad \text{or} \quad \mathbf{60' = 1^{\circ}}$$

One **second, 1″**, is $\frac{1}{60}$ of a minute.

$$\mathbf{1'' = \left(\frac{1}{60}\right)' = \left(\frac{1}{3600}\right)^{\circ}} \quad \text{or} \quad \mathbf{60'' = 1'} \text{ and } \mathbf{3600'' = 1^{\circ}}$$

The measure 12° 42′ 38″ represents 12 degrees, 42 minutes, 38 seconds.

Classroom Example 3
Perform each calculation.
(a) 28° 35′ + 63° 52′
(b) 180° − 117° 29′

Answers:
(a) 92° 27′ (b) 62° 31′

EXAMPLE 3 Calculating with Degrees, Minutes, and Seconds

Perform each calculation.

(a) 51° 29′ + 32° 46′ **(b)** 90° − 73° 12′

SOLUTION

(a)
$$\begin{array}{r} 51^{\circ}\,29' \\ +\;32^{\circ}\,46' \\ \hline 83^{\circ}\,75' \end{array}$$
Add degrees and minutes separately.

The sum 83° 75′ can be rewritten as follows.

$$\begin{aligned} &83^{\circ}\,75' \\ &= 83^{\circ} + 1^{\circ}\,15' && 75' = 60' + 15' = 1^{\circ}\,15' \\ &= 84^{\circ}\,15' && \text{Add.} \end{aligned}$$

(b)
$$\begin{array}{r} 90^{\circ} \\ -\;73^{\circ}\,12' \\ \hline \end{array} \quad \text{can be written} \quad \begin{array}{r} 89^{\circ}\,60' \\ -\;73^{\circ}\,12' \\ \hline 16^{\circ}\,48' \end{array}$$
Write 90° as 89° 60′.

✔ **Now Try Exercises 41 and 45.**

An alternative way to measure angles involves decimal degrees. For example,

$$12.4238^{\circ} \quad \text{represents} \quad 12\frac{4238}{10{,}000}^{\circ}.$$

Classroom Example 4
(a) Convert 105° 20′ 32″ to decimal degrees to the nearest thousandth.
(b) Convert 85.263° to degrees, minutes, and seconds to the nearest second.

Answers:
(a) 105.342° (b) 85° 15′ 47″

EXAMPLE 4 Converting between Angle Measures

(a) Convert 74° 08′ 14″ to decimal degrees to the nearest thousandth.

(b) Convert 34.817° to degrees, minutes, and seconds to the nearest second.

SOLUTION

(a) 74° 08′ 14″

$$= 74^{\circ} + \left(\frac{8}{60}\right)^{\circ} + \left(\frac{14}{3600}\right)^{\circ} \qquad 08' \cdot \frac{1^{\circ}}{60'} = \left(\frac{8}{60}\right)^{\circ} \text{ and } 14'' \cdot \frac{1^{\circ}}{3600''} = \left(\frac{14}{3600}\right)^{\circ}$$

$$\approx 74^{\circ} + 0.1333^{\circ} + 0.0039^{\circ} \qquad \text{Divide to express the fractions as decimals.}$$

$$\approx 74.137^{\circ} \qquad \text{Add and round to the nearest thousandth.}$$

```
NORMAL FLOAT AUTO REAL DEGREE MP
74°8'14"
                    74.13722222
74°8'14"
                         74.137
34.817▸DMS
                    34°49'1.2"
```

This screen shows how the TI-84 Plus performs the conversions in **Example 4.** The ▶DMS option is found in the ANGLE Menu.

(b) $34.817°$

$$\begin{aligned}
&= 34° + 0.817° && \text{Write as a sum.}\\
&= 34° + 0.817(60') && 0.817° \cdot \tfrac{60'}{1°} = 0.817(60')\\
&= 34° + 49.02' && \text{Multiply.}\\
&= 34° + 49' + 0.02' && \text{Write } 49.02' \text{ as a sum.}\\
&= 34° + 49' + 0.02(60'') && 0.02' \cdot \tfrac{60''}{1'} = 0.02(60'')\\
&= 34° + 49' + 1.2'' && \text{Multiply.}\\
&\approx 34°\,49'\,01'' && \text{Approximate to the nearest second.}
\end{aligned}$$

✔ **Now Try Exercises 61 and 71.**

Standard Position An angle is in **standard position** if its vertex is at the origin and its initial side lies on the positive x-axis. The angles in **Figures 8(a) and 8(b)** are in standard position. An angle in standard position is said to lie in the quadrant in which its terminal side lies. An acute angle is in quadrant I (**Figure 8(a)**) and an obtuse angle is in quadrant II (**Figure 8(b)**). **Figure 8(c)** shows ranges of angle measures for each quadrant when $0° < \theta < 360°$.

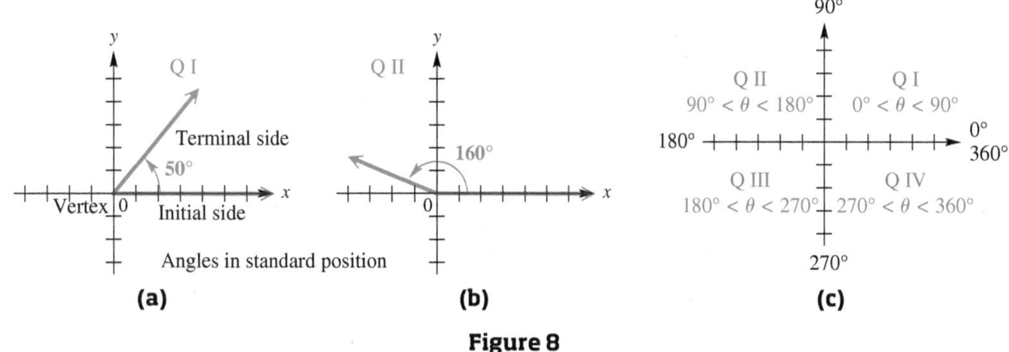

Figure 8

Quadrantal Angles

Angles in standard position whose terminal sides lie on the x-axis or y-axis, such as angles with measures 90°, 180°, 270°, and so on, are **quadrantal angles.**

Coterminal Angles A complete rotation of a ray results in an angle measuring 360°. By continuing the rotation, angles of measure larger than 360° can be produced. The angles in **Figure 9** with measures 60° and 420° have the same initial side and the same terminal side, but different amounts of rotation. Such angles are **coterminal angles.** ***Their measures differ by a multiple of* 360°.** As shown in **Figure 10,** angles with measures 110° and 830° are coterminal.

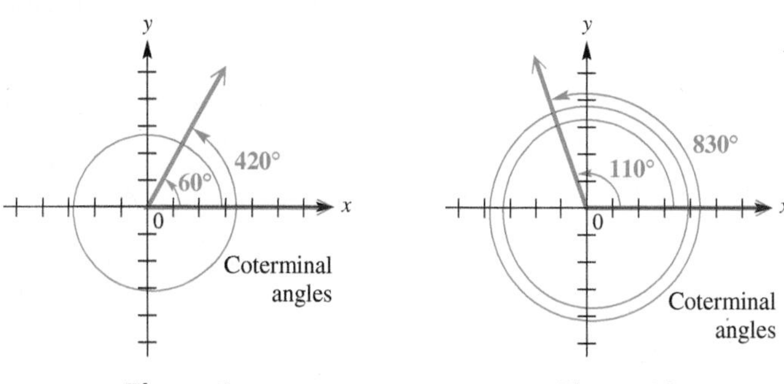

Figure 9 **Figure 10**

Classroom Example 5
Find the angle of least positive measure that is coterminal with each angle.

(a) $1106°$ (b) $-150°$
(c) $-603°$

Answers:
(a) $26°$ (b) $210°$ (c) $117°$

EXAMPLE 5 Finding Measures of Coterminal Angles

Find the angle of least positive measure that is coterminal with each angle.

(a) $908°$ **(b)** $-75°$ **(c)** $-800°$

SOLUTION

(a) Subtract $360°$ as many times as needed to obtain an angle with measure greater than $0°$ but less than $360°$.

$$908° - 2 \cdot 360° = 188° \quad \text{Multiply } 2 \cdot 360°\text{. Then subtract.}$$

An angle of $188°$ is coterminal with an angle of $908°$. See **Figure 11.**

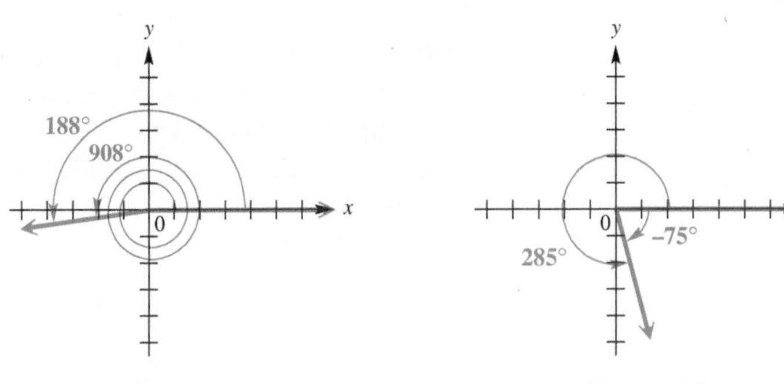

Figure 11 **Figure 12**

(b) Add $360°$ to the given negative angle measure to obtain the angle of least *positive* measure. See **Figure 12.**

$$-75° + 360° = 285°$$

(c) The least integer multiple of $360°$ greater than $800°$ is

$$3 \cdot 360° = 1080°.$$

Add $1080°$ to $-800°$ to obtain

$$-800° + 1080° = 280°.$$

✔ **Now Try Exercises 81, 91, and 95.**

Teaching Tip Remind students that the set of integers is
$\{\ldots, -3, -2, -1, 0, 1, 2, 3, \ldots\}$.

Sometimes it is necessary to find an expression that will generate all angles coterminal with a given angle. For example, we can obtain any angle coterminal with $60°$ by adding an integer multiple of $360°$ to $60°$. Let n represent any integer. Then the following expression represents all such coterminal angles.

$$60° + \boldsymbol{n} \cdot \mathbf{360°} \quad \text{Angles coterminal with } 60°$$

The table below shows a few possibilities.

Examples of Angles Coterminal with 60°

Value of n	Angle Coterminal with 60°
2	$60° + 2 \cdot 360° = 780°$
1	$60° + 1 \cdot 360° = 420°$
0	$60° + 0 \cdot 360° = 60°$ (the angle itself)
-1	$60° + (-1) \cdot 360° = -300°$
-2	$60° + (-2) \cdot 360° = -660°$

This table shows some examples of coterminal quadrantal angles.

Examples of Coterminal Quadrantal Angles

Quadrantal Angle θ	Coterminal with θ
$0°$	$\pm 360°, \pm 720°$
$90°$	$-630°, -270°, 450°$
$180°$	$-180°, 540°, 900°$
$270°$	$-450°, -90°, 630°$

Classroom Example 6
A wheel makes 270 revolutions per min. Through how many degrees will a point on the edge of the wheel move in 5 sec?

Answer: 8100°

EXAMPLE 6 Analyzing Revolutions of a Disk Drive

A constant angular velocity disk drive spins a disk at a constant speed. Suppose a disk makes 480 revolutions per min. Through how many degrees will a point on the edge of the disk move in 2 sec?

SOLUTION The disk revolves 480 times in 1 min, or $\frac{480}{60}$ times $=$ 8 times per sec (because 60 sec $=$ 1 min). In 2 sec, the disk will revolve $2 \cdot 8 = 16$ times. Each revolution is 360°, so in 2 sec a point on the edge of the disk will revolve

$$16 \cdot 360° = 5760°.$$

A unit analysis expression can also be used.

$$\frac{480 \text{ rev}}{1 \text{ min}} \times \frac{1 \text{ min}}{60 \text{ sec}} \times \frac{360°}{1 \text{ rev}} \times 2 \text{ sec} = 5760°$$ Divide out common units.

✓ **Now Try Exercise 123.**

Complete solutions to all exercises denoted with a green ■ are available in the eText.

1.1 Exercises

Teaching Tip If students are reading the text or watching videos before coming to class, use the **Concept Preview** exercises to assess their readiness for the class lecture or activities.

1. $\frac{1}{360}$ 2. 90°
3. 180° 4. 45°
5. 90° 6. $\frac{1}{60}$
7. $\frac{1}{60}$ 8. 12.5°
9. 55° 15′ 10. $n \cdot 360°$
11. (a) 60° (b) 150°
12. (a) 30° (b) 120°
13. (a) 45° (b) 135°
14. (a) 0° (b) 90°
15. (a) 36° (b) 126°
16. (a) 80° (b) 170°
17. (a) 89° (b) 179°
18. (a) 1° (b) 91°
19. (a) 75° 40′ (b) 165° 40′
20. (a) 50° 10′ (b) 140° 10′
21. (a) 69° 49′ 30″ (b) 159° 49′ 30″

CONCEPT PREVIEW *Fill in the blank(s) to correctly complete each sentence.*

1. One degree, written 1°, represents ____ of a complete rotation.

2. If the measure of an angle is $x°$, its complement can be expressed as ____ $- x°$.

3. If the measure of an angle is $x°$, its supplement can be expressed as ____ $- x°$.

4. The measure of an angle that is its own complement is ____.

5. The measure of an angle that is its own supplement is ____.

6. One minute, written 1′, is ____ of a degree.

7. One second, written 1″, is ____ of a minute.

8. 12° 30′ written in decimal degrees is ____.

9. 55.25° written in degrees and minutes is ____.

10. If n represents any integer, then an expression representing all angles coterminal with 45° is 45° + ____.

Find the measure of **(a)** *the complement and* **(b)** *the supplement of an angle with the given measure. See* ***Examples 1 and 3.***

11. 30° **12.** 60° **13.** 45° **14.** 90°

15. 54° **16.** 10° **17.** 1° **18.** 89°

19. 14° 20′ **20.** 39° 50′ **21.** 20° 10′ 30″ **22.** 50° 40′ 50″

22. (a) 39° 19′ 10″
(b) 129° 19′ 10″

23. 70°; 110° 24. 150°; 30°
25. 30°; 60° 26. 55°; 35°
27. 40°; 140° 28. 90°; 90°
29. 107°; 73° 30. 80°; 100°
31. 69°; 21° 32. 40°; 50°

33. 150° 34. 142° 30′
35. 7° 30′ 36. 22° 30′
37. 130° 38. 125°

39. 83° 59′ 40. 158° 47′
41. 179° 19′ 42. 143° 20′
43. −23° 49′ 44. −26° 25′
45. 38° 32′ 46. 72° 47′
47. 60° 34′ 48. 55° 09′
49. 17° 01′ 49″ 50. 53° 41′ 13″
51. 30° 27′ 52. 59° 59′

53. 35.5° 54. 82.5°
55. 112.25° 56. 133.75°
57. −60.2° 58. −70.8°
59. 20.91° 60. 38.705°
61. 91.598° 62. 34.860°
63. 274.316° 64. 165.853°

65. 39° 15′ 00″ 66. 46° 45′ 00″
67. 126° 45′ 36″ 68. 174° 15′ 18″
69. −18° 30′ 54″ 70. −25° 29′ 06″
71. 31° 25′ 47″ 72. 59° 05′ 07″
73. 89° 54′ 01″ 74. 102° 22′ 38″
75. 178° 35′ 58″ 76. 122° 41′ 07″

77. 392° 78. 446°
79. 386° 30′ 80. 418° 40′
81. 320° 82. 262°
83. 234° 30′ 84. 156° 40′
85. 1° 86. 181°
87. 359° 88. 179°
89. 179° 90. 339°
91. 130° 92. 280°
93. 240° 94. 160°
95. 120° 96. 200°

In Exercises 97–100, answers may vary.

97. 450°, 810°; −270°, −630°
98. 540°, 900°; −180°, −540°
99. 360°, 720°; −360°, −720°
100. 630°, 990°; −90°, −450°

101. $30° + n \cdot 360°$
102. $45° + n \cdot 360°$
103. $135° + n \cdot 360°$
104. $225° + n \cdot 360°$

Find the measure of each marked angle. ***See Example 2.***

23. 24.

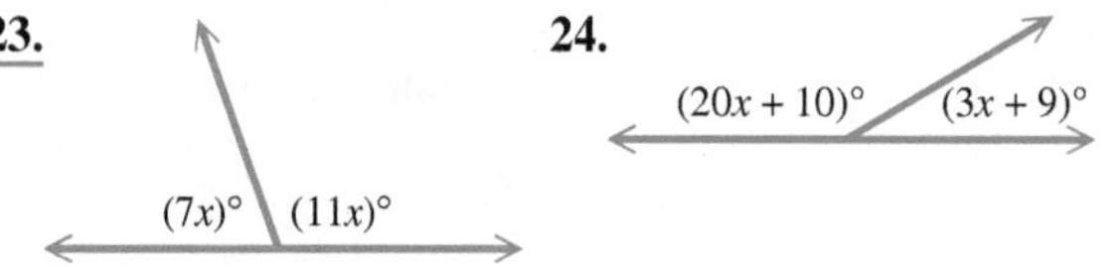

25.

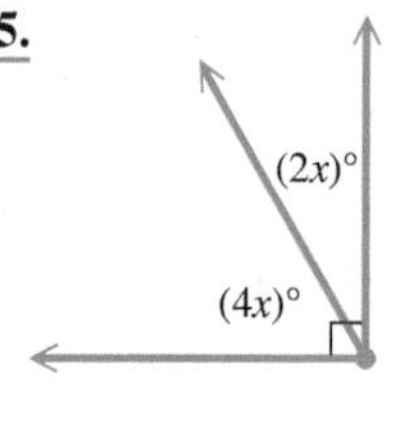

26. 27. 28.

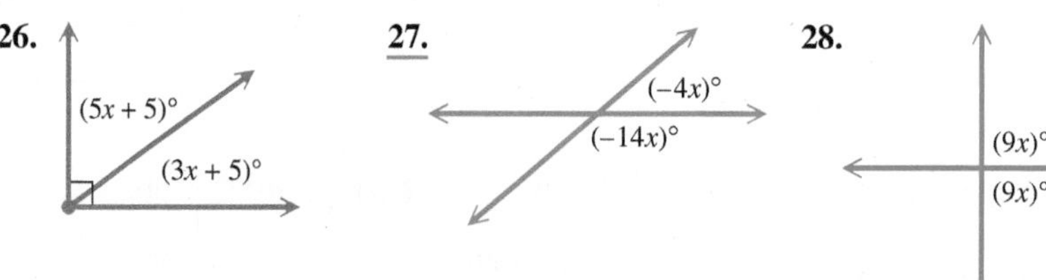

29. supplementary angles with measures $10x + 7$ and $7x + 3$ degrees

30. supplementary angles with measures $6x - 4$ and $8x - 12$ degrees

31. complementary angles with measures $9x + 6$ and $3x$ degrees

32. complementary angles with measures $3x - 5$ and $6x - 40$ degrees

Find the measure of the smaller angle formed by the hands of a clock at the following times.

33. 34.

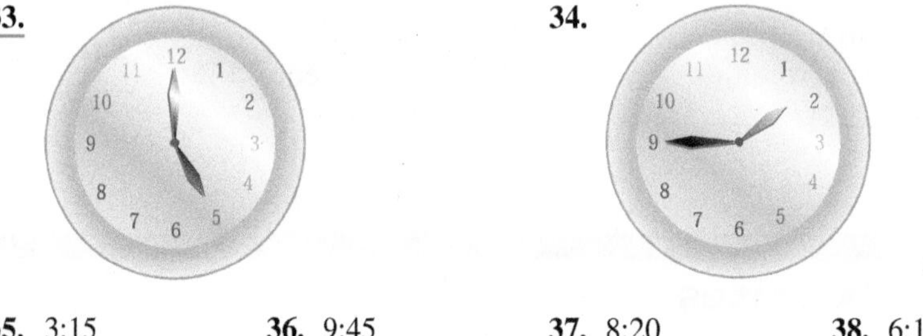

35. 3:15 36. 9:45 37. 8:20 38. 6:10

Perform each calculation. ***See Example 3.***

39. 62° 18′ + 21° 41′ 40. 75° 15′ + 83° 32′ 41. 97° 42′ + 81° 37′

42. 110° 25′ + 32° 55′ 43. 47° 29′ − 71° 18′ 44. 47° 23′ − 73° 48′

45. 90° − 51° 28′ 46. 90° − 17° 13′ 47. 180° − 119° 26′

48. 180° − 124° 51′ 49. 90° − 72° 58′ 11″ 50. 90° − 36° 18′ 47″

51. 26° 20′ + 18° 17′ − 14° 10′ 52. 55° 30′ + 12° 44′ − 8° 15′

Convert each angle measure to decimal degrees. If applicable, round to the nearest thousandth of a degree. ***See Example 4(a).***

53. 35° 30′ 54. 82° 30′ 55. 112° 15′ 56. 133° 45′

57. −60° 12′ 58. −70° 48′ 59. 20° 54′ 36″ 60. 38° 42′ 18″

61. 91° 35′ 54″ 62. 34° 51′ 35″ 63. 274° 18′ 59″ 64. 165° 51′ 09″

Convert each angle measure to degrees, minutes, and seconds. If applicable, round to the nearest second. ***See Example 4(b).***

65. 39.25° 66. 46.75° 67. 126.76° 68. 174.255°

69. −18.515° 70. −25.485° 71. 31.4296° 72. 59.0854°

73. 89.9004° 74. 102.3771° 75. 178.5994° 76. 122.6853°

105. $-90° + n \cdot 360°$

106. $-180° + n \cdot 360°$

107. $0° + n \cdot 360°$, or $n \cdot 360°$

108. $360° + n \cdot 360°$, or $n \cdot 360°$

109. 0° and 360° are coterminal angles.

110. C and D

Angles other than those given are possible in Exercises 111–122.

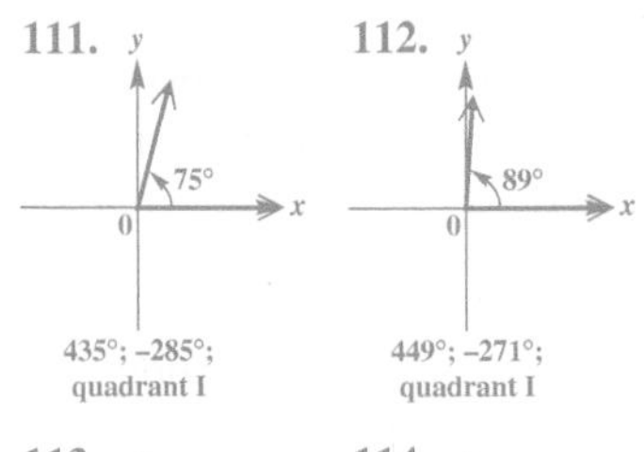

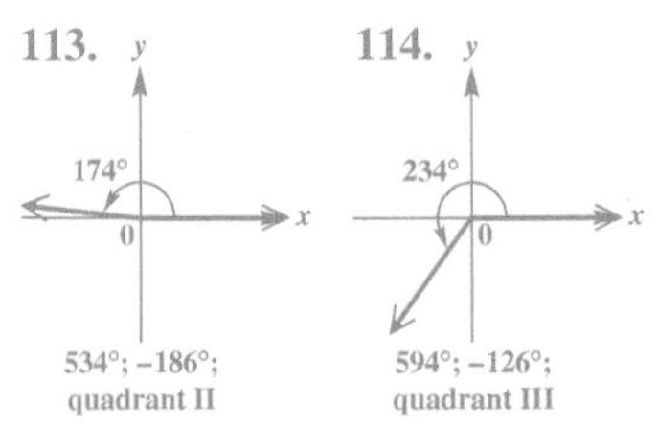

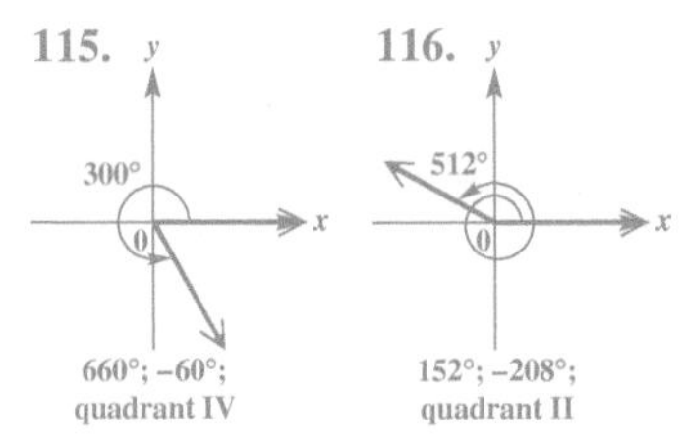

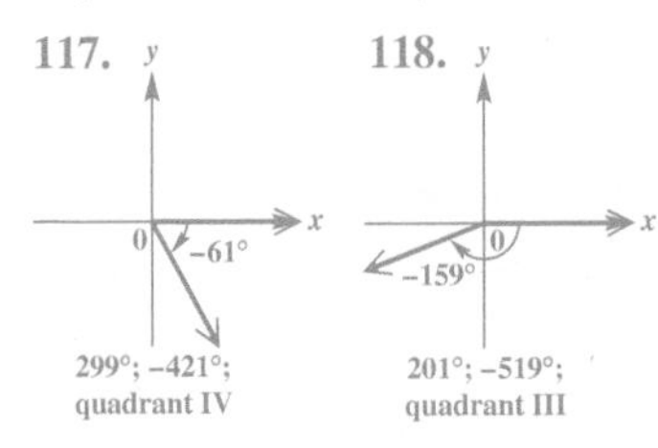

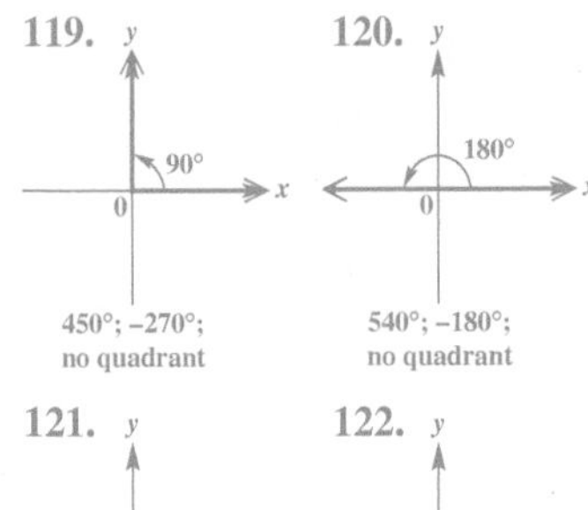

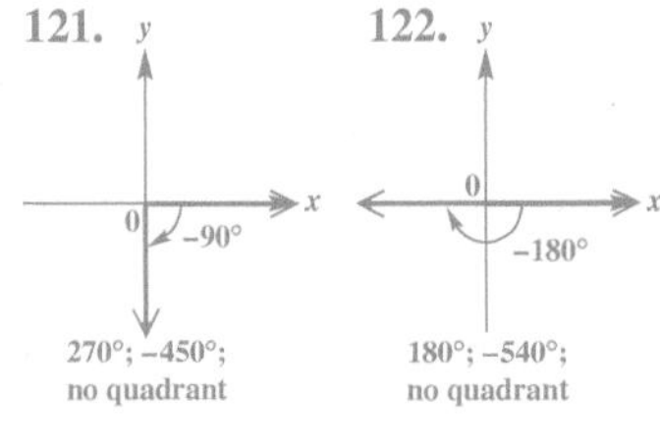

123. $\frac{3}{4}$ **124.** 1.5

125. 1800° **126.** 12,000°

127. 12.5 rotations per hr

Find the angle of least positive measure (not equal to the given measure) that is coterminal with each angle. ***See Example 5.***

77. 32°	**78.** 86°	**79.** 26° 30′	**80.** 58° 40′
81. −40°	**82.** −98°	**83.** −125° 30′	**84.** −203° 20′
85. 361°	**86.** 541°	**87.** −361°	**88.** −541°
89. 539°	**90.** 699°	**91.** 850°	**92.** 1000°
93. 5280°	**94.** 8440°	**95.** −5280°	**96.** −8440°

Give two positive and two negative angles that are coterminal with the given quadrantal angle.

97. 90°	**98.** 180°	**99.** 0°	**100.** 270°

Write an expression that generates all angles coterminal with each angle. Let n represent any integer.

101. 30°	**102.** 45°	**103.** 135°	**104.** 225°
105. −90°	**106.** −180°	**107.** 0°	**108.** 360°

109. Why do the answers to **Exercises 107 and 108** give the same set of angles?

110. ***Concept Check*** Which two of the following are not coterminal with $r°$?

A. $360° + r°$ **B.** $r° - 360°$ **C.** $360° - r°$ **D.** $r° + 180°$

Concept Check *Sketch each angle in standard position. Draw an arrow representing the correct amount of rotation. Find the measure of two other angles, one positive and one negative, that are coterminal with the given angle. Give the quadrant of each angle, if applicable.*

111. 75°	**112.** 89°	**113.** 174°	**114.** 234°
115. 300°	**116.** 512°	**117.** −61°	**118.** −159°
119. 90°	**120.** 180°	**121.** −90°	**122.** −180°

Solve each problem. ***See Example 6.***

123. ***Revolutions of a Turntable*** A turntable in a shop makes 45 revolutions per min. How many revolutions does it make per second?

124. ***Revolutions of a Windmill*** A windmill makes 90 revolutions per min. How many revolutions does it make per second?

125. ***Rotating Tire*** A tire is rotating 600 times per min. Through how many degrees does a point on the edge of the tire move in $\frac{1}{2}$ sec?

126. ***Rotating Airplane Propeller*** An airplane propeller rotates 1000 times per min. Find the number of degrees that a point on the edge of the propeller will rotate in 2 sec.

127. ***Rotating Pulley*** A pulley rotates through 75° in 1 min. How many rotations does the pulley make in 1 hr?

128. 5′, or 0.08°
129. 4 sec
130. 36°

128. *Surveying* One student in a surveying class measures an angle as 74.25°, while another student measures the same angle as 74° 20′. Find the difference between these measurements, both to the nearest minute and to the nearest hundredth of a degree.

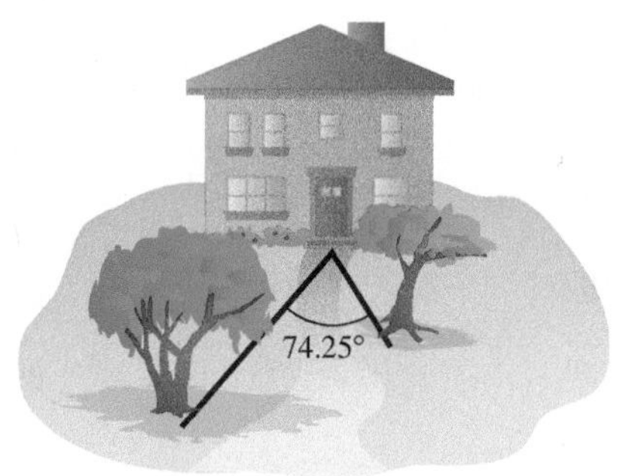

129. *Viewing Field of a Telescope* As a consequence of Earth's rotation, celestial objects such as the moon and the stars appear to move across the sky, rising in the east and setting in the west. As a result, if a telescope on Earth remains stationary while viewing a celestial object, the object will slowly move outside the viewing field of the telescope. For this reason, a motor is often attached to telescopes so that the telescope rotates at the same rate as Earth. Determine how long it should take the motor to turn the telescope through an angle of 1 min in a direction perpendicular to Earth's axis.

130. *Angle Measure of a Star on the American Flag* Determine the measure of the angle in each point of the five-pointed star appearing on the American flag. (*Hint:* Inscribe the star in a circle, and use the following theorem from geometry: *An angle whose vertex lies on the circumference of a circle is equal to half the central angle that cuts off the same arc.* See the figure.)

1.2 Angle Relationships and Similar Triangles

- Geometric Properties
- Triangles

Geometric Properties In **Figure 13,** we extended the sides of angle NMP to form another angle, RMQ. The pair of angles NMP and RMQ are **vertical angles.** Another pair of vertical angles, NMQ and PMR, are also formed. Vertical angles have the following important property.

Q R M N P

Vertical angles

Figure 13

Vertical Angles

Vertical angles have equal measures.

Parallel lines are lines that lie in the same plane and do not intersect. **Figure 14** shows parallel lines m and n. When a line q intersects two parallel lines, q is called a **transversal.** In **Figure 14,** the transversal intersecting the parallel lines forms eight angles, indicated by numbers.

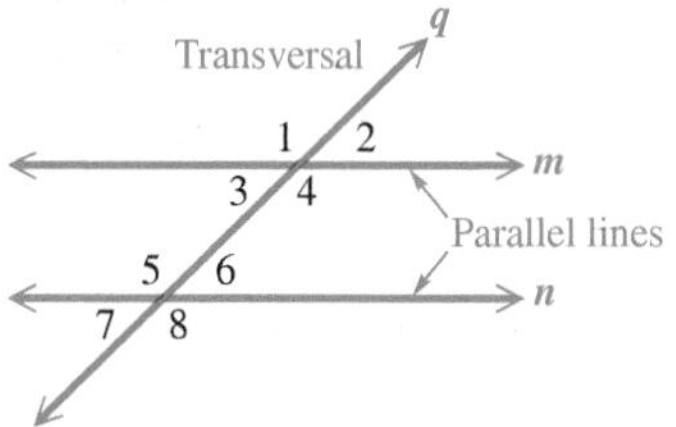

Figure 14

We learn in geometry that the degree measures of angles 1 through 8 in **Figure 14** possess some special properties. The following table gives the names of these angles and rules about their measures.

Angle Pairs of Parallel Lines Intersected by a Transversal

Name	Sketch	Rule
Alternate interior angles	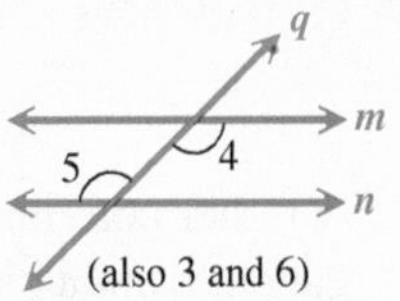 q, m, n, 5, 4 (also 3 and 6)	Angle measures are equal.
Alternate exterior angles	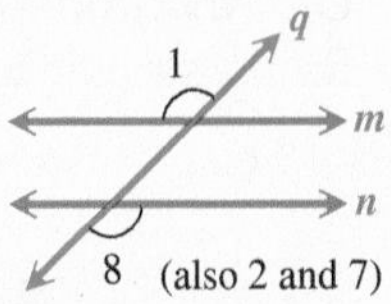 q, 1, m, n, 8 (also 2 and 7)	Angle measures are equal.
Interior angles on the same side of a transversal	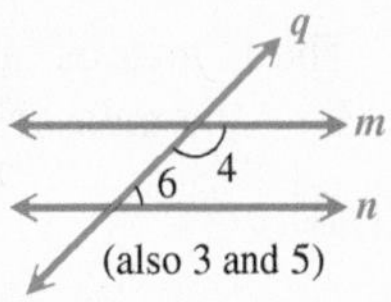 q, m, 6, 4, n (also 3 and 5)	Angle measures add to 180°.
Corresponding angles	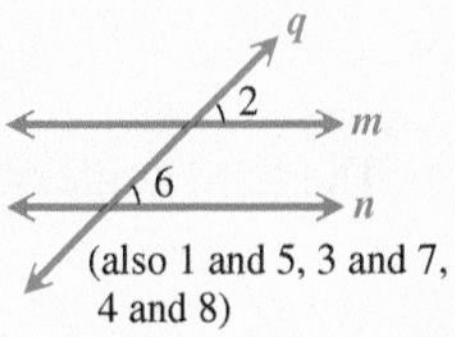 q, 2, m, 6, n (also 1 and 5, 3 and 7, 4 and 8)	Angle measures are equal.

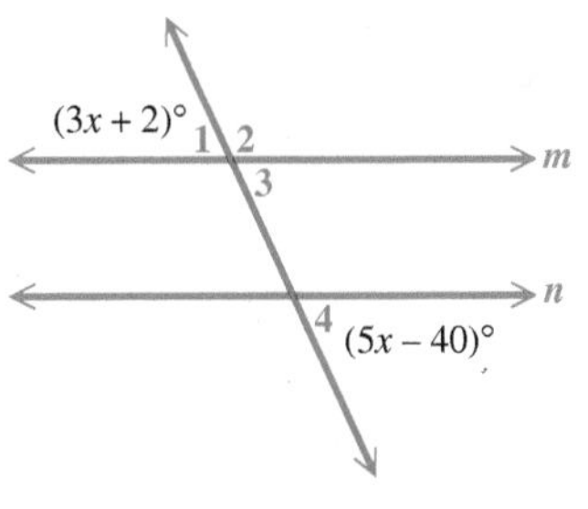

Figure 15

EXAMPLE 1 Finding Angle Measures

Find the measures of angles 1, 2, 3, and 4 in **Figure 15,** given that lines m and n are parallel.

SOLUTION Angles 1 and 4 are alternate exterior angles, so they are equal.

$$3x + 2 = 5x - 40 \quad \text{Alternate exterior angles have equal measures.}$$

$$42 = 2x \quad \text{Subtract } 3x \text{ and add 40.}$$

$$21 = x \quad \text{Divide by 2.}$$

Angle 1 has measure

$$3x + 2$$

$$= 3 \cdot 21 + 2 \quad \text{Substitute 21 for } x.$$

$$= 65°. \quad \text{Multiply, and then add.}$$

Angle 4 has measure

$$5x - 40$$

$$= 5 \cdot 21 - 40 \quad \text{Substitute 21 for } x.$$

$$= 65°. \quad \text{Multiply, and then subtract.}$$

Angle 2 is the supplement of a 65° angle, so it has measure

$$180° - 65° = 115°.$$

Angle 3 is a vertical angle to angle 1, so its measure is also 65°. (There are other ways to determine these measures.)

✔ **Now Try Exercises 11 and 19.**

Classroom Example 1
Find the measures of angles 1, 2, 3, and 4 in the figure, given that lines m and n are parallel.

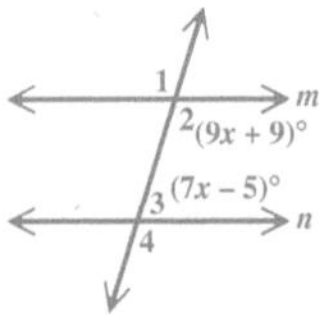

m and *n* are parallel.

Answer: angles 1, 2, and 4: 108°; angle 3: 72°

Triangles An important property of triangles, first proved by Greek geometers, deals with the sum of the measures of the angles of any triangle.

(a)

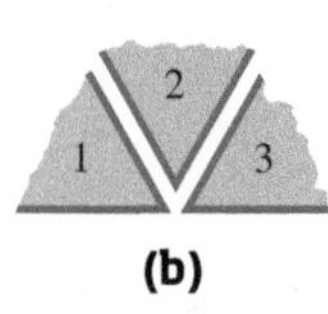

(b)

Figure 16

Angle Sum of a Triangle

***The sum of the measures of the angles of any triangle is* 180°.**

A rather convincing argument for the truth of this statement uses any size triangle cut from a piece of paper. Tear each corner from the triangle, as suggested in **Figure 16(a).** We should be able to rearrange the pieces so that the three angles form a straight angle, which has measure 180°, as shown in **Figure 16(b).**

EXAMPLE 2 Applying the Angle Sum of a Triangle Property

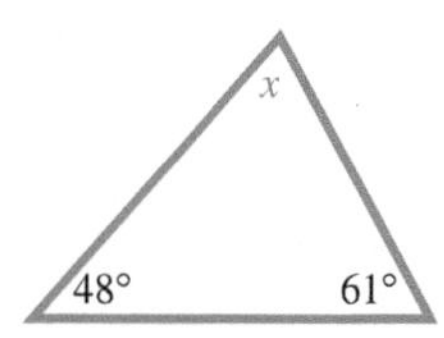

Figure 17

The measures of two of the angles of a triangle are 48° and 61°. See **Figure 17.** Find the measure of the third angle, x.

SOLUTION

$$48° + 61° + x = 180° \quad \text{The sum of the angles is 180°.}$$

$$109° + x = 180° \quad \text{Add.}$$

$$x = 71° \quad \text{Subtract 109°.}$$

The third angle of the triangle measures 71°.

✔ **Now Try Exercises 13 and 23.**

Classroom Example 2
The measures of two of the angles of a triangle are 33° and 26°. Find the measure of the third angle.

Answer: 121°

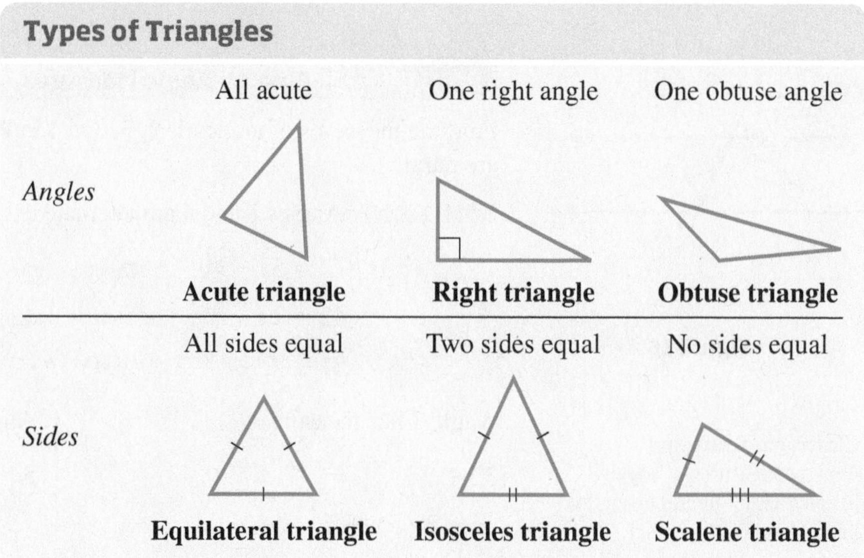

Teaching Tip Give students several similar triangles that are oriented differently. Have them

(a) visualize how they can be lined up (i.e., slide, rotation, reflection), and

(b) name the similarity (i.e., triangle ABC is similar to triangle DEF).

Similar triangles are triangles of exactly the same shape but not necessarily the same size. **Figure 18** on the next page shows three pairs of similar triangles. The two triangles in **Figure 18(c)** have not only the same shape but also the same size. Triangles that are both the same size and the same shape are **congruent triangles.** If two triangles are congruent, then it is possible to pick one of them up and place it on top of the other so that they coincide.

If two triangles are congruent, then they must be similar. However, two similar triangles need not be congruent.

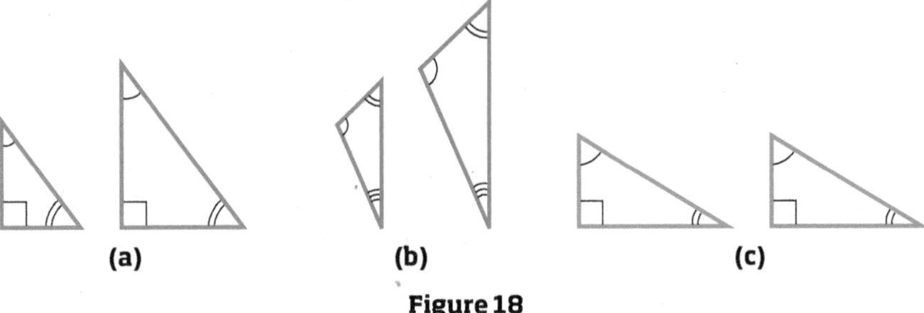

Figure 18

The triangular supports for a child's swing set are congruent (and thus similar) triangles, machine-produced with exactly the same dimensions each time. These supports are just one example of similar triangles. The supports of a long bridge, all the same shape but increasing in size toward the center of the bridge, are examples of similar (but not congruent) figures. See the photo.

Consider the correspondence between triangles ABC and DEF in **Figure 19.**

Angle A corresponds to angle D.
Angle B corresponds to angle E.
Angle C corresponds to angle F.
Side AB corresponds to side DE.
Side BC corresponds to side EF.
Side AC corresponds to side DF.

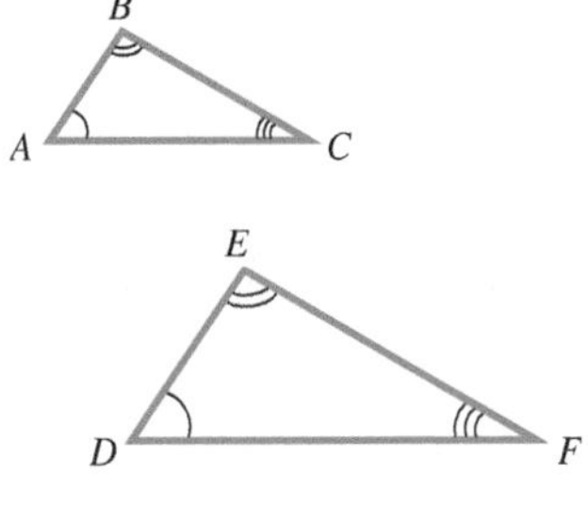

Figure 19

The small arcs found at the angles in **Figure 19** denote the corresponding angles in the triangles.

Conditions for Similar Triangles

Triangle ABC is similar to triangle DEF if the following conditions hold.

1. Corresponding angles have the same measure.
2. Corresponding sides are proportional. (That is, the ratios of their corresponding sides are equal.)

Classroom Example 3
In the figure, triangles DEF and GHI are similar. Find all unknown angle measures.

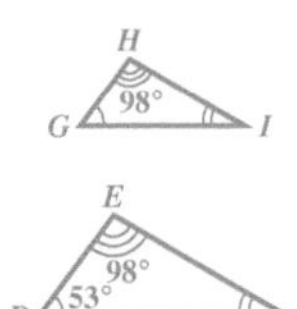

Answer: $G = 53°$; $I = F = 29°$

EXAMPLE 3 Finding Angle Measures in Similar Triangles

In **Figure 20,** triangles ABC and NMP are similar. Find all unknown angle measures.

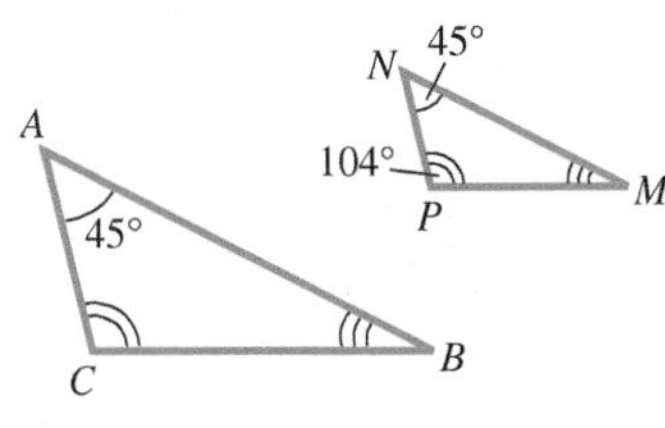

Figure 20

SOLUTION First, we find the measure of angle M using the angle sum property of a triangle.

$$104° + 45° + M = 180° \quad \text{The sum of the angles is } 180°.$$

$$149° + M = 180° \quad \text{Add.}$$

$$M = 31° \quad \text{Subtract } 149°.$$

The triangles are similar, so corresponding angles have the same measure. Because C corresponds to P and P measures 104°, angle C also measures 104°. Angles B and M correspond, so B measures 31°.

✔ **Now Try Exercise 49.**

Classroom Example 4
Given that triangle MNP and triangle QSR are similar, find the lengths of the unknown sides of triangle QSR.

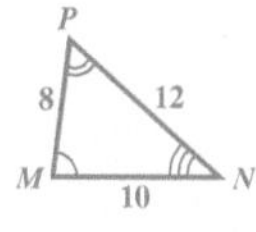

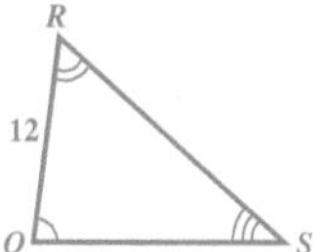

Answer: $QS = 15$; $RS = 18$

EXAMPLE 4 Finding Side Lengths in Similar Triangles

Given that triangle ABC and triangle DFE in **Figure 21** are similar, find the lengths of the unknown sides of triangle DFE.

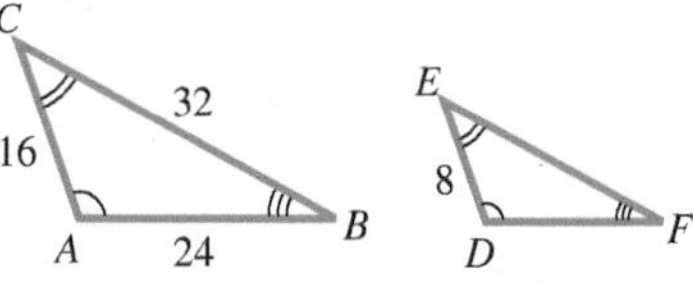

Figure 21

SOLUTION Similar triangles have corresponding sides in proportion. Use this fact to find the unknown side lengths in triangle DFE.

Side DF of triangle DFE corresponds to side AB of triangle ABC, and sides DE and AC correspond. This leads to the following proportion.

$$\frac{8}{16} = \frac{DF}{24}$$

Recall this property of proportions from algebra.

$$\textbf{If } \frac{a}{b} = \frac{c}{d}, \textbf{ then } ad = bc.$$

We use this property to solve the equation for DF.

$$\frac{8}{16} = \frac{DF}{24} \quad \text{Corresponding sides are proportional.}$$

$$8 \cdot 24 = 16 \cdot DF \quad \text{If } \tfrac{a}{b} = \tfrac{c}{d}, \text{ then } ad = bc.$$

$$192 = 16 \cdot DF \quad \text{Multiply.}$$

$$12 = DF \quad \text{Divide by 16.}$$

Teaching Tip Remind students that it is INCORRECT to "reduce across the equals symbol." In the first proportion, do not reduce 8 and 24, and in the second, do not reduce 8 and 32.

Explain also that other forms of these proportions exist.

Side DF has length 12.

Side EF corresponds to CB. This leads to another proportion.

$$\frac{8}{16} = \frac{EF}{32} \quad \text{Corresponding sides are proportional.}$$

$$8 \cdot 32 = 16 \cdot EF \quad \text{If } \tfrac{a}{b} = \tfrac{c}{d}, \text{ then } ad = bc.$$

$$16 = EF \quad \text{Solve for } EF.$$

Side EF has length 16.

✔ **Now Try Exercise 55.**

Classroom Example 5
Joey wants to know the height of a tree in a park near his home. The tree casts a 38-ft shadow at the same time that Joey, who is 63 in. tall, casts a 42-in. shadow. Find the height of the tree.

Answer: 57 ft

EXAMPLE 5 Finding the Height of a Flagpole

Workers must measure the height of a building flagpole. They find that at the instant when the shadow of the building is 18 m long, the shadow of the flagpole is 27 m long. The building is 10 m high. Find the height of the flagpole.

SOLUTION **Figure 22** shows the information given in the problem. The two triangles are similar, so corresponding sides are in proportion.

$$\frac{MN}{10} = \frac{27}{18} \quad \text{Corresponding sides are proportional.}$$

$$\frac{MN}{10} = \frac{3}{2} \quad \text{Write } \tfrac{27}{18} \text{ in lowest terms.}$$

$$MN \cdot 2 = 10 \cdot 3 \quad \text{Property of proportions}$$

$$MN = 15 \quad \text{Solve for } MN.$$

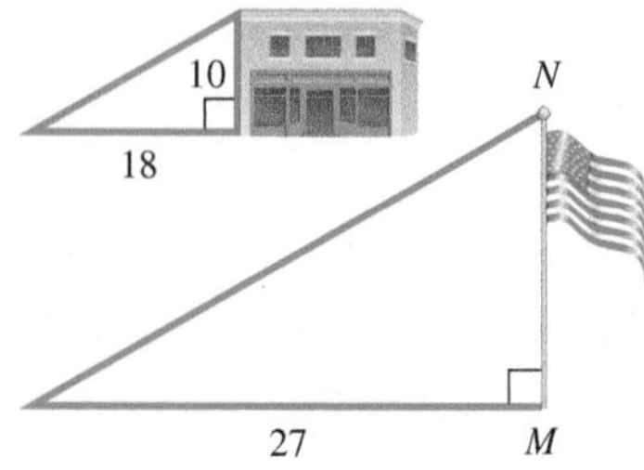

Figure 22

The flagpole is 15 m high.

✔ **Now Try Exercise 59.**

1.2 Exercises

1. 180°
2. right; two
3. three
4. sides; angles

Answers are given in numerical order in Exercises 5 and 6.

5. 49°; 49°; 131°; 131°; 49°; 49°; 131°
6. 60°; 120°; 60°; 60°; 65°; 60°; 65°; 55°; 55°; 55°
7. *A* and *P*; *B* and *Q*; *C* and *R*; *AC* and *PR*; *BC* and *QR*; *AB* and *PQ*
8. *A* and *P*; *C* and *R*; *B* and *Q*; *AC* and *PR*; *CB* and *RQ*; *AB* and *PQ*
9. *A* and *C*; *E* and *D*; *ABE* and *CBD*; *EB* and *DB*; *AB* and *CB*; *AE* and *CD*
10. *H* and *F*; *K* and *E*; *HGK* and *FGE*; *HK* and *FE*; *GK* and *GE*; *HG* and *FG*
11. 51°; 51°
12. 139°; 139°
13. 50°; 60°; 70°
14. 20°; 30°; 130°

CONCEPT PREVIEW *Fill in the blank(s) to correctly complete each sentence.*

1. The sum of the measures of the angles of any triangle is ______.
2. An isosceles right triangle has one ______ angle and ______ equal sides.
3. An equilateral triangle has ______ equal sides.
4. If two triangles are similar, then their corresponding ______ are proportional and their corresponding ______ have equal measure.

CONCEPT PREVIEW *In each figure, find the measures of the numbered angles, given that lines m and n are parallel.*

5.

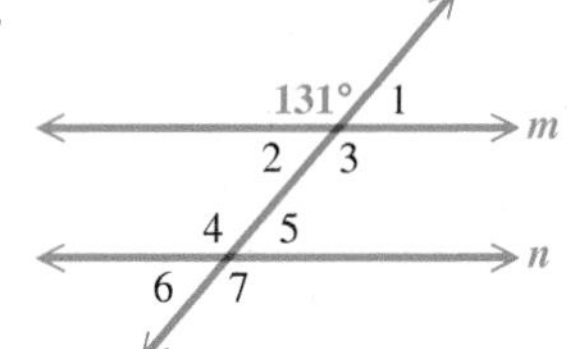

6.

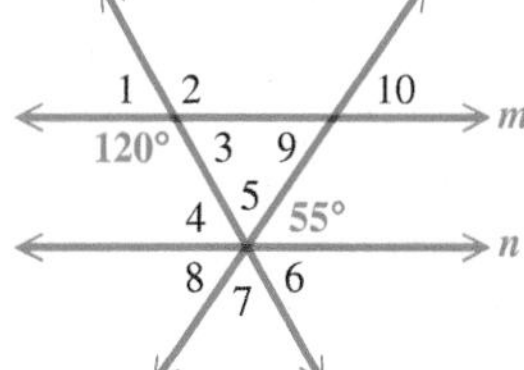

CONCEPT PREVIEW *Name the corresponding angles and the corresponding sides of each pair of similar triangles.*

7.

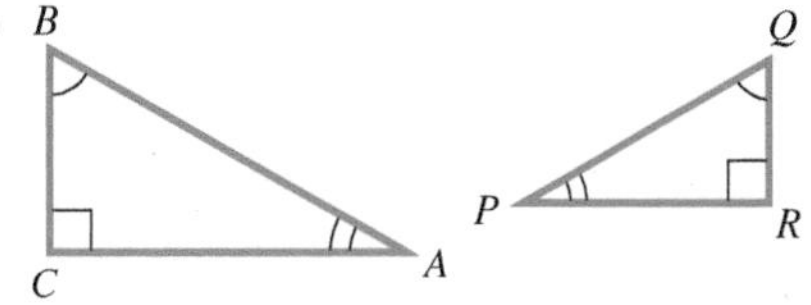

8.

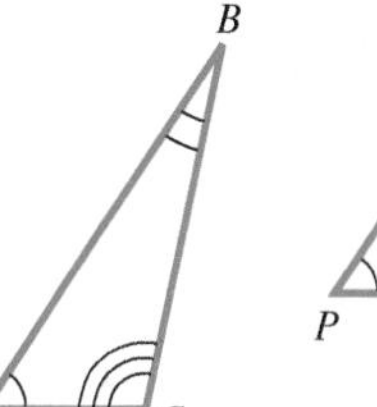

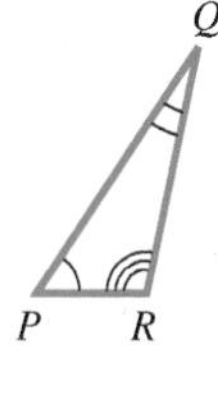

9. (*EA* is parallel to *CD*.)

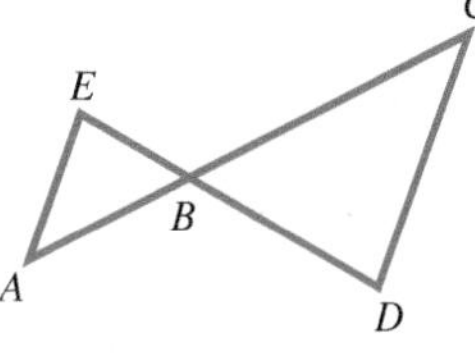

10. (*HK* is parallel to *EF*.)

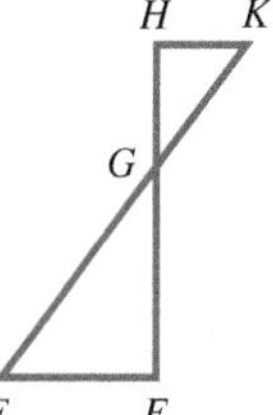

Find the measure of each marked angle. In Exercises 19–22, m and n are parallel. ***See Examples 1 and 2.***

11.

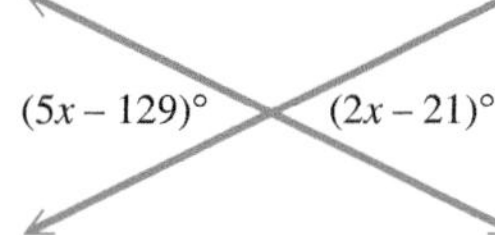

12.

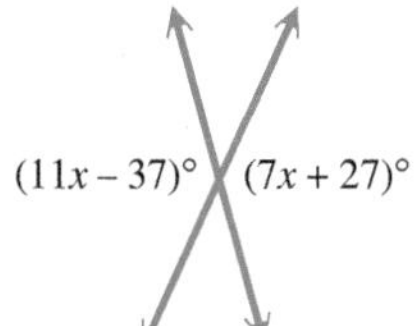

13.

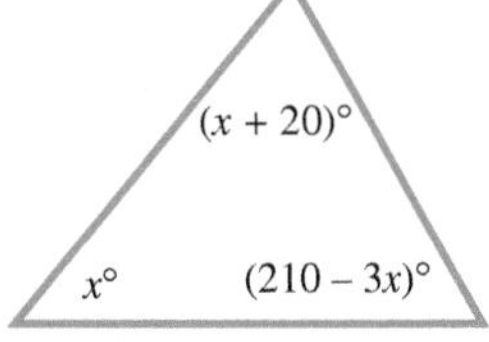

14.

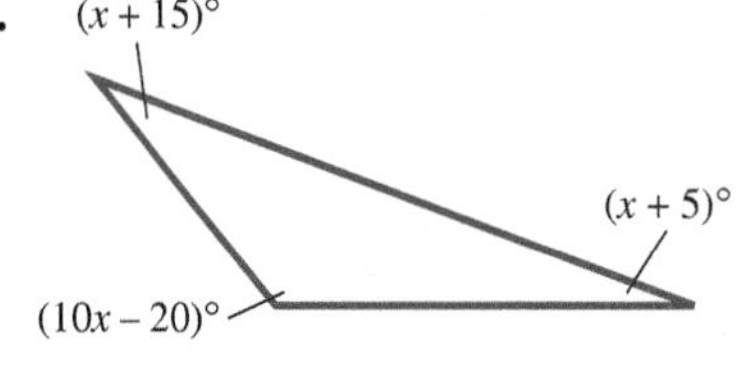

15. 60°; 60°; 60°
16. 60°; 60°; 60°
17. 45°; 75°; 120°
18. 35°; 20°; 55°
19. 49°; 49°
20. 117°; 117°
21. 48°; 132°
22. 141°; 141°

23. 91°
24. 47°
25. 2° 29′
26. 1° 32′
27. 25.4°
28. 100.7°
29. 22° 29′ 34″
30. 66° 06′ 37″

31. no
32. no

33. right; scalene
34. obtuse; scalene
35. acute; equilateral
36. acute; isosceles
37. right; scalene
38. obtuse; isosceles

15.

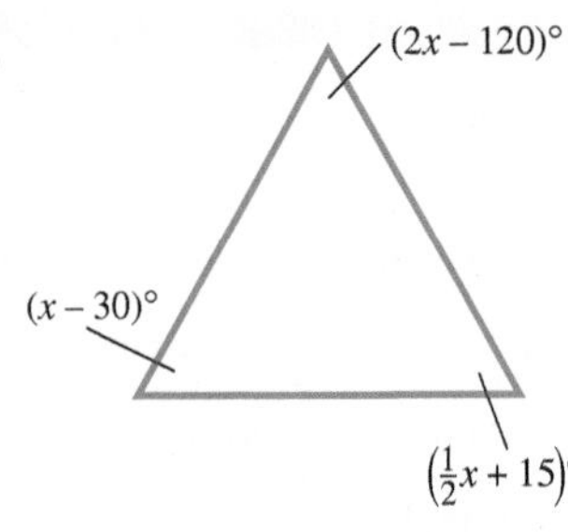

16.

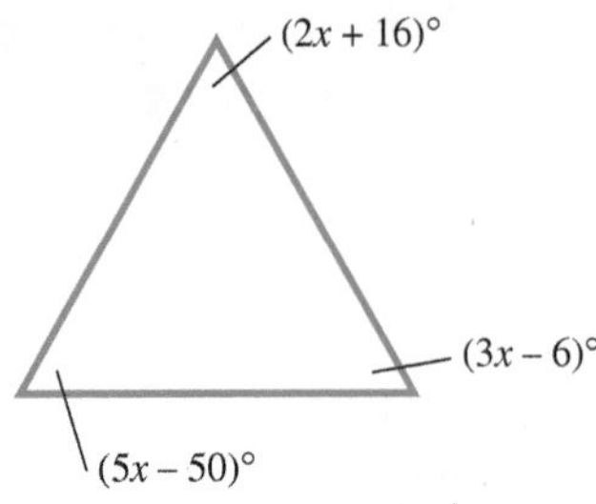

17.

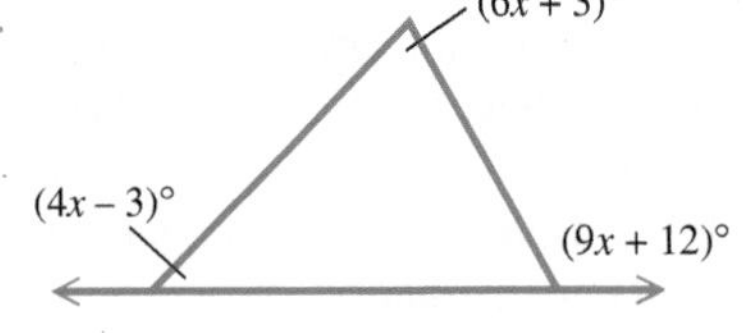

18.

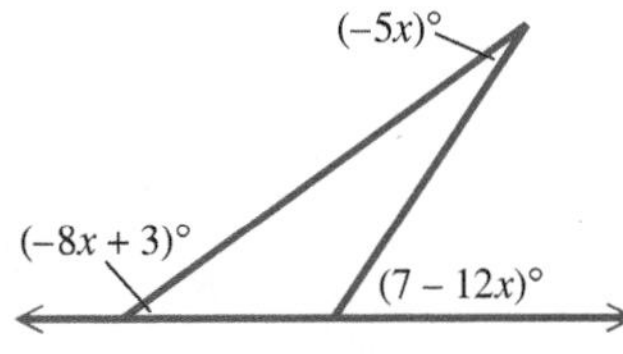

19.

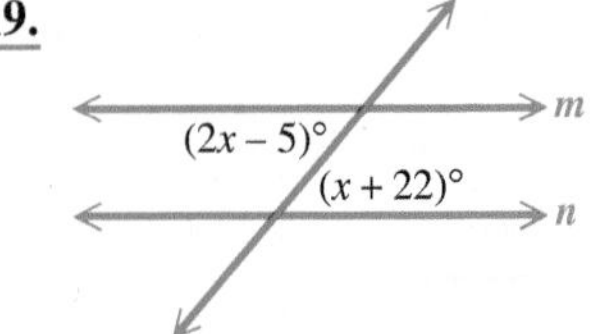

20.

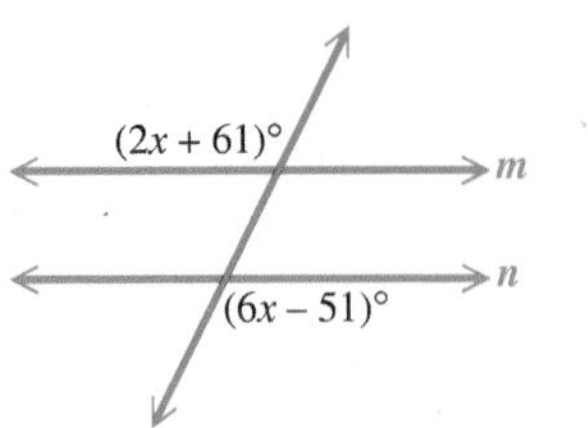

21.

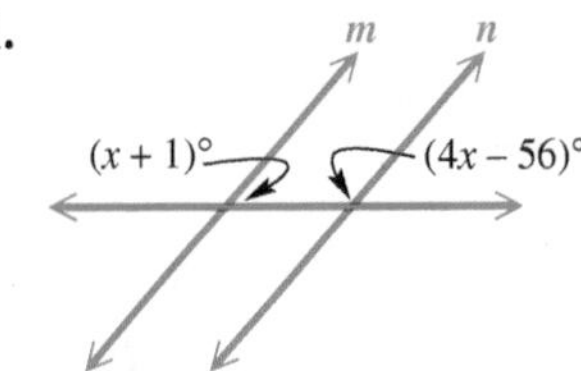

22.

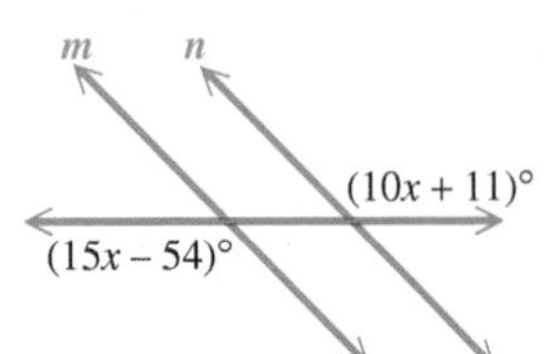

The measures of two angles of a triangle are given. Find the measure of the third angle. ***See Example 2.***

23. 37°, 52°
24. 29°, 104°
25. 147° 12′, 30° 19′
26. 136° 50′, 41° 38′
27. 74.2°, 80.4°
28. 29.6°, 49.7°
29. 51° 20′ 14″, 106° 10′ 12″
30. 17° 41′ 13″, 96° 12′ 10″

31. *Concept Check* Can a triangle have angles of measures 85° and 100°?

32. *Concept Check* Can a triangle have two obtuse angles?

Concept Check Classify each triangle as acute, right, *or* obtuse. *Also classify each as* equilateral, isosceles, *or* scalene. ***See the discussion following Example 2.***

33.

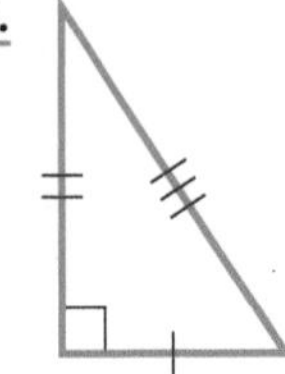

34.

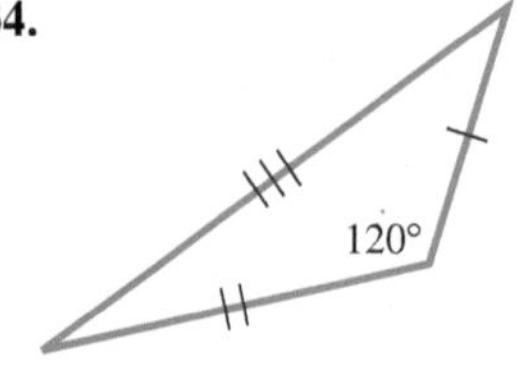

35.

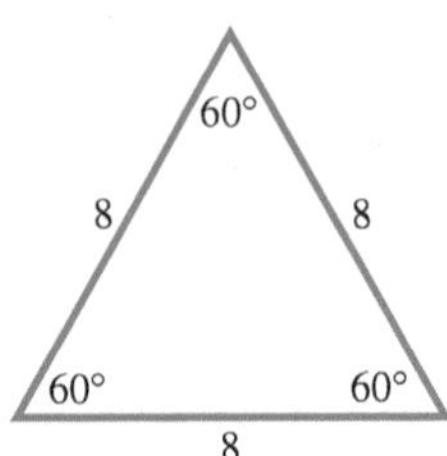

36.

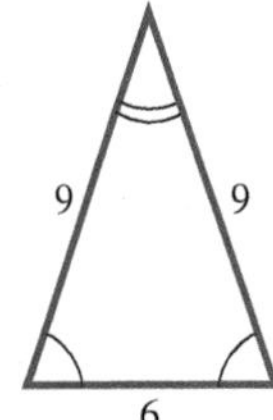

37.

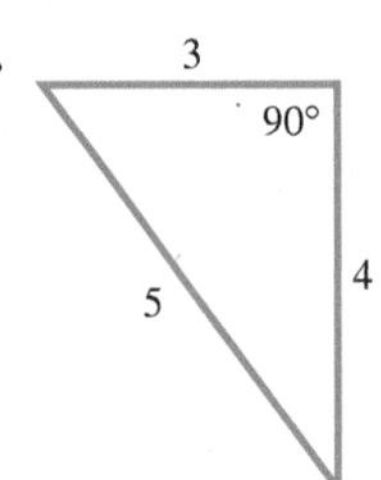

38.

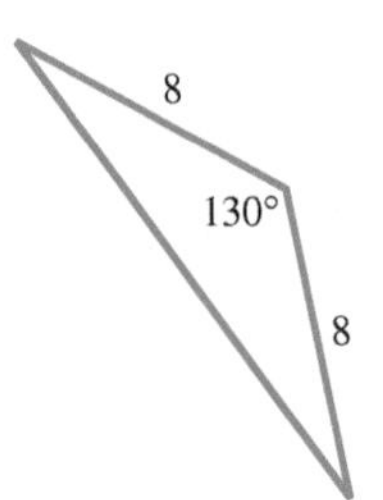

39. right; isosceles
40. right; scalene
41. obtuse; scalene
42. acute; equilateral
43. acute; isosceles
44. right; scalene

45. Angles 1, 2, and 3 form a straight angle on line m and, therefore, sum to 180°. It follows that the sum of the measures of the angles of triangle PQR is 180° because the angles marked 1 are alternate interior angles whose measures are equal, as are the angles marked 2.
46. Connect the right end of the semicircle to the point where the arc crosses the semicircle. The triangle is equilateral, and therefore each of its angles measures 60°.

47. $Q = 42°$; $B = R = 48°$
48. $P = 78°$; $M = 46°$; $A = N = 56°$
49. $B = 106°$; $A = M = 44°$
50. $T = 74°$; $Y = 28°$; $Z = W = 78°$

39.

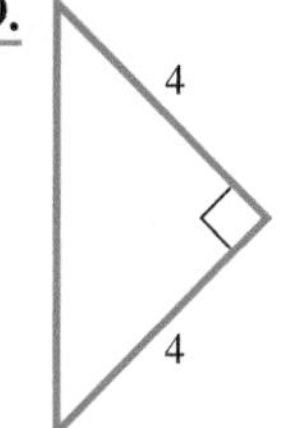

40.

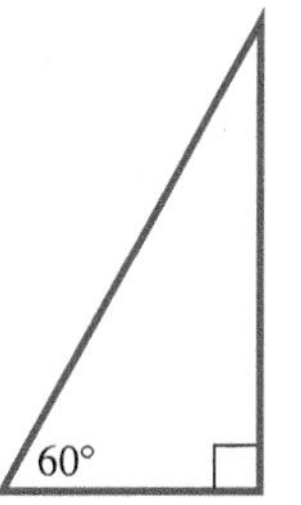

41.

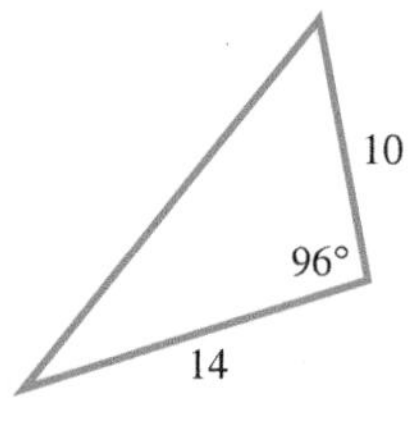

42.

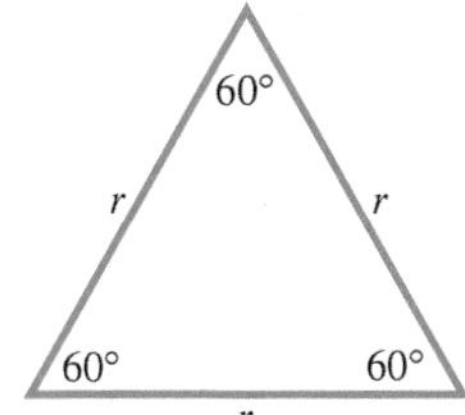

43.

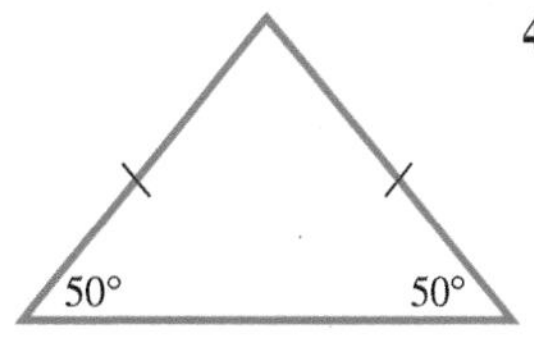

44.

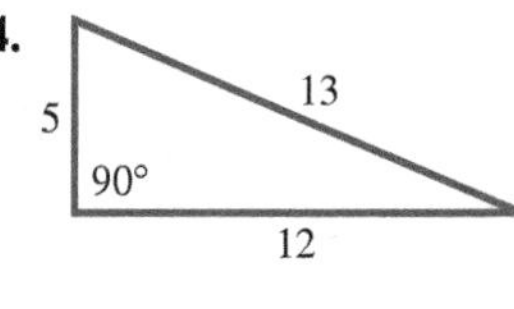

45. ***Angle Sum of a Triangle*** Use this figure to discuss why the measures of the angles of a triangle must add up to the same sum as the measure of a straight angle.

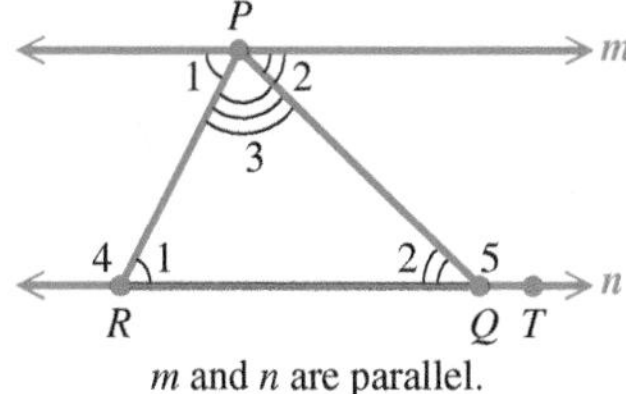

46. ***Carpentry Technique*** The following technique is used by carpenters to draw a 60° angle with a straightedge and a compass. Why does this technique work? (*Source:* Hamilton, J. E. and M. S. Hamilton, *Math to Build On*, Construction Trades Press.)

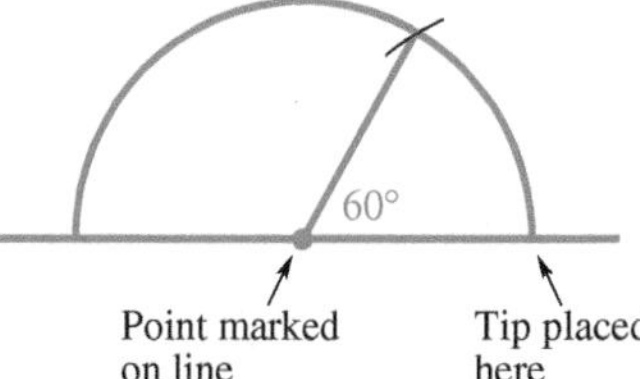

"Draw a straight line segment, and mark a point near the midpoint. Now place the tip on the marked point, and draw a semicircle. Without changing the setting of the compass, place the tip at the right intersection of the line and the semicircle, and then mark a small arc across the semicircle. Finally, draw a line segment from the marked point on the original segment to the point where the arc crosses the semicircle. This will form a 60° angle with the original segment."

Find all unknown angle measures in each pair of similar triangles. ***See Example 3.***

47.

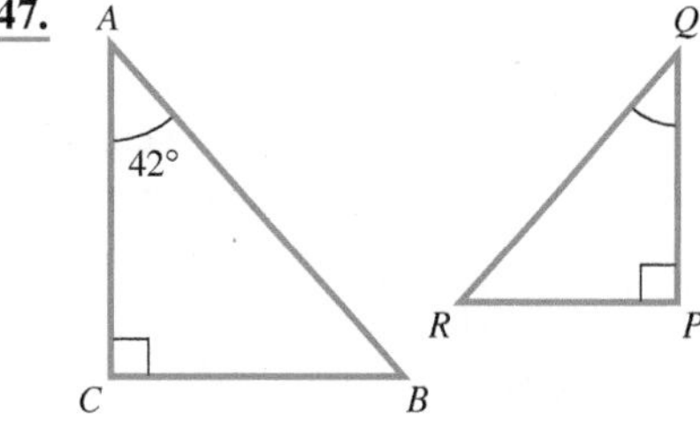

48.

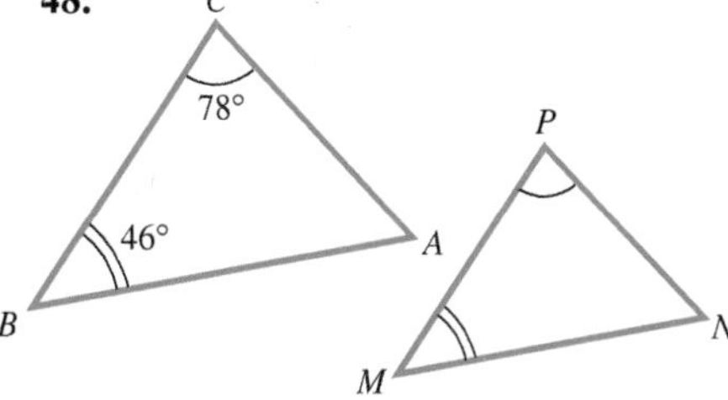

49.

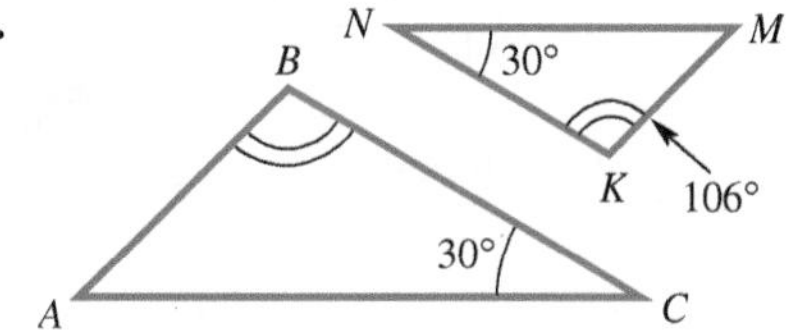

50.

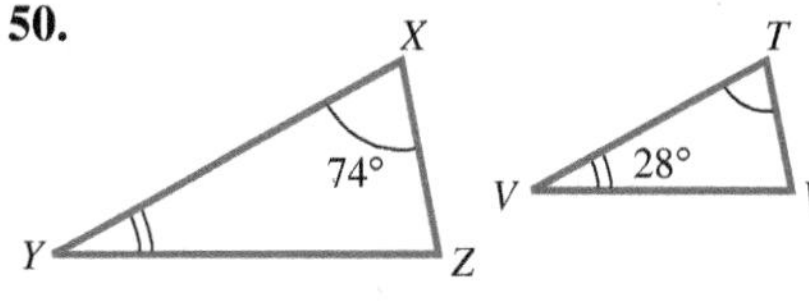

51. $X = M = 52°$
52. $T = 20°$; $V = 64°$; $R = U = 96°$
53. $a = 20$; $b = 15$
54. $a = 30$; $b = 60$
55. $a = 6$; $b = 7.5$
56. $a = 2$
57. $x = 6$
58. $m = 18$
59. 30 m
60. 108 ft
61. 500 m; 700 m
62. 14 m

51.

52.

Find the unknown side lengths in each pair of similar triangles. ***See Example 4.***

53.

54.

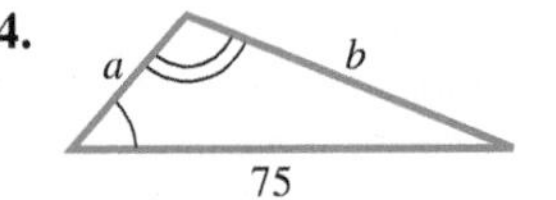

55.

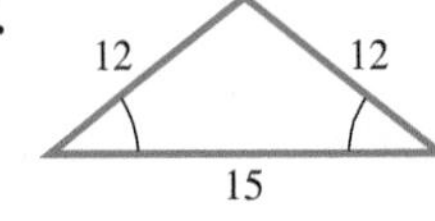

56.

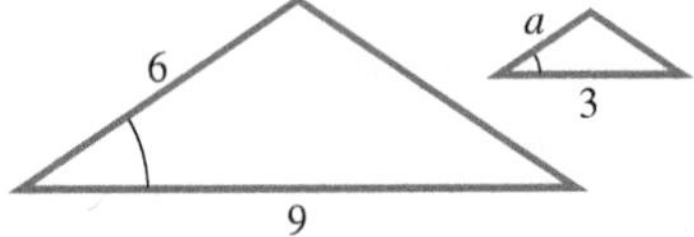

57.

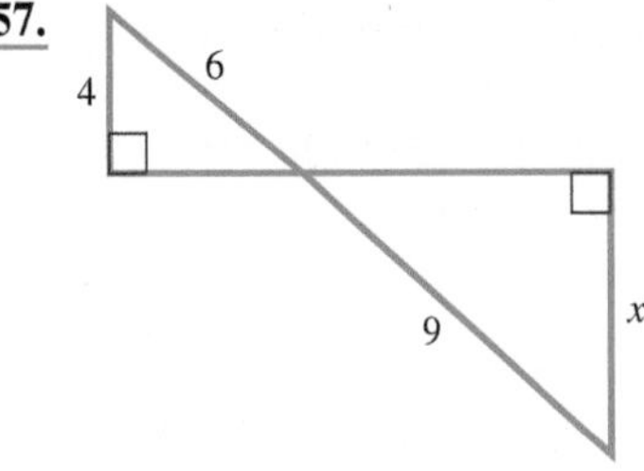

58.

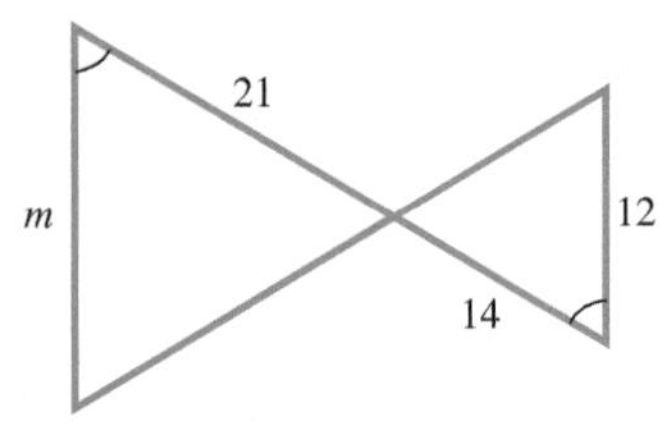

Solve each problem. ***See Example 5.***

59. ***Height of a Tree*** A tree casts a shadow 45 m long. At the same time, the shadow cast by a vertical 2-m stick is 3 m long. Find the height of the tree.

60. ***Height of a Lookout Tower*** A forest fire lookout tower casts a shadow 180 ft long at the same time that the shadow of a 9-ft truck is 15 ft long. Find the height of the tower.

61. ***Lengths of Sides of a Triangle*** On a photograph of a triangular piece of land, the lengths of the three sides are 4 cm, 5 cm, and 7 cm, respectively. The shortest side of the actual piece of land is 400 m long. Find the lengths of the other two sides.

62. ***Height of a Lighthouse*** The Biloxi lighthouse in the figure casts a shadow 28 m long at 7 A.M. At the same time, the shadow of the lighthouse keeper, who is 1.75 m tall, is 3.5 m long. How tall is the lighthouse?

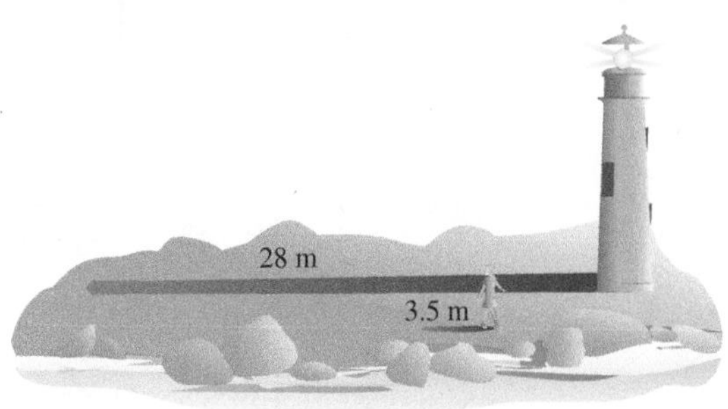

NOT TO SCALE

63. 112.5 ft 64. $506\frac{2}{3}$ ft

65. $x = 110$ 66. $y = 12$

67. $c \approx 111.1$ 68. $m \approx 85.3$

69. (a) 236,000 mi
(b) no

70. (a) 5,438,000 mi
(b) yes

63. *Height of a Building* A house is 15 ft tall. Its shadow is 40 ft long at the same time that the shadow of a nearby building is 300 ft long. Find the height of the building.

64. *Height of a Carving of Lincoln* Assume that Lincoln was $6\frac{1}{3}$ ft tall and his head $\frac{3}{4}$ ft long. Knowing that the carved head of Lincoln at Mt. Rushmore is 60 ft tall, find how tall his entire body would be if it were carved into the mountain.

In each figure, there are two similar triangles. Find the unknown measurement. Give approximations to the nearest tenth.

65.

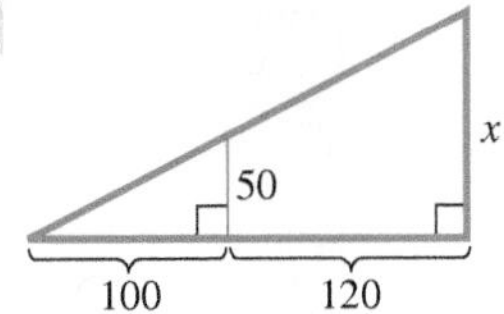

66.

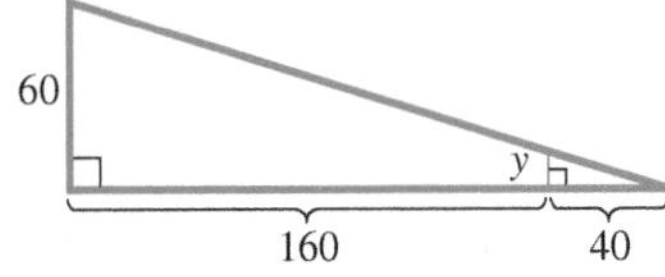

67.

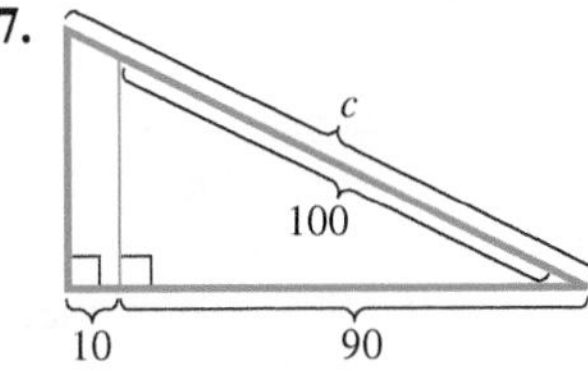

68.

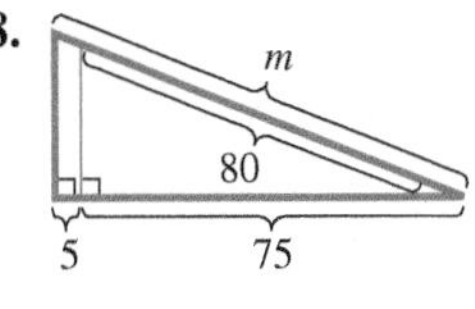

Solve each problem.

69. *Solar Eclipse on Earth* The sun has a diameter of about 865,000 mi with a maximum distance from Earth's surface of about 94,500,000 mi. The moon has a smaller diameter of 2159 mi. For a total solar eclipse to occur, the moon must pass between Earth and the sun. The moon must also be close enough to Earth for the moon's **umbra** (shadow) to reach the surface of Earth. (*Source:* Karttunen, H., P. Kröger, H. Oja, M. Putannen, and K. Donners, Editors, *Fundamental Astronomy,* Fourth Edition, Springer-Verlag.)

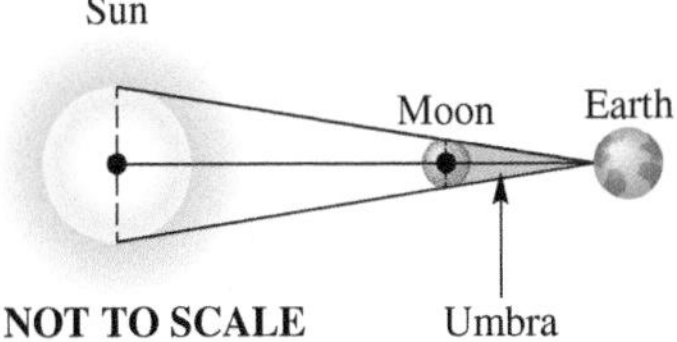

(a) Calculate the maximum distance, to the nearest thousand miles, that the moon can be from Earth and still have a total solar eclipse occur. (*Hint:* Use similar triangles.)

(b) The closest approach of the moon to Earth's surface was 225,745 mi and the farthest was 251,978 mi. (*Source: World Almanac and Book of Facts.*) Can a total solar eclipse occur every time the moon is between Earth and the sun?

70. *Solar Eclipse on Neptune* (**Refer to Exercise 69.**) The sun's distance from Neptune is approximately 2,800,000,000 mi (2.8 billion mi). The largest moon of Neptune is Triton, with a diameter of approximately 1680 mi. (*Source: World Almanac and Book of Facts.*)

(a) Calculate the maximum distance, to the nearest thousand miles, that Triton can be from Neptune for a total eclipse of the sun to occur on Neptune. (*Hint:* Use similar triangles.)

(b) Triton is approximately 220,000 mi from Neptune. Is it possible for Triton to cause a total eclipse on Neptune?

71. (a) 2900 mi
(b) no
72. (a) 1,830,000 mi
(b) yes
73. (a) $\frac{1}{4}$
(b) 30 arc degrees

71. *Solar Eclipse on Mars* **(Refer to Exercise 69.)** The sun's distance from the surface of Mars is approximately 142,000,000 mi. One of Mars' two moons, Phobos, has a maximum diameter of 17.4 mi. (*Source: World Almanac and Book of Facts.*)

(a) Calculate the maximum distance, to the nearest hundred miles, that the moon Phobos can be from Mars for a total eclipse of the sun to occur on Mars.

(b) Phobos is approximately 5800 mi from Mars. Is it possible for Phobos to cause a total eclipse on Mars?

72. *Solar Eclipse on Jupiter* **(Refer to Exercise 69.)** The sun's distance from the surface of Jupiter is approximately 484,000,000 mi. One of Jupiter's moons, Ganymede, has a diameter of 3270 mi. (*Source: World Almanac and Book of Facts.*)

(a) Calculate the maximum distance, to the nearest thousand miles, that the moon Ganymede can be from Jupiter for a total eclipse of the sun to occur on Jupiter.

(b) Ganymede is approximately 665,000 mi from Jupiter. Is it possible for Ganymede to cause a total eclipse on Jupiter?

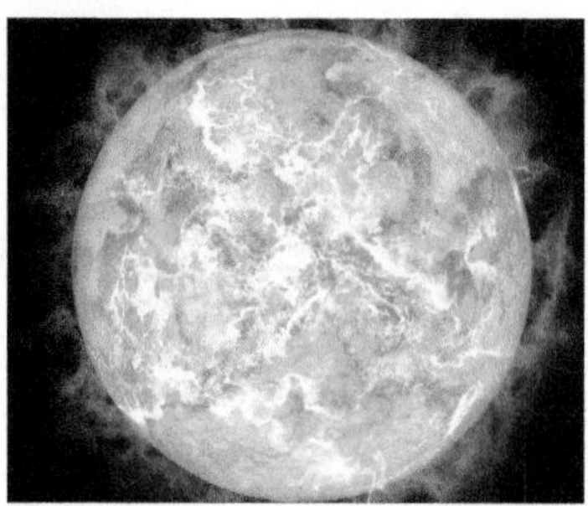

73. *Sizes and Distances in the Sky* Astronomers use degrees, minutes, and seconds to measure sizes and distances in the sky along an arc from the horizon to the zenith point directly overhead. An adult observer on Earth can judge distances in the sky using his or her hand at arm's length. An outstretched hand will be about 20 arc degrees wide from the tip of the thumb to the tip of the little finger. A clenched fist at arm's length measures about 10 arc degrees, and a thumb corresponds to about 2 arc degrees. (*Source:* Levy, D. H., *Skywatching,* The Nature Company.)

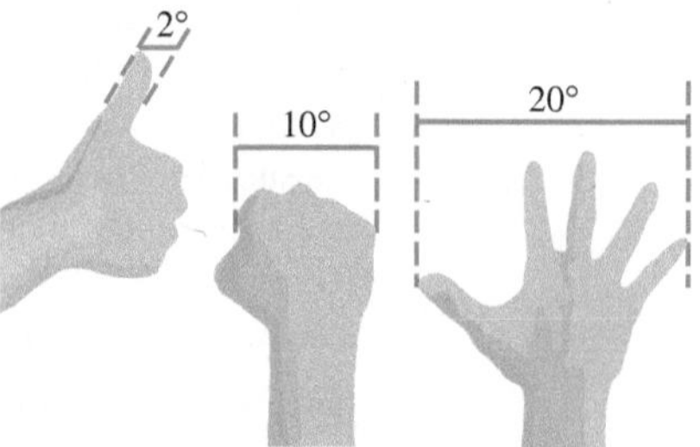

(a) The apparent size of the moon is about 31 arc minutes. Approximately what part of your thumb would cover the moon?

(b) If an outstretched hand plus a fist cover the distance between two bright stars, about how far apart in arc degrees are the stars?

74. *Estimates of Heights* There is a relatively simple way to make a reasonable estimate of a vertical height.

Step 1 Hold a 1-ft ruler vertically at arm's length and approach the object to be measured.

Step 2 Stop when one end of the ruler lines up with the top of the object and the other end with its base.

Step 3 Now pace off the distance to the object, taking normal strides. The number of paces will be the approximate height of the object in feet.

74. **(a)** The ratios of corresponding sides of similar triangles CAG and HAD are equal and $HD = 1$ ft.
(b) The ratios of corresponding sides of similar triangles AGE and ADB are equal.
(c) $BD = EF = 1$ pace
(d) From parts (a)–(c),

$$CG = \frac{AG}{AD} = \frac{EG}{BD} = \frac{EG}{1} = EG.$$

(e) The height of the tree in feet is (approximately) the number of paces.

Furnish the reasons in parts (a)–(d), which refer to the figure. (Assume that the length of one pace is EF.) Then answer the question in part (e).

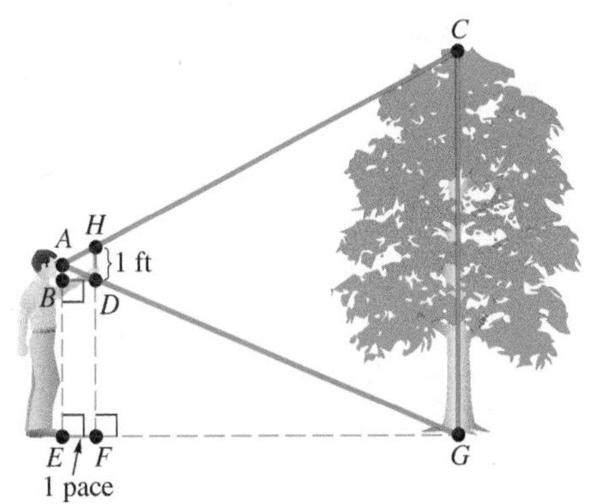

Reasons

(a) $CG = \dfrac{CG}{1} = \dfrac{AG}{AD}$ ____________________

(b) $\dfrac{AG}{AD} = \dfrac{EG}{BD}$ ____________________

(c) $\dfrac{EG}{EF} = \dfrac{EG}{BD} = \dfrac{EG}{1}$ ____________________

(d) CG ft $= EG$ paces ____________________

(e) What is the height of the tree in feet?

Chapter 1 Quiz (Sections 1.1–1.2)

[1.1]
1. **(a)** 71° **(b)** 161°
2. 65°; 115°
3. 26°; 64°

[1.2]
4. 20°; 24°; 136°
5. 130°; 50°

[1.1]
6. **(a)** 77.2025° **(b)** 22° 01′ 30″
7. **(a)** 50° **(b)** 300°
(c) 170° **(d)** 417°

1. Find the measure of **(a)** the complement and **(b)** the supplement of an angle measuring 19°.

Find the measure of each unknown angle.

2.

$(3x + 5)°$ $(5x + 15)°$

3.

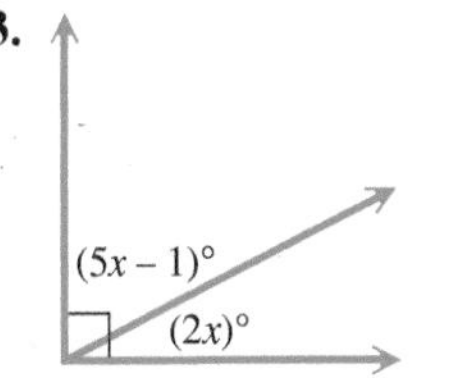

4.

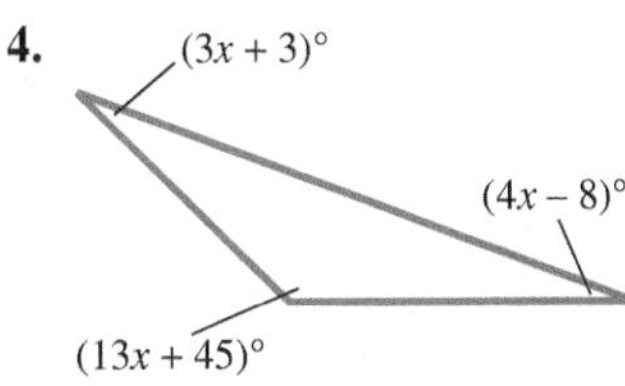

5.

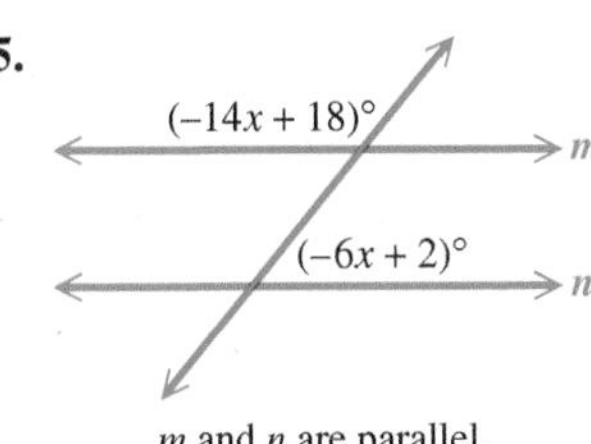

Solve each problem.

6. Perform each conversion.

(a) 77° 12′ 09″ to decimal degrees **(b)** 22.0250° to degrees, minutes, seconds

7. Find the angle of least positive measure (not equal to the given measure) that is coterminal with each angle.

(a) 410° **(b)** −60° **(c)** 890° **(d)** 57°

8. 1800°

[1.2]
9. 10 ft
10. (a) $x = 12$; $y = 10$
(b) $x = 5$

8. *Rotating Flywheel* A flywheel rotates 300 times per min. Through how many degrees does a point on the edge of the flywheel move in 1 sec?

9. *Length of a Shadow* If a vertical antenna 45 ft tall casts a shadow 15 ft long, how long would the shadow of a 30-ft pole be at the same time and place?

10. Find the values of x and y.

(a)

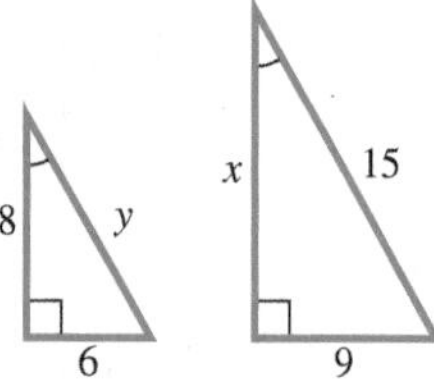

(b)

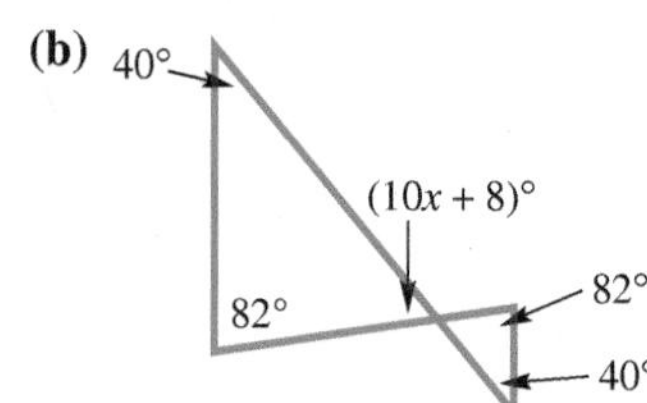

1.3 Trigonometric Functions

- The Pythagorean Theorem and the Distance Formula
- Trigonometric Functions
- Quadrantal Angles

The Pythagorean Theorem and the Distance Formula The distance between any two points in a plane can be found by using a formula derived from the **Pythagorean theorem.**

Pythagorean Theorem

In a right triangle, the sum of the squares of the lengths of the legs is equal to the square of the length of the hypotenuse.

$$a^2 + b^2 = c^2$$

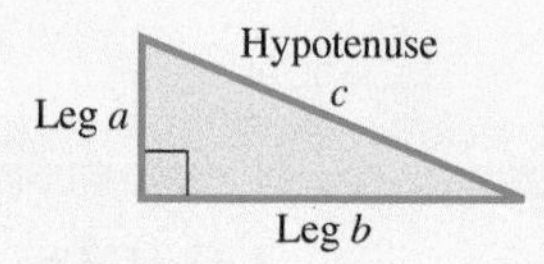

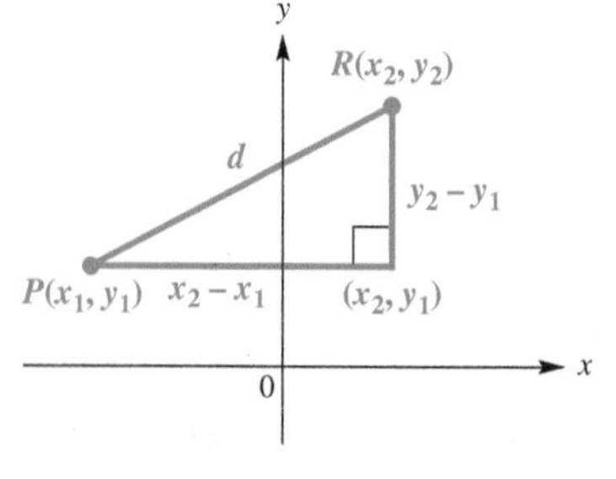

Figure 23

To find the distance between two points $P(x_1, y_1)$ and $R(x_2, y_2)$, draw a line segment connecting the points, as shown in **Figure 23.** Complete a right triangle by drawing a line through (x_1, y_1) parallel to the x-axis and a line through (x_2, y_2) parallel to the y-axis. The ordered pair at the right angle is (x_2, y_1).

The horizontal side of the right triangle in **Figure 23** has length $x_2 - x_1$, while the vertical side has length $y_2 - y_1$. If d represents the distance between the two original points, then by the Pythagorean theorem,

$$d^2 = (x_2 - x_1)^2 + (y_2 - y_1)^2.$$

Solving for d, we obtain the **distance formula.**

Distance Formula

Suppose that $P(x_1, y_1)$ and $R(x_2, y_2)$ are two points in a coordinate plane. The distance between P and R, written $d(P, R)$, is given by the following formula.

$$d(P, R) = \sqrt{(x_2 - x_1)^2 + (y_2 - y_1)^2}$$

That is, the distance between two points in a coordinate plane is the square root of the sum of the square of the difference between their x-coordinates and the square of the difference between their y-coordinates.

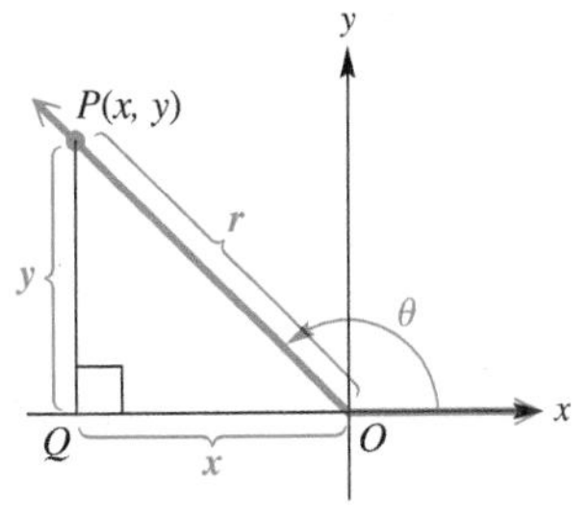

Figure 24

Trigonometric Functions To define the six **trigonometric functions,** we start with an angle θ in standard position and choose any point P having coordinates (x, y) on the terminal side of angle θ. (The point P must not be the vertex of the angle.) See **Figure 24.** A perpendicular from P to the x-axis at point Q determines a right triangle, having vertices at O, P, and Q. We find the distance r from $P(x, y)$ to the origin, $(0, 0)$, using the distance formula.

$$d(O, P) = \sqrt{(x_2 - x_1)^2 + (y_2 - y_1)^2} \quad \text{Distance formula}$$

$$r = \sqrt{(x - 0)^2 + (y - 0)^2} \quad \text{Substitute } (x, y) \text{ for } (x_2, y_2) \text{ and } (0, 0) \text{ for } (x_1, y_1).$$

$$r = \sqrt{x^2 + y^2} \quad \text{Subtract.}$$

Notice that $r > 0$ because this is the undirected distance.

The six trigonometric functions of angle θ are

sine, cosine, tangent, cotangent, secant, and **cosecant,**

abbreviated **sin, cos, tan, cot, sec,** and **csc.**

Trigonometric Functions

Let (x, y) be a point other than the origin on the terminal side of an angle θ in standard position. The distance from the point to the origin is $r = \sqrt{x^2 + y^2}$. The six trigonometric functions of θ are defined as follows.

$$\sin \theta = \frac{y}{r} \qquad \cos \theta = \frac{x}{r} \qquad \tan \theta = \frac{y}{x} \quad (x \neq 0)$$

$$\csc \theta = \frac{r}{y} \quad (y \neq 0) \qquad \sec \theta = \frac{r}{x} \quad (x \neq 0) \qquad \cot \theta = \frac{x}{y} \quad (y \neq 0)$$

Classroom Example 1
The terminal side of an angle θ in standard position passes through the point $(12, 5)$. Find the values of the six trigonometric functions of angle θ.

Answer:
$\sin \theta = \frac{5}{13}$; $\cos \theta = \frac{12}{13}$;
$\tan \theta = \frac{5}{12}$; $\cot \theta = \frac{12}{5}$;
$\sec \theta = \frac{13}{12}$; $\csc \theta = \frac{13}{5}$

EXAMPLE 1 Finding Function Values of an Angle

The terminal side of an angle θ in standard position passes through the point $(8, 15)$. Find the values of the six trigonometric functions of angle θ.

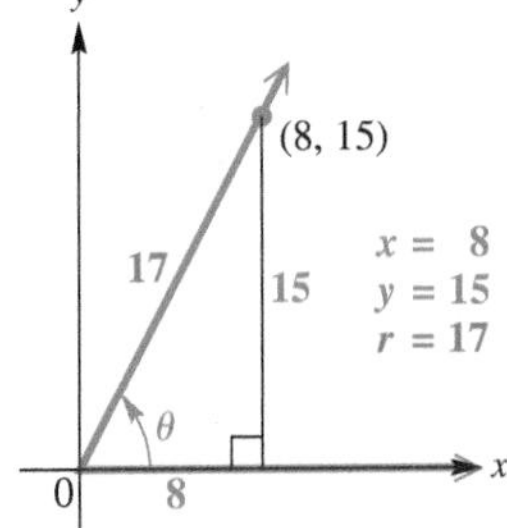

Figure 25

SOLUTION **Figure 25** shows angle θ and the triangle formed by dropping a perpendicular from the point $(8, 15)$ to the x-axis. The point $(8, 15)$ is 8 units to the right of the y-axis and 15 units above the x-axis, so $x = 8$ and $y = 15$. Now use $r = \sqrt{x^2 + y^2}$.

$$r = \sqrt{8^2 + 15^2} = \sqrt{64 + 225} = \sqrt{289} = 17$$

We can now use these values for x, y, and r to find the values of the six trigonometric functions of angle θ.

$$\sin \theta = \frac{y}{r} = \frac{15}{17} \qquad \cos \theta = \frac{x}{r} = \frac{8}{17} \qquad \tan \theta = \frac{y}{x} = \frac{15}{8}$$

$$\csc \theta = \frac{r}{y} = \frac{17}{15} \qquad \sec \theta = \frac{r}{x} = \frac{17}{8} \qquad \cot \theta = \frac{x}{y} = \frac{8}{15}$$

✔ **Now Try Exercise 13.**

Classroom Example 2
The terminal side of an angle θ in standard position passes through the point $(8, -6)$. Find the values of the six trigonometric functions of angle θ.

Answer:
$\sin\theta = -\frac{3}{5}$; $\cos\theta = \frac{4}{5}$;
$\tan\theta = -\frac{3}{4}$; $\cot\theta = -\frac{4}{3}$;
$\sec\theta = \frac{5}{4}$; $\csc\theta = -\frac{5}{3}$

EXAMPLE 2 Finding Function Values of an Angle

The terminal side of an angle θ in standard position passes through the point $(-3, -4)$. Find the values of the six trigonometric functions of angle θ.

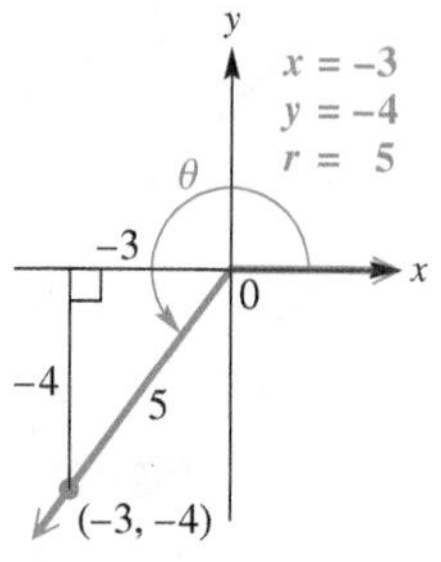

Figure 26

SOLUTION As shown in **Figure 26,** $x = -3$ and $y = -4$.

$$r = \sqrt{(-3)^2 + (-4)^2} \quad r = \sqrt{x^2 + y^2}$$

$$r = \sqrt{25} \quad \text{Simplify the radicand.}$$

$$r = 5 \quad r > 0$$

Now we use the definitions of the trigonometric functions.

$$\sin\theta = \frac{-4}{5} = -\frac{4}{5} \qquad \cos\theta = \frac{-3}{5} = -\frac{3}{5} \qquad \tan\theta = \frac{-4}{-3} = \frac{4}{3}$$

$$\csc\theta = \frac{5}{-4} = -\frac{5}{4} \qquad \sec\theta = \frac{5}{-3} = -\frac{5}{3} \qquad \cot\theta = \frac{-3}{-4} = \frac{3}{4}$$

✔ Now Try Exercise 17.

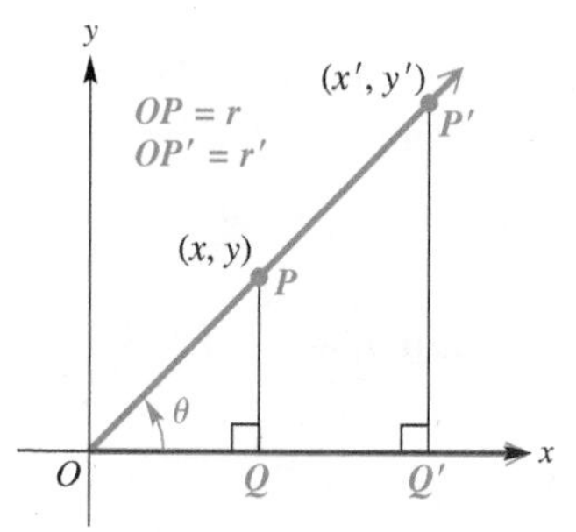

Figure 27

We can find the six trigonometric functions using *any* point other than the origin on the terminal side of an angle. To see why any point can be used, refer to **Figure 27,** which shows an angle θ and two distinct points on its terminal side. Point P has coordinates (x, y), and point P' (read **"P-prime"**) has coordinates (x', y'). Let r be the length of the hypotenuse of triangle OPQ, and let r' be the length of the hypotenuse of triangle $OP'Q'$. Because corresponding sides of similar triangles are proportional, we have

$$\frac{y}{r} = \frac{y'}{r'}. \quad \text{Corresponding sides are proportional.}$$

Thus $\sin\theta = \frac{y}{r}$ is the same no matter which point is used to find it. A similar result holds for the other five trigonometric functions.

We can also find the trigonometric function values of an angle if we know the equation of the line coinciding with the terminal ray. Recall from algebra that the graph of the equation

$$Ax + By = 0 \quad \text{Linear equation in two variables}$$

is a line that passes through the origin $(0, 0)$. If we restrict x to have only nonpositive or only nonnegative values, we obtain as the graph a ray with endpoint at the origin. For example, the graph of $x + 2y = 0$, $x \geq 0$, shown in **Figure 28,** is a ray that can serve as the terminal side of an angle θ in standard position. By choosing a point on the ray, we can find the trigonometric function values of the angle.

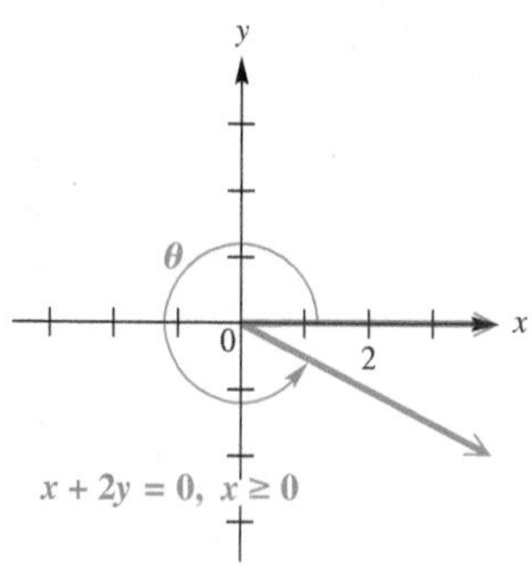

Figure 28

EXAMPLE 3 Finding Function Values of an Angle

Find the six trigonometric function values of an angle θ in standard position, if the terminal side of θ is defined by $x + 2y = 0$, $x \geq 0$.

SOLUTION The angle is shown in **Figure 29.** We can use *any* point except $(0, 0)$ on the terminal side of θ to find the trigonometric function values. We choose $x = 2$ and find the corresponding y-value.

$$x + 2y = 0, \quad x \geq 0$$

$$2 + 2y = 0 \qquad \text{Let } x = 2.$$

$$2y = -2 \qquad \text{Subtract 2.}$$

$$y = -1 \qquad \text{Divide by 2.}$$

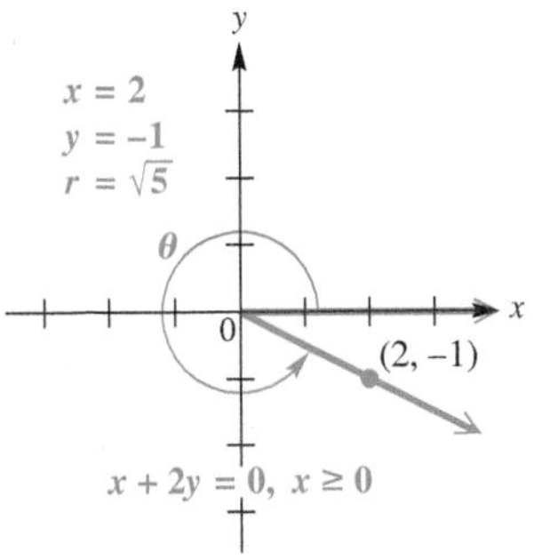

Figure 29

The point $(2, -1)$ lies on the terminal side, and so $r = \sqrt{2^2 + (-1)^2} = \sqrt{5}$. Now we use the definitions of the trigonometric functions.

$$\sin\theta = \frac{y}{r} = \frac{-1}{\sqrt{5}} = \frac{-1}{\sqrt{5}} \cdot \frac{\sqrt{5}}{\sqrt{5}} = -\frac{\sqrt{5}}{5}$$

$$\cos\theta = \frac{x}{r} = \frac{2}{\sqrt{5}} = \frac{2}{\sqrt{5}} \cdot \frac{\sqrt{5}}{\sqrt{5}} = \frac{2\sqrt{5}}{5}$$

Multiply by $\frac{\sqrt{5}}{\sqrt{5}}$, a form of 1, to rationalize the denominators.

$$\tan\theta = \frac{y}{x} = \frac{-1}{2} = -\frac{1}{2}$$

$$\csc\theta = \frac{r}{y} = \frac{\sqrt{5}}{-1} = -\sqrt{5} \qquad \sec\theta = \frac{r}{x} = \frac{\sqrt{5}}{2} \qquad \cot\theta = \frac{x}{y} = \frac{2}{-1} = -2$$

✔ **Now Try Exercise 51.**

Classroom Example 3
Find the six trigonometric function values of an angle θ in standard position, if the terminal side of θ is defined by $3x - 2y = 0$, $x \leq 0$.

Answer:

$\sin\theta = -\frac{3\sqrt{13}}{13}$;

$\cos\theta = -\frac{2\sqrt{13}}{13}$;

$\tan\theta = \frac{3}{2}$; $\cot\theta = \frac{2}{3}$;

$\sec\theta = -\frac{\sqrt{13}}{2}$; $\csc\theta = -\frac{\sqrt{13}}{3}$

Recall that when the equation of a line is written in the form

$$y = mx + b, \qquad \text{Slope-intercept form}$$

the coefficient m of x gives the slope of the line. In **Example 3,** the equation $x + 2y = 0$ can be written as $y = -\frac{1}{2}x$, so the slope of this line is $-\frac{1}{2}$. Notice that $\tan\theta = -\frac{1}{2}$.

In general, it is true that m = tan θ.

NOTE The trigonometric function values we found in **Examples 1–3** are *exact.* If we were to use a calculator to approximate these values, the decimal results would not be acceptable if exact values were required.

Quadrantal Angles If the terminal side of an angle in standard position lies along the y-axis, any point on this terminal side has x-coordinate 0. Similarly, an angle with terminal side on the x-axis has y-coordinate 0 for any point on the terminal side. Because the values of x and y appear in the denominators of some trigonometric functions, and because a fraction is undefined if its denominator is 0, some trigonometric function values of quadrantal angles (i.e., those with terminal side on an axis) are undefined.

When determining trigonometric function values of quadrantal angles, **Figure 30** can help find the ratios. Because *any* point on the terminal side can be used, it is convenient to choose the point one unit from the origin, with $r = 1$. (Later we will extend this idea to the *unit circle.*)

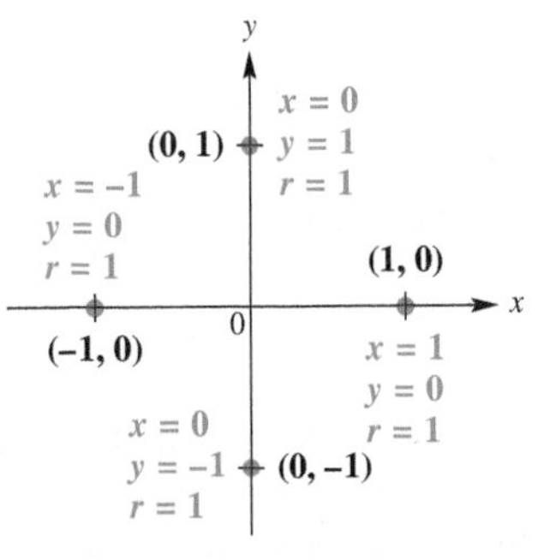

Figure 30

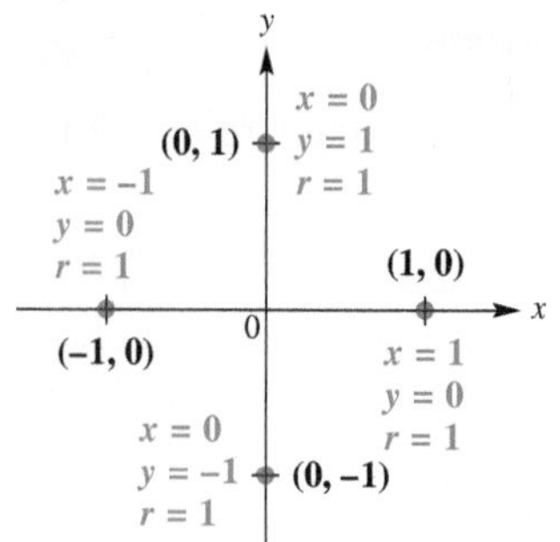

Figure 30 (repeated)

```
NORMAL FLOAT AUTO REAL DEGREE MP
sin(90)
                                1
cos(90)
                                0
tan(90)
                            Error
```

A calculator in degree mode returns the correct values for sin 90° and cos 90°. The screen shows an ERROR message for tan 90°, because 90° is not in the domain of the tangent function.

Classroom Example 4
Find the values of the six trigonometric functions for each angle.

(a) an angle of 360°

(b) an angle θ in standard position with terminal side passing through $(0, -5)$

Answers:

(a) $\sin\theta = 0$; $\cos\theta = 1$; $\tan\theta = 0$; $\cot\theta$: undefined; $\sec\theta = 1$; $\csc\theta$: undefined

(b) $\sin\theta = -1$; $\cos\theta = 0$; $\tan\theta$: undefined; $\cot\theta = 0$; $\sec\theta$: undefined; $\csc\theta = -1$

To find the function values of a quadrantal angle, determine the position of the terminal side, choose the one of these four points that lies on this terminal side, and then use the definitions involving x, y, and r.

EXAMPLE 4 Finding Function Values of Quadrantal Angles

Find the values of the six trigonometric functions for each angle.

(a) an angle of 90°

(b) an angle θ in standard position with terminal side passing through $(-3, 0)$

SOLUTION

(a) Figure 31 shows that the terminal side passes through $(0, 1)$. So $x = 0$, $y = 1$, and $r = 1$. Thus, we have the following.

$$\sin 90° = \frac{1}{1} = 1 \qquad \cos 90° = \frac{0}{1} = 0 \qquad \tan 90° = \frac{1}{0} \quad \text{Undefined}$$

$$\csc 90° = \frac{1}{1} = 1 \qquad \sec 90° = \frac{1}{0} \quad \text{Undefined} \qquad \cot 90° = \frac{0}{1} = 0$$

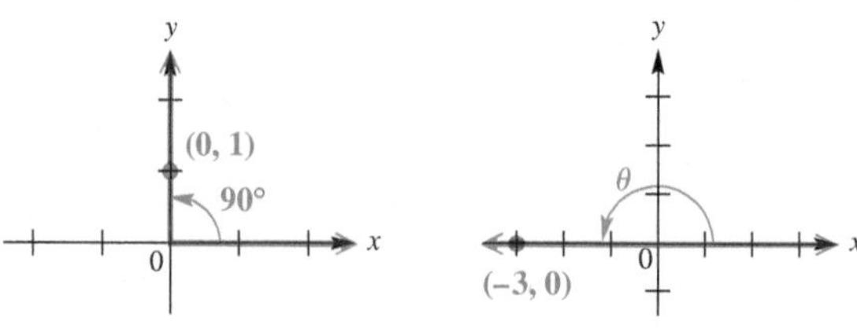

Figure 31 **Figure 32**

(b) Figure 32 shows the angle. Here, $x = -3$, $y = 0$, and $r = 3$, so the trigonometric functions have the following values.

$$\sin\theta = \frac{0}{3} = 0 \qquad \cos\theta = \frac{-3}{3} = -1 \qquad \tan\theta = \frac{0}{-3} = 0$$

$$\csc\theta = \frac{3}{0} \quad \text{Undefined} \qquad \sec\theta = \frac{3}{-3} = -1 \qquad \cot\theta = \frac{-3}{0} \quad \text{Undefined}$$

Verify that these values can also be found using the point $(-1, 0)$.

✔ **Now Try Exercises 23, 67, 69, and 71.**

The conditions under which the trigonometric function values of quadrantal angles are undefined are summarized here.

Conditions for Undefined Function Values

Identify the terminal side of a quadrantal angle.

- If the terminal side of the quadrantal angle lies along the y-axis, then the tangent and secant functions are undefined.
- If the terminal side of the quadrantal angle lies along the x-axis, then the cotangent and cosecant functions are undefined.

The function values of some commonly used quadrantal angles, 0°, 90°, 180°, 270°, and 360°, are summarized in the table on the next page. They can be determined when needed by using **Figure 30** and the method of **Example 4(a).**

For other quadrantal angles such as $-90°$, $-270°$, and $450°$, first determine the coterminal angle that lies between $0°$ and $360°$, and then refer to the table entries for that particular angle. For example, the function values of a $-90°$ angle would correspond to those of a $270°$ angle.

Function Values of Quadrantal Angles

θ	$\sin \theta$	$\cos \theta$	$\tan \theta$	$\cot \theta$	$\sec \theta$	$\csc \theta$
0°	0	1	0	Undefined	1	Undefined
90°	1	0	Undefined	0	Undefined	1
180°	0	−1	0	Undefined	−1	Undefined
270°	−1	0	Undefined	0	Undefined	−1
360°	0	1	0	Undefined	1	Undefined

The values given in this table can be found with a calculator that has trigonometric function keys. ***Make sure the calculator is set to degree mode.***

NORMAL FLOAT AUTO REAL DEGREE MP
MATHPRINT CLASSIC
NORMAL SCI ENG
FLOAT 0 1 2 3 4 5 6 7 8 9
RADIAN DEGREE
FUNCTION PARAMETRIC POLAR SEQ
THICK DOT-THICK THIN DOT-THIN
SEQUENTIAL SIMUL
REAL a+bi re^(θi)
FULL HORIZONTAL GRAPH-TABLE

TI-84 Plus

Figure 33

CAUTION ***One of the most common errors involving calculators in trigonometry occurs when the calculator is set for radian measure, rather than degree measure.*** Be sure to set your calculator to degree mode. See **Figure 33.**

1.3 Exercises

1. hypotenuse 2. origin
3. same 4. 0
5. positive; negative
6. $\cot \theta$; $\csc \theta$

7. $3\sqrt{2}$ 8. $-\frac{\sqrt{2}}{2}$
9. $-\frac{\sqrt{2}}{2}$ 10. 1

In Exercises 11–30 and 51–62, we give, in order, sine, cosine, tangent, cotangent, secant, and cosecant.

11.

12.

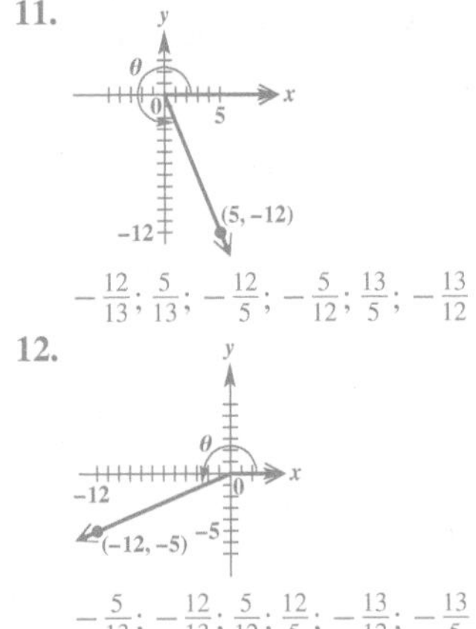

11. $-\frac{12}{13}$; $\frac{5}{13}$; $-\frac{12}{5}$; $-\frac{5}{12}$; $\frac{13}{5}$; $-\frac{13}{12}$

12. $-\frac{5}{13}$; $-\frac{12}{13}$; $\frac{5}{12}$; $\frac{12}{5}$; $-\frac{13}{12}$; $-\frac{13}{5}$

Answer graphs are not included for Exercises 13–30. They are similar to the ones in Exercises 11 and 12.

CONCEPT PREVIEW *Fill in the blank(s) to correctly complete each sentence.*

1. The Pythagorean theorem for right triangles states that the sum of the squares of the lengths of the legs is equal to the square of the ________.

2. In the definitions of the sine, cosine, secant, and cosecant functions, r is interpreted geometrically as the distance from a given point (x, y) on the terminal side of an angle θ in standard position to the ________.

3. For any nonquadrantal angle θ, $\sin \theta$ and $\csc \theta$ will have the ________ sign.
(same/opposite)

4. If $\cot \theta$ is undefined, then $\tan \theta =$ ________.

5. If the terminal side of an angle θ lies in quadrant III, then the values of $\tan \theta$ and $\cot \theta$ are ________, and all other trigonometric function values are
(positive/negative)
________.
(positive/negative)

6. If a quadrantal angle θ is coterminal with $0°$ or $180°$, then the trigonometric functions ________ and ________ are undefined.

CONCEPT PREVIEW *The terminal side of an angle θ in standard position passes through the point $(-3, -3)$. Use the figure to find the following values. Rationalize denominators when applicable.*

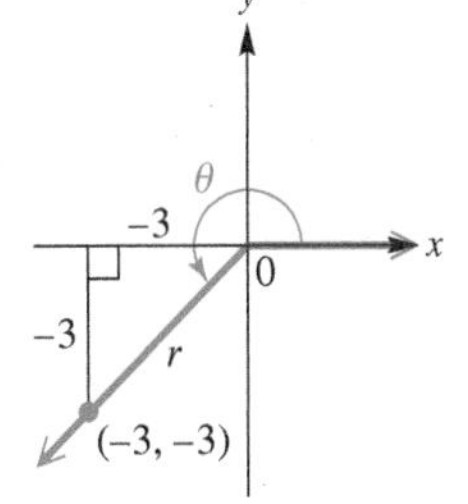

7. r **8.** $\sin \theta$

9. $\cos \theta$ **10.** $\tan \theta$

13. $\frac{4}{5}$; $\frac{3}{5}$; $\frac{4}{3}$; $\frac{3}{4}$; $\frac{5}{3}$; $\frac{5}{4}$

14. $-\frac{3}{5}$; $-\frac{4}{5}$; $\frac{3}{4}$; $\frac{4}{3}$; $-\frac{5}{4}$; $-\frac{5}{3}$

15. $\frac{15}{17}$; $-\frac{8}{17}$; $-\frac{15}{8}$; $-\frac{8}{15}$; $-\frac{17}{8}$; $\frac{17}{15}$

16. $-\frac{8}{17}$; $\frac{15}{17}$; $-\frac{8}{15}$; $-\frac{15}{8}$; $\frac{17}{15}$; $-\frac{17}{8}$

17. $-\frac{24}{25}$; $-\frac{7}{25}$; $\frac{24}{7}$; $\frac{7}{24}$; $-\frac{25}{7}$; $-\frac{25}{24}$

18. $-\frac{7}{25}$; $-\frac{24}{25}$; $\frac{7}{24}$; $\frac{24}{7}$; $-\frac{25}{24}$; $-\frac{25}{7}$

19. 1; 0; undefined; 0; undefined; 1

20. 1; 0; undefined; 0; undefined; 1

21. 0; -1; 0; undefined; -1; undefined

22. 0; -1; 0; undefined; -1; undefined

23. -1; 0; undefined; 0; undefined; -1

24. -1; 0; undefined; 0; undefined; -1

25. $\frac{\sqrt{3}}{2}$; $\frac{1}{2}$; $\sqrt{3}$; $\frac{\sqrt{3}}{3}$; 2; $\frac{2\sqrt{3}}{3}$

26. $\frac{\sqrt{3}}{2}$; $-\frac{1}{2}$; $-\sqrt{3}$; $-\frac{\sqrt{3}}{3}$; -2; $\frac{2\sqrt{3}}{3}$

27. $\frac{\sqrt{2}}{2}$; $\frac{\sqrt{2}}{2}$; 1; 1; $\sqrt{2}$; $\sqrt{2}$

28. $-\frac{\sqrt{2}}{2}$; $-\frac{\sqrt{2}}{2}$; 1; 1; $-\sqrt{2}$; $-\sqrt{2}$

29. $-\frac{1}{2}$; $-\frac{\sqrt{3}}{2}$; $\frac{\sqrt{3}}{3}$; $\sqrt{3}$; $-\frac{2\sqrt{3}}{3}$; -2

30. $\frac{1}{2}$; $-\frac{\sqrt{3}}{2}$; $-\frac{\sqrt{3}}{3}$; $-\sqrt{3}$; $-\frac{2\sqrt{3}}{3}$; 2

31. negative 32. negative
33. negative 34. negative
35. positive 36. negative
37. positive 38. negative
39. negative 40. negative
41. positive 42. positive
43. negative 44. negative
45. positive 46. positive
47. positive 48. positive
49. positive 50. positive

Answer graphs are not included for Exercises 51–62.

51. $-\frac{2\sqrt{5}}{5}$; $\frac{\sqrt{5}}{5}$; -2; $-\frac{1}{2}$; $\sqrt{5}$; $-\frac{\sqrt{5}}{2}$

52. $-\frac{3\sqrt{34}}{34}$; $\frac{5\sqrt{34}}{34}$; $-\frac{3}{5}$; $-\frac{5}{3}$; $\frac{\sqrt{34}}{5}$; $-\frac{\sqrt{34}}{3}$

53. $\frac{6\sqrt{37}}{37}$; $-\frac{\sqrt{37}}{37}$; -6; $-\frac{1}{6}$; $-\sqrt{37}$; $\frac{\sqrt{37}}{6}$

Sketch an angle θ in standard position such that θ has the least positive measure, and the given point is on the terminal side of θ. Then find the values of the six trigonometric functions for each angle. Rationalize denominators when applicable. ***See Examples 1, 2, and 4.***

11. $(5, -12)$ **12.** $(-12, -5)$ **13.** $(3, 4)$ **14.** $(-4, -3)$

15. $(-8, 15)$ **16.** $(15, -8)$ **17.** $(-7, -24)$ **18.** $(-24, -7)$

19. $(0, 2)$ **20.** $(0, 5)$ **21.** $(-4, 0)$ **22.** $(-5, 0)$

23. $(0, -4)$ **24.** $(0, -3)$ **25.** $(1, \sqrt{3})$ **26.** $(-1, \sqrt{3})$

27. $(\sqrt{2}, \sqrt{2})$ **28.** $(-\sqrt{2}, -\sqrt{2})$ **29.** $(-2\sqrt{3}, -2)$ **30.** $(-2\sqrt{3}, 2)$

Concept Check *Suppose that the point (x, y) is in the indicated quadrant. Determine whether the given ratio is positive or negative. Recall that $r = \sqrt{x^2 + y^2}$. (Hint: Drawing a sketch may help.)*

31. II, $\frac{x}{r}$ **32.** III, $\frac{y}{r}$ **33.** IV, $\frac{y}{x}$ **34.** IV, $\frac{x}{y}$ **35.** II, $\frac{y}{r}$

36. III, $\frac{x}{r}$ **37.** IV, $\frac{x}{r}$ **38.** IV, $\frac{y}{r}$ **39.** II, $\frac{x}{y}$ **40.** II, $\frac{y}{x}$

41. III, $\frac{y}{x}$ **42.** III, $\frac{x}{y}$ **43.** III, $\frac{r}{x}$ **44.** III, $\frac{r}{y}$ **45.** I, $\frac{x}{y}$

46. I, $\frac{y}{x}$ **47.** I, $\frac{y}{r}$ **48.** I, $\frac{x}{r}$ **49.** I, $\frac{r}{x}$ **50.** I, $\frac{r}{y}$

An equation of the terminal side of an angle θ in standard position is given with a restriction on x. Sketch the least positive such angle θ, and find the values of the six trigonometric functions of θ. ***See Example 3.***

51. $2x + y = 0, x \geq 0$ **52.** $3x + 5y = 0, x \geq 0$ **53.** $-6x - y = 0, x \leq 0$

54. $-5x - 3y = 0, x \leq 0$ **55.** $-4x + 7y = 0, x \leq 0$ **56.** $6x - 5y = 0, x \geq 0$

57. $x + y = 0, x \geq 0$ **58.** $x - y = 0, x \geq 0$ **59.** $-\sqrt{3}x + y = 0, x \leq 0$

60. $\sqrt{3}x + y = 0, x \leq 0$ **61.** $x = 0, y \geq 0$ **62.** $y = 0, x \leq 0$

Find the indicated function value. If it is undefined, say so. ***See Example 4.***

63. $\cos 90°$ **64.** $\sin 90°$ **65.** $\tan 180°$ **66.** $\cot 90°$

67. $\sec 180°$ **68.** $\csc 270°$ **69.** $\sin(-270°)$ **70.** $\cos(-90°)$

71. $\cot 540°$ **72.** $\tan 450°$ **73.** $\csc(-450°)$ **74.** $\sec(-540°)$

75. $\sin 1800°$ **76.** $\cos 1800°$ **77.** $\csc 1800°$

78. $\cot 1800°$ **79.** $\sec 1800°$ **80.** $\tan 1800°$

81. $\cos(-900°)$ **82.** $\sin(-900°)$ **83.** $\tan(-900°)$

84. How can the answer to **Exercise 83** be given once the answers to **Exercises 81 and 82** have been determined?

Use trigonometric function values of quadrantal angles to evaluate each expression. An expression such as $\cot^2 90°$ means $(\cot 90°)^2$, which is equal to $0^2 = 0$.

85. $\cos 90° + 3 \sin 270°$ **86.** $\tan 0° - 6 \sin 90°$

87. $3 \sec 180° - 5 \tan 360°$ **88.** $4 \csc 270° + 3 \cos 180°$

89. $\tan 360° + 4 \sin 180° + 5 \cos^2 180°$ **90.** $5 \sin^2 90° + 2 \cos^2 270° - \tan 360°$

91. $\sin^2 180° + \cos^2 180°$ **92.** $\sin^2 360° + \cos^2 360°$

54. $\frac{5\sqrt{34}}{34}$; $-\frac{3\sqrt{34}}{34}$; $-\frac{5}{3}$; $-\frac{3}{5}$; $-\frac{\sqrt{34}}{3}$; $\frac{\sqrt{34}}{5}$

55. $-\frac{4\sqrt{65}}{65}$; $-\frac{7\sqrt{65}}{65}$; $\frac{4}{7}$; $\frac{7}{4}$; $-\frac{\sqrt{65}}{7}$; $-\frac{\sqrt{65}}{4}$

56. $\frac{6\sqrt{61}}{61}$; $\frac{5\sqrt{61}}{61}$; $\frac{6}{5}$; $\frac{5}{6}$; $\frac{\sqrt{61}}{5}$; $\frac{\sqrt{61}}{6}$

57. $-\frac{\sqrt{2}}{2}$; $\frac{\sqrt{2}}{2}$; -1; -1; $\sqrt{2}$; $-\sqrt{2}$

58. $\frac{\sqrt{2}}{2}$; $\frac{\sqrt{2}}{2}$; 1; 1; $\sqrt{2}$; $\sqrt{2}$

59. $-\frac{\sqrt{3}}{2}$; $-\frac{1}{2}$; $\sqrt{3}$; $\frac{\sqrt{3}}{3}$; -2; $-\frac{2\sqrt{3}}{3}$

60. $\frac{\sqrt{3}}{2}$; $-\frac{1}{2}$; $-\sqrt{3}$; $-\frac{\sqrt{3}}{3}$; -2; $\frac{2\sqrt{3}}{3}$

61. 1; 0; undefined; 0; undefined; 1

62. 0; -1; 0; undefined; -1; undefined

63. 0	64. 1
65. 0	66. 0
67. -1	68. -1
69. 1	70. 0
71. undefined	72. undefined
73. -1	74. -1
75. 0	76. 1
77. undefined	78. undefined
79. 1	80. 0
81. -1	82. 0
83. 0	

84. Once $\sin(-900^\circ)$ and $\cos(-900^\circ)$ are found, $\tan(-900^\circ)$ can be found by division as follows.

$$\tan(-900^\circ) = \frac{\sin(-900^\circ)}{\cos(-900^\circ)} = \frac{0}{-1} = 0$$

85. -3	86. -6
87. -3	88. -7
89. 5	90. 5
91. 1	92. 1
93. 0	94. 3
95. 0	96. -7
97. 1	98. 1
99. 0	100. 0
101. 0	102. -1
103. undefined	104. undefined
105. 0	106. 1
107. undefined	108. undefined

109. They are equal.
110. They are equal.
111. They are negatives of each other.
112. They are equal.

113. 0.940; 0.342
114. 40° 115. 35° 116. 45°
117. decrease; increase
118. decrease; decrease

93. $\sec^2 180^\circ - 3\sin^2 360^\circ + \cos 180^\circ$ **94.** $2\sec 0^\circ + 4\cot^2 90^\circ + \cos 360^\circ$

95. $-2\sin^4 0^\circ + 3\tan^2 0^\circ$ **96.** $-3\sin^4 90^\circ + 4\cos^3 180^\circ$

97. $\sin^2(-90^\circ) + \cos^2(-90^\circ)$ **98.** $\cos^2(-180^\circ) + \sin^2(-180^\circ)$

If n is an integer, $n \cdot 180^\circ$ *represents an integer multiple of* 180°, $(2n+1) \cdot 90^\circ$ *represents an odd integer multiple of* 90°, *and so on. Determine whether each expression is equal to* 0, 1, *or* -1, *or is* undefined.

99. $\cos[(2n+1) \cdot 90^\circ]$ **100.** $\sin[n \cdot 180^\circ]$ **101.** $\tan[n \cdot 180^\circ]$

102. $\sin[270^\circ + n \cdot 360^\circ]$ **103.** $\tan[(2n+1) \cdot 90^\circ]$ **104.** $\cot[n \cdot 180^\circ]$

105. $\cot[(2n+1) \cdot 90^\circ]$ **106.** $\cos[n \cdot 360^\circ]$

107. $\sec[(2n+1) \cdot 90^\circ]$ **108.** $\csc[n \cdot 180^\circ]$

Concept Check In later chapters we will study trigonometric functions of angles other than quadrantal angles, such as 15°, 30°, 60°, 75°, *and so on. To prepare for some important concepts, provide conjectures in each exercise. Use a calculator set to degree mode.*

109. The angles 15° and 75° are complementary. Determine sin 15° and cos 75°. Make a conjecture about the sines and cosines of complementary angles, and test this hypothesis with other pairs of complementary angles.

110. The angles 25° and 65° are complementary. Determine tan 25° and cot 65°. Make a conjecture about the tangents and cotangents of complementary angles, and test this hypothesis with other pairs of complementary angles.

111. Determine sin 10° and $\sin(-10^\circ)$. Make a conjecture about the sine of an angle and the sine of its negative, and test this hypothesis with other angles.

112. Determine cos 20° and $\cos(-20^\circ)$. Make a conjecture about the cosine of an angle and the cosine of its negative, and test this hypothesis with other angles.

Set a TI graphing calculator to parametric and degree modes. Use the window values shown in the first screen, and enter the equations shown in the second screen. The corresponding graph in the third screen is a circle of radius 1. *Trace to move a short distance around the circle. In the third screen, the point on the circle corresponds to the angle* T = 25°. *Because* $r = 1$, cos 25° *is* X = 0.90630779 *and* sin 25° *is* Y = 0.42261826.

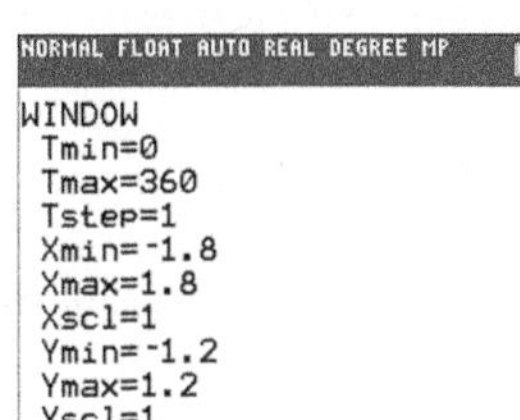

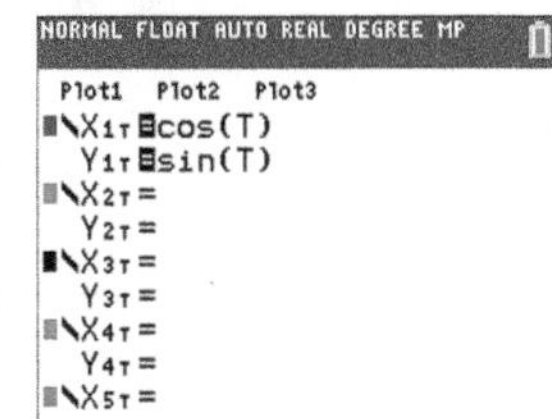

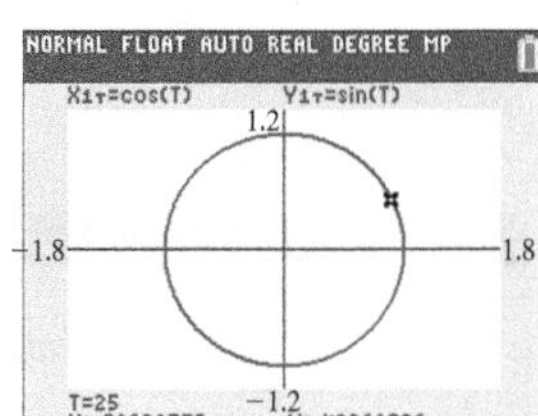

Use this information to answer each question.

113. Use the right- and left-arrow keys to move to the point corresponding to 20° (T = 20). Approximate cos 20° and sin 20° to the nearest thousandth.

114. For what angle T, $0^\circ \le T \le 90^\circ$, is $\cos T \approx 0.766$?

115. For what angle T, $0^\circ \le T \le 90^\circ$, is $\sin T \approx 0.574$?

116. For what angle T, $0^\circ \le T \le 90^\circ$ does cos T equal sin T?

117. As T increases from 0° to 90°, does the cosine increase or decrease? What about the sine?

118. As T increases from 90° to 180°, does the cosine increase or decrease? What about the sine?

1.4 Using the Definitions of the Trigonometric Functions

- Reciprocal Identities
- Signs and Ranges of Function Values
- Pythagorean Identities
- Quotient Identities

Identities are equations that are true for all values of the variables for which all expressions are defined.

$$(x+y)^2 = x^2 + 2xy + y^2 \qquad 2(x+3) = 2x + 6 \qquad \text{Identities}$$

Reciprocal Identities

Recall the definition of a reciprocal.

The **reciprocal** of a nonzero number x is $\frac{1}{x}$.

Examples: The reciprocal of 2 is $\frac{1}{2}$, and the reciprocal of $\frac{8}{11}$ is $\frac{11}{8}$. There is no reciprocal for 0 because $\frac{1}{0}$ is undefined.

The definitions of the trigonometric functions in the previous section were written so that functions in the same column were reciprocals of each other. Because $\sin\theta = \frac{y}{r}$ and $\csc\theta = \frac{r}{y}$,

$$\sin\theta = \frac{1}{\csc\theta} \quad \text{and} \quad \csc\theta = \frac{1}{\sin\theta}, \quad \text{provided } \sin\theta \neq 0.$$

Teaching Tip Students may be tempted to associate secant with sine and cosecant with cosine. Note this common misconception.

Also, $\cos\theta$ and $\sec\theta$ are reciprocals, as are $\tan\theta$ and $\cot\theta$. The **reciprocal identities** hold for any angle θ that does not lead to a 0 denominator.

Reciprocal Identities

For all angles θ for which both functions are defined, the following identities hold.

$$\sin\theta = \frac{1}{\csc\theta} \qquad \cos\theta = \frac{1}{\sec\theta} \qquad \tan\theta = \frac{1}{\cot\theta}$$

$$\csc\theta = \frac{1}{\sin\theta} \qquad \sec\theta = \frac{1}{\cos\theta} \qquad \cot\theta = \frac{1}{\tan\theta}$$

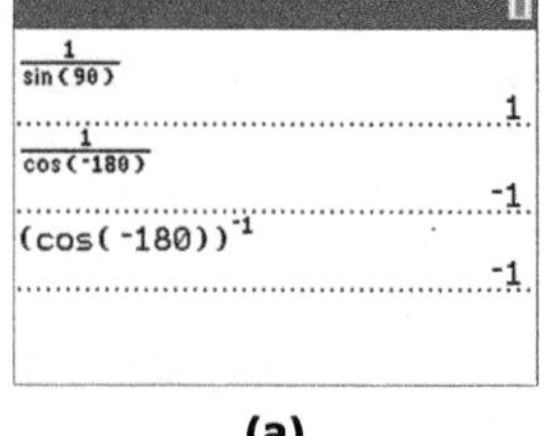

(a)

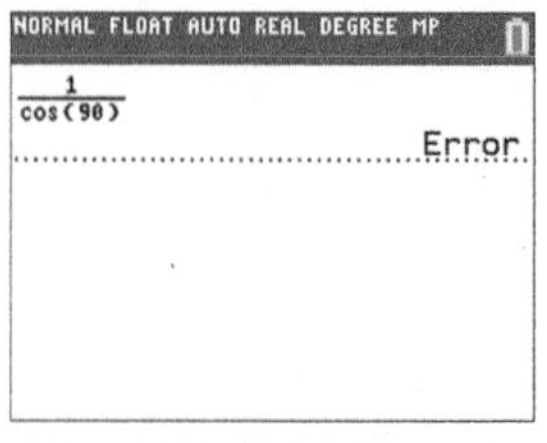

(b)

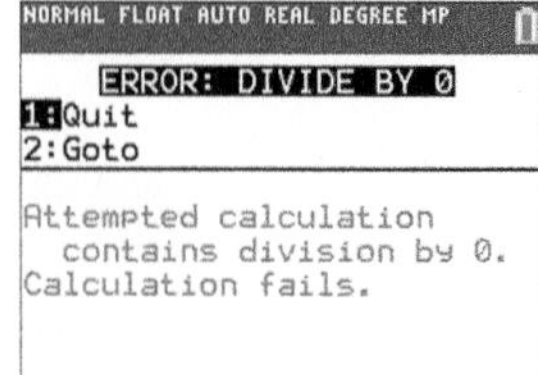

(c)

Figure 34

The screen in **Figure 34(a)** shows that $\csc 90° = 1$ and $\sec(-180°) = -1$ using appropriate reciprocal identities. The third entry uses the reciprocal function key x^{-1} to evaluate $\sec(-180°)$. **Figure 34(b)** shows that attempting to find $\sec 90°$ by entering $\frac{1}{\cos 90°}$ produces an ERROR message, indicating that the reciprocal is undefined. See **Figure 34(c).** ■

CAUTION ***Be sure not to use the inverse trigonometric function keys to find reciprocal function values.*** For example, consider the following.

$$\cos^{-1}(-180°) \neq (\cos(-180°))^{-1}$$

This is the *inverse cosine function,* which will be discussed later in the text.

This is the *reciprocal function,* which correctly evaluates $\sec(-180°)$, as seen in **Figure 34(a).**

$$(\cos(-180°))^{-1} = \frac{1}{\cos(-180°)} = \sec(-180°)$$

The reciprocal identities can be written in different forms. For example,

$$\sin\theta = \frac{1}{\csc\theta} \text{ is equivalent to } \csc\theta = \frac{1}{\sin\theta} \text{ and } (\sin\theta)(\csc\theta) = 1.$$

Classroom Example 1
Find each function value.
(a) $\tan\theta$, given that $\cot\theta = 4$
(b) $\sec\theta$, given that
$\cos\theta = -\frac{2}{\sqrt{20}}$

Answers:
(a) $\frac{1}{4}$ (b) $-\sqrt{5}$

EXAMPLE 1 Using the Reciprocal Identities

Find each function value.

(a) $\cos\theta$, given that $\sec\theta = \frac{5}{3}$ **(b)** $\sin\theta$, given that $\csc\theta = -\frac{\sqrt{12}}{2}$

SOLUTION

(a) We use the fact that $\cos\theta$ is the reciprocal of $\sec\theta$.

$$\cos\theta = \frac{1}{\sec\theta} = \frac{1}{\frac{5}{3}} = 1 \div \frac{5}{3} = 1 \cdot \frac{3}{5} = \frac{3}{5} \quad \text{Simplify the complex fraction.}$$

(b)
$$\sin\theta = \frac{1}{\csc\theta} \quad \sin\theta \text{ is the reciprocal of } \csc\theta.$$
$$= \frac{1}{-\frac{\sqrt{12}}{2}} \quad \text{Substitute } \csc\theta = -\frac{\sqrt{12}}{2}.$$
$$= -\frac{2}{\sqrt{12}} \quad \text{Simplify the complex fraction as in part (a).}$$
$$= -\frac{2}{2\sqrt{3}} \quad \sqrt{12} = \sqrt{4\cdot 3} = 2\sqrt{3}$$
$$= -\frac{1}{\sqrt{3}} \quad \text{Divide out the common factor 2.}$$
$$= -\frac{1}{\sqrt{3}} \cdot \frac{\sqrt{3}}{\sqrt{3}} \quad \text{Rationalize the denominator.}$$
$$= -\frac{\sqrt{3}}{3} \quad \text{Multiply.}$$

✔ **Now Try Exercises 11 and 19.**

Teaching Tip Some students use the sentence "All Students Take Calculus" to remember which of the three basic functions are positive in each quadrant. A indicates "all" in quadrant I, S represents "sine" in quadrant II, T represents "tangent" in quadrant III, and C stands for "cosine" in quadrant IV.

Signs and Ranges of Function Values In the definitions of the trigonometric functions, r is the distance from the origin to the point (x, y). This distance is undirected, so $r > 0$. If we choose a point (x, y) in quadrant I, then both x and y will be positive, and the values of all six functions will be positive.

A point (x, y) in quadrant II satisfies $x < 0$ and $y > 0$. This makes the values of sine and cosecant positive for quadrant II angles, while the other four functions take on negative values. Similar results can be obtained for the other quadrants.

This important information is summarized here.

Signs of Trigonometric Function Values

θ in Quadrant	$\sin\theta$	$\cos\theta$	$\tan\theta$	$\cot\theta$	$\sec\theta$	$\csc\theta$
I	+	+	+	+	+	+
II	+	−	−	−	−	+
III	−	−	+	+	−	−
IV	−	+	−	−	+	−

$x < 0, y > 0, r > 0$
II
Sine and cosecant positive

$x > 0, y > 0, r > 0$
I
All functions positive

$x < 0, y < 0, r > 0$
III
Tangent and cotangent positive

$x > 0, y < 0, r > 0$
IV
Cosine and secant positive

Classroom Example 2
Determine the signs of the trigonometric functions of an angle in standard position with the given measure.

(a) 54° (b) 260° (c) -60°

Answers:

(a) All values are positive.

(b) Tangent and cotangent are positive. All others are negative.

(c) Cosine and secant are positive. All others are negative.

EXAMPLE 2 Determining Signs of Functions of Nonquadrantal Angles

Determine the signs of the trigonometric functions of an angle in standard position with the given measure.

(a) 87° **(b)** 300° **(c)** -200°

SOLUTION

(a) An angle of 87° is in the first quadrant, with x, y, and r all positive, so all of its trigonometric function values are positive.

(b) A 300° angle is in quadrant IV, so the cosine and secant are positive, while the sine, cosecant, tangent, and cotangent are negative.

(c) A -200° angle is in quadrant II. The sine and cosecant are positive, and all other function values are negative.

✔ **Now Try Exercises 27, 29, and 33.**

NOTE Because numbers that are reciprocals always have the same sign, the sign of a function value automatically determines the sign of the reciprocal function value.

Classroom Example 3
Identify the quadrant (or possible quadrants) of an angle θ that satisfies the given conditions.

(a) $\tan\theta > 0$, $\csc\theta < 0$

(b) $\sin\theta > 0$, $\csc\theta > 0$

Answers: (a) III (b) I or II

EXAMPLE 3 Identifying the Quadrant of an Angle

Identify the quadrant (or possible quadrants) of an angle θ that satisfies the given conditions.

(a) $\sin\theta > 0$, $\tan\theta < 0$ **(b)** $\cos\theta < 0$, $\sec\theta < 0$

SOLUTION

(a) Because $\sin\theta > 0$ in quadrants I and II and $\tan\theta < 0$ in quadrants II and IV, both conditions are met only in quadrant II.

(b) The cosine and secant functions are both negative in quadrants II and III, so in this case θ could be in either of these two quadrants.

✔ **Now Try Exercises 43 and 49.**

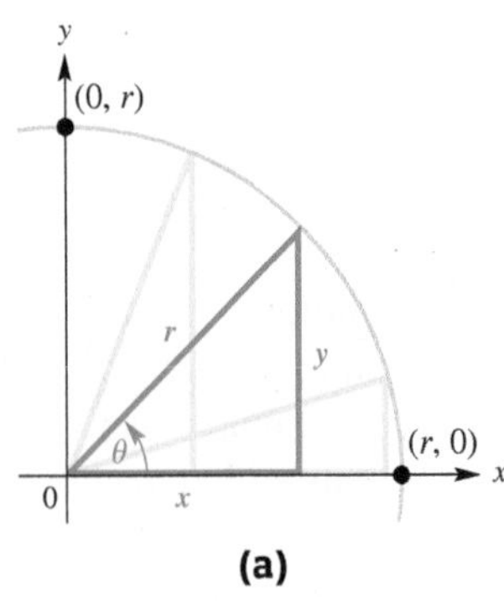

(a)

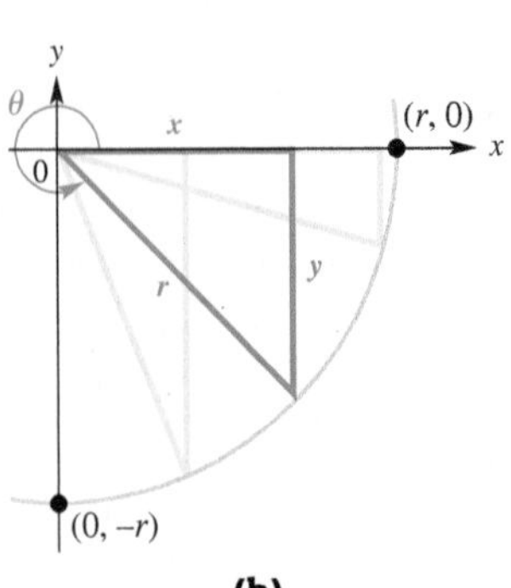

(b)

Figure 35

Figure 35(a) shows an angle θ as it increases in measure from near 0° toward 90°. In each case, the value of r is the same. As the measure of the angle increases, y increases but never exceeds r, so $y \le r$. Dividing both sides by the positive number r gives $\frac{y}{r} \le 1$.

In a similar way, angles in quadrant IV as in **Figure 35(b)** suggest that

$$-1 \le \frac{y}{r},$$

so $$-1 \le \frac{y}{r} \le 1$$

and $$\mathbf{-1 \le \sin\theta \le 1.} \qquad \frac{y}{r} = \sin\theta \text{ for any angle } \theta.$$

Similarly, $$\mathbf{-1 \le \cos\theta \le 1.}$$

The tangent of an angle is defined as $\frac{y}{x}$. It is possible that $x < y$, $x = y$, or $x > y$. Thus, $\frac{y}{x}$ can take any value, so **tan θ can be any real number, as can cot θ.**

The functions sec θ and csc θ are reciprocals of the functions cos θ and sin θ, respectively, making

$$\sec\theta \le -1 \quad \text{or} \quad \sec\theta \ge 1 \quad \text{and} \quad \csc\theta \le -1 \quad \text{or} \quad \csc\theta \ge 1.$$

In summary, the ranges of the trigonometric functions are as follows.

Ranges of Trigonometric Functions

Trigonometric Function of θ	Range (Set-Builder Notation)	Range (Interval Notation)
$\sin\theta$, $\cos\theta$	$\{y \mid \lvert y\rvert \le 1\}$	$[-1, 1]$
$\tan\theta$, $\cot\theta$	$\{y \mid y \text{ is a real number}\}$	$(-\infty, \infty)$
$\sec\theta$, $\csc\theta$	$\{y \mid \lvert y\rvert \ge 1\}$	$(-\infty, -1] \cup [1, \infty)$

Classroom Example 4
Determine whether each statement is *possible* or *impossible*.
(a) $\cot\theta = -0.999$
(b) $\cos\theta = -1.7$
(c) $\csc\theta = 0$

Answers:
(a) possible
(b) impossible
(c) impossible

EXAMPLE 4 Determining Whether a Value Is in the Range of a Trigonometric Function

Determine whether each statement is *possible* or *impossible*.

(a) $\sin\theta = 2.5$ **(b)** $\tan\theta = 110.47$ **(c)** $\sec\theta = 0.6$

SOLUTION

(a) For any value of θ, we know that $-1 \le \sin\theta \le 1$. Because $2.5 > 1$, it is impossible to find a value of θ that satisfies $\sin\theta = 2.5$.

(b) The tangent function can take on any real number value. Thus, $\tan\theta = 110.47$ is possible.

(c) Because $\lvert\sec\theta\rvert \ge 1$ for all θ for which the secant is defined, the statement $\sec\theta = 0.6$ is impossible.

✔ **Now Try Exercises 53, 57, and 59.**

The six trigonometric functions are defined in terms of x, y, and r, where the Pythagorean theorem shows that $r^2 = x^2 + y^2$ and $r > 0$. With these relationships, knowing the value of only one function and the quadrant in which the angle lies makes it possible to find the values of the other trigonometric functions.

Classroom Example 5
Suppose that angle θ is in quadrant III and $\tan\theta = \frac{8}{5}$. Find the values of the five remaining trigonometric functions.

Answer:
$\sin\theta = -\frac{8\sqrt{89}}{89}$; $\cos\theta = -\frac{5\sqrt{89}}{89}$;
$\csc\theta = -\frac{\sqrt{89}}{8}$; $\sec\theta = -\frac{\sqrt{89}}{5}$;
$\cot\theta = \frac{5}{8}$

EXAMPLE 5 Finding All Function Values Given One Value and the Quadrant

Suppose that angle θ is in quadrant II and $\sin\theta = \frac{2}{3}$. Find the values of the five remaining trigonometric functions.

SOLUTION Choose any point on the terminal side of angle θ. For simplicity, since $\sin\theta = \frac{y}{r}$, choose the point with $r = 3$.

$$\sin\theta = \frac{2}{3} \quad \text{Given value}$$

$$\frac{y}{r} = \frac{2}{3} \quad \text{Substitute } \tfrac{y}{r} \text{ for } \sin\theta.$$

Because $\frac{y}{r} = \frac{2}{3}$ and $r = 3$, it follows that $y = 2$. We must find the value of x.

$$x^2 + y^2 = r^2 \quad \text{Pythagorean theorem}$$
$$x^2 + 2^2 = 3^2 \quad \text{Substitute.}$$
$$x^2 + 4 = 9 \quad \text{Apply exponents.}$$
$$x^2 = 5 \quad \text{Subtract 4.}$$

Remember *both* roots.

$$x = \sqrt{5} \quad \text{or} \quad x = -\sqrt{5} \quad \text{Square root property: If } x^2 = k, \text{ then } x = \sqrt{k} \text{ or } x = -\sqrt{k}.$$

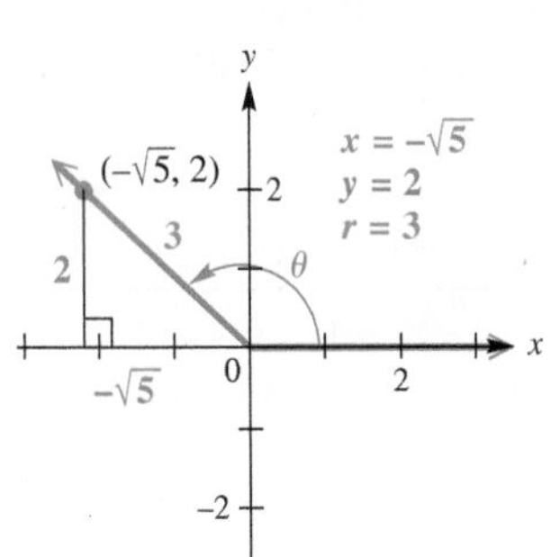

Figure 36

Because θ is in quadrant II, x must be negative. Choose $x = -\sqrt{5}$ so that the point $\left(-\sqrt{5}, 2\right)$ is on the terminal side of θ. See **Figure 36.**

$$\cos\theta = \frac{x}{r} = \frac{-\sqrt{5}}{3} = -\frac{\sqrt{5}}{3}$$

$$\sec\theta = \frac{r}{x} = \frac{3}{-\sqrt{5}} = -\frac{3}{\sqrt{5}} \cdot \frac{\sqrt{5}}{\sqrt{5}} = -\frac{3\sqrt{5}}{5}$$

These have rationalized denominators.

$$\tan\theta = \frac{y}{x} = \frac{2}{-\sqrt{5}} = -\frac{2}{\sqrt{5}} \cdot \frac{\sqrt{5}}{\sqrt{5}} = -\frac{2\sqrt{5}}{5}$$

$$\cot\theta = \frac{x}{y} = \frac{-\sqrt{5}}{2} = -\frac{\sqrt{5}}{2}$$

$$\csc\theta = \frac{r}{y} = \frac{3}{2}$$

✔ **Now Try Exercise 75.**

Pythagorean Identities

We now derive three new identities.

$$x^2 + y^2 = r^2 \quad \text{Pythagorean theorem}$$
$$\frac{x^2}{r^2} + \frac{y^2}{r^2} = \frac{r^2}{r^2} \quad \text{Divide by } r^2.$$
$$\left(\frac{x}{r}\right)^2 + \left(\frac{y}{r}\right)^2 = 1 \quad \text{Power rule for exponents; } \frac{a^m}{b^m} = \left(\frac{a}{b}\right)^m$$

$(\cos\theta)^2$ and $\cos^2\theta$ are equivalent forms.

$$(\cos\theta)^2 + (\sin\theta)^2 = 1 \quad \cos\theta = \frac{x}{r},\ \sin\theta = \frac{y}{r}$$
$$\mathbf{\sin^2\theta + \cos^2\theta = 1} \quad \text{Apply exponents; commutative property}$$

Starting again with $x^2 + y^2 = r^2$ and dividing through by x^2 gives the following.

$$\frac{x^2}{x^2} + \frac{y^2}{x^2} = \frac{r^2}{x^2} \quad \text{Divide by } x^2.$$
$$1 + \left(\frac{y}{x}\right)^2 = \left(\frac{r}{x}\right)^2 \quad \text{Power rule for exponents}$$
$$1 + (\tan\theta)^2 = (\sec\theta)^2 \quad \tan\theta = \frac{y}{x},\ \sec\theta = \frac{r}{x}$$
$$\mathbf{\tan^2\theta + 1 = \sec^2\theta} \quad \text{Apply exponents; commutative property}$$

Similarly, dividing through by y^2 leads to another identity.

$$\mathbf{1 + \cot^2\theta = \csc^2\theta}$$

These three identities are the **Pythagorean identities** because the original equation that led to them, $x^2 + y^2 = r^2$, comes from the Pythagorean theorem.

Teaching Tip Point out the importance of the exponents in the Pythagorean identities. Emphasize that without the exponents, the resulting equations are not identities.

Pythagorean Identities

For all angles θ for which the function values are defined, the following identities hold.

$$\sin^2\theta + \cos^2\theta = 1 \qquad \tan^2\theta + 1 = \sec^2\theta \qquad 1 + \cot^2\theta = \csc^2\theta$$

We give only one form of each identity. However, algebraic transformations produce equivalent forms. For example, by subtracting $\sin^2\theta$ from both sides of $\sin^2\theta + \cos^2\theta = 1$, we obtain an equivalent identity.

$$\cos^2\theta = 1 - \sin^2\theta \quad \text{Alternative form}$$

It is important to be able to transform these identities quickly and also to recognize their equivalent forms.

LOOKING AHEAD TO CALCULUS
The reciprocal, Pythagorean, and quotient identities are used in calculus to find derivatives and integrals of trigonometric functions. A standard technique of integration called **trigonometric substitution** relies on the Pythagorean identities.

Quotient Identities Consider the quotient of the functions $\sin\theta$ and $\cos\theta$, for $\cos\theta \neq 0$.

$$\frac{\sin\theta}{\cos\theta} = \frac{\frac{y}{r}}{\frac{x}{r}} = \frac{y}{r} \div \frac{x}{r} = \frac{y}{r} \cdot \frac{r}{x} = \frac{y}{x} = \tan\theta$$

Similarly, $\frac{\cos\theta}{\sin\theta} = \cot\theta$, for $\sin\theta \neq 0$. Thus, we have the **quotient identities.**

Quotient Identities

For all angles θ for which the denominators are not zero, the following identities hold.

$$\frac{\sin\theta}{\cos\theta} = \tan\theta \qquad \frac{\cos\theta}{\sin\theta} = \cot\theta$$

Classroom Example 6
Find $\cos\theta$ and $\tan\theta$, given that $\sin\theta = -\frac{\sqrt{2}}{3}$ and $\cos\theta > 0$.

Answer:
$\cos\theta = \frac{\sqrt{7}}{3}$; $\tan\theta = -\frac{\sqrt{14}}{7}$

EXAMPLE 6 Using Identities to Find Function Values

Find $\sin\theta$ and $\tan\theta$, given that $\cos\theta = -\frac{\sqrt{3}}{4}$ and $\sin\theta > 0$.

SOLUTION Start with the Pythagorean identity that includes $\cos\theta$.

$$\sin^2\theta + \cos^2\theta = 1 \quad \text{Pythagorean identity}$$

$$\sin^2\theta + \left(-\frac{\sqrt{3}}{4}\right)^2 = 1 \quad \text{Replace } \cos\theta \text{ with } -\frac{\sqrt{3}}{4}.$$

$$\sin^2\theta + \frac{3}{16} = 1 \quad \text{Square } -\frac{\sqrt{3}}{4}.$$

$$\sin^2\theta = \frac{13}{16} \quad \text{Subtract } \frac{3}{16}.$$

$$\sin\theta = \pm\frac{\sqrt{13}}{4} \quad \text{Take square roots.}$$

Choose the correct sign here.

$$\sin\theta = \frac{\sqrt{13}}{4} \quad \text{Choose the positive square root because } \sin\theta \text{ is positive.}$$

Teaching Tip In **Example 6,** the answer can be checked by solving

$$\left(-\sqrt{3}\right)^2 + y^2 = 4^2$$

for y and then applying the trigonometric function definitions.

To find $\tan \theta$, use the values of $\cos \theta$ and $\sin \theta$ and the quotient identity for $\tan \theta$.

$$\tan \theta = \frac{\sin \theta}{\cos \theta} = \frac{\frac{\sqrt{13}}{4}}{-\frac{\sqrt{3}}{4}} = \frac{\sqrt{13}}{4}\left(-\frac{4}{\sqrt{3}}\right) = -\frac{\sqrt{13}}{\sqrt{3}}$$

$$= -\frac{\sqrt{13}}{\sqrt{3}} \cdot \frac{\sqrt{3}}{\sqrt{3}} = -\frac{\sqrt{39}}{3} \qquad \text{Rationalize the denominator.}$$

Now Try Exercise 79.

CAUTION ***In exercises like Examples 5 and 6, be careful to choose the correct sign when square roots are taken.*** Refer as needed to the diagrams preceding **Example 2** that summarize the signs of the functions.

Classroom Example 7
Find $\sin \theta$ and $\cos \theta$, given that $\cot \theta = -\frac{7}{24}$ and θ is in quadrant II.

Answer: $\sin \theta = \frac{24}{25}$; $\cos \theta = -\frac{7}{25}$

EXAMPLE 7 Using Identities to Find Function Values

Find $\sin \theta$ and $\cos \theta$, given that $\tan \theta = \frac{4}{3}$ and θ is in quadrant III.

SOLUTION Because θ is in quadrant III, $\sin \theta$ and $\cos \theta$ will both be negative. It is tempting to say that since $\tan \theta = \frac{\sin \theta}{\cos \theta}$ and $\tan \theta = \frac{4}{3}$, then $\sin \theta = -4$ and $\cos \theta = -3$. This is *incorrect,* however—both $\sin \theta$ and $\cos \theta$ must be in the interval $[-1, 1]$.

We use the Pythagorean identity $\tan^2 \theta + 1 = \sec^2 \theta$ to find $\sec \theta$, and then the reciprocal identity $\cos \theta = \frac{1}{\sec \theta}$ to find $\cos \theta$.

$$\tan^2 \theta + 1 = \sec^2 \theta \qquad \text{Pythagorean identity}$$

$$\left(\frac{4}{3}\right)^2 + 1 = \sec^2 \theta \qquad \tan \theta = \tfrac{4}{3}$$

$$\frac{16}{9} + 1 = \sec^2 \theta \qquad \text{Square } \tfrac{4}{3}.$$

$$\frac{25}{9} = \sec^2 \theta \qquad \text{Add.}$$

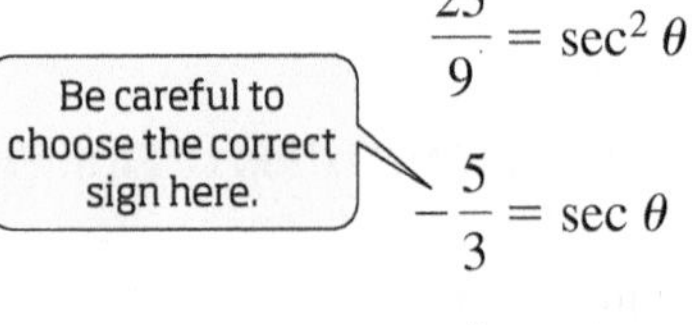

$$-\frac{5}{3} = \sec \theta \qquad \text{Choose the negative square root because } \sec \theta \text{ is negative when } \theta \text{ is in quadrant III.}$$

$$-\frac{3}{5} = \cos \theta \qquad \text{Secant and cosine are reciprocals.}$$

Now we use this value of $\cos \theta$ to find $\sin \theta$.

$$\sin^2 \theta = 1 - \cos^2 \theta \qquad \text{Pythagorean identity (alternative form)}$$

$$\sin^2 \theta = 1 - \left(-\frac{3}{5}\right)^2 \qquad \cos \theta = -\tfrac{3}{5}$$

$$\sin^2 \theta = 1 - \frac{9}{25} \qquad \text{Square } -\tfrac{3}{5}.$$

$$\sin^2 \theta = \frac{16}{25} \qquad \text{Subtract.}$$

Again, be careful.

$$\sin \theta = -\frac{4}{5} \qquad \text{Choose the negative square root.}$$

Now Try Exercise 77.

Teaching Tip When students use the alternative method described in the Note, they often make errors involving quadrant placement and signs. Warn them about this.

NOTE **Example 7** can also be worked by sketching θ in standard position in quadrant III, finding r to be 5, and then using the definitions of $\sin\theta$ and $\cos\theta$ in terms of x, y, and r. See **Figure 37.**

When using this method, be sure to choose the correct signs for x and y as determined by the quadrant in which the terminal side of θ lies. This is analogous to choosing the correct signs after applying the Pythagorean identities.

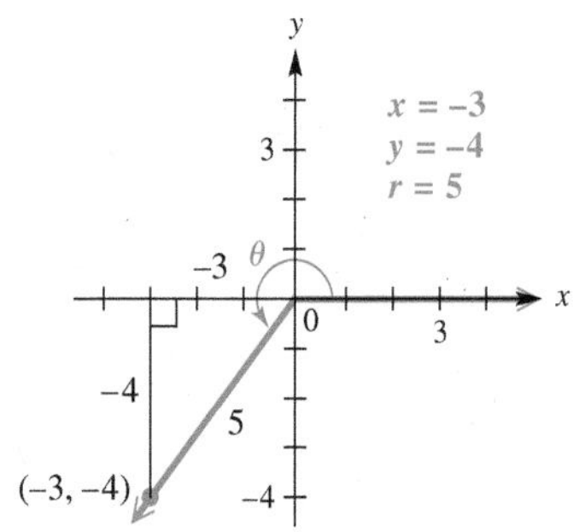

Figure 37

1.4 Exercises

1. cos θ; sec θ 2. tan θ; cot θ
3. sin θ; csc θ 4. I

5. possible 6. impossible
7. impossible 8. impossible
9. possible 10. impossible

11. $\frac{3}{2}$ 12. $\frac{8}{5}$
13. $-\frac{7}{3}$ 14. $-\frac{43}{8}$
15. $\frac{1}{5}$ 16. $\frac{1}{18}$
17. $-\frac{2}{5}$ 18. $-\frac{7}{11}$
19. $\frac{\sqrt{2}}{2}$ 20. $\frac{\sqrt{6}}{4}$
21. −0.4 22. −100
23. 0.8 24. 0.125

25. Because $-1 \le \cos\theta \le 1$, it is not possible that $\cos\theta = \frac{3}{2}$.

26. Because $\cot 90° = 0$, $\frac{1}{\cot 90°}$ and consequently $\tan 90°$ are undefined.

27. All are positive.
28. All are positive.
29. Tangent and cotangent are positive. All others are negative.
30. Tangent and cotangent are positive. All others are negative.
31. Sine and cosecant are positive. All others are negative.
32. Sine and cosecant are positive. All others are negative.
33. Cosine and secant are positive. All others are negative.
34. Cosine and secant are positive. All others are negative.

CONCEPT PREVIEW *Fill in the blank(s) to correctly complete each sentence.*

1. Given $\cos\theta = \frac{1}{\sec\theta}$, two equivalent forms of this identity are $\sec\theta = \frac{1}{\_\_\_\_}$ and $\cos\theta \cdot$ ______ = 1.

2. Given $\tan\theta = \frac{1}{\cot\theta}$, two equivalent forms of this identity are $\cot\theta = \frac{1}{\_\_\_\_}$ and $\tan\theta \cdot$ ______ = 1.

3. For an angle θ measuring 105°, the trigonometric functions ______ and ______ are positive, and the remaining trigonometric functions are negative.

4. If $\sin\theta > 0$ and $\tan\theta > 0$, then θ is in quadrant ______.

CONCEPT PREVIEW *Determine whether each statement is* possible *or* impossible.

5. $\sin\theta = \frac{1}{2}$, $\csc\theta = 2$ **6.** $\tan\theta = 2$, $\cot\theta = -2$ **7.** $\sin\theta > 0$, $\csc\theta < 0$

8. $\cos\theta = 1.5$ **9.** $\cot\theta = -1.5$ **10.** $\sin^2\theta + \cos^2\theta = 2$

Use the appropriate reciprocal identity to find each function value. Rationalize denominators when applicable. ***See Example 1.***

11. $\sec\theta$, given that $\cos\theta = \frac{2}{3}$ **12.** $\sec\theta$, given that $\cos\theta = \frac{5}{8}$

13. $\csc\theta$, given that $\sin\theta = -\frac{3}{7}$ **14.** $\csc\theta$, given that $\sin\theta = -\frac{8}{43}$

15. $\cot\theta$, given that $\tan\theta = 5$ **16.** $\cot\theta$, given that $\tan\theta = 18$

17. $\cos\theta$, given that $\sec\theta = -\frac{5}{2}$ **18.** $\cos\theta$, given that $\sec\theta = -\frac{11}{7}$

19. $\sin\theta$, given that $\csc\theta = \frac{\sqrt{8}}{2}$ **20.** $\sin\theta$, given that $\csc\theta = \frac{\sqrt{24}}{3}$

21. $\tan\theta$, given that $\cot\theta = -2.5$ **22.** $\tan\theta$, given that $\cot\theta = -0.01$

23. $\sin\theta$, given that $\csc\theta = 1.25$ **24.** $\cos\theta$, given that $\sec\theta = 8$

25. *Concept Check* What is **wrong** with the following item that appears on a trigonometry test?

"*Find* $\sec\theta$, *given that* $\cos\theta = \frac{3}{2}$."

26. *Concept Check* What is **wrong** with the statement $\tan 90° = \frac{1}{\cot 90°}$?

35. Sine and cosecant are positive. All others are negative.
36. Cosine and secant are positive. All others are negative.
37. All are positive.
38. All are positive.

39. I, II **40.** I, IV
41. I **42.** I
43. II **44.** III
45. I **46.** I
47. III **48.** II
49. III, IV **50.** II, IV

51. $\cos\theta$ and $\sec\theta$ are reciprocal functions, and $\sin\theta$ and $\csc\theta$ are reciprocal functions. The pairs have the same sign for each quadrant.
52. Because $\tan\theta$ and $\cot\theta$ are reciprocal functions, they have the same sign for all angles θ for which both are defined.

53. impossible **54.** impossible
55. possible **56.** possible
57. possible **58.** possible
59. impossible **60.** impossible
61. possible **62.** possible
63. possible **64.** possible

65. $-\frac{4}{5}$ **66.** $-\frac{3}{5}$
67. $-\frac{\sqrt{5}}{2}$ **68.** $-\frac{4}{3}$
69. $-\frac{\sqrt{3}}{3}$ **70.** $\sqrt{3}$
71. 1.05 **72.** -0.75

In Exercises 73–84, we give, in order, sine, cosine, tangent, cotangent, secant, and cosecant.
73. $\frac{15}{17}; -\frac{8}{17}; -\frac{15}{8}; -\frac{8}{15}; -\frac{17}{8}; \frac{17}{15}$
74. $-\frac{4}{5}; -\frac{3}{5}; \frac{4}{3}; \frac{3}{4}; -\frac{5}{3}; -\frac{5}{4}$
75. $\frac{\sqrt{5}}{7}; \frac{2\sqrt{11}}{7}; \frac{\sqrt{55}}{22}; \frac{2\sqrt{55}}{5}; \frac{7\sqrt{11}}{22}; \frac{7\sqrt{5}}{5}$
76. $-\frac{\sqrt{3}}{2}; -\frac{1}{2}; \sqrt{3}; \frac{\sqrt{3}}{3}; -2; -\frac{2\sqrt{3}}{3}$
77. $\frac{8\sqrt{67}}{67}; \frac{\sqrt{201}}{67}; \frac{8\sqrt{3}}{3}; \frac{\sqrt{3}}{8}; \frac{\sqrt{201}}{3}; \frac{\sqrt{67}}{8}$
78. $\frac{1}{2}; -\frac{\sqrt{3}}{2}; -\frac{\sqrt{3}}{3}; -\sqrt{3}; -\frac{2\sqrt{3}}{3}; 2$
79. $\frac{\sqrt{2}}{6}; -\frac{\sqrt{34}}{6}; -\frac{\sqrt{17}}{17}; -\sqrt{17}; -\frac{3\sqrt{34}}{17}; 3\sqrt{2}$

Determine the signs of the trigonometric functions of an angle in standard position with the given measure. ***See Example 2.***

27. 74° **28.** 84° **29.** 218° **30.** 195°
31. 178° **32.** 125° **33.** −80° **34.** −15°
35. 855° **36.** 1005° **37.** −345° **38.** −640°

Identify the quadrant (or possible quadrants) of an angle θ that satisfies the given conditions. ***See Example 3.***

39. $\sin\theta > 0,\ \csc\theta > 0$ **40.** $\cos\theta > 0,\ \sec\theta > 0$ **41.** $\cos\theta > 0,\ \sin\theta > 0$
42. $\sin\theta > 0,\ \tan\theta > 0$ **43.** $\tan\theta < 0,\ \cos\theta < 0$ **44.** $\cos\theta < 0,\ \sin\theta < 0$
45. $\sec\theta > 0,\ \csc\theta > 0$ **46.** $\csc\theta > 0,\ \cot\theta > 0$ **47.** $\sec\theta < 0,\ \csc\theta < 0$
48. $\cot\theta < 0,\ \sec\theta < 0$ **49.** $\sin\theta < 0,\ \csc\theta < 0$ **50.** $\tan\theta < 0,\ \cot\theta < 0$

51. Why are the answers to **Exercises 41 and 45** the same?

52. Why is there no angle θ that satisfies $\tan\theta > 0$, $\cot\theta < 0$?

Determine whether each statement is possible *or* impossible. ***See Example 4.***

53. $\sin\theta = 2$ **54.** $\sin\theta = 3$ **55.** $\cos\theta = -0.96$
56. $\cos\theta = -0.56$ **57.** $\tan\theta = 0.93$ **58.** $\cot\theta = 0.93$
59. $\sec\theta = -0.3$ **60.** $\sec\theta = -0.9$ **61.** $\csc\theta = 100$
62. $\csc\theta = -100$ **63.** $\cot\theta = -4$ **64.** $\cot\theta = -6$

Use identities to solve each of the following. Rationalize denominators when applicable. ***See Examples 5–7.***

65. Find $\cos\theta$, given that $\sin\theta = \frac{3}{5}$ and θ is in quadrant II.

66. Find $\sin\theta$, given that $\cos\theta = \frac{4}{5}$ and θ is in quadrant IV.

67. Find $\csc\theta$, given that $\cot\theta = -\frac{1}{2}$ and θ is in quadrant IV.

68. Find $\sec\theta$, given that $\tan\theta = \frac{\sqrt{7}}{3}$ and θ is in quadrant III.

69. Find $\tan\theta$, given that $\sin\theta = \frac{1}{2}$ and θ is in quadrant II.

70. Find $\cot\theta$, given that $\csc\theta = -2$ and θ is in quadrant III.

71. Find $\cot\theta$, given that $\csc\theta = -1.45$ and θ is in quadrant III.

72. Find $\tan\theta$, given that $\sin\theta = 0.6$ and θ is in quadrant II.

Give all six trigonometric function values for each angle θ. Rationalize denominators when applicable. ***See Examples 5–7.***

73. $\tan\theta = -\frac{15}{8}$, and θ is in quadrant II **74.** $\cos\theta = -\frac{3}{5}$, and θ is in quadrant III

75. $\sin\theta = \frac{\sqrt{5}}{7}$, and θ is in quadrant I **76.** $\tan\theta = \sqrt{3}$, and θ is in quadrant III

77. $\cot\theta = \frac{\sqrt{3}}{8}$, and θ is in quadrant I **78.** $\csc\theta = 2$, and θ is in quadrant II

79. $\sin\theta = \frac{\sqrt{2}}{6}$, and $\cos\theta < 0$ **80.** $\cos\theta = \frac{\sqrt{5}}{8}$, and $\tan\theta < 0$

81. $\sec\theta = -4$, and $\sin\theta > 0$ **82.** $\csc\theta = -3$, and $\cos\theta > 0$

83. $\sin\theta = 1$ **84.** $\cos\theta = 1$

80. $-\frac{\sqrt{59}}{8}; \frac{\sqrt{5}}{8}; -\frac{\sqrt{295}}{5}; -\frac{\sqrt{295}}{59}; \frac{8\sqrt{5}}{5}; -\frac{8\sqrt{59}}{59}$

81. $\frac{\sqrt{15}}{4}; -\frac{1}{4}; -\sqrt{15}; -\frac{\sqrt{15}}{15}; -4; \frac{4\sqrt{15}}{15}$

82. $-\frac{1}{3}; \frac{2\sqrt{2}}{3}; -\frac{\sqrt{2}}{4}; -2\sqrt{2}; \frac{3\sqrt{2}}{4}; -3$

83. 1; 0; undefined; 0; undefined; 1

84. 0; 1; 0; undefined; 1; undefined

87. This statement is false. For example, $\sin 180° + \cos 180° = 0 + (-1) = -1 \neq 1$.

88. This statement is false. 2 is not in the range of the sine function.

89. negative 90. negative
91. positive 92. positive
93. negative 94. negative
95. negative 96. negative

97. positive 98. positive
99. negative 100. negative
101. positive 102. positive
103. negative 104. negative

105. 2° 106. 5°
107. 3° 108. 1°

109. Quadrant II is the only quadrant in which the cosine is negative and the sine is positive.

110. III

Work each problem.

85. Derive the identity $1 + \cot^2 \theta = \csc^2 \theta$ by dividing $x^2 + y^2 = r^2$ by y^2.

86. Derive the quotient identity $\frac{\cos \theta}{\sin \theta} = \cot \theta$.

87. ***Concept Check*** *True* or *false*: For all angles θ, $\sin \theta + \cos \theta = 1$. If the statement is false, give an example showing why.

88. ***Concept Check*** *True* or *false*: Since $\cot \theta = \frac{\cos \theta}{\sin \theta}$, if $\cot \theta = \frac{1}{2}$ with θ in quadrant I, then $\cos \theta = 1$ and $\sin \theta = 2$. If the statement is false, explain why.

Concept Check *Suppose that* $90° < \theta < 180°$. *Find the sign of each function value.*

89. $\sin 2\theta$ 90. $\csc 2\theta$ 91. $\tan \frac{\theta}{2}$ 92. $\cot \frac{\theta}{2}$

93. $\cot(\theta + 180°)$ 94. $\tan(\theta + 180°)$ 95. $\cos(-\theta)$ 96. $\sec(-\theta)$

Concept Check *Suppose that* $-90° < \theta < 90°$. *Find the sign of each function value.*

97. $\cos \frac{\theta}{2}$ 98. $\sec \frac{\theta}{2}$ 99. $\sec(\theta + 180°)$ 100. $\cos(\theta + 180°)$

101. $\sec(-\theta)$ 102. $\cos(-\theta)$ 103. $\cos(\theta - 180°)$ 104. $\sec(\theta - 180°)$

Concept Check *Find a solution for each equation.*

105. $\tan(3\theta - 4°) = \frac{1}{\cot(5\theta - 8°)}$

106. $\cos(6\theta + 5°) = \frac{1}{\sec(4\theta + 15°)}$

107. $\sin(4\theta + 2°) \csc(3\theta + 5°) = 1$

108. $\sec(2\theta + 6°) \cos(5\theta + 3°) = 1$

109. ***Concept Check*** The screen below was obtained with the calculator in degree mode. Use it to justify that an angle of 14,879° is a quadrant II angle.

```
NORMAL FLOAT AUTO REAL DEGREE MP
cos(14879)
                    -.4848096202
sin(14879)
                     .8746197071
```

110. ***Concept Check*** The screen below was obtained with the calculator in degree mode. In which quadrant does a 1294° angle lie?

```
NORMAL FLOAT AUTO REAL DEGREE MP
tan(1294)
                     .6745085168
sin(1294)
                    -.5591929035
```

Chapter 1 Test Prep

Key Terms

1.1 line
line segment (or segment)
ray
endpoint of a ray
angle
side of an angle
vertex of an angle
initial side
terminal side
positive angle
negative angle
degree
acute angle
right angle
obtuse angle
straight angle
complementary angles (complements)
supplementary angles (supplements)
minute
second
angle in standard position
quadrantal angle
coterminal angles

1.2 vertical angles
parallel lines
transversal
similar triangles
congruent triangles

1.3 sine (sin)
cosine (cos)
tangent (tan)
cotangent (cot)
secant (sec)
cosecant (csc)
degree mode

1.4 reciprocal

New Symbols

⏋	right angle symbol (for a right triangle)	$^\circ$	degree
θ	Greek letter theta	$'$	minute
		$''$	second

Quick Review

Concepts	Examples

1.1 Angles

Types of Angles
Two positive angles with a sum of 90° are **complementary angles.**

70° and $90^\circ - 70^\circ = 20^\circ$ are complementary.

Two positive angles with a sum of 180° are **supplementary angles.**

70° and $180^\circ - 70^\circ = 110^\circ$ are supplementary.

1 degree = 60 minutes $(\mathbf{1^\circ = 60'})$

1 minute = 60 seconds $(\mathbf{1' = 60''})$

$$15^\circ\, 30'\, 45''$$

$$= 15^\circ + \frac{30}{60}^\circ + \frac{45}{3600}^\circ \qquad 30' \cdot \frac{1^\circ}{60'} = \frac{30}{60}^\circ \text{ and } 45'' \cdot \frac{1^\circ}{3600''} = \frac{45}{3600}^\circ.$$

$$= 15.5125^\circ \qquad \text{Decimal degrees}$$

Coterminal angles have measures that differ by a multiple of 360°. Their terminal sides coincide when in standard position.

The acute angle θ in the figure is in standard position. If θ measures 46°, find the measure of a positive and a negative coterminal angle.

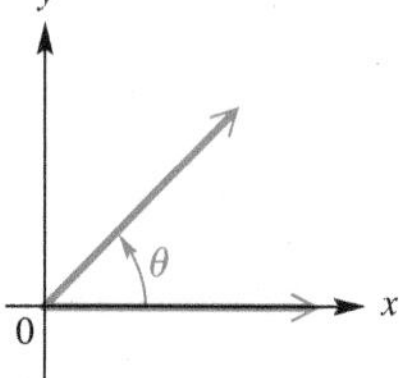

$$46^\circ + 360^\circ = 406^\circ$$

$$46^\circ - 360^\circ = -314^\circ$$

1.2 Angle Relationships and Similar Triangles

Vertical angles have equal measures.

When a transversal intersects two parallel lines, the following angles formed have equal measure:

- **alternate interior angles,**
- **alternate exterior angles,** and
- **corresponding angles.**

Interior angles on the same side of a transversal are supplementary.

Find the measures of angles 1, 2, 3, and 4.

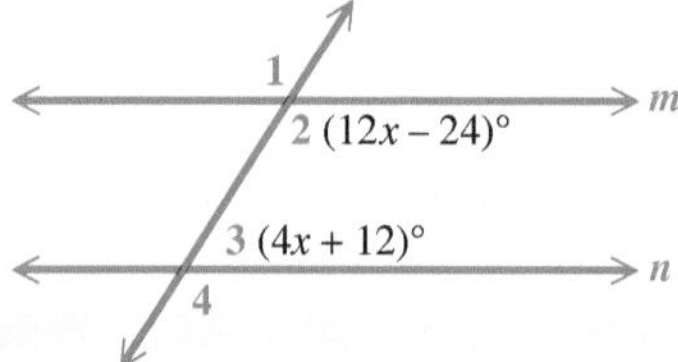

m and n are parallel lines.

$$12x - 24 + 4x + 12 = 180$$
$$16x - 12 = 180 \qquad \text{Interior angles on the same side of a transversal are supplementary.}$$
$$x = 12$$

Angle 2 has measure $12 \cdot 12 - 24 = 120^\circ$.

Angle 3 has measure $4 \cdot 12 + 12 = 60^\circ$.

Angle 1 is a vertical angle to angle 2, so its measure is 120°.

Angle 4 corresponds to angle 2, so its measure is 120°.

Angle Sum of a Triangle
The sum of the measures of the angles of any triangle is 180°.

The measures of two angles of a triangle are $42^\circ\, 20'$ and $35^\circ\, 10'$. Find the measure of the third angle, x.

$$42^\circ\, 20' + 35^\circ\, 10' + x = 180^\circ \qquad \text{The sum of the angles is } 180^\circ.$$
$$77^\circ\, 30' + x = 180^\circ$$
$$x = 102^\circ\, 30'$$

Concepts	Examples
Similar triangles have corresponding angles with the same measures and corresponding sides proportional. **Congruent triangles** are the same size and the same shape.	Find the unknown side length.

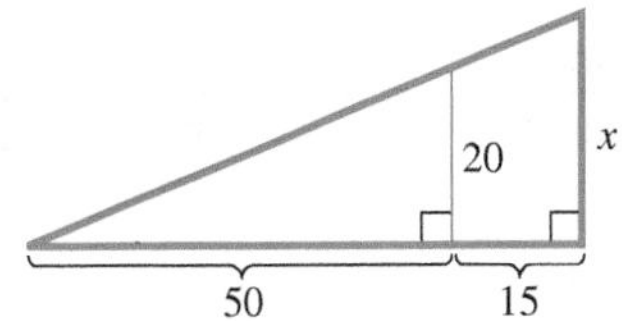

$$\frac{x}{20} = \frac{50 + 15}{50}$$ Corresponding sides of similar triangles are proportional.

$$50x = 1300$$ Property of proportions

$$x = 26$$ Divide by 50.

1.3 Trigonometric Functions

Trigonometric Functions

Let (x, y) be a point other than the origin on the terminal side of an angle θ in standard position. The distance from the point to the origin is

$$r = \sqrt{x^2 + y^2}.$$

The six trigonometric functions of θ are defined as follows.

$$\sin\theta = \frac{y}{r} \qquad \cos\theta = \frac{x}{r} \qquad \tan\theta = \frac{y}{x}\ (x \neq 0)$$

$$\csc\theta = \frac{r}{y}\ (y \neq 0) \quad \sec\theta = \frac{r}{x}\ (x \neq 0) \quad \cot\theta = \frac{x}{y}\ (y \neq 0)$$

See the summary table of trigonometric function values for quadrantal angles in this section.

If the point $(-2, 3)$ is on the terminal side of an angle θ in standard position, find the values of the six trigonometric functions of θ.

Here $x = -2$ and $y = 3$, so

$$r = \sqrt{(-2)^2 + 3^2} = \sqrt{4 + 9} = \sqrt{13}.$$

$$\sin\theta = \frac{3\sqrt{13}}{13} \qquad \cos\theta = -\frac{2\sqrt{13}}{13} \qquad \tan\theta = -\frac{3}{2}$$

$$\csc\theta = \frac{\sqrt{13}}{3} \qquad \sec\theta = -\frac{\sqrt{13}}{2} \qquad \cot\theta = -\frac{2}{3}$$

1.4 Using the Definitions of the Trigonometric Functions

Reciprocal Identities

$$\sin\theta = \frac{1}{\csc\theta} \qquad \cos\theta = \frac{1}{\sec\theta} \qquad \tan\theta = \frac{1}{\cot\theta}$$

$$\csc\theta = \frac{1}{\sin\theta} \qquad \sec\theta = \frac{1}{\cos\theta} \qquad \cot\theta = \frac{1}{\tan\theta}$$

Pythagorean Identities

$$\sin^2\theta + \cos^2\theta = 1 \qquad \tan^2\theta + 1 = \sec^2\theta$$

$$1 + \cot^2\theta = \csc^2\theta$$

If $\cot\theta = -\frac{2}{3}$, find $\tan\theta$.

$$\tan\theta = \frac{1}{\cot\theta} = \frac{1}{-\frac{2}{3}} = -\frac{3}{2}$$

Find $\sin\theta$ and $\tan\theta$, given that $\cos\theta = \frac{\sqrt{3}}{5}$ and $\sin\theta < 0$.

$$\sin^2\theta + \cos^2\theta = 1$$ Pythagorean identity

$$\sin^2\theta + \left(\frac{\sqrt{3}}{5}\right)^2 = 1$$ Replace $\cos\theta$ with $\frac{\sqrt{3}}{5}$.

$$\sin^2\theta + \frac{3}{25} = 1$$ Square $\frac{\sqrt{3}}{5}$.

$$\sin^2\theta = \frac{22}{25}$$ Subtract $\frac{3}{25}$.

$$\sin\theta = -\frac{\sqrt{22}}{5}$$ Choose the negative root.

Concepts	Examples
Quotient Identities $\dfrac{\sin\theta}{\cos\theta} = \tan\theta \qquad \dfrac{\cos\theta}{\sin\theta} = \cot\theta$	To find $\tan\theta$, use the values of $\sin\theta$ and $\cos\theta$ from the preceding page and the quotient identity $\tan\theta = \frac{\sin\theta}{\cos\theta}$. $\tan\theta = \dfrac{\sin\theta}{\cos\theta} = \dfrac{-\frac{\sqrt{22}}{5}}{\frac{\sqrt{3}}{5}} = -\dfrac{\sqrt{22}}{\sqrt{3}} \cdot \dfrac{\sqrt{3}}{\sqrt{3}} = -\dfrac{\sqrt{66}}{3}$ Simplify the complex fraction, and rationalize the denominator.
Signs of the Trigonometric Functions 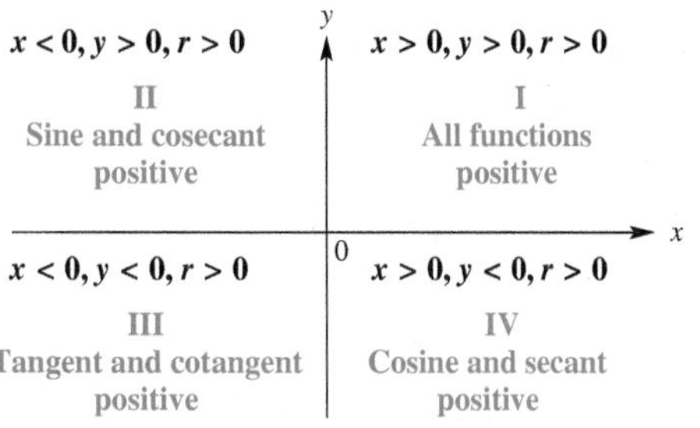	Identify the quadrant(s) of any angle θ that satisfies $\sin\theta < 0$, $\tan\theta > 0$. Because $\sin\theta < 0$ in quadrants III and IV, and $\tan\theta > 0$ in quadrants I and III, both conditions are met only in quadrant III.

Chapter 1 Review Exercises

1. complement: 55°; supplement: 145°
2. 309°
3. 186° 4. 72°
5. 9360° 6. 1280°
7. 119.134° 8. 47.420°
9. 275° 06′ 02″
10. −61° 30′ 12″
11. 40°; 60°; 80°
12. 58°; 58°

1. Give the measures of the complement and the supplement of an angle measuring 35°.

Find the angle of least positive measure that is coterminal with each angle.

2. −51° **3.** −174° **4.** 792°

Work each problem.

5. *Rotating Propeller* The propeller of a speedboat rotates 650 times per min. Through how many degrees does a point on the edge of the propeller rotate in 2.4 sec?

6. *Rotating Pulley* A pulley is rotating 320 times per min. Through how many degrees does a point on the edge of the pulley move in $\frac{2}{3}$ sec?

Convert decimal degrees to degrees, minutes, seconds, and convert degrees, minutes, seconds to decimal degrees. If applicable, round to the nearest second or the nearest thousandth of a degree.

7. 119° 08′ 03″ **8.** 47° 25′ 11″ **9.** 275.1005° **10.** −61.5034°

Find the measure of each marked angle.

11.

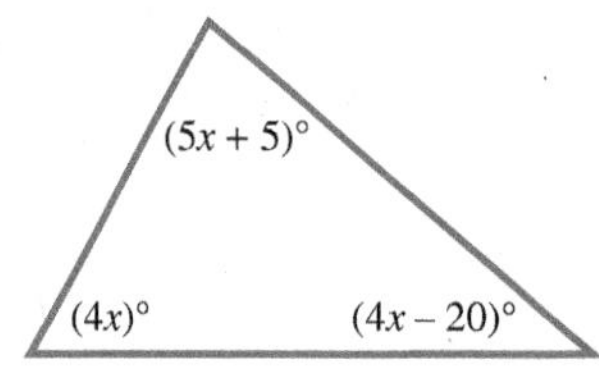

12.

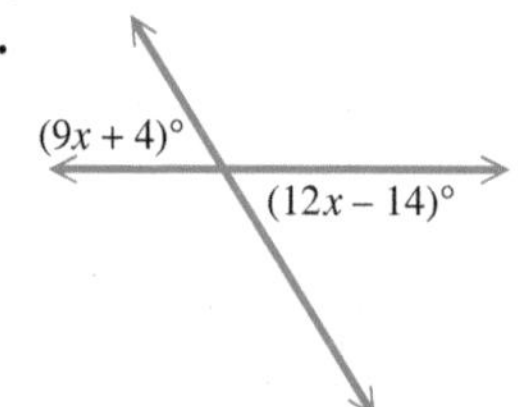

13. 105°; 105°
14. $x = 30°$; $y = 30°$
15. 0.25 km
16. $\theta = \beta - \alpha$
17. $N = 12°$; $R = 82°$; $M = 86°$
18. $V = 41°$; $Z = 32°$; $Y = U = 107°$
19. $p = 7$; $q = 7$
20. $m = 45$; $n = 60$
21. $k = 14$
22. $r \approx 15.4$
23. 12 ft

13.

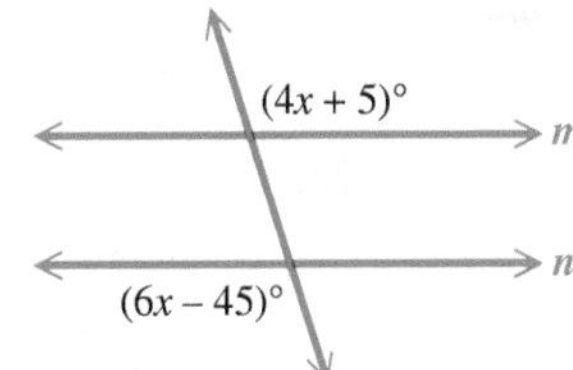

m and *n* are parallel.

14.

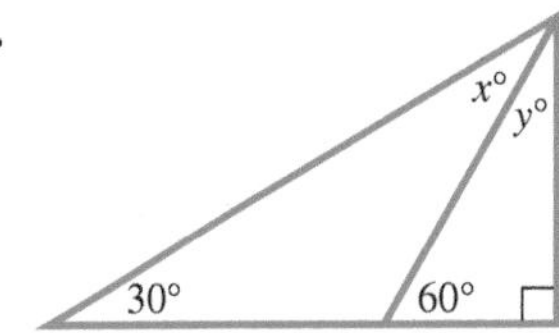

Solve each problem.

15. ***Length of a Road*** A camera is located on a satellite with its lens positioned at C in the figure. Length PC represents the distance from the lens to the film PQ, and BA represents a straight road on the ground. Use the measurements given in the figure to find the length of the road. (*Source:* Kastner, B., *Space Mathematics*, NASA.)

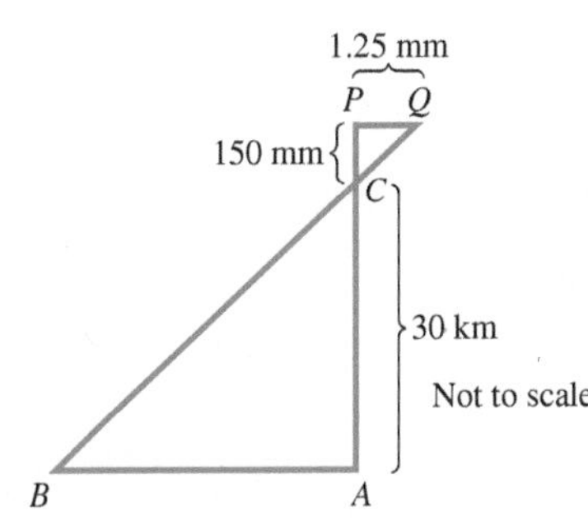

16. Express θ in terms of α and β.

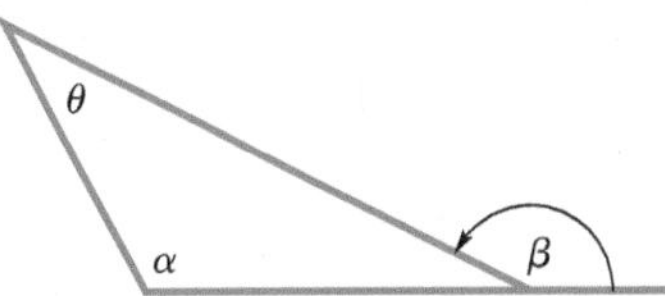

Find all unknown angle measures in each pair of similar triangles.

17.

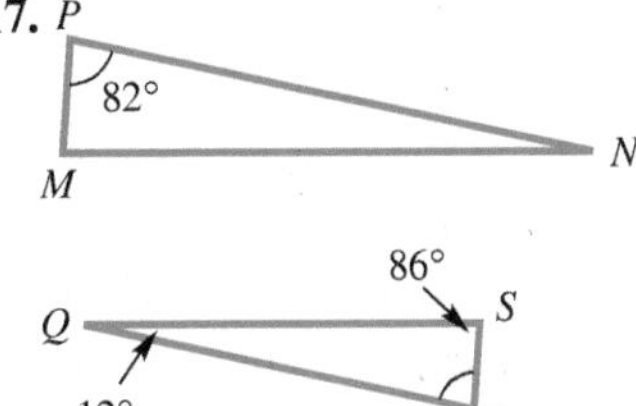

18.

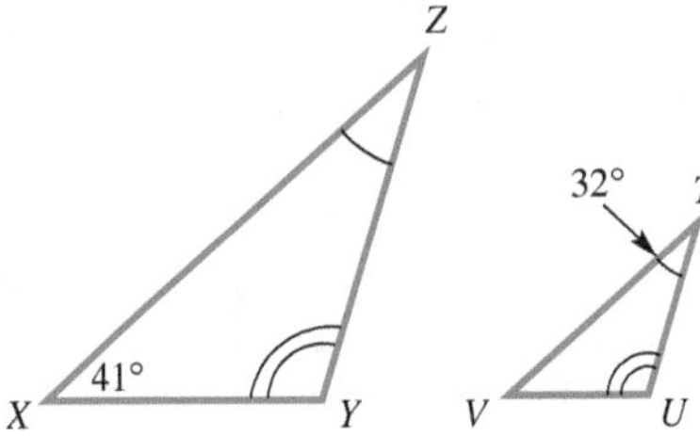

Find the unknown side lengths in each pair of similar triangles.

19.

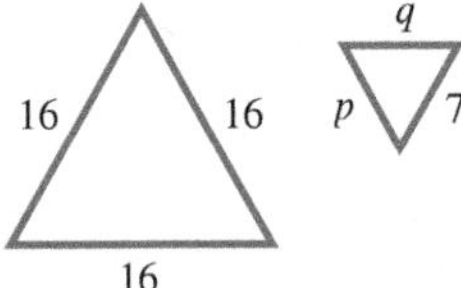

20.

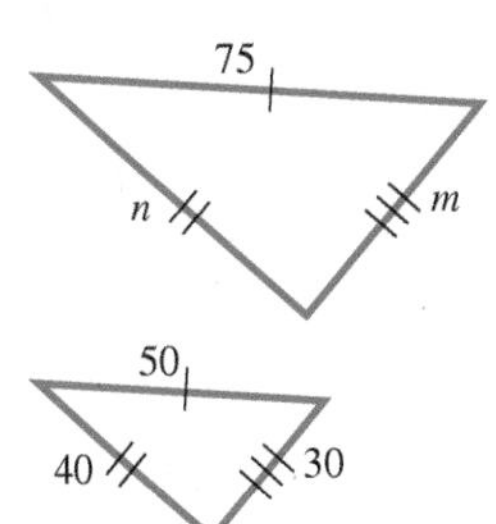

In each figure, there are two similar triangles. Find the unknown measurement.

21.

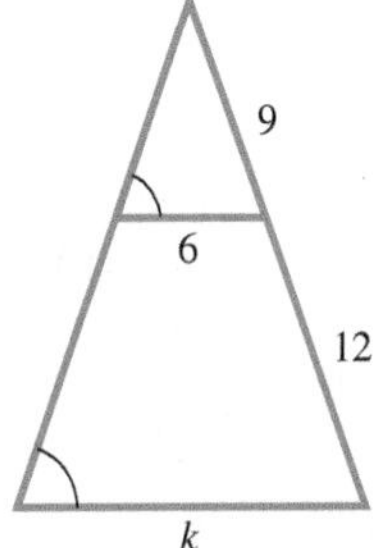

22.

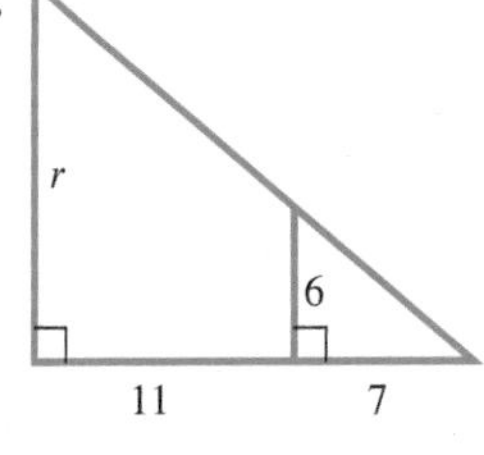

23. ***Length of a Shadow*** If a tree 20 ft tall casts a shadow 8 ft long, how long would the shadow of a 30-ft tree be at the same time and place?

In Exercises 24–35, we give, in order, sine, cosine, tangent, cotangent, secant, and cosecant.

24. $-\frac{\sqrt{2}}{2}; -\frac{\sqrt{2}}{2}; 1; 1; -\sqrt{2}; -\sqrt{2}$

25. $-\frac{\sqrt{3}}{2}; \frac{1}{2}; -\sqrt{3}; -\frac{\sqrt{3}}{3}; 2; -\frac{2\sqrt{3}}{3}$

26. 0; -1; 0; undefined; -1; undefined

27. $-\frac{4}{5}; \frac{3}{5}; -\frac{4}{3}; -\frac{3}{4}; \frac{5}{3}; -\frac{5}{4}$

28. $-\frac{2\sqrt{85}}{85}; \frac{9\sqrt{85}}{85}; -\frac{2}{9}; -\frac{9}{2}; \frac{\sqrt{85}}{9}; -\frac{\sqrt{85}}{2}$

29. $\frac{15}{17}; -\frac{8}{17}; -\frac{15}{8}; -\frac{8}{15}; -\frac{17}{8}; \frac{17}{15}$

30. $-\frac{5\sqrt{26}}{26}; \frac{\sqrt{26}}{26}; -5; -\frac{1}{5}; \sqrt{26}; -\frac{\sqrt{26}}{5}$

31. $-\frac{1}{2}; \frac{\sqrt{3}}{2}; -\frac{\sqrt{3}}{3}; -\sqrt{3}; \frac{2\sqrt{3}}{3}; -2$

32. $\frac{\sqrt{2}}{2}; -\frac{\sqrt{2}}{2}; -1; -1; -\sqrt{2}; \sqrt{2}$

33.

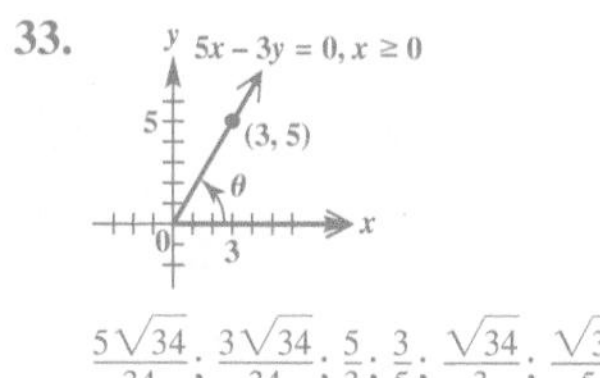

$\frac{5\sqrt{34}}{34}; \frac{3\sqrt{34}}{34}; \frac{5}{3}; \frac{3}{5}; \frac{\sqrt{34}}{3}; \frac{\sqrt{34}}{5}$

34.

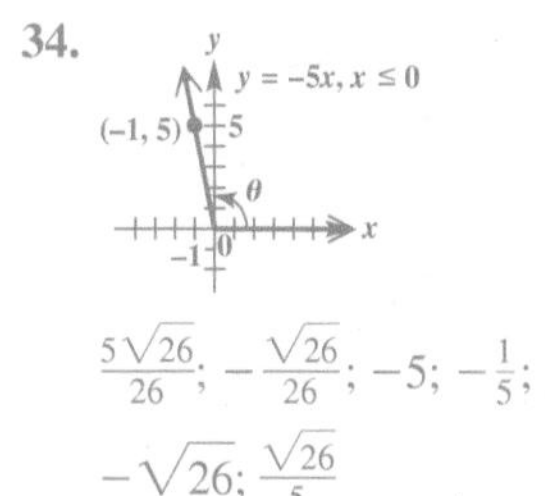

$\frac{5\sqrt{26}}{26}; -\frac{\sqrt{26}}{26}; -5; -\frac{1}{5}; -\sqrt{26}; \frac{\sqrt{26}}{5}$

35.

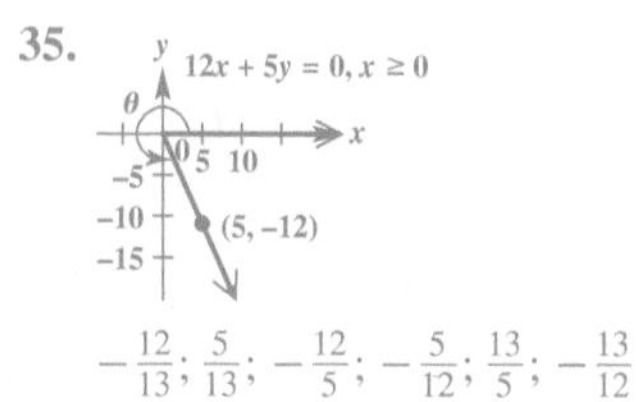

$-\frac{12}{13}; \frac{5}{13}; -\frac{12}{5}; -\frac{5}{12}; \frac{13}{5}; -\frac{13}{12}$

In Exercises 36–37 and 39–44, we give, in order, sine, cosine, tangent, cotangent, secant, and cosecant.

36. 0; -1; 0; undefined; -1; undefined

37. -1; 0; undefined; 0; undefined; -1

Find the six trigonometric function values for each angle. Rationalize denominators when applicable.

24. **25.** **26.**

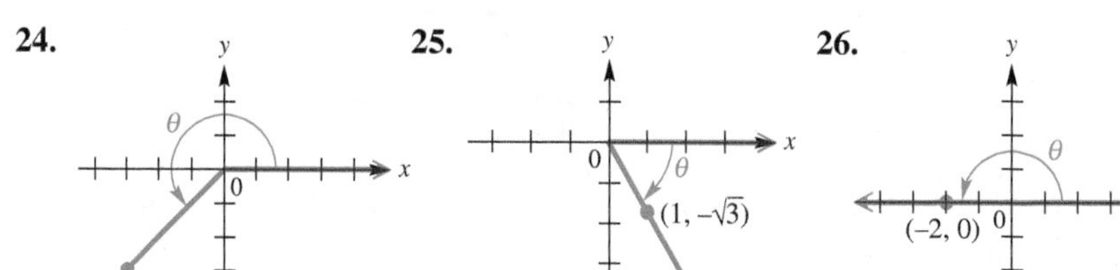

Find the values of the six trigonometric functions for an angle in standard position having each given point on its terminal side. Rationalize denominators when applicable.

27. $(3, -4)$ **28.** $(9, -2)$ **29.** $(-8, 15)$

30. $(1, -5)$ **31.** $\left(6\sqrt{3}, -6\right)$ **32.** $\left(-2\sqrt{2}, 2\sqrt{2}\right)$

An equation of the terminal side of an angle θ in standard position is given with a restriction on x. Sketch the least positive such angle θ, and find the values of the six trigonometric functions of θ.

33. $5x - 3y = 0,\ x \geq 0$ **34.** $y = -5x,\ x \leq 0$ **35.** $12x + 5y = 0,\ x \geq 0$

Complete the table with the appropriate function values of the given quadrantal angles. If the value is undefined, say so.

	θ	$\sin\theta$	$\cos\theta$	$\tan\theta$	$\cot\theta$	$\sec\theta$	$\csc\theta$
36.	180°						
37.	−90°						

38. ***Concept Check*** If the terminal side of a quadrantal angle lies along the y-axis, which of its trigonometric functions are undefined?

Give all six trigonometric function values for each angle θ. Rationalize denominators when applicable.

39. $\cos\theta = -\frac{5}{8}$, and θ is in quadrant III **40.** $\sin\theta = \frac{\sqrt{3}}{5}$, and $\cos\theta < 0$

41. $\sec\theta = -\sqrt{5}$, and θ is in quadrant II **42.** $\tan\theta = 2$, and θ is in quadrant III

43. $\sec\theta = \frac{5}{4}$, and θ is in quadrant IV **44.** $\sin\theta = -\frac{2}{5}$, and θ is in quadrant III

45. Decide whether each statement is *possible* or *impossible*.

(a) $\sec\theta = -\frac{2}{3}$ **(b)** $\tan\theta = 1.4$ **(c)** $\cos\theta = 5$

46. ***Concept Check*** If, for some particular angle θ, $\sin\theta < 0$ and $\cos\theta > 0$, in what quadrant must θ lie? What is the sign of $\tan\theta$?

Solve each problem.

47. ***Swimmer in Distress*** A lifeguard located 20 yd from the water spots a swimmer in distress. The swimmer is 30 yd from shore and 100 yd east of the lifeguard. Suppose the lifeguard runs and then swims to the swimmer in a direct line, as shown in the figure. How far east from his original position will he enter the water? (*Hint:* Find the value of x in the sketch.)

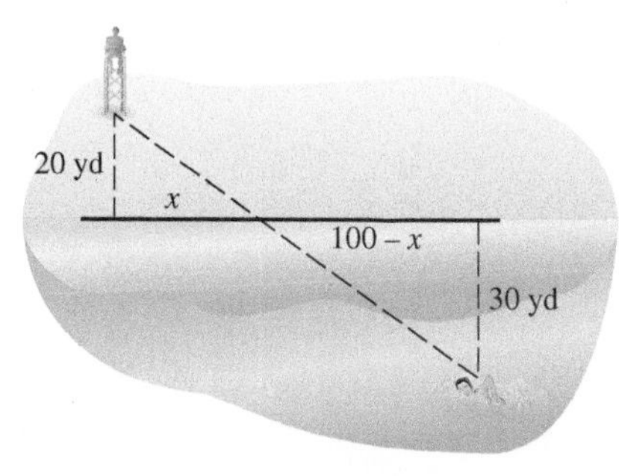

38. tangent and secant

39. $-\frac{\sqrt{39}}{8}; -\frac{5}{8}; \frac{\sqrt{39}}{5}; \frac{5\sqrt{39}}{39}; -\frac{8}{5}; -\frac{8\sqrt{39}}{39}$

40. $\frac{\sqrt{3}}{5}; -\frac{\sqrt{22}}{5}; -\frac{\sqrt{66}}{22}; -\frac{\sqrt{66}}{3}; -\frac{5\sqrt{22}}{22}; \frac{5\sqrt{3}}{3}$

41. $\frac{2\sqrt{5}}{5}; -\frac{\sqrt{5}}{5}; -2; -\frac{1}{2}; -\sqrt{5}; \frac{\sqrt{5}}{2}$

42. $-\frac{2\sqrt{5}}{5}; -\frac{\sqrt{5}}{5}; 2; \frac{1}{2}; -\sqrt{5}; -\frac{\sqrt{5}}{2}$

43. $-\frac{3}{5}; \frac{4}{5}; -\frac{3}{4}; -\frac{4}{3}; \frac{5}{4}; -\frac{5}{3}$

44. $-\frac{2}{5}; -\frac{\sqrt{21}}{5}; \frac{2\sqrt{21}}{21}; \frac{\sqrt{21}}{2}; -\frac{5\sqrt{21}}{21}; -\frac{5}{2}$

45. (a) impossible (b) possible (c) impossible

46. IV; –

47. 40 yd 48. 50″
49. 9500 ft 50. 13,500 ft

48. ***Angle through Which the Celestial North Pole Moves*** At present, the north star Polaris is located very near the celestial north pole. However, because Earth is inclined 23.5°, the moon's gravitational pull on Earth is uneven. As a result, Earth slowly precesses (moves in) like a spinning top, and the direction of the celestial north pole traces out a circular path once every 26,000 yr. See the figure. For example, in approximately A.D. 14,000 the star Vega—not the star Polaris—will be located at the celestial north pole. As viewed from the center C of this circular path, calculate the angle (to the nearest second) through which the celestial north pole moves each year. (*Source:* Zeilik, M., S. Gregory, and E. Smith, *Introductory Astronomy and Astrophysics*, Second Edition, Saunders College Publishers.)

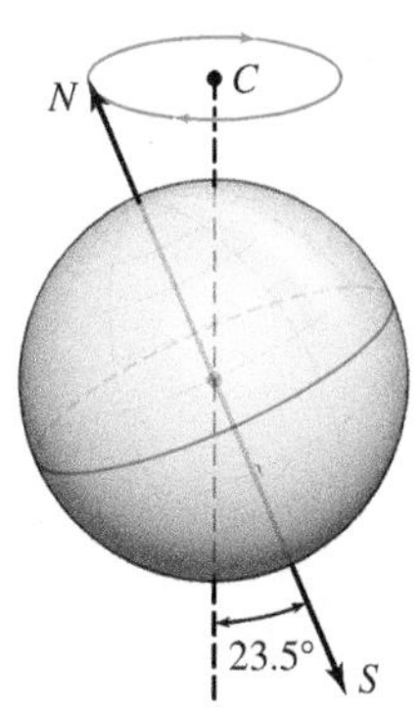

49. ***Depth of a Crater on the Moon*** The depths of unknown craters on the moon can be approximated by comparing the lengths of their shadows to the shadows of nearby craters with known depths. The crater Aristillus is 11,000 ft deep, and its shadow was measured as 1.5 mm on a photograph. Its companion crater, Autolycus, had a shadow of 1.3 mm on the same photograph. Use similar triangles to determine the depth of the crater Autolycus to the nearest hundred feet. (*Source:* Webb, T., *Celestial Objects for Common Telescopes,* Dover Publications.)

50. ***Height of a Lunar Peak*** The lunar mountain peak Huygens has a height of 21,000 ft. The shadow of Huygens on a photograph was 2.8 mm, while the nearby mountain Bradley had a shadow of 1.8 mm on the same photograph. Calculate the height of Bradley. (*Source:* Webb, T., *Celestial Objects for Common Telescopes,* Dover Publications.)

Chapter 1 Test

[1.1]
1. 23°; 113°
2. 145°; 35°
3. 20°; 70°

[1.2]
4. 130°; 130°
5. 110°; 110°
6. 20°; 30°; 130°
7. 60°; 40°; 100°

[1.1]
8. 74.31°
9. 45° 12′ 09″

10. (a) 30° (b) 280° (c) 90°

1. Give the measures of the complement and the supplement of an angle measuring 67°.

Find the measure of each marked angle.

2.

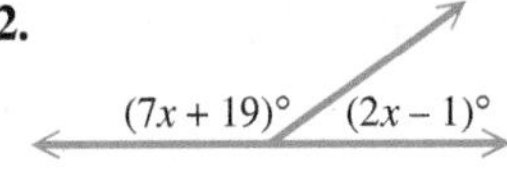

3.

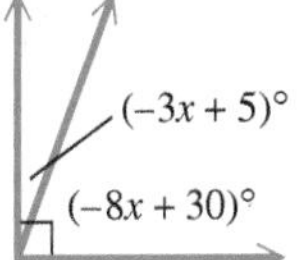

4.

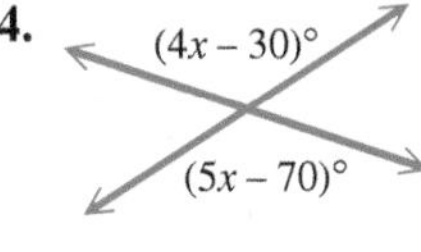

5.

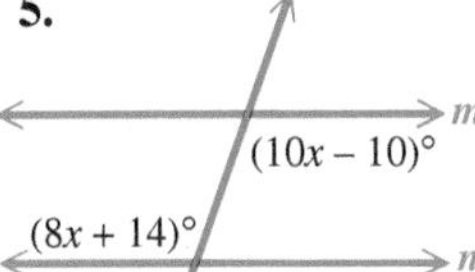

m and *n* are parallel.

6.

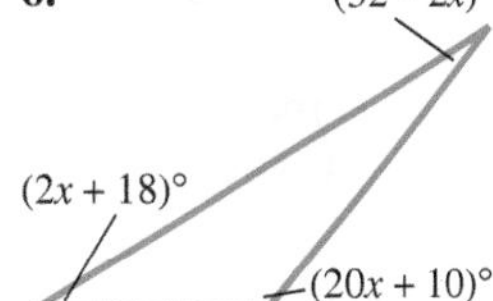

7.

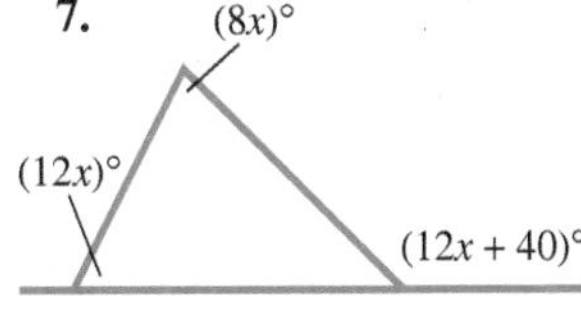

Perform each conversion.

8. 74° 18′ 36″ to decimal degrees

9. 45.2025° to degrees, minutes, seconds

Solve each problem.

10. Find the angle of least positive measure that is coterminal with each angle.

(a) 390° (b) −80° (c) 810°

11. **Rotating Tire** A tire rotates 450 times per min. Through how many degrees does a point on the edge of the tire move in 1 sec?

12. **Length of a Shadow** If a vertical pole 30 ft tall casts a shadow 8 ft long, how long would the shadow of a 40-ft pole be at the same time and place?

13. Find the unknown side lengths in this pair of similar triangles.

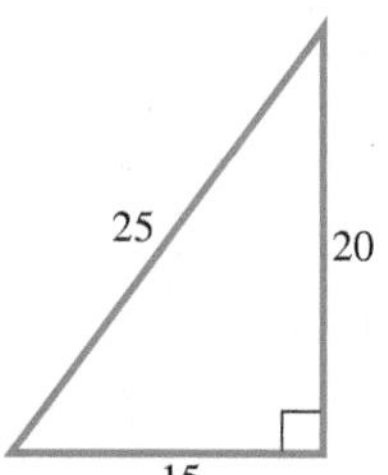

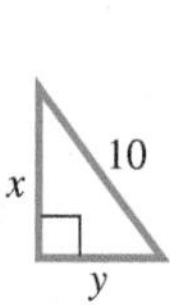

Sketch an angle θ in standard position such that θ has the least positive measure, and the given point is on the terminal side of θ. Then find the values of the six trigonometric functions for the angle. If any of these are undefined, say so.

14. $(2, -7)$

15. $(0, -2)$

Work each problem.

16. Draw a sketch of an angle in standard position having the line with the equation $3x - 4y = 0$, $x \le 0$, as its terminal side. Indicate the angle of least positive measure θ, and find the values of the six trigonometric functions of θ.

17. Complete the table with the appropriate function values of the given quadrantal angles. If the value is undefined, say so.

θ	$\sin\theta$	$\cos\theta$	$\tan\theta$	$\cot\theta$	$\sec\theta$	$\csc\theta$
90°						
−360°						
630°						

18. If the terminal side of a quadrantal angle lies along the negative x-axis, which two of its trigonometric function values are undefined?

19. Identify the possible quadrant(s) in which θ must lie under the given conditions.
(a) $\cos\theta > 0$, $\tan\theta > 0$ (b) $\sin\theta < 0$, $\csc\theta < 0$ (c) $\cot\theta > 0$, $\cos\theta < 0$

20. Decide whether each statement is *possible* or *impossible*.
(a) $\sin\theta = 1.5$ (b) $\sec\theta = 4$ (c) $\tan\theta = 10{,}000$

21. Find the value of $\sec\theta$ if $\cos\theta = -\frac{7}{12}$.

22. Find the five remaining trigonometric function values of θ if $\sin\theta = \frac{3}{7}$ and θ is in quadrant II.

---

11. 2700°

[1.2]

12. $10\frac{2}{3}$ ft, or 10 ft, 8 in.

13. $x = 8$; $y = 6$

[1.3]

14.

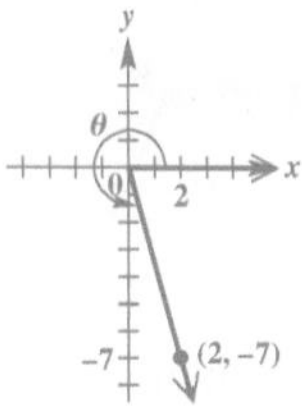

$\sin\theta = -\frac{7\sqrt{53}}{53}$; $\cos\theta = \frac{2\sqrt{53}}{53}$; $\tan\theta = -\frac{7}{2}$; $\cot\theta = -\frac{2}{7}$; $\sec\theta = \frac{\sqrt{53}}{2}$; $\csc\theta = -\frac{\sqrt{53}}{7}$

15.

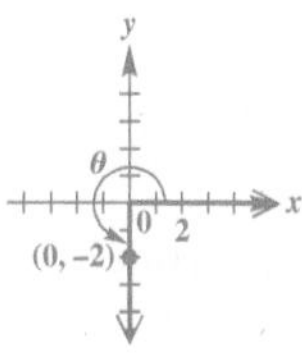

$\sin\theta = -1$; $\cos\theta = 0$; $\tan\theta$ is undefined; $\cot\theta = 0$; $\sec\theta$ is undefined; $\csc\theta = -1$

16.

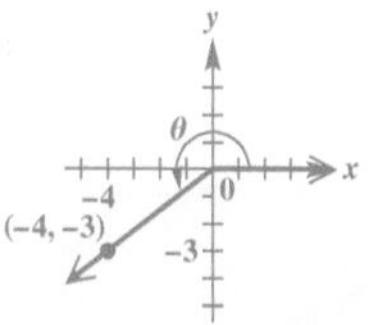

$\sin\theta = -\frac{3}{5}$; $\cos\theta = -\frac{4}{5}$; $\tan\theta = \frac{3}{4}$; $\cot\theta = \frac{4}{3}$; $\sec\theta = -\frac{5}{4}$; $\csc\theta = -\frac{5}{3}$

17. row 1: 1, 0, undefined, 0, undefined, 1;
row 2: 0, 1, 0, undefined, 1, undefined;
row 3: −1, 0, undefined, 0, undefined, −1

18. cosecant and cotangent

[1.4]

19. (a) I (b) III, IV (c) III

20. (a) impossible (b) possible (c) possible

21. $\sec\theta = -\frac{12}{7}$

22. $\cos\theta = -\frac{2\sqrt{10}}{7}$; $\tan\theta = -\frac{3\sqrt{10}}{20}$; $\cot\theta = -\frac{2\sqrt{10}}{3}$; $\sec\theta = -\frac{7\sqrt{10}}{20}$; $\csc\theta = \frac{7}{3}$

2 Acute Angles and Right Triangles

Trigonometry is used in safe roadway design to provide sufficient visibility around curves as well as a smooth-flowing, comfortable ride.

2.1 Trigonometric Functions of Acute Angles

- Right-Triangle-Based Definitions of the Trigonometric Functions
- Cofunctions
- How Function Values Change as Angles Change
- Trigonometric Function Values of Special Angles

Right-Triangle-Based Definitions of the Trigonometric Functions

Angles in standard position can be used to define the trigonometric functions. There is also another way to approach them: as ratios of the lengths of the sides of right triangles.

Figure 1 shows an acute angle A in standard position. The definitions of the trigonometric function values of angle A require x, y, and r. As drawn in **Figure 1,** x and y are the lengths of the two legs of the right triangle ABC, and r is the length of the hypotenuse.

The side of length y is the **side opposite** angle A, and the side of length x is the **side adjacent** to angle A. We use the lengths of these sides to replace x and y in the definitions of the trigonometric functions, and the length of the hypotenuse to replace r, to obtain the following right-triangle-based definitions. In the definitions, we use the standard abbreviations for the sine, cosine, tangent, cosecant, secant, and cotangent functions.

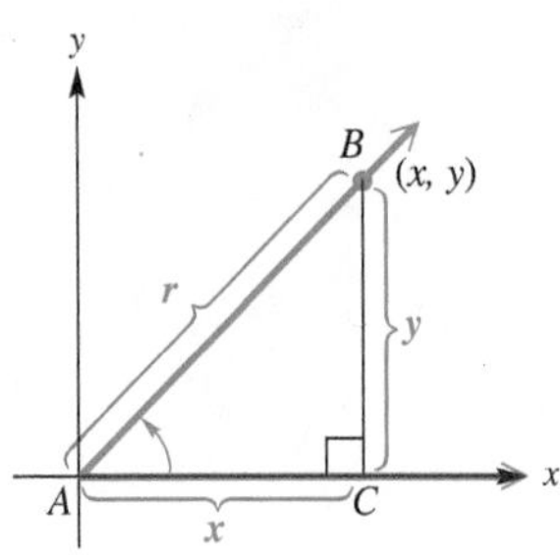

Figure 1

Right-Triangle-Based Definitions of Trigonometric Functions

Let A represent any acute angle in standard position.

$$\sin A = \frac{y}{r} = \frac{\text{side opposite } A}{\text{hypotenuse}} \qquad \csc A = \frac{r}{y} = \frac{\text{hypotenuse}}{\text{side opposite } A}$$

$$\cos A = \frac{x}{r} = \frac{\text{side adjacent to } A}{\text{hypotenuse}} \qquad \sec A = \frac{r}{x} = \frac{\text{hypotenuse}}{\text{side adjacent to } A}$$

$$\tan A = \frac{y}{x} = \frac{\text{side opposite } A}{\text{side adjacent to } A} \qquad \cot A = \frac{x}{y} = \frac{\text{side adjacent to } A}{\text{side opposite } A}$$

Classroom Example 1
Find the sine, cosine, and tangent values for angles A and B in the right triangle.

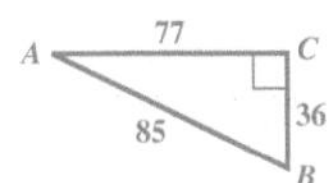

Answer:
$\sin A = \frac{36}{85}$; $\cos A = \frac{77}{85}$;
$\tan A = \frac{36}{77}$; $\sin B = \frac{77}{85}$;
$\cos B = \frac{36}{85}$; $\tan B = \frac{77}{36}$

NOTE We will sometimes shorten wording like "side opposite A" to just "side opposite" when the meaning is obvious.

EXAMPLE 1 Finding Trigonometric Function Values of an Acute Angle

Find the sine, cosine, and tangent values for angles A and B in the right triangle in **Figure 2.**

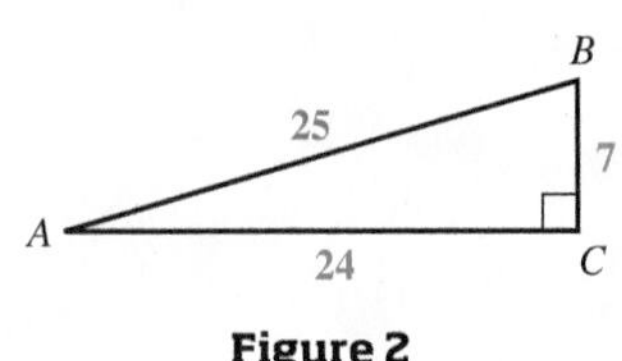

Figure 2

SOLUTION The length of the side opposite angle A is 7, the length of the side adjacent to angle A is 24, and the length of the hypotenuse is 25.

$$\sin A = \frac{\text{side opposite}}{\text{hypotenuse}} = \frac{7}{25} \qquad \cos A = \frac{\text{side adjacent}}{\text{hypotenuse}} = \frac{24}{25} \qquad \tan A = \frac{\text{side opposite}}{\text{side adjacent}} = \frac{7}{24}$$

The length of the side opposite angle B is 24, and the length of the side adjacent to angle B is 7.

$$\sin B = \frac{24}{25} \qquad \cos B = \frac{7}{25} \qquad \tan B = \frac{24}{7}$$

Use the right-triangle-based definitions of the trigonometric functions.

✔ Now Try Exercise 7.

Teaching Tip Introduce the mnemonic **sohcahtoa** to help students remember that

"**s**ine is **o**pposite over **h**ypotenuse, **c**osine is **a**djacent over **h**ypotenuse, and **t**angent is **o**pposite over **a**djacent."

NOTE The cosecant, secant, and cotangent ratios are reciprocals of the sine, cosine, and tangent values, respectively, so in **Example 1** we have

$$\csc A = \frac{25}{7} \qquad \sec A = \frac{25}{24} \qquad \cot A = \frac{24}{7}$$

$$\csc B = \frac{25}{24} \qquad \sec B = \frac{25}{7} \qquad \text{and} \qquad \cot B = \frac{7}{24}.$$

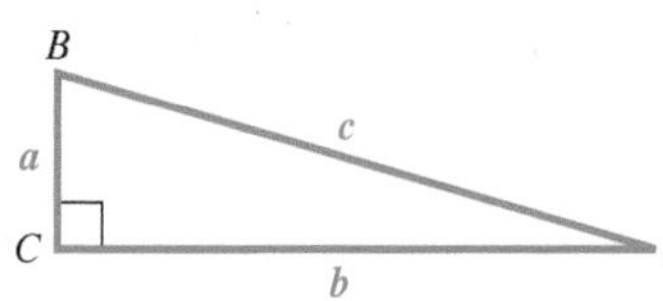

Whenever we use A, B, and C to name angles in a right triangle, C will be the right angle.

Figure 3

Cofunctions

Figure 3 shows a right triangle with acute angles A and B and a right angle at C. The length of the side opposite angle A is a, and the length of the side opposite angle B is b. The length of the hypotenuse is c. By the preceding definitions, $\sin A = \frac{a}{c}$. Also, $\cos B = \frac{a}{c}$. Thus, we have the following.

$$\mathbf{\sin A = \frac{a}{c} = \cos B}$$

Similarly, $\mathbf{\tan A = \frac{a}{b} = \cot B}$ and $\mathbf{\sec A = \frac{c}{b} = \csc B}$.

In any right triangle, the sum of the two acute angles is 90°, so they are *complementary*. In **Figure 3,** A and B are thus complementary, and we have established that $\sin A = \cos B$. This can also be written as follows.

$$\sin A = \cos(90° - A) \qquad B = 90° - A$$

This is an example of a more general relationship between **cofunction** pairs.

sine, cosine
tangent, cotangent } Cofunction pairs
secant, cosecant

Cofunction Identities

For any acute angle A, the following hold.

$$\mathbf{\sin A = \cos(90° - A) \quad \sec A = \csc(90° - A) \quad \tan A = \cot(90° - A)}$$

$$\mathbf{\cos A = \sin(90° - A) \quad \csc A = \sec(90° - A) \quad \cot A = \tan(90° - A)}$$

The cofunction identities state the following.

Cofunction values of complementary angles are equal.

Classroom Example 2
Write each function in terms of its cofunction.
(a) sin 9° (b) cot 76°
(c) csc 45°

Answers:
(a) cos 81° (b) tan 14°
(c) sec 45°

EXAMPLE 2 Writing Functions in Terms of Cofunctions

Write each function in terms of its cofunction.

(a) cos 52° **(b)** tan 71° **(c)** sec 24°

SOLUTION

(a)

Cofunctions

$$\cos 52° = \sin(90° - 52°) = \sin 38° \qquad \cos A = \sin(90° - A)$$

Complementary angles

(b) $\tan 71° = \cot(90° - 71°) = \cot 19°$ **(c)** $\sec 24° = \csc 66°$

✔ **Now Try Exercises 25 and 27.**

Classroom Example 3
Find one solution for each equation. Assume all angles involved are acute angles.

(a) $\cot(\theta - 8°) = \tan(4\theta + 13°)$

(b) $\sec(5\theta + 14°) = \csc(2\theta - 8°)$

Answers:
(a) $17°$ (b) $12°$

EXAMPLE 3 Solving Equations Using Cofunction Identities

Find one solution for each equation. Assume all angles involved are acute angles.

(a) $\cos(\theta + 4°) = \sin(3\theta + 2°)$ **(b)** $\tan(2\theta - 18°) = \cot(\theta + 18°)$

SOLUTION

(a) Sine and cosine are cofunctions, so $\cos(\theta + 4°) = \sin(3\theta + 2°)$ is true if the sum of the angles is $90°$.

$$(\theta + 4°) + (3\theta + 2°) = 90° \quad \text{Complementary angles}$$
$$4\theta + 6° = 90° \quad \text{Combine like terms.}$$
$$4\theta = 84° \quad \text{Subtract } 6° \text{ from each side.}$$
$$\theta = 21° \quad \text{Divide by 4.}$$

(b) Tangent and cotangent are cofunctions.

$$(2\theta - 18°) + (\theta + 18°) = 90° \quad \text{Complementary angles}$$
$$3\theta = 90° \quad \text{Combine like terms.}$$
$$\theta = 30° \quad \text{Divide by 3.}$$

✓ **Now Try Exercises 31 and 33.**

How Function Values Change as Angles Change **Figure 4** shows three right triangles. From left to right, the length of each hypotenuse is the same, but angle A increases in measure. As angle A increases in measure from $0°$ to $90°$, the length of the side opposite angle A also increases.

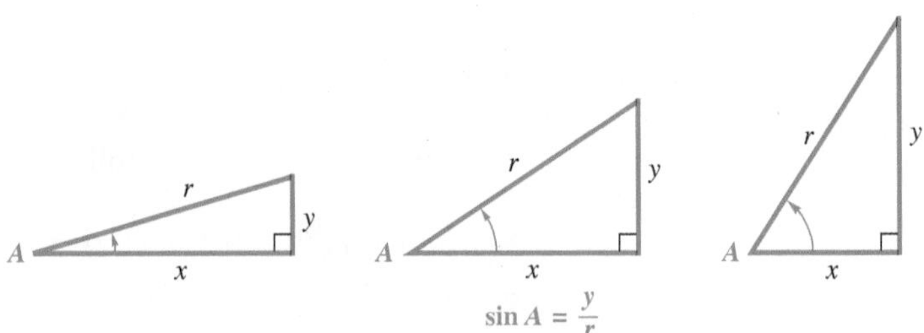

As A increases, y increases. Because r is fixed, $\sin A$ increases.

Figure 4

In the ratio

$$\sin A = \frac{\text{side opposite}}{\text{hypotenuse}} = \frac{y}{r},$$

as angle A increases, the numerator of this fraction also increases, while the denominator is fixed. Therefore, $\sin A$ *increases* as A increases from $0°$ to $90°$.

As angle A increases from $0°$ to $90°$, the length of the side adjacent to A decreases. Because r is fixed, the ratio $\frac{x}{r}$ decreases. This ratio gives $\cos A$, showing that the values of cosine *decrease* as the angle measure changes from $0°$ to $90°$. Finally, increasing A from $0°$ to $90°$ causes y to increase and x to decrease, making the values of $\frac{y}{x} = \tan A$ increase.

A similar discussion shows that as A increases from $0°$ to $90°$, the values of $\sec A$ increase, while the values of $\cot A$ and $\csc A$ decrease.

Classroom Example 4 Determine whether each statement is *true* or *false*.

(a) $\tan 25^\circ < \tan 23^\circ$

(b) $\csc 44^\circ < \csc 40^\circ$

Answers:

(a) false (b) true

EXAMPLE 4 Comparing Function Values of Acute Angles

Determine whether each statement is *true* or *false*.

(a) $\sin 21^\circ > \sin 18^\circ$ **(b)** $\sec 56^\circ \leq \sec 49^\circ$

SOLUTION

(a) In the interval from 0° to 90°, as the angle increases, so does the sine of the angle. This makes $\sin 21^\circ > \sin 18^\circ$ a true statement.

(b) For fixed r, increasing an angle from 0° to 90° causes x to decrease. Therefore, $\sec \theta = \frac{r}{x}$ increases. The statement $\sec 56^\circ \leq \sec 49^\circ$ is false.

Now Try Exercises 41 and 47.

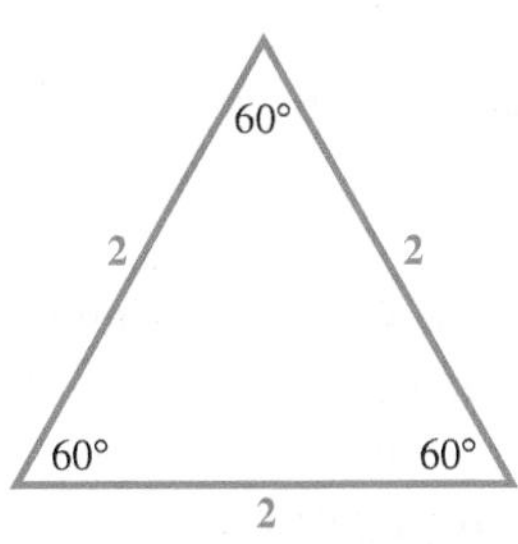

Equilateral triangle

(a)

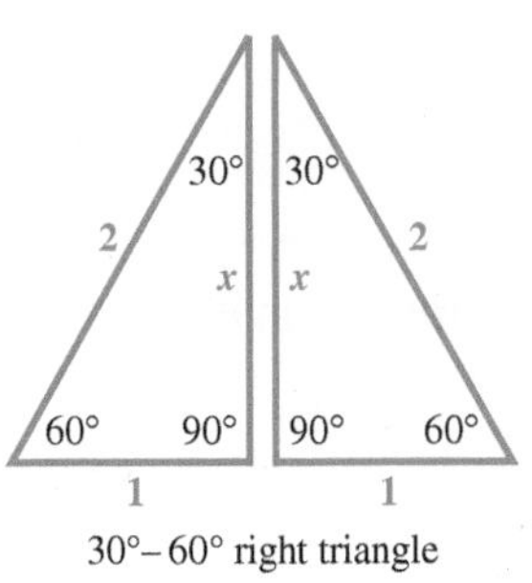

30°–60° right triangle

(b)

Figure 5

Trigonometric Function Values of Special Angles Certain special angles, such as 30°, 45°, and 60°, occur so often in trigonometry and in more advanced mathematics that they deserve special study. We start with an equilateral triangle, a triangle with all sides of equal length. Each angle of such a triangle measures 60°. Although the results we will obtain are independent of the length, for convenience we choose the length of each side to be 2 units. See **Figure 5(a).**

Bisecting one angle of this equilateral triangle leads to two right triangles, each of which has angles of 30°, 60°, and 90°, as shown in **Figure 5(b).** An angle bisector of an equilateral triangle also bisects the opposite side. Thus the shorter leg has length 1. Let x represent the length of the longer leg.

$$2^2 = 1^2 + x^2 \quad \text{Pythagorean theorem}$$

$$4 = 1 + x^2 \quad \text{Apply the exponents.}$$

$$3 = x^2 \quad \text{Subtract 1 from each side.}$$

$$\sqrt{3} = x \quad \text{Square root property; choose the positive root.}$$

Figure 6 summarizes our results using a 30°–60° right triangle. As shown in the figure, the side opposite the 30° angle has length 1. For the 30° angle,

$$\text{hypotenuse} = 2, \quad \text{side opposite} = 1, \quad \text{side adjacent} = \sqrt{3}.$$

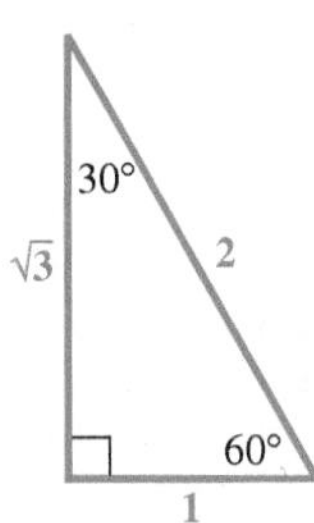

Figure 6

Now we use the definitions of the trigonometric functions.

$$\sin 30^\circ = \frac{\text{side opposite}}{\text{hypotenuse}} = \frac{1}{2}$$

$$\cos 30^\circ = \frac{\text{side adjacent}}{\text{hypotenuse}} = \frac{\sqrt{3}}{2}$$

$$\tan 30^\circ = \frac{\text{side opposite}}{\text{side adjacent}} = \frac{1}{\sqrt{3}} = \frac{1}{\sqrt{3}} \cdot \frac{\sqrt{3}}{\sqrt{3}} = \frac{\sqrt{3}}{3}$$

$$\csc 30^\circ = \frac{2}{1} = 2$$

Rationalize the denominators.

$$\sec 30^\circ = \frac{2}{\sqrt{3}} = \frac{2}{\sqrt{3}} \cdot \frac{\sqrt{3}}{\sqrt{3}} = \frac{2\sqrt{3}}{3}$$

$$\cot 30^\circ = \frac{\sqrt{3}}{1} = \sqrt{3}$$

Teaching Tip Tell students that a good way to obtain one of these function values is to draw the appropriate "famous" right triangle as shown in **Figure 6 or 7** and then use the right-triangle-based definition of the function value.

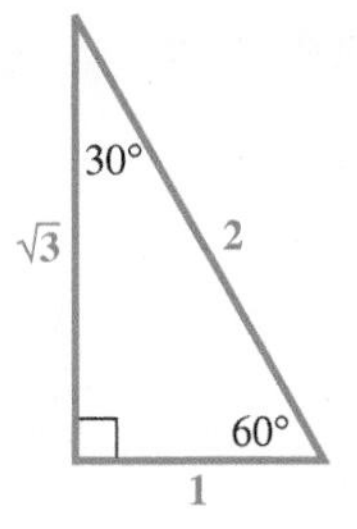

Figure 6 (repeated)

EXAMPLE 5 Finding Trigonometric Function Values for 60°

Find the six trigonometric function values for a 60° angle.

SOLUTION Refer to **Figure 6** to find the following ratios.

$$\sin 60° = \frac{\sqrt{3}}{2} \qquad \cos 60° = \frac{1}{2} \qquad \tan 60° = \frac{\sqrt{3}}{1} = \sqrt{3}$$

$$\csc 60° = \frac{2}{\sqrt{3}} = \frac{2\sqrt{3}}{3} \qquad \sec 60° = \frac{2}{1} = 2 \qquad \cot 60° = \frac{1}{\sqrt{3}} = \frac{\sqrt{3}}{3}$$

✔ **Now Try Exercises 49, 51, and 53.**

NOTE The results in **Example 5** can also be found using the fact that cofunction values of the complementary angles 60° and 30° are equal.

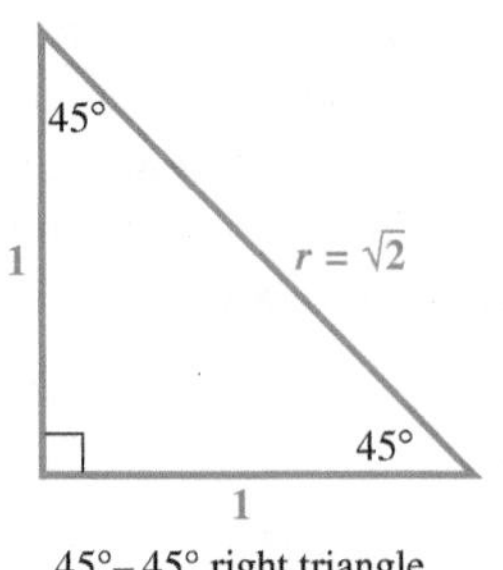

45°–45° right triangle

Figure 7

We find the values of the trigonometric functions for 45° by starting with a 45°–45° right triangle, as shown in **Figure 7.** This triangle is isosceles. For simplicity, we choose the lengths of the equal sides to be 1 unit. (As before, the results are independent of the length of the equal sides.) If *r* represents the length of the hypotenuse, then we can find its value using the Pythagorean theorem.

$$1^2 + 1^2 = r^2 \quad \text{Pythagorean theorem}$$

$$2 = r^2 \quad \text{Simplify.}$$

$$\sqrt{2} = r \quad \text{Choose the positive root.}$$

Now we use the measures indicated on the 45°–45° right triangle in **Figure 7.**

$$\sin 45° = \frac{1}{\sqrt{2}} = \frac{\sqrt{2}}{2} \qquad \cos 45° = \frac{1}{\sqrt{2}} = \frac{\sqrt{2}}{2} \qquad \tan 45° = \frac{1}{1} = 1$$

$$\csc 45° = \frac{\sqrt{2}}{1} = \sqrt{2} \qquad \sec 45° = \frac{\sqrt{2}}{1} = \sqrt{2} \qquad \cot 45° = \frac{1}{1} = 1$$

Classroom Example 5
Find the six trigonometric function values for a 30° angle.

Answer:
$\sin 30° = \frac{1}{2}$; $\cos 30° = \frac{\sqrt{3}}{2}$; $\tan 30° = \frac{\sqrt{3}}{3}$; $\csc 30° = 2$; $\sec 30° = \frac{2\sqrt{3}}{3}$; $\cot 30° = \sqrt{3}$

Function values for 30°, 45°, and 60° are summarized in the table that follows.

Function Values of Special Angles

θ	$\sin\theta$	$\cos\theta$	$\tan\theta$	$\cot\theta$	$\sec\theta$	$\csc\theta$
30°	$\frac{1}{2}$	$\frac{\sqrt{3}}{2}$	$\frac{\sqrt{3}}{3}$	$\sqrt{3}$	$\frac{2\sqrt{3}}{3}$	2
45°	$\frac{\sqrt{2}}{2}$	$\frac{\sqrt{2}}{2}$	1	1	$\sqrt{2}$	$\sqrt{2}$
60°	$\frac{\sqrt{3}}{2}$	$\frac{1}{2}$	$\sqrt{3}$	$\frac{\sqrt{3}}{3}$	2	$\frac{2\sqrt{3}}{3}$

NOTE You will be able to reproduce this table quickly if you learn the values of sin 30°, sin 45°, and sin 60°. Then you can complete the rest of the table using the reciprocal, cofunction, and quotient identities.

Teaching Tip If the TI-84 Plus calculator is in radian mode, using the degree symbol from the ANGLE menu will override the radian mode.

A calculator can find trigonometric function values at the touch of a key. So why do we spend so much time finding values for special angles? We do this because a calculator gives only *approximate* values in most cases instead of *exact* values. A scientific calculator gives the following approximation for tan 30°.

$$\tan 30° \approx 0.57735027 \quad \approx \text{ means "is approximately equal to."}$$

Earlier, however, we found the exact value.

$$\tan 30° = \frac{\sqrt{3}}{3} \quad \text{Exact value}$$

Figure 8 shows mode display options for the TI-84 Plus. **Figure 9** displays the output when evaluating the tangent, sine, and cosine of 30°. (The calculator should be in degree mode to enter angle measure in degrees.)

Figure 8

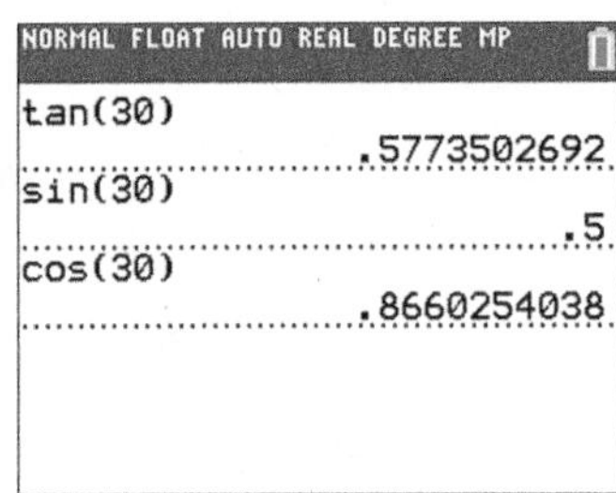

Figure 9

2.1 Exercises

1. C 2. H
3. B 4. G
5. E 6. A

In Exercises 7–10, we give, in order, sine, cosine, and tangent.

7. $\frac{21}{29}; \frac{20}{29}; \frac{21}{20}$ 8. $\frac{45}{53}; \frac{28}{53}; \frac{45}{28}$
9. $\frac{n}{p}; \frac{m}{p}; \frac{n}{m}$ 10. $\frac{k}{z}; \frac{y}{z}; \frac{k}{y}$

CONCEPT PREVIEW *Match each trigonometric function in Column I with its value in Column II. Choices may be used once, more than once, or not at all.*

I		II		
1. sin 30°	**2.** cos 45°	**A.** $\sqrt{3}$	**B.** 1	**C.** $\frac{1}{2}$
3. tan 45°	**4.** sec 60°	**D.** $\frac{\sqrt{3}}{2}$	**E.** $\frac{2\sqrt{3}}{3}$	**F.** $\frac{\sqrt{3}}{3}$
5. csc 60°	**6.** cot 30°	**G.** 2	**H.** $\frac{\sqrt{2}}{2}$	**I.** $\sqrt{2}$

Find exact values or expressions for sin *A*, cos *A*, *and* tan *A*. ***See Example 1.***

7.

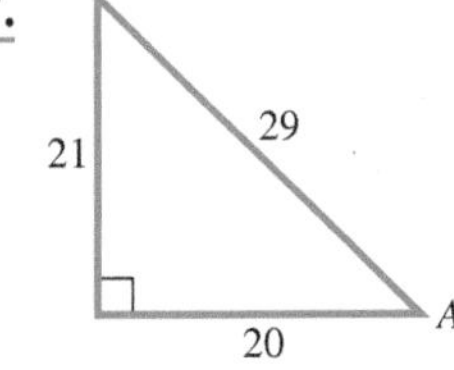

8.

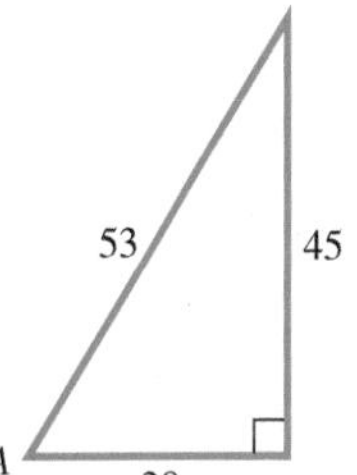

9.

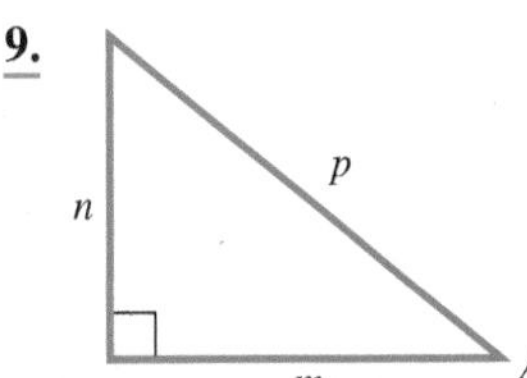

10.

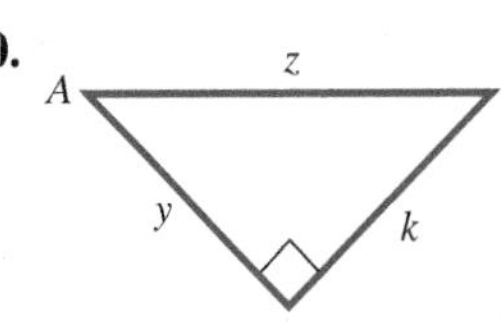

In Exercises 11–19, we give, in order, the unknown side, sine, cosine, tangent, cotangent, secant, and cosecant.

11. $c = 13$; $\frac{12}{13}$; $\frac{5}{13}$; $\frac{12}{5}$; $\frac{5}{12}$; $\frac{13}{5}$; $\frac{13}{12}$

12. $c = 5$; $\frac{4}{5}$; $\frac{3}{5}$; $\frac{4}{3}$; $\frac{3}{4}$; $\frac{5}{3}$; $\frac{5}{4}$

13. $b = \sqrt{13}$; $\frac{\sqrt{13}}{7}$; $\frac{6}{7}$; $\frac{\sqrt{13}}{6}$; $\frac{6\sqrt{13}}{13}$; $\frac{7}{6}$; $\frac{7\sqrt{13}}{13}$

14. $a = \sqrt{95}$; $\frac{7}{12}$; $\frac{\sqrt{95}}{12}$; $\frac{7\sqrt{95}}{95}$; $\frac{\sqrt{95}}{7}$; $\frac{12\sqrt{95}}{95}$; $\frac{12}{7}$

15. $b = \sqrt{91}$; $\frac{\sqrt{91}}{10}$; $\frac{3}{10}$; $\frac{\sqrt{91}}{3}$; $\frac{3\sqrt{91}}{91}$; $\frac{10}{3}$; $\frac{10\sqrt{91}}{91}$

16. $a = \sqrt{57}$; $\frac{8}{11}$; $\frac{\sqrt{57}}{11}$; $\frac{8\sqrt{57}}{57}$; $\frac{\sqrt{57}}{8}$; $\frac{11\sqrt{57}}{57}$; $\frac{11}{8}$

17. $b = \sqrt{3}$; $\frac{\sqrt{3}}{2}$; $\frac{1}{2}$; $\sqrt{3}$; $\frac{\sqrt{3}}{3}$; 2; $\frac{2\sqrt{3}}{3}$

18. $b = \sqrt{2}$; $\frac{\sqrt{2}}{2}$; $\frac{\sqrt{2}}{2}$; 1; 1; $\sqrt{2}$; $\sqrt{2}$

19. $a = \sqrt{21}$; $\frac{2}{5}$; $\frac{\sqrt{21}}{5}$; $\frac{2\sqrt{21}}{21}$; $\frac{\sqrt{21}}{2}$; $\frac{5\sqrt{21}}{21}$; $\frac{5}{2}$

20. $\sin A = \cos(90° - A)$; $\cos A = \sin(90° - A)$; $\tan A = \cot(90° - A)$; $\cot A = \tan(90° - A)$; $\sec A = \csc(90° - A)$; $\csc A = \sec(90° - A)$

21. sin 60° 22. cos 45°
23. sec 30° 24. tan 17°
25. csc 51° 26. cot 64.6°
27. cos 51.3° 28. $\sin(70° - \theta)$
29. $\csc(75° - \theta)$

30. They are always the same.

31. 40° 32. 40°
33. 20° 34. 20°
35. 12° 36. 12°
37. 35° 38. 15°
39. 18° 40. 35°

41. true 42. true
43. false 44. false
45. true 46. false
47. true 48. false

49. $\frac{\sqrt{3}}{3}$ 50. $\sqrt{3}$
51. $\frac{1}{2}$ 52. $\frac{\sqrt{3}}{2}$

Suppose ABC is a right triangle with sides of lengths a, b, and c and right angle at C.

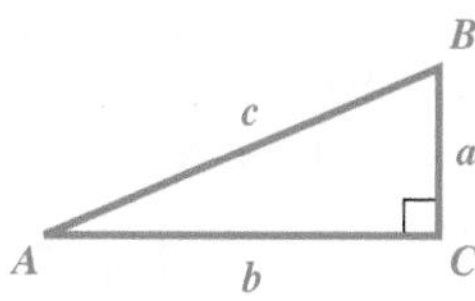

Use the Pythagorean theorem to find the unknown side length. Then find exact values of the six trigonometric functions for angle B. Rationalize denominators when applicable. **See Example 1.**

11. $a = 5, b = 12$ **12.** $a = 3, b = 4$ **13.** $a = 6, c = 7$

14. $b = 7, c = 12$ **15.** $a = 3, c = 10$ **16.** $b = 8, c = 11$

17. $a = 1, c = 2$ **18.** $a = \sqrt{2}, c = 2$ **19.** $b = 2, c = 5$

20. ***Concept Check*** Give the six cofunction identities.

Write each function in terms of its cofunction. Assume that all angles labeled θ are acute angles. **See Example 2.**

21. cos 30° **22.** sin 45° **23.** csc 60°

24. cot 73° **25.** sec 39° **26.** tan 25.4°

27. sin 38.7° **28.** $\cos(\theta + 20°)$ **29.** $\sec(\theta + 15°)$

30. ***Concept Check*** With a calculator, evaluate $\sin(90° - \theta)$ and $\cos \theta$ for various values of θ. (Check values greater than 90° and less than 0°.) Comment on the results.

Find one solution for each equation. Assume that all angles involved are acute angles. **See Example 3.**

31. $\tan \alpha = \cot(\alpha + 10°)$ **32.** $\cos \theta = \sin(2\theta - 30°)$

33. $\sin(2\theta + 10°) = \cos(3\theta - 20°)$ **34.** $\sec(\beta + 10°) = \csc(2\beta + 20°)$

35. $\tan(3B + 4°) = \cot(5B - 10°)$ **36.** $\cot(5\theta + 2°) = \tan(2\theta + 4°)$

37. $\sin(\theta - 20°) = \cos(2\theta + 5°)$ **38.** $\cos(2\theta + 50°) = \sin(2\theta - 20°)$

39. $\sec(3\beta + 10°) = \csc(\beta + 8°)$ **40.** $\csc(\beta + 40°) = \sec(\beta - 20°)$

Determine whether each statement is true *or* false. **See Example 4.**

41. sin 50° > sin 40° **42.** tan 28° ≤ tan 40°

43. sin 46° < cos 46° (*Hint*: cos 46° = sin 44°) **44.** cos 28° < sin 28° (*Hint*: sin 28° = cos 62°)

45. tan 41° < cot 41° **46.** cot 30° < tan 40°

47. sec 60° > sec 30° **48.** csc 20° < csc 30°

Give the exact value of each expression. **See Example 5.**

49. tan 30° **50.** cot 30° **51.** sin 30° **52.** cos 30°

53. sec 30° **54.** csc 30° **55.** csc 45° **56.** sec 45°

57. cos 45° **58.** cot 45° **59.** tan 45° **60.** sin 45°

61. sin 60° **62.** cos 60° **63.** tan 60° **64.** csc 60°

53. $\frac{2\sqrt{3}}{3}$ 54. 2

55. $\sqrt{2}$ 56. $\sqrt{2}$

57. $\frac{\sqrt{2}}{2}$ 58. 1

59. 1 60. $\frac{\sqrt{2}}{2}$

61. $\frac{\sqrt{3}}{2}$ 62. $\frac{1}{2}$

63. $\sqrt{3}$ 64. $\frac{2\sqrt{3}}{3}$

65. 60°

66. The student gave an *approximation*. The *exact* value of sin 45° is $\frac{\sqrt{2}}{2}$.

67. $y = \frac{\sqrt{3}}{3}x$ 68. $y = \sqrt{3}x$

69. 60° 70. 30°

71. (a) 60° (b) k (c) $k\sqrt{3}$ (d) 2; $\sqrt{3}$; 30°; 60°

72. (a) 45° (b) $k\sqrt{2}$ (c) $\sqrt{2}$

73. $x = \frac{9\sqrt{3}}{2}$; $y = \frac{9}{2}$; $z = \frac{3\sqrt{3}}{2}$; $w = 3\sqrt{3}$

74. $a = 12$; $b = 12\sqrt{3}$; $d = 12\sqrt{3}$; $c = 12\sqrt{6}$

75. $p = 15$; $r = 15\sqrt{2}$; $q = 5\sqrt{6}$; $t = 10\sqrt{6}$

76. $m = \frac{7\sqrt{3}}{3}$; $a = \frac{14\sqrt{3}}{3}$; $n = \frac{14\sqrt{3}}{3}$; $q = \frac{14\sqrt{6}}{3}$

Concept Check *Work each problem.*

65. What value of A between 0° and 90° will produce the output shown on the graphing calculator screen?

```
NORMAL FLOAT FRAC REAL DEGREE MP
√3/2
                    .8660254038
sin(A)
                    .8660254038
```

66. A student was asked to give the exact value of sin 45°. Using a calculator, he gave the answer 0.7071067812. Explain why the teacher did not give him credit.

67. Find the equation of the line that passes through the origin and makes a 30° angle with the x-axis.

68. Find the equation of the line that passes through the origin and makes a 60° angle with the x-axis.

69. What angle does the line $y = \sqrt{3}x$ make with the positive x-axis?

70. What angle does the line $y = \frac{\sqrt{3}}{3}x$ make with the positive x-axis?

71. Consider an equilateral triangle with each side having length $2k$.

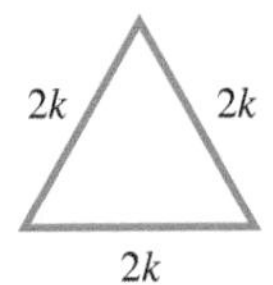

(a) What is the measure of each angle?

(b) Label one angle A. Drop a perpendicular from A to the side opposite A. Two 30° angles are formed at A, and two right triangles are formed. What is the length of the sides opposite the 30° angles?

(c) What is the length of the perpendicular in part (b)?

(d) From the results of parts (a)–(c), complete the following statement:

In a 30°–60° right triangle, the hypotenuse is always _____ times as long as the shorter leg, and the longer leg has a length that is _____ times as long as that of the shorter leg. Also, the shorter leg is opposite the _____ angle, and the longer leg is opposite the _____ angle.

72. Consider a square with each side of length k.

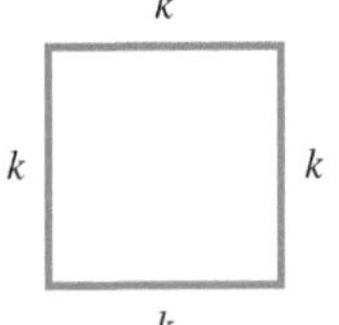

(a) Draw a diagonal of the square. What is the measure of each angle formed by a side of the square and this diagonal?

(b) What is the length of the diagonal?

(c) From the results of parts (a) and (b), complete the following statement:

In a 45°–45° right triangle, the hypotenuse has a length that is _____ times as long as either leg.

Find the exact value of the variables in each figure.

73.

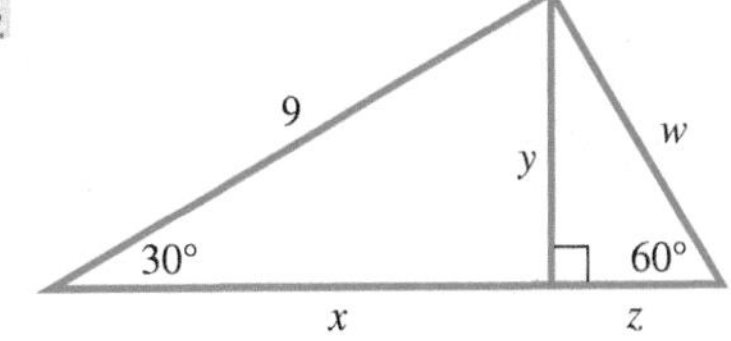

74.

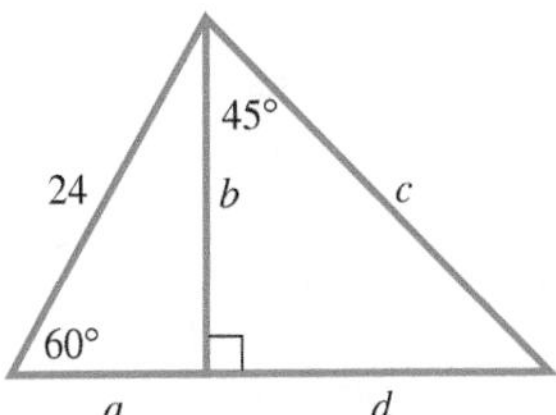

75.

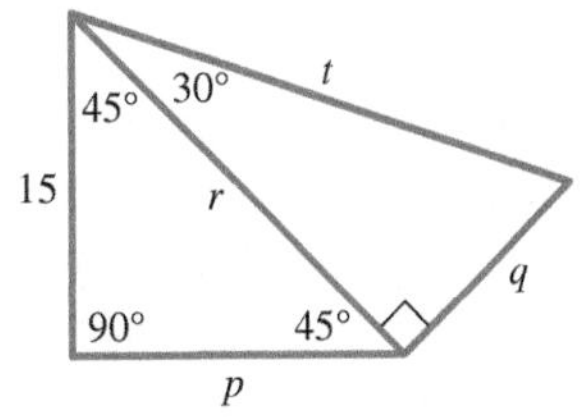

76.

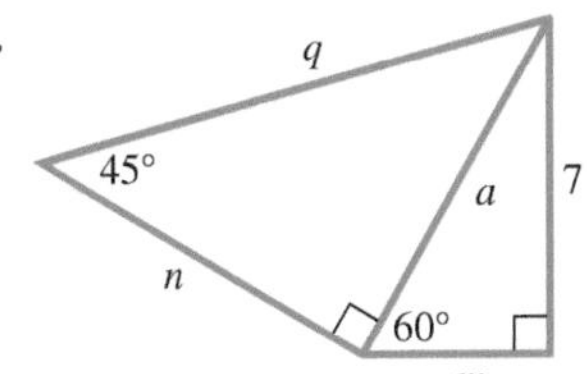

77. $\mathcal{A} = \frac{s^2}{2}$ 78. $\mathcal{A} = \frac{s^2\sqrt{3}}{4}$

79. $\left(\frac{\sqrt{2}}{2}, \frac{\sqrt{2}}{2}\right)$; 45°

80. yes

81. 82.

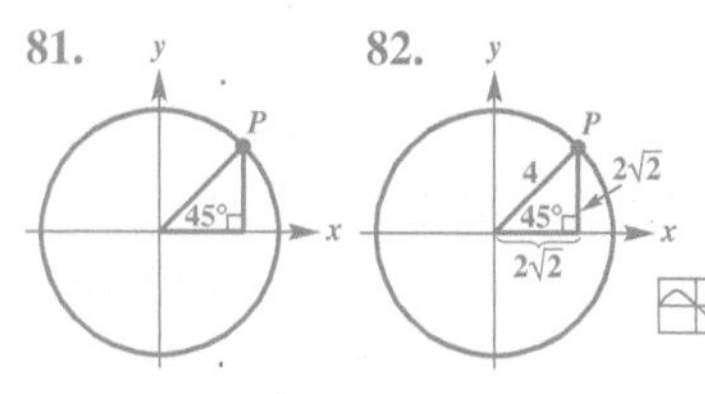

83. the legs; $\left(2\sqrt{2}, 2\sqrt{2}\right)$

84. $\left(1, \sqrt{3}\right)$

Find a formula for the area of each figure in terms of s.

77.

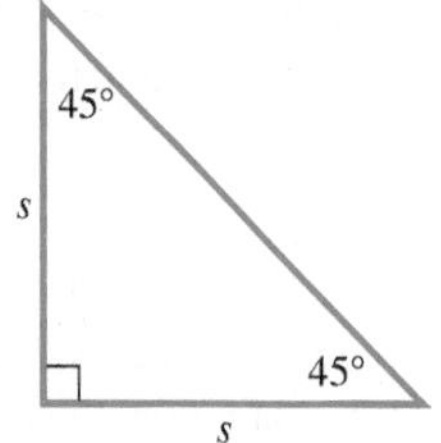

78.

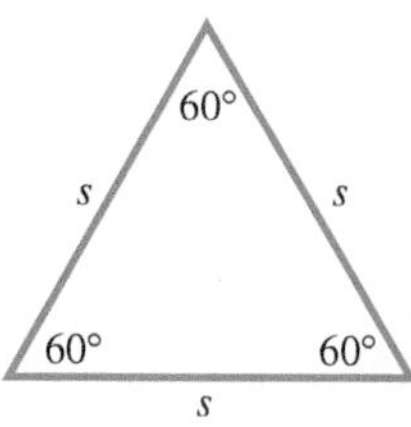

79. With a graphing calculator, find the coordinates of the point of intersection of the graphs of $y = x$ and $y = \sqrt{1 - x^2}$. These coordinates are the cosine and sine of what angle between 0° and 90°?

80. *Concept Check* Suppose we know the length of one side and one acute angle of a 30°–60° right triangle. Is it possible to determine the measures of all the sides and angles of the triangle?

Relating Concepts

For individual or collaborative investigation *(Exercises 81-84)*

The figure shows a 45° central angle in a circle with radius 4 units. To find the coordinates of point P on the circle, ***work Exercises 81–84 in order.***

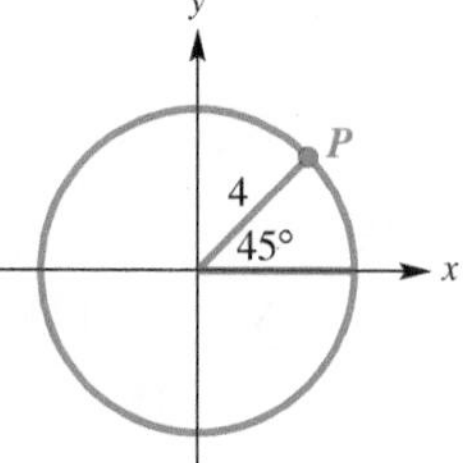

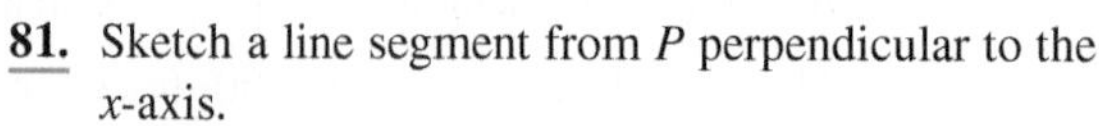

81. Sketch a line segment from P perpendicular to the x-axis.

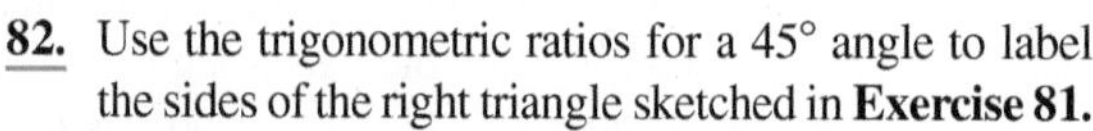

82. Use the trigonometric ratios for a 45° angle to label the sides of the right triangle sketched in **Exercise 81.**

83. Which sides of the right triangle give the coordinates of point P? What are the coordinates of P?

84. The figure at the right shows a 60° central angle in a circle of radius 2 units. Follow the same procedure as in **Exercises 81–83** to find the coordinates of P in the figure.

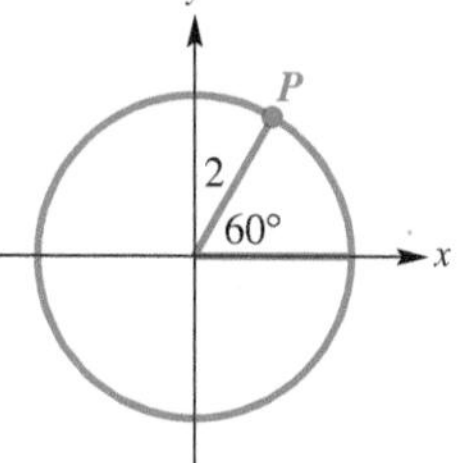

2.2 Trigonometric Functions of Non-Acute Angles

- Reference Angles
- Special Angles as Reference Angles
- Determination of Angle Measures with Special Reference Angles

Reference Angles Associated with every nonquadrantal angle in standard position is an acute angle called its *reference angle*. A **reference angle** for an angle θ, written θ', is the acute angle made by the terminal side of angle θ and the x-axis.

NOTE Reference angles are always positive and are between 0° and 90°.

Figure 10 shows several angles θ (each less than one complete counterclockwise revolution) in quadrants II, III, and IV, respectively, with the reference angle θ' also shown. In quadrant I, θ and θ' are the same. If an angle θ is negative or has measure greater than 360°, its reference angle is found by first finding its coterminal angle that is between 0° and 360°, and then using the diagrams in **Figure 10.**

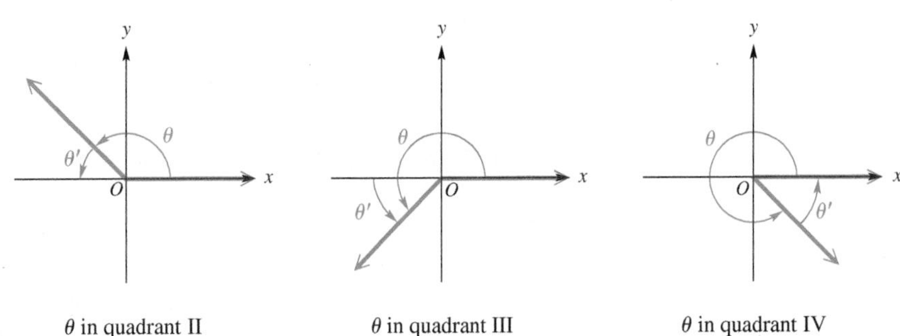

Figure 10

CAUTION A common error is to find the reference angle by using the terminal side of θ and the y-axis. ***The reference angle is always found with reference to the x-axis.***

Classroom Example 1
Find the reference angle for each angle.
(a) 294° (b) 883°

Answers:
(a) 66° (b) 17°

EXAMPLE 1 Finding Reference Angles

Find the reference angle for each angle.

(a) 218° **(b)** 1387°

SOLUTION

(a) As shown in **Figure 11(a),** the positive acute angle made by the terminal side of this angle and the x-axis is

$$218^\circ - 180^\circ = 38^\circ.$$

For $\theta = 218^\circ$, the reference angle $\theta' = 38^\circ$.

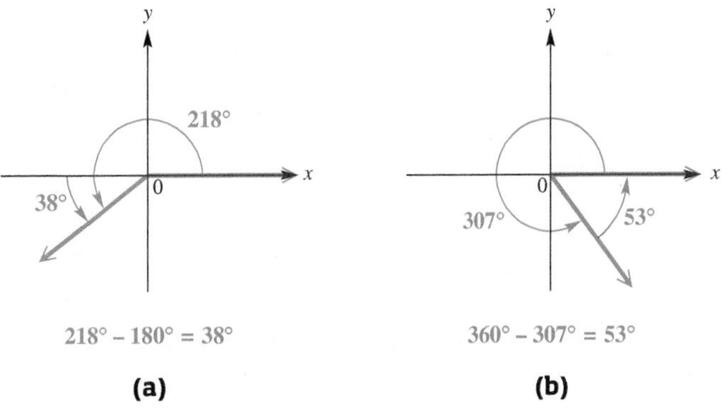

Figure 11

(b) First find a coterminal angle between 0° and 360°. Divide 1387° by 360° to obtain a quotient of about 3.9. Begin by subtracting 360° three times (because of the whole number 3 in 3.9).

$$\begin{aligned} &1387^\circ - 3 \cdot 360^\circ \\ &= 1387^\circ - 1080^\circ \quad \text{Multiply.} \\ &= 307^\circ \quad \text{Subtract.} \end{aligned}$$

The reference angle for 307° (and thus for 1387°) is $360^\circ - 307^\circ = 53^\circ$. See **Figure 11(b).**

✔ **Now Try Exercises 5 and 9.**

The preceding example suggests the following table for finding the reference angle θ' for any angle θ between 0° and 360°.

Reference Angle θ' for θ, where $0° < \theta < 360°$*

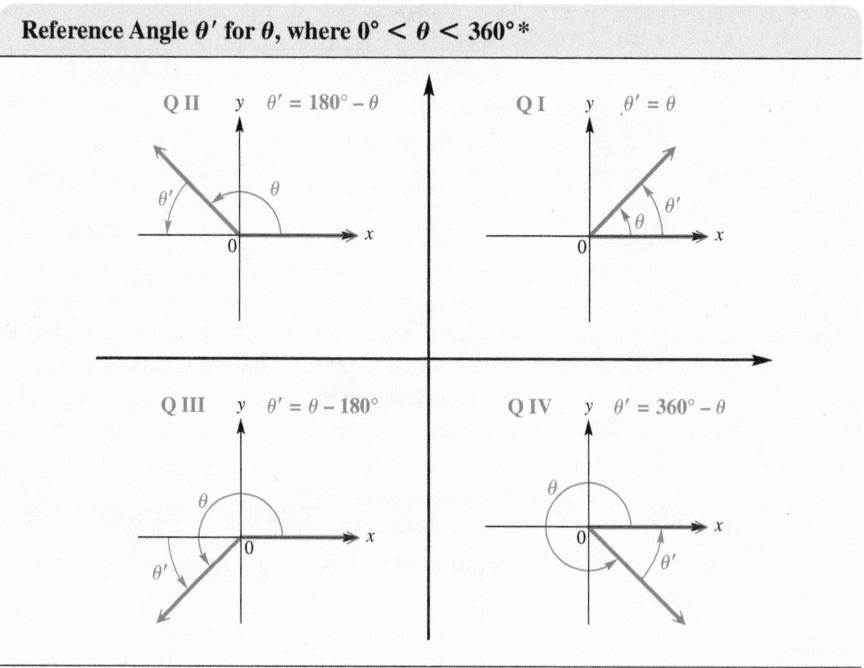

Special Angles as Reference Angles We can now find exact trigonometric function values of angles with reference angles of 30°, 45°, or 60°.

EXAMPLE 2 Finding Trigonometric Function Values of a Quadrant III Angle

Find exact values of the six trigonometric functions of 210°.

SOLUTION An angle of 210° is shown in **Figure 12.** The reference angle is

$$210° - 180° = 30°.$$

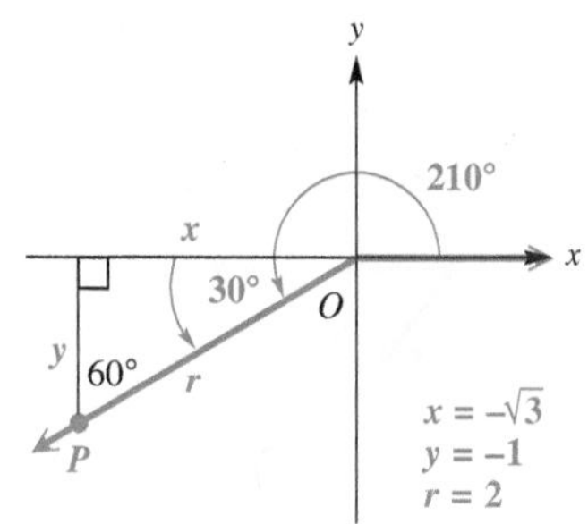

Figure 12

To find the trigonometric function values of 210°, choose point P on the terminal side of the angle so that the distance from the origin O to P is 2. (Any positive number would work, but 2 is most convenient.) By the results from 30°–60° right triangles, the coordinates of point P become $(-\sqrt{3}, -1)$, with $x = -\sqrt{3}$, $y = -1$, and $r = 2$. Then, by the definitions of the trigonometric functions, we obtain the following.

$$\sin 210° = \frac{-1}{2} = -\frac{1}{2} \qquad \csc 210° = \frac{2}{-1} = -2$$

$$\cos 210° = \frac{-\sqrt{3}}{2} = -\frac{\sqrt{3}}{2} \qquad \sec 210° = \frac{2}{-\sqrt{3}} = -\frac{2\sqrt{3}}{3}$$

Rationalize denominators as needed.

$$\tan 210° = \frac{-1}{-\sqrt{3}} = \frac{\sqrt{3}}{3} \qquad \cot 210° = \frac{-\sqrt{3}}{-1} = \sqrt{3}$$

Now Try Exercise 19.

Classroom Example 2
Find the values of the six trigonometric functions of 225°.

Answer:

$\sin 225° = -\frac{\sqrt{2}}{2}$;
$\cos 225° = -\frac{\sqrt{2}}{2}$;
$\tan 225° = 1$;
$\csc 225° = -\sqrt{2}$;
$\sec 225° = -\sqrt{2}$;
$\cot 225° = 1$

*The authors would like to thank Bethany Vaughn and Theresa Matick, of Vincennes Lincoln High School, for their suggestions concerning this table.

Teaching Tip Review the signs of each trigonometric function by quadrant. As mentioned previously, the sentence "**A**ll **S**tudents **T**ake **C**alculus" may help some students remember signs of the basic functions by quadrant. (You could tell them that you graduated from **A**ll **S**tate **T**eachers' **C**ollege.)

Notice in **Example 2** that the trigonometric function values of 210° correspond in absolute value to those of its reference angle 30°. The signs are different for the sine, cosine, secant, and cosecant functions because 210° is a quadrant III angle. These results suggest a shortcut for finding the trigonometric function values of a non-acute angle, using the reference angle.

In **Example 2,** the reference angle for 210° is 30°. Using the trigonometric function values of 30°, and choosing the correct signs for a quadrant III angle, we obtain the same results.

We determine the values of the trigonometric functions for any nonquadrantal angle θ as follows. Keep in mind that all function values are positive when the terminal side is in Quadrant I, the sine and cosecant are positive in Quadrant II, the tangent and cotangent are positive in Quadrant III, and the cosine and secant are positive in Quadrant IV. In other cases, the function values are negative.

Finding Trigonometric Function Values for Any Nonquadrantal Angle θ

Step 1 If $\theta > 360°$, or if $\theta < 0°$, then find a coterminal angle by adding or subtracting 360° as many times as needed to obtain an angle greater than 0° but less than 360°.

Step 2 Find the reference angle θ'.

Step 3 Find the trigonometric function values for reference angle θ'.

Step 4 Determine the correct signs for the values found in Step 3. (Use the table of signs given earlier in the text or the paragraph above, if necessary.) This gives the values of the trigonometric functions for angle θ.

NOTE To avoid sign errors when finding the trigonometric function values of an angle, sketch it in standard position. Include a reference triangle complete with appropriate values for x, y, and r as done in **Figure 12.**

Classroom Example 3
Find the exact value of each expression.

(a) $\sin(-150°)$ (b) $\cot 780°$

Answers:
(a) $-\frac{1}{2}$ (b) $\frac{\sqrt{3}}{3}$

EXAMPLE 3 Finding Trigonometric Function Values Using Reference Angles

Find the exact value of each expression.

(a) $\cos(-240°)$ **(b)** $\tan 675°$

SOLUTION

(a) Because an angle of $-240°$ is coterminal with an angle of

$$-240° + 360° = 120°,$$

the reference angle is $180° - 120° = 60°$, as shown in **Figure 13(a).** The cosine is negative in quadrant II.

$$\begin{aligned} &\cos(-240°) \\ &= \cos 120° \quad \longleftarrow \text{Coterminal angle} \\ &= -\cos 60° \quad \longleftarrow \text{Reference angle} \\ &= -\frac{1}{2} \quad \text{Evaluate.} \end{aligned}$$

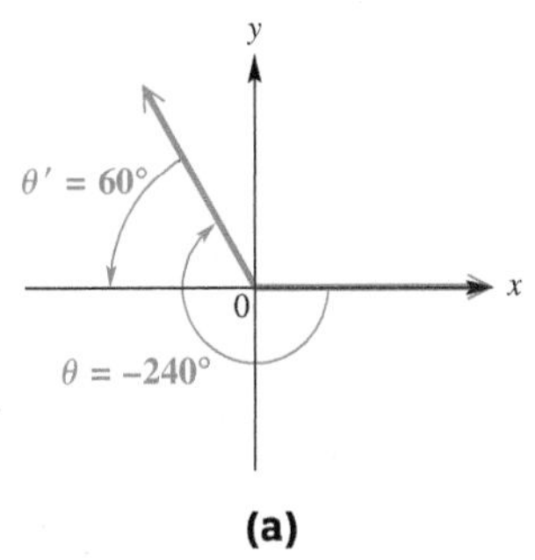

Figure 13

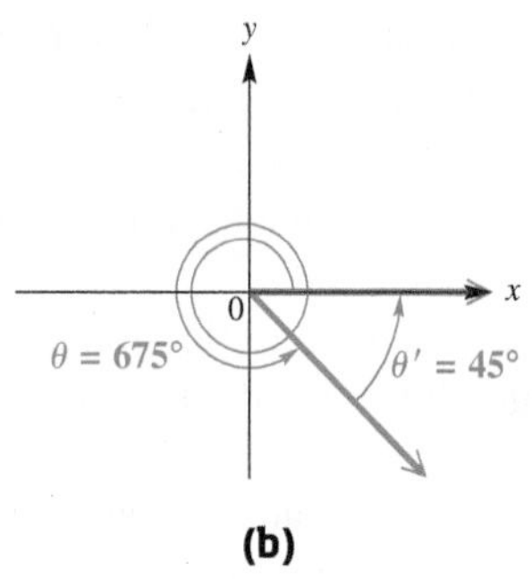

(b)

Figure 13

(b) Subtract 360° to find an angle between 0° and 360° coterminal with 675°.

$$675° - 360° = 315°$$

As shown in **Figure 13(b),** the reference angle is $360° - 315° = 45°$. An angle of 315° is in quadrant IV, so the tangent will be negative.

$\tan 675°$

$= \tan 315°$ Coterminal angle

$= -\tan 45°$ Reference angle; quadrant-based sign choice

$= -1$ Evaluate. ✔ **Now Try Exercises 37 and 39.**

Classroom Example 4
Evaluate.

$\sin^2 45° + 3\cos^2 135° - 2\tan 225°$

Answer: 0

EXAMPLE 4 Using Function Values of Special Angles

Evaluate $\cos 120° + 2\sin^2 60° - \tan^2 30°$

SOLUTION Use the procedure explained earlier to determine $\cos 120° = -\frac{1}{2}$. Then use the values $\cos 120° = -\frac{1}{2}$, $\sin 60° = \frac{\sqrt{3}}{2}$, and $\tan 30° = \frac{\sqrt{3}}{3}$.

$\cos 120° + 2\sin^2 60° - \tan^2 30°$

$= -\frac{1}{2} + 2\left(\frac{\sqrt{3}}{2}\right)^2 - \left(\frac{\sqrt{3}}{3}\right)^2$ Substitute values.

$= -\frac{1}{2} + 2\left(\frac{3}{4}\right) - \frac{3}{9}$ Apply the exponents.

$= \frac{2}{3}$ Simplify.

✔ **Now Try Exercise 47.**

Classroom Example 5
Evaluate each function by first expressing it in terms of a function of an angle between 0° and 360°.
(a) sin 585° **(b)** cot(−930°)

Answers:
(a) $\sin 585° = \sin 225° = -\frac{\sqrt{2}}{2}$
(b) $\cot(-930°) = \cot 150° = -\sqrt{3}$

EXAMPLE 5 Using Coterminal Angles to Find Function Values

Evaluate each function by first expressing it in terms of a function of an angle between 0° and 360°.

(a) $\cos 780°$ **(b)** $\cot(-405°)$

SOLUTION

(a) Subtract 360° as many times as necessary to obtain an angle between 0° and 360°, which gives the following.

$\cos 780°$

$= \cos(780° - 2 \cdot 360°)$ Subtract 720°, which is $2 \cdot 360°$.

$= \cos 60°$ Multiply first and then subtract.

$= \frac{1}{2}$ Evaluate.

(b) Add 360° twice to obtain $-405° + 2(360°) = 315°$, which is located in quadrant IV and has reference angle 45°. The cotangent will be negative.

$$\cot(-405°) = \cot 315° = -\cot 45° = -1$$

✔ **Now Try Exercises 27 and 31.**

Determination of Angle Measures with Special Reference Angles

The ideas discussed in this section can be used "in reverse" to find the measures of certain angles, given a trigonometric function value and an interval in which the angle must lie. We are most often interested in the interval $[0°, 360°)$.

Classroom Example 6
Find all values of θ, if θ is in the interval $[0°, 360°)$ and $\sin\theta = -\frac{\sqrt{3}}{2}$.

Answer: 240°; 300°

Teaching Tip In **Example 6**, relate cosine to the value of the x-coordinate, mentioning that the value of x does not change if a point is reflected from quadrant II to quadrant III.

EXAMPLE 6 Finding Angle Measures Given an Interval and a Function Value

Find all values of θ, if θ is in the interval $[0°, 360°)$ and $\cos\theta = -\frac{\sqrt{2}}{2}$.

SOLUTION The value of $\cos\theta$ is negative, so θ may lie in either quadrant II or III. Because the absolute value of $\cos\theta$ is $\frac{\sqrt{2}}{2}$, the reference angle θ' must be 45°. The two possible angles θ are sketched in **Figure 14.**

$180° - 45° = 135°$ Quadrant II angle θ (from **Figure 14 (a)**)

$180° + 45° = 225°$ Quadrant III angle θ (from **Figure 14 (b)**)

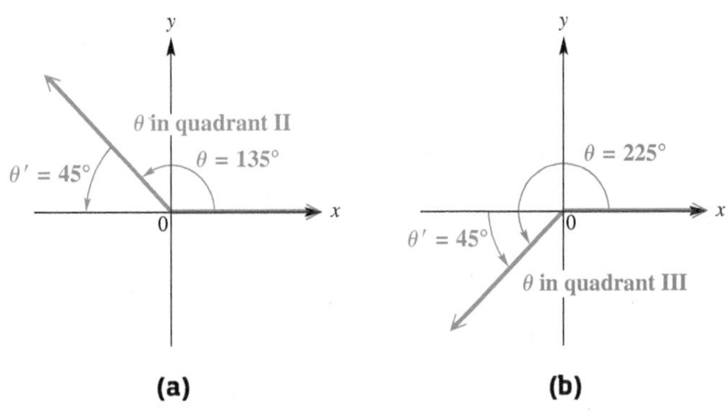

Figure 14

Now Try Exercise 61.

2.2 Exercises

1. negative; III; 60°; $-\frac{\sqrt{3}}{2}$
2. positive; I; 30°; $\frac{\sqrt{3}}{2}$
3. positive; III; 30°; $\frac{\sqrt{3}}{3}$
4. negative; II; 45°; $-\sqrt{2}$

CONCEPT PREVIEW *Fill in the blanks to correctly complete each sentence.*

1. The value of sin 240° is ______ (positive/negative) because 240° is in quadrant ____. The reference angle is ____, and the *exact* value of sin 240° is ____.

2. The value of cos 390° is ______ (positive/negative) because 390° is in quadrant ____. The reference angle is ____, and the *exact* value of cos 390° is ____.

3. The value of $\tan(-150°)$ is ______ (positive/negative) because $-150°$ is in quadrant ____. The reference angle is ____, and the *exact* value of $\tan(-150°)$ is ____.

4. The value of sec 135° is ______ (positive/negative) because 135° is in quadrant ____. The reference angle is ____, and the *exact* value of sec 135° is ____.

5. C
6. F
7. A
8. B
9. D
10. B

11. $\frac{\sqrt{3}}{3}$; $\sqrt{3}$
12. $\frac{\sqrt{2}}{2}$; $\frac{\sqrt{2}}{2}$; $\sqrt{2}$; $\sqrt{2}$
13. $\frac{\sqrt{3}}{2}$; $\frac{\sqrt{3}}{3}$; $\frac{2\sqrt{3}}{3}$
14. $-\frac{1}{2}$; $-\frac{\sqrt{3}}{3}$; -2
15. -1; -1
16. $\frac{1}{2}$; $-\sqrt{3}$; $-\frac{2\sqrt{3}}{3}$
17. $-\frac{\sqrt{3}}{2}$; $-\frac{2\sqrt{3}}{3}$
18. $\sqrt{3}$; $\frac{\sqrt{3}}{3}$

In Exercises 19–36, we give, in order, sine, cosine, tangent, cotangent, secant, and cosecant.

19. $-\frac{\sqrt{3}}{2}$; $\frac{1}{2}$; $-\sqrt{3}$; $-\frac{\sqrt{3}}{3}$; 2; $-\frac{2\sqrt{3}}{3}$
20. $-\frac{\sqrt{2}}{2}$; $\frac{\sqrt{2}}{2}$; -1; -1; $\sqrt{2}$; $-\sqrt{2}$
21. $\frac{\sqrt{2}}{2}$; $\frac{\sqrt{2}}{2}$; 1; 1; $\sqrt{2}$; $\sqrt{2}$
22. $\frac{\sqrt{3}}{2}$; $\frac{1}{2}$; $\sqrt{3}$; $\frac{\sqrt{3}}{3}$; 2; $\frac{2\sqrt{3}}{3}$
23. $\frac{\sqrt{3}}{2}$; $-\frac{1}{2}$; $-\sqrt{3}$; $-\frac{\sqrt{3}}{3}$; -2; $\frac{2\sqrt{3}}{3}$
24. $\frac{\sqrt{2}}{2}$; $-\frac{\sqrt{2}}{2}$; -1; -1; $-\sqrt{2}$; $\sqrt{2}$
25. $-\frac{1}{2}$; $-\frac{\sqrt{3}}{2}$; $\frac{\sqrt{3}}{3}$; $\sqrt{3}$; $-\frac{2\sqrt{3}}{3}$; -2
26. $\frac{1}{2}$; $\frac{\sqrt{3}}{2}$; $\frac{\sqrt{3}}{3}$; $\sqrt{3}$; $\frac{2\sqrt{3}}{3}$; 2
27. $-\frac{\sqrt{2}}{2}$; $-\frac{\sqrt{2}}{2}$; 1; 1; $-\sqrt{2}$; $-\sqrt{2}$
28. $\frac{\sqrt{3}}{2}$; $\frac{1}{2}$; $\sqrt{3}$; $\frac{\sqrt{3}}{3}$; 2; $\frac{2\sqrt{3}}{3}$
29. $\frac{\sqrt{3}}{2}$; $\frac{1}{2}$; $\sqrt{3}$; $\frac{\sqrt{3}}{3}$; 2; $\frac{2\sqrt{3}}{3}$
30. $-\frac{1}{2}$; $\frac{\sqrt{3}}{2}$; $-\frac{\sqrt{3}}{3}$; $-\sqrt{3}$; $\frac{2\sqrt{3}}{3}$; -2
31. $-\frac{1}{2}$; $-\frac{\sqrt{3}}{2}$; $\frac{\sqrt{3}}{3}$; $\sqrt{3}$; $-\frac{2\sqrt{3}}{3}$; -2
32. $\frac{\sqrt{3}}{2}$; $\frac{1}{2}$; $\sqrt{3}$; $\frac{\sqrt{3}}{3}$; 2; $\frac{2\sqrt{3}}{3}$
33. $\frac{1}{2}$; $-\frac{\sqrt{3}}{2}$; $-\frac{\sqrt{3}}{3}$; $-\sqrt{3}$; $-\frac{2\sqrt{3}}{3}$; 2

Concept Check *Match each angle in Column I with its reference angle in Column II. Choices may be used once, more than once, or not at all.* ***See Example 1.***

I		II	
5. 98°	**6.** 212°	**A.** 45°	**B.** 60°
7. −135°	**8.** −60°	**C.** 82°	**D.** 30°
9. 750°	**10.** 480°	**E.** 38°	**F.** 32°

Complete the table with exact trigonometric function values. Do not use a calculator. ***See Examples 2 and 3.***

	θ	$\sin\theta$	$\cos\theta$	$\tan\theta$	$\cot\theta$	$\sec\theta$	$\csc\theta$
11.	30°	$\frac{1}{2}$	$\frac{\sqrt{3}}{2}$			$\frac{2\sqrt{3}}{3}$	2
12.	45°			1	1		
13.	60°		$\frac{1}{2}$	$\sqrt{3}$		2	
14.	120°	$\frac{\sqrt{3}}{2}$		$-\sqrt{3}$			$\frac{2\sqrt{3}}{3}$
15.	135°	$\frac{\sqrt{2}}{2}$	$-\frac{\sqrt{2}}{2}$			$-\sqrt{2}$	$\sqrt{2}$
16.	150°		$-\frac{\sqrt{3}}{2}$	$-\frac{\sqrt{3}}{3}$			2
17.	210°	$-\frac{1}{2}$		$\frac{\sqrt{3}}{3}$	$\sqrt{3}$		−2
18.	240°	$-\frac{\sqrt{3}}{2}$	$-\frac{1}{2}$			−2	$-\frac{2\sqrt{3}}{3}$

Find exact values of the six trigonometric functions of each angle. Rationalize denominators when applicable. ***See Examples 2, 3, and 5.***

19. 300° **20.** 315° **21.** 405° **22.** 420° **23.** 480° **24.** 495°

25. 570° **26.** 750° **27.** 1305° **28.** 1500° **29.** −300° **30.** −390°

31. −510° **32.** −1020° **33.** −1290° **34.** −855° **35.** −1860° **36.** −2205°

Find the exact value of each expression. ***See Example 3.***

37. $\sin 1305°$ **38.** $\sin 1500°$ **39.** $\cos(-510°)$ **40.** $\tan(-1020°)$

41. $\csc(-855°)$ **42.** $\sec(-495°)$ **43.** $\tan 3015°$ **44.** $\cot 2280°$

Evaluate each expression. ***See Example 4.***

45. $\sin^2 120° + \cos^2 120°$

46. $\sin^2 225° + \cos^2 225°$

47. $2\tan^2 120° + 3\sin^2 150° - \cos^2 180°$

48. $\cot^2 135° - \sin 30° + 4\tan 45°$

49. $\sin^2 225° - \cos^2 270° + \tan^2 60°$

50. $\cot^2 90° - \sec^2 180° + \csc^2 135°$

51. $\cos^2 60° + \sec^2 150° - \csc^2 210°$

52. $\cot^2 135° + \tan^4 60° - \sin^4 180°$

34. $-\frac{\sqrt{2}}{2}; -\frac{\sqrt{2}}{2}; 1; 1; -\sqrt{2}; -\sqrt{2}$

35. $-\frac{\sqrt{3}}{2}; \frac{1}{2}; -\sqrt{3}; -\frac{\sqrt{3}}{3}; 2; -\frac{2\sqrt{3}}{3}$

36. $-\frac{\sqrt{2}}{2}; \frac{\sqrt{2}}{2}; -1; -1; \sqrt{2}; -\sqrt{2}$

37. $-\frac{\sqrt{2}}{2}$ 38. $\frac{\sqrt{3}}{2}$

39. $-\frac{\sqrt{3}}{2}$ 40. $\sqrt{3}$

41. $-\sqrt{2}$ 42. $-\sqrt{2}$

43. -1 44. $-\frac{\sqrt{3}}{3}$

45. 1 46. 1

47. $\frac{23}{4}$ 48. $\frac{9}{2}$

49. $\frac{7}{2}$ 50. 1

51. $-\frac{29}{12}$ 52. 10

53. false; $0 \neq \frac{\sqrt{3}+1}{2}$

54. false; $\frac{1+\sqrt{3}}{2} \neq 1$

55. false; $\frac{1}{2} \neq \sqrt{3}$

56. true

57. true 58. true

59. false; $0 \neq \sqrt{2}$

60. true

61. 30°; 150° 62. 30°; 330°

63. 120°; 300° 64. 135°; 225°

65. 45°; 315° 66. 120°; 300°

67. 210°; 330° 68. 240°; 300°

69. 30°; 210° 70. 120°; 240°

71. 225°; 315° 72. 135°; 315°

73. $(-3\sqrt{3}, 3)$

74. $(-5\sqrt{2}, -5\sqrt{2})$

75. yes 76. no

77. positive 78. positive

79. positive 80. negative

81. negative 82. negative

83. When an integer multiple of 360° is added to θ, the resulting angle is coterminal with θ. The sine values of coterminal angles are equal.

84. See the answer to Exercise 83, and replace "sine" with "cosine."

85. 0.9 86. -0.4

87. 45°; 225° 88. 135°; 315°

Determine whether each statement is true *or* false. *If false, tell why.* ***See Example 4.***

53. $\cos(30° + 60°) = \cos 30° + \cos 60°$

54. $\sin 30° + \sin 60° = \sin(30° + 60°)$

55. $\cos 60° = 2 \cos 30°$

56. $\cos 60° = 2 \cos^2 30° - 1$

57. $\sin^2 45° + \cos^2 45° = 1$

58. $\tan^2 60° + 1 = \sec^2 60°$

59. $\cos(2 \cdot 45°) = 2 \cos 45°$

60. $\sin(2 \cdot 30°) = 2 \sin 30° \cdot \cos 30°$

Find all values of θ, if θ is in the interval $[0°, 360°)$ and has the given function value. ***See Example 6.***

61. $\sin \theta = \frac{1}{2}$

62. $\cos \theta = \frac{\sqrt{3}}{2}$

63. $\tan \theta = -\sqrt{3}$

64. $\sec \theta = -\sqrt{2}$

65. $\cos \theta = \frac{\sqrt{2}}{2}$

66. $\cot \theta = -\frac{\sqrt{3}}{3}$

67. $\csc \theta = -2$

68. $\sin \theta = -\frac{\sqrt{3}}{2}$

69. $\tan \theta = \frac{\sqrt{3}}{3}$

70. $\cos \theta = -\frac{1}{2}$

71. $\csc \theta = -\sqrt{2}$

72. $\cot \theta = -1$

Concept Check *Find the coordinates of the point P on the circumference of each circle. (Hint: Sketch x- and y-axes, and interpret so that the angle is in standard position.)*

73.

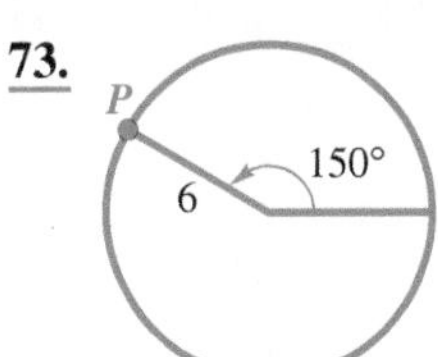

74.

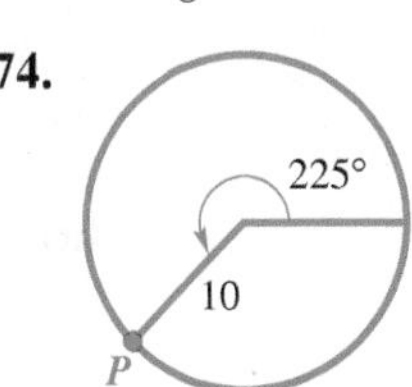

75. ***Concept Check*** Does there exist an angle θ with the function values $\cos \theta = 0.6$ and $\sin \theta = -0.8$?

76. ***Concept Check*** Does there exist an angle θ with the function values $\cos \theta = \frac{2}{3}$ and $\sin \theta = \frac{3}{4}$?

Suppose θ is in the interval $(90°, 180°)$. Find the sign of each of the following.

77. $\cos \frac{\theta}{2}$

78. $\sin \frac{\theta}{2}$

79. $\sec(\theta + 180°)$

80. $\cot(\theta + 180°)$

81. $\sin(-\theta)$

82. $\cos(-\theta)$

Concept Check *Work each problem.*

83. Why is $\sin \theta = \sin(\theta + n \cdot 360°)$ true for any angle θ and any integer n?

84. Why is $\cos \theta = \cos(\theta + n \cdot 360°)$ true for any angle θ and any integer n?

85. Without using a calculator, determine which of the following numbers is closest to $\sin 115°$: -0.9, -0.1, 0, 0.1, or 0.9.

86. Without using a calculator, determine which of the following numbers is closest to $\cos 115°$: -0.6, -0.4, 0, 0.4, or 0.6.

87. For what angles θ between 0° and 360° is $\cos \theta = \sin \theta$ true?

88. For what angles θ between 0° and 360° is $\cos \theta = -\sin \theta$ true?

2.3 Approximations of Trigonometric Function Values

- Calculator Approximations of Trigonometric Function Values
- Calculator Approximations of Angle Measures
- An Application

Calculator Approximations of Trigonometric Function Values We learned how to find exact function values for special angles and for angles having special reference angles earlier in this chapter. In this section we investigate how calculators provide approximations for function values of angles that do not satisfy these conditions. (Of course, they can also be used to find exact values such as $\cos(-240°)$ and $\tan 675°$, as seen in **Figure 15.**)

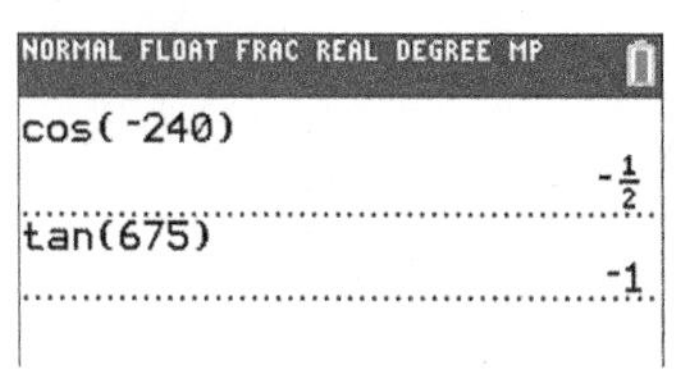

Degree mode

Figure 15

CAUTION It is important to remember that ***when we are evaluating trigonometric functions of angles given in degrees, the calculator must be in degree mode.*** An easy way to check this is to enter sin 90. If the displayed answer is 1, then the calculator is in degree mode.

Also remember that if the angle or the reference angle is not a special or quadrantal angle, then the value given by the calculator is an *approximation*. And even if the angle or reference angle *is* a special angle, the value given by the calculator will often be an approximation.

Teaching Tip Caution students *not* to use the $\sin^{-1}$, $\cos^{-1}$, and $\tan^{-1}$ keys when evaluating reciprocal functions with a calculator.

EXAMPLE 1 Finding Function Values with a Calculator

Approximate the value of each expression.

(a) $\sin 49° 12'$ **(b)** $\sec 97.977°$ **(c)** $\dfrac{1}{\cot 51.4283°}$ **(d)** $\sin(-246°)$

SOLUTION See **Figure 16.** We give values to eight decimal places below.

```
NORMAL FLOAT FRAC REAL DEGREE MP
sin(49°12')
                    .7569950557
1/cos(97.977)
                   -7.205879213
tan(51.4283)
                    1.253948151
sin(-246)
                    .9135454576
```

Degree mode

Figure 16

(a) We *may* begin by converting 49° 12′ to decimal degrees.

$$49°\,12' = 49\frac{12°}{60} = 49.2°$$

However, some calculators allow direct entry of degrees, minutes, and seconds. (The method of entry varies among models.) Entering either $\sin(49°\,12')$ or $\sin 49.2°$ gives the same approximation.

$$\sin 49°\,12' = \sin 49.2° \approx 0.75699506$$

(b) There are no dedicated calculator keys for the secant, cosecant, and cotangent functions. However, we can use reciprocal identities to evaluate them. Recall that $\sec\theta = \frac{1}{\cos\theta}$ for all angles θ, where $\cos\theta \neq 0$. Therefore, we use the reciprocal of the cosine function to evaluate the secant function.

$$\sec 97.977° = \frac{1}{\cos 97.977°} \approx -7.20587921$$

(c) Use the reciprocal identity $\frac{1}{\cot\theta} = \tan\theta$ to simplify the expression first.

$$\frac{1}{\cot 51.4283°} = \tan 51.4283° \approx 1.25394815$$

(d) $\sin(-246°) \approx 0.91354546$

✔ **Now Try Exercises 11, 13, 17, and 21.**

Classroom Example 1
Approximate the value of each expression.

(a) $\tan 68°\,43'$
(b) $\cos 193.622°$
(c) $\csc 35.8471°$
(d) $\sec(-287°)$

Answers:
(a) 2.56707352
(b) −0.97187064
(c) 1.70757967
(d) 3.42030362

Calculator Approximations of Angle Measures To find the measure of an angle having a certain trigonometric function value, calculators have three *inverse functions* (denoted $\sin^{-1}$, $\cos^{-1}$, and $\tan^{-1}$). ***If x is an appropriate number, then $\sin^{-1} x$, $\cos^{-1} x$, or $\tan^{-1} x$ gives the measure of an angle whose sine, cosine, or tangent, respectively, is x.*** For applications in this chapter, these functions will return angles in quadrant I.

EXAMPLE 2 Using Inverse Trigonometric Functions to Find Angles

Find an angle θ in the interval $[0°, 90°)$ that satisfies each condition.

(a) $\sin \theta = 0.96770915$ **(b)** $\sec \theta = 1.0545829$

SOLUTION

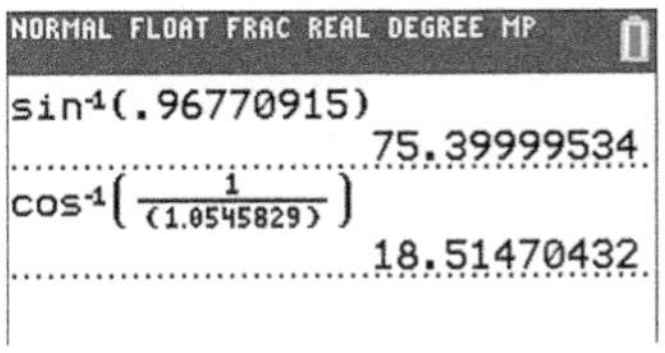

NORMAL FLOAT FRAC REAL DEGREE MP
sin⁻¹(.96770915)
75.39999534
cos⁻¹(1/(1.0545829))
18.51470432

Degree mode

Figure 17

(a) Using degree mode and the inverse sine function, we find that an angle θ having sine value 0.96770915 is 75.399995°. (There are infinitely many such angles, but the calculator gives only this one.)

$$\theta = \sin^{-1} 0.96770915 \approx 75.399995°$$

See **Figure 17.**

(b) Use the identity $\cos \theta = \frac{1}{\sec \theta}$. If $\sec \theta = 1.0545829$, then

$$\cos \theta = \frac{1}{1.0545829}.$$

Now, find θ using the inverse cosine function. See **Figure 17.**

$$\theta = \cos^{-1}\left(\frac{1}{1.0545829}\right) \approx 18.514704°$$

Now Try Exercises 31 and 35.

Classroom Example 2
Use a calculator to find an angle θ in the interval $[0°, 90°)$ that satisfies each condition.

(a) $\cos \theta = 0.92118541$

(b) $\cot \theta = 1.4466474$

Answers:

(a) 22.899999°

(b) 34.654300°

CAUTION Compare **Examples 1(b) and 2(b).**

- To determine the secant of an angle, as in **Example 1(b),** we find the *reciprocal of the cosine* of the angle.
- To determine an angle with a given secant value, as in **Example 2(b),** we find the *inverse cosine of the reciprocal* of the value.

An Application

EXAMPLE 3 Finding Grade Resistance

When an automobile travels uphill or downhill on a highway, it experiences a force due to gravity. This force F in pounds is the **grade resistance** and is modeled by the equation

$$F = W \sin \theta,$$

where θ is the grade and W is the weight of the automobile. If the automobile is moving uphill, then $\theta > 0°$; if downhill, then $\theta < 0°$. See **Figure 18.** (*Source:* Mannering, F. and W. Kilareski, *Principles of Highway Engineering and Traffic Analysis,* Second Edition, John Wiley and Sons.)

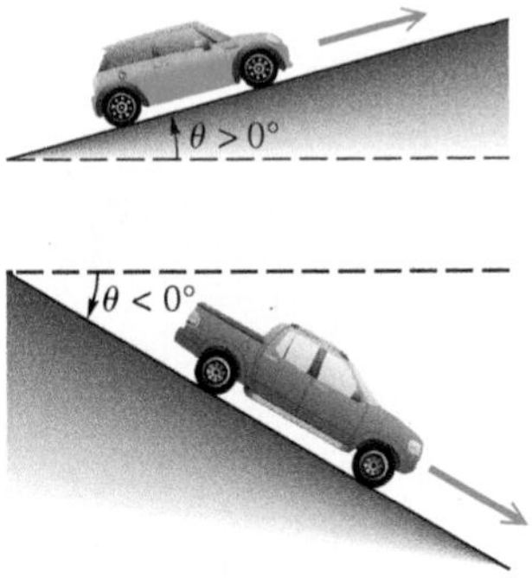

$\theta > 0°$

$\theta < 0°$

Figure 18

Classroom Example 3
Refer to **Example 3.**

(a) Calculate F to the nearest 10 lb for a 5500-lb car traveling an uphill grade with $\theta = 3.9°$.

(b) Calculate F to the nearest 10 lb for a 2800-lb car traveling a downhill grade with $\theta = -4.8°$.

(c) A 2400-lb car traveling uphill has a grade resistance of 288 lb. What is the angle of the grade?

Answers:
(a) 370 lb (b) −230 lb
(c) 6.89°

(a) Calculate F to the nearest 10 lb for a 2500-lb car traveling an uphill grade with $\theta = 2.5°$.

(b) Calculate F to the nearest 10 lb for a 5000-lb truck traveling a downhill grade with $\theta = -6.1°$.

(c) Calculate F for $\theta = 0°$ and $\theta = 90°$. Do these answers agree with intuition?

SOLUTION

(a)

$F = W \sin \theta$ Given model for grade resistance

$F = 2500 \sin 2.5°$ Substitute given values.

$F \approx 110$ lb Evaluate.

(b) $F = W \sin \theta = 5000 \sin(-6.1°) \approx -530$ lb

F is negative because the truck is moving downhill.

(c) $F = W \sin \theta = W \sin 0° = W(0) = 0$ lb

$F = W \sin \theta = W \sin 90° = W(1) = W$ lb

This agrees with intuition because if $\theta = 0°$, then there is level ground and gravity does not cause the vehicle to roll. If θ were 90°, the road would be vertical and the full weight of the vehicle would be pulled downward by gravity, so $F = W$.

✔ **Now Try Exercises 69 and 71.**

2.3 Exercises

1. J 2. B
3. E 4. F
5. D 6. I
7. H 8. A
9. G 10. C

In Exercises 11–28, the number of decimal places may vary depending on the calculator used. We show six places.

11. 0.625243 12. 0.750111
13. 1.027349 14. 1.776815
15. 15.055723 16. 1.841771
17. 0.740805 18. −5.729742
19. 1.483014 20. 1.907415
21. $\tan 23.4° \approx 0.432739$
22. $\cos 14.8° \approx 0.966823$
23. $\cot 77° \approx 0.230868$
24. $\tan 33° \approx 0.649408$
25. $\tan 4.72° \approx 0.082566$
26. $\sin 3.69° \approx 0.064358$
27. $\cos 51° \approx 0.629320$
28. $\tan 22° \approx 0.404026$

CONCEPT PREVIEW *Match each trigonometric function value or angle in Column I with its appropriate approximation in Column II.*

I		II	
1. $\sin 83°$	**2.** $\cos^{-1} 0.45$	**A.** 88.09084757°	**B.** 63.25631605°
3. $\tan 16°$	**4.** $\cot 27°$	**C.** 1.909152433°	**D.** 17.45760312°
5. $\sin^{-1} 0.30$	**6.** $\sec 18°$	**E.** 0.2867453858	**F.** 1.962610506
7. $\csc 80°$	**8.** $\tan^{-1} 30$	**G.** 14.47751219°	**H.** 1.015426612
9. $\csc^{-1} 4$	**10.** $\cot^{-1} 30$	**I.** 1.051462224	**J.** 0.9925461516

Use a calculator to approximate the value of each expression. Give answers to six decimal places. In Exercises 21–28, simplify the expression before using the calculator. ***See Example 1.***

11. $\sin 38° 42'$ **12.** $\cos 41° 24'$ **13.** $\sec 13° 15'$

14. $\csc 145° 45'$ **15.** $\cot 183° 48'$ **16.** $\tan 421° 30'$

17. $\sin(-312° 12')$ **18.** $\tan(-80° 06')$ **19.** $\csc(-317° 36')$

20. $\cot(-512° 20')$ **21.** $\dfrac{1}{\cot 23.4°}$ **22.** $\dfrac{1}{\sec 14.8°}$

23. $\dfrac{\cos 77°}{\sin 77°}$ **24.** $\dfrac{\sin 33°}{\cos 33°}$ **25.** $\cot(90° - 4.72°)$

26. $\cos(90° - 3.69°)$ **27.** $\dfrac{1}{\csc(90° - 51°)}$ **28.** $\dfrac{1}{\tan(90° - 22°)}$

29. 55.845496° **30.** 81.168073°
31. 16.166641° **32.** 57.997172°
33. 38.491580° **34.** 46.173582°
35. 68.673241° **36.** 30.502748°
37. 45.526434° **38.** 31.199998°
39. 12.227282° **40.** 77.831359°

41. The calculator is not in degree mode.
42. Find an angle coterminal with 2000° that is within the capability of the calculator. Then find the cosine of that coterminal angle.
43. 56° **44.** 0.3746065934

45. 1 **46.** −1
47. 1 **48.** 0
49. 0 **50.** 1

51. false **52.** false
53. true **54.** true
55. false **56.** false
57. false **58.** false
59. true **60.** true
61. true **62.** false

63. 68°; 112° **64.** 32°; 148°
65. 44°; 316° **66.** 84°; 276°
67. 51°; 231° **68.** 35°; 215°

Find a value of θ in the interval $[0°, 90°)$ that satisfies each statement. Write each answer in decimal degrees to six decimal places. ***See Example 2.***

29. $\tan\theta = 1.4739716$	**30.** $\tan\theta = 6.4358841$	**31.** $\sin\theta = 0.27843196$
32. $\sin\theta = 0.84802194$	**33.** $\cot\theta = 1.2575516$	**34.** $\csc\theta = 1.3861147$
35. $\sec\theta = 2.7496222$	**36.** $\sec\theta = 1.1606249$	**37.** $\cos\theta = 0.70058013$
38. $\cos\theta = 0.85536428$	**39.** $\csc\theta = 4.7216543$	**40.** $\cot\theta = 0.21563481$

Concept Check Answer each question.

41. A student, wishing to use a calculator to verify the value of sin 30°, enters the information correctly but gets a display of −0.98803162. He knows that the display should be 0.5, and he also knows that his calculator is in good working order. What might the problem be?

42. At one time, a certain make of calculator did not allow the input of angles outside of a particular interval when finding trigonometric function values. For example, trying to find cos 2000° using the methods of this section gave an error message, despite the fact that cos 2000° can be evaluated. How would we use this calculator to find cos 2000°?

43. What value of A, to the nearest degree, between 0° and 90° will produce the output in the graphing calculator screen?

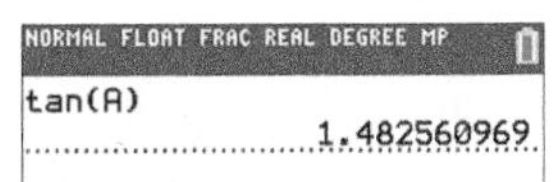

44. What value of A will produce the output (in degrees) in the graphing calculator screen? Give as many decimal places as shown on the calculator.

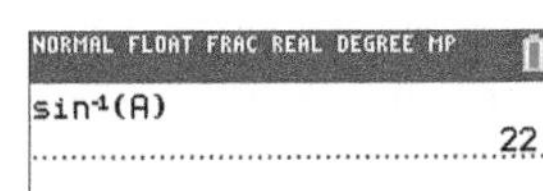

Use a calculator to evaluate each expression.

45. $\sin 35° \cos 55° + \cos 35° \sin 55°$

46. $\cos 100° \cos 80° - \sin 100° \sin 80°$

47. $\sin^2 36° + \cos^2 36°$

48. $2 \sin 25° 13' \cos 25° 13' - \sin 50° 26'$

49. $\cos 75° 29' \cos 14° 31' - \sin 75° 29' \sin 14° 31'$

50. $\sin 28° 14' \cos 61° 46' + \cos 28° 14' \sin 61° 46'$

Use a calculator to decide whether each statement is true *or* false. *It may be that a true statement will lead to results that differ in the last decimal place due to rounding error.*

51. $\sin 10° + \sin 10° = \sin 20°$

52. $\cos 40° = 2 \cos 20°$

53. $\sin 50° = 2 \sin 25° \cos 25°$

54. $\cos 70° = 2 \cos^2 35° - 1$

55. $\cos 40° = 1 - 2 \sin^2 80°$

56. $2 \cos 38° 22' = \cos 76° 44'$

57. $\sin 39° 48' + \cos 39° 48' = 1$

58. $\frac{1}{2} \sin 40° = \sin\left[\frac{1}{2}(40°)\right]$

59. $1 + \cot^2 42.5° = \csc^2 42.5°$

60. $\tan^2 72° 25' + 1 = \sec^2 72° 25'$

61. $\cos(30° + 20°) = \cos 30° \cos 20° - \sin 30° \sin 20°$

62. $\cos(30° + 20°) = \cos 30° + \cos 20°$

Find two angles in the interval $[0°, 360°)$ that satisfy each of the following. Round answers to the nearest degree.

63. $\sin\theta = 0.92718385$

64. $\sin\theta = 0.52991926$

65. $\cos\theta = 0.71933980$

66. $\cos\theta = 0.10452846$

67. $\tan\theta = 1.2348971$

68. $\tan\theta = 0.70020753$

69. 70 lb 70. -100 lb
71. $-2.9°$ 72. $2.9°$
73. 2500 lb 74. 2800 lb
75. A 2200-lb car on a 2° uphill grade has greater grade resistance.
76. $\sin\theta$: 0.0000, 0.0087, 0.0175, 0.0262, 0.0349, 0.0436, 0.0523, 0.0610, 0.0698

$\tan\theta$: 0.0000, 0.0087, 0.0175, 0.0262, 0.0349, 0.0437, 0.0524, 0.0612, 0.0699

$\frac{\pi\theta}{180}$: 0.0000, 0.0087, 0.0175, 0.0262, 0.0349, 0.0436, 0.0524, 0.0611, 0.0698

(a) If θ is small, then $\sin\theta \approx \tan\theta \approx \frac{\pi\theta}{180}$.

(b) $F = W\sin\theta \approx W\tan\theta \approx \frac{W\pi\theta}{180}$

(c) 80 lb
(d) 118 lb

77. 703 ft 78. 1701 ft
79. R would decrease; 644 ft, 1559 ft

(Modeling) Grade Resistance *Solve each problem.* **See Example 3.**

69. Find the grade resistance, to the nearest ten pounds, for a 2100-lb car traveling on a 1.8° uphill grade.

70. Find the grade resistance, to the nearest ten pounds, for a 2400-lb car traveling on a $-2.4°$ downhill grade.

71. A 2600-lb car traveling downhill has a grade resistance of -130 lb. Find the angle of the grade to the nearest tenth of a degree.

72. A 3000-lb car traveling uphill has a grade resistance of 150 lb. Find the angle of the grade to the nearest tenth of a degree.

73. A car traveling on a 2.7° uphill grade has a grade resistance of 120 lb. Determine the weight of the car to the nearest hundred pounds.

74. A car traveling on a $-3°$ downhill grade has a grade resistance of -145 lb. Determine the weight of the car to the nearest hundred pounds.

75. Which has the greater grade resistance: a 2200-lb car on a 2° uphill grade or a 2000-lb car on a 2.2° uphill grade?

76. Complete the table for values of $\sin\theta$, $\tan\theta$, and $\frac{\pi\theta}{180}$ to four decimal places.

θ	0°	0.5°	1°	1.5°	2°	2.5°	3°	3.5°	4°
$\sin\theta$									
$\tan\theta$									
$\frac{\pi\theta}{180}$									

(a) How do $\sin\theta$, $\tan\theta$, and $\frac{\pi\theta}{180}$ compare for small grades θ?

(b) Highway grades are usually small. Give two approximations of the grade resistance $F = W\sin\theta$ that do not use the sine function.

(c) A stretch of highway has a 4-ft vertical rise for every 100 ft of horizontal run. Use an approximation from part (b) to estimate the grade resistance, to the nearest pound, for a 2000-lb car on this stretch of highway.

(d) Without evaluating a trigonometric function, estimate the grade resistance, to the nearest pound, for an 1800-lb car on a stretch of highway that has a 3.75° grade.

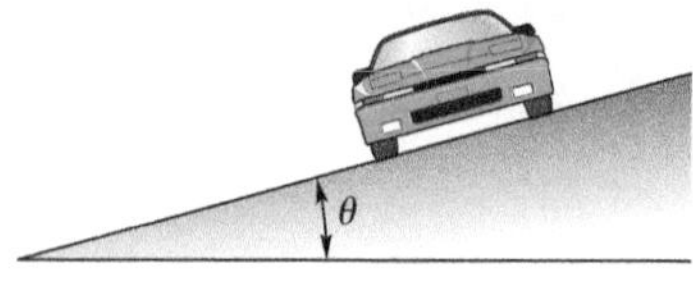

(Modeling) Design of Highway Curves *When highway curves are designed, the outside of the curve is often slightly elevated or inclined above the inside of the curve. See the figure. This inclination is the* ***superelevation.*** *For safety reasons, it is important that both the curve's radius and superelevation be correct for a given speed limit. If an automobile is traveling at velocity V (in feet per second), the safe radius R, in feet, for a curve with superelevation θ is modeled by the formula*

$$R = \frac{V^2}{g(f + \tan\theta)},$$

where f and g are constants. (*Source:* Mannering, F. and W. Kilareski, *Principles of Highway Engineering and Traffic Analysis,* Second Edition, John Wiley and Sons.)

77. A roadway is being designed for automobiles traveling at 45 mph. If $\theta = 3°$, $g = 32.2$, and $f = 0.14$, calculate R to the nearest foot. (*Hint:* 45 mph = 66 ft per sec)

78. Determine the radius of the curve, to the nearest foot, if the speed in **Exercise 77** is increased to 70 mph.

79. How would increasing angle θ affect the results? Verify your answer by repeating **Exercises 77 and 78** with $\theta = 4°$.

80. 55 mph

81. (a) 2×10^8 m per sec
(b) 2×10^8 m per sec

82. (a) 19° (b) 50°

83. 48.7° 84. 7.9°

85. 155 ft

80. Refer to **Exercise 77** and use the same values for f and g. A highway curve has radius $R = 1150$ ft and a superelevation of $\theta = 2.1°$. What should the speed limit (in miles per hour) be for this curve?

(Modeling) Speed of Light When a light ray travels from one medium, such as air, to another medium, such as water or glass, the speed of the light changes, and the light ray is bent, or ***refracted,*** *at the boundary between the two media. (This is why objects under water appear to be in a different position from where they really are.) It can be shown in physics that these changes are related by* ***Snell's law***

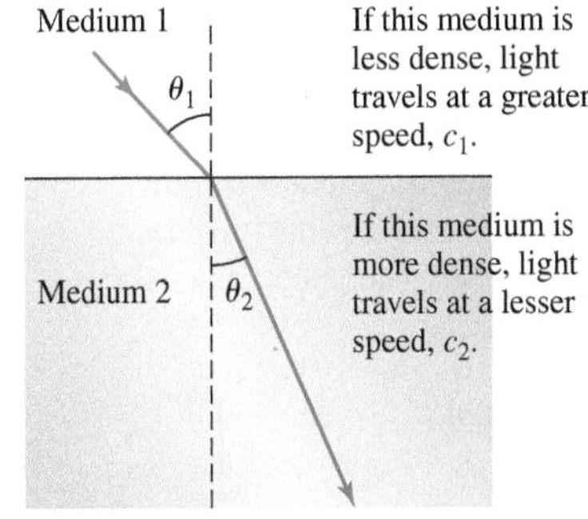

$$\frac{c_1}{c_2} = \frac{\sin \theta_1}{\sin \theta_2},$$

where c_1 is the speed of light in the first medium, c_2 is the speed of light in the second medium, and θ_1 and θ_2 are the angles shown in the figure. In Exercises 81 and 82, assume that $c_1 = 3 \times 10^8$ m per sec.

81. Find the speed of light in the second medium for each of the following.

(a) $\theta_1 = 46°, \theta_2 = 31°$ **(b)** $\theta_1 = 39°, \theta_2 = 28°$

82. Find θ_2 for each of the following values of θ_1 and c_2. Round to the nearest degree.

(a) $\theta_1 = 40°, c_2 = 1.5 \times 10^8$ m per sec **(b)** $\theta_1 = 62°, c_2 = 2.6 \times 10^8$ m per sec

(Modeling) Fish's View of the World The figure shows a fish's view of the world above the surface of the water. (Source: Walker, J., "The Amateur Scientist," *Scientific American.) Suppose that a light ray comes from the horizon, enters the water, and strikes the fish's eye.*

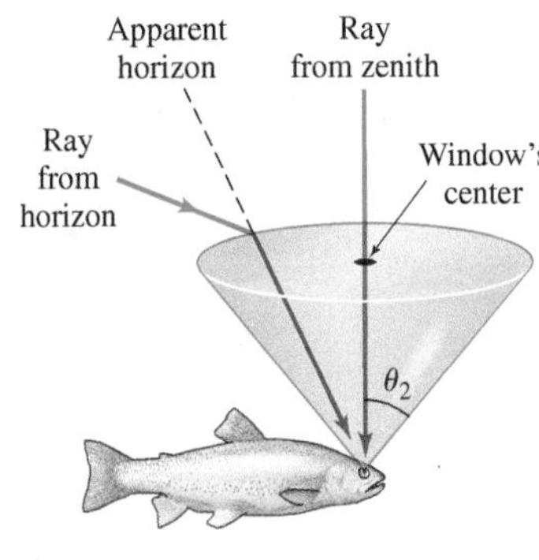

83. Assume that this ray gives a value of 90° for angle θ_1 in the formula for Snell's law. (In a practical situation, this angle would probably be a little less than 90°.) The speed of light in water is about 2.254×10^8 m per sec. Find angle θ_2 to the nearest tenth.

84. Refer to **Exercise 83.** Suppose an object is located at a true angle of 29.6° above the horizon. Find the apparent angle above the horizon to a fish.

(Modeling) Braking Distance If aerodynamic resistance is ignored, the braking distance D (in feet) for an automobile to change its velocity from V_1 to V_2 (feet per second) can be modeled using the following equation.

$$D = \frac{1.05(V_1{}^2 - V_2{}^2)}{64.4(K_1 + K_2 + \sin \theta)}$$

K_1 is a constant determined by the efficiency of the brakes and tires, K_2 is a constant determined by the rolling resistance of the automobile, and θ is the grade of the highway. (Source: Mannering, F. and W. Kilareski, *Principles of Highway Engineering and Traffic Analysis,* Second Edition, John Wiley and Sons.*)*

85. Compute the number of feet, to the nearest unit, required to slow a car from 55 mph to 30 mph while traveling uphill with a grade of $\theta = 3.5°$. Let $K_1 = 0.4$ and $K_2 = 0.02$. (*Hint:* Change miles per hour to feet per second.)

86. 194 ft
87. Negative values of θ require greater distances for slowing down than positive values.
88. 78 mph

89. A: 69 mph; B: 66 mph
90. To find the reduced speed r, we use the fact that $\cos \theta = \frac{r}{a}$, which leads to $r = a \cos \theta$. This confirms the "cosine effect" that reduces the radar reading.

91. 550 ft **92.** 369 ft

86. Repeat **Exercise 85** with $\theta = -2°$.

87. How is braking distance affected by grade θ?

88. An automobile is traveling at 90 mph on a highway with a downhill grade of $\theta = -3.5°$. The driver sees a stalled truck in the road 200 ft away and immediately applies the brakes. Assuming that a collision cannot be avoided, how fast (in miles per hour, to the nearest unit) is the car traveling when it hits the truck? (Use the same values for K_1 and K_2 as in **Exercise 85.**)

*(Modeling) **Measuring Speed by Radar*** *Any offset between a stationary radar gun and a moving target creates a "cosine effect" that reduces the radar reading by the cosine of the angle between the gun and the vehicle. That is, the radar speed reading is the product of the actual speed and the cosine of the angle. (Source:* Fischetti, M., "Working Knowledge," *Scientific American.)*

89. Find the radar readings, to the nearest unit, for Auto A and Auto B shown in the figure.

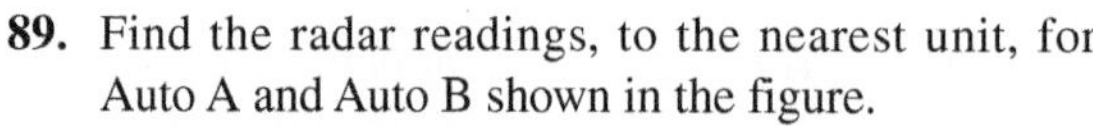

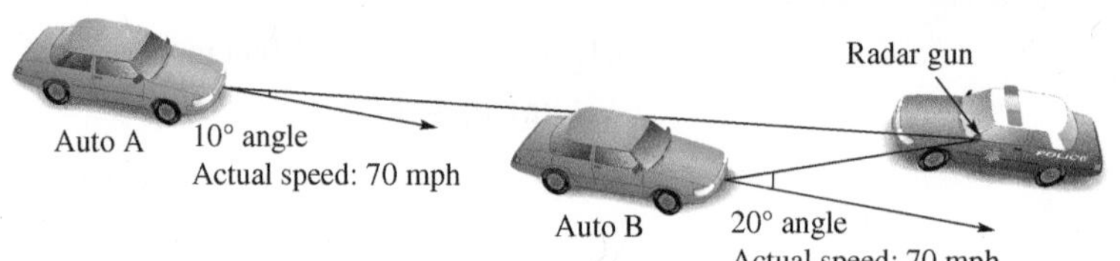

90. The speed reported by a radar gun is reduced by the cosine of angle θ, shown in the figure, where r represents reduced speed and a represents actual speed. Use the figure to show why this "cosine effect" occurs.

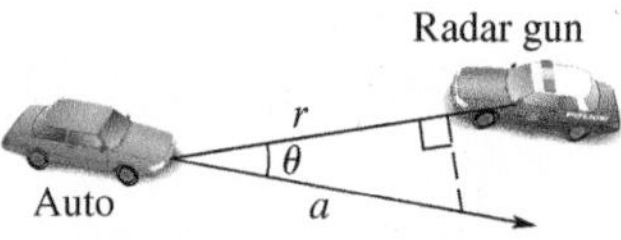

*(Modeling) **Length of a Sag Curve*** *When a highway goes downhill and then uphill, it has a* ***sag curve.*** *Sag curves are designed so that at night, headlights shine sufficiently far down the road to allow a safe stopping distance. See the figure. S and L are in feet.*

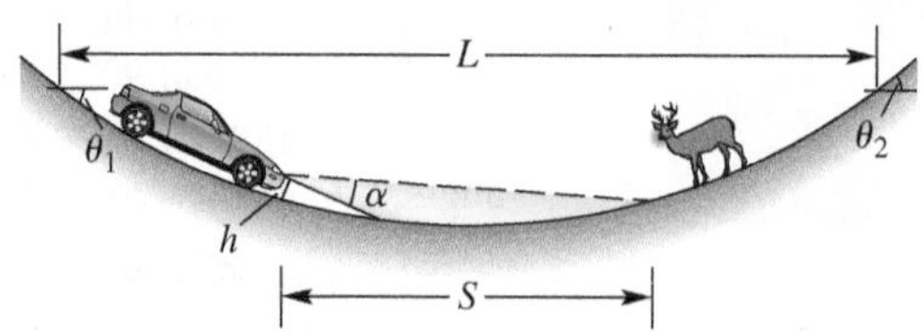

The minimum length L of a sag curve is determined by the height h of the car's headlights above the pavement, the downhill grade $\theta_1 < 0°$, the uphill grade $\theta_2 > 0°$, and the safe stopping distance S for a given speed limit. In addition, L is dependent on the vertical alignment of the headlights. Headlights are usually pointed upward at a slight angle α above the horizontal of the car. Using these quantities, for a 55 mph speed limit, L can be modeled by the formula

$$L = \frac{(\theta_2 - \theta_1)S^2}{200(h + S \tan \alpha)},$$

where $S < L$. (Source: Mannering, F. and W. Kilareski, *Principles of Highway Engineering and Traffic Analysis,* Second Edition, John Wiley and Sons.*)*

91. Compute length L, to the nearest foot, if $h = 1.9$ ft, $\alpha = 0.9°$, $\theta_1 = -3°$, $\theta_2 = 4°$, and $S = 336$ ft.

92. Repeat **Exercise 91** with $\alpha = 1.5°$.

Chapter 2 Quiz (Sections 2.1-2.3)

[2.1]

1. $\sin A = \frac{3}{5}$; $\cos A = \frac{4}{5}$; $\tan A = \frac{3}{4}$; $\cot A = \frac{4}{3}$; $\sec A = \frac{5}{4}$; $\csc A = \frac{5}{3}$

2.

θ	$\sin\theta$	$\cos\theta$	$\tan\theta$
30°	$\frac{1}{2}$	$\frac{\sqrt{3}}{2}$	$\frac{\sqrt{3}}{3}$
45°	$\frac{\sqrt{2}}{2}$	$\frac{\sqrt{2}}{2}$	1
60°	$\frac{\sqrt{3}}{2}$	$\frac{1}{2}$	$\sqrt{3}$

θ	$\cot\theta$	$\sec\theta$	$\csc\theta$
30°	$\sqrt{3}$	$\frac{2\sqrt{3}}{3}$	2
45°	1	$\sqrt{2}$	$\sqrt{2}$
60°	$\frac{\sqrt{3}}{3}$	2	$\frac{2\sqrt{3}}{3}$

3. $w = 18$; $x = 18\sqrt{3}$; $y = 18$; $z = 18\sqrt{2}$

4. $\mathscr{A} = 3x^2 \sin\theta$

[2.2]

In Exercises 5–7, we give, in order, sine, cosine, tangent, cotangent, secant, and cosecant.

5. $\frac{\sqrt{2}}{2}$; $-\frac{\sqrt{2}}{2}$; -1; -1; $-\sqrt{2}$; $\sqrt{2}$

6. $-\frac{1}{2}$; $-\frac{\sqrt{3}}{2}$; $\frac{\sqrt{3}}{3}$; $\sqrt{3}$; $-\frac{2\sqrt{3}}{3}$; -2

7. $-\frac{\sqrt{3}}{2}$; $\frac{1}{2}$; $-\sqrt{3}$; $-\frac{\sqrt{3}}{3}$; 2; $-\frac{2\sqrt{3}}{3}$

8. 60°; 120°

9. 135°; 225°

[2.3]

10. 0.673013

11. −1.181763

12. 69.497888°

13. 24.777233°

[2.1–2.3]

14. false

15. true

Solve each problem.

1. Find exact values of the six trigonometric functions for angle A in the right triangle.

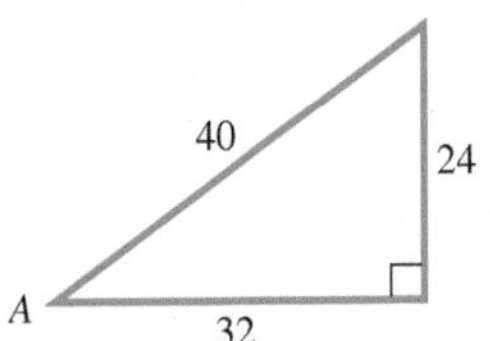

2. Complete the table with exact trigonometric function values.

θ	$\sin\theta$	$\cos\theta$	$\tan\theta$	$\cot\theta$	$\sec\theta$	$\csc\theta$
30°						
45°						
60°						

3. Find the exact value of each variable in the figure.

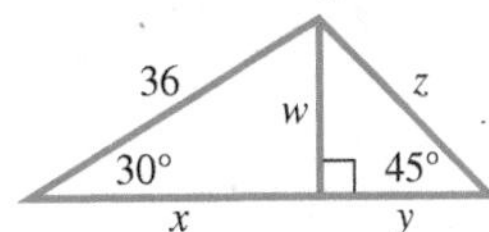

4. *Area of a Solar Cell* A solar cell converts the energy of sunlight directly into electrical energy. The amount of energy a cell produces depends on its area. Suppose a solar cell is hexagonal, as shown in the first figure on the right. Express its area $\mathscr{A}$ in terms of $\sin\theta$ and any side x. (*Hint:* Consider one of the six equilateral triangles from the hexagon. See the second figure on the right.) (*Source:* Kastner, B., *Space Mathematics,* NASA.)

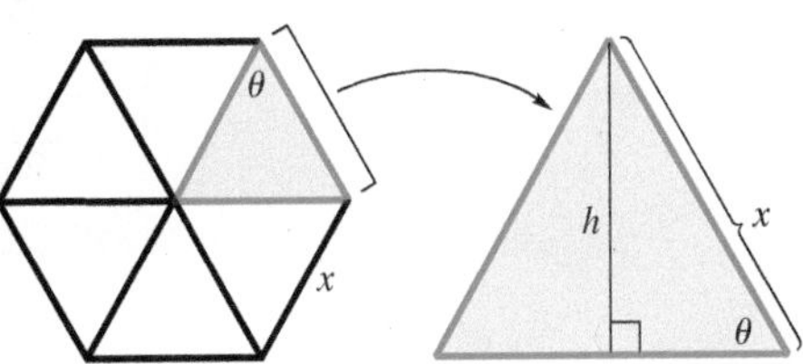

Find exact values of the six trigonometric functions for each angle. Rationalize denominators when applicable.

5. 135° **6.** −150° **7.** 1020°

Find all values of θ, if θ is in the interval [0°, 360°) and has the given function value.

8. $\sin\theta = \frac{\sqrt{3}}{2}$ **9.** $\sec\theta = -\sqrt{2}$

Use a calculator to approximate the value of each expression. Give answers to six decimal places.

10. $\sin 42^\circ 18'$ **11.** $\sec(-212^\circ 12')$

Find a value of θ in the interval [0°, 90°) that satisfies each statement. Write each answer in decimal degrees to six decimal places.

12. $\tan\theta = 2.6743210$ **13.** $\csc\theta = 2.3861147$

Determine whether each statement is true or false.

14. $\sin(60^\circ + 30^\circ) = \sin 60^\circ + \sin 30^\circ$ **15.** $\tan(90^\circ - 35^\circ) = \cot 35^\circ$

2.4 Solutions and Applications of Right Triangles

- Historical Background
- Significant Digits
- Solving Triangles
- Angles of Elevation or Depression

Historical Background The beginnings of trigonometry can be traced back to antiquity. **Figure 19** shows the Babylonian tablet **Plimpton 322,** which provides a table of secant values. The Greek mathematicians Hipparchus and Claudius Ptolemy developed a table of chords, which gives values of sines of angles between 0° and 90° in increments of 15 minutes. Until the advent of scientific calculators in the late 20th century, tables were used to find function values that we now obtain with the stroke of a key.

Applications of *spherical trigonometry* accompanied the study of astronomy for these ancient civilizations. Until the mid-20th century, spherical trigonometry was studied in undergraduate courses. See **Figure 20**.

An introduction to applications of the *plane trigonometry* studied in this text involves applying the ratios to sides of objects that take the shape of right triangles.

Plimpton 322

Figure 19

Figure 20

Significant Digits A number that represents the result of counting, or a number that results from theoretical work and is not the result of measurement, is an **exact number.** There are 50 states in the United States. In this statement, 50 is an exact number.

Most values obtained for trigonometric applications are measured values that are *not* exact. Suppose we quickly measure a room as 15 ft by 18 ft. See **Figure 21.** To calculate the length of a diagonal of the room, we can use the Pythagorean theorem.

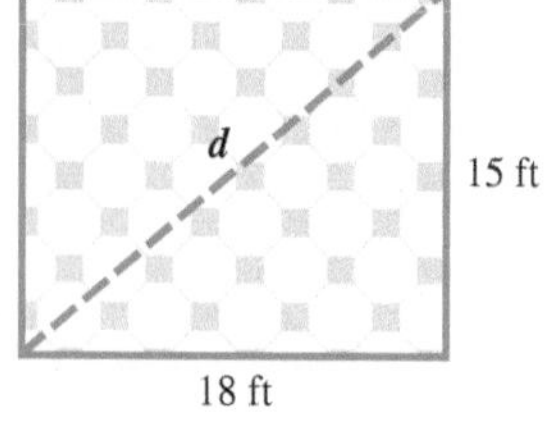

Figure 21

$d^2 = 15^2 + 18^2$ Pythagorean theorem

$d^2 = 549$ Apply the exponents and add.

$d = \sqrt{549}$ Square root property;

$d \approx 23.430749$ Choose the positive root.

Teaching Tip Make the analogy that this discussion is similar to the idea that a chain is only as strong as its weakest link.

Should this answer be given as the length of the diagonal of the room? Of course not. The number 23.430749 contains six decimal places, while the original data of 15 ft and 18 ft are accurate only to the nearest foot. In practice, the results of a calculation can be no more accurate than the least accurate number in the calculation. Thus, we should indicate that the diagonal of the 15-by-18-ft room is approximately 23 ft.

Teaching Tip To help students identify significant digits, write numbers using scientific notation. For example,

$$408 = 4.08 \times 10^2$$
$$6.700 = 6.700 \times 10^0$$
$$0.0025 = 2.5 \times 10^{-3}$$
$$7300 = 7.3 \times 10^3.$$

The number of digits in the first factor of a number written in scientific notation represents the number of significant digits.

If a wall measured to the nearest foot is 18 ft long, this actually means that the wall has length between 17.5 ft and 18.5 ft. If the wall is measured more accurately as 18.3 ft long, then its length is really between 18.25 ft and 18.35 ft. The results of physical measurement are only approximately accurate and depend on the precision of the measuring instrument as well as the aptness of the observer. The digits obtained by actual measurement are **significant digits.** The measurement 18 ft is said to have two significant digits; 18.3 ft has three significant digits.

In the following numbers, the significant digits are identified in color.

408   21.5   18.00   6.700   0.0025   0.09810   7300

Notice the following.

- 18.00 has four significant digits. The zeros in this number represent measured digits accurate to the nearest hundredth.
- The number 0.0025 has only two significant digits, 2 and 5, because the zeros here are used only to locate the decimal point.
- The number 7300 causes some confusion because it is impossible to determine whether the zeros are measured values. The number 7300 may have two, three, or four significant digits. When presented with this situation, we assume that the zeros are not significant, unless the context of the problem indicates otherwise.

To determine the number of significant digits for answers in applications of angle measure, use the following table.

Significant Digits for Angles

Angle Measure to Nearest	Examples	Write Answer to This Number of Significant Digits
Degree	62°, 36°	two
Ten minutes, or nearest tenth of a degree	52° 30′, 60.4°	three
Minute, or nearest hundredth of a degree	81° 48′, 71.25°	four
Ten seconds, or nearest thousandth of a degree	10° 52′ 20″, 21.264°	five

To perform calculations with measured numbers, start by identifying the number with the least number of significant digits. Round the final answer to the same number of significant digits as this number. ***Remember that the answer is no more accurate than the least accurate number in the calculation.***

Solving Triangles ***To solve a triangle means to find the measures of all the angles and sides of the triangle.*** As shown in **Figure 22,** we use a to represent the length of the side opposite angle A, b for the length of the side opposite angle B, and so on. In a right triangle, the letter c is reserved for the hypotenuse.

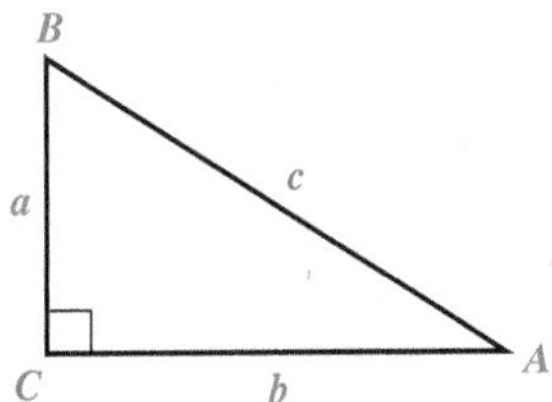

When we are solving triangles, a labeled sketch is an important aid.

Figure 22

EXAMPLE 1 Solving a Right Triangle Given an Angle and a Side

Solve right triangle ABC, if $A = 34° 30'$ and $c = 12.7$ in.

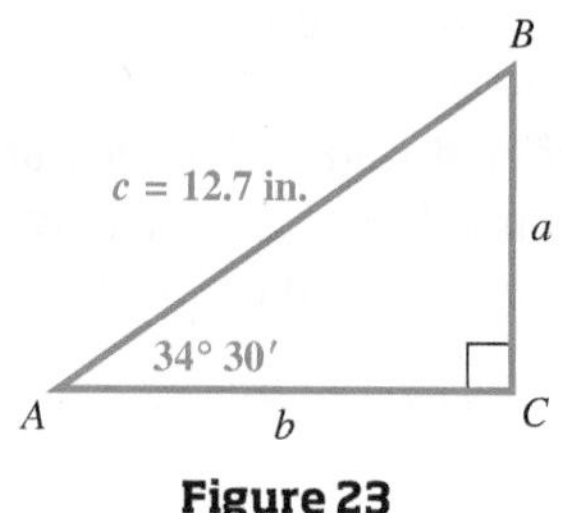

Figure 23

SOLUTION To solve the triangle, find the measures of the remaining sides and angles. See **Figure 23.** To find the value of a, use a trigonometric function involving the known values of angle A and side c. Because the sine of angle A is given by the quotient of the side opposite A and the hypotenuse, use sin A.

$$\sin A = \frac{a}{c} \qquad \sin A = \frac{\text{side opposite}}{\text{hypotenuse}}$$

$$\sin 34° 30' = \frac{a}{12.7} \qquad A = 34° 30',\ c = 12.7$$

$$a = 12.7 \sin 34° 30' \qquad \text{Multiply by 12.7 and rewrite.}$$

$$a = 12.7 \sin 34.5° \qquad \text{Convert to decimal degrees.}$$

$$a \approx 12.7(0.56640624) \qquad \text{Use a calculator.}$$

$$a \approx 7.19 \text{ in.} \qquad \text{Three significant digits}$$

Classroom Example 1
Solve right triangle ABC, if $B = 28° 40'$ and $a = 25.3$ cm.

Answer:
$b = 13.8$ cm; $c = 28.8$ cm; $A = 61° 20'$

Assuming that $34° 30'$ is given to the nearest ten minutes, we rounded the answer to three significant digits.

To find the value of b, we could substitute the value of a just calculated and the given value of c in the Pythagorean theorem. It is better, however, to use the information given in the problem rather than a result just calculated. If an error is made in finding a, then b also would be incorrect. And, rounding more than once may cause the result to be less accurate. To find b, use cos A.

Teaching Tip When solving triangles, encourage students to refer to a sketch of the problem to see if their answer is reasonable.

$$\cos A = \frac{b}{c} \qquad \cos A = \frac{\text{side adjacent}}{\text{hypotenuse}}$$

$$\cos 34° 30' = \frac{b}{12.7} \qquad A = 34° 30',\ c = 12.7$$

$$b = 12.7 \cos 34° 30' \qquad \text{Multiply by 12.7 and rewrite.}$$

$$b \approx 10.5 \text{ in.} \qquad \text{Three significant digits}$$

Once b is found, the Pythagorean theorem can be used to verify the results.

All that remains to solve triangle ABC is to find the measure of angle B.

$$A + B = 90° \qquad A \text{ and } B \text{ are complementary angles.}$$

$$34° 30' + B = 90° \qquad A = 34° 30'$$

$$B = 89° 60' - 34° 30' \qquad \text{Rewrite } 90°. \text{ Subtract } 34° 30'.$$

$$B = 55° 30' \qquad \text{Subtract degrees and minutes separately.}$$

✔ **Now Try Exercise 25.**

LOOKING AHEAD TO CALCULUS
The derivatives of the **parametric equations** $x = f(t)$ and $y = g(t)$ often represent the rate of change of physical quantities, such as velocities. When x and y are related by an equation, the derivatives are **related rates** because a change in one causes a related change in the other. Determining these rates in calculus often requires solving a right triangle.

NOTE In **Example 1,** we could have found the measure of angle B first and then used the trigonometric function values of B to find the lengths of the unknown sides. A right triangle can usually be solved in several ways, each producing the correct answer.

To maintain accuracy, always use given information as much as possible, and avoid rounding in intermediate steps.

EXAMPLE 2 Solving a Right Triangle Given Two Sides

Solve right triangle ABC, if $a = 29.43$ cm and $c = 53.58$ cm.

B, $a = 29.43$ cm, $c = 53.58$ cm, C, b, A

Figure 24

SOLUTION We draw a sketch showing the given information, as in **Figure 24.** One way to begin is to find angle A using the sine function.

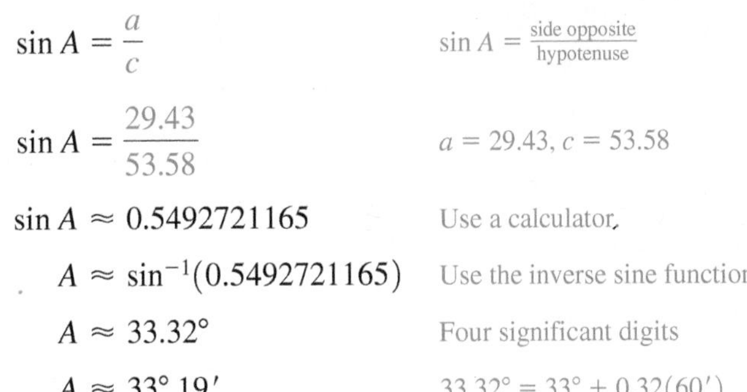

$$\sin A = \frac{a}{c} \qquad \sin A = \frac{\text{side opposite}}{\text{hypotenuse}}$$

$$\sin A = \frac{29.43}{53.58} \qquad a = 29.43,\ c = 53.58$$

$$\sin A \approx 0.5492721165 \qquad \text{Use a calculator.}$$

$$A \approx \sin^{-1}(0.5492721165) \qquad \text{Use the inverse sine function.}$$

$$A \approx 33.32^\circ \qquad \text{Four significant digits}$$

$$A \approx 33^\circ\, 19' \qquad 33.32^\circ = 33^\circ + 0.32(60')$$

Classroom Example 2
Solve right triangle ABC, if $a = 44.25$ cm and $b = 55.87$ cm.

Answer:
$A = 38^\circ\, 23'$; $B = 51^\circ\, 37'$; $c = 71.27$ cm

The measure of B is approximately

$$90^\circ - 33^\circ\, 19' = 56^\circ\, 41'. \qquad 90^\circ = 89^\circ\, 60'$$

Teaching Tip Remind students to use the appropriate **trigonometric function** key to find a **length** and the appropriate **inverse trigonometric function** key to find an **angle.**

We now find b from the Pythagorean theorem.

$$a^2 + b^2 = c^2 \qquad \text{Pythagorean theorem}$$

$$29.43^2 + b^2 = 53.58^2 \qquad a = 29.43,\ c = 53.58$$

$$b^2 = 53.58^2 - 29.43^2 \qquad \text{Subtract } 29.43^2.$$

$$b = \sqrt{2004.6915} \qquad \text{Simplify on the right; square root property}$$

$$b \approx 44.77 \text{ cm}$$

Choose the positive square root.

✔ **Now Try Exercise 35.**

Angles of Elevation or Depression In applications of right triangles, the **angle of elevation** from point X to point Y (above X) is the acute angle formed by ray XY and a horizontal ray with endpoint at X. See **Figure 25(a).** The **angle of depression** from point X to point Y (below X) is the acute angle formed by ray XY and a horizontal ray with endpoint X. See **Figure 25(b).**

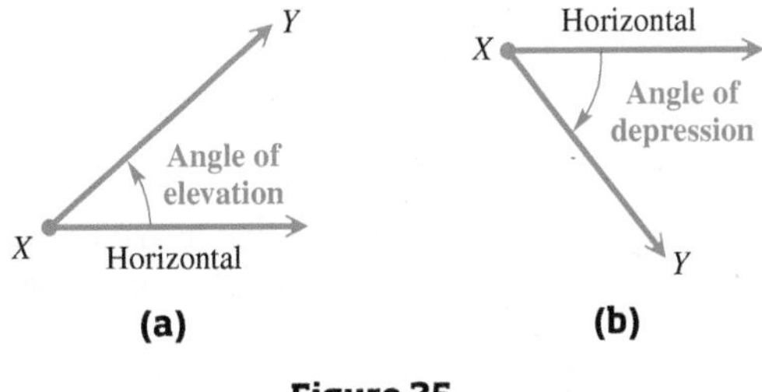

Figure 25

CAUTION Be careful when interpreting the angle of depression. ***Both the angle of elevation and the angle of depression are measured between the line of sight and a horizontal line.***

George Polya (1887–1985)

Polya, a native of Budapest, Hungary, wrote more than 250 papers and a number of books. He proposed a general outline for solving applied problems in his classic book *How to Solve It.*

To solve applied trigonometry problems, follow the same procedure as solving a triangle. ***Drawing a sketch and labeling it correctly in Step 1 is crucial.***

Solving an Applied Trigonometry Problem

Step 1 Draw a sketch, and label it with the given information. Label the quantity to be found with a variable.

Step 2 Use the sketch to write an equation relating the given quantities to the variable.

Step 3 Solve the equation, and check that the answer makes sense.

EXAMPLE 3 Finding a Length Given the Angle of Elevation

At a point A, 123 ft from the base of a flagpole, the angle of elevation to the top of the flagpole is $26° 40'$. Find the height of the flagpole.

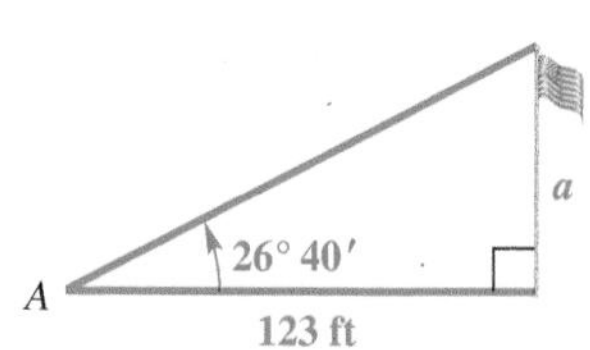

Figure 26

SOLUTION

Step 1 See **Figure 26.** The length of the side adjacent to A is known, and the length of the side opposite A must be found. We will call it a.

Step 2 The tangent ratio involves the given values. Write an equation.

$$\tan A = \frac{\text{side opposite}}{\text{side adjacent}} \quad \text{Tangent ratio}$$

$$\tan 26° 40' = \frac{a}{123} \quad A = 26° 40';\ \text{side adjacent} = 123$$

Step 3

$$a = 123 \tan 26° 40' \quad \text{Multiply by 123 and rewrite.}$$

$$a \approx 123(0.50221888) \quad \text{Use a calculator.}$$

$$a \approx 61.8 \text{ ft} \quad \text{Three significant digits}$$

The height of the flagpole is 61.8 ft.

Now Try Exercise 53.

Classroom Example 3
The angle of elevation from a point on the ground 15.5 m from the base of a tree to the top of the tree is 60.4°. Find the height of the tree.

Answer: 27.3 m

Classroom Example 4
Find the angle of depression from the top of a 97-ft building to the base of a building across the street located 42 ft away.

Answer: 67°

EXAMPLE 4 Finding an Angle of Depression

From the top of a 210-ft cliff, David observes a lighthouse that is 430 ft offshore. Find the angle of depression from the top of the cliff to the base of the lighthouse.

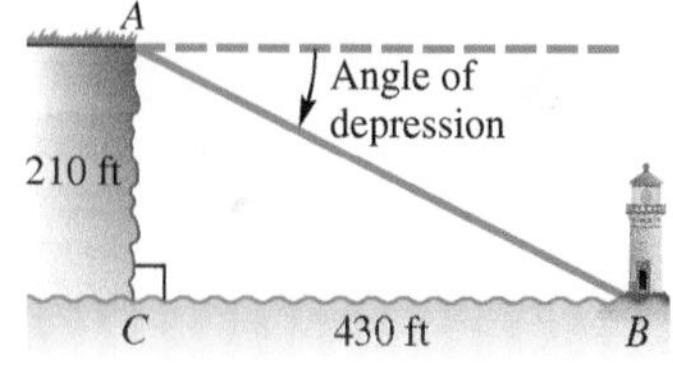

Figure 27

SOLUTION As shown in **Figure 27,** the angle of depression is measured from a horizontal line down to the base of the lighthouse. The angle of depression and angle B, in the right triangle shown, are alternate interior angles whose measures are equal. We use the tangent ratio to solve for angle B.

$$\tan B = \frac{210}{430} \quad \text{Tangent ratio}$$

$$B = \tan^{-1}\left(\frac{210}{430}\right) \quad \text{Use the inverse tangent function.}$$

$$B \approx 26° \quad \text{Two significant digits}$$

Now Try Exercise 55.

2.4 Exercises

1. B 2. D
3. A 4. F
5. C 6. E

7. 23.825 to 23.835
8. 28,999.5 to 29,000.5
9. 8958.5 to 8959.5
10. No. Points scored in basketball are exact numbers and cannot take on values that are not whole numbers.
11. 0.05 12. 0.5

Note to instructor: While most of the measures resulting from solving triangles in this chapter are approximations, for convenience we use = rather than ≈ in the answers.

13. $B = 53° 40'$; $a = 571$ m; $b = 777$ m
14. $Y = 42.2°$; $x = 66.4$ cm; $y = 60.2$ cm

CONCEPT PREVIEW *Match each equation in Column I with the appropriate right triangle in Column II. In each case, the goal is to find the value of x.*

I

1. $x = 5 \cot 38°$
2. $x = 5 \cos 38°$
3. $x = 5 \tan 38°$
4. $x = 5 \csc 38°$
5. $x = 5 \sin 38°$
6. $x = 5 \sec 38°$

II

A.

B.
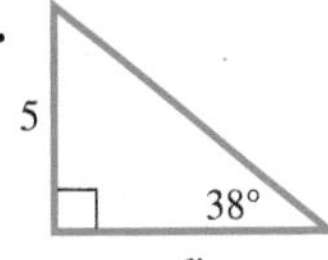

C.
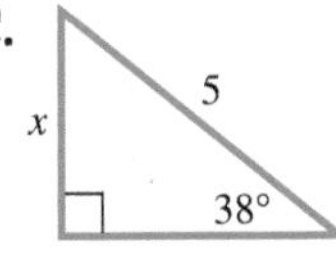

D.
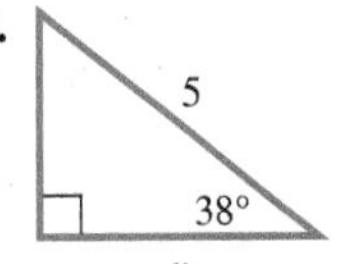

E.
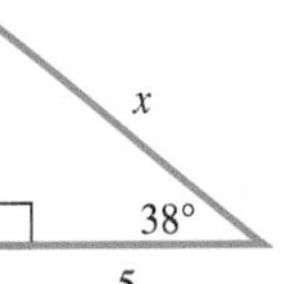

F.
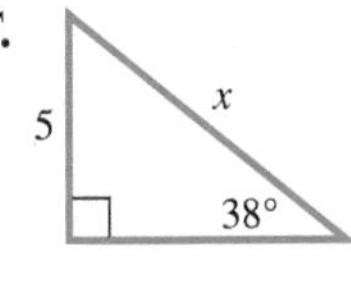

Concept Check Refer to the discussion of accuracy and significant digits in this section to answer the following.

7. *Lake Ponchartrain Causeway* The world's longest bridge over a body of water (continuous) is the causeway that joins the north and south shores of Lake Ponchartrain, a saltwater lake that lies north of New Orleans, Louisiana. It consists of two parallel spans. The longer of the spans measures 23.83 mi. State the range represented by this number. (*Source:* www.worldheritage.org)

8. *Mt. Everest* When Mt. Everest was first surveyed, the surveyors obtained a height of 29,000 ft to the nearest foot. State the range represented by this number. (The surveyors thought no one would believe a measurement of 29,000 ft, so they reported it as 29,002.) (*Source:* Dunham, W., *The Mathematical Universe*, John Wiley and Sons.)

9. *Vehicular Tunnel* The E. Johnson Memorial Tunnel in Colorado, which measures 8959 ft, is one of the longest land vehicular tunnels in the United States. What is the range of this number? (*Source: World Almanac and Book of Facts.*)

10. *WNBA Scorer* Women's National Basketball Association player Maya Moore of the Minnesota Lynx received the 2014 award for the most points scored, 812. Is it appropriate to consider this number between 811.5 and 812.5? Why or why not? (*Source:* www.wnba.com)

11. If h is the actual height of a building and the height is measured as 58.6 ft, then $|h - 58.6| \leq$ ____.

12. If w is the actual weight of a car and the weight is measured as 1542 lb, then $|w - 1542| \leq$ ____.

Solve each right triangle. When two sides are given, give angles in degrees and minutes. ***See Examples 1 and 2.***

13.
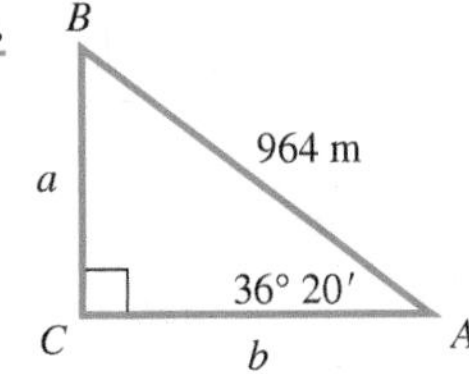

14.
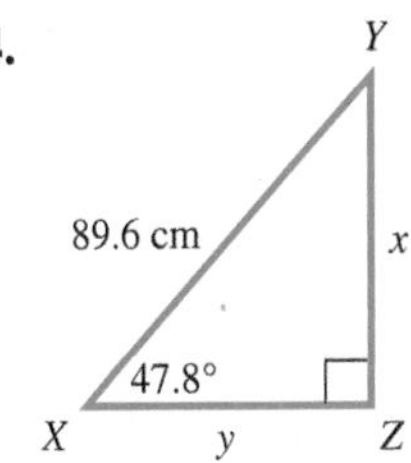

15. $M = 38.8°$; $n = 154$ m; $p = 198$ m
16. $B = 58° \, 20'$; $c = 68.4$ km; $b = 58.2$ km
17. $A = 47.9108°$; $c = 84.816$ cm; $a = 62.942$ cm
18. $A = 21.4858°$; $b = 3330.68$ m; $a = 1311.04$ m
19. $A = 37° \, 40'$; $B = 52° \, 20'$; $c = 20.5$ ft
20. $A = 18° \, 20'$; $B = 71° \, 40'$; $b = 14.5$ m

15.

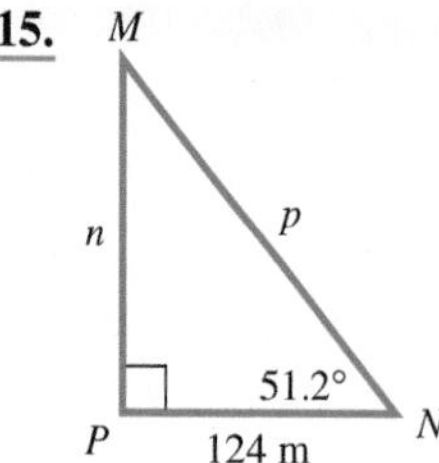

16.

B
c
35.9 km
31° 40′
A
b
C

17.

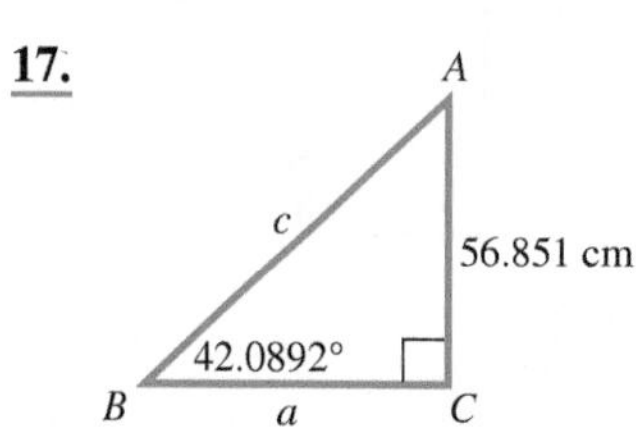

18.

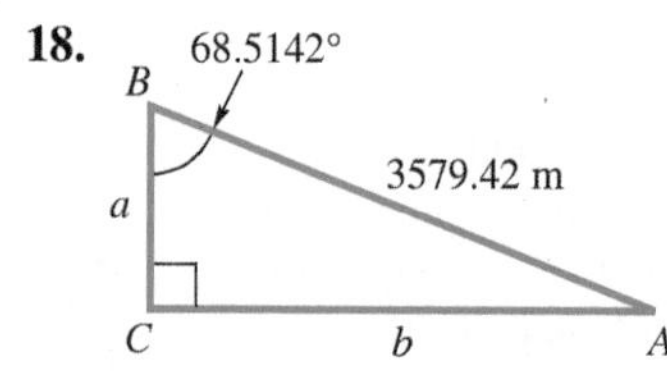

19.

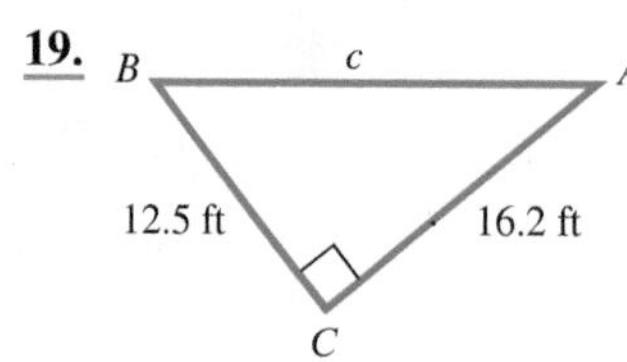

20.

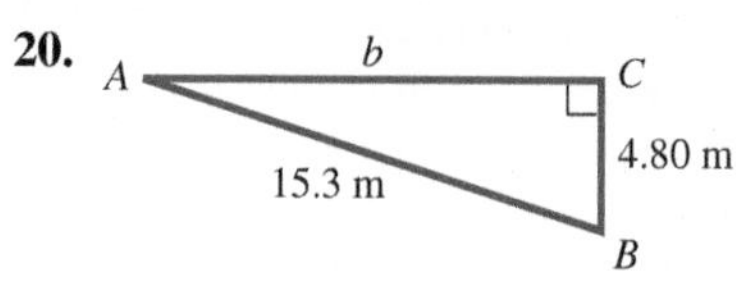

21. No. Given three angles (the two acute angles and the right angle), there are infinitely many similar triangles satisfying the conditions.
22. It is easiest to find the remaining acute angle by simply subtracting the given angle measure from 90°.
23. If we are given one side and one acute angle, an appropriate trigonometric function of that angle will make it possible to solve for one of the remaining sides. Then the complete solution can follow.
24. If we are given two sides, the Pythagorean theorem can be used to find the remaining side. Alternatively, an inverse trigonometric function can be used to find one of the acute angles. Then the complete solution can follow.

Concept Check Answer each question.

21. Can a right triangle be solved if we are given measures of its two acute angles and no side lengths? Why or why not?

22. If we are given an acute angle and a side in a right triangle, what unknown part of the triangle requires the least work to find?

23. Why can we always solve a right triangle if we know the measures of one side and one acute angle?

24. Why can we always solve a right triangle if we know the lengths of two sides?

25. $B = 62.0°$; $a = 8.17$ ft; $b = 15.4$ ft
26. $A = 44.0°$; $a = 20.6$ m; $b = 21.4$ m
27. $A = 17.0°$; $a = 39.1$ in.; $c = 134$ in.
28. $B = 27.5°$; $b = 6.61$ m; $c = 14.3$ m
29. $B = 29.0°$; $a = 70.7$ cm; $c = 80.9$ cm
30. $A = 38.3°$; $b = 35.6$ ft; $c = 45.3$ ft
31. $A = 36°$; $B = 54°$; $b = 18$ m
32. $A = 51°$; $B = 39°$; $a = 40$ ft
33. $c = 85.9$ yd; $A = 62° \, 50'$; $B = 27° \, 10'$
34. $c = 1080$ m; $A = 63° \, 00'$; $B = 27° \, 00'$
35. $b = 42.3$ cm; $A = 24° \, 10'$; $B = 65° \, 50'$

Solve each right triangle. In each case, $C = 90°$. If angle information is given in degrees and minutes, give answers in the same way. If angle information is given in decimal degrees, do likewise in answers. When two sides are given, give angles in degrees and minutes. ***See Examples 1 and 2.***

25. $A = 28.0°$, $c = 17.4$ ft

26. $B = 46.0°$, $c = 29.7$ m

27. $B = 73.0°$, $b = 128$ in.

28. $A = 62.5°$, $a = 12.7$ m

29. $A = 61.0°$, $b = 39.2$ cm

30. $B = 51.7°$, $a = 28.1$ ft

31. $a = 13$ m, $c = 22$ m

32. $b = 32$ ft, $c = 51$ ft

33. $a = 76.4$ yd, $b = 39.3$ yd

34. $a = 958$ m, $b = 489$ m

35. $a = 18.9$ cm, $c = 46.3$ cm

36. $b = 219$ m, $c = 647$ m

37. $A = 53°24'$, $c = 387.1$ ft

38. $A = 13°47'$, $c = 1285$ m

39. $B = 39°09'$, $c = 0.6231$ m

40. $B = 82°51'$, $c = 4.825$ cm

Concept Check Answer each question.

41. What is the meaning of the term *angle of elevation*?

42. Can an angle of elevation be more than 90°?

36. $a = 609$ m; $A = 70° 10'$; $B = 19° 50'$

37. $B = 36° 36'$; $a = 310.8$ ft; $b = 230.8$ ft

38. $B = 76° 13'$; $a = 306.2$ m; $b = 1248$ m

39. $A = 50° 51'$; $a = 0.4832$ m; $b = 0.3934$ m

40. $A = 7° 09'$; $b = 4.787$ cm; $a = 0.6006$ cm

41. If B is a point above point A, as shown in the figure, then the angle of elevation from A to B is the acute angle formed by the horizontal line through A and the line of sight from A to B.

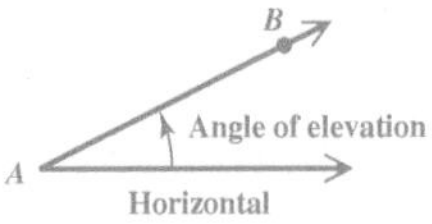

42. No. An angle of elevation (and also an angle of depression) must be an acute angle.

43. Angles DAB and ABC are alternate interior angles formed by the transversal AB intersecting parallel lines AD and BC. Therefore, they have the same measure.

44. An angle of depression (and also an angle of elevation) must be measured from the *horizontal* to the line of sight. Angle CAB is formed by the *vertical* line AC and the line of sight.

45. 9.35 m
46. 33.4 m
47. 128 ft
48. 864,900 mi
49. 26.92 in.
50. 134.7 cm

43. Why does the angle of depression DAB in the figure have the same measure as the angle of elevation ABC?

44. Why is angle CAB *not* an angle of depression in the figure for **Exercise 43?**

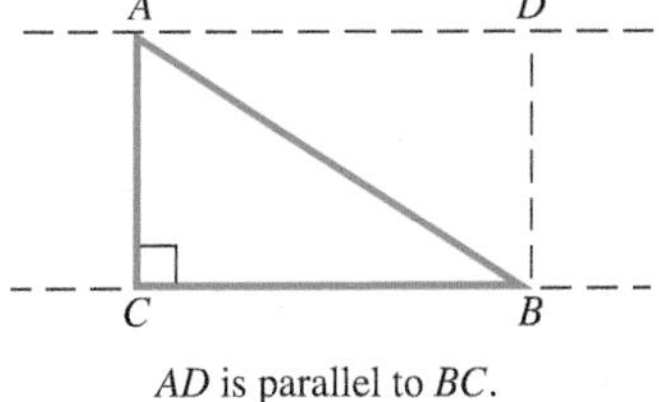

AD is parallel to BC.

Solve each problem. ***See Examples 1–4.***

45. ***Height of a Ladder on a Wall*** A 13.5-m fire truck ladder is leaning against a wall. Find the distance d the ladder goes up the wall (above the top of the fire truck) if the ladder makes an angle of 43° 50′ with the horizontal.

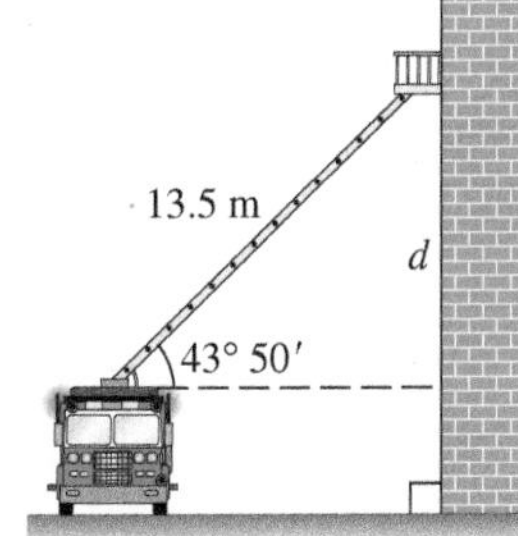

46. ***Distance across a Lake*** To find the distance RS across a lake, a surveyor lays off length $RT = 53.1$ m, so that angle $T = 32° 10'$ and angle $S = 57° 50'$. Find length RS.

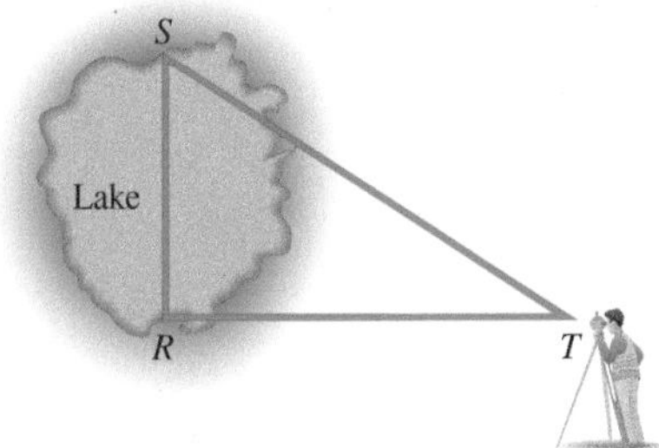

47. ***Height of a Building*** From a window 30.0 ft above the street, the angle of elevation to the top of the building across the street is 50.0° and the angle of depression to the base of this building is 20.0°. Find the height of the building across the street.

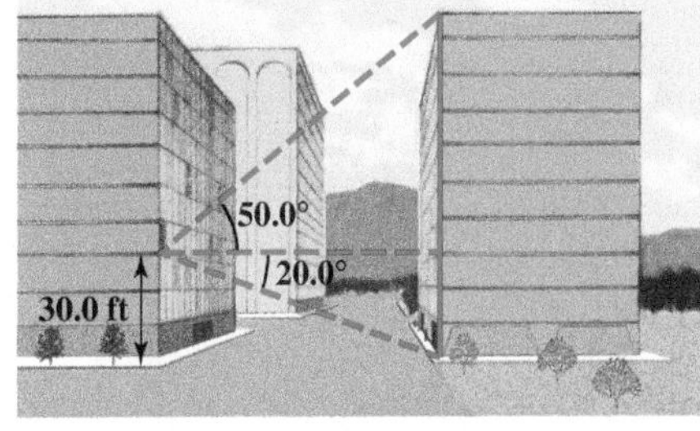

48. ***Diameter of the Sun*** To determine the diameter of the sun, an astronomer might sight with a **transit** (a device used by surveyors for measuring angles) first to one edge of the sun and then to the other, estimating that the included angle equals 32′. Assuming that the distance d from Earth to the sun is 92,919,800 mi, approximate the diameter of the sun.

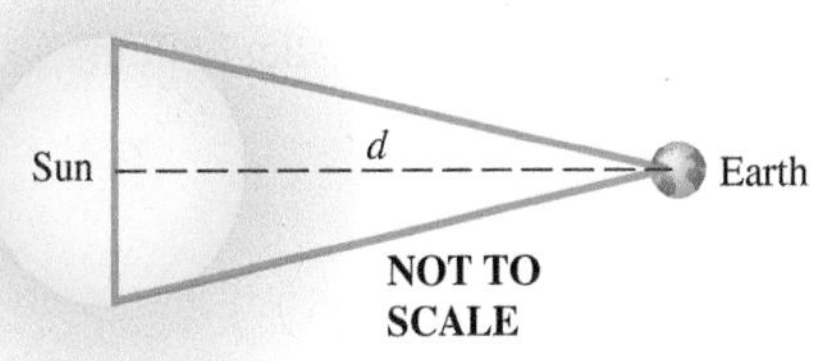

49. ***Side Lengths of a Triangle*** The length of the base of an isosceles triangle is 42.36 in. Each base angle is 38.12°. Find the length of each of the two equal sides of the triangle. (*Hint:* Divide the triangle into two right triangles.)

50. ***Altitude of a Triangle*** Find the altitude of an isosceles triangle having base 184.2 cm if the angle opposite the base is 68° 44′.

51. 28.0 m 52. 469 m
53. 13.3 ft 54. 42,600 ft
55. 37° 35′ 56. 146 m

Solve each problem. ***See Examples 3 and 4.***

51. *Height of a Tower* The shadow of a vertical tower is 40.6 m long when the angle of elevation of the sun is 34.6°. Find the height of the tower.

52. *Distance from the Ground to the Top of a Building* The angle of depression from the top of a building to a point on the ground is 32° 30′. How far is the point on the ground from the top of the building if the building is 252 m high?

53. *Length of a Shadow* Suppose that the angle of elevation of the sun is 23.4°. Find the length of the shadow cast by a person who is 5.75 ft tall.

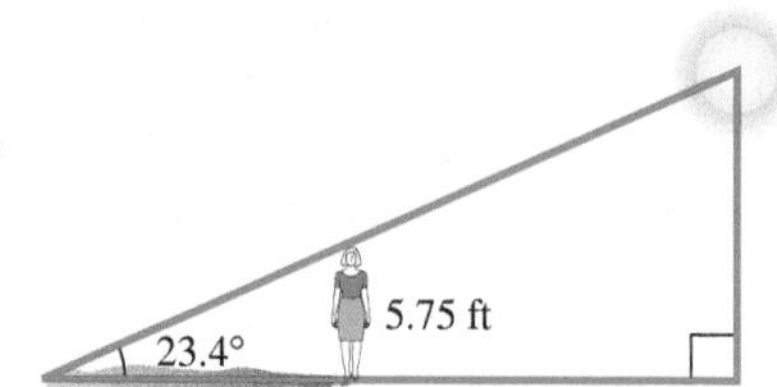

54. *Airplane Distance* An airplane is flying 10,500 ft above level ground. The angle of depression from the plane to the base of a tree is 13° 50′. How far horizontally must the plane fly to be directly over the tree?

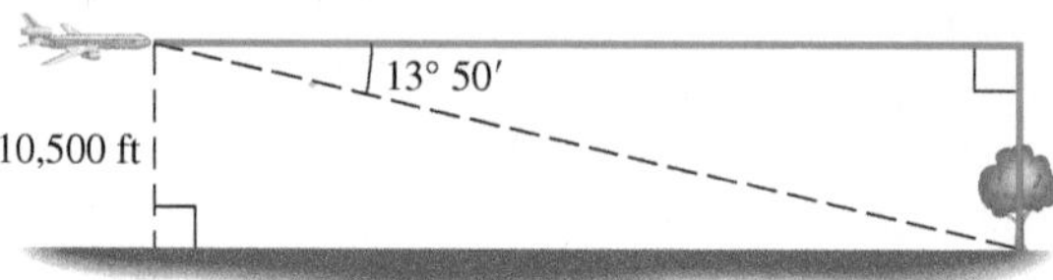

55. *Angle of Depression of a Light* A company safety committee has recommended that a floodlight be mounted in a parking lot so as to illuminate the employee exit, as shown in the figure. Find the angle of depression of the light to the nearest minute.

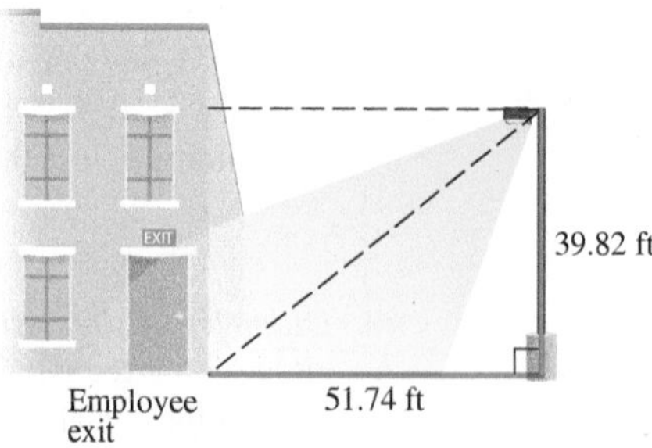

56. *Height of a Building* The angle of elevation from the top of a small building to the top of a nearby taller building is 46° 40′, and the angle of depression to the bottom is 14° 10′. If the shorter building is 28.0 m high, find the height of the taller building.

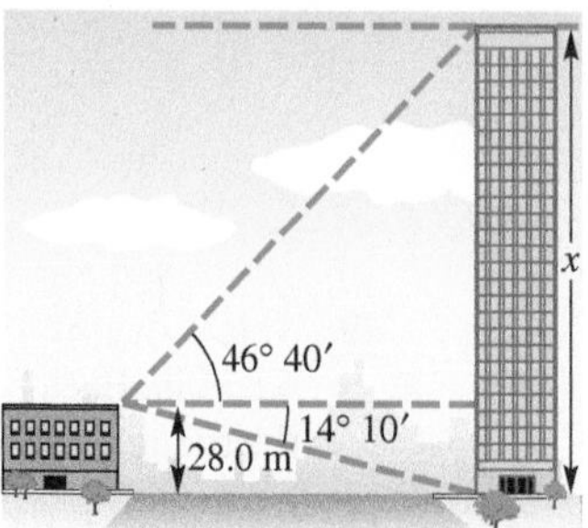

57. 42.18° 58. 63.39°
59. 22° 60. 583 ft
61. (a) 29,000 ft
(b) shorter
62. 34.0 mi

57. *Angle of Elevation of the Sun* The length of the shadow of a building 34.09 m tall is 37.62 m. Find the angle of elevation of the sun to the nearest hundredth of a degree.

58. *Angle of Elevation of the Sun* The length of the shadow of a flagpole 55.20 ft tall is 27.65 ft. Find the angle of elevation of the sun to the nearest hundredth of a degree.

59. *Angle of Elevation of the Pyramid of the Sun* The Pyramid of the Sun is in the ancient Mexican city of Teotihuacan. The base is a square with sides about 700 ft long. The height of the pyramid is about 200 ft. Find the angle of elevation θ of the edge indicated in the figure to two significant digits. (*Hint:* The base of the triangle in the figure is half the diagonal of the square base of the pyramid.) (*Source*: www.britannica.com)

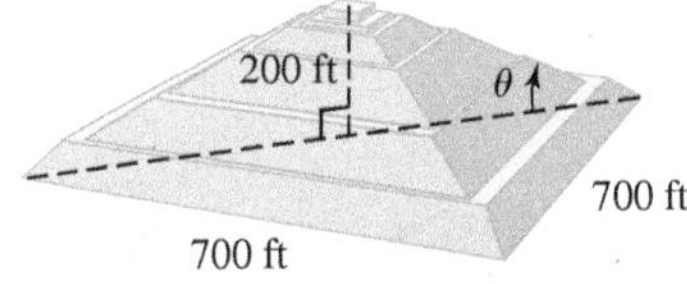

60. *Cloud Ceiling* The U.S. Weather Bureau defines a **cloud ceiling** as the altitude of the lowest clouds that cover more than half the sky. To determine a cloud ceiling, a powerful searchlight projects a circle of light vertically on the bottom of the cloud. An observer sights the circle of light in the crosshairs of a tube called a **clinometer.** A pendant hanging vertically from the tube and resting on a protractor gives the angle of elevation. Find the cloud ceiling if the searchlight is located 1000 ft from the observer and the angle of elevation is 30.0° as measured with a clinometer at eye-height 6 ft. (Assume three significant digits.)

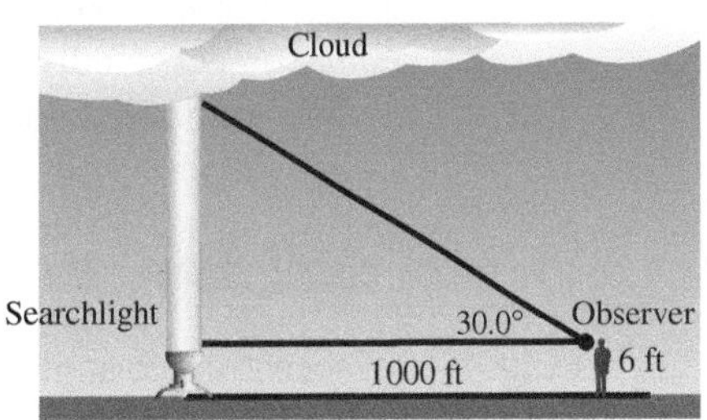

61. *Height of Mt. Everest* The highest mountain peak in the world is Mt. Everest, located in the Himalayas. The height of this enormous mountain was determined in 1856 by surveyors using trigonometry long before it was first climbed in 1953. This difficult measurement had to be done from a great distance. At an altitude of 14,545 ft on a different mountain, the straight-line distance to the peak of Mt. Everest is 27.0134 mi and its angle of elevation is $\theta = 5.82°$. (*Source:* Dunham, W., *The Mathematical Universe*, John Wiley and Sons.)

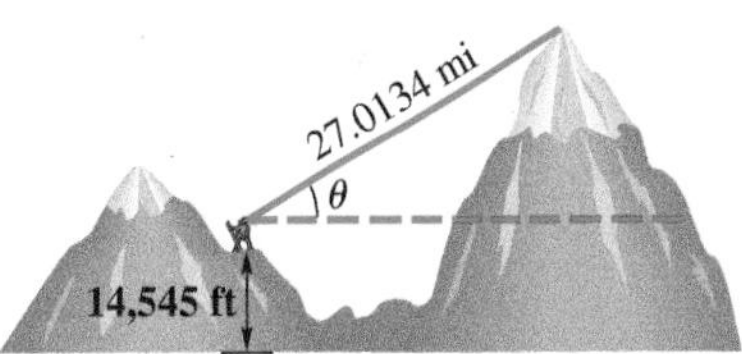

(a) Approximate the height (in feet) of Mt. Everest.

(b) In the actual measurement, Mt. Everest was over 100 mi away and the curvature of Earth had to be taken into account. Would the curvature of Earth make the peak appear taller or shorter than it actually is?

62. *Error in Measurement* A degree may seem like a very small unit, but an error of one degree in measuring an angle may be very significant. For example, suppose a laser beam directed toward the visible center of the moon misses its assigned target by 30.0″. How far is it (in miles) from its assigned target? Take the distance from the surface of Earth to that of the moon to be 234,000 mi. (*Source: A Sourcebook of Applications of School Mathematics* by Donald Bushaw et al.)

2.5 Further Applications of Right Triangles

- Historical Background
- Bearing
- Further Applications

Historical Background

Johann Müller, known as Regiomontanus (see **Figure 28**), was a fifteenth-century German astronomer whose best known book is *On Triangles of Every Kind.* He used the recently-invented printing process of Gutenberg to promote his research. In his excellent book *Trigonometric Delights* *, Eli Maor writes:

Regiomontanus

Figure 28

> Regiomontanus was the first publisher of mathematical and astronomical books for commercial use. In 1474 he printed his *Ephemerides*, tables listing the position of the sun, moon, and planets for each day from 1475 to 1506. This work brought him great acclaim; Christopher Columbus had a copy of it on his fourth voyage to the New World and used it to predict the famous lunar eclipse of February 29, 1504. The hostile natives had for some time refused Columbus's men food and water, and he warned them that God would punish them and take away the moon's light. His admonition was at first ridiculed, but when at the appointed hour the eclipse began, the terrified natives immediately repented and fell into submission.

Bearing

We now investigate navigation problems. **Bearing** refers to the direction of motion of an object, such as a ship or airplane, or the direction of a second object at a distance relative to the ship or airplane.

We introduce two methods of measuring bearing.

Expressing Bearing (Method 1)

When a single angle is given, it is understood that bearing is measured in a clockwise direction from due north.

Several sample bearings using Method 1 are shown in **Figure 29.**

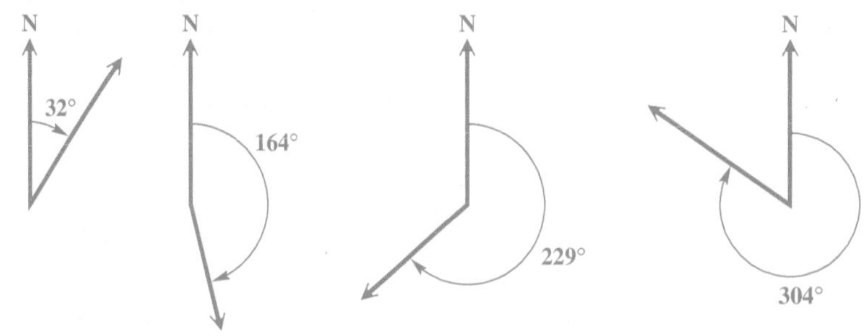

Bearings of 32°, 164°, 229°, and 304°

Figure 29

Teaching Tip Emphasize the need to draw accurate sketches describing each application problem. Encourage students to use their sketches to check that their answers are reasonable.

CAUTION ***A correctly labeled sketch is crucial*** when solving applications like those that follow. Some of the necessary information is often not directly stated in the problem and can be determined only from the sketch.

*Excerpt from *Trigonometric Delights* by Eli Maor, copyright ©1998 by Princeton University Press. Used by permission of Princeton University Press.

EXAMPLE 1 Solving a Problem Involving Bearing (Method 1)

Radar stations A and B are on an east-west line, 3.7 km apart. Station A detects a plane at C, on a bearing of 61°. Station B simultaneously detects the same plane, on a bearing of 331°. Find the distance from A to C.

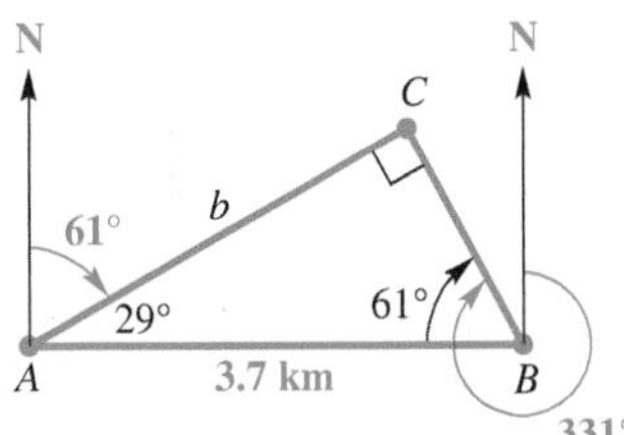

Figure 30

SOLUTION Begin with a sketch showing the given information. See **Figure 30.** A line drawn due north is perpendicular to an east-west line, so right angles are formed at A and B. Angles CBA and CAB can be found as follows.

$$\angle CBA = 331° - 270° = 61° \quad \text{and} \quad \angle CAB = 90° - 61° = 29°$$

A right triangle is formed. The distance from A to C, denoted b in the figure, can be found using the cosine function for angle CAB.

$$\cos 29° = \frac{b}{3.7} \qquad \text{Cosine ratio}$$

$$b = 3.7 \cos 29° \qquad \text{Multiply by 3.7 and rewrite.}$$

$$b \approx 3.2 \text{ km} \qquad \text{Two significant digits}$$

✔ **Now Try Exercise 23.**

Classroom Example 1
Radar stations A and B are on an east-west line, 8.6 km apart. Station A detects a plane at C, on a bearing of 53°. Station B simultaneously detects the same plane, on a bearing of 323°. Find the distance from B to C.

Answer: 5.2 km

Expressing Bearing (Method 2)

Start with a north-south line and use an acute angle to show the direction, either east or west, from this line.

Figure 31 shows several sample bearings using this method. Either N or S always comes first, followed by an acute angle, and then E or W.

N
42°
N 42° E

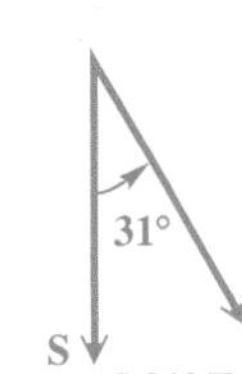

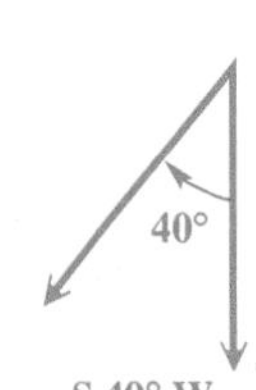

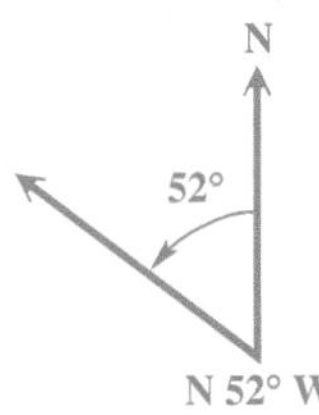

Figure 31

Classroom Example 2
A ship leaves port and sails on a bearing of S 34° W for 2.5 hr. It then turns and sails on a bearing of N 56° W for 3.0 hr. If the ship's rate is 18 knots (nautical miles per hour), find the distance that the ship is from port.

Answer: 70 nautical mi

EXAMPLE 2 Solving a Problem Involving Bearing (Method 2)

A ship leaves port and sails on a bearing of N 47° E for 3.5 hr. It then turns and sails on a bearing of S 43° E for 4.0 hr. If the ship's rate is 22 knots (nautical miles per hour), find the distance that the ship is from port.

SOLUTION Draw and label a sketch as in **Figure 32.** Choose a point C on a bearing of N 47° E from port at point A. Then choose a point B on a bearing of S 43° E from point C. Because north-south lines are parallel, angle ACD measures 47° by alternate interior angles. The measure of angle ACB is

$$47° + 43° = 90°,$$

making triangle ABC a right triangle.

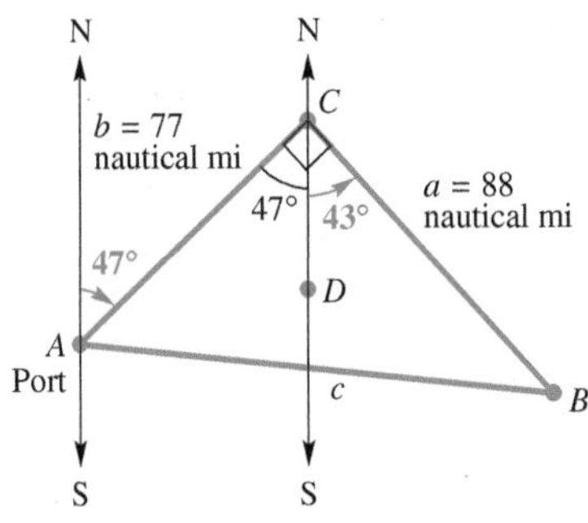

Figure 32

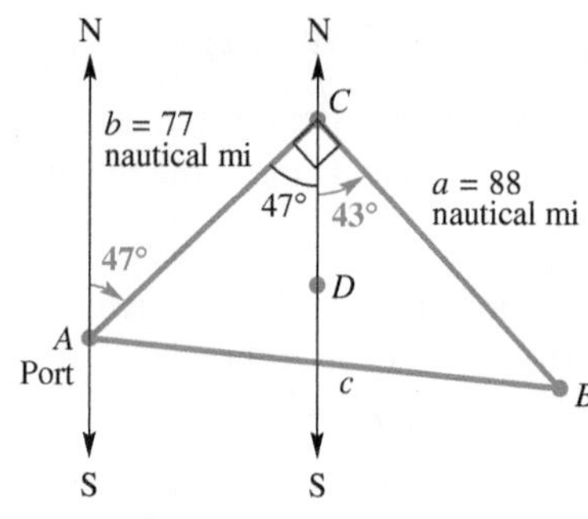

Figure 32 (repeated)

Use the formula relating distance, rate, and time to find the distances in **Figure 32** from A to C and from C to B.

$$b = 22 \times 3.5 = 77 \text{ nautical mi}$$
$$a = 22 \times 4.0 = 88 \text{ nautical mi}$$
Distance = rate × time

Now find c, the distance from port at point A to the ship at point B.

$$a^2 + b^2 = c^2 \quad \text{Pythagorean theorem}$$
$$88^2 + 77^2 = c^2 \quad a = 88,\ b = 77$$
$$c = \sqrt{88^2 + 77^2} \quad \text{If } a^2 + b^2 = c^2 \text{ and } c > 0, \text{ then } c = \sqrt{a^2 + b^2}.$$
$$c \approx 120 \text{ nautical mi} \quad \text{Two significant digits}$$

✔ **Now Try Exercise 29.**

Further Applications

Classroom Example 3
Repeat **Example 3** for $\theta = 2°\,41'\,38''$ and $b = 3.5000$ cm.

Answers:
(a) 74.427 cm
(b) 0.0076765 cm

EXAMPLE 3 Using Trigonometry to Measure a Distance

The **subtense bar method** is a method that surveyors use to determine a small distance d between two points P and Q. The subtense bar with length b is centered at Q and situated perpendicular to the line of sight between P and Q. See **Figure 33.** Angle θ is measured, and then the distance d can be determined.

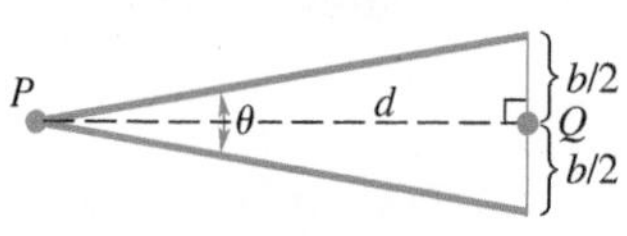

Figure 33

(a) Find d when $\theta = 1°\,23'\,12''$ and $b = 2.0000$ cm.

(b) How much change would there be in the value of d if θ measured 1″ larger?

SOLUTION

(a) From **Figure 33,** we obtain the following.

$$\cot \frac{\theta}{2} = \frac{d}{\frac{b}{2}} \quad \text{Cotangent ratio}$$

$$d = \frac{b}{2} \cot \frac{\theta}{2} \quad \text{Multiply and rewrite.}$$

Let $b = 2$. To evaluate $\frac{\theta}{2}$, we change θ to decimal degrees.

$$1°\,23'\,12'' \approx 1.386666667°$$

Use $\cot \theta = \frac{1}{\tan \theta}$ to evaluate.

Then $$d = \frac{2}{2} \cot \frac{1.386666667°}{2} \approx 82.634110 \text{ cm.}$$

(b) If θ is 1″ larger, then $\theta = 1°\,23'\,13'' \approx 1.386944444°$.

$$d = \frac{2}{2} \cot \frac{1.386944444°}{2} \approx 82.617558 \text{ cm}$$

The difference is $82.634110 - 82.617558 = 0.016552$ cm.

✔ **Now Try Exercise 41.**

Classroom Example 4
Marla needs to find the height of a building. From a given point on the ground, she finds that the angle of elevation to the top of the building is 74.2°. She then walks back 35 ft. From the second point, the angle of elevation to the top of the building is 51.8°. Find the height of the building to the nearest foot.

Answer: 69 ft

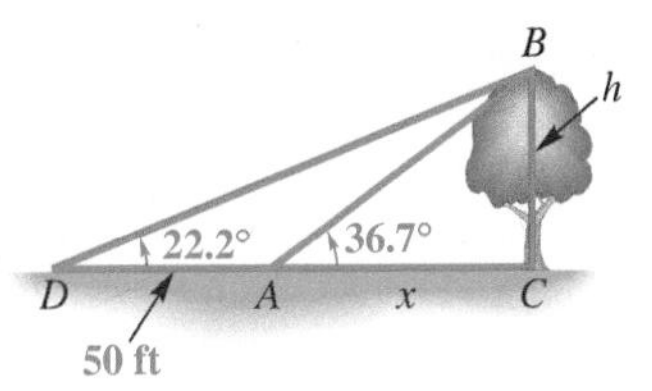

Figure 34

EXAMPLE 4 Solving a Problem Involving Angles of Elevation

Francisco needs to know the height of a tree. From a given point on the ground, he finds that the angle of elevation to the top of the tree is 36.7°. He then moves back 50 ft. From the second point, the angle of elevation to the top of the tree is 22.2°. See **Figure 34.** Find the height of the tree to the nearest foot.

ALGEBRAIC SOLUTION

Figure 34 shows two unknowns: x, the distance from the center of the trunk of the tree to the point where the first observation was made, and h, the height of the tree. See **Figure 35** in the Graphing Calculator Solution. Because nothing is given about the length of the hypotenuse of either triangle ABC or triangle BCD, we use a ratio that does not involve the hypotenuse—namely, the tangent.

In triangle ABC, $\tan 36.7° = \frac{h}{x}$ or $h = x \tan 36.7°$.

In triangle BCD, $\tan 22.2° = \frac{h}{50 + x}$ or $h = (50 + x) \tan 22.2°$.

Each expression equals h, so the expressions must be equal.

$$x \tan 36.7° = (50 + x) \tan 22.2° \quad \text{Equate expressions for } h.$$

$$x \tan 36.7° = 50 \tan 22.2° + x \tan 22.2° \quad \text{Distributive property}$$

$$x \tan 36.7° - x \tan 22.2° = 50 \tan 22.2° \quad \text{Write the } x\text{-terms on one side.}$$

$$x(\tan 36.7° - \tan 22.2°) = 50 \tan 22.2° \quad \text{Factor out } x.$$

$$x = \frac{50 \tan 22.2°}{\tan 36.7° - \tan 22.2°} \quad \text{Divide by the coefficient of } x.$$

We saw above that $h = x \tan 36.7°$. Substitute for x.

$$h = \left(\frac{50 \tan 22.2°}{\tan 36.7° - \tan 22.2°}\right) \tan 36.7°$$

Use a calculator.

$$\tan 36.7° = 0.74537703 \quad \text{and} \quad \tan 22.2° = 0.40809244$$

Thus,

$$\tan 36.7° - \tan 22.2° = 0.74537703 - 0.40809244 = 0.33728459$$

and
$$h = \left(\frac{50(0.40809244)}{0.33728459}\right) 0.74537703 \approx 45.$$

To the nearest foot, the height of the tree is 45 ft.

GRAPHING CALCULATOR SOLUTION*

In **Figure 35,** we have superimposed **Figure 34** on coordinate axes with the origin at D. By definition, the tangent of the angle between the x-axis and the graph of a line with equation $y = mx + b$ is the slope of the line, m. For line DB, $m = \tan 22.2°$. Because b equals 0, the equation of line DB is

$$y_1 = (\tan 22.2°)x.$$

The equation of line AB is

$$y_2 = (\tan 36.7°)x + b.$$

Because $b \neq 0$ here, we use the point $A(50, 0)$ and the point-slope form to find the equation.

$$y_2 - y_0 = m(x - x_0) \quad \text{Point-slope form}$$

$$y_2 - 0 = m(x - 50) \quad x_0 = 50,\ y_0 = 0$$

$$y_2 = \tan 36.7°(x - 50)$$

Lines y_1 and y_2 are graphed in **Figure 36.** The y-coordinate of the point of intersection of the graphs gives the length of BC, or h. Thus, $h \approx 45$.

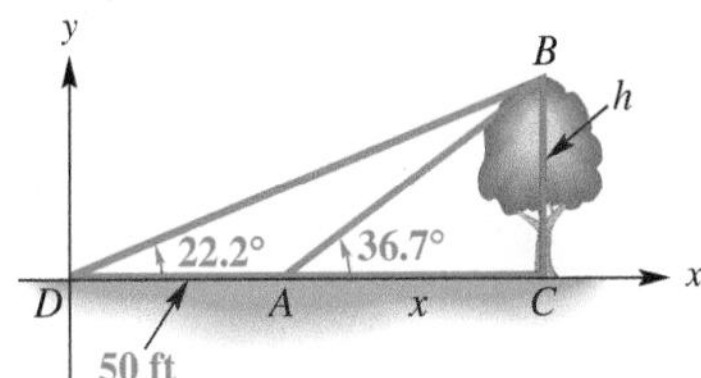

Figure 35

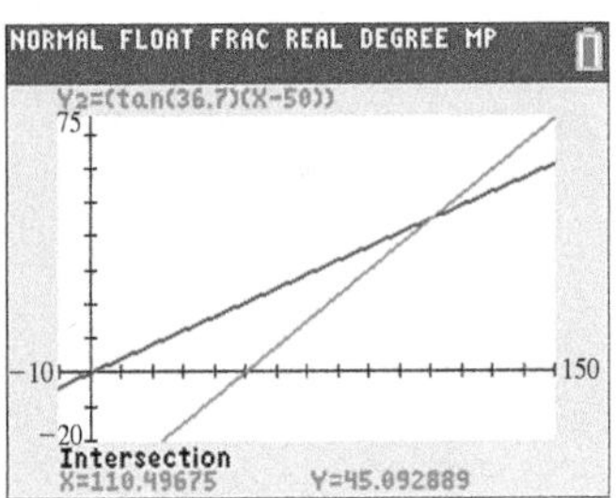

Figure 36

✔ Now Try Exercise 31.

*Source: Reprinted with permission from *The Mathematics Teacher,* copyright 1995 by the National Council of Teachers of Mathematics.

2.5 Exercises

1. C 2. D
3. A 4. G
5. B 6. H
7. F 8. J
9. I 10. E

11. 270°; N 90° W, or S 90° W
12. 90°; N 90°E, or S 90° E
13. 0°; N 0°E, or N 0° W
14. 180°; S 0° E, or S 0° W
15. 315°; N 45° W
16. 225°; S 45° W
17. 135°; S 45° E
18. 45°; N 45° E

19. 220 mi

CONCEPT PREVIEW *Match the measure of bearing in Column I with the appropriate graph in Column II.*

I

1. 20°
2. S 70° W
3. 160°
4. S 20° W
5. N 70° W
6. 340°
7. 110°
8. 270°
9. 180°
10. N 70° E

II

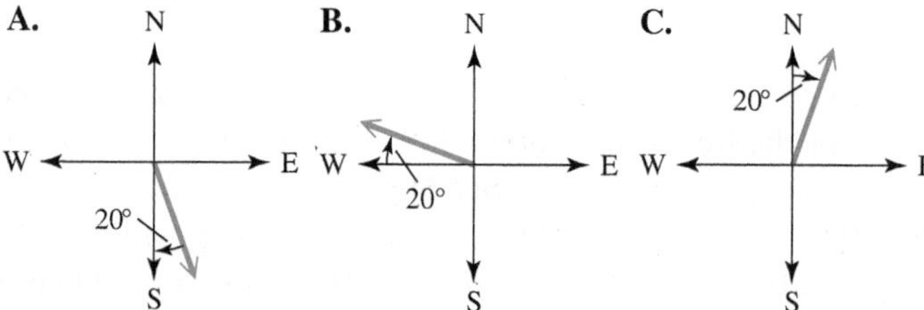

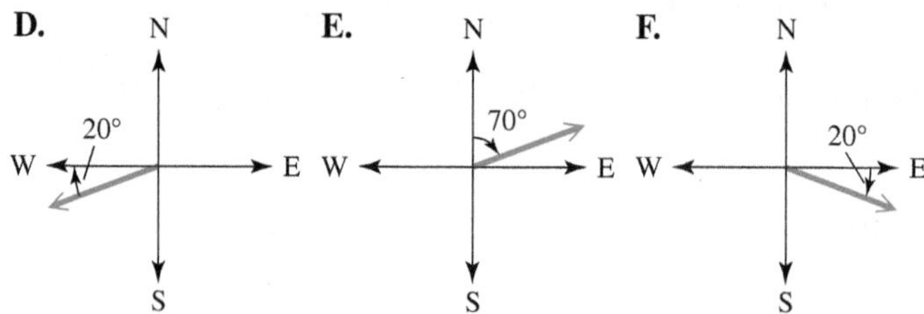

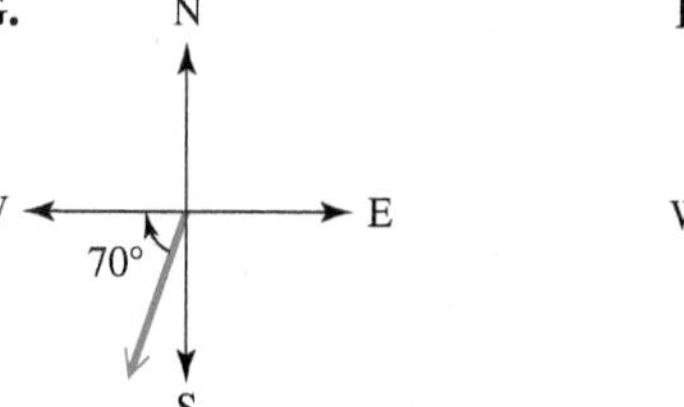

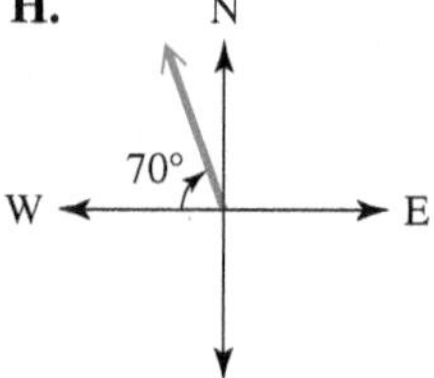

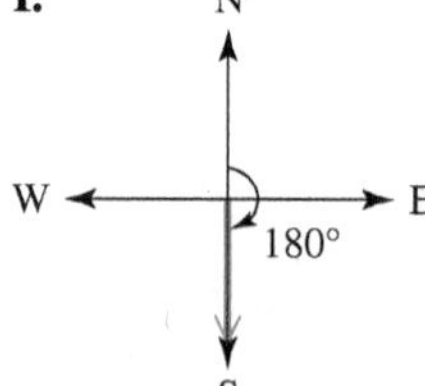

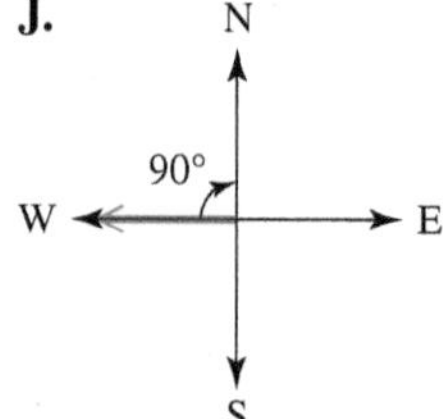

The two methods of expressing bearing can be interpreted using a rectangular coordinate system. Suppose that an observer for a radar station is located at the origin of a coordinate system. Find the bearing of an airplane located at each point. Express the bearing using both methods.

11. $(-4, 0)$ **12.** $(5, 0)$ **13.** $(0, 4)$ **14.** $(0, -2)$

15. $(-5, 5)$ **16.** $(-3, -3)$ **17.** $(2, -2)$ **18.** $(2, 2)$

Solve each problem. ***See Examples 1 and 2.***

19. ***Distance Flown by a Plane*** A plane flies 1.3 hr at 110 mph on a bearing of 38°. It then turns and flies 1.5 hr at the same speed on a bearing of 128°. How far is the plane from its starting point?

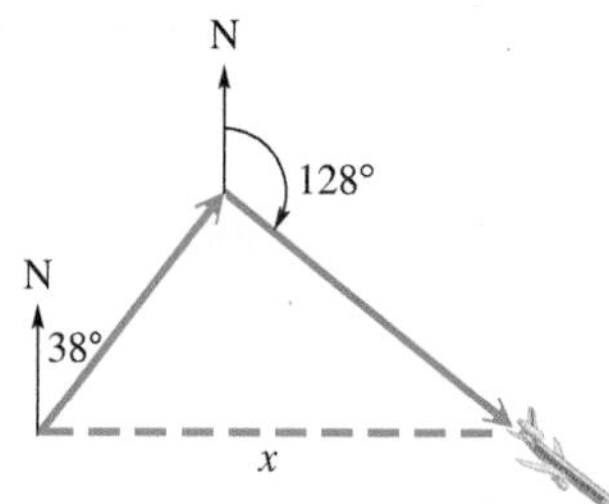

20. 150 km
21. 47 nautical mi
22. 70 nautical mi
23. 2203 ft 24. 5856 m
25. 148 mi 26. 2.01 mi
27. 430 mi 28. 350 mi
29. 140 mi 30. 130 mi

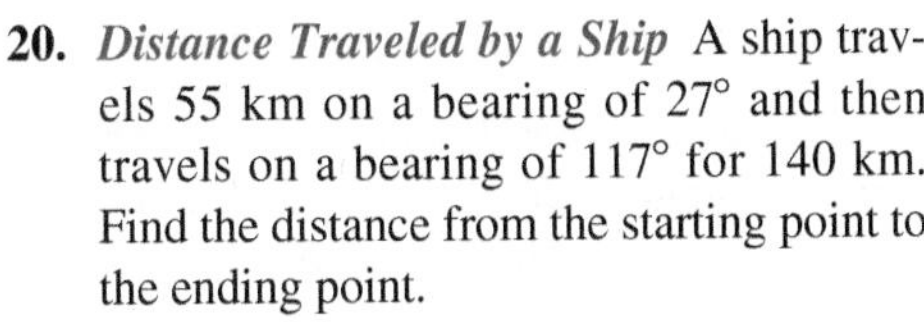

20. ***Distance Traveled by a Ship*** A ship travels 55 km on a bearing of 27° and then travels on a bearing of 117° for 140 km. Find the distance from the starting point to the ending point.

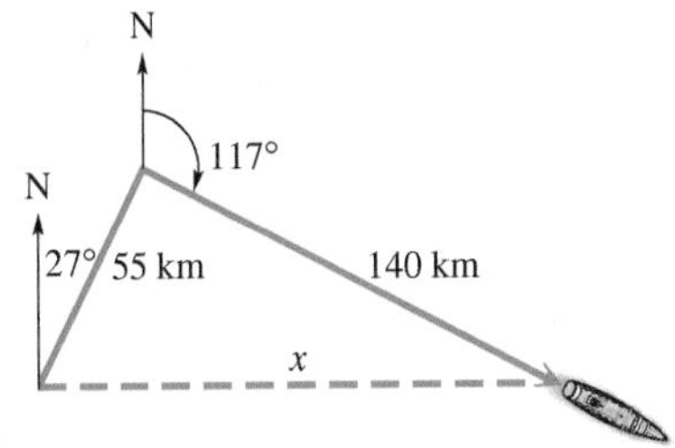

21. ***Distance between Two Ships*** Two ships leave a port at the same time. The first ship sails on a bearing of 40° at 18 knots (nautical miles per hour) and the second on a bearing of 130° at 26 knots. How far apart are they after 1.5 hr?

22. ***Distance between Two Ships*** Two ships leave a port at the same time. The first ship sails on a bearing of 52° at 17 knots and the second on a bearing of 322° at 22 knots. How far apart are they after 2.5 hr?

23. ***Distance between Two Docks*** Two docks are located on an east-west line 2587 ft apart. From dock A, the bearing of a coral reef is 58° 22′. From dock B, the bearing of the coral reef is 328° 22′. Find the distance from dock A to the coral reef.

24. ***Distance between Two Lighthouses*** Two lighthouses are located on a north-south line. From lighthouse A, the bearing of a ship 3742 m away is 129° 43′. From lighthouse B, the bearing of the ship is 39° 43′. Find the distance between the lighthouses.

25. ***Distance between Two Ships*** A ship leaves its home port and sails on a bearing of S 61°50′ E. Another ship leaves the same port at the same time and sails on a bearing of N 28°10′ E. If the first ship sails at 24.0 mph and the second sails at 28.0 mph, find the distance between the two ships after 4 hr.

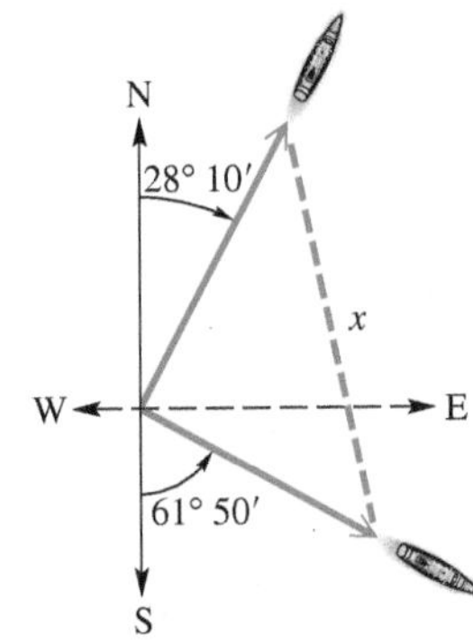

26. ***Distance between Transmitters*** Radio direction finders are set up at two points A and B, which are 2.50 mi apart on an east-west line. From A, it is found that the bearing of a signal from a radio transmitter is N 36°20′ E, and from B the bearing of the same signal is N 53°40′ W. Find the distance of the transmitter from B.

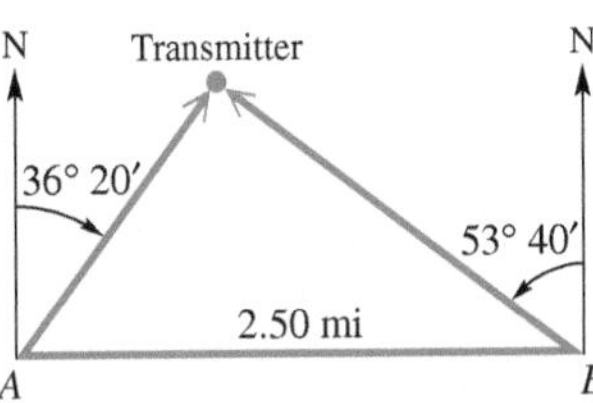

27. ***Flying Distance*** The bearing from A to C is S 52° E. The bearing from A to B is N 84° E. The bearing from B to C is S 38° W. A plane flying at 250 mph takes 2.4 hr to go from A to B. Find the distance from A to C.

28. ***Flying Distance*** The bearing from A to C is N 64° W. The bearing from A to B is S 82° W. The bearing from B to C is N 26° E. A plane flying at 350 mph takes 1.8 hr to go from A to B. Find the distance from B to C.

29. ***Distance between Two Cities*** The bearing from Winston-Salem, North Carolina, to Danville, Virginia, is N 42° E. The bearing from Danville to Goldsboro, North Carolina, is S 48° E. A car traveling at 65 mph takes 1.1 hr to go from Winston-Salem to Danville and 1.8 hr to go from Danville to Goldsboro. Find the distance from Winston-Salem to Goldsboro.

30. ***Distance between Two Cities*** The bearing from Atlanta to Macon is S 27° E, and the bearing from Macon to Augusta is N 63° E. An automobile traveling at 62 mph needs $1\frac{1}{4}$ hr to go from Atlanta to Macon and $1\frac{3}{4}$ hr to go from Macon to Augusta. Find the distance from Atlanta to Augusta.

31. 114 ft
32. 147 m
33. 5.18 m
34. 2.47 km
35. 433 ft
36. 448 m
37. 10.8 ft
38. $A = 35.987°$, or $35° 59' 10''$; $B = 54.013°$, or $54° 00' 50''$

Solve each problem. ***See Examples 3 and 4.***

31. ***Height of a Pyramid*** The angle of elevation from a point on the ground to the top of a pyramid is $35° 30'$. The angle of elevation from a point 135 ft farther back to the top of the pyramid is $21° 10'$. Find the height of the pyramid.

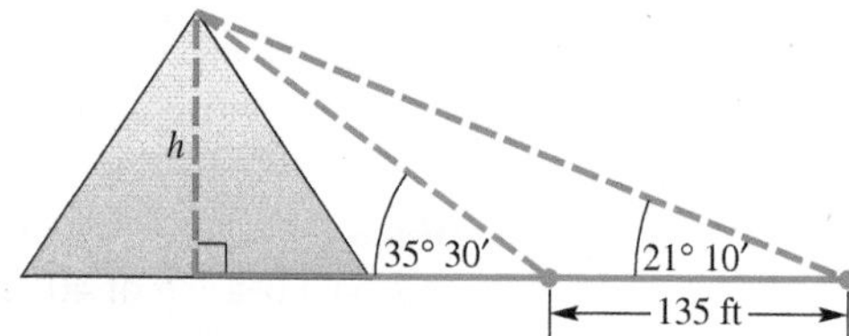

32. ***Distance between a Whale and a Lighthouse*** A whale researcher is watching a whale approach directly toward a lighthouse as she observes from the top of this lighthouse. When she first begins watching the whale, the angle of depression to the whale is $15° 50'$. Just as the whale turns away from the lighthouse, the angle of depression is $35° 40'$. If the height of the lighthouse is 68.7 m, find the distance traveled by the whale as it approached the lighthouse.

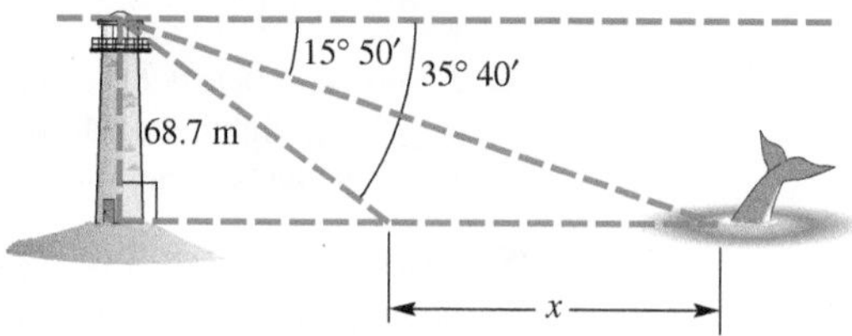

33. ***Height of an Antenna*** A scanner antenna is on top of the center of a house. The angle of elevation from a point 28.0 m from the center of the house to the top of the antenna is $27° 10'$, and the angle of elevation to the bottom of the antenna is $18° 10'$. Find the height of the antenna.

34. ***Height of Mt. Whitney*** The angle of elevation from Lone Pine to the top of Mt. Whitney is $10° 50'$. A hiker, traveling 7.00 km from Lone Pine along a straight, level road toward Mt. Whitney, finds the angle of elevation to be $22° 40'$. Find the height of the top of Mt. Whitney above the level of the road.

35. Find h as indicated in the figure.

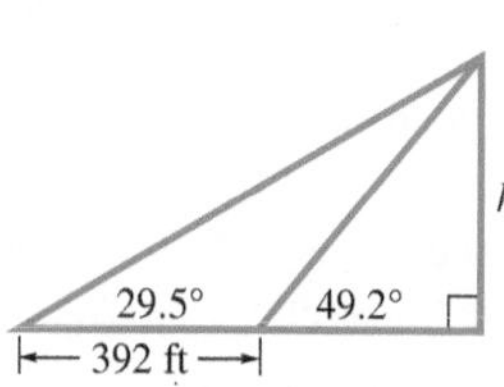

36. Find h as indicated in the figure.

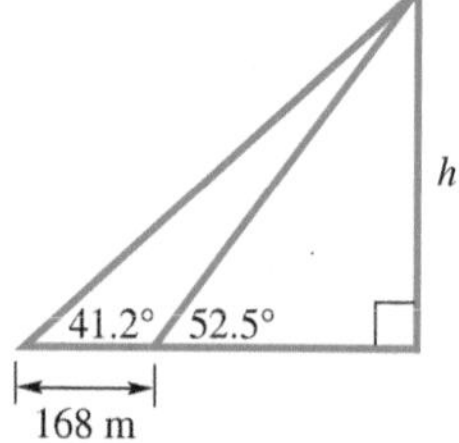

37. ***Distance of a Plant from a Fence*** In one area, the lowest angle of elevation of the sun in winter is $23° 20'$. Find the minimum distance x that a plant needing full sun can be placed from a fence 4.65 ft high.

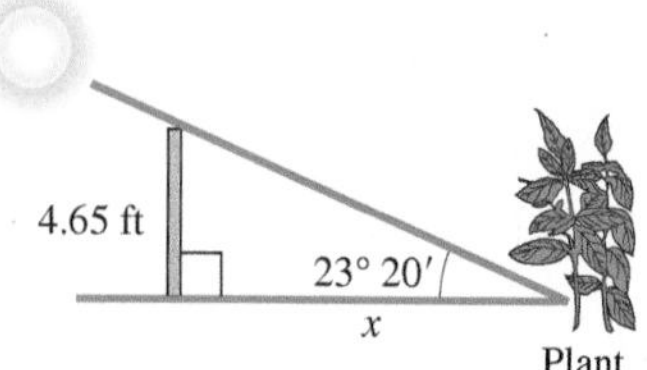

38. ***Distance through a Tunnel*** A tunnel is to be built from A to B. Both A and B are visible from C. If AC is 1.4923 mi and BC is 1.0837 mi, and if C is 90°, find the measures of angles A and B.

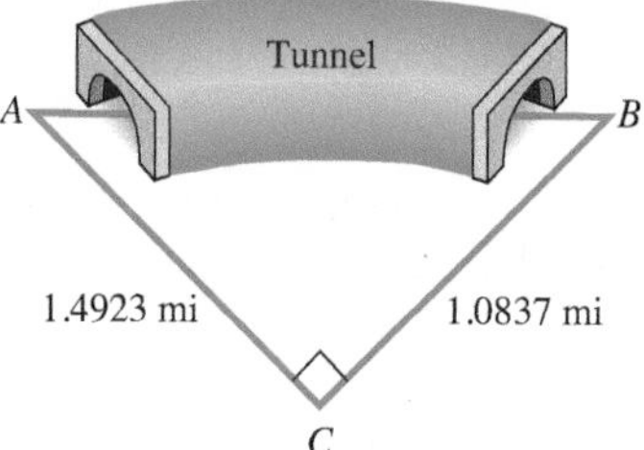

39. 1.95 mi 40. 84.7 m

41. (a) $d = \frac{b}{2}\left(\cot \frac{\alpha}{2} + \cot \frac{\beta}{2}\right)$
(b) 345.4 cm

42. (a) 67.00 ft; 67.14 ft; 66.84 ft; D increases and then decreases.
(b) 64.40 ft; 67.14 ft; 69.93 ft; D increases.
(c) v; The shot-putter should concentrate on achieving as large a value of v as possible.

39. ***Height of a Plane above Earth*** Find the minimum height h above the surface of Earth so that a pilot at point A in the figure can see an object on the horizon at C, 125 mi away. Assume 4.00×10^3 mi as the radius of Earth.

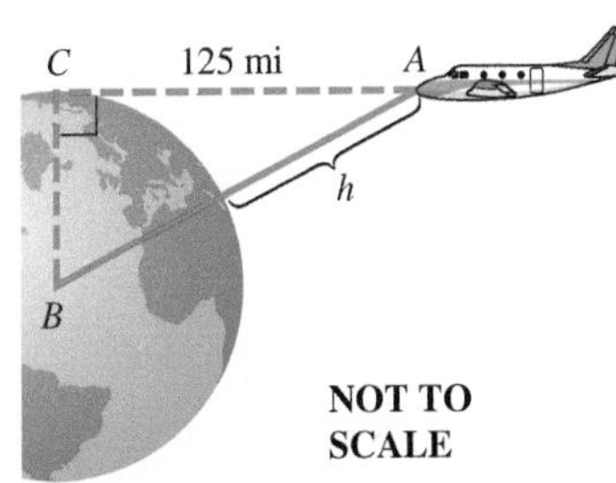

40. ***Length of a Side of a Piece of Land*** A piece of land has the shape shown in the figure. Find the length x.

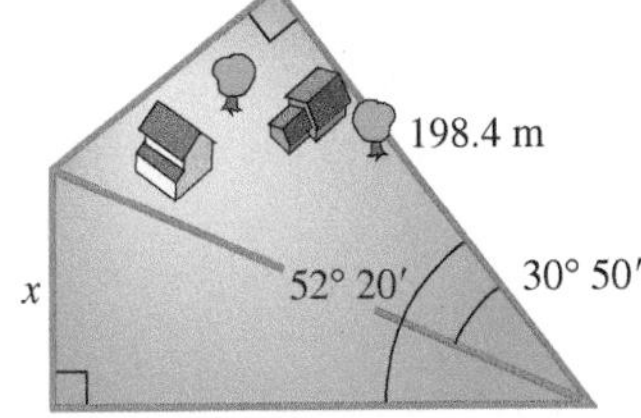

41. ***(Modeling) Distance between Two Points*** A variation of the subtense bar method that surveyors use to determine larger distances d between two points P and Q is shown in the figure. The subtense bar with length b is placed between points P and Q so that the bar is centered on and perpendicular to the line of sight between P and Q. Angles α and β are measured from points P and Q, respectively. (*Source:* Mueller, I. and K. Ramsayer, *Introduction to Surveying*, Frederick Ungar Publishing Co.)

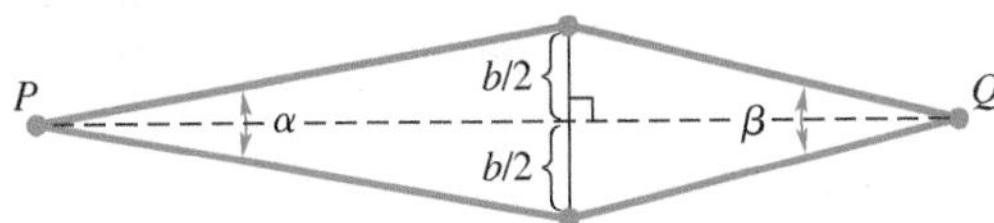

(a) Find a formula for d involving α, β, and b.

(b) Use the formula from part (a) to determine d if $\alpha = 37' \, 48''$, $\beta = 42' \, 03''$, and $b = 2.000$ cm.

42. ***(Modeling) Distance of a Shot Put*** A shot-putter trying to improve performance may wonder whether there is an optimal angle to aim for, or whether the velocity (speed) at which the ball is thrown is more important. The figure shows the path of a steel ball thrown by a shot-putter. The distance D depends on initial velocity v, height h, and angle θ when the ball is released.

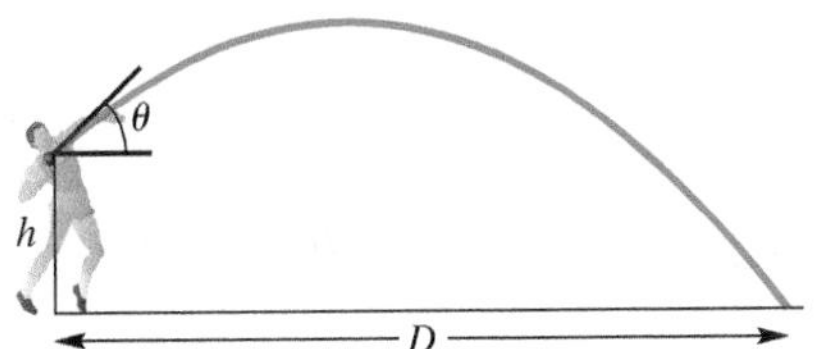

One model developed for this situation gives D as

$$D = \frac{v^2 \sin \theta \cos \theta + v \cos \theta \sqrt{(v \sin \theta)^2 + 64h}}{32}.$$

Typical ranges for the variables are v: 33–46 ft per sec; h: 6–8 ft; and θ: 40°–45°. (*Source:* Kreighbaum, E. and K. Barthels, *Biomechanics*, Allyn & Bacon.)

(a) To see how angle θ affects distance D, let $v = 44$ ft per sec and $h = 7$ ft. Calculate D, to the nearest hundredth, for $\theta = 40°$, 42°, and 45°. How does distance D change as θ increases?

(b) To see how velocity v affects distance D, let $h = 7$ and $\theta = 42°$. Calculate D, to the nearest hundredth, for $v = 43$, 44, and 45 ft per sec. How does distance D change as v increases?

(c) Which affects distance D more, v or θ? What should the shot-putter do to improve performance?

43. (a) 320 ft
(b) $R\left(1 - \cos \frac{\theta}{2}\right)$

44. (a) 23 ft (b) 48 ft
(c) As the speed limit increases, more land needs to be cleared inside the curve.

45. $y = (\tan 35°)(x - 25)$

46. $y = (\tan 15°)(x - 5)$

49. $y = \frac{\sqrt{3}}{3}x,\ x \le 0$

50. $y = -\sqrt{3}x,\ x \ge 0$

43. *(Modeling) Highway Curves* A basic highway curve connecting two straight sections of road may be circular. In the figure, the points P and S mark the beginning and end of the curve. Let Q be the point of intersection where the two straight sections of highway leading into the curve would meet if extended. The radius of the curve is R, and the central angle θ denotes how many degrees the curve turns. (*Source:* Mannering, F. and W. Kilareski, *Principles of Highway Engineering and Traffic Analysis*, Second Edition, John Wiley and Sons.)

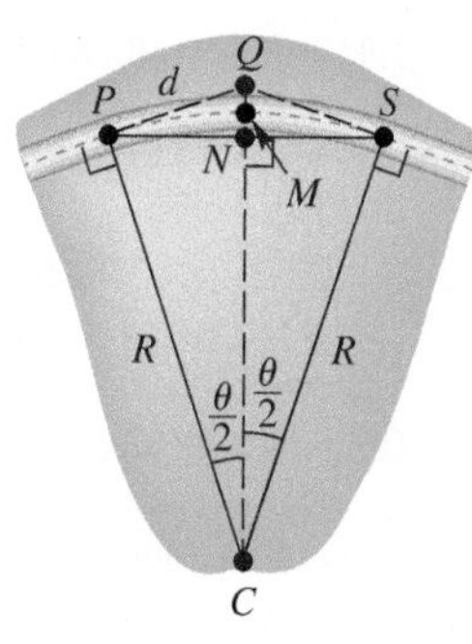

(a) If $R = 965$ ft and $\theta = 37°$, find the distance d between P and Q.

(b) Find an expression in terms of R and θ for the distance between points M and N.

44. *(Modeling) Stopping Distance on a Curve* Refer to **Exercise 43.** When an automobile travels along a circular curve, objects like trees and buildings situated on the inside of the curve can obstruct the driver's vision. These obstructions prevent the driver from seeing sufficiently far down the highway to ensure a safe stopping distance. In the figure, the *minimum* distance d that should be cleared on the inside of the highway is modeled by the equation

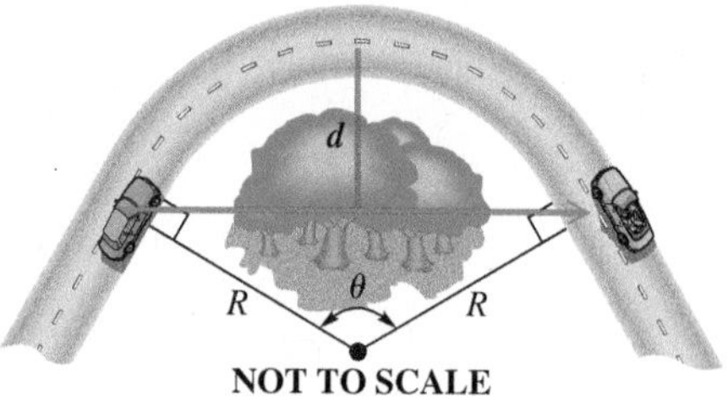

$$d = R\left(1 - \cos \frac{\theta}{2}\right).$$

(*Source:* Mannering, F. and W. Kilareski, *Principles of Highway Engineering and Traffic Analysis,* Second Edition, John Wiley and Sons.)

(a) It can be shown that if θ is measured in degrees, then $\theta \approx \frac{57.3S}{R}$, where S is the safe stopping distance for the given speed limit. Compute d to the nearest foot for a 55 mph speed limit if $S = 336$ ft and $R = 600$ ft.

(b) Compute d to the nearest foot for a 65 mph speed limit given $S = 485$ ft and $R = 600$ ft.

(c) How does the speed limit affect the amount of land that should be cleared on the inside of the curve?

The figure to the right indicates that the equation of a line passing through the point $(a, 0)$ and making an angle θ with the x-axis is

$$y = (\tan \theta)(x - a).$$

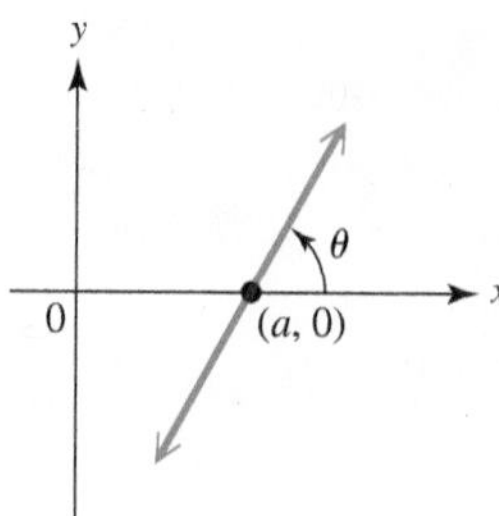

45. Find an equation of the line passing through the point $(25, 0)$ that makes an angle of 35° with the x-axis.

46. Find an equation of the line passing through the point $(5, 0)$ that makes an angle of 15° with the x-axis.

47. Show that a line bisecting the first and third quadrants satisfies the equation given in the instructions.

48. Show that a line bisecting the second and fourth quadrants satisfies the equation given in the instructions.

49. The ray $y = x,\ x \ge 0$, contains the origin and all points in the coordinate system whose bearing is 45°. Determine an equation of a ray consisting of the origin and all points whose bearing is 240°.

50. Repeat **Exercise 49** for a bearing of 150°.

Chapter 2 Test Prep

Key Terms

2.1 side opposite side adjacent cofunctions	2.2 reference angle 2.4 exact number significant digits	angle of elevation angle of depression	2.5 bearing

Quick Review

Concepts | **Examples**

2.1 Trigonometric Functions of Acute Angles

Right-Triangle-Based Definitions of Trigonometric Functions

Let A represent any acute angle in standard position.

$$\sin A = \frac{y}{r} = \frac{\textbf{side opposite}}{\textbf{hypotenuse}} \qquad \csc A = \frac{r}{y} = \frac{\textbf{hypotenuse}}{\textbf{side opposite}}$$

$$\cos A = \frac{x}{r} = \frac{\textbf{side adjacent}}{\textbf{hypotenuse}} \qquad \sec A = \frac{r}{x} = \frac{\textbf{hypotenuse}}{\textbf{side adjacent}}$$

$$\tan A = \frac{y}{x} = \frac{\textbf{side opposite}}{\textbf{side adjacent}} \qquad \cot A = \frac{x}{y} = \frac{\textbf{side adjacent}}{\textbf{side opposite}}$$

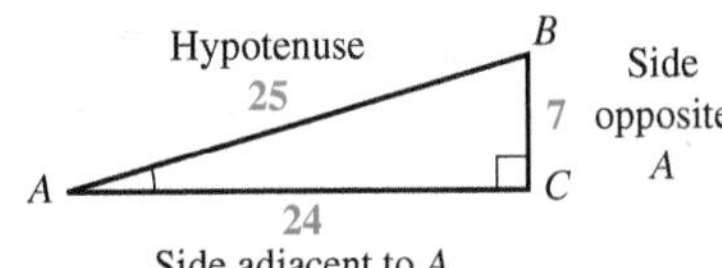

$$\sin A = \frac{7}{25} \quad \cos A = \frac{24}{25} \quad \tan A = \frac{7}{24}$$

$$\csc A = \frac{25}{7} \quad \sec A = \frac{25}{24} \quad \cot A = \frac{24}{7}$$

Cofunction Identities

For any acute angle A, cofunction values of complementary angles are equal.

$$\sin A = \cos(90° - A) \qquad \cos A = \sin(90° - A)$$
$$\sec A = \csc(90° - A) \qquad \csc A = \sec(90° - A)$$
$$\tan A = \cot(90° - A) \qquad \cot A = \tan(90° - A)$$

$$\sin 55° = \cos(90° - 55°) = \cos 35°$$
$$\sec 48° = \csc(90° - 48°) = \csc 42°$$
$$\tan 72° = \cot(90° - 72°) = \cot 18°$$

Function Values of Special Angles

θ	$\sin\theta$	$\cos\theta$	$\tan\theta$	$\cot\theta$	$\sec\theta$	$\csc\theta$
30°	$\frac{1}{2}$	$\frac{\sqrt{3}}{2}$	$\frac{\sqrt{3}}{3}$	$\sqrt{3}$	$\frac{2\sqrt{3}}{3}$	2
45°	$\frac{\sqrt{2}}{2}$	$\frac{\sqrt{2}}{2}$	1	1	$\sqrt{2}$	$\sqrt{2}$
60°	$\frac{\sqrt{3}}{2}$	$\frac{1}{2}$	$\sqrt{3}$	$\frac{\sqrt{3}}{3}$	2	$\frac{2\sqrt{3}}{3}$

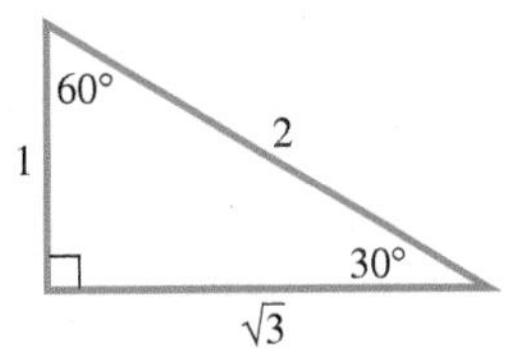

30°–60° right triangle

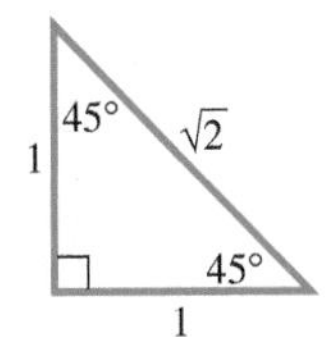

45°–45° right triangle

Concepts	Examples

2.2 Trigonometric Functions of Non-Acute Angles

Reference Angle θ' for θ in $(0°, 360°)$

θ in Quadrant	I	II	III	IV
θ' is	θ	$180° - \theta$	$\theta - 180°$	$360° - \theta$

Quadrant I: For $\theta = 25°$, $\theta' = 25°$

Quadrant II: For $\theta = 152°$, $\theta' = 28°$

Quadrant III: For $\theta = 200°$, $\theta' = 20°$

Quadrant IV: For $\theta = 320°$, $\theta' = 40°$

Finding Trigonometric Function Values for Any Nonquadrantal Angle θ

Step 1 Add or subtract 360° as many times as needed to obtain an angle greater than 0° but less than 360°.

Step 2 Find the reference angle θ'.

Step 3 Find the trigonometric function values for θ'.

Step 4 Determine the correct signs for the values found in Step 3.

Find sin 1050°.

$$1050° - 2(360°) = 330° \quad \text{Coterminal angle in quadrant IV}$$

The reference angle for 330° is $\theta' = 30°$.

$$\sin 1050°$$
$$= -\sin 30° \quad \text{Sine is negative in quadrant IV.}$$
$$= -\frac{1}{2} \quad \sin 30° = \frac{1}{2}$$

2.3 Approximations of Trigonometric Function Values

To approximate a trigonometric function value of an angle in degrees, make sure the calculator is in degree mode.

Approximate each value.

$$\cos 50° \, 15' = \cos 50.25° \approx 0.63943900$$

$$\csc 32.5° = \frac{1}{\sin 32.5°} \approx 1.86115900 \quad \csc\theta = \frac{1}{\sin\theta}$$

To find the corresponding angle measure given a trigonometric function value, use an appropriate inverse function.

Find an angle θ in the interval $[0°, 90°)$ that satisfies each condition in color.

$$\cos\theta \approx 0.73677482$$
$$\theta \approx \cos^{-1}(0.73677482)$$
$$\theta \approx 42.542600°$$

$$\csc\theta \approx 1.04766792$$
$$\sin\theta \approx \frac{1}{1.04766792} \quad \sin\theta = \frac{1}{\csc\theta}$$
$$\theta \approx \sin^{-1}\left(\frac{1}{1.04766792}\right)$$
$$\theta \approx 72.65°$$

2.4 Solutions and Applications of Right Triangles

Solving an Applied Trigonometry Problem

Step 1 Draw a sketch, and label it with the given information. Label the quantity to be found with a variable.

Find the angle of elevation of the sun if a 48.6-ft flagpole casts a shadow 63.1 ft long.

Step 1 See the sketch. We must find θ.

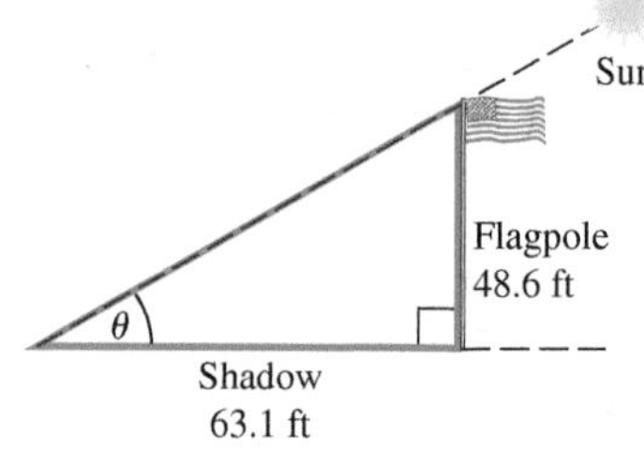

Concepts	Examples
Step 2 Use the sketch to write an equation relating the given quantities to the variable.	***Step 2*** $\tan\theta = \dfrac{48.6}{63.1}$ $\tan\theta \approx 0.770206$
Step 3 Solve the equation, and check that the answer makes sense.	***Step 3*** $\theta = \tan^{-1} 0.770206$ $\theta \approx 37.6°$ The angle of elevation rounded to three significant digits is 37.6°, or 37° 40′.

2.5 Further Applications of Right Triangles

Expressing Bearing

Method 1 When a single angle is given, bearing is measured in a clockwise direction from due north.

Method 2 Start with a north-south line and use an acute angle to show direction, either east or west, from this line.

Example: 220°

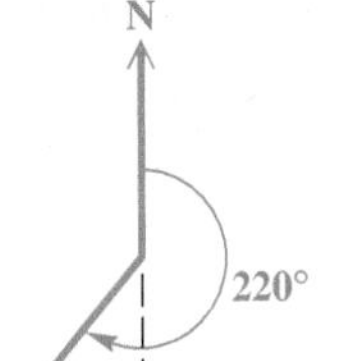

Example: S 40° W

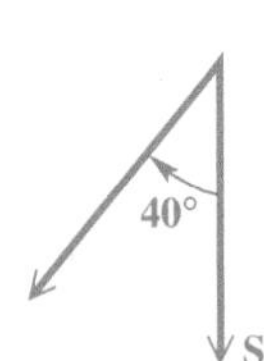

Chapter 2 Review Exercises

In Exercises 1, 2, and 13–16, we give, in order, sine, cosine, tangent, cotangent, secant, and cosecant.

1. $\frac{60}{61}; \frac{11}{61}; \frac{60}{11}; \frac{11}{60}; \frac{61}{11}; \frac{61}{60}$
2. $\frac{20}{29}; \frac{21}{29}; \frac{20}{21}; \frac{21}{20}; \frac{29}{21}; \frac{29}{20}$
3. 10°
4. 10°
5. 7°
6. 30°
7. true
8. false; For $0° \le \theta \le 90°$, as the angle increases, $\cos\theta$ decreases.
9. true
10. false; For $0° < \theta \le 90°$, as the angle increases, $\csc\theta$ decreases.
11. $\cos A = \frac{b}{c}$ and $\sin B = \frac{b}{c}$, so $\cos A = \sin B$. (This is an example of equality of cofunctions of complementary angles.)

Find exact values of the six trigonometric functions for each angle A.

1.

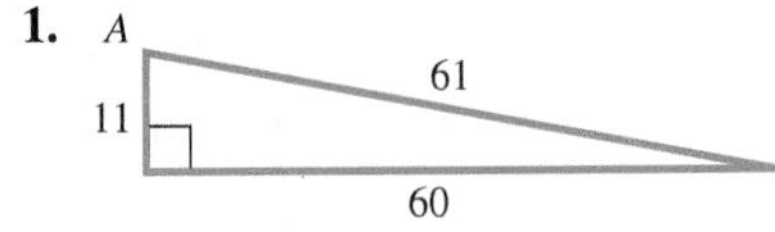

2.

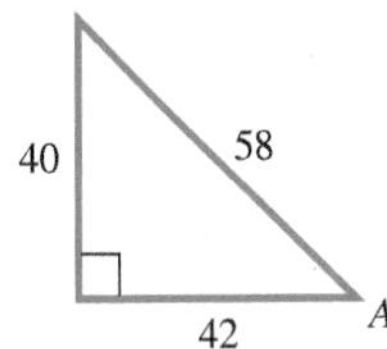

Find one solution for each equation. Assume that all angles involved are acute angles.

3. $\sin 4\beta = \cos 5\beta$

4. $\sec(2\theta + 10°) = \csc(4\theta + 20°)$

5. $\tan(5x + 11°) = \cot(6x + 2°)$

6. $\cos\left(\frac{3\theta}{5} + 11°\right) = \sin\left(\frac{7\theta}{10} + 40°\right)$

Determine whether each statement is true *or* false. *If false, tell why.*

7. $\sin 46° < \sin 58°$

8. $\cos 47° < \cos 58°$

9. $\tan 60° \ge \cot 40°$

10. $\csc 22° \le \csc 68°$

11. Explain why, in the figure, the cosine of angle A is equal to the sine of angle B.

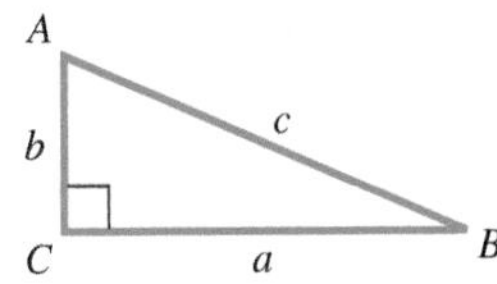

12. D

13. $-\frac{\sqrt{3}}{2}; \frac{1}{2}; -\sqrt{3}; -\frac{\sqrt{3}}{3}; 2; -\frac{2\sqrt{3}}{3}$

14. $\frac{\sqrt{3}}{2}; -\frac{1}{2}; -\sqrt{3}; -\frac{\sqrt{3}}{3}; -2; \frac{2\sqrt{3}}{3}$

15. $-\frac{1}{2}; \frac{\sqrt{3}}{2}; -\frac{\sqrt{3}}{3}; -\sqrt{3}; \frac{2\sqrt{3}}{3}; -2$

16. $\frac{\sqrt{2}}{2}; -\frac{\sqrt{2}}{2}; -1; -1; -\sqrt{2}; \sqrt{2}$

17. 120°; 240° 18. 210°; 330°
19. 150°; 210° 20. 135°; 315°

21. $3 - \frac{2\sqrt{3}}{3}$ 22. 1

23. $\frac{5}{2}$

24. (a) $-\frac{\sqrt{2}}{2}; -\frac{\sqrt{2}}{2}; 1$
(b) $-\frac{\sqrt{3}}{2}; \frac{1}{2}; -\sqrt{3}$

25. −1.356342 26. 0.953717
27. 1.021034 28. −0.715930
29. 0.208344 30. 1.936213

31. 55.673870° 32. 41.635092°
33. 12.733938° 34. 37.695528°
35. 63.008286° 36. 5.999827°

37. 47°; 133° 38. 54°; 234°

39. false; $1.4088321 \neq 1$
40. true
41. true
42. false; $1.3382612 \neq 0.99452190$

43. No, this will result in an angle having tangent equal to 25. The function $\tan^{-1}$ is not the reciprocal of the tangent (cotangent) but is, rather, the *inverse tangent function*. To find cot 25°, the student must find the *reciprocal* of tan 25°.
44. We must find an angle having secant equal to 10, which means that its cosine is $\frac{1}{10}$. Find $\cos^{-1}\left(\frac{1}{10}\right) \approx 84.26082952°$.

---

12. Which one of the following cannot be *exactly* determined using the methods of this chapter?

A. cos 135° **B.** cot(−45°) **C.** sin 300° **D.** tan 140°

Find exact values of the six trigonometric functions for each angle. Do not use a calculator. Rationalize denominators when applicable.

13. 1020° **14.** 120° **15.** −1470° **16.** −225°

Find all values of θ, if θ is in the interval $[0°, 360°)$ and θ has the given function value.

17. $\cos\theta = -\frac{1}{2}$ **18.** $\sin\theta = -\frac{1}{2}$

19. $\sec\theta = -\frac{2\sqrt{3}}{3}$ **20.** $\cot\theta = -1$

Evaluate each expression. Give exact values.

21. $\tan^2 120° - 2\cot 240°$ **22.** $\cos 60° + 2\sin^2 30°$ **23.** $\sec^2 300° - 2\cos^2 150°$

24. Find the sine, cosine, and tangent function values for each angle.

(a)

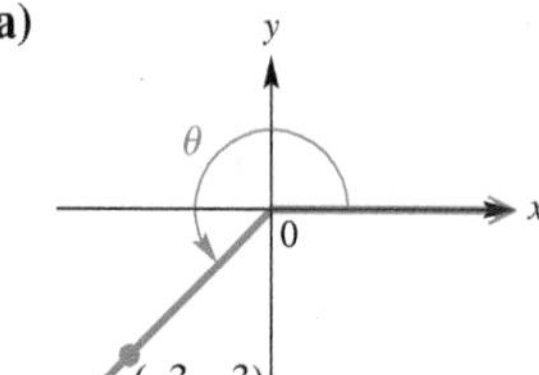

(b)

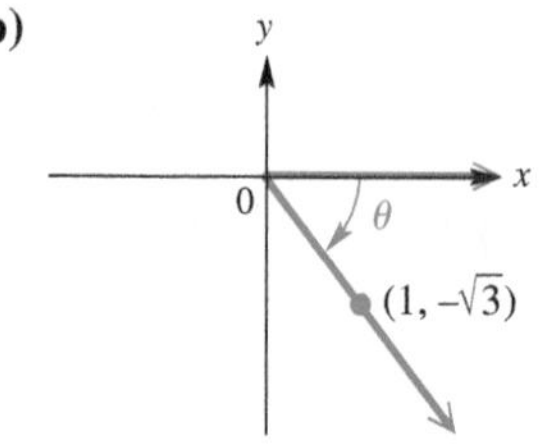

Use a calculator to approximate the value of each expression. Give answers to six decimal places.

25. sec 222° 30′ **26.** sin 72° 30′ **27.** csc 78° 21′

28. cot 305.6° **29.** tan 11.7689° **30.** sec 58.9041°

Use a calculator to find each value of θ, where θ is in the interval $[0°, 90°)$. Give answers in decimal degrees to six decimal places.

31. $\sin\theta = 0.82584121$ **32.** $\cot\theta = 1.1249386$ **33.** $\cos\theta = 0.97540415$

34. $\sec\theta = 1.2637891$ **35.** $\tan\theta = 1.9633124$ **36.** $\csc\theta = 9.5670466$

Find two angles in the interval $[0°, 360°)$ that satisfy each of the following. Round answers to the nearest degree.

37. $\sin\theta = 0.73135370$ **38.** $\tan\theta = 1.3763819$

Determine whether each statement is true *or* false. *If false, tell why. Use a calculator for Exercises 39 and 42.*

39. $\sin 50° + \sin 40° = \sin 90°$ **40.** $1 + \tan^2 60° = \sec^2 60°$

41. $\sin 240° = 2\sin 120° \cdot \cos 120°$ **42.** $\sin 42° + \sin 42° = \sin 84°$

43. A student wants to use a calculator to find the value of cot 25°. However, instead of entering $\frac{1}{\tan 25}$, he enters $\tan^{-1} 25$. Assuming the calculator is in degree mode, will this produce the correct answer? Explain.

44. Explain the process for using a calculator to find $\sec^{-1} 10$.

45. $B = 31° 30'$; $a = 638$; $b = 391$
46. $A = 19° 25'$; $B = 70° 35'$; $c = 390.3$
47. $B = 50.28°$; $a = 32.38$ m; $c = 50.66$ m
48. $A = 42° 07'$; $c = 402.5$ m; $a = 270.0$ m

49. 137 ft
50. $r = 13$; $\theta = 23°$
51. 73.7 ft
52. 20.4 m

Solve each right triangle. In Exercise 46, give angles to the nearest minute. In Exercises 47 and 48, label the triangle ABC as in Exercises 45 and 46.

45. 

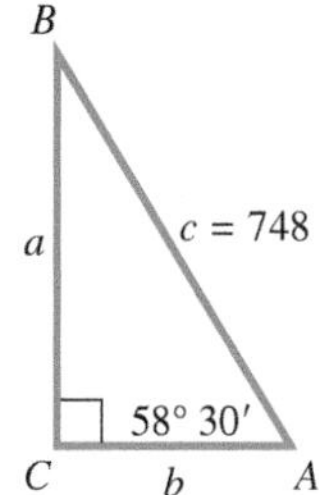

46.

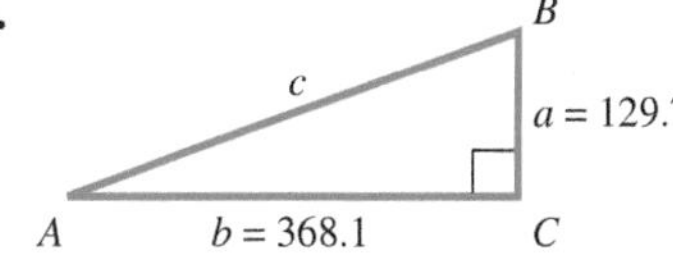

47. $A = 39.72°$, $b = 38.97$ m

48. $B = 47° 53'$, $b = 298.6$ m

Solve each problem. (Source for Exercises 49 and 50: Parker, M., Editor, *She Does Math,* Mathematical Association of America.*)*

49. ***Height of a Tree*** A civil engineer must determine the vertical height of the tree shown in the figure. The given angle was measured with a **clinometer.** Find the height of the leaning tree to the nearest whole number.

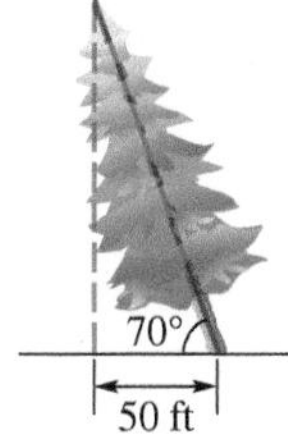

This is a picture of one type of clinometer, called an Abney hand level and clinometer. (Courtesy of Keuffel & Esser Co.)

50. ***(Modeling) Double Vision*** To correct mild double vision, a small amount of prism is added to a patient's eyeglasses. The amount of light shift this causes is measured in **prism diopters.** A patient needs 12 prism diopters horizontally and 5 prism diopters vertically. A prism that corrects for both requirements should have length r and be set at angle θ. Find the values of r and θ in the figure.

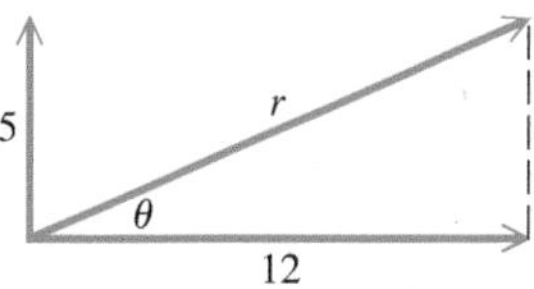

51. ***Height of a Tower*** The angle of elevation from a point 93.2 ft from the base of a tower to the top of the tower is $38° 20'$. Find the height of the tower.

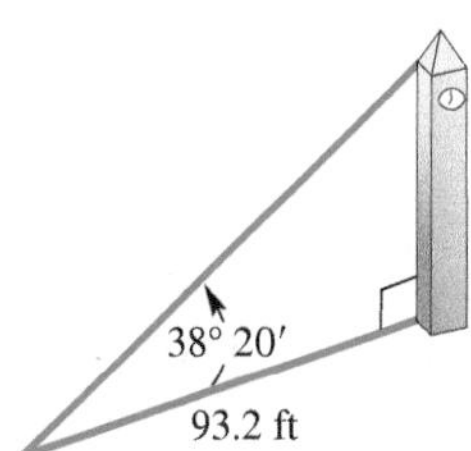

52. ***Height of a Tower*** The angle of depression from a television tower to a point on the ground 36.0 m from the bottom of the tower is 29.5°. Find the height of the tower.

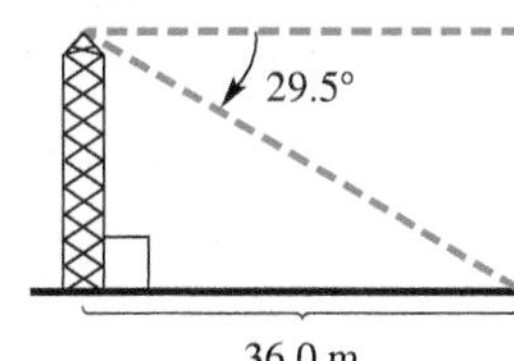

53. 18.75 cm 54. 50.24 m
55. 1200 m 56. 110 km
57. 140 mi
58. $h = k(\tan B - \tan A)$
59. One possible answer:
Find the value of x.

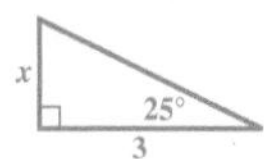

60. One possible answer:
Find the value of θ.

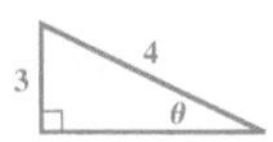

61. (a) 716 mi (b) 1104 mi
62. (a) $x_Q = x_P + d \sin \theta$, $y_Q = y_P + d \cos \theta$
(b) (181.34, 523.02)

53. *Length of a Diagonal* One side of a rectangle measures 15.24 cm. The angle between the diagonal and that side is 35.65°. Find the length of the diagonal.

54. *Length of Sides of an Isosceles Triangle* An isosceles triangle has a base of length 49.28 m. The angle opposite the base is 58.746°. Find the length of each of the two equal sides.

55. *Distance between Two Points* The bearing of point B from point C is 254°. The bearing of point A from point C is 344°. The bearing of point A from point B is 32°. If the distance from A to C is 780 m, find the distance from A to B.

56. *Distance a Ship Sails* The bearing from point A to point B is S 55° E, and the bearing from point B to point C is N 35° E. If a ship sails from A to B, a distance of 81 km, and then from B to C, a distance of 74 km, how far is it from A to C?

57. *Distance between Two Points* Two cars leave an intersection at the same time. One heads due south at 55 mph. The other travels due west. After 2 hr, the bearing of the car headed west from the car headed south is 324°. How far apart are they at that time?

58. Find a formula for h in terms of k, A, and B. Assume $A < B$.

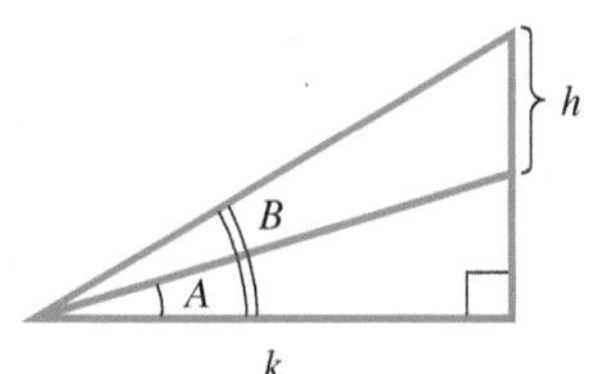

59. Create a right triangle problem whose solution is $3 \tan 25°$.

60. Create a right triangle problem whose solution can be found by evaluating θ if $\sin \theta = \frac{3}{4}$.

61. ***(Modeling) Height of a Satellite*** Artificial satellites that orbit Earth often use VHF signals to communicate with the ground. VHF signals travel in straight lines. The height h of the satellite above Earth and the time T that the satellite can communicate with a fixed location on the ground are related by the model

$$h = R\left(\frac{1}{\cos \frac{180T}{P}} - 1\right),$$

where $R = 3955$ mi is the radius of Earth and P is the period for the satellite to orbit Earth. (*Source:* Schlosser, W., T. Schmidt-Kaler, and E. Milone, *Challenges of Astronomy,* Springer-Verlag.)

(a) Find h to the nearest mile when $T = 25$ min and $P = 140$ min. (Evaluate the cosine function in degree mode.)

(b) What is the value of h to the nearest mile if T is increased to 30 min?

62. ***(Modeling) Fundamental Surveying Problem*** The first fundamental problem of surveying is to determine the coordinates of a point Q given the coordinates of a point P, the distance between P and Q, and the bearing θ from P to Q. See the figure. (*Source:* Mueller, I. and K. Ramsayer, *Introduction to Surveying,* Frederick Ungar Publishing Co.)

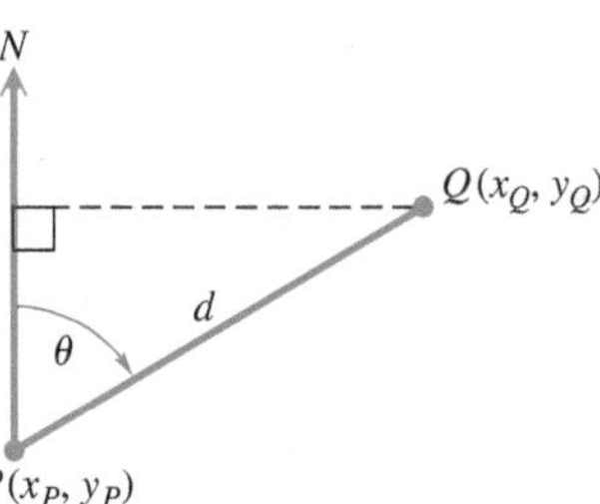

(a) Find a formula for the coordinates (x_Q, y_Q) of the point Q given θ, the coordinates (x_P, y_P) of P, and the distance d between P and Q.

(b) Use the formula found in part (a) to determine the coordinates (x_Q, y_Q) if $(x_P, y_P) = (123.62, 337.95)$, $\theta = 17°19'22''$, and $d = 193.86$ ft.

Chapter 2 Test

Solve each problem.

1. Find exact values of the six trigonometric functions for angle A in the right triangle.

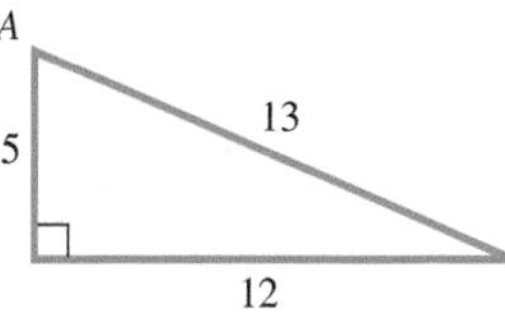

2. Find the exact value of each variable in the figure.

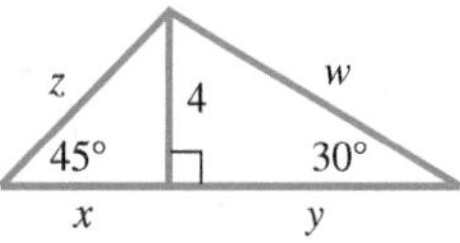

3. Find a solution for $\sin(\theta + 15°) = \cos(2\theta + 30°)$.

4. Determine whether each statement is *true* or *false*. If false, tell why.

(a) $\sin 24° < \sin 48°$ **(b)** $\cos 24° < \cos 48°$

(c) $\cos(60° + 30°) = \cos 60° \cdot \cos 30° - \sin 60° \cdot \sin 30°$

Find exact values of the six trigonometric functions for each angle. Rationalize denominators when applicable.

5. $240°$ **6.** $-135°$ **7.** $990°$

Find all values of θ, if θ is in the interval $[0°, 360°)$ and has the given function value.

8. $\cos\theta = -\dfrac{\sqrt{2}}{2}$ **9.** $\csc\theta = -\dfrac{2\sqrt{3}}{3}$ **10.** $\tan\theta = 1$

Solve each problem.

11. How would we find $\cot\theta$ using a calculator, if $\tan\theta = 1.6778490$? Evaluate $\cot\theta$.

12. Use a calculator to approximate the value of each expression. Give answers to six decimal places.

(a) $\sin 78° 21'$ **(b)** $\tan 117.689°$ **(c)** $\sec 58.9041°$

13. Find the value of θ in the interval $[0°, 90°]$ in decimal degrees, if

$$\sin\theta = 0.27843196.$$

Give the answer to six decimal places.

14. Solve the right triangle.

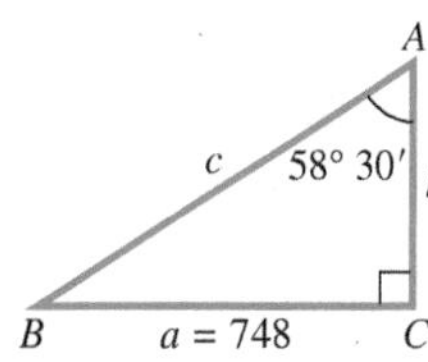

15. ***Antenna Mast Guy Wire*** A guy wire 77.4 m long is attached to the top of an antenna mast that is 71.3 m high. Find the angle that the wire makes with the ground.

[2.1]

1. $\sin A = \frac{12}{13}$; $\cos A = \frac{5}{13}$; $\tan A = \frac{12}{5}$; $\cot A = \frac{5}{12}$; $\sec A = \frac{13}{5}$; $\csc A = \frac{13}{12}$

2. $x = 4$; $y = 4\sqrt{3}$; $z = 4\sqrt{2}$; $w = 8$

3. $15°$

[2.1, 2.2]

4. **(a)** true

(b) false; For $0° \le \theta \le 90°$, as the angle increases, $\cos\theta$ decreases.

(c) true

In Exercises 5–7, we give, in order, sine, cosine, tangent, cotangent, secant, and cosecant.

[2.2]

5. $-\frac{\sqrt{3}}{2}$; $-\frac{1}{2}$; $\sqrt{3}$; $\frac{\sqrt{3}}{3}$; -2; $-\frac{2\sqrt{3}}{3}$

6. $-\frac{\sqrt{2}}{2}$; $-\frac{\sqrt{2}}{2}$; 1; 1; $-\sqrt{2}$; $-\sqrt{2}$

7. -1; 0; undefined; 0; undefined; -1

8. 135°; 225°

9. 240°; 300°

10. 45°; 225°

[2.3]

11. Take the reciprocal of $\tan\theta$ to obtain $\cot\theta = 0.59600119$.

12. **(a)** 0.979399

(b) -1.905608

(c) 1.936213

13. 16.166641°

[2.4]

14. $B = 31° 30'$; $c = 877$; $b = 458$

15. 67.1°, or 67° 10′

16. 15.5 ft
17. 8800 ft

[2.5]
18. 72 nautical mi
19. 92 km
20. 448 m

16. ***Height of a Flagpole*** To measure the height of a flagpole, Jan Marie found that the angle of elevation from a point 24.7 ft from the base to the top is $32° 10'$. What is the height of the flagpole?

17. ***Altitude of a Mountain*** The highest point in Texas is Guadalupe Peak. The angle of depression from the top of this peak to a small miner's cabin at an approximate elevation of 2000 ft is 26°. The cabin is located 14,000 ft horizontally from a point directly under the top of the mountain. Find the altitude of the top of the mountain to the nearest hundred feet.

18. ***Distance between Two Points*** Two ships leave a port at the same time. The first ship sails on a bearing of 32° at 16 knots (nautical miles per hour) and the second on a bearing of 122° at 24 knots. How far apart are they after 2.5 hr?

19. ***Distance of a Ship from a Pier*** A ship leaves a pier on a bearing of S 62° E and travels for 75 km. It then turns and continues on a bearing of N 28° E for 53 km. How far is the ship from the pier?

20. Find h as indicated in the figure.

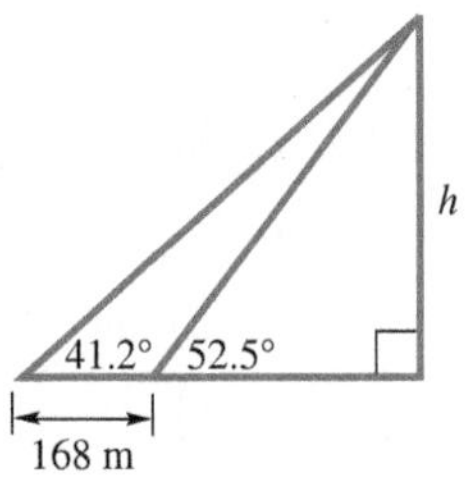

3 Radian Measure and the Unit Circle

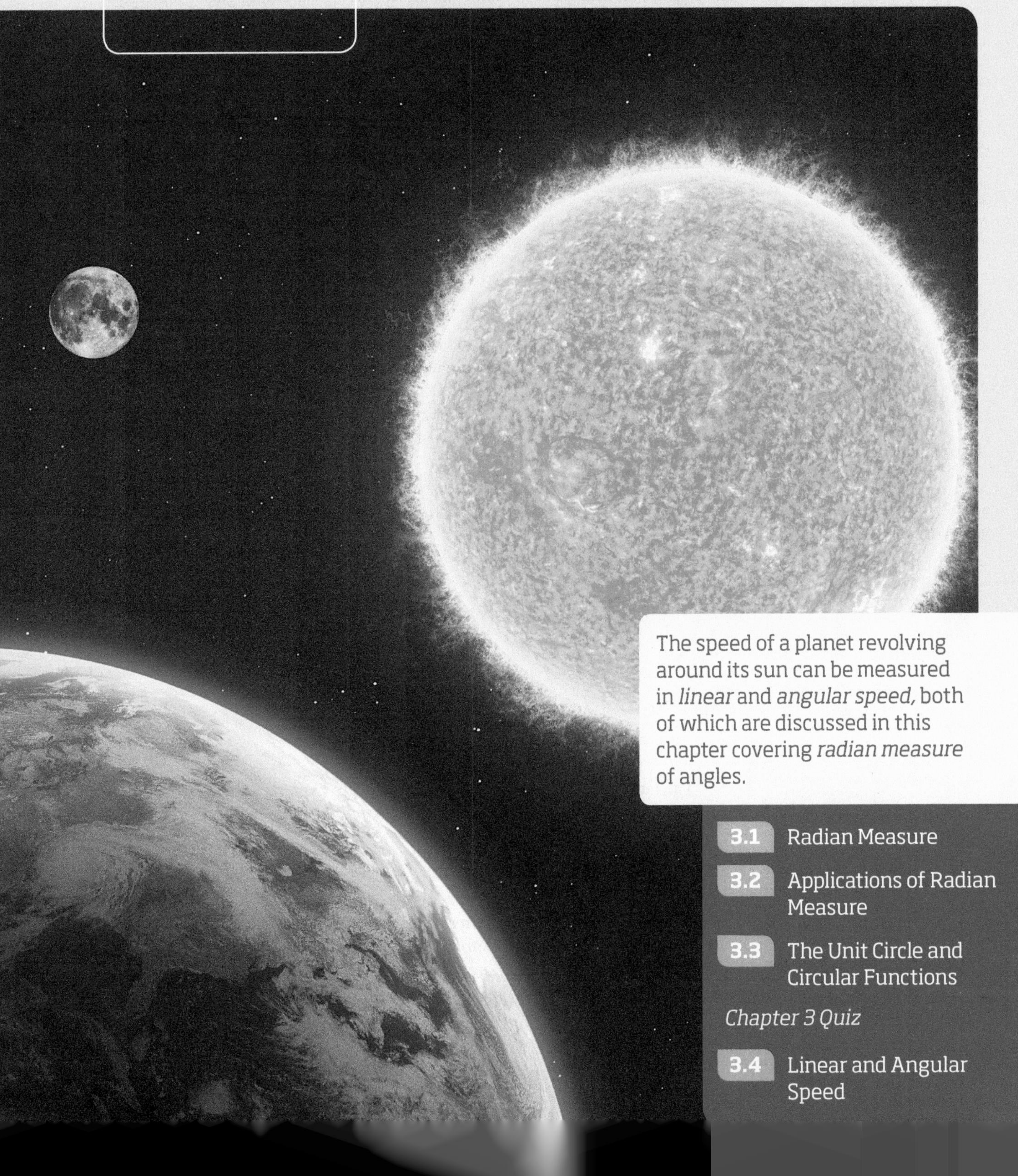

The speed of a planet revolving around its sun can be measured in *linear* and *angular speed,* both of which are discussed in this chapter covering *radian measure* of angles.

3.1 Radian Measure

- Radian Measure
- Conversions between Degrees and Radians
- Trigonometric Function Values of Angles in Radians

Radian Measure We have seen that angles can be measured in degrees. In more theoretical work in mathematics, *radian measure* of angles is preferred. Radian measure enables us to treat the trigonometric functions as functions with domains of *real numbers,* rather than angles.

Figure 1 shows an angle θ in standard position, along with a circle of radius r. The vertex of θ is at the center of the circle. Because angle θ intercepts an arc on the circle equal in length to the radius of the circle, we say that angle θ has a measure of *1 radian.*

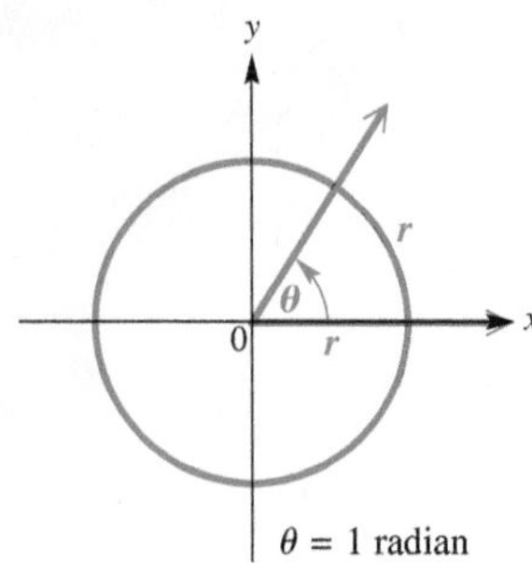

Figure 1

Radian

An angle with its vertex at the center of a circle that intercepts an arc on the circle equal in length to the radius of the circle has a measure of **1 radian.**

Teaching Tip Students may resist the idea of using radian measure. Explain that radians have wide applications in angular motion problems, as well as engineering and science.

It follows that an angle of measure 2 radians intercepts an arc equal in length to twice the radius of the circle, an angle of measure $\frac{1}{2}$ radian intercepts an arc equal in length to half the radius of the circle, and so on. ***In general, if θ is a central angle of a circle of radius r, and θ intercepts an arc of length s, then the radian measure of θ is $\frac{s}{r}$. See*** **Figure 2.**

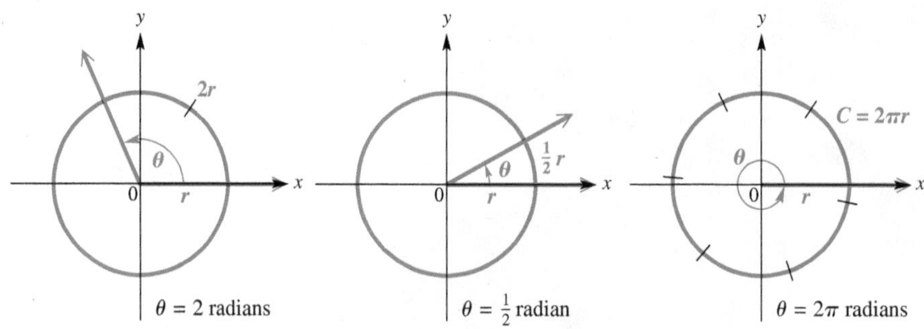

Figure 2

The ratio $\frac{s}{r}$ is a pure number, where s and r are expressed in the same units. ***Thus, "radians" is not a unit of measure like feet or centimeters.***

Conversions between Degrees and Radians The **circumference** of a circle—the distance around the circle—is given by $C = 2\pi r$, where r is the radius of the circle. The formula $C = 2\pi r$ shows that the radius can be measured off 2π times around a circle. Therefore, an angle of 360°, which corresponds to a complete circle, intercepts an arc equal in length to 2π times the radius of the circle. Thus, an angle of 360° has a measure of 2π radians.

$$360° = 2\pi \text{ radians}$$

Teaching Tip Emphasize the difference between an angle of a degrees (written $a°$) and an angle of a radians (sometimes written a^r). Tell students that although radian measures are often given exactly in terms of π, they can also be approximated using decimals.

An angle of 180° is half the size of an angle of 360°, so an angle of 180° has half the radian measure of an angle of 360°.

$$180° = \frac{1}{2}(2\pi) \text{ radians} = \pi \text{ radians} \quad \text{Degree/radian relationship}$$

We can use the relationship $180° = \pi$ radians to develop a method for converting between degrees and radians as follows.

$$180° = \pi \text{ radians} \quad \text{Degree/radian relationship}$$

$$1° = \frac{\pi}{180} \text{ radian} \quad \text{Divide by 180.} \qquad \text{or} \qquad 1 \text{ radian} = \frac{180°}{\pi} \quad \text{Divide by } \pi.$$

Classroom Example 1
Convert each degree measure to radians.
(a) 108° (b) −135° (c) 325.7°

Answers:
(a) $\frac{3\pi}{5}$ radians (b) $-\frac{3\pi}{4}$ radians
(c) 5.685 radians

NOTE Replacing π with its approximate integer value 3 in the fractions above and simplifying gives a couple of facts to help recall the relationship between degrees and radians. Remember that these are only approximations.

$$1° \approx \frac{1}{60} \text{ radian} \quad \text{and} \quad 1 \text{ radian} \approx 60°$$

Classroom Example 2
Convert each radian measure to degrees.
(a) $\frac{11\pi}{12}$ (b) $-\frac{7\pi}{6}$ (c) −2.92

Answers:
(a) 165° (b) −210°
(c) −167.3°, or −167° 18′

Converting between Degrees and Radians

- Multiply a degree measure by $\frac{\pi}{180}$ radian and simplify to convert to radians.
- Multiply a radian measure by $\frac{180°}{\pi}$ and simplify to convert to degrees.

EXAMPLE 1 Converting Degrees to Radians

Convert each degree measure to radians.

(a) 45° (b) −270° (c) 249.8°

SOLUTION

(a) $45° = 45\left(\frac{\pi}{180} \text{ radian}\right) = \frac{\pi}{4} \text{ radian}$ Multiply by $\frac{\pi}{180}$ radian.

(b) $-270° = -270\left(\frac{\pi}{180} \text{ radian}\right) = -\frac{3\pi}{2} \text{ radians}$ Multiply by $\frac{\pi}{180}$ radian. Write in lowest terms.

(c) $249.8° = 249.8\left(\frac{\pi}{180} \text{ radian}\right) \approx 4.360 \text{ radians}$ Nearest thousandth

✔ Now Try Exercises 11, 17, and 47.

```
NORMAL FLOAT AUTO REAL RADIAN MP
45°
                    .7853981634
-270°
                    -4.71238898
249.8°
                    4.359832471
```

This radian mode screen shows TI-84 Plus conversions for **Example 1.** Verify that the first two results are *approximations* for the *exact* values of $\frac{\pi}{4}$ and $-\frac{3\pi}{2}$.

EXAMPLE 2 Converting Radians to Degrees

Convert each radian measure to degrees.

(a) $\frac{9\pi}{4}$ (b) $-\frac{5\pi}{6}$ (c) 4.25

SOLUTION

(a) $\frac{9\pi}{4} \text{ radians} = \frac{9\pi}{4}\left(\frac{180°}{\pi}\right) = 405°$ Multiply by $\frac{180°}{\pi}$.

(b) $-\frac{5\pi}{6} \text{ radians} = -\frac{5\pi}{6}\left(\frac{180°}{\pi}\right) = -150°$ Multiply by $\frac{180°}{\pi}$.

(c) $4.25 \text{ radians} = 4.25\left(\frac{180°}{\pi}\right) \approx 243.5°, \text{ or } 243° \, 30'$ $0.50706(60') \approx 30'$

✔ Now Try Exercises 31, 35, and 59.

```
NORMAL FLOAT AUTO REAL DEGREE MP
(9π/4)r
                    405
(-5π/6)r
                    -150
4.25r▸DMS
          243°30'25.427"
```

This degree mode screen shows how a TI-84 Plus calculator converts the radian measures in **Example 2** to degree measures.

NOTE Another way to convert a radian measure that is a rational multiple of π, such as $\frac{9\pi}{4}$, to degrees is to substitute $180°$ for π. In **Example 2(a),** doing this would give the following.

$$\frac{9\pi}{4} \text{ radians} = \frac{9(180°)}{4} = 405°$$

One of the most important facts to remember when working with angles and their measures is summarized in the following statement.

Agreement on Angle Measurement Units

If no unit of angle measure is specified, then the angle is understood to be measured in radians.

For example, **Figure 3(a)** shows an angle of 30°, and **Figure 3(b)** shows an angle of 30 (which means 30 radians). An angle with measure 30 radians is coterminal with an angle of approximately 279°.

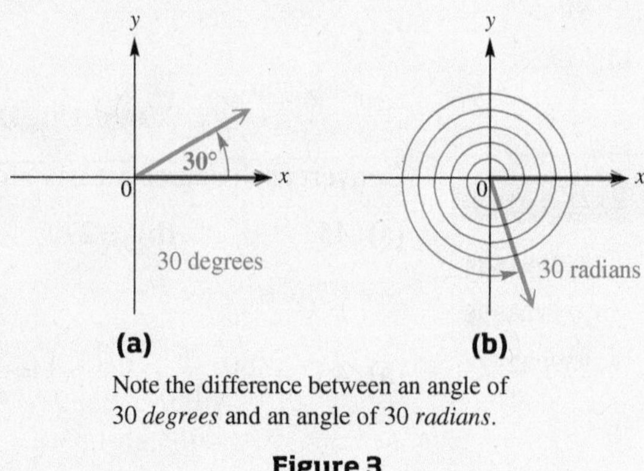

(a) **(b)**

Note the difference between an angle of 30 *degrees* and an angle of 30 *radians*.

Figure 3

The following table and **Figure 4** on the next page give some equivalent angle measures in degrees and radians. Keep in mind that

$$\mathbf{180° = \pi \text{ radians.}}$$

Equivalent Angle Measures

Degrees	Radians		Degrees	Radians	
	Exact	Approximate		Exact	Approximate
0°	0	0	90°	$\frac{\pi}{2}$	1.57
30°	$\frac{\pi}{6}$	0.52	180°	π	3.14
45°	$\frac{\pi}{4}$	0.79	270°	$\frac{3\pi}{2}$	4.71
60°	$\frac{\pi}{3}$	1.05	360°	2π	6.28

These exact values are *rational multiples of* π.

LOOKING AHEAD TO CALCULUS

In calculus, radian measure is much easier to work with than degree measure. If x is measured in radians, then the derivative of $f(x) = \sin x$ is

$$f'(x) = \cos x.$$

However, if x is measured in degrees, then the derivative of $f(x) = \sin x$ is

$$f'(x) = \frac{\pi}{180}\cos x.$$

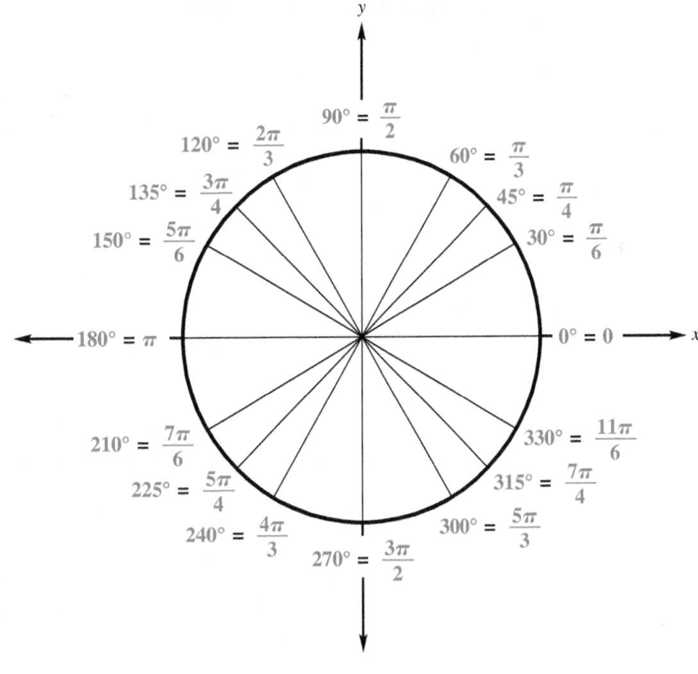

Figure 4

Learn the equivalences in* Figure 4*. They appear often in trigonometry.

Trigonometric Function Values of Angles in Radians Trigonometric function values for angles measured in radians can be found by first converting radian measure to degrees. *(Try to skip this intermediate step as soon as possible, however, and find the function values directly from radian measure.)*

Classroom Example 3
Find each function value.
(a) $\cos \frac{5\pi}{6}$ (b) $\cot \frac{3\pi}{4}$
(c) $\sec \frac{5\pi}{3}$
Answers:
(a) $-\frac{\sqrt{3}}{2}$ (b) -1
(c) 2

EXAMPLE 3 Finding Function Values of Angles in Radian Measure

Find each function value.

(a) $\tan \frac{2\pi}{3}$ **(b)** $\sin \frac{3\pi}{2}$ **(c)** $\cos\left(-\frac{4\pi}{3}\right)$

SOLUTION

(a) First convert $\frac{2\pi}{3}$ radians to degrees.

$$\tan \frac{2\pi}{3} = \tan\left(\frac{2\pi}{3} \cdot \frac{180°}{\pi}\right)$$ Multiply by $\frac{180°}{\pi}$ to convert radians to degrees.

Consider the reference angle.

$$= \tan 120°$$ Multiply.

$$= -\sqrt{3}$$ $\tan 120° = -\tan 60° = -\sqrt{3}$

(b) $\sin \frac{3\pi}{2} = \sin 270° = -1$ $\frac{3\pi}{2}$ radians $= 270°$

(c) $$\cos\left(-\frac{4\pi}{3}\right) = \cos\left(-\frac{4\pi}{3} \cdot \frac{180°}{\pi}\right)$$ Convert radians to degrees.

$$= \cos(-240°)$$ Multiply.

$$= -\frac{1}{2}$$ $\cos(-240°) = -\cos 60° = -\frac{1}{2}$

Now Try Exercises 69, 79, and 83.

3.1 Exercises

1. radius
2. 2π; π
3. $\frac{\pi}{180}$
4. $\frac{180°}{\pi}$
5. 1
6. 2
7. 3
8. -1
9. -3
10. -2
11. $\frac{\pi}{3}$
12. $\frac{\pi}{6}$
13. $\frac{\pi}{2}$
14. $\frac{2\pi}{3}$
15. $\frac{5\pi}{6}$
16. $\frac{3\pi}{2}$
17. $-\frac{5\pi}{3}$
18. $-\frac{7\pi}{4}$
19. $\frac{5\pi}{2}$
20. $\frac{8\pi}{3}$
21. 10π
22. 20π
23. 0
24. π
25. -5π
26. -10π
27. Radian measure provides a method for measuring angles in which the central angle, θ, of a circle is the ratio of the intercepted arc, s, to the radius of the circle, r. That is, $\theta = \frac{s}{r}$.
28. If the radius of a circle is 1, then the measure of a central angle, θ, in radians, is equal to the arc length, s. That is, $\theta = \frac{s}{1} = s$. Therefore, an angle of radian measure t in standard position will intercept an arc of length t on the circle.
29. 60°
30. 480°
31. 315°
32. 120°
33. 330°
34. 675°
35. $-30°$
36. $-288°$
37. 126°
38. 132°
39. $-48°$
40. $-63°$
41. 153°
42. 66°
43. $-900°$
44. 2700°

CONCEPT PREVIEW *Fill in the blank(s) to correctly complete each sentence.*

1. An angle with its vertex at the center of a circle that intercepts an arc on the circle equal in length to the ________ of the circle has measure 1 radian.

2. $360° =$ ______ radians, and $180° =$ ______ radians.

3. To convert to radians, multiply a degree measure by ______ radian and simplify.

4. To convert to degrees, multiply a radian measure by ______ and simplify.

CONCEPT PREVIEW *Each angle θ is an integer (e.g., 0, ± 1, ± 2, . . .) when measured in radians. Give the radian measure of the angle. (It helps to remember that $\pi \approx 3$.)*

5. 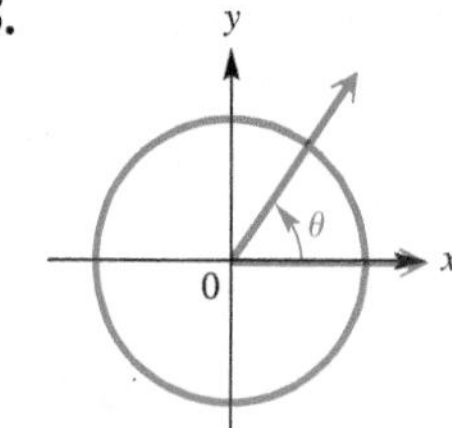

6. 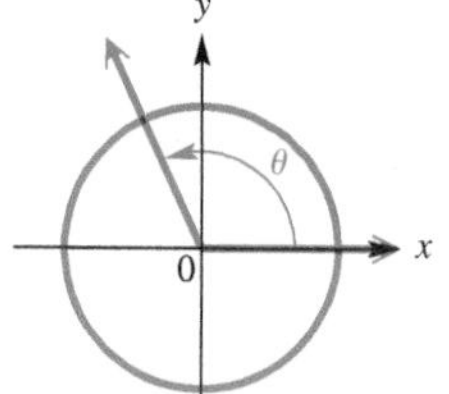

7. 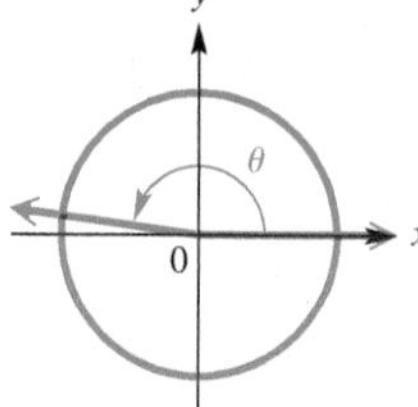

8. 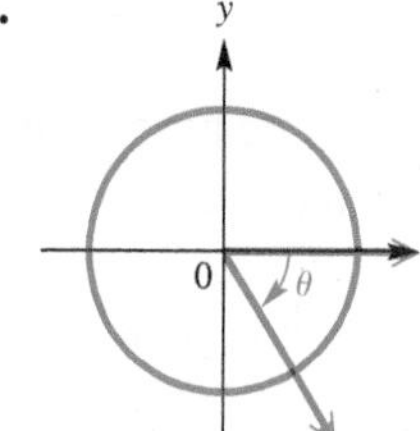

9. 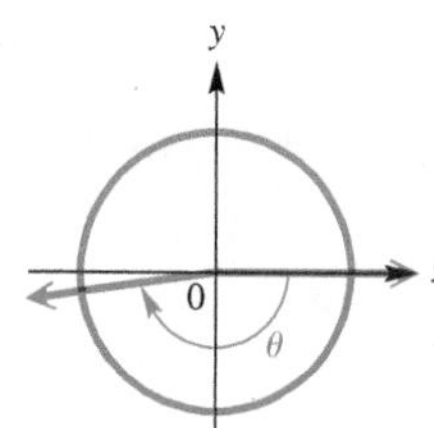

10. 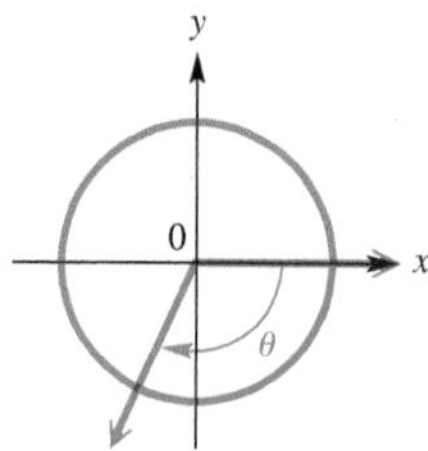

Convert each degree measure to radians. Leave answers as multiples of π. ***See Examples 1(a) and 1(b).***

11. 60°	12. 30°	13. 90°	14. 120°
15. 150°	16. 270°	17. $-300°$	18. $-315°$
19. 450°	20. 480°	21. 1800°	22. 3600°
23. 0°	24. 180°	25. $-900°$	26. $-1800°$

27. *Concept Check* Explain the meaning of radian measure.

28. *Concept Check* Explain why an angle of radian measure t in standard position intercepts an arc of length t on a circle of radius 1.

Convert each radian measure to degrees. ***See Examples 2(a) and 2(b).***

29. $\frac{\pi}{3}$	30. $\frac{8\pi}{3}$	31. $\frac{7\pi}{4}$	32. $\frac{2\pi}{3}$
33. $\frac{11\pi}{6}$	34. $\frac{15\pi}{4}$	35. $-\frac{\pi}{6}$	36. $-\frac{8\pi}{5}$
37. $\frac{7\pi}{10}$	38. $\frac{11\pi}{15}$	39. $-\frac{4\pi}{15}$	40. $-\frac{7\pi}{20}$
41. $\frac{17\pi}{20}$	42. $\frac{11\pi}{30}$	43. -5π	44. 15π

45. 0.681 46. 1.292
47. 0.742 48. 4.623
49. 2.429 50. 3.051
51. 1.122 52. 1.484
53. 0.985 54. 2.140
55. −0.832 56. −0.402

57. 114° 35′ 58. 286° 29′
59. 99° 42′ 60. 175° 20′
61. 19° 35′ 62. 564° 14′
63. −287° 06′ 64. −198° 55′

65. In the expression "sin 30," 30 means 30 radians; $\sin 30° = \frac{1}{2}$, while $\sin 30 \approx -0.9880$.

66. A central angle measuring 1 radian intercepts an arc length equal to the radius of the circle. It is approximately 60°.

67. $\frac{\sqrt{3}}{2}$ 68. $\frac{\sqrt{3}}{2}$
69. 1 70. $\frac{\sqrt{3}}{3}$
71. $\frac{2\sqrt{3}}{3}$ 72. $\sqrt{2}$
73. 1 74. 1
75. $-\sqrt{3}$ 76. $-\frac{\sqrt{3}}{3}$
77. $\frac{1}{2}$ 78. $-\frac{\sqrt{3}}{3}$
79. −1 80. −1
81. $-\frac{\sqrt{3}}{2}$ 82. $\frac{\sqrt{3}}{3}$
83. $\frac{1}{2}$ 84. $\frac{\sqrt{3}}{2}$
85. $\sqrt{3}$ 86. $-\frac{2\sqrt{3}}{3}$

87. We begin the answers with the blank next to 30°, and then proceed counterclockwise from there: $\frac{\pi}{6}$; 45; $\frac{\pi}{3}$; 120; 135; $\frac{5\pi}{6}$; π; $\frac{7\pi}{6}$; $\frac{5\pi}{4}$; 240; 300; $\frac{7\pi}{4}$; $\frac{11\pi}{6}$.

88. $\frac{\pi^2}{180}$ radian

89. 3π; $-\pi$ (Answers may vary.)

90. $\frac{\pi}{2} + 2n\pi$

Convert each degree measure to radians. If applicable, round to the nearest thousandth. ***See Example 1(c).***

45. 39° 46. 74° 47. 42.5° 48. 264.9°

49. 139° 10′ 50. 174° 50′ 51. 64.29° 52. 85.04°

53. 56° 25′ 54. 122° 37′ 55. −47.69° 56. −23.01°

Convert each radian measure to degrees. Write answers to the nearest minute. ***See Example 2(c).***

57. 2 58. 5 59. 1.74 60. 3.06

61. 0.3417 62. 9.84763 63. −5.01095 64. −3.47189

65. ***Concept Check*** The value of sin 30 is not $\frac{1}{2}$. Why is this true?

66. ***Concept Check*** What is meant by an angle of one radian?

Find each exact function value. ***See Example 3.***

67. $\sin \frac{\pi}{3}$ 68. $\cos \frac{\pi}{6}$ 69. $\tan \frac{\pi}{4}$ 70. $\cot \frac{\pi}{3}$

71. $\sec \frac{\pi}{6}$ 72. $\csc \frac{\pi}{4}$ 73. $\sin \frac{\pi}{2}$ 74. $\csc \frac{\pi}{2}$

75. $\tan \frac{5\pi}{3}$ 76. $\cot \frac{2\pi}{3}$ 77. $\sin \frac{5\pi}{6}$ 78. $\tan \frac{5\pi}{6}$

79. $\cos 3\pi$ 80. $\sec \pi$ 81. $\sin\left(-\frac{8\pi}{3}\right)$ 82. $\cot\left(-\frac{2\pi}{3}\right)$

83. $\sin\left(-\frac{7\pi}{6}\right)$ 84. $\cos\left(-\frac{\pi}{6}\right)$ 85. $\tan\left(-\frac{14\pi}{3}\right)$ 86. $\csc\left(-\frac{13\pi}{3}\right)$

87. ***Concept Check*** The figure shows the same angles measured in both degrees and radians. Complete the missing measures.

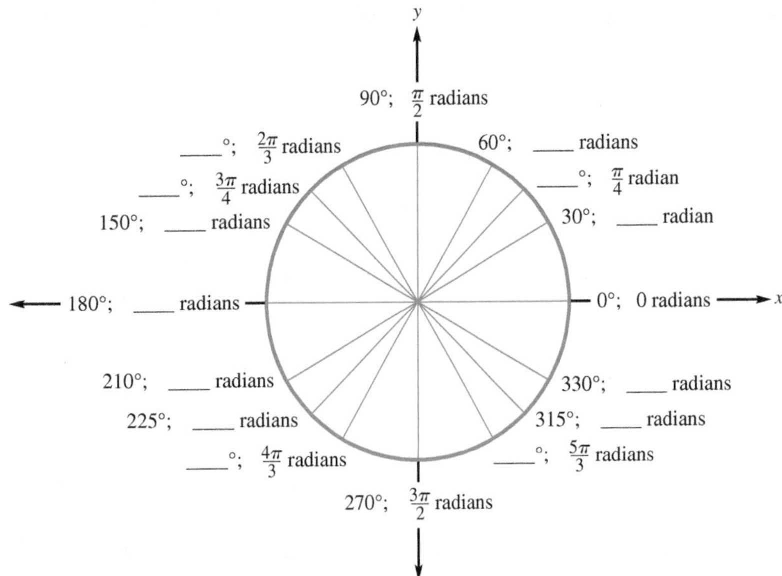

88. ***Concept Check*** What is the exact radian measure of an angle measuring π degrees?

89. ***Concept Check*** Find two angles, one positive and one negative, that are coterminal with an angle of π radians.

90. ***Concept Check*** Give an expression that generates all angles coterminal with an angle of $\frac{\pi}{2}$ radians. Let n represent any integer.

91. (a) 4π (b) $\frac{2\pi}{3}$
92. (a) 24π (b) 6π
93. (a) 5π (b) $\frac{8\pi}{3}$
94. (a) 16π (b) 60π
95. 24π
96. (a) $\frac{\pi}{200}$
(b) 3.15°; 0.055 radian

Solve each problem.

91. *Rotating Hour Hand on a Clock* Through how many radians does the hour hand on a clock rotate in **(a)** 24 hr and **(b)** 4 hr?

92. *Rotating Minute Hand on a Clock* Through how many radians does the minute hand on a clock rotate in **(a)** 12 hr and **(b)** 3 hr?

93. *Orbits of a Space Vehicle* A space vehicle is orbiting Earth in a circular orbit. What radian measure corresponds to **(a)** 2.5 orbits and **(b)** $\frac{4}{3}$ orbits?

94. *Rotating Pulley* A circular pulley is rotating about its center. Through how many radians does it turn in **(a)** 8 rotations and **(b)** 30 rotations?

95. *Revolutions of a Carousel* A stationary horse on a carousel makes 12 complete revolutions. Through what radian measure angle does the horse revolve?

96. *Railroad Engineering* Some engineers use the term **grade** to represent $\frac{1}{100}$ of a right angle and express grade as a percent. For example, an angle of 0.9° would be referred to as a 1% grade. (*Source*: Hay, W., *Railroad Engineering,* John Wiley and Sons.)

(a) By what number should we multiply a grade (disregarding the % symbol) to convert it to radians?

(b) In a rapid-transit rail system, the maximum grade allowed between two stations is 3.5%. Express this angle in degrees and in radians.

3.2 Applications of Radian Measure

- Arc Length on a Circle
- Area of a Sector of a Circle

Arc Length on a Circle The formula for finding the length of an arc of a circle follows directly from the definition of an angle θ in radians, where $\theta = \frac{s}{r}$.

In **Figure 5,** we see that angle QOP has measure 1 radian and intercepts an arc of length r on the circle. We also see that angle ROT has measure θ radians and intercepts an arc of length s on the circle. From plane geometry, we know that the lengths of the arcs are proportional to the measures of their central angles.

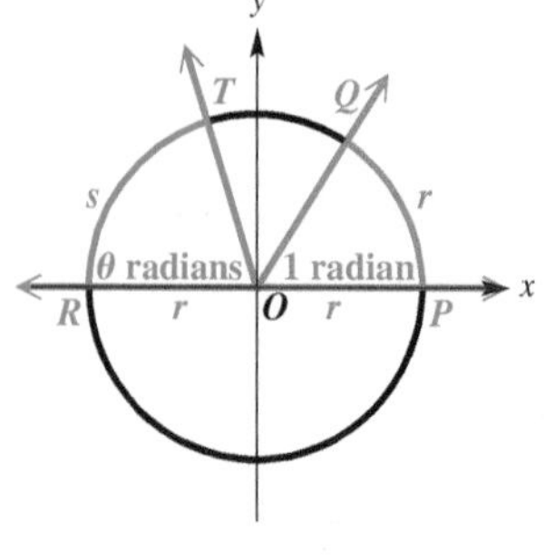

Figure 5

$$\frac{s}{r} = \frac{\theta}{1} \quad \text{Set up a proportion.}$$

Multiplying each side by r gives

$$s = r\theta. \quad \text{Solve for } s.$$

Arc Length

The length s of the arc intercepted on a circle of radius r by a central angle of measure θ radians is given by the product of the radius and the radian measure of the angle.

$$s = r\theta, \quad \textbf{where } \theta \textbf{ is in radians}$$

CAUTION ***When the formula $s = r\theta$ is applied, the value of θ MUST be expressed in radians, not degrees.***

Classroom Example 1
A circle has radius 25.60 cm. Find the length of the arc intercepted by a central angle having each of the following measures.

(a) $\frac{7\pi}{8}$ radians (b) 54°

Answers:
(a) 22.4π cm ≈ 70.37 cm
(b) 7.68π cm ≈ 24.13 cm

Teaching Tip To help students appreciate radian measure, show them that in degrees, the formula for arc length is

$$s = \frac{\theta}{360}(2\pi r) = \frac{\theta \pi r}{180}.$$

(See **Exercise 22.**)

EXAMPLE 1 Finding Arc Length Using $s = r\theta$

A circle has radius 18.20 cm. Find the length of the arc intercepted by a central angle having each of the following measures.

(a) $\frac{3\pi}{8}$ radians **(b)** 144°

SOLUTION

(a) As shown in **Figure 6,** $r = 18.20$ cm and $\theta = \frac{3\pi}{8}$.

$s = r\theta$ Arc length formula

$s = 18.20\left(\frac{3\pi}{8}\right)$ Let $r = 18.20$ and $\theta = \frac{3\pi}{8}$.

$s \approx 21.44$ cm Use a calculator.

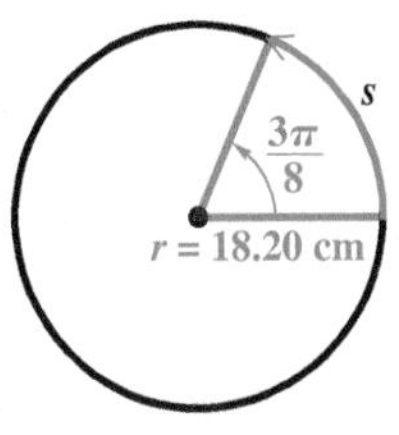

Figure 6

(b) The formula $s = r\theta$ requires that θ be measured in radians. First, convert θ to radians by multiplying 144° by $\frac{\pi}{180}$ radian.

$$144° = 144\left(\frac{\pi}{180}\right) = \frac{4\pi}{5} \text{ radians} \quad \text{Convert from degrees to radians.}$$

The length s is found using $s = r\theta$.

$$s = r\theta = 18.20\left(\frac{4\pi}{5}\right) \approx 45.74 \text{ cm} \quad \text{Let } r = 18.20 \text{ and } \theta = \frac{4\pi}{5}.$$

Be sure to use radians for θ in $s = r\theta$.

✔ Now Try Exercises 13 and 17.

Classroom Example 2
Erie, Pennsylvania, is approximately due north of Columbia, South Carolina. The latitude of Erie is 42° N, and that of Columbia is 34° N. The radius of Earth is 6400 km. Find the north-south distance between the two cities.

Answer: 890 km

EXAMPLE 2 Finding the Distance between Two Cities

Latitude gives the measure of a central angle with vertex at Earth's center whose initial side goes through the equator and whose terminal side goes through the given location. Reno, Nevada, is approximately due north of Los Angeles. The latitude of Reno is 40° N, and that of Los Angeles is 34° N. (The N in 34° N means *north* of the equator.) The radius of Earth is 6400 km. Find the north-south distance between the two cities.

SOLUTION As shown in **Figure 7,** the central angle between Reno and Los Angeles is

$$40° - 34° = 6°.$$

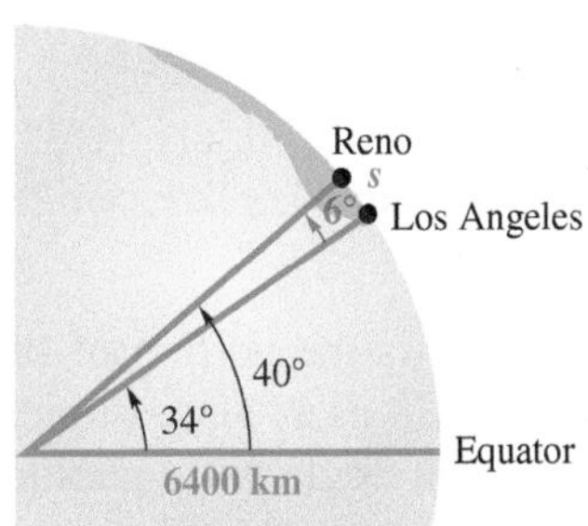

Figure 7

The distance between the two cities can be found using the formula $s = r\theta$, after 6° is converted to radians.

$$6° = 6\left(\frac{\pi}{180}\right) = \frac{\pi}{30} \text{ radian}$$

The distance between the two cities is given by s.

$$s = r\theta = 6400\left(\frac{\pi}{30}\right) \approx 670 \text{ km} \quad \text{Let } r = 6400 \text{ and } \theta = \frac{\pi}{30}.$$

✔ Now Try Exercise 23.

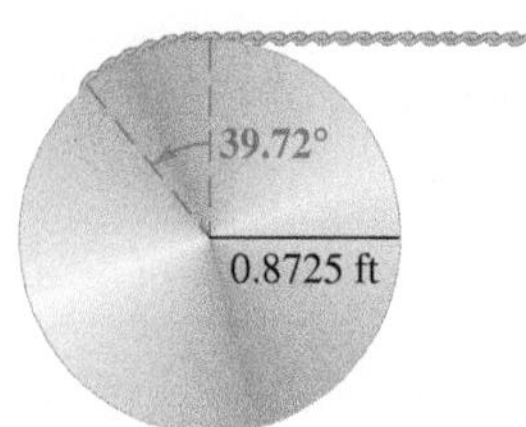

Figure 8

Classroom Example 3
Repeat **Example 3** if the radius of the drum is 0.327 m and the drum is rotated through an angle of 132.6°.

Answer: 0.757 m

EXAMPLE 3 Finding a Length Using $s = r\theta$

A rope is being wound around a drum with radius 0.8725 ft. (See **Figure 8.**) How much rope will be wound around the drum if the drum is rotated through an angle of 39.72°?

SOLUTION The length of rope wound around the drum is the arc length for a circle of radius 0.8725 ft and a central angle of 39.72°. Use the formula $s = r\theta$, with the angle converted to radian measure. The length of the rope wound around the drum is approximated by s.

Convert to radian measure.

$$s = r\theta = 0.8725\left[39.72\left(\frac{\pi}{180}\right)\right] \approx 0.6049 \text{ ft}$$

✔ Now Try Exercise 35(a).

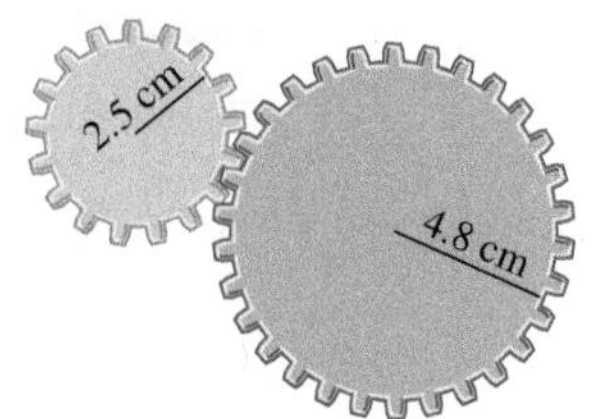

Figure 9

Classroom Example 4
Repeat **Example 4** if the radii of the gears are 3.6 in. and 5.4 in. and the smaller gear rotates through 150°.

Answer: 100°

EXAMPLE 4 Finding an Angle Measure Using $s = r\theta$

Two gears are adjusted so that the smaller gear drives the larger one, as shown in **Figure 9.** If the smaller gear rotates through an angle of 225°, through how many degrees will the larger gear rotate?

SOLUTION First find the radian measure of the angle of rotation for the smaller gear, and then find the arc length on the smaller gear. This arc length will correspond to the arc length of the motion of the larger gear. Because $225° = \frac{5\pi}{4}$ radians, for the smaller gear we have arc length

$$s = r\theta = 2.5\left(\frac{5\pi}{4}\right) = \frac{12.5\pi}{4} = \frac{25\pi}{8} \text{ cm.}$$

The tips of the two mating gear teeth must move at the same linear speed, or the teeth will break. So we must have "equal arc lengths in equal times." An arc with this length s on the larger gear corresponds to an angle measure θ, in radians, where $s = r\theta$.

$$s = r\theta \quad \text{Arc length formula}$$

$$\frac{25\pi}{8} = 4.8\theta \quad \text{Let } s = \tfrac{25\pi}{8} \text{ and } r = 4.8 \text{ (for the larger gear).}$$

$$\frac{125\pi}{192} = \theta \quad 4.8 = \tfrac{48}{10} = \tfrac{24}{5}\text{; Multiply by } \tfrac{5}{24} \text{ to solve for } \theta.$$

Converting θ back to degrees shows that the larger gear rotates through

$$\frac{125\pi}{192}\left(\frac{180°}{\pi}\right) \approx 117°. \quad \text{Convert } \theta = \tfrac{125\pi}{192} \text{ to degrees.}$$

✔ Now Try Exercise 29.

Area of a Sector of a Circle A **sector of a circle** is the portion of the interior of a circle intercepted by a central angle. Think of it as a "piece of pie." See **Figure 10.** A complete circle can be thought of as an angle with measure 2π radians. If a central angle for a sector has measure θ radians, then the sector makes up the fraction $\frac{\theta}{2\pi}$ of a complete circle. The area $\mathcal{A}$ of a complete circle with radius r is $\mathcal{A} = \pi r^2$. Therefore, we have the following.

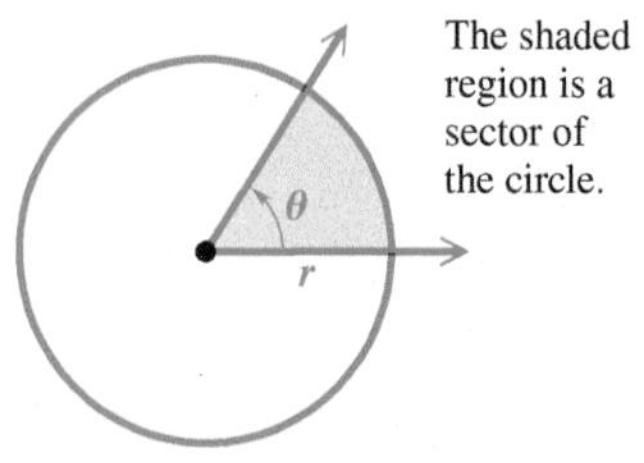

The shaded region is a sector of the circle.

Figure 10

$$\text{Area } \mathcal{A} \text{ of a sector} = \frac{\theta}{2\pi}(\pi r^2) = \frac{1}{2}r^2\theta, \quad \text{where } \theta \text{ is in radians.}$$

Teaching Tip To find the arc length or area of a sector, you may first find the ratio

$$\frac{\theta}{2\pi}, \quad \text{or} \quad \frac{\theta}{360^\circ}.$$

Then, the arc length is this fraction of the circumference $(2\pi r)$, and the sector area is this fraction of the total area (πr^2).

Area of a Sector

The area $\mathscr{A}$ of a sector of a circle of radius r and central angle θ is given by the following formula.

$$\mathscr{A} = \frac{1}{2}r^2\theta, \quad \text{where } \theta \text{ is in radians}$$

CAUTION ***As in the formula for arc length, the value of θ must be in radians when this formula is used to find the area of a sector.***

EXAMPLE 5 Finding the Area of a Sector-Shaped Field

A center-pivot irrigation system provides water to a sector-shaped field with the measures shown in **Figure 11.** Find the area of the field.

Center-pivot irrigation system

Figure 11

SOLUTION First, convert 15° to radians.

$$15^\circ = 15\left(\frac{\pi}{180}\right) = \frac{\pi}{12} \text{ radian} \qquad \text{Convert to radians.}$$

Now find the area of a sector of a circle.

$$\mathscr{A} = \frac{1}{2}r^2\theta \qquad \text{Formula for area of a sector}$$

$$\mathscr{A} = \frac{1}{2}(321)^2\left(\frac{\pi}{12}\right) \qquad \text{Let } r = 321 \text{ and } \theta = \tfrac{\pi}{12}.$$

$$\mathscr{A} \approx 13{,}500 \text{ m}^2 \qquad \text{Multiply.}$$

✔ **Now Try Exercise 57.**

Classroom Example 5
Find the area of a sector of a circle having radius 15.20 ft and central angle 108.0°.
Answer: 217.8 ft^2

3.2 Exercises

1. 2π 2. 20π
3. 8 4. 8

CONCEPT PREVIEW *Find the exact length of each arc intercepted by the given central angle.*

1.

$\frac{\pi}{2}$, 4

2.

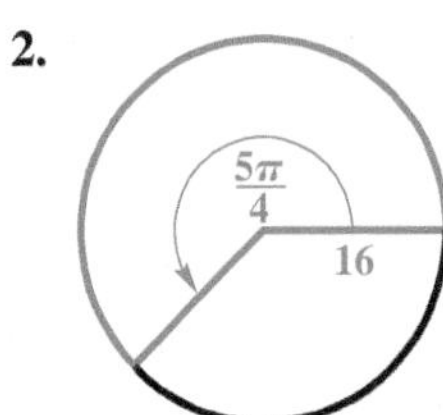

CONCEPT PREVIEW *Find the radius of each circle.*

3.

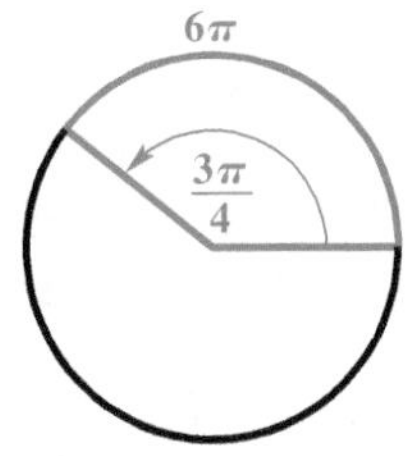

4.

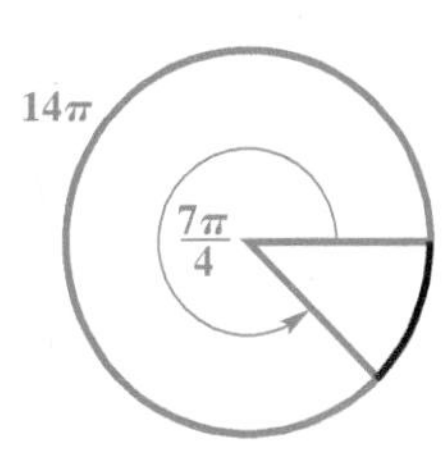

5. 1 6. 2

7. 6π 8. 75π

9. 1.5 10. 1

11. 60° 12. 240°

13. 25.8 cm 14. 3.08 cm
15. 3.61 ft 16. 11.9 mi
17. 5.05 m 18. 169 cm
19. 55.3 in. 20. 71.4 ft

21. The length is doubled.

22. $s = \frac{\pi r \theta}{180}$

CONCEPT PREVIEW *Find the measure of each central angle (in radians).*

5.

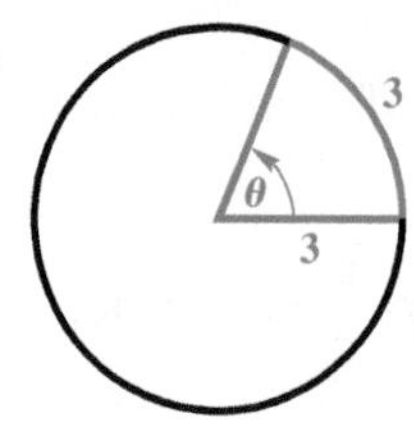

6.

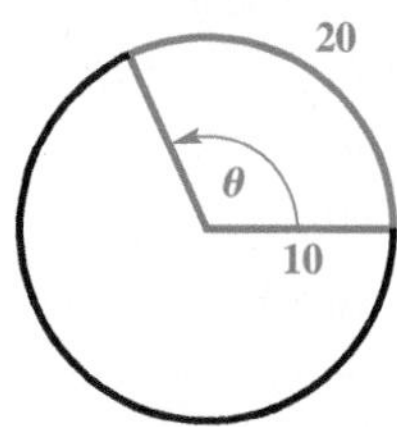

CONCEPT PREVIEW *Find the area of each sector.*

7.

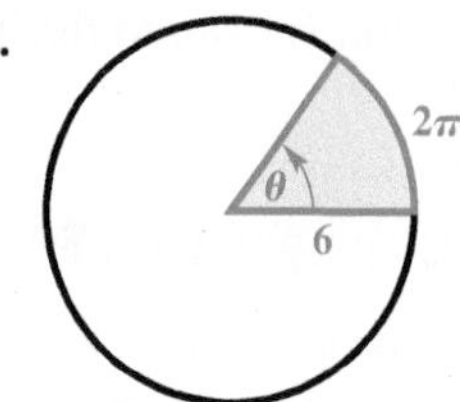

8.

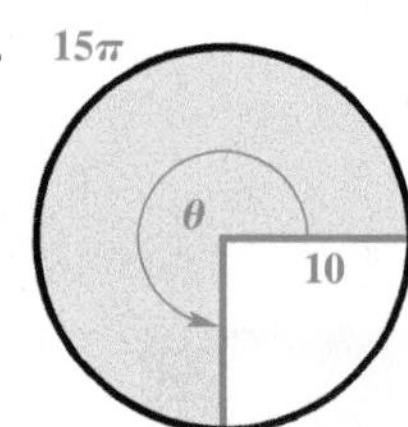

CONCEPT PREVIEW *Find the measure (in radians) of each central angle. The number inside the sector is the area.*

9.

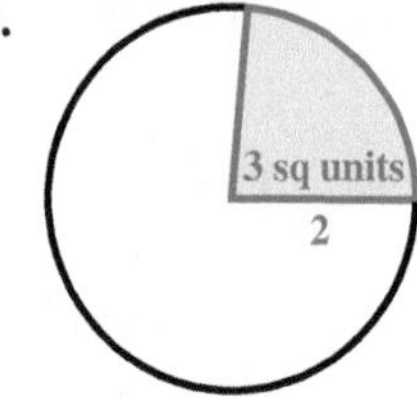

10.

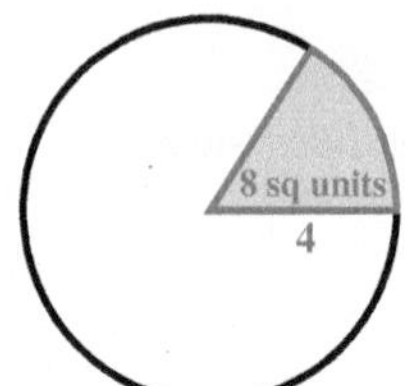

CONCEPT PREVIEW *Find the measure (in degrees) of each central angle. The number inside the sector is the area.*

11.

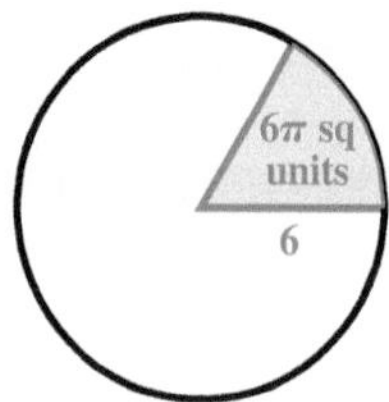

12.

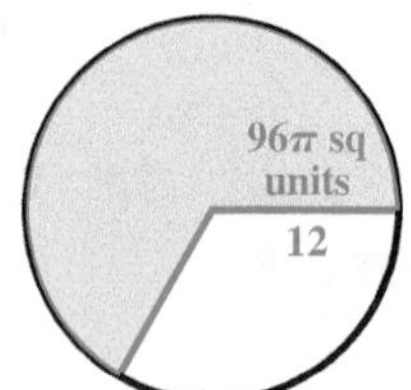

Unless otherwise directed, give calculator approximations in answers in the rest of this exercise set.

Find the length to three significant digits of each arc intercepted by a central angle θ in a circle of radius r. ***See Example 1.***

13. $r = 12.3$ cm, $\theta = \frac{2\pi}{3}$ radians

14. $r = 0.892$ cm, $\theta = \frac{11\pi}{10}$ radians

15. $r = 1.38$ ft, $\theta = \frac{5\pi}{6}$ radians

16. $r = 3.24$ mi, $\theta = \frac{7\pi}{6}$ radians

17. $r = 4.82$ m, $\theta = 60°$

18. $r = 71.9$ cm, $\theta = 135°$

19. $r = 15.1$ in., $\theta = 210°$

20. $r = 12.4$ ft, $\theta = 330°$

21. *Concept Check* If the radius of a circle is doubled, how is the length of the arc intercepted by a fixed central angle changed?

22. *Concept Check* Radian measure simplifies many formulas, such as the formula for arc length, $s = r\theta$. Give the corresponding formula when θ is measured in degrees instead of radians.

23. 3500 km	24. 1500 km
25. 5900 km	26. 8800 km
27. 44° N	28. 43° N
29. 156°	30. 213°
31. 38.5°	32. 82.3°
33. 18.7 cm	34. 29.2 in.
35. (a) 11.6 in.	(b) 37° 05′

Distance between Cities *Find the distance in kilometers between each pair of cities, assuming they lie on the same north-south line. Assume that the radius of Earth is* 6400 km. ***See Example 2.***

23. Panama City, Panama, 9° N, and Pittsburgh, Pennsylvania, 40° N

24. Farmersville, California, 36° N, and Penticton, British Columbia, 49° N

25. New York City, New York, 41° N, and Lima, Peru, 12° S

26. Halifax, Nova Scotia, 45° N, and Buenos Aires, Argentina, 34° S

27. *Latitude of Madison* Madison, South Dakota, and Dallas, Texas, are 1200 km apart and lie on the same north-south line. The latitude of Dallas is 33° N. What is the latitude of Madison?

28. *Latitude of Toronto* Charleston, South Carolina, and Toronto, Canada, are 1100 km apart and lie on the same north-south line. The latitude of Charleston is 33° N. What is the latitude of Toronto?

Work each problem. ***See Examples 3 and 4.***

29. *Gear Movement* Two gears are adjusted so that the smaller gear drives the larger one, as shown in the figure. If the smaller gear rotates through an angle of 300°, through how many degrees does the larger gear rotate?

30. *Gear Movement* Repeat **Exercise 29** for gear radii of 4.8 in. and 7.1 in. and for an angle of 315° for the smaller gear.

31. *Rotating Wheels* The rotation of the smaller wheel in the figure causes the larger wheel to rotate. Through how many degrees does the larger wheel rotate if the smaller one rotates through 60.0°?

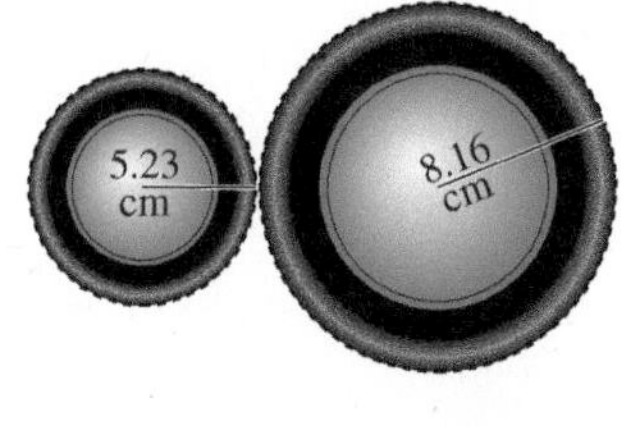

32. *Rotating Wheels* Repeat **Exercise 31** for wheel radii of 6.84 in. and 12.46 in. and an angle of 150.0° for the smaller wheel.

33. *Rotating Wheels* Find the radius of the larger wheel in the figure if the smaller wheel rotates 80.0° when the larger wheel rotates 50.0°.

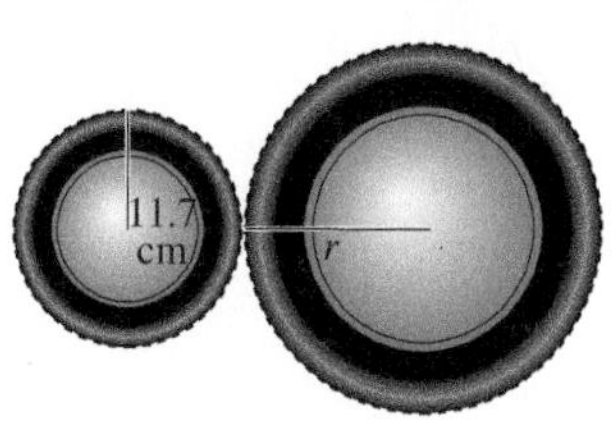

34. *Rotating Wheels* Repeat **Exercise 33** if the smaller wheel of radius 14.6 in. rotates 120.0° when the larger wheel rotates 60.0°.

35. *Pulley Raising a Weight* Refer to the figure.

(a) How many inches will the weight in the figure rise if the pulley is rotated through an angle of 71° 50′?

(b) Through what angle, to the nearest minute, must the pulley be rotated to raise the weight 6 in.?

36. 12.7 cm
37. 146 in.
38. (a) 39,616 rotations
(b) 62.9 mi; yes

39. 3π in. 40. 4π in.
41. 27π in. 42. 39π in.

43. 0.20 km 44. 0.12 mi
45. 840 ft 46. 360 m

47. 1116.1 m^2
48. 3744.8 km^2
49. 706.9 ft^2
50. 10,602.9 yd^2
51. 114.0 cm^2
52. 365.3 m^2
53. 1885.0 mi^2
54. 19,085.2 km^2

36. ***Pulley Raising a Weight*** Find the radius of the pulley in the figure if a rotation of 51.6° raises the weight 11.4 cm.

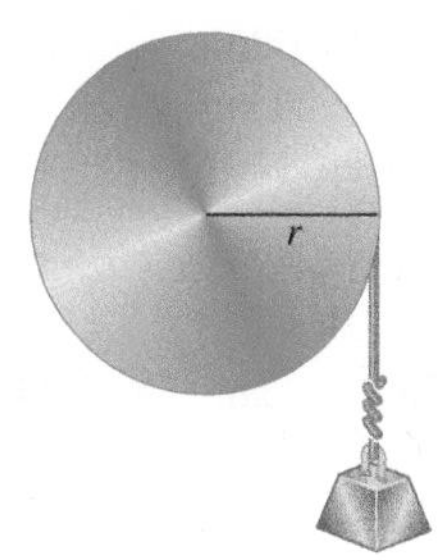

37. ***Bicycle Chain Drive*** The figure shows the chain drive of a bicycle. How far will the bicycle move if the pedals are rotated through 180.0°? Assume the radius of the bicycle wheel is 13.6 in.

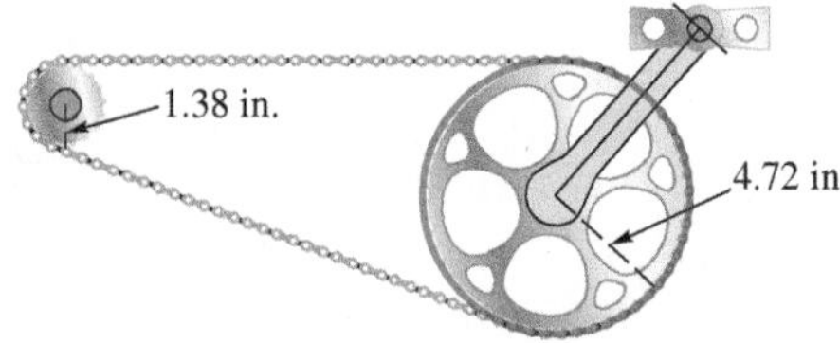

38. ***Car Speedometer*** The speedometer of Terry's Honda CR-V is designed to be accurate with tires of radius 14 in.

(a) Find the number of rotations of a tire in 1 hr if the car is driven at 55 mph.

(b) Suppose that oversize tires of radius 16 in. are placed on the car. If the car is now driven for 1 hr with the speedometer reading 55 mph, how far has the car gone? If the speed limit is 55 mph, does Terry deserve a speeding ticket?

Suppose the tip of the minute hand of a clock is 3 in. *from the center of the clock. For each duration, determine the distance traveled by the tip of the minute hand. Leave answers as multiples of* π.

39. 30 min **40.** 40 min

41. 4.5 hr **42.** $6\frac{1}{2}$ hr

*If a central angle is very small, there is little difference in length between an arc and the inscribed chord. See the figure. Approximate each of the following lengths by finding the necessary arc length. (Note: When a central angle intercepts an arc, the arc is said to **subtend** the angle.)*

Arc length ≈ length of inscribed chord

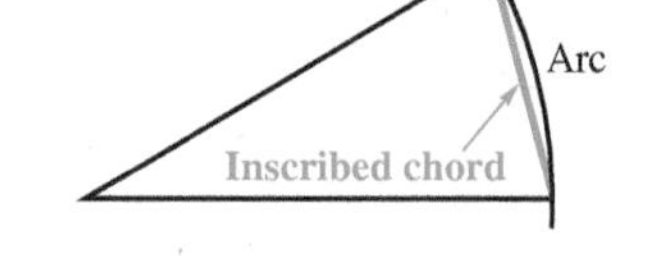

43. ***Length of a Train*** A railroad track in the desert is 3.5 km away. A train on the track subtends (horizontally) an angle of 3° 20′. Find the length of the train.

44. Repeat **Exercise 43** for a railroad track 2.7 mi away and a train that subtends an angle of 2° 30′.

45. ***Distance to a Boat*** The mast of a boat is 32.0 ft high. If it subtends an angle of 2° 11′, how far away is it?

46. Repeat **Exercise 45** for a boat mast 11.0 m high that subtends an angle of 1° 45′.

Find the area of a sector of a circle having radius r and central angle θ. Express answers to the nearest tenth. ***See Example 5.***

47. $r = 29.2$ m, $\theta = \frac{5\pi}{6}$ radians **48.** $r = 59.8$ km, $\theta = \frac{2\pi}{3}$ radians

49. $r = 30.0$ ft, $\theta = \frac{\pi}{2}$ radians **50.** $r = 90.0$ yd, $\theta = \frac{5\pi}{6}$ radians

51. $r = 12.7$ cm, $\theta = 81°$ **52.** $r = 18.3$ m, $\theta = 125°$

53. $r = 40.0$ mi, $\theta = 135°$ **54.** $r = 90.0$ km, $\theta = 270°$

55. 3.6
56. 16 m
57. 8060 yd^2
58. The new area is double the original area.; yes
59. 20 in.
60. 64°
61. (a) $13\frac{1}{3}°$; $\frac{2\pi}{27}$ (b) 478 ft (c) 17.7 ft (d) 672 ft^2
62. 75.4 in.^{2}
63. (a) 140 ft (b) 102 ft (c) 622 ft^2
64. 550 m; 1800 m

Work each problem. ***See Example 5.***

55. *Angle Measure* Find the measure (in radians) of a central angle of a sector of area 16 in.^{2} in a circle of radius 3.0 in.

56. *Radius Length* Find the radius of a circle in which a central angle of $\frac{\pi}{6}$ radian determines a sector of area 64 m^2.

57. *Irrigation Area* A center-pivot irrigation system provides water to a sector-shaped field as shown in the figure. Find the area of the field if $\theta = 40.0°$ and $r = 152$ yd.

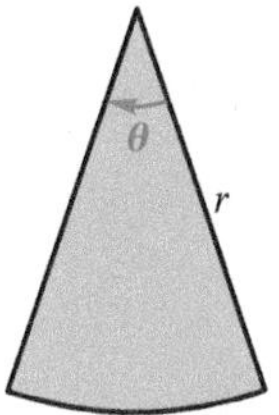

58. *Irrigation Area* Suppose that in **Exercise 57** the angle is halved and the radius length is doubled. How does the new area compare to the original area? Does this result hold in general for any values of θ and r?

59. *Arc Length* A circular sector has an area of 50 in.^{2}. The radius of the circle is 5 in. What is the arc length of the sector?

60. *Angle Measure* In a circle, a sector has an area of 16 cm^2 and an arc length of 6.0 cm. What is the measure of the central angle in degrees?

61. *Measures of a Structure* The figure illustrates Medicine Wheel, a Native American structure in northern Wyoming. There are 27 aboriginal spokes in the wheel, all equally spaced.

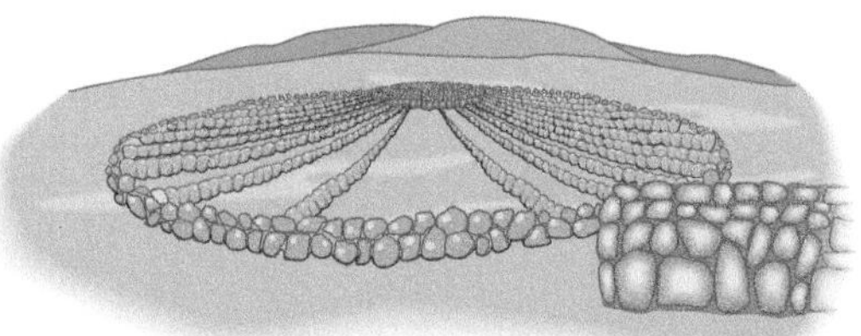

(a) Find the measure of each central angle in degrees and in radians in terms of π.

(b) If the radius of the wheel is 76.0 ft, find the circumference.

(c) Find the length of each arc intercepted by consecutive pairs of spokes.

(d) Find the area of each sector formed by consecutive spokes.

62. *Area Cleaned by a Windshield Wiper* The Ford Model A, built from 1928 to 1931, had a single windshield wiper on the driver's side. The total arm and blade was 10 in. long and rotated back and forth through an angle of 95°. The shaded region in the figure is the portion of the windshield cleaned by the 7-in. wiper blade. Find the area of the region cleaned to the nearest tenth.

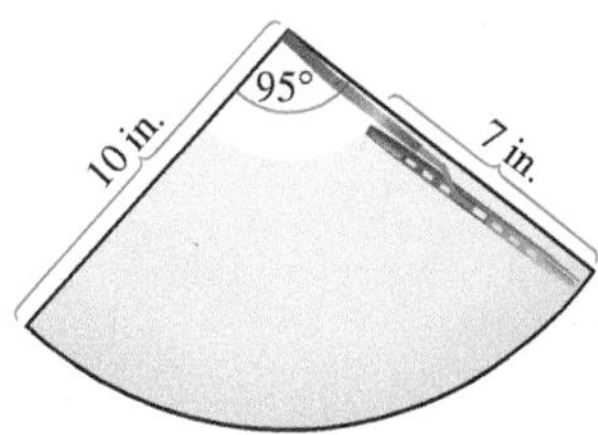

63. *Circular Railroad Curves* In the United States, circular railroad curves are designated by the **degree of curvature,** the central angle subtended by a chord of 100 ft. Suppose a portion of track has curvature 42.0°. (*Source:* Hay, W., *Railroad Engineering,* John Wiley and Sons.)

(a) What is the radius of the curve?

(b) What is the length of the arc determined by the 100-ft chord?

(c) What is the area of the portion of the circle bounded by the arc and the 100-ft chord?

64. *Land Required for a Solar-Power Plant* A 300-megawatt solar-power plant requires approximately 950,000 m^2 of land area to collect the required amount of energy from sunlight. If this land area is circular, what is its radius? If this land area is a 35° sector of a circle, what is its radius?

65. 1900 yd^2 66. 1.15 mi
67. radius: 3950 mi; circumference: 24,800 mi
68. (a) 0°
(b) Answers will vary.

65. *Area of a Lot* A frequent problem in surveying city lots and rural lands adjacent to curves of highways and railways is that of finding the area when one or more of the boundary lines is the arc of a circle. Find the area (to two significant digits) of the lot shown in the figure. (*Source:* Anderson, J. and E. Michael, *Introduction to Surveying,* McGraw-Hill.)

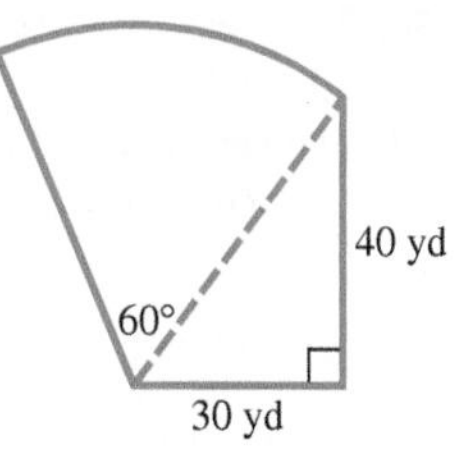

66. *Nautical Miles* **Nautical miles** are used by ships and airplanes. They are different from **statute miles,** where 1 mi = 5280 ft. A nautical mile is defined to be the arc length along the equator intercepted by a central angle AOB of $1'$, as illustrated in the figure. If the equatorial radius of Earth is 3963 mi, use the arc length formula to approximate the number of statute miles in 1 nautical mile. Round the answer to two decimal places.

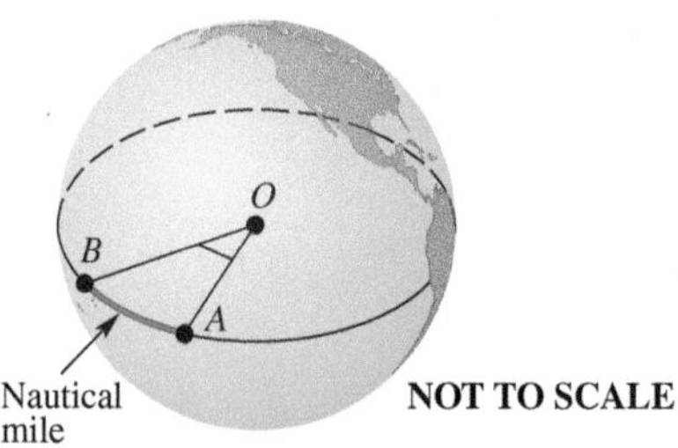

67. *Circumference of Earth* The first accurate estimate of the distance around Earth was done by the Greek astronomer Eratosthenes (276–195 B.C.), who noted that the noontime position of the sun at the summer solstice in the city of Syene differed by 7° 12′ from its noontime position in the city of Alexandria. (See the figure.) The distance between these two cities is 496 mi. Use the arc length formula to estimate the radius of Earth. Then find the circumference of Earth. (*Source:* Zeilik, M., *Introductory Astronomy and Astrophysics,* Third Edition, Saunders College Publishers.)

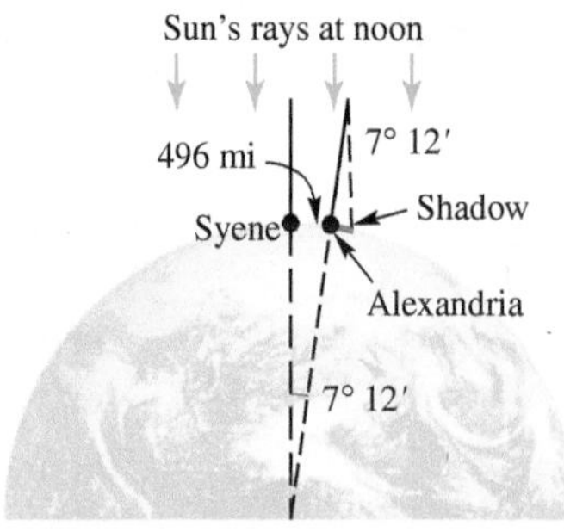

68. *Longitude* **Longitude** is the angular distance (expressed in degrees) East or West of the prime meridian, which goes from the North Pole to the South Pole through Greenwich, England. Arcs of 1° longitude are 110 km apart at the equator, and therefore 15° arcs subtend 15(110) km, or 1650 km, at the equator.

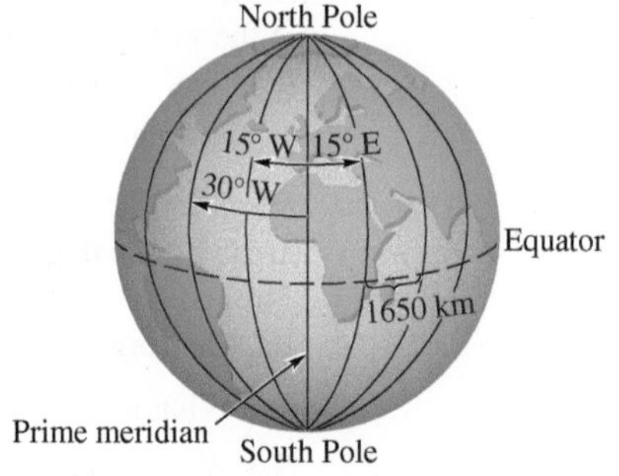

Because Earth rotates 15° per hr, longitude is found by taking the difference between time zones multiplied by 15°. For example, if it is 12 noon where we are (in the United States) and 5 P.M. in Greenwich, we are located at longitude 5(15°), or 75° W.

(a) What is the longitude at Greenwich, England?

(b) Use time zones to determine the longitude where you live.

69. The area is quadrupled.

70. $\mathcal{A} = \frac{\pi r^2 \theta}{360}$

71. $V = \frac{r^2 \theta h}{2}$ (θ in radians)

72. $V = \frac{1}{2}\theta(r_1^2 - r_2^2)h$ (θ in radians)

73. $r = \frac{L}{\theta}$

74. $h = r \cos \frac{\theta}{2}$

75. $d = r\left(1 - \cos \frac{\theta}{2}\right)$

76. $d = \frac{L}{\theta}\left(1 - \cos \frac{\theta}{2}\right)$

69. ***Concept Check*** If the radius of a circle is doubled and the central angle of a sector is unchanged, how is the area of the sector changed?

70. ***Concept Check*** Give the formula for the area of a sector when the angle is measured in degrees.

Volume of a Solid *Multiply the area of the base by the height to find a formula for the volume V of each solid.*

71.

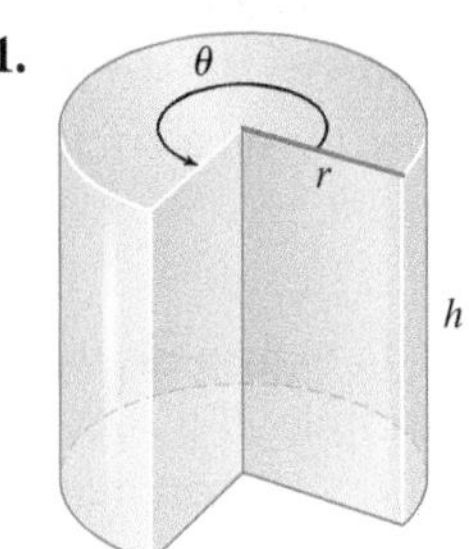

72.

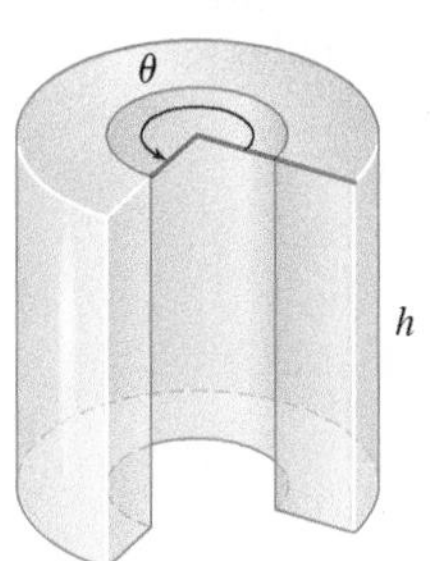

Outside radius is r_1, inside radius is r_2.

Relating Concepts

For individual or collaborative investigation *(Exercises 73–76)*

(Modeling) Measuring Paper Curl *Manufacturers of paper determine its quality by its curl. The curl of a sheet of paper is measured by holding it at the center of one edge and comparing the arc formed by the free end to arcs on a chart lying flat on a table. Each arc in the chart corresponds to a number d that gives the depth of the arc. See the figure. (Source:* Tabakovic, H., J. Paullet, and R. Bertram, "Measuring the Curl of Paper," *The College Mathematics Journal,* Vol. 30, No. 4.*)*

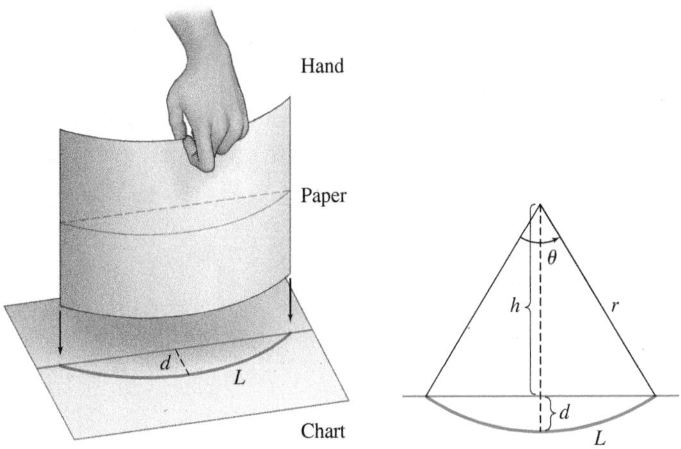

To produce the chart, it is necessary to find a function that relates d to the length of arc L. ***Work Exercises 73–76 in order,*** *to determine that function. Refer to the figure on the right.*

73. Express L in terms of r and θ, and then solve for r.

74. Use a right triangle to relate r, h, and θ. Solve for h.

75. Express d in terms of r and h. Then substitute the answer from **Exercise 74** for h. Factor out r.

76. Use the answer from **Exercise 73** to substitute for r in the result from **Exercise 75.** This result is a formula that gives d for specific values of θ.

3.3 The Unit Circle and Circular Functions

- Circular Functions
- Values of the Circular Functions
- Determining a Number with a Given Circular Function Value
- Applications of Circular Functions
- Function Values as Lengths of Line Segments

We have defined the six trigonometric functions in such a way that the domain of each function was a set of *angles* in standard position. These angles can be measured in degrees or in radians. In advanced courses, such as calculus, it is necessary to modify the trigonometric functions so that their domains consist of *real numbers* rather than angles. We do this by using the relationship between an angle θ and an arc of length s on a circle.

Circular Functions

In **Figure 12,** we start at the point $(1, 0)$ and measure an arc of length s along the circle. If $s > 0$, then the arc is measured in a counterclockwise direction, and if $s < 0$, then the direction is clockwise. (If $s = 0$, then no arc is measured.) Let the endpoint of this arc be at the point (x, y). The circle in **Figure 12** is the **unit circle**—it has center at the origin and radius 1 unit (hence the name *unit circle*). Recall from algebra that the equation of this circle is

$$x^2 + y^2 = 1. \quad \text{The unit circle}$$

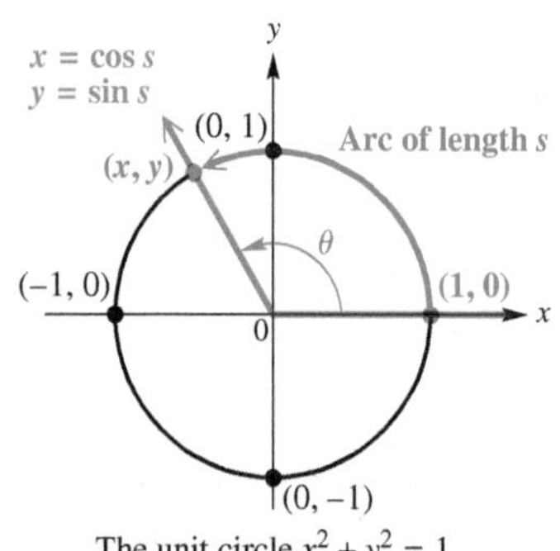

The unit circle $x^2 + y^2 = 1$

Figure 12

The radian measure of θ is related to the arc length s. For θ measured in radians and for r and s measured in the same linear units, we know that

$$s = r\theta.$$

When the radius has measure 1 unit, the formula $s = r\theta$ becomes $s = \theta$. Thus, the trigonometric functions of angle θ in radians found by choosing a point (x, y) on the unit circle can be rewritten as functions of the arc length s, a real number. When interpreted this way, they are called **circular functions.**

LOOKING AHEAD TO CALCULUS

If you plan to study calculus, you must become very familiar with radian measure. In calculus, the trigonometric or circular functions are always understood to have real number domains.

Circular Functions

The following functions are defined for any real number s represented by a directed arc on the unit circle.

$$\sin s = y \qquad \cos s = x \qquad \tan s = \frac{y}{x} \quad (x \neq 0)$$

$$\csc s = \frac{1}{y} \quad (y \neq 0) \qquad \sec s = \frac{1}{x} \quad (x \neq 0) \qquad \cot s = \frac{x}{y} \quad (y \neq 0)$$

Teaching Tip This is a good opportunity to review the concept of symmetry of a graph with respect to the axes and to the origin. Draw the unit circle and discuss the relative positions of the points (a, b), $(-a, b)$, $(-a, -b)$, and $(a, -b)$.

The unit circle is symmetric with respect to the x-axis, the y-axis, and the origin. If a point (a, b) lies on the unit circle, so do $(a, -b)$, $(-a, b)$, and $(-a, -b)$. Furthermore, each of these points has a *reference arc* of equal magnitude. For a point on the unit circle, its **reference arc** is the shortest arc from the point itself to the nearest point on the x-axis. (This concept is analogous to the reference angle concept.) Using the concept of symmetry makes determining sines and cosines of the real numbers identified in **Figure 13*** a relatively simple procedure if we know the coordinates of the points labeled in quadrant I.

*The authors thank Professor Marvel Townsend of the University of Florida for her suggestion to include **Figure 13.**

Teaching Tip Draw the unit circle and trace a point slowly counterclockwise, starting at $(1, 0)$. Students should see that the y-value (sine function)

- starts at 0
- increases to 1 (at $\frac{\pi}{2}$ radians)
- decreases to 0 (at π radians)
- decreases again to -1 (at $\frac{3\pi}{2}$ radians)
- returns to 0 after one full revolution (2π radians).

This analysis will help illustrate the range associated with the sine and cosine functions.

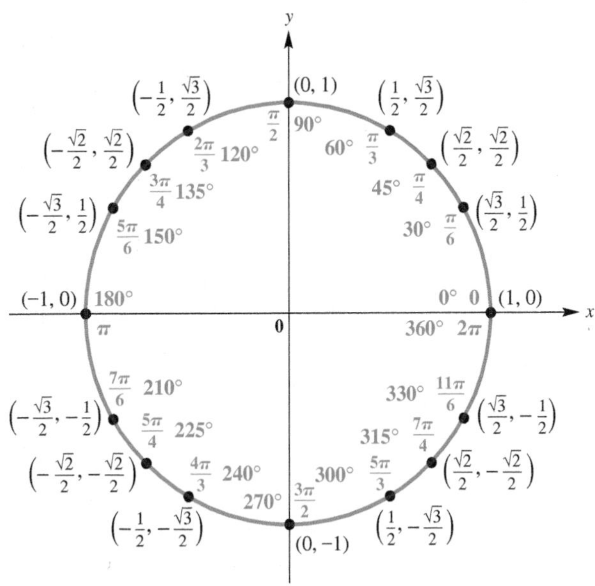

Figure 13

For example, the quadrant I real number $\frac{\pi}{3}$ is associated with the point $\left(\frac{1}{2}, \frac{\sqrt{3}}{2}\right)$ on the unit circle. Therefore, we can use symmetry to identify the coordinates of points having $\frac{\pi}{3}$ as reference arc.

Symmetry and Function Values for Real Numbers with Reference Arc $\frac{\pi}{3}$

s	Quadrant of s	Symmetry Type and Corresponding Point	$\cos s$	$\sin s$
$\frac{\pi}{3}$	I	not applicable; $\left(\frac{1}{2}, \frac{\sqrt{3}}{2}\right)$	$\frac{1}{2}$	$\frac{\sqrt{3}}{2}$
$\pi - \frac{\pi}{3} = \frac{2\pi}{3}$	II	y-axis; $\left(-\frac{1}{2}, \frac{\sqrt{3}}{2}\right)$	$-\frac{1}{2}$	$\frac{\sqrt{3}}{2}$
$\pi + \frac{\pi}{3} = \frac{4\pi}{3}$	III	origin; $\left(-\frac{1}{2}, -\frac{\sqrt{3}}{2}\right)$	$-\frac{1}{2}$	$-\frac{\sqrt{3}}{2}$
$2\pi - \frac{\pi}{3} = \frac{5\pi}{3}$	IV	x-axis; $\left(\frac{1}{2}, -\frac{\sqrt{3}}{2}\right)$	$\frac{1}{2}$	$-\frac{\sqrt{3}}{2}$

NOTE Because $\cos s = x$ and $\sin s = y$, we can replace x and y in the equation of the unit circle $x^2 + y^2 = 1$ and obtain the following.

$$\cos^2 s + \sin^2 s = 1 \quad \text{Pythagorean identity}$$

The ordered pair (x, y) represents a point on the unit circle, and therefore

$$-1 \leq x \leq 1 \quad \text{and} \quad -1 \leq y \leq 1,$$

$$-1 \leq \cos s \leq 1 \quad \text{and} \quad -1 \leq \sin s \leq 1.$$

For any value of s, both $\sin s$ and $\cos s$ exist, so the domain of these functions is the set of all real numbers.

For tan s, defined as $\frac{y}{x}$, x must not equal 0. The only way x can equal 0 is when the arc length s is $\frac{\pi}{2}, -\frac{\pi}{2}, \frac{3\pi}{2}, -\frac{3\pi}{2}$, and so on. To avoid a 0 denominator, the domain of the tangent function must be restricted to those values of s that satisfy

$$s \neq (2n + 1)\frac{\pi}{2}, \quad \textbf{where } n \textbf{ is any integer.}$$

The definition of secant also has x in the denominator, so the domain of secant is the same as the domain of tangent. Both cotangent and cosecant are defined with a denominator of y. To guarantee that $y \neq 0$, the domain of these functions must be the set of all values of s that satisfy

$$s \neq n\pi, \quad \textbf{where } n \textbf{ is any integer.}$$

Domains of the Circular Functions

The domains of the circular functions are as follows.

Sine and Cosine Functions: $(-\infty, \infty)$

Tangent and Secant Functions:

$$\{s \mid s \neq (2n + 1)\frac{\pi}{2}, \quad \textbf{where } n \textbf{ is any integer}\}$$

Cotangent and Cosecant Functions:

$$\{s \mid s \neq n\pi, \quad \textbf{where } n \textbf{ is any integer}\}$$

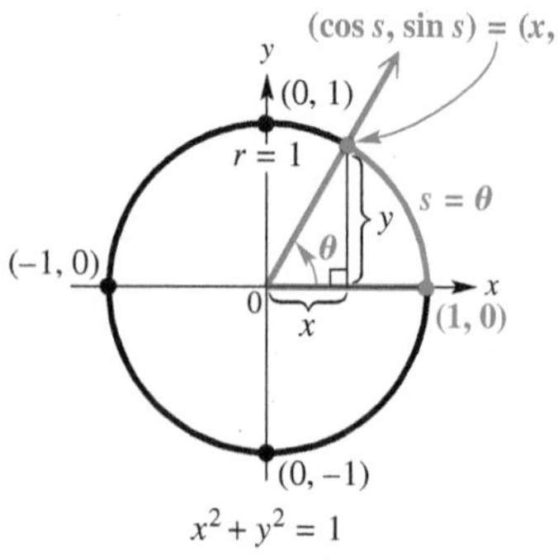

Figure 14

Values of the Circular Functions The circular functions of real numbers correspond to the trigonometric functions of angles measured in radians. Let us assume that angle θ is in standard position, superimposed on the unit circle. See **Figure 14.** Suppose that θ is the *radian* measure of this angle. Using the arc length formula

$$s = r\theta \quad \text{with } r = 1, \quad \text{we have} \quad s = \theta.$$

Thus, the length of the intercepted arc is the real number that corresponds to the radian measure of θ. We use the trigonometric function definitions to obtain the following.

$$\sin\theta = \frac{y}{r} = \frac{y}{1} = y = \sin s, \quad \cos\theta = \frac{x}{r} = \frac{x}{1} = x = \cos s, \quad \text{and so on.}$$

As shown here, the trigonometric functions and the circular functions lead to the same function values, provided that we think of the angles as being in radian measure. This leads to the following important result.

Evaluating a Circular Function

Circular function values of real numbers are obtained in the same manner as trigonometric function values of angles measured in radians. This applies both to methods of finding exact values (such as reference angle analysis) and to calculator approximations. ***Calculators must be in radian mode when they are used to find circular function values.***

Classroom Example 1
Find the exact values of $\sin(-3\pi)$, $\cos(-3\pi)$, and $\cot(-3\pi)$.

Answer:
$\sin(-3\pi) = 0$;
$\cos(-3\pi) = -1$;
$\cot(-3\pi)$ is undefined.

Teaching Tip Point out the importance of being able to sketch the unit circle as in **Figure 15.** This will enable students to more easily determine function values of multiples of $\frac{\pi}{2}$.

EXAMPLE 1 Finding Exact Circular Function Values

Find the exact values of $\sin \frac{3\pi}{2}$, $\cos \frac{3\pi}{2}$, and $\tan \frac{3\pi}{2}$.

SOLUTION Evaluating a circular function at the real number $\frac{3\pi}{2}$ is equivalent to evaluating it at $\frac{3\pi}{2}$ radians. An angle of $\frac{3\pi}{2}$ radians intersects the unit circle at the point $(0, -1)$, as shown in **Figure 15.** Because

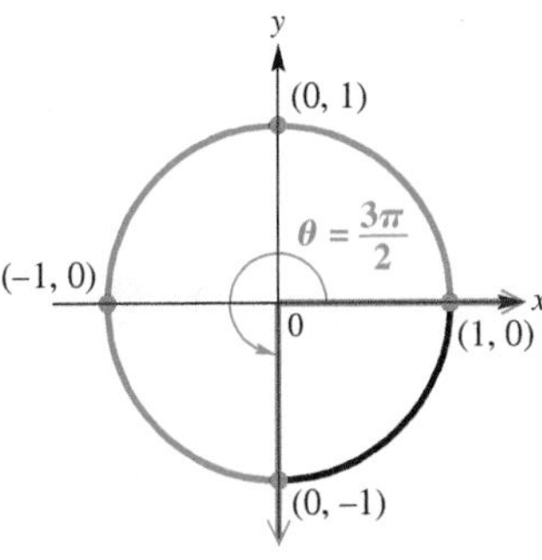

Figure 15

$$\sin s = y, \quad \cos s = x, \quad \text{and} \quad \tan s = \frac{y}{x},$$

it follows that

$$\sin \frac{3\pi}{2} = -1, \quad \cos \frac{3\pi}{2} = 0, \quad \text{and} \quad \tan \frac{3\pi}{2} \text{ is undefined.}$$

✔ **Now Try Exercises 11 and 13.**

Classroom Example 2
Find each exact function value using the specified method.
(a) Use **Figure 13** to find the exact values of $\sin \frac{4\pi}{3}$ and $\cos \frac{4\pi}{3}$.
(b) Use **Figure 13** and the definition of the tangent to find the exact value of $\tan\left(-\frac{9\pi}{4}\right)$.
(c) Use reference angles and radian-to-degree conversion to find the exact value of $\sin \frac{11\pi}{6}$.

Answers:
(a) $\sin \frac{4\pi}{3} = -\frac{\sqrt{3}}{2}$;
$\cos \frac{4\pi}{3} = -\frac{1}{2}$
(b) -1
(c) $-\frac{1}{2}$

EXAMPLE 2 Finding Exact Circular Function Values

Find each exact function value using the specified method.

(a) Use **Figure 13** to find the exact values of $\cos \frac{7\pi}{4}$ and $\sin \frac{7\pi}{4}$.

(b) Use **Figure 13** and the definition of the tangent to find the exact value of $\tan\left(-\frac{5\pi}{3}\right)$.

(c) Use reference angles and radian-to-degree conversion to find the exact value of $\cos \frac{2\pi}{3}$.

SOLUTION

(a) In **Figure 13,** we see that the real number $\frac{7\pi}{4}$ corresponds to the unit circle point $\left(\frac{\sqrt{2}}{2}, -\frac{\sqrt{2}}{2}\right)$.

$$\cos \frac{7\pi}{4} = \frac{\sqrt{2}}{2} \quad \text{and} \quad \sin \frac{7\pi}{4} = -\frac{\sqrt{2}}{2}$$

(b) Moving around the unit circle $\frac{5\pi}{3}$ units in the *negative* direction yields the same ending point as moving around $\frac{\pi}{3}$ units in the positive direction. Thus, $-\frac{5\pi}{3}$ corresponds to $\left(\frac{1}{2}, \frac{\sqrt{3}}{2}\right)$.

$$\tan\left(-\frac{5\pi}{3}\right) = \tan \frac{\pi}{3} = \frac{\frac{\sqrt{3}}{2}}{\frac{1}{2}} = \frac{\sqrt{3}}{2} \div \frac{1}{2} = \frac{\sqrt{3}}{2} \cdot \frac{2}{1} = \sqrt{3}$$

$\tan s = \frac{y}{x}$

Simplify this complex fraction.

(c) An angle of $\frac{2\pi}{3}$ radians corresponds to an angle of 120°. In standard position, 120° lies in quadrant II with a reference angle of 60°.

Cosine is negative in quadrant II.

$$\cos \frac{2\pi}{3} = \cos 120^\circ = -\cos 60^\circ = -\frac{1}{2}$$

Reference angle

✔ **Now Try Exercises 17, 23, 27, and 31.**

EXAMPLE 3 Approximating Circular Function Values

Find a calculator approximation for each circular function value.

(a) $\cos 1.85$ **(b)** $\cos 0.5149$ **(c)** $\cot 1.3209$ **(d)** $\sec(-2.9234)$

SOLUTION

```
NORMAL FIX4 AUTO REAL RADIAN MP
cos(1.85)
                    -.2756
cos(.5149)
                     .8703
1/tan(1.3209)
                     .2552
1/cos(-2.9234)
                   -1.0243
```

Radian mode

This is how the TI-84 Plus calculator displays the results of **Example 3,** fixed to four decimal places.

(a) $\cos 1.85 \approx -0.2756$ Use a calculator in radian mode.

(b) $\cos 0.5149 \approx 0.8703$ Use a calculator in radian mode.

(c) As before, to find cotangent, secant, and cosecant function values, we must use the appropriate reciprocal functions. To find $\cot 1.3209$, first find $\tan 1.3209$ and then find the reciprocal.

$$\cot 1.3209 = \frac{1}{\tan 1.3209} \approx 0.2552$$ Tangent and cotangent are reciprocals.

(d) $$\sec(-2.9234) = \frac{1}{\cos(-2.9234)} \approx -1.0243$$ Cosine and secant are reciprocals.

✔ **Now Try Exercises 33, 39, and 43.**

Classroom Example 3
Find a calculator approximation for each circular function value.
(a) $\sin 3.42$ (b) $\tan 0.8234$
(c) $\sec 5.6041$ (d) $\csc(-2.7335)$

Answers:
(a) -0.2748 (b) 1.0790
(c) 1.2851 (d) -2.5198

CAUTION ***Remember, when used to find a circular function value of a real number, a calculator must be in radian mode.***

Determining a Number with a Given Circular Function Value We can reverse the process of **Example 3** and use a calculator to determine an angle measure, given a trigonometric function value of the angle. ***Remember that the keys marked*** $\sin^{-1}$, $\cos^{-1}$, ***and*** $\tan^{-1}$ ***do not represent reciprocal functions. They enable us to find inverse function values.***

For reasons explained in a later chapter, the following statements are true.

- For all x in $[-1, 1]$, a calculator in radian mode returns a single value in $\left[-\frac{\pi}{2}, \frac{\pi}{2}\right]$ for $\sin^{-1} x$.
- For all x in $[-1, 1]$, a calculator in radian mode returns a single value in $[0, \pi]$ for $\cos^{-1} x$.
- For all real numbers x, a calculator in radian mode returns a single value in $\left(-\frac{\pi}{2}, \frac{\pi}{2}\right)$ for $\tan^{-1} x$.

Classroom Example 4
Find each value as specified.
(a) Approximate the value of s in the interval $\left[0, \frac{\pi}{2}\right]$ if $\sin s = 0.3210$.
(b) Find the exact value of s in the interval $\left[\frac{3\pi}{2}, 2\pi\right]$ if $\tan s = -\frac{\sqrt{3}}{3}$.

Answers:
(a) 0.3268 (b) $\frac{11\pi}{6}$

EXAMPLE 4 Finding Numbers Given Circular Function Values

Find each value as specified.

(a) Approximate the value of s in the interval $\left[0, \frac{\pi}{2}\right]$ if $\cos s = 0.9685$.

(b) Find the exact value of s in the interval $\left[\pi, \frac{3\pi}{2}\right]$ if $\tan s = 1$.

SOLUTION

(a) Because we are given a cosine value and want to determine the real number in $\left[0, \frac{\pi}{2}\right]$ that has this cosine value, we use the *inverse cosine* function of a calculator. With the calculator in radian mode, we find s as follows.

$$s = \cos^{-1}(0.9685) \approx 0.2517$$

See **Figure 16.** The screen indicates that the real number in $\left[0, \frac{\pi}{2}\right]$ having cosine equal to 0.9685 is 0.2517.

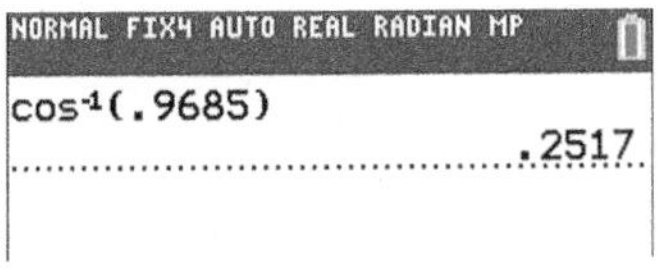

```
NORMAL FIX4 AUTO REAL RADIAN MP
cos⁻¹(.9685)
                         .2517
```

Radian mode

Figure 16

(b) Recall that $\tan \frac{\pi}{4} = 1$, and in quadrant III tan s is positive.

$$\tan\left(\pi + \frac{\pi}{4}\right) = \tan \frac{5\pi}{4} = 1$$

Thus, $s = \frac{5\pi}{4}$. See **Figure 17.**

Now Try Exercises 63 and 71.

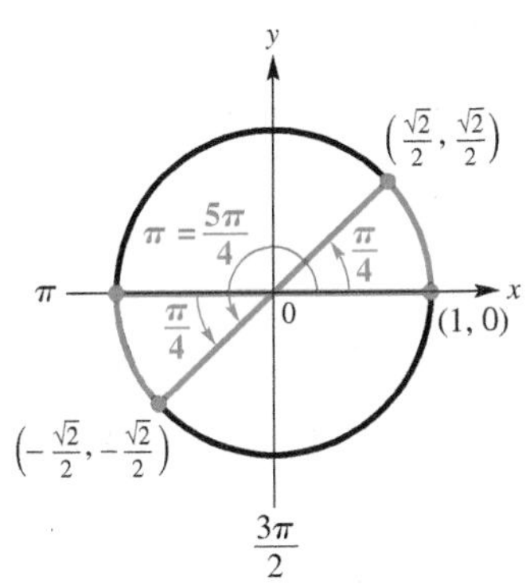

Figure 17

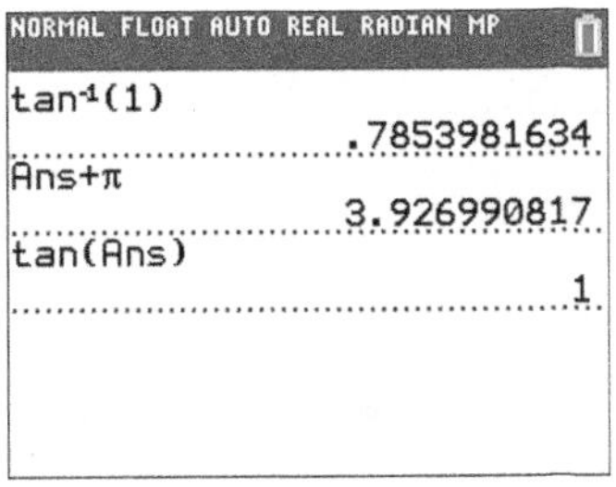

```
NORMAL FLOAT AUTO REAL RADIAN MP
tan⁻¹(1)
                  .7853981634
Ans+π
                  3.926990817
tan(Ans)
                            1
```

This screen supports the result in **Example 4(b)** with calculator approximations.

Applications of Circular Functions

EXAMPLE 5 Modeling the Angle of Elevation of the Sun

The angle of elevation θ of the sun in the sky at any latitude L is calculated with the formula

$$\sin \theta = \cos D \cos L \cos \omega + \sin D \sin L,$$

where $\theta = 0$ corresponds to sunrise and $\theta = \frac{\pi}{2}$ occurs if the sun is directly overhead. The Greek letter ω (lowercase *omega*) is the number of radians that Earth has rotated through since noon, when $\omega = 0$. D is the declination of the sun, which varies because Earth is tilted on its axis. (*Source*: Winter, C., R. Sizmann, and L. L. Vant-Hull, Editors, *Solar Power Plants,* Springer-Verlag.)

Sacramento, California, has latitude $L = 38.5°$, or 0.6720 radian. Find the angle of elevation θ of the sun at 3 P.M. on February 29, 2012, where at that time $D \approx -0.1425$ and $\omega \approx 0.7854$.

Classroom Example 5
Repeat **Example 5** for Sitka, Alaska, which has latitude 57.1°.

Answer: 15.1°

SOLUTION Use the given formula for $\sin \theta$.

$$\sin \theta = \cos D \cos L \cos \omega + \sin D \sin L$$

$$\sin \theta = \cos(-0.1425) \cos(0.6720) \cos(0.7854) + \sin(-0.1425) \sin(0.6720)$$

Let $D = -0.1425$, $L = 0.6720$, and $\omega = 0.7854$.

$$\sin \theta \approx 0.4593426188$$

$$\theta \approx 0.4773 \text{ radian, or } 27.3°$$ Use inverse sine.

Now Try Exercise 89.

Function Values as Lengths of Line Segments

The diagram shown in **Figure 18** illustrates a correspondence that ties together the right triangle ratio definitions of the trigonometric functions and the unit circle interpretation. The arc SR is the first-quadrant portion of the unit circle, and the standard-position angle POQ is designated θ. By definition, the coordinates of P are $(\cos \theta, \sin \theta)$. The six trigonometric functions of θ can be interpreted as lengths of line segments found in **Figure 18.**

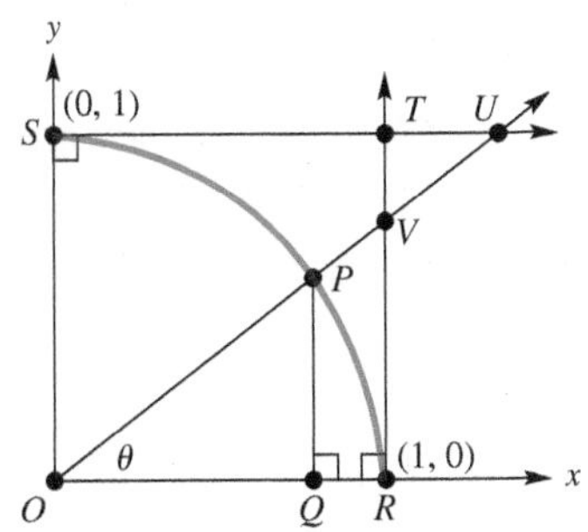

Figure 18

For $\cos \theta$ and $\sin \theta$, use right triangle POQ and right triangle ratios.

$$\cos \theta = \frac{\text{side adjacent to } \theta}{\text{hypotenuse}} = \frac{OQ}{OP} = \frac{OQ}{1} = OQ$$

$$\sin \theta = \frac{\text{side opposite } \theta}{\text{hypotenuse}} = \frac{PQ}{OP} = \frac{PQ}{1} = PQ$$

For $\tan \theta$ and $\sec \theta$, use right triangle VOR in **Figure 18** (repeated below in the margin) and right triangle ratios.

$$\tan \boldsymbol{\theta} = \frac{\text{side opposite } \theta}{\text{side adjacent to } \theta} = \frac{VR}{OR} = \frac{VR}{1} = \boldsymbol{VR}$$

$$\sec \boldsymbol{\theta} = \frac{\text{hypotenuse}}{\text{side adjacent to } \theta} = \frac{OV}{OR} = \frac{OV}{1} = \boldsymbol{OV}$$

For $\csc \theta$ and $\cot \theta$, first note that US and OR are parallel. Thus angle SUO is equal to θ because it is an alternate interior angle to angle POQ, which is equal to θ. Use right triangle USO and right triangle ratios.

$$\csc SUO = \csc \boldsymbol{\theta} = \frac{\text{hypotenuse}}{\text{side opposite } \theta} = \frac{OU}{OS} = \frac{OU}{1} = \boldsymbol{OU}$$

$$\cot SUO = \cot \boldsymbol{\theta} = \frac{\text{side adjacent to } \theta}{\text{side opposite } \theta} = \frac{US}{OS} = \frac{US}{1} = \boldsymbol{US}$$

Figure 19 uses color to illustrate the results found above.

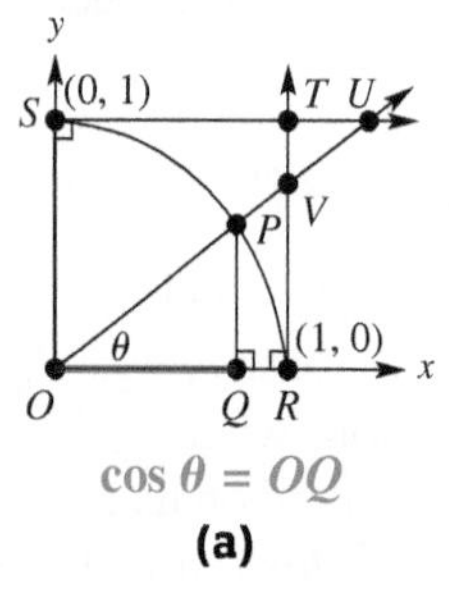

$\cos \theta = OQ$
(a)

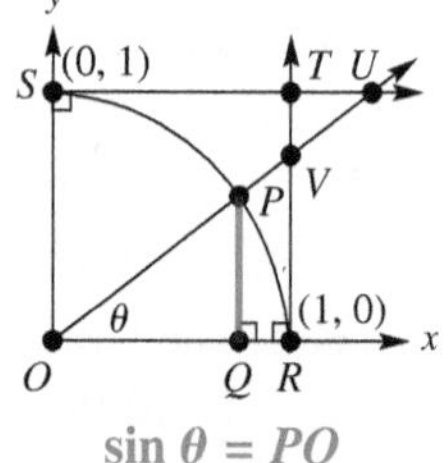

$\sin \theta = PQ$
(b)

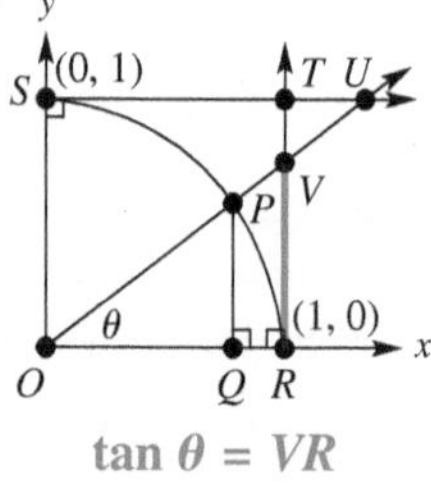

$\tan \theta = VR$
(c)

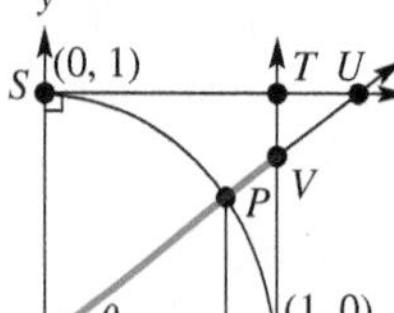

$\sec \theta = OV$
(d)

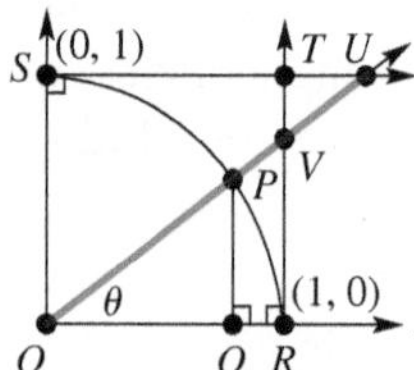

$\csc \theta = OU$
(e)

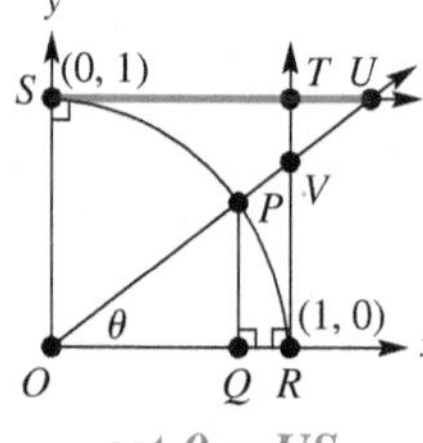

$\cot \theta = US$
(f)

Figure 19

Classroom Example 6
Repeat **Example 6,** but suppose that angle TVU measures 45°.

Answer:

$OQ = \frac{\sqrt{2}}{2}$ $\quad PQ = \frac{\sqrt{2}}{2}$

$VR = 1$ $\quad OV = \sqrt{2}$

$OU = \sqrt{2}$ $\quad US = 1$

EXAMPLE 6 Finding Lengths of Line Segments

Figure 18 is repeated in the margin. Suppose that angle TVU measures 60°. Find the exact lengths of segments OQ, PQ, VR, OV, OU, and US.

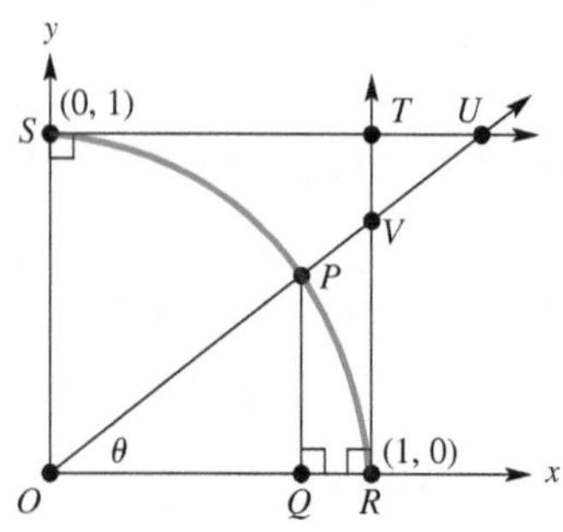

Figure 18 (repeated)

SOLUTION Angle TVU has the same measure as angle OVR because they are vertical angles. Therefore, angle OVR measures 60°. Because it is one of the acute angles in right triangle VOR, θ must be its complement, measuring 30°.

$$OQ = \cos 30° = \frac{\sqrt{3}}{2} \qquad OV = \sec 30° = \frac{2\sqrt{3}}{3}$$

$$PQ = \sin 30° = \frac{1}{2} \qquad OU = \csc 30° = 2$$

$$VR = \tan 30° = \frac{\sqrt{3}}{3} \qquad US = \cot 30° = \sqrt{3}$$

Use the equations found in **Figure 19,** with $\theta = 30°$.

✔ **Now Try Exercise 93.**

3.3 Exercises

1. Counterclockwise from 0 radians, the coordinates are $(1, 0)$, $\left(\frac{\sqrt{3}}{2}, \frac{1}{2}\right)$, $\left(\frac{\sqrt{2}}{2}, \frac{\sqrt{2}}{2}\right)$, $\left(\frac{1}{2}, \frac{\sqrt{3}}{2}\right)$, and $(0, 1)$.
2. 1
3. $\frac{\sqrt{2}}{2}$
4. $\frac{\sqrt{3}}{2}$
5. 1
6. $\frac{\pi}{3}$
7. $\sin\theta = \frac{\sqrt{2}}{2}$; $\cos\theta = \frac{\sqrt{2}}{2}$; $\tan\theta = 1$; $\cot\theta = 1$; $\sec\theta = \sqrt{2}$; $\csc\theta = \sqrt{2}$
8. $\sin\theta = \frac{8}{17}$; $\cos\theta = -\frac{15}{17}$; $\tan\theta = -\frac{8}{15}$; $\cot\theta = -\frac{15}{8}$; $\sec\theta = -\frac{17}{15}$; $\csc\theta = \frac{17}{8}$
9. $\sin\theta = -\frac{12}{13}$; $\cos\theta = \frac{5}{13}$; $\tan\theta = -\frac{12}{5}$; $\cot\theta = -\frac{5}{12}$; $\sec\theta = \frac{13}{5}$; $\csc\theta = -\frac{13}{12}$
10. $\sin\theta = -\frac{1}{2}$; $\cos\theta = -\frac{\sqrt{3}}{2}$; $\tan\theta = \frac{\sqrt{3}}{3}$; $\cot\theta = \sqrt{3}$; $\sec\theta = -\frac{2\sqrt{3}}{3}$; $\csc\theta = -2$
11. (a) 1 (b) 0 (c) undefined
12. (a) 0 (b) -1 (c) 0
13. (a) 0 (b) 1 (c) 0
14. (a) 0 (b) -1 (c) 0
15. (a) 0 (b) -1 (c) 0
16. (a) 1 (b) 0 (c) undefined
17. $-\frac{1}{2}$
18. $\frac{1}{2}$
19. -1
20. -2
21. -2
22. $-\sqrt{3}$
23. $-\frac{1}{2}$
24. $\sqrt{3}$

CONCEPT PREVIEW *Fill in the blanks to complete the coordinates for each point indicated in the first quadrant of the unit circle in Exercise 1. Then use it to find each exact circular function value in Exercises 2–5, and work Exercise 6.*

1.

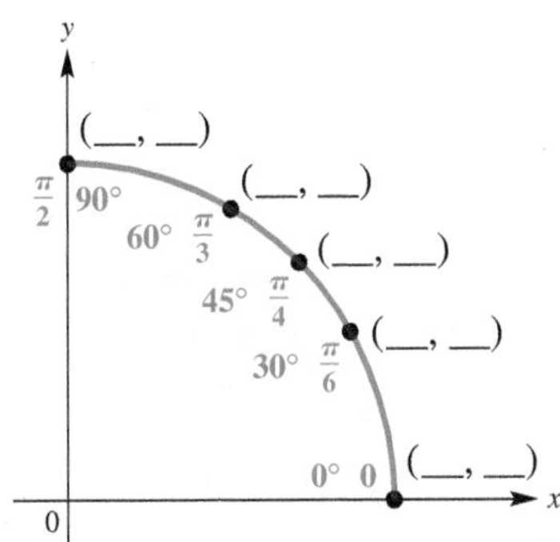

2. $\cos 0$

3. $\sin \frac{\pi}{4}$

4. $\sin \frac{\pi}{3}$

5. $\tan \frac{\pi}{4}$

6. Find s in the interval $\left[0, \frac{\pi}{2}\right]$ if $\cos s = \frac{1}{2}$.

CONCEPT PREVIEW *Each figure shows an angle θ in standard position with its terminal side intersecting the unit circle. Evaluate the six circular function values of θ.*

7.

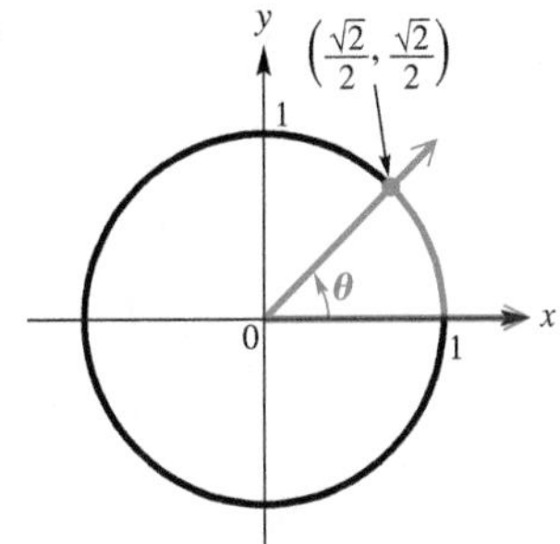

8.

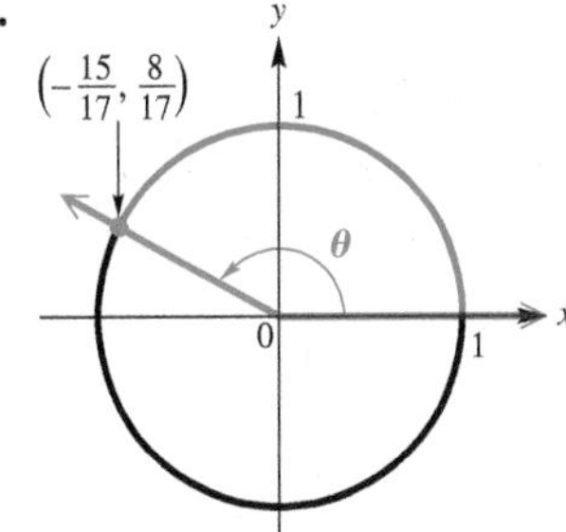

9.

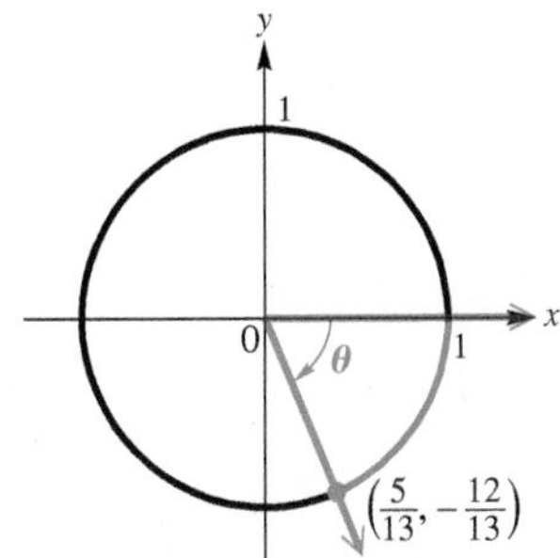

10.

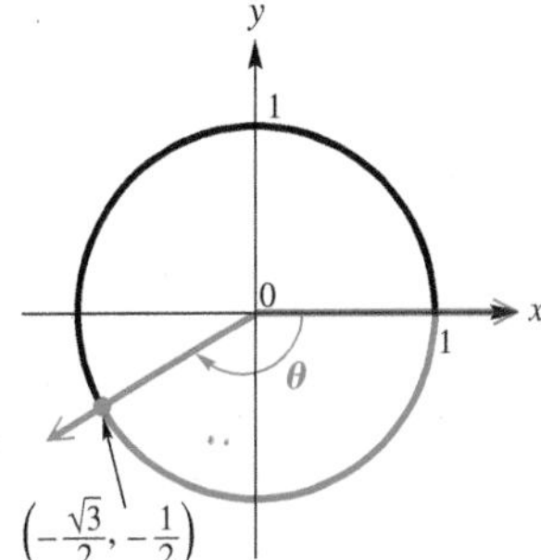

Find the exact values of ***(a)*** $\sin s$, ***(b)*** $\cos s$, *and* ***(c)*** $\tan s$ *for each real number* s. ***See Example 1.***

11. $s = \frac{\pi}{2}$

12. $s = \pi$

13. $s = 2\pi$

14. $s = 3\pi$

15. $s = -\pi$

16. $s = -\frac{3\pi}{2}$

Find each exact function value. ***See Example 2.***

17. $\sin \frac{7\pi}{6}$

18. $\cos \frac{5\pi}{3}$

19. $\tan \frac{3\pi}{4}$

20. $\sec \frac{2\pi}{3}$

21. $\csc \frac{11\pi}{6}$

22. $\cot \frac{5\pi}{6}$

23. $\cos\left(-\frac{4\pi}{3}\right)$

24. $\tan\left(-\frac{17\pi}{3}\right)$

25. $\frac{\sqrt{2}}{2}$ 26. $-\sqrt{2}$
27. $\frac{\sqrt{3}}{2}$ 28. $-\frac{1}{2}$
29. $\frac{2\sqrt{3}}{3}$ 30. $\frac{2\sqrt{3}}{3}$
31. $-\frac{\sqrt{3}}{3}$ 32. $-\frac{\sqrt{2}}{2}$

25. $\cos \frac{7\pi}{4}$ **26.** $\sec \frac{5\pi}{4}$ **27.** $\sin\left(-\frac{4\pi}{3}\right)$ **28.** $\sin\left(-\frac{5\pi}{6}\right)$

29. $\sec \frac{23\pi}{6}$ **30.** $\csc \frac{13\pi}{3}$ **31.** $\tan \frac{5\pi}{6}$ **32.** $\cos \frac{3\pi}{4}$

33. 0.5736 34. 0.7314
35. 0.4068 36. 0.5397
37. 1.2065 38. 0.1944
39. 14.3338 40. 1.0170
41. −1.0460 42. −2.1291
43. −3.8665 44. 1.1848

Find a calculator approximation to four decimal places for each circular function value. ***See Example 3.***

33. $\sin 0.6109$ **34.** $\sin 0.8203$ **35.** $\cos(-1.1519)$

36. $\cos(-5.2825)$ **37.** $\tan 4.0203$ **38.** $\tan 6.4752$

39. $\csc(-9.4946)$ **40.** $\csc 1.3875$ **41.** $\sec 2.8440$

42. $\sec(-8.3429)$ **43.** $\cot 6.0301$ **44.** $\cot 3.8426$

45. 0.7 46. 0.8
47. 0.9 48. −0.8
49. −0.6 50. −1.0
51. 2.3 or 4.0 52. 4.4 or 5.0
53. 0.8 or 2.4 54. 1.3 or 5.0

Concept Check *The figure displays a unit circle and an angle of* 1 *radian. The tick marks on the circle are spaced at every two-tenths radian. Use the figure to estimate each value.*

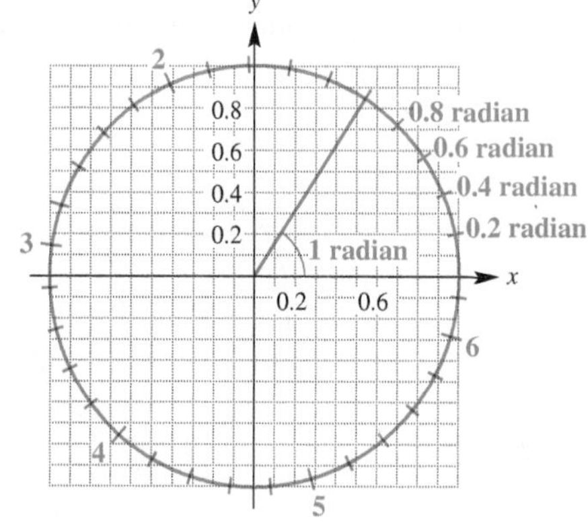

45. $\cos 0.8$ **46.** $\cos 0.6$ **47.** $\sin 2$

48. $\sin 5.4$ **49.** $\sin 3.8$ **50.** $\cos 3.2$

51. a positive angle whose cosine is -0.65

52. a positive angle whose sine is -0.95

53. a positive angle whose sine is 0.7

54. a positive angle whose cosine is 0.3

55. negative 56. negative
57. negative 58. positive
59. positive 60. negative

Concept Check *Without using a calculator, decide whether each function value is positive or negative. (Hint: Consider the radian measures of the quadrantal angles, and remember that* $\pi \approx 3.14$.*)*

55. $\cos 2$ **56.** $\sin(-1)$ **57.** $\sin 5$

58. $\cos 6$ **59.** $\tan 6.29$ **60.** $\tan(-6.29)$

61. 0.2095 62. 0.6720
63. 1.4426 64. 1.2799
65. 0.3887 66. 1.3634

Find the approximate value of s, to four decimal places, in the interval $\left[0, \frac{\pi}{2}\right]$ *that makes each statement true.* ***See Example 4(a).***

61. $\tan s = 0.2126$ **62.** $\cos s = 0.7826$ **63.** $\sin s = 0.9918$

64. $\cot s = 0.2994$ **65.** $\sec s = 1.0806$ **66.** $\csc s = 1.0219$

67. $\frac{5\pi}{6}$ 68. $\frac{2\pi}{3}$
69. $\frac{4\pi}{3}$ 70. $\frac{7\pi}{6}$
71. $\frac{7\pi}{4}$ 72. $\frac{11\pi}{6}$
73. $\frac{4\pi}{3}, \frac{5\pi}{3}$ 74. $\frac{2\pi}{3}, \frac{4\pi}{3}$
75. $\frac{\pi}{4}, \frac{3\pi}{4}, \frac{5\pi}{4}, \frac{7\pi}{4}$
76. $\frac{\pi}{3}, \frac{2\pi}{3}, \frac{4\pi}{3}, \frac{5\pi}{3}$
77. $-\frac{11\pi}{6}, -\frac{7\pi}{6}, -\frac{5\pi}{6}, -\frac{\pi}{6}, \frac{\pi}{6}, \frac{5\pi}{6}$
78. $-\frac{3\pi}{4}, -\frac{\pi}{4}, \frac{\pi}{4}, \frac{3\pi}{4}$
79. $(-0.8011, 0.5985)$
80. $(-0.9668, -0.2555)$
81. $(0.4385, -0.8987)$
82. $(-0.7259, 0.6878)$
83. I 84. IV
85. II 86. III
87. 0.9428 88. 2.8040
89. (a) 32.4° (b) Answers may vary.
90. 8.6 hr; 15.4 hr

Find the exact value of s in the given interval that has the given circular function value. ***See Example 4(b).***

67. $\left[\frac{\pi}{2}, \pi\right]$; $\sin s = \frac{1}{2}$

68. $\left[\frac{\pi}{2}, \pi\right]$; $\cos s = -\frac{1}{2}$

69. $\left[\pi, \frac{3\pi}{2}\right]$; $\tan s = \sqrt{3}$

70. $\left[\pi, \frac{3\pi}{2}\right]$; $\sin s = -\frac{1}{2}$

71. $\left[\frac{3\pi}{2}, 2\pi\right]$; $\tan s = -1$

72. $\left[\frac{3\pi}{2}, 2\pi\right]$; $\cos s = \frac{\sqrt{3}}{2}$

Find the exact values of s in the given interval that satisfy the given condition.

73. $[0, 2\pi)$; $\sin s = -\frac{\sqrt{3}}{2}$

74. $[0, 2\pi)$; $\cos s = -\frac{1}{2}$

75. $[0, 2\pi)$; $\cos^2 s = \frac{1}{2}$

76. $[0, 2\pi)$; $\tan^2 s = 3$

77. $[-2\pi, \pi)$; $3 \tan^2 s = 1$

78. $[-\pi, \pi)$; $\sin^2 s = \frac{1}{2}$

Suppose an arc of length s lies on the unit circle $x^2 + y^2 = 1$, starting at the point $(1, 0)$ and terminating at the point (x, y). (See **Figure 12.**) *Use a calculator to find the approximate coordinates for (x, y) to four decimal places. (Hint: $x = \cos s$ and $y = \sin s$.)*

79. $s = 2.5$ **80.** $s = 3.4$ **81.** $s = -7.4$ **82.** $s = -3.9$

Concept Check For each value of s, use a calculator to find sin s and cos s, and then use the results to decide in which quadrant an angle of s radians lies.

83. $s = 51$ **84.** $s = 49$ **85.** $s = 65$ **86.** $s = 79$

Concept Check Each graphing calculator screen shows a point on the unit circle. Find the length, to four decimal places, of the shortest arc of the circle from $(1, 0)$ to the point.

87. $x^2 + y^2 = 1$

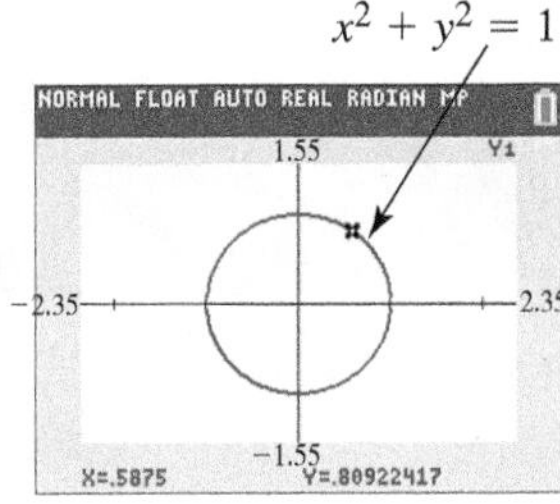

88. $x^2 + y^2 = 1$

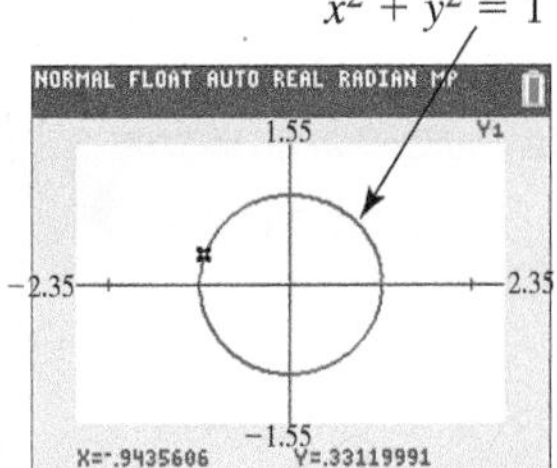

(Modeling) Solve each problem. ***See Example 5.***

89. *Elevation of the Sun* Refer to **Example 5.**

(a) Repeat the example for New Orleans, which has latitude $L = 30°$.

(b) Compare the answers. Do they agree with intuition?

90. *Length of a Day* The number of daylight hours H at any location can be calculated using the formula

$$\cos(0.1309H) = -\tan D \tan L,$$

where D and L are defined as in **Example 5.** Use this trigonometric equation to calculate the shortest and longest days in Minneapolis, Minnesota, if its latitude $L = 44.88°$, the shortest day occurs when $D = -23.44°$, and the longest day occurs when $D = 23.44°$. Remember to convert degrees to radians. Round the answer to the nearest tenth. (*Source:* Winter, C., R. Sizmann, and L. L. Vant-Hull, Editors, *Solar Power Plants,* Springer-Verlag.)

91. (a) 30° (b) 60° (c) 75° (d) 86° (e) 86° (f) 60°

92. (a) 1° (b) 19° (c) 53° (d) 58° (e) 48° (f) 12°

93. (a) $\frac{1}{2}$ (b) $\frac{\sqrt{3}}{2}$ (c) $\sqrt{3}$ (d) 2 (e) $\frac{2\sqrt{3}}{3}$ (f) $\frac{\sqrt{3}}{3}$

94. (a) 0.7880 (b) 0.6157 (c) 0.7813 (d) 1.269 (e) 1.624 (f) 1.280

91. ***Maximum Temperatures*** Because the values of the circular functions repeat every 2π, they may be used to describe phenomena that repeat periodically. For example, the maximum afternoon temperature in a given city might be modeled by

$$t = 60 - 30\cos\left(\frac{\pi}{6}x\right),$$

where t represents the maximum afternoon temperature in degrees Fahrenheit in month x, with $x = 0$ representing January, $x = 1$ representing February, and so on. Find the maximum afternoon temperature, to the nearest degree, for each of the following months.

(a) January **(b)** April **(c)** May

(d) June **(e)** August **(f)** October

92. ***Temperature in Fairbanks*** Suppose the temperature in Fairbanks is modeled by

$$T(x) = 37\sin\left[\frac{2\pi}{365}(x - 101)\right] + 25,$$

where $T(x)$ is the temperature in degrees Fahrenheit on day x, with $x = 1$ corresponding to January 1 and $x = 365$ corresponding to December 31. Use a calculator to estimate the temperature, to the nearest degree, on the following days. (*Source:* Lando, B. and C. Lando, "Is the Graph of Temperature Variation a Sine Curve?," *The Mathematics Teacher,* vol. 70.)

(a) March 1 (day 60) **(b)** April 1 (day 91) **(c)** Day 150

(d) June 15 **(e)** September 1 **(f)** October 31

Refer to **Figures 18 and 19,** *and work each problem.* ***See Example 6.***

93. Suppose that angle θ measures 60°. Find the exact length of each segment.

(a) OQ **(b)** PQ **(c)** VR

(d) OV **(e)** OU **(f)** US

94. Repeat **Exercise 93** for $\theta = 38°$. Give lengths as approximations to four significant digits.

Chapter 3 Quiz (Sections 3.1–3.3)

[3.1]
1. $\frac{5\pi}{4}$ 2. $-\frac{11\pi}{6}$
3. 300° 4. −210°

[3.2]
5. 1.5 6. 67,500 in.^{2}

[3.3]
7. $\frac{\sqrt{2}}{2}$ 8. $-\frac{1}{2}$
9. 0 10. $\frac{2\pi}{3}$

Convert each degree measure to radians.

1. 225° **2.** −330°

Convert each radian measure to degrees.

3. $\frac{5\pi}{3}$ **4.** $-\frac{7\pi}{6}$

A central angle of a circle with radius 300 in. *intercepts an arc of* 450 in. *(These measures are accurate to the nearest inch.) Find each measure.*

5. the radian measure of the angle **6.** the area of the sector

Find each exact circular function value.

7. $\cos\frac{7\pi}{4}$ **8.** $\sin\left(-\frac{5\pi}{6}\right)$ **9.** $\tan 3\pi$

10. Find the exact value of s in the interval $\left[\frac{\pi}{2}, \pi\right]$ if $\sin s = \frac{\sqrt{3}}{2}$.

3.4 Linear and Angular Speed

- Linear Speed
- Angular Speed

Linear Speed

There are situations when we need to know how fast a point on a circular disk is moving or how fast the central angle of such a disk is changing. Some examples occur with machinery involving gears or pulleys or the speed of a car around a curved portion of highway.

Suppose that point P moves at a constant speed along a circle of radius r and center O. See **Figure 20.** The measure of how fast the position of P is changing is the **linear speed.** If v represents linear speed, then

$$\text{speed} = \frac{\text{distance}}{\text{time}}, \quad \text{or} \quad v = \frac{s}{t},$$

where s is the length of the arc traced by point P at time t. (This formula is just a restatement of $r = \frac{d}{t}$ with s as distance, v as rate (speed), and t as time.)

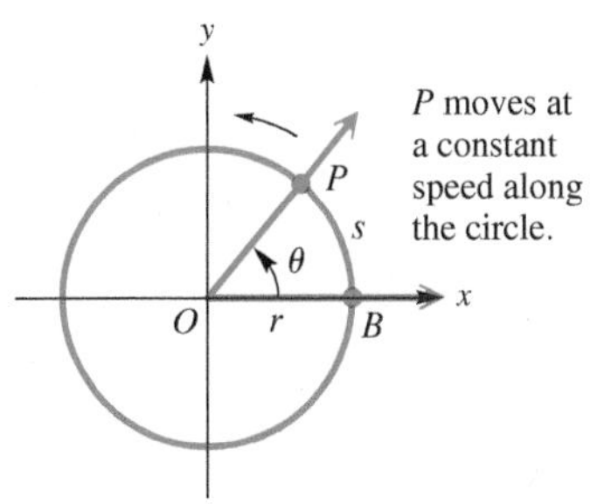

Figure 20

Angular Speed

Refer to **Figure 20.** As point P in the figure moves along the circle, ray OP rotates around the origin. Because ray OP is the terminal side of angle POB, the measure of the angle changes as P moves along the circle. The measure of how fast angle POB is changing is its **angular speed.** Angular speed, symbolized ω, is given as

$$\omega = \frac{\theta}{t}, \quad \textbf{where } \theta \textbf{ is in radians.}$$

Here θ is the measure of angle POB at time t. ***As with earlier formulas in this chapter, θ must be measured in radians, with ω expressed in radians per unit of time.***

Formulas for Angular and Linear Speed

Angular Speed ω	Linear Speed v
$\omega = \frac{\theta}{t}$	$v = \frac{s}{t}$
(ω in radians per unit time t, θ in radians)	$v = \frac{r\theta}{t}$
	$v = r\omega$

The length s of the arc intercepted on a circle of radius r by a central angle of measure θ radians is $s = r\theta$. Using this formula, the formula for linear speed, $v = \frac{s}{t}$, can be written in several useful forms.

$$v = \frac{s}{t} \qquad \text{Formula for linear speed}$$

$$v = \frac{r\theta}{t} \qquad s = r\theta$$

$$v = r \cdot \frac{\theta}{t} \qquad \frac{ab}{c} = a \cdot \frac{b}{c}$$

$$v = r\omega \qquad \omega = \frac{\theta}{t}$$

As an example of linear and angular speeds, consider the following. The human joint that can be flexed the fastest is the wrist, which can rotate through 90°, or $\frac{\pi}{2}$ radians, in 0.045 sec while holding a tennis racket. The angular speed of a human wrist swinging a tennis racket is

$$\omega = \frac{\theta}{t} \qquad \text{Formula for angular speed}$$

$$\omega = \frac{\frac{\pi}{2}}{0.045} \qquad \text{Let } \theta = \frac{\pi}{2} \text{ and } t = 0.045.$$

$$\omega \approx 35 \text{ radians per sec.} \qquad \text{Use a calculator.}$$

If the radius (distance) from the tip of the racket to the wrist joint is 2 ft, then the speed at the tip of the racket is

$v = r\omega$ Formula for linear speed

$v \approx 2(35)$ Let $r = 2$ and $\omega = 35$.

$v = 70$ ft per sec, or about 48 mph. Use a calculator.

In a tennis serve the arm rotates at the shoulder, so the final speed of the racket is considerably greater. (*Source:* Cooper, J. and R. Glassow, *Kinesiology,* Second Edition, C.V. Mosby.)

Classroom Example 1
Suppose that P is on a circle with radius 15 in., and ray OP is rotating with angular speed $\frac{\pi}{12}$ radian per sec.

(a) Find the angle generated by P in 10 sec.

(b) Find the distance traveled by P along the circle in 10 sec.

(c) Find the linear speed of P in inches per second.

Answers:

(a) $\frac{5\pi}{6}$ radians (b) $\frac{25\pi}{2}$ in.

(c) $\frac{5\pi}{4}$ in. per sec

Teaching Tip Mention that, while problems involving linear and angular speed can be solved using the formula for circumference, the formulas presented in this section are much more efficient.

EXAMPLE 1 Using Linear and Angular Speed Formulas

Suppose that point P is on a circle with radius 10 cm, and ray OP is rotating with angular speed $\frac{\pi}{18}$ radian per sec.

(a) Find the angle generated by P in 6 sec.

(b) Find the distance traveled by P along the circle in 6 sec.

(c) Find the linear speed of P in centimeters per second.

SOLUTION

(a) Solve for θ in the angular speed formula $\omega = \frac{\theta}{t}$, and substitute the known quantities $\omega = \frac{\pi}{18}$ radian per sec and $t = 6$ sec.

$\theta = \omega t$ Angular speed formula solved for θ

$\theta = \frac{\pi}{18}(6)$ Let $\omega = \frac{\pi}{18}$ and $t = 6$.

$\theta = \frac{\pi}{3}$ radians Multiply.

(b) To find the distance traveled by P, use the arc length formula $s = r\theta$ with $r = 10$ cm and, from part (a), $\theta = \frac{\pi}{3}$ radians.

$s = r\theta = 10\left(\frac{\pi}{3}\right) = \frac{10\pi}{3}$ cm Let $r = 10$ and $\theta = \frac{\pi}{3}$.

(c) Use the formula for linear speed with $r = 10$ cm and $\omega = \frac{\pi}{18}$ radians per sec.

$v = r\omega = 10\left(\frac{\pi}{18}\right) = \frac{5\pi}{9}$ cm per sec Linear speed formula

Now Try Exercise 7.

Classroom Example 2
Repeat **Example 2** for a belt that runs a pulley of radius 5 in. at 120 revolutions per min.

Answers:

(a) 4π radians per sec

(b) $20\pi \approx 63$ in. per sec

EXAMPLE 2 Finding Angular Speed of a Pulley and Linear Speed of a Belt

A belt runs a pulley of radius 6 cm at 80 revolutions per min. See **Figure 21.**

(a) Find the angular speed of the pulley in radians per second.

(b) Find the linear speed of the belt in centimeters per second.

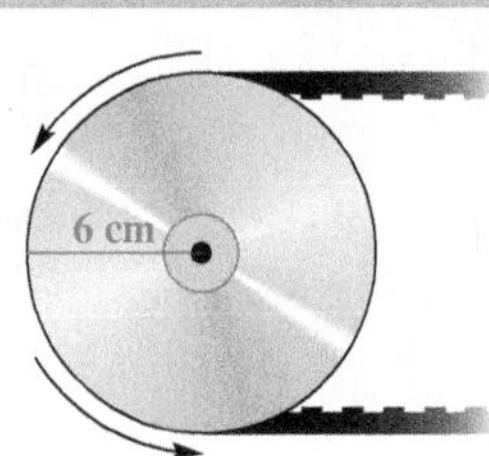

Figure 21

SOLUTION

(a) The angular speed 80 revolutions per min can be converted to radians per second using the following facts.

$$1 \text{ revolution} = 2\pi \text{ radians} \quad \text{and} \quad 1 \text{ min} = 60 \text{ sec}$$

We multiply by the corresponding **unit fractions.** Here, just as with the unit circle, the word *unit* means 1, so multiplying by a unit fraction is equivalent to multiplying by 1. We divide out common units in the same way that we divide out common factors.

$$\omega = \frac{80 \text{ revolutions}}{1 \text{ min}} \cdot \frac{2\pi \text{ radians}}{1 \text{ revolution}} \cdot \frac{1 \text{ min}}{60 \text{ sec}}$$

$$\omega = \frac{160\pi \text{ radians}}{60 \text{ sec}} \qquad \text{Multiply. Divide out common units.}$$

$$\omega = \frac{8\pi}{3} \text{ radians per sec} \qquad \text{Angular speed}$$

(b) The linear speed v of the belt will be the same as that of a point on the circumference of the pulley.

$$v = r\omega = 6\left(\frac{8\pi}{3}\right) = 16\pi \approx 50 \text{ cm per sec} \qquad \text{Linear speed}$$

✔ **Now Try Exercise 47.**

EXAMPLE 3 Finding Linear Speed and Distance Traveled by a Satellite

A satellite traveling in a circular orbit 1600 km above the surface of Earth takes 2 hr to make an orbit. The radius of Earth is approximately 6400 km. See **Figure 22.**

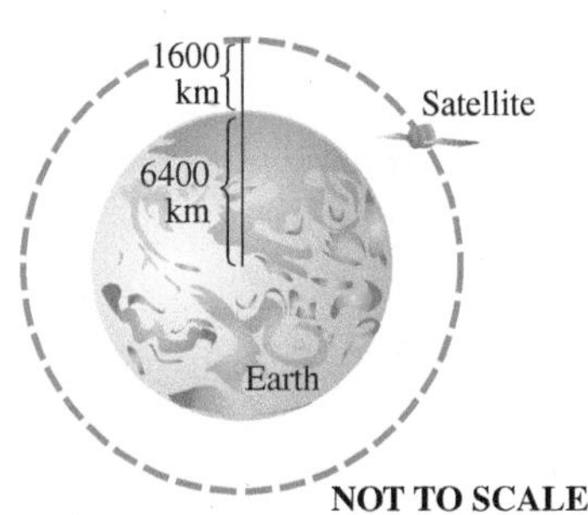

Figure 22

(a) Approximate the linear speed of the satellite in kilometers per hour.

(b) Approximate the distance the satellite travels in 4.5 hr.

SOLUTION

(a) The distance of the satellite from the center of Earth is approximately

$$r = 1600 + 6400 = 8000 \text{ km}.$$

The angular speed 1 orbit per 2 hr can be converted to radians per hour using the fact that 1 orbit $= 2\pi$ radians.

Unit fraction: $\frac{2\pi \text{ radians}}{1 \text{ orbit}} = 1$

$$\omega = \frac{1 \text{ orbit}}{2 \text{ hr}} \cdot \frac{2\pi \text{ radians}}{1 \text{ orbit}} = \pi \text{ radians per hr} \qquad \text{Angular speed}$$

We now use the formula for linear speed with $r = 8000$ km and $\omega = \pi$ radians per hr.

$$v = r\omega = 8000\pi \approx 25{,}000 \text{ km per hr} \qquad \text{Linear speed}$$

(b) To approximate the distance traveled by the satellite, we use $s = vt$.

This is similar to the distance formula $d = rt$.

$$s = vt \qquad \text{Formula for arc length}$$

$$s = 8000\pi(4.5) \qquad \text{Let } v = 8000\pi \text{ and } t = 4.5.$$

$$s \approx 110{,}000 \text{ km} \qquad \text{Multiply. Approximate to two significant digits.}$$

✔ **Now Try Exercise 45.**

Classroom Example 3
A satellite traveling in a circular orbit 1800 km above the surface of Earth takes 2.5 hr to make an orbit.
(a) Approximate the linear speed of the satellite in kilometers per hour.
(b) Approximate the distance the satellite travels in 3.5 hr.

Answers:
(a) $6560\pi \approx 21{,}000$ km per hr
(b) $22{,}960\pi \approx 72{,}000$ km

3.4 Exercises

1. linear speed (or linear velocity)
2. angular speed (or angular velocity)
3. 2π
4. π
5. 2π
6. 5π
7. (a) $\frac{\pi}{2}$ radians (b) 10π cm (c) $\frac{5\pi}{3}$ cm per sec
8. (a) $\frac{2\pi}{5}$ radians (b) 12π cm (c) 3π cm per sec
9. (a) 3π radians (b) 24π in. (c) $\frac{8\pi}{3}$ in. per min
10. (a) 40π radians (b) 480π ft (c) 96π ft per min
11. 2π radians
12. $\frac{5\pi}{4}$ radians
13. 7.4 radians
14. 6.9 radians
15. $\frac{3\pi}{32}$ radian per sec
16. $\frac{\pi}{25}$ radian per sec
17. 0.1803 radian per sec
18. 2.078 radians per sec
19. $\frac{6}{5}$ min
20. 9 min

CONCEPT PREVIEW *Fill in the blank to correctly complete each sentence. As necessary, refer to the figure that shows point P moving at a constant speed along the unit circle.*

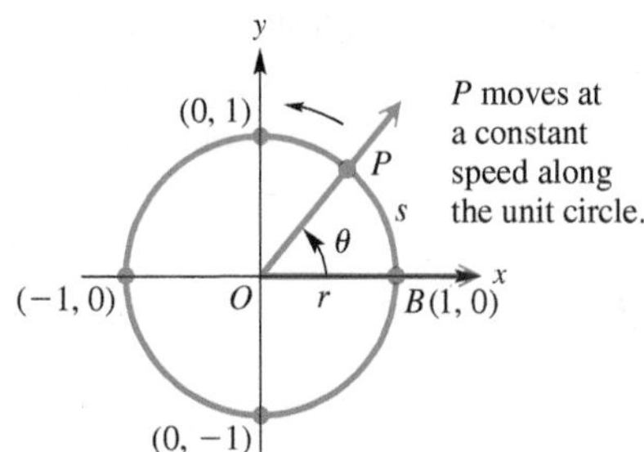

1. The measure of how fast the position of point P is changing is the ________.
2. The measure of how fast angle POB is changing is the ________.
3. If the angular speed of point P is 1 radian per sec, then P will move around the entire unit circle in ______ sec.
4. If the angular speed of point P is π radians per sec, then the linear speed is ______ unit(s) per sec.
5. An angular speed of 1 revolution per min on the unit circle is equivalent to an angular speed, ω, of ______ radians per min.
6. If P is rotating with angular speed $\frac{\pi}{2}$ radians per sec, then the distance traveled by P in 10 sec is ______ units.

Suppose that point P is on a circle with radius r, and ray OP is rotating with angular speed ω. Use the given values of r, ω, and t to do the following. ***See Example 1.***

(a) Find the angle generated by P in time t.

(b) Find the distance traveled by P along the circle in time t.

(c) Find the linear speed of P.

7. $r = 20$ cm, $\omega = \frac{\pi}{12}$ radian per sec, $t = 6$ sec
8. $r = 30$ cm, $\omega = \frac{\pi}{10}$ radian per sec, $t = 4$ sec
9. $r = 8$ in., $\omega = \frac{\pi}{3}$ radians per min, $t = 9$ min
10. $r = 12$ ft, $\omega = 8\pi$ radians per min, $t = 5$ min

Use the formula $\omega = \frac{\theta}{t}$ to find the value of the missing variable.

11. $\omega = \frac{2\pi}{3}$ radians per sec, $t = 3$ sec
12. $\omega = \frac{\pi}{4}$ radian per min, $t = 5$ min
13. $\omega = 0.91$ radian per min, $t = 8.1$ min
14. $\omega = 4.3$ radians per min, $t = 1.6$ min
15. $\theta = \frac{3\pi}{4}$ radians, $t = 8$ sec
16. $\theta = \frac{2\pi}{5}$ radians, $t = 10$ sec
17. $\theta = 3.871$ radians, $t = 21.47$ sec
18. $\theta = 5.225$ radians, $t = 2.515$ sec
19. $\theta = \frac{2\pi}{9}$ radian, $\omega = \frac{5\pi}{27}$ radian per min
20. $\theta = \frac{3\pi}{8}$ radians, $\omega = \frac{\pi}{24}$ radian per min

21. 8π m per sec
22. $\frac{72\pi}{5}$ cm per sec
23. $\frac{9}{5}$ radians per sec
24. 6 radians per sec
25. $\frac{8}{\pi}$ m
26. 66.85 cm
27. 18π cm
28. $\frac{216\pi}{5}$ yd
29. 12 sec
30. 4 sec
31. $\frac{3\pi}{32}$ radian per sec
32. $\frac{\pi}{18}$ radian per sec
33. $\frac{\pi}{6}$ radian per hr
34. $\frac{\pi}{30}$ radian per sec
35. $\frac{\pi}{30}$ radian per min
36. 600π radians per min
37. $\frac{7\pi}{30}$ cm per min
38. $\frac{14\pi}{15}$ mm per sec
39. 168π m per min
40. 1260π cm per min
41. 1500π m per min
42. $112{,}880\pi$ cm per min
43. 16.6 mph

Use the formula $v = r\omega$ to find the value of the missing variable.

21. $r = 12$ m, $\omega = \frac{2\pi}{3}$ radians per sec

22. $r = 8$ cm, $\omega = \frac{9\pi}{5}$ radians per sec

23. $v = 9$ m per sec, $r = 5$ m

24. $v = 18$ ft per sec, $r = 3$ ft

25. $v = 12$ m per sec, $\omega = \frac{3\pi}{2}$ radians per sec

26. $v = 24.93$ cm per sec, $\omega = 0.3729$ radian per sec

The formula $\omega = \frac{\theta}{t}$ can be rewritten as $\theta = \omega t$. Substituting ωt for θ converts $s = r\theta$ to $s = r\omega t$. Use the formula $s = r\omega t$ to find the value of the missing variable.

27. $r = 6$ cm, $\omega = \frac{\pi}{3}$ radians per sec, $t = 9$ sec

28. $r = 9$ yd, $\omega = \frac{2\pi}{5}$ radians per sec, $t = 12$ sec

29. $s = 6\pi$ cm, $r = 2$ cm, $\omega = \frac{\pi}{4}$ radian per sec

30. $s = \frac{12\pi}{5}$ m, $r = \frac{3}{2}$ m, $\omega = \frac{2\pi}{5}$ radians per sec

31. $s = \frac{3\pi}{4}$ km, $r = 2$ km, $t = 4$ sec

32. $s = \frac{8\pi}{9}$ m, $r = \frac{4}{3}$ m, $t = 12$ sec

Find the angular speed ω for each of the following.

33. the hour hand of a clock

34. the second hand of a clock

35. the minute hand of a clock

36. a gear revolving 300 times per min

Find the linear speed v for each of the following.

37. the tip of the minute hand of a clock, if the hand is 7 cm long

38. the tip of the second hand of a clock, if the hand is 28 mm long

39. a point on the edge of a flywheel of radius 2 m, rotating 42 times per min

40. a point on the tread of a tire of radius 18 cm, rotating 35 times per min

41. the tip of a propeller 3 m long, rotating 500 times per min (*Hint*: $r = 1.5$ m)

42. a point on the edge of a gyroscope of radius 83 cm, rotating 680 times per min

Solve each problem. ***See Examples 1–3.***

43. ***Speed of a Bicycle*** The tires of a bicycle have radius 13.0 in. and are turning at the rate of 215 revolutions per min. See the figure. How fast is the bicycle traveling in miles per hour? (*Hint:* 5280 ft = 1 mi)

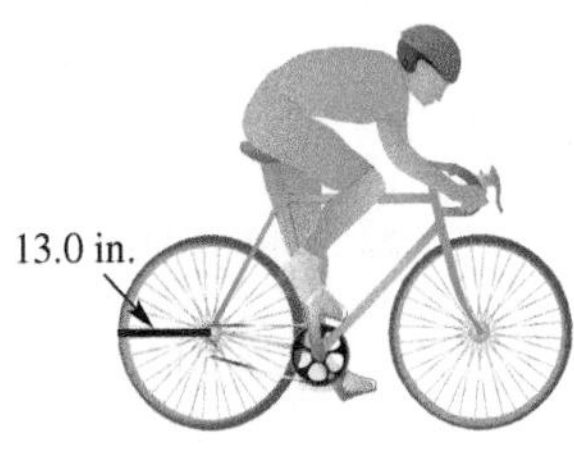

44. 24.62 hr
45. (a) $\frac{2\pi}{365}$ radian
(b) $\frac{\pi}{4380}$ radian per hr
(c) 67,000 mph
46. (a) $\frac{\pi}{12}$ radian per hr
(b) 0
(c) 1700 km per hr
(d) 1200 km per hr
47. (a) 3.1 cm per sec
(b) 0.24 radian per sec
48. larger pulley: $\frac{25\pi}{18}$ radians per sec; smaller pulley: $\frac{125\pi}{48}$ radians per sec
49. 3.73 cm 50. 29 sec
51. 523.6 radians per sec
52. 125 ft per sec

44. ***Hours in a Martian Day*** Mars rotates on its axis at the rate of about 0.2552 radian per hr. Approximately how many hours are in a Martian day (or *sol*)? (*Source: World Almanac and Book of Facts.*)

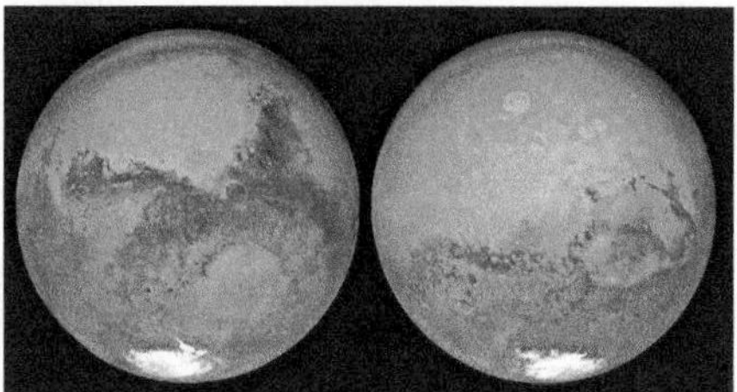

Opposite sides of Mars

45. ***Angular and Linear Speeds of Earth*** The orbit of Earth about the sun is almost circular. Assume that the orbit is a circle with radius 93,000,000 mi. Its angular and linear speeds are used in designing solar-power facilities.

93,000,000 mi
Earth
θ
Sun
NOT TO SCALE

(a) Assume that a year is 365 days, and find the angle formed by Earth's movement in one day.

(b) Give the angular speed in radians per hour.

(c) Find the approximate linear speed of Earth in miles per hour.

46. ***Angular and Linear Speeds of Earth*** Earth revolves on its axis once every 24 hr. Assuming that Earth's radius is 6400 km, find the following.

(a) angular speed of Earth in radians per hour

(b) linear speed at the North Pole or South Pole

(c) approximate linear speed at Quito, Ecuador, a city on the equator

(d) approximate linear speed at Salem, Oregon (halfway from the equator to the North Pole)

47. ***Speeds of a Pulley and a Belt*** The pulley shown has a radius of 12.96 cm. Suppose it takes 18 sec for 56 cm of belt to go around the pulley.

12.96 cm

(a) Find the linear speed of the belt in centimeters per second.

(b) Find the angular speed of the pulley in radians per second.

48. ***Angular Speeds of Pulleys*** The two pulleys in the figure have radii of 15 cm and 8 cm, respectively. The larger pulley rotates 25 times in 36 sec. Find the angular speed of each pulley in radians per second.

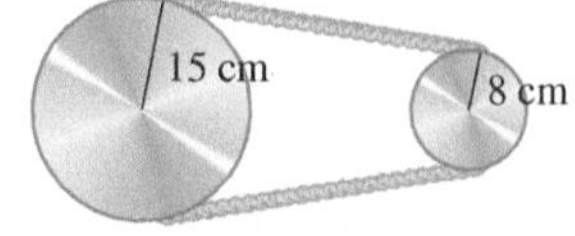

49. ***Radius of a Spool of Thread*** A thread is being pulled off a spool at the rate of 59.4 cm per sec. Find the radius of the spool if it makes 152 revolutions per min.

50. ***Time to Move along a Railroad Track*** A railroad track is laid along the arc of a circle of radius 1800 ft. The circular part of the track subtends a central angle of 40°. How long (in seconds) will it take a point on the front of a train traveling 30.0 mph to go around this portion of the track?

51. ***Angular Speed of a Motor Propeller*** The propeller of a 90-horsepower outboard motor at full throttle rotates at exactly 5000 revolutions per min. Find the angular speed of the propeller in radians per second.

52. ***Linear Speed of a Golf Club*** The shoulder joint can rotate at 25.0 radians per sec. If a golfer's arm is straight and the distance from the shoulder to the club head is 5.00 ft, find the linear speed of the club head from shoulder rotation. (*Source*: Cooper, J. and R. Glassow, *Kinesiology*, Second Edition, C.V. Mosby.)

Chapter 3 Test Prep

Key Terms

3.1 radian circumference **3.2** latitude sector of a circle subtend	degree of curvature nautical mile statute mile longitude	**3.3** unit circle circular functions reference arc	**3.4** linear speed v angular speed ω unit fraction

Quick Review

Concepts	Examples

3.1 Radian Measure

An angle with its vertex at the center of a circle that intercepts an arc on the circle equal in length to the radius of the circle has a measure of **1 radian.**

$$180° = \pi \text{ radians} \quad \text{Degree/Radian Relationship}$$

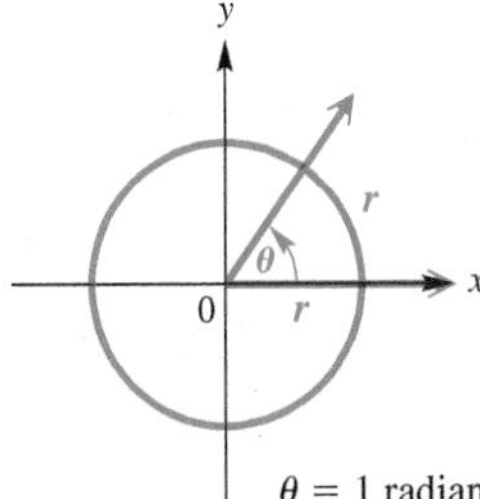

Converting between Degrees and Radians

- Multiply a degree measure by $\frac{\pi}{180}$ radian and simplify to convert to radians.
- Multiply a radian measure by $\frac{180°}{\pi}$ and simplify to convert to degrees.

Convert 135° to radians.

$$135° = 135\left(\frac{\pi}{180}\text{ radian}\right) = \frac{3\pi}{4}\text{ radians}$$

Convert $-\frac{5\pi}{3}$ radians to degrees.

$$-\frac{5\pi}{3}\text{ radians} = -\frac{5\pi}{3}\left(\frac{180°}{\pi}\right) = -300°$$

3.2 Applications of Radian Measure

Arc Length

The length s of the arc intercepted on a circle of radius r by a central angle of measure θ radians is given by the product of the radius and the radian measure of the angle.

$$s = r\theta, \quad \text{where } \theta \text{ is in radians}$$

Find the central angle θ in the figure.

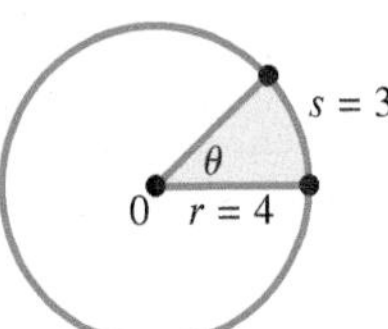

$$\theta = \frac{s}{r} = \frac{3}{4}\text{ radian}$$

Area of a Sector

The area $\mathcal{A}$ of a sector of a circle of radius r and central angle θ is given by the following formula.

$$\mathcal{A} = \frac{1}{2}r^2\theta, \quad \text{where } \theta \text{ is in radians}$$

Find the area $\mathcal{A}$ of the sector in the figure above.

$$\mathcal{A} = \frac{1}{2}(4)^2\left(\frac{3}{4}\right) = 6 \text{ sq units}$$

Concepts	Examples

3.3 The Unit Circle and Circular Functions

Circular Functions

Start at the point $(1, 0)$ on the unit circle $x^2 + y^2 = 1$ and measure off an arc of length $|s|$ along the circle, moving counterclockwise if s is positive and clockwise if s is negative. Let the endpoint of the arc be at the point (x, y). The six circular functions of s are defined as follows. (Assume that no denominators are 0.)

$$\sin s = y \qquad \cos s = x \qquad \tan s = \frac{y}{x}$$

$$\csc s = \frac{1}{y} \qquad \sec s = \frac{1}{x} \qquad \cot s = \frac{x}{y}$$

The Unit Circle

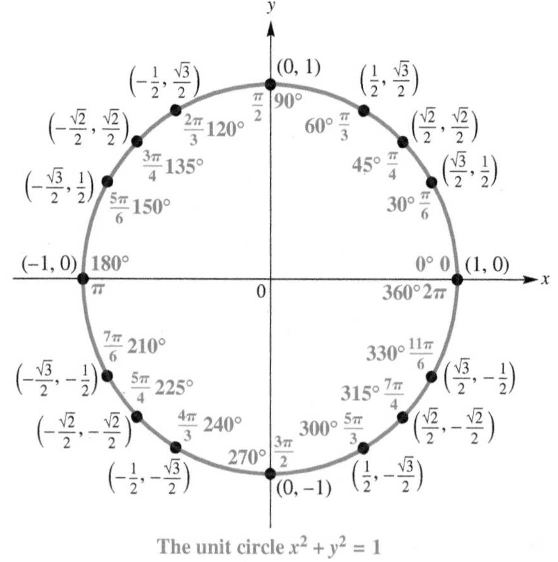

The unit circle $x^2 + y^2 = 1$

Use the unit circle to find each value.

$$\sin \frac{5\pi}{6} = \frac{1}{2}$$

$$\cos \frac{3\pi}{2} = 0$$

$$\tan \frac{\pi}{4} = \frac{\frac{\sqrt{2}}{2}}{\frac{\sqrt{2}}{2}} = 1$$

$$\csc \frac{7\pi}{4} = \frac{1}{-\frac{\sqrt{2}}{2}} = -\sqrt{2}$$

$$\sec \frac{7\pi}{6} = \frac{1}{-\frac{\sqrt{3}}{2}} = -\frac{2\sqrt{3}}{3}$$

$$\cot \frac{\pi}{3} = \frac{\frac{1}{2}}{\frac{\sqrt{3}}{2}} = \frac{\sqrt{3}}{3}$$

$$\sin 0 = 0$$

$$\cos \frac{\pi}{2} = 0$$

Find the exact value of s in $\left[0, \frac{\pi}{2}\right]$ if $\cos s = \frac{\sqrt{3}}{2}$.

In $\left[0, \frac{\pi}{2}\right]$, the arc length $s = \frac{\pi}{6}$ is associated with the point $\left(\frac{\sqrt{3}}{2}, \frac{1}{2}\right)$. The first coordinate is

$$\cos s = \cos \frac{\pi}{6} = \frac{\sqrt{3}}{2}.$$

Thus we have $s = \frac{\pi}{6}$.

3.4 Linear and Angular Speed

Formulas for Angular and Linear Speed

Angular Speed ω	Linear Speed v
$\omega = \frac{\theta}{t}$	$v = \frac{s}{t}$
(ω in radians per unit time t, θ in radians)	$v = \frac{r\theta}{t}$
	$v = r\omega$

A belt runs a machine pulley of radius 8 in. at 60 revolutions per min.

(a) Find the angular speed ω in radians per minute.

$$\omega = \frac{60 \text{ revolutions}}{1 \text{ min}} \cdot \frac{2\pi \text{ radians}}{1 \text{ revolution}}$$

$$\omega = 120\pi \text{ radians per min}$$

(b) Find the linear speed v in inches per minute.

$$v = r\omega$$

$$v = 8(120\pi)$$

$$v = 960\pi \text{ in. per min}$$

Chapter 3 Review Exercises

1. A central angle of a circle that intercepts an arc of length 2 times the radius of the circle has a measure of 2 radians.
2. (a) II (b) III (c) III (d) I
3. Three of many possible answers are $1 + 2\pi$, $1 + 4\pi$, and $1 + 6\pi$.
4. $\frac{\pi}{6} + 2n\pi$

5. $\frac{\pi}{4}$ 6. $\frac{2\pi}{3}$
7. $\frac{35\pi}{36}$ 8. $\frac{11\pi}{6}$
9. $\frac{40\pi}{9}$ 10. $\frac{17\pi}{3}$

11. 225° 12. 162°
13. 480° 14. 216°
15. −110° 16. −756°
17. π in. 18. $\frac{4\pi}{3}$ in.
19. 12π in. 20. 32π in.
21. 35.8 cm 22. 8.77 cm
23. 49.06° 24. $\frac{\pi}{3}$ radians
25. 273 m^2 26. 2263 in.^{2}
27. 2156 mi 28. $\mathcal{A} = \frac{s^2}{2\theta}$
29. (a) $\frac{\pi}{3}$ radians (b) 2π in.

Concept Check *Work each problem.*

1. What is the meaning of "an angle with measure 2 radians"?
2. Consider each angle in standard position having the given radian measure. In what quadrant does the terminal side lie?
   (a) 3 (b) 4 (c) −2 (d) 7
3. Find three angles coterminal with an angle of 1 radian.
4. Give an expression that generates all angles coterminal with an angle of $\frac{\pi}{6}$ radian. Let n represent any integer.

Convert each degree measure to radians. Leave answers as multiples of π.

5. 45° 6. 120° 7. 175° 8. 330° 9. 800° 10. 1020°

Convert each radian measure to degrees.

11. $\frac{5\pi}{4}$ 12. $\frac{9\pi}{10}$ 13. $\frac{8\pi}{3}$ 14. $\frac{6\pi}{5}$ 15. $-\frac{11\pi}{18}$ 16. $-\frac{21\pi}{5}$

Suppose the tip of the minute hand of a clock is 2 in. *from the center of the clock. For each duration, determine the distance traveled by the tip of the minute hand. Leave answers as multiples of π.*

17. 15 min 18. 20 min 19. 3 hr 20. 8 hr

Solve each problem. Use a calculator as necessary.

21. ***Arc Length*** The radius of a circle is 15.2 cm. Find the length of an arc of the circle intercepted by a central angle of $\frac{3\pi}{4}$ radians.
22. ***Arc Length*** Find the length of an arc intercepted by a central angle of 0.769 radian on a circle with radius 11.4 cm.
23. ***Angle Measure*** Find the measure (in degrees) of a central angle that intercepts an arc of length 7.683 cm in a circle of radius 8.973 cm.
24. ***Angle Measure*** Find the measure (in radians) of a central angle whose sector has area $\frac{50\pi}{3}$ cm^2 in a circle of radius 10 cm.
25. ***Area of a Sector*** Find the area of a sector of a circle having a central angle of 21° 40′ in a circle of radius 38.0 m.
26. ***Area of a Sector*** A central angle of $\frac{7\pi}{4}$ radians forms a sector of a circle. Find the area of the sector if the radius of the circle is 28.69 in.
27. ***Diameter of the Moon*** The distance to the moon is approximately 238,900 mi. Use the arc length formula to estimate the diameter d of the moon if angle θ in the figure is measured to be 0.5170°.

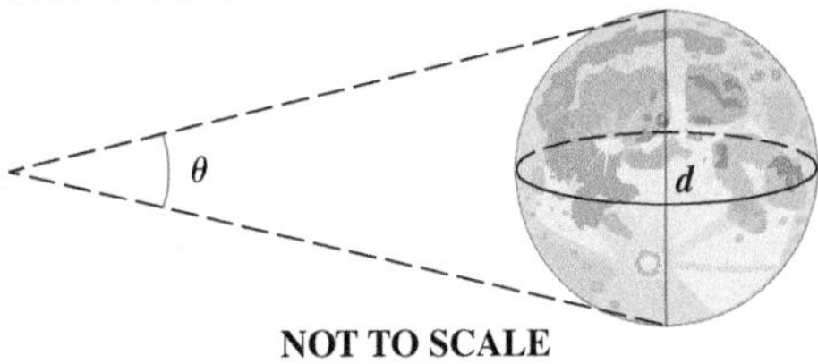

28. ***Concept Check*** Using $s = r\theta$ and $\mathcal{A} = \frac{1}{2}r^2\theta$, express $\mathcal{A}$ in terms of s and θ.
29. ***Concept Check*** The hour hand of a wall clock measures 6 in. from its tip to the center of the clock.
   (a) Through what angle (in radians) does the hour hand pass between 1 o'clock and 3 o'clock?
   (b) What distance does the tip of the hour hand travel during the time period from 1 o'clock to 3 o'clock?

30. Because $\theta_1 = \frac{s}{r}$, if the radius were doubled, then the angle would be halved—that is,

$$\theta_2 = \frac{s}{2r} = \frac{1}{2} \cdot \frac{s}{r} = \frac{1}{2}\theta_1.$$

31. 4500 km **32.** 12,000 km

33. $\frac{3}{4}$; 1.5 sq units

34. $\frac{1}{2}$; 16 sq units

35. $\sqrt{3}$ **36.** $-\frac{1}{2}$

37. $-\frac{1}{2}$ **38.** $-\sqrt{3}$

39. 2 **40.** undefined

41. tan 1 **42.** tan 1

43. sin 2

44. (a) A (b) C (c) B

45. 0.8660 **46.** 2.7976

47. 0.9703 **48.** −11.4266

49. 1.9513 **50.** −1.0080

51. 0.3898 **52.** 1.3265

53. 0.5148 **54.** 0.9424

55. 1.1054 **56.** 1.3497

57. $\frac{\pi}{4}$ **58.** $\frac{2\pi}{3}$

59. $\frac{7\pi}{6}$ **60.** $\frac{11\pi}{6}$

30. ***Concept Check*** What would happen to the central angle for a given arc length of a circle if the circle's radius were doubled? (Assume everything else is unchanged.)

Distance between Cities *Assume that the radius of Earth is* 6400 km.

31. Find the distance in kilometers between cities on a north-south line that are on latitudes 28° N and 12° S, respectively.

32. Two cities on the equator have longitudes of 72° E and 35° W, respectively. Find the distance between the cities.

Concept Check *Find the measure of each central angle θ (in radians) and the area of each sector.*

33. 

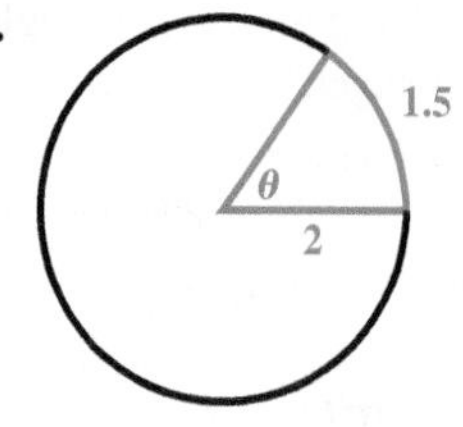

34.

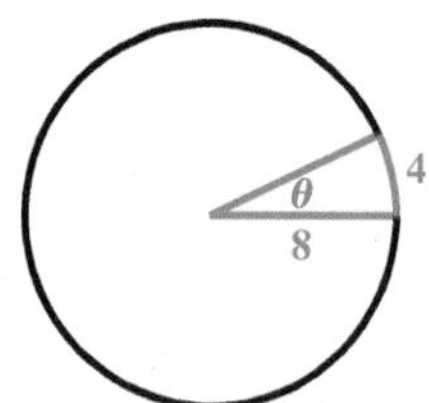

Find each exact function value.

35. $\tan \frac{\pi}{3}$ **36.** $\cos \frac{2\pi}{3}$ **37.** $\sin\left(-\frac{5\pi}{6}\right)$

38. $\tan\left(-\frac{7\pi}{3}\right)$ **39.** $\csc\left(-\frac{11\pi}{6}\right)$ **40.** $\cot(-13\pi)$

Concept Check *Without using a calculator, determine which of the two values is greater.*

41. tan 1 or tan 2 **42.** sin 1 or tan 1 **43.** cos 2 or sin 2

44. ***Concept Check*** Match each domain in Column II with the appropriate circular function pair in Column I.

I	II
(a) sine and cosine	**A.** $(-\infty, \infty)$
(b) tangent and secant	**B.** $\{s \mid s \neq n\pi$, where n is any integer$\}$
(c) cotangent and cosecant	**C.** $\{s \mid s \neq (2n+1)\frac{\pi}{2}$, where n is any integer$\}$

Find a calculator approximation to four decimal places for each circular function value.

45. sin 1.0472 **46.** tan 1.2275 **47.** cos(−0.2443)

48. cot 3.0543 **49.** sec 7.3159 **50.** csc 4.8386

Find the approximate value of s, to four decimal places, in the interval $\left[0, \frac{\pi}{2}\right]$ *that makes each statement true.*

51. $\cos s = 0.9250$ **52.** $\tan s = 4.0112$ **53.** $\sin s = 0.4924$

54. $\csc s = 1.2361$ **55.** $\cot s = 0.5022$ **56.** $\sec s = 4.5600$

Find the exact value of s in the given interval that has the given circular function value.

57. $\left[0, \frac{\pi}{2}\right]$; $\cos s = \frac{\sqrt{2}}{2}$ **58.** $\left[\frac{\pi}{2}, \pi\right]$; $\tan s = -\sqrt{3}$

59. $\left[\pi, \frac{3\pi}{2}\right]$; $\sec s = -\frac{2\sqrt{3}}{3}$ **60.** $\left[\frac{3\pi}{2}, 2\pi\right]$; $\sin s = -\frac{1}{2}$

61. (a) 20π radians
(b) 300π cm
(c) 10π cm per sec

62. (a) $\frac{\pi}{3}$ radians (b) 15π ft
(c) $\frac{5\pi}{4}$ ft per min

63. 1260π cm per sec

64. (a) 47 ft
(b) $\frac{\pi}{36}$ radian per sec

65. 5 in.

66. (a) 0; The face of the moon is not visible.
(b) $\frac{1}{2}$; Half the face of the moon is visible.
(c) 1; The face of the moon is completely visible.
(d) $\frac{1}{2}$; Half the face of the moon is visible.

Suppose that point P is on a circle with radius r, and ray OP is rotating with angular speed ω. Use the given values of r, ω, and t to do the following.

(a) Find the angle generated by P in time t.

(b) Find the distance traveled by P along the circle in time t.

(c) Find the linear speed of P.

61. $r = 15$ cm, $\omega = \frac{2\pi}{3}$ radians per sec, $t = 30$ sec

62. $r = 45$ ft, $\omega = \frac{\pi}{36}$ radian per min, $t = 12$ min

Solve each problem.

63. ***Linear Speed of a Flywheel*** Find the linear speed of a point on the edge of a flywheel of radius 7 cm if the flywheel is rotating 90 times per sec.

64. ***Angular Speed of a Ferris Wheel*** A Ferris wheel has radius 25 ft. A person takes a seat, and then the wheel turns $\frac{5\pi}{6}$ radians.

(a) How far is the person above the ground to the nearest foot?

(b) If it takes 30 sec for the wheel to turn $\frac{5\pi}{6}$ radians, what is the angular speed of the wheel?

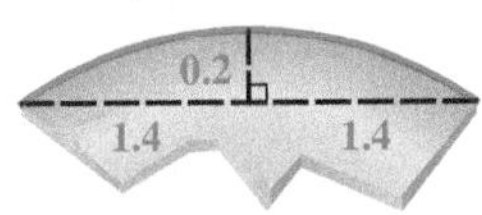

Figure A

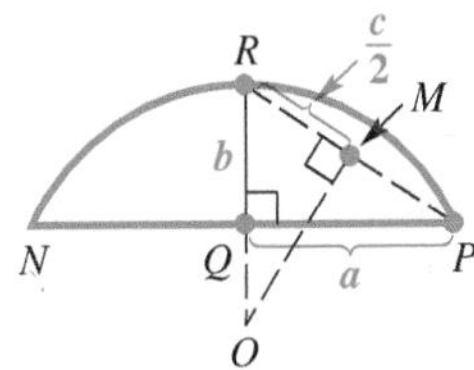

Figure B

65. ***(Modeling) Archaeology*** An archaeology professor believes that an unearthed fragment is a piece of the edge of a circular ceremonial plate and uses a formula that will give the radius of the original plate using measurements from the fragment, shown in **Figure A.** Measurements are in inches.

In **Figure B,** a is $\frac{1}{2}$ the length of chord NP, and b is the distance from the midpoint of chord NP to the circle. According to the formula, the radius r of the circle, OR, is given by

$$r = \frac{a^2 + b^2}{2b}.$$

What is the radius of the original plate from which the fragment came?

66. ***(Modeling) Phase Angle of the Moon*** Because the moon orbits Earth, we observe different phases of the moon during the period of a month. In the figure, t is the **phase angle.**

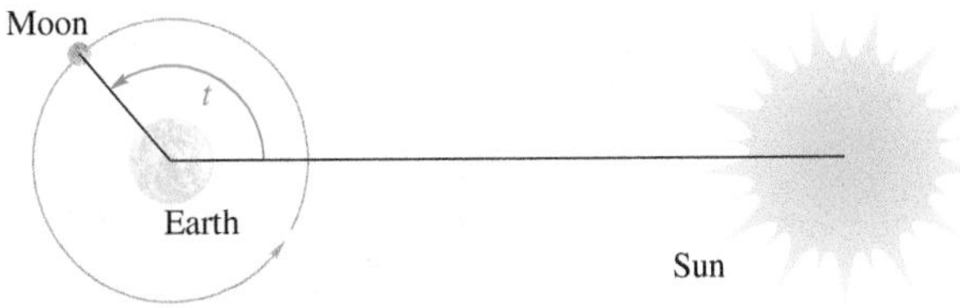

The **phase** F of the moon is modeled by

$$F(t) = \frac{1}{2}(1 - \cos t)$$

and gives the fraction of the moon's face that is illuminated by the sun. (*Source:* Duffet-Smith, P., *Practical Astronomy with Your Calculator,* Cambridge University Press.) Evaluate each expression and interpret the result.

(a) $F(0)$ **(b)** $F\left(\frac{\pi}{2}\right)$ **(c)** $F(\pi)$ **(d)** $F\left(\frac{3\pi}{2}\right)$

Chapter 3 Test

Convert each degree measure to radians.

1. 120° **2.** −45° **3.** 5° (to the nearest thousandth)

Convert each radian measure to degrees.

4. $\frac{3\pi}{4}$ **5.** $-\frac{7\pi}{6}$ **6.** 4 (to the nearest minute)

7. A central angle of a circle with radius 150 cm intercepts an arc of 200 cm. Find each measure.

(a) the radian measure of the angle **(b)** the area of a sector with that central angle

8. ***Rotation of Gas Gauge Arrow*** The arrow on a car's gasoline gauge is $\frac{1}{2}$ in. long. See the figure. Through what angle does the arrow rotate when it moves 1 in. on the gauge?

Find each exact function value.

9. $\sin \frac{3\pi}{4}$ **10.** $\cos\left(-\frac{7\pi}{6}\right)$ **11.** $\tan \frac{3\pi}{2}$

12. $\sec \frac{8\pi}{3}$ **13.** $\tan \pi$ **14.** $\cos \frac{3\pi}{2}$

15. Determine the six exact circular function values of s in the figure.

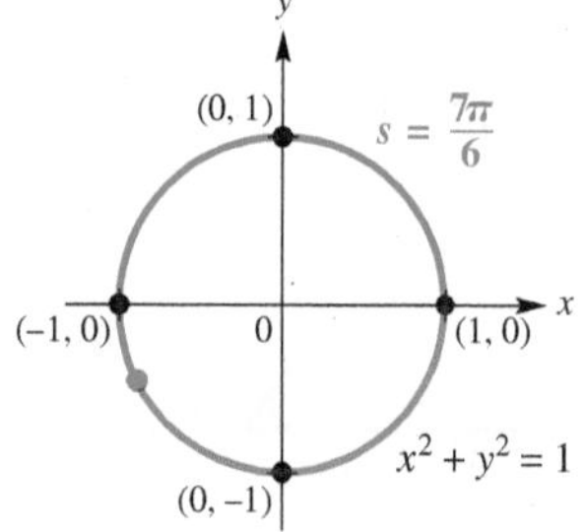

16. Give the domains of the six circular functions.

17. **(a)** Use a calculator to approximate s in the interval $\left[0, \frac{\pi}{2}\right]$ if $\sin s = 0.8258$.

(b) Find the exact value of s in the interval $\left[0, \frac{\pi}{2}\right]$ if $\cos s = \frac{1}{2}$.

18. ***Angular and Linear Speed of a Point*** Suppose that point P is on a circle with radius 60 cm, and ray OP is rotating with angular speed $\frac{\pi}{12}$ radian per sec.

(a) Find the angle generated by P in 8 sec.

(b) Find the distance traveled by P along the circle in 8 sec.

(c) Find the linear speed of P.

19. ***Orbital Speed of Jupiter*** It takes Jupiter 11.86 yr to complete one orbit around the sun. See the figure. If Jupiter's average distance from the sun is 483,800,000 mi, find its orbital speed (speed along its orbital path) in miles per second. (*Source: World Almanac and Book of Facts.*)

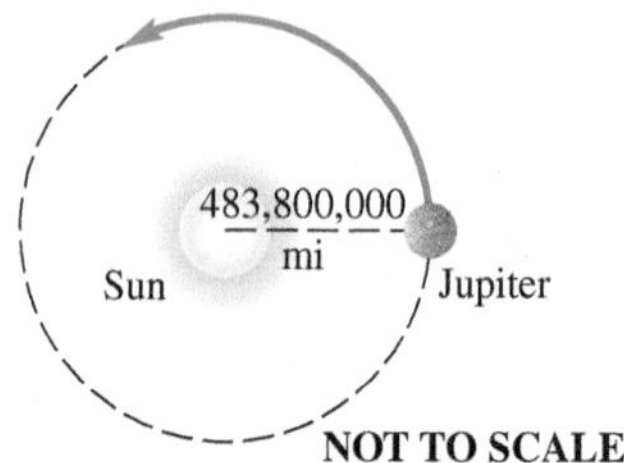

20. ***Ferris Wheel*** A Ferris wheel has radius 50.0 ft. A person takes a seat, and then the wheel turns $\frac{2\pi}{3}$ radians.

(a) How far is the person above the ground?

(b) If it takes 30 sec for the wheel to turn $\frac{2\pi}{3}$ radians, what is the angular speed of the wheel?

[3.1]

1. $\frac{2\pi}{3}$ 2. $-\frac{\pi}{4}$
3. 0.087 4. 135°
5. −210° 6. 229° 11′

[3.2]

7. (a) $\frac{4}{3}$ (b) 15,000 cm^2
8. 2 radians

[3.3]

9. $\frac{\sqrt{2}}{2}$ 10. $-\frac{\sqrt{3}}{2}$
11. undefined 12. −2
13. 0 14. 0
15. $\sin \frac{7\pi}{6} = -\frac{1}{2}$; $\cos \frac{7\pi}{6} = -\frac{\sqrt{3}}{2}$; $\tan \frac{7\pi}{6} = \frac{\sqrt{3}}{3}$; $\csc \frac{7\pi}{6} = -2$; $\sec \frac{7\pi}{6} = -\frac{2\sqrt{3}}{3}$; $\cot \frac{7\pi}{6} = \sqrt{3}$
16. sine and cosine: $(-\infty, \infty)$; tangent and secant: $\{s \mid s \neq (2n+1)\frac{\pi}{2}$, where n is any integer$\}$; cotangent and cosecant: $\{s \mid s \neq n\pi$, where n is any integer$\}$
17. (a) 0.9716 (b) $\frac{\pi}{3}$

[3.4]

18. (a) $\frac{2\pi}{3}$ radians (b) 40π cm (c) 5π cm per sec
19. 8.127 mi per sec
20. (a) 75 ft (b) $\frac{\pi}{45}$ radian per sec

4 Graphs of the Circular Functions

Phenomena that repeat in a regular pattern, such as average monthly temperature, fractional part of the moon's illumination, and high and low tides, can be modeled by *periodic functions*.

4.1 Graphs of the Sine and Cosine Functions

- Periodic Functions
- Graph of the Sine Function
- Graph of the Cosine Function
- Techniques for Graphing, Amplitude, and Period
- Connecting Graphs with Equations
- A Trigonometric Model

Periodic Functions Phenomena that repeat with a predictable pattern, such as tides, phases of the moon, and hours of daylight, can be modeled by sine and cosine functions. These functions are *periodic*. The periodic graph in **Figure 1** represents a normal heartbeat.

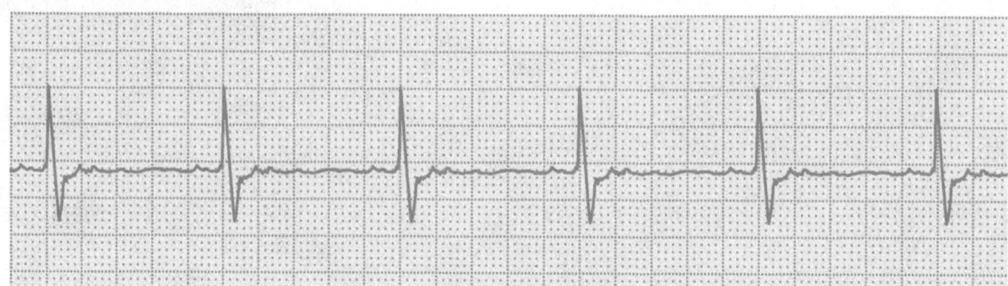

Figure 1

Periodic functions are defined as follows.

LOOKING AHEAD TO CALCULUS

Periodic functions are used throughout calculus, so it is important to know their characteristics. One use of these functions is to describe the location of a point in the plane using **polar coordinates,** an alternative to rectangular coordinates.

Periodic Function

A **periodic function** is a function f such that

$$f(x) = f(x + np),$$

for every real number x in the domain of f, every integer n, and some positive real number p. The least possible positive value of p is the **period** of the function.

The circumference of the unit circle is 2π, so the least value of p for which the sine and cosine functions repeat is 2π. ***Therefore, the sine and cosine functions are periodic functions with period 2π.*** For every positive integer n,

$$\sin x = \sin(x + n \cdot 2\pi) \quad \text{and} \quad \cos x = \cos(x + n \cdot 2\pi).$$

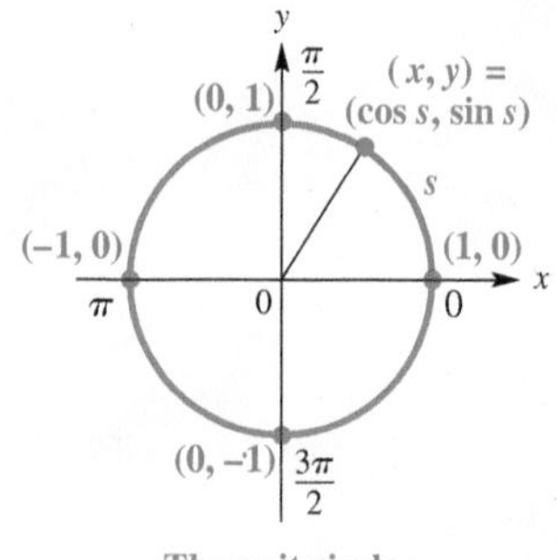

Figure 2

Graph of the Sine Function We have seen that for a real number s, the point on the unit circle corresponding to s has coordinates $(\cos s, \sin s)$. See **Figure 2.** Trace along the circle to verify the results shown in the table.

As s Increases from	$\sin s$	$\cos s$
0 to $\frac{\pi}{2}$	Increases from 0 to 1	Decreases from 1 to 0
$\frac{\pi}{2}$ to π	Decreases from 1 to 0	Decreases from 0 to -1
π to $\frac{3\pi}{2}$	Decreases from 0 to -1	Increases from -1 to 0
$\frac{3\pi}{2}$ to 2π	Increases from -1 to 0	Increases from 0 to 1

To avoid confusion when graphing the sine function, we use x rather than s. This corresponds to the letters in the xy-coordinate system. Selecting key values of x and finding the corresponding values of $\sin x$ leads to the table in **Figure 3.**

Teaching Tip In preparation for the material presented in this chapter, have students review **Appendix C** on Functions and **Appendix D** on Graphing Techniques.

To obtain the traditional graph in **Figure 3,** we plot the points from the table, use symmetry, and join them with a smooth curve. Because $y = \sin x$ is periodic with period 2π and has domain $(-\infty, \infty)$, the graph continues in the same pattern in both directions. This graph is a **sine wave,** or **sinusoid.**

Sine Function $f(x) = \sin x$

Domain: $(-\infty, \infty)$ Range: $[-1, 1]$

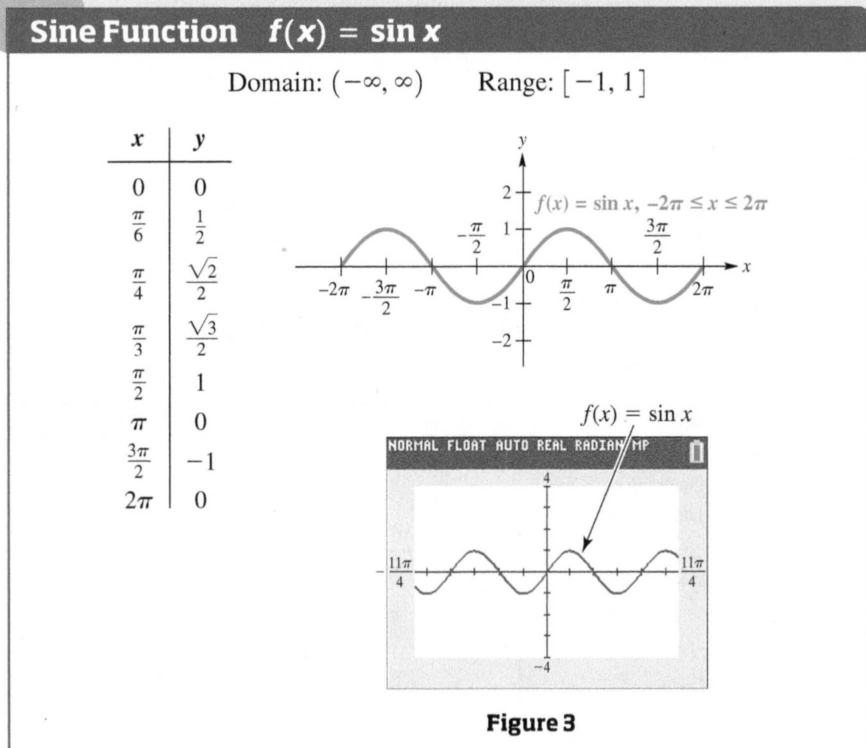

x	y
0	0
$\frac{\pi}{6}$	$\frac{1}{2}$
$\frac{\pi}{4}$	$\frac{\sqrt{2}}{2}$
$\frac{\pi}{3}$	$\frac{\sqrt{3}}{2}$
$\frac{\pi}{2}$	1
π	0
$\frac{3\pi}{2}$	-1
2π	0

Figure 3

- The graph is continuous over its entire domain, $(-\infty, \infty)$.
- Its x-intercepts have x-values of the form $n\pi$, where n is an integer.
- Its period is 2π.
- The graph is symmetric with respect to the origin, so the function is an odd function. For all x in the domain, $\sin(-x) = -\sin x$.

NOTE A function f is an **odd function** if for all x in the domain of f,

$$f(-x) = -f(x).$$

The graph of an odd function is symmetric with respect to the origin. This means that if (x, y) belongs to the function, then $(-x, -y)$ also belongs to the function. For example, $\left(\frac{\pi}{2}, 1\right)$ and $\left(-\frac{\pi}{2}, -1\right)$ are points on the graph of $y = \sin x$, illustrating the property $\sin(-x) = -\sin x$.

The sine function is related to the unit circle. **Its domain consists of real numbers corresponding to angle measures (or arc lengths) on the unit circle. Its range corresponds to y-coordinates (or sine values) on the unit circle.**

Consider the unit circle in **Figure 2** and assume that the line from the origin to some point on the circle is part of the pedal of a bicycle, with a foot placed on the circle itself. As the pedal is rotated from 0 radians on the horizontal axis through various angles, the angle (or arc length) giving the pedal's location and its corresponding height from the horizontal axis given by $\sin x$ are used to create points on the sine graph. See **Figure 4** on the next page.

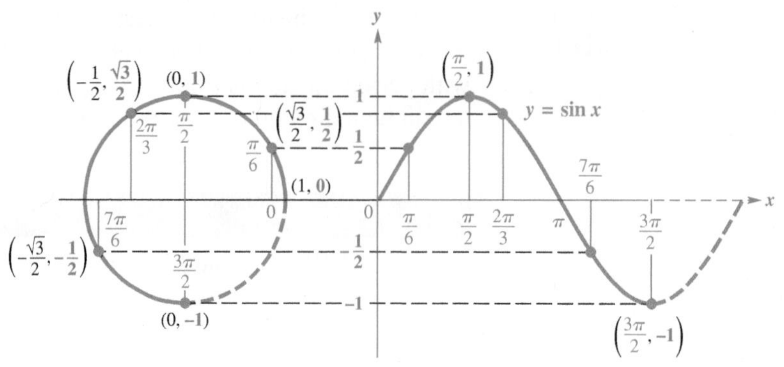

Figure 4

> **LOOKING AHEAD TO CALCULUS**
> The discussion of the derivative of a function in calculus shows that for the sine function, the slope of the tangent line at any point x is given by $\cos x$. For example, look at the graph of $y = \sin x$ and notice that a tangent line at $x = \pm\frac{\pi}{2}, \pm\frac{3\pi}{2}, \pm\frac{5\pi}{2}, \ldots$ will be horizontal and thus have slope 0. Now look at the graph of $y = \cos x$ and see that for these values, $\cos x = 0$.

Graph of the Cosine Function ***The graph of $y = \cos x$ in* Figure 5 *is the graph of the sine function shifted, or translated, $\frac{\pi}{2}$ units to the left.***

Cosine Function $f(x) = \cos x$

Domain: $(-\infty, \infty)$ Range: $[-1, 1]$

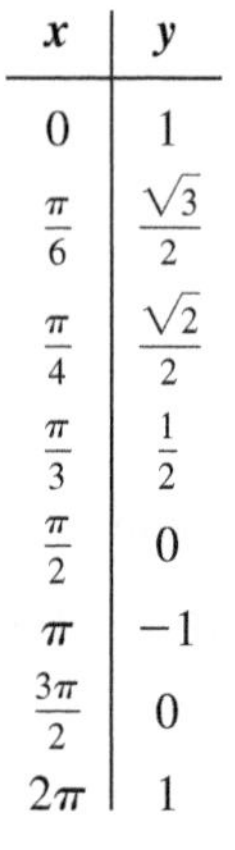

x	y
0	1
$\frac{\pi}{6}$	$\frac{\sqrt{3}}{2}$
$\frac{\pi}{4}$	$\frac{\sqrt{2}}{2}$
$\frac{\pi}{3}$	$\frac{1}{2}$
$\frac{\pi}{2}$	0
π	-1
$\frac{3\pi}{2}$	0
2π	1

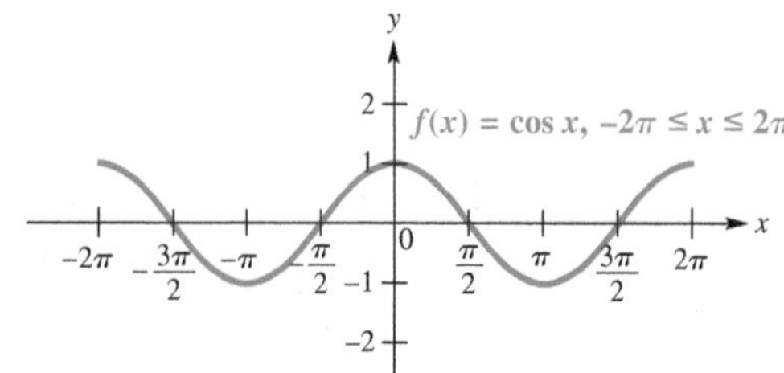

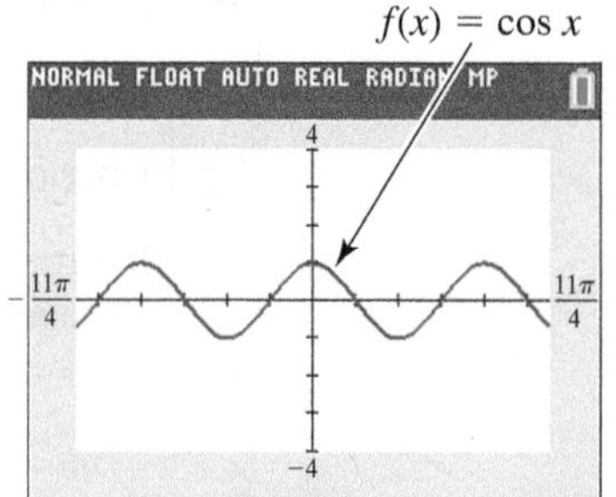

Figure 5

- The graph is continuous over its entire domain, $(-\infty, \infty)$.
- Its x-intercepts have x-values of the form $(2n + 1)\frac{\pi}{2}$, where n is an integer.
- Its period is 2π.
- The graph is symmetric with respect to the y-axis, so the function is an even function. For all x in the domain, $\cos(-x) = \cos x$.

NOTE A function f is an **even function** if for all x in the domain of f,

$$f(-x) = f(x).$$

The graph of an even function is symmetric with respect to the y-axis. This means that if (x, y) belongs to the function, then $(-x, y)$ also belongs to the function. For example, $\left(\frac{\pi}{2}, 0\right)$ and $\left(-\frac{\pi}{2}, 0\right)$ are points on the graph of $y = \cos x$, illustrating the property $\cos(-x) = \cos x$.

The calculator graphs of $f(x) = \sin x$ in **Figure 3** and $f(x) = \cos x$ in **Figure 5** are shown in the ZTrig viewing window

$$\left[-\frac{11\pi}{4}, \frac{11\pi}{4}\right] \text{ by } [-4, 4] \quad \left(\tfrac{11\pi}{4} \approx 8.639379797\right)$$

of the TI-84 Plus calculator, with Xscl $= \frac{\pi}{2}$ and Yscl $= 1$. (Other models have different trigonometry viewing windows.) ■

Techniques for Graphing, Amplitude, and Period The examples that follow show graphs that are "stretched" or "compressed" (shrunk) either vertically, horizontally, or both when compared with the graphs of $y = \sin x$ or $y = \cos x$.

Classroom Example 1

Graph $y = \frac{1}{2} \sin x$, and compare to the graph of $y = \sin x$.

Answer:

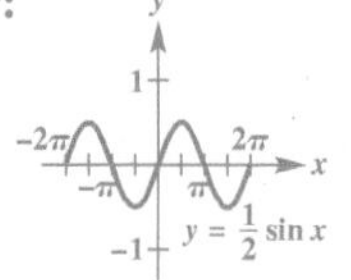

The only change in the graph is the range, which becomes $\left[-\frac{1}{2}, \frac{1}{2}\right]$.

EXAMPLE 1 Graphing $y = a \sin x$

Graph $y = 2 \sin x$, and compare to the graph of $y = \sin x$.

SOLUTION For a given value of x, the value of y is twice what it would be for $y = \sin x$. See the table of values. The change in the graph is the range, which becomes $[-2, 2]$. See **Figure 6,** which also includes a graph of $y = \sin x$.

x	0	$\frac{\pi}{2}$	π	$\frac{3\pi}{2}$	2π
$\sin x$	0	1	0	−1	0
$2 \sin x$	0	2	0	−2	0

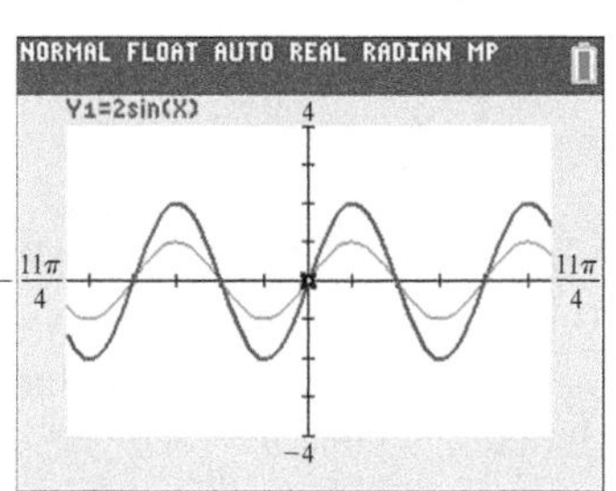

The graph of $y = 2 \sin x$ is shown in blue, and that of $y = \sin x$ is shown in red. Compare to **Figure 6.**

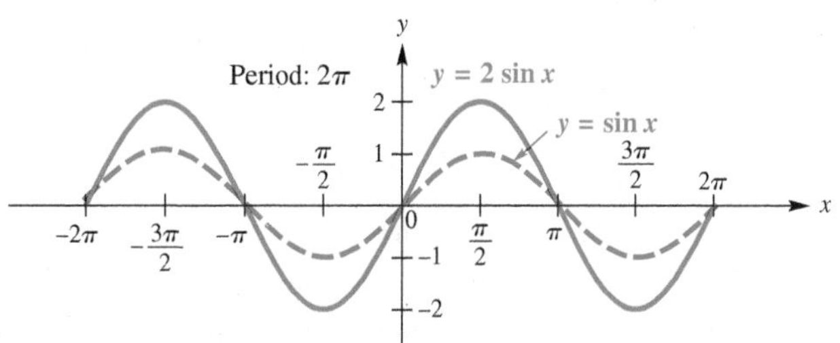

Figure 6

The **amplitude** of a periodic function is half the difference between the maximum and minimum values. It describes the height of the graph both above and below a horizontal line passing through the "middle" of the graph. Thus, for the basic sine function $y = \sin x$ (and also for the basic cosine function $y = \cos x$), the amplitude is computed as follows.

$$\frac{1}{2}[1 - (-1)] = \frac{1}{2}(2) = 1 \quad \text{Amplitude of } y = \sin x$$

For $y = 2 \sin x$, the amplitude is

$$\frac{1}{2}[2 - (-2)] = \frac{1}{2}(4) = 2. \quad \text{Amplitude of } y = 2 \sin x$$

We can think of the graph of $y = a \sin x$ as a vertical stretching of the graph of $y = \sin x$ when $a > 1$ and a vertical shrinking when $0 < a < 1$.

✔ **Now Try Exercise 15.**

Amplitude

The graph of $\boldsymbol{y = a \sin x}$ or $\boldsymbol{y = a \cos x}$, with $a \neq 0$, will have the same shape as the graph of $y = \sin x$ or $y = \cos x$, respectively, except with range $[-|a|, |a|]$. The amplitude is $|a|$.

Teaching Tip Students may think that the period of $y = \sin 2x$ is 4π. Explain that a factor of 2 causes values of the argument to *increase twice as fast*, thereby shortening the length of the period to half its original value.

While the coefficient a in $y = a \sin x$ or $y = a \cos x$ affects the amplitude of the graph, the coefficient of x in the argument affects the period. Consider $y = \sin 2x$. We can complete a table of values for the interval $[0, 2\pi]$.

x	0	$\frac{\pi}{4}$	$\frac{\pi}{2}$	$\frac{3\pi}{4}$	π	$\frac{5\pi}{4}$	$\frac{3\pi}{2}$	$\frac{7\pi}{4}$	2π
sin 2x	0	1	0	−1	0	1	0	−1	0

Note that one complete cycle occurs in π units, not 2π units. Therefore, the period here is π, which equals $\frac{2\pi}{2}$. Now consider $y = \sin 4x$. Look at the next table.

x	0	$\frac{\pi}{8}$	$\frac{\pi}{4}$	$\frac{3\pi}{8}$	$\frac{\pi}{2}$	$\frac{5\pi}{8}$	$\frac{3\pi}{4}$	$\frac{7\pi}{8}$	π
sin 4x	0	1	0	−1	0	1	0	−1	0

These values suggest that one complete cycle is achieved in $\frac{\pi}{2}$ or $\frac{2\pi}{4}$ units, which is reasonable because

$$\sin\left(4 \cdot \frac{\pi}{2}\right) = \sin 2\pi = 0.$$

In general, the graph of a function of the form $y = \sin bx$ or $y = \cos bx$, for $b > 0$, will have a period different from 2π when $b \neq 1$.

To see why this is so, remember that the values of $\sin bx$ or $\cos bx$ will take on all possible values as bx ranges from 0 to 2π. Therefore, to find the period of either of these functions, we must solve the following three-part inequality.

$$0 \leq bx \leq 2\pi \quad \text{bx ranges from 0 to 2π.}$$

$$0 \leq x \leq \frac{2\pi}{b} \quad \text{Divide each part by the positive number b.}$$

Thus, the period is $\frac{2\pi}{b}$. By dividing the interval $\left[0, \frac{2\pi}{b}\right]$ into four equal parts, we obtain the values for which $\sin bx$ or $\cos bx$ is -1, 0, or 1. These values will give minimum points, x-intercepts, and maximum points on the graph. (If a function has $b < 0$, then identities can be used to rewrite the function so that $b > 0$.)

NOTE One method to divide an interval into four equal parts is as follows.

Step 1 Find the midpoint of the interval by adding the x-values of the endpoints and dividing by 2.

Step 2 Find the quarter points (the midpoints of the two intervals found in Step 1) using the same procedure.

Classroom Example 2

Graph $y = \sin \frac{1}{3}x$, and compare to the graph of $y = \sin x$.

Answer:

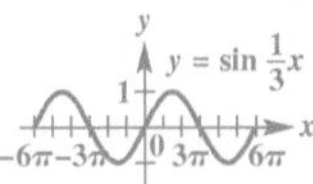

The period changes from 2π to 6π.

EXAMPLE 2 Graphing $y = \sin bx$

Graph $y = \sin 2x$, and compare to the graph of $y = \sin x$.

SOLUTION In this function the coefficient of x is 2, so $b = 2$ and the period is $\frac{2\pi}{2} = \pi$. Therefore, the graph will complete one period over the interval $[0, \pi]$.

We can divide the interval $[0, \pi]$ into four equal parts by first finding its midpoint: $\frac{1}{2}(0 + \pi) = \frac{\pi}{2}$. The quarter points are found next by determining the midpoints of the two intervals $\left[0, \frac{\pi}{2}\right]$ and $\left[\frac{\pi}{2}, \pi\right]$.

$$\frac{1}{2}\left(\frac{\pi}{2} + \pi\right) = \frac{1}{2}\left(\frac{3\pi}{2}\right) = \frac{3\pi}{4}$$

$$\frac{1}{2}\left(0 + \frac{\pi}{2}\right) = \frac{\pi}{4} \quad \text{and} \quad \frac{1}{2}\left(\frac{\pi}{2} + \pi\right) = \frac{3\pi}{4} \quad \text{Quarter points}$$

The interval $[0, \pi]$ is divided into four equal parts using these x-values.

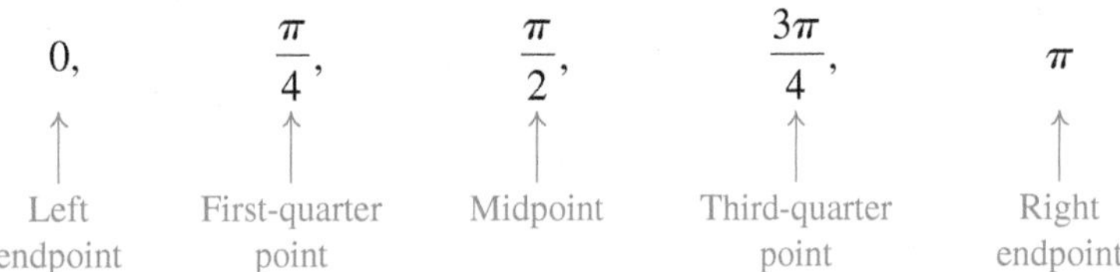

We plot the points from the table of values given at the top of the previous page, and join them with a smooth sinusoidal curve. More of the graph can be sketched by repeating this cycle, as shown in **Figure 7.** The amplitude is not changed.

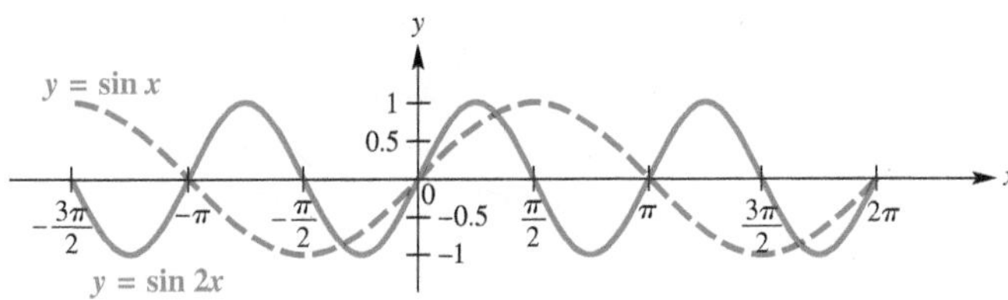

Figure 7

We can think of the graph of $y = \sin bx$ as a horizontal stretching of the graph of $y = \sin x$ when $0 < b < 1$ and a horizontal shrinking when $b > 1$.

Now Try Exercise 27.

Period

For $b > 0$, the graph of $y = \sin bx$ will resemble that of $y = \sin x$, but with period $\frac{2\pi}{b}$. Also, the graph of $y = \cos bx$ will resemble that of $y = \cos x$, but with period $\frac{2\pi}{b}$.

Classroom Example 3

Graph $y = \cos \frac{1}{2}x$ over one period.

Answer:

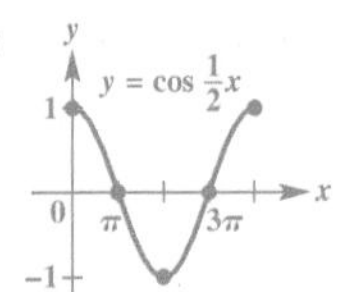

EXAMPLE 3 Graphing $y = \cos bx$

Graph $y = \cos \frac{2}{3}x$ over one period.

SOLUTION The period is

$$\frac{2\pi}{\frac{2}{3}} = 2\pi \div \frac{2}{3} = 2\pi \cdot \frac{3}{2} = 3\pi.$$

To divide by a fraction, multiply by its reciprocal.

We divide the interval $[0, 3\pi]$ into four equal parts to obtain the x-values $0, \frac{3\pi}{4}, \frac{3\pi}{2}, \frac{9\pi}{4}$, and 3π that yield minimum points, maximum points, and x-intercepts. We use these values to obtain a table of key points for one period.

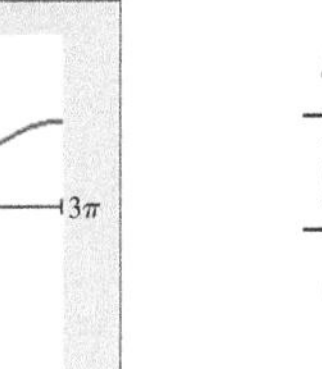

This screen shows a graph of the function in **Example 3.** By choosing Xscl $= \frac{3\pi}{4}$, the x-intercepts, maxima, and minima coincide with tick marks on the x-axis.

x	0	$\frac{3\pi}{4}$	$\frac{3\pi}{2}$	$\frac{9\pi}{4}$	3π
$\frac{2}{3}x$	0	$\frac{\pi}{2}$	π	$\frac{3\pi}{2}$	2π
$\cos \frac{2}{3}x$	1	0	-1	0	1

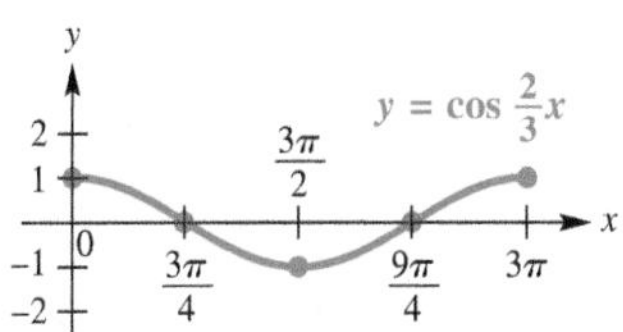

Figure 8

The amplitude is 1 because the maximum value is 1, the minimum value is -1, and $\frac{1}{2}[1 - (-1)] = \frac{1}{2}(2) = 1$. We plot these points and join them with a smooth curve. The graph is shown in **Figure 8.**

Now Try Exercise 25.

NOTE Look at the middle row of the table in **Example 3.** Dividing $\left[0, \frac{2\pi}{b}\right]$ into four equal parts gives the values $0, \frac{\pi}{2}, \pi, \frac{3\pi}{2}$, and 2π for this row, resulting here in values of -1, 0, or 1. These values lead to key points on the graph, which can be plotted and joined with a smooth sinusoidal curve.

Guidelines for Sketching Graphs of Sine and Cosine Functions

To graph $y = a \sin bx$ or $y = a \cos bx$, with $b > 0$, follow these steps.

Step 1 Find the period, $\frac{2\pi}{b}$. Start at 0 on the x-axis, and lay off a distance of $\frac{2\pi}{b}$.

Step 2 Divide the interval into four equal parts. (See the Note preceding **Example 2.**)

Step 3 Evaluate the function for each of the five x-values resulting from Step 2. The points will be maximum points, minimum points, and x-intercepts.

Step 4 Plot the points found in Step 3, and join them with a sinusoidal curve having amplitude $|a|$.

Step 5 Draw the graph over additional periods as needed.

Classroom Example 4
Graph $y = -3 \sin 2x$ over one period using the guidelines preceding **Example 4.**

Answer:

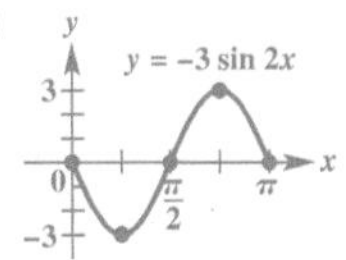

EXAMPLE 4 Graphing $y = a \sin bx$

Graph $y = -2 \sin 3x$ over one period using the preceding guidelines.

SOLUTION

Step 1 For this function, $b = 3$, so the period is $\frac{2\pi}{3}$. The function will be graphed over the interval $\left[0, \frac{2\pi}{3}\right]$.

Step 2 Divide the interval $\left[0, \frac{2\pi}{3}\right]$ into four equal parts to obtain the x-values $0, \frac{\pi}{6}, \frac{\pi}{3}, \frac{\pi}{2}$, and $\frac{2\pi}{3}$.

Step 3 Make a table of values determined by the x-values from Step 2.

x	0	$\frac{\pi}{6}$	$\frac{\pi}{3}$	$\frac{\pi}{2}$	$\frac{2\pi}{3}$
$3x$	0	$\frac{\pi}{2}$	π	$\frac{3\pi}{2}$	2π
$\sin 3x$	0	1	0	-1	0
$-2 \sin 3x$	0	-2	0	2	0

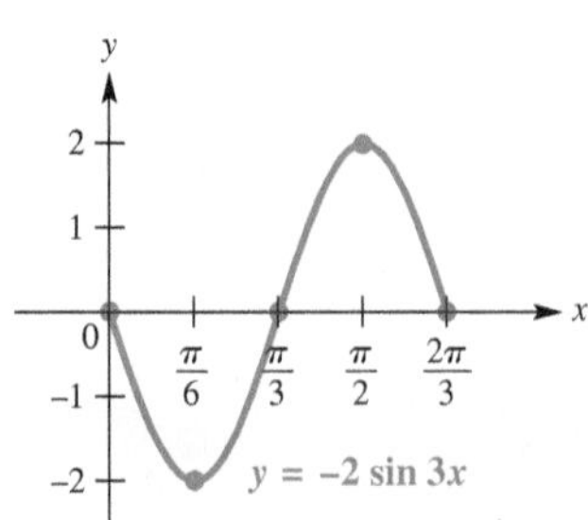

Figure 9

Step 4 Plot the points $(0, 0)$, $\left(\frac{\pi}{6}, -2\right)$, $\left(\frac{\pi}{3}, 0\right)$, $\left(\frac{\pi}{2}, 2\right)$, and $\left(\frac{2\pi}{3}, 0\right)$, and join them with a sinusoidal curve having amplitude 2. See **Figure 9.**

Step 5 The graph can be extended by repeating the cycle.

Notice that when a is negative, the graph of $y = a \sin bx$ is a reflection across the x-axis of the graph of $y = |a| \sin bx$.

Now Try Exercise 29.

Classroom Example 5

Graph $y = 2 \cos \frac{\pi}{2}x$ over one period.

Answer:

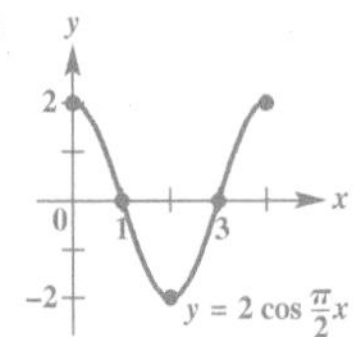

EXAMPLE 5 Graphing $y = a \cos bx$ (Where b Is a Multiple of π)

Graph $y = -3 \cos \pi x$ over one period.

SOLUTION

Step 1 Here $b = \pi$ and the period is $\frac{2\pi}{\pi} = 2$, so we will graph the function over the interval $[0, 2]$.

Step 2 Dividing $[0, 2]$ into four equal parts yields the x-values $0, \frac{1}{2}, 1, \frac{3}{2}$, and 2.

Step 3 Make a table using these x-values.

x	0	$\frac{1}{2}$	1	$\frac{3}{2}$	2
πx	0	$\frac{\pi}{2}$	π	$\frac{3\pi}{2}$	2π
$\cos \pi x$	1	0	-1	0	1
$-3 \cos \pi x$	-3	0	3	0	-3

Step 4 Plot the points $(0, -3)$, $\left(\frac{1}{2}, 0\right)$, $(1, 3)$, $\left(\frac{3}{2}, 0\right)$, and $(2, -3)$, and join them with a sinusoidal curve having amplitude $|-3| = 3$. See **Figure 10.**

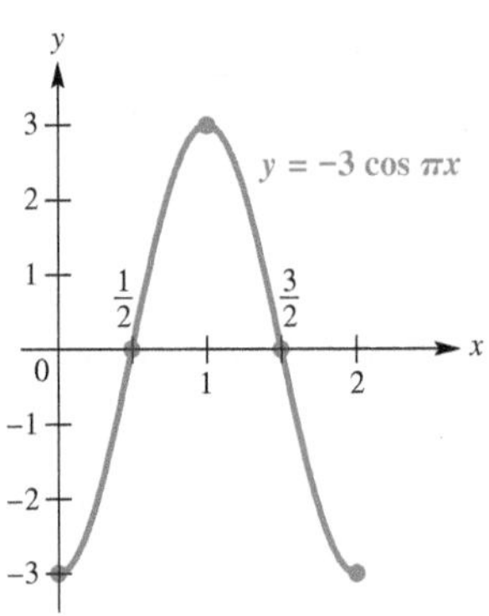

Figure 10

Step 5 The graph can be extended by repeating the cycle.

Notice that when b is an integer multiple of π, the first coordinates of the x-intercepts of the graph are rational numbers.

✔ **Now Try Exercise 37.**

Connecting Graphs with Equations

Classroom Example 6

Determine an equation of the form $y = a \cos bx$ or $y = a \sin bx$, where $b > 0$, for the given graph.

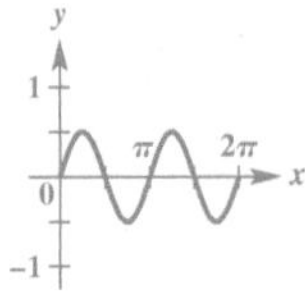

Answer: $y = \frac{1}{2} \sin 2x$

EXAMPLE 6 Determining an Equation for a Graph

Determine an equation of the form $y = a \cos bx$ or $y = a \sin bx$, where $b > 0$, for the given graph.

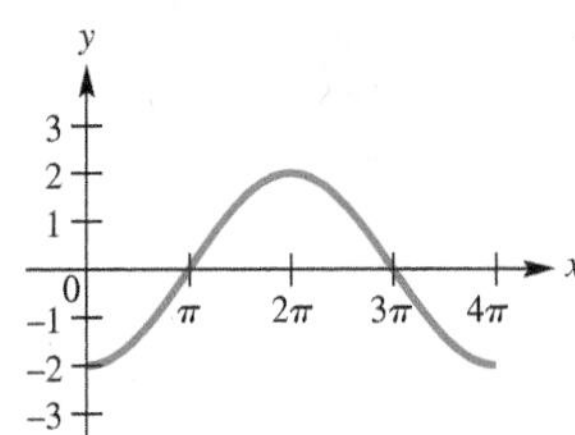

SOLUTION This graph is that of a cosine function that is reflected across its horizontal axis, the x-axis. The amplitude is half the distance between the maximum and minimum values.

$$\frac{1}{2}[2 - (-2)] = \frac{1}{2}(4) = 2 \quad \text{The amplitude } |a| \text{ is 2.}$$

Because the graph completes a cycle on the interval $[0, 4\pi]$, the period is 4π. We use this fact to solve for b.

$$4\pi = \frac{2\pi}{b} \quad \text{Period} = \frac{2\pi}{b}$$

$$4\pi b = 2\pi \quad \text{Multiply each side by } b.$$

$$b = \frac{1}{2} \quad \text{Divide each side by } 4\pi.$$

An equation for the graph is

$$y = -2 \cos \frac{1}{2}x.$$

x-axis reflection (points to -2); Horizontal stretch (points to $\frac{1}{2}$)

✔ **Now Try Exercise 41.**

Classroom Example 7
The average temperature (in °F) in Phoenix, Arizona, can be approximated by the function

$$f(x) = 19 \sin\left[\frac{\pi}{6}(x - 4.3)\right] + 74,$$

where x is the month and $x = 1$ corresponds to January, $x = 2$ to February, and so on.

(a) To observe the graph over a two-year interval, graph f in the window $[0, 25]$ by $[0, 125]$.

(b) According to this model, what is the average temperature during the month of October?

(c) What would be an approximation for the average annual temperature in Phoenix?

Answers:

(a)

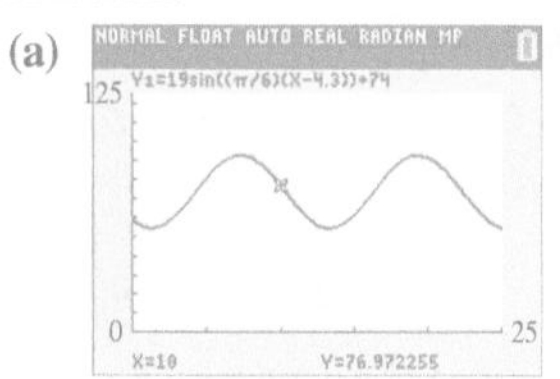

(b) 77°F (c) 74°F

A Trigonometric Model Sine and cosine functions may be used to model many real-life phenomena that repeat their values in a cyclical, or periodic, manner. Average temperature in a certain geographic location is one such example.

EXAMPLE 7 Interpreting a Sine Function Model

The average temperature (in °F) at Mould Bay, Canada, can be approximated by the function

$$f(x) = 34 \sin\left[\frac{\pi}{6}(x - 4.3)\right],$$

where x is the month and $x = 1$ corresponds to January, $x = 2$ to February, and so on.

(a) To observe the graph over a two-year interval, graph f in the window $[0, 25]$ by $[-45, 45]$.

(b) According to this model, what is the average temperature during the month of May?

(c) What would be an approximation for the average annual temperature at Mould Bay?

SOLUTION

(a) The graph of $f(x) = 34 \sin\left[\frac{\pi}{6}(x - 4.3)\right]$ is shown in **Figure 11.** Its amplitude is 34, and the period is

$$\frac{2\pi}{\frac{\pi}{6}} = 2\pi \div \frac{\pi}{6} = 2\pi \cdot \frac{6}{\pi} = 12. \quad \text{Simplify the complex fraction.}$$

Function f has a period of 12 months, or 1 year, which agrees with the changing of the seasons.

(b) May is the fifth month, so the average temperature during May is

$$f(5) = 34 \sin\left[\frac{\pi}{6}(5 - 4.3)\right] \approx 12°\text{F}. \quad \text{Let } x = 5 \text{ in the given function.}$$

See the display at the bottom of the screen in **Figure 11.**

(c) From the graph, it appears that the average annual temperature is about 0°F because the graph is centered vertically about the line $y = 0$.

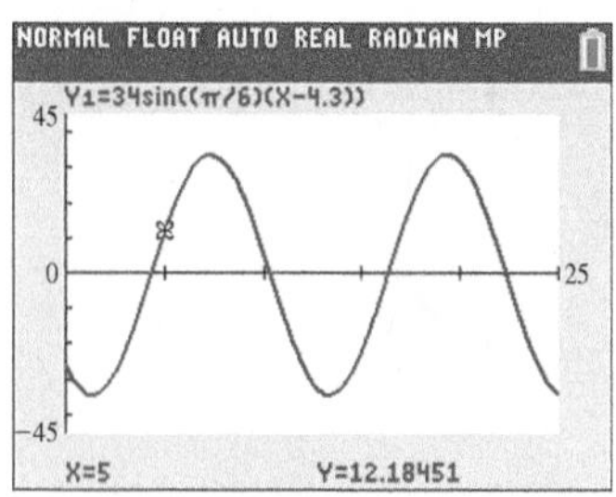

Figure 11

✔ **Now Try Exercise 57.**

4.1 Exercises

1. 1; 2π
2. rises; falls

CONCEPT PREVIEW *Fill in the blank(s) to correctly complete each sentence.*

1. The amplitude of the graphs of the sine and cosine functions is ______, and the period of each is ______.

2. For the x-values 0 to $\frac{\pi}{2}$, the graph of the sine function ______ (rises/falls) and that of the cosine function ______ (rises/falls).

3. $n\pi$
4. $(-\infty, \infty)$; $[-1, 1]$
5. $\frac{\pi}{2}$
6. -1; 1
7. E
8. D
9. B
10. A
11. F
12. C

13.
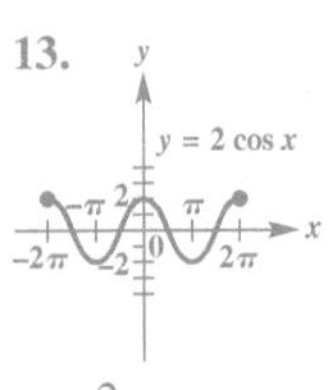

2

14. 3

15.
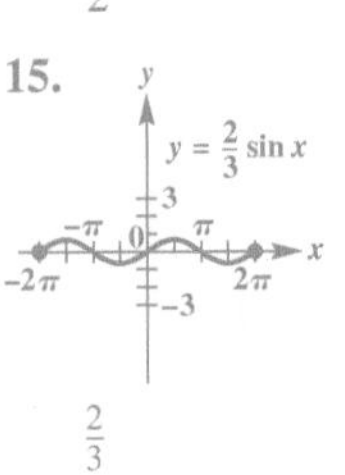

$\frac{2}{3}$

16.
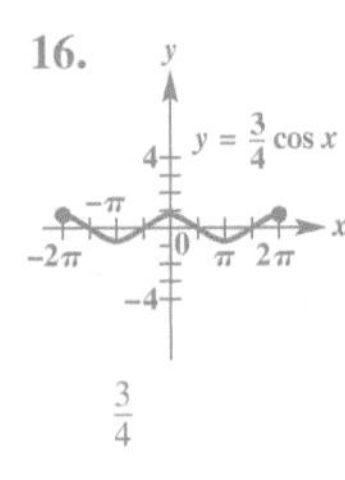

$\frac{3}{4}$

17. 18.
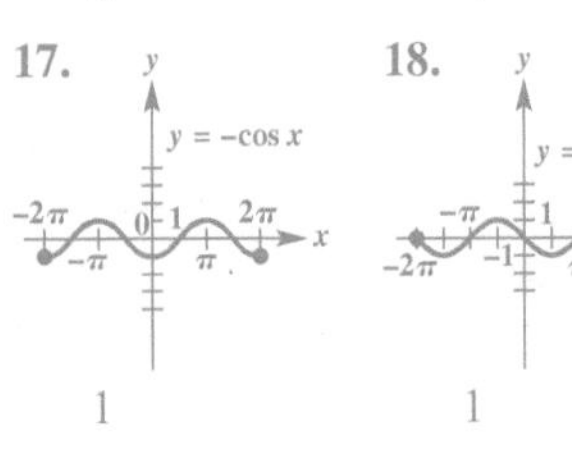

1 1

19. 20.
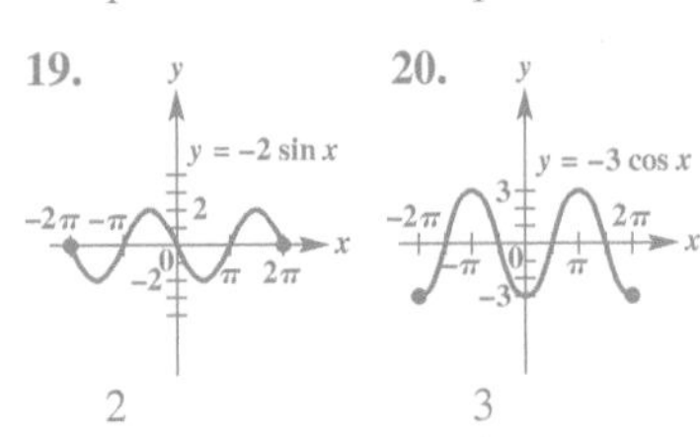

2 3

21.
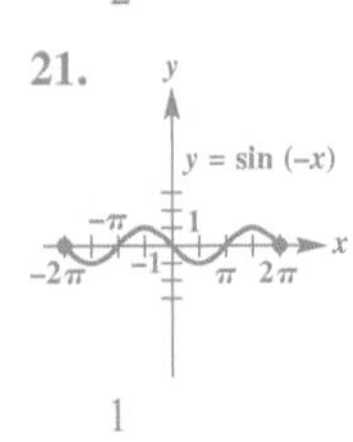

1

22. Because $\sin(-x) = -\sin x$ for all x, the graphs are the same.

23. 24. 25. 26.
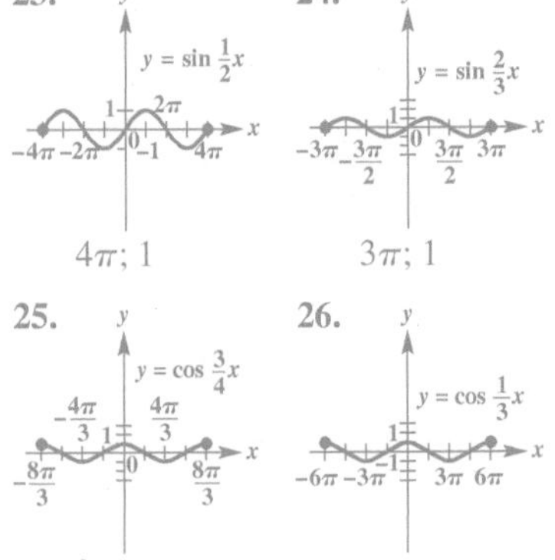

23. 4π; 1
24. 3π; 1
25. $\frac{8\pi}{3}$; 1
26. 6π; 1

3. The graph of the sine function crosses the x-axis for all numbers of the form ______, where n is an integer.

4. The domain of both the sine and cosine functions (in interval form) is ______, and the range is ______.

5. The least positive number x for which $\cos x = 0$ is ______.

6. On the interval $[\pi, 2\pi]$, the function values of the cosine function increase from ______ to ______.

Concept Check Match each function with its graph in choices A–F.

7. $y = -\sin x$ **8.** $y = -\cos x$ **9.** $y = \sin 2x$

10. $y = \cos 2x$ **11.** $y = 2 \sin x$ **12.** $y = 2 \cos x$

A.
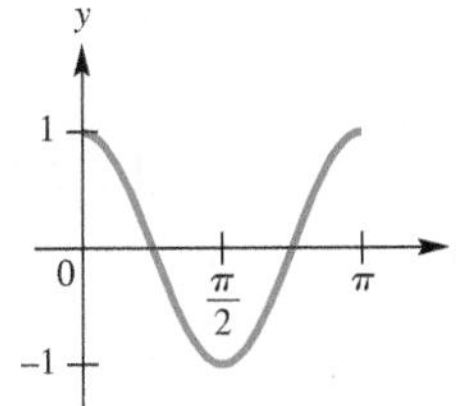

B.
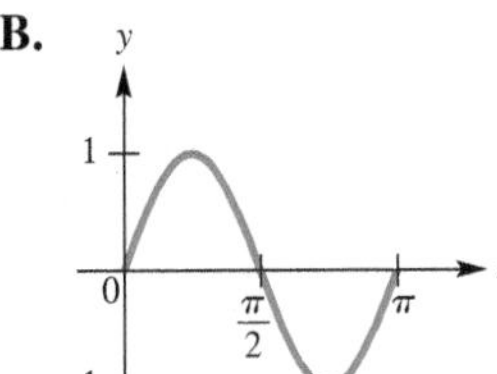

C.
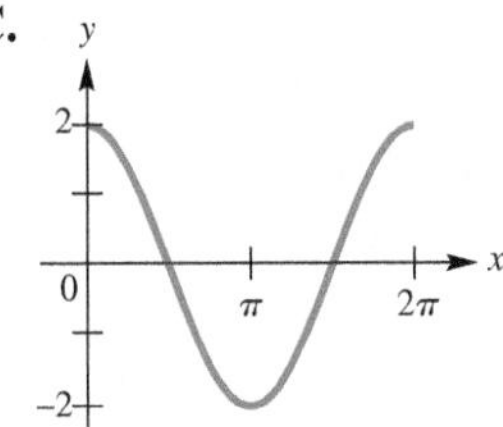

D.
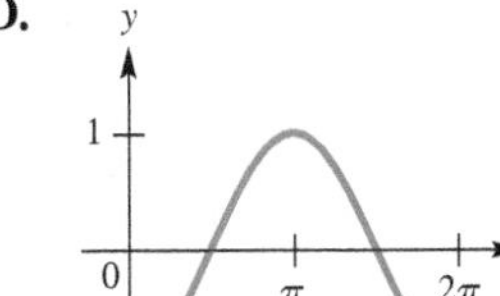

E.
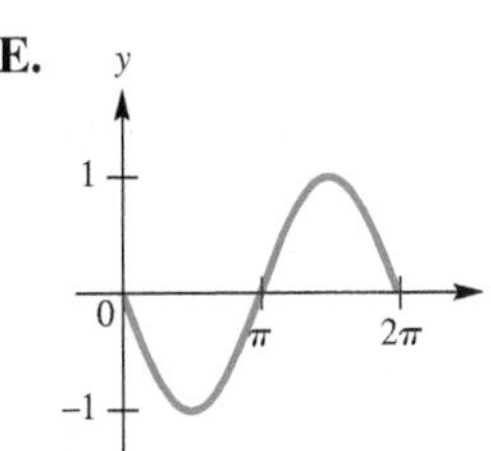

F.
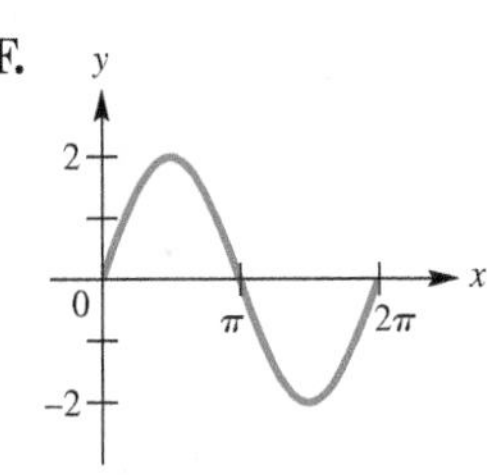

Graph each function over the interval $[-2\pi, 2\pi]$. Give the amplitude. ***See Example 1.***

13. $y = 2 \cos x$ **14.** $y = 3 \sin x$ **15.** $y = \frac{2}{3} \sin x$

16. $y = \frac{3}{4} \cos x$ **17.** $y = -\cos x$ **18.** $y = -\sin x$

19. $y = -2 \sin x$ **20.** $y = -3 \cos x$ **21.** $y = \sin(-x)$

22. ***Concept Check*** In **Exercise 21,** why is the graph the same as that of $y = -\sin x$?

Graph each function over a two-period interval. Give the period and amplitude. ***See Examples 2–5.***

23. $y = \sin \frac{1}{2}x$ **24.** $y = \sin \frac{2}{3}x$ **25.** $y = \cos \frac{3}{4}x$

26. $y = \cos \frac{1}{3}x$ **27.** $y = \sin 3x$ **28.** $y = \cos 2x$

29. $y = 2 \sin \frac{1}{4}x$ **30.** $y = 3 \sin 2x$ **31.** $y = -2 \cos 3x$

32. $y = -5 \cos 2x$ **33.** $y = \cos \pi x$ **34.** $y = -\sin \pi x$

35. $y = -2 \sin 2\pi x$ **36.** $y = 3 \cos 2\pi x$ **37.** $y = \frac{1}{2} \cos \frac{\pi}{2}x$

38. $y = -\frac{2}{3} \sin \frac{\pi}{4}x$ **39.** $y = \pi \sin \pi x$ **40.** $y = -\pi \cos \pi x$

27. 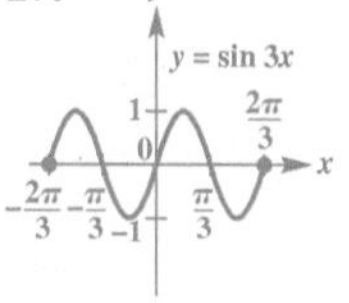 $y = \sin 3x$

$\frac{2\pi}{3}$; 1

28. 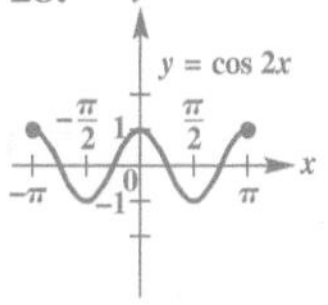 $y = \cos 2x$

π; 1

29. 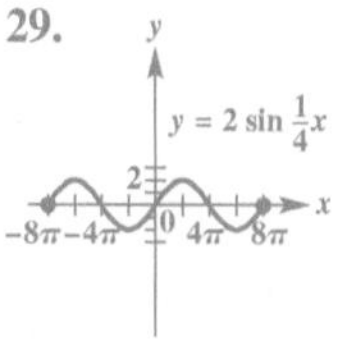 $y = 2 \sin \frac{1}{4}x$

8π; 2

30.

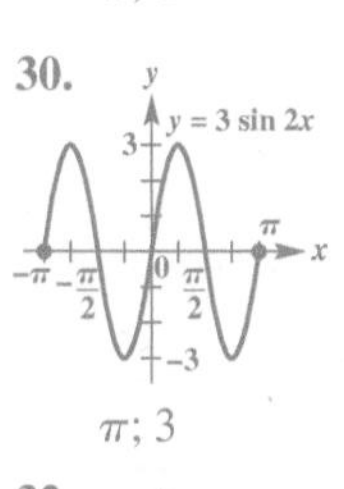

π; 3

31.

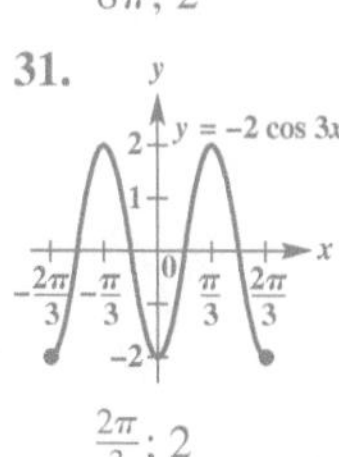

$\frac{2\pi}{3}$; 2

32.

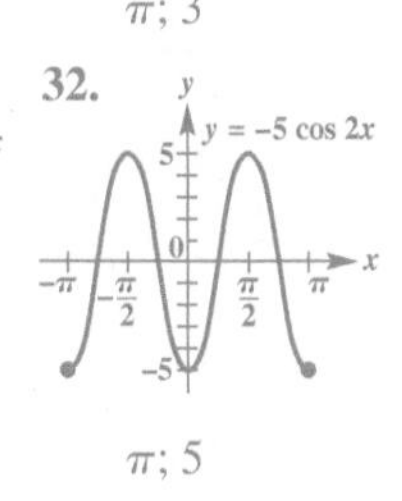

π; 5

33.

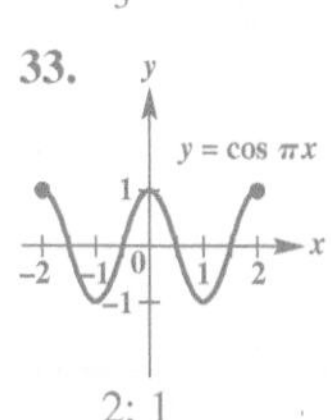

2; 1

34.

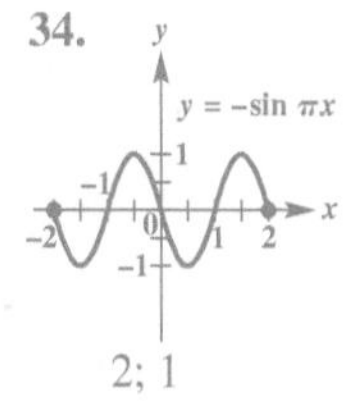

2; 1

35.

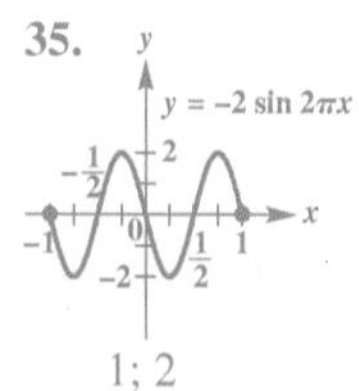

1; 2

36. 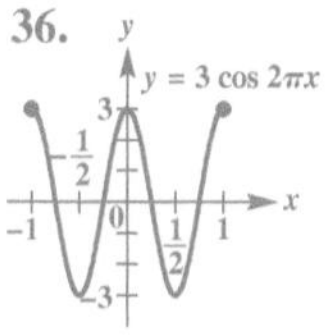 $y = 3 \cos 2\pi x$

1; 3

37. 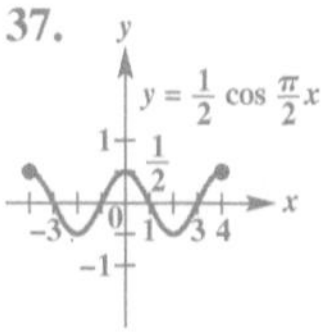 $y = \frac{1}{2} \cos \frac{\pi}{2}x$

4; $\frac{1}{2}$

38.

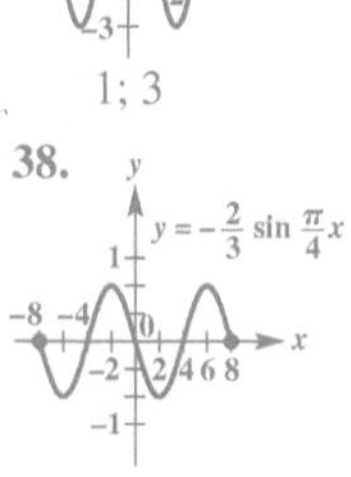

8; $\frac{2}{3}$

39.

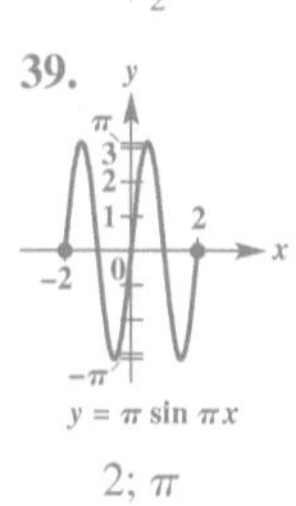

2; π

40.

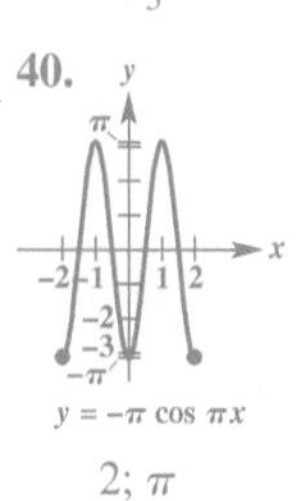

2; π

41. $y = 2 \cos 2x$
42. $y = -2 \sin 2x$
43. $y = -3 \cos \frac{1}{2}x$
44. $y = 3 \cos \frac{1}{2}x$
45. $y = 3 \sin 4x$
46. $y = -3 \cos 4x$

Connecting Graphs with Equations *Determine an equation of the form* $y = a \cos bx$ *or* $y = a \sin bx$*, where* $b > 0$*, for the given graph.* ***See Example 6.***

41.

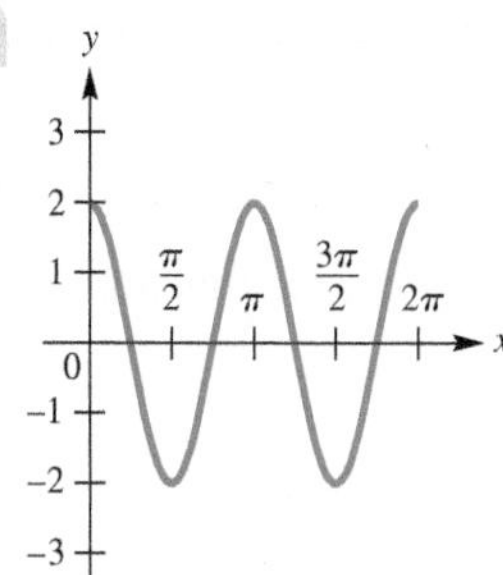

42.

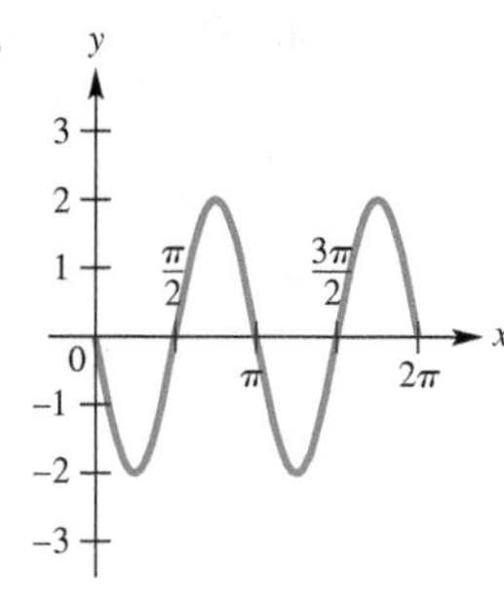

43.

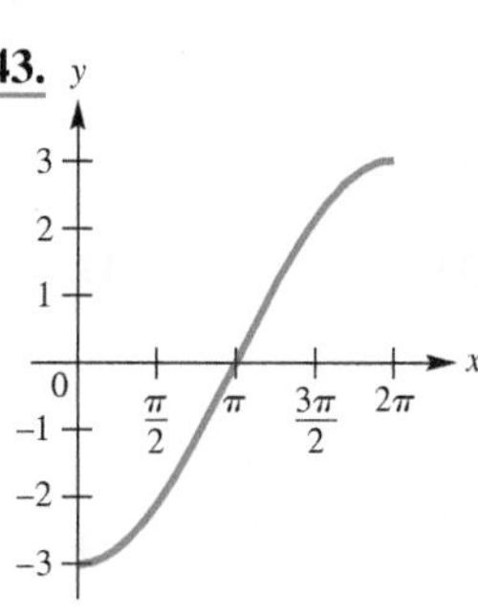

44.

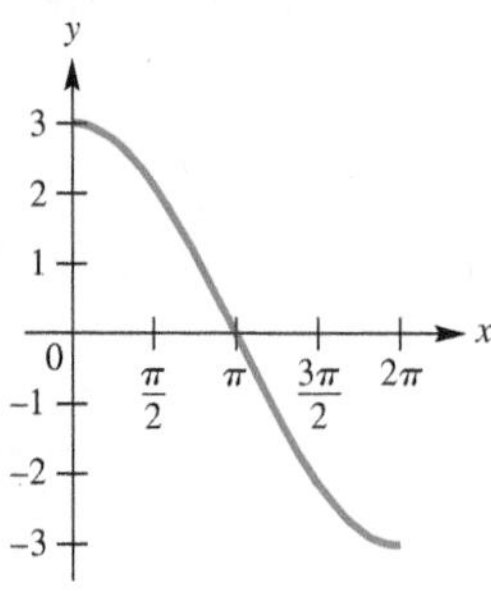

45.

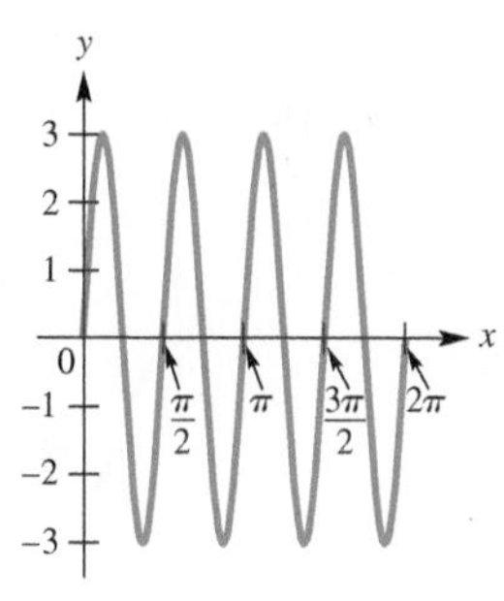

46.

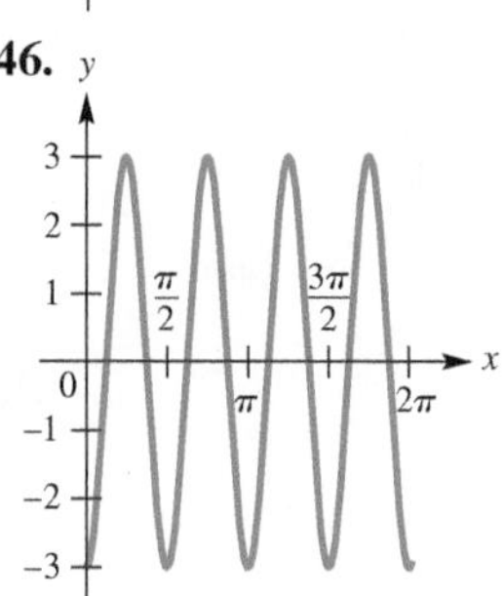

(Modeling) Solve each problem.

47. ***Average Annual Temperature*** Scientists believe that the average annual temperature in a given location is periodic. The average temperature at a given place during a given season fluctuates as time goes on, from colder to warmer, and back to colder. The graph shows an idealized description of the temperature (in °F) for approximately the last 150 thousand years of a particular location.

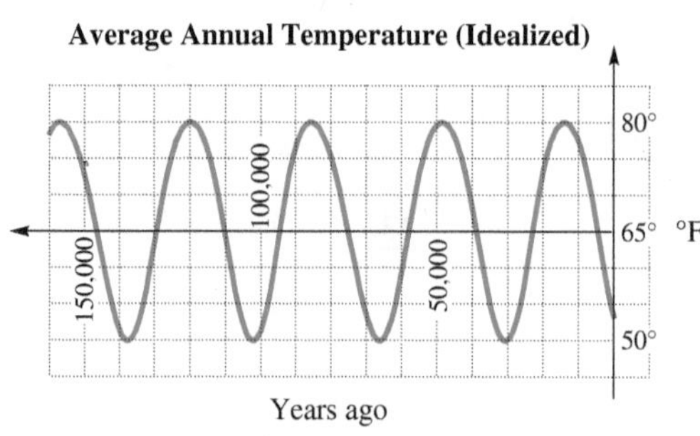

(a) Find the highest and lowest temperatures recorded.

(b) Use these two numbers to find the amplitude.

(c) Find the period of the function.

(d) What is the trend of the temperature now?

48. ***Blood Pressure Variation*** The graph gives the variation in blood pressure for a typical person. **Systolic** and **diastolic pressures** are the upper and lower limits of the periodic changes in pressure that produce the pulse. The length of time between peaks is the period of the pulse.

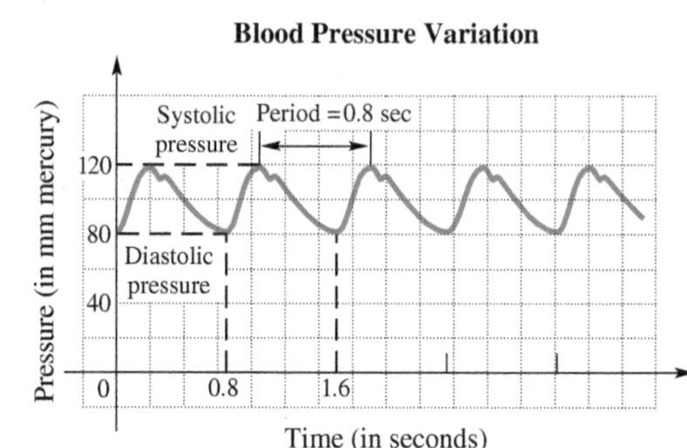

(a) Find the systolic and diastolic pressures.

(b) Find the amplitude of the graph.

(c) Find the pulse rate (the number of pulse beats in 1 min) for this person.

47. (a) 80°F; 50°F
(b) 15
(c) 35,000 yr
(d) downward
48. (a) 120; 80 (b) 20
(c) 75
49. 24 hr
50. $\frac{2.6-0.2}{2} = 1.2$
51. 6:00 P.M.; 0.2 ft
52. 7:19 P.M.; 0 ft
53. 3:18 A.M.; 2.4 ft
54. (a) 2 hr (b) 1 yr
55. (a)

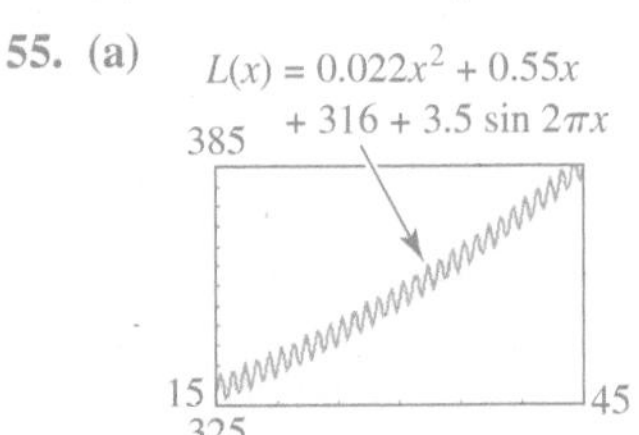

(b) maxima: $x = \frac{1}{4}, \frac{5}{4}, \frac{9}{4}, \ldots$;
minima: $x = \frac{3}{4}, \frac{7}{4}, \frac{11}{4}, \ldots$
(c) The quadratic function provides the general increasing nature of the level, while the sine function provides the fluctuations as the years go by.

(Modeling) Tides for Kahului Harbor The chart shows the tides for Kahului Harbor (on the island of Maui, Hawaii). To identify high and low tides and times for other Maui areas, the following adjustments must be made.

Hana:	High, +40 min, −0.1 ft; Low, +18 min, −0.2 ft	Makena:	High, +1:21, −0.5 ft; Low, +1:09, −0.2 ft
Maalaea:	High, +1:52, −0.1 ft; Low, +1:19, −0.2 ft	Lahaina:	High, +1:18, −0.2 ft; Low, +1:01, −0.1 ft

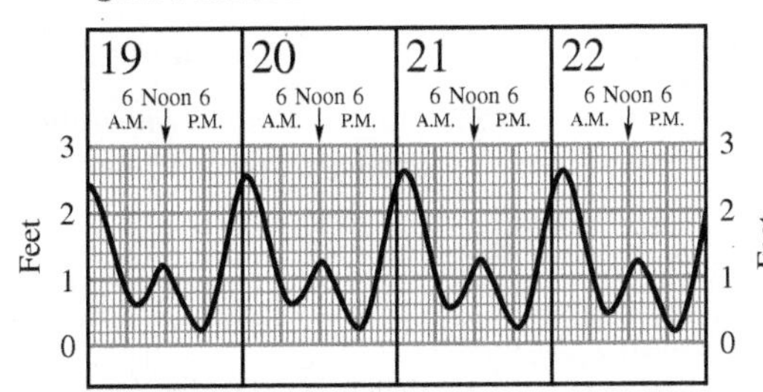

Source: Maui News. Original chart prepared by Edward K. Noda and Associates.

Use the graph to approximate each answer.

49. The graph is an example of a periodic function. What is the period (in hours)?

50. What is the amplitude?

51. At what time on January 20 was low tide at Kahului? What was the height then?

52. Repeat **Exercise 51** for Maalaea.

53. At what time on January 22 was high tide at Lahaina? What was the height then?

(Modeling) Solve each problem.

54. *Activity of a Nocturnal Animal* Many activities of living organisms are periodic. For example, the graph at the right below shows the time that a certain nocturnal animal begins its evening activity.

(a) Find the amplitude of this graph.

(b) Find the period.

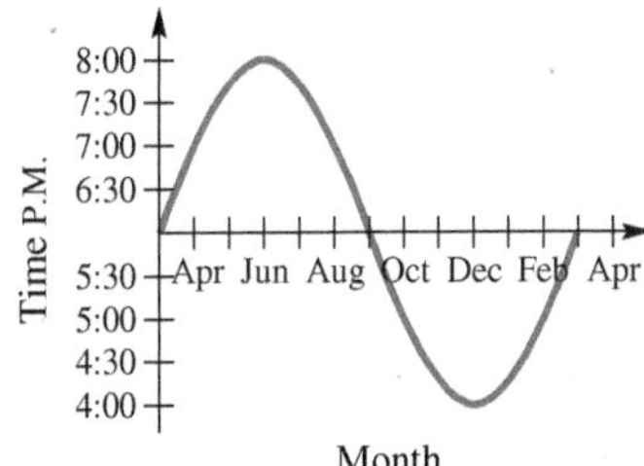

55. *Atmospheric Carbon Dioxide* At Mauna Loa, Hawaii, atmospheric carbon dioxide levels in parts per million (ppm) were measured regularly, beginning in 1958. The function

$$L(x) = 0.022x^2 + 0.55x + 316 + 3.5 \sin 2\pi x$$

can be used to model these levels, where x is in years and $x = 0$ corresponds to 1960. (*Source*: Nilsson, A., *Greenhouse Earth*, John Wiley and Sons.)

(a) Graph L in the window $[15, 45]$ by $[325, 385]$.

(b) When do the seasonal maximum and minimum carbon dioxide levels occur?

(c) L is the sum of a quadratic function and a sine function. What is the significance of each of these functions?

56. (a)

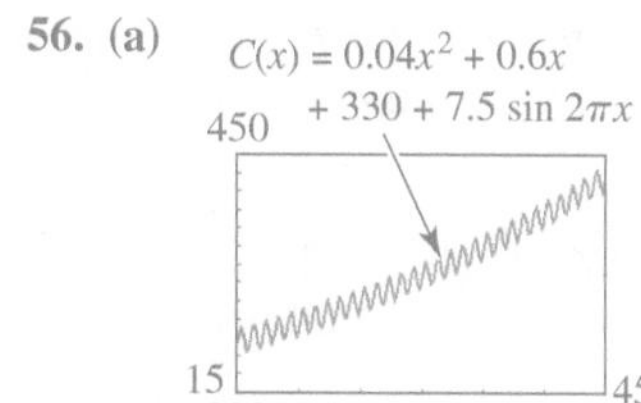

(b) The coefficient of $\sin 2\pi x$, 7.5, is greater than the one in Exercise 55.

57. (a) 31°F (b) 38°F (c) 57°F (d) 58°F (e) 37°F (f) 16°F

58. (a) 1.998 watts per m^2 (b) -46.461 watts per m^2 (c) 46.478 watts per m^2 (d) Answers may vary. A possible answer is $N = 82.5$. (Because N represents a day number, which should be a natural number, we might interpret day 82.5 as noon on the 82nd day.)

59. 1; 240°, or $\frac{4\pi}{3}$

60. 1; 120°, or $\frac{2\pi}{3}$

61. No. For $b > 0$, $b \neq 1$, the graph of $y = \sin bx$ has amplitude 1 and period $\frac{2\pi}{b}$, while that of $y = b \sin x$ has amplitude b and period 2π.

62. No. For $b > 0$, $b \neq 1$, the graph of $y = \cos bx$ has amplitude 1 and period $\frac{2\pi}{b}$, while that of $y = b \cos x$ has amplitude b and period 2π.

56. ***Atmospheric Carbon Dioxide*** Refer to **Exercise 55.** The carbon dioxide content in the atmosphere at Barrow, Alaska, in parts per million (ppm) can be modeled by the function

$$C(x) = 0.04x^2 + 0.6x + 330 + 7.5 \sin 2\pi x,$$

where $x = 0$ corresponds to 1970. (*Source*: Zeilik, M. and S. Gregory, *Introductory Astronomy and Astrophysics,* Brooks/Cole.)

(a) Graph C in the window $[5, 50]$ by $[320, 450]$.

(b) What part of the function causes the amplitude of the oscillations in the graph of C to be larger than the amplitude of the oscillations in the graph of L in **Exercise 55,** which models Hawaii?

57. ***Average Daily Temperature*** The temperature in Anchorage, Alaska, can be approximated by the function

$$T(x) = 37 + 21 \sin\left[\frac{2\pi}{365}(x - 91)\right],$$

where $T(x)$ is the temperature in degrees Fahrenheit on day x, with $x = 1$ corresponding to January 1 and $x = 365$ corresponding to December 31. Use a calculator to estimate the temperature on the following days. (*Source*: *World Almanac and Book of Facts*.)

(a) March 15 (day 74) **(b)** April 5 (day 95) **(c)** Day 200

(d) June 25 **(e)** October 1 **(f)** December 31

58. ***Fluctuation in the Solar Constant*** The **solar constant** S is the amount of energy per unit area that reaches Earth's atmosphere from the sun. It is equal to 1367 watts per m^2 but varies slightly throughout the seasons. This fluctuation ΔS in S can be calculated using the formula

$$\Delta S = 0.034S \sin\left[\frac{2\pi(82.5 - N)}{365.25}\right].$$

In this formula, N is the day number covering a four-year period, where $N = 1$ corresponds to January 1 of a leap year and $N = 1461$ corresponds to December 31 of the fourth year. (*Source*: Winter, C., R. Sizmann, and L. L.Vant-Hull, Editors, *Solar Power Plants,* Springer-Verlag.)

(a) Calculate ΔS for $N = 80$, which is the spring equinox in the first year.

(b) Calculate ΔS for $N = 1268$, which is the summer solstice in the fourth year.

(c) What is the maximum value of ΔS?

(d) Find a value for N where ΔS is equal to 0.

Musical Sound Waves *Pure sounds produce single sine waves on an oscilloscope. Find the amplitude and period of each sine wave graph. On the vertical scale, each square represents* 0.5. *On the horizontal scale, each square represents* 30° *or* $\frac{\pi}{6}$.

59.

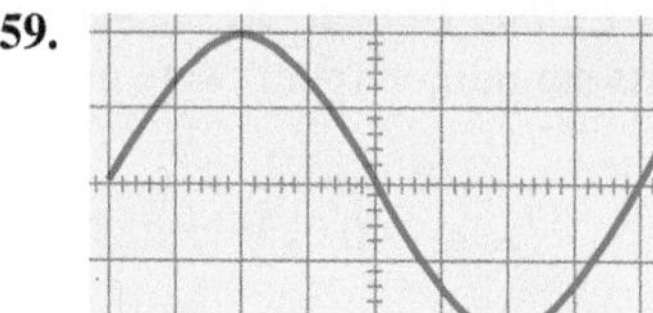

60.

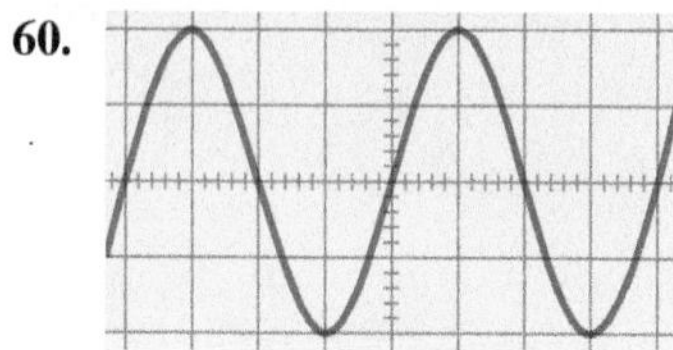

61. ***Concept Check*** Compare the graphs of $y = \sin 2x$ and $y = 2 \sin x$ over the interval $[0, 2\pi]$. Can we say that, in general, $\sin bx = b \sin x$ for $b > 0$? Explain.

62. ***Concept Check*** Compare the graphs of $y = \cos 3x$ and $y = 3 \cos x$ over the interval $[0, 2\pi]$. Can we say that, in general, $\cos bx = b \cos x$ for $b > 0$? Explain.

Relating Concepts

For individual or collaborative investigation *(Exercises 63–66)*

Connecting the Unit Circle and Sine Graph *Using a TI-84 Plus calculator, adjust the settings to correspond to the following screens.*

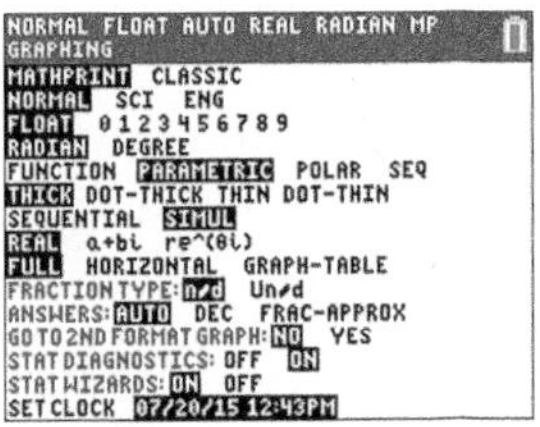

MODE

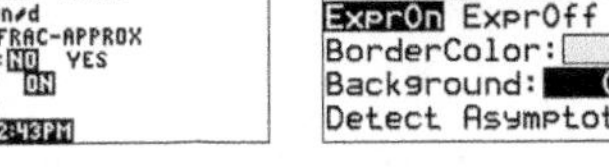

FORMAT

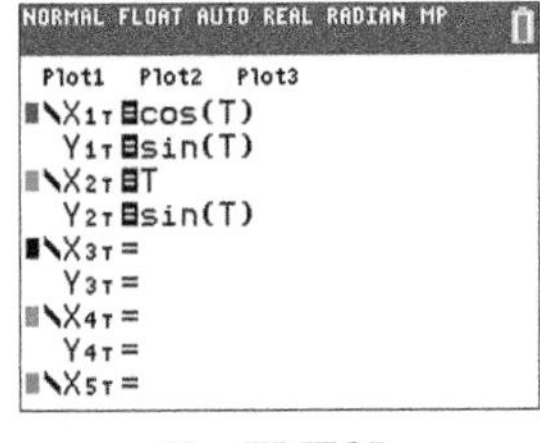

Y = EDITOR

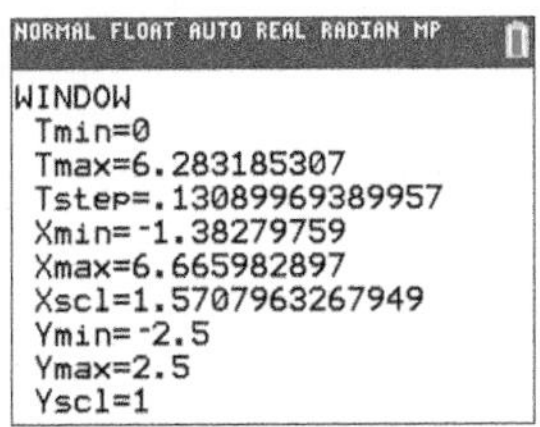

Graph the two equations (which are in ***parametric form****), and watch as the unit circle and the sine function are graphed simultaneously. Press the* TRACE *key once to obtain the screen shown on the left below. Then press the up-arrow key to obtain the screen shown on the right below. The screen on the left gives a unit circle interpretation of* $\cos 0 = 1$ *and* $\sin 0 = 0$. *The screen on the right gives a rectangular coordinate graph interpretation of* $\sin 0 = 0$.

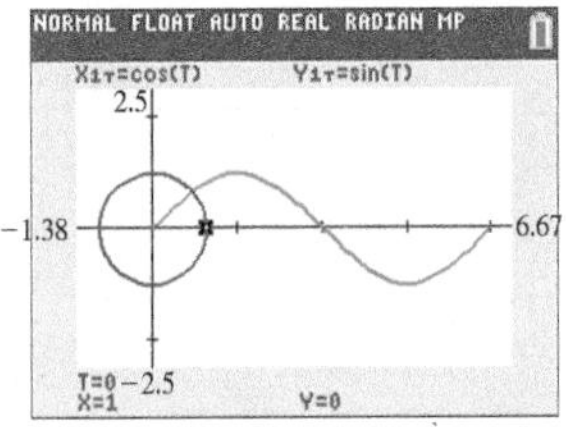

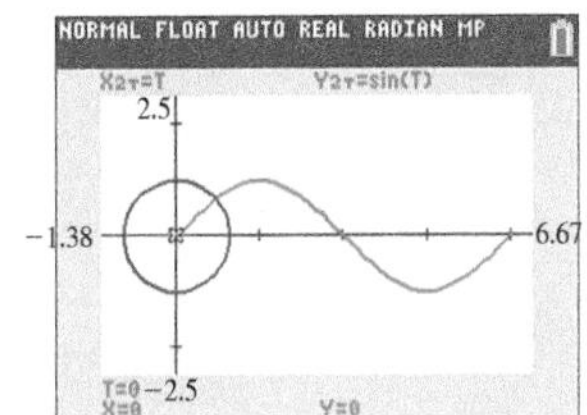

63. $X = -0.4161468$, $Y = 0.90929743$; X is cos 2 and Y is sin 2.
64. $X = 2$, $Y = 0.90929743$; $\sin 2 = 0.90929743$
65. $X = 2$, $Y = -0.4161468$; $\cos 2 = -0.4161468$
66. For an arc length T on the unit circle, $X = \cos T$ and $Y = \sin T$.

63. On the unit circle graph, let $T = 2$. Find X and Y, and interpret their values.

64. On the sine graph, let $T = 2$. What values of X and Y are displayed? Interpret these values with an equation in X and Y.

65. Now go back and redefine Y_{2T} as $\cos(T)$. Graph both equations. On the cosine graph, let $T = 2$. What values of X and Y are displayed? Interpret these values with an equation in X and Y.

66. Explain the relationship between the coordinates of the unit circle and the coordinates of the sine and cosine graphs.

4.2 Translations of the Graphs of the Sine and Cosine Functions

- Horizontal Translations
- Vertical Translations
- Combinations of Translations
- A Trigonometric Model

Horizontal Translations The graph of the function

$$y = f(x - d)$$

is translated *horizontally* compared to the graph of $y = f(x)$. The translation is d units to the right if $d > 0$ and $|d|$ units to the left if $d < 0$. See **Figure 12.**

With circular functions, a horizontal translation is a **phase shift.** In the function $y = f(x - d)$, the expression $x - d$ is the **argument.**

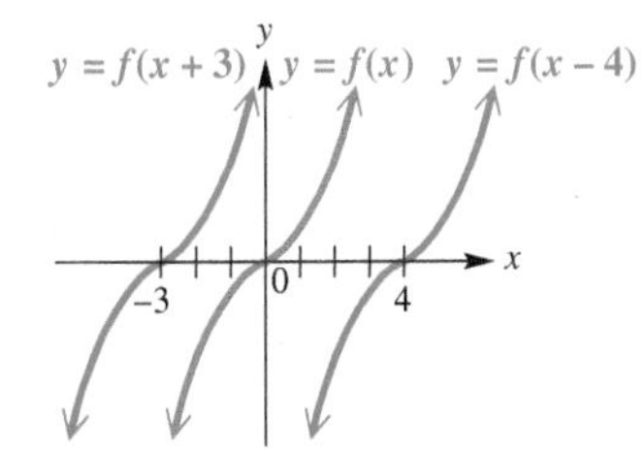

Horizontal translations of $y = f(x)$

Figure 12

Classroom Example 1

Graph $y = \sin\left(x + \frac{3\pi}{4}\right)$ over one period.

Answer:

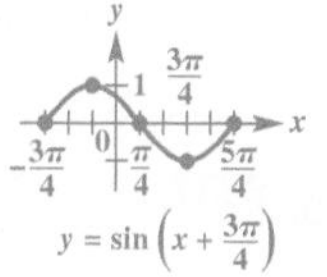

EXAMPLE 1 Graphing $y = \sin(x - d)$

Graph $y = \sin\left(x - \frac{\pi}{3}\right)$ over one period.

SOLUTION ***Method 1*** For the argument $x - \frac{\pi}{3}$ to result in all possible values throughout one period, it must take on all values between 0 and 2π, inclusive. To find an interval of one period, we solve the following three-part inequality.

$$0 \le x - \frac{\pi}{3} \le 2\pi \quad \text{Three-part inequality}$$

$$\frac{\pi}{3} \le x \le \frac{7\pi}{3} \quad \text{Add } \frac{\pi}{3} \text{ to each part.}$$

Use the method described in the previous section to divide the interval $\left[\frac{\pi}{3}, \frac{7\pi}{3}\right]$ into four equal parts, obtaining the following x-values.

$$\frac{\pi}{3}, \frac{5\pi}{6}, \frac{4\pi}{3}, \frac{11\pi}{6}, \frac{7\pi}{3}$$ These are *key* x-values.

A table of values using these x-values follows.

x	$\frac{\pi}{3}$	$\frac{5\pi}{6}$	$\frac{4\pi}{3}$	$\frac{11\pi}{6}$	$\frac{7\pi}{3}$
$x - \frac{\pi}{3}$	0	$\frac{\pi}{2}$	π	$\frac{3\pi}{2}$	2π
$\sin\left(x - \frac{\pi}{3}\right)$	0	1	0	-1	0

We join the corresponding points with a smooth curve to obtain the solid blue graph shown in **Figure 13.** The period is 2π, and the amplitude is 1.

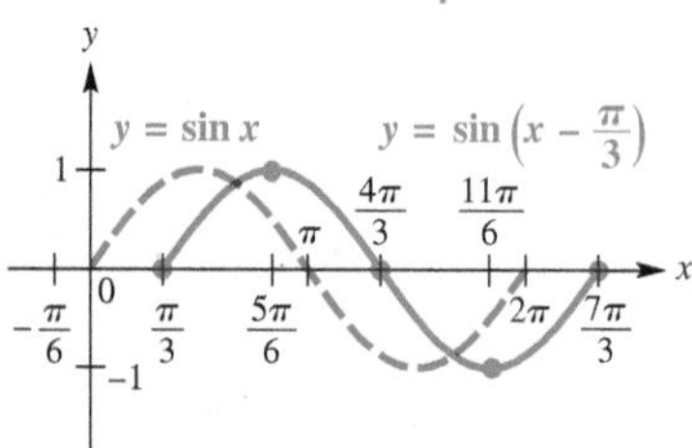

The graph can be extended through additional periods by repeating the given portion of the graph, as necessary.

Figure 13

Method 2 We can also graph $y = \sin\left(x - \frac{\pi}{3}\right)$ by using a horizontal translation of the graph of $y = \sin x$. The argument $x - \frac{\pi}{3}$ indicates that the graph will be translated $\frac{\pi}{3}$ units to the *right* (the phase shift) compared to the graph of $y = \sin x$. See **Figure 13.**

To graph a function using this method, first graph the basic circular function, and then graph the desired function using the appropriate translation.

✔ **Now Try Exercise 39.**

CAUTION In **Example 1,** the argument of the function is $\left(x - \frac{\pi}{3}\right)$. The parentheses are important here. If the function had been

$$y = \sin x - \frac{\pi}{3},$$

the graph would be that of $y = \sin x$ translated *vertically* $\frac{\pi}{3}$ units down.

Classroom Example 2

Graph $y = -2\cos\left(x - \frac{\pi}{3}\right)$ over one period.

Answer:

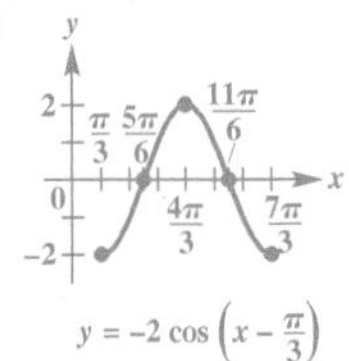

EXAMPLE 2 Graphing $y = a\cos(x - d)$

Graph $y = 3\cos\left(x + \frac{\pi}{4}\right)$ over one period.

SOLUTION *Method 1* We first solve the following three-part inequality.

$$0 \leq x + \frac{\pi}{4} \leq 2\pi \quad \text{Three-part inequality}$$

$$-\frac{\pi}{4} \leq x \leq \frac{7\pi}{4} \quad \text{Subtract } \tfrac{\pi}{4} \text{ from each part.}$$

Dividing this interval into four equal parts gives the following x-values. We use them to make a table of key points.

$$-\frac{\pi}{4}, \frac{\pi}{4}, \frac{3\pi}{4}, \frac{5\pi}{4}, \frac{7\pi}{4} \quad \text{Key } x\text{-values}$$

x	$-\frac{\pi}{4}$	$\frac{\pi}{4}$	$\frac{3\pi}{4}$	$\frac{5\pi}{4}$	$\frac{7\pi}{4}$
$x + \frac{\pi}{4}$	0	$\frac{\pi}{2}$	π	$\frac{3\pi}{2}$	2π
$\cos\left(x + \frac{\pi}{4}\right)$	1	0	-1	0	1
$3\cos\left(x + \frac{\pi}{4}\right)$	3	0	-3	0	3

These x-values lead to maximum points, minimum points, and x-intercepts.

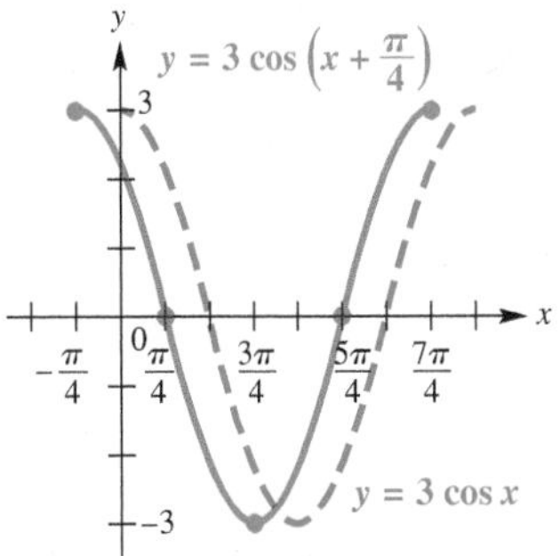

Figure 14

We join the corresponding points with a smooth curve to obtain the solid blue graph shown in **Figure 14.** The period is 2π, and the amplitude is 3.

Method 2 Write $y = 3\cos\left(x + \frac{\pi}{4}\right)$ in the form $y = a\cos(x - d)$.

$$y = 3\cos\left(x + \frac{\pi}{4}\right), \quad \text{or} \quad y = 3\cos\left[x - \left(-\frac{\pi}{4}\right)\right] \quad \text{Rewrite to subtract } -\tfrac{\pi}{4}.$$

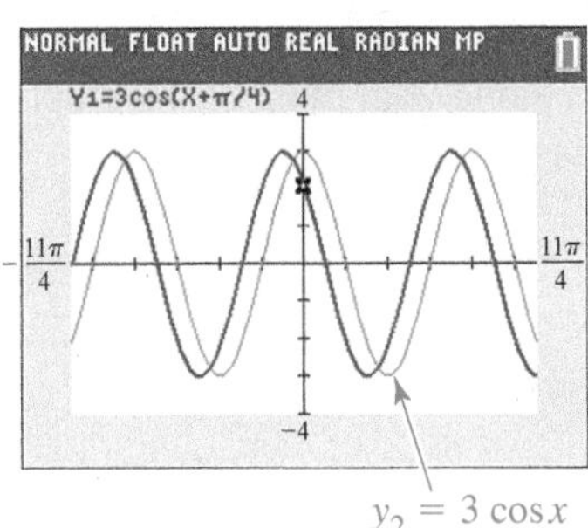

$y_2 = 3\cos x$

This result shows that $d = -\frac{\pi}{4}$. Because $-\frac{\pi}{4}$ is negative, the phase shift is $\left|-\frac{\pi}{4}\right| = \frac{\pi}{4}$ unit to the left. The graph is the same as that of $y = 3\cos x$ (the red graph in the calculator screen shown in the margin), except that it is translated $\frac{\pi}{4}$ unit to the left (the blue graph).

Now Try Exercise 41.

Classroom Example 3

Graph $y = \frac{3}{2}\cos(2x - \pi)$ over two periods.

Answer:

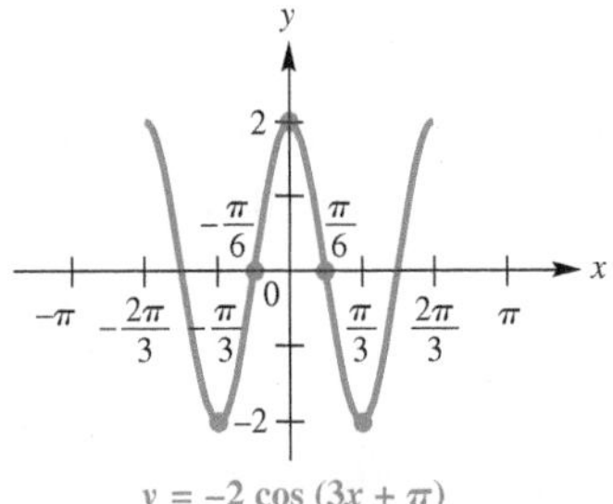

Figure 15

EXAMPLE 3 Graphing $y = a\cos[b(x - d)]$

Graph $y = -2\cos(3x + \pi)$ over two periods.

SOLUTION *Method 1* We first solve the three-part inequality

$$0 \leq 3x + \pi \leq 2\pi$$

to find the interval $\left[-\frac{\pi}{3}, \frac{\pi}{3}\right]$. Dividing this interval into four equal parts gives the points $\left(-\frac{\pi}{3}, -2\right)$, $\left(-\frac{\pi}{6}, 0\right)$, $(0, 2)$, $\left(\frac{\pi}{6}, 0\right)$, and $\left(\frac{\pi}{3}, -2\right)$. We plot these points and join them with a smooth curve. By graphing an additional half period to the left and to the right, we obtain the graph shown in **Figure 15.**

Method 2 First write the equation in the form $y = a\cos[b(x - d)]$.

$$y = -2\cos(3x + \pi), \quad \text{or} \quad y = -2\cos\left[3\left(x + \frac{\pi}{3}\right)\right] \quad \text{Rewrite by factoring out 3.}$$

Then $a = -2$, $b = 3$, and $d = -\frac{\pi}{3}$. The amplitude is $|-2| = 2$, and the period is $\frac{2\pi}{3}$ (because the value of b is 3). The phase shift is $\left|-\frac{\pi}{3}\right| = \frac{\pi}{3}$ units to the left compared to the graph of $y = -2\cos 3x$. Again, see **Figure 15.**

Now Try Exercise 47.

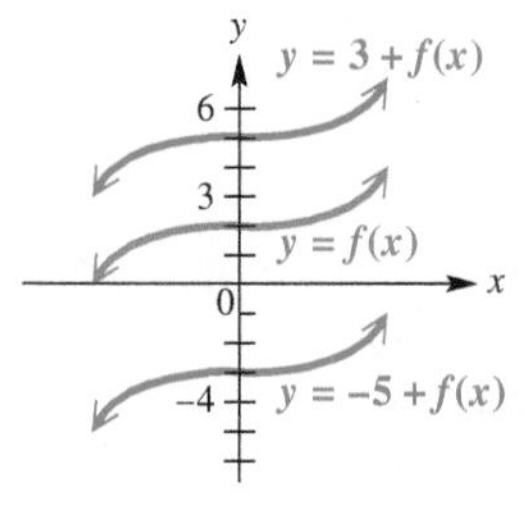

Vertical translations of $y = f(x)$

Figure 16

Vertical Translations

The graph of a function of the form

$$y = c + f(x)$$

is translated *vertically* compared to the graph of $y = f(x)$. See **Figure 16.** The translation is c units up if $c > 0$ and is $|c|$ units down if $c < 0$.

Classroom Example 4
Graph $y = -2 + 3 \cos 2x$ over two periods.

Answer:

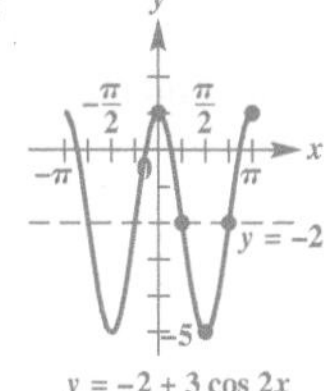

EXAMPLE 4 Graphing $y = c + a \cos bx$

Graph $y = 3 - 2 \cos 3x$ over two periods.

SOLUTION We use Method 1. The values of y will be 3 greater than the corresponding values of y in $y = -2 \cos 3x$. This means that the graph of $y = 3 - 2 \cos 3x$ is the same as the graph of $y = -2 \cos 3x$, vertically translated 3 units up. The period of $y = -2 \cos 3x$ is $\frac{2\pi}{3}$, so the key points have these x-values.

$$0, \quad \frac{\pi}{6}, \quad \frac{\pi}{3}, \quad \frac{\pi}{2}, \quad \frac{2\pi}{3} \qquad \text{Key } x\text{-values}$$

Use these x-values to make a table of points.

x	0	$\frac{\pi}{6}$	$\frac{\pi}{3}$	$\frac{\pi}{2}$	$\frac{2\pi}{3}$
$\cos 3x$	1	0	−1	0	1
$2 \cos 3x$	2	0	−2	0	2
$3 - 2 \cos 3x$	1	3	5	3	1

The key points are shown on the graph in **Figure 17,** along with more of the graph, which is sketched using the fact that the function is periodic.

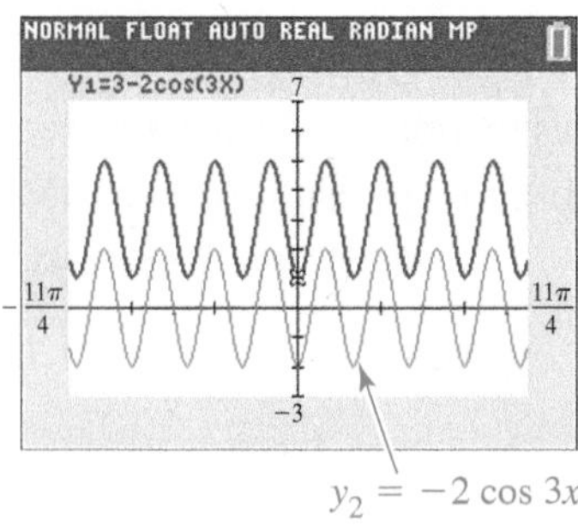

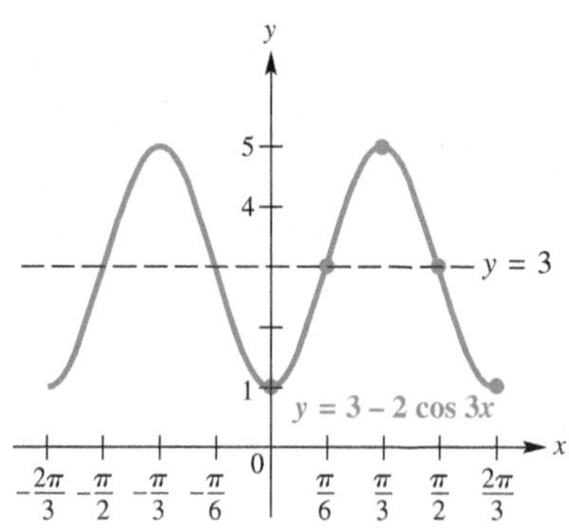

Figure 17

✔ Now Try Exercise 51.

Teaching Tip In **Example 4,** students may find it easiest to sketch the graph of

$$y = c + f(x)$$

by setting up a "temporary axis" at $y = c$. See **Figure 17,** for example.

CAUTION If we use ***Method 2*** to graph the function $y = 3 - 2 \cos 3x$ in **Example 4,** we must ***first*** graph

$$y = -2 \cos 3x$$

and ***then*** apply the vertical translation 3 units up. ***To begin,*** use the fact that $a = -2$ and $b = 3$ to determine that the amplitude is 2, the period is $\frac{2\pi}{3}$, and the graph is the reflection of the graph of $y = 2 \cos 3x$ across the x-axis. ***Then,*** because $c = 3$, translate the graph of $y = -2 \cos 3x$ up 3 units. See **Figure 17.**

If the vertical translation is applied first, then the reflection must be across the line $y = 3$, not across the x-axis.

Combinations of Translations

Further Guidelines for Sketching Graphs of Sine and Cosine Functions

To graph $y = c + a \sin[b(x - d)]$ or $y = c + a \cos[b(x - d)]$, with $b > 0$, follow these steps.

Method 1

Step 1 Find an interval whose length is one period $\frac{2\pi}{b}$ by solving the three-part inequality $0 \le b(x - d) \le 2\pi$.

Step 2 Divide the interval into four equal parts to obtain five key x-values.

Step 3 Evaluate the function for each of the five x-values resulting from Step 2. The points will be maximum points, minimum points, and points that intersect the line $y = c$ ("middle" points of the wave).

Step 4 Plot the points found in Step 3, and join them with a sinusoidal curve having amplitude $|a|$.

Step 5 Draw the graph over additional periods, as needed.

Method 2

Step 1 Graph $y = a \sin bx$ or $y = a \cos bx$. The amplitude of the function is $|a|$, and the period is $\frac{2\pi}{b}$.

Step 2 Use translations to graph the desired function. The vertical translation is c units up if $c > 0$ and $|c|$ units down if $c < 0$. The horizontal translation (phase shift) is d units to the right if $d > 0$ and $|d|$ units to the left if $d < 0$.

Teaching Tip Give several examples of sinusoidal graphs, and ask students to determine their corresponding equations. Point out that any given graph may have several different corresponding equations, all of which are valid.

Many computer graphing utilities have animation features that enable us to incrementally change the values of a, b, c, and d in the sinusoidal graphs of

$$y = c + a \sin b(x - d)$$

and $y = c + a \cos b(x - d)$.

This provides an excellent visual image of how these parameters affect sinusoidal graphs.

Classroom Example 5
Graph $y = 4 - 2 \sin(3x - \pi)$ over two periods.

Answer:

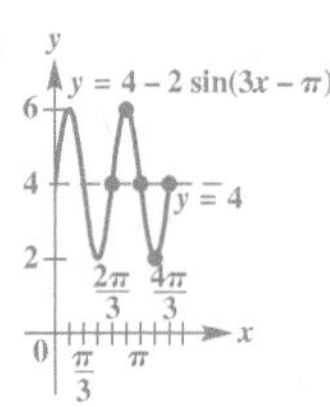

EXAMPLE 5 Graphing $y = c + a \sin[b(x - d)]$

Graph $y = -1 + 2 \sin(4x + \pi)$ over two periods.

SOLUTION We use Method 1. We must first write the expression on the right side of the equation in the form $c + a \sin[b(x - d)]$.

$$y = -1 + 2\sin(4x + \pi), \quad \text{or} \quad y = -1 + 2\sin\left[4\left(x + \frac{\pi}{4}\right)\right] \quad \text{Rewrite by factoring out 4.}$$

Step 1 Find an interval whose length is one period.

$$0 \le 4\left(x + \frac{\pi}{4}\right) \le 2\pi \quad \text{Three-part inequality}$$

$$0 \le x + \frac{\pi}{4} \le \frac{\pi}{2} \quad \text{Divide each part by 4.}$$

$$-\frac{\pi}{4} \le x \le \frac{\pi}{4} \quad \text{Subtract } \tfrac{\pi}{4} \text{ from each part.}$$

Step 2 Divide the interval $\left[-\frac{\pi}{4}, \frac{\pi}{4}\right]$ into four equal parts to obtain these x-values.

$$-\frac{\pi}{4}, \quad -\frac{\pi}{8}, \quad 0, \quad \frac{\pi}{8}, \quad \frac{\pi}{4} \quad \text{Key } x\text{-values}$$

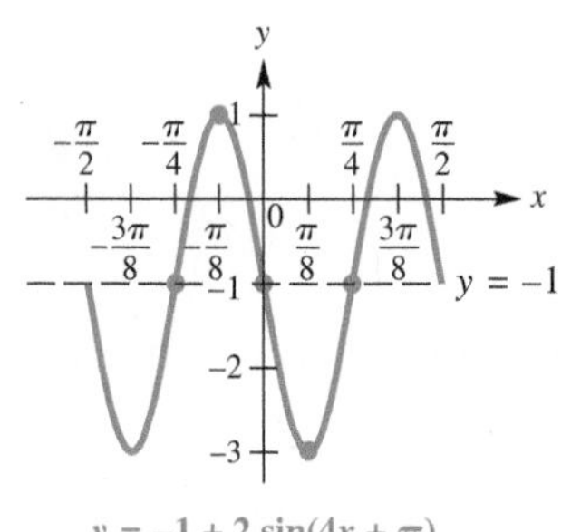

$y = -1 + 2\sin(4x + \pi)$

Figure 18

Step 3 Make a table of values.

x	$-\frac{\pi}{4}$	$-\frac{\pi}{8}$	0	$\frac{\pi}{8}$	$\frac{\pi}{4}$
$x + \frac{\pi}{4}$	0	$\frac{\pi}{8}$	$\frac{\pi}{4}$	$\frac{3\pi}{8}$	$\frac{\pi}{2}$
$4(x + \frac{\pi}{4})$	0	$\frac{\pi}{2}$	π	$\frac{3\pi}{2}$	2π
$\sin[4(x + \frac{\pi}{4})]$	0	1	0	-1	0
$2\sin[4(x + \frac{\pi}{4})]$	0	2	0	-2	0
$-1 + 2\sin(4x + \pi)$	-1	1	-1	-3	-1

Steps 4 and 5 Plot the points found in the table and join them with a sinusoidal curve. **Figure 18** shows the graph, extended to the right and left to include two full periods.

✔ **Now Try Exercise 57.**

A Trigonometric Model For natural phenomena that occur in periodic patterns (such as seasonal temperatures, phases of the moon, heights of tides) a sinusoidal function will provide a good approximation of a set of data points.

EXAMPLE 6 Modeling Temperature with a Sine Function

The maximum average monthly temperature in New Orleans, Louisiana, is 83°F, and the minimum is 53°F. The table shows the average monthly temperatures. The scatter diagram for a two-year interval in **Figure 19** strongly suggests that the temperatures can be modeled with a sine curve.

Month	°F	Month	°F
Jan	53	July	83
Feb	56	Aug	83
Mar	62	Sept	79
Apr	68	Oct	70
May	76	Nov	61
June	81	Dec	55

Source: World Almanac and Book of Facts.

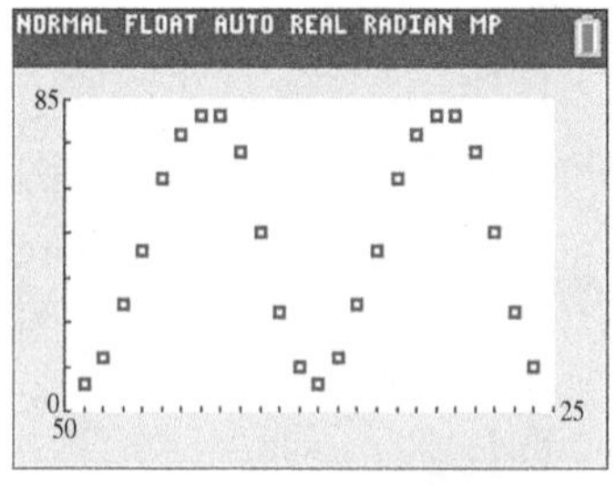

Figure 19

(a) Using only the maximum and minimum temperatures, determine a function of the form

$$f(x) = a\sin[b(x - d)] + c, \quad \text{where } a, b, c, \text{ and } d \text{ are constants,}$$

that models the average monthly temperature in New Orleans. Let x represent the month, with January corresponding to $x = 1$.

(b) On the same coordinate axes, graph f for a two-year period together with the actual data values found in the table.

(c) Use the **sine regression** feature of a graphing calculator to determine a second model for these data.

Classroom Example 6
Repeat **Example 6** for the following table, which gives the mean maximum temperatures in degrees Celsius for Sydney, Australia.

Month	°C	Month	°C
Jan	25.9	July	16.3
Feb	25.8	Aug	17.8
Mar	24.8	Sept	20.0
Apr	22.4	Oct	22.1
May	19.5	Nov	23.6
June	17.0	Dec	25.2

Source: www.bom.gov.au

Classroom Example 6
Answers:

(a) $f(x) = 4.8 \sin\left[\frac{\pi}{6}(x+2)\right] + 21.1$

(b)

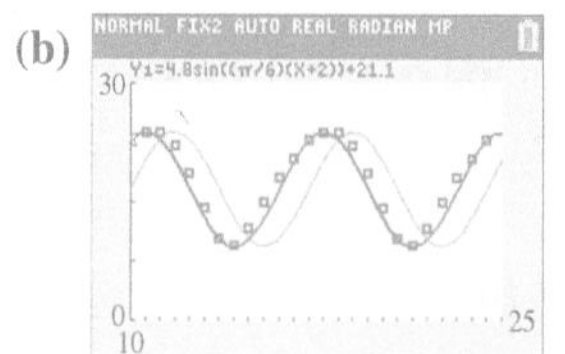

(c) $f(x) = 4.67 \sin(0.53x + 0.97) + 21.67$

SOLUTION

(a) We use the maximum and minimum average monthly temperatures to find the amplitude a.

$$a = \frac{83 - 53}{2} = 15 \quad \text{Amplitude } a$$

The average of the maximum and minimum temperatures is a good choice for c. The average is

$$\frac{83 + 53}{2} = 68. \quad \text{Vertical translation } c$$

Because temperatures repeat every 12 months, b can be found as follows.

$$12 = \frac{2\pi}{b} \quad \text{Period} = \frac{2\pi}{b}$$

$$b = \frac{\pi}{6} \quad \text{Solve for } b.$$

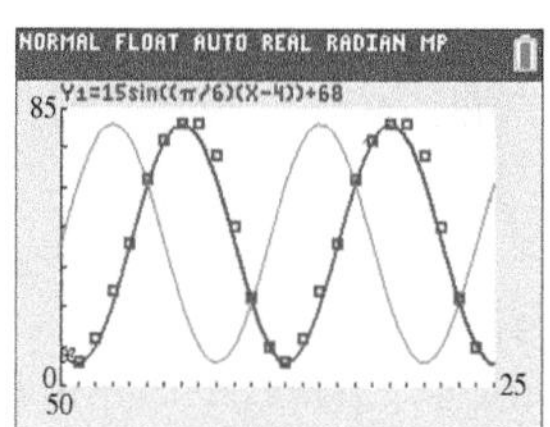

Figure 20

The coldest month is January, when $x = 1$, and the hottest month is July, when $x = 7$. A good choice for d is 4 because April, when $x = 4$, is located at the midpoint between January and July. Also, notice that the average monthly temperature in April is 68°F, which is the value of the vertical translation, c. The average monthly temperature in New Orleans is modeled closely by the following equation.

$$f(x) = a \sin[b(x - d)] + c$$

$$f(x) = 15 \sin\left[\frac{\pi}{6}(x - 4)\right] + 68 \quad \text{Substitute for } a, b, c, \text{ and } d.$$

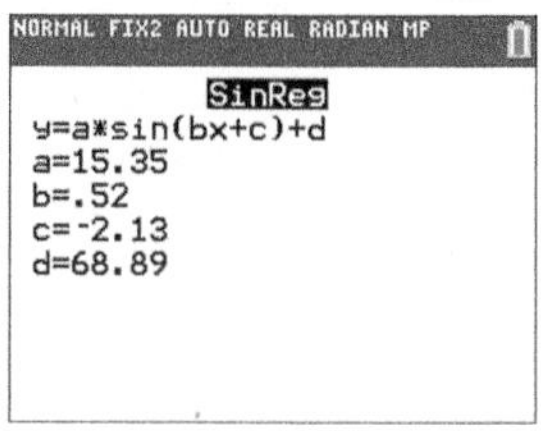

(a)

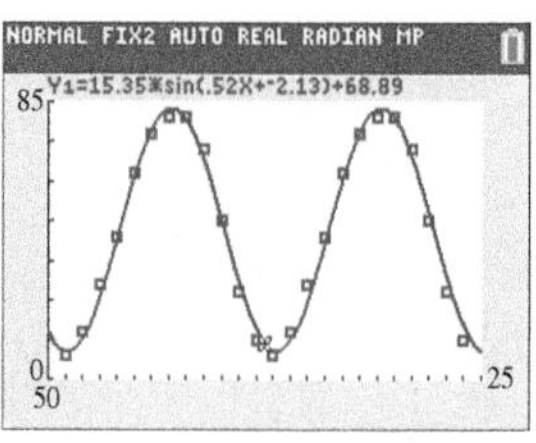

(b)

Figure 21

(b) **Figure 20** shows two iterations of the data points from the table, along with the graph of $y = 15 \sin\left[\frac{\pi}{6}(x - 4)\right] + 68$. The graph of $y = 15 \sin \frac{\pi}{6}x + 68$ is shown for comparison.

(c) We used the given data for a two-year period and the sine regression capability of a graphing calculator to produce the model

$$f(x) = 15.35 \sin(0.52x - 2.13) + 68.89$$

described in **Figure 21(a).** Its graph along with the data points is shown in **Figure 21(b).**

✔ **Now Try Exercise 61.**

4.2 Exercises

1. $\frac{\pi}{4}$; left 2. $\frac{\pi}{6}$; right

CONCEPT PREVIEW *Fill in the blank(s) to correctly complete each sentence.*

1. The graph of $y = \sin\left(x + \frac{\pi}{4}\right)$ is obtained by shifting the graph of $y = \sin x$ ______ unit(s) to the ______.
(right/left)

2. The graph of $y = \cos\left(x - \frac{\pi}{6}\right)$ is obtained by shifting the graph of $y = \cos x$ ______ unit(s) to the ______.
(right/left)

3. 4
4. 3; x
5. 6; up
6. 5; down
7. $\frac{\pi}{5}$; left; 5; 3; up
8. $\frac{\pi}{6}$; right; 3; 2; down

9. D
10. G
11. H
12. A
13. B
14. E
15. I
16. C

17. The graph of $y = \sin x + 1$ is obtained by shifting the graph of $y = \sin x$ up 1 unit. The graph of $y = \sin(x + 1)$ is obtained by shifting the graph of $y = \sin x$ to the left 1 unit.
18. $y = \sin x + 1$

19. B
20. D
21. C
22. A

23. right
24. left

25. $y = -1 + \sin x$
26. $y = 2 + \cos x$
27. $y = \cos\left(x - \frac{\pi}{3}\right)$
28. $y = \cos\left(x - \frac{\pi}{6}\right)$

29. 2; 2π; none; π to the left
30. 3; 2π; none; $\frac{\pi}{2}$ to the left
31. $\frac{1}{4}$; 4π; none; π to the left
32. $\frac{1}{2}$; 4π; none; 2π to the left
33. 3; 4; none; $\frac{1}{2}$ to the right
34. 1; 2; none; $\frac{1}{3}$ to the right
35. 1; $\frac{2\pi}{3}$; up 2; $\frac{\pi}{15}$ to the right
36. $\frac{1}{2}$; π; down 1; $\frac{3\pi}{2}$ to the right

37.

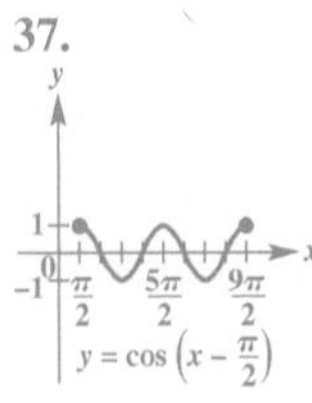

$y = \cos\left(x - \frac{\pi}{2}\right)$

38.

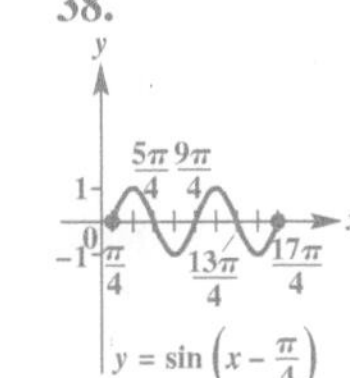

$y = \sin\left(x - \frac{\pi}{4}\right)$

39.

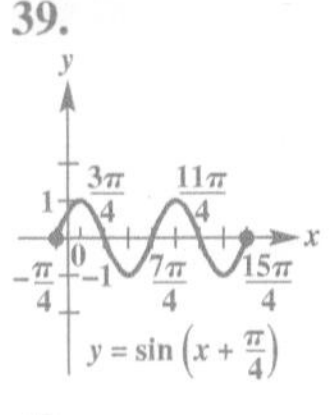

$y = \sin\left(x + \frac{\pi}{4}\right)$

40.

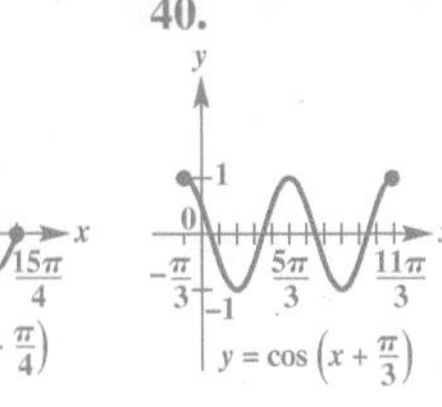

$y = \cos\left(x + \frac{\pi}{3}\right)$

41. 42.

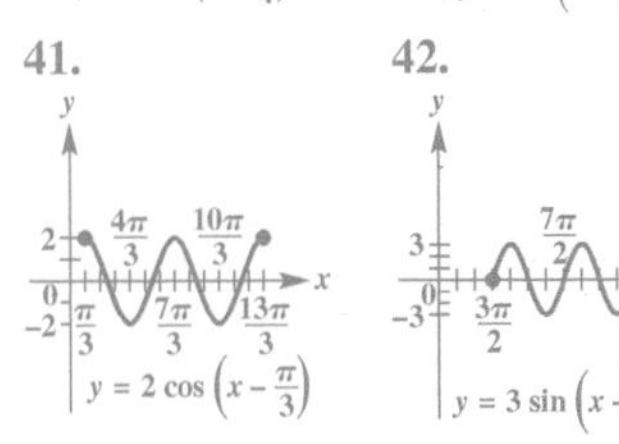

$y = 2\cos\left(x - \frac{\pi}{3}\right)$ $y = 3\sin\left(x - \frac{3\pi}{2}\right)$

3. The graph of $y = 4 \sin x$ is obtained by stretching the graph of $y = \sin x$ vertically by a factor of ______.

4. The graph of $y = -3 \sin x$ is obtained by stretching the graph of $y = \sin x$ by a factor of ______ and reflecting across the ______-axis.

5. The graph of $y = 6 + 3 \sin x$ is obtained by shifting the graph of $y = 3 \sin x$ ______ unit(s) __________ (up/down).

6. The graph of $y = -5 + 2 \cos x$ is obtained by shifting the graph of $y = 2 \cos x$ ______ unit(s) __________ (up/down).

7. The graph of $y = 3 + 5 \cos\left(x + \frac{\pi}{5}\right)$ is obtained by shifting the graph of $y = \cos x$ ______ unit(s) horizontally to the __________ (right/left), stretching it vertically by a factor of ______, and then shifting it ______ unit(s) vertically __________ (up/down).

8. Repeat **Exercise 7** for $y = -2 + 3 \cos\left(x - \frac{\pi}{6}\right)$.

Concept Check Match each function with its graph in choices A–I. (One choice will not be used.)

9. $y = \sin\left(x - \frac{\pi}{4}\right)$ **10.** $y = \sin\left(x + \frac{\pi}{4}\right)$ **11.** $y = \cos\left(x - \frac{\pi}{4}\right)$

12. $y = \cos\left(x + \frac{\pi}{4}\right)$ **13.** $y = 1 + \sin x$ **14.** $y = -1 + \sin x$

15. $y = 1 + \cos x$ **16.** $y = -1 + \cos x$

A.

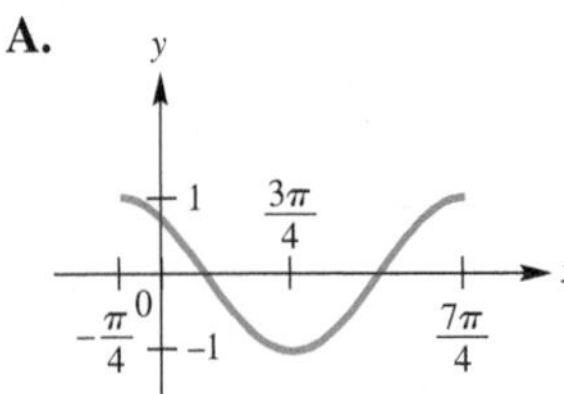

B.

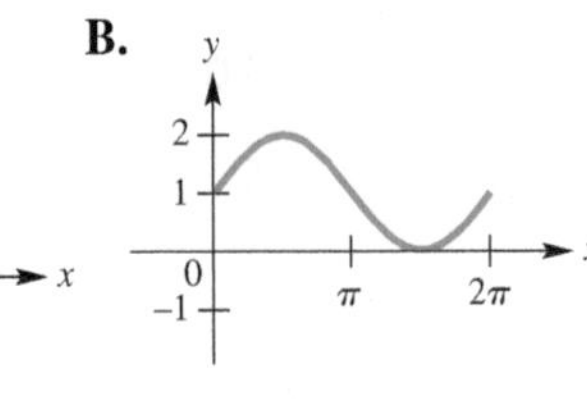

C.

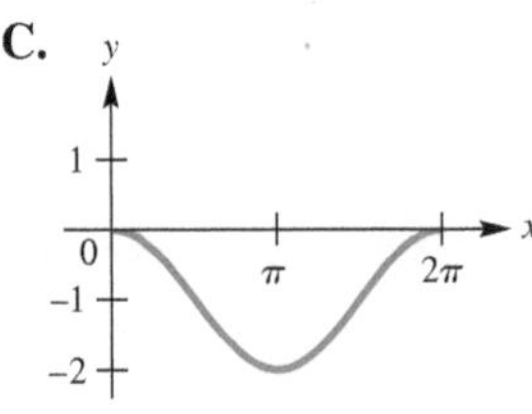

D.

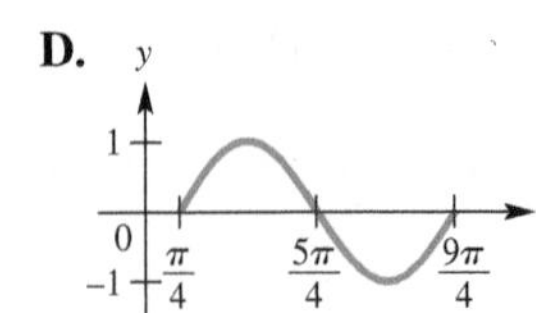

E.

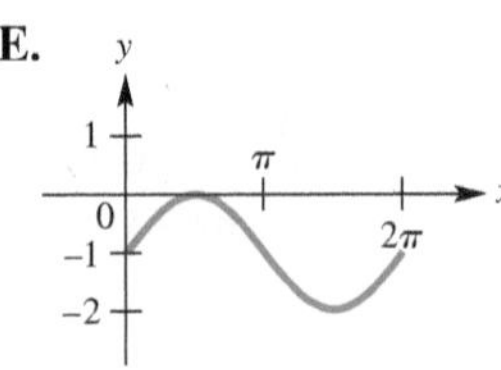

F.

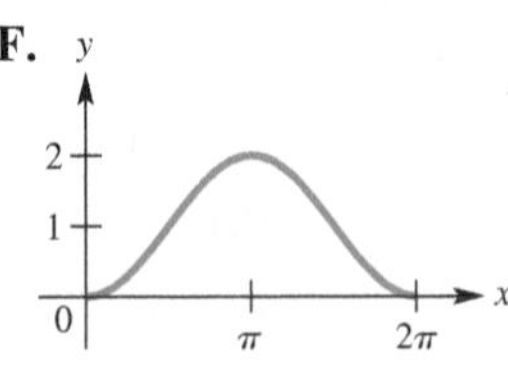

G.

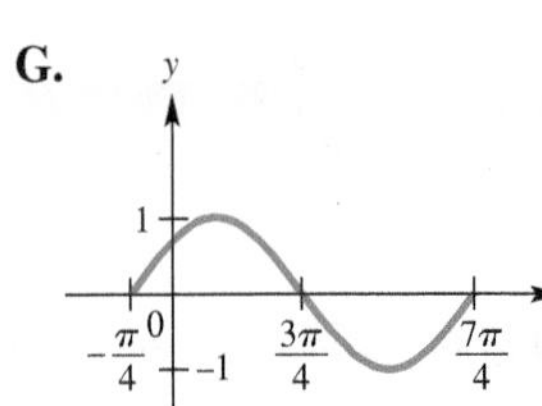

H.

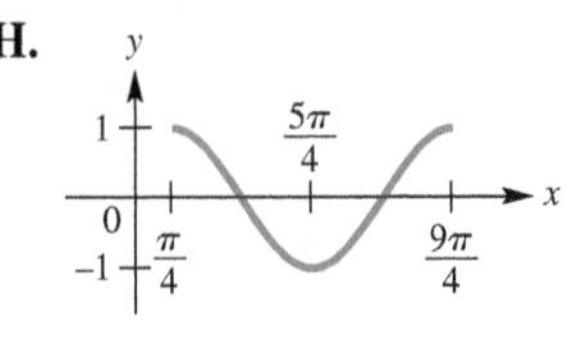

I.

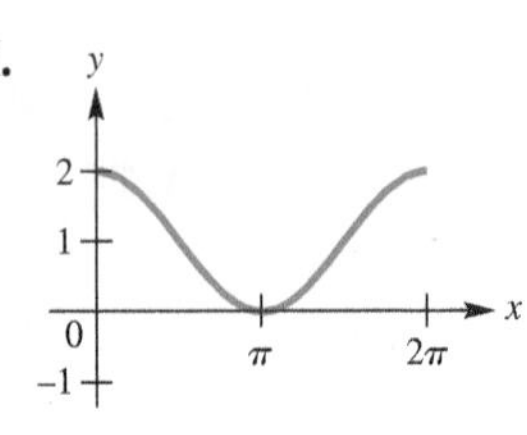

17. The graphs of $y = \sin x + 1$ and $y = \sin(x + 1)$ are **NOT** the same. Explain why this is so.

18. *Concept Check* Refer to **Exercise 17.** Which one of the two graphs is the same as that of $y = 1 + \sin x$?

43. $y = \frac{3}{2}\sin\left[2\left(x + \frac{\pi}{4}\right)\right]$

44. $y = -\frac{1}{2}\cos\left[4\left(x + \frac{\pi}{2}\right)\right]$

45. 46.

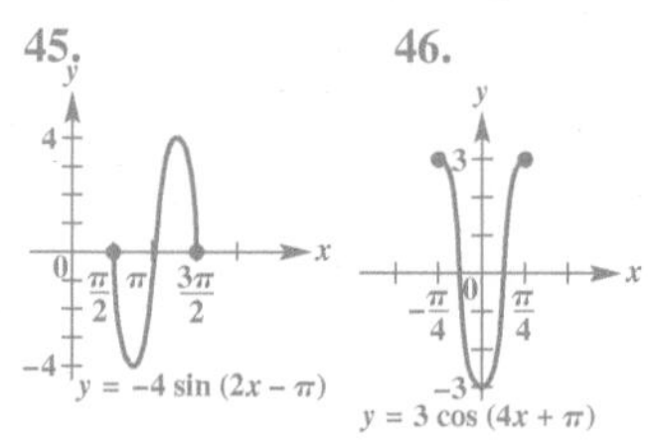

47. 48.

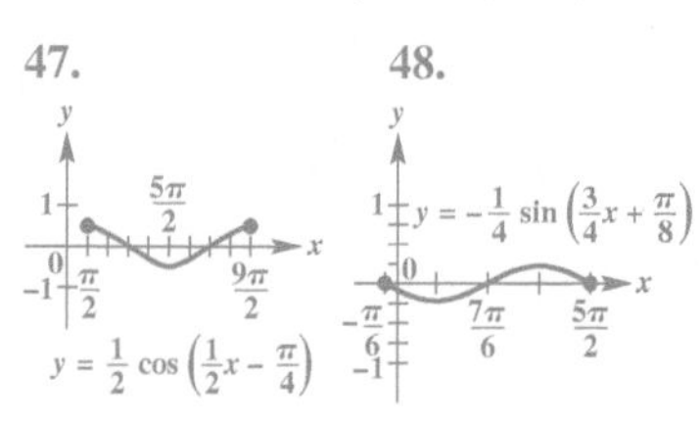

49. $y = -3 + 2\sin x$

50.

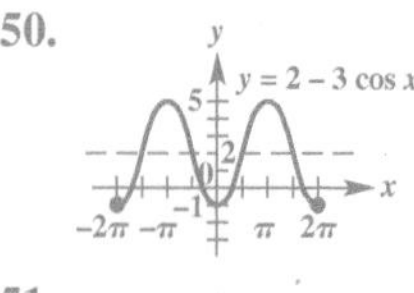

51.

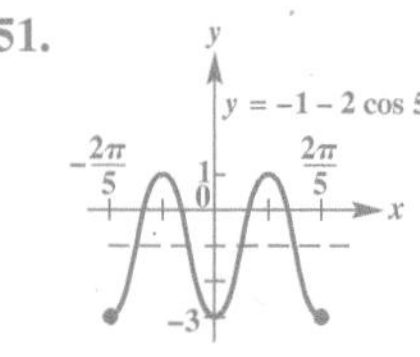

52.

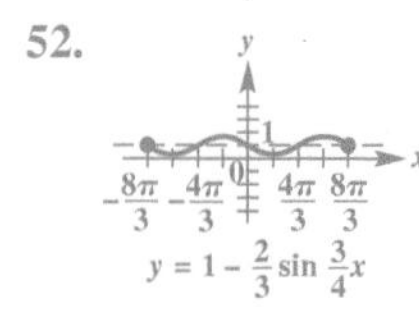

53.

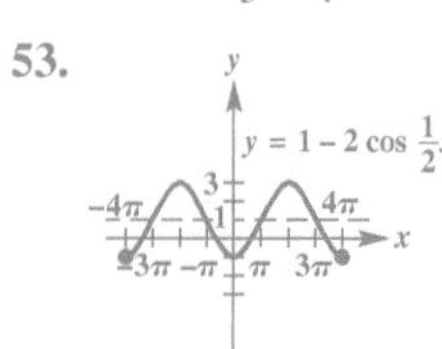

54.

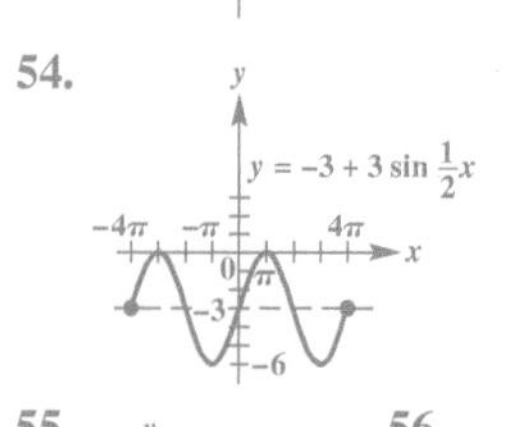

55. 56.

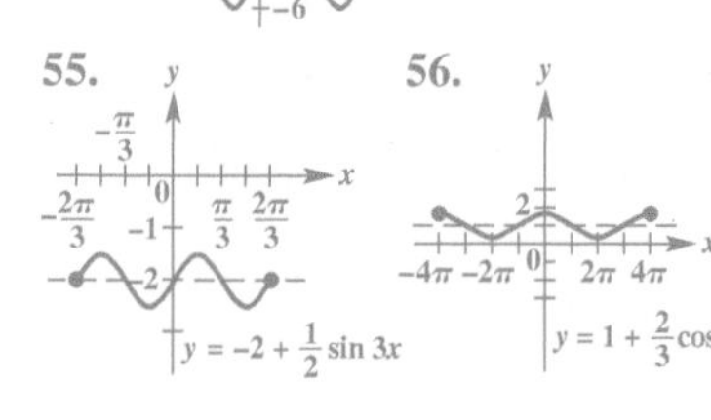

Concept Check *Match each function in Column I with the appropriate description in Column II.*

I	II
19. $y = 3\sin(2x - 4)$	**A.** amplitude $= 2$, period $= \frac{\pi}{2}$, phase shift $= \frac{3}{4}$
20. $y = 2\sin(3x - 4)$	**B.** amplitude $= 3$, period $= \pi$, phase shift $= 2$
21. $y = -4\sin(3x - 2)$	**C.** amplitude $= 4$, period $= \frac{2\pi}{3}$, phase shift $= \frac{2}{3}$
22. $y = -2\sin(4x - 3)$	**D.** amplitude $= 2$, period $= \frac{2\pi}{3}$, phase shift $= \frac{4}{3}$

Concept Check *Fill in each blank with the word* right *or the word* left.

23. If the graph of $y = \cos x$ is translated $\frac{\pi}{2}$ units horizontally to the ________, it will coincide with the graph of $y = \sin x$.

24. If the graph of $y = \sin x$ is translated $\frac{\pi}{2}$ units horizontally to the ________, it will coincide with the graph of $y = \cos x$.

Connecting Graphs with Equations *Each function graphed is of the form $y = c + \cos x$, $y = c + \sin x$, $y = \cos(x - d)$, or $y = \sin(x - d)$, where d is the least possible positive value. Determine an equation of the graph.*

25.

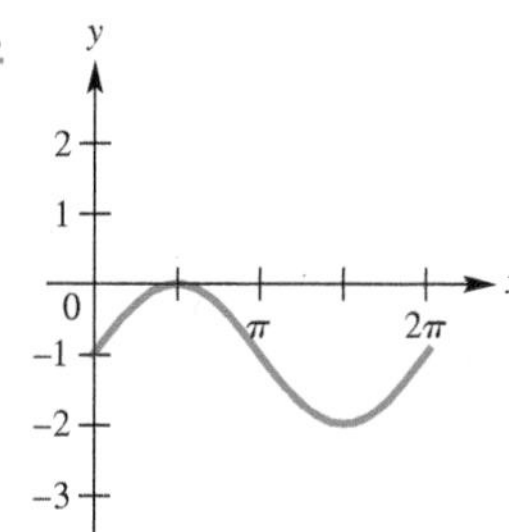

26.

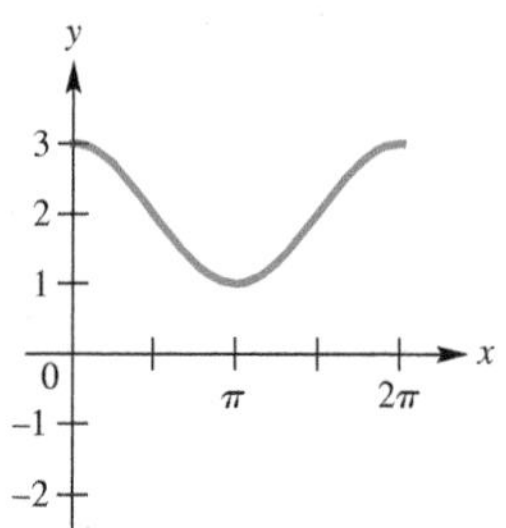

27.

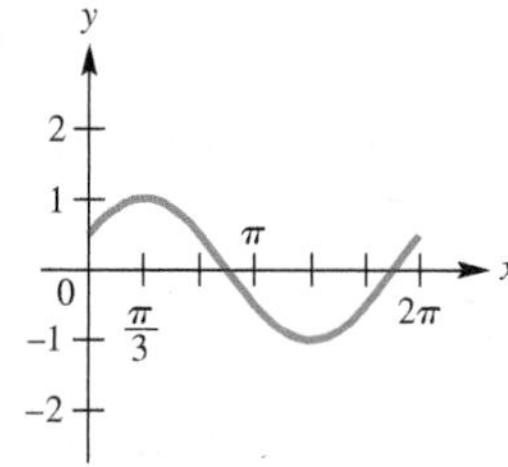

28.

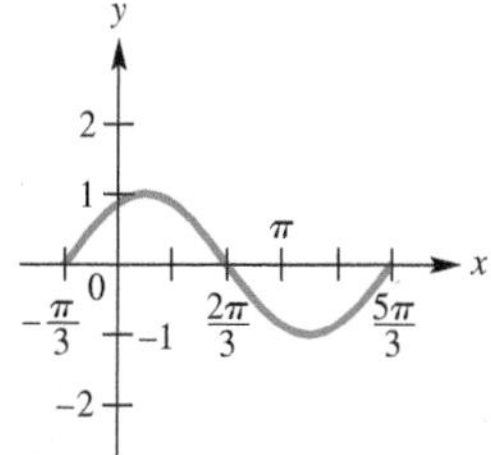

Find the amplitude, the period, any vertical translation, and any phase shift of the graph of each function. ***See Examples 1–5.***

29. $y = 2\sin(x + \pi)$

30. $y = 3\sin\left(x + \frac{\pi}{2}\right)$

31. $y = -\frac{1}{4}\cos\left(\frac{1}{2}x + \frac{\pi}{2}\right)$

32. $y = -\frac{1}{2}\sin\left(\frac{1}{2}x + \pi\right)$

33. $y = 3\cos\left[\frac{\pi}{2}\left(x - \frac{1}{2}\right)\right]$

34. $y = -\cos\left[\pi\left(x - \frac{1}{3}\right)\right]$

35. $y = 2 - \sin\left(3x - \frac{\pi}{5}\right)$

36. $y = -1 + \frac{1}{2}\cos(2x - 3\pi)$

57. **58.**

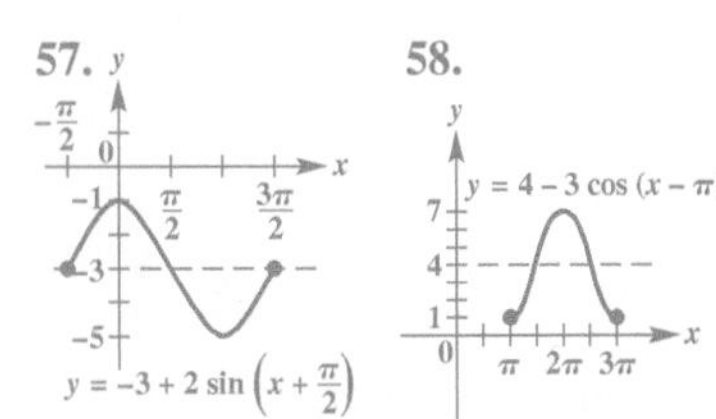

59.

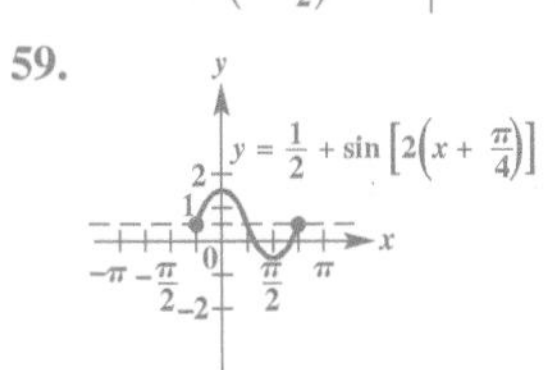

60.

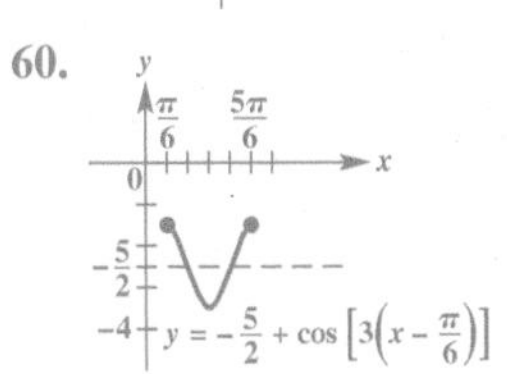

61. (a)

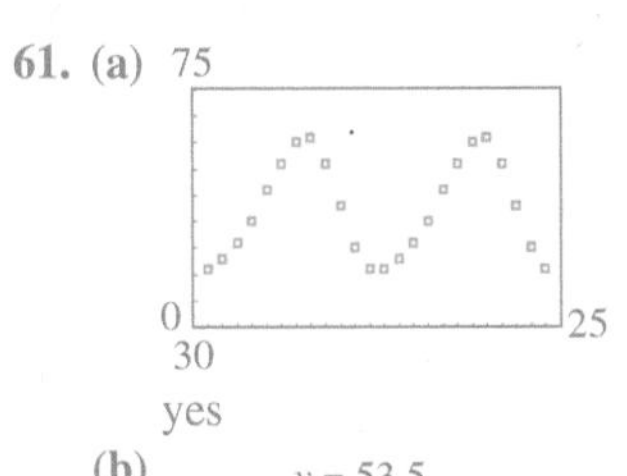

yes

(b)

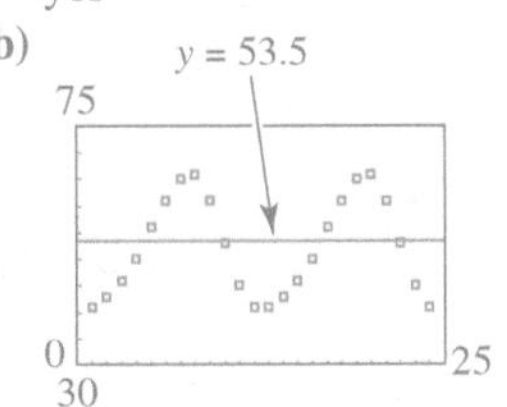

It represents the average annual temperature.

(c) 12.5; 12; 4.5

(d) $f(x) = 12.5\sin\left[\frac{\pi}{6}(x - 4.5)\right] + 53.5$

(e)

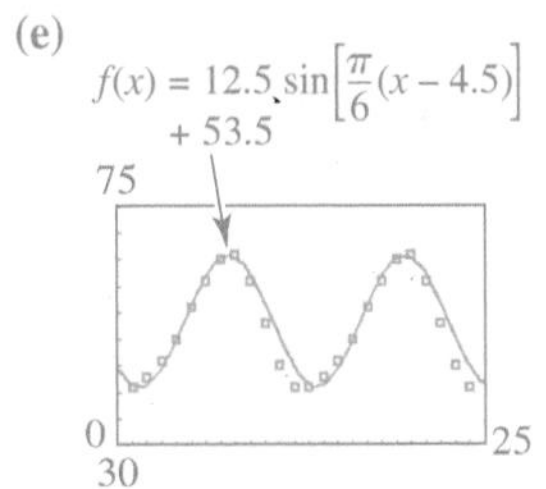

The function gives a good model for the data.

(f)

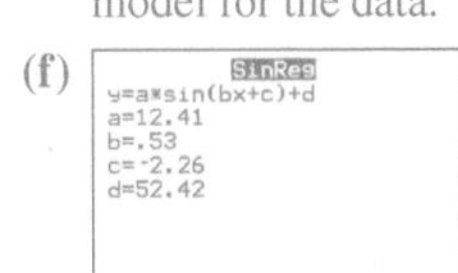

TI-84 Plus fixed to the nearest hundredth

Graph each function over a two-period interval. ***See Examples 1 and 2.***

37. $y = \cos\left(x - \frac{\pi}{2}\right)$ **38.** $y = \sin\left(x - \frac{\pi}{4}\right)$ **39.** $y = \sin\left(x + \frac{\pi}{4}\right)$

40. $y = \cos\left(x + \frac{\pi}{3}\right)$ **41.** $y = 2\cos\left(x - \frac{\pi}{3}\right)$ **42.** $y = 3\sin\left(x - \frac{3\pi}{2}\right)$

Graph each function over a one-period interval. ***See Example 3.***

43. $y = \frac{3}{2}\sin\left[2\left(x + \frac{\pi}{4}\right)\right]$ **44.** $y = -\frac{1}{2}\cos\left[4\left(x + \frac{\pi}{2}\right)\right]$

45. $y = -4\sin(2x - \pi)$ **46.** $y = 3\cos(4x + \pi)$

47. $y = \frac{1}{2}\cos\left(\frac{1}{2}x - \frac{\pi}{4}\right)$ **48.** $y = -\frac{1}{4}\sin\left(\frac{3}{4}x + \frac{\pi}{8}\right)$

Graph each function over a two-period interval. ***See Example 4.***

49. $y = -3 + 2\sin x$ **50.** $y = 2 - 3\cos x$ **51.** $y = -1 - 2\cos 5x$

52. $y = 1 - \frac{2}{3}\sin\frac{3}{4}x$ **53.** $y = 1 - 2\cos\frac{1}{2}x$ **54.** $y = -3 + 3\sin\frac{1}{2}x$

55. $y = -2 + \frac{1}{2}\sin 3x$ **56.** $y = 1 + \frac{2}{3}\cos\frac{1}{2}x$

Graph each function over a one-period interval. ***See Example 5.***

57. $y = -3 + 2\sin\left(x + \frac{\pi}{2}\right)$ **58.** $y = 4 - 3\cos(x - \pi)$

59. $y = \frac{1}{2} + \sin\left[2\left(x + \frac{\pi}{4}\right)\right]$ **60.** $y = -\frac{5}{2} + \cos\left[3\left(x - \frac{\pi}{6}\right)\right]$

(Modeling) Solve each problem. ***See Example 6.***

61. ***Average Monthly Temperature*** The average monthly temperature (in °F) in Seattle, Washington, is shown in the table.

Month	°F	Month	°F
Jan	41	July	65
Feb	43	Aug	66
Mar	46	Sept	61
Apr	50	Oct	53
May	56	Nov	45
June	61	Dec	41

Source: World Almanac and Book of Facts.

(a) Plot the average monthly temperature over a two-year period, letting $x = 1$ correspond to January of the first year. Do the data seem to indicate a translated sine graph?

(b) The highest average monthly temperature is 66°F in August, and the lowest average monthly temperature is 41°F in January. Their average is 53.5°F. Graph the data together with the line $y = 53.5$. What does this line represent with regard to temperature in Seattle?

(c) Approximate the amplitude, period, and phase shift of the translated sine wave.

(d) Determine a function of the form $f(x) = a\sin[b(x - d)] + c$, where a, b, c, and d are constants, that models the data.

(e) Graph f together with the data on the same coordinate axes. How well does f model the given data?

(f) Use the sine regression capability of a graphing calculator to find the equation of a sine curve that fits these data.

62. (a) 73.5°F

(b) See the graph in part (d).

(c) $f(x) = 19.5 \cos\left[\frac{\pi}{6}(x - 7)\right] + 73.5$

(d) $f(x) = 19.5 \cos\left[\frac{\pi}{6}(x - 7)\right] + 73.5$

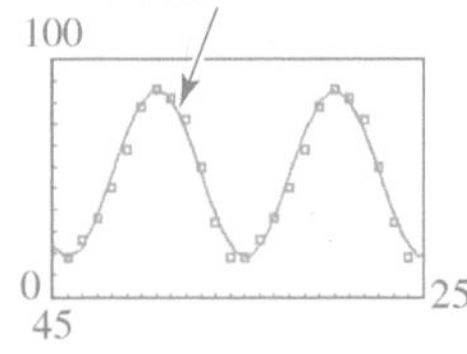

The function gives a good model for the data.

(e)

TI-84 Plus fixed to the nearest hundredth

In the answers to Exercises 63–66, we give the model and one graph of the data and equation.

63. (a) See the graph in part (c).

(b) $y = 12.28 \sin(0.52x + 1.06) + 63.96$

(c)

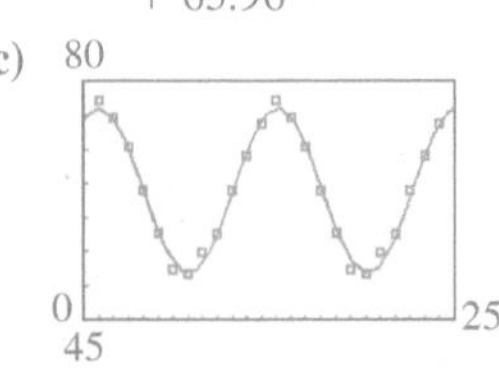

64. (a) See the graph in part (c).

(b) $y = 12.95 \sin(0.52x + 1.05) + 72.50$

(c)

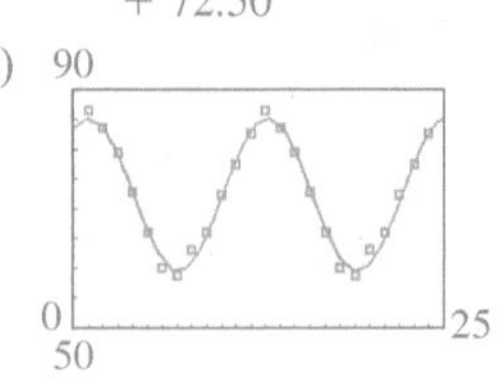

65. (a) See the graph in part (c).

(b) $y = 0.49 \sin(0.21x + 0.41) + 0.52$

(c)

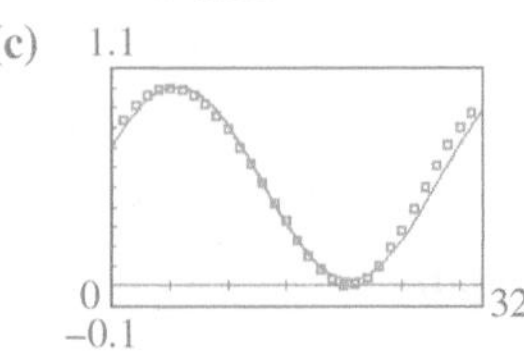

66. (a) See the graph in part (c).

(b) $y = 0.50 \sin(0.21x + 2.36) + 0.48$

(c)

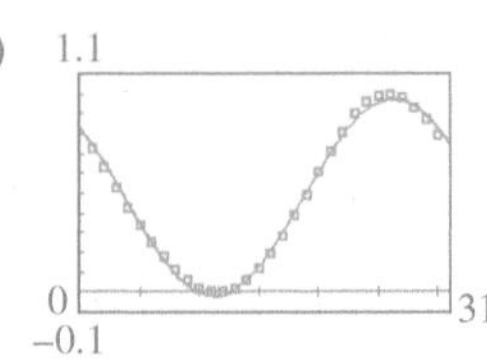

62. ***Average Monthly Temperature*** The average monthly temperature (in °F) in Phoenix, Arizona, is shown in the table.

(a) Predict the average annual temperature.

(b) Plot the average monthly temperature over a two-year period, letting $x = 1$ correspond to January of the first year.

(c) Determine a function of the form $f(x) = a \cos[b(x - d)] + c$, where a, b, c, and d are constants, that models the data.

(d) Graph f together with the data on the same coordinate axes. How well does f model the data?

(e) Use the sine regression capability of a graphing calculator to find the equation of a sine curve that fits these data (two years).

Month	°F	Month	°F
Jan	54	July	93
Feb	58	Aug	91
Mar	63	Sept	86
Apr	70	Oct	75
May	79	Nov	62
June	89	Dec	54

Source: World Almanac and Book of Facts.

(Modeling) Monthly Temperatures *A set of temperature data (in °F) is given for a particular location. (Source:* www.weatherbase.com*)*

(a) Plot the data over a two-year interval.

(b) Use sine regression to determine a model for the two-year interval. Let $x = 1$ represent January of the first year.

(c) Graph the equation from part (b) together with the data on the same coordinate axes.

63. Average Monthly Temperature, Buenos Aires, Argentina

Jan	Feb	Mar	Apr	May	Jun	Jul	Aug	Sept	Oct	Nov	Dec
77.2	74.7	70.5	63.9	57.7	52.2	51.6	54.9	57.6	63.9	69.1	73.8

64. Average High Temperature, Buenos Aires, Argentina

Jan	Feb	Mar	Apr	May	Jun	Jul	Aug	Sept	Oct	Nov	Dec
86.7	83.7	79.5	72.9	66.2	60.1	58.8	63.1	66.0	72.5	77.5	82.6

(Modeling) Fractional Part of the Moon Illuminated *The tables give the fractional part of the moon that is illuminated during the month indicated. (Source:* http://aa.usno.navy.mil*)*

(a) Plot the data for the month.

(b) Use sine regression to determine a model for the data.

(c) Graph the equation from part (b) together with the data on the same coordinate axes.

65. January 2015

Day	1	2	3	4	5	6	7	8	9	10	11	12	13	14	15	16
Fraction	0.84	0.91	0.96	0.99	1.00	0.99	0.96	0.92	0.86	0.79	0.70	0.62	0.52	0.42	0.33	0.23

Day	17	18	19	20	21	22	23	24	25	26	27	28	29	30	31
Fraction	0.15	0.08	0.03	0.00	0.01	0.04	0.10	0.19	0.28	0.39	0.50	0.61	0.71	0.80	0.87

66. November 2015

Day	1	2	3	4	5	6	7	8	9	10	11	12	13	14
Fraction	0.73	0.63	0.53	0.43	0.34	0.25	0.18	0.11	0.06	0.02	0.00	0.00	0.02	0.06

Day	15	16	17	18	19	20	21	22	23	24	25	26	27	28	29	30
Fraction	0.12	0.19	0.28	0.39	0.49	0.61	0.71	0.81	0.90	0.96	0.99	1.00	0.98	0.93	0.87	0.79

Chapter 4 Quiz (Sections 4.1-4.2)

1. Give the amplitude, period, vertical translation, and phase shift of the function $y = 3 - 4\sin\left(2x + \frac{\pi}{2}\right)$.

Graph each function over a two-period interval. Give the period and amplitude.

2. $y = -4\sin x$
3. $y = -\dfrac{1}{2}\cos 2x$
4. $y = 3\sin \pi x$
5. $y = -2\cos\left(x + \dfrac{\pi}{4}\right)$
6. $y = 2 + \sin(2x - \pi)$
7. $y = -1 + \dfrac{1}{2}\sin x$

Connecting Graphs with Equations *Each function graphed is of the form $y = a\cos bx$ or $y = a\sin bx$, where $b > 0$. Determine an equation of the graph.*

8.

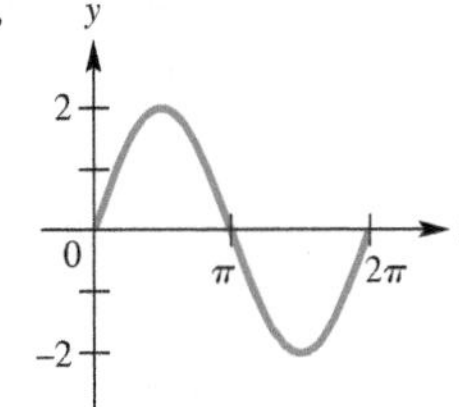

9.

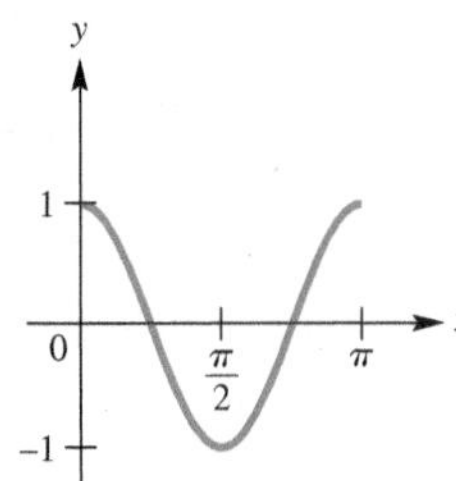

10.

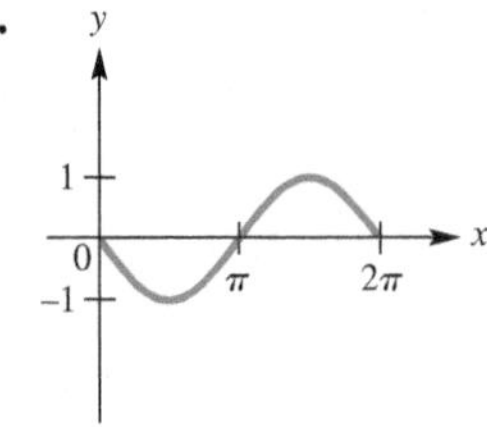

(Modeling) Average Monthly Temperature *The average temperature (in °F) at a certain location can be approximated by the function*

$$f(x) = 12\sin\left[\frac{\pi}{6}(x - 3.9)\right] + 72,$$

where $x = 1$ represents January, $x = 2$ represents February, and so on.

11. What is the average temperature in April?
12. What is the lowest average monthly temperature? What is the highest?

[4.1, 4.2]
1. 4; π; 3 up; $\frac{\pi}{4}$ to the left

[4.1]
2. 2π; 4

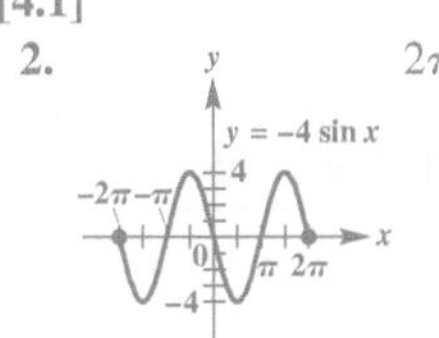

3. π; $\frac{1}{2}$

4. 2; 3

[4.2]
5. 2π; 2

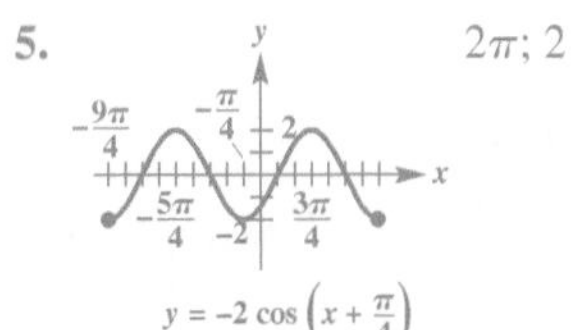

6. π; 1

7. 2π; $\frac{1}{2}$

[4.1]
8. $y = 2\sin x$
9. $y = \cos 2x$
10. $y = -\sin x$

[4.1, 4.2]
11. 73°F
12. 60°F; 84°F

4.3 Graphs of the Tangent and Cotangent Functions

- Graph of the Tangent Function
- Graph of the Cotangent Function
- Techniques for Graphing
- Connecting Graphs with Equations

Graph of the Tangent Function Consider the table of selected points accompanying the graph of the tangent function in **Figure 22** on the next page. These points include special values between $-\frac{\pi}{2}$ and $\frac{\pi}{2}$. The tangent function is undefined for odd multiples of $\frac{\pi}{2}$ and, thus, has *vertical asymptotes* for such values. A **vertical asymptote** is a vertical line that the graph approaches but does not intersect. As the x-values get closer and closer to the line, the function values increase or decrease without bound. Furthermore, because

$$\tan(-x) = -\tan x, \quad \text{Odd function}$$

the graph of the tangent function is symmetric with respect to the origin.

x	$y = \tan x$
$-\frac{\pi}{3}$	$-\sqrt{3} \approx -1.7$
$-\frac{\pi}{4}$	-1
$-\frac{\pi}{6}$	$-\frac{\sqrt{3}}{3} \approx -0.6$
0	0
$\frac{\pi}{6}$	$\frac{\sqrt{3}}{3} \approx 0.6$
$\frac{\pi}{4}$	1
$\frac{\pi}{3}$	$\sqrt{3} \approx 1.7$

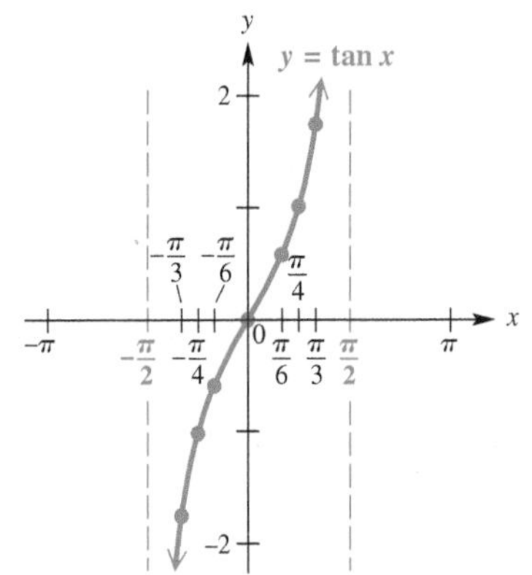

Figure 22

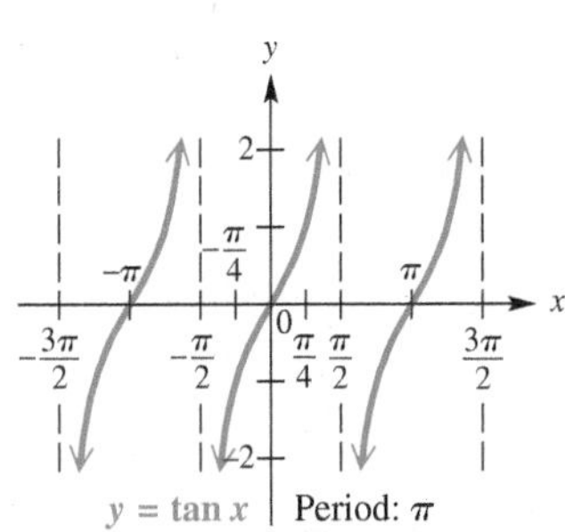

The graph continues in this pattern.

Figure 23

The tangent function has period π. Because $\tan x = \frac{\sin x}{\cos x}$, tangent values are 0 when sine values are 0, and are undefined when cosine values are 0. As x-values increase from $-\frac{\pi}{2}$ to $\frac{\pi}{2}$, tangent values range from $-\infty$ to ∞ and increase throughout the interval. Those same values are repeated as x increases from $\frac{\pi}{2}$ to $\frac{3\pi}{2}$, from $\frac{3\pi}{2}$ to $\frac{5\pi}{2}$, and so on. The graph of $y = \tan x$ from $-\frac{3\pi}{2}$ to $\frac{3\pi}{2}$ is shown in **Figure 23.**

Tangent Function $f(x) = \tan x$

Domain: $\left\{x \,|\, x \neq (2n + 1)\frac{\pi}{2}\text{, where } n \text{ is any integer}\right\}$ Range: $(-\infty, \infty)$

x	y
$-\frac{\pi}{2}$	undefined
$-\frac{\pi}{4}$	-1
0	0
$\frac{\pi}{4}$	1
$\frac{\pi}{2}$	undefined

$f(x) = \tan x,\ -\frac{\pi}{2} < x < \frac{\pi}{2}$

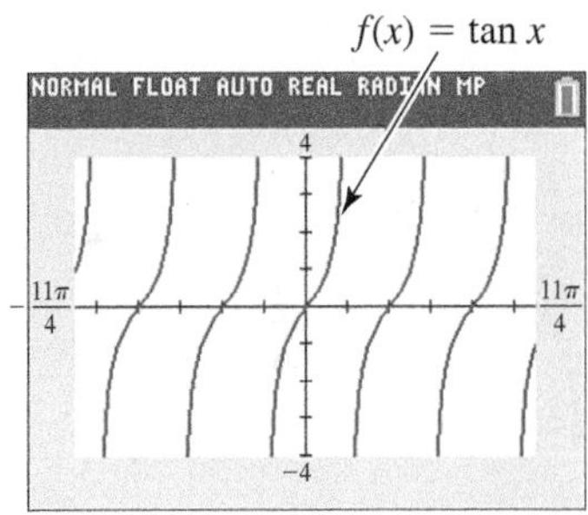

Figure 24

- The graph is discontinuous at values of x of the form $x = (2n + 1)\frac{\pi}{2}$ and has vertical asymptotes at these values.
- Its x-intercepts have x-values of the form $n\pi$.
- Its period is π.
- There are no minimum or maximum values, so its graph has no amplitude.
- The graph is symmetric with respect to the origin, so the function is an odd function. For all x in the domain, $\tan(-x) = -\tan x$.

Graph of the Cotangent Function A similar analysis for selected points between 0 and π for the graph of the cotangent function yields the graph in **Figure 25** on the next page. Here the vertical asymptotes are at x-values that are integer multiples of π. Because

$$\cot(-x) = -\cot x, \quad \text{Odd function}$$

this graph is also symmetric with respect to the origin.

x	$y = \cot x$
$\frac{\pi}{6}$	$\sqrt{3} \approx 1.7$
$\frac{\pi}{4}$	1
$\frac{\pi}{3}$	$\frac{\sqrt{3}}{3} \approx 0.6$
$\frac{\pi}{2}$	0
$\frac{2\pi}{3}$	$-\frac{\sqrt{3}}{3} \approx -0.6$
$\frac{3\pi}{4}$	-1
$\frac{5\pi}{6}$	$-\sqrt{3} \approx -1.7$

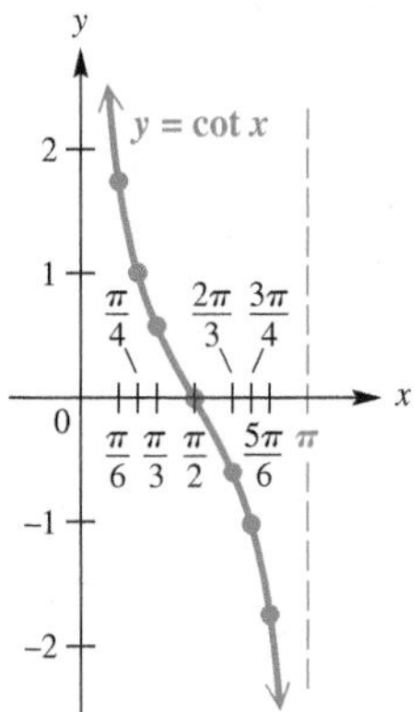

Figure 25

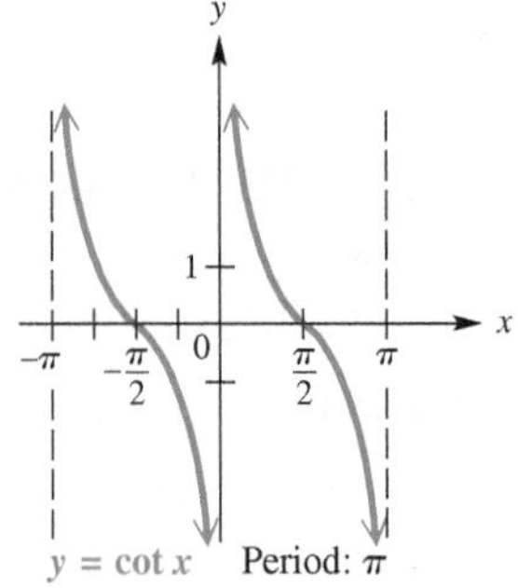

y = cot x Period: π

The graph continues in this pattern.

Figure 26

The cotangent function also has period π. Cotangent values are 0 when cosine values are 0, and are undefined when sine values are 0. As x-values increase from 0 to π, cotangent values range from ∞ to $-\infty$ and decrease throughout the interval. Those same values are repeated as x increases from π to 2π, from 2π to 3π, and so on. The graph of $y = \cot x$ from $-\pi$ to π is shown in **Figure 26.**

Cotangent Function $f(x) = \cot x$

Domain: $\{x \mid x \neq n\pi$, where n is any integer$\}$ Range: $(-\infty, \infty)$

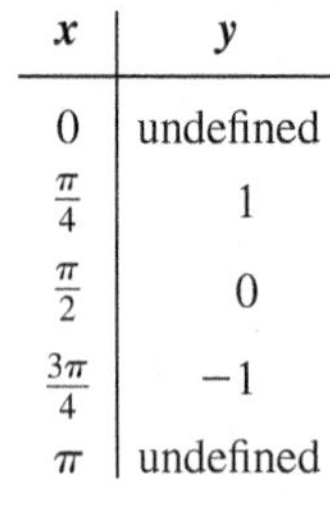

x	y
0	undefined
$\frac{\pi}{4}$	1
$\frac{\pi}{2}$	0
$\frac{3\pi}{4}$	-1
π	undefined

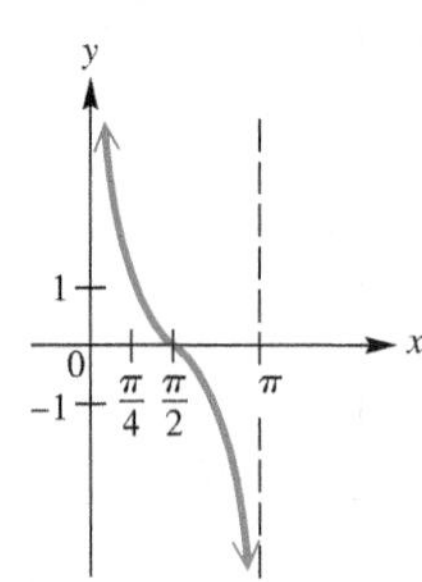

$f(x) = \cot x,\ 0 < x < \pi$

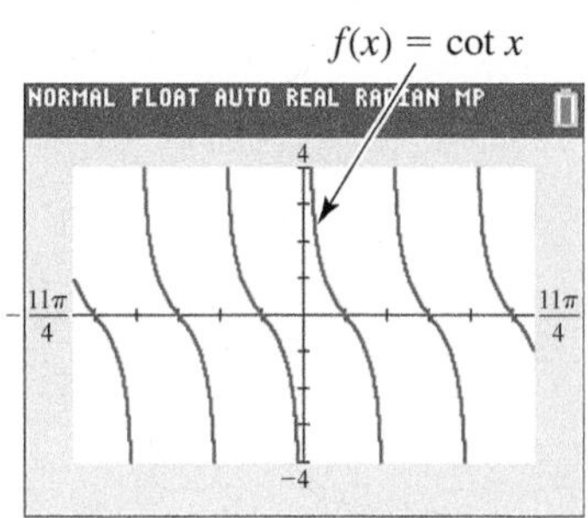

Figure 27

- The graph is discontinuous at values of x of the form $x = n\pi$ and has vertical asymptotes at these values.
- Its x-intercepts have x-values of the form $(2n + 1)\frac{\pi}{2}$.
- Its period is π.
- There are no minimum or maximum values, so its graph has no amplitude.
- The graph is symmetric with respect to the origin, so the function is an odd function. For all x in the domain, $\cot(-x) = -\cot x$.

The tangent function can be graphed directly with a graphing calculator, using the tangent key. To graph the cotangent function, however, we must use one of the identities

$$\cot x = \frac{1}{\tan x} \quad \text{or} \quad \cot x = \frac{\cos x}{\sin x}$$

because graphing calculators generally do not have cotangent keys. ■

Techniques for Graphing

Guidelines for Sketching Graphs of Tangent and Cotangent Functions

To graph $y = a \tan bx$ or $y = a \cot bx$, with $b > 0$, follow these steps.

Step 1 Determine the period, $\frac{\pi}{b}$. To locate two adjacent vertical asymptotes, solve the following equations for x:

For $y = a \tan bx$: $bx = -\frac{\pi}{2}$ and $bx = \frac{\pi}{2}$.

For $y = a \cot bx$: $bx = 0$ and $bx = \pi$.

Step 2 Sketch the two vertical asymptotes found in Step 1.

Step 3 Divide the interval formed by the vertical asymptotes into four equal parts.

Step 4 Evaluate the function for the first-quarter point, midpoint, and third-quarter point, using the x-values found in Step 3.

Step 5 Join the points with a smooth curve, approaching the vertical asymptotes. Indicate additional asymptotes and periods of the graph as necessary.

Teaching Tip Emphasize to students that the period of the tangent and cotangent functions is $\frac{\pi}{b}$, not $\frac{2\pi}{b}$.

EXAMPLE 1 **Graphing $y = \tan bx$**

Graph $y = \tan 2x$.

SOLUTION

Step 1 The period of this function is $\frac{\pi}{2}$. To locate two adjacent vertical asymptotes, solve $2x = -\frac{\pi}{2}$ and $2x = \frac{\pi}{2}$ (because this is a tangent function). The two asymptotes have equations $x = -\frac{\pi}{4}$ and $x = \frac{\pi}{4}$.

Step 2 Sketch the two vertical asymptotes $x = \pm\frac{\pi}{4}$, as shown in **Figure 28.**

Step 3 Divide the interval $\left(-\frac{\pi}{4}, \frac{\pi}{4}\right)$ into four equal parts to find key x-values.

first-quarter value: $-\frac{\pi}{8}$, middle value: 0, third-quarter value: $\frac{\pi}{8}$ Key x-values

Step 4 Evaluate the function for the x-values found in Step 3.

x	$-\frac{\pi}{8}$	0	$\frac{\pi}{8}$
$2x$	$-\frac{\pi}{4}$	0	$\frac{\pi}{4}$
$\tan 2x$	-1	0	1

Another period has been graphed, one half period to the left and one half period to the right.

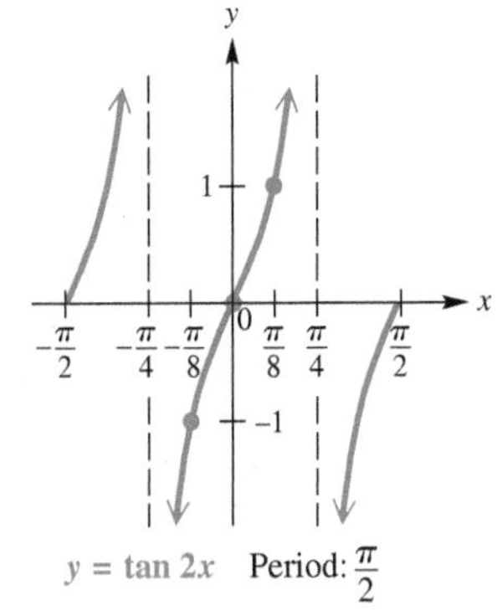

Figure 28

Step 5 Join these points with a smooth curve, approaching the vertical asymptotes. See **Figure 28.**

✔ **Now Try Exercise 13.**

Classroom Example 1
Graph $y = \tan \frac{2}{3}x$.

Answer:

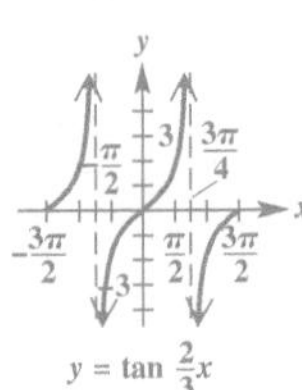

EXAMPLE 2 Graphing $y = a \tan bx$

Graph $y = -3 \tan \frac{1}{2}x$.

SOLUTION The period is $\frac{\pi}{\frac{1}{2}} = \pi \div \frac{1}{2} = \pi \cdot \frac{2}{1} = 2\pi$. Adjacent asymptotes are at $x = -\pi$ and $x = \pi$. Dividing the interval $(-\pi, \pi)$ into four equal parts gives key x-values of $-\frac{\pi}{2}$, 0, and $\frac{\pi}{2}$. Evaluating the function at these x-values gives the following key points.

$$\left(-\frac{\pi}{2}, 3\right),\quad (0, 0),\quad \left(\frac{\pi}{2}, -3\right) \qquad \text{Key points}$$

By plotting these points and joining them with a smooth curve, we obtain the graph shown in **Figure 29.** Because the coefficient -3 is negative, the graph is reflected across the x-axis compared to the graph of $y = 3 \tan \frac{1}{2}x$.

Figure 29

Now Try Exercise 21.

Classroom Example 2
Graph $y = -\frac{1}{2} \tan 2x$.

Answer:

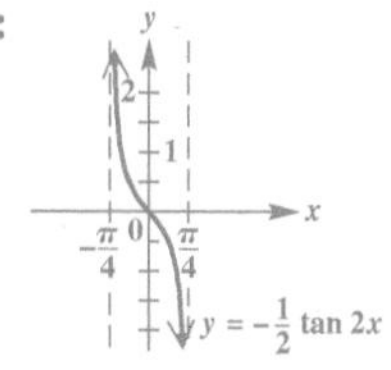

NOTE The function $y = -3 \tan \frac{1}{2}x$ in **Example 2,** graphed in **Figure 29,** has a graph that compares to the graph of $y = \tan x$ as follows.

1. The period is larger because $b = \frac{1}{2}$, and $\frac{1}{2} < 1$.
2. The graph is stretched vertically because $a = -3$, and $|-3| > 1$.
3. Each branch of the graph falls from left to right (that is, the function decreases) between each pair of adjacent asymptotes because $a = -3$, and $-3 < 0$. When $a < 0$, the graph is reflected across the x-axis compared to the graph of $y = |a| \tan bx$.

Classroom Example 3
Graph $y = 3 \cot \frac{1}{2}x$.

Answer:

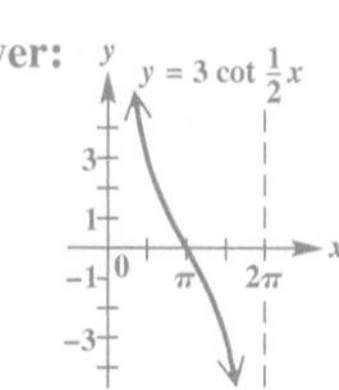

EXAMPLE 3 Graphing $y = a \cot bx$

Graph $y = \frac{1}{2} \cot 2x$.

SOLUTION Because this function involves the cotangent, we can locate two adjacent asymptotes by solving the equations $2x = 0$ and $2x = \pi$. The lines $x = 0$ (the y-axis) and $x = \frac{\pi}{2}$ are two such asymptotes. We divide the interval $\left(0, \frac{\pi}{2}\right)$ into four equal parts, obtaining key x-values of $\frac{\pi}{8}$, $\frac{\pi}{4}$, and $\frac{3\pi}{8}$. Evaluating the function at these x-values gives the key points $\left(\frac{\pi}{8}, \frac{1}{2}\right)$, $\left(\frac{\pi}{4}, 0\right)$, $\left(\frac{3\pi}{8}, -\frac{1}{2}\right)$. We plot these points and join them with a smooth curve approaching the asymptotes to obtain the graph shown in **Figure 30.**

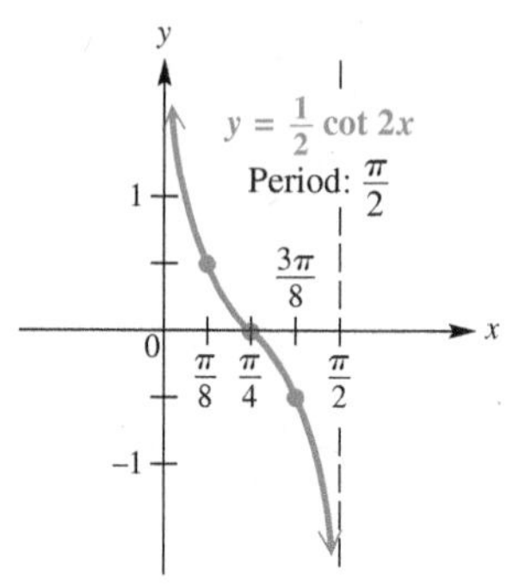

Figure 30

Now Try Exercise 23.

Like the other circular functions, the graphs of the tangent and cotangent functions may be translated horizontally and vertically.

EXAMPLE 4 Graphing $y = c + \tan x$

Graph $y = 2 + \tan x$.

ANALYTIC SOLUTION

Every value of y for this function will be 2 units more than the corresponding value of y in $y = \tan x$, causing the graph of $y = 2 + \tan x$ to be translated 2 units up compared to the graph of $y = \tan x$. See **Figure 31.**

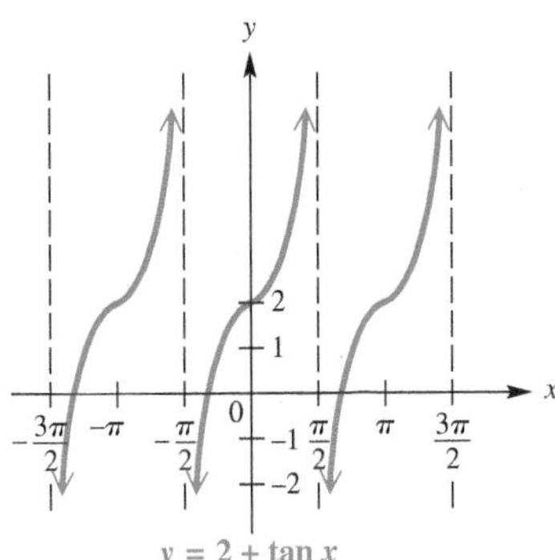

Figure 31

Three periods of the function are shown in **Figure 31.** Because the period of $y = 2 + \tan x$ is π, additional asymptotes and periods of the function can be drawn by repeating the basic graph every π units on the x-axis to the left or to the right of the graph shown.

GRAPHING CALCULATOR SOLUTION

Observe **Figures 32 and 33.** In these figures

$$y_2 = \tan x$$

is the red graph and

$$y_1 = 2 + \tan x$$

is the blue graph. Notice that for the arbitrarily-chosen value of $\frac{\pi}{4}$ (approximately 0.78539816), the difference in the y-values is

$$y_1 - y_2 = 3 - 1 = 2.$$

This illustrates the vertical translation 2 units up.

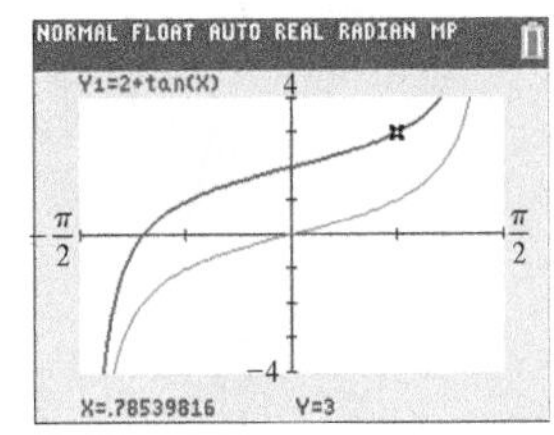

Figure 32

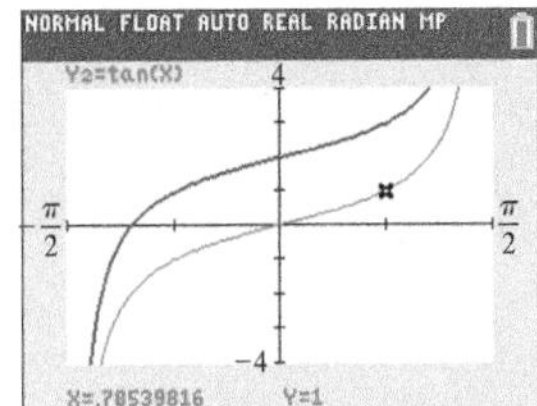

Figure 33

✓ **Now Try Exercise 29.**

Classroom Example 4

Graph $y = -3 + \tan x$.

Answer:

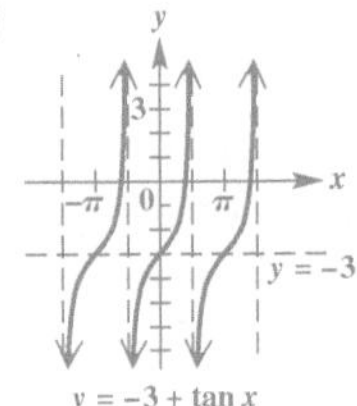

EXAMPLE 5 Graphing $y = c + a\cot(x - d)$

Graph $y = -2 - \cot\left(x - \frac{\pi}{4}\right)$.

SOLUTION Here $b = 1$, so the period is π. The negative sign in front of the cotangent will cause the graph to be reflected across the x-axis, and the argument $\left(x - \frac{\pi}{4}\right)$ indicates a phase shift (horizontal shift) $\frac{\pi}{4}$ unit to the right. Because $c = -2$, the graph will then be translated 2 units down. To locate adjacent asymptotes, because this function involves the cotangent, we solve the following.

$$x - \frac{\pi}{4} = 0 \quad \text{and} \quad x - \frac{\pi}{4} = \pi$$

$$x = \frac{\pi}{4} \quad \text{and} \quad x = \frac{5\pi}{4} \quad \text{Add } \tfrac{\pi}{4}.$$

Dividing the interval $\left(\frac{\pi}{4}, \frac{5\pi}{4}\right)$ into four equal parts and evaluating the function at the three key x-values within the interval give these points.

$$\left(\frac{\pi}{2}, -3\right), \quad \left(\frac{3\pi}{4}, -2\right), \quad (\pi, -1) \quad \text{Key points}$$

We join these points with a smooth curve. This period of the graph, along with the one in the domain interval $\left(-\frac{3\pi}{4}, \frac{\pi}{4}\right)$, is shown in **Figure 34** on the next page.

Classroom Example 5

Graph $y = 3 + \cot\left(x + \frac{\pi}{2}\right)$.

Answer:

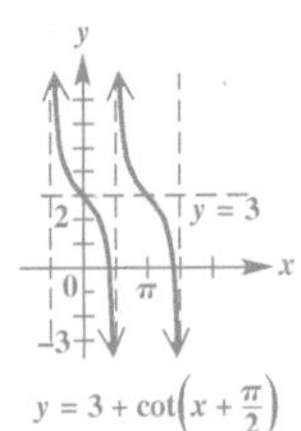

$y = -2 - \cot\left(x - \frac{\pi}{4}\right)$

Figure 34

✔ Now Try Exercise 37.

Connecting Graphs with Equations

Classroom Example 6
Determine an equation for each graph.

(a)

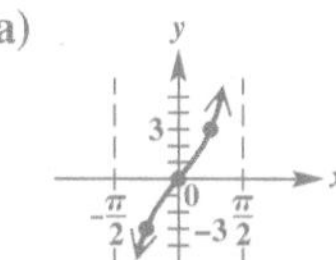

(b)

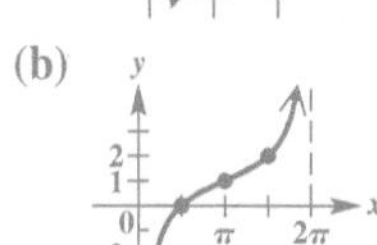

Answers:
(a) $y = 3 \tan x$
(b) $y = 1 - \cot \frac{1}{2}x$

EXAMPLE 6 Determining an Equation for a Graph

Determine an equation for each graph.

(a)

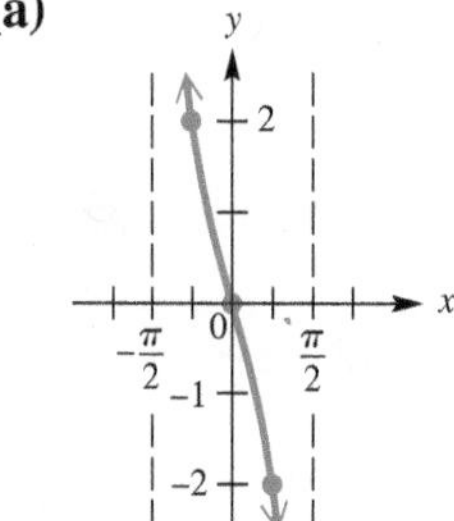

(b)

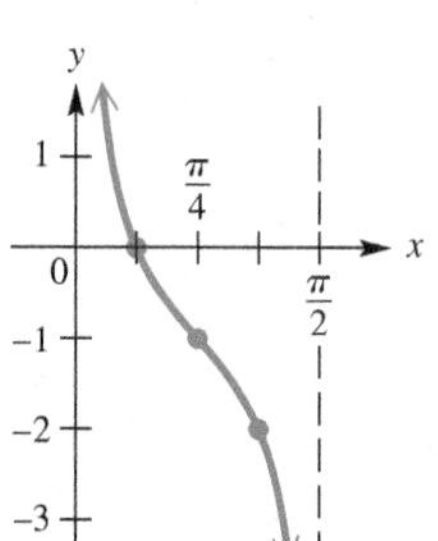

SOLUTION

(a) This graph is that of $y = \tan x$ but reflected across the x-axis and stretched vertically by a factor of 2. Therefore, an equation for this graph is

$$y = -2 \tan x.$$

Vertical stretch
x-axis reflection

(b) This is the graph of a cotangent function, but the period is $\frac{\pi}{2}$ rather than π. Therefore, the coefficient of x is 2. This graph is vertically translated 1 unit down compared to the graph of $y = \cot 2x$. An equation for this graph is

$$y = -1 + \cot 2x.$$

Period is $\frac{\pi}{2}$.
Vertical translation 1 unit down

✔ Now Try Exercises 39 and 43.

NOTE Because the circular functions are periodic, there are infinitely many equations that correspond to each graph in **Example 6.** Confirm that both

$$y = -1 - \cot(-2x) \quad \text{and} \quad y = -1 - \tan\left(2x - \frac{\pi}{2}\right)$$

are equations for the graph in **Example 6(b).** When writing the equation from a graph, it is practical to write the simplest form. Therefore, we choose values of b where $b > 0$ and write the function without a phase shift when possible.

4.3 Exercises

1. π
2. $\frac{\pi}{2}$
3. increases
4. decreases
5. $-\frac{\pi}{2}$
6. $-\pi$
7. C
8. A
9. B
10. D
11. F
12. E

13. 14.

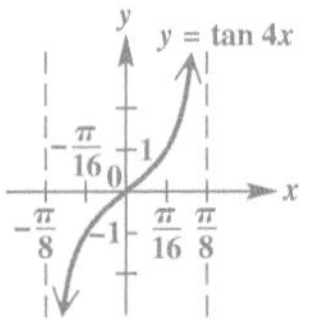

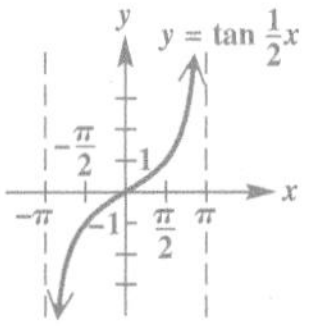

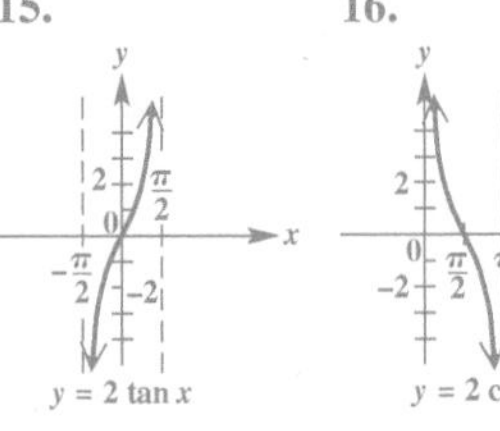

15. 16.

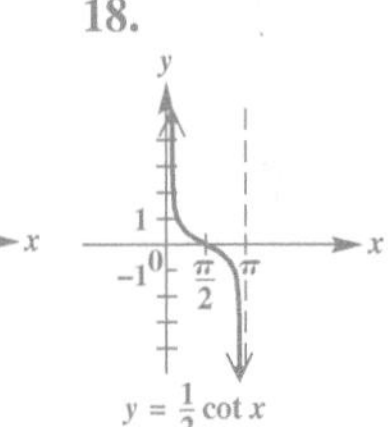

17. 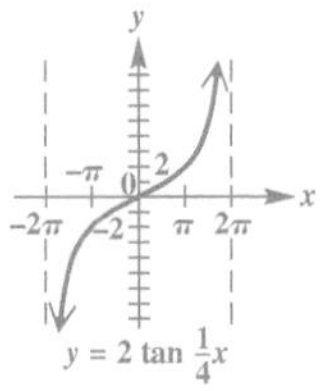 18.

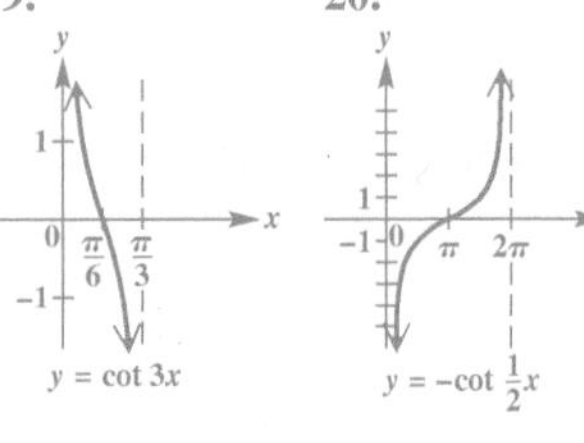

19. 20.

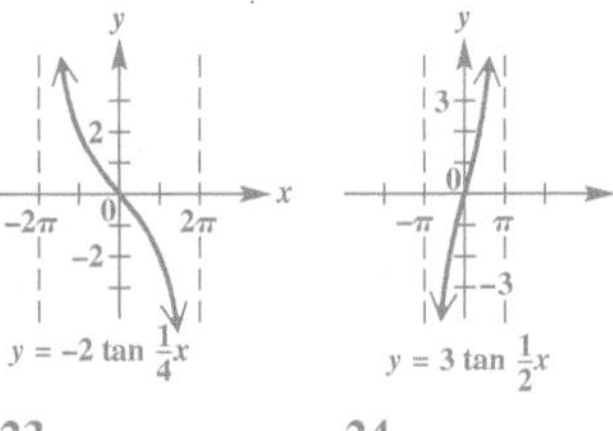

21. 22.

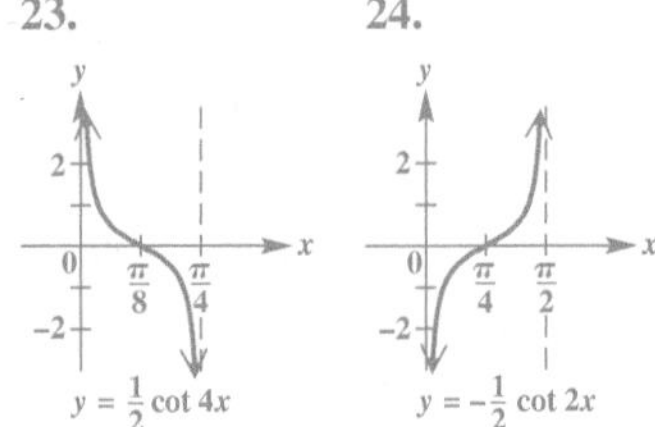

23. 24.

$y = \frac{1}{2}\cot 4x$ $y = -\frac{1}{2}\cot 2x$

CONCEPT PREVIEW *Fill in the blank to correctly complete each sentence.*

1. The least positive value x for which $\tan x = 0$ is ____.
2. The least positive value x for which $\cot x = 0$ is ____.
3. Between any two successive vertical asymptotes, the graph of $y = \tan x$ ____________.
(increases/decreases)
4. Between any two successive vertical asymptotes, the graph of $y = \cot x$ ____________.
(increases/decreases)
5. The negative value k with the greatest value for which $x = k$ is a vertical asymptote of the graph of $y = \tan x$ is ____.
6. The negative value k with the greatest value for which $x = k$ is a vertical asymptote of the graph of $y = \cot x$ is ____.

Concept Check Match each function with its graph from choices A–F.

7. $y = -\tan x$
8. $y = -\cot x$
9. $y = \tan\left(x - \frac{\pi}{4}\right)$
10. $y = \cot\left(x - \frac{\pi}{4}\right)$
11. $y = \cot\left(x + \frac{\pi}{4}\right)$
12. $y = \tan\left(x + \frac{\pi}{4}\right)$

A.

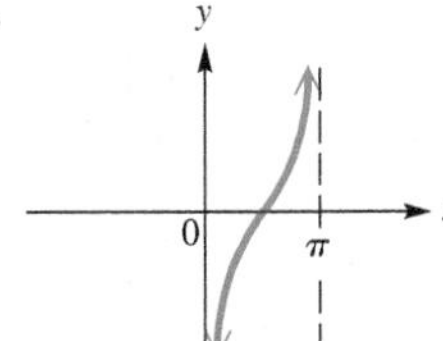

B.

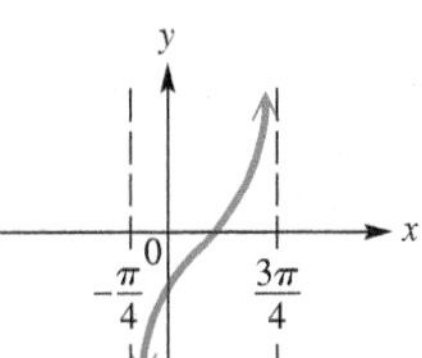

C.

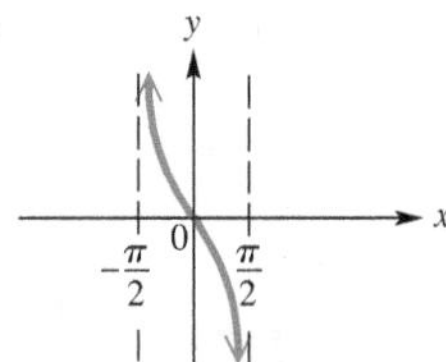

D.

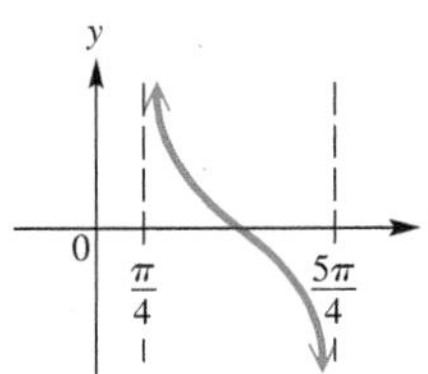

E.

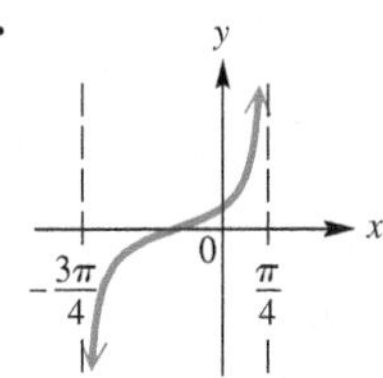

F.

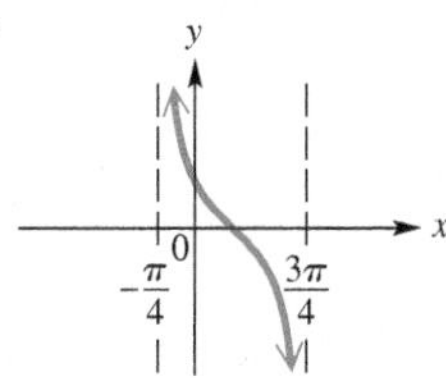

Graph each function over a one-period interval. ***See Examples 1–3.***

13. $y = \tan 4x$
14. $y = \tan \frac{1}{2}x$
15. $y = 2\tan x$
16. $y = 2\cot x$
17. $y = 2\tan\frac{1}{4}x$
18. $y = \frac{1}{2}\cot x$
19. $y = \cot 3x$
20. $y = -\cot\frac{1}{2}x$
21. $y = -2\tan\frac{1}{4}x$
22. $y = 3\tan\frac{1}{2}x$
23. $y = \frac{1}{2}\cot 4x$
24. $y = -\frac{1}{2}\cot 2x$

Graph each function over a two-period interval. ***See Examples 4 and 5.***

25. $y = \tan(2x - \pi)$
26. $y = \tan\left(\frac{x}{2} + \pi\right)$
27. $y = \cot\left(3x + \frac{\pi}{4}\right)$
28. $y = \cot\left(2x - \frac{3\pi}{2}\right)$
29. $y = 1 + \tan x$
30. $y = 1 - \tan x$

25.

$y = \tan(2x - \pi)$

26.

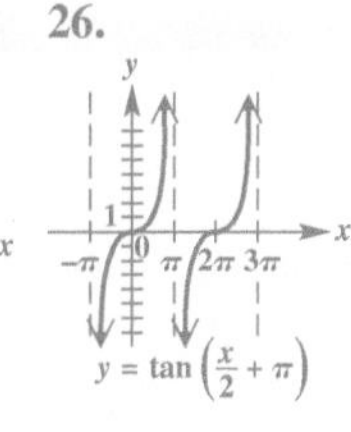

$y = \tan\left(\frac{x}{2} + \pi\right)$

27.

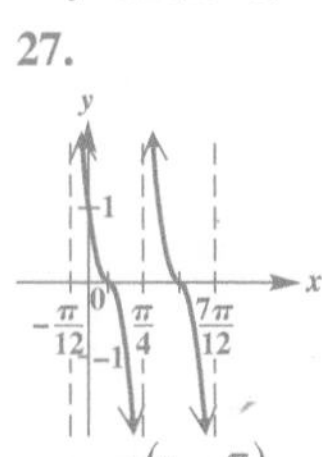

$y = \cot\left(3x + \frac{\pi}{4}\right)$

28.

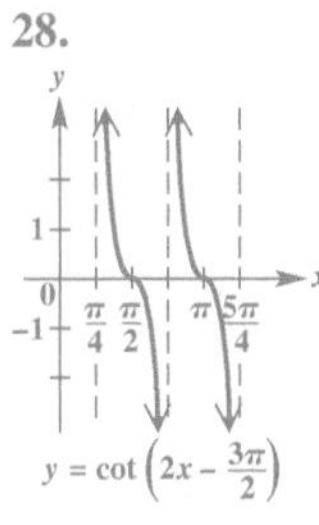

$y = \cot\left(2x - \frac{3\pi}{2}\right)$

29. 30.

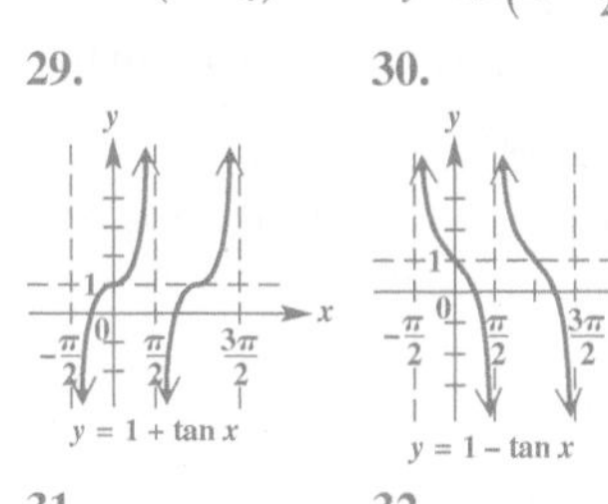

$y = 1 + \tan x$ $\quad$ $y = 1 - \tan x$

31. 32.

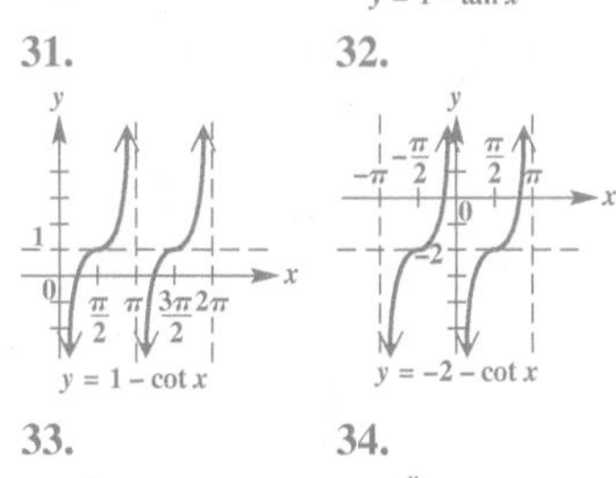

$y = 1 - \cot x$ $\quad$ $y = -2 - \cot x$

33. 34.

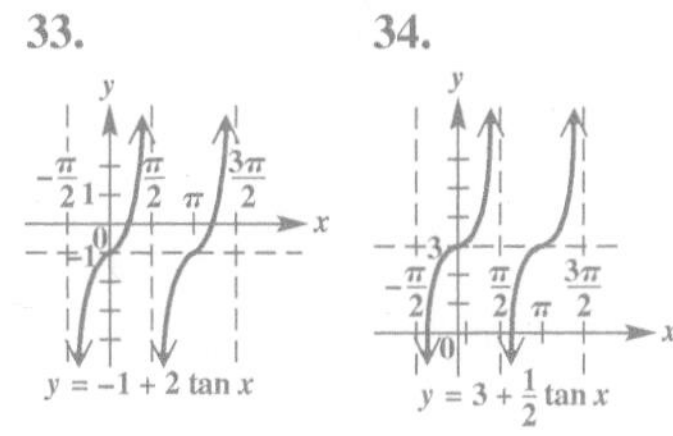

$y = -1 + 2\tan x$ $\quad$ $y = 3 + \frac{1}{2}\tan x$

35. 36.

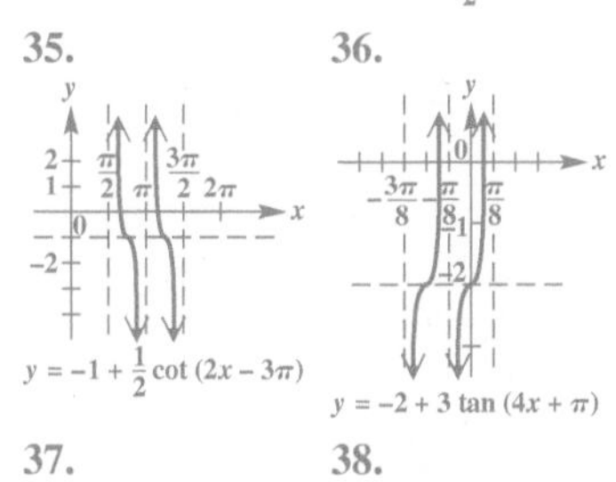

$y = -1 + \frac{1}{2}\cot(2x - 3\pi)$ $\quad$ $y = -2 + 3\tan(4x + \pi)$

37. 38.

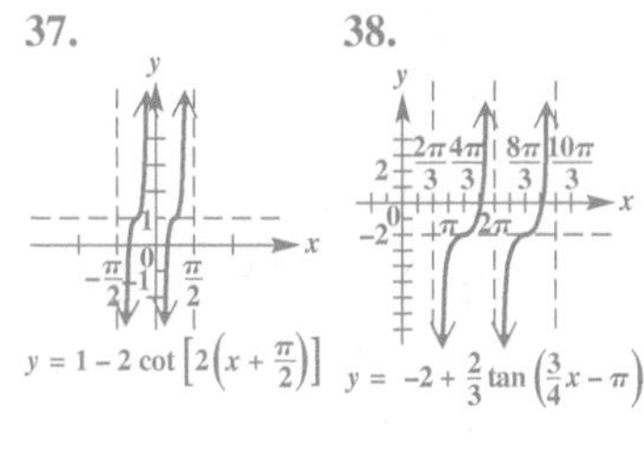

$y = 1 - 2\cot\left[2\left(x + \frac{\pi}{2}\right)\right]$ $\quad$ $y = -2 + \frac{2}{3}\tan\left(\frac{3}{4}x - \pi\right)$

31. $y = 1 - \cot x$

32. $y = -2 - \cot x$

33. $y = -1 + 2\tan x$

34. $y = 3 + \frac{1}{2}\tan x$

35. $y = -1 + \frac{1}{2}\cot(2x - 3\pi)$

36. $y = -2 + 3\tan(4x + \pi)$

37. $y = 1 - 2\cot\left[2\left(x + \frac{\pi}{2}\right)\right]$

38. $y = -2 + \frac{2}{3}\tan\left(\frac{3}{4}x - \pi\right)$

Connecting Graphs with Equations *Determine the simplest form of an equation for each graph. Choose $b > 0$, and include no phase shifts. (Midpoints and quarter-points are identified by dots.)* ***See Example 6.***

39.

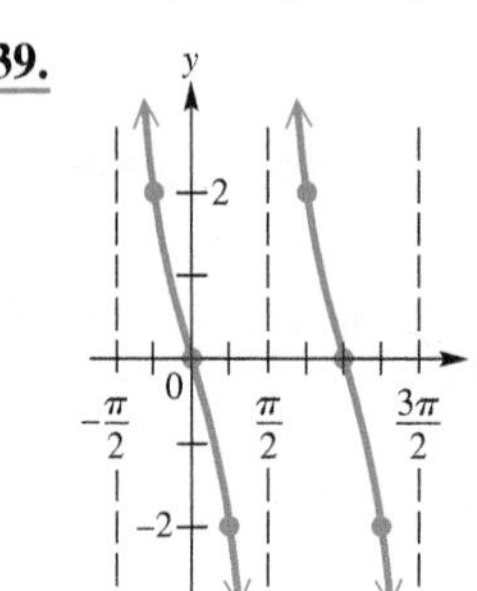

40.

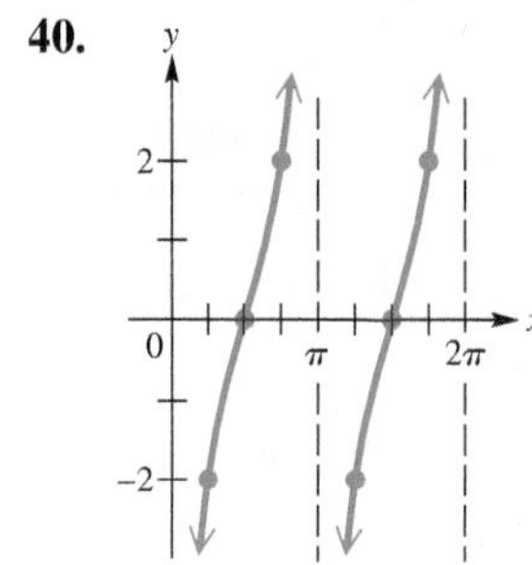

41.

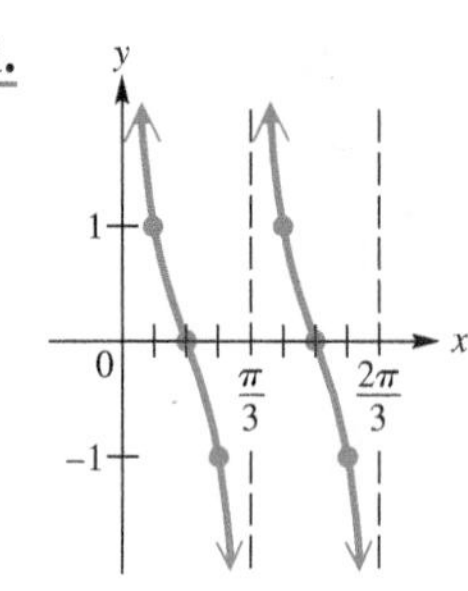

42.

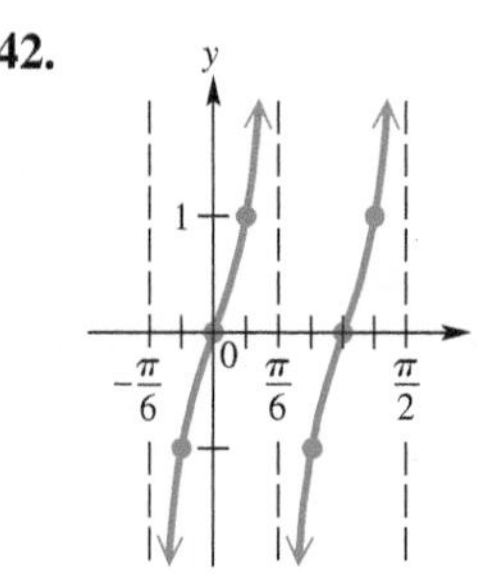

43.

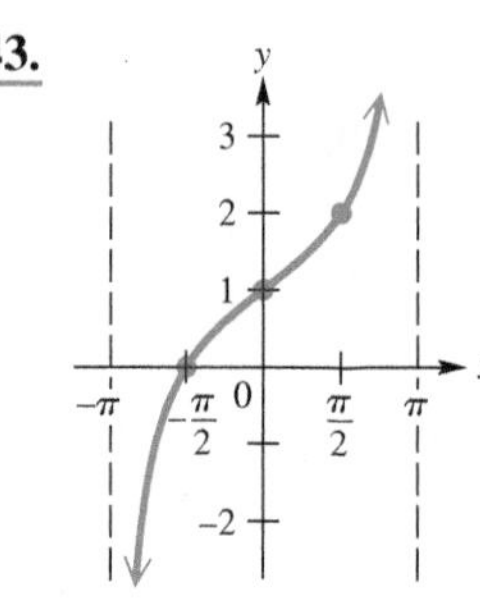

44.

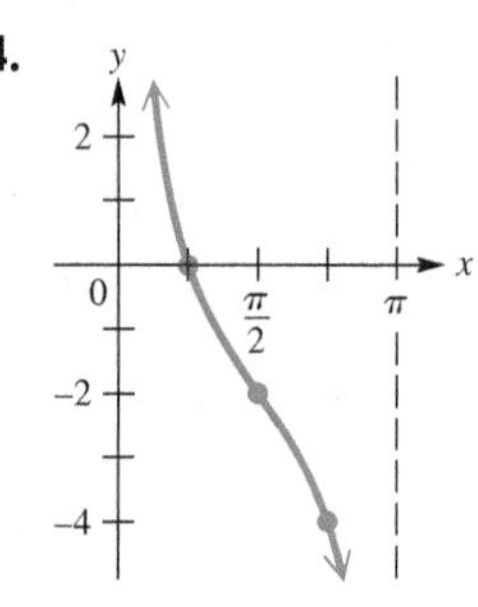

Concept Check *Decide whether each statement is* true *or* false. *If* false, *explain why.*

45. The least positive number k for which $x = k$ is an asymptote for the tangent function is $\frac{\pi}{2}$.

46. The least positive number k for which $x = k$ is an asymptote for the cotangent function is $\frac{\pi}{2}$.

47. The graph of $y = \tan x$ in **Figure 23** suggests that $\tan(-x) = \tan x$ for all x in the domain of $\tan x$.

48. The graph of $y = \cot x$ in **Figure 26** suggests that $\cot(-x) = -\cot x$ for all x in the domain of $\cot x$.

Work each exercise.

49. ***Concept Check*** If c is any number, then how many solutions does the equation $c = \tan x$ have in the interval $(-2\pi, 2\pi]$?

50. ***Concept Check*** Consider the function defined by $f(x) = -4\tan(2x + \pi)$. What is the domain of f? What is its range?

51. Show that $\tan(-x) = -\tan x$ by writing $\tan(-x)$ as $\frac{\sin(-x)}{\cos(-x)}$ and then using the relationships for $\sin(-x)$ and $\cos(-x)$.

52. Show that $\cot(-x) = -\cot x$ by writing $\cot(-x)$ as $\frac{\cos(-x)}{\sin(-x)}$ and then using the relationships for $\cos(-x)$ and $\sin(-x)$.

39. $y = -2 \tan x$
40. $y = -2 \cot x$
41. $y = \cot 3x$
42. $y = \tan 3x$
43. $y = 1 + \tan \frac{1}{2}x$
44. $y = -2 + 2 \cot x$

45. true
46. false; The least such k is π.
47. false; $\tan(-x) = -\tan x$ for all x in the domain.
48. true

49. four
50. domain: $\left\{x \mid x \neq (2n + 1)\frac{\pi}{4}, \text{ where } n \text{ is any integer}\right\}$; range: $(-\infty, \infty)$

53. 0 m 54. -2.9 m
55. 12.3 m
56. It leads to $\tan \frac{\pi}{2}$, which is undefined.

57. π 58. $\frac{5\pi}{4}$
59. $x = \frac{5\pi}{4} + n\pi$ 60. $(0.32, 0)$
61. $(3.46, 0)$
62. $\{x \mid x = 0.32 + n\pi\}$

(Modeling) Distance of a Rotating Beacon A rotating beacon is located at point A next to a long wall. The beacon is 4 m *from the wall. The distance d is given by*

$$d = 4 \tan 2\pi t,$$

where t is time measured in seconds since the beacon started rotating. (When $t = 0$, the beacon is aimed at point R. When the beacon is aimed to the right of R, the value of d is positive; d is negative when the beacon is aimed to the left of R.) Find d for each time. Round to the nearest tenth if applicable.

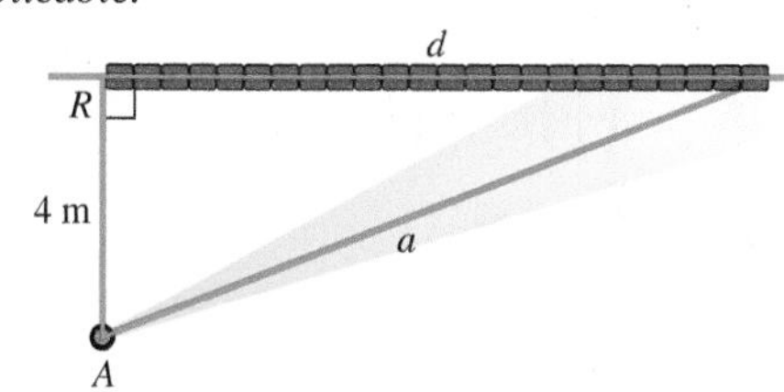

53. $t = 0$

54. $t = 0.4$

55. $t = 1.2$

56. Why is 0.25 a meaningless value for t?

Relating Concepts

For individual or collaborative investigation *(Exercises 57–62)*

Consider the following function from **Example 5. Work these exercises in order.**

$$y = -2 - \cot\left(x - \frac{\pi}{4}\right)$$

57. What is the least positive number for which $y = \cot x$ is undefined?

58. Let k represent the number found in **Exercise 57.** Set $x - \frac{\pi}{4}$ equal to k, and solve to find a positive number for which $\cot\left(x - \frac{\pi}{4}\right)$ is undefined.

59. Based on the answer in **Exercise 58** and the fact that the cotangent function has period π, give the general form of the equations of the asymptotes of the graph of $y = -2 - \cot\left(x - \frac{\pi}{4}\right)$. Let n represent any integer.

60. Use the capabilities of a calculator to find the x-intercept with least positive x-value of the graph of this function. Round to the nearest hundredth.

61. Use the fact that the period of this function is π to find the next positive x-intercept. Round to the nearest hundredth.

62. Give the solution set of the equation $-2 - \cot\left(x - \frac{\pi}{4}\right) = 0$ over all real numbers. Let n represent any integer.

4.4 Graphs of the Secant and Cosecant Functions

- Graph of the Secant Function
- Graph of the Cosecant Function
- Techniques for Graphing
- Connecting Graphs with Equations
- Addition of Ordinates

Graph of the Secant Function Consider the table of selected points accompanying the graph of the secant function in **Figure 35** on the next page. These points include special values from $-\pi$ to π. The secant function is undefined for odd multiples of $\frac{\pi}{2}$ and thus, like the tangent function, has vertical asymptotes for such values. Furthermore, because

$$\sec(-x) = \sec x, \quad \text{Even function}$$

the graph of the secant function is symmetric with respect to the y-axis.

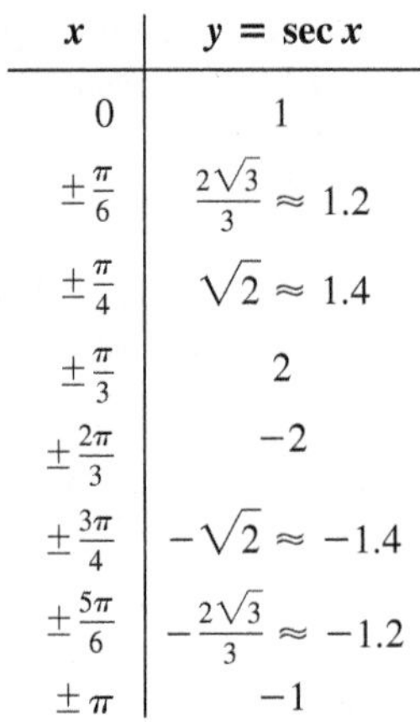

x	$y = \sec x$
0	1
$\pm\frac{\pi}{6}$	$\frac{2\sqrt{3}}{3} \approx 1.2$
$\pm\frac{\pi}{4}$	$\sqrt{2} \approx 1.4$
$\pm\frac{\pi}{3}$	2
$\pm\frac{2\pi}{3}$	-2
$\pm\frac{3\pi}{4}$	$-\sqrt{2} \approx -1.4$
$\pm\frac{5\pi}{6}$	$-\frac{2\sqrt{3}}{3} \approx -1.2$
$\pm\pi$	-1

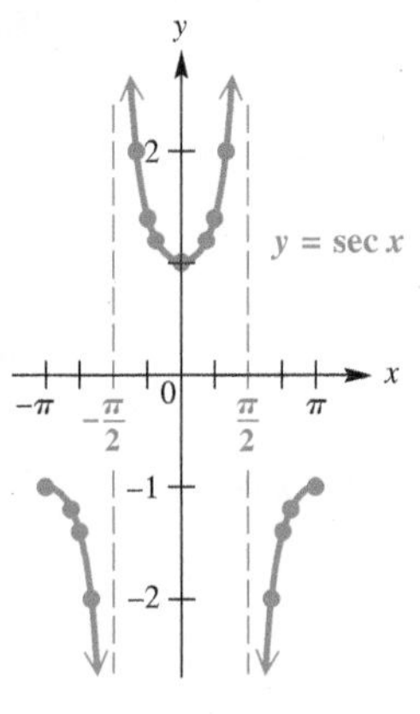

Figure 35

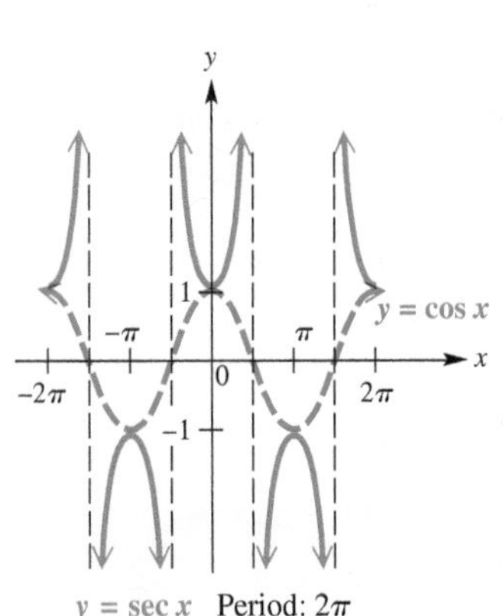

$y = \sec x$ Period: 2π

Figure 36

Because secant values are reciprocals of corresponding cosine values, the period of the secant function is 2π, the same as for $y = \cos x$. When $\cos x = 1$, the value of $\sec x$ is also 1. Likewise, when $\cos x = -1$, $\sec x = -1$. For all x, $-1 \leq \cos x \leq 1$, and thus, $|\sec x| \geq 1$ for all x in its domain. **Figure 36** shows how the graphs of $y = \cos x$ and $y = \sec x$ are related.

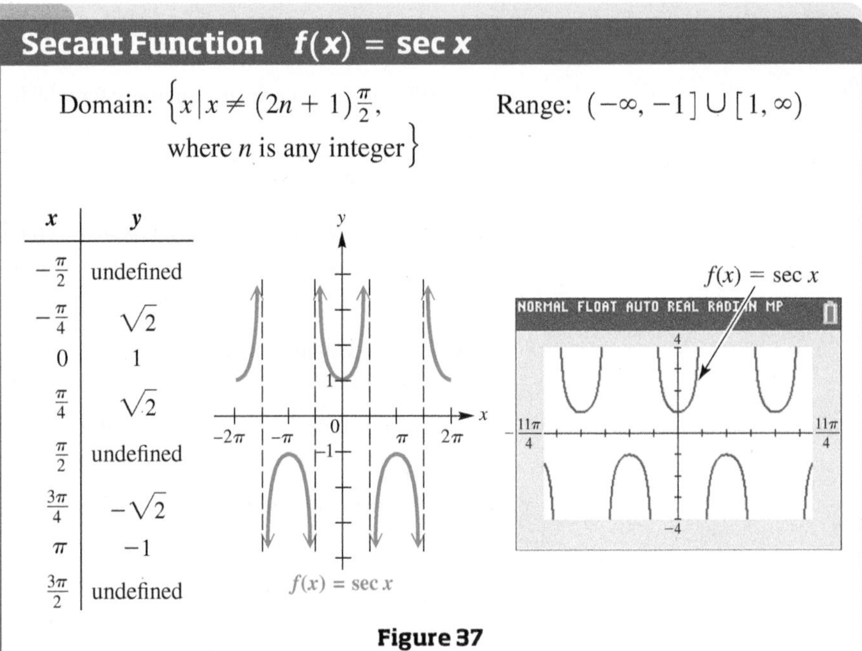

Secant Function $f(x) = \sec x$

Domain: $\left\{x \mid x \neq (2n+1)\frac{\pi}{2},\right.$ where n is any integer$\left.\right\}$ Range: $(-\infty, -1] \cup [1, \infty)$

x	y
$-\frac{\pi}{2}$	undefined
$-\frac{\pi}{4}$	$\sqrt{2}$
0	1
$\frac{\pi}{4}$	$\sqrt{2}$
$\frac{\pi}{2}$	undefined
$\frac{3\pi}{4}$	$-\sqrt{2}$
π	-1
$\frac{3\pi}{2}$	undefined

Figure 37

- The graph is discontinuous at values of x of the form $x = (2n+1)\frac{\pi}{2}$ and has vertical asymptotes at these values.
- There are no x-intercepts.
- Its period is 2π.
- There are no minimum or maximum values, so its graph has no amplitude.
- The graph is symmetric with respect to the y-axis, so the function is an even function. For all x in the domain, $\sec(-x) = \sec x$.

As we shall see, locating the vertical asymptotes for the graph of a function involving the secant (as well as the cosecant) is helpful in sketching its graph.

Graph of the Cosecant function A similar analysis for selected points between $-\pi$ and π for the graph of the cosecant function yields the graph in **Figure 38.** The vertical asymptotes are at x-values that are integer multiples of π. This graph is symmetric with respect to the origin because

$$\csc(-x) = -\csc x. \quad \text{Odd function}$$

x	$y = \csc x$	x	$y = \csc x$
$\frac{\pi}{6}$	2	$-\frac{\pi}{6}$	-2
$\frac{\pi}{4}$	$\sqrt{2} \approx 1.4$	$-\frac{\pi}{4}$	$-\sqrt{2} \approx -1.4$
$\frac{\pi}{3}$	$\frac{2\sqrt{3}}{3} \approx 1.2$	$-\frac{\pi}{3}$	$-\frac{2\sqrt{3}}{3} \approx -1.2$
$\frac{\pi}{2}$	1	$-\frac{\pi}{2}$	-1
$\frac{2\pi}{3}$	$\frac{2\sqrt{3}}{3} \approx 1.2$	$-\frac{2\pi}{3}$	$-\frac{2\sqrt{3}}{3} \approx -1.2$
$\frac{3\pi}{4}$	$\sqrt{2} \approx 1.4$	$-\frac{3\pi}{4}$	$-\sqrt{2} \approx -1.4$
$\frac{5\pi}{6}$	2	$-\frac{5\pi}{6}$	-2

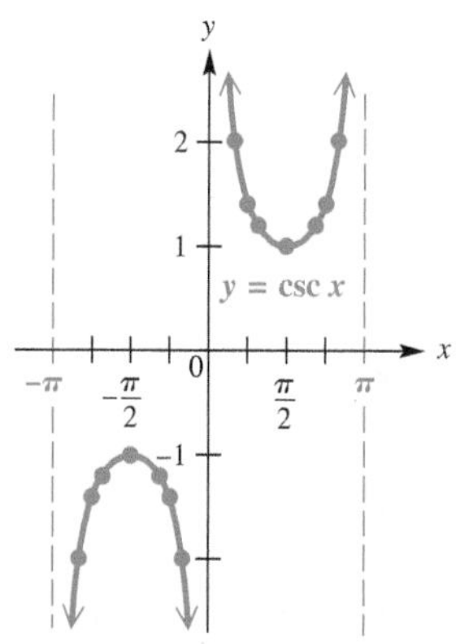

Figure 38

$y = \csc x$ Period: 2π

Figure 39

Because cosecant values are reciprocals of corresponding sine values, the period of the cosecant function is 2π, the same as for $y = \sin x$. When $\sin x = 1$, the value of $\csc x$ is also 1. Likewise, when $\sin x = -1$, $\csc x = -1$. For all x, $-1 \le \sin x \le 1$, and thus $|\csc x| \ge 1$ for all x in its domain. **Figure 39** shows how the graphs of $y = \sin x$ and $y = \csc x$ are related.

Cosecant Function $f(x) = \csc x$

Domain: $\{x \mid x \ne n\pi$, where n is any integer$\}$ Range: $(-\infty, -1] \cup [1, \infty)$

x	y
0	undefined
$\frac{\pi}{6}$	2
$\frac{\pi}{3}$	$\frac{2\sqrt{3}}{3}$
$\frac{\pi}{2}$	1
$\frac{2\pi}{3}$	$\frac{2\sqrt{3}}{3}$
π	undefined
$\frac{3\pi}{2}$	-1
2π	undefined

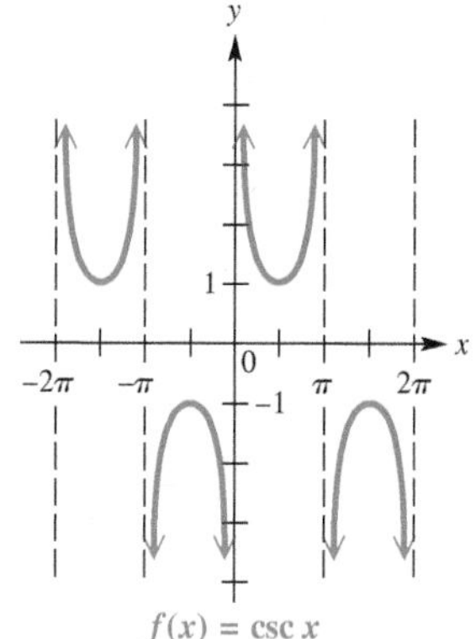

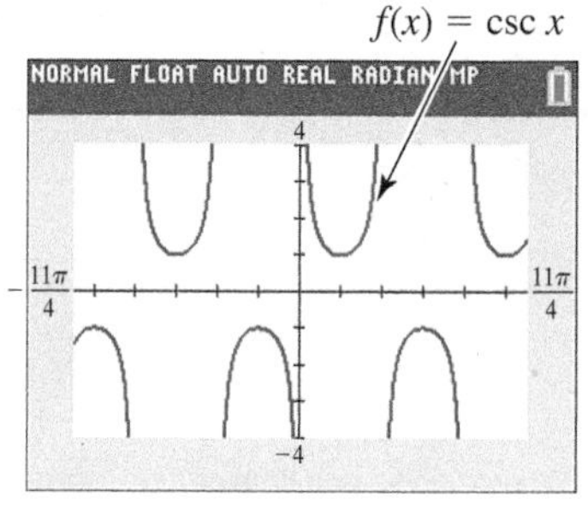

Figure 40

- The graph is discontinuous at values of x of the form $x = n\pi$ and has vertical asymptotes at these values.
- There are no x-intercepts.
- Its period is 2π.
- There are no minimum or maximum values, so its graph has no amplitude.
- The graph is symmetric with respect to the origin, so the function is an odd function. For all x in the domain, $\csc(-x) = -\csc x$.

Calculators do not have keys for the cosecant and secant functions. To graph them with a graphing calculator, use

$$\csc x = \frac{1}{\sin x} \quad \text{and} \quad \sec x = \frac{1}{\cos x}. \quad \text{Reciprocal identities}$$

Techniques for Graphing

Guidelines for Sketching Graphs of Cosecant and Secant Functions

To graph $\boldsymbol{y = a \csc bx}$ or $\boldsymbol{y = a \sec bx}$, with $b > 0$, follow these steps.

Step 1 Graph the corresponding reciprocal function as a guide, using a dashed curve.

To Graph	Use as a Guide
$y = a \csc bx$	$y = a \sin bx$
$y = a \sec bx$	$y = a \cos bx$

Step 2 Sketch the vertical asymptotes. They will have equations of the form $x = k$, where k corresponds to an x-intercept of the graph of the guide function.

Step 3 Sketch the graph of the desired function by drawing the typical U-shaped branches between the adjacent asymptotes. The branches will be above the graph of the guide function when the guide function values are positive and below the graph of the guide function when the guide function values are negative. The graph will resemble those in **Figures 37 and 40** in the function boxes given earlier in this section.

Like graphs of the sine and cosine functions, graphs of the secant and cosecant functions may be translated vertically and horizontally. The period of both basic functions is 2π.

Classroom Example 1
Graph $y = 3 \sec 2x$.

Answer:

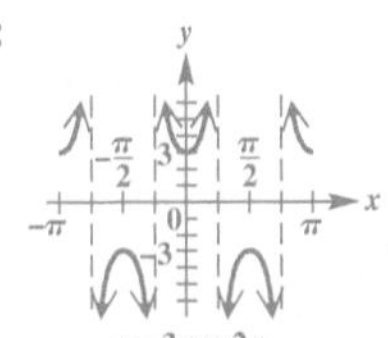

EXAMPLE 1 Graphing $y = a \sec bx$

Graph $y = 2 \sec \frac{1}{2}x$.

SOLUTION

Step 1 This function involves the secant, so the corresponding reciprocal function will involve the cosine. The guide function to graph is

$$y = 2 \cos \frac{1}{2}x.$$

Using the guidelines given earlier, we find that this guide function has amplitude 2 and that one period of the graph lies along the interval that satisfies the following inequality.

$$0 \le \frac{1}{2}x \le 2\pi$$

$$0 \le x \le 4\pi \quad \text{Multiply each part by 2.}$$

Dividing the interval $[0, 4\pi]$ into four equal parts gives these key points.

$$(0, 2), \quad (\pi, 0), \quad (2\pi, -2), \quad (3\pi, 0), \quad (4\pi, 2) \quad \text{Key points}$$

These key points are plotted and joined with a dashed red curve to indicate that this graph is only a guide. An additional period is graphed as shown in **Figure 41(a).**

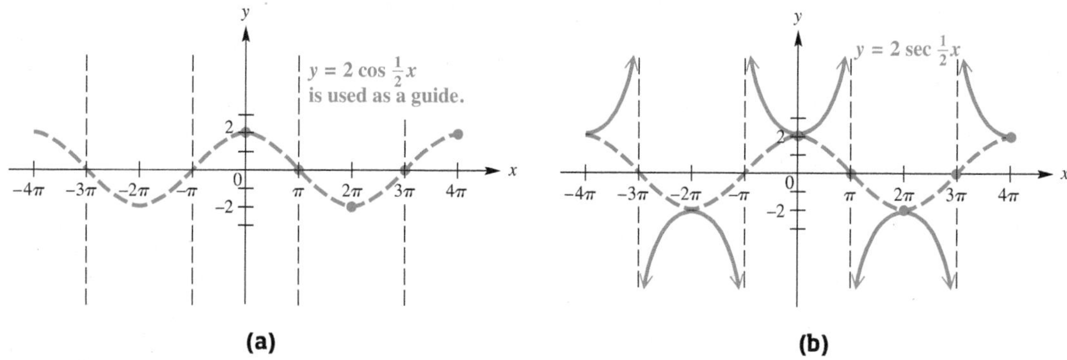

Figure 41

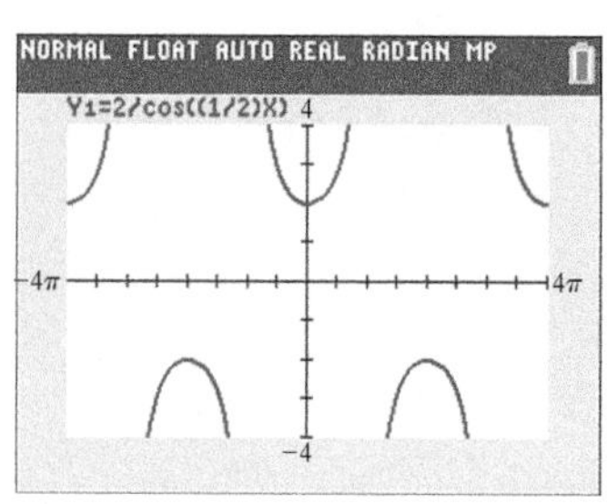

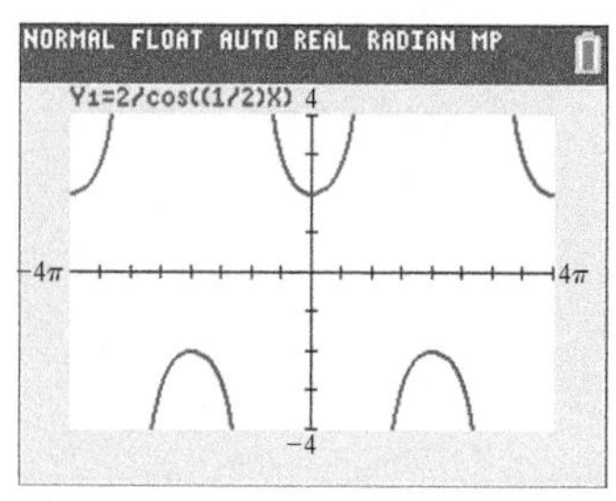

This is a calculator graph of the function in **Example 1.**

Step 2 Sketch the vertical asymptotes as shown in **Figure 41(a).** These occur at x-values for which the guide function equals 0, such as

$$x = -3\pi, \quad x = -\pi, \quad x = \pi, \quad x = 3\pi.$$

Step 3 Sketch the graph of $y = 2\sec\frac{1}{2}x$ by drawing typical U-shaped branches, approaching the asymptotes. See the solid blue graph in **Figure 41(b).**

✔ **Now Try Exercise 11.**

Classroom Example 2

Graph $y = \frac{1}{2}\csc\left(x + \frac{\pi}{4}\right)$.

Answer:

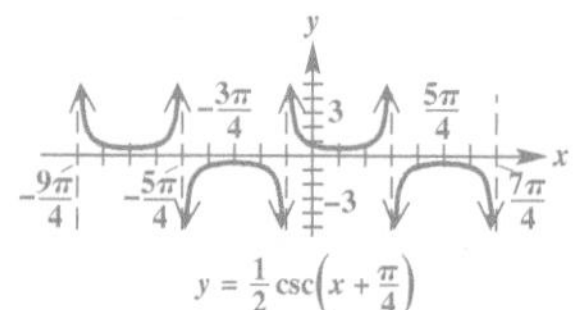

EXAMPLE 2 Graphing $y = a\csc(x - d)$

Graph $y = \frac{3}{2}\csc\left(x - \frac{\pi}{2}\right)$.

SOLUTION

Step 1 Graph the corresponding reciprocal function

$$y = \frac{3}{2}\sin\left(x - \frac{\pi}{2}\right),$$

shown as a red dashed curve in **Figure 42.**

Step 2 Sketch the vertical asymptotes through the x-intercepts of the graph of $y = \frac{3}{2}\sin\left(x - \frac{\pi}{2}\right)$. These x-values have the form $(2n + 1)\frac{\pi}{2}$, where n is any integer. See the black dashed lines in **Figure 42.**

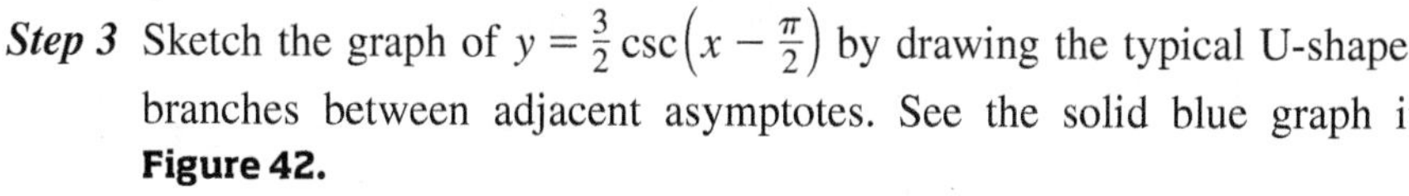

Step 3 Sketch the graph of $y = \frac{3}{2}\csc\left(x - \frac{\pi}{2}\right)$ by drawing the typical U-shaped branches between adjacent asymptotes. See the solid blue graph in **Figure 42.**

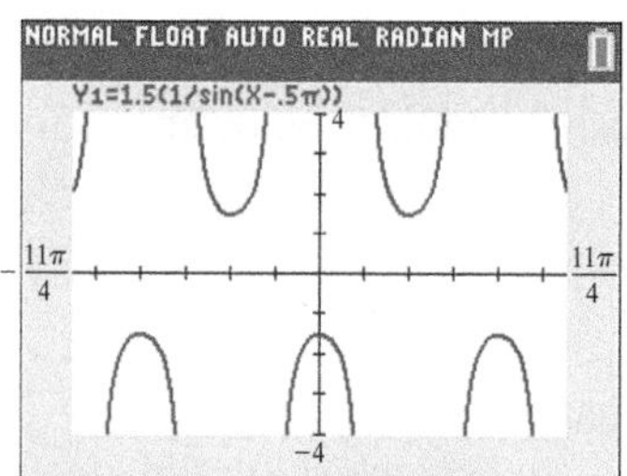

This is a calculator graph of the function in **Example 2.** (The use of decimal equivalents when defining y_1 eliminates the need for some parentheses.)

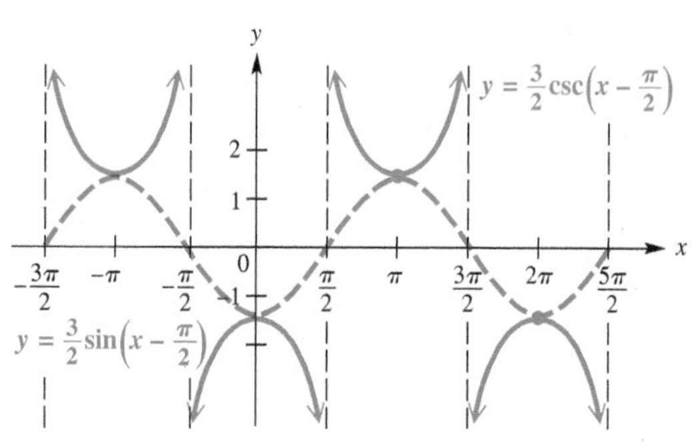

Figure 42

✔ **Now Try Exercise 13.**

Connecting Graphs with Equations

Classroom Example 3
Determine an equation for each graph.

(a)

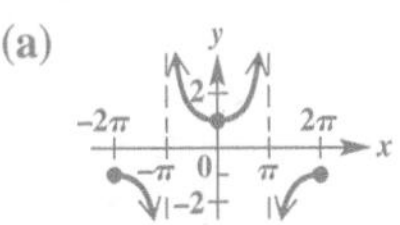

(b)

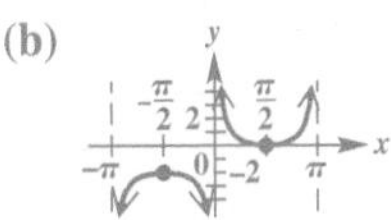

Answers:

(a) $y = \sec \frac{1}{2}x$

(b) $y = -1 + \csc x$

EXAMPLE 3 Determining an Equation for a Graph

Determine an equation for each graph.

(a)

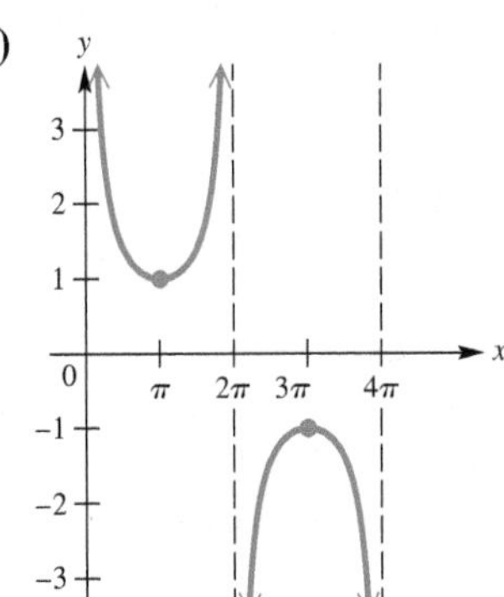

(b)

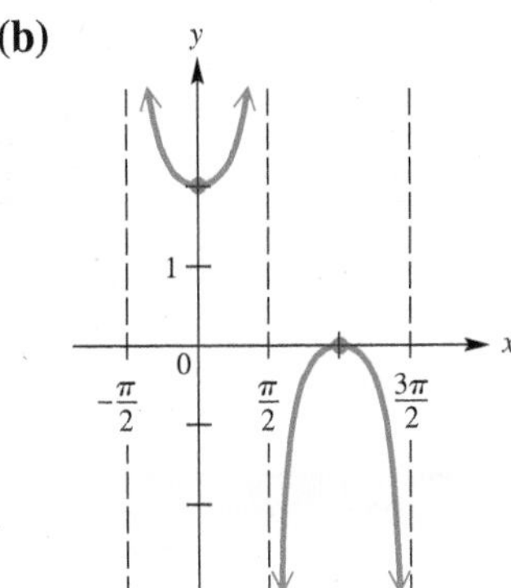

SOLUTION

(a) This graph is that of a cosecant function that is stretched horizontally having period 4π. If $y = \csc bx$, where $b > 0$, then we must have $b = \frac{1}{2}$ and

$$y = \csc \frac{1}{2}x. \qquad \frac{2\pi}{\frac{1}{2}} = 4\pi$$

Horizontal stretch

(b) This is the graph of $y = \sec x$, translated 1 unit up. An equation is

$$y = 1 + \sec x.$$

Vertical translation

✓ **Now Try Exercises 25 and 27.**

Classroom Example 4
Repeat **Example 4** with the value $\frac{\pi}{5}$ for x.

Answer: For $x = \frac{\pi}{5}$ (approximately 0.62831853) the y-coordinates are 0.80901699, 0.58778525, and 1.3968022. The sum of the first two y-coordinates is equal to the third (with slight discrepancy due to roundoff).

Addition of Ordinates A function formed by combining two other functions, such as

$$y_3 = y_1 + y_2,$$

has historically been graphed using a method known as **addition of ordinates.** (The x-value of a point is sometimes called its **abscissa,** while its y-value is called its **ordinate.**)

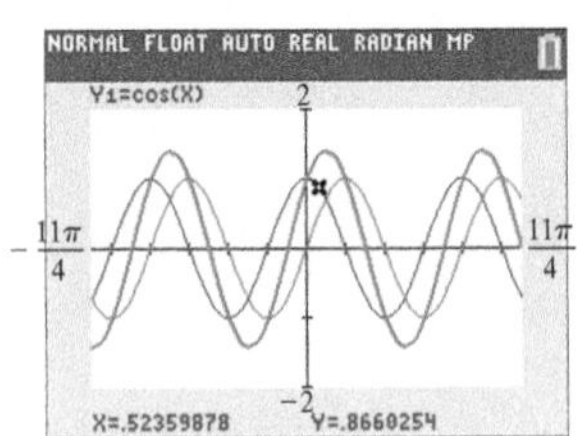

Figure 43

EXAMPLE 4 Illustrating Addition of Ordinates

Use the functions $y_1 = \cos x$ and $y_2 = \sin x$ to illustrate addition of ordinates for

$$y_3 = \cos x + \sin x$$

with the value $\frac{\pi}{6}$ for x.

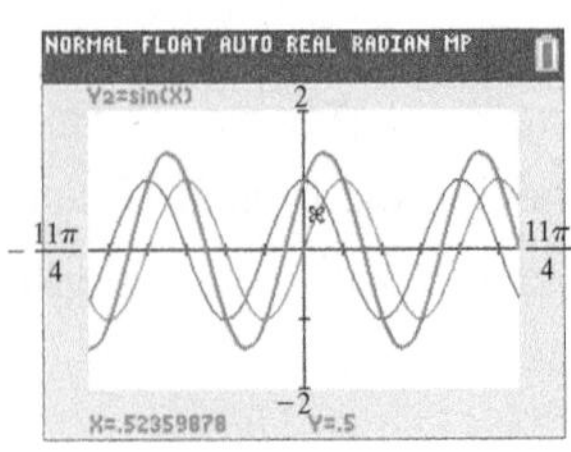

Figure 44

SOLUTION In **Figures 43-45,** $y_1 = \cos x$ is graphed in blue, $y_2 = \sin x$ is graphed in red, and their sum, $y_1 + y_2 = \cos x + \sin x$, is graphed as y_3 in green. If the ordinates (y-values) for $x = \frac{\pi}{6}$ (approximately 0.52359878) in **Figures 43 and 44** are added, their sum is found in **Figure 45.** Verify that

$$0.8660254 + 0.5 = 1.3660254.$$

(This would occur for *any* value of x.)

✓ **Now Try Exercise 43.**

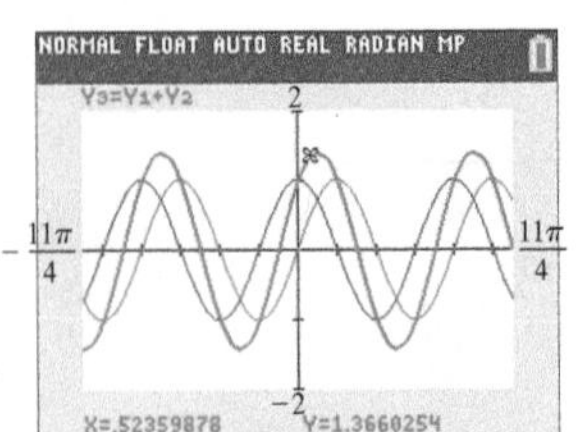

Figure 45

4.4 Exercises

1. A 2. B
3. D 4. F
5. C 6. E

7. B 8. C
9. D 10. A

11. 12.

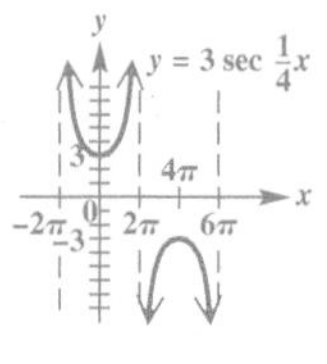

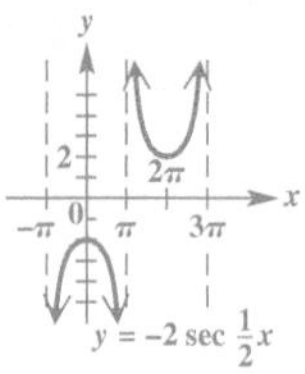

$y = 3 \sec \frac{1}{4}x$ $y = -2 \sec \frac{1}{2}x$

13. 14.

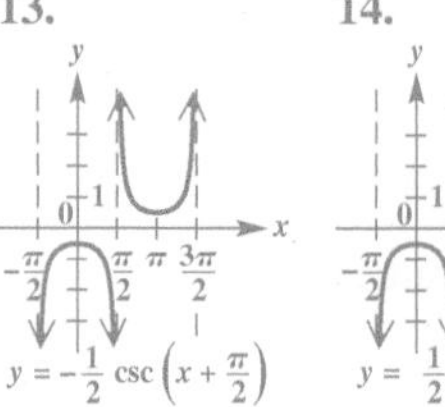

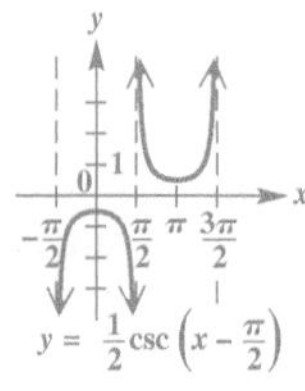

15. 16.

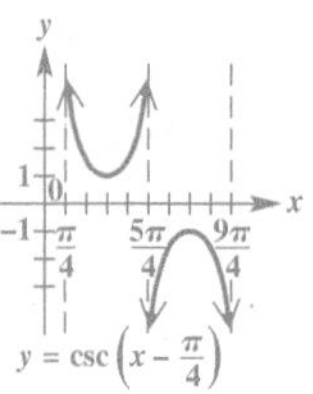

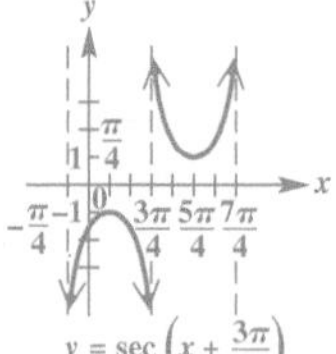

17. 18.

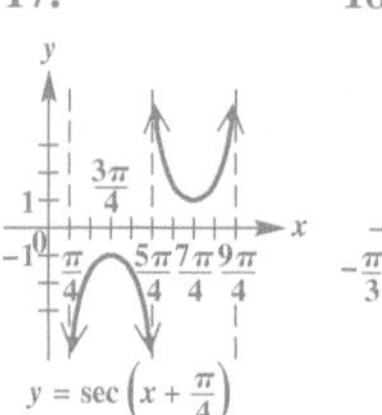

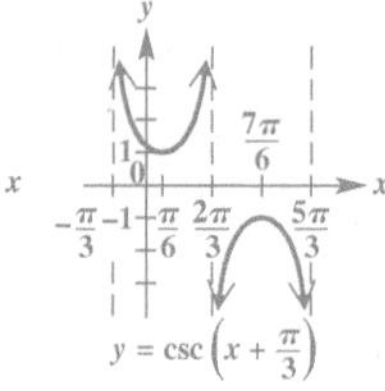

19. 20.

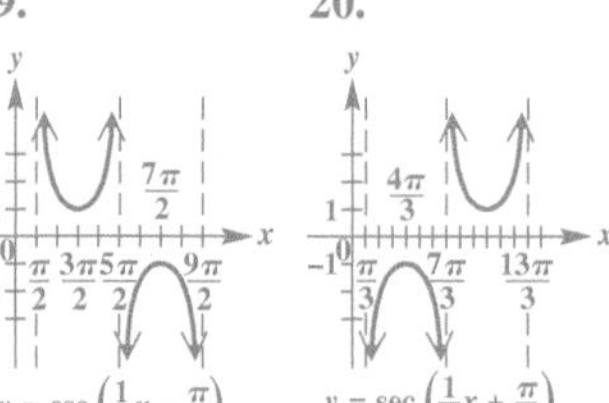

21. 22.

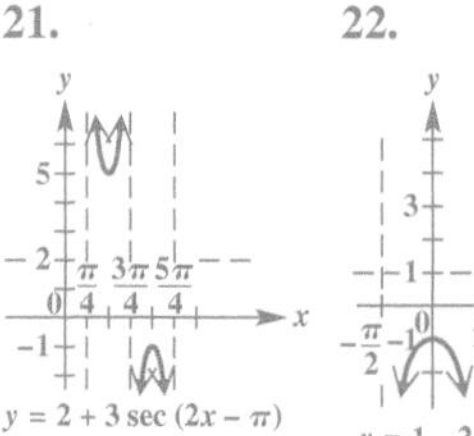

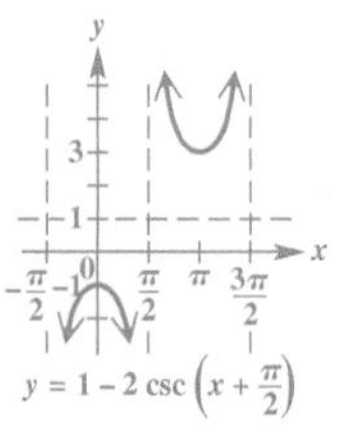

CONCEPT PREVIEW *Match each description in Column I with the correct value in Column II. Refer to the basic graphs as needed.*

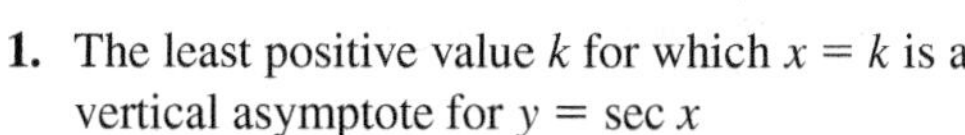

I	II
1. The least positive value k for which $x = k$ is a vertical asymptote for $y = \sec x$	**A.** $\frac{\pi}{2}$
2. The least positive value k for which $x = k$ is a vertical asymptote for $y = \csc x$	**B.** π
3. The least positive value that is in the range of $y = \sec x$	**C.** $-\pi$
4. The greatest negative value that is in the range of $y = \csc x$	**D.** 1
5. The greatest negative value of x for which $\sec x = -1$	**E.** $\frac{3\pi}{2}$
6. The least positive value of x for which $\csc x = -1$	**F.** -1

Concept Check Match each function with its graph from choices A–D.

7. $y = -\csc x$ **8.** $y = -\sec x$ **9.** $y = \sec\left(x - \frac{\pi}{2}\right)$ **10.** $y = \csc\left(x + \frac{\pi}{2}\right)$

A.

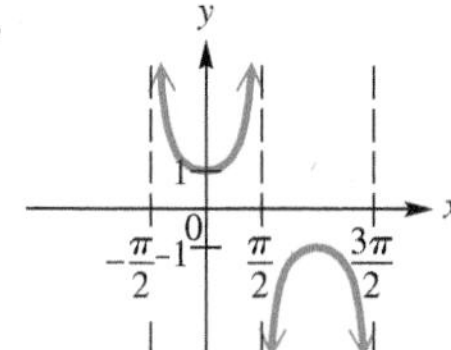

B.

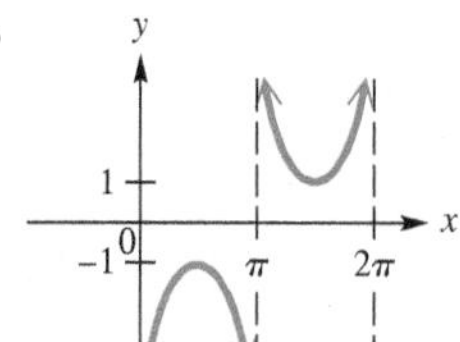

C.

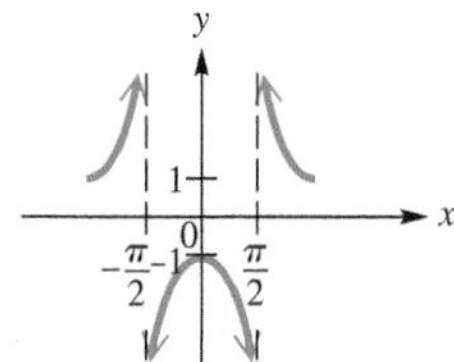

D.

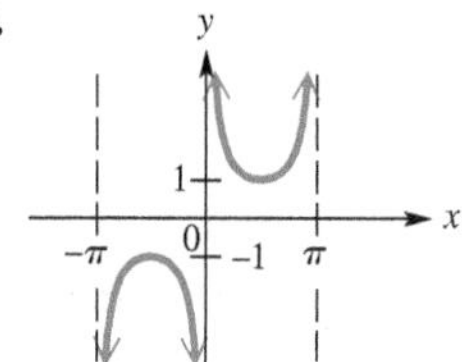

Graph each function over a one-period interval. ***See Examples 1 and 2.***

11. $y = 3 \sec \frac{1}{4}x$ **12.** $y = -2 \sec \frac{1}{2}x$ **13.** $y = -\frac{1}{2} \csc\left(x + \frac{\pi}{2}\right)$

14. $y = \frac{1}{2} \csc\left(x - \frac{\pi}{2}\right)$ **15.** $y = \csc\left(x - \frac{\pi}{4}\right)$ **16.** $y = \sec\left(x + \frac{3\pi}{4}\right)$

17. $y = \sec\left(x + \frac{\pi}{4}\right)$ **18.** $y = \csc\left(x + \frac{\pi}{3}\right)$

19. $y = \csc\left(\frac{1}{2}x - \frac{\pi}{4}\right)$ **20.** $y = \sec\left(\frac{1}{2}x + \frac{\pi}{3}\right)$

21. $y = 2 + 3 \sec(2x - \pi)$ **22.** $y = 1 - 2 \csc\left(x + \frac{\pi}{2}\right)$

23. $y = 1 - \frac{1}{2} \csc\left(x - \frac{3\pi}{4}\right)$ **24.** $y = 2 + \frac{1}{4} \sec\left(\frac{1}{2}x - \pi\right)$

23. $y = 1 - \frac{1}{2}\csc\left(x - \frac{3\pi}{4}\right)$

24. $y = 2 + \frac{1}{4}\sec\left(\frac{1}{2}x - \pi\right)$

25. $y = \sec 4x$ 26. $y = \sec 2x$

27. $y = -2 + \csc x$

28. $y = 1 + \csc x$

29. $y = -1 - \sec x$

30. $y = 1 - \csc \frac{1}{2}x$

31. true

32. false; Secant values are undefined when $x = (2n + 1)\frac{\pi}{2}$, while cosecant values are undefined when $x = n\pi$.

33. true 34. true

35. none

36. domain: $\left\{x \mid x \neq \frac{n\pi}{4}\text{, where } n \text{ is any integer}\right\}$; range: $(-\infty, -2] \cup [2, \infty)$

39. 4 m 40. 6.3 m

41. 63.7 m 42. undefined

Connecting Graphs with Equations *Determine an equation for each graph.* ***See Example 3.***

25.

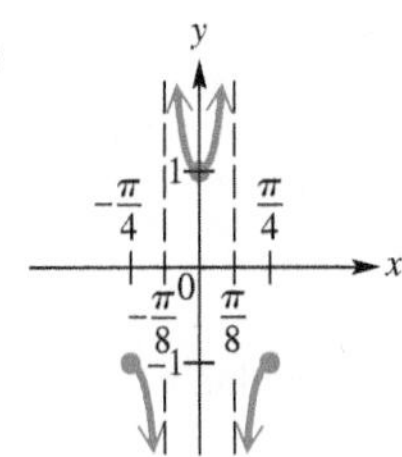

26.

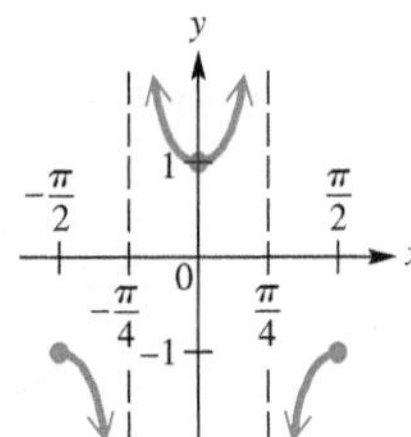

27.

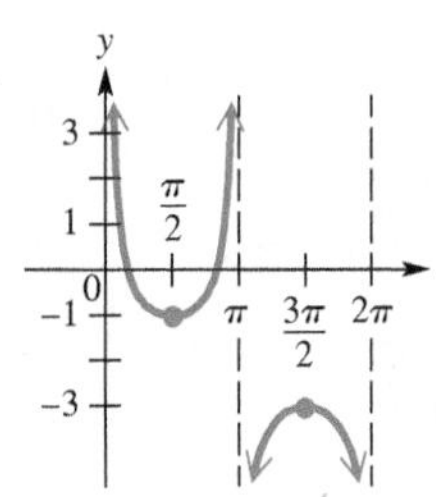

28.

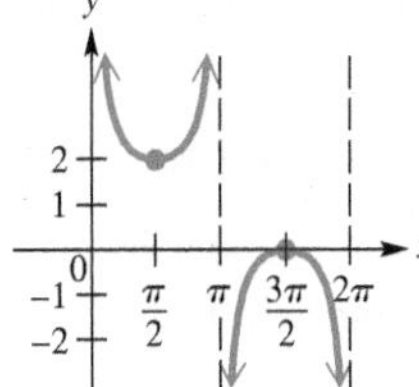

29.

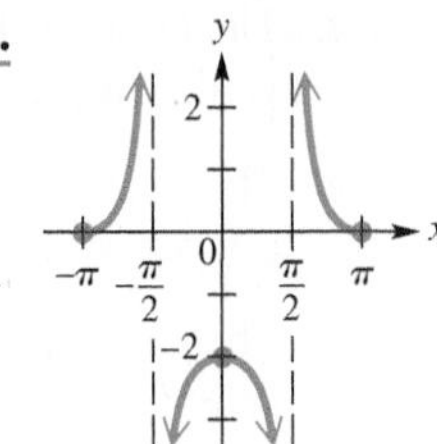

30.

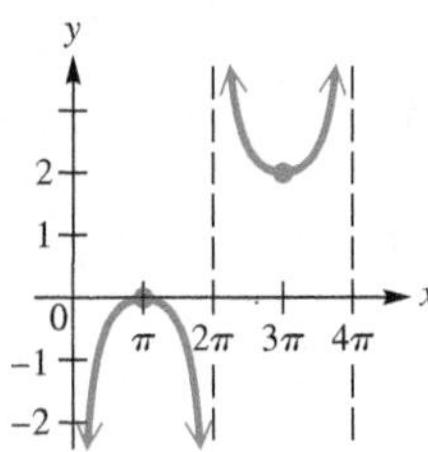

Concept Check *Decide whether each statement is* true *or* false. *If* false, *explain why.*

31. The tangent and secant functions are undefined for the same values.

32. The secant and cosecant functions are undefined for the same values.

33. The graph of $y = \sec x$ in **Figure 37** suggests that $\sec(-x) = \sec x$ for all x in the domain of $\sec x$.

34. The graph of $y = \csc x$ in **Figure 40** suggests that $\csc(-x) = -\csc x$ for all x in the domain of $\csc x$.

Work each problem.

35. ***Concept Check*** If c is any number such that $-1 < c < 1$, then how many solutions does the equation $c = \sec x$ have over the entire domain of the secant function?

36. ***Concept Check*** Consider the function $g(x) = -2\csc(4x + \pi)$. What is the domain of g? What is its range?

37. Show that $\sec(-x) = \sec x$ by writing $\sec(-x)$ as $\frac{1}{\cos(-x)}$ and then using the relationship between $\cos(-x)$ and $\cos x$.

38. Show that $\csc(-x) = -\csc x$ by writing $\csc(-x)$ as $\frac{1}{\sin(-x)}$ and then using the relationship between $\sin(-x)$ and $\sin x$.

(Modeling) Distance of a Rotating Beacon *The distance a in the figure (repeated from the exercise set in the previous section) is given by*

$$a = 4|\sec 2\pi t|.$$

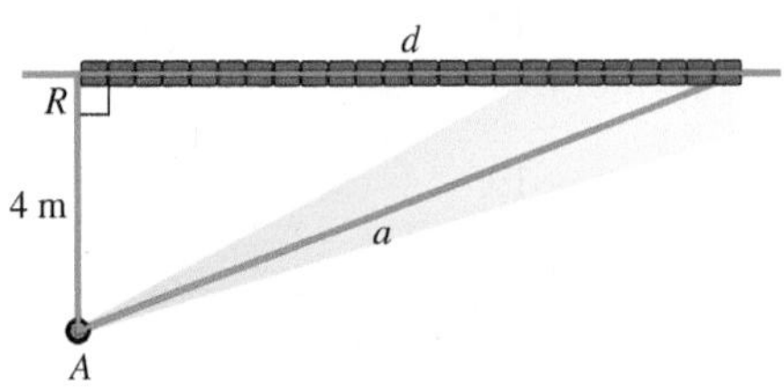

Find the value of a for each time t. Round to the nearest tenth if applicable.

39. $t = 0$ **40.** $t = 0.86$ **41.** $t = 1.24$ **42.** $t = 0.25$

Concepts	Examples

4.4 Graphs of the Secant and Cosecant Functions

Secant and Cosecant Functions

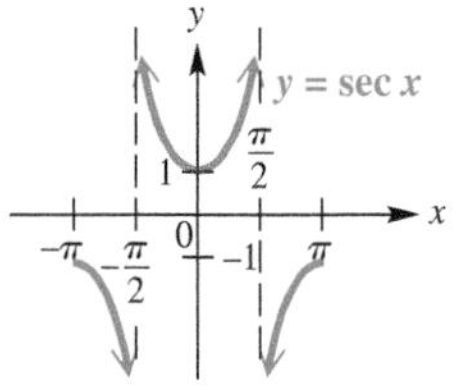

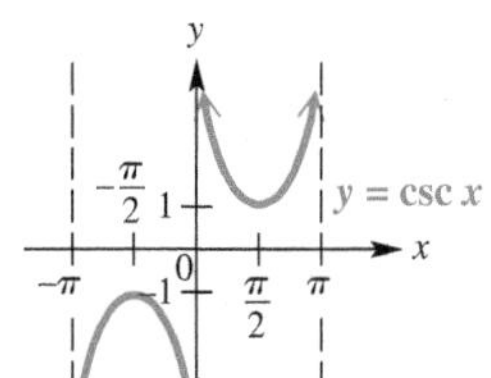

Domain: $\{x \mid x \neq (2n+1)\frac{\pi}{2},$ where n is any integer$\}$

Range: $(-\infty, -1] \cup [1, \infty)$

Period: 2π

Domain: $\{x \mid x \neq n\pi,$ where n is any integer$\}$

Range: $(-\infty, -1] \cup [1, \infty)$

Period: 2π

Graph $y = \sec\left(x + \frac{\pi}{4}\right)$ over a one-period interval.

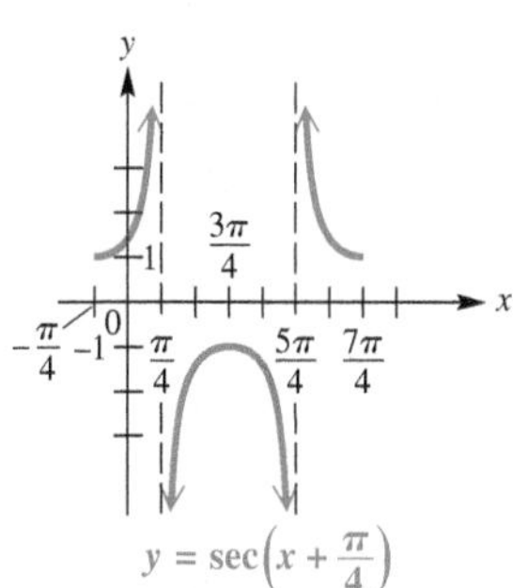

period: 2π

phase shift: $\frac{\pi}{4}$ unit left

domain: $\left\{x \mid x \neq \frac{\pi}{4} + n\pi,\right.$ where n is any integer$\left.\right\}$

range: $(-\infty, -1] \cup [1, \infty)$

4.5 Harmonic Motion

Simple Harmonic Motion

The position of a point oscillating about an equilibrium position at time t is modeled by either

$$s(t) = a \cos \omega t \quad \text{or} \quad s(t) = a \sin \omega t,$$

where a and ω are constants, with $\omega > 0$. The amplitude of the motion is $|a|$, the period is $\frac{2\pi}{\omega}$, and the frequency is $\frac{\omega}{2\pi}$ oscillations per time unit.

A spring oscillates according to

$$s(t) = -5 \cos 6t,$$

where t is in seconds and $s(t)$ is in inches. Find the amplitude, period, and frequency.

$$\text{amplitude} = |-5| = 5 \text{ in.} \quad \text{period} = \frac{2\pi}{6} = \frac{\pi}{3} \text{ sec}$$

$$\text{frequency} = \frac{3}{\pi} \text{ oscillation per sec}$$

Chapter 4 Review Exercises

1. B 2. D
3. sine, cosine, tangent, cotangent
4. secant, cosecant, tangent, cotangent
5. 2; 2π; none; none
6. not applicable; $\frac{\pi}{3}$; none; none
7. $\frac{1}{2}$; $\frac{2\pi}{3}$; none; none
8. 2; $\frac{2\pi}{5}$; none; none
9. 2; 8π; 1 up; none
10. $\frac{1}{4}$; 3π; 3 up; none

1. *Concept Check* Which one of the following statements is true about the graph of $y = 4 \sin 2x$?

   A. It has amplitude 2 and period $\frac{\pi}{2}$. **B.** It has amplitude 4 and period π.

   C. Its range is $[0, 4]$. **D.** Its range is $[-4, 0]$.

2. *Concept Check* Which one of the following statements is false about the graph of $y = -3 \cos \frac{1}{2}x$?

   A. Its range is $[-3, 3]$. **B.** Its domain is $(-\infty, \infty)$.

   C. Its amplitude is 3, and its period is 4π. **D.** Its amplitude is -3, and its period is π.

3. *Concept Check* Which of the basic circular functions can have y-value $\frac{1}{2}$?

4. *Concept Check* Which of the basic circular functions can have y-value 2?

11. 3; 2π; none; $\frac{\pi}{2}$ to the left
12. 1; 2π; none; $\frac{3\pi}{4}$ to the right
13. not applicable; π; none; $\frac{\pi}{8}$ to the right
14. not applicable; 2; none; 2 to the right
15. not applicable; $\frac{\pi}{3}$; none; $\frac{\pi}{9}$ to the right
16. not applicable; 2π; none; $\frac{3\pi}{2}$ to the left

17. tangent 18. sine
19. cosine 20. cosecant
21. cotangent 22. secant

23. By the given condition, $f(x) = f(x + 10n)$ for all integers n. If we let $n = 2$, then $25 = 5 + 10 \cdot 2$, and therefore $f(5) = f(25)$. Thus $f(25) = 3$.
24. By the given condition, $f(x) = f(x + \pi n)$ for all integers n. If we let $n = -2$, then $-\frac{4\pi}{5} = \frac{6\pi}{5} + \pi(-2)$, and therefore $f\left(\frac{6\pi}{5}\right) = f\left(-\frac{4\pi}{5}\right)$. Thus $f\left(-\frac{4\pi}{5}\right) = 1$.

25.

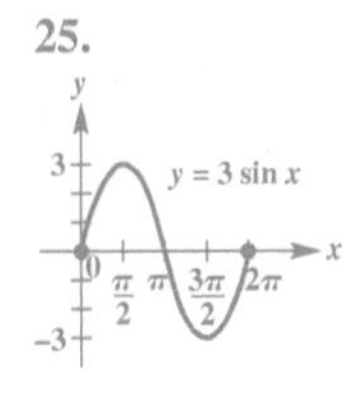

26.

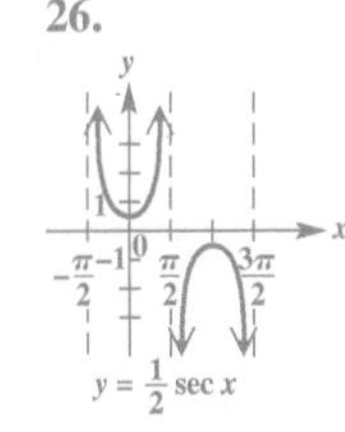

27.

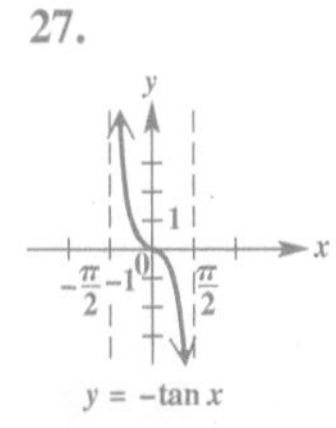

28.

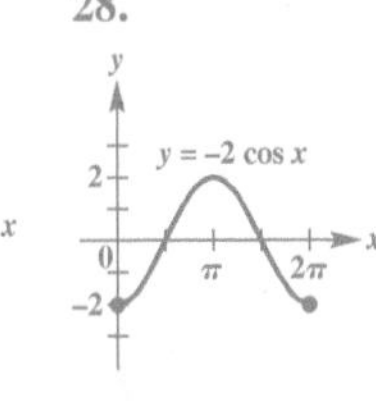

29.

y = 2 + cot x

30.

y = −1 + csc x

31.

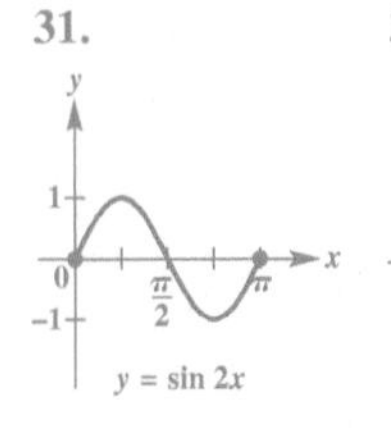

32.

y = tan 3x

For each function, give the amplitude, period, vertical translation, and phase shift, as applicable.

5. $y = 2 \sin x$
6. $y = \tan 3x$
7. $y = -\frac{1}{2} \cos 3x$
8. $y = 2 \sin 5x$
9. $y = 1 + 2 \sin \frac{1}{4}x$
10. $y = 3 - \frac{1}{4} \cos \frac{2}{3}x$
11. $y = 3 \cos\left(x + \frac{\pi}{2}\right)$
12. $y = -\sin\left(x - \frac{3\pi}{4}\right)$
13. $y = \frac{1}{2} \csc\left(2x - \frac{\pi}{4}\right)$
14. $y = 2 \sec(\pi x - 2\pi)$
15. $y = \frac{1}{3} \tan\left(3x - \frac{\pi}{3}\right)$
16. $y = \cot\left(\frac{x}{2} + \frac{3\pi}{4}\right)$

Concept Check *Identify the circular function that satisfies each description.*

17. period is π; x-intercepts have x-values of the form $n\pi$, where n is any integer
18. period is 2π; graph passes through the origin
19. period is 2π; graph passes through the point $\left(\frac{\pi}{2}, 0\right)$
20. period is 2π; domain is $\{x \mid x \neq n\pi$, where n is any integer$\}$
21. period is π; function is decreasing on the interval $(0, \pi)$
22. period is 2π; has vertical asymptotes of the form $x = (2n + 1)\frac{\pi}{2}$, where n is any integer

Provide a short explanation.

23. Suppose that f defines a sine function with period 10 and $f(5) = 3$. Explain why $f(25) = 3$.
24. Suppose that f defines a sine function with period π and $f\left(\frac{6\pi}{5}\right) = 1$. Explain why $f\left(-\frac{4\pi}{5}\right) = 1$.

Graph each function over a one-period interval.

25. $y = 3 \sin x$
26. $y = \frac{1}{2} \sec x$
27. $y = -\tan x$
28. $y = -2 \cos x$
29. $y = 2 + \cot x$
30. $y = -1 + \csc x$
31. $y = \sin 2x$
32. $y = \tan 3x$
33. $y = 3 \cos 2x$
34. $y = \frac{1}{2} \cot 3x$
35. $y = \cos\left(x - \frac{\pi}{4}\right)$
36. $y = \tan\left(x - \frac{\pi}{2}\right)$
37. $y = \sec\left(2x + \frac{\pi}{3}\right)$
38. $y = \sin\left(3x + \frac{\pi}{2}\right)$
39. $y = 1 + 2 \cos 3x$
40. $y = -1 - 3 \sin 2x$
41. $y = 2 \sin \pi x$
42. $y = -\frac{1}{2} \cos(\pi x - \pi)$

(Modeling) Monthly Temperatures *A set of temperature data (in °F) is given for a particular location. (Source: www.weatherbase.com)*

(a) Plot the data over a two-year interval.

(b) Use sine regression to determine a model for the two-year interval. Let $x = 1$ represent January of the first year.

(c) Graph the equation from part (b) together with the data on the same coordinate axes.

43. Average Monthly Temperature, Auckland, New Zealand

Jan	Feb	Mar	Apr	May	Jun	Jul	Aug	Sept	Oct	Nov	Dec
67.6	68.5	65.8	61.3	57.2	53.2	51.6	52.9	55.4	58.1	61.2	64.9

33. 34.

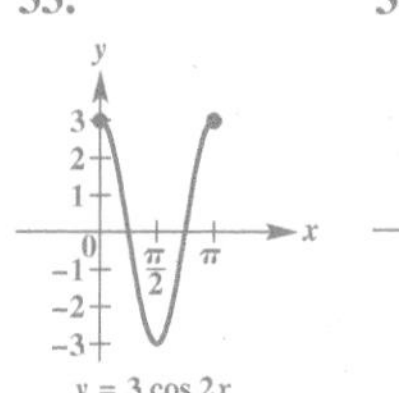

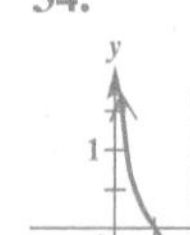

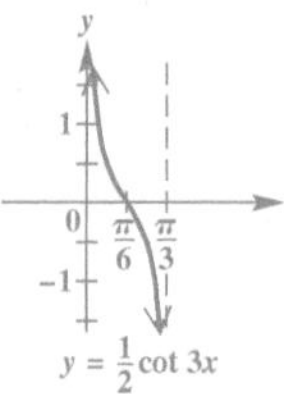

35. 36.

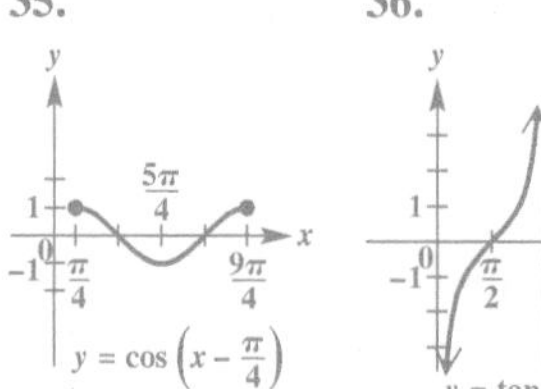

37. 38.

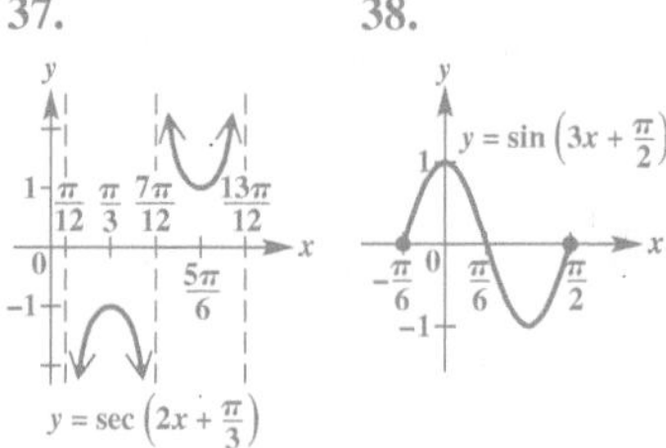

39. 40.

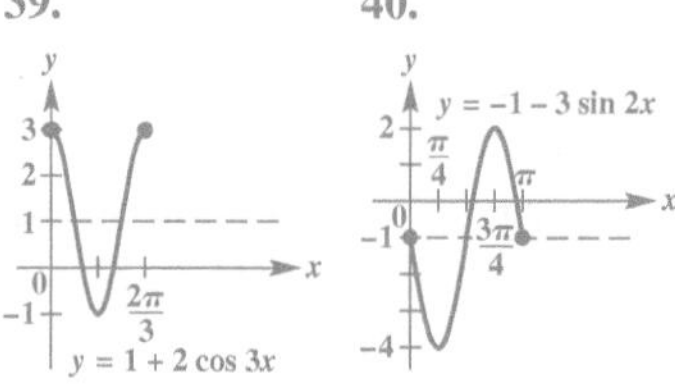

41. 42.

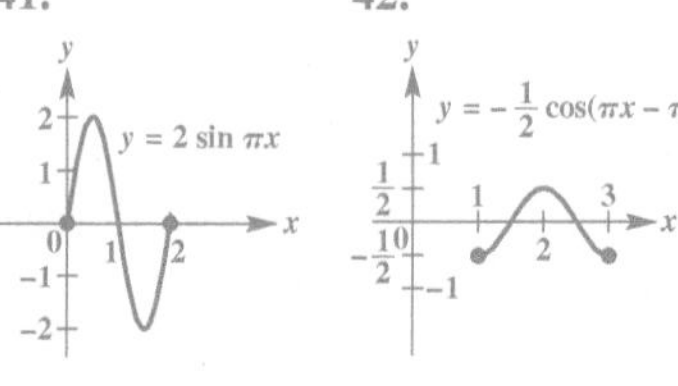

43. (a) See the graph in part (c).

(b) $y = 8.02 \sin(0.52x + 0.84) + 59.83$

(c)

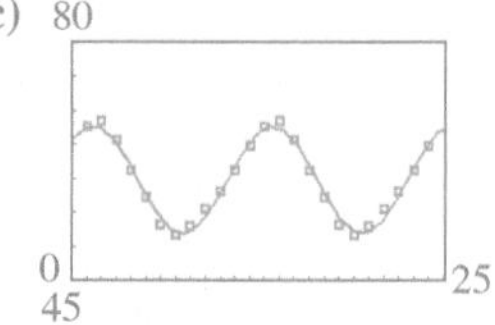

44. (a) See the graph in part (c).

(b) $y = 7.64 \sin(0.52x + 0.86) + 53.53$

(c)

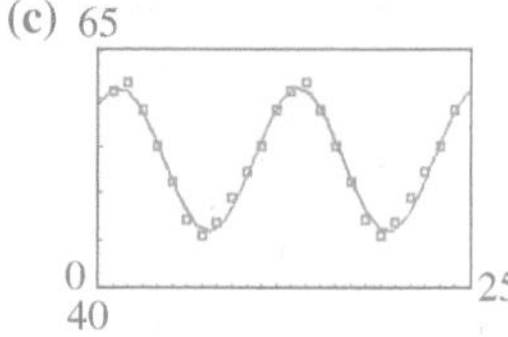

45. $y = 1 - \sin x$

46. $y = \frac{1}{2} \cos 2x$

47. $y = 2 \tan \frac{1}{2}x$

44. Average Low Temperature, Auckland, New Zealand

Jan	Feb	Mar	Apr	May	Jun	Jul	Aug	Sept	Oct	Nov	Dec
60.8	61.7	58.8	54.9	51.1	47.1	45.5	46.8	49.5	52.2	55.0	58.8

Connecting Graphs with Equations Determine the simplest form of an equation for each graph. Choose $b > 0$, and include no phase shifts.

45.

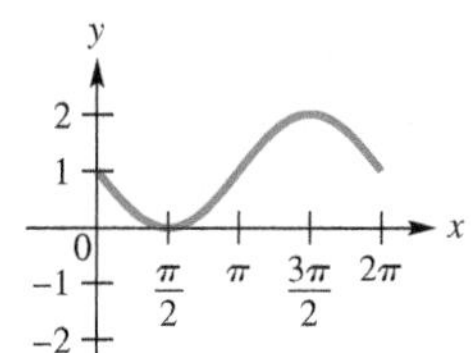

46.

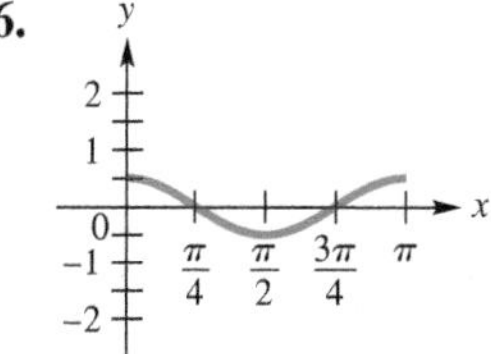

47.

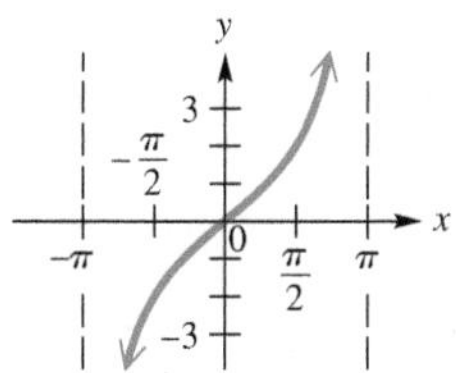

48.

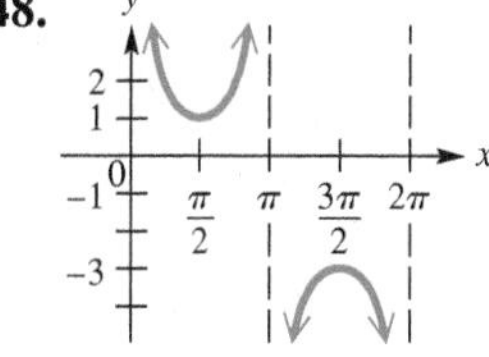

Solve each problem.

49. ***Viewing Angle to an Object*** Suppose that a person whose eyes are h_1 feet from the ground is standing d feet from an object h_2 feet tall, where $h_2 > h_1$. Let θ be the angle of elevation to the top of the object. See the figure.

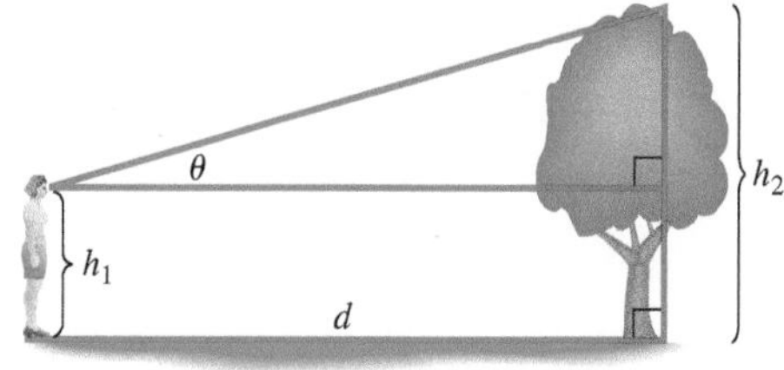

(a) Show that $d = (h_2 - h_1) \cot \theta$.

(b) Let $h_2 = 55$ and $h_1 = 5$. Graph d for the interval $0 < \theta \le \frac{\pi}{2}$.

50. ***(Modeling) Tides*** The figure shows a function f that models the tides in feet at Clearwater Beach, Florida, x hours after midnight. (*Source*: Pentcheff, D., *WWW Tide and Current Predictor*.)

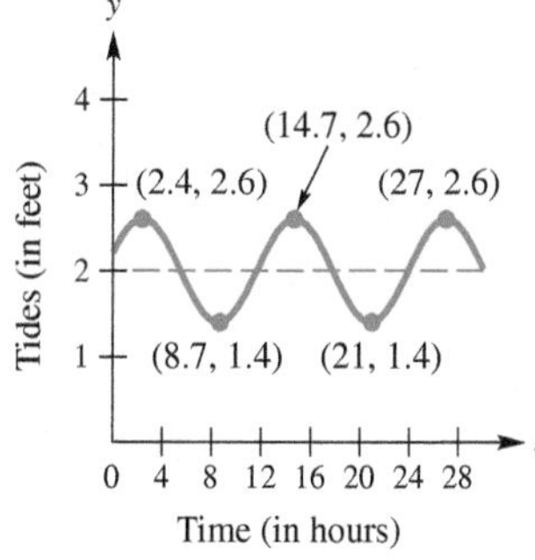

(a) Find the time between high tides.

(b) What is the difference in water levels between high tide and low tide?

(c) The tides can be modeled by

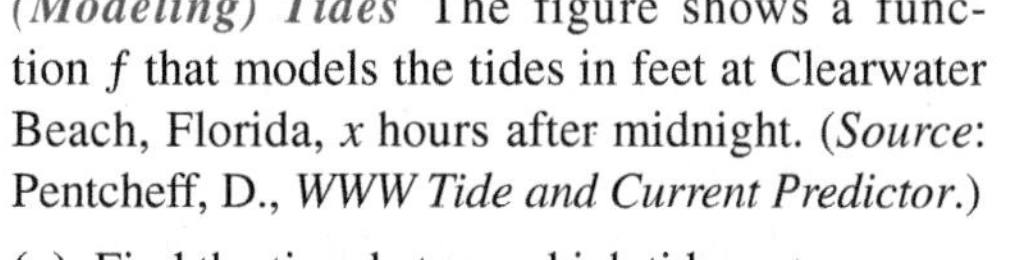

$$f(x) = 0.6 \cos[0.511(x - 2.4)] + 2.$$

Estimate the tides, to the nearest hundredth, when $x = 10$.

51. ***(Modeling) Maximum Temperatures*** The maximum afternoon temperature (in °F) in a given city can be modeled by

$$t = 60 - 30 \cos \frac{x\pi}{6},$$

where t represents the maximum afternoon temperature in month x, with $x = 0$ representing January, $x = 1$ representing February, and so on. Find the maximum afternoon temperature, to the nearest degree, for each month.

(a) January **(b)** April **(c)** May

(d) June **(e)** August **(f)** October

48. $y = -1 + 2 \csc x$

49. **(b)**

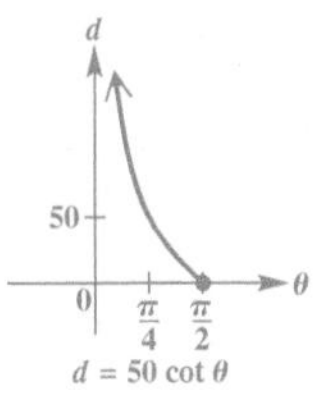

50. **(a)** 12.3 hr **(b)** 1.2 ft
(c) 1.56 ft

51. **(a)** 30°F **(b)** 60°F
(c) 75°F **(d)** 86°F
(e) 86°F **(f)** 60°F

52. **(a)** See the graph in part (c).
(b) $f(x) = 25.5 \sin\left[\frac{\pi}{6}(x - 4)\right] + 47.5$
(c) $f(x) = 25.5 \sin\left[\frac{\pi}{6}(x - 4)\right] + 47.5$

80

0 25

0

The function gives an excellent model for the data.

(d)

SinReg
y=a*sin(bx+c)+d
a=24.96
b=.52
c=-2.15
d=48.90

TI-84 Plus fixed to the nearest hundredth

53. **(a)** 100 **(b)** 258
(c) 122 **(d)** 296

54. **(a)** about 20 yr
(b) maximum: about 150,000; minimum: about 5000

55. amplitude: 4; period: 2; frequency: $\frac{1}{2}$ cycle per sec

56. amplitude: 3; period: π; frequency: $\frac{1}{\pi}$ cycle per sec

52. *(Modeling) Average Monthly Temperature* The average monthly temperature (in °F) in Chicago, Illinois, is shown in the table.

Month	°F	Month	°F
Jan	22	July	73
Feb	27	Aug	72
Mar	37	Sept	64
Apr	48	Oct	52
May	59	Nov	39
June	68	Dec	27

Source: World Almanac and Book of Facts.

(a) Plot the average monthly temperature over a two-year period. Let $x = 1$ correspond to January of the first year.

(b) To model the data, determine a function of the form $f(x) = a \sin[b(x - d)] + c$, where a, b, c, and d are constants.

(c) Graph f together with the data on the same coordinate axes. How well does f model the data?

(d) Use the sine regression capability of a graphing calculator to find the equation of a sine curve of the form $y = a \sin(bx + c) + d$ that fits these data.

53. *(Modeling) Pollution Trends* The amount of pollution in the air is lower after heavy spring rains and higher after periods of little rain. In addition to this seasonal fluctuation, the long-term trend is upward. An idealized graph of this situation is shown in the figure.

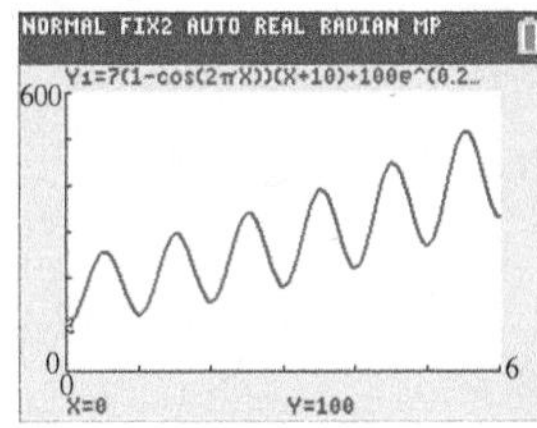

Circular functions can be used to model the fluctuating part of the pollution levels. Powers of the number e (e is the base of the natural logarithm; $e \approx 2.718282$) can be used to model long-term growth. The pollution level in a certain area might be given by

$$y = 7(1 - \cos 2\pi x)(x + 10) + 100e^{0.2x},$$

where x is the time in years, with $x = 0$ representing January 1 of the base year. July 1 of the same year would be represented by $x = 0.5$, October 1 of the following year would be represented by $x = 1.75$, and so on. Find the pollution levels on each date.

(a) January 1, base year (See the screen.) **(b)** July 1, base year

(c) January 1, following year **(d)** July 1, following year

54. *(Modeling) Lynx and Hare Populations* The figure shows the populations of lynx and hares in Canada for the years 1847–1903. The hares are food for the lynx. An increase in hare population causes an increase in lynx population some time later. The increasing lynx population then causes a decline in hare population. The two graphs have the same period.

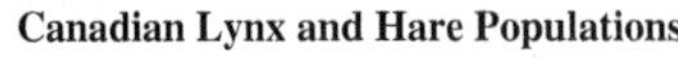

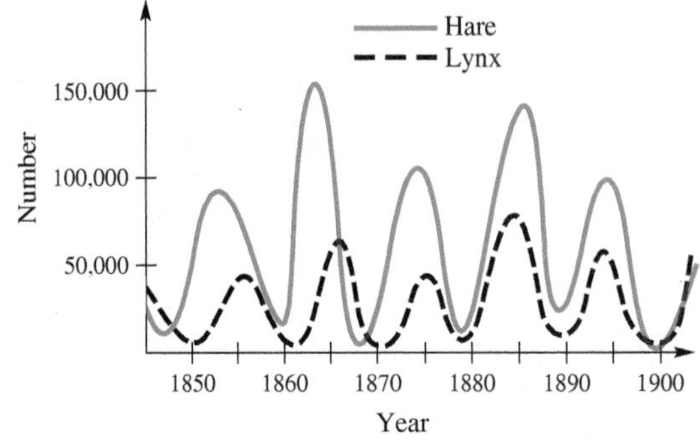

(a) Estimate the length of one period.

(b) Estimate the maximum and minimum hare populations.

An object in simple harmonic motion has position function $s(t)$ inches from an equilibrium point, where t is the time in seconds. Find the amplitude, period, and frequency.

55. $s(t) = 4 \sin \pi t$ **56.** $s(t) = 3 \cos 2t$

57. In **Exercise 55,** what does the frequency represent? Find the position of the object relative to the equilibrium point at 1.5 sec, 2 sec, and 3.25 sec.

58. In **Exercise 56,** what does the period represent? What does the amplitude represent?

57. The frequency is the number of cycles in one unit of time; -4; 0; $-2\sqrt{2}$

58. The period is the time to complete one cycle. The amplitude is the maximum distance (on either side) from the equilibrium point.

Chapter 4 Test

1. Identify each of the following basic circular function graphs.

(a)

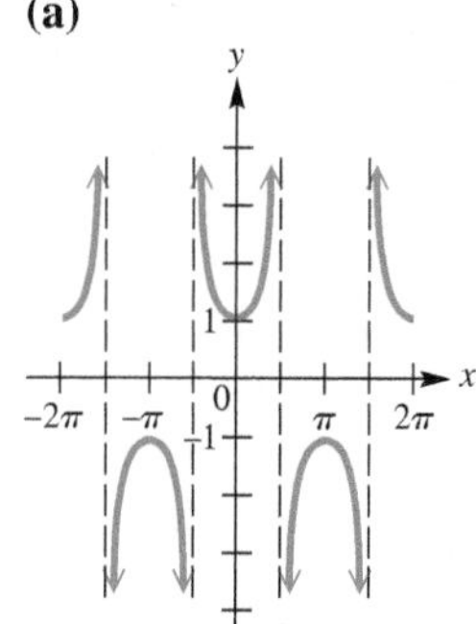

(b)

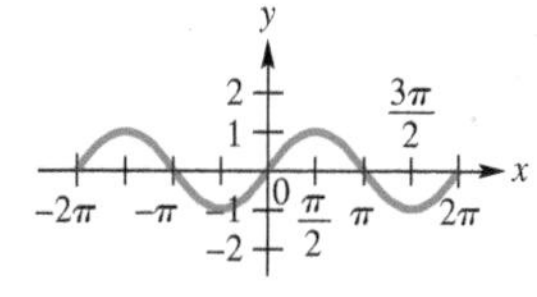

(c)

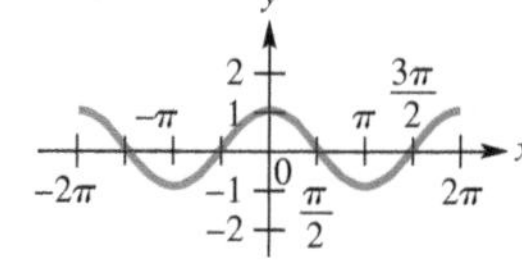

(d)

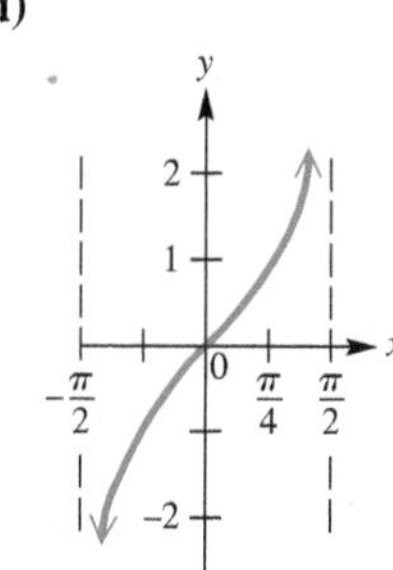

(e)

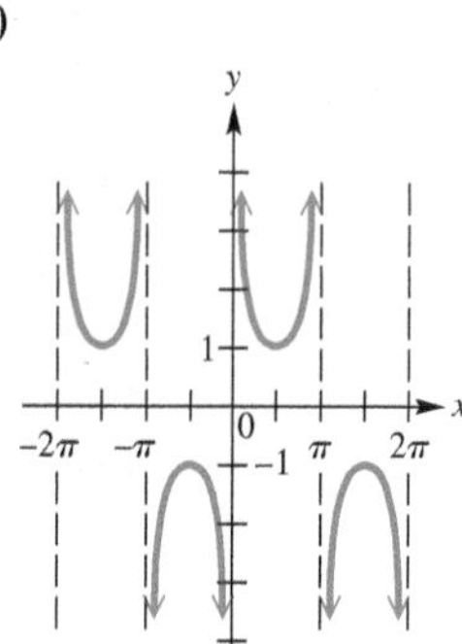

(f)

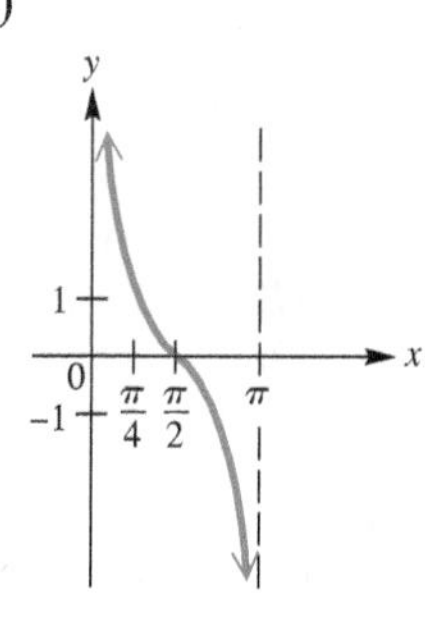

2. ***Connecting Graphs with Equations*** Determine the simplest form of an equation for each graph. Choose $b > 0$, and include no phase shifts.

(a)

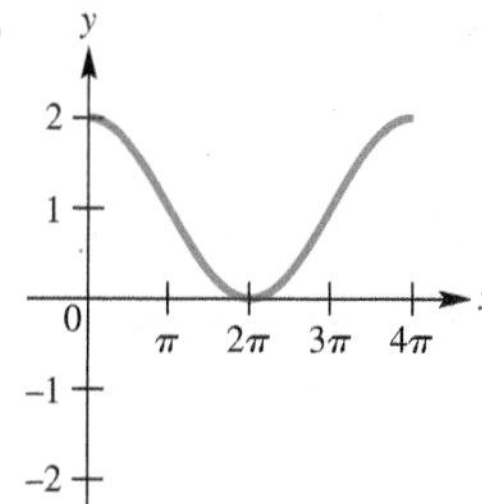

(b)

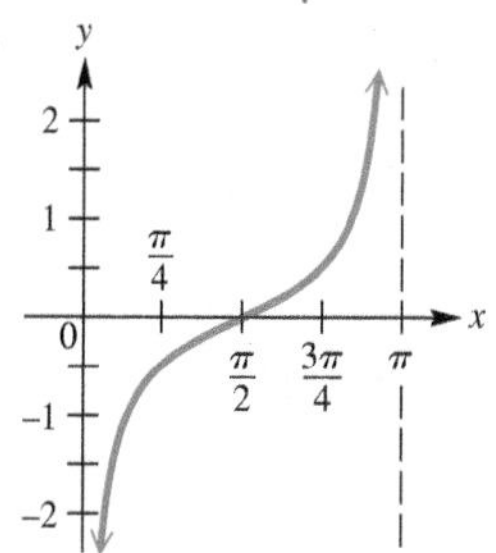

3. Answer each question.

(a) What is the domain of the cosine function?

(b) What is the range of the sine function?

(c) What is the least positive value for which the tangent function is undefined?

(d) What is the range of the secant function?

[4.1–4.4]

1. (a) $y = \sec x$ (b) $y = \sin x$ (c) $y = \cos x$ (d) $y = \tan x$ (e) $y = \csc x$ (f) $y = \cot x$

2. (a) $y = 1 + \cos \frac{1}{2}x$ (b) $y = -\frac{1}{2} \cot x$

[4.1, 4.3, 4.4]

3. (a) $(-\infty, \infty)$ (b) $[-1, 1]$ (c) $\frac{\pi}{2}$ (d) $(-\infty, -1] \cup [1, \infty)$

[4.2]

4. (a) π (b) 6 (c) $[-3, 9]$ (d) $(0, -3)$ (e) $\frac{\pi}{4}$ to the left $\left(\text{that is, } -\frac{\pi}{4}\right)$

5.

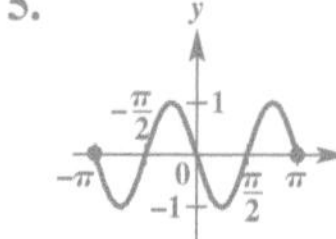

[4.1]

6.

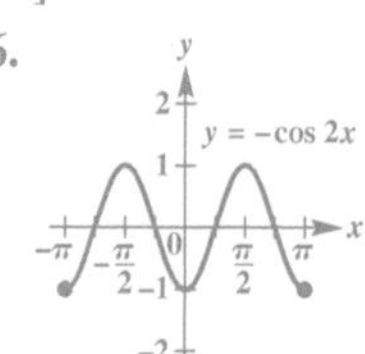

[4.2]

7.

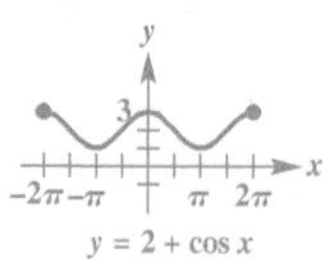

8.

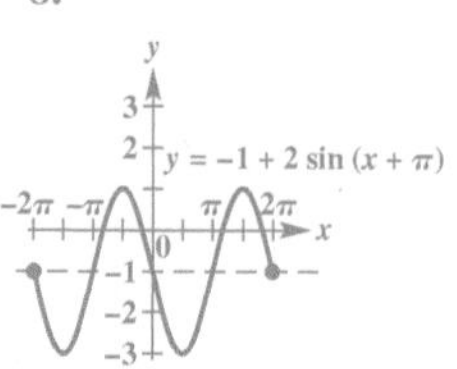

[4.3]

9. 10.

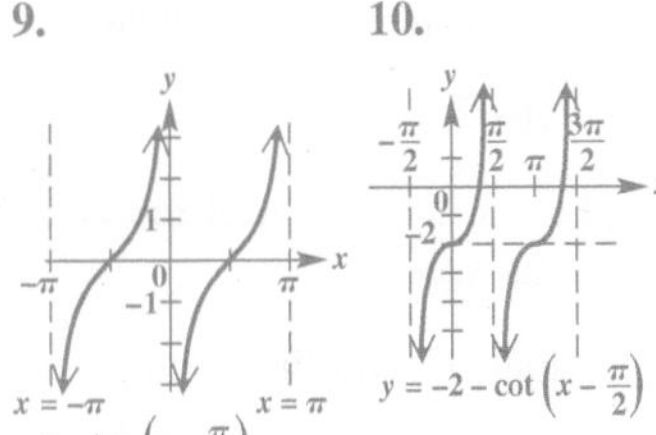

[4.4]

11. 12.

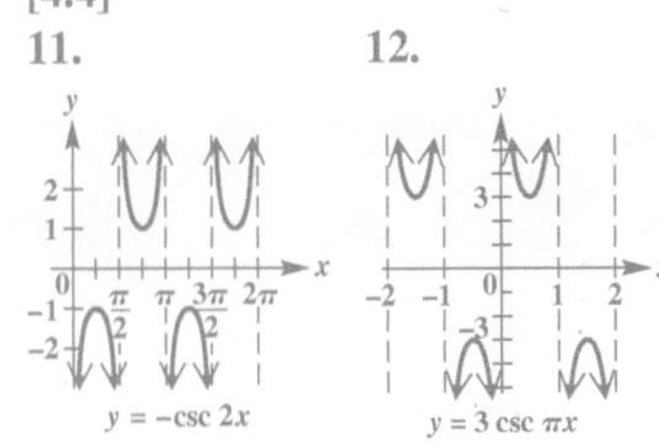

[4.1, 4.2]

13. (a)

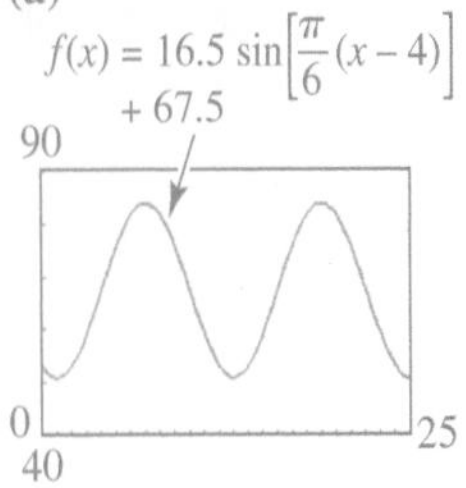

(b) 16.5; 12; 4 to the right; 67.5 up

(c) 53°F

(d) 51°F in January; 84°F in July

(e) 67.5°F; This is the vertical translation.

[4.5]

14. (a) 4 in. (b) after $\frac{1}{8}$ sec

(c) 4 cycles per sec; $\frac{1}{4}$ sec

[4.3, 4.4]

15. Both the tangent and secant functions are (by definition) undefined when $\cos x = 0$ and are defined for all other x-values, so their domains are the same. Both the cotangent and cosecant functions are (by definition) undefined when $\sin x = 0$ and are defined for all other x-values, so their domains are the same.

4. Consider the function $y = 3 - 6 \sin\left(2x + \frac{\pi}{2}\right)$.

(a) What is its period?

(b) What is the amplitude of its graph?

(c) What is its range?

(d) What is the y-intercept of its graph?

(e) What is its phase shift?

Graph each function over a two-period interval. Identify asymptotes when applicable.

5. $y = \sin(2x + \pi)$

6. $y = -\cos 2x$

7. $y = 2 + \cos x$

8. $y = -1 + 2 \sin(x + \pi)$

9. $y = \tan\left(x - \frac{\pi}{2}\right)$

10. $y = -2 - \cot\left(x - \frac{\pi}{2}\right)$

11. $y = -\csc 2x$

12. $y = 3 \csc \pi x$

(Modeling) Solve each problem.

13. ***Average Monthly Temperature*** The average monthly temperature (in °F) in San Antonio, Texas, can be modeled by

$$f(x) = 16.5 \sin\left[\frac{\pi}{6}(x-4)\right] + 67.5,$$

where x is the month and $x = 1$ corresponds to January. (*Source: World Almanac and Book of Facts.*)

(a) Graph f in the window $[0, 25]$ by $[40, 90]$.

(b) Determine the amplitude, period, phase shift, and vertical translation of f.

(c) What is the average monthly temperature for the month of December?

(d) Determine the minimum and maximum average monthly temperatures and the months when they occur.

(e) What would be an approximation for the average annual temperature in San Antonio? How is this related to the vertical translation of the sine function in the formula for f?

14. ***Spring Motion*** The position of a weight attached to a spring is

$$s(t) = -4 \cos 8\pi t \text{ inches after } t \text{ seconds.}$$

(a) Find the maximum height that the weight rises above the equilibrium position of $s(t) = 0$.

(b) When does the weight first reach its maximum height if $t \geq 0$?

(c) What are the frequency and period?

15. Explain why the domains of the tangent and secant functions are the same, and then give a similar explanation for the cotangent and cosecant functions.

5 Trigonometric Identities

Electricity that passes through wires to homes and businesses alternates its direction on those wires and is modeled by *sine* and *cosine functions.*

5.1 Fundamental Identities

- Fundamental Identities
- Uses of the Fundamental Identities

Fundamental Identities

Recall that a function is **even** if $f(-x) = f(x)$ for all x in the domain of f, and a function is **odd** if $f(-x) = -f(x)$ for all x in the domain of f. We have used graphs to classify the trigonometric functions as even or odd. We can also use **Figure 1** to do this.

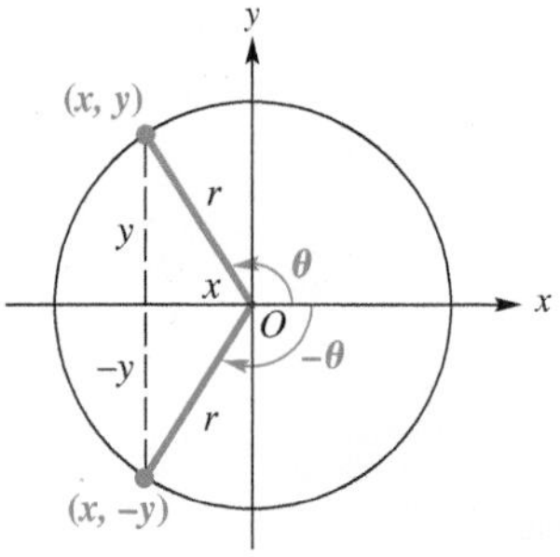

As suggested by the circle in **Figure 1,** an angle θ having the point (x, y) on its terminal side has a corresponding angle $-\theta$ with the point $(x, -y)$ on its terminal side.

$\sin(-\theta) = -\frac{y}{r} = -\sin\theta$

Figure 1

From the definition of sine, we see that $\sin(-\theta)$ and $\sin\theta$ are negatives of each other. That is,

$$\sin(-\theta) = \frac{-y}{r} \quad \text{and} \quad \sin\theta = \frac{y}{r},$$

so $$\boldsymbol{\sin(-\theta) = -\sin\theta}$$ Sine is an odd function.

Teaching Tip To help students distinguish between identities and conditional equations, provide several examples. Ask students to identify each example as one of these.

This is an example of an **identity,** an equation that is satisfied by *every* value in the domain of its variable. Some examples from algebra follow.

$$\begin{aligned} x^2 - y^2 &= (x + y)(x - y) \\ x(x + y) &= x^2 + xy \\ x^2 + 2xy + y^2 &= (x + y)^2 \end{aligned}$$ Identities

Figure 1 shows an angle θ in quadrant II, but the same result holds for θ in any quadrant. The figure also suggests the following identity for cosine.

$$\cos(-\theta) = \frac{x}{r} \quad \text{and} \quad \cos\theta = \frac{x}{r}$$

$$\boldsymbol{\cos(-\theta) = \cos\theta}$$ Cosine is an even function.

We use the identities for $\sin(-\theta)$ and $\cos(-\theta)$ to find $\tan(-\theta)$ in terms of $\tan\theta$.

$$\tan(-\theta) = \frac{\sin(-\theta)}{\cos(-\theta)} = \frac{-\sin\theta}{\cos\theta} = -\frac{\sin\theta}{\cos\theta}$$

$$\boldsymbol{\tan(-\theta) = -\tan\theta}$$ Tangent is an odd function.

The reciprocal identities are used to determine that cosecant and cotangent are odd functions and secant is an even function. These **even-odd identities** together with the reciprocal, quotient, and Pythagorean identities make up the **fundamental identities.**

NOTE In trigonometric identities, θ can represent an angle in degrees or radians, or a real number.

Teaching Tip Encourage students to memorize the identities presented in this section as well as subsequent sections. Point out that numerical values can be used to help check whether or not an identity was recalled correctly.

Fundamental Identities

Reciprocal Identities

$$\cot \theta = \frac{1}{\tan \theta} \qquad \sec \theta = \frac{1}{\cos \theta} \qquad \csc \theta = \frac{1}{\sin \theta}$$

Quotient Identities

$$\tan \theta = \frac{\sin \theta}{\cos \theta} \qquad \cot \theta = \frac{\cos \theta}{\sin \theta}$$

Pythagorean Identities

$$\sin^2 \theta + \cos^2 \theta = 1 \qquad \tan^2 \theta + 1 = \sec^2 \theta \qquad 1 + \cot^2 \theta = \csc^2 \theta$$

Even-Odd Identities

$$\sin(-\theta) = -\sin \theta \qquad \cos(-\theta) = \cos \theta \qquad \tan(-\theta) = -\tan \theta$$

$$\csc(-\theta) = -\csc \theta \qquad \sec(-\theta) = \sec \theta \qquad \cot(-\theta) = -\cot \theta$$

NOTE We will also use alternative forms of the fundamental identities. *For example, two other forms of* $\sin^2 \theta + \cos^2 \theta = 1$ *are*

$$\sin^2 \theta = 1 - \cos^2 \theta \quad \textit{and} \quad \cos^2 \theta = 1 - \sin^2 \theta.$$

Uses of the Fundamental Identities We can use these identities to find the values of other trigonometric functions from the value of a given trigonometric function.

Classroom Example 1

If $\cos \theta = \frac{5}{8}$ and θ is in quadrant IV, find each function value.

(a) $\sin \theta$ (b) $\tan \theta$

(c) $\sec(-\theta)$

Answers:

(a) $-\frac{\sqrt{39}}{8}$ (b) $-\frac{\sqrt{39}}{5}$

(c) $\frac{8}{5}$

EXAMPLE 1 Finding Trigonometric Function Values Given One Value and the Quadrant

If $\tan \theta = -\frac{5}{3}$ and θ is in quadrant II, find each function value.

(a) $\sec \theta$ (b) $\sin \theta$ (c) $\cot(-\theta)$

SOLUTION

(a) We use an identity that relates the tangent and secant functions.

$$\tan^2 \theta + 1 = \sec^2 \theta \qquad \text{Pythagorean identity}$$

$$\left(-\frac{5}{3}\right)^2 + 1 = \sec^2 \theta \qquad \tan \theta = -\frac{5}{3}$$

$$\frac{25}{9} + 1 = \sec^2 \theta \qquad \text{Square } -\frac{5}{3}.$$

$$\frac{34}{9} = \sec^2 \theta \qquad \text{Add; } 1 = \frac{9}{9}$$

$$-\sqrt{\frac{34}{9}} = \sec \theta \qquad \text{Take the negative square root because } \theta \text{ is in quadrant II.}$$

Choose the correct sign.

$$\sec \theta = -\frac{\sqrt{34}}{3} \qquad \text{Simplify the radical: } -\sqrt{\frac{34}{9}} = -\frac{\sqrt{34}}{\sqrt{9}} = -\frac{\sqrt{34}}{3}, \text{ and rewrite.}$$

Teaching Tip Warn students that the given information in **Example 1,** $\tan \theta = -\frac{5}{3} = \frac{-5}{3}$, does not mean that $\sin \theta = -5$ and $\cos \theta = 3$. Ask them why these values cannot be correct.

(b)

$$\tan\theta = \frac{\sin\theta}{\cos\theta}$$ Quotient identity

$$\cos\theta\tan\theta = \sin\theta$$ Multiply each side by $\cos\theta$.

$$\left(\frac{1}{\sec\theta}\right)\tan\theta = \sin\theta$$ Reciprocal identity

$$\left(-\frac{3\sqrt{34}}{34}\right)\left(-\frac{5}{3}\right) = \sin\theta$$ $\tan\theta = -\frac{5}{3}$, and from part (a), $\frac{1}{\sec\theta} = \frac{1}{-\frac{\sqrt{34}}{3}} = -\frac{3}{\sqrt{34}} = -\frac{3}{\sqrt{34}}\cdot\frac{\sqrt{34}}{\sqrt{34}} = -\frac{3\sqrt{34}}{34}$.

$$\sin\theta = \frac{5\sqrt{34}}{34}$$ Multiply and rewrite.

(c)

$$\cot(-\theta) = \frac{1}{\tan(-\theta)}$$ Reciprocal identity

$$\cot(-\theta) = \frac{1}{-\tan\theta}$$ Even-odd identity

$$\cot(-\theta) = \frac{1}{-\left(-\frac{5}{3}\right)}$$ $\tan\theta = -\frac{5}{3}$

$$\cot(-\theta) = \frac{3}{5}$$ $\frac{1}{-\left(-\frac{5}{3}\right)} = 1 \div \frac{5}{3} = 1\cdot\frac{3}{5} = \frac{3}{5}$

✔ **Now Try Exercises 11, 19, and 31.**

CAUTION ***When taking the square root, be sure to choose the sign based on the quadrant of θ and the function being evaluated.***

Classroom Example 2
Write $\tan\theta$ in terms of $\cos\theta$.

Answer:
$\tan\theta = \frac{\pm\sqrt{1-\cos^2\theta}}{\cos\theta}$

EXAMPLE 2 Writing One Trignometric Function in Terms of Another

Write $\cos x$ in terms of $\tan x$.

SOLUTION By identities, $\sec x$ is related to both $\cos x$ and $\tan x$.

$$1 + \tan^2 x = \sec^2 x$$ Pythagorean identity

$$\frac{1}{1+\tan^2 x} = \frac{1}{\sec^2 x}$$ Take reciprocals.

$$\frac{1}{1+\tan^2 x} = \cos^2 x$$ The reciprocal of $\sec^2 x$ is $\cos^2 x$.

$$\pm\sqrt{\frac{1}{1+\tan^2 x}} = \cos x$$ Take the square root of each side.

Remember both the positive and negative roots.

$$\cos x = \frac{\pm 1}{\sqrt{1+\tan^2 x}}$$ Quotient rule for radicals: $\sqrt[n]{\frac{a}{b}} = \frac{\sqrt[n]{a}}{\sqrt[n]{b}}$; rewrite.

$$\cos x = \frac{\pm\sqrt{1+\tan^2 x}}{1+\tan^2 x}$$ Rationalize the denominator.

The choice of the + sign or the − sign is made depending on the quadrant of x.

✔ **Now Try Exercise 47.**

$y_1 = \sin^2 x + \cos^2 x$
$y_2 = 1$

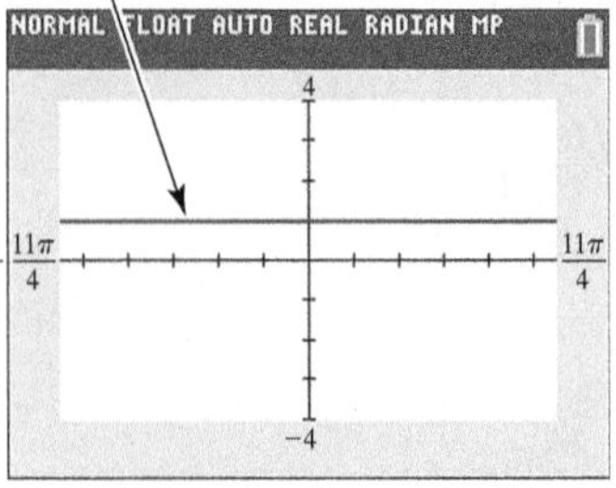

NORMAL FLOAT AUTO REAL RADIAN MP

4, −4, $-\frac{11\pi}{4}$, $\frac{11\pi}{4}$

With an identity, there should be no difference between the two graphs.

Figure 2

Figure 2 supports the identity $\sin^2 x + \cos^2 x = 1$. ■

The functions tan θ, cot θ, sec θ, and csc θ can easily be expressed in terms of sin θ, cos θ, or both. We make such substitutions in an expression to simplify it.

Classroom Example 3

Write 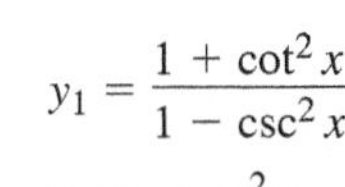 $\frac{1 + \tan^2 \theta}{1 - \sec^2 \theta}$ in terms of sin θ and cos θ, and then simplify the expression so that no quotients appear.

Answer: $-\csc^2 \theta$

EXAMPLE 3 Rewriting an Expression in Terms of Sine and Cosine

Write $\frac{1 + \cot^2 \theta}{1 - \csc^2 \theta}$ in terms of sin θ and cos θ, and then simplify the expression so that no quotients appear.

SOLUTION

$$\frac{1 + \cot^2 \theta}{1 - \csc^2 \theta} \quad \text{Given expression}$$

$$= \frac{1 + \frac{\cos^2 \theta}{\sin^2 \theta}}{1 - \frac{1}{\sin^2 \theta}} \quad \text{Quotient identities}$$

$$= \frac{\left(1 + \frac{\cos^2 \theta}{\sin^2 \theta}\right)\sin^2 \theta}{\left(1 - \frac{1}{\sin^2 \theta}\right)\sin^2 \theta} \quad \text{Simplify the complex fraction by multiplying both numerator and denominator by the LCD.}$$

$$= \frac{\sin^2 \theta + \cos^2 \theta}{\sin^2 \theta - 1} \quad \text{Distributive property: } (b + c)a = ba + ca$$

$$= \frac{1}{-\cos^2 \theta} \quad \text{Pythagorean identities}$$

$$= -\sec^2 \theta \quad \text{Reciprocal identity}$$

Now Try Exercise 59.

$y_1 = \frac{1 + \cot^2 x}{1 - \csc^2 x}$

$y_2 = -\sec^2 x$

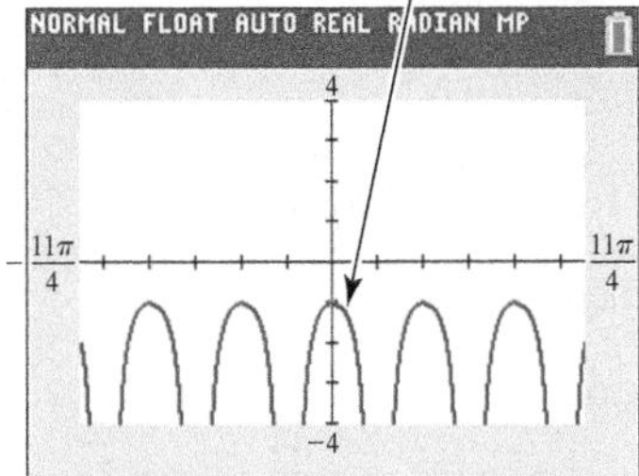

The graph supports the result in **Example 3.** The graphs of y_1 and y_2 coincide.

CAUTION ***When working with trigonometric expressions and identities, be sure to write the argument of the function.*** For example, we would *not* write $\sin^2 + \cos^2 = 1$. An argument such as θ is necessary to write this correctly as $\sin^2 \theta + \cos^2 \theta = 1$.

5.1 Exercises

1. B 2. D
3. E 4. C
5. A

CONCEPT PREVIEW *For each expression in Column I, choose the expression from Column II that completes an identity.*

I	II
1. $\frac{\cos x}{\sin x} =$ ____	**A.** $\sin^2 x + \cos^2 x$
2. $\tan x =$ ____	**B.** $\cot x$
3. $\cos(-x) =$ ____	**C.** $\sec^2 x$
4. $\tan^2 x + 1 =$ ____	**D.** $\frac{\sin x}{\cos x}$
5. $1 =$ ____	**E.** $\cos x$

6. -2.6
7. -0.65 8. 0.625
9. -0.75 10. $\frac{2}{3}$
11. $\frac{\sqrt{7}}{4}$ 12. $\frac{\sqrt{11}}{6}$
13. $-\frac{5\sqrt{26}}{26}$ 14. $-\frac{3\sqrt{10}}{10}$
15. $-\frac{2\sqrt{5}}{5}$ 16. $-\frac{\sqrt{33}}{6}$
17. $-\frac{\sqrt{15}}{5}$ 18. $-\frac{\sqrt{77}}{11}$
19. $-\frac{\sqrt{105}}{11}$ 20. $-\frac{3\sqrt{5}}{7}$
21. $-\frac{4}{9}$ 22. $-\frac{5}{8}$
23. $\sin\theta$ is the reciprocal of $\csc\theta$ and therefore has the same sign.
24. There is no number or angle whose cosine is 3.
25. $f(-x) = \frac{-\sin x}{-x} = \frac{\sin x}{x} = f(x)$; even
26. $f(-x) = -x\cos x = -f(x)$; odd
27. $f(x) = \sec x$; even
28. $f(x) = \csc x$; odd
29. $f(x) = \cot x$; odd
30. $f(x) = \tan x$; odd
31. $\cos\theta = -\frac{\sqrt{5}}{3}$; $\tan\theta = -\frac{2\sqrt{5}}{5}$; $\cot\theta = -\frac{\sqrt{5}}{2}$; $\sec\theta = -\frac{3\sqrt{5}}{5}$; $\csc\theta = \frac{3}{2}$
32. $\sin\theta = \frac{2\sqrt{6}}{5}$; $\tan\theta = 2\sqrt{6}$; $\cot\theta = \frac{\sqrt{6}}{12}$; $\sec\theta = 5$; $\csc\theta = \frac{5\sqrt{6}}{12}$
33. $\sin\theta = -\frac{\sqrt{17}}{17}$; $\cos\theta = \frac{4\sqrt{17}}{17}$; $\cot\theta = -4$; $\sec\theta = \frac{\sqrt{17}}{4}$; $\csc\theta = -\sqrt{17}$
34. $\sin\theta = -\frac{2}{5}$; $\cos\theta = -\frac{\sqrt{21}}{5}$; $\tan\theta = \frac{2\sqrt{21}}{21}$; $\cot\theta = \frac{\sqrt{21}}{2}$; $\sec\theta = -\frac{5\sqrt{21}}{21}$
35. $\sin\theta = \frac{3}{5}$; $\cos\theta = \frac{4}{5}$; $\tan\theta = \frac{3}{4}$; $\sec\theta = \frac{5}{4}$; $\csc\theta = \frac{5}{3}$

CONCEPT PREVIEW *Use identities to correctly complete each sentence.*

6. If $\tan\theta = 2.6$, then $\tan(-\theta) =$ ______.
7. If $\cos\theta = -0.65$, then $\cos(-\theta) =$ ______.
8. If $\tan\theta = 1.6$, then $\cot\theta =$ ______.
9. If $\cos\theta = 0.8$ and $\sin\theta = 0.6$, then $\tan(-\theta) =$ ______.
10. If $\sin\theta = \frac{2}{3}$, then $-\sin(-\theta) =$ ______.

Find $\sin\theta$. ***See Example 1.***

11. $\cos\theta = \frac{3}{4}$, θ in quadrant I
12. $\cos\theta = \frac{5}{6}$, θ in quadrant I
13. $\cot\theta = -\frac{1}{5}$, θ in quadrant IV
14. $\cot\theta = -\frac{1}{3}$, θ in quadrant IV
15. $\cos(-\theta) = \frac{\sqrt{5}}{5}$, $\tan\theta < 0$
16. $\cos(-\theta) = \frac{\sqrt{3}}{6}$, $\cot\theta < 0$
17. $\tan\theta = -\frac{\sqrt{6}}{2}$, $\cos\theta > 0$
18. $\tan\theta = -\frac{\sqrt{7}}{2}$, $\sec\theta > 0$
19. $\sec\theta = \frac{11}{4}$, $\cot\theta < 0$
20. $\sec\theta = \frac{7}{2}$, $\tan\theta < 0$
21. $\csc\theta = -\frac{9}{4}$
22. $\csc\theta = -\frac{8}{5}$
23. Why is it unnecessary to give the quadrant of θ in **Exercises 21 and 22?**
24. *Concept Check* What is **WRONG** with the statement of this problem?

Find $\cos(-\theta)$ *if* $\cos\theta = 3$.

Concept Check Find $f(-x)$ *to determine whether each function is* even *or* odd.

25. $f(x) = \dfrac{\sin x}{x}$
26. $f(x) = x\cos x$

Concept Check Identify the basic trigonometric function graphed and determine whether it is even *or* odd.

27.

28.

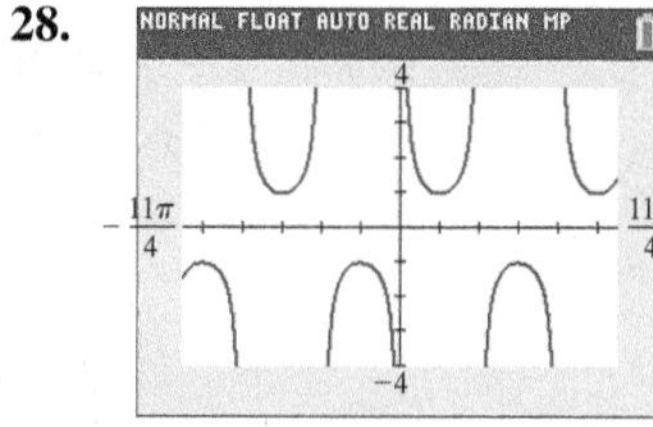

29.

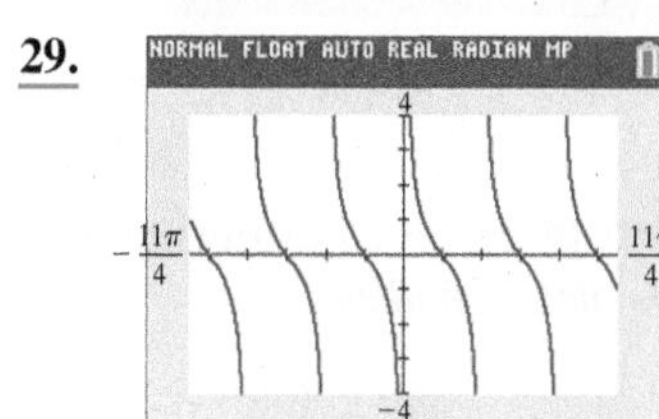

30.

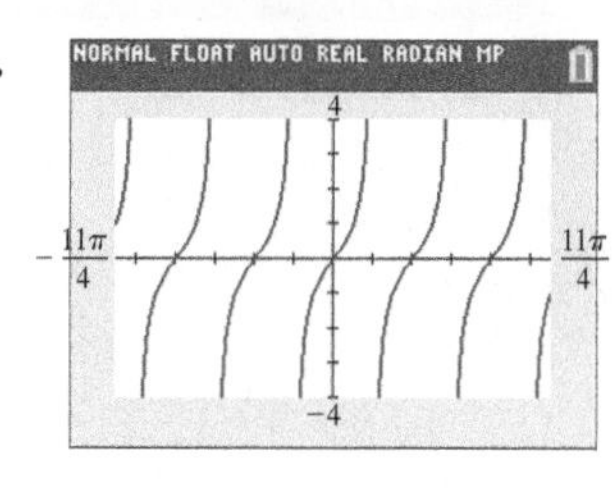

Find the remaining five trigonometric functions of θ. ***See Example 1.***

31. $\sin\theta = \frac{2}{3}$, θ in quadrant II
32. $\cos\theta = \frac{1}{5}$, θ in quadrant I
33. $\tan\theta = -\frac{1}{4}$, θ in quadrant IV
34. $\csc\theta = -\frac{5}{2}$, θ in quadrant III
35. $\cot\theta = \frac{4}{3}$, $\sin\theta > 0$
36. $\sin\theta = -\frac{4}{5}$, $\cos\theta < 0$
37. $\sec\theta = \frac{4}{3}$, $\sin\theta < 0$
38. $\cos\theta = -\frac{1}{4}$, $\sin\theta > 0$

36. $\cos\theta = -\frac{3}{5}$; $\tan\theta = \frac{4}{3}$;
$\cot\theta = \frac{3}{4}$; $\sec\theta = -\frac{5}{3}$;
$\csc\theta = -\frac{5}{4}$

37. $\sin\theta = -\frac{\sqrt{7}}{4}$; $\cos\theta = \frac{3}{4}$;
$\tan\theta = -\frac{\sqrt{7}}{3}$; $\cot\theta = -\frac{3\sqrt{7}}{7}$;
$\csc\theta = -\frac{4\sqrt{7}}{7}$

38. $\sin\theta = \frac{\sqrt{15}}{4}$; $\tan\theta = -\sqrt{15}$;
$\cot\theta = -\frac{\sqrt{15}}{15}$; $\sec\theta = -4$;
$\csc\theta = \frac{4\sqrt{15}}{15}$

39. C
40. A
41. E
42. D
43. B

44. The student has forgotten to write the arguments of the identity; for example,

$1 + \cot^2 \boldsymbol{\theta} = \csc^2 \boldsymbol{\theta}$.

45. $\sin\theta = \frac{\pm\sqrt{2x+1}}{x+1}$

46. $\tan\theta = \frac{\pm 2\sqrt{2x+4}}{x}$

47. $\sin x = \pm\sqrt{1-\cos^2 x}$

48. $\cot x = \frac{\pm\sqrt{1-\sin^2 x}}{\sin x}$

49. $\tan x = \pm\sqrt{\sec^2 x - 1}$

50. $\cot x = \pm\sqrt{\csc^2 x - 1}$

51. $\csc x = \frac{\pm\sqrt{1-\cos^2 x}}{1-\cos^2 x}$

52. $\sec x = \frac{\pm\sqrt{1-\sin^2 x}}{1-\sin^2 x}$

In Exercises 53–78, there may be more than one possible answer.

53. $\cos\theta$
54. $\sin\theta$
55. 1
56. 1
57. $\cot\theta$
58. $\tan\theta$
59. $\cos^2\theta$
60. $\csc^2\theta$
61. $\sec\theta - \cos\theta$
62. $\tan^2\theta$
63. $-\cot\theta + 1$
64. $\tan\theta + 1$
65. $\sin^2\theta\cos^2\theta$
66. $\sin^2\theta\cos^2\theta$
67. $\tan\theta\sin\theta$
68. $\cot\theta\cos\theta$
69. $\cot\theta - \tan\theta$
70. $\tan\theta - \cot\theta$
71. $\cos^2\theta$
72. $-\sin^2\theta$
73. $\tan^2\theta$
74. $\tan^4\theta$
75. $-\sec\theta$
76. $-\sin\theta$
77. $\sec^2\theta$
78. $-\tan^2\theta$

Concept Check For each expression in Column I, choose the expression from Column II that completes an identity. One or both expressions may need to be rewritten.

I	II
39. $-\tan x \cos x =$ _____	**A.** $\frac{\sin^2 x}{\cos^2 x}$
40. $\sec^2 x - 1 =$ _____	**B.** $\frac{1}{\sec^2 x}$
41. $\frac{\sec x}{\csc x} =$ _____	**C.** $\sin(-x)$
42. $1 + \sin^2 x =$ _____	**D.** $\csc^2 x - \cot^2 x + \sin^2 x$
43. $\cos^2 x =$ _____	**E.** $\tan x$

44. A student writes "$1 + \cot^2 = \csc^2$." Comment on this student's work.

45. ***Concept Check*** Suppose that $\cos\theta = \frac{x}{x+1}$. Find an expression in x for $\sin\theta$.

46. ***Concept Check*** Suppose that $\sec\theta = \frac{x+4}{x}$. Find an expression in x for $\tan\theta$.

Perform each transformation. ***See Example 2.***

47. Write $\sin x$ in terms of $\cos x$.

48. Write $\cot x$ in terms of $\sin x$.

49. Write $\tan x$ in terms of $\sec x$.

50. Write $\cot x$ in terms of $\csc x$.

51. Write $\csc x$ in terms of $\cos x$.

52. Write $\sec x$ in terms of $\sin x$.

Write each expression in terms of sine and cosine, and then simplify the expression so that no quotients appear and all functions are of θ only. ***See Example 3.***

53. $\cot\theta\sin\theta$

54. $\tan\theta\cos\theta$

55. $\sec\theta\cot\theta\sin\theta$

56. $\csc\theta\cos\theta\tan\theta$

57. $\cos\theta\csc\theta$

58. $\sin\theta\sec\theta$

59. $\sin^2\theta(\csc^2\theta - 1)$

60. $\cot^2\theta(1 + \tan^2\theta)$

61. $(1 - \cos\theta)(1 + \sec\theta)$

62. $(\sec\theta - 1)(\sec\theta + 1)$

63. $\frac{1 + \tan(-\theta)}{\tan(-\theta)}$

64. $\frac{1 + \cot\theta}{\cot\theta}$

65. $\frac{1 - \cos^2(-\theta)}{1 + \tan^2(-\theta)}$

66. $\frac{1 - \sin^2(-\theta)}{1 + \cot^2(-\theta)}$

67. $\sec\theta - \cos\theta$

68. $\csc\theta - \sin\theta$

69. $(\sec\theta + \csc\theta)(\cos\theta - \sin\theta)$

70. $(\sin\theta - \cos\theta)(\csc\theta + \sec\theta)$

71. $\sin\theta(\csc\theta - \sin\theta)$

72. $\cos\theta(\cos\theta - \sec\theta)$

73. $\frac{1 + \tan^2\theta}{1 + \cot^2\theta}$

74. $\frac{\sec^2\theta - 1}{\csc^2\theta - 1}$

75. $\frac{\csc\theta}{\cot(-\theta)}$

76. $\frac{\tan(-\theta)}{\sec\theta}$

77. $\sin^2(-\theta) + \tan^2(-\theta) + \cos^2(-\theta)$

78. $-\sec^2(-\theta) + \sin^2(-\theta) + \cos^2(-\theta)$

79. $\frac{25\sqrt{6}-60}{12}$; $\frac{-25\sqrt{6}-60}{12}$

80. $\frac{2\sqrt{2}+8}{9}$; $\frac{-2\sqrt{2}+8}{9}$

81. identity
82. not an identity
83. not an identity
84. identity

85. $y = -\sin 2x$
86. It is the negative of $y = \sin 2x$.
87. $y = \cos 4x$
88. It is the same function.
89. (a) $y = -\sin 4x$
(b) $y = \cos 2x$
(c) $y = 5 \sin 3x$
90. Students who ignore negative signs will enjoy graphing cosine and secant functions containing a negative coefficient of x in the argument, because it can be ignored and the graph will still be correct.

Work each problem.

79. Let $\cos x = \frac{1}{5}$. Find all possible values of $\frac{\sec x - \tan x}{\sin x}$.

80. Let $\csc x = -3$. Find all possible values of $\frac{\sin x + \cos x}{\sec x}$.

Use a graphing calculator to make a conjecture about whether each equation is an identity.

81. $\cos 2x \doteq 1 - 2\sin^2 x$

82. $2 \sin x = \sin 2x$

83. $\sin x = \sqrt{1 - \cos^2 x}$

84. $\cos 2x = \cos^2 x - \sin^2 x$

Relating Concepts

For individual or collaborative investigation *(Exercises 85–90)*

Previously we graphed functions of the form

$$y = c + a \cdot f[b(x - d)]$$

with the assumption that $b > 0$. To see what happens when $b < 0$, ***work Exercises 85–90 in order.***

85. Use an even-odd identity to write $y = \sin(-2x)$ as a function of $2x$.

86. How is the answer to **Exercise 85** related to $y = \sin 2x$?

87. Use an even-odd identity to write $y = \cos(-4x)$ as a function of $4x$.

88. How is the answer to **Exercise 87** related to $y = \cos 4x$?

89. Use the results from **Exercises 85–88** to rewrite the following with a positive value of b.

(a) $y = \sin(-4x)$ **(b)** $y = \cos(-2x)$ **(c)** $y = -5 \sin(-3x)$

90. Write a short response to this statement, which is often used by one of the authors of this text in trigonometry classes:

Students who tend to ignore negative signs should enjoy graphing functions involving the cosine and the secant.

5.2 Verifying Trigonometric Identities

- Strategies
- Verifying Identities by Working with One Side
- Verifying Identities by Working with Both Sides

Strategies One of the skills required for more advanced work in mathematics, especially in calculus, is the ability to use identities to write expressions in alternative forms. We develop this skill by using the fundamental identities to verify that a trigonometric equation is an identity (for those values of the variable for which it is defined).

CAUTION ***The procedure for verifying identities is not the same as that for solving equations.*** Techniques used in solving equations, such as adding the same term to each side, and multiplying each side by the same term, should *not* be used when working with identities.

LOOKING AHEAD TO CALCULUS

Trigonometric identities are used in calculus to simplify trigonometric expressions, determine derivatives of trigonometric functions, and change the form of some integrals.

Teaching Tip There is no substitute for experience when it comes to verifying identities. Guide students through several examples, giving hints such as

"Apply a reciprocal identity"

or

"Use a different form of the Pythagorean identity $\sin^2\theta + \cos^2\theta = 1$."

Hints for Verifying Identities

1. ***Learn the fundamental identities.*** Whenever you see either side of a fundamental identity, the other side should come to mind. ***Also, be aware of equivalent forms of the fundamental identities.*** For example,

$$\sin^2\theta = 1 - \cos^2\theta \quad \text{is an alternative form of} \quad \sin^2\theta + \cos^2\theta = 1.$$

2. ***Try to rewrite the more complicated side*** of the equation so that it is identical to the simpler side.

3. ***It is sometimes helpful to express all trigonometric functions in the equation in terms of sine and cosine*** and then simplify the result.

4. ***Usually, any factoring or indicated algebraic operations should be performed.*** These *algebraic* identities are often used in verifying trigonometric identities.

$$x^2 + 2xy + y^2 = (x + y)^2$$
$$x^2 - 2xy + y^2 = (x - y)^2$$
$$x^3 - y^3 = (x - y)(x^2 + xy + y^2)$$
$$x^3 + y^3 = (x + y)(x^2 - xy + y^2)$$
$$x^2 - y^2 = (x + y)(x - y)$$

For example, the expression

$$\sin^2 x + 2\sin x + 1 \quad \text{can be factored as} \quad (\sin x + 1)^2.$$

The sum or difference of two trigonometric expressions can be found in the same way as any other rational expression. For example,

$$\frac{1}{\sin\theta} + \frac{1}{\cos\theta}$$

$$= \frac{1 \cdot \cos\theta}{\sin\theta\cos\theta} + \frac{1 \cdot \sin\theta}{\cos\theta\sin\theta} \qquad \text{Write with the LCD.}$$

$$= \frac{\cos\theta + \sin\theta}{\sin\theta\cos\theta}. \qquad \frac{a}{c} + \frac{b}{c} = \frac{a+b}{c}$$

5. ***When selecting substitutions, keep in mind the side that is not changing, because it represents the goal.*** For example, to verify that the equation

$$\tan^2 x + 1 = \frac{1}{\cos^2 x}$$

is an identity, think of an identity that relates $\tan x$ to $\cos x$. In this case, because $\sec x = \frac{1}{\cos x}$ and $\sec^2 x = \tan^2 x + 1$, the secant function is the best link between the two sides.

6. If an expression contains $1 + \sin x$, ***multiplying both numerator and denominator*** by $1 - \sin x$ would give $1 - \sin^2 x$, which could be replaced with $\cos^2 x$. Similar procedures apply for $1 - \sin x$, $1 + \cos x$, and $1 - \cos x$.

Verifying Identities by Working with One Side Avoid the temptation to use algebraic properties of equations to verify identities.

One strategy is to work with one side and rewrite it to match the other side.

Classroom Example 1
Verify that the following equation is an identity.

$\sec\theta(\sin\theta + \cos\theta) = 1 + \tan\theta$

Answer: The verification appears in the *Solutions to Classroom Examples.*

EXAMPLE 1 Verifying an Identity (Working with One Side)

Verify that the following equation is an identity.

$$\cot\theta + 1 = \csc\theta(\cos\theta + \sin\theta)$$

SOLUTION We use the fundamental identities to rewrite one side of the equation so that it is identical to the other side. The right side is more complicated, so we work with it, as suggested in Hint 2, and use Hint 3 to change all functions to expressions involving sine or cosine.

Steps	Reasons
(Right side of given equation) $\csc\theta(\cos\theta + \sin\theta) = \frac{1}{\sin\theta}(\cos\theta + \sin\theta)$	$\csc\theta = \frac{1}{\sin\theta}$
$= \frac{\cos\theta}{\sin\theta} + \frac{\sin\theta}{\sin\theta}$	Distributive property: $a(b + c) = ab + ac$
$= \cot\theta + 1$ (Left side of given equation)	$\frac{\cos\theta}{\sin\theta} = \cot\theta$; $\frac{\sin\theta}{\sin\theta} = 1$

For $\theta = x$,
$y_1 = \cot x + 1$
$y_2 = \csc x(\cos x + \sin x)$

NORMAL FLOAT AUTO REAL RADIAN MP

The graphs coincide, which supports the conclusion in **Example 1.**

The given equation is an identity. The right side of the equation is identical to the left side.

✔ **Now Try Exercise 45.**

Classroom Example 2
Verify that the following equation is an identity.

$\cot^2 x(\tan^2 x + 1) = \csc^2 x$

Answer: The verification appears in the *Solutions to Classroom Examples.*

EXAMPLE 2 Verifying an Identity (Working with One Side)

Verify that the following equation is an identity.

$$\tan^2 x(1 + \cot^2 x) = \frac{1}{1 - \sin^2 x}$$

SOLUTION We work with the more complicated left side, as suggested in Hint 2. Again, we use the fundamental identities.

Steps	Reasons
(Left side of given equation) $\tan^2 x(1 + \cot^2 x) = \tan^2 x + \tan^2 x\cot^2 x$	Distributive property
$= \tan^2 x + \tan^2 x \cdot \frac{1}{\tan^2 x}$	$\cot^2 x = \frac{1}{\tan^2 x}$
$= \tan^2 x + 1$	$\tan^2 x \cdot \frac{1}{\tan^2 x} = 1$
$= \sec^2 x$	Pythagorean identity
$= \frac{1}{\cos^2 x}$	$\sec^2 x = \frac{1}{\cos^2 x}$
$= \frac{1}{1 - \sin^2 x}$ (Right side of given equation)	Pythagorean identity

$y_1 = \tan^2 x(1 + \cot^2 x)$
$y_2 = \frac{1}{1 - \sin^2 x}$

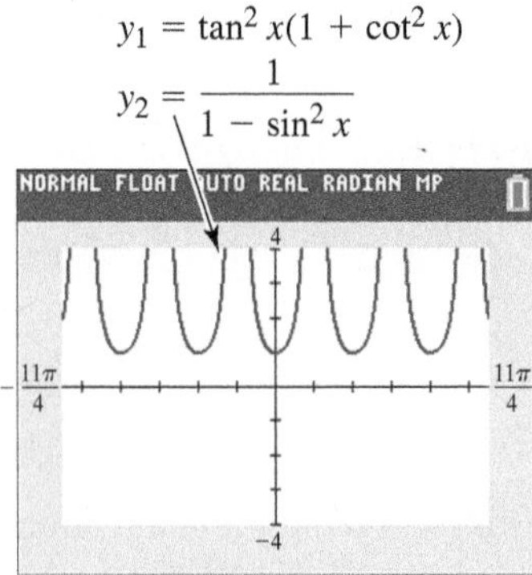

The screen supports the conclusion in **Example 2.**

Because the left side of the equation is identical to the right side, the given equation is an identity.

✔ **Now Try Exercise 49.**

Classroom Example 3
Verify that the following equation is an identity.

$$\frac{\tan^2 t}{\sec^2 t} = (1 + \cos t)(1 - \cos t)$$

Answer: The verification appears in the *Solutions to Classroom Examples.*

EXAMPLE 3 Verifying an Identity (Working with One Side)

Verify that the following equation is an identity.

$$\frac{\tan t - \cot t}{\sin t \cos t} = \sec^2 t - \csc^2 t$$

SOLUTION We transform the more complicated left side to match the right side.

$$\frac{\tan t - \cot t}{\sin t \cos t} = \frac{\tan t}{\sin t \cos t} - \frac{\cot t}{\sin t \cos t} \qquad \frac{a-b}{c} = \frac{a}{c} - \frac{b}{c}$$

$$= \tan t \cdot \frac{1}{\sin t \cos t} - \cot t \cdot \frac{1}{\sin t \cos t} \qquad \frac{a}{b} = a \cdot \frac{1}{b}$$

$$= \frac{\sin t}{\cos t} \cdot \frac{1}{\sin t \cos t} - \frac{\cos t}{\sin t} \cdot \frac{1}{\sin t \cos t} \qquad \tan t = \frac{\sin t}{\cos t};\ \cot t = \frac{\cos t}{\sin t}$$

$$= \frac{1}{\cos^2 t} - \frac{1}{\sin^2 t} \qquad \text{Multiply.}$$

$$= \sec^2 t - \csc^2 t \qquad \frac{1}{\cos^2 t} = \sec^2 t;\ \frac{1}{\sin^2 t} = \csc^2 t$$

Hint 3 about writing all trigonometric functions in terms of sine and cosine was used in the third line of the solution.

✔ **Now Try Exercise 53.**

Classroom Example 4
Verify that the following equation is an identity.

$$\frac{\sec x + \tan x}{\sin x} = \frac{\csc x}{\sec x - \tan x}$$

Answer: The verification appears in the *Solutions to Classroom Examples.*

EXAMPLE 4 Verifying an Identity (Working with One Side)

Verify that the following equation is an identity.

$$\frac{\cos x}{1 - \sin x} = \frac{1 + \sin x}{\cos x}$$

SOLUTION We work on the right side, using Hint 6 in the list given earlier to multiply the numerator and denominator on the right by $1 - \sin x$.

$$\frac{1 + \sin x}{\cos x} = \frac{(1 + \sin x)(1 - \sin x)}{\cos x(1 - \sin x)} \qquad \text{Multiply by 1 in the form } \frac{1 - \sin x}{1 - \sin x}.$$

$$= \frac{1 - \sin^2 x}{\cos x(1 - \sin x)} \qquad (x + y)(x - y) = x^2 - y^2$$

$$= \frac{\cos^2 x}{\cos x(1 - \sin x)} \qquad 1 - \sin^2 x = \cos^2 x$$

$$= \frac{\cos x \cdot \cos x}{\cos x(1 - \sin x)} \qquad a^2 = a \cdot a$$

$$= \frac{\cos x}{1 - \sin x} \qquad \text{Write in lowest terms.}$$

✔ **Now Try Exercise 59.**

Teaching Tip Show that the identity in **Example 4** can also be verified by multiplying the numerator and denominator on the left side by $1 + \sin x$.

Verifying Identities by Working with Both Sides If both sides of an identity appear to be equally complex, the identity can be verified by working independently on the left side and on the right side, until each side is changed into some common third result. ***Each step, on each side, must be reversible.*** With all steps reversible, the procedure is as shown in the margin. The left side leads to a common third expression, which leads back to the right side.

left = right
↘↖ ↙↗
common third expression

NOTE Working with both sides is often a good alternative for identities that are difficult. In practice, if working with one side does not seem to be effective, switch to the other side. Somewhere along the way it may happen that the same expression occurs on both sides.

Classroom Example 5 Verify that the following equation is an identity.

$$\frac{\cot\alpha - \csc\alpha}{\cot\alpha + \csc\alpha} = \frac{1 - 2\cos\alpha + \cos^2\alpha}{-\sin^2\alpha}$$

Answer: The verification appears in the *Solutions to Classroom Examples.*

EXAMPLE 5 Verifying an Identity (Working with Both Sides)

Verify that the following equation is an identity.

$$\frac{\sec\alpha + \tan\alpha}{\sec\alpha - \tan\alpha} = \frac{1 + 2\sin\alpha + \sin^2\alpha}{\cos^2\alpha}$$

SOLUTION Both sides appear equally complex, so we verify the identity by changing each side into a common third expression. We work first on the left, multiplying the numerator and denominator by $\cos\alpha$.

$$\underbrace{\frac{\sec\alpha + \tan\alpha}{\sec\alpha - \tan\alpha}}_{\text{Left side of given equation}} = \frac{(\sec\alpha + \tan\alpha)\cos\alpha}{(\sec\alpha - \tan\alpha)\cos\alpha} \quad \text{Multiply by 1 in the form } \tfrac{\cos\alpha}{\cos\alpha}.$$

$$= \frac{\sec\alpha\cos\alpha + \tan\alpha\cos\alpha}{\sec\alpha\cos\alpha - \tan\alpha\cos\alpha} \quad \text{Distributive property}$$

$$= \frac{1 + \tan\alpha\cos\alpha}{1 - \tan\alpha\cos\alpha} \quad \sec\alpha\cos\alpha = 1$$

$$= \frac{1 + \frac{\sin\alpha}{\cos\alpha}\cdot\cos\alpha}{1 - \frac{\sin\alpha}{\cos\alpha}\cdot\cos\alpha} \quad \tan\alpha = \tfrac{\sin\alpha}{\cos\alpha}$$

$$= \frac{1 + \sin\alpha}{1 - \sin\alpha} \quad \text{Simplify.}$$

On the right side of the original equation, we begin by factoring.

$$\underbrace{\frac{1 + 2\sin\alpha + \sin^2\alpha}{\cos^2\alpha}}_{\text{Right side of given equation}} = \frac{(1 + \sin\alpha)^2}{\cos^2\alpha} \quad \text{Factor the numerator; } x^2 + 2xy + y^2 = (x + y)^2.$$

$$= \frac{(1 + \sin\alpha)^2}{1 - \sin^2\alpha} \quad \cos^2\alpha = 1 - \sin^2\alpha$$

$$= \frac{(1 + \sin\alpha)^2}{(1 + \sin\alpha)(1 - \sin\alpha)} \quad \text{Factor the denominator; } x^2 - y^2 = (x + y)(x - y).$$

$$= \frac{1 + \sin\alpha}{1 - \sin\alpha} \quad \text{Write in lowest terms.}$$

We have shown that

$$\overbrace{\frac{\sec\alpha + \tan\alpha}{\sec\alpha - \tan\alpha}}^{\text{Left side of given equation}} = \overbrace{\frac{1 + \sin\alpha}{1 - \sin\alpha}}^{\text{Common third expression}} = \overbrace{\frac{1 + 2\sin\alpha + \sin^2\alpha}{\cos^2\alpha}}^{\text{Right side of given equation}},$$

and thus have verified that the given equation is an identity.

Now Try Exercise 75.

CAUTION Use the method of **Example 5** *only* if the steps are reversible.

There are usually several ways to verify a given identity. Another way to begin verifying the identity in **Example 5** is to work on the left as follows.

$$\underbrace{\frac{\sec \alpha + \tan \alpha}{\sec \alpha - \tan \alpha}}_{\text{Left side of given equation in Example 5}} = \frac{\frac{1}{\cos \alpha} + \frac{\sin \alpha}{\cos \alpha}}{\frac{1}{\cos \alpha} - \frac{\sin \alpha}{\cos \alpha}} \quad \text{Fundamental identities}$$

$$= \frac{\frac{1 + \sin \alpha}{\cos \alpha}}{\frac{1 - \sin \alpha}{\cos \alpha}} \quad \text{Add and subtract fractions.}$$

$$= \frac{1 + \sin \alpha}{\cos \alpha} \div \frac{1 - \sin \alpha}{\cos \alpha} \quad \text{Simplify the complex fraction. Use the definition of division.}$$

$$= \frac{1 + \sin \alpha}{\cos \alpha} \cdot \frac{\cos \alpha}{1 - \sin \alpha} \quad \text{Multiply by the reciprocal.}$$

$$= \frac{1 + \sin \alpha}{1 - \sin \alpha} \quad \text{Multiply and write in lowest terms.}$$

Compare this with the result shown in **Example 5** for the right side to see that the two sides indeed agree.

Classroom Example 6
FM radio stations broadcast at higher frequencies than AM stations. For a certain classical music FM station, the energy stored in the inductor is given by

$$L(t) = 5 \cos^2 620{,}000{,}000t$$

and the energy in the capacitor is given by

$$C(t) = 5 \sin^2 620{,}000{,}000t.$$

Write a simplified expression for $E(t)$, the total energy in the circuit.

Answer: 5

EXAMPLE 6 Applying a Pythagorean Identity to Electronics

Tuners in radios select a radio station by adjusting the frequency. A tuner may contain an inductor L and a capacitor C, as illustrated in **Figure 3.** The energy stored in the inductor at time t is given by

$$L(t) = k \sin^2 2\pi Ft$$

and the energy stored in the capacitor is given by

$$C(t) = k \cos^2 2\pi Ft,$$

where F is the frequency of the radio station and k is a constant. The total energy E in the circuit is given by

$$E(t) = L(t) + C(t).$$

Show that E is a constant function. (*Source*: Weidner, R. and R. Sells, *Elementary Classical Physics,* Vol. 2, Allyn & Bacon.)

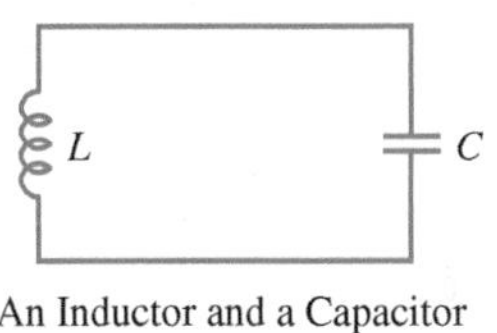

An Inductor and a Capacitor

Figure 3

SOLUTION

$$\begin{aligned} E(t) &= L(t) + C(t) && \text{Given equation} \\ &= k \sin^2 2\pi Ft + k \cos^2 2\pi Ft && \text{Substitute.} \\ &= k[\sin^2 2\pi Ft + \cos^2 2\pi Ft] && \text{Factor out } k. \\ &= k(1) && \sin^2\theta + \cos^2\theta = 1 \text{ (Here } \theta = 2\pi Ft.) \\ &= k && \text{Identity property} \end{aligned}$$

Because k is a constant, $E(t)$ is a constant function. ✔ **Now Try Exercise 105.**

5.2 Exercises

To the student: **Exercises 1–44** are designed for practice in using the fundamental identities and applying algebraic techniques to trigonometric expressions. These skills are essential in verifying the identities that follow.

CONCEPT PREVIEW *Match each expression in Column I with its correct factorization in Column II.*

I	II
1. $x^2 - y^2$	**A.** $(x + y)(x^2 - xy + y^2)$
2. $x^3 - y^3$	**B.** $(x + y)(x - y)$
3. $x^3 + y^3$	**C.** $(x + y)^2$
4. $x^2 + 2xy + y^2$	**D.** $(x - y)(x^2 + xy + y^2)$

CONCEPT PREVIEW *Fill in the blank(s) to correctly complete each fundamental identity.*

5. $\sin^2 \theta + \cos^2 \theta =$ ______ **6.** $\tan^2 \theta + 1 =$ ______

7. $\sin(-\theta) =$ ______ **8.** $\sec(-\theta) =$ ______

9. $\tan \theta = \dfrac{1}{\_\_\_\_} = \dfrac{\sin \theta}{\_\_\_\_}$ **10.** $\cot \theta = \dfrac{1}{\_\_\_\_} = \dfrac{\cos \theta}{\_\_\_\_}$

Perform each indicated operation and simplify the result so that there are no quotients.

11. $\cot \theta + \dfrac{1}{\cot \theta}$ **12.** $\dfrac{\sec x}{\csc x} + \dfrac{\csc x}{\sec x}$ **13.** $\tan x(\cot x + \csc x)$

14. $\cos \beta(\sec \beta + \csc \beta)$ **15.** $\dfrac{1}{\csc^2 \theta} + \dfrac{1}{\sec^2 \theta}$ **16.** $\dfrac{\cos x}{\sec x} + \dfrac{\sin x}{\csc x}$

17. $(\sin \alpha - \cos \alpha)^2$ **18.** $(\tan x + \cot x)^2$ **19.** $(1 + \sin t)^2 + \cos^2 t$

20. $(1 + \tan \theta)^2 - 2 \tan \theta$ **21.** $\dfrac{1}{1 + \cos x} - \dfrac{1}{1 - \cos x}$ **22.** $\dfrac{1}{\sin \alpha - 1} - \dfrac{1}{\sin \alpha + 1}$

Factor each trigonometric expression.

23. $\sin^2 \theta - 1$ **24.** $\sec^2 \theta - 1$

25. $(\sin x + 1)^2 - (\sin x - 1)^2$ **26.** $(\tan x + \cot x)^2 - (\tan x - \cot x)^2$

27. $2 \sin^2 x + 3 \sin x + 1$ **28.** $4 \tan^2 \beta + \tan \beta - 3$

29. $\cos^4 x + 2 \cos^2 x + 1$ **30.** $\cot^4 x + 3 \cot^2 x + 2$

31. $\sin^3 x - \cos^3 x$ **32.** $\sin^3 \alpha + \cos^3 \alpha$

Each expression simplifies to a constant, a single function, or a power of a function. Use fundamental identities to simplify each expression.

33. $\tan \theta \cos \theta$ **34.** $\cot \alpha \sin \alpha$ **35.** $\sec r \cos r$

36. $\cot t \tan t$ **37.** $\dfrac{\sin \beta \tan \beta}{\cos \beta}$ **38.** $\dfrac{\csc \theta \sec \theta}{\cot \theta}$

39. $\sec^2 x - 1$ **40.** $\csc^2 t - 1$ **41.** $\dfrac{\sin^2 x}{\cos^2 x} + \sin x \csc x$

42. $\dfrac{1}{\tan^2 \alpha} + \cot \alpha \tan \alpha$ **43.** $1 - \dfrac{1}{\csc^2 x}$ **44.** $1 - \dfrac{1}{\sec^2 x}$

1. B **2.** D
3. A **4.** C

5. 1 **6.** $\sec^2 \theta$
7. $-\sin \theta$ **8.** $\sec \theta$
9. $\cot \theta$; $\cos \theta$ **10.** $\tan \theta$; $\sin \theta$

11. $\csc \theta \sec \theta$
12. $\csc x \sec x$
13. $1 + \sec x$
14. $1 + \cot \beta$
15. 1 **16.** 1
17. $1 - 2 \sin \alpha \cos \alpha$
18. $\sec^2 x + \csc^2 x$
19. $2 + 2 \sin t$
20. $\sec^2 \theta$
21. $-2 \cot x \csc x$
22. $-2 \sec^2 \alpha$

23. $(\sin \theta + 1)(\sin \theta - 1)$
24. $(\sec \theta + 1)(\sec \theta - 1)$
25. $4 \sin x$ **26.** 4
27. $(2 \sin x + 1)(\sin x + 1)$
28. $(4 \tan \beta - 3)(\tan \beta + 1)$
29. $(\cos^2 x + 1)^2$
30. $(\cot^2 x + 2)(\cot^2 x + 1)$, or $\csc^2 x(\cot^2 x + 2)$
31. $(\sin x - \cos x)(1 + \sin x \cos x)$
32. $(\sin \alpha + \cos \alpha)(1 - \sin \alpha \cos \alpha)$

33. $\sin \theta$ **34.** $\cos \alpha$
35. 1 **36.** 1
37. $\tan^2 \beta$ **38.** $\sec^2 \theta$
39. $\tan^2 x$ **40.** $\cot^2 t$
41. $\sec^2 x$ **42.** $\csc^2 \alpha$
43. $\cos^2 x$ **44.** $\sin^2 x$

Verify that each equation is an identity. ***See Examples 1–5.***

45. $\dfrac{\cot \theta}{\csc \theta} = \cos \theta$

46. $\dfrac{\tan \alpha}{\sec \alpha} = \sin \alpha$

47. $\dfrac{1 - \sin^2 \beta}{\cos \beta} = \cos \beta$

48. $\dfrac{\tan^2 \alpha + 1}{\sec \alpha} = \sec \alpha$

49. $\cos^2 \theta(\tan^2 \theta + 1) = 1$

50. $\sin^2 \beta(1 + \cot^2 \beta) = 1$

51. $\cot \theta + \tan \theta = \sec \theta \csc \theta$

52. $\sin^2 \alpha + \tan^2 \alpha + \cos^2 \alpha = \sec^2 \alpha$

53. $\dfrac{\cos \alpha}{\sec \alpha} + \dfrac{\sin \alpha}{\csc \alpha} = \sec^2 \alpha - \tan^2 \alpha$

54. $\dfrac{\sin^2 \theta}{\cos \theta} = \sec \theta - \cos \theta$

55. $\sin^4 \theta - \cos^4 \theta = 2 \sin^2 \theta - 1$

56. $\sec^4 x - \sec^2 x = \tan^4 x + \tan^2 x$

57. $\dfrac{1 - \cos x}{1 + \cos x} = (\cot x - \csc x)^2$

58. $(\sec \alpha - \tan \alpha)^2 = \dfrac{1 - \sin \alpha}{1 + \sin \alpha}$

59. $\dfrac{\cos \theta + 1}{\tan^2 \theta} = \dfrac{\cos \theta}{\sec \theta - 1}$

60. $\dfrac{(\sec \theta - \tan \theta)^2 + 1}{\sec \theta \csc \theta - \tan \theta \csc \theta} = 2 \tan \theta$

61. $\dfrac{1}{1 - \sin \theta} + \dfrac{1}{1 + \sin \theta} = 2 \sec^2 \theta$

62. $\dfrac{1}{\sec \alpha - \tan \alpha} = \sec \alpha + \tan \alpha$

63. $\dfrac{\cot \alpha + 1}{\cot \alpha - 1} = \dfrac{1 + \tan \alpha}{1 - \tan \alpha}$

64. $\dfrac{\csc \theta + \cot \theta}{\tan \theta + \sin \theta} = \cot \theta \csc \theta$

65. $\dfrac{\cos \theta}{\sin \theta \cot \theta} = 1$

66. $\sin^2 \theta(1 + \cot^2 \theta) - 1 = 0$

67. $\dfrac{\sec^4 \theta - \tan^4 \theta}{\sec^2 \theta + \tan^2 \theta} = \sec^2 \theta - \tan^2 \theta$

68. $\dfrac{\sin^4 \alpha - \cos^4 \alpha}{\sin^2 \alpha - \cos^2 \alpha} = 1$

69. $\dfrac{\tan^2 t - 1}{\sec^2 t} = \dfrac{\tan t - \cot t}{\tan t + \cot t}$

70. $\dfrac{\cot^2 t - 1}{1 + \cot^2 t} = 1 - 2 \sin^2 t$

71. $\sin^2 \alpha \sec^2 \alpha + \sin^2 \alpha \csc^2 \alpha = \sec^2 \alpha$

72. $\tan^2 \alpha \sin^2 \alpha = \tan^2 \alpha + \cos^2 \alpha - 1$

73. $\dfrac{\tan x}{1 + \cos x} + \dfrac{\sin x}{1 - \cos x} = \cot x + \sec x \csc x$

74. $\dfrac{\sin \theta}{1 - \cos \theta} - \dfrac{\sin \theta \cos \theta}{1 + \cos \theta} = \csc \theta(1 + \cos^2 \theta)$

75. $\dfrac{1 + \cos x}{1 - \cos x} - \dfrac{1 - \cos x}{1 + \cos x} = 4 \cot x \csc x$

76. $\dfrac{1 + \sin \theta}{1 - \sin \theta} - \dfrac{1 - \sin \theta}{1 + \sin \theta} = 4 \tan \theta \sec \theta$

77. $\dfrac{1 - \sin \theta}{1 + \sin \theta} = \sec^2 \theta - 2 \sec \theta \tan \theta + \tan^2 \theta$

78. $\sin \theta + \cos \theta = \dfrac{\sin \theta}{1 - \cot \theta} + \dfrac{\cos \theta}{1 - \tan \theta}$

79. $\dfrac{-1}{\tan \alpha - \sec \alpha} + \dfrac{-1}{\tan \alpha + \sec \alpha} = 2 \tan \alpha$

80. $(1 + \sin x + \cos x)^2 = 2(1 + \sin x)(1 + \cos x)$

81. $(1 - \cos^2 \alpha)(1 + \cos^2 \alpha) = 2 \sin^2 \alpha - \sin^4 \alpha$

82. $(\sec \alpha + \csc \alpha)(\cos \alpha - \sin \alpha) = \cot \alpha - \tan \alpha$

83. $\dfrac{1 - \cos x}{1 + \cos x} = \csc^2 x - 2 \csc x \cot x + \cot^2 x$

89. $(\sec\theta + \tan\theta)(1 - \sin\theta) = \cos\theta$

90. $(\csc\theta + \cot\theta)(\sec\theta - 1) = \tan\theta$

91. $\frac{\cos\theta + 1}{\sin\theta + \tan\theta} = \cot\theta$

92. $\tan\theta\sin\theta + \cos\theta = \sec\theta$

93. identity
94. identity
95. not an identity
96. not an identity

101. (a) $I = k(1 - \sin^2\theta)$
(b) When $\theta = 0$, $\cos\theta = 1$, its maximum value. Thus, $\cos^2\theta$ will be a maximum and, as a result, I will be maximized if k is a positive constant.

102. (a) $P = 16k\cos^2 2\pi t$
(b) $P = 16k(1 - \sin^2 2\pi t)$

84. $\dfrac{1 - \cos\theta}{1 + \cos\theta} = 2\csc^2\theta - 2\csc\theta\cot\theta - 1$

85. $(2\sin x + \cos x)^2 + (2\cos x - \sin x)^2 = 5$

86. $\sin^2 x(1 + \cot x) + \cos^2 x(1 - \tan x) + \cot^2 x = \csc^2 x$

87. $\sec x - \cos x + \csc x - \sin x - \sin x\tan x = \cos x\cot x$

88. $\sin^3\theta + \cos^3\theta = (\cos\theta + \sin\theta)(1 - \cos\theta\sin\theta)$

Graph each expression and use the graph to make a conjecture, predicting what might be an identity. Then verify your conjecture algebraically.

89. $(\sec\theta + \tan\theta)(1 - \sin\theta)$

90. $(\csc\theta + \cot\theta)(\sec\theta - 1)$

91. $\dfrac{\cos\theta + 1}{\sin\theta + \tan\theta}$

92. $\tan\theta\sin\theta + \cos\theta$

Graph the expressions on each side of the equals symbol to determine whether the equation might be an identity. (Note: Use a domain whose length is at least 2π.) If the equation looks like an identity, then verify it algebraically. ***See Example 1.***

93. $\dfrac{2 + 5\cos x}{\sin x} = 2\csc x + 5\cot x$

94. $1 + \cot^2 x = \dfrac{\sec^2 x}{\sec^2 x - 1}$

95. $\dfrac{\tan x - \cot x}{\tan x + \cot x} = 2\sin^2 x$

96. $\dfrac{1}{1 + \sin x} + \dfrac{1}{1 - \sin x} = \sec^2 x$

By substituting a number for t, show that the equation is not an identity.

97. $\sin(\csc t) = 1$

98. $\sqrt{\cos^2 t} = \cos t$

99. $\csc t = \sqrt{1 + \cot^2 t}$

100. $\cos t = \sqrt{1 - \sin^2 t}$

(Modeling) Work each problem.

101. ***Intensity of a Lamp*** According to **Lambert's law,** the intensity of light from a single source on a flat surface at point P is given by

$$I = k\cos^2\theta,$$

where k is a constant. (*Source:* Winter, C., *Solar Power Plants,* Springer-Verlag.)

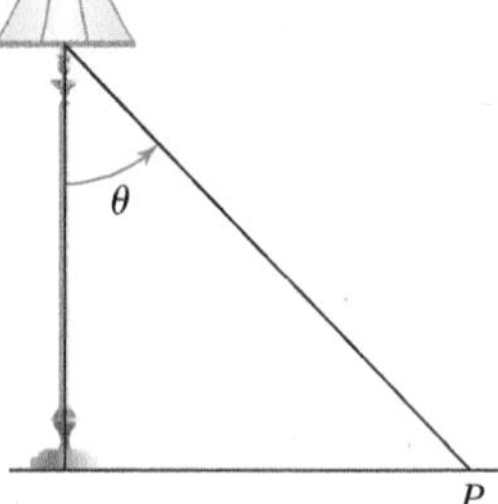

(a) Write I in terms of the sine function.

(b) Why does the maximum value of I occur when $\theta = 0$?

102. ***Oscillating Spring*** The distance or displacement y of a weight attached to an oscillating spring from its natural position is modeled by

$$y = 4\cos 2\pi t,$$

where t is time in seconds. Potential energy is the energy of position and is given by

$$P = ky^2,$$

where k is a constant. The weight has the greatest potential energy when the spring is stretched the most. (*Source*: Weidner, R. and R. Sells, *Elementary Classical Physics,* Vol. 2, Allyn & Bacon.)

(a) Write an expression for P that involves the cosine function.

(b) Use a fundamental identity to write P in terms of $\sin 2\pi t$.

103. 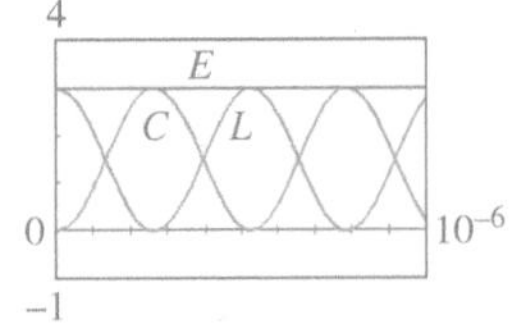

The sum of L and C equals 3.

104. 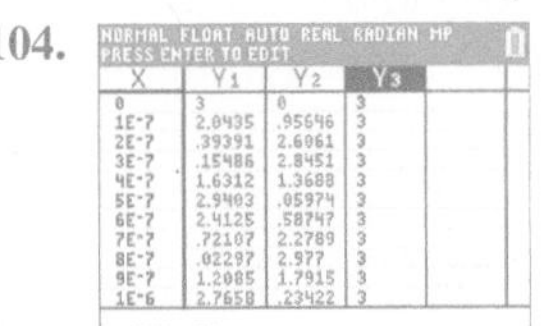

NORMAL FLOAT AUTO REAL RADIAN MP
PRESS ENTER TO EDIT

X	Y1	Y2	Y3
0	3	0	3
1E-7	2.0435	.95646	3
2E-7	.39391	2.6061	3
3E-7	.15486	2.8451	3
4E-7	1.6312	1.3688	3
5E-7	2.9403	.05974	3
6E-7	2.4125	.58747	3
7E-7	.72107	2.2789	3
8E-7	.02297	2.977	3
9E-7	1.2085	1.7915	3
1E-6	2.7658	.23422	3

Y3=Y1+Y2

Let $y_1 = L(t)$, $y_2 = C(t)$, and $y_3 = E(t)$. $y_3 = 3$ for all inputs.

105. $E(t) = 3$

*(Modeling) Radio Tuners See **Example 6.** Let the energy stored in the inductor be given by*

$$L(t) = 3\cos^2 6{,}000{,}000t$$

and let the energy stored in the capacitor be given by

$$C(t) = 3\sin^2 6{,}000{,}000t,$$

where t is time in seconds. The total energy E in the circuit is given by

$$E(t) = L(t) + C(t).$$

103. Graph L, C, and E in the window $[0, 10^{-6}]$ by $[-1, 4]$, with Xscl $= 10^{-7}$ and Yscl $= 1$. Interpret the graph.

104. Make a table of values for L, C, and E starting at $t = 0$, incrementing by 10^{-7}. Interpret the results.

105. Use a fundamental identity to derive a simplified expression for $E(t)$.

5.3 Sum and Difference Identities for Cosine

- Difference Identity for Cosine
- Sum Identity for Cosine
- Cofunction Identities
- Applications of the Sum and Difference Identities
- Verifying an Identity

Difference Identity for Cosine Several examples presented earlier should have convinced you by now that

$$\cos(A - B) \text{ does not equal } \cos A - \cos B.$$

For example, if $A = \frac{\pi}{2}$ and $B = 0$, then

$$\cos(A - B) = \cos\left(\frac{\pi}{2} - 0\right) = \cos\frac{\pi}{2} = 0,$$

while $$\cos A - \cos B = \cos\frac{\pi}{2} - \cos 0 = 0 - 1 = -1.$$

To derive a formula for $\cos(A - B)$, we start by locating angles A and B in standard position on a unit circle, with $B < A$. Let S and Q be the points where the terminal sides of angles A and B, respectively, intersect the circle. Let P be the point $(1, 0)$, and locate point R on the unit circle so that angle POR equals the difference $A - B$. See **Figure 4.**

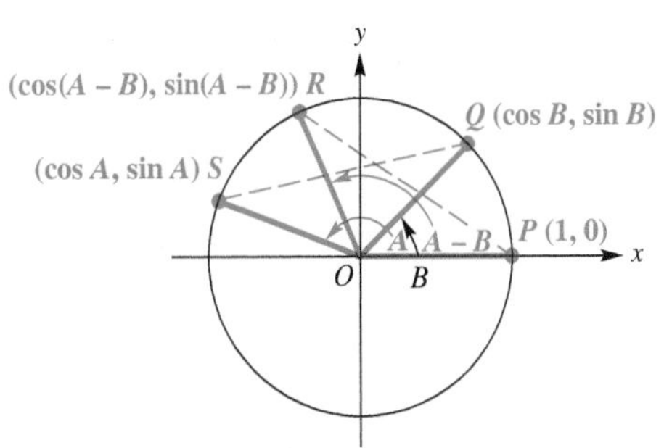

Figure 4

Because point Q is on the unit circle, the x-coordinate of Q is the cosine of angle B, while the y-coordinate of Q is the sine of angle B.

$$Q \text{ has coordinates } (\cos B, \sin B).$$

In the same way,

$$S \text{ has coordinates } (\cos A, \sin A),$$

and $$R \text{ has coordinates } (\cos(A-B), \sin(A-B)).$$

Angle SOQ also equals $A - B$. The central angles SOQ and POR are equal, so chords PR and SQ are equal. Because $PR = SQ$, by the distance formula,

$$\sqrt{[\cos(A-B)-1]^2 + [\sin(A-B)-0]^2}$$
$$= \sqrt{(\cos A - \cos B)^2 + (\sin A - \sin B)^2}.$$

$d = \sqrt{(x_2 - x_1)^2 + (y_2 - y_1)^2}$

Square each side of this equation. Then square each expression, remembering that for any values of x and y, $(x - y)^2 = x^2 - 2xy + y^2$.

$$[\cos(A-B)-1]^2 + [\sin(A-B)-0]^2$$
$$= (\cos A - \cos B)^2 + (\sin A - \sin B)^2$$
$$\cos^2(A-B) - 2\cos(A-B) + 1 + \sin^2(A-B)$$
$$= \cos^2 A - 2\cos A \cos B + \cos^2 B + \sin^2 A - 2\sin A \sin B + \sin^2 B$$

For any value of x, $\sin^2 x + \cos^2 x = 1$, so we can rewrite the equation.

$$2 - 2\cos(A-B) = 2 - 2\cos A \cos B - 2\sin A \sin B$$

Use $\sin^2 x + \cos^2 x = 1$ three times and add like terms.

$$\mathbf{\cos(A-B) = \cos A \cos B + \sin A \sin B}$$

Subtract 2, and then divide by -2.

This is the identity for $\cos(A - B)$. Although **Figure 4** shows angles A and B in the second and first quadrants, respectively, this result is the same for any values of these angles.

Sum Identity for Cosine To find a similar expression for $\cos(A + B)$, rewrite $A + B$ as $A - (-B)$ and use the identity for $\cos(A - B)$.

$\cos(A+B) = \cos[A-(-B)]$	Definition of subtraction
$= \cos A \cos(-B) + \sin A \sin(-B)$	Cosine difference identity
$= \cos A \cos B + \sin A(-\sin B)$	Even-odd identities
$\mathbf{\cos(A+B) = \cos A \cos B - \sin A \sin B}$	Multiply.

Cosine of a Sum or Difference

$$\cos(A+B) = \cos A \cos B - \sin A \sin B$$
$$\cos(A-B) = \cos A \cos B + \sin A \sin B$$

These identities are important in calculus and useful in certain applications. For example, the method shown in **Example 1** can be applied to find an exact value for $\cos 15°$.

Classroom Example 1
Find the *exact* value of each expression.

(a) $\cos(-75^\circ)$ (b) $\cos \frac{17\pi}{12}$

(c) $\cos 173^\circ \cos 83^\circ + \sin 173^\circ \sin 83^\circ$

Answers:

(a) $\frac{\sqrt{6}-\sqrt{2}}{4}$ (b) $\frac{\sqrt{2}-\sqrt{6}}{4}$

(c) 0

Teaching Tip Write the following equivalences so that students can see how to express multiples of $\frac{\pi}{12}$, like the one in **Example 1(b).**

$$\frac{\pi}{6}=\frac{2\pi}{12},\quad \frac{\pi}{4}=\frac{3\pi}{12},\quad \frac{\pi}{3}=\frac{4\pi}{12}$$

```
NORMAL FLOAT AUTO REAL RADIAN MP
cos(5π/12)
                    .2588190451
(√6-√2)/4
                    .2588190451
```

This screen supports the solution in **Example 1(b)** by showing that the decimal approximations for $\cos \frac{5\pi}{12}$ and $\frac{\sqrt{6}-\sqrt{2}}{4}$ agree.

EXAMPLE 1 Finding Exact Cosine Function Values

Find the *exact* value of each expression.

(a) $\cos 15^\circ$ **(b)** $\cos \frac{5\pi}{12}$ **(c)** $\cos 87^\circ \cos 93^\circ - \sin 87^\circ \sin 93^\circ$

SOLUTION

(a) To find $\cos 15^\circ$, we write 15° as the sum or difference of two angles with known function values, such as 45° and 30°, because

$$15^\circ = 45^\circ - 30^\circ. \quad \text{(We could also use } 60^\circ - 45^\circ.)$$

Then we use the cosine difference identity.

$\cos 15^\circ$

$= \cos(45^\circ - 30^\circ)$ — $15^\circ = 45^\circ - 30^\circ$

$= \cos 45^\circ \cos 30^\circ + \sin 45^\circ \sin 30^\circ$ — Cosine difference identity

$= \frac{\sqrt{2}}{2} \cdot \frac{\sqrt{3}}{2} + \frac{\sqrt{2}}{2} \cdot \frac{1}{2}$ — Substitute known values.

$= \frac{\sqrt{6}+\sqrt{2}}{4}$ — Multiply, and then add fractions.

(b) $\cos \frac{5\pi}{12}$

$= \cos\left(\frac{\pi}{6} + \frac{\pi}{4}\right)$ — $\frac{\pi}{6} = \frac{2\pi}{12}$ and $\frac{\pi}{4} = \frac{3\pi}{12}$

$= \cos \frac{\pi}{6} \cos \frac{\pi}{4} - \sin \frac{\pi}{6} \sin \frac{\pi}{4}$ — Cosine sum identity

$= \frac{\sqrt{3}}{2} \cdot \frac{\sqrt{2}}{2} - \frac{1}{2} \cdot \frac{\sqrt{2}}{2}$ — Substitute known values.

$= \frac{\sqrt{6}-\sqrt{2}}{4}$ — Multiply, and then subtract fractions.

(c) $\cos 87^\circ \cos 93^\circ - \sin 87^\circ \sin 93^\circ$

$= \cos(87^\circ + 93^\circ)$ — Cosine sum identity

$= \cos 180^\circ$ — Add.

$= -1$ — $\cos 180^\circ = -1$

✔ **Now Try Exercises 9, 13, and 17.**

Cofunction Identities We can use the identity for the cosine of the difference of two angles and the fundamental identities to derive *cofunction identities,* presented previously for values of θ in the interval $[0^\circ, 90^\circ]$. For example, substituting 90° for A and θ for B in the identity for $\cos(A - B)$ gives the following.

$\mathbf{\cos(90^\circ - \theta)} = \cos 90^\circ \cos \theta + \sin 90^\circ \sin \theta$ — Cosine difference identity

$= 0 \cdot \cos \theta + 1 \cdot \sin \theta$ — $\cos 90^\circ = 0$ and $\sin 90^\circ = 1$

$= \mathbf{\sin \theta}$ — Simplify.

This result is true for *any* value of θ because the identity for $\cos(A - B)$ is true for any values of A and B.

Teaching Tip Mention that these identities state that the trigonometric function of an angle is the same as the cofunction of its complement. Verify the cofunction identities for acute angles using complementary angles in a right triangle along with the right-triangle-based definitions of the trigonometric functions. Emphasize that these identities apply to *any* angle θ, not just to acute angles.

Cofunction Identities

The following identities hold for any angle θ for which the functions are defined.

$$\cos(90° - \theta) = \sin\theta \qquad \cot(90° - \theta) = \tan\theta$$
$$\sin(90° - \theta) = \cos\theta \qquad \sec(90° - \theta) = \csc\theta$$
$$\tan(90° - \theta) = \cot\theta \qquad \csc(90° - \theta) = \sec\theta$$

The same identities can be obtained for a real number domain by replacing 90° with $\frac{\pi}{2}$.

NOTE Because trigonometric (circular) functions are periodic, the solutions in **Example 2** are not unique. We give only one of infinitely many possibilities.

Classroom Example 2
Find one value of θ or x that satisfies each of the following.

(a) $\sec\theta = \csc 62°$

(b) $\tan\theta = \cot(-54°)$

(c) $\cos x = \sin\frac{7\pi}{6}$

Answers:

(a) $\theta = 28°$ (b) $\theta = 144°$

(c) $x = -\frac{2\pi}{3}$

Teaching Tip Support the results from **Example 2** using a calculator.

EXAMPLE 2 Using Cofunction Identities to Find θ

Find one value of θ or x that satisfies each of the following.

(a) $\cot\theta = \tan 25°$ **(b)** $\sin\theta = \cos(-30°)$ **(c)** $\csc\frac{3\pi}{4} = \sec x$

SOLUTION

(a) Because tangent and cotangent are cofunctions, $\tan(90° - \theta) = \cot\theta$.

$$\cot\theta = \tan 25°$$
$$\tan(90° - \theta) = \tan 25° \quad \text{Cofunction identity}$$
$$90° - \theta = 25° \quad \text{Set angle measures equal.}$$
$$\theta = 65° \quad \text{Solve for } \theta.$$

(b)
$$\sin\theta = \cos(-30°)$$
$$\cos(90° - \theta) = \cos(-30°) \quad \text{Cofunction identity}$$
$$90° - \theta = -30° \quad \text{Set angle measures equal.}$$
$$\theta = 120° \quad \text{Solve for } \theta.$$

(c)
$$\csc\frac{3\pi}{4} = \sec x$$
$$\csc\frac{3\pi}{4} = \csc\left(\frac{\pi}{2} - x\right) \quad \text{Cofunction identity}$$
$$\frac{3\pi}{4} = \frac{\pi}{2} - x \quad \text{Set angle measures equal.}$$
$$x = -\frac{\pi}{4} \quad \text{Solve for } x;\ \frac{\pi}{2} - \frac{3\pi}{4} = \frac{2\pi}{4} - \frac{3\pi}{4} = -\frac{\pi}{4}.$$

✔ **Now Try Exercises 37 and 41.**

Applications of the Sum and Difference Identities If either angle A or angle B in the identities for $\cos(A + B)$ and $\cos(A - B)$ is a quadrantal angle, then the identity allows us to write the expression in terms of a single function of A or B.

Classroom Example 3
Write $\cos(90° + \theta)$ as a trigonometric function of θ alone.

Answer: $-\sin\theta$

EXAMPLE 3 Reducing cos(A − B) to a Function of a Single Variable

Write $\cos(180° - \theta)$ as a trigonometric function of θ alone.

SOLUTION $\cos(180° - \theta)$

$= \cos 180° \cos\theta + \sin 180° \sin\theta$ Cosine difference identity

$= (-1)\cos\theta + (0)\sin\theta$ $\cos 180° = -1$ and $\sin 180° = 0$

$= -\cos\theta$ Simplify.

✔ **Now Try Exercise 49.**

Classroom Example 4
Suppose that $\cos s = \frac{15}{17}$, $\sin t = -\frac{24}{25}$, and both s and t are in quadrant IV. Find $\cos(s - t)$.

Answer: $\frac{297}{425}$

EXAMPLE 4 Finding cos(s + t) Given Information about s and t

Suppose that $\sin s = \frac{3}{5}$, $\cos t = -\frac{12}{13}$, and both s and t are in quadrant II. Find $\cos(s + t)$.

SOLUTION By the cosine sum identity,

$$\cos(s + t) = \cos s \cos t - \sin s \sin t.$$

The values of $\sin s$ and $\cos t$ are given, so we can find $\cos(s + t)$ if we know the values of $\cos s$ and $\sin t$. There are two ways to do this.

Method 1 We use angles in standard position. To find $\cos s$ and $\sin t$, we sketch two reference triangles in the second quadrant, one with $\sin s = \frac{3}{5}$ and the other with $\cos t = -\frac{12}{13}$. Notice that for angle t, we use -12 to denote the length of the side that lies along the x-axis. See **Figure 5.**

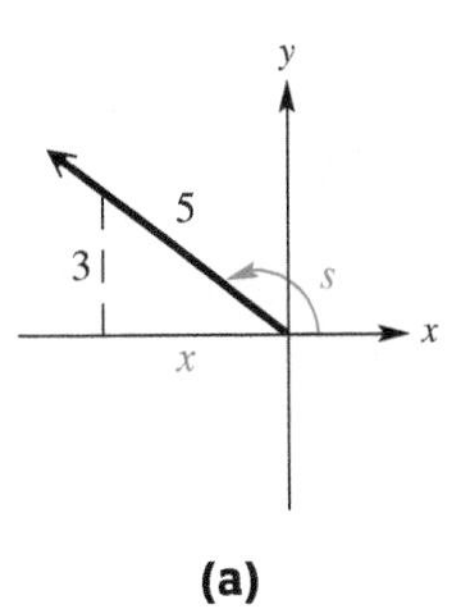

(a)

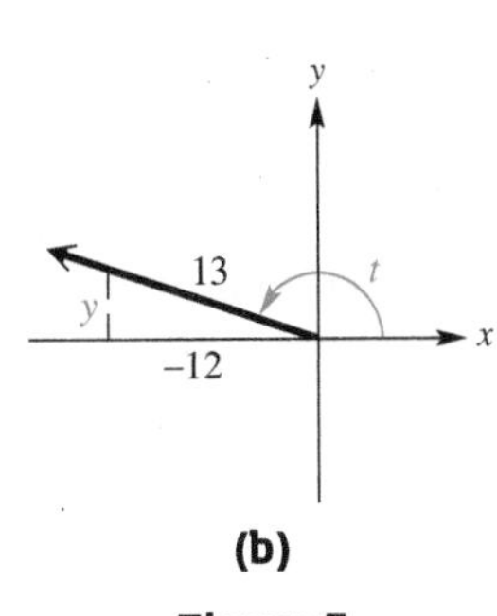

(b)

Figure 5

In **Figure 5(a),** $y = 3$ and $r = 5$. We must find x.

$x^2 + y^2 = r^2$ Pythagorean theorem

$x^2 + 3^2 = 5^2$ Substitute.

$x^2 = 16$ Isolate x^2.

$x = -4$ Choose the negative square root here.

Thus, $\cos s = \frac{x}{r} = -\frac{4}{5}$.

In **Figure 5(b),** $x = -12$ and $r = 13$. We must find y.

$x^2 + y^2 = r^2$ Pythagorean theorem

$(-12)^2 + y^2 = 13^2$ Substitute.

$y^2 = 25$ Isolate y^2.

$y = 5$ Choose the positive square root here.

Thus, $\sin t = \frac{y}{r} = \frac{5}{13}$.

Now we can find $\cos(s + t)$.

$\cos(s + t) = \cos s \cos t - \sin s \sin t$ Cosine sum identity (1)

$= -\frac{4}{5}\left(-\frac{12}{13}\right) - \frac{3}{5} \cdot \frac{5}{13}$ Substitute.

$= \frac{48}{65} - \frac{15}{65}$ Multiply.

$\cos(s + t) = \frac{33}{65}$ Subtract.

Method 2 We use Pythagorean identities here. To find cos s, recall that $\sin^2 s + \cos^2 s = 1$, where s is in quadrant II.

$$\left(\frac{3}{5}\right)^2 + \cos^2 s = 1 \quad \sin s = \frac{3}{5}$$

$$\frac{9}{25} + \cos^2 s = 1 \quad \text{Square } \frac{3}{5}.$$

$$\cos^2 s = \frac{16}{25} \quad \text{Subtract } \frac{9}{25}.$$

$$\cos s = -\frac{4}{5}$$ cos $s < 0$ because s is in quadrant II.

To find sin t, we use $\sin^2 t + \cos^2 t = 1$, where t is in quadrant II.

$$\sin^2 t + \left(-\frac{12}{13}\right)^2 = 1 \quad \cos t = -\frac{12}{13}$$

$$\sin^2 t + \frac{144}{169} = 1 \quad \text{Square } -\frac{12}{13}.$$

$$\sin^2 t = \frac{25}{169} \quad \text{Subtract } \frac{144}{169}.$$

$$\sin t = \frac{5}{13}$$ sin $t > 0$ because t is in quadrant II.

From this point, the problem is solved using the same steps beginning with the equation marked (1) in Method 1 on the previous page. The result is

$$\cos(s + t) = \frac{33}{65}. \quad \text{Same result as in Method 1}$$

Now Try Exercise 51.

Classroom Example 5
See **Example 5.** Because household current is supplied at different voltages in different countries, international travelers often carry electrical adapters to connect items they have brought from home to a power source. The voltage V in a typical European 220-volt outlet can be expressed by the function

$$V(t) = 311 \sin \omega t.$$

(a) European generators rotate at precisely 50 cycles per sec. Determine ω for these electric generators.

(b) What is the maximum voltage in the outlet?

(c) Determine the least positive value of ϕ in radians so that the graph of

$$V(t) = 311 \cos(\omega t + \phi)$$

is the same as the graph of

$$V(t) = 311 \sin \omega t.$$

Answers:

(a) 100π radians per sec

(b) 311 volts

(c) $\phi = \frac{3\pi}{2}$

EXAMPLE 5 Applying the Cosine Difference Identity to Voltage

Common household electric current is called **alternating current** because the current alternates direction within the wires. The voltage V in a typical 115-volt outlet can be expressed by the function

$$V(t) = 163 \sin \omega t,$$

where ω is the angular speed (in radians per second) of the rotating generator at the electrical plant and t is time in seconds. (*Source*: Bell, D., *Fundamentals of Electric Circuits,* Fourth Edition, Prentice-Hall.)

(a) It is essential for electric generators to rotate at precisely 60 cycles per sec so household appliances and computers will function properly. Determine ω for these electric generators.

(b) Graph V in the window $[0, 0.05]$ by $[-200, 200]$.

(c) Determine a value of ϕ so that the graph of

$$V(t) = 163 \cos(\omega t - \phi)$$

is the same as the graph of

$$V(t) = 163 \sin \omega t.$$

SOLUTION

(a) We convert 60 cycles per sec to radians per second as follows.

$$\omega = \frac{60 \text{ cycles}}{1 \text{ sec}} \cdot \frac{2\pi \text{ radians}}{1 \text{ cycle}} = 120\pi \text{ radians per sec.}$$

(b)

$$V(t) = 163 \sin \omega t$$

$$V(t) = 163 \sin 120\pi t \quad \text{From part (a), } \omega = 120\pi \text{ radians per sec.}$$

For $x = t$,
$V(t) = 163 \sin 120\pi t$

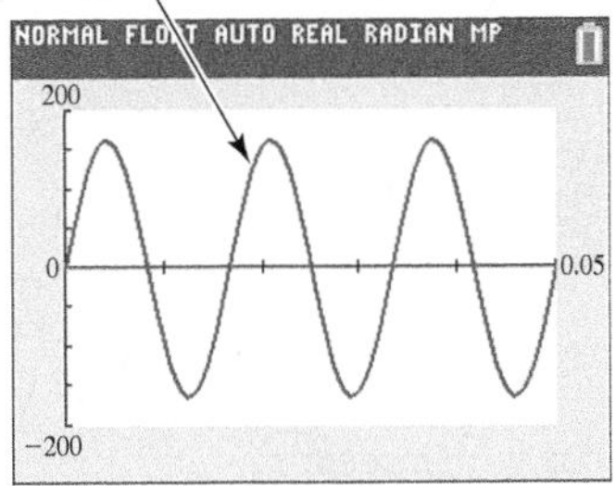

Figure 6

Because the amplitude of the function $V(t)$ is 163, an appropriate interval for the range is $[-200, 200]$, as shown in the graph in **Figure 6.**

(c) Use the even-odd identity for cosine and a cofunction identity.

$$\cos\left(x - \frac{\pi}{2}\right) = \cos\left[-\left(\frac{\pi}{2} - x\right)\right] = \cos\left(\frac{\pi}{2} - x\right) = \sin x$$

Therefore, if $\phi = \frac{\pi}{2}$, then

$$V(t) = 163 \cos(\omega t - \phi)$$

$$V(t) = 163 \cos\left(\omega t - \frac{\pi}{2}\right)$$

$$V(t) = 163 \sin \omega t.$$

✔ **Now Try Exercise 75.**

Verifying an Identity

Classroom Example 6
Verify that the following equation is an identity.

$$\sec(\pi + x) = -\sec x$$

Answer: The verification appears in the *Solutions to Classroom Examples.*

EXAMPLE 6 Verifying an Identity

Verify that the following equation is an identity.

$$\sec\left(\frac{3\pi}{2} - x\right) = -\csc x$$

SOLUTION We work with the more complicated left side.

$$\sec\left(\frac{3\pi}{2} - x\right) = \frac{1}{\cos\left(\frac{3\pi}{2} - x\right)} \quad \text{Reciprocal identity}$$

$$= \frac{1}{\cos \frac{3\pi}{2} \cos x + \sin \frac{3\pi}{2} \sin x} \quad \text{Cosine difference identity}$$

$$= \frac{1}{0 \cdot \cos x + (-1)\sin x} \quad \cos \frac{3\pi}{2} = 0 \text{ and } \sin \frac{3\pi}{2} = -1$$

$$= \frac{1}{-\sin x} \quad \text{Simplify.}$$

$$= -\csc x \quad \text{Reciprocal identity}$$

The left side is identical to the right side, so the given equation is an identity.

✔ **Now Try Exercise 67.**

5.3 Exercises

1. F 2. A
3. E 4. G
5. E 6. C
7. H 8. B

9. $\frac{\sqrt{6}-\sqrt{2}}{4}$ 10. $\frac{\sqrt{6}+\sqrt{2}}{4}$
11. $\frac{\sqrt{2}-\sqrt{6}}{4}$ 12. $\frac{\sqrt{2}-\sqrt{6}}{4}$
13. $\frac{\sqrt{2}-\sqrt{6}}{4}$ 14. $\frac{\sqrt{6}+\sqrt{2}}{4}$
15. $\frac{\sqrt{6}+\sqrt{2}}{4}$ 16. $\frac{\sqrt{2}-\sqrt{6}}{4}$
17. 0 18. -1

19. cot 3° 20. cos 75°
21. $\sin \frac{5\pi}{12}$ 22. $\cos \frac{\pi}{10}$
23. sec 75° 36′
24. $\cos(-52° 14')$
25. $\cos\left(-\frac{\pi}{8}\right)$
26. $\tan\left(-\frac{2\pi}{5}\right)$
27. $\csc(-56° 42')$
28. $\cot(-84° 03')$
29. $\tan(-86.9814°)$
30. $\cos(-8.0142°)$

31. tan 32. cos
33. cos 34. tan
35. csc 36. tan

CONCEPT PREVIEW *Match each expression in Column I with the correct expression in Column II to form an identity. Choices may be used once, more than once, or not at all.*

I	II
1. $\cos(x + y) =$ ____	**A.** $\cos x \cos y + \sin x \sin y$
2. $\cos(x - y) =$ ____	**B.** $\tan x$
3. $\cos\left(\frac{\pi}{2} - x\right) =$ ____	**C.** $-\cos x$
4. $\sin\left(\frac{\pi}{2} - x\right) =$ ____	**D.** $-\sin x$
5. $\cos\left(x - \frac{\pi}{2}\right) =$ ____	**E.** $\sin x$
6. $\sin\left(x - \frac{\pi}{2}\right) =$ ____	**F.** $\cos x \cos y - \sin x \sin y$
7. $\tan\left(\frac{\pi}{2} - x\right) =$ ____	**G.** $\cos x$
8. $\cot\left(\frac{\pi}{2} - x\right) =$ ____	**H.** $\cot x$

Find the exact value of each expression. (Do not use a calculator.) **See Example 1.**

9. cos 75° **10.** $\cos(-15°)$

11. $\cos(-105°)$ (*Hint*: $-105° = -60° + (-45°)$) **12.** cos 105° (*Hint*: $105° = 60° + 45°$)

13. $\cos \frac{7\pi}{12}$ **14.** $\cos \frac{\pi}{12}$

15. $\cos\left(-\frac{\pi}{12}\right)$ **16.** $\cos\left(-\frac{7\pi}{12}\right)$

17. $\cos 40° \cos 50° - \sin 40° \sin 50°$ **18.** $\cos \frac{7\pi}{9} \cos \frac{2\pi}{9} - \sin \frac{7\pi}{9} \sin \frac{2\pi}{9}$

Write each function value in terms of the cofunction of a complementary angle. **See Example 2.**

19. tan 87° **20.** sin 15° **21.** $\cos \frac{\pi}{12}$ **22.** $\sin \frac{2\pi}{5}$

23. csc 14° 24′ **24.** sin 142° 14′ **25.** $\sin \frac{5\pi}{8}$ **26.** $\cot \frac{9\pi}{10}$

27. sec 146° 42′ **28.** tan 174° 03′ **29.** cot 176.9814° **30.** sin 98.0142°

Use identities to fill in each blank with the appropriate trigonometric function name. **See Example 2.**

31. $\cot \frac{\pi}{3} =$ ____ $\frac{\pi}{6}$ **32.** $\sin \frac{2\pi}{3} =$ ____ $\left(-\frac{\pi}{6}\right)$

33. ____ 33° = sin 57° **34.** ____ 72° = cot 18°

35. $\cos 70° = \frac{1}{\text{\_\_\_\_ } 20°}$ **36.** $\tan 24° = \frac{1}{\text{\_\_\_\_ } 66°}$

For Exercises 37–42, other answers are possible. We give the most obvious one.

37. $15°$ **38.** $20°$
39. $-\frac{\pi}{6}$ **40.** $\frac{5\pi}{12}$
41. $20°$ **42.** $40°$

43. $\cos\theta$ **44.** $\sin\theta$
45. $-\cos\theta$ **46.** $-\sin\theta$
47. $\cos\theta$ **48.** $-\sin\theta$
49. $-\cos\theta$ **50.** $\sin\theta$

51. $\frac{16}{65}$; $-\frac{56}{65}$ **52.** $-\frac{36}{85}$; $\frac{84}{85}$

53. $\frac{4-6\sqrt{6}}{25}$; $\frac{4+6\sqrt{6}}{25}$

54. $\frac{-2\sqrt{10}+2}{9}$; $\frac{-2\sqrt{10}-2}{9}$

55. $\frac{2\sqrt{638}-\sqrt{30}}{56}$; $\frac{2\sqrt{638}+\sqrt{30}}{56}$

56. $\frac{\sqrt{62}-\sqrt{70}}{24}$; $\frac{\sqrt{62}+\sqrt{70}}{24}$

57. true **58.** false
59. false **60.** true
61. true **62.** true
63. true **64.** true
65. false **66.** false

Find one value of θ or x that satisfies each of the following. ***See Example 2.***

37. $\tan\theta = \cot(45° + 2\theta)$

38. $\sin\theta = \cos(2\theta + 30°)$

39. $\sec x = \csc\frac{2\pi}{3}$

40. $\cos x = \sin\frac{\pi}{12}$

41. $\sin(3\theta - 15°) = \cos(\theta + 25°)$

42. $\cot(\theta - 10°) = \tan(2\theta - 20°)$

Use the identities for the cosine of a sum or difference to write each expression as a trigonometric function of θ alone. ***See Example 3.***

43. $\cos(0° - \theta)$

44. $\cos(90° - \theta)$

45. $\cos(\theta - 180°)$

46. $\cos(\theta - 270°)$

47. $\cos(0° + \theta)$

48. $\cos(90° + \theta)$

49. $\cos(180° + \theta)$

50. $\cos(270° + \theta)$

Find $\cos(s + t)$ *and* $\cos(s - t)$. ***See Example 4.***

51. $\sin s = \frac{3}{5}$ and $\sin t = -\frac{12}{13}$, s in quadrant I and t in quadrant III

52. $\cos s = -\frac{8}{17}$ and $\cos t = -\frac{3}{5}$, s and t in quadrant III

53. $\cos s = -\frac{1}{5}$ and $\sin t = \frac{3}{5}$, s and t in quadrant II

54. $\sin s = \frac{2}{3}$ and $\sin t = -\frac{1}{3}$, s in quadrant II and t in quadrant IV

55. $\sin s = \frac{\sqrt{5}}{7}$ and $\sin t = \frac{\sqrt{6}}{8}$, s and t in quadrant I

56. $\cos s = \frac{\sqrt{2}}{4}$ and $\sin t = -\frac{\sqrt{5}}{6}$, s and t in quadrant IV

Concept Check Determine whether each statement is true *or* false.

57. $\cos 42° = \cos(30° + 12°)$

58. $\cos(-24°) = \cos 16° - \cos 40°$

59. $\cos 74° = \cos 60° \cos 14° + \sin 60° \sin 14°$

60. $\cos 140° = \cos 60° \cos 80° - \sin 60° \sin 80°$

61. $\cos\frac{\pi}{3} = \cos\frac{\pi}{12}\cos\frac{\pi}{4} - \sin\frac{\pi}{12}\sin\frac{\pi}{4}$

62. $\cos\frac{2\pi}{3} = \cos\frac{11\pi}{12}\cos\frac{\pi}{4} + \sin\frac{11\pi}{12}\sin\frac{\pi}{4}$

63. $\cos 70° \cos 20° - \sin 70° \sin 20° = 0$

64. $\cos 85° \cos 40° + \sin 85° \sin 40° = \frac{\sqrt{2}}{2}$

65. $\tan\left(x - \frac{\pi}{2}\right) = \cot x$

66. $\sin\left(x - \frac{\pi}{2}\right) = \cos x$

Verify that each equation is an identity. (Hint: $\cos 2x = \cos(x + x)$.*)* ***See Example 6.***

67. $\cos\left(\frac{\pi}{2} + x\right) = -\sin x$

68. $\sec(\pi - x) = -\sec x$

69. $\cos 2x = \cos^2 x - \sin^2 x$

70. $1 + \cos 2x - \cos^2 x = \cos^2 x$

71. $\cos 2x = 1 - 2\sin^2 x$

72. $\cos 2x = 2\cos^2 x - 1$

73. $\cos 2x = \frac{\cot^2 x - 1}{\cot^2 x + 1}$

74. $\sec 2x = \frac{\cot^2 x + 1}{\cot^2 x - 1}$

75. (a) 3
(b) 163 and -163
(c) no

76. (a) For $x = t$,
$P(t) = \frac{0.4}{10}\cos\left[\frac{20\pi}{4.9} - 1026t\right]$

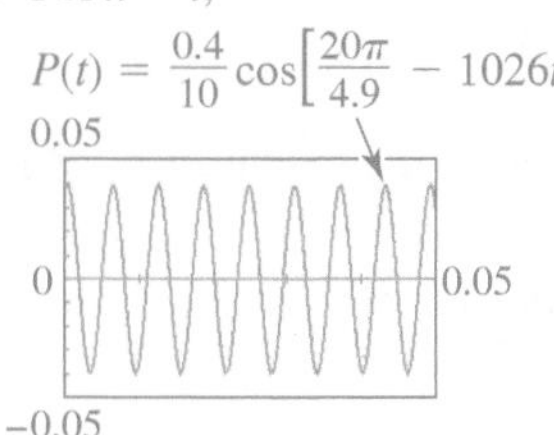

The pressure P is oscillating.

(b) For $x = r$,
$P(r) = \frac{3}{r}\cos\left[\frac{2\pi r}{4.9} - 10{,}260\right]$

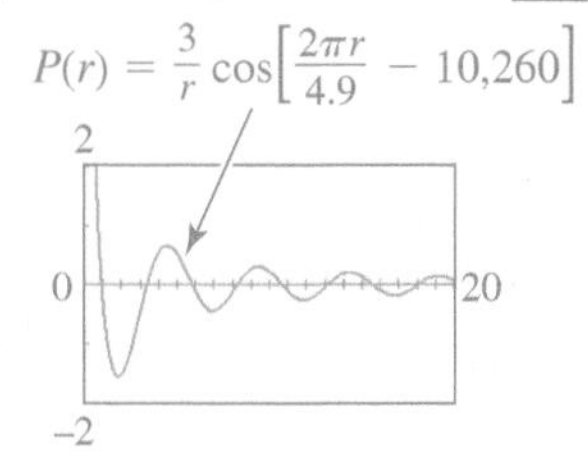

The pressure oscillates and amplitude decreases as r increases.

(c) $P = \frac{a}{n\lambda}\cos ct$

(Modeling) Solve each problem. ***See Example 5.***

75. ***Electric Current*** The voltage V in a typical 115-volt outlet can be expressed by the function

$$V(t) = 163 \sin 120\pi t,$$

where 120π is the angular speed (in radians per second) of the rotating generator at an electrical power plant, and t is time in seconds. (*Source:* Bell, D., *Fundamentals of Electric Circuits,* Fourth Edition, Prentice-Hall.)

(a) How many times does the current oscillate in 0.05 sec?

(b) What are the maximum and minimum voltages in this outlet?

(c) Is the voltage always equal to 115 volts?

76. ***Sound Waves*** Sound is a result of waves applying pressure to a person's eardrum. For a pure sound wave radiating outward in a spherical shape, the trigonometric function

$$P = \frac{a}{r}\cos\left(\frac{2\pi r}{\lambda} - ct\right)$$

can be used to model the sound pressure at a radius of r feet from the source, where t is time in seconds, λ is length of the sound wave in feet, c is speed of sound in feet per second, and a is maximum sound pressure at the source measured in pounds per square foot. (*Source:* Beranek, L., *Noise and Vibration Control,* Institute of Noise Control Engineering, Washington, D.C.) Let $\lambda = 4.9$ ft and $c = 1026$ ft per sec.

(a) Let $a = 0.4$ lb per ft^2. Graph the sound pressure at distance $r = 10$ ft from its source in the window $[0, 0.05]$ by $[-0.05, 0.05]$. Describe P at this distance.

(b) Now let $a = 3$ and $t = 10$. Graph the sound pressure in the window $[0, 20]$ by $[-2, 2]$. What happens to pressure P as radius r increases?

(c) Suppose a person stands at a radius r so that $r = n\lambda$, where n is a positive integer. Use the difference identity for cosine to simplify P in this situation.

Relating Concepts

For individual or collaborative investigation *(Exercises 77–82)*

(This discussion applies to functions of both angles and real numbers.) The result of **Example 3** in this section can be written as an identity.

$$\cos(180° - \theta) = -\cos\theta$$

This is an example of a **reduction formula,** which is an identity that *reduces* a function of a quadrantal angle plus or minus θ to a function of θ alone. Another example of a reduction formula is

$$\cos(270° + \theta) = \sin\theta.$$

Here is an interesting method for quickly determining a reduction formula for a trigonometric function f of the form

$$f(Q \pm \theta), \quad \text{where } Q \text{ is a quadrantal angle.}$$

There are two cases to consider, and in each case, think of θ as a small positive angle in order to determine the quadrant in which $Q \pm \theta$ will lie.

77. $\cos(90° + \theta) = -\sin \theta$
78. $\cos(270° - \theta) = -\sin \theta$
79. $\cos(180° + \theta) = -\cos \theta$
80. $\cos(270° + \theta) = \sin \theta$
81. $\sin(180° + \theta) = -\sin \theta$
82. $\tan(270° - \theta) = \cot \theta$

Case 1 **Q is a quadrantal angle whose terminal side lies along the x-axis.**

Determine the quadrant in which $Q \pm \theta$ will lie for a small positive angle θ. If the given function f is positive in that quadrant, use a $+$ sign on the reduced form. If f is negative in that quadrant, use a $-$ sign. The reduced form will have that sign, f as the function, and θ as the argument.

Example:

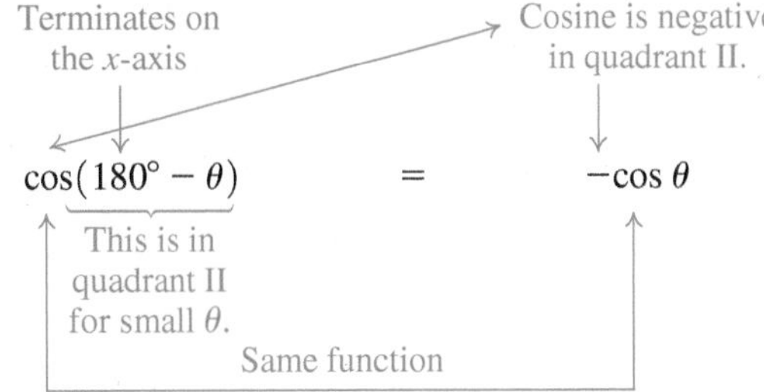

Case 2 **Q is a quadrantal angle whose terminal side lies along the y-axis.**

Determine the quadrant in which $Q \pm \theta$ will lie for a small positive angle θ. If the given function f is positive in that quadrant, use a $+$ sign on the reduced form. If f is negative in that quadrant, use a $-$ sign. The reduced form will have that sign, the ***cofunction of f*** as the function, and θ as the argument.

Example:

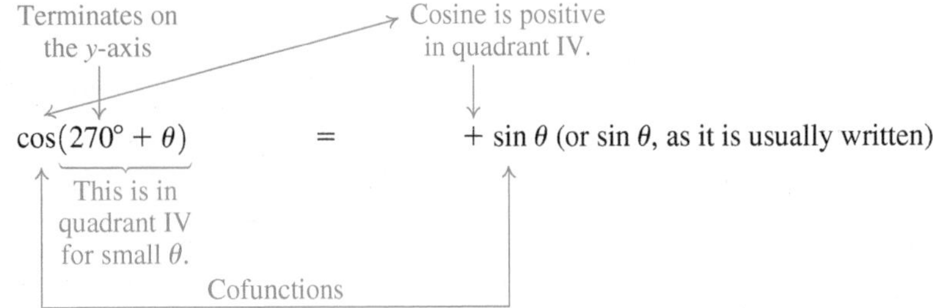

Use these ideas to write a reduction formula for each of the following.

77. $\cos(90° + \theta)$ **78.** $\cos(270° - \theta)$ **79.** $\cos(180° + \theta)$

80. $\cos(270° + \theta)$ **81.** $\sin(180° + \theta)$ **82.** $\tan(270° - \theta)$

5.4 Sum and Difference Identities for Sine and Tangent

- Sum and Difference Identities for Sine
- Sum and Difference Identities for Tangent
- Applications of the Sum and Difference Identities
- Verifying an Identity

Sum and Difference Identities for Sine

We can use the cosine sum and difference identities from the previous section to derive similar identities for sine and tangent. In $\sin \theta = \cos(90° - \theta)$, replace θ with $A + B$.

$$\sin(A + B) = \cos[90° - (A + B)] \quad \text{Cofunction identity}$$

$$= \cos[(90° - A) - B] \quad \text{Distribute negative sign and regroup.}$$

$$= \cos(90° - A)\cos B + \sin(90° - A)\sin B \quad \text{Cosine difference identity}$$

$$\mathbf{\sin(A + B) = \sin A \cos B + \cos A \sin B} \quad \text{Cofunction identities}$$

Now we write $\sin(A - B)$ as $\sin[A + (-B)]$ and use the identity just found for $\sin(A + B)$.

$$\sin(A - B) = \sin[A + (-B)] \quad \text{Definition of subtraction}$$

$$= \sin A \cos(-B) + \cos A \sin(-B) \quad \text{Sine sum identity}$$

$$\mathbf{\sin(A - B) = \sin A \cos B - \cos A \sin B} \quad \text{Even-odd identities}$$

Sine of a Sum or Difference

$$\sin(A + B) = \sin A \cos B + \cos A \sin B$$

$$\sin(A - B) = \sin A \cos B - \cos A \sin B$$

Sum and Difference Identities for Tangent We can derive the identity for $\tan(A + B)$ as follows.

$$\tan(A + B) = \frac{\sin(A + B)}{\cos(A + B)} \quad \text{Fundamental identity}$$

We express this result in terms of the tangent function.

$$= \frac{\sin A \cos B + \cos A \sin B}{\cos A \cos B - \sin A \sin B} \quad \text{Sum identities}$$

$$= \frac{\dfrac{\sin A \cos B + \cos A \sin B}{1}}{\dfrac{\cos A \cos B - \sin A \sin B}{1}} \cdot \frac{\dfrac{1}{\cos A \cos B}}{\dfrac{1}{\cos A \cos B}} \quad \text{Multiply by 1, where } 1 = \frac{\frac{1}{\cos A \cos B}}{\frac{1}{\cos A \cos B}}.$$

$$= \frac{\dfrac{\sin A \cos B}{\cos A \cos B} + \dfrac{\cos A \sin B}{\cos A \cos B}}{\dfrac{\cos A \cos B}{\cos A \cos B} - \dfrac{\sin A \sin B}{\cos A \cos B}} \quad \text{Multiply numerators. Multiply denominators.}$$

$$= \frac{\dfrac{\sin A}{\cos A} + \dfrac{\sin B}{\cos B}}{1 - \dfrac{\sin A}{\cos A} \cdot \dfrac{\sin B}{\cos B}} \quad \text{Simplify.}$$

$$\mathbf{\tan(A + B) = \frac{\tan A + \tan B}{1 - \tan A \tan B}} \quad \frac{\sin \theta}{\cos \theta} = \tan \theta$$

We can replace B with $-B$ and use the fact that $\tan(-B) = -\tan B$ to obtain the identity for the tangent of the difference of two angles, as seen below.

Tangent of a Sum or Difference

$$\tan(A + B) = \frac{\tan A + \tan B}{1 - \tan A \tan B} \qquad \tan(A - B) = \frac{\tan A - \tan B}{1 + \tan A \tan B}$$

Applications of the Sum and Difference Identities

Classroom Example 1
Find the *exact* value of each expression.

(a) $\sin(-15°)$

(b) $\tan \frac{13\pi}{12}$

(c) $\frac{\tan 100° - \tan 70°}{1 + \tan 100° \tan 70°}$

Answers:

(a) $\frac{\sqrt{2} - \sqrt{6}}{4}$ (b) $2 - \sqrt{3}$

(c) $\frac{\sqrt{3}}{3}$

EXAMPLE 1 Finding Exact Sine and Tangent Function Values

Find the *exact* value of each expression.

(a) $\sin 75°$ **(b)** $\tan \frac{7\pi}{12}$ **(c)** $\sin 40° \cos 160° - \cos 40° \sin 160°$

SOLUTION

(a) $\sin 75°$

$= \sin(45° + 30°)$ $\quad 75° = 45° + 30°$

$= \sin 45° \cos 30° + \cos 45° \sin 30°$ $\quad$ Sine sum identity

$= \frac{\sqrt{2}}{2} \cdot \frac{\sqrt{3}}{2} + \frac{\sqrt{2}}{2} \cdot \frac{1}{2}$ $\quad$ Substitute known values.

$= \frac{\sqrt{6} + \sqrt{2}}{4}$ $\quad$ Multiply, and then add fractions.

(b) $\tan \frac{7\pi}{12}$

$= \tan\left(\frac{\pi}{3} + \frac{\pi}{4}\right)$ $\quad \frac{\pi}{3} = \frac{4\pi}{12}$ and $\frac{\pi}{4} = \frac{3\pi}{12}$

$= \frac{\tan \frac{\pi}{3} + \tan \frac{\pi}{4}}{1 - \tan \frac{\pi}{3} \tan \frac{\pi}{4}}$ $\quad$ Tangent sum identity

$= \frac{\sqrt{3} + 1}{1 - \sqrt{3} \cdot 1}$ $\quad$ Substitute known values.

$= \frac{\sqrt{3} + 1}{1 - \sqrt{3}} \cdot \frac{1 + \sqrt{3}}{1 + \sqrt{3}}$ $\quad$ Rationalize the denominator.

$= \frac{\sqrt{3} + 3 + 1 + \sqrt{3}}{1 - 3}$ $\quad (a + b)(c + d) = ac + ad + bc + bd;$ $(x - y)(x + y) = x^2 - y^2$

$= \frac{4 + 2\sqrt{3}}{-2}$ $\quad$ Combine like terms.

$= \frac{2(2 + \sqrt{3})}{2(-1)}$ $\quad$ Factor out 2.

Factor first. Then divide out the common factor.

$= -2 - \sqrt{3}$ $\quad$ Write in lowest terms.

(c) $\sin 40° \cos 160° - \cos 40° \sin 160°$

$= \sin(40° - 160°)$ $\quad$ Sine difference identity

$= \sin(-120°)$ $\quad$ Subtract.

$= -\sin 120°$ $\quad$ Even-odd identity

$= -\frac{\sqrt{3}}{2}$ $\quad$ Substitute the known value.

✔ **Now Try Exercises 9, 15, and 25.**

Classroom Example 2
Write each function as an expression involving functions of θ alone.

(a) $\sin(120° + \theta)$

(b) $\tan(\theta + 3\pi)$

(c) $\sin(\theta - 270°)$

Answers:

(a) $\frac{\sqrt{3}\cos\theta - \sin\theta}{2}$

(b) $\tan\theta$ (c) $\cos\theta$

EXAMPLE 2 Writing Functions as Expressions Involving Functions of θ

Write each function as an expression involving functions of θ alone.

(a) $\sin(30° + \theta)$ **(b)** $\tan(45° - \theta)$ **(c)** $\sin(180° - \theta)$

SOLUTION

(a) $\sin(30° + \theta)$

$= \sin 30° \cos\theta + \cos 30° \sin\theta$ Sine sum identity

$= \frac{1}{2}\cos\theta + \frac{\sqrt{3}}{2}\sin\theta$ $\sin 30° = \frac{1}{2}$ and $\cos 30° = \frac{\sqrt{3}}{2}$

$= \frac{\cos\theta + \sqrt{3}\sin\theta}{2}$ $\frac{a}{b} \cdot c = \frac{ac}{b}$; Add fractions.

(b) $\tan(45° - \theta)$

$= \frac{\tan 45° - \tan\theta}{1 + \tan 45° \tan\theta}$ Tangent difference identity

$= \frac{1 - \tan\theta}{1 + 1 \cdot \tan\theta}$ $\tan 45° = 1$

$= \frac{1 - \tan\theta}{1 + \tan\theta}$ Multiply.

(c) $\sin(180° - \theta)$

$= \sin 180° \cos\theta - \cos 180° \sin\theta$ Sine difference identity

$= 0 \cdot \cos\theta - (-1)\sin\theta$ $\sin 180° = 0$ and $\cos 180° = -1$

$= \sin\theta$ Simplify.

✔ **Now Try Exercises 33, 39, and 43.**

Classroom Example 3
Suppose that A and B are angles in standard position, such that $\cos A = -\frac{7}{25}$, $\pi < A < \frac{3\pi}{2}$, and $\sin B = -\frac{3}{5}$, $\frac{3\pi}{2} < B < 2\pi$.

Find each of the following.

(a) $\sin(A - B)$ (b) $\tan(A - B)$

(c) the quadrant of $A - B$

Answers:

(a) $-\frac{117}{125}$ (b) $-\frac{117}{44}$

(c) quadrant IV

EXAMPLE 3 Finding Function Values and the Quadrant of $A + B$

Suppose that A and B are angles in standard position such that $\sin A = \frac{4}{5}$, $\frac{\pi}{2} < A < \pi$, and $\cos B = -\frac{5}{13}$, $\pi < B < \frac{3\pi}{2}$. Find each of the following.

(a) $\sin(A + B)$ **(b)** $\tan(A + B)$ **(c)** the quadrant of $A + B$

SOLUTION

(a) The identity for $\sin(A + B)$ involves $\sin A$, $\cos A$, $\sin B$, and $\cos B$. We are given values of $\sin A$ and $\cos B$. We must find values of $\cos A$ and $\sin B$.

$\sin^2 A + \cos^2 A = 1$ Fundamental identity

$\left(\frac{4}{5}\right)^2 + \cos^2 A = 1$ $\sin A = \frac{4}{5}$

$\frac{16}{25} + \cos^2 A = 1$ Square $\frac{4}{5}$.

$\cos^2 A = \frac{9}{25}$ Subtract $\frac{16}{25}$.

Pay attention to signs. → $\cos A = -\frac{3}{5}$ Take square roots. Because A is in quadrant II, $\cos A < 0$.

In the same way, $\sin B = -\frac{12}{13}$. Now find $\sin(A + B)$.

$$\begin{aligned}
\sin(A+B) &= \sin A\cos B + \cos A \sin B && \text{Sine sum identity}\\
&= \frac{4}{5}\left(-\frac{5}{13}\right) + \left(-\frac{3}{5}\right)\left(-\frac{12}{13}\right) && \text{Substitute the given values for } \sin A \text{ and } \cos B \text{ and the values found for } \cos A \text{ and } \sin B.\\
&= -\frac{20}{65} + \frac{36}{65} && \text{Multiply.}\\
\sin(A+B) &= \frac{16}{65} && \text{Add.}
\end{aligned}$$

(b) To find $\tan(A + B)$, use the values of sine and cosine from part (a), $\sin A = \frac{4}{5}$, $\cos A = -\frac{3}{5}$, $\sin B = -\frac{12}{13}$, and $\cos B = -\frac{5}{13}$, to obtain $\tan A$ and $\tan B$.

$$\begin{aligned}
\tan A &= \frac{\sin A}{\cos A}\\
&= \frac{\frac{4}{5}}{-\frac{3}{5}}\\
&= \frac{4}{5} \div \left(-\frac{3}{5}\right)\\
&= \frac{4}{5}\cdot\left(-\frac{5}{3}\right)\\
\tan A &= -\frac{4}{3}
\end{aligned}$$

$$\begin{aligned}
\tan B &= \frac{\sin B}{\cos B}\\
&= \frac{-\frac{12}{13}}{-\frac{5}{13}}\\
&= -\frac{12}{13} \div \left(-\frac{5}{13}\right)\\
&= -\frac{12}{13}\cdot\left(-\frac{13}{5}\right)\\
\tan B &= \frac{12}{5}
\end{aligned}$$

$$\begin{aligned}
\tan(A+B) &= \frac{\tan A + \tan B}{1 - \tan A \tan B} && \text{Tangent sum identity}\\
&= \frac{\left(-\frac{4}{3}\right) + \frac{12}{5}}{1 - \left(-\frac{4}{3}\right)\left(\frac{12}{5}\right)} && \text{Substitute.}\\
&= \frac{\frac{16}{15}}{1 + \frac{48}{15}} && \text{Perform the indicated operations.}\\
&= \frac{\frac{16}{15}}{\frac{63}{15}} && \text{Add terms in the denominator.}\\
&= \frac{16}{15} \div \frac{63}{15} && \text{Simplify the complex fraction.}\\
&= \frac{16}{15}\cdot\frac{15}{63} && \text{Definition of division}\\
\tan(A+B) &= \frac{16}{63} && \text{Multiply.}
\end{aligned}$$

(c) $\sin(A + B) = \frac{16}{65}$ and $\tan(A + B) = \frac{16}{63}$ See parts (a) and (b).

Both are positive. Therefore, $A + B$ must be in quadrant I, because it is the only quadrant in which both sine and tangent are positive.

✔ **Now Try Exercise 51.**

Verifying an Identity

Classroom Example 4
Verify that the following equation is an identity.

$$\tan\left(\frac{\pi}{4}+t\right)+\tan\left(\frac{\pi}{4}-t\right)=\frac{2\sec^2 t}{1-\tan^2 t}$$

Answer: The verification appears in the *Solutions to Classroom Examples.*

EXAMPLE 4 Verifying an Identity

Verify that the equation is an identity.

$$\sin\left(\frac{\pi}{6}+\theta\right)+\cos\left(\frac{\pi}{3}+\theta\right)=\cos\theta$$

SOLUTION Work on the left side, using the sine and cosine sum identities.

$$\sin\left(\frac{\pi}{6}+\theta\right)+\cos\left(\frac{\pi}{3}+\theta\right)$$

$$=\left(\sin\frac{\pi}{6}\cos\theta+\cos\frac{\pi}{6}\sin\theta\right)+\left(\cos\frac{\pi}{3}\cos\theta-\sin\frac{\pi}{3}\sin\theta\right)$$

Sine sum identity; cosine sum identity

$$=\left(\frac{1}{2}\cos\theta+\frac{\sqrt{3}}{2}\sin\theta\right)+\left(\frac{1}{2}\cos\theta-\frac{\sqrt{3}}{2}\sin\theta\right)$$

$\sin\frac{\pi}{6}=\frac{1}{2}$; $\cos\frac{\pi}{6}=\frac{\sqrt{3}}{2}$; $\cos\frac{\pi}{3}=\frac{1}{2}$; $\sin\frac{\pi}{3}=\frac{\sqrt{3}}{2}$

$$=\frac{1}{2}\cos\theta+\frac{1}{2}\cos\theta \quad \text{Simplify.}$$

$$=\cos\theta \quad \text{Add.}$$

✔ Now Try Exercise 63.

5.4 Exercises

1. D 2. A
3. B 4. C

5. C 6. B
7. A 8. D

9. $\frac{\sqrt{6}-\sqrt{2}}{4}$ 10. $\frac{-\sqrt{6}-\sqrt{2}}{4}$
11. $-2+\sqrt{3}$ 12. $-2-\sqrt{3}$
13. $\frac{\sqrt{6}+\sqrt{2}}{4}$ 14. $\frac{\sqrt{2}-\sqrt{6}}{4}$
15. $2-\sqrt{3}$ 16. $2+\sqrt{3}$
17. $\frac{\sqrt{6}+\sqrt{2}}{4}$ 18. $\frac{\sqrt{6}-\sqrt{2}}{4}$

CONCEPT PREVIEW *Match each expression in Column I with the correct expression in Column II to form an identity.*

I	II
1. $\sin(A+B)$	**A.** $\sin A\cos B-\cos A\sin B$
2. $\sin(A-B)$	**B.** $\frac{\tan A+\tan B}{1-\tan A\tan B}$
3. $\tan(A+B)$	**C.** $\frac{\tan A-\tan B}{1+\tan A\tan B}$
4. $\tan(A-B)$	**D.** $\sin A\cos B+\cos A\sin B$

CONCEPT PREVIEW *Match each expression in Column I with its equivalent expression in Column II.*

I	II
5. $\sin 60°\cos 45°+\cos 60°\sin 45°$	**A.** $\tan\frac{7\pi}{12}$
6. $\sin 60°\cos 45°-\cos 60°\sin 45°$	**B.** $\sin 15°$
7. $\frac{\tan\frac{\pi}{3}+\tan\frac{\pi}{4}}{1-\tan\frac{\pi}{3}\tan\frac{\pi}{4}}$	**C.** $\sin 105°$
8. $\frac{\tan\frac{\pi}{3}-\tan\frac{\pi}{4}}{1+\tan\frac{\pi}{3}\tan\frac{\pi}{4}}$	**D.** $\tan\frac{\pi}{12}$

Find the exact value of each expression. **See Example 1.**

9. $\sin 165°$ **10.** $\sin 255°$ **11.** $\tan 165°$ **12.** $\tan 285°$

19. $\frac{-\sqrt{6}-\sqrt{2}}{4}$ 20. $\frac{-\sqrt{2}-\sqrt{6}}{4}$
21. $-2-\sqrt{3}$ 22. $2+\sqrt{3}$
23. $-2+\sqrt{3}$ 24. $\frac{\sqrt{6}-\sqrt{2}}{4}$
25. $\frac{\sqrt{2}}{2}$ 26. 1
27. 1 28. 1
29. -1 30. -1
31. 0 32. $-\sqrt{3}$
33. $\frac{\sqrt{3}\cos\theta-\sin\theta}{2}$
34. $\frac{\sqrt{3}\cos\theta+\sin\theta}{2}$
35. $\frac{\cos\theta-\sqrt{3}\sin\theta}{2}$
36. $\frac{\sqrt{2}(\cos\theta+\sin\theta)}{2}$
37. $\frac{\sqrt{2}(\sin x-\cos x)}{2}$
38. $\frac{\sqrt{2}(\cos\theta+\sin\theta)}{2}$
39. $\frac{\sqrt{3}\tan\theta+1}{\sqrt{3}-\tan\theta}$
40. $\frac{1+\tan x}{1-\tan x}$
41. $\frac{\sqrt{2}(\cos x+\sin x)}{2}$
42. $\frac{\sqrt{2}(\cos x+\sin x)}{2}$
43. $-\cos\theta$ 44. $\tan\theta$
45. $-\tan x$ 46. $-\sin x$
47. $-\tan x$
48. $\tan 270°$ is undefined.
49. Cotangent, secant, and cosecant formulas can be written using their reciprocal functions: tangent, cosine, and sine.
51. (a) $\frac{63}{65}$ (b) $\frac{63}{16}$ (c) I
52. (a) $-\frac{63}{65}$ (b) $-\frac{63}{16}$ (c) IV
53. (a) $\frac{77}{85}$ (b) $-\frac{77}{36}$ (c) II
54. (a) $-\frac{36}{85}$ (b) $\frac{36}{77}$ (c) III
55. (a) $\frac{4\sqrt{2}+\sqrt{5}}{9}$ (b) $\frac{-\sqrt{5}-\sqrt{2}}{2}$ (c) II
56. (a) $\frac{-8\sqrt{6}-3}{25}$ (b) $\frac{\sqrt{6}+6}{4}$ (c) III
57. $\sin\left(\frac{\pi}{2}+\theta\right)=\cos\theta$
58. $\sin\left(\frac{3\pi}{2}+\theta\right)=-\cos\theta$
59. $\tan\left(\frac{\pi}{2}+\theta\right)=-\cot\theta$
60. $\tan\left(\frac{\pi}{2}-\theta\right)=\cot\theta$

13. $\sin\frac{5\pi}{12}$ **14.** $\sin\frac{13\pi}{12}$ **15.** $\tan\frac{\pi}{12}$ **16.** $\tan\frac{5\pi}{12}$

17. $\sin\frac{7\pi}{12}$ **18.** $\sin\frac{\pi}{12}$ **19.** $\sin\left(-\frac{7\pi}{12}\right)$ **20.** $\sin\left(-\frac{5\pi}{12}\right)$

21. $\tan\left(-\frac{5\pi}{12}\right)$ **22.** $\tan\left(-\frac{7\pi}{12}\right)$ **23.** $\tan\frac{11\pi}{12}$ **24.** $\sin\left(-\frac{13\pi}{12}\right)$

25. $\sin 76°\cos 31°-\cos 76°\sin 31°$ **26.** $\sin 40°\cos 50°+\cos 40°\sin 50°$

27. $\sin\frac{\pi}{5}\cos\frac{3\pi}{10}+\cos\frac{\pi}{5}\sin\frac{3\pi}{10}$ **28.** $\sin\frac{5\pi}{9}\cos\frac{\pi}{18}-\cos\frac{5\pi}{9}\sin\frac{\pi}{18}$

29. $\frac{\tan 80°+\tan 55°}{1-\tan 80°\tan 55°}$ **30.** $\frac{\tan 80°-\tan(-55°)}{1+\tan 80°\tan(-55°)}$

31. $\frac{\tan\frac{5\pi}{9}+\tan\frac{4\pi}{9}}{1-\tan\frac{5\pi}{9}\tan\frac{4\pi}{9}}$ **32.** $\frac{\tan\frac{5\pi}{12}+\tan\frac{\pi}{4}}{1-\tan\frac{5\pi}{12}\tan\frac{\pi}{4}}$

Write each function as an expression involving functions of θ or x alone. ***See Example 2.***

33. $\cos(30°+\theta)$ **34.** $\cos(\theta-30°)$ **35.** $\cos(60°+\theta)$

36. $\cos(45°-\theta)$ **37.** $\cos\left(\frac{3\pi}{4}-x\right)$ **38.** $\sin(45°+\theta)$

39. $\tan(\theta+30°)$ **40.** $\tan\left(\frac{\pi}{4}+x\right)$ **41.** $\sin\left(\frac{\pi}{4}+x\right)$

42. $\sin\left(\frac{3\pi}{4}-x\right)$ **43.** $\sin(270°-\theta)$ **44.** $\tan(180°+\theta)$

45. $\tan(2\pi-x)$ **46.** $\sin(\pi+x)$ **47.** $\tan(\pi-x)$

48. Why is it not possible to use the method of **Example 2** to find a formula for $\tan(270°-\theta)$?

49. Why is it that standard trigonometry texts usually do not develop formulas for the cotangent, secant, and cosecant of the sum and difference of two numbers or angles?

50. Show that if A, B, and C are the angles of a triangle, then

$$\sin(A+B+C)=0.$$

Use the given information to find ***(a)*** $\sin(s+t)$, ***(b)*** $\tan(s+t)$, *and* ***(c)*** *the quadrant of* $s+t$. ***See Example 3.***

51. $\cos s=\frac{3}{5}$ and $\sin t=\frac{5}{13}$, s and t in quadrant I

52. $\sin s=\frac{3}{5}$ and $\sin t=-\frac{12}{13}$, s in quadrant I and t in quadrant III

53. $\cos s=-\frac{8}{17}$ and $\cos t=-\frac{3}{5}$, s and t in quadrant III

54. $\cos s=-\frac{15}{17}$ and $\sin t=\frac{4}{5}$, s in quadrant II and t in quadrant I

55. $\sin s=\frac{2}{3}$ and $\sin t=-\frac{1}{3}$, s in quadrant II and t in quadrant IV

56. $\cos s=-\frac{1}{5}$ and $\sin t=\frac{3}{5}$, s and t in quadrant II

Graph each expression and use the graph to make a conjecture, predicting what might be an identity. Then verify your conjecture algebraically.

57. $\sin\left(\frac{\pi}{2}+\theta\right)$ **58.** $\sin\left(\frac{3\pi}{2}+\theta\right)$ **59.** $\tan\left(\frac{\pi}{2}+\theta\right)$ **60.** $\tan\left(\frac{\pi}{2}-\theta\right)$

71. (a) 425 lb (c) 0°
72. (a) 408 lb (b) 46.1°
73. $-20 \cos \frac{\pi t}{4}$
74. (a)
For $x = t$,
$V = V_1 + V_2$
$= 30 \sin 120\pi t + 40 \cos 120\pi t$

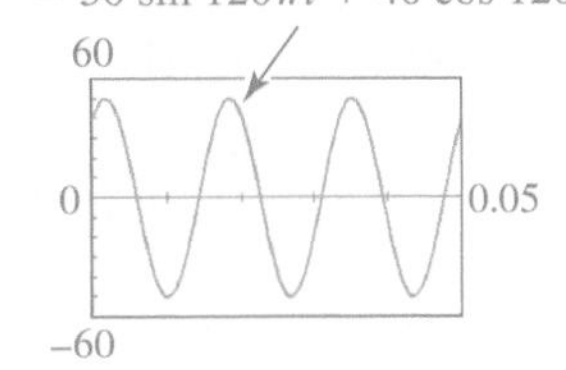

(b) $a = 50$; $\phi \approx -5.356$

Verify that each equation is an identity. ***See Example 4.***

61. $\sin 2x = 2 \sin x \cos x$ (*Hint*: $\sin 2x = \sin(x + x)$)

62. $\sin(x + y) + \sin(x - y) = 2 \sin x \cos y$

63. $\sin\left(\frac{7\pi}{6} + x\right) - \cos\left(\frac{2\pi}{3} + x\right) = 0$

64. $\tan(x - y) - \tan(y - x) = \frac{2(\tan x - \tan y)}{1 + \tan x \tan y}$

65. $\frac{\cos(\alpha - \beta)}{\cos \alpha \sin \beta} = \tan \alpha + \cot \beta$

66. $\frac{\sin(s + t)}{\cos s \cos t} = \tan s + \tan t$

67. $\frac{\sin(x - y)}{\sin(x + y)} = \frac{\tan x - \tan y}{\tan x + \tan y}$

68. $\frac{\sin(x + y)}{\cos(x - y)} = \frac{\cot x + \cot y}{1 + \cot x \cot y}$

69. $\frac{\sin(s - t)}{\sin t} + \frac{\cos(s - t)}{\cos t} = \frac{\sin s}{\sin t \cos t}$

70. $\frac{\tan(\alpha + \beta) - \tan \beta}{1 + \tan(\alpha + \beta) \tan \beta} = \tan \alpha$

(Modeling) Solve each problem.

71. ***Back Stress*** If a person bends at the waist with a straight back making an angle of θ degrees with the horizontal, then the force F exerted on the back muscles can be modeled by the equation

$$F = \frac{0.6W \sin(\theta + 90°)}{\sin 12°},$$

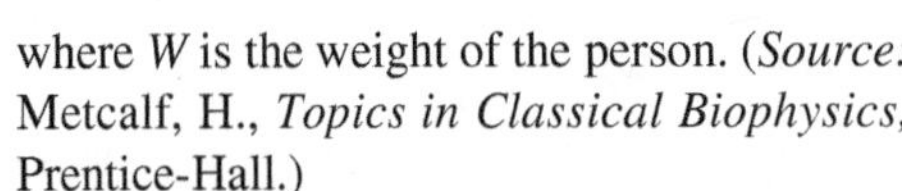

where W is the weight of the person. (*Source:* Metcalf, H., *Topics in Classical Biophysics,* Prentice-Hall.)

(a) Calculate force F, to the nearest pound, for $W = 170$ lb and $\theta = 30°$.

(b) Use an identity to show that F is approximately equal to $2.9W \cos \theta$.

(c) For what value of θ is F maximum?

72. ***Back Stress*** Refer to **Exercise 71.**

(a) Suppose a 200-lb person bends at the waist so that $\theta = 45°$. Calculate the force, to the nearest pound, exerted on the person's back muscles.

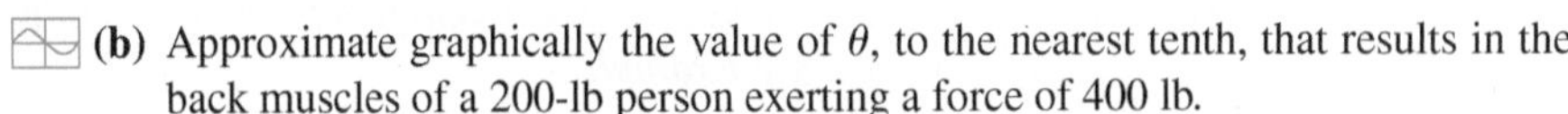

(b) Approximate graphically the value of θ, to the nearest tenth, that results in the back muscles of a 200-lb person exerting a force of 400 lb.

73. ***Voltage*** A coil of wire rotating in a magnetic field induces a voltage

$$E = 20 \sin\left(\frac{\pi t}{4} - \frac{\pi}{2}\right).$$

Use an identity from this section to express this in terms of $\cos \frac{\pi t}{4}$.

74. ***Voltage of a Circuit*** When the two voltages

$$V_1 = 30 \sin 120\pi t \quad \text{and} \quad V_2 = 40 \cos 120\pi t$$

are applied to the same circuit, the resulting voltage V will be equal to their sum. (*Source*: Bell, D., *Fundamentals of Electric Circuits,* Second Edition, Reston Publishing Company.)

(a) Graph the sum in the window $[0, 0.05]$ by $[-60, 60]$.

(b) Use the graph to estimate values for a and ϕ so that $V = a \sin(120\pi t + \phi)$.

(c) Use identities to verify that the expression for V in part (b) is valid.

75. $y' = y \cos R - z \sin R$
76. $z' = z \cos R + y \sin R$

77. $180° - \beta$
78. $\theta = \beta - \alpha$
79. $\tan \theta = \frac{\tan \beta - \tan \alpha}{1 + \tan \beta \tan \alpha}$

81. 18.4° 82. 80.8°

(Modeling) Roll of a Spacecraft *The figure on the left below shows the three quantities that determine the motion of a spacecraft. A conventional three-dimensional spacecraft coordinate system is shown on the right.*

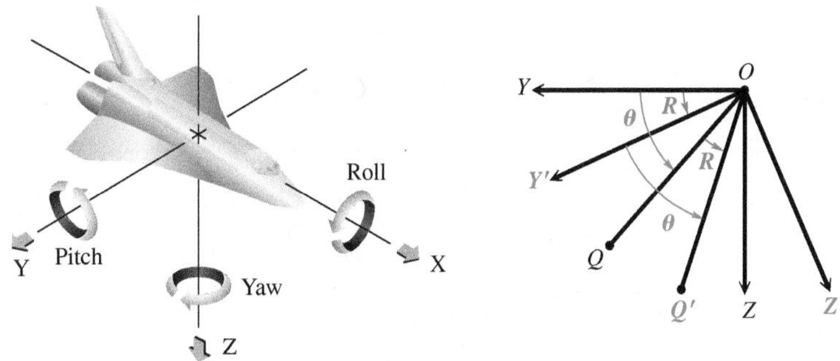

Angle $YOQ = \theta$ *and* $OQ = r$. *The coordinates of* Q *are* (x, y, z), *where*

$$y = r \cos \theta \quad \text{and} \quad z = r \sin \theta.$$

When the spacecraft performs a rotation, it is necessary to find the coordinates in the spacecraft system after the rotation takes place. For example, suppose the spacecraft undergoes roll through angle R. *The coordinates* (x, y, z) *of point* Q *become* (x', y', z'), *the coordinates of the corresponding point* Q'. *In the new reference system,* $OQ' = r$ *and, because the roll is around the x-axis and angle* $Y'OQ' = YOQ = \theta$,

$$x' = x, \quad y' = r \cos(\theta + R), \quad \text{and} \quad z' = r \sin(\theta + R).$$

(*Source: Kastner, B.,* Space Mathematics, *NASA.*)

75. Write y' in terms of y, R, and z.

76. Write z' in terms of y, R, and z.

Relating Concepts

For individual or collaborative investigation *(Exercises 77–82)*

Refer to the figure on the left below. By the definition of $\tan \theta$,

$m = \tan \theta$, *where* m *is the slope and* θ *is the angle of inclination of the line.*

The following exercises, which depend on properties of triangles, refer to triangle ABC in the figure on the right below. ***Work Exercises 77–82 in order.*** *Assume that all angles are measured in degrees.*

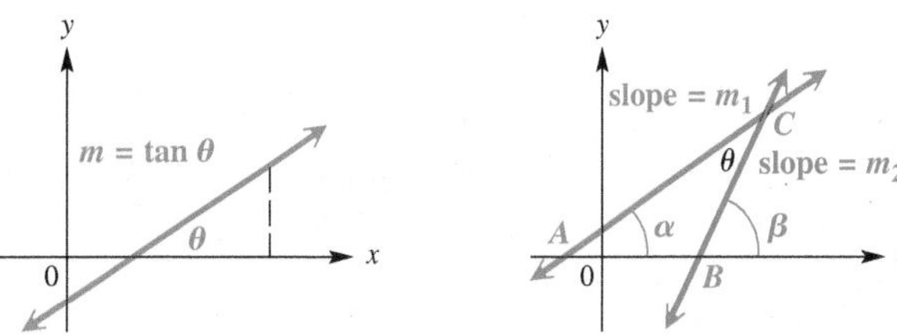

77. In terms of β, what is the measure of angle ABC?

78. Use the fact that the sum of the angles in a triangle is 180° to express θ in terms of α and β.

79. Apply the formula for $\tan(A - B)$ to obtain an expression for $\tan \theta$ in terms of $\tan \alpha$ and $\tan \beta$.

80. Replace $\tan \alpha$ with m_1 and $\tan \beta$ with m_2 to obtain $\tan \theta = \frac{m_2 - m_1}{1 + m_1 m_2}$.

Use the result from ***Exercise 80*** *to find the acute angle between each pair of lines. (Note that the tangent of the angle will be positive.) Use a calculator and round to the nearest tenth of a degree.*

81. $x + y = 9, \quad 2x + y = -1$

82. $5x - 2y + 4 = 0, \quad 3x + 5y = 6$

Chapter 5 Quiz (Sections 5.1–5.4)

[5.1]
1. $\cos\theta = \frac{24}{25}$; $\tan\theta = -\frac{7}{24}$; $\cot\theta = -\frac{24}{7}$; $\sec\theta = \frac{25}{24}$; $\csc\theta = -\frac{25}{7}$
2. $\frac{\cos^2 x + 1}{\sin^2 x}$

[5.4]
3. $\frac{-\sqrt{6} - \sqrt{2}}{4}$

[5.3]
4. $-\cos\theta$

[5.3, 5.4]
5. (a) $-\frac{16}{65}$ (b) $-\frac{63}{65}$ (c) III

[5.4]
6. $\frac{-1 + \tan x}{1 + \tan x}$

1. If $\sin\theta = -\frac{7}{25}$ and θ is in quadrant IV, find the remaining five trigonometric function values of θ.

2. Express $\cot^2 x + \csc^2 x$ in terms of $\sin x$ and $\cos x$, and simplify.

3. Find the exact value of $\sin\left(-\frac{7\pi}{12}\right)$.

4. Express $\cos(180° - \theta)$ as a function of θ alone.

5. If $\cos A = \frac{3}{5}$, $\sin B = -\frac{5}{13}$, $0 < A < \frac{\pi}{2}$, and $\pi < B < \frac{3\pi}{2}$, find each of the following.

(a) $\cos(A + B)$ **(b)** $\sin(A + B)$ **(c)** the quadrant of $A + B$

6. Express $\tan\left(\frac{3\pi}{4} + x\right)$ as a function of x alone.

Verify that each equation is an identity.

7. $\dfrac{1 + \sin\theta}{\cot^2\theta} = \dfrac{\sin\theta}{\csc\theta - 1}$

8. $\sin\left(\dfrac{\pi}{3} + \theta\right) - \sin\left(\dfrac{\pi}{3} - \theta\right) = \sin\theta$

9. $\dfrac{\sin^2\theta - \cos^2\theta}{\sin^4\theta - \cos^4\theta} = 1$

10. $\dfrac{\cos(x + y) + \cos(x - y)}{\sin(x - y) + \sin(x + y)} = \cot x$

5.5 Double-Angle Identities

- Double-Angle Identities
- An Application
- Product-to-Sum and Sum-to-Product Identities

Double-Angle Identities When $A = B$ in the identities for the sum of two angles, the **double-angle identities** result. To derive an expression for $\cos 2A$, we let $B = A$ in the identity $\cos(A + B) = \cos A \cos B - \sin A \sin B$.

$$\cos 2A = \cos(A + A) \qquad 2A = A + A$$
$$= \cos A \cos A - \sin A \sin A \qquad \text{Cosine sum identity}$$
$$\mathbf{\cos 2A = \cos^2 A - \sin^2 A} \qquad a \cdot a = a^2$$

Two other useful forms of this identity can be obtained by substituting

$$\cos^2 A = 1 - \sin^2 A \quad \text{or} \quad \sin^2 A = 1 - \cos^2 A.$$

Replacing $\cos^2 A$ with the expression $1 - \sin^2 A$ gives the following.

$$\cos 2A = \cos^2 A - \sin^2 A \qquad \text{Double-angle identity from above}$$
$$= (1 - \sin^2 A) - \sin^2 A \qquad \text{Fundamental identity}$$
$$\mathbf{\cos 2A = 1 - 2\sin^2 A} \qquad \text{Subtract.}$$

Replacing $\sin^2 A$ with $1 - \cos^2 A$ gives a third form.

$$\cos 2A = \cos^2 A - \sin^2 A \qquad \text{Double-angle identity from above}$$
$$= \cos^2 A - (1 - \cos^2 A) \qquad \text{Fundamental identity}$$
$$= \cos^2 A - 1 + \cos^2 A \qquad \text{Distributive property}$$
$$\mathbf{\cos 2A = 2\cos^2 A - 1} \qquad \text{Add.}$$

Teaching Tip Students might find it helpful to see each formula illustrated with a concrete example that they can check. For instance, show that

$$\cos 60° = \cos(2 \cdot 30°)$$
$$= \cos^2 30° - \sin^2 30°.$$

LOOKING AHEAD TO CALCULUS

The identities

$$\cos 2A = 1 - 2\sin^2 A$$

and $$\cos 2A = 2\cos^2 A - 1$$

can be rewritten as

$$\sin^2 A = \frac{1}{2}(1 - \cos 2A)$$

and $$\cos^2 A = \frac{1}{2}(1 + \cos 2A).$$

These identities are used to integrate the functions

$$f(A) = \sin^2 A$$

and $$g(A) = \cos^2 A.$$

We find $\sin 2A$ using $\sin(A + B) = \sin A \cos B + \cos A \sin B$, with $B = A$.

$$\sin 2A = \sin(A + A) \qquad 2A = A + A$$

$$= \sin A \cos A + \cos A \sin A \qquad \text{Sine sum identity}$$

$$\mathbf{\sin 2A = 2 \sin A \cos A} \qquad \text{Add.}$$

Using the identity for $\tan(A + B)$, we find $\tan 2A$.

$$\tan 2A = \tan(A + A) \qquad 2A = A + A$$

$$= \frac{\tan A + \tan A}{1 - \tan A \tan A} \qquad \text{Tangent sum identity}$$

$$\mathbf{\tan 2A = \frac{2\tan A}{1 - \tan^2 A}} \qquad \text{Simplify.}$$

NOTE In general, for a trigonometric function f,

$$f(2A) \neq 2f(A).$$

Teaching Tip Students often make the error of writing $\cos 2A$ as $2\cos A$. Refer them to the Note above.

Double-Angle Identities

$$\mathbf{\cos 2A = \cos^2 A - \sin^2 A} \qquad \mathbf{\cos 2A = 1 - 2\sin^2 A}$$

$$\mathbf{\cos 2A = 2\cos^2 A - 1} \qquad \mathbf{\sin 2A = 2\sin A \cos A}$$

$$\mathbf{\tan 2A = \frac{2\tan A}{1 - \tan^2 A}}$$

Classroom Example 1

Given $\sin\theta = \frac{8}{17}$ and $\cos\theta < 0$, find $\sin 2\theta$, $\cos 2\theta$, and $\tan 2\theta$.

Answer: $\sin 2\theta = -\frac{240}{289}$; $\cos 2\theta = \frac{161}{289}$; $\tan 2\theta = -\frac{240}{161}$

EXAMPLE 1 Finding Function Values of 2θ Given Information about θ

Given $\cos\theta = \frac{3}{5}$ and $\sin\theta < 0$, find $\sin 2\theta$, $\cos 2\theta$, and $\tan 2\theta$.

SOLUTION To find $\sin 2\theta$, we must first find the value of $\sin\theta$.

$$\sin^2\theta + \cos^2\theta = 1 \qquad \text{Pythagorean identity}$$

$$\sin^2\theta + \left(\frac{3}{5}\right)^2 = 1 \qquad \cos\theta = \tfrac{3}{5}$$

$$\sin^2\theta = \frac{16}{25} \qquad \left(\tfrac{3}{5}\right)^2 = \tfrac{9}{25};\ \text{Subtract } \tfrac{9}{25}.$$

Pay attention to signs here.

$$\sin\theta = -\frac{4}{5} \qquad \text{Take square roots. Choose the negative square root because } \sin\theta < 0.$$

Now use the double-angle identity for sine.

$$\sin 2\theta = 2\sin\theta\cos\theta = 2\left(-\frac{4}{5}\right)\left(\frac{3}{5}\right) = -\frac{24}{25} \qquad \sin\theta = -\tfrac{4}{5} \text{ and } \cos\theta = \tfrac{3}{5}$$

Now we find $\cos 2\theta$, using the first of the double-angle identities for cosine.

Any of the three forms may be used.

$$\cos 2\theta = \cos^2\theta - \sin^2\theta = \frac{9}{25} - \frac{16}{25} = -\frac{7}{25} \qquad \cos\theta = \tfrac{3}{5} \text{ and } \left(\tfrac{3}{5}\right)^2 = \tfrac{9}{25};\ \sin\theta = -\tfrac{4}{5} \text{ and } \left(-\tfrac{4}{5}\right)^2 = \tfrac{16}{25}$$

The value of $\tan 2\theta$ can be found in either of two ways. We can use the double-angle identity and the fact that $\tan\theta = \frac{\sin\theta}{\cos\theta} = \frac{-\frac{4}{5}}{\frac{3}{5}} = -\frac{4}{5} \div \frac{3}{5} = -\frac{4}{5}\cdot\frac{5}{3} = -\frac{4}{3}$.

$$\tan 2\theta = \frac{2\tan\theta}{1-\tan^2\theta} = \frac{2\left(-\frac{4}{3}\right)}{1-\left(-\frac{4}{3}\right)^2} = \frac{-\frac{8}{3}}{-\frac{7}{9}} = \frac{24}{7}$$

Alternatively, we can find $\tan 2\theta$ by finding the quotient of $\sin 2\theta$ and $\cos 2\theta$.

$$\tan 2\theta = \frac{\sin 2\theta}{\cos 2\theta} = \frac{-\frac{24}{25}}{-\frac{7}{25}} = \frac{24}{7} \quad \text{Same result as above}$$

Now Try Exercise 11.

Classroom Example 2
Find the values of the six trigonometric functions of θ given $\cos 2\theta = -\frac{12}{13}$ and $180° < \theta < 270°$.

Answer:

$\sin\theta = -\frac{5\sqrt{26}}{26}$; $\cos\theta = -\frac{\sqrt{26}}{26}$;
$\tan\theta = 5$; $\cot\theta = \frac{1}{5}$;
$\sec\theta = -\sqrt{26}$; $\csc\theta = -\frac{\sqrt{26}}{5}$

Teaching Tip Just as

$$\cos(A+B) \neq \cos A + \cos B,$$

caution students not to simplify $\cos 2A = \frac{4}{5}$ in **Example 2** as $\cos A = \frac{2}{5}$. Show that

$$\cos 2A = 2\cos^2 A - 1$$

can also be used to work **Example 2.**

EXAMPLE 2 Finding Function Values of θ Given Information about 2θ

Find the values of the six trigonometric functions of θ given $\cos 2\theta = \frac{4}{5}$ and $90° < \theta < 180°$.

SOLUTION We must obtain a trigonometric function value of θ alone.

$$\cos 2\theta = 1 - 2\sin^2\theta \quad \text{Double-angle identity}$$

$$\frac{4}{5} = 1 - 2\sin^2\theta \quad \cos 2\theta = \frac{4}{5}$$

$$-\frac{1}{5} = -2\sin^2\theta \quad \text{Subtract 1 from each side.}$$

$$\frac{1}{10} = \sin^2\theta \quad \text{Multiply by } -\frac{1}{2}.$$

$$\sin\theta = \sqrt{\frac{1}{10}} \quad \text{Take square roots. Choose the positive square root because } \theta \text{ terminates in quadrant II.}$$

$$\sin\theta = \frac{1}{\sqrt{10}}\cdot\frac{\sqrt{10}}{\sqrt{10}} \quad \text{Quotient rule for radicals; rationalize the denominator.}$$

$$\sin\theta = \frac{\sqrt{10}}{10} \quad \sqrt{a}\cdot\sqrt{a} = a$$

Now find values of $\cos\theta$ and $\tan\theta$ by sketching and labeling a right triangle in quadrant II. Because $\sin\theta = \frac{1}{\sqrt{10}}$, the triangle in **Figure 7** is labeled accordingly. The Pythagorean theorem is used to find the remaining leg.

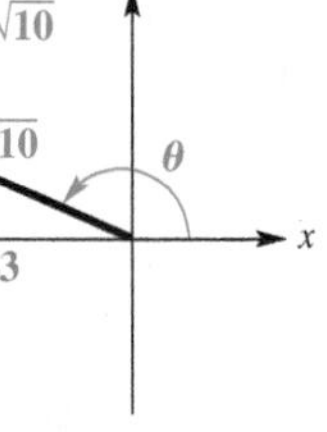

Figure 7

$$\cos\theta = \frac{-3}{\sqrt{10}} = -\frac{3\sqrt{10}}{10} \quad \text{and} \quad \tan\theta = \frac{1}{-3} = -\frac{1}{3} \quad \cos\theta = \frac{x}{r} \text{ and } \tan\theta = \frac{y}{x}$$

We find the other three functions using reciprocals.

$$\csc\theta = \frac{1}{\sin\theta} = \sqrt{10}, \quad \sec\theta = \frac{1}{\cos\theta} = -\frac{\sqrt{10}}{3}, \quad \cot\theta = \frac{1}{\tan\theta} = -3$$

Now Try Exercise 15.

Classroom Example 3
Verify that the following equation is an identity.

$$\cos^4 \beta - \sin^4 \beta = \cos 2\beta$$

Answer: The verification appears in the *Solutions to Classroom Examples.*

EXAMPLE 3 Verifying an Identity

Verify that the following equation is an identity.

$$\cot x \sin 2x = 1 + \cos 2x$$

SOLUTION We start by working on the left side, writing all functions in terms of sine and cosine and then simplifying the result.

$$\cot x \sin 2x = \frac{\cos x}{\sin x} \cdot \sin 2x \qquad \text{Quotient identity}$$

$$= \frac{\cos x}{\sin x}(2 \sin x \cos x) \qquad \text{Double-angle identity}$$

$$= 2 \cos^2 x \qquad \text{Multiply.}$$

$$= 1 + \cos 2x \qquad \cos 2x = 2\cos^2 x - 1,\text{ so } 2\cos^2 x = 1 + \cos 2x.$$

Be able to recognize alternative forms of identities.

✓ **Now Try Exercise 17.**

Classroom Example 4
Simplify each expression.
(a) $2 \cos^2 5x - 1$
(b) $\sin 165° \cos 165°$

Answers:
(a) $\cos 10x$ (b) $-\frac{1}{4}$

EXAMPLE 4 Simplifying Expressions Using Double-Angle Identities

Simplify each expression.

(a) $\cos^2 7x - \sin^2 7x$ **(b)** $\sin 15° \cos 15°$

SOLUTION

(a) This expression suggests one of the double-angle identities for cosine: $\cos 2A = \cos^2 A - \sin^2 A$. Substitute $7x$ for A.

$$\cos^2 7x - \sin^2 7x = \cos 2(7x) = \cos 14x$$

(b) If the expression $\sin 15° \cos 15°$ were

$$2 \sin 15° \cos 15°,$$

we could apply the identity for $\sin 2A$ directly because $\sin 2A = 2 \sin A \cos A$.

$$\sin 15° \cos 15°$$

$$= \frac{1}{2}(2) \sin 15° \cos 15° \qquad \text{Multiply by 1 in the form } \tfrac{1}{2}(2).$$

This is not an obvious way to begin, but it is indeed valid.

$$= \frac{1}{2}(2 \sin 15° \cos 15°) \qquad \text{Associative property}$$

$$= \frac{1}{2}\sin(2 \cdot 15°) \qquad 2 \sin A \cos A = \sin 2A,\text{ with } A = 15°$$

$$= \frac{1}{2}\sin 30° \qquad \text{Multiply.}$$

$$= \frac{1}{2} \cdot \frac{1}{2} \qquad \sin 30° = \tfrac{1}{2}$$

$$= \frac{1}{4} \qquad \text{Multiply.}$$

✓ **Now Try Exercises 37 and 39.**

Identities involving larger multiples of the variable can be derived by repeated use of the double-angle identities and other identities.

Classroom Example 5
Write $\cos 3x$ in terms of $\cos x$.

Answer:
$\cos 3x = 4\cos^3 x - 3\cos x$

EXAMPLE 5 **Deriving a Multiple-Angle Identity**

Write $\sin 3x$ in terms of $\sin x$.

SOLUTION

$$\begin{aligned}
&\sin 3x \\
&= \sin(2x + x) && \text{Use the simple fact that } 3 = 2 + 1 \text{ here.} \quad 3x = 2x + x \\
&= \sin 2x \cos x + \cos 2x \sin x && \text{Sine sum identity} \\
&= (2\sin x \cos x)\cos x + (\cos^2 x - \sin^2 x)\sin x && \text{Double-angle identities} \\
&= 2\sin x \cos^2 x + \cos^2 x \sin x - \sin^3 x && \text{Multiply.} \\
&= 2\sin x(1 - \sin^2 x) + (1 - \sin^2 x)\sin x - \sin^3 x && \cos^2 x = 1 - \sin^2 x \\
&= 2\sin x - 2\sin^3 x + \sin x - \sin^3 x - \sin^3 x && \text{Distributive property} \\
&= 3\sin x - 4\sin^3 x && \text{Combine like terms.}
\end{aligned}$$

✔ **Now Try Exercise 49.**

An Application

Classroom Example 6
Refer to **Example 6.** The voltage V in a typical European outlet can be expressed by the function

$$V(t) = 311 \sin 100\pi t.$$

Find the maximum wattage of a European light bulb with $R = 686$.

Answer: 141 watts

EXAMPLE 6 **Determining Wattage Consumption**

If a toaster is plugged into a common household outlet, the wattage consumed is not constant. Instead, it varies at a high frequency according to the model

$$W = \frac{V^2}{R},$$

where V is the voltage and R is a constant that measures the resistance of the toaster in ohms. (*Source*: Bell, D., *Fundamentals of Electric Circuits,* Fourth Edition, Prentice-Hall.)

Graph the wattage W consumed by a toaster with $R = 15$ and $V = 163 \sin 120\pi t$ in the window $[0, 0.05]$ by $[-500, 2000]$. How many oscillations are there?

SOLUTION Substituting the given values into the wattage equation gives

$$W = \frac{V^2}{R} = \frac{(163 \sin 120\pi t)^2}{15}.$$

To determine the range of W, we note that $\sin 120\pi t$ has maximum value 1, so the expression for W has maximum value $\frac{163^2}{15} \approx 1771$. The minimum value is 0. The graph in **Figure 8** shows that there are six oscillations.

For $x = t$,
$W(t) = \frac{(163 \sin 120\pi t)^2}{15}$

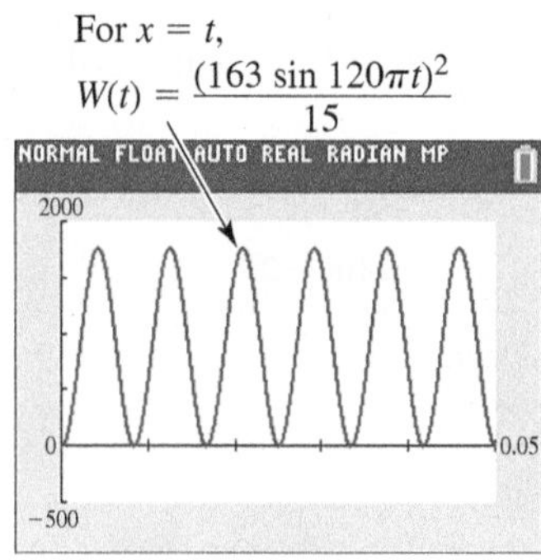

Figure 8

✔ **Now Try Exercise 69.**

Product-to-Sum and Sum-to-Product Identities We can add the corresponding sides of the identities for $\cos(A + B)$ and $\cos(A - B)$ to derive a product-to-sum identity that is useful in calculus.

$$\begin{aligned}
\cos(A + B) &= \cos A \cos B - \sin A \sin B \\
\cos(A - B) &= \cos A \cos B + \sin A \sin B \\
\hline
\cos(A + B) + \cos(A - B) &= 2\cos A \cos B \qquad \text{Add.}
\end{aligned}$$

$$\mathbf{\cos A \cos B = \frac{1}{2}[\cos(A + B) + \cos(A - B)]}$$

LOOKING AHEAD TO CALCULUS

The product-to-sum identities are used in calculus to find **integrals** of functions that are products of trigonometric functions. The classic calculus text by Earl Swokowski includes the following example:

Evaluate $\int \cos 5x \cos 3x \, dx$.

The first solution line reads: "We may write

$\cos 5x \cos 3x = \frac{1}{2}[\cos 8x + \cos 2x]$."

Similarly, subtracting $\cos(A + B)$ from $\cos(A - B)$ gives

$$\sin A \sin B = \frac{1}{2}[\cos(A - B) - \cos(A + B)].$$

Using the identities for $\sin(A + B)$ and $\sin(A - B)$ in the same way, we obtain two more identities. Those and the previous ones are now summarized.

Product-to-Sum Identities

$$\cos A \cos B = \frac{1}{2}[\cos(A + B) + \cos(A - B)]$$

$$\sin A \sin B = \frac{1}{2}[\cos(A - B) - \cos(A + B)]$$

$$\sin A \cos B = \frac{1}{2}[\sin(A + B) + \sin(A - B)]$$

$$\cos A \sin B = \frac{1}{2}[\sin(A + B) - \sin(A - B)]$$

Classroom Example 7
Write 6 sin 40° sin 15° as the sum or difference of two functions.

Answer: 3 cos 25° − 3 cos 55°

EXAMPLE 7 Using a Product-to-Sum Identity

Write 4 cos 75° sin 25° as the sum or difference of two functions.

SOLUTION

$4 \cos 75° \sin 25°$

$= 4\left[\frac{1}{2}(\sin(75° + 25°) - \sin(75° - 25°))\right]$ Use the identity for $\cos A \sin B$, with $A = 75°$ and $B = 25°$.

$= 2 \sin 100° - 2 \sin 50°$ Simplify.

Now Try Exercise 57.

We can transform the product-to-sum identities into equivalent useful forms—the sum-to-product identities—using substitution. Consider the product-to-sum identity for $\sin A \cos B$.

$$\sin A \cos B = \frac{1}{2}[\sin(A + B) + \sin(A - B)]$$ Product-to-sum identity

Let $u = A + B$ and $v = A - B$. Use substitution variables to write the product-to-sum identity in terms of u and v.

Then $u + v = 2A$ and $u - v = 2B$,

so $A = \frac{u + v}{2}$ and $B = \frac{u - v}{2}$.

$$\sin\left(\frac{u + v}{2}\right)\cos\left(\frac{u - v}{2}\right) = \frac{1}{2}(\sin u + \sin v)$$ Substitute.

$$\sin u + \sin v = 2 \sin\left(\frac{u + v}{2}\right)\cos\left(\frac{u - v}{2}\right)$$ Multiply by 2. Interchange sides.

The other three sum-to-product identities are derived using the same substitutions into the other three product-to-sum formulas.

Sum-to-Product Identities

$$\sin A + \sin B = 2\sin\left(\frac{A+B}{2}\right)\cos\left(\frac{A-B}{2}\right)$$

$$\sin A - \sin B = 2\cos\left(\frac{A+B}{2}\right)\sin\left(\frac{A-B}{2}\right)$$

$$\cos A + \cos B = 2\cos\left(\frac{A+B}{2}\right)\cos\left(\frac{A-B}{2}\right)$$

$$\cos A - \cos B = -2\sin\left(\frac{A+B}{2}\right)\sin\left(\frac{A-B}{2}\right)$$

Classroom Example 8
Write $\cos 3\theta + \cos 7\theta$ as a product of two functions.

Answer: $2\cos 5\theta \cos 2\theta$

EXAMPLE 8 Using a Sum-to-Product Identity

Write $\sin 2\theta - \sin 4\theta$ as a product of two functions.

SOLUTION $\sin 2\theta - \sin 4\theta$

$= 2\cos\left(\frac{2\theta + 4\theta}{2}\right)\sin\left(\frac{2\theta - 4\theta}{2}\right)$ — Use the identity for $\sin A - \sin B$, with $A = 2\theta$ and $B = 4\theta$.

$= 2\cos\frac{6\theta}{2}\sin\left(\frac{-2\theta}{2}\right)$ — Simplify the numerators.

$= 2\cos 3\theta \sin(-\theta)$ — Divide.

$= -2\cos 3\theta \sin\theta$ — $\sin(-\theta) = -\sin\theta$

✔ **Now Try Exercise 63.**

5.5 Exercises

1. C 2. E
3. B 4. A
5. F 6. D

7. $\cos 2\theta = \frac{17}{25}$; $\sin 2\theta = -\frac{4\sqrt{21}}{25}$
8. $\cos 2\theta = \frac{119}{169}$; $\sin 2\theta = -\frac{120}{169}$
9. $\cos 2x = -\frac{3}{5}$; $\sin 2x = \frac{4}{5}$
10. $\cos 2x = -\frac{8}{17}$; $\sin 2x = \frac{15}{17}$
11. $\cos 2\theta = \frac{39}{49}$; $\sin 2\theta = -\frac{4\sqrt{55}}{49}$
12. $\cos 2\theta = -\frac{19}{25}$; $\sin 2\theta = \frac{2\sqrt{66}}{25}$
13. $\cos\theta = \frac{2\sqrt{5}}{5}$; $\sin\theta = \frac{\sqrt{5}}{5}$
14. $\cos\theta = -\frac{\sqrt{14}}{4}$; $\sin\theta = -\frac{\sqrt{2}}{4}$

CONCEPT PREVIEW *Match each expression in Column I with its value in Column II.*

I		II	
1. $2\cos^2 15° - 1$	2. $\frac{2\tan 15°}{1 - \tan^2 15°}$	A. $\frac{1}{2}$	B. $\frac{\sqrt{2}}{2}$
3. $2\sin 22.5° \cos 22.5°$	4. $\cos^2\frac{\pi}{6} - \sin^2\frac{\pi}{6}$	C. $\frac{\sqrt{3}}{2}$	D. $-\sqrt{3}$
5. $4\sin\frac{\pi}{3}\cos\frac{\pi}{3}$	6. $\frac{2\tan\frac{\pi}{3}}{1 - \tan^2\frac{\pi}{3}}$	E. $\frac{\sqrt{3}}{3}$	F. $\sqrt{3}$

Find values of the sine and cosine functions for each angle measure. ***See Examples 1 and 2.***

7. 2θ, given $\sin\theta = \frac{2}{5}$ and $\cos\theta < 0$
8. 2θ, given $\cos\theta = -\frac{12}{13}$ and $\sin\theta > 0$
9. $2x$, given $\tan x = 2$ and $\cos x > 0$
10. $2x$, given $\tan x = \frac{5}{3}$ and $\sin x < 0$
11. 2θ, given $\sin\theta = -\frac{\sqrt{5}}{7}$ and $\cos\theta > 0$
12. 2θ, given $\cos\theta = \frac{\sqrt{3}}{5}$ and $\sin\theta > 0$
13. θ, given $\cos 2\theta = \frac{3}{5}$ and θ terminates in quadrant I
14. θ, given $\cos 2\theta = \frac{3}{4}$ and θ terminates in quadrant III

15. $\cos\theta = -\frac{\sqrt{42}}{12}$; $\sin\theta = \frac{\sqrt{102}}{12}$

16. $\cos\theta = -\frac{\sqrt{30}}{6}$; $\sin\theta = \frac{\sqrt{6}}{6}$

37. $\frac{\sqrt{3}}{2}$ 38. $\frac{\sqrt{3}}{3}$

39. $\frac{\sqrt{3}}{2}$ 40. $\frac{\sqrt{2}}{2}$

41. $-\frac{\sqrt{2}}{2}$ 42. $\frac{\sqrt{2}}{4}$

43. $\frac{1}{2}\tan 102°$ 44. $\frac{1}{4}\tan 68°$

45. $\frac{1}{4}\cos 94.2°$ 46. $\frac{1}{16}\sin 59°$

47. $-\cos\frac{4\pi}{5}$ 48. $\cos 4x$

49. $\sin 4x = 4\sin x\cos^3 x - 4\sin^3 x\cos x$

50. $\cos 3x = 4\cos^3 x - 3\cos x$

51. $\tan 3x = \frac{3\tan x - \tan^3 x}{1 - 3\tan^2 x}$

52. $\cos 4x = 8\cos^4 x - 8\cos^2 x + 1$

53. $\cos^4 x - \sin^4 x = \cos 2x$

54. $\frac{4\tan x\cos^2 x - 2\tan x}{1 - \tan^2 x} = \sin 2x$

55. $\frac{2\tan x}{2 - \sec^2 x} = \tan 2x$

56. $\frac{\cot^2 x - 1}{2\cot x} = \cot 2x$

57. $\sin 160° - \sin 44°$

58. $\sin 225° + \sin 55°$

59. $\sin\frac{\pi}{2} - \sin\frac{\pi}{6}$

60. $\frac{5}{2}\cos 5x + \frac{5}{2}\cos x$

61. $3\cos x - 3\cos 9x$

62. $4\cos 2x - 4\cos 16x$

15. θ, given $\cos 2\theta = -\frac{5}{12}$ and $90° < \theta < 180°$

16. θ, given $\cos 2\theta = \frac{2}{3}$ and $90° < \theta < 180°$

Verify that each equation is an identity. ***See Example 3.***

17. $(\sin x + \cos x)^2 = \sin 2x + 1$

18. $\sec 2x = \dfrac{\sec^2 x + \sec^4 x}{2 + \sec^2 x - \sec^4 x}$

19. $(\cos 2x + \sin 2x)^2 = 1 + \sin 4x$

20. $(\cos 2x - \sin 2x)^2 = 1 - \sin 4x$

21. $\tan 8\theta - \tan 8\theta\tan^2 4\theta = 2\tan 4\theta$

22. $\sin 2x = \dfrac{2\tan x}{1 + \tan^2 x}$

23. $\cos 2\theta = \dfrac{2 - \sec^2\theta}{\sec^2\theta}$

24. $\tan 2\theta = \dfrac{-2\tan\theta}{\sec^2\theta - 2}$

25. $\sin 4x = 4\sin x\cos x\cos 2x$

26. $\dfrac{1 + \cos 2x}{\sin 2x} = \cot x$

27. $\dfrac{2\cos 2\theta}{\sin 2\theta} = \cot\theta - \tan\theta$

28. $\cot 4\theta = \dfrac{1 - \tan^2 2\theta}{2\tan 2\theta}$

29. $\tan x + \cot x = 2\csc 2x$

30. $\cos 2x = \dfrac{1 - \tan^2 x}{1 + \tan^2 x}$

31. $1 + \tan x\tan 2x = \sec 2x$

32. $\dfrac{\cot A - \tan A}{\cot A + \tan A} = \cos 2A$

33. $\sin 2A\cos 2A = \sin 2A - 4\sin^3 A\cos A$

34. $\sin 4x = 4\sin x\cos x - 8\sin^3 x\cos x$

35. $\tan(\theta - 45°) + \tan(\theta + 45°) = 2\tan 2\theta$

36. $\cot\theta\tan(\theta + \pi) - \sin(\pi - \theta)\cos\left(\dfrac{\pi}{2} - \theta\right) = \cos^2\theta$

Simplify each expression. ***See Example 4.***

37. $\cos^2 15° - \sin^2 15°$

38. $\dfrac{2\tan 15°}{1 - \tan^2 15°}$

39. $1 - 2\sin^2 15°$

40. $1 - 2\sin^2 22\dfrac{1}{2}°$

41. $2\cos^2 67\dfrac{1}{2}° - 1$

42. $\cos^2\dfrac{\pi}{8} - \dfrac{1}{2}$

43. $\dfrac{\tan 51°}{1 - \tan^2 51°}$

44. $\dfrac{\tan 34°}{2(1 - \tan^2 34°)}$

45. $\dfrac{1}{4} - \dfrac{1}{2}\sin^2 47.1°$

46. $\dfrac{1}{8}\sin 29.5°\cos 29.5°$

47. $\sin^2\dfrac{2\pi}{5} - \cos^2\dfrac{2\pi}{5}$

48. $\cos^2 2x - \sin^2 2x$

Express each function as a trigonometric function of x. ***See Example 5.***

49. $\sin 4x$ **50.** $\cos 3x$ **51.** $\tan 3x$ **52.** $\cos 4x$

Graph each expression and use the graph to make a conjecture, predicting what might be an identity. Then verify your conjecture algebraically.

53. $\cos^4 x - \sin^4 x$

54. $\dfrac{4\tan x\cos^2 x - 2\tan x}{1 - \tan^2 x}$

55. $\dfrac{2\tan x}{2 - \sec^2 x}$

56. $\dfrac{\cot^2 x - 1}{2\cot x}$

Write each expression as a sum or difference of trigonometric functions. ***See Example 7.***

57. $2\sin 58°\cos 102°$

58. $2\cos 85°\sin 140°$

59. $2\sin\dfrac{\pi}{6}\cos\dfrac{\pi}{3}$

60. $5\cos 3x\cos 2x$

61. $6\sin 4x\sin 5x$

62. $8\sin 7x\sin 9x$

63. $-2 \sin 3x \sin x$
64. $2 \cos 6.5x \cos 1.5x$
65. $-2 \sin 11.5° \cos 36.5°$
66. $2 \cos 98.5° \sin 3.5°$
67. $2 \cos 6x \cos 2x$
68. $2 \cos 6x \sin 3x$

69. $a = -885.6$; $c = 885.6$; $\omega = 240\pi$
70. (a) For $x = t$, $W = VI = (163 \sin 120\pi t)(1.23 \sin 120\pi t)$

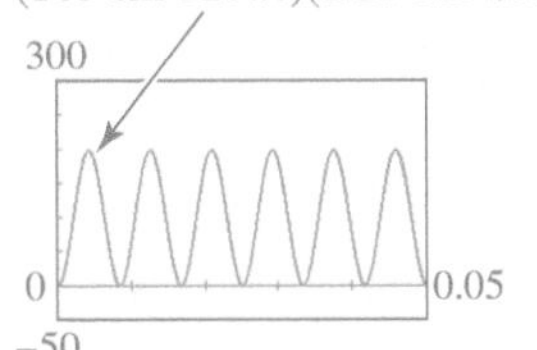

(b) maximum: 200.49 watts; minimum: 0 watts
(c) $a = -100.245$; $\omega = 240\pi$; $c = 100.245$
(e) 100 watts

Write each expression as a product of trigonometric functions. ***See Example 8.***

63. $\cos 4x - \cos 2x$	**64.** $\cos 5x + \cos 8x$	**65.** $\sin 25° + \sin(-48°)$
66. $\sin 102° - \sin 95°$	**67.** $\cos 4x + \cos 8x$	**68.** $\sin 9x - \sin 3x$

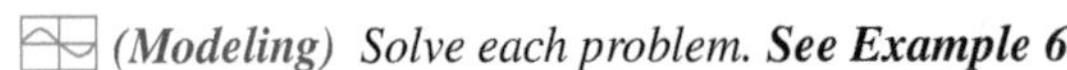

(Modeling) Solve each problem. ***See Example 6.***

69. ***Wattage Consumption*** Use the identity $\cos 2\theta = 1 - 2 \sin^2 \theta$ to determine values of a, c, and ω so that the equation

$$W = \frac{(163 \sin 120\pi t)^2}{15} \quad \text{becomes} \quad W = a \cos(\omega t) + c.$$

Round to the nearest tenth as necessary. Check by graphing both expressions for W on the same coordinate axes.

70. ***Amperage, Wattage, and Voltage*** Amperage is a measure of the amount of electricity that is moving through a circuit, whereas voltage is a measure of the force pushing the electricity. The wattage W consumed by an electrical device can be determined by calculating the product of the amperage I and voltage V. (*Source*: Wilcox, G. and C. Hesselberth, *Electricity for Engineering Technology*, Allyn & Bacon.)

(a) A household circuit has voltage

$$V = 163 \sin 120\pi t$$

when an incandescent light bulb is turned on with amperage

$$I = 1.23 \sin 120\pi t.$$

Graph the wattage $W = VI$ consumed by the light bulb in the window $[0, 0.05]$ by $[-50, 300]$.

(b) Determine the maximum and minimum wattages used by the light bulb.

(c) Use identities to determine values for a, c, and ω so that $W = a \cos(\omega t) + c$.

(d) Check by graphing both expressions for W on the same coordinate axes.

(e) Use the graph to estimate the average wattage used by the light. For how many watts (to the nearest integer) would this incandescent light bulb be rated?

5.6 Half-Angle Identities

- Half-Angle Identities
- Applications of the Half-Angle Identities
- Verifying an Identity

Half-Angle Identities From alternative forms of the identity for $\cos 2A$, we derive identities for $\sin \frac{A}{2}$, $\cos \frac{A}{2}$, and $\tan \frac{A}{2}$, known as **half-angle identities.**

We derive the identity for $\sin \frac{A}{2}$ as follows.

$$\cos 2x = 1 - 2 \sin^2 x \qquad \text{Cosine double-angle identity}$$

$$2 \sin^2 x = 1 - \cos 2x \qquad \text{Add } 2 \sin^2 x \text{ and subtract } \cos 2x.$$

Remember both the positive and negative square roots.

$$\sin x = \pm\sqrt{\frac{1 - \cos 2x}{2}} \qquad \text{Divide by 2 and take square roots.}$$

$$\mathbf{\sin \frac{A}{2} = \pm\sqrt{\frac{1 - \cos A}{2}}} \qquad \text{Let } 2x = A, \text{ so } x = \tfrac{A}{2}. \text{ Substitute.}$$

The $\pm$ symbol indicates that the appropriate sign is chosen depending on the quadrant of $\frac{A}{2}$. For example, if $\frac{A}{2}$ is a quadrant III angle, we choose the negative sign because the sine function is negative in quadrant III.

We derive the identity for $\cos \frac{A}{2}$ using another double-angle identity.

$$\cos 2x = 2\cos^2 x - 1 \qquad \text{Cosine double-angle identity}$$

$$1 + \cos 2x = 2\cos^2 x \qquad \text{Add 1.}$$

$$\cos^2 x = \frac{1 + \cos 2x}{2} \qquad \text{Rewrite and divide by 2.}$$

$$\cos x = \pm\sqrt{\frac{1 + \cos 2x}{2}} \qquad \text{Take square roots.}$$

$$\mathbf{\cos \frac{A}{2} = \pm\sqrt{\frac{1 + \cos A}{2}}} \qquad \text{Replace } x \text{ with } \tfrac{A}{2}.$$

An identity for $\tan \frac{A}{2}$ comes from the identities for $\sin \frac{A}{2}$ and $\cos \frac{A}{2}$.

$$\tan \frac{A}{2} = \frac{\sin \frac{A}{2}}{\cos \frac{A}{2}} = \frac{\pm\sqrt{\frac{1 - \cos A}{2}}}{\pm\sqrt{\frac{1 + \cos A}{2}}} = \pm\sqrt{\frac{1 - \cos A}{1 + \cos A}}$$

We derive an alternative identity for $\tan \frac{A}{2}$ using double-angle identities.

$$\tan \frac{A}{2} = \frac{\sin \frac{A}{2}}{\cos \frac{A}{2}} = \frac{2\sin \frac{A}{2}\cos \frac{A}{2}}{2\cos^2 \frac{A}{2}} \qquad \text{Multiply by } 2\cos \tfrac{A}{2} \text{ in numerator and denominator.}$$

$$= \frac{\sin 2\left(\frac{A}{2}\right)}{1 + \cos 2\left(\frac{A}{2}\right)} \qquad \text{Double-angle identities}$$

$$\tan \frac{A}{2} = \frac{\sin A}{1 + \cos A} \qquad \text{Simplify.}$$

Teaching Tip Show students how to derive

$$\tan \frac{A}{2} = \frac{1 - \cos A}{\sin A}$$

by multiplying the numerator and denominator of

$$\tan \frac{A}{2} = \frac{\sin A}{1 + \cos A}$$

by $1 - \cos A$.

From the identity $\tan \frac{A}{2} = \frac{\sin A}{1 + \cos A}$, we can also derive an equivalent identity.

$$\tan \frac{A}{2} = \frac{1 - \cos A}{\sin A}$$

Half-Angle Identities

In the following identities, the $\pm$ symbol indicates that the sign is chosen based on the function under consideration and the quadrant of $\frac{A}{2}$.

$$\cos \frac{A}{2} = \pm\sqrt{\frac{1 + \cos A}{2}} \qquad \sin \frac{A}{2} = \pm\sqrt{\frac{1 - \cos A}{2}}$$

$$\tan \frac{A}{2} = \pm\sqrt{\frac{1 - \cos A}{1 + \cos A}} \qquad \tan \frac{A}{2} = \frac{\sin A}{1 + \cos A} \qquad \tan \frac{A}{2} = \frac{1 - \cos A}{\sin A}$$

Teaching Tip Point out that the first identity for $\tan \frac{A}{2}$ follows directly from the cosine and sine half-angle identities. However, the other two identities for $\tan \frac{A}{2}$ have the advantage of requiring no sign choice.

Three of these identities require a sign choice. When using these identities, select the plus or minus sign according to the quadrant in which $\frac{A}{2}$ terminates. For example, if an angle $A = 324°$, then $\frac{A}{2} = 162°$, which lies in quadrant II. So when $A = 324°$, $\cos \frac{A}{2}$ and $\tan \frac{A}{2}$ are negative, and $\sin \frac{A}{2}$ is positive.

Classroom Example 1
Find the exact value of sin 22.5° using the half-angle identity for sine.

Answer: $\frac{\sqrt{2-\sqrt{2}}}{2}$

Teaching Tip Have students compare the value of cos 15° in **Example 1** to the value

$$\frac{\sqrt{6}+\sqrt{2}}{4}$$

found earlier in this chapter, where we used the identity for the cosine of the difference of two angles. Although the expressions look completely different, they are equal, as suggested by a calculator approximation for both, 0.96592583.

Classroom Example 2
Find the exact value of tan 75° using the identity $\tan\frac{A}{2}=\frac{\sin A}{1+\cos A}$.

Answer: $2+\sqrt{3}$

Classroom Example 3
Given $\cos s=-\frac{3}{7}$, with $\pi<s<\frac{3\pi}{2}$, find $\sin\frac{s}{2}$, $\cos\frac{s}{2}$, and $\tan\frac{s}{2}$.

Answer:
$\sin\frac{s}{2}=\frac{\sqrt{35}}{7}$; $\cos\frac{s}{2}=-\frac{\sqrt{14}}{7}$; $\tan\frac{s}{2}=-\frac{\sqrt{10}}{2}$

Applications of the Half-Angle Identities

EXAMPLE 1 Using a Half-Angle Identity to Find an Exact Value

Find the exact value of cos 15° using the half-angle identity for cosine.

SOLUTION
$$\cos 15^\circ=\cos\frac{30^\circ}{2}=\sqrt{\frac{1+\cos 30^\circ}{2}}$$
Choose the positive square root.

$$=\sqrt{\frac{1+\frac{\sqrt{3}}{2}}{2}}=\sqrt{\frac{\left(1+\frac{\sqrt{3}}{2}\right)\cdot 2}{2\cdot 2}}=\frac{\sqrt{2+\sqrt{3}}}{2}$$
Simplify the radicals.

✔ **Now Try Exercise 11.**

EXAMPLE 2 Using a Half-Angle Identity to Find an Exact Value

Find the exact value of tan 22.5° using the identity $\tan\frac{A}{2}=\frac{\sin A}{1+\cos A}$.

SOLUTION Because $22.5^\circ=\frac{45^\circ}{2}$, replace A with 45°.

$$\tan 22.5^\circ=\tan\frac{45^\circ}{2}=\frac{\sin 45^\circ}{1+\cos 45^\circ}=\frac{\frac{\sqrt{2}}{2}}{1+\frac{\sqrt{2}}{2}}=\frac{\frac{\sqrt{2}}{2}}{1+\frac{\sqrt{2}}{2}}\cdot\frac{2}{2}$$

$$=\frac{\sqrt{2}}{2+\sqrt{2}}=\frac{\sqrt{2}}{2+\sqrt{2}}\cdot\frac{2-\sqrt{2}}{2-\sqrt{2}}=\frac{2\sqrt{2}-2}{2}$$
Rationalize the denominator.

Factor first, and then divide out the common factor.

$$=\frac{2\left(\sqrt{2}-1\right)}{2}=\sqrt{2}-1$$

✔ **Now Try Exercise 13.**

EXAMPLE 3 Finding Function Values of $\frac{s}{2}$ Given Information about s

Given $\cos s=\frac{2}{3}$, with $\frac{3\pi}{2}<s<2\pi$, find $\sin\frac{s}{2}$, $\cos\frac{s}{2}$, and $\tan\frac{s}{2}$.

SOLUTION The angle associated with $\frac{s}{2}$ terminates in quadrant II because

$$\frac{3\pi}{2}<s<2\pi \quad\text{and}\quad \frac{3\pi}{4}<\frac{s}{2}<\pi. \qquad \text{Divide by 2.}$$

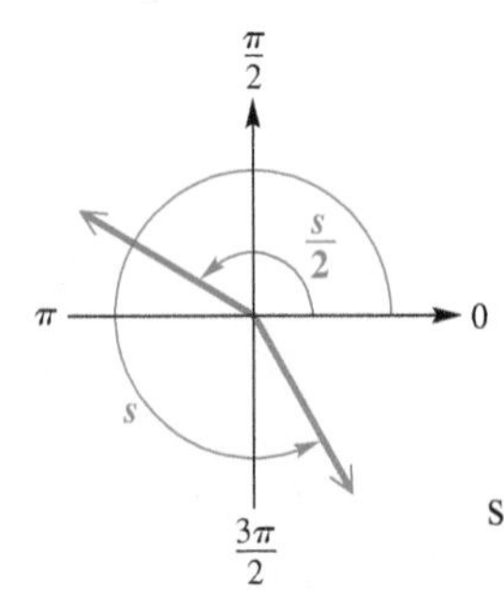

Figure 9

See **Figure 9.** In quadrant II, the values of $\cos\frac{s}{2}$ and $\tan\frac{s}{2}$ are negative and the value of $\sin\frac{s}{2}$ is positive. Use the appropriate half-angle identities and simplify.

$$\sin\frac{s}{2}=\sqrt{\frac{1-\frac{2}{3}}{2}}=\sqrt{\frac{1}{6}}=\frac{\sqrt{1}}{\sqrt{6}}\cdot\frac{\sqrt{6}}{\sqrt{6}}=\frac{\sqrt{6}}{6}$$

$$\cos\frac{s}{2}=-\sqrt{\frac{1+\frac{2}{3}}{2}}=-\sqrt{\frac{5}{6}}=-\frac{\sqrt{5}}{\sqrt{6}}\cdot\frac{\sqrt{6}}{\sqrt{6}}=-\frac{\sqrt{30}}{6}$$

Rationalize all denominators.

$$\tan\frac{s}{2}=\frac{\sin\frac{s}{2}}{\cos\frac{s}{2}}=\frac{\frac{\sqrt{6}}{6}}{-\frac{\sqrt{30}}{6}}=\frac{\sqrt{6}}{-\sqrt{30}}=-\frac{\sqrt{6}}{\sqrt{30}}\cdot\frac{\sqrt{30}}{\sqrt{30}}=-\frac{\sqrt{180}}{30}=-\frac{6\sqrt{5}}{6\cdot 5}=-\frac{\sqrt{5}}{5}$$

Notice that it is not necessary to use a half-angle identity for $\tan\frac{s}{2}$ once we find $\sin\frac{s}{2}$ and $\cos\frac{s}{2}$. However, using this identity provides an excellent check.

✔ **Now Try Exercise 19.**

Classroom Example 4
Simplify each expression.

(a) $\pm\sqrt{\dfrac{1-\cos 8x}{2}}$

(b) $\pm\sqrt{\dfrac{1-\cos 9\beta}{1+\cos 9\beta}}$

Answers:
(a) $\sin 4x$ (b) $\tan\frac{9\beta}{2}$

EXAMPLE 4 Simplifying Expressions Using Half-Angle Identities

Simplify each expression.

(a) $\pm\sqrt{\dfrac{1+\cos 12x}{2}}$ **(b)** $\dfrac{1-\cos 5\alpha}{\sin 5\alpha}$

SOLUTION

(a) This matches part of the identity for $\cos\frac{A}{2}$. Replace A with $12x$.

$$\cos\frac{A}{2} = \pm\sqrt{\frac{1+\cos A}{2}} = \pm\sqrt{\frac{1+\cos 12x}{2}} = \cos\frac{12x}{2} = \cos 6x$$

(b) Use the identity $\tan\frac{A}{2} = \frac{1-\cos A}{\sin A}$ with $A = 5\alpha$.

$$\frac{1-\cos 5\alpha}{\sin 5\alpha} = \tan\frac{5\alpha}{2}$$

✔ **Now Try Exercises 37 and 39.**

Verifying an Identity

Classroom Example 5
Verify that the following equation is an identity.

$$\tan^2\left(\frac{x}{2}\right) = \frac{\sec x + \cos x - 2}{\sec x - \cos x}$$

Answer: The verification appears in the *Solutions to Classroom Examples*.

EXAMPLE 5 Verifying an Identity

Verify that the following equation is an identity.

$$\left(\sin\frac{x}{2} + \cos\frac{x}{2}\right)^2 = 1 + \sin x$$

SOLUTION We work on the more complicated left side.

$$\left(\sin\frac{x}{2} + \cos\frac{x}{2}\right)^2$$

Remember the middle term when squaring a binomial.

$$= \sin^2\frac{x}{2} + 2\sin\frac{x}{2}\cos\frac{x}{2} + \cos^2\frac{x}{2} \quad (x+y)^2 = x^2 + 2xy + y^2$$

$$= 1 + 2\sin\frac{x}{2}\cos\frac{x}{2} \quad \sin^2\tfrac{x}{2} + \cos^2\tfrac{x}{2} = 1$$

$$= 1 + \sin 2\left(\frac{x}{2}\right) \quad 2\sin\tfrac{x}{2}\cos\tfrac{x}{2} = \sin 2\left(\tfrac{x}{2}\right)$$

$$= 1 + \sin x \quad \text{Multiply.}$$

✔ **Now Try Exercise 47.**

5.6 Exercises

1. − 2. +
3. + 4. −

CONCEPT PREVIEW *Determine whether the positive or negative square root should be selected.*

1. $\sin 195° = \pm\sqrt{\dfrac{1-\cos 390°}{2}}$

2. $\cos 58° = \pm\sqrt{\dfrac{1+\cos 116°}{2}}$

3. $\tan 225° = \pm\sqrt{\dfrac{1-\cos 450°}{1+\cos 450°}}$

4. $\sin(-10°) = \pm\sqrt{\dfrac{1-\cos(-20°)}{2}}$

5. C 6. A
7. D 8. E
9. F 10. B

11. $\frac{\sqrt{2+\sqrt{2}}}{2}$ 12. $-\frac{\sqrt{2-\sqrt{3}}}{2}$

13. $2-\sqrt{3}$ 14. $-\frac{\sqrt{2+\sqrt{3}}}{2}$

15. $-\frac{\sqrt{2+\sqrt{3}}}{2}$ 16. $\frac{\sqrt{2-\sqrt{3}}}{2}$

17. Because

$$\sin 7.5° = \sin\left(\frac{1}{2} \cdot \frac{30°}{2}\right),$$

first use the half-angle identity for sine and then use the half-angle identity for cosine within that calculation. The exact value is

$$\frac{\sqrt{2-\sqrt{2+\sqrt{3}}}}{2}.$$

19. $\frac{\sqrt{10}}{4}$ 20. $\frac{\sqrt{13}}{4}$

21. 3 22. $-\frac{\sqrt{5}}{5}$

23. $\frac{\sqrt{50-10\sqrt{5}}}{10}$ 24. $\frac{\sqrt{50-15\sqrt{10}}}{10}$

25. $-\sqrt{7}$ 26. $\frac{\sqrt{5}}{5}$

27. $\frac{\sqrt{5}}{5}$ 28. $-\frac{\sqrt{3}}{2}$

29. $-\frac{\sqrt{42}}{12}$ 30. $-\frac{\sqrt{6}}{6}$

31. 0.127 32. 2.65

33. sin 20° 34. cos 38°
35. tan 73.5° 36. cot 82.5°
37. tan 29.87° 38. tan 79.1°

CONCEPT PREVIEW *Match each expression in Column I with its value in Column II.*

I		II	
5. sin 15°	**6.** tan 15°	**A.** $2-\sqrt{3}$	**B.** $\frac{\sqrt{2-\sqrt{2}}}{2}$
7. $\cos\frac{\pi}{8}$	**8.** $\tan\left(-\frac{\pi}{8}\right)$	**C.** $\frac{\sqrt{2-\sqrt{3}}}{2}$	**D.** $\frac{\sqrt{2+\sqrt{2}}}{2}$
9. tan 67.5°	**10.** cos 67.5°	**E.** $1-\sqrt{2}$	**F.** $1+\sqrt{2}$

Use a half-angle identity to find each exact value. ***See Examples 1 and 2.***

11. sin 67.5° **12.** sin 195° **13.** tan 195°

14. cos 195° **15.** cos 165° **16.** sin 165°

17. Explain how to use identities from this section to find the exact value of sin 7.5°.

18. The half-angle identity

$$\tan\frac{A}{2} = \pm\sqrt{\frac{1-\cos A}{1+\cos A}}$$

can be used to find $\tan 22.5° = \sqrt{3-2\sqrt{2}}$, and the half-angle identity

$$\tan\frac{A}{2} = \frac{\sin A}{1+\cos A}$$

can be used to find $\tan 22.5° = \sqrt{2}-1$. Show that these answers are the same, without using a calculator. (*Hint*: If $a > 0$ and $b > 0$ and $a^2 = b^2$, then $a = b$.)

Use the given information to find each of the following. ***See Example 3.***

19. $\cos\frac{x}{2}$, given $\cos x = \frac{1}{4}$, with $0 < x < \frac{\pi}{2}$

20. $\sin\frac{x}{2}$, given $\cos x = -\frac{5}{8}$, with $\frac{\pi}{2} < x < \pi$

21. $\tan\frac{\theta}{2}$, given $\sin\theta = \frac{3}{5}$, with $90° < \theta < 180°$

22. $\cos\frac{\theta}{2}$, given $\sin\theta = -\frac{4}{5}$, with $180° < \theta < 270°$

23. $\sin\frac{x}{2}$, given $\tan x = 2$, with $0 < x < \frac{\pi}{2}$

24. $\cos\frac{x}{2}$, given $\cot x = -3$, with $\frac{\pi}{2} < x < \pi$

25. $\tan\frac{\theta}{2}$, given $\tan\theta = \frac{\sqrt{7}}{3}$, with $180° < \theta < 270°$

26. $\cot\frac{\theta}{2}$, given $\tan\theta = -\frac{\sqrt{5}}{2}$, with $90° < \theta < 180°$

27. $\sin\theta$, given $\cos 2\theta = \frac{3}{5}$ and θ terminates in quadrant I

28. $\cos\theta$, given $\cos 2\theta = \frac{1}{2}$ and θ terminates in quadrant II

29. $\cos x$, given $\cos 2x = -\frac{5}{12}$, with $\frac{\pi}{2} < x < \pi$

30. $\sin x$, given $\cos 2x = \frac{2}{3}$, with $\pi < x < \frac{3\pi}{2}$

31. *Concept Check* If $\cos x \approx 0.9682$ and $\sin x = 0.250$, then $\tan\frac{x}{2} \approx$ ______.

32. *Concept Check* If $\cos x = -0.750$ and $\sin x \approx 0.6614$, then $\tan\frac{x}{2} \approx$ ______.

Simplify each expression. ***See Example 4.***

33. $\sqrt{\frac{1-\cos 40°}{2}}$ **34.** $\sqrt{\frac{1+\cos 76°}{2}}$ **35.** $\sqrt{\frac{1-\cos 147°}{1+\cos 147°}}$

36. $\sqrt{\frac{1+\cos 165°}{1-\cos 165°}}$ **37.** $\frac{1-\cos 59.74°}{\sin 59.74°}$ **38.** $\frac{\sin 158.2°}{1+\cos 158.2°}$

39. $\cos 9x$ 40. $\cos 10\alpha$

41. $\tan 4\theta$ 42. $\tan \frac{5A}{2}$

43. $\cos \frac{x}{8}$ 44. $\sin \frac{3\theta}{10}$

54. $\cot \frac{A}{2} = \frac{1 + \cos A}{\sin A}$

55. $\frac{\sin x}{1 + \cos x} = \tan \frac{x}{2}$

56. $\frac{1 - \cos x}{\sin x} = \tan \frac{x}{2}$

57. $\frac{\tan \frac{x}{2} + \cot \frac{x}{2}}{\cot \frac{x}{2} - \tan \frac{x}{2}} = \sec x$

58. $1 - 8 \sin^2 \frac{x}{2} \cos^2 \frac{x}{2} = \cos 2x$

59. $106°$ 60. $84°$

61. 2 62. 3.9

39. $\pm\sqrt{\dfrac{1 + \cos 18x}{2}}$ 40. $\pm\sqrt{\dfrac{1 + \cos 20\alpha}{2}}$ 41. $\pm\sqrt{\dfrac{1 - \cos 8\theta}{1 + \cos 8\theta}}$

42. $\pm\sqrt{\dfrac{1 - \cos 5A}{1 + \cos 5A}}$ 43. $\pm\sqrt{\dfrac{1 + \cos \frac{x}{4}}{2}}$ 44. $\pm\sqrt{\dfrac{1 - \cos \frac{3\theta}{5}}{2}}$

Verify that each equation is an identity. ***See Example 5.***

45. $\sec^2 \dfrac{x}{2} = \dfrac{2}{1 + \cos x}$

46. $\cot^2 \dfrac{x}{2} = \dfrac{(1 + \cos x)^2}{\sin^2 x}$

47. $\sin^2 \dfrac{x}{2} = \dfrac{\tan x - \sin x}{2 \tan x}$

48. $\dfrac{\sin 2x}{2 \sin x} = \cos^2 \dfrac{x}{2} - \sin^2 \dfrac{x}{2}$

49. $\dfrac{2}{1 + \cos x} - \tan^2 \dfrac{x}{2} = 1$

50. $\tan \dfrac{\theta}{2} = \csc \theta - \cot \theta$

51. $1 - \tan^2 \dfrac{\theta}{2} = \dfrac{2 \cos \theta}{1 + \cos \theta}$

52. $\cos x = \dfrac{1 - \tan^2 \frac{x}{2}}{1 + \tan^2 \frac{x}{2}}$

53. Use the half-angle identity

$$\tan \frac{A}{2} = \frac{\sin A}{1 + \cos A}$$

to derive the equivalent identity

$$\tan \frac{A}{2} = \frac{1 - \cos A}{\sin A}$$

by multiplying both the numerator and the denominator by $1 - \cos A$.

54. Use the identity $\tan \frac{A}{2} = \frac{\sin A}{1 + \cos A}$ to determine an identity for $\cot \frac{A}{2}$.

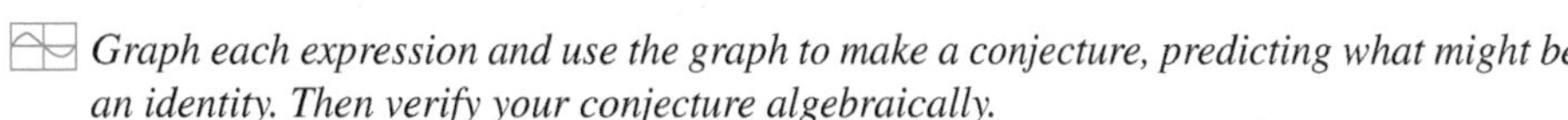

Graph each expression and use the graph to make a conjecture, predicting what might be an identity. Then verify your conjecture algebraically.

55. $\dfrac{\sin x}{1 + \cos x}$

56. $\dfrac{1 - \cos x}{\sin x}$

57. $\dfrac{\tan \frac{x}{2} + \cot \frac{x}{2}}{\cot \frac{x}{2} - \tan \frac{x}{2}}$

58. $1 - 8 \sin^2 \dfrac{x}{2} \cos^2 \dfrac{x}{2}$

(Modeling) ***Mach Number*** *An airplane flying faster than the speed of sound sends out sound waves that form a cone, as shown in the figure. The cone intersects the ground to form a* ***hyperbola.*** *As this hyperbola passes over a particular point on the ground, a sonic boom is heard at that point. If θ is the angle at the vertex of the cone, then*

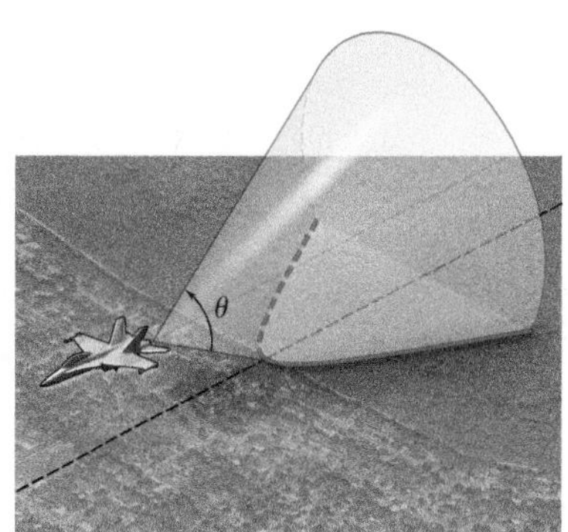

$$\sin \frac{\theta}{2} = \frac{1}{m},$$

where m is the Mach number for the speed of the plane. (We assume $m > 1$.) The Mach number is the ratio of the speed of the plane to the speed of sound. Thus, a speed of Mach 1.4 means that the plane is flying at 1.4 times the speed of sound.

In each of the following exercises, θ or m is given. Find the other value (θ to the nearest degree and m to the nearest tenth as applicable).

59. $m = \dfrac{5}{4}$ 60. $m = \dfrac{3}{2}$ 61. $\theta = 60°$ 62. $\theta = 30°$

63. (a) $\cos \frac{\theta}{2} = \frac{R - b}{R}$

(b) $\tan \frac{\theta}{4} = \frac{b}{50}$

64. 54°

65. $\frac{\sqrt{10 + 2\sqrt{5}}}{4}$

66. $\frac{(\sqrt{10 + 2\sqrt{5}})(-5 + 3\sqrt{5})}{20}$

67. $\frac{(\sqrt{10 + 2\sqrt{5}})(\sqrt{5} + 1)}{4}$

68. $\frac{(\sqrt{10 + 2\sqrt{5}})(5 - \sqrt{5})}{10}$

69. $1 + \sqrt{5}$

70. $\frac{\sqrt{5} - 1}{4}$

71. $\frac{\sqrt{10 + 2\sqrt{5}}}{4}$

72. $\frac{(\sqrt{10 + 2\sqrt{5}})(\sqrt{5} + 1)}{4}$

73. $\frac{(\sqrt{10 + 2\sqrt{5}})(-5 + 3\sqrt{5})}{20}$

74. $\frac{(\sqrt{10 + 2\sqrt{5}})(5 - \sqrt{5})}{10}$

75. $1 + \sqrt{5}$

76. $\frac{\sqrt{5} - 1}{4}$

77. They are both radii of the circle.

78. It is the supplement of a 30° angle.

79. Their sum is 180° − 150° = 30°, and they are equal.

80. $2 + \sqrt{3}$

82. $\frac{\sqrt{6} + \sqrt{2}}{4}$

83. $\frac{\sqrt{6} - \sqrt{2}}{4}$

84. $2 - \sqrt{3}$

Solve each problem.

63. *(Modeling) Railroad Curves* In the United States, circular railroad curves are designated by the **degree of curvature,** the central angle subtended by a chord of 100 ft. See the figure. (*Source:* Hay, W. W., *Railroad Engineering,* John Wiley and Sons.)

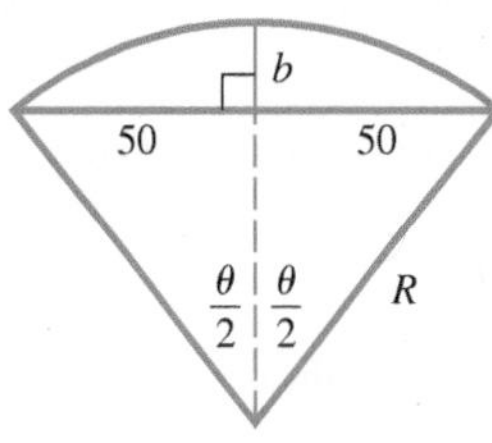

(a) Use the figure to write an expression for $\cos \frac{\theta}{2}$.

(b) Use the result of part (a) and the half-angle identity $\tan \frac{A}{2} = \frac{1 - \cos A}{\sin A}$ to write an expression for $\tan \frac{\theta}{4}$.

64. In **Exercise 63,** if $b = 12$, what is the measure of angle θ to the nearest degree?

Advanced methods of trigonometry can be used to find the following exact *value.*

$$\sin 18° = \frac{\sqrt{5} - 1}{4}$$

(See Hobson's A Treatise on Plane Trigonometry.) *Use this value and identities to find each exact value. Support answers with calculator approximations if desired.*

65. cos 18° **66.** tan 18° **67.** cot 18° **68.** sec 18°

69. csc 18° **70.** cos 72° **71.** sin 72° **72.** tan 72°

73. cot 72° **74.** csc 72° **75.** sec 72° **76.** sin 162°

Relating Concepts

For individual or collaborative investigation *(Exercises 77–84)*

These exercises use results from plane geometry to obtain exact values of the trigonometric functions of 15°.

Start with a right triangle ACB having a 60° angle at A and a 30° angle at B. Let the hypotenuse of this triangle have length 2. Extend side BC and draw a semicircle with diameter along BC extended, center at B, and radius AB. Draw segment AE. (See the figure.) Any angle inscribed in a semicircle is a right angle, so triangle EAD is a right triangle. ***Work Exercises 77–84 in order.***

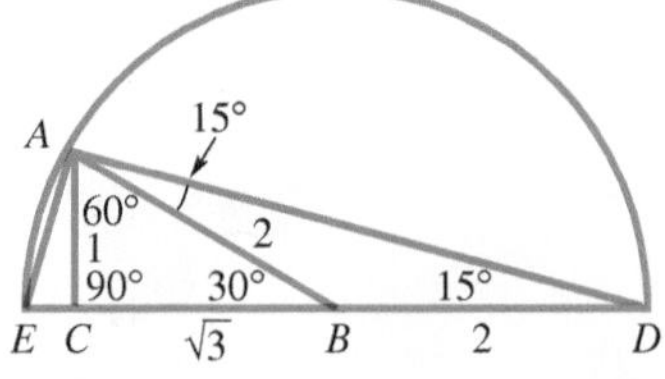

77. Why is $AB = BD$ true? Conclude that triangle ABD is isosceles.

78. Why does angle ABD have measure 150°?

79. Why do angles DAB and ADB both have measures of 15°?

80. What is the length DC?

81. Use the Pythagorean theorem to show that the length AD is $\sqrt{6} + \sqrt{2}$.

82. Use angle ADB of triangle EAD to find cos 15°.

83. Show that AE has length $\sqrt{6} - \sqrt{2}$ and find sin 15°.

84. Use triangle ACD to find tan 15°.

Summary Exercises on Verifying Trigonometric Identities

Teaching Tip Remind students that if they have trouble verifying an identity, they should move on to the next one. Their brains may continue to work on the original one subconsciously, and they may be able to come back and finish it at a later time.

These summary exercises provide practice with the various types of trigonometric identities presented in this chapter. Verify that each equation is an identity.

1. $\tan\theta + \cot\theta = \sec\theta\csc\theta$

2. $\csc\theta\cos^2\theta + \sin\theta = \csc\theta$

3. $\tan\frac{x}{2} = \csc x - \cot x$

4. $\sec(\pi - x) = -\sec x$

5. $\frac{\sin t}{1+\cos t} = \frac{1-\cos t}{\sin t}$

6. $\frac{1-\sin t}{\cos t} = \frac{1}{\sec t + \tan t}$

7. $\sin 2\theta = \frac{2\tan\theta}{1+\tan^2\theta}$

8. $\frac{2}{1+\cos x} - \tan^2\frac{x}{2} = 1$

9. $\cot\theta - \tan\theta = \frac{2\cos^2\theta - 1}{\sin\theta\cos\theta}$

10. $\frac{1}{\sec t - 1} + \frac{1}{\sec t + 1} = 2\cot t\csc t$

11. $\frac{\sin(x+y)}{\cos(x-y)} = \frac{\cot x + \cot y}{1+\cot x\cot y}$

12. $1 - \tan^2\frac{\theta}{2} = \frac{2\cos\theta}{1+\cos\theta}$

13. $\frac{\sin\theta + \tan\theta}{1+\cos\theta} = \tan\theta$

14. $\csc^4 x - \cot^4 x = \frac{1+\cos^2 x}{1-\cos^2 x}$

15. $\cos x = \frac{1-\tan^2\frac{x}{2}}{1+\tan^2\frac{x}{2}}$

16. $\cos 2x = \frac{2-\sec^2 x}{\sec^2 x}$

17. $\frac{\tan^2 t + 1}{\tan t\csc^2 t} = \tan t$

18. $\frac{\sin s}{1+\cos s} + \frac{1+\cos s}{\sin s} = 2\csc s$

19. $\tan 4\theta = \frac{2\tan 2\theta}{2-\sec^2 2\theta}$

20. $\tan\left(\frac{x}{2} + \frac{\pi}{4}\right) = \sec x + \tan x$

21. $\frac{\cot s - \tan s}{\cos s + \sin s} = \frac{\cos s - \sin s}{\sin s\cos s}$

22. $\frac{\tan\theta - \cot\theta}{\tan\theta + \cot\theta} = 1 - 2\cos^2\theta$

23. $\frac{\tan(x+y) - \tan y}{1+\tan(x+y)\tan y} = \tan x$

24. $2\cos^2\frac{x}{2}\tan x = \tan x + \sin x$

25. $\frac{\cos^4 x - \sin^4 x}{\cos^2 x} = 1 - \tan^2 x$

26. $\frac{\csc t + 1}{\csc t - 1} = (\sec t + \tan t)^2$

27. $\frac{2(\sin x - \sin^3 x)}{\cos x} = \sin 2x$

28. $\frac{1}{2}\cot\frac{x}{2} - \frac{1}{2}\tan\frac{x}{2} = \cot x$

29. $\frac{\cos(x+y) + \cos(y-x)}{\sin(x+y) - \sin(y-x)} = \cot x$

30. $\sin(60° - x) - \sin(60° + x) = -\sin x$

31. $\sin(60° + x) + \sin(60° - x) = \sqrt{3}\cos x$

32. $\sin x + \sin 3x + \sin 5x + \sin 7x = 4\cos x\cos 2x\sin 4x$

33. $\sin^3\theta + \cos^3\theta + \sin\theta\cos^2\theta + \sin^2\theta\cos\theta = \sin\theta + \cos\theta$

34. $\frac{\cos x + \sin x}{\cos x - \sin x} - \frac{\cos x - \sin x}{\cos x + \sin x} = 2\tan 2x$

Chapter 5 Test Prep

Quick Review

Concepts	Examples

5.1 Fundamental Identities

Reciprocal Identities

$$\cot\theta = \frac{1}{\tan\theta} \qquad \sec\theta = \frac{1}{\cos\theta} \qquad \csc\theta = \frac{1}{\sin\theta}$$

Quotient Identities

$$\tan\theta = \frac{\sin\theta}{\cos\theta} \qquad \cot\theta = \frac{\cos\theta}{\sin\theta}$$

Pythagorean Identities

$$\sin^2\theta + \cos^2\theta = 1 \qquad \tan^2\theta + 1 = \sec^2\theta$$

$$1 + \cot^2\theta = \csc^2\theta$$

Even-Odd Identities

$$\sin(-\theta) = -\sin\theta \quad \cos(-\theta) = \cos\theta \quad \tan(-\theta) = -\tan\theta$$

$$\csc(-\theta) = -\csc\theta \quad \sec(-\theta) = \sec\theta \quad \cot(-\theta) = -\cot\theta$$

If θ is in quadrant IV and $\sin\theta = -\frac{3}{5}$, find $\csc\theta$, $\cos\theta$, and $\sin(-\theta)$.

$$\csc\theta = \frac{1}{\sin\theta} = \frac{1}{-\frac{3}{5}} = -\frac{5}{3}$$ Reciprocal identity

$$\sin^2\theta + \cos^2\theta = 1$$ Pythagorean identity

$$\left(-\frac{3}{5}\right)^2 + \cos^2\theta = 1$$ Substitute.

$$\cos^2\theta = \frac{16}{25}$$ $\left(-\frac{3}{5}\right)^2 = \frac{9}{25}$; Subtract $\frac{9}{25}$.

$$\cos\theta = +\sqrt{\frac{16}{25}}$$ $\cos\theta$ is positive in quadrant IV.

$$\cos\theta = \frac{4}{5}$$

$$\sin(-\theta) = -\sin\theta = -\left(-\frac{3}{5}\right) = \frac{3}{5}$$

Even-odd identity

5.2 Verifying Trigonometric Identities

See the box titled Hints for Verifying Identities in **Section 5.2.**

5.3 Sum and Difference Identities for Cosine

5.4 Sum and Difference Identities for Sine and Tangent

Cofunction Identities

$$\cos(90° - \theta) = \sin\theta \qquad \cot(90° - \theta) = \tan\theta$$

$$\sin(90° - \theta) = \cos\theta \qquad \sec(90° - \theta) = \csc\theta$$

$$\tan(90° - \theta) = \cot\theta \qquad \csc(90° - \theta) = \sec\theta$$

Find one value of θ such that $\tan\theta = \cot 78°$.

$$\tan\theta = \cot 78°$$

$$\cot(90° - \theta) = \cot 78°$$ Cofunction identity

$$90° - \theta = 78°$$ Set angles equal.

$$\theta = 12°$$ Solve for θ.

Sum and Difference Identities

$$\cos(A - B) = \cos A\cos B + \sin A\sin B$$

$$\cos(A + B) = \cos A\cos B - \sin A\sin B$$

$$\sin(A + B) = \sin A\cos B + \cos A\sin B$$

$$\sin(A - B) = \sin A\cos B - \cos A\sin B$$

$$\tan(A + B) = \frac{\tan A + \tan B}{1 - \tan A\tan B}$$

$$\tan(A - B) = \frac{\tan A - \tan B}{1 + \tan A\tan B}$$

Find the exact value of $\cos(-15°)$.

$$\cos(-15°)$$

$$= \cos(30° - 45°)$$ $-15° = 30° - 45°$

$$= \cos 30°\cos 45° + \sin 30°\sin 45°$$ Cosine difference identity

$$= \frac{\sqrt{3}}{2}\cdot\frac{\sqrt{2}}{2} + \frac{1}{2}\cdot\frac{\sqrt{2}}{2}$$ Substitute known values.

$$= \frac{\sqrt{6} + \sqrt{2}}{4}$$ Simplify.

Concepts	Examples

5.5 Double-Angle Identities

Double-Angle Identities

$$\cos 2A = \cos^2 A - \sin^2 A \qquad \cos 2A = 1 - 2\sin^2 A$$

$$\cos 2A = 2\cos^2 A - 1 \qquad \sin 2A = 2\sin A\cos A$$

$$\tan 2A = \frac{2\tan A}{1-\tan^2 A}$$

Given $\cos\theta = -\frac{5}{13}$ and $\sin\theta > 0$, find $\sin 2\theta$.

Sketch a triangle in quadrant II because $\cos\theta < 0$ and $\sin\theta > 0$. Use it to find that $\sin\theta = \frac{12}{13}$.

$$\sin 2\theta = 2\sin\theta\cos\theta$$
$$= 2\left(\frac{12}{13}\right)\left(-\frac{5}{13}\right)$$
$$= -\frac{120}{169}$$

Product-to-Sum Identities

$$\cos A\cos B = \frac{1}{2}[\cos(A+B) + \cos(A-B)]$$

$$\sin A\sin B = \frac{1}{2}[\cos(A-B) - \cos(A+B)]$$

$$\sin A\cos B = \frac{1}{2}[\sin(A+B) + \sin(A-B)]$$

$$\cos A\sin B = \frac{1}{2}[\sin(A+B) - \sin(A-B)]$$

Write $\sin(-\theta)\sin 2\theta$ as the difference of two functions.

$$\sin(-\theta)\sin 2\theta$$
$$= \frac{1}{2}[\cos(-\theta-2\theta) - \cos(-\theta+2\theta)]$$
$$= \frac{1}{2}[\cos(-3\theta) - \cos\theta]$$
$$= \frac{1}{2}\cos(-3\theta) - \frac{1}{2}\cos\theta$$
$$= \frac{1}{2}\cos 3\theta - \frac{1}{2}\cos\theta$$

Sum-to-Product Identities

$$\sin A + \sin B = 2\sin\left(\frac{A+B}{2}\right)\cos\left(\frac{A-B}{2}\right)$$

$$\sin A - \sin B = 2\cos\left(\frac{A+B}{2}\right)\sin\left(\frac{A-B}{2}\right)$$

$$\cos A + \cos B = 2\cos\left(\frac{A+B}{2}\right)\cos\left(\frac{A-B}{2}\right)$$

$$\cos A - \cos B = -2\sin\left(\frac{A+B}{2}\right)\sin\left(\frac{A-B}{2}\right)$$

Write $\cos\theta + \cos 3\theta$ as a product of two functions.

$$\cos\theta + \cos 3\theta$$
$$= 2\cos\left(\frac{\theta+3\theta}{2}\right)\cos\left(\frac{\theta-3\theta}{2}\right)$$
$$= 2\cos\left(\frac{4\theta}{2}\right)\cos\left(\frac{-2\theta}{2}\right)$$
$$= 2\cos 2\theta\cos(-\theta)$$
$$= 2\cos 2\theta\cos\theta$$

5.6 Half-Angle Identities

Half-Angle Identities

$$\cos\frac{A}{2} = \pm\sqrt{\frac{1+\cos A}{2}} \qquad \sin\frac{A}{2} = \pm\sqrt{\frac{1-\cos A}{2}}$$

$$\tan\frac{A}{2} = \pm\sqrt{\frac{1-\cos A}{1+\cos A}} \qquad \tan\frac{A}{2} = \frac{\sin A}{1+\cos A}$$

$$\tan\frac{A}{2} = \frac{1-\cos A}{\sin A}$$

(In the identities involving radicals, the sign is chosen based on the function under consideration and the quadrant of $\frac{A}{2}$.)

Find the exact value of $\tan 67.5°$.

We choose the last form with $A = 135°$.

$$\tan 67.5° = \tan\frac{135°}{2} = \frac{1-\cos 135°}{\sin 135°} = \frac{1-\left(-\frac{\sqrt{2}}{2}\right)}{\frac{\sqrt{2}}{2}}$$

$$= \frac{1+\frac{\sqrt{2}}{2}}{\frac{\sqrt{2}}{2}}\cdot\frac{2}{2} = \frac{2+\sqrt{2}}{\sqrt{2}} = \sqrt{2}+1$$

Rationalize the denominator and simplify.

Chapter 5 Review Exercises

Concept Check For each expression in Column I, choose the expression from Column II that completes an identity.

I		II	
1. $\sec x =$ ____	2. $\csc x =$ ____	A. $\frac{1}{\sin x}$	B. $\frac{1}{\cos x}$
3. $\tan x =$ ____	4. $\cot x =$ ____	C. $\frac{\sin x}{\cos x}$	D. $\frac{1}{\cot^2 x}$
5. $\tan^2 x =$ ____	6. $\sec^2 x =$ ____	E. $\frac{1}{\cos^2 x}$	F. $\frac{\cos x}{\sin x}$

Use identities to write each expression in terms of $\sin \theta$ *and* $\cos \theta$, *and then simplify so that no quotients appear and all functions are of* θ *only.*

7. $\sec^2 \theta - \tan^2 \theta$ **8.** $\frac{\cot(-\theta)}{\sec(-\theta)}$ **9.** $\tan^2 \theta(1 + \cot^2 \theta)$

10. $\csc \theta - \sin \theta$ **11.** $\tan \theta - \sec \theta \csc \theta$ **12.** $\csc^2 \theta + \sec^2 \theta$

Work each problem.

13. Use the trigonometric identities to find $\sin x$, $\tan x$, and $\cot(-x)$, given $\cos x = \frac{3}{5}$ and x in quadrant IV.

14. Given $\tan x = -\frac{5}{4}$, where $\frac{\pi}{2} < x < \pi$, use the trigonometric identities to find $\cot x$, $\csc x$, and $\sec x$.

15. Find the exact values of the six trigonometric functions of 165°.

16. Find the exact values of $\sin x$, $\cos x$, and $\tan x$, for $x = \frac{\pi}{12}$, using
(a) difference identities **(b)** half-angle identities.

Concept Check For each expression in Column I, use an identity to choose an expression from Column II with the same value. Choices may be used once, more than once, or not at all.

I		II	
17. $\cos 210°$	18. $\sin 35°$	A. $\sin(-35°)$	B. $\cos 55°$
19. $\tan(-35°)$	20. $-\sin 35°$	C. $\sqrt{\frac{1 + \cos 150°}{2}}$	D. $2 \sin 150° \cos 150°$
21. $\cos 35°$	22. $\cos 75°$	E. $\cot(-35°)$	F. $\cos^2 150° - \sin^2 150°$
23. $\sin 75°$	24. $\sin 300°$	G. $\cos(-35°)$	H. $\cot 125°$
25. $\cos 300°$	26. $\cos(-55°)$	I. $\cos 150° \cos 60° - \sin 150° \sin 60°$	
		J. $\sin 15° \cos 60° + \cos 15° \sin 60°$	

Use the given information to find $\sin(x + y)$, $\cos(x - y)$, $\tan(x + y)$, *and the quadrant of* $x + y$.

27. $\sin x = -\frac{3}{5}$, $\cos y = -\frac{7}{25}$, x and y in quadrant III

28. $\sin x = \frac{3}{5}$, $\cos y = \frac{24}{25}$, x in quadrant I, y in quadrant IV

29. $\sin x = -\frac{1}{2}$, $\cos y = -\frac{2}{5}$, x and y in quadrant III

30. $\sin y = -\frac{2}{3}$, $\cos x = -\frac{1}{5}$, x in quadrant II, y in quadrant III

---

1. B 2. A
3. C 4. F
5. D 6. E

In Exercises 7–12, there may be more than one possible answer.

7. 1 8. $-\cos^2 \theta \csc \theta$
9. $\sec^2 \theta$ 10. $\cos \theta \cot \theta$
11. $-\cot \theta$ 12. $\csc^2 \theta \sec^2 \theta$

13. $\sin x = -\frac{4}{5}$; $\tan x = -\frac{4}{3}$; $\cot(-x) = \frac{3}{4}$

14. $\cot x = -\frac{4}{5}$; $\csc x = \frac{\sqrt{41}}{5}$; $\sec x = -\frac{\sqrt{41}}{4}$

15. $\sin 165° = \frac{\sqrt{6} - \sqrt{2}}{4}$; $\cos 165° = \frac{-\sqrt{6} - \sqrt{2}}{4}$; $\tan 165° = -2 + \sqrt{3}$; $\csc 165° = \sqrt{6} + \sqrt{2}$; $\sec 165° = -\sqrt{6} + \sqrt{2}$; $\cot 165° = -2 - \sqrt{3}$

16. (a) $\sin \frac{\pi}{12} = \frac{\sqrt{6} - \sqrt{2}}{4}$; $\cos \frac{\pi}{12} = \frac{\sqrt{6} + \sqrt{2}}{4}$; $\tan \frac{\pi}{12} = 2 - \sqrt{3}$
(b) $\sin \frac{\pi}{12} = \frac{\sqrt{2 - \sqrt{3}}}{2}$; $\cos \frac{\pi}{12} = \frac{\sqrt{2 + \sqrt{3}}}{2}$; $\tan \frac{\pi}{12} = 2 - \sqrt{3}$

17. I 18. B
19. H 20. A
21. G 22. C
23. J 24. D
25. F 26. B

27. $\frac{117}{125}$; $\frac{4}{5}$; $-\frac{117}{44}$; II
28. $\frac{44}{125}$; $\frac{3}{5}$; $\frac{44}{117}$; I
29. $\frac{2 + 3\sqrt{7}}{10}$; $\frac{2\sqrt{3} + \sqrt{21}}{10}$; $\frac{-25\sqrt{3} - 8\sqrt{21}}{9}$; II
30. $\frac{2 - 2\sqrt{30}}{15}$; $\frac{\sqrt{5} - 4\sqrt{6}}{15}$; $\frac{18\sqrt{6} - 50\sqrt{5}}{91}$; IV

31. $\frac{4 - 9\sqrt{11}}{50}$; $\frac{12\sqrt{11} - 3}{50}$; $\frac{\sqrt{11} - 16}{21}$; IV

32. $\frac{\sqrt{231} - 2}{18}$; $\frac{\sqrt{77} - 2\sqrt{3}}{18}$; $\frac{8\sqrt{77} - 81\sqrt{3}}{65}$; II

33. $\sin \theta = \frac{\sqrt{14}}{4}$; $\cos \theta = \frac{\sqrt{2}}{4}$

34. $\sin B = -\frac{\sqrt{7}}{4}$; $\cos B = \frac{3}{4}$

35. $\sin 2x = \frac{3}{5}$; $\cos 2x = -\frac{4}{5}$

36. $\sin 2y = -\frac{24}{25}$; $\cos 2y = -\frac{7}{25}$

37. $\frac{1}{2}$ 38. $\frac{\sqrt{14}}{4}$

39. $\frac{\sqrt{5} - 1}{2}$ 40. $\frac{\sqrt{6}}{3}$

41. 0.5 42. -0.96

43. $\frac{\sin 2x + \sin x}{\cos x - \cos 2x} = \cot \frac{x}{2}$

44. $\frac{1 - \cos 2x}{\sin 2x} = \tan x$

45. $\frac{\sin x}{1 - \cos x} = \cot \frac{x}{2}$

46. $\frac{\cos x \sin 2x}{1 + \cos 2x} = \sin x$

47. $\frac{2(\sin x - \sin^3 x)}{\cos x} = \sin 2x$

48. $\csc x - \cot x = \tan \frac{x}{2}$

31. $\sin x = \frac{1}{10}$, $\cos y = \frac{4}{5}$, x in quadrant I, y in quadrant IV

32. $\cos x = \frac{2}{9}$, $\sin y = -\frac{1}{2}$, x in quadrant IV, y in quadrant III

Find values of the sine and cosine functions for each angle measure.

33. θ, given $\cos 2\theta = -\frac{3}{4}$, $90° < 2\theta < 180°$

34. B, given $\cos 2B = \frac{1}{8}$, $540° < 2B < 720°$

35. $2x$, given $\tan x = 3$, $\sin x < 0$

36. $2y$, given $\sec y = -\frac{5}{3}$, $\sin y > 0$

Use the given information to find each of the following.

37. $\cos \frac{\theta}{2}$, given $\cos \theta = -\frac{1}{2}$, $90° < \theta < 180°$

38. $\sin \frac{A}{2}$, given $\cos A = -\frac{3}{4}$, $90° < A < 180°$

39. $\tan x$, given $\tan 2x = 2$, $\pi < x < \frac{3\pi}{2}$

40. $\sin y$, given $\cos 2y = -\frac{1}{3}$, $\frac{\pi}{2} < y < \pi$

41. $\tan \frac{x}{2}$, given $\sin x = 0.8$, $0 < x < \frac{\pi}{2}$

42. $\sin 2x$, given $\sin x = 0.6$, $\frac{\pi}{2} < x < \pi$

Graph each expression and use the graph to make a conjecture, predicting what might be an identity. Then verify your conjecture algebraically.

43. $\frac{\sin 2x + \sin x}{\cos x - \cos 2x}$

44. $\frac{1 - \cos 2x}{\sin 2x}$

45. $\frac{\sin x}{1 - \cos x}$

46. $\frac{\cos x \sin 2x}{1 + \cos 2x}$

47. $\frac{2(\sin x - \sin^3 x)}{\cos x}$

48. $\csc x - \cot x$

Verify that each equation is an identity.

49. $\sin^2 x - \sin^2 y = \cos^2 y - \cos^2 x$

50. $2 \cos^3 x - \cos x = \frac{\cos^2 x - \sin^2 x}{\sec x}$

51. $\frac{\sin^2 x}{2 - 2 \cos x} = \cos^2 \frac{x}{2}$

52. $\frac{\sin 2x}{\sin x} = \frac{2}{\sec x}$

53. $2 \cos A - \sec A = \cos A - \frac{\tan A}{\csc A}$

54. $\frac{2 \tan B}{\sin 2B} = \sec^2 B$

55. $1 + \tan^2 \alpha = 2 \tan \alpha \csc 2\alpha$

56. $\frac{2 \cot x}{\tan 2x} = \csc^2 x - 2$

57. $\tan \theta \sin 2\theta = 2 - 2 \cos^2 \theta$

58. $\csc A \sin 2A - \sec A = \cos 2A \sec A$

59. $2 \tan x \csc 2x - \tan^2 x = 1$

60. $2 \cos^2 \theta - 1 = \frac{1 - \tan^2 \theta}{1 + \tan^2 \theta}$

61. $\tan \theta \cos^2 \theta = \frac{2 \tan \theta \cos^2 \theta - \tan \theta}{1 - \tan^2 \theta}$

62. $\sec^2 \alpha - 1 = \frac{\sec 2\alpha - 1}{\sec 2\alpha + 1}$

63. $\frac{\sin^2 x - \cos^2 x}{\csc x} = 2 \sin^3 x - \sin x$

64. $\sin^3 \theta = \sin \theta - \cos^2 \theta \sin \theta$

65. $\tan 4\theta = \frac{2 \tan 2\theta}{2 - \sec^2 2\theta}$

66. $2 \cos^2 \frac{x}{2} \tan x = \tan x + \sin x$

67. $\tan\left(\frac{x}{2} + \frac{\pi}{4}\right) = \sec x + \tan x$

68. $\frac{1}{2} \cot \frac{x}{2} - \frac{1}{2} \tan \frac{x}{2} = \cot x$

69. $-\cot \frac{x}{2} = \frac{\sin 2x + \sin x}{\cos 2x - \cos x}$

70. $\frac{\sin 3t + \sin 2t}{\sin 3t - \sin 2t} = \frac{\tan \frac{5t}{2}}{\tan \frac{t}{2}}$

71. (a) $D = \frac{v^2 \sin 2\theta}{32}$
(b) 35 ft

72. (a) $\frac{1}{\omega}$

(Modeling) Solve each problem.

71. ***Distance Traveled by an Object*** The distance D of an object thrown (or projected) from height h (in feet) at angle θ with initial velocity v is illustrated in the figure and modeled by the formula

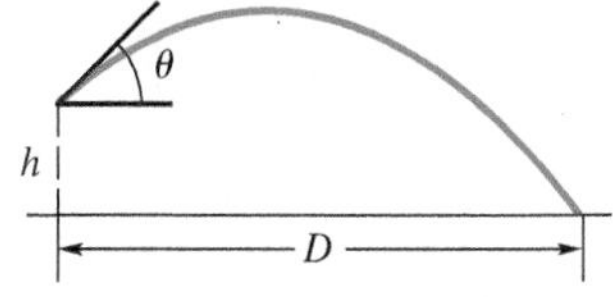

$$D = \frac{v^2 \sin \theta \cos \theta + v \cos \theta \sqrt{(v \sin \theta)^2 + 64h}}{32}.$$

(*Source*: Kreighbaum, E. and K. Barthels, *Biomechanics,* Allyn & Bacon.)

(a) Find D when $h = 0$—that is, when the object is projected from the ground.

(b) Suppose a car driving over loose gravel kicks up a small stone at a velocity of 36 ft per sec (about 25 mph) and an angle $\theta = 30°$. How far, to the nearest foot, will the stone travel?

72. ***Amperage, Wattage, and Voltage*** Suppose that for an electric heater, voltage is given by $V = a \sin 2\pi\omega t$ and amperage by $I = b \sin 2\pi\omega t$, where t is time in seconds.

(a) Find the period of the graph for the voltage.

(b) Show that the graph of the wattage $W = VI$ has half the period of the voltage.

Chapter 5 Test

[5.1]
1. $\sin \theta = -\frac{7}{25}$; $\tan \theta = -\frac{7}{24}$; $\cot \theta = -\frac{24}{7}$; $\sec \theta = \frac{25}{24}$; $\csc \theta = -\frac{25}{7}$
2. $\cos \theta$
3. -1

[5.3]
4. $\frac{\sqrt{6} - \sqrt{2}}{4}$

[5.3, 5.4]
5. (a) $-\sin \theta$ (b) $\tan x$

[5.6]
6. $-\frac{\sqrt{2 - \sqrt{2}}}{2}$
7. $\cot \frac{1}{2}x - \cot x = \csc x$

[5.3, 5.4]
8. (a) $\frac{33}{65}$ (b) $-\frac{56}{65}$ (c) $\frac{63}{16}$ (d) II

[5.5, 5.6]
9. (a) $-\frac{7}{25}$ (b) $-\frac{24}{25}$ (c) $\frac{24}{7}$ (d) $\frac{\sqrt{5}}{5}$ (e) 2

[5.3]
15. (a) $V = 163 \cos\left(\frac{\pi}{2} - \omega t\right)$
(b) 163 volts; $\frac{1}{240}$ sec

1. If $\cos \theta = \frac{24}{25}$ and θ is in quadrant IV, find the other five trigonometric functions of θ.

2. Express $\sec \theta - \sin \theta \tan \theta$ as a single function of θ.

3. Express $\tan^2 x - \sec^2 x$ in terms of $\sin x$ and $\cos x$, and simplify.

4. Find the exact value of $\cos \frac{5\pi}{12}$.

5. Express **(a)** $\cos(270° - \theta)$ and **(b)** $\tan(\pi + x)$ as functions of θ or x alone.

6. Use a half-angle identity to find the exact value of $\sin(-22.5°)$.

7. Graph $y = \cot \frac{1}{2}x - \cot x$ and use the graph to make a conjecture, predicting what might be an identity. Then verify your conjecture algebraically.

8. Given that $\sin A = \frac{5}{13}$, $\cos B = -\frac{3}{5}$, A is a quadrant I angle, and B is a quadrant II angle, find each of the following.

(a) $\sin(A + B)$ **(b)** $\cos(A + B)$ **(c)** $\tan(A - B)$ **(d)** the quadrant of $A + B$

9. Given that $\cos \theta = -\frac{3}{5}$ and $90° < \theta < 180°$, find each of the following.

(a) $\cos 2\theta$ **(b)** $\sin 2\theta$ **(c)** $\tan 2\theta$ **(d)** $\cos \frac{\theta}{2}$ **(e)** $\tan \frac{\theta}{2}$

Verify that each equation is an identity.

10. $\sec^2 B = \frac{1}{1 - \sin^2 B}$

11. $\cos 2A = \frac{\cot A - \tan A}{\csc A \sec A}$

12. $\frac{\sin 2x}{\cos 2x + 1} = \tan x$

13. $\tan^2 x - \sin^2 x = (\tan x \sin x)^2$

14. $\frac{\tan x - \cot x}{\tan x + \cot x} = 2 \sin^2 x - 1$

15. ***(Modeling) Voltage*** The voltage in common household current is expressed as $V = 163 \sin \omega t$, where ω is the angular speed (in radians per second) of the generator at an electrical plant and t is time (in seconds).

(a) Use an identity to express V in terms of cosine.

(b) If $\omega = 120\pi$, what is the maximum voltage? Give the least positive value of t when the maximum voltage occurs.

6 Inverse Circular Functions and Trigonometric Equations

Sound waves, such as those initiated by musical instruments, travel in sinusoidal patterns that can be graphed as sine or cosine functions and described by *trigonometric equations.*

6.1 Inverse Circular Functions

- Inverse Functions
- Inverse Sine Function
- Inverse Cosine Function
- Inverse Tangent Function
- Other Inverse Circular Functions
- Inverse Function Values

Inverse Functions

We now review some basic concepts from algebra. ***For a function f, every element x in the domain corresponds to one and only one element y, or $f(x)$, in the range.*** This means the following:

1. If point (a, b) lies on the graph of f, then there is no other point on the graph that has a as first coordinate.

2. Other points may have b as second coordinate, however, because the definition of function allows range elements to be used more than once.

If a function is defined so that ***each range element is used only once,*** then it is a **one-to-one function.** For example, the function

$$f(x) = x^3 \text{ is a one-to-one function}$$

because every real number has exactly one real cube root. However,

$$g(x) = x^2 \text{ is not a one-to-one function}$$

because $g(2) = 4$ and $g(-2) = 4$. There are two domain elements, 2 and -2, that correspond to the range element 4.

The **horizontal line test** helps determine graphically whether a function is one-to-one.

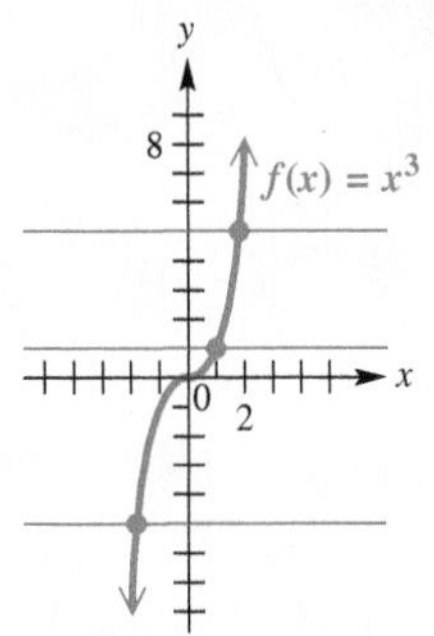

$f(x) = x^3$ is a one-to-one function. It satisfies the conditions of the horizontal line test.

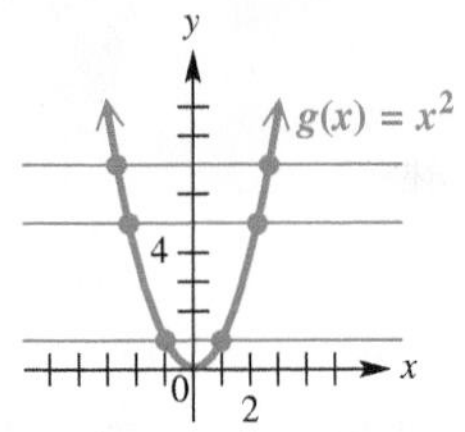

$g(x) = x^2$ is not one-to-one. It does not satisfy the conditions of the horizontal line test.

Figure 1

Horizontal Line Test

A function is one-to-one if every horizontal line intersects the graph of the function at most once.

This test is applied to the graphs of $f(x) = x^3$ and $g(x) = x^2$ in **Figure 1.**

By interchanging the components of the ordered pairs of a one-to-one function f, we obtain a new set of ordered pairs that satisfies the definition of a function. This new function is the *inverse function,* or *inverse,* of f.

Inverse Function

The **inverse function** of a one-to-one function f is defined as follows.

$$f^{-1} = \{(y, x) \mid (x, y) \text{ belongs to } f\}$$

The special notation used for inverse functions is f^{-1} (read "**f-inverse**"). It represents the function created by interchanging the input (domain) and the output (range) of a one-to-one function.

CAUTION ***Do not confuse the -1 in f^{-1} with a negative exponent.*** The symbol $f^{-1}(x)$ represents the inverse function of f, *not* $\frac{1}{f(x)}$.

The following statements summarize the concepts of inverse functions.

Summary of Inverse Functions

1. In a one-to-one function, each x-value corresponds to only one y-value and each y-value corresponds to only one x-value.
2. If a function f is one-to-one, then f has an inverse function f^{-1}.
3. The domain of f is the range of f^{-1}, and the range of f is the domain of f^{-1}. That is, if the point (a, b) lies on the graph of f, then the point (b, a) lies on the graph of f^{-1}.
4. The graphs of f and f^{-1} are reflections of each other across the line $y = x$.
5. To find $f^{-1}(x)$ for $f(x)$, follow these steps.

   Step 1 Replace $f(x)$ with y and interchange x and y.

   Step 2 Solve for y.

   Step 3 Replace y with $f^{-1}(x)$.

Figure 2 illustrates some of these concepts.

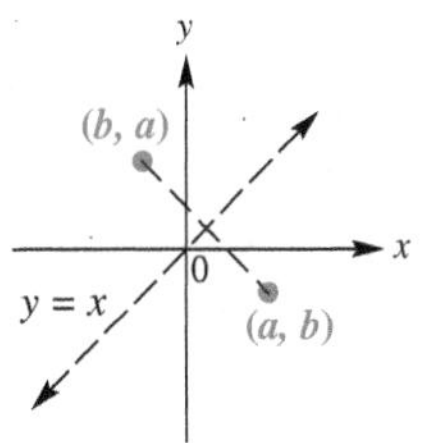

(b, a) is the reflection of (a, b) across the line $y = x$.

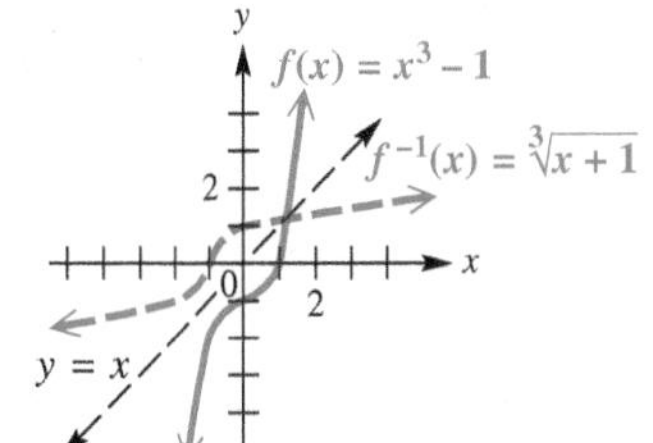

The graph of f^{-1} is the reflection of the graph of f across the line $y = x$.

Figure 2

If the domain of $g(x) = x^2$ is restricted so that $x \geq 0$, then it is a one-to-one function.

Figure 3

We often restrict the domain of a function that is not one-to-one to make it one-to-one without changing the range. We saw in **Figure 1** that $g(x) = x^2$, with its natural domain $(-\infty, \infty)$, is not one-to-one. However, if we restrict its domain to the set of nonnegative numbers $[0, \infty)$, we obtain a new function f that is one-to-one and has the same range as g, $[0, \infty)$. See **Figure 3.**

NOTE We could have restricted the domain of $g(x) = x^2$ to $(-\infty, 0]$ to obtain a different one-to-one function. For the trigonometric functions, such choices are based on general agreement by mathematicians.

LOOKING AHEAD TO CALCULUS

The **inverse circular functions** are used in calculus to solve certain types of related-rates problems and to integrate certain rational functions.

Inverse Sine Function Refer to the graph of the sine function in **Figure 4** on the next page. Applying the horizontal line test, we see that $y = \sin x$ does not define a one-to-one function. If we restrict the domain to the interval $\left[-\frac{\pi}{2}, \frac{\pi}{2}\right]$, which is the part of the graph in **Figure 4** shown in color, this restricted function is one-to-one and has an inverse function. The range of $y = \sin x$ is $[-1, 1]$, so the domain of the inverse function will be $[-1, 1]$, and its range will be $\left[-\frac{\pi}{2}, \frac{\pi}{2}\right]$.

Teaching Tip Mention that the interval $\left[-\frac{\pi}{2}, \frac{\pi}{2}\right]$ contains exactly enough of the graph of the sine function to include all possible values of y. Other intervals could also be used, but this interval is an accepted convention that is adopted by mathematicians and used by scientific calculators and graphing calculators.

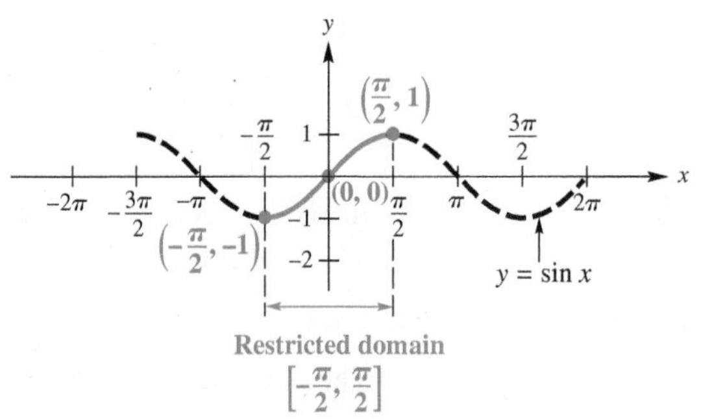

Figure 4

Reflecting the graph of $y = \sin x$ on the restricted domain, shown in **Figure 5(a),** across the line $y = x$ gives the graph of the inverse function, shown in **Figure 5(b).** Some key points are labeled on the graph. The equation of the inverse of $y = \sin x$ is found by interchanging x and y to obtain

$$x = \sin y.$$

This equation is solved for y by writing

$$y = \sin^{-1} x \quad \text{(read "inverse sine of } x\text{")}.$$

As **Figure 5(b)** shows, the domain of $y = \sin^{-1} x$ is $[-1, 1]$, while the restricted domain of $y = \sin x$, $\left[-\frac{\pi}{2}, \frac{\pi}{2}\right]$, is the range of $y = \sin^{-1} x$. ***An alternative notation for* $\sin^{-1} x$ *is* arcsin x.**

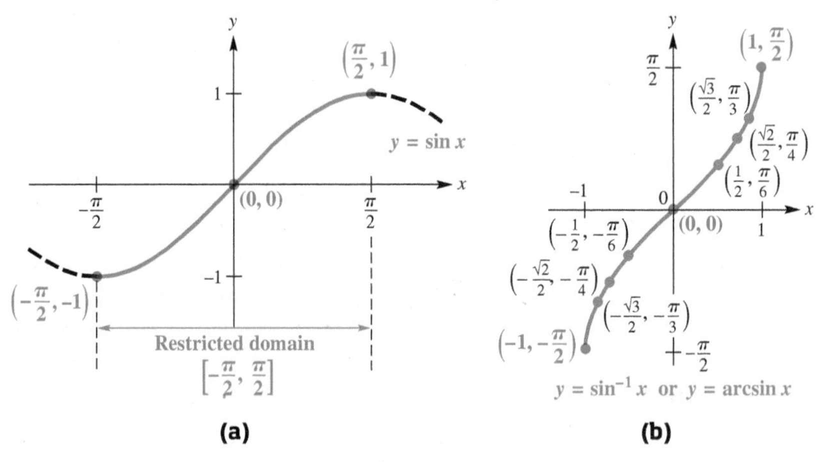

(a) **(b)**

Figure 5

Teaching Tip In problems like **Example 1(a),** use phrases such as

"y is a value *in radians* between $-\frac{\pi}{2}$ and $\frac{\pi}{2}$ whose sine is equal to $\frac{1}{2}$"

to help students understand the meaning of inverse sine. Illustrate the answer graphically by graphing $y = \sin x$ and $y = \frac{1}{2}$ in the interval $\left[-\frac{\pi}{2}, \frac{\pi}{2}\right]$, and then determine the point of intersection.

Inverse Sine Function

$y = \sin^{-1} x$ or $y = \arcsin x$ means that $x = \sin y$, for $-\frac{\pi}{2} \le y \le \frac{\pi}{2}$.

We can think of* $y = \sin^{-1} x$ *or* $y = \arcsin x$ *as
"y is the number (angle) in the interval* $\left[-\frac{\pi}{2}, \frac{\pi}{2}\right]$ *whose sine is x."

Thus, we can write $y = \sin^{-1} x$ as $\sin y = x$ to evaluate it. We must pay close attention to the domain and range intervals.

EXAMPLE 1 Finding Inverse Sine Values

Find the value of each real number y if it exists.

(a) $y = \arcsin \frac{1}{2}$ **(b)** $y = \sin^{-1}(-1)$ **(c)** $y = \sin^{-1}(-2)$

ALGEBRAIC SOLUTION

(a) The graph of the function defined by $y = \arcsin x$ (**Figure 5(b)**) includes the point $\left(\frac{1}{2}, \frac{\pi}{6}\right)$. Therefore, $\arcsin \frac{1}{2} = \frac{\pi}{6}$.

Alternatively, we can think of $y = \arcsin \frac{1}{2}$ as "y is the number in $\left[-\frac{\pi}{2}, \frac{\pi}{2}\right]$ whose sine is $\frac{1}{2}$." Then we can write the given equation as $\sin y = \frac{1}{2}$. Because $\sin \frac{\pi}{6} = \frac{1}{2}$ and $\frac{\pi}{6}$ is in the range of the arcsine function, $y = \frac{\pi}{6}$.

(b) Writing the equation $y = \sin^{-1}(-1)$ in the form $\sin y = -1$ shows that $y = -\frac{\pi}{2}$. Notice that the point $\left(-1, -\frac{\pi}{2}\right)$ is on the graph of $y = \sin^{-1} x$.

(c) Because -2 is not in the domain of the inverse sine function, $\sin^{-1}(-2)$ does not exist.

GRAPHING CALCULATOR SOLUTION

(a)–(c) We graph the equation $y_1 = \sin^{-1} x$ and find the points with x-values $\frac{1}{2} = 0.5$ and -1. For these two x-values, **Figure 6** indicates that $y = \frac{\pi}{6} \approx 0.52359878$ and $y = -\frac{\pi}{2} \approx -1.570796$.

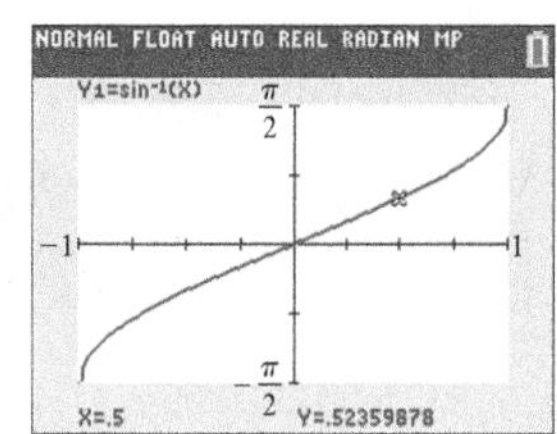

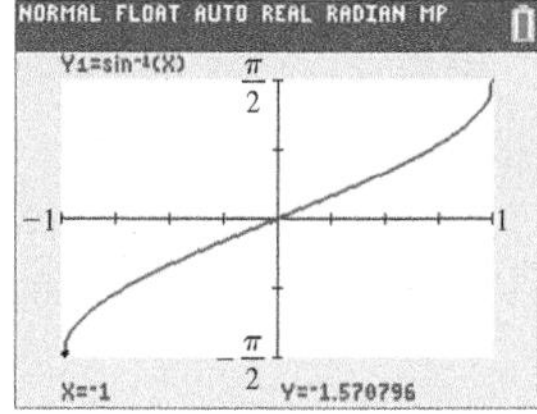

Figure 6

Because $\sin^{-1}(-2)$ does not exist, a calculator will give an error message for this input.

✔ **Now Try Exercises 13, 21, and 25.**

Classroom Example 1
Find the value of each real number y if it exists.

(a) $y = \arcsin \frac{\sqrt{3}}{2}$

(b) $y = \sin^{-1}\left(-\frac{1}{2}\right)$

(c) $y = \sin^{-1} \sqrt{2}$

Answers:

(a) $\frac{\pi}{3}$ (b) $-\frac{\pi}{6}$

(c) $\sin^{-1}\sqrt{2}$ does not exist.

CAUTION In **Example 1(b),** it is tempting to give the value of $\sin^{-1}(-1)$ as $\frac{3\pi}{2}$ because $\sin \frac{3\pi}{2} = -1$. However, $\frac{3\pi}{2}$ is not in the range of the inverse sine function. ***Be certain that the number given for an inverse function value is in the range of the particular inverse function being considered.***

Inverse Sine Function $y = \sin^{-1} x$ or $y = \arcsin x$

Domain: $[-1, 1]$ Range: $\left[-\frac{\pi}{2}, \frac{\pi}{2}\right]$

x	y
-1	$-\frac{\pi}{2}$
$-\frac{\sqrt{2}}{2}$	$-\frac{\pi}{4}$
0	0
$\frac{\sqrt{2}}{2}$	$\frac{\pi}{4}$
1	$\frac{\pi}{2}$

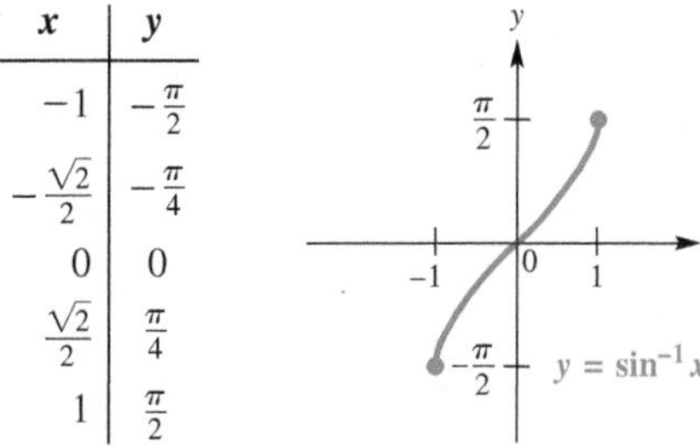

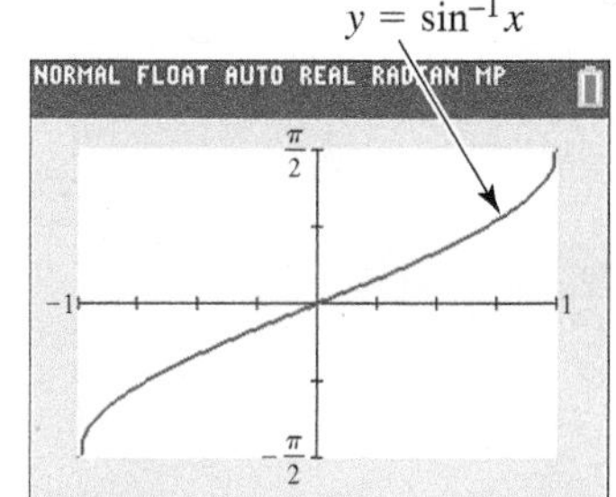

Figure 7

- The inverse sine function is increasing on the open interval $(-1, 1)$ and continuous on its domain $[-1, 1]$.
- Its x- and y-intercepts are both $(0, 0)$.
- Its graph is symmetric with respect to the origin, so the function is an odd function. For all x in the domain, $\sin^{-1}(-x) = -\sin^{-1} x$.

Inverse Cosine Function The function

$$y = \cos^{-1} x \quad \text{or} \quad y = \arccos x$$

is defined by restricting the domain of the function $y = \cos x$ to the interval $[0, \pi]$ as in **Figure 8.** This restricted function, which is the part of the graph in **Figure 8** shown in color, is one-to-one and has an inverse function. The inverse function, $y = \cos^{-1} x$, is found by interchanging the roles of x and y. Reflecting the graph of $y = \cos x$ across the line $y = x$ gives the graph of the inverse function shown in **Figure 9.** Some key points are shown on the graph.

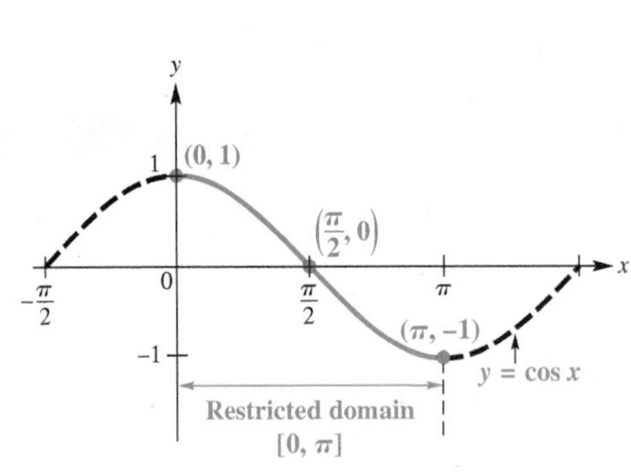

Figure 8

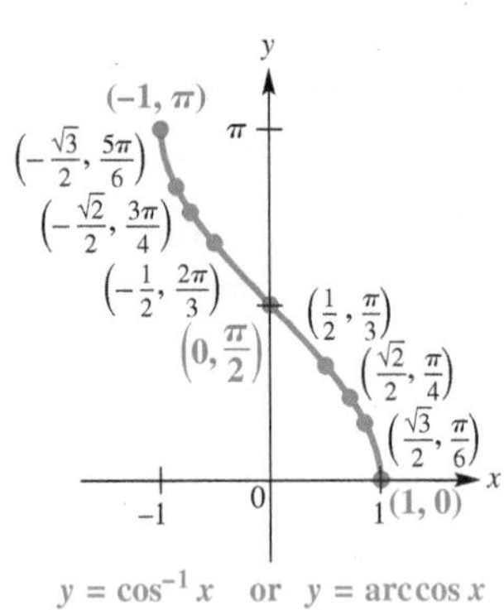

Figure 9

Classroom Example 2
Find the value of each real number y if it exists.

(a) $y = \arccos 0$

(b) $y = \cos^{-1} \frac{1}{2}$

Answers:
(a) $\frac{\pi}{2}$ (b) $\frac{\pi}{3}$

Inverse Cosine Function

$y = \cos^{-1} x$ or $y = \arccos x$ means that $x = \cos y$, for $0 \le y \le \pi$.

We can think of $y = \cos^{-1} x$ or $y = \arccos x$ as

"y is the number (angle) in the interval $[0, \pi]$ whose cosine is x."

EXAMPLE 2 Finding Inverse Cosine Values

Find the value of each real number y if it exists.

(a) $y = \arccos 1$ **(b)** $y = \cos^{-1}\left(-\frac{\sqrt{2}}{2}\right)$

SOLUTION

(a) Because the point $(1, 0)$ lies on the graph of $y = \arccos x$ in **Figure 9,** the value of y, or arccos 1, is 0. Alternatively, we can think of $y = \arccos 1$ as

"y is the number in $[0, \pi]$ whose cosine is 1," or $\cos y = 1$.

Thus, $y = 0$, since $\cos 0 = 1$ and 0 is in the range of the arccosine function.

(b) We must find the value of y that satisfies

$$\cos y = -\frac{\sqrt{2}}{2}, \quad \text{where } y \text{ is in the interval } [0, \pi],$$

which is the range of the function $y = \cos^{-1} x$. The only value for y that satisfies these conditions is $\frac{3\pi}{4}$. Again, this can be verified from the graph in **Figure 9.**

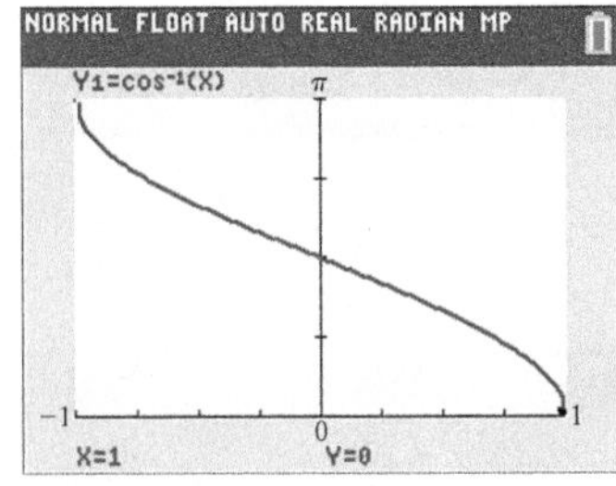

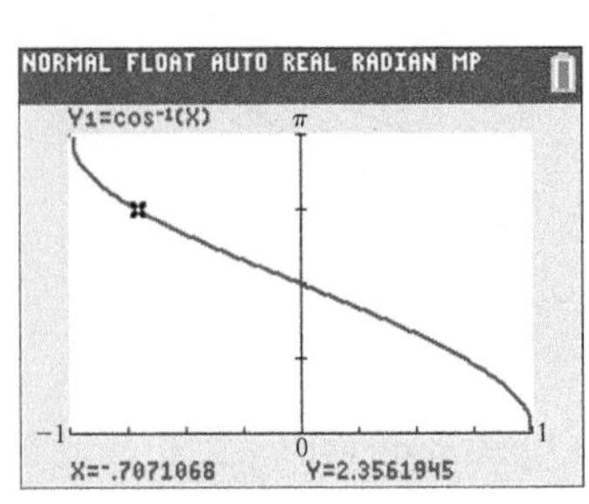

These screens support the results of **Example 2** because

$$-\frac{\sqrt{2}}{2} \approx -0.7071068$$

and $\frac{3\pi}{4} \approx 2.3561945.$

✓ **Now Try Exercises 15 and 23.**

Inverse Cosine Function $y = \cos^{-1} x$ or $y = \arccos x$

Domain: $[-1, 1]$ Range: $[0, \pi]$

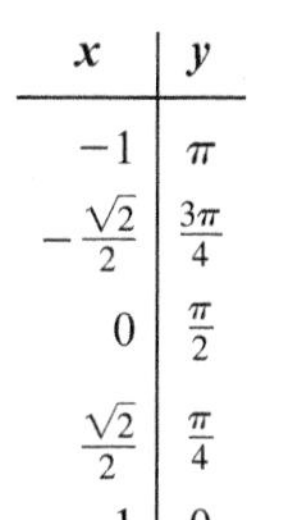

x	y
-1	π
$-\frac{\sqrt{2}}{2}$	$\frac{3\pi}{4}$
0	$\frac{\pi}{2}$
$\frac{\sqrt{2}}{2}$	$\frac{\pi}{4}$
1	0

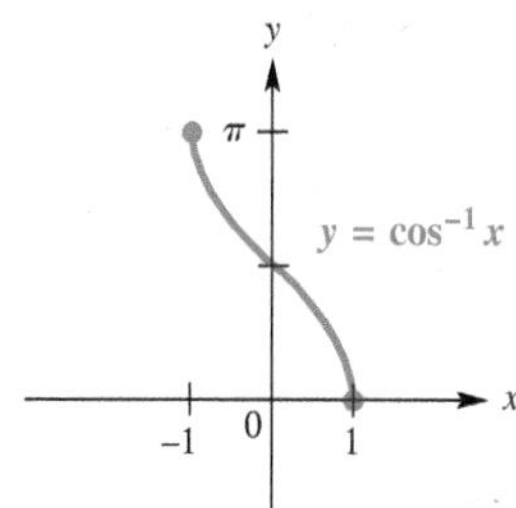

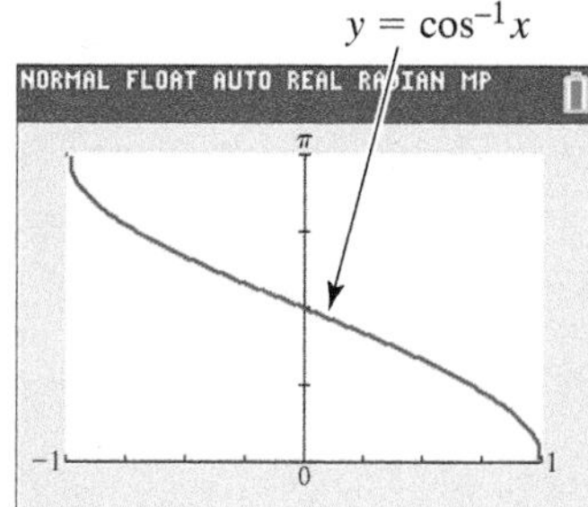

Figure 10

- The inverse cosine function is decreasing on the open interval $(-1, 1)$ and continuous on its domain $[-1, 1]$.
- Its x-intercept is $(1, 0)$ and its y-intercept is $\left(0, \frac{\pi}{2}\right)$.
- Its graph is not symmetric with respect to either the y-axis or the origin.

Inverse Tangent Function Restricting the domain of the function $y = \tan x$ to the open interval $\left(-\frac{\pi}{2}, \frac{\pi}{2}\right)$ yields a one-to-one function. By interchanging the roles of x and y, we obtain the inverse tangent function given by

$$y = \tan^{-1} x \quad \text{or} \quad y = \arctan x.$$

Figure 11 shows the graph of the restricted tangent function. **Figure 12** gives the graph of $y = \tan^{-1} x$.

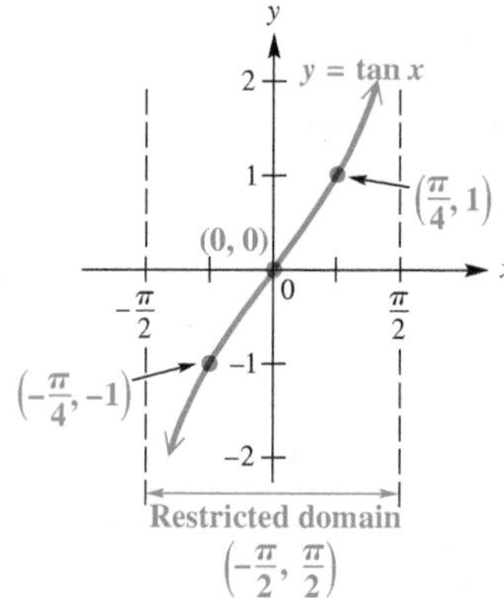

Figure 11

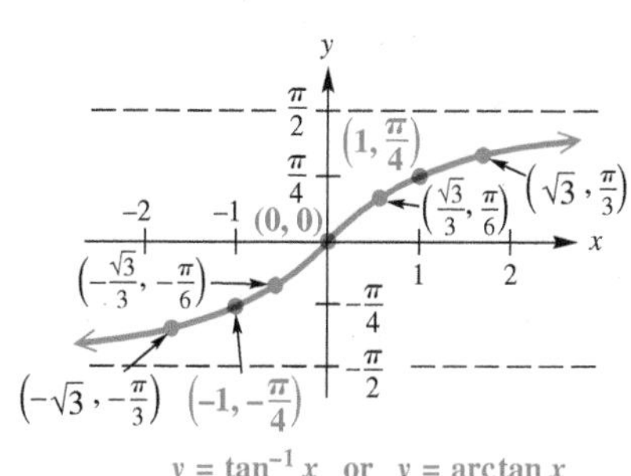

Figure 12

Inverse Tangent Function

$y = \tan^{-1} x$ or $y = \arctan x$ means that $x = \tan y$, for $-\frac{\pi}{2} < y < \frac{\pi}{2}$.

We can think of $y = \tan^{-1} x$ or $y = \arctan x$ as
"y is the number (angle) in the interval $\left(-\frac{\pi}{2}, \frac{\pi}{2}\right)$ whose tangent is x."

We summarize this discussion about the inverse tangent function as follows.

Inverse Tangent Function $y = \tan^{-1} x$ or $y = \arctan x$

Domain: $(-\infty, \infty)$ Range: $\left(-\frac{\pi}{2}, \frac{\pi}{2}\right)$

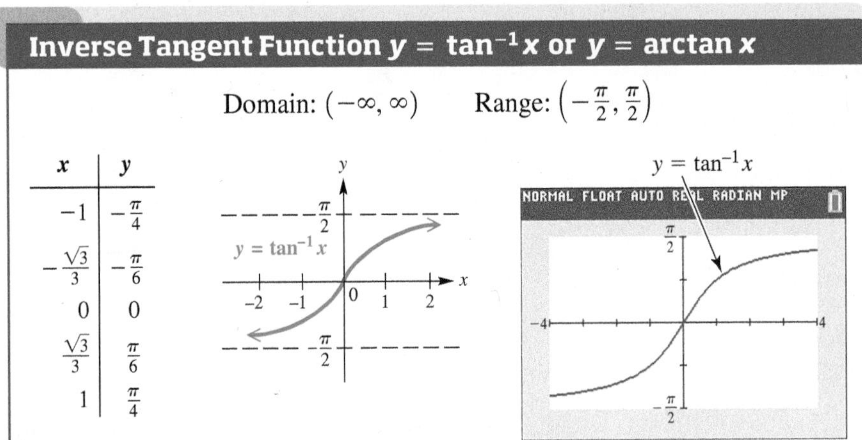

x	y
-1	$-\frac{\pi}{4}$
$-\frac{\sqrt{3}}{3}$	$-\frac{\pi}{6}$
0	0
$\frac{\sqrt{3}}{3}$	$\frac{\pi}{6}$
1	$\frac{\pi}{4}$

Figure 13

- The inverse tangent function is increasing on $(-\infty, \infty)$ and continuous on its domain $(-\infty, \infty)$.
- Its x- and y-intercepts are both $(0, 0)$.
- Its graph is symmetric with respect to the origin, so the function is an odd function. For all x in the domain, $\tan^{-1}(-x) = -\tan^{-1} x$.
- The lines $y = \frac{\pi}{2}$ and $y = -\frac{\pi}{2}$ are horizontal asymptotes.

Other Inverse Circular Functions The other three inverse trigonometric functions are defined similarly. Their graphs are shown in **Figure 14.**

Inverse Cotangent, Secant, and Cosecant Functions*

$y = \cot^{-1} x$ or $y = \text{arccot}\, x$ means that $x = \cot y$, for $0 < y < \pi$.

$y = \sec^{-1} x$ or $y = \text{arcsec}\, x$ means that $x = \sec y$, for $0 \le y \le \pi$, $y \neq \frac{\pi}{2}$.

$y = \csc^{-1} x$ or $y = \text{arccsc}\, x$ means that $x = \csc y$, for $-\frac{\pi}{2} \le y \le \frac{\pi}{2}$, $y \neq 0$.

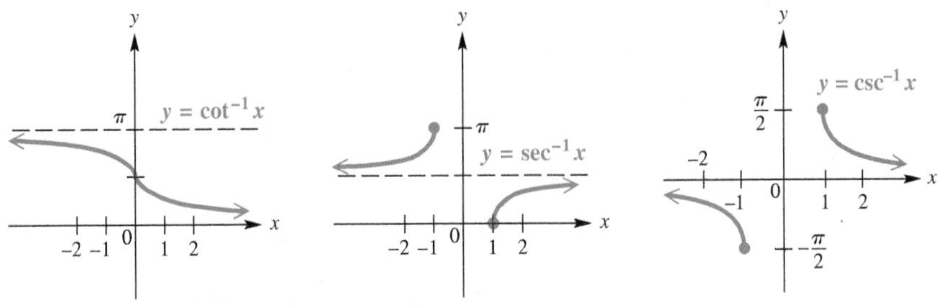

Figure 14

*The inverse secant and inverse cosecant functions are sometimes defined with different ranges. We use intervals that match those of the inverse cosine and inverse sine functions, respectively (except for one missing point).

The table gives all six inverse circular functions with their domains and ranges.

Summary of Inverse Circular Functions

Inverse Function	Domain	Range	
		Interval	**Quadrants of the Unit Circle**
$y = \sin^{-1} x$	$[-1, 1]$	$\left[-\frac{\pi}{2}, \frac{\pi}{2}\right]$	I and IV
$y = \cos^{-1} x$	$[-1, 1]$	$[0, \pi]$	I and II
$y = \tan^{-1} x$	$(-\infty, \infty)$	$\left(-\frac{\pi}{2}, \frac{\pi}{2}\right)$	I and IV
$y = \cot^{-1} x$	$(-\infty, \infty)$	$(0, \pi)$	I and II
$y = \sec^{-1} x$	$(-\infty, -1] \cup [1, \infty)$	$\left[0, \frac{\pi}{2}\right) \cup \left(\frac{\pi}{2}, \pi\right]$	I and II
$y = \csc^{-1} x$	$(-\infty, -1] \cup [1, \infty)$	$\left[-\frac{\pi}{2}, 0\right) \cup \left(0, \frac{\pi}{2}\right]$	I and IV

Inverse Function Values The inverse circular functions are formally defined with real number ranges. However, there are times when it may be convenient to find degree-measured angles equivalent to these real number values. It is also often convenient to think in terms of the unit circle and choose the inverse function values on the basis of the quadrants given in the preceding table.

Classroom Example 3

Find the *degree measure* of θ if it exists.

(a) $\theta = \arctan \sqrt{3}$

(b) $\theta = \csc^{-1}\left(-\sqrt{2}\right)$

Answers:

(a) $60°$ (b) $-45°$

EXAMPLE 3 Finding Inverse Function Values (Degree-Measured Angles)

Find the *degree measure* of θ if it exists.

(a) $\theta = \arctan 1$ **(b)** $\theta = \sec^{-1} 2$

SOLUTION

(a) Here θ must be in $(-90°, 90°)$, but because 1 is positive, θ must be in quadrant I. The alternative statement, $\tan \theta = 1$, leads to $\theta = 45°$.

(b) Write the equation as $\sec \theta = 2$. For $\sec^{-1} x$, θ is in quadrant I or II. Because 2 is positive, θ is in quadrant I and $\theta = 60°$, since $\sec 60° = 2$. Note that $60°$ (the degree equivalent of $\frac{\pi}{3}$) is in the range of the inverse secant function.

Now Try Exercises 37 and 45.

The inverse trigonometric function keys on a calculator give correct results for the inverse sine, inverse cosine, and inverse tangent functions.

$$\sin^{-1} 0.5 = 30°, \quad \sin^{-1}(-0.5) = -30°,$$
$$\tan^{-1}(-1) = -45°, \quad \text{and} \quad \cos^{-1}(-0.5) = 120°$$

Degree mode

However, finding $\cot^{-1} x$, $\sec^{-1} x$, and $\csc^{-1} x$ with a calculator is not as straightforward because these functions must first be expressed in terms of $\tan^{-1} x$, $\cos^{-1} x$, and $\sin^{-1} x$, respectively. If $y = \sec^{-1} x$, for example, then we have $\sec y = x$, which must be written in terms of cosine as follows.

$$\text{If} \quad \sec y = x, \quad \text{then} \quad \frac{1}{\cos y} = x, \quad \text{or} \quad \cos y = \frac{1}{x}, \quad \text{and} \quad y = \cos^{-1} \frac{1}{x}.$$

Use the following to evaluate these inverse functions on a calculator.

$$\sec^{-1} x \textit{ is evaluated as } \cos^{-1}\frac{1}{x}; \quad \csc^{-1} x \textit{ is evaluated as } \sin^{-1}\frac{1}{x};$$

$$\cot^{-1} x \textit{ is evaluated as } \begin{cases} \tan^{-1}\frac{1}{x} & \text{if } x > 0 \\ 180° + \tan^{-1}\frac{1}{x} & \text{if } x < 0. \end{cases}$$ Degree mode

Classroom Example 4
Use a calculator to approximate each value.

(a) Find y in radians if $y = \sec^{-1}(-4)$.

(b) Find θ in degrees if $\theta = \text{arccot}(-0.2528)$.

Answers:

(a) 1.823476582

(b) 104.1871349°

EXAMPLE 4 Finding Inverse Function Values with a Calculator

Use a calculator to approximate each value.

(a) Find y in radians if $y = \csc^{-1}(-3)$.

(b) Find θ in degrees if $\theta = \text{arccot}(-0.3541)$.

SOLUTION

(a) With the calculator in radian mode, enter $\csc^{-1}(-3)$ as $\sin^{-1}\left(\frac{1}{-3}\right)$ to obtain $y \approx -0.3398369095$. See **Figure 15(a).**

(b) A calculator in degree mode gives the inverse tangent value of a negative number as a quadrant IV angle. The restriction on the range of arccotangent implies that θ must be in quadrant II.

$$\text{arccot}(-0.3541) \quad \text{is entered as} \quad \tan^{-1}\left(\frac{1}{-0.3541}\right) + 180°.$$

As shown in **Figure 15(b),**

$$\theta \approx 109.4990544°.$$

✔ **Now Try Exercises 53 and 65.**

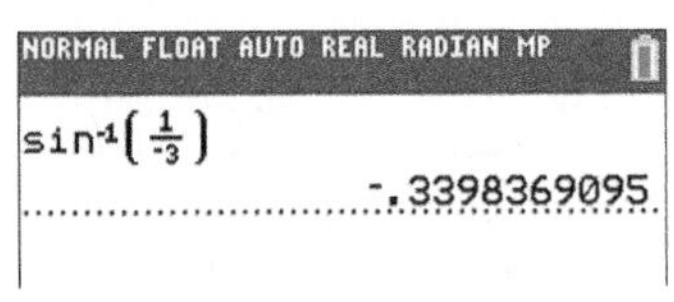

(a)

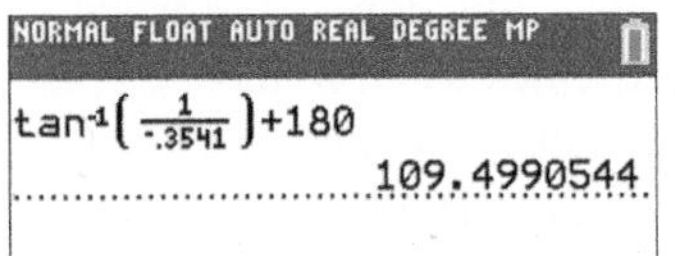

(b)

Figure 15

CAUTION ***Be careful when using a calculator to evaluate the inverse cotangent of a negative quantity.*** Enter the inverse tangent of the *reciprocal* of the negative quantity, which returns an angle in quadrant IV. Because inverse cotangent is negative in quadrant II, adjust the calculator result by adding π or 180° accordingly. (Note that $\cot^{-1} 0 = \frac{\pi}{2}$ or 90°.)

Classroom Example 5
Evaluate each expression without using a calculator.

(a) $\cos\left(\sin^{-1}\frac{2}{3}\right)$

(b) $\sec\left(\cot^{-1}\left(-\frac{15}{8}\right)\right)$

Answers:

(a) $\frac{\sqrt{5}}{3}$ (b) $-\frac{17}{15}$

EXAMPLE 5 Finding Function Values Using Definitions of the Trigonometric Functions

Evaluate each expression without using a calculator.

(a) $\sin\left(\tan^{-1}\frac{3}{2}\right)$ **(b)** $\tan\left(\cos^{-1}\left(-\frac{5}{13}\right)\right)$

SOLUTION

(a) Let $\theta = \tan^{-1}\frac{3}{2}$, and thus $\tan\theta = \frac{3}{2}$. The inverse tangent function yields values only in quadrants I and IV, and because $\frac{3}{2}$ is positive, θ is in quadrant I. Sketch θ in quadrant I, and label a triangle, as shown in **Figure 16** on the next page. By the Pythagorean theorem, the hypotenuse is $\sqrt{13}$. The value of sine is the quotient of the side opposite and the hypotenuse.

$$\sin\left(\tan^{-1}\frac{3}{2}\right) = \sin\theta = \frac{3}{\sqrt{13}} = \frac{3}{\sqrt{13}} \cdot \frac{\sqrt{13}}{\sqrt{13}} = \frac{3\sqrt{13}}{13}$$

Rationalize the denominator.

15. $\left\{\frac{3\pi}{4}, \frac{7\pi}{4}\right\}$ 16. $\left\{\frac{\pi}{2}\right\}$

17. $\left\{\frac{\pi}{6}, \frac{5\pi}{6}\right\}$ 18. $\left\{\frac{\pi}{3}, \frac{5\pi}{3}\right\}$

19. $\emptyset$ 20. $\emptyset$

21. $\left\{\frac{\pi}{4}, \frac{2\pi}{3}, \frac{5\pi}{4}, \frac{5\pi}{3}\right\}$

22. $\left\{\frac{\pi}{4}, \frac{3\pi}{4}, \frac{7\pi}{6}, \frac{11\pi}{6}\right\}$

23. $\{\pi\}$

24. $\left\{\frac{\pi}{6}, \frac{\pi}{2}, \frac{3\pi}{2}, \frac{11\pi}{6}\right\}$

25. $\left\{\frac{7\pi}{6}, \frac{3\pi}{2}, \frac{11\pi}{6}\right\}$

26. $\left\{0, \frac{2\pi}{3}, \frac{4\pi}{3}\right\}$

27. $\{30°, 210°, 240°, 300°\}$
28. $\{0°, 45°, 225°\}$
29. $\{90°, 210°, 330°\}$
30. $\{60°, 135°, 240°, 315°\}$
31. $\{45°, 135°, 225°, 315°\}$
32. $\{0°, 180°\}$
33. $\{45°, 225°\}$ 34. $\{90°, 270°\}$
35. $\{0°, 30°, 150°, 180°\}$
36. $\{0°, 90°, 180°, 270°\}$
37. $\{0°, 45°, 135°, 180°, 225°, 315°\}$
38. $\{45°, 135°, 225°, 315°\}$
39. $\{53.6°, 126.4°, 187.9°, 352.1°\}$
40. $\{78.0°, 282.0°\}$
41. $\{149.6°, 329.6°, 106.3°, 286.3°\}$
42. $\{38.4°, 218.4°, 104.8°, 284.8°\}$
43. $\emptyset$ 44. $\emptyset$
45. $\{57.7°, 159.2°\}$
46. $\{114.3°, 335.7°\}$

47. $\{180° + 360°n$, where n is any integer$\}$

48. $\{135° + 180°n$, where n is any integer$\}$

49. $\left\{\frac{\pi}{3} + 2n\pi, \frac{2\pi}{3} + 2n\pi\right.$, where n is any integer$\left.\right\}$

50. $\left\{\frac{5\pi}{6} + n\pi\right.$, where n is any integer$\left.\right\}$

51. $\{19.5° + 360°n, 160.5° + 360°n, 210° + 360°n, 330° + 360°n$, where n is any integer$\}$

52. $\{90° + 360°n, 221.8° + 360°n, 318.2° + 360°n$, where n is any integer$\}$

53. $\left\{\frac{\pi}{3} + 2n\pi, \pi + 2n\pi, \frac{5\pi}{3} + 2n\pi\right.$, where n is any integer$\left.\right\}$

13. ***Concept Check*** Suppose that in solving an equation over the interval $[0°, 360°)$, we reach the step $\sin\theta = -\frac{1}{2}$. Why is $-30°$ not a correct answer?

14. ***Concept Check*** Lindsay solved the equation $\sin x = 1 - \cos x$ by squaring each side to obtain

$$\sin^2 x = 1 - 2\cos x + \cos^2 x.$$

Several steps later, using correct algebra, she concluded that the solution set for solutions over the interval $[0, 2\pi)$ is $\left\{0, \frac{\pi}{2}, \frac{3\pi}{2}\right\}$. Explain why this is not correct.

*Solve each equation for exact solutions over the interval $[0, 2\pi)$. **See Examples 1–3.***

15. $2\cot x + 1 = -1$
16. $\sin x + 2 = 3$
17. $2\sin x + 3 = 4$
18. $2\sec x + 1 = \sec x + 3$
19. $\tan^2 x + 3 = 0$
20. $\sec^2 x + 2 = -1$
21. $(\cot x - 1)\left(\sqrt{3}\cot x + 1\right) = 0$
22. $(\csc x + 2)\left(\csc x - \sqrt{2}\right) = 0$
23. $\cos^2 x + 2\cos x + 1 = 0$
24. $2\cos^2 x - \sqrt{3}\cos x = 0$
25. $-2\sin^2 x = 3\sin x + 1$
26. $2\cos^2 x - \cos x = 1$

*Solve each equation for solutions over the interval $[0°, 360°)$. Give solutions to the nearest tenth as appropriate. **See Examples 2–5.***

27. $\left(\cot\theta - \sqrt{3}\right)\left(2\sin\theta + \sqrt{3}\right) = 0$
28. $(\tan\theta - 1)(\cos\theta - 1) = 0$
29. $2\sin\theta - 1 = \csc\theta$
30. $\tan\theta + 1 = \sqrt{3} + \sqrt{3}\cot\theta$
31. $\tan\theta - \cot\theta = 0$
32. $\cos^2\theta = \sin^2\theta + 1$
33. $\csc^2\theta - 2\cot\theta = 0$
34. $\sin^2\theta\cos\theta = \cos\theta$
35. $2\tan^2\theta\sin\theta - \tan^2\theta = 0$
36. $\sin^2\theta\cos^2\theta = 0$
37. $\sec^2\theta\tan\theta = 2\tan\theta$
38. $\cos^2\theta - \sin^2\theta = 0$
39. $9\sin^2\theta - 6\sin\theta = 1$
40. $4\cos^2\theta + 4\cos\theta = 1$
41. $\tan^2\theta + 4\tan\theta + 2 = 0$
42. $3\cot^2\theta - 3\cot\theta - 1 = 0$
43. $\sin^2\theta - 2\sin\theta + 3 = 0$
44. $2\cos^2\theta + 2\cos\theta + 1 = 0$
45. $\cot\theta + 2\csc\theta = 3$
46. $2\sin\theta = 1 - 2\cos\theta$

*Solve each equation (x in radians and θ in degrees) for all exact solutions where appropriate. Round approximate answers in radians to four decimal places and approximate answers in degrees to the nearest tenth. Write answers using the least possible nonnegative angle measures. **See Examples 1–5.***

47. $\cos\theta + 1 = 0$
48. $\tan\theta + 1 = 0$
49. $3\csc x - 2\sqrt{3} = 0$
50. $\cot x + \sqrt{3} = 0$
51. $6\sin^2\theta + \sin\theta = 1$
52. $3\sin^2\theta - \sin\theta = 2$
53. $2\cos^2 x + \cos x - 1 = 0$
54. $4\cos^2 x - 1 = 0$
55. $\sin\theta\cos\theta - \sin\theta = 0$
56. $\tan\theta\csc\theta - \sqrt{3}\csc\theta = 0$
57. $\sin x(3\sin x - 1) = 1$
58. $\tan x(\tan x - 2) = 5$

54. $\left\{\frac{\pi}{3} + n\pi, \frac{2\pi}{3} + n\pi\text{, where } n \text{ is any integer}\right\}$

55. $\{180°n$, where n is any integer$\}$

56. $\{60° + 180°n$, where n is any integer$\}$

57. $\{0.8751 + 2n\pi, 2.2665 + 2n\pi, 3.5908 + 2n\pi, 5.8340 + 2n\pi$, where n is any integer$\}$

58. $\{1.2886 + n\pi, 2.1747 + n\pi$, where n is any integer$\}$

59. $\{33.6° + 360°n, 326.4° + 360°n$, where n is any integer$\}$

60. $\{71.6° + 180°n, 135° + 180°n$, where n is any integer$\}$

61. $\{45° + 180°n, 108.4° + 180°n$, where n is any integer$\}$

62. $\{33.7° + 180°n, 135° + 180°n$, where n is any integer$\}$

63. $\{0.6806, 1.4159\}$

64. $\{0, 0.3760\}$

65. (a) 0.00164 and 0.00355 (b) $[0.00164, 0.00355]$ (c) outward

66. 14°

67. (a) $\frac{1}{4}$ sec (b) $\frac{1}{6}$ sec (c) 0.21 sec

68. (a) 2 sec (b) $3\frac{1}{3}$ sec

59. $5 + 5\tan^2\theta = 6\sec\theta$

60. $\sec^2\theta = 2\tan\theta + 4$

61. $\dfrac{2\tan\theta}{3 - \tan^2\theta} = 1$

62. $\dfrac{2\cot^2\theta}{\cot\theta + 3} = 1$

The following equations cannot be solved by algebraic methods. Use a graphing calculator to find all solutions over the interval $[0, 2\pi)$. Express solutions to four decimal places.

63. $x^2 + \sin x - x^3 - \cos x = 0$

64. $x^3 - \cos^2 x = \frac{1}{2}x - 1$

(Modeling) Solve each problem.

65. ***Pressure on the Eardrum*** **See Example 6.** No musical instrument can generate a true pure tone. A pure tone has a unique, constant frequency and amplitude that sounds rather dull and uninteresting. The pressures caused by pure tones on the eardrum are sinusoidal. The change in pressure P in pounds per square foot on a person's eardrum from a pure tone at time t in seconds can be modeled using the equation

$$P = A\sin(2\pi ft + \phi),$$

where f is the frequency in cycles per second, and ϕ is the phase angle. When P is positive, there is an increase in pressure and the eardrum is pushed inward. When P is negative, there is a decrease in pressure and the eardrum is pushed outward. (*Source*: Roederer, J., *Introduction to the Physics and Psychophysics of Music*, Second Edition, Springer-Verlag.)

(a) Determine algebraically the values of t for which $P = 0$ over $[0, 0.005]$.

(b) From a graph and the answer in part (a), determine the interval for which $P \le 0$ over $[0, 0.005]$.

(c) Would an eardrum hearing this tone be vibrating outward or inward when $P < 0$?

66. ***Accident Reconstruction*** To reconstruct accidents in which a vehicle vaults into the air after hitting an obstruction, the model

$$0.342D\cos\theta + h\cos^2\theta = \frac{16D^2}{V_0^{\,2}}$$

can be used. V_0 is velocity in feet per second of the vehicle when it hits the obstruction, D is distance (in feet) from the obstruction to the landing point, and h is the difference in height (in feet) between landing point and takeoff point. Angle θ is the takeoff angle, the angle between the horizontal and the path of the vehicle. Find θ to the nearest degree if $V_0 = 60$, $D = 80$, and $h = 2$.

67. ***Electromotive Force*** In an electric circuit, suppose that the electromotive force in volts at t seconds can be modeled by

$$V = \cos 2\pi t.$$

Find the least value of t where $0 \le t \le \frac{1}{2}$ for each value of V.

(a) $V = 0$ **(b)** $V = 0.5$ **(c)** $V = 0.25$

68. ***Voltage Induced by a Coil of Wire*** A coil of wire rotating in a magnetic field induces a voltage modeled by

$$E = 20\sin\left(\frac{\pi t}{4} - \frac{\pi}{2}\right),$$

where t is time in seconds. Find the least positive time to produce each voltage.

(a) 0 **(b)** $10\sqrt{3}$

6.3 Trigonometric Equations II

- Equations with Half-Angles
- Equations with Multiple Angles
- An Application

In this section, we discuss trigonometric equations that involve functions of half-angles and multiple angles. Solving these equations often requires adjusting solution intervals to fit given domains.

Teaching Tip As a slight variation of the problem in **Example 1,** replace $\frac{x}{2}$ with u, solve $2 \sin u = 1$ for u, and then multiply the solutions by 2 to find x.

Equations with Half-Angles

EXAMPLE 1 Solving an Equation with a Half-Angle

Solve $2 \sin \frac{x}{2} = 1$

(a) over the interval $[0, 2\pi)$ **(b)** for all solutions.

Classroom Example 1

Solve $2 \cos \frac{x}{2} - \sqrt{2} = 0$

(a) over the interval $[0, 2\pi)$

(b) for all solutions.

Answers:

(a) $\left\{\frac{\pi}{2}\right\}$

(b) $\left\{\frac{\pi}{2} + 4n\pi, \frac{7\pi}{2} + 4n\pi,\right.$ where n is any integer$\left.\right\}$

SOLUTION

(a) To solve over the interval $[0, 2\pi)$, we must have

$$0 \leq x < 2\pi.$$

The corresponding inequality for $\frac{x}{2}$ is

$$0 \leq \frac{x}{2} < \pi. \quad \text{Divide by 2.}$$

To find all values of $\frac{x}{2}$ over the interval $[0, \pi)$ that satisfy the given equation, first solve for $\sin \frac{x}{2}$.

$$2 \sin \frac{x}{2} = 1 \quad \text{Original equation}$$

$$\sin \frac{x}{2} = \frac{1}{2} \quad \text{Divide by 2.}$$

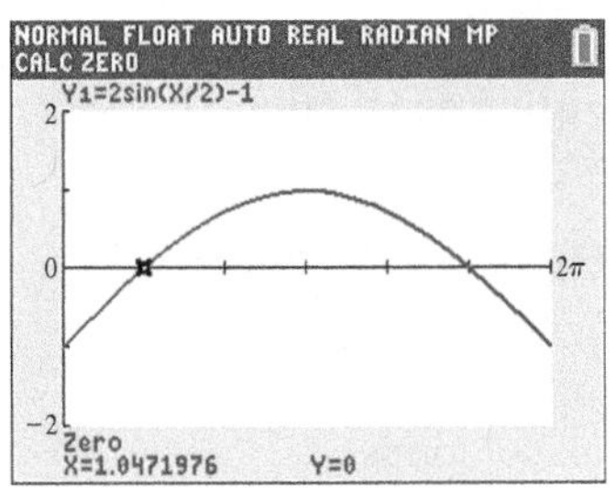

The x-intercepts correspond to the solutions found in **Example 1(a).** Using Xscl $= \frac{\pi}{3}$ makes it possible to support the exact solutions by counting the tick marks from 0 on the graph.

The two numbers over the interval $[0, \pi)$ with sine value $\frac{1}{2}$ are $\frac{\pi}{6}$ and $\frac{5\pi}{6}$.

$$\frac{x}{2} = \frac{\pi}{6} \quad \text{or} \quad \frac{x}{2} = \frac{5\pi}{6} \quad \text{Definition of inverse sine}$$

$$x = \frac{\pi}{3} \quad \text{or} \quad x = \frac{5\pi}{3} \quad \text{Multiply by 2.}$$

The solution set over the given interval is $\left\{\frac{\pi}{3}, \frac{5\pi}{3}\right\}$.

(b) The argument $\frac{x}{2}$ in the expression $\sin \frac{x}{2}$ can also be written $\frac{1}{2}x$ to see that the value of b in $\sin bx$ is $\frac{1}{2}$. From earlier work we know that the period is $\frac{2\pi}{b}$, so we replace b with $\frac{1}{2}$ in this expression and perform the calculation. Here the period is

$$\frac{2\pi}{\frac{1}{2}} = 2\pi \div \frac{1}{2} = 2\pi \cdot 2 = 4\pi.$$

All solutions are found by adding integer multiples of 4π.

$$\left\{\frac{\pi}{3} + 4n\pi, \frac{5\pi}{3} + 4n\pi, \text{ where } n \text{ is any integer}\right\}$$

✔ **Now Try Exercises 25 and 39.**

Equations with Multiple Angles

Classroom Example 2
Solve $\cos 2x = \sin x$ over the interval $[0, 2\pi)$.

Answer: $\left\{\frac{\pi}{6}, \frac{5\pi}{6}, \frac{3\pi}{2}\right\}$

EXAMPLE 2 Solving an Equation Using a Double-Angle Identity

Solve $\cos 2x = \cos x$ over the interval $[0, 2\pi)$.

SOLUTION First convert $\cos 2x$ to a function of x alone. Use the identity $\cos 2x = 2\cos^2 x - 1$ so that the equation involves only $\cos x$. Then factor.

$$\cos 2x = \cos x \quad \text{Original equation}$$
$$2\cos^2 x - 1 = \cos x \quad \text{Cosine double-angle identity}$$
$$2\cos^2 x - \cos x - 1 = 0 \quad \text{Subtract } \cos x.$$
$$(2\cos x + 1)(\cos x - 1) = 0 \quad \text{Factor.}$$
$$2\cos x + 1 = 0 \quad \text{or} \quad \cos x - 1 = 0 \quad \text{Zero-factor property}$$
$$\cos x = -\frac{1}{2} \quad \text{or} \quad \cos x = 1 \quad \text{Solve each equation for } \cos x.$$

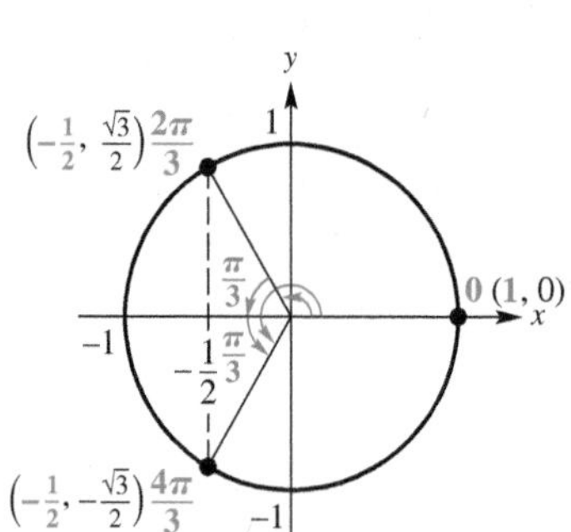

Figure 28

If we use the unit circle to analyze these results, we recognize that a radian-measured angle having cosine $-\frac{1}{2}$ must be in quadrant II or III with reference angle $\frac{\pi}{3}$. Another possibility is that it has a value of 1 at 0 radians. We can use **Figure 28** to determine that solutions over the required interval are as follows.

$$x = \frac{2\pi}{3} \quad \text{or} \quad x = \frac{4\pi}{3} \quad \text{or} \quad x = 0$$

The solution set is $\left\{0, \frac{2\pi}{3}, \frac{4\pi}{3}\right\}$. ✔ **Now Try Exercise 27.**

CAUTION Because 2 is not a factor of $\cos 2x$, $\frac{\cos 2x}{2} \neq \cos x$. In **Example 2,** we changed $\cos 2x$ to a function of x alone using an identity.

Classroom Example 3
Solve $2\cos^2\theta - 2\sin^2\theta + 1 = 0$
(a) over the interval $[0, 360°)$
(b) for all solutions.

Answers:
(a) $\{60°, 120°, 240°, 300°\}$
(b) $\{60° + 180°n, 120° + 180°n$, where n is any integer$\}$

EXAMPLE 3 Solving an Equation Using a Double-Angle Identity

Solve $4\sin\theta\cos\theta = \sqrt{3}$

(a) over the interval $[0°, 360°)$ **(b)** for all solutions.

SOLUTION

(a)
$$4\sin\theta\cos\theta = \sqrt{3} \quad \text{Original equation}$$
$$2(2\sin\theta\cos\theta) = \sqrt{3} \quad 4 = 2 \cdot 2$$
$$2\sin 2\theta = \sqrt{3} \quad \text{Sine double-angle identity}$$
$$\sin 2\theta = \frac{\sqrt{3}}{2} \quad \text{Divide by 2.}$$

From the given interval $0° \le \theta < 360°$, the corresponding interval for 2θ is $0° \le 2\theta < 720°$. Because the sine is positive in quadrants I and II, solutions over this interval are as follows.

$$2\theta = 60°, 120°, 420°, 480°, \quad \text{Reference angle is } 60°.$$
$$\text{or} \quad \theta = 30°, 60°, 210°, 240° \quad \text{Divide by 2.}$$

The final two solutions for 2θ were found by adding 360° to 60° and 120°, respectively, which gives the solution set $\{30°, 60°, 210°, 240°\}$.

Teaching Tip Students should be aware that

$$y = \sin ax$$

may have as many as $2a$ solutions from 0° to 360°.

(b) All angles 2θ that are solutions of the equation $\sin 2\theta = \frac{\sqrt{3}}{2}$ are found by adding integer multiples of 360° to the basic solution angles, 60° and 120°.

$$2\theta = 60° + 360°n \quad \text{and} \quad 2\theta = 120° + 360°n \qquad \text{Add integer multiples of } 360°.$$

$$\theta = 30° + 180°n \quad \text{and} \quad \theta = 60° + 180°n \qquad \text{Divide by 2.}$$

All solutions are given by the following set, where 180° represents the period of $\sin 2\theta$.

$$\{30° + 180°n, 60° + 180°n, \text{ where } n \text{ is any integer}\}$$

✔ **Now Try Exercises 23 and 47.**

NOTE Solving an equation by squaring both sides may introduce extraneous values. We use this method in **Example 4,** and all proposed solutions must be checked.

Classroom Example 4
Solve $\cot 2x - \csc 2x = 1$ over the interval $[0, 2\pi)$.

Answer: $\left\{\frac{3\pi}{4}, \frac{7\pi}{4}\right\}$

EXAMPLE 4 Solving an Equation with a Multiple Angle

Solve $\tan 3x + \sec 3x = 2$ over the interval $[0, 2\pi)$.

SOLUTION The tangent and secant functions are related by the identity $1 + \tan^2 x = \sec^2 x$. One way to begin is to express the left side in terms of secant.

$$\tan 3x + \sec 3x = 2$$

$$\tan 3x = 2 - \sec 3x \qquad \text{Subtract } \sec 3x.$$

$$(\tan 3x)^2 = (2 - \sec 3x)^2 \qquad \text{Square each side.}$$

$$\tan^2 3x = 4 - 4\sec 3x + \sec^2 3x \qquad (x-y)^2 = x^2 - 2xy + y^2$$

$$\sec^2 3x - 1 = 4 - 4\sec 3x + \sec^2 3x \qquad \text{Replace } \tan^2 3x \text{ with } \sec^2 3x - 1.$$

$$4\sec 3x = 5 \qquad \text{Simplify.}$$

$$\sec 3x = \frac{5}{4} \qquad \text{Divide by 4.}$$

$$\frac{1}{\cos 3x} = \frac{5}{4} \qquad \sec\theta = \frac{1}{\cos\theta}$$

$$\cos 3x = \frac{4}{5} \qquad \text{Use reciprocals.}$$

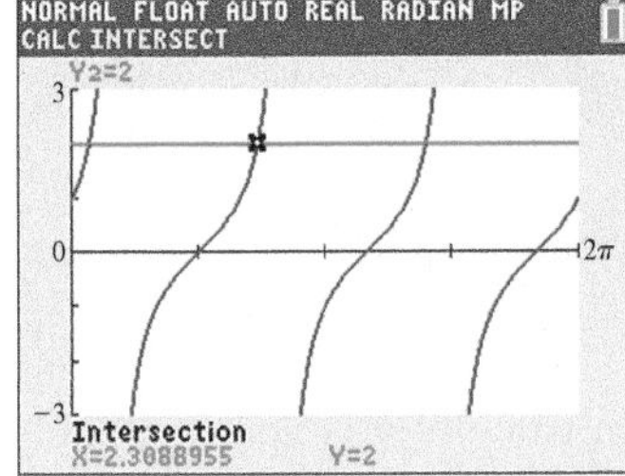

The screen shows the graphs of

$$y_1 = \tan 3x + \sec 3x$$

and

$$y_2 = 2.$$

One solution is approximately 2.3089. An advantage of using a graphing calculator is that extraneous values do not appear.

Multiply each term of the inequality $0 \le x < 2\pi$ by 3 to find the interval for $3x$: $[0, 6\pi)$. Use a calculator and the fact that cosine is positive in quadrants I and IV.

$$3x \approx 0.6435, 5.6397, 6.9267, 11.9229, 13.2099, 18.2061 \qquad \text{These numbers have cosine value equal to } \tfrac{4}{5}.$$

$$x \approx 0.2145, 1.8799, 2.3089, 3.9743, 4.4033, 6.0687 \qquad \text{Divide by 3.}$$

Both sides of the equation were squared, so each proposed solution must be checked. Verify by substitution in the given equation that the solution set is

$$\{0.2145, 2.3089, 4.4033\}.$$

✔ **Now Try Exercise 53.**

An Application A piano string can vibrate at more than one frequency when it is struck. It produces a complex wave that can mathematically be modeled by a sum of several pure tones. When a piano key with a frequency of f_1 is played, the corresponding string vibrates not only at f_1 but also at the higher frequencies of $2f_1, 3f_1, 4f_1, \ldots, nf_1$. f_1 is the **fundamental frequency** of the string, and higher frequencies are the **upper harmonics.** The human ear will hear the sum of these frequencies as one complex tone. (*Source*: Roederer, J., *Introduction to the Physics and Psychophysics of Music,* Second Edition, Springer-Verlag.)

EXAMPLE 5 Analyzing Pressures of Upper Harmonics

Suppose that the A key above middle C is played on a piano. Its fundamental frequency is $f_1 = 440$ Hz, and its associated pressure is expressed as

$$P_1 = 0.002 \sin 880\pi t.$$

Classroom Example 5
Repeat **Example 5,** but in each case, replace the word "maximum" with "minimum."

Answers:
(a) -0.00317
(b) 0.00208, 0.00436, 0.00663, 0.00890

The string will also vibrate at

$$f_2 = 880, \quad f_3 = 1320, \quad f_4 = 1760, \quad f_5 = 2200, \ldots \text{ Hz.}$$

The corresponding pressures of these upper harmonics are as follows.

$$P_2 = \frac{0.002}{2} \sin 1760\pi t, \qquad P_3 = \frac{0.002}{3} \sin 2640\pi t,$$

$$P_4 = \frac{0.002}{4} \sin 3520\pi t, \qquad \text{and} \qquad P_5 = \frac{0.002}{5} \sin 4400\pi t$$

The graph of $P = P_1 + P_2 + P_3 + P_4 + P_5$ can be found by entering each P_i as a separate function y_i and graphing their sum. The graph, shown in **Figure 29,** is "saw-toothed."

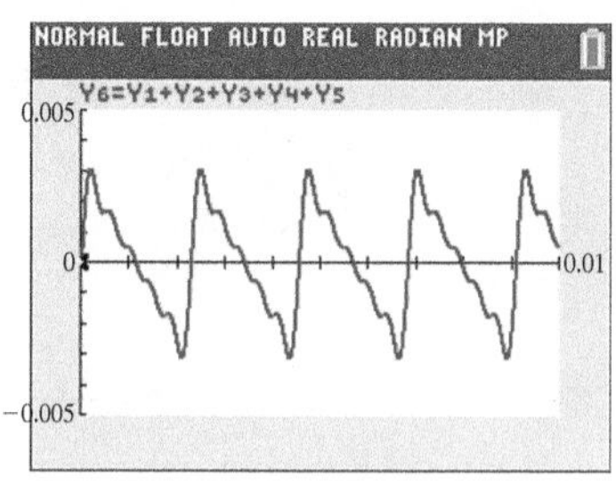

Figure 29

(a) Approximate the maximum value of P.

(b) At what values of $t = x$ does this maximum occur over $[0, 0.01]$?

SOLUTION

(a) A graphing calculator shows that the maximum value of P is approximately 0.00317. See **Figure 30.**

(b) The maximum occurs at

$$t = x \approx 0.000191, 0.00246, 0.00474, 0.00701, \text{ and } 0.00928.$$

Figure 30 shows how the second value is found. The other values are found similarly.

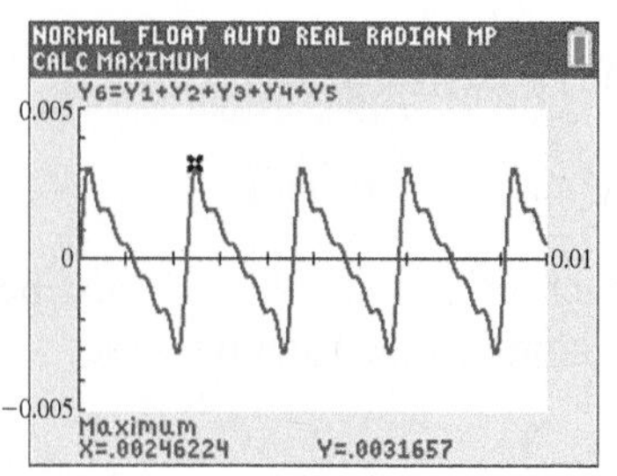

Figure 30

✔ **Now Try Exercise 57.**

6.3 Exercises

1. $\left\{\frac{\pi}{6}, \frac{5\pi}{6}, \frac{7\pi}{6}, \frac{11\pi}{6}\right\}$
2. $\left\{\frac{\pi}{12}, \frac{11\pi}{12}, \frac{13\pi}{12}, \frac{23\pi}{12}\right\}$
3. $\left\{\frac{7\pi}{12}, \frac{11\pi}{12}, \frac{19\pi}{12}, \frac{23\pi}{12}\right\}$
4. $\left\{\frac{2\pi}{3}, \frac{5\pi}{6}, \frac{5\pi}{3}, \frac{11\pi}{6}\right\}$
5. $\left\{\frac{\pi}{2}, \frac{3\pi}{2}\right\}$
6. $\left\{\frac{\pi}{4}, \frac{3\pi}{4}, \frac{5\pi}{4}, \frac{7\pi}{4}\right\}$
7. $\{0°\}$ 8. $\varnothing$
9. $\{240°\}$ 10. $\{270°\}$
11. $\varnothing$ 12. $\varnothing$
13. $\left\{\frac{\pi}{3}, \pi, \frac{4\pi}{3}\right\}$ 14. $\left\{\frac{\pi}{8}, \frac{5\pi}{6}, \frac{5\pi}{4}\right\}$
15. $\{60°, 210°, 240°, 310°\}$
16. $\{135°, 180°, 225°, 270°\}$
17. $\left\{\frac{\pi}{12}, \frac{11\pi}{12}, \frac{13\pi}{12}, \frac{23\pi}{12}\right\}$
18. $\left\{\frac{\pi}{3}, \frac{2\pi}{3}, \frac{4\pi}{3}, \frac{5\pi}{3}\right\}$
19. $\{90°, 210°, 330°\}$
20. $\{0°, 60°, 120°, 180°, 240°, 300°\}$
21. $\left\{\frac{\pi}{18}, \frac{7\pi}{18}, \frac{13\pi}{18}, \frac{19\pi}{18}, \frac{25\pi}{18}, \frac{31\pi}{18}\right\}$
22. $\left\{\frac{\pi}{18}, \frac{7\pi}{18}, \frac{13\pi}{18}, \frac{19\pi}{18}, \frac{25\pi}{18}, \frac{31\pi}{18}\right\}$
23. $\{67.5°, 112.5°, 247.5°, 292.5°\}$
24. $\{15°, 75°, 195°, 255°\}$
25. $\left\{\frac{\pi}{2}, \frac{3\pi}{2}\right\}$
26. $\left\{0, \frac{\pi}{4}, \frac{\pi}{2}, \frac{3\pi}{4}, \pi, \frac{5\pi}{4}, \frac{3\pi}{2}, \frac{7\pi}{4}\right\}$
27. $\left\{0, \frac{\pi}{3}, \pi, \frac{5\pi}{3}\right\}$ 28. $\left\{0, \frac{2\pi}{3}, \frac{4\pi}{3}\right\}$
29. $\varnothing$ 30. $\varnothing$
31. $\{180°\}$ 32. $\{0°\}$
33. $\left\{\frac{\pi}{3}, \pi, \frac{5\pi}{3}\right\}$
34. $\left\{\frac{\pi}{12}, \frac{5\pi}{12}, \frac{13\pi}{12}, \frac{17\pi}{12}\right\}$
35. $\left\{\frac{\pi}{12} + \frac{2n\pi}{3}, \frac{\pi}{4} + \frac{2n\pi}{3}, \text{ where } n \text{ is any integer}\right\}$
36. $\left\{\frac{5\pi}{12} + n\pi, \frac{7\pi}{12} + n\pi, \text{ where } n \text{ is any integer}\right\}$
37. $\{720°n, \text{ where } n \text{ is any integer}\}$
38. $\{180° + 720°n, \text{ where } n \text{ is any integer}\}$
39. $\left\{\frac{2\pi}{3} + 4n\pi, \frac{4\pi}{3} + 4n\pi, \text{ where } n \text{ is any integer}\right\}$

CONCEPT PREVIEW *Refer to* ***Exercises 1–6*** *in the previous section, and use those results to solve each equation over the interval* $[0, 2\pi)$.

1. $\cos 2x = \frac{1}{2}$
2. $\cos 2x = \frac{\sqrt{3}}{2}$
3. $\sin 2x = -\frac{1}{2}$
4. $\sin 2x = -\frac{\sqrt{3}}{2}$
5. $\cos 2x = -1$
6. $\cos 2x = 0$

CONCEPT PREVIEW *Refer to* ***Exercises 7–12*** *in the previous section, and use those results to solve each equation over the interval* $[0°, 360°)$.

7. $\sin \frac{\theta}{2} = 0$
8. $\sin \frac{\theta}{2} = -1$
9. $\cos \frac{\theta}{2} = -\frac{1}{2}$
10. $\cos \frac{\theta}{2} = -\frac{\sqrt{2}}{2}$
11. $\sin \frac{\theta}{2} = -\frac{\sqrt{2}}{2}$
12. $\sin \frac{\theta}{2} = -\frac{\sqrt{3}}{2}$

Concept Check Answer each question.

13. Suppose solving a trigonometric equation for solutions over the interval $[0, 2\pi)$ leads to $2x = \frac{2\pi}{3}, 2\pi, \frac{8\pi}{3}$. What are the corresponding values of x?
14. Suppose solving a trigonometric equation for solutions over the interval $[0, 2\pi)$ leads to $\frac{1}{2}x = \frac{\pi}{16}, \frac{5\pi}{12}, \frac{5\pi}{8}$. What are the corresponding values of x?
15. Suppose solving a trigonometric equation for solutions over the interval $[0°, 360°)$ leads to $3\theta = 180°, 630°, 720°, 930°$. What are the corresponding values of θ?
16. Suppose solving a trigonometric equation for solutions over the interval $[0°, 360°)$ leads to $\frac{1}{3}\theta = 45°, 60°, 75°, 90°$. What are the corresponding values of θ?

Solve each equation in x for exact solutions over the interval $[0, 2\pi)$ *and each equation in θ for exact solutions over the interval* $[0°, 360°)$. ***See Examples 1–4.***

17. $2 \cos 2x = \sqrt{3}$
18. $2 \cos 2x = -1$
19. $\sin 3\theta = -1$
20. $\sin 3\theta = 0$
21. $3 \tan 3x = \sqrt{3}$
22. $\cot 3x = \sqrt{3}$
23. $\sqrt{2} \cos 2\theta = -1$
24. $2\sqrt{3} \sin 2\theta = \sqrt{3}$
25. $\sin \frac{x}{2} = \sqrt{2} - \sin \frac{x}{2}$
26. $\tan 4x = 0$
27. $\sin x = \sin 2x$
28. $\cos 2x - \cos x = 0$
29. $8 \sec^2 \frac{x}{2} = 4$
30. $\sin^2 \frac{x}{2} - 2 = 0$
31. $\sin \frac{\theta}{2} = \csc \frac{\theta}{2}$
32. $\sec \frac{\theta}{2} = \cos \frac{\theta}{2}$
33. $\cos 2x + \cos x = 0$
34. $\sin x \cos x = \frac{1}{4}$

Solve each equation (x in radians and θ in degrees) for all exact solutions where appropriate. Round approximate answers in radians to four decimal places and approximate answers in degrees to the nearest tenth. Write answers using the least possible nonnegative angle measures. ***See Examples 1–4.***

35. $\sqrt{2} \sin 3x - 1 = 0$
36. $-2 \cos 2x = \sqrt{3}$
37. $\cos \frac{\theta}{2} = 1$
38. $\sin \frac{\theta}{2} = 1$
39. $2\sqrt{3} \sin \frac{x}{2} = 3$
40. $2\sqrt{3} \cos \frac{x}{2} = -3$

41. $2 \sin \theta = 2 \cos 2\theta$ **42.** $\cos \theta - 1 = \cos 2\theta$ **43.** $1 - \sin x = \cos 2x$

44. $\sin 2x = 2 \cos^2 x$ **45.** $3 \csc^2 \frac{x}{2} = 2 \sec x$ **46.** $\cos x = \sin^2 \frac{x}{2}$

47. $2 - \sin 2\theta = 4 \sin 2\theta$ **48.** $4 \cos 2\theta = 8 \sin \theta \cos \theta$

49. $2 \cos^2 2\theta = 1 - \cos 2\theta$ **50.** $\sin \theta - \sin 2\theta = 0$

Solve each equation for solutions over the interval $[0, 2\pi)$. Write solutions as exact values or to four decimal places, as appropriate. ***See Example 4.***

51. $\sin \frac{x}{2} - \cos \frac{x}{2} = 0$ **52.** $\sin \frac{x}{2} + \cos \frac{x}{2} = 1$

53. $\tan 2x + \sec 2x = 3$ **54.** $\tan 2x - \sec 2x = 2$

The following equations cannot be solved by algebraic methods. Use a graphing calculator to find all solutions over the interval $[0, 2\pi)$. Express solutions to four decimal places.

55. $2 \sin 2x - x^3 + 1 = 0$ **56.** $3 \cos \frac{x}{2} + \sqrt{x} - 2 = -\frac{1}{2}x + 2$

(Modeling) Solve each problem. ***See Example 5.***

57. ***Pressure of a Plucked String*** If a string with a fundamental frequency of 110 Hz is plucked in the middle, it will vibrate at the odd harmonics of 110, 330, 550, . . . Hz but not at the even harmonics of 220, 440, 660, . . . Hz. The resulting pressure P caused by the string is graphed below and can be modeled by the following equation.

$$P = 0.003 \sin 220\pi t + \frac{0.003}{3} \sin 660\pi t + \frac{0.003}{5} \sin 1100\pi t + \frac{0.003}{7} \sin 1540\pi t$$

(*Source:* Benade, A., *Fundamentals of Musical Acoustics,* Dover Publications. Roederer, J., *Introduction to the Physics and Psychophysics of Music,* Second Edition, Springer-Verlag.)

(a) Duplicate the graph shown here.

(b) Describe the shape of the sound wave that is produced.

(c) At lower frequencies, the inner ear will hear a tone only when the eardrum is moving outward. This occurs when P is negative. Determine the times over the interval $[0, 0.03]$ when this will occur.

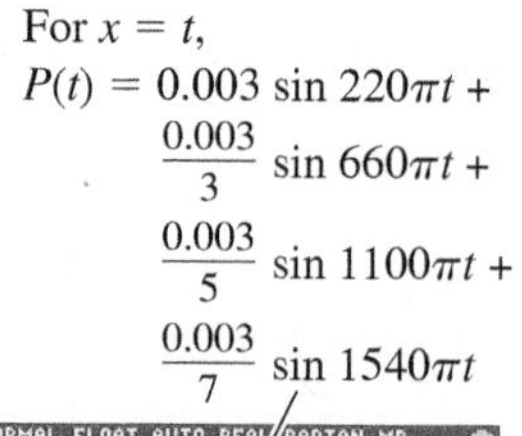

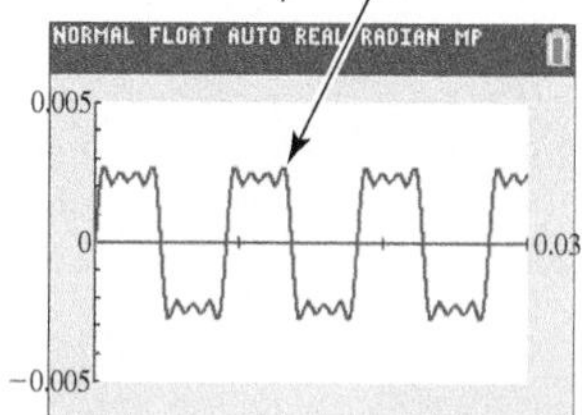

58. ***Hearing Beats in Music*** Musicians sometimes tune instruments by playing the same tone on two different instruments and listening for a phenomenon known as **beats.** Beats occur when two tones vary in frequency by only a few hertz. When the two instruments are in tune, the beats disappear. The ear hears beats because the pressure slowly rises and falls as a result of this slight variation in the frequency. (*Source:* Pierce, J., *The Science of Musical Sound,* Scientific American Books.)

(a) Consider the two tones with frequencies of 220 Hz and 223 Hz and pressures $P_1 = 0.005 \sin 440\pi t$ and $P_2 = 0.005 \sin 446\pi t$, respectively. A graph of the pressure $P = P_1 + P_2$ felt by an eardrum over the 1-sec interval $[0.15, 1.15]$ is shown here. How many beats are there in 1 sec?

(b) Repeat part (a) with frequencies of 220 and 216 Hz.

(c) Determine a simple way to find the number of beats per second if the frequency of each tone is given.

For $x = t$,
$P(t) = 0.005 \sin 440\pi t + 0.005 \sin 446\pi t$

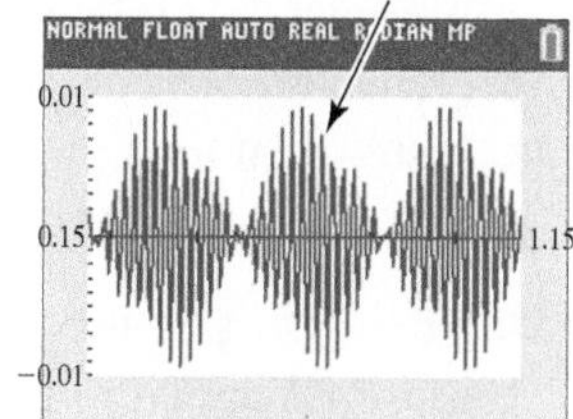

---

40. $\left\{\frac{5\pi}{3} + 4n\pi, \frac{7\pi}{3} + 4n\pi\text{, where } n \text{ is any integer}\right\}$

41. $\{30° + 360°n, 150° + 360°n, 270° + 360°n\text{, where } n \text{ is any integer}\}$

42. $\{60° + 360°n, 90° + 180°n, 300° + 360°n\text{, where } n \text{ is any integer}\}$

43. $\left\{n\pi, \frac{\pi}{6} + 2n\pi, \frac{5\pi}{6} + 2n\pi\text{, where } n \text{ is any integer}\right\}$

44. $\left\{\frac{\pi}{2} + n\pi, \frac{\pi}{4} + n\pi\text{, where } n \text{ is any integer}\right\}$

45. $\{1.3181 + 2n\pi, 4.9651 + 2n\pi\text{, where } n \text{ is any integer}\}$

46. $\{1.2310 + 2n\pi, 5.0522 + 2n\pi\text{, where } n \text{ is any integer}\}$

47. $\{11.8° + 180°n, 78.2° + 180°n\text{, where } n \text{ is any integer}\}$

48. $\{22.5° + 180°n, 112.5° + 180°n\text{, where } n \text{ is any integer}\}$

49. $\{30° + 180°n, 90° + 180°n, 150° + 180°n\text{, where } n \text{ is any integer}\}$

50. $\{180°n, 60° + 360°n, 300° + 360°n\text{, where } n \text{ is any integer}\}$

51. $\left\{\frac{\pi}{2}\right\}$ **52.** $\{0, \pi\}$

53. $\{0.4636, 3.6052\}$

54. $\{1.8925, 5.0341\}$

55. $\{1.2802\}$

56. $\{0.6919, 2.0820\}$

57. (a) See the graph in the text column.
(b) The graph is periodic, and the wave has "jagged square" tops and bottoms.
(c) This will occur when t is in one of these intervals: $(0.0045, 0.0091)$, $(0.0136, 0.0182)$, $(0.0227, 0.0273)$.

58. (a) 3 beats per sec
(b) For $x = t$,
$P(t) = 0.005 \sin 440\pi t + 0.005 \sin 432\pi t$

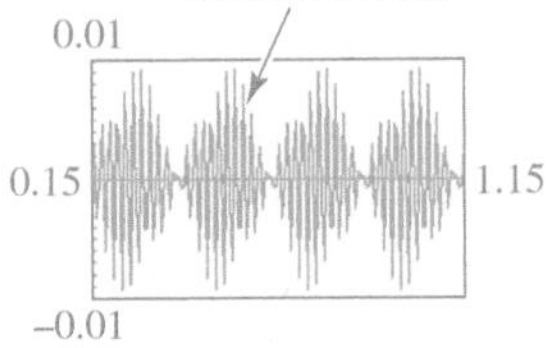

0.01
0.15 1.15
−0.01

4 beats per sec
(c) The number of beats is equal to the absolute value of the difference in the frequencies of the two tones.

59. (a) For $x = t$,
$P(t) = \frac{1}{2}\sin[2\pi(220)t] + \frac{1}{3}\sin[2\pi(330)t] + \frac{1}{4}\sin[2\pi(440)t]$

1
0 0.03
−1

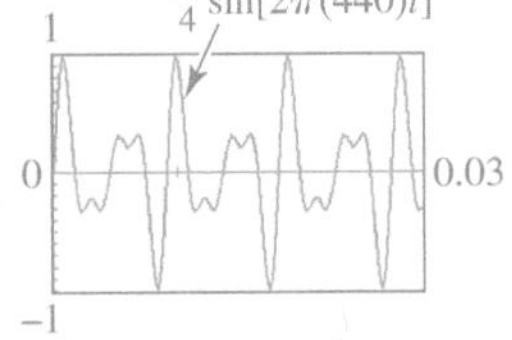

(b) 0.0007576, 0.009847, 0.01894, 0.02803
(c) 110 Hz
(d) For $x = t$,
$P(t) = \sin[2\pi(110)t] + \frac{1}{2}\sin[2\pi(220)t] + \frac{1}{3}\sin[2\pi(330)t] + \frac{1}{4}\sin[2\pi(440)t]$

2
0 0.03
−2

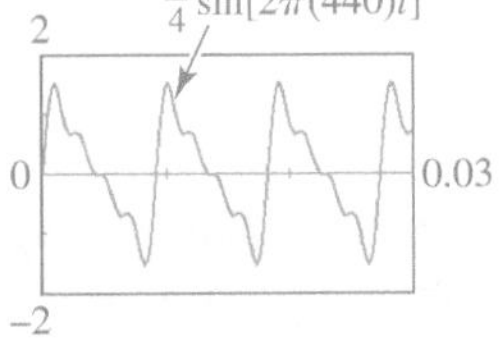

60. (a) 91.25 days after March 21, on June 20
(b) 273.8 days after March 21, on December 19
(c) 228.7 days after March 21, on November 4, and again after 318.8 days, on February 2

61. (a) when $x = 7$ (during July)
(b) when $x = 2.3$ (during February) and when $x = 11.7$ (during November)

62. (a) when $x = 4$ (during April) and when $x = 10$ (during October)
(b) when $x = 2.2$ (during February) and when $x = 11.8$ (during November)

63. 0.001 sec 64. 0.0007 sec
65. 0.004 sec 66. 0.0014 sec

59. ***Hearing Difference Tones*** When a musical instrument creates a tone of 110 Hz, it also creates tones at 220, 330, 440, 550, 660, . . . Hz. A small speaker cannot reproduce the 110-Hz vibration but it can reproduce the higher frequencies, which are the **upper harmonics.** The low tones can still be heard because the speaker produces **difference tones** of the upper harmonics. The difference between consecutive frequencies is 110 Hz, and this difference tone will be heard by a listener. (*Source*: Benade, A., *Fundamentals of Musical Acoustics,* Dover Publications.)

(a) In the window $[0, 0.03]$ by $[-1, 1]$, graph the upper harmonics represented by the pressure

$$P = \frac{1}{2}\sin[2\pi(220)t] + \frac{1}{3}\sin[2\pi(330)t] + \frac{1}{4}\sin[2\pi(440)t].$$

(b) Estimate all t-coordinates where P is maximum.

(c) What does a person hear in addition to the frequencies of 220, 330, and 440 Hz?

(d) Graph the pressure produced by a speaker that can vibrate at 110 Hz and above.

60. ***Daylight Hours in New Orleans*** The seasonal variation in length of daylight can be modeled by a sine function. For example, the daily number of hours of daylight in New Orleans is given by

$$h = \frac{35}{3} + \frac{7}{3}\sin\frac{2\pi x}{365},$$

where x is the number of days after March 21 (disregarding leap year). (*Source:* Bushaw, D., et al., *A Sourcebook of Applications of School Mathematics*, Mathematical Association of America.)

(a) On what date will there be about 14 hr of daylight?

(b) What date has the least number of hours of daylight?

(c) When will there be about 10 hr of daylight?

61. ***Average Monthly Temperature in Vancouver*** The following function approximates average monthly temperature y (in °F) in Vancouver, Canada. Here x represents the month, where $x = 1$ corresponds to January, $x = 2$ corresponds to February, and so on. (*Source:* www.weather.com)

$$f(x) = 14\sin\left[\frac{\pi}{6}(x - 4)\right] + 50$$

When is the average monthly temperature **(a)** 64°F **(b)** 39°F?

62. ***Average Monthly Temperature in Phoenix*** The following function approximates average monthly temperature y (in °F) in Phoenix, Arizona. Here x represents the month, where $x = 1$ corresponds to January, $x = 2$ corresponds to February, and so on. (*Source:* www.weather.com)

$$f(x) = 19.5\cos\left[\frac{\pi}{6}(x - 7)\right] + 70.5$$

When is the average monthly temperature **(a)** 70.5°F **(b)** 55°F?

(Modeling) Alternating Electric Current *The study of alternating electric current requires solving equations of the form*

$$i = I_{max}\sin 2\pi ft,$$

for time t in seconds, where i is instantaneous current in amperes, I_{max} is maximum current in amperes, and f is the number of cycles per second. (*Source:* Hannon, R. H., *Basic Technical Mathematics with Calculus,* W. B. Saunders Company.) *Find the least positive value of t, given the following data.*

63. $i = 40, I_{max} = 100, f = 60$

64. $i = 50, I_{max} = 100, f = 120$

65. $i = I_{max}, f = 60$

66. $i = \frac{1}{2}I_{max}, f = 60$

Chapter 6 Quiz (Sections 6.1-6.3)

[6.1]

1.

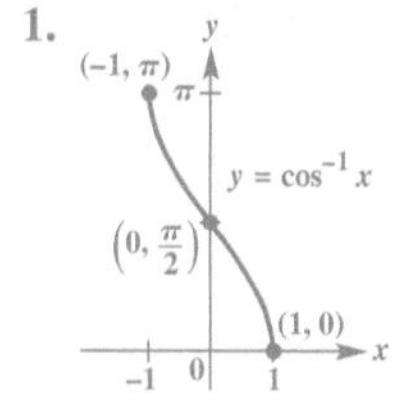

$[-1, 1]$; $[0, \pi]$

2. (a) $-\frac{\pi}{4}$ (b) $\frac{\pi}{3}$ (c) $\frac{5\pi}{6}$

3. (a) 22.568922° (b) 137.431085°

4. (a) $\frac{5\sqrt{41}}{41}$ (b) $\frac{\sqrt{3}}{2}$

[6.2]

5. $\{60°, 120°\}$

6. $\{60°, 180°, 300°\}$

7. (a) 0 sec (b) 0.20 sec

8. $\{0.6089, 1.3424, 3.7505, 4.4840\}$

[6.3]

9. $\left\{\frac{\pi}{6}, \frac{2\pi}{3}, \frac{7\pi}{6}, \frac{5\pi}{3}\right\}$

10. $\left\{\frac{5\pi}{3} + 4n\pi, \frac{7\pi}{3} + 4n\pi, \text{ where } n \text{ is any integer}\right\}$

1. Graph $y = \cos^{-1} x$, and indicate the coordinates of three points on the graph. Give the domain and range.

2. Find the exact value of each real number y. Do not use a calculator.

(a) $y = \sin^{-1}\left(-\frac{\sqrt{2}}{2}\right)$ **(b)** $y = \tan^{-1} \sqrt{3}$ **(c)** $y = \sec^{-1}\left(-\frac{2\sqrt{3}}{3}\right)$

3. Use a calculator to approximate each value in decimal degrees.

(a) $\theta = \arccos 0.92341853$ **(b)** $\theta = \cot^{-1}(-1.0886767)$

4. Evaluate each expression without using a calculator.

(a) $\cos\left(\tan^{-1} \frac{4}{5}\right)$ **(b)** $\sin\left(\cos^{-1}\left(-\frac{1}{2}\right) + \tan^{-1}(-\sqrt{3})\right)$

Solve each equation for exact solutions over the interval $[0°, 360°)$.

5. $2 \sin \theta - \sqrt{3} = 0$ **6.** $\cos \theta + 1 = 2 \sin^2 \theta$

7. ***(Modeling) Electromotive Force*** In an electric circuit, suppose that

$$V = \cos 2\pi t$$

models the electromotive force in volts at t seconds. Find the least value of t where $0 \le t \le \frac{1}{2}$ for each value of V.

(a) $V = 1$ **(b)** $V = 0.30$

Solve each equation for solutions over the interval $[0, 2\pi)$. *Round approximate answers to four decimal places.*

8. $\tan^2 x - 5 \tan x + 3 = 0$ **9.** $3 \cot 2x - \sqrt{3} = 0$

10. Solve $\cos \frac{x}{2} + \sqrt{3} = -\cos \frac{x}{2}$, giving all solutions in radians.

6.4 Equations Involving Inverse Trigonometric Functions

- Solution for x in Terms of y Using Inverse Functions
- Solution of Inverse Trigonometric Equations

Solution for x in Terms of y Using Inverse Functions

EXAMPLE 1 Solving an Equation for a Specified Variable

Solve $y = 3 \cos 2x$ for x, where x is restricted to the interval $\left[0, \frac{\pi}{2}\right]$.

SOLUTION We want to isolate $\cos 2x$ on one side of the equation so that we can solve for $2x$, and then for x.

$$y = 3 \cos 2x \quad \text{Our goal is to isolate } x.$$

$$\frac{y}{3} = \cos 2x \quad \text{Divide by 3.}$$

$$2x = \arccos \frac{y}{3} \quad \text{Definition of arccosine}$$

$$x = \frac{1}{2} \arccos \frac{y}{3} \quad \text{Multiply by } \frac{1}{2}.$$

An equivalent form of this answer is $x = \frac{1}{2} \cos^{-1} \frac{y}{3}$.

Classroom Example 1
Solve $y = 4 \tan 3x$ for x, where x is restricted to the interval $\left(-\frac{\pi}{6}, \frac{\pi}{6}\right)$.

Answer: $x = \frac{1}{3} \arctan \frac{y}{4}$

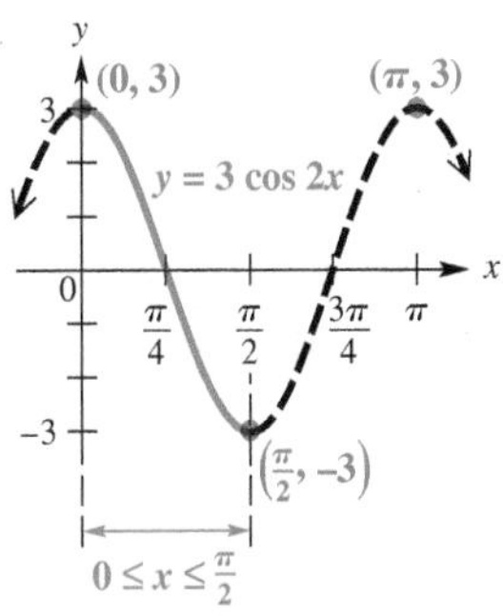

Figure 31

Because the function $y = 3 \cos 2x$ is periodic, with period π, there are infinitely many domain values (x-values) that will result in a given range value (y-value). For example, the x-values 0 and π both correspond to the y-value 3. See **Figure 31.** The restriction $0 \leq x \leq \frac{\pi}{2}$ given in the original problem ensures that this function is one-to-one, and, correspondingly, that

$$x = \frac{1}{2} \operatorname{arccos} \frac{y}{3}$$

has a one-to-one relationship. Thus, each y-value in $[-3, 3]$ substituted into this equation will lead to a single x-value.

✔ **Now Try Exercise 9.**

Solution of Inverse Trigonometric Equations

Classroom Example 2
Solve $3 \arctan x = \pi$.
Answer: $\{\sqrt{3}\}$

EXAMPLE 2 Solving an Equation Involving an Inverse Trigonometric Function

Solve $2 \arcsin x = \pi$.

SOLUTION First solve for $\arcsin x$, and then for x.

$$2 \arcsin x = \pi \quad \text{Original equation}$$

$$\arcsin x = \frac{\pi}{2} \quad \text{Divide by 2.}$$

$$x = \sin \frac{\pi}{2} \quad \text{Definition of arcsine}$$

$$x = 1 \quad \arcsin 1 = \tfrac{\pi}{2}$$

CHECK

$$2 \arcsin x = \pi \quad \text{Original equation}$$

$$2 \arcsin 1 \stackrel{?}{=} \pi \quad \text{Let } x = 1.$$

$$2\left(\frac{\pi}{2}\right) \stackrel{?}{=} \pi \quad \text{Substitute the inverse value.}$$

$$\pi = \pi \ \checkmark \quad \text{True}$$

The solution set is $\{1\}$.

✔ **Now Try Exercise 27.**

Classroom Example 3
Solve $\sec^{-1} x = \csc^{-1} 2$.
Answer: $\left\{\frac{2\sqrt{3}}{3}\right\}$

EXAMPLE 3 Solving an Equation Involving Inverse Trigonometric Functions

Solve $\cos^{-1} x = \sin^{-1} \frac{1}{2}$.

SOLUTION Let $\sin^{-1} \frac{1}{2} = u$. Then $\sin u = \frac{1}{2}$, and for u in quadrant I we have the following.

$$\cos^{-1} x = \sin^{-1} \frac{1}{2} \quad \text{Original equation}$$

$$\cos^{-1} x = u \quad \text{Substitute.}$$

$$\cos u = x \quad \text{Alternative form}$$

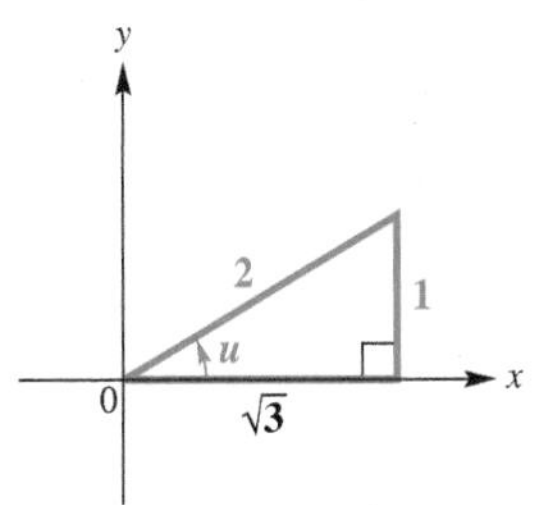

Figure 32

Sketch a triangle and label it using the facts that u is in quadrant I and $\sin u = \frac{1}{2}$. See **Figure 32.** Because $x = \cos u$, we have $x = \frac{\sqrt{3}}{2}$. The solution set is $\left\{\frac{\sqrt{3}}{2}\right\}$.

✔ **Now Try Exercise 35.**

Classroom Example 4
Solve $\arccos x - \arcsin x = \pi$.

Answer: $\left\{-\frac{\sqrt{2}}{2}\right\}$

EXAMPLE 4 Solving an Inverse Trigonometric Equation Using an Identity

Solve $\arcsin x - \arccos x = \frac{\pi}{6}$.

SOLUTION Isolate one inverse function on one side of the equation.

$$\arcsin x - \arccos x = \frac{\pi}{6} \quad \text{Original equation}$$

$$\arcsin x = \arccos x + \frac{\pi}{6} \quad \text{Add } \arccos x. \quad (1)$$

$$x = \sin\left(\arccos x + \frac{\pi}{6}\right) \quad \text{Definition of arcsine}$$

Let $u = \arccos x$. The arccosine function yields angles in quadrants I and II, so $0 \le u \le \pi$ by definition.

$$x = \sin\left(u + \frac{\pi}{6}\right) \quad \text{Substitute.}$$

$$x = \sin u \cos \frac{\pi}{6} + \cos u \sin \frac{\pi}{6} \quad \text{Sine sum identity} \quad (2)$$

Use equation (1) and the definition of the arcsine function.

$$-\frac{\pi}{2} \le \arccos x + \frac{\pi}{6} \le \frac{\pi}{2} \quad \text{Range of arcsine is } \left[-\frac{\pi}{2}, \frac{\pi}{2}\right].$$

$$-\frac{2\pi}{3} \le \arccos x \le \frac{\pi}{3} \quad \text{Subtract } \frac{\pi}{6} \text{ from each part.}$$

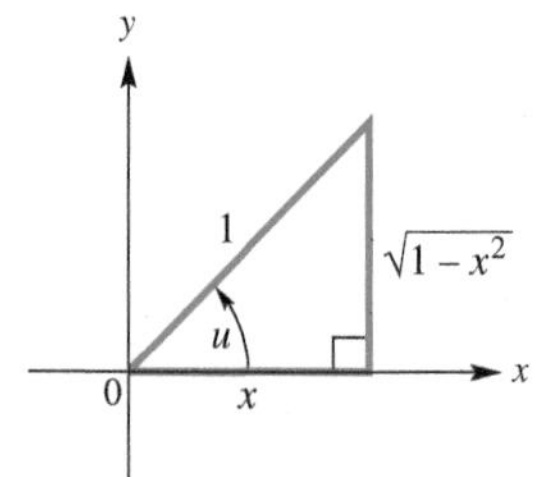

Figure 33

Because both $0 \le \arccos x \le \pi$ **and** $-\frac{2\pi}{3} \le \arccos x \le \frac{\pi}{3}$, the intersection yields $0 \le \arccos x \le \frac{\pi}{3}$. This places u in quadrant I, and we can sketch the triangle in **Figure 33.** From this triangle we find that $\sin u = \sqrt{1 - x^2}$. Now substitute into equation (2) using $\sin u = \sqrt{1 - x^2}$, $\sin \frac{\pi}{6} = \frac{1}{2}$, $\cos \frac{\pi}{6} = \frac{\sqrt{3}}{2}$, and $\cos u = x$.

$$x = \sin u \cos \frac{\pi}{6} + \cos u \sin \frac{\pi}{6} \quad (2)$$

$$x = \left(\sqrt{1 - x^2}\right)\frac{\sqrt{3}}{2} + x \cdot \frac{1}{2} \quad \text{Substitute.}$$

$$2x = \left(\sqrt{1 - x^2}\right)\sqrt{3} + x \quad \text{Multiply by 2.}$$

$$x = \left(\sqrt{3}\right)\sqrt{1 - x^2} \quad \text{Subtract } x\text{; commutative property}$$

Square each factor.

$$x^2 = 3(1 - x^2) \quad \text{Square each side; } (ab)^2 = a^2b^2$$

$$x^2 = 3 - 3x^2 \quad \text{Distributive property}$$

$$x^2 = \frac{3}{4} \quad \text{Add } 3x^2\text{. Divide by 4.}$$

Choose the positive square root, $x > 0$.

$$x = \sqrt{\frac{3}{4}} \quad \text{Take the square root on each side.}$$

$$x = \frac{\sqrt{3}}{2} \quad \text{Quotient rule: } \sqrt[n]{\frac{a}{b}} = \frac{\sqrt[n]{a}}{\sqrt[n]{b}}$$

CHECK A check is necessary because we squared each side when solving the equation.

$$\arcsin x - \arccos x = \frac{\pi}{6} \quad \text{Original equation}$$

$$\arcsin \frac{\sqrt{3}}{2} - \arccos \frac{\sqrt{3}}{2} \stackrel{?}{=} \frac{\pi}{6} \quad \text{Let } x = \frac{\sqrt{3}}{2}.$$

$$\frac{\pi}{3} - \frac{\pi}{6} \stackrel{?}{=} \frac{\pi}{6} \quad \text{Substitute inverse values.}$$

$$\frac{\pi}{6} = \frac{\pi}{6} \quad ✓ \text{ True}$$

The solution set is $\left\{\frac{\sqrt{3}}{2}\right\}$.

✓ **Now Try Exercise 37.**

6.4 Exercises

1. C 2. A
3. C 4. C
5. A 6. B

7. $x = \arccos \frac{y}{5}$
8. $x = \arcsin 4y$
9. $x = \frac{1}{2} \arctan \frac{y}{3}$
10. $x = 2 \arcsin \frac{y}{3}$
11. $x = 4 \arccos \frac{y}{6}$
12. $x = 3 \arcsin(-y)$
13. $x = \frac{1}{5} \arccos\left(-\frac{y}{2}\right)$
14. $x = \frac{1}{5} \operatorname{arccot} \frac{y}{3}$

CONCEPT PREVIEW *Answer each question.*

1. Which one of the following equations has solution 0?

A. $\arctan 1 = x$ **B.** $\arccos 0 = x$ **C.** $\arcsin 0 = x$

2. Which one of the following equations has solution $\frac{\pi}{4}$?

A. $\arcsin \frac{\sqrt{2}}{2} = x$ **B.** $\arccos\left(-\frac{\sqrt{2}}{2}\right) = x$ **C.** $\arctan \frac{\sqrt{3}}{3} = x$

3. Which one of the following equations has solution $\frac{3\pi}{4}$?

A. $\arctan 1 = x$ **B.** $\arcsin \frac{\sqrt{2}}{2} = x$ **C.** $\arccos\left(-\frac{\sqrt{2}}{2}\right) = x$

4. Which one of the following equations has solution $-\frac{\pi}{6}$?

A. $\arctan \frac{\sqrt{3}}{3} = x$ **B.** $\arccos\left(-\frac{1}{2}\right) = x$ **C.** $\arcsin\left(-\frac{1}{2}\right) = x$

5. Which one of the following equations has solution π?

A. $\arccos(-1) = x$ **B.** $\arccos 1 = x$ **C.** $\arcsin(-1) = x$

6. Which one of the following equations has solution $-\frac{\pi}{2}$?

A. $\arctan(-1) = x$ **B.** $\arcsin(-1) = x$ **C.** $\arccos(-1) = x$

Solve each equation for x, where x is restricted to the given interval. ***See Example 1.***

7. $y = 5 \cos x$, for x in $[0, \pi]$

8. $y = \frac{1}{4} \sin x$, for x in $\left[-\frac{\pi}{2}, \frac{\pi}{2}\right]$

9. $y = 3 \tan 2x$, for x in $\left(-\frac{\pi}{4}, \frac{\pi}{4}\right)$

10. $y = 3 \sin \frac{x}{2}$, for x in $[-\pi, \pi]$

11. $y = 6 \cos \frac{x}{4}$, for x in $[0, 4\pi]$

12. $y = -\sin \frac{x}{3}$, for x in $\left[-\frac{3\pi}{2}, \frac{3\pi}{2}\right]$

13. $y = -2 \cos 5x$, for x in $\left[0, \frac{\pi}{5}\right]$

14. $y = 3 \cot 5x$, for x in $\left(0, \frac{\pi}{5}\right)$

15. $x = \arcsin(y + 2)$
16. $x = \text{arccot}(y - 1)$
17. $x = \arcsin \frac{y+4}{2}$
18. $x = \arccos \frac{y-4}{3}$
19. $x = \frac{1}{3} \text{ arccot } 2y$
20. $x = \text{arcsec } 12y$
21. $x = -3 + \arccos y$
22. $x = \frac{1}{2}(1 + \arctan y)$
23. $x = \frac{1}{2} \sec^{-1}\left(\frac{y - \sqrt{2}}{3}\right)$
24. $x = 2 \csc^{-1}\left(\frac{y + \sqrt{3}}{2}\right)$
25. The argument of the sine function is x, not $x - 2$. To solve for x, first add 2 and then use the definition of arcsine.
26. There is no number that has cosine value of 2, so $\cos^{-1} 2$ is not defined.
27. $\left\{-\frac{\sqrt{2}}{2}\right\}$ 28. $\left\{-\frac{\sqrt{3}}{2}\right\}$
29. $\left\{-2\sqrt{2}\right\}$ 30. $\varnothing$
31. $\{\pi - 3\}$ 32. $\left\{\frac{3\sqrt{3} + 2\pi}{6}\right\}$
33. $\left\{\frac{3}{5}\right\}$ 34. $\left\{\frac{12}{5}\right\}$
35. $\left\{\frac{4}{5}\right\}$ 36. $\left\{\frac{3}{4}\right\}$
37. $\{0\}$ 38. $\left\{\frac{\sqrt{3}}{2}\right\}$
39. $\left\{\frac{1}{2}\right\}$ 40. $\varnothing$
41. $\left\{-\frac{1}{2}\right\}$ 42. $\left\{\frac{\sqrt{5}}{5}\right\}$
43. $\{0\}$ 44. $\{0\}$
45. $y_1 = \sin^{-1} x - \cos^{-1} x - \frac{\pi}{6}$

15. $y = \sin x - 2$, for x in $\left[-\frac{\pi}{2}, \frac{\pi}{2}\right]$

16. $y = \cot x + 1$, for x in $(0, \pi)$

17. $y = -4 + 2 \sin x$, for x in $\left[-\frac{\pi}{2}, \frac{\pi}{2}\right]$

18. $y = 4 + 3 \cos x$, for x in $[0, \pi]$

19. $y = \frac{1}{2} \cot 3x$, for x in $\left(0, \frac{\pi}{3}\right)$

20. $y = \frac{1}{12} \sec x$, for x in $\left[0, \frac{\pi}{2}\right) \cup \left(\frac{\pi}{2}, \pi\right]$

21. $y = \cos(x + 3)$, for x in $[-3, \pi - 3]$

22. $y = \tan(2x - 1)$, for x in $\left(\frac{1}{2} - \frac{\pi}{4}, \frac{1}{2} + \frac{\pi}{4}\right)$

23. $y = \sqrt{2} + 3 \sec 2x$, for x in $\left[0, \frac{\pi}{4}\right) \cup \left(\frac{\pi}{4}, \frac{\pi}{2}\right]$

24. $y = -\sqrt{3} + 2 \csc \frac{x}{2}$, for x in $[-\pi, 0) \cup (0, \pi]$

25. Refer to **Exercise 15.** A student solving this equation wrote $y = \sin(x - 2)$ as the first step, inserting parentheses as shown. Explain why this is incorrect.

26. Explain why the equation $\sin^{-1} x = \cos^{-1} 2$ cannot have a solution. (No work is required.)

Solve each equation for exact solutions. ***See Examples 2 and 3.***

27. $-4 \arcsin x = \pi$

28. $6 \arccos x = 5\pi$

29. $\frac{4}{3} \cos^{-1} \frac{x}{4} = \pi$

30. $4 \tan^{-1} x = -3\pi$

31. $2 \arccos\left(\frac{x}{3} - \frac{\pi}{3}\right) = 2\pi$

32. $6 \arccos\left(x - \frac{\pi}{3}\right) = \pi$

33. $\arcsin x = \arctan \frac{3}{4}$

34. $\arctan x = \arccos \frac{5}{13}$

35. $\cos^{-1} x = \sin^{-1} \frac{3}{5}$

36. $\cot^{-1} x = \tan^{-1} \frac{4}{3}$

Solve each equation for exact solutions. ***See Example 4.***

37. $\sin^{-1} x - \tan^{-1} 1 = -\frac{\pi}{4}$

38. $\sin^{-1} x + \tan^{-1} \sqrt{3} = \frac{2\pi}{3}$

39. $\arccos x + 2 \arcsin \frac{\sqrt{3}}{2} = \pi$

40. $\arccos x + 2 \arcsin \frac{\sqrt{3}}{2} = \frac{\pi}{3}$

41. $\arcsin 2x + \arccos x = \frac{\pi}{6}$

42. $\arcsin 2x + \arcsin x = \frac{\pi}{2}$

43. $\cos^{-1} x + \tan^{-1} x = \frac{\pi}{2}$

44. $\sin^{-1} x + \tan^{-1} x = 0$

Use a graphing calculator in each of the following.

45. Provide graphical support for the solution in **Example 4** by showing that the graph of

$$y = \sin^{-1} x - \cos^{-1} x - \frac{\pi}{6} \quad \text{has a zero of} \quad \frac{\sqrt{3}}{2} \approx 0.8660254.$$

46. Provide graphical support for the solution in **Example 4** by showing that the x-coordinate of the point of intersection of the graphs of

$$y_1 = \sin^{-1} x - \cos^{-1} x \quad \text{and} \quad y_2 = \frac{\pi}{6} \quad \text{is} \quad \frac{\sqrt{3}}{2} \approx 0.8660254.$$

The following equations cannot be solved by algebraic methods. Use a graphing calculator to find all solutions over the interval $[0, 6]$. Express solutions to four decimal places.

47. $(\arctan x)^3 - x + 2 = 0$

48. $\pi \sin^{-1}(0.2x) - 3 = -\sqrt{x}$

(Modeling) Solve each problem.

49. ***Tone Heard by a Listener*** When two sources located at different positions produce the same pure tone, the human ear will often hear one sound that is equal to the sum of the individual tones. Because the sources are at different locations, they will have different phase angles ϕ. If two speakers located at different positions produce pure tones $P_1 = A_1 \sin(2\pi ft + \phi_1)$ and $P_2 = A_2 \sin(2\pi ft + \phi_2)$, where $-\frac{\pi}{4} \le \phi_1, \phi_2 \le \frac{\pi}{4}$, then the resulting tone heard by a listener can be written as $P = A \sin(2\pi ft + \phi)$, where

$$A = \sqrt{(A_1 \cos \phi_1 + A_2 \cos \phi_2)^2 + (A_1 \sin \phi_1 + A_2 \sin \phi_2)^2}$$

$$\text{and} \quad \phi = \arctan\left(\frac{A_1 \sin \phi_1 + A_2 \sin \phi_2}{A_1 \cos \phi_1 + A_2 \cos \phi_2}\right).$$

(*Source*: Fletcher, N. and T. Rossing, *The Physics of Musical Instruments,* Second Edition, Springer-Verlag.)

(a) Calculate A and ϕ if $A_1 = 0.0012$, $\phi_1 = 0.052$, $A_2 = 0.004$, and $\phi_2 = 0.61$. Also, if $f = 220$, find an expression for

$$P = A \sin(2\pi ft + \phi).$$

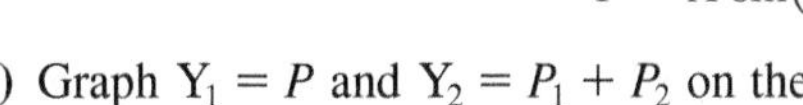

(b) Graph $Y_1 = P$ and $Y_2 = P_1 + P_2$ on the same coordinate axes over the interval $[0, 0.01]$. Are the two graphs the same?

50. ***Tone Heard by a Listener*** Repeat **Exercise 49.** Use $A_1 = 0.0025$, $\phi_1 = \frac{\pi}{7}$, $A_2 = 0.001$, $\phi_2 = \frac{\pi}{6}$, and $f = 300$.

51. ***Depth of Field*** When a large-view camera is used to take a picture of an object that is not parallel to the film, the lens board should be tilted so that the planes containing the subject, the lens board, and the film intersect in a line. This gives the best "depth of field." See the figure. (*Source*: Bushaw, D., et al., *A Sourcebook of Applications of School Mathematics,* Mathematical Association of America.)

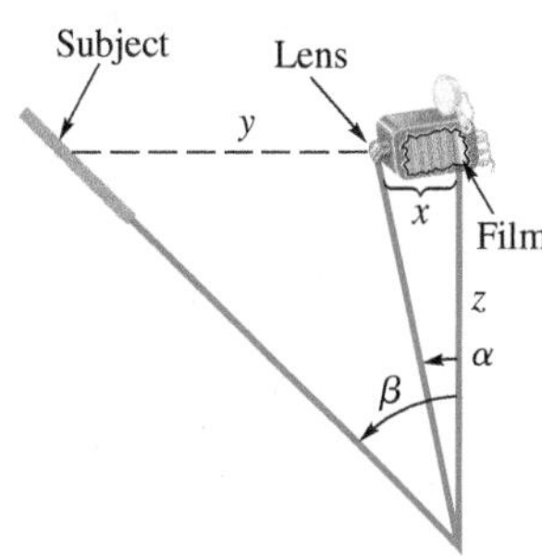

(a) Write two equations, one relating α, x, and z, and the other relating β, x, y, and z.

(b) Eliminate z from the equations in part (a) to get one equation relating α, β, x, and y.

(c) Solve the equation from part (b) for α.

(d) Solve the equation from part (b) for β.

46.

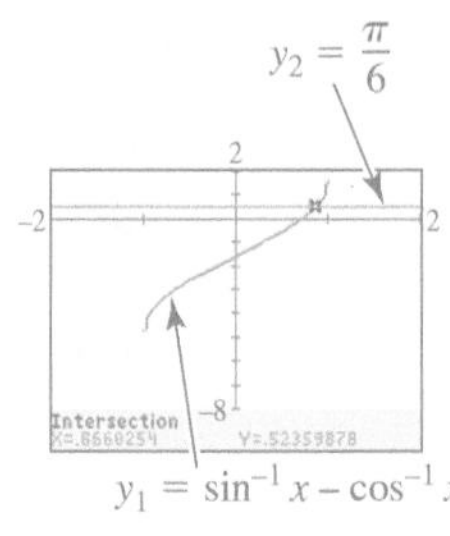

47. $\{4.4622\}$ **48.** $\{2.2824\}$

49. (a) $A \approx 0.00506$, $\phi \approx 0.484$; $P = 0.00506 \sin(440\pi t + 0.484)$

(b)

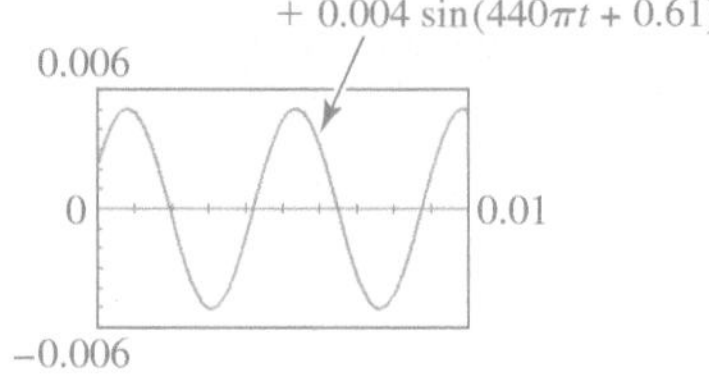

The two graphs are the same.

50. (a) $A \approx 0.0035$, $\phi \approx 0.470$; $P = 0.0035 \sin(600\pi t + 0.47)$

(b)

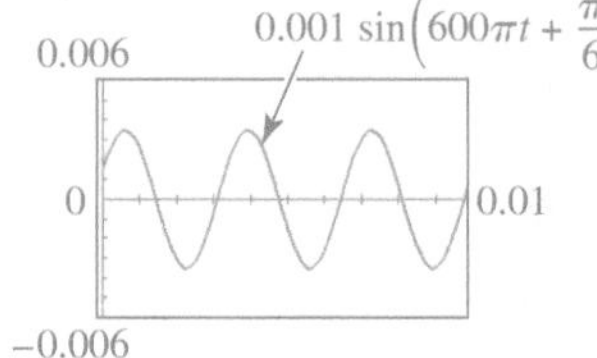

The two graphs are the same.

51. (a) $\tan \alpha = \frac{x}{z}$; $\tan \beta = \frac{x + y}{z}$

(b) $\frac{x}{\tan \alpha} = \frac{x + y}{\tan \beta}$

(c) $\alpha = \arctan\left(\frac{x \tan \beta}{x + y}\right)$

(d) $\beta = \arctan\left(\frac{(x + y) \tan \alpha}{x}\right)$

52. (a) $x = \sin u,\ -\frac{\pi}{2} \le u \le \frac{\pi}{2}$

(b)

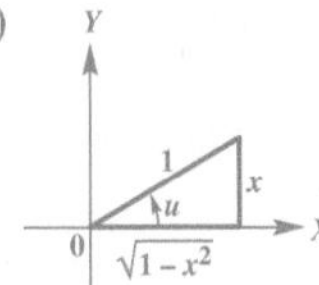

(c) $\tan u = \frac{x\sqrt{1-x^2}}{1-x^2}$

(d) $u = \arctan \frac{x\sqrt{1-x^2}}{1-x^2}$

53. (a) $t = \frac{1}{2\pi f} \arcsin \frac{E}{E_{max}}$

(b) 0.00068 sec

54. (b) (i) 0.94 or 4.26
(ii) 0.60 or 6.64
(c) (i) 0.54
(ii) 0.61

55. (a) $t = \frac{3}{4\pi} \arcsin 3y$

(b) 0.27 sec

52. ***Programming Language for Inverse Functions*** In some programming languages, the only inverse trigonometric function available is arctangent. The other inverse trigonometric functions can be expressed in terms of arctangent.

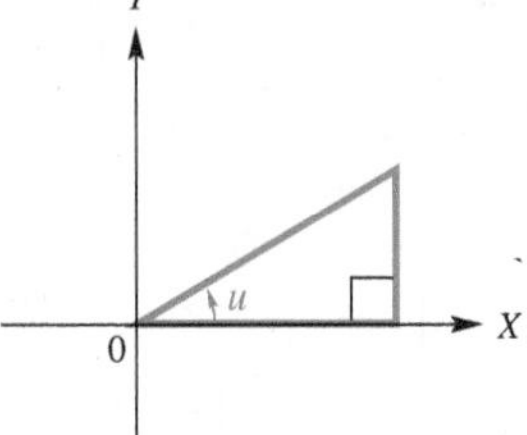

(a) Let $u = \arcsin x$. Solve the equation for x in terms of u.

(b) Use the result of part (a) to label the three sides of the triangle in the figure in terms of x.

(c) Use the triangle from part (b) to write an equation for $\tan u$ in terms of x.

(d) Solve the equation from part (c) for u.

53. ***Alternating Electric Current*** In the study of alternating electric current, instantaneous voltage is modeled by

$$E = E_{max} \sin 2\pi f t,$$

where f is the number of cycles per second, E_{max} is the maximum voltage, and t is time in seconds.

(a) Solve the equation for t.

(b) Find the least positive value of t if $E_{max} = 12$, $E = 5$, and $f = 100$. Use a calculator and round to two significant digits.

54. ***Viewing Angle of an Observer*** While visiting a museum, an observer views a painting that is 3 ft high and hangs 6 ft above the ground. See the figure. Assume her eyes are 5 ft above the ground, and let x be the distance from the spot where she is standing to the wall displaying the painting.

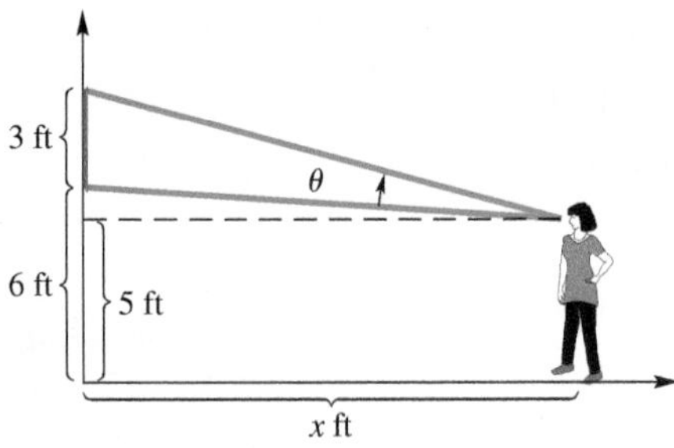

(a) Show that θ, the viewing angle subtended by the painting, is given by

$$\theta = \tan^{-1}\left(\frac{4}{x}\right) - \tan^{-1}\left(\frac{1}{x}\right).$$

(b) Find the value of x to the nearest hundredth for each value of θ.

(i) $\theta = \frac{\pi}{6}$ **(ii)** $\theta = \frac{\pi}{8}$

(c) Find the value of θ to the nearest hundredth for each value of x.

(i) $x = 4$ **(ii)** $x = 3$

55. ***Movement of an Arm*** In the equation below, t is time (in seconds) and y is the angle formed by a rhythmically moving arm.

$$y = \frac{1}{3} \sin \frac{4\pi t}{3}$$

(a) Solve the equation for t.

(b) At what time, to the nearest hundredth of a second, does the arm first form an angle of 0.3 radian?

Chapter 6 Test Prep

Key Terms

6.1 one-to-one function
inverse function

New Symbols

f^{-1}	inverse of function f	$\cot^{-1} x$ ($\text{arccot}\, x$)	inverse cotangent of x
$\sin^{-1} x$ ($\arcsin x$)	inverse sine of x	$\sec^{-1} x$ ($\text{arcsec}\, x$)	inverse secant of x
$\cos^{-1} x$ ($\arccos x$)	inverse cosine of x	$\csc^{-1} x$ ($\text{arccsc}\, x$)	inverse cosecant of x
$\tan^{-1} x$ ($\arctan x$)	inverse tangent of x		

Quick Review

Concepts

6.1 Inverse Circular Functions

Inverse Function	Domain	Range: Interval	Range: Quadrants of the Unit Circle
$y = \sin^{-1} x$	$[-1, 1]$	$\left[-\frac{\pi}{2}, \frac{\pi}{2}\right]$	I and IV
$y = \cos^{-1} x$	$[-1, 1]$	$[0, \pi]$	I and II
$y = \tan^{-1} x$	$(-\infty, \infty)$	$\left(-\frac{\pi}{2}, \frac{\pi}{2}\right)$	I and IV
$y = \cot^{-1} x$	$(-\infty, \infty)$	$(0, \pi)$	I and II
$y = \sec^{-1} x$	$(-\infty, -1] \cup [1, \infty)$	$\left[0, \frac{\pi}{2}\right) \cup \left(\frac{\pi}{2}, \pi\right]$	I and II
$y = \csc^{-1} x$	$(-\infty, -1] \cup [1, \infty)$	$\left[-\frac{\pi}{2}, 0\right) \cup \left(0, \frac{\pi}{2}\right]$	I and IV

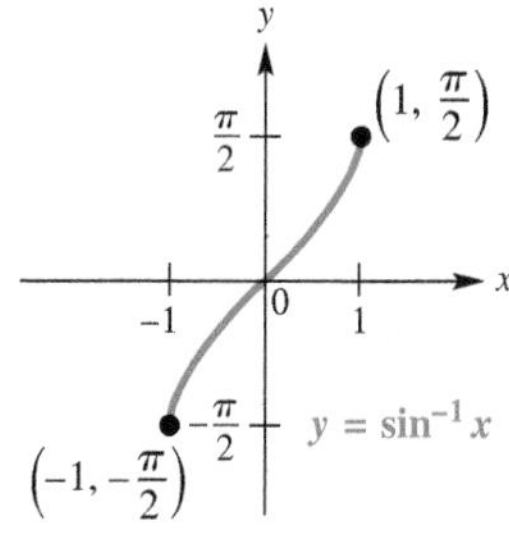

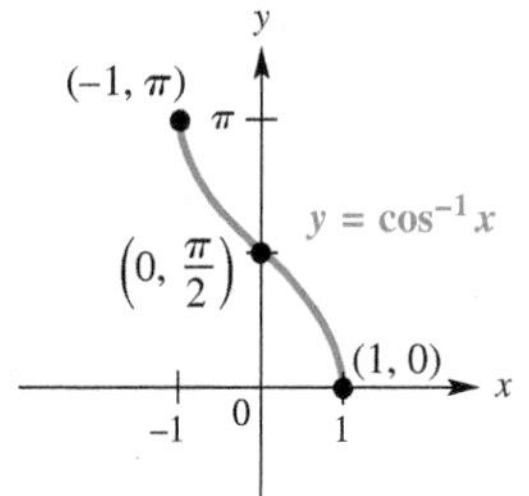

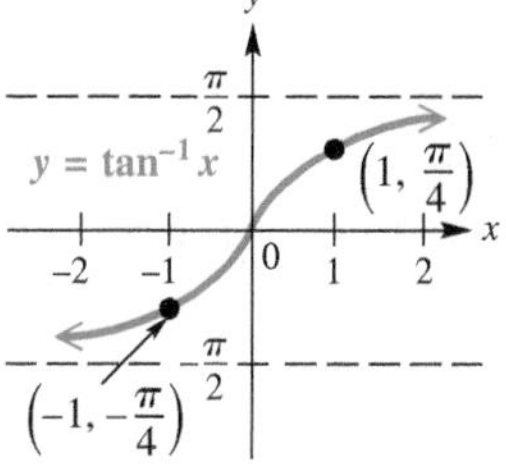

See the section for graphs of the other inverse circular (trigonometric) functions.

Examples

Evaluate $y = \cos^{-1} 0$.

Write $y = \cos^{-1} 0$ as $\cos y = 0$. Then

$$y = \frac{\pi}{2}$$

because $\cos \frac{\pi}{2} = 0$ and $\frac{\pi}{2}$ is in the range of $\cos^{-1} x$.

Use a calculator to find y in radians if $y = \sec^{-1}(-3)$.

With the calculator in radian mode, enter $\sec^{-1}(-3)$ as $\cos^{-1}\left(\frac{1}{-3}\right)$ to obtain

$$y \approx 1.9106332.$$

Evaluate $\sin\left(\tan^{-1}\left(-\frac{3}{4}\right)\right)$.

Let $u = \tan^{-1}\left(-\frac{3}{4}\right)$. Then $\tan u = -\frac{3}{4}$. Because $\tan u$ is negative when u is in quadrant IV, sketch a triangle as shown.

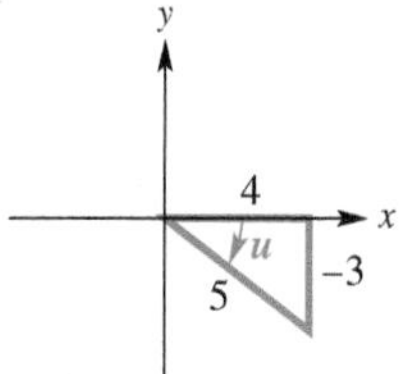

We want $\sin\left(\tan^{-1}\left(-\frac{3}{4}\right)\right) = \sin u$. From the triangle, we have the following.

$$\sin u = -\frac{3}{5}$$

Concepts	Examples

6.2 Trigonometric Equations I

6.3 Trigonometric Equations II

Solving a Trigonometric Equation

1. Decide whether the equation is linear or quadratic in form in order to determine the solution method.
2. If only one trigonometric function is present, solve the equation for that function.
3. If more than one trigonometric function is present, rewrite the equation so that one side equals 0. Then try to factor and apply the zero-factor property.
4. If the equation is quadratic in form, but not factorable, use the quadratic formula. Check that solutions are in the desired interval.
5. Try using identities to change the form of the equation. It may be helpful to square each side of the equation first. In this case, check for extraneous solutions.

Solve $\tan\theta + \sqrt{3} = 2\sqrt{3}$ over the interval $[0°, 360°)$.

$\tan\theta + \sqrt{3} = 2\sqrt{3}$ Original equation

$\tan\theta = \sqrt{3}$ Subtract $\sqrt{3}$.

$\theta = 60°$ Definition of inverse tangent

Another solution over $[0°, 360°)$ is

$$\theta = 60° + 180° = 240°.$$

The solution set is $\{60°, 240°\}$.

Solve $2\cos^2 x = 1$ for all solutions.

$2\cos^2 x = 1$ Original equation

$2\cos^2 x - 1 = 0$ Subtract 1.

$\cos 2x = 0$ Cosine double-angle identity

$2x = \frac{\pi}{2} + 2n\pi$ and $2x = \frac{3\pi}{2} + 2n\pi$

Add integer multiples of 2π.

$x = \frac{\pi}{4} + n\pi$ and $x = \frac{3\pi}{4} + n\pi$

Divide by 2.

The solution set, where π is the period of $\cos 2x$, is

$$\left\{\frac{\pi}{4} + n\pi, \frac{3\pi}{4} + n\pi, \text{ where } n \text{ is any integer}\right\}.$$

6.4 Equations Involving Inverse Trigonometric Functions

We solve equations of the form $y = f(x)$, where $f(x)$ involves a trigonometric function, using inverse trigonometric functions.

Solve $y = 2\sin 3x$ for x, where x is restricted to the interval $\left[-\frac{\pi}{6}, \frac{\pi}{6}\right]$.

$y = 2\sin 3x$ Original equation

$\frac{y}{2} = \sin 3x$ Divide by 2.

$3x = \arcsin\frac{y}{2}$ Definition of arcsine

$x = \frac{1}{3}\arcsin\frac{y}{2}$ Multiply by $\frac{1}{3}$.

Techniques introduced in this section also show how to solve equations that involve inverse functions.

Solve.

$4\tan^{-1} x = \pi$ Original equation

$\tan^{-1} x = \frac{\pi}{4}$ Divide by 4.

$x = \tan\frac{\pi}{4}$ Definition of arctangent

$x = 1$ Evaluate.

The solution set is $\{1\}$.

Chapter 6 Review Exercises

1. Graph the inverse sine, cosine, and tangent functions, indicating the coordinates of three points on each graph. Give the domain and range for each.

Concept Check Determine whether each statement is true *or* false. *If false, tell why.*

2. The ranges of the inverse tangent and inverse cotangent functions are the same.
3. It is true that $\sin \frac{11\pi}{6} = -\frac{1}{2}$, and therefore $\arcsin\left(-\frac{1}{2}\right) = \frac{11\pi}{6}$.
4. For all x, $\tan(\tan^{-1} x) = x$.

Find the exact value of each real number y. Do not use a calculator.

5. $y = \sin^{-1} \frac{\sqrt{2}}{2}$
6. $y = \arccos\left(-\frac{1}{2}\right)$
7. $y = \tan^{-1}\left(-\sqrt{3}\right)$
8. $y = \arcsin(-1)$
9. $y = \cos^{-1}\left(-\frac{\sqrt{2}}{2}\right)$
10. $y = \arctan \frac{\sqrt{3}}{3}$
11. $y = \sec^{-1}(-2)$
12. $y = \operatorname{arccsc} \frac{2\sqrt{3}}{3}$
13. $y = \operatorname{arccot}(-1)$

Give the degree measure of θ. Do not use a calculator.

14. $\theta = \arccos \frac{1}{2}$
15. $\theta = \arcsin\left(-\frac{\sqrt{3}}{2}\right)$
16. $\theta = \tan^{-1} 0$

Use a calculator to approximate each value in decimal degrees.

17. $\theta = \arctan 1.7804675$
18. $\theta = \sin^{-1}(-0.66045320)$
19. $\theta = \cos^{-1} 0.80396577$
20. $\theta = \cot^{-1} 4.5046388$
21. $\theta = \operatorname{arcsec} 3.4723155$
22. $\theta = \csc^{-1} 7.4890096$

Evaluate each expression without using a calculator.

23. $\cos(\arccos(-1))$
24. $\sin\left(\arcsin\left(-\frac{\sqrt{3}}{2}\right)\right)$
25. $\arccos\left(\cos \frac{3\pi}{4}\right)$
26. $\operatorname{arcsec}(\sec \pi)$
27. $\tan^{-1}\left(\tan \frac{\pi}{4}\right)$
28. $\cos^{-1}(\cos 0)$
29. $\sin\left(\arccos \frac{3}{4}\right)$
30. $\cos(\arctan 3)$
31. $\cos(\csc^{-1}(-2))$
32. $\sec\left(2 \sin^{-1}\left(-\frac{1}{3}\right)\right)$
33. $\tan\left(\arcsin \frac{3}{5} + \arccos \frac{5}{7}\right)$

Write each trigonometric expression as an algebraic expression in u, for $u > 0$.

34. $\cos\left(\arctan \frac{u}{\sqrt{1 - u^2}}\right)$
35. $\tan\left(\operatorname{arcsec} \frac{\sqrt{u^2 + 1}}{u}\right)$

Solve each equation for exact solutions over the interval $[0, 2\pi)$ *where appropriate. Give approximate solutions to four decimal places.*

36. $\sin^2 x = 1$
37. $2 \tan x - 1 = 0$
38. $3 \sin^2 x - 5 \sin x + 2 = 0$
39. $\tan x = \cot x$
40. $\sec^2 2x = 2$
41. $\tan^2 2x - 1 = 0$

1.

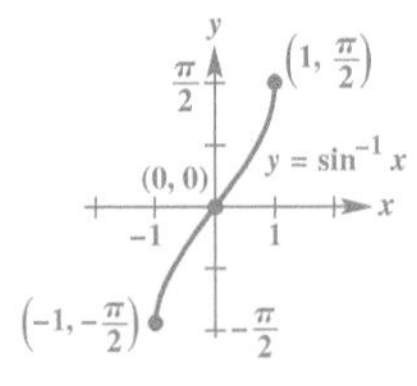

$[-1, 1]$; $\left[-\frac{\pi}{2}, \frac{\pi}{2}\right]$

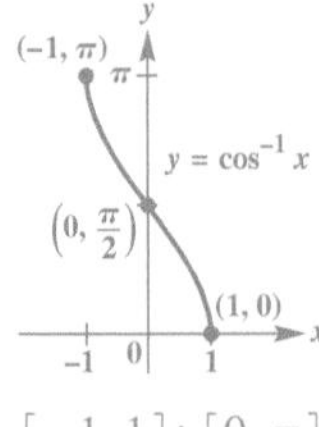

$[-1, 1]$; $[0, \pi]$

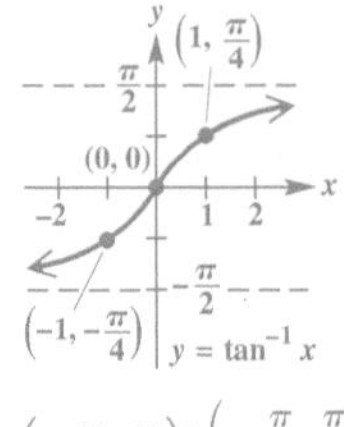

$(-\infty, \infty)$; $\left(-\frac{\pi}{2}, \frac{\pi}{2}\right)$

2. false; The range of the inverse tangent function is $\left(-\frac{\pi}{2}, \frac{\pi}{2}\right)$, while that of the inverse cotangent is $(0, \pi)$.
3. false; $\arcsin\left(-\frac{1}{2}\right) = -\frac{\pi}{6}$, not $\frac{11\pi}{6}$.
4. true
5. $\frac{\pi}{4}$
6. $\frac{2\pi}{3}$
7. $-\frac{\pi}{3}$
8. $-\frac{\pi}{2}$
9. $\frac{3\pi}{4}$
10. $\frac{\pi}{6}$
11. $\frac{2\pi}{3}$
12. $\frac{\pi}{3}$
13. $\frac{3\pi}{4}$
14. $60°$
15. $-60°$
16. $0°$
17. $60.67924514°$
18. $-41.33444556°$
19. $36.4895081°$
20. $12.51631252°$
21. $73.26220613°$
22. $7.673567973°$
23. -1
24. $-\frac{\sqrt{3}}{2}$
25. $\frac{3\pi}{4}$
26. π

27. $\frac{\pi}{4}$ 28. 0

29. $\frac{\sqrt{7}}{4}$ 30. $\frac{\sqrt{10}}{10}$

31. $\frac{\sqrt{3}}{2}$ 32. $\frac{9}{7}$

33. $\frac{294 + 125\sqrt{6}}{92}$

34. $\sqrt{1 - u^2}$

35. $\frac{1}{u}$

36. $\left\{\frac{\pi}{2}, \frac{3\pi}{2}\right\}$

37. $\{0.4636, 3.6052\}$

38. $\left\{0.7297, \frac{\pi}{2}, 2.4119\right\}$

39. $\left\{\frac{\pi}{4}, \frac{3\pi}{4}, \frac{5\pi}{4}, \frac{7\pi}{4}\right\}$

40. $\left\{\frac{\pi}{8}, \frac{3\pi}{8}, \frac{5\pi}{8}, \frac{7\pi}{8}, \frac{9\pi}{8}, \frac{11\pi}{8}, \frac{13\pi}{8}, \frac{15\pi}{8}\right\}$

41. $\left\{\frac{\pi}{8}, \frac{3\pi}{8}, \frac{5\pi}{8}, \frac{7\pi}{8}, \frac{9\pi}{8}, \frac{11\pi}{8}, \frac{13\pi}{8}, \frac{15\pi}{8}\right\}$

42. $\{2n\pi$, where n is any integer$\}$

43. $\left\{\frac{\pi}{3} + 2n\pi, \pi + 2n\pi, \frac{5\pi}{3} + 2n\pi\right.$, where n is any integer$\left.\right\}$

44. $\left\{\frac{\pi}{6} + n\pi, \frac{\pi}{3} + n\pi\right.$, where n is any integer$\left.\right\}$

45. $\{270°\}$

46. $\{45°, 153.4°, 225°, 333.4°\}$

47. $\{45°, 90°, 225°, 270°\}$

48. $\{15°, 75°, 195°, 255°\}$

49. $\{70.5°, 180°, 289.5°\}$

50. $\{53.5°, 118.4°, 233.5°, 298.4°\}$

51. $\{300° + 720°n, 420° + 720°n$, where n is any integer$\}$

52. $\{30° + 360°n, 150° + 360°n, 270° + 360°n$, where n is any integer$\}$

53. $\{180° + 360°n$, where n is any integer$\}$

54. $\{-1\}$

55. $\emptyset$ 56. $\left\{\frac{3\sqrt{5}}{7}\right\}$

57. $\left\{-\frac{1}{2}\right\}$ 58. $x = 2 \arccos \frac{y}{3}$

59. $x = \arcsin 2y$

60. $x = \arcsin\left(\frac{5y + 3}{4}\right)$

61. $x = \left(\frac{1}{3} \arctan 2y\right) - \frac{2}{3}$

62. $t = \frac{50}{\pi} \arccos\left(\frac{d - 550}{450}\right)$

Solve each equation for all exact solutions, in radians.

42. $\sec \frac{x}{2} = \cos \frac{x}{2}$ **43.** $\cos 2x + \cos x = 0$ **44.** $4 \sin x \cos x = \sqrt{3}$

Solve each equation for exact solutions over the interval $[0°\ 360°)$ where appropriate. Give approximate solutions to the nearest tenth of a degree.

45. $\sin^2 \theta + 3 \sin \theta + 2 = 0$ **46.** $2 \tan^2 \theta = \tan \theta + 1$

47. $\sin 2\theta = \cos 2\theta + 1$ **48.** $2 \sin 2\theta = 1$

49. $3 \cos^2 \theta + 2 \cos \theta - 1 = 0$ **50.** $5 \cot^2 \theta - \cot \theta - 2 = 0$

Solve each equation for all exact solutions, in degrees.

51. $2\sqrt{3} \cos \frac{\theta}{2} = -3$ **52.** $\sin \theta - \cos 2\theta = 0$ **53.** $\tan \theta - \sec \theta = 1$

Solve each equation for x.

54. $4\pi - 4 \cot^{-1} x = \pi$ **55.** $\frac{4}{3} \arctan \frac{x}{2} = \pi$

56. $\arccos x = \arcsin \frac{2}{7}$ **57.** $\arccos x + \arctan 1 = \frac{11\pi}{12}$

58. $y = 3 \cos \frac{x}{2}$, for x in $[0, 2\pi]$ **59.** $y = \frac{1}{2} \sin x$, for x in $\left[-\frac{\pi}{2}, \frac{\pi}{2}\right]$

60. $y = \frac{4}{5} \sin x - \frac{3}{5}$, for x in $\left[-\frac{\pi}{2}, \frac{\pi}{2}\right]$

61. $y = \frac{1}{2} \tan(3x + 2)$, for x in $\left(-\frac{2}{3} - \frac{\pi}{6}, -\frac{2}{3} + \frac{\pi}{6}\right)$

62. Solve $d = 550 + 450 \cos\left(\frac{\pi}{50}t\right)$ for t, where t is in the interval $[0, 50]$.

(Modeling) Solve each problem.

63. ***Viewing Angle of an Observer*** A 10-ft-wide chalkboard is situated 5 ft from the left wall of a classroom. See the figure. A student sitting next to the wall x feet from the front of the classroom has a viewing angle of θ radians.

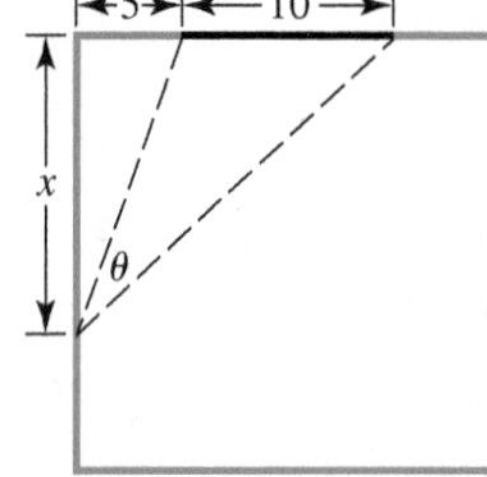

(a) Show that the value of θ is given by

$$y_1 = \tan^{-1}\left(\frac{15}{x}\right) - \tan^{-1}\left(\frac{5}{x}\right).$$

(b) Graph y_1 with a graphing calculator to estimate the value of x that maximizes the viewing angle.

64. ***Snell's Law*** Snell's law states that

$$\frac{c_1}{c_2} = \frac{\sin \theta_1}{\sin \theta_2},$$

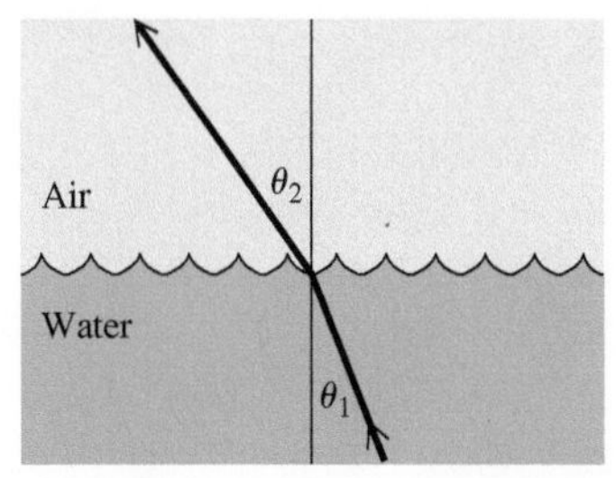

where c_1 is the speed of light in one medium, c_2 is the speed of light in a second medium, and θ_1 and θ_2 are the angles shown in the figure. Suppose a light is shining up through water into the air as in the figure. As θ_1 increases, θ_2 approaches 90°, at which point no light will emerge from the water. Assume the ratio $\frac{c_1}{c_2}$ in this case is 0.752. For what value of θ_1, to the nearest tenth, does $\theta_2 = 90°$? This value of θ_1 is the **critical angle** for water.

63. (b)

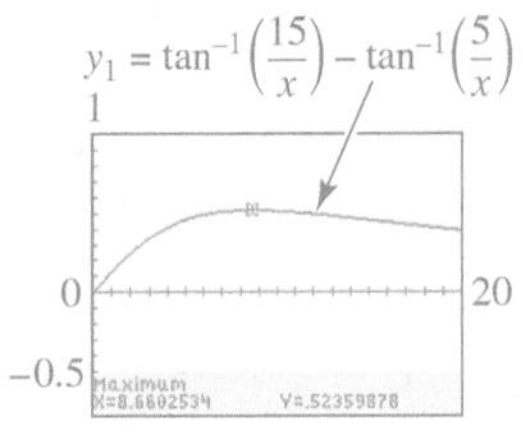

8.6602534 ft; There may be a discrepancy in the final digits.

64. 48.8°

65. No light will emerge from the water.

66. (a) 42.2° (b) 90° (c) 48.0°

65. *Snell's Law* Refer to **Exercise 64.** What happens when θ_1 is greater than the critical angle?

66. *British Nautical Mile* The British nautical mile is defined as the length of a minute of arc of a meridian. Because Earth is flat at its poles, the nautical mile, in feet, is given by

$$L = 6077 - 31 \cos 2\theta,$$

where θ is the latitude in degrees. See the figure. (*Source:* Bushaw, D., et al., *A Sourcebook of Applications of School Mathematics,* National Council of Teachers of Mathematics.) Give answers to the nearest tenth if applicable.

A nautical mile is the length on any of the meridians cut by a central angle of measure 1 minute.

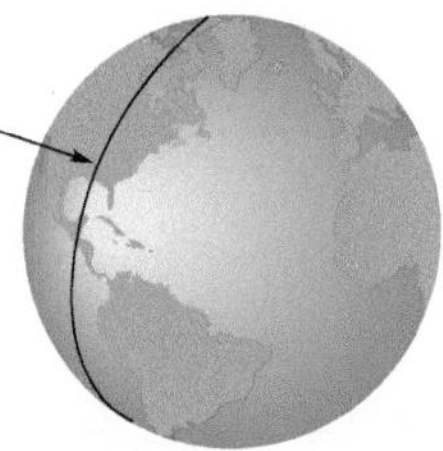

(a) Find the latitude between 0° and 90° at which the nautical mile is 6074 ft.

(b) At what latitude between 0° and 180° is the nautical mile 6108 ft?

(c) In the United States, the nautical mile is defined everywhere as 6080.2 ft. At what latitude between 0° and 90° does this agree with the British nautical mile?

Chapter 6 Test

[6.1]

1.

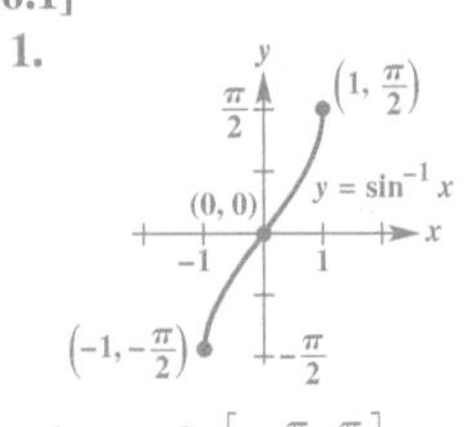

$[-1, 1]$; $\left[-\frac{\pi}{2}, \frac{\pi}{2}\right]$

2. (a) $\frac{2\pi}{3}$ (b) $-\frac{\pi}{3}$ (c) 0 (d) $\frac{2\pi}{3}$

3. (a) 30° (b) −45° (c) 135° (d) −60°

4. (a) 43.97° (b) 22.72° (c) 125.47°

5. (a) $\frac{\sqrt{5}}{3}$ (b) $\frac{4\sqrt{2}}{9}$

6. No number has a sine value of 3.

7. $\arcsin\left(\sin \frac{5\pi}{6}\right) = \arcsin \frac{1}{2}$ $= \frac{\pi}{6} \neq \frac{5\pi}{6}$

8. $\frac{u\sqrt{1-u^2}}{1-u^2}$

[6.2, 6.3]

9. {30°, 330°}

10. {90°, 270°}

11. {18.4°, 135°, 198.4°, 315°}

1. Graph $y = \sin^{-1} x$, and indicate the coordinates of three points on the graph. Give the domain and range.

2. Find the exact value of each real number y. Do not use a calculator.

(a) $y = \arccos\left(-\frac{1}{2}\right)$ **(b)** $y = \sin^{-1}\left(-\frac{\sqrt{3}}{2}\right)$

(c) $y = \tan^{-1} 0$ **(d)** $y = \operatorname{arcsec}(-2)$

3. Give the degree measure of θ. Do not use a calculator.

(a) $\theta = \arccos \frac{\sqrt{3}}{2}$ **(b)** $\theta = \tan^{-1}(-1)$

(c) $\theta = \cot^{-1}(-1)$ **(d)** $\theta = \csc^{-1}\left(-\frac{2\sqrt{3}}{3}\right)$

4. Use a calculator to approximate each value in decimal degrees to the nearest hundredth.

(a) $\sin^{-1} 0.69431882$ **(b)** $\sec^{-1} 1.0840880$

(c) $\cot^{-1}(-0.7125586)$

5. Evaluate each expression without using a calculator.

(a) $\cos\left(\arcsin \frac{2}{3}\right)$ **(b)** $\sin\left(2 \cos^{-1} \frac{1}{3}\right)$

6. Explain why $\sin^{-1} 3$ is not defined.

7. Explain why $\arcsin\left(\sin \frac{5\pi}{6}\right) \neq \frac{5\pi}{6}$.

8. Write $\tan(\arcsin u)$ as an algebraic expression in u, for $u > 0$.

Solve each equation for exact solutions over the interval $[0°, 360°)$ *where appropriate. Give approximate solutions to the nearest tenth of a degree.*

9. $-3 \sec \theta + 2\sqrt{3} = 0$ **10.** $\sin^2 \theta = \cos^2 \theta + 1$ **11.** $\csc^2 \theta - 2 \cot \theta = 4$

12. $\left\{0, \frac{2\pi}{3}, \frac{4\pi}{3}\right\}$

13. $\left\{\frac{\pi}{12}, \frac{7\pi}{12}, \frac{3\pi}{4}, \frac{5\pi}{4}, \frac{17\pi}{12}, \frac{23\pi}{12}\right\}$

14. $\{0.3649, 1.2059, 3.5065, 4.3475\}$

15. $\{90° + 180°n$, where n is any integer$\}$

16. $\left\{\frac{2\pi}{3} + 4n\pi, \frac{4\pi}{3} + 4n\pi\right.$, where n is any integer$\left.\right\}$

17. $\left\{\frac{\pi}{2} + 2n\pi\right.$, where n is any integer$\left.\right\}$

[6.4]

18. (a) $x = \frac{1}{3} \arccos y$

(b) $x = \operatorname{arccot}\left(\frac{y-4}{3}\right)$

19. (a) $\left\{\frac{4}{5}\right\}$ (b) $\left\{\frac{\sqrt{3}}{3}\right\}$

[6.3]

20. P first reaches its maximum at approximately 2.5×10^{-4}. The maximum is approximately 0.003166.

Solve each equation for exact solutions over the interval $[0, 2\pi)$ where appropriate. Give approximate solutions to four decimal places.

12. $\cos x = \cos 2x$ **13.** $\sqrt{2} \cos 3x - 1 = 0$ **14.** $\sin x \cos x = \frac{1}{3}$

Solve each equation for all exact solutions in radians (for x) or in degrees (for θ). Write answers using the least possible nonnegative angle measures.

15. $\sin^2 \theta = -\cos 2\theta$ **16.** $2\sqrt{3} \sin \frac{x}{2} = 3$ **17.** $\csc x - \cot x = 1$

Work each problem.

18. Solve each equation for x, where x is restricted to the given interval.

(a) $y = \cos 3x$, for x in $\left[0, \frac{\pi}{3}\right]$ **(b)** $y = 4 + 3 \cot x$, for x in $(0, \pi)$

19. Solve each equation for exact solutions.

(a) $\arcsin x = \arctan \frac{4}{3}$ **(b)** $\operatorname{arccot} x + 2 \arcsin \frac{\sqrt{3}}{2} = \pi$

20. ***Upper Harmonics Pressures*** Suppose that the E key above middle C is played on a piano, and its fundamental frequency is $f_1 = 330$ Hz. Its associated pressure is expressed as

$$P_1 = 0.002 \sin 660\pi t.$$

The pressures associated with the next four frequencies are $P_2 = \frac{0.002}{2} \sin 1320\pi t$, $P_3 = \frac{0.002}{3} \sin 1980\pi t$, $P_4 = \frac{0.002}{4} \sin 2640\pi t$, and $P_5 = \frac{0.002}{5} \sin 3300\pi t$. Duplicate the graph shown below of

$$P = P_1 + P_2 + P_3 + P_4 + P_5.$$

Approximate the maximum value of P to four significant digits and the least positive value of t for which P reaches this maximum.

For $x = t$,
$y_6 = P_1 + P_2 + P_3 + P_4 + P_5 = P$

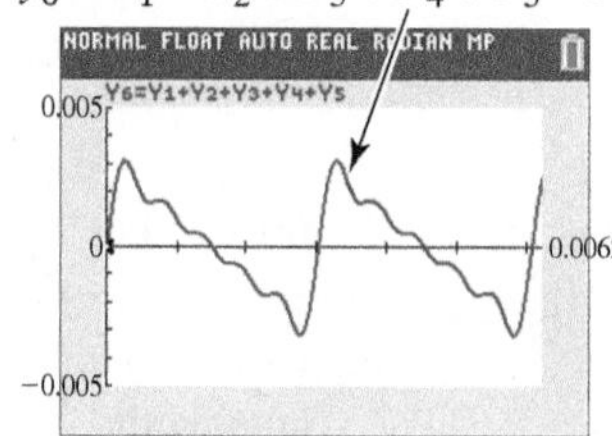

7 Applications of Trigonometry and Vectors

Surveyors use a method known as *triangulation* to measure distances when direct measurements cannot be made due to obstructions in the line of sight.

7.1 Oblique Triangles and the Law of Sines

- Congruency and Oblique Triangles
- Derivation of the Law of Sines
- Solutions of SAA and ASA Triangles (Case 1)
- Area of a Triangle

Congruency and Oblique Triangles We now turn our attention to solving triangles that are *not* right triangles. To do this we develop new relationships, or laws, that exist between the sides and angles of any triangle. The congruence axioms assist in this process. ***Recall that two triangles are congruent if their corresponding sides and angles are equal.***

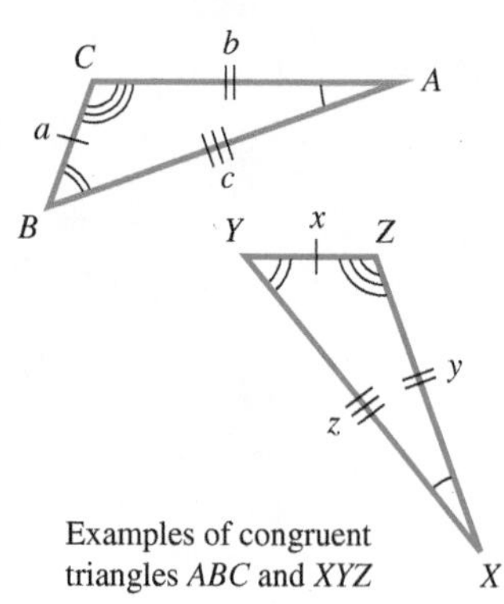

Examples of congruent triangles *ABC* and *XYZ*

Congruence Axioms

Side-Angle-Side (SAS)	If two sides and the included angle of one triangle are equal, respectively, to two sides and the included angle of a second triangle, then the triangles are congruent.
Angle-Side-Angle (ASA)	If two angles and the included side of one triangle are equal, respectively, to two angles and the included side of a second triangle, then the triangles are congruent.
Side-Side-Side (SSS)	If three sides of one triangle are equal, respectively, to three sides of a second triangle, then the triangles are congruent.

If a side and *any* two angles are given **(SAA),** the third angle can be determined by the angle sum formula

$$A + B + C = 180°.$$

Then the ASA axiom can be applied. Whenever SAS, ASA, or SSS is given, the triangle is unique.

A triangle that is not a right triangle is an **oblique triangle.** ***Recall that a triangle can be solved—that is, the measures of the three sides and three angles can be found—if at least one side and any other two measures are known.***

Data Required for Solving Oblique Triangles

There are four possible cases.

Case 1 One side and two angles are known (SAA or ASA).

Case 2 Two sides and one angle not included between the two sides are known (SSA). This case may lead to more than one triangle.

Case 3 Two sides and the angle included between the two sides are known (SAS).

Case 4 Three sides are known (SSS).

NOTE ***If we know three angles of a triangle, we cannot find unique side lengths because AAA assures us only of similarity, not congruence.*** For example, there are infinitely many triangles *ABC* of different sizes with $A = 35°$, $B = 65°$, and $C = 80°$.

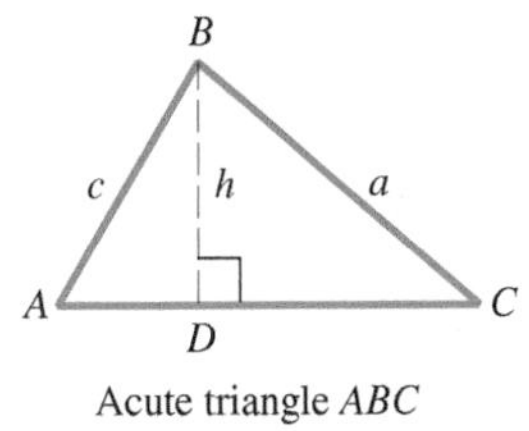

Acute triangle ABC

(a)

Obtuse triangle ABC

(b)

We label oblique triangles as we did right triangles: side a opposite angle A, side b opposite angle B, and side c opposite angle C.

Figure 1

Solving a triangle with given information matching Case 1 or Case 2 requires using the *law of sines,* while solving a triangle with given information matching Case 3 or Case 4 requires using the *law of cosines.*

Derivation of the Law of Sines To derive the law of sines, we start with an oblique triangle, such as the **acute triangle** in **Figure 1(a)** or the **obtuse triangle** in **Figure 1(b).** This discussion applies to both triangles. First, construct the perpendicular from B to side AC (or its extension). Let h be the length of this perpendicular. Then c is the hypotenuse of right triangle ADB, and a is the hypotenuse of right triangle BDC.

In triangle ADB, $\sin A = \frac{h}{c}$, or $h = c \sin A$.

In triangle BDC, $\sin C = \frac{h}{a}$, or $h = a \sin C$.

Because $h = c \sin A$ and $h = a \sin C$, we set these two expressions equal.

$$a \sin C = c \sin A$$

$$\frac{a}{\sin A} = \frac{c}{\sin C} \quad \text{Divide each side by } \sin A \sin C.$$

In a similar way, by constructing perpendicular lines from the other vertices, we can show that these two equations are also true.

$$\frac{a}{\sin A} = \frac{b}{\sin B} \quad \text{and} \quad \frac{b}{\sin B} = \frac{c}{\sin C}$$

This discussion proves the following theorem.

Teaching Tip The law of sines says that in a triangle, the ratio of the length of any side to the sine of its opposite angle is constant. (This constant is the diameter of the circumscribed circle of the triangle. See **Exercise 61** in this section.)

Law of Sines

In any triangle ABC, with sides a, b, and c, the following hold.

$$\frac{a}{\sin A} = \frac{b}{\sin B}, \quad \frac{a}{\sin A} = \frac{c}{\sin C}, \quad \text{and} \quad \frac{b}{\sin B} = \frac{c}{\sin C}$$

This can be written in compact form as follows.

$$\frac{a}{\sin A} = \frac{b}{\sin B} = \frac{c}{\sin C}$$

That is, according to the law of sines, the lengths of the sides in a triangle are proportional to the sines of the measures of the angles opposite them.

In practice we can also use an alternative form of the law of sines.

$$\frac{\sin A}{a} = \frac{\sin B}{b} = \frac{\sin C}{c} \quad \text{Alternative form of the law of sines}$$

NOTE When using the law of sines, a good strategy is to select a form that has the unknown variable in the numerator and where all other variables are known. This makes computation easier.

Solutions of SAA and ASA Triangles (Case 1)

Classroom Example 1
Solve triangle ABC if $A = 28.8°$, $C = 102.6°$, and $c = 25.3$ in.

Answer: $a = 12.5$ in., $B = 48.6°$, $b = 19.4$ in.

EXAMPLE 1 Applying the Law of Sines (SAA)

Solve triangle ABC if $A = 32.0°$, $B = 81.8°$, and $a = 42.9$ cm.

SOLUTION Start by drawing a triangle, roughly to scale, and labeling the given parts as in **Figure 2.** The values of A, B, and a are known, so use the form of the law of sines that involves these variables, and then solve for b.

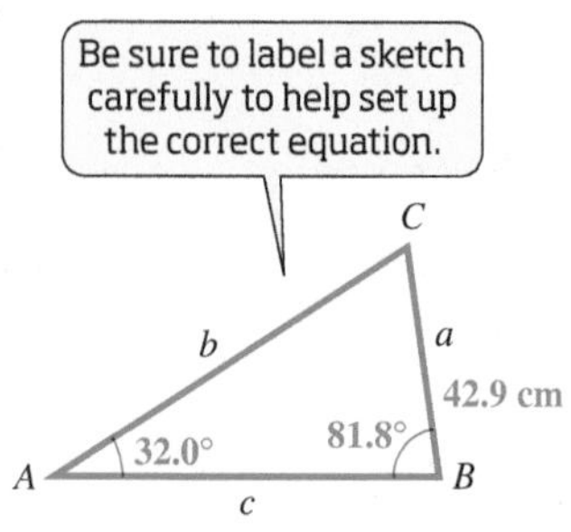

Figure 2

$$\frac{a}{\sin A} = \frac{b}{\sin B}$$ Choose a form of the law of sines that has the unknown variable in the numerator.

$$\frac{42.9}{\sin 32.0°} = \frac{b}{\sin 81.8°}$$ Substitute the given values.

$$b = \frac{42.9 \sin 81.8°}{\sin 32.0°}$$ Multiply by sin 81.8° and rewrite.

$$b \approx 80.1 \text{ cm}$$ Approximate with a calculator.

To find C, use the fact that the sum of the angles of any triangle is 180°.

$$A + B + C = 180°$$ Angle sum formula

$$C = 180° - A - B$$ Solve for C.

$$C = 180° - 32.0° - 81.8°$$ Substitute.

$$C = 66.2°$$ Subtract.

Teaching Tip Encourage students to draw a scale version of the given triangle to help solve it, and then check answers for reasonableness.

Now use the law of sines to find c. (The Pythagorean theorem does not apply because this is not a right triangle.)

$$\frac{a}{\sin A} = \frac{c}{\sin C}$$ Law of sines

$$\frac{42.9}{\sin 32.0°} = \frac{c}{\sin 66.2°}$$ Substitute known values.

$$c = \frac{42.9 \sin 66.2°}{\sin 32.0°}$$ Multiply by sin 66.2° and rewrite.

$$c \approx 74.1 \text{ cm}$$ Approximate with a calculator.

✔ **Now Try Exercise 17.**

CAUTION Whenever possible, use given values in solving triangles, rather than values obtained in intermediate steps, to avoid rounding errors.

Classroom Example 2
Repeat **Example 2** if $C = 117.2°$, $A = 28.8°$, and $b = 75.6$ ft.

Answer: 65.1 ft

EXAMPLE 2 Applying the Law of Sines (ASA)

An engineer wishes to measure the distance across a river. See **Figure 3.** He determines that $C = 112.90°$, $A = 31.10°$, and $b = 347.6$ ft. Find the distance a.

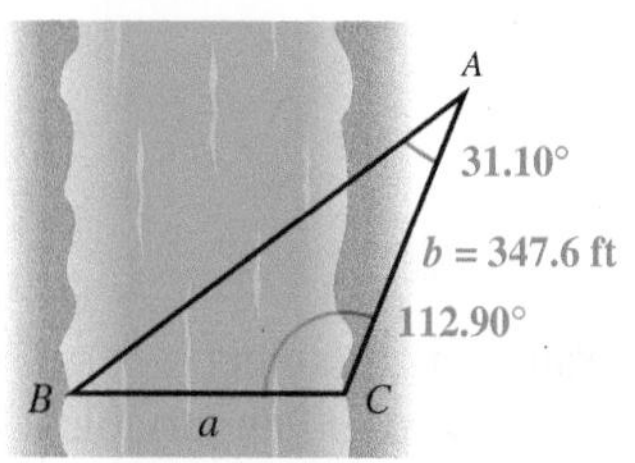

Figure 3

SOLUTION To use the law of sines, one side and the angle opposite it must be known. Here b is the only side whose length is given, so angle B must be found before the law of sines can be used.

$$B = 180° - A - C$$ Angle sum formula, solved for B

$$B = 180° - 31.10° - 112.90°$$ Substitute the given values.

$$B = 36.00°$$ Subtract.

Now use the form of the law of sines involving A, B, and b to find side a.

Solve for a.

$$\frac{a}{\sin A} = \frac{b}{\sin B}$$ Law of sines

$$\frac{a}{\sin 31.10°} = \frac{347.6}{\sin 36.00°}$$ Substitute known values.

$$a = \frac{347.6 \sin 31.10°}{\sin 36.00°}$$ Multiply by sin 31.10°.

$$a \approx 305.5 \text{ ft}$$ Use a calculator.

Now Try Exercise 33.

Recall that **bearing** is used in navigation to refer to direction of motion or direction of a distant object relative to current course. We consider two methods for expressing bearing.

Method 1

When a single angle is given, such as 220°, this bearing is measured in a clockwise direction from north.

Example: 220°

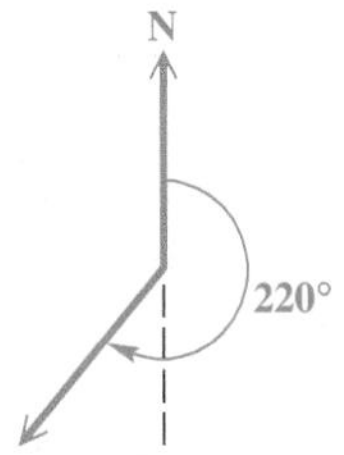

Method 2

Start with a north-south line and use an acute angle to show direction, either east or west, from this line.

Example: S 40° W

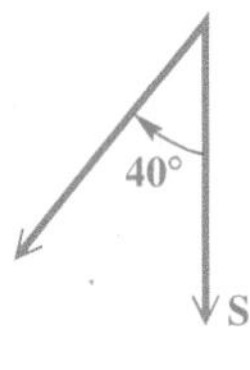

Classroom Example 3
The bearing of a lighthouse from a ship was found to be N 52° W. After the ship sailed 5.8 km due south, the new bearing was N 23° W. Find the distance, to the nearest tenth kilometer, between the ship and the lighthouse at each location.

Answer: first location: 4.7 km; second location: 9.4 km

EXAMPLE 3 Applying the Law of Sines (ASA)

Two ranger stations are on an east-west line 110 mi apart. A forest fire is located on a bearing of N 42° E from the western station at A and a bearing of N 15° E from the eastern station at B. To the nearest ten miles, how far is the fire from the western station?

SOLUTION **Figure 4** shows the two ranger stations at points A and B and the fire at point C. Angle BAC measures $90° - 42° = 48°$, obtuse angle B measures $90° + 15° = 105°$, and the third angle, C, measures $180° - 105° - 48° = 27°$. We use the law of sines to find side b.

Solve for b.

$$\frac{b}{\sin B} = \frac{c}{\sin C}$$ Law of sines

$$\frac{b}{\sin 105°} = \frac{110}{\sin 27°}$$ Substitute known values.

$$b = \frac{110 \sin 105°}{\sin 27°}$$ Multiply by sin 105°.

$$b \approx 230 \text{ mi}$$ Use a calculator and give two significant digits.

Now Try Exercise 35.

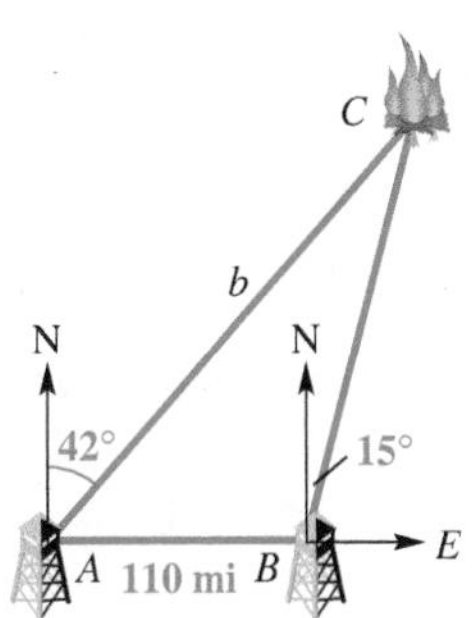

Figure 4

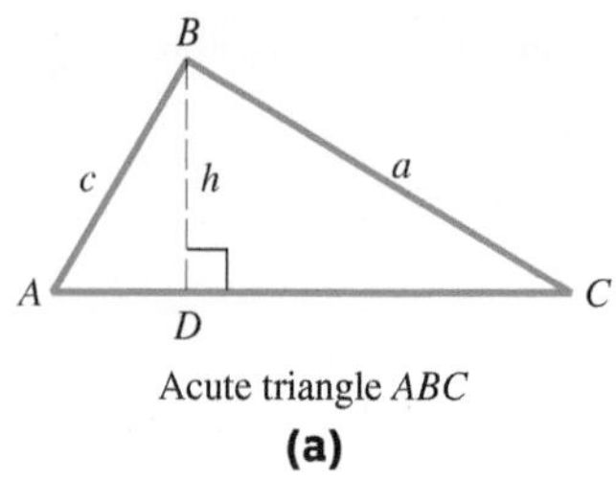

Acute triangle ABC
(a)

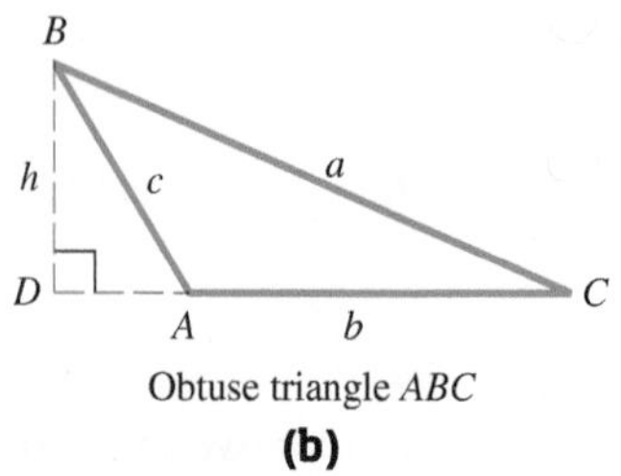

Obtuse triangle ABC
(b)

Figure 5

Area of a Triangle A familiar formula for the area of a triangle is

$$\mathcal{A} = \frac{1}{2}bh, \quad \text{where } \mathcal{A} \text{ represents area, } b \text{ base, and } h \text{ height.}$$

This formula cannot always be used easily because in practice, h is often unknown. To find another formula, refer to acute triangle ABC in **Figure 5(a)** or obtuse triangle ABC in **Figure 5(b).**

A perpendicular has been drawn from B to the base of the triangle (or the extension of the base). Consider right triangle ADB in either figure.

$$\sin A = \frac{h}{c}, \quad \text{or} \quad h = c \sin A$$

Substitute into the formula for the area of a triangle.

$$\mathcal{A} = \frac{1}{2}bh = \frac{1}{2}bc \sin A$$

Any other pair of sides and the angle between them could have been used.

Teaching Tip Emphasize that in each of these formulas, two sides and an included angle are used.

Area of a Triangle (SAS)

In any triangle ABC, the area $\mathcal{A}$ is given by the following formulas.

$$\mathcal{A} = \frac{1}{2}bc \sin A, \quad \mathcal{A} = \frac{1}{2}ab \sin C, \quad \text{and} \quad \mathcal{A} = \frac{1}{2}ac \sin B$$

That is, the area is half the product of the lengths of two sides and the sine of the angle included between them.

NOTE If the included angle measures 90°, its sine is 1 and the formula becomes the familiar $\mathcal{A} = \frac{1}{2}bh$.

Classroom Example 4
Find the area of triangle DEF in the figure.

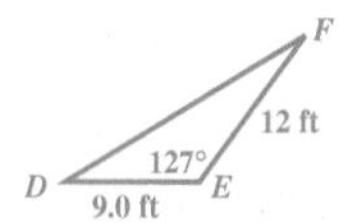

Answer: 43 ft²

EXAMPLE 4 Finding the Area of a Triangle (SAS)

Find the area of triangle ABC in **Figure 6.**

SOLUTION Substitute $B = 55° \, 10'$, $a = 34.0$ ft, and $c = 42.0$ ft into the area formula.

$$\mathcal{A} = \frac{1}{2}ac \sin B = \frac{1}{2}(34.0)(42.0) \sin 55° \, 10' \approx 586 \text{ ft}^2$$

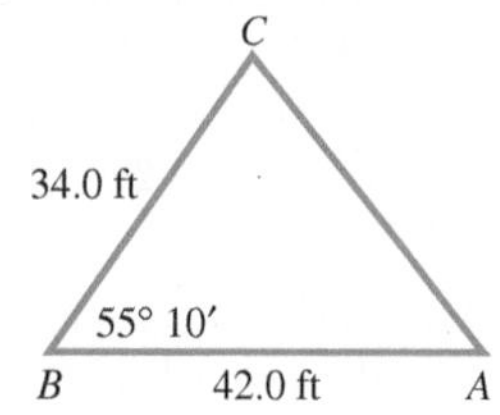

Figure 6

✔ **Now Try Exercise 51.**

EXAMPLE 5 Finding the Area of a Triangle (ASA)

Find the area of triangle ABC in **Figure 7.**

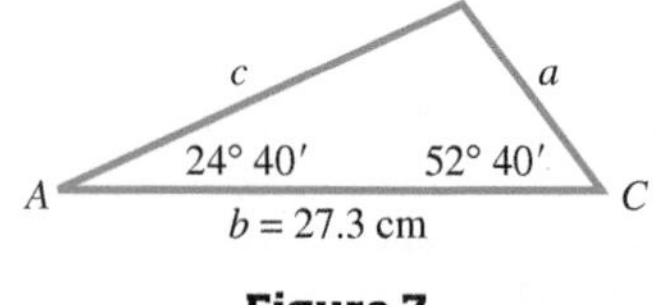

Figure 7

SOLUTION Before the area formula can be used, we must find side a or c.

First find remaining angle B.

$$180° = A + B + C \quad \text{Angle sum formula}$$

$$B = 180° - 24° \, 40' - 52° \, 40' \quad \text{Substitute and solve for } B.$$

$$B = 102° \, 40' \quad \text{Subtract.}$$

Classroom Example 5
Find the area of triangle ABC if $B = 58° 10'$, $a = 32.5$ cm, and $C = 73° 30'$.

Answer: 576 cm²

Next use the law of sines to find side a.

Solve for a.

$$\frac{a}{\sin A} = \frac{b}{\sin B} \quad \text{Law of sines}$$

$$\frac{a}{\sin 24° 40'} = \frac{27.3}{\sin 102° 40'} \quad \text{Substitute known values.}$$

$$a = \frac{27.3 \sin 24° 40'}{\sin 102° 40'} \quad \text{Multiply by } \sin 24° 40'.$$

$$a \approx 11.7 \text{ cm} \quad \text{Use a calculator.}$$

Now that we know two sides, a and b, and their included angle C, we find the area.

$$\mathcal{A} = \frac{1}{2}ab \sin C \approx \frac{1}{2}(11.7)(27.3) \sin 52° 40' \approx 127 \text{ cm}^2$$

11.7 is an approximation. In practice, use the calculator value.

✔ **Now Try Exercise 57.**

7.1 Exercises

Note to instructor: Although most of the measurements resulting from solving triangles in this chapter are approximations, for convenience we use = rather than ≈ in the answers.

1. oblique 2. side
3. angles 4. $\sin B$; $\sin C$
5. a; b; c 6. $\sin C$
7. The law of sines may be used.
8. The law of sines may be used.
9. There is not sufficient information to use the law of sines.
10. There is not sufficient information to use the law of sines.
11. $\sqrt{3}$ 12. $10\sqrt{2}$

CONCEPT PREVIEW *Fill in the blank(s) to correctly complete each sentence.*

1. A triangle that is not a right triangle is a(n) ______ triangle.

2. The measures of the three sides and three angles of a triangle can be found if at least one ______ and any other two measures are known.

3. If we know three ______ of a triangle, we cannot find a unique solution for the triangle.

4. In the law of sines, $\frac{a}{\sin A} = \frac{b}{\_\_\_} = \frac{c}{\_\_\_}$.

5. An alternative form of the law of sines is $\frac{\sin A}{\_\_\_} = \frac{\sin B}{\_\_\_} = \frac{\sin C}{\_\_\_}$.

6. For any triangle ABC, its area can be found using the formula $\mathcal{A} = \frac{1}{2}ab$ ______.

CONCEPT PREVIEW *Consider each case and determine whether there is sufficient information to solve the triangle using the law of sines.*

7. Two angles and the side included between them are known.

8. Two angles and a side opposite one of them are known.

9. Two sides and the angle included between them are known.

10. Three sides are known.

Find the length of each side labeled a. Do not use a calculator.

11.

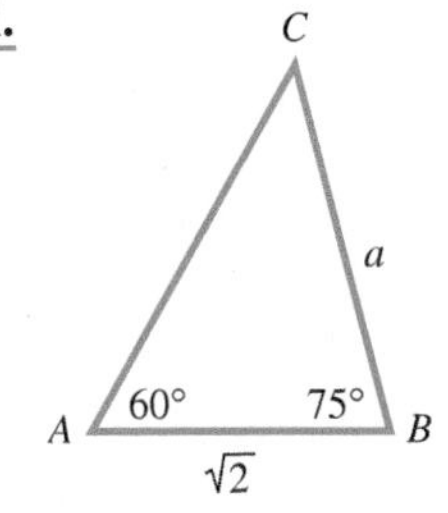

12.

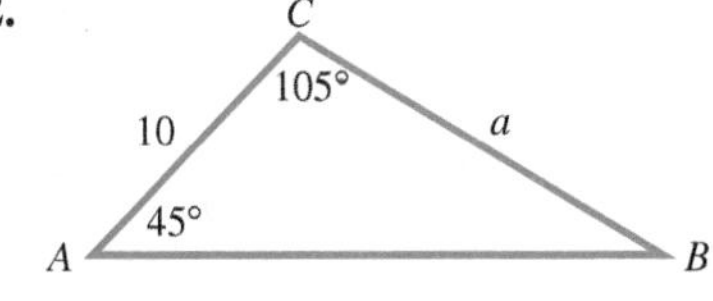

Determine the remaining sides and angles of each triangle ABC. **See Example 1.**

13.

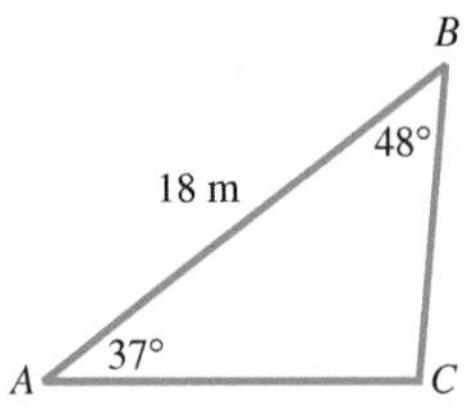

14.

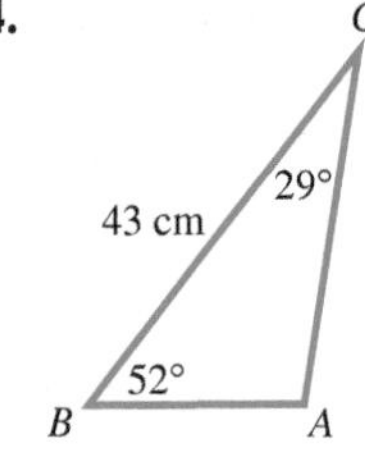

15.

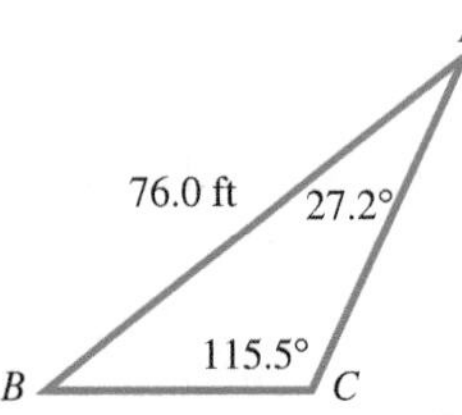

16.

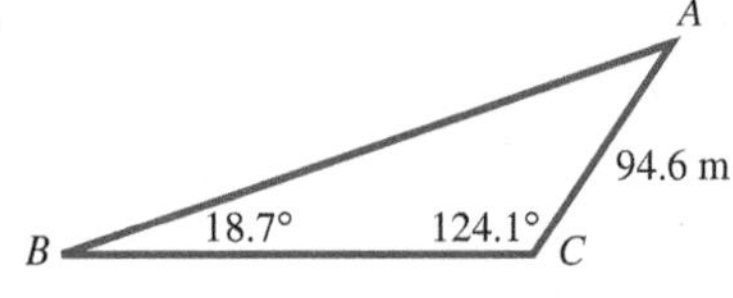

17. $A = 68.41°$, $B = 54.23°$, $a = 12.75$ ft

18. $C = 74.08°$, $B = 69.38°$, $c = 45.38$ m

19. $A = 87.2°$, $b = 75.9$ yd, $C = 74.3°$

20. $B = 38° 40'$, $a = 19.7$ cm, $C = 91° 40'$

21. $B = 20° 50'$, $C = 103° 10'$, $AC = 132$ ft

22. $A = 35.3°$, $B = 52.8°$, $AC = 675$ ft

23. $A = 39.70°$, $C = 30.35°$, $b = 39.74$ m

24. $C = 71.83°$, $B = 42.57°$, $a = 2.614$ cm

25. $B = 42.88°$, $C = 102.40°$, $b = 3974$ ft

26. $C = 50.15°$, $A = 106.1°$, $c = 3726$ yd

27. $A = 39° 54'$, $a = 268.7$ m, $B = 42° 32'$

28. $C = 79° 18'$, $c = 39.81$ mm, $A = 32° 57'$

Concept Check Answer each question.

29. Why can the law of sines not be used to solve a triangle if we are given only the lengths of the three sides of the triangle?

30. In **Example 1,** we begin (as seen there) by solving for b and C. Why is it a better idea to solve for c by using a and $\sin A$ than by using b and $\sin B$?

31. Eli Maor, a perceptive trigonometry student, makes this statement: "If we know *any* two angles and one side of a triangle, then the triangle is uniquely determined." Why is this true? Refer to the congruence axioms given in this section.

32. In a triangle, if a is twice as long as b, is A necessarily twice as large as B?

Solve each problem. **See Examples 2 and 3.**

33. ***Distance across a River*** To find the distance AB across a river, a surveyor laid off a distance $BC = 354$ m on one side of the river. It is found that $B = 112° 10'$ and $C = 15° 20'$. Find AB. See the figure.

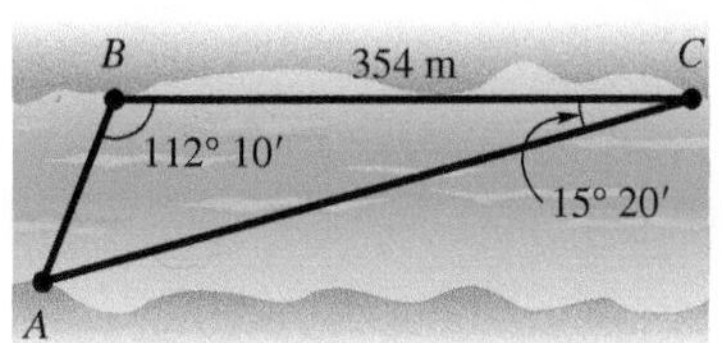

34. ***Distance across a Canyon*** To determine the distance RS across a deep canyon, Rhonda lays off a distance $TR = 582$ yd. She then finds that $T = 32° 50'$ and $R = 102° 20'$. Find RS. See the figure.

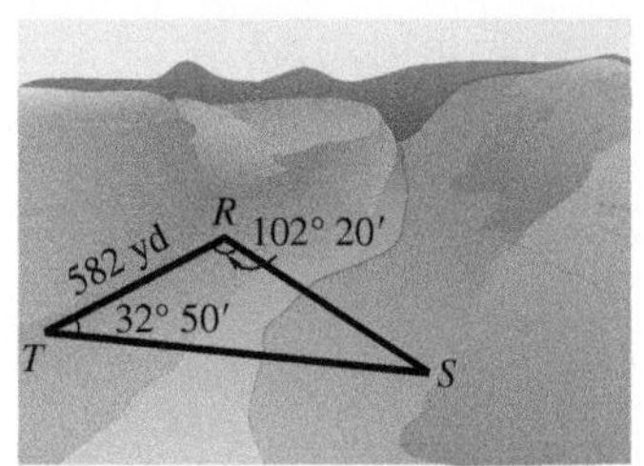

13. $C = 95°$, $b = 13$ m, $a = 11$ m
14. $A = 99°$, $b = 34$ cm, $c = 21$ cm
15. $B = 37.3°$, $a = 38.5$ ft, $b = 51.0$ ft
16. $A = 37.2°$, $a = 178$ m, $c = 244$ m
17. $C = 57.36°$, $b = 11.13$ ft, $c = 11.55$ ft
18. $A = 36.54°$, $b = 44.17$ m, $a = 28.10$ m
19. $B = 18.5°$, $a = 239$ yd, $c = 230$ yd
20. $A = 49° 40'$, $b = 16.1$ cm, $c = 25.8$ cm
21. $A = 56° 00'$, $AB = 361$ ft, $BC = 308$ ft
22. $C = 91.9°$, $BC = 490$ ft, $AB = 847$ ft
23. $B = 110.0°$, $a = 27.01$ m, $c = 21.36$ m
24. $A = 65.60°$, $b = 1.942$ cm, $c = 2.727$ cm
25. $A = 34.72°$, $a = 3326$ ft, $c = 5704$ ft
26. $B = 23.75°$, $a = 4663$ yd, $b = 1955$ yd
27. $C = 97° 34'$, $b = 283.2$ m, $c = 415.2$ m
28. $B = 67° 45'$, $b = 37.50$ mm, $a = 22.04$ mm
29. To use the law of sines, we must know an angle measure, the length of the side opposite it, and at least one other angle measure or side length.
30. It is best to use given values, when possible, to avoid accumulating round-off errors.
31. If two angles and a side are given, the third angle can be determined using the angle sum formula. Then the ASA congruence axiom can be applied. This triangle is uniquely determined because there is only one possible triangle that meets these initial conditions.
32. no
33. 118 m **34.** 448 yd

35. 17.8 km 36. 1.93 mi
37. first location: 5.1 mi; second location: 7.2 mi
38. 293.4 m
39. 0.49 mi 40. 10.4 in.
41. 111° 42. 12
43. The distance is 419,000 km, which compares favorably to the actual value.

35. *Distance a Ship Travels* A ship is sailing due north. At a certain point the bearing of a lighthouse 12.5 km away is N 38.8° E. Later on, the captain notices that the bearing of the lighthouse has become S 44.2° E. How far did the ship travel between the two observations of the lighthouse?

36. *Distance between Radio Direction Finders* Radio direction finders are placed at points A and B, which are 3.46 mi apart on an east-west line, with A west of B. From A the bearing of a certain radio transmitter is 47.7°, and from B the bearing is 302.5°. Find the distance of the transmitter from A.

37. *Distance between a Ship and a Lighthouse* The bearing of a lighthouse from a ship was found to be N 37° E. After the ship sailed 2.5 mi due south, the new bearing was N 25° E. Find the distance between the ship and the lighthouse at each location.

38. *Distance across a River* Standing on one bank of a river flowing north, Mark notices a tree on the opposite bank at a bearing of 115.45°. Lisa is on the same bank as Mark, but 428.3 m away. She notices that the bearing of the tree is 45.47°. The two banks are parallel. What is the distance across the river?

39. *Height of a Balloon* A balloonist is directly above a straight road 1.5 mi long that joins two villages. She finds that the town closer to her is at an angle of depression of 35°, and the farther town is at an angle of depression of 31°. How high above the ground is the balloon?

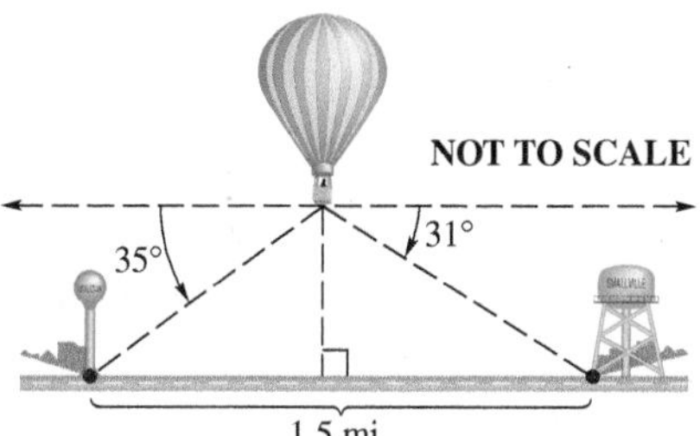

40. *Measurement of a Folding Chair* A folding chair is to have a seat 12.0 in. deep with angles as shown in the figure. How far down from the seat should the crossing legs be joined? (Find length x in the figure.)

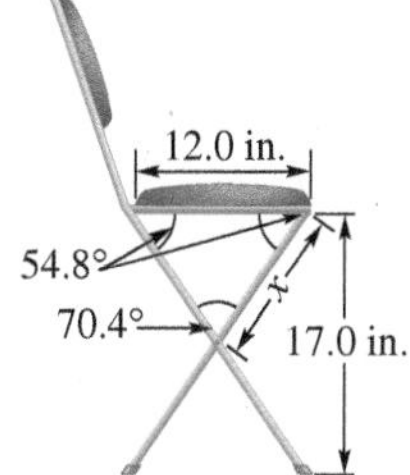

41. *Angle Formed by Radii of Gears* Three gears are arranged as shown in the figure. Find angle θ.

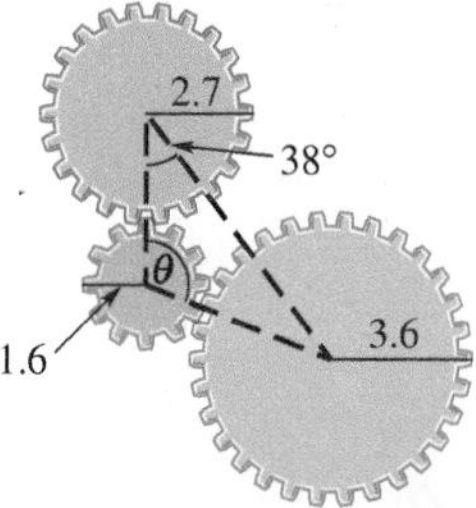

42. *Distance between Atoms* Three atoms with atomic radii of 2.0, 3.0, and 4.5 are arranged as in the figure. Find the distance between the centers of atoms A and C.

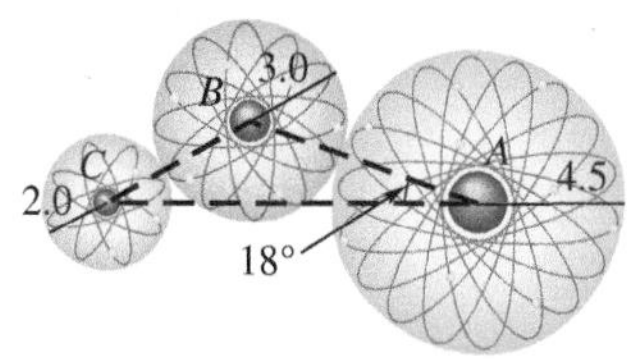

43. *Distance to the Moon* The moon is a relatively close celestial object, so its distance can be measured directly by taking two different photographs at precisely the same time from two different locations. The moon will have a different angle of elevation at each location. On April 29, 1976, at 11:35 A.M., the lunar angles of elevation during a partial solar eclipse at Bochum in upper Germany and at Donaueschingen in lower Germany were measured as 52.6997° and 52.7430°, respectively. The two cities are 398 km apart.

Calculate the distance to the moon, to the nearest thousand kilometers, from Bochum on this day, and compare it with the actual value of 406,000 km. Disregard the curvature of Earth in this calculation. (*Source:* Scholosser, W., T. Schmidt-Kaler, and E. Milone, *Challenges of Astronomy,* Springer-Verlag.)

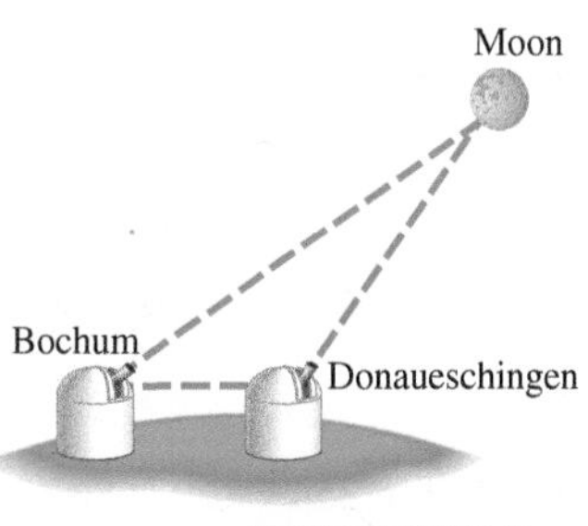

44. 10,285 ft
45. 6600 ft **46.** 5100 ft
47. $\frac{\sqrt{3}}{2}$ sq unit **48.** $\sqrt{3}$ sq units
49. $\frac{\sqrt{2}}{2}$ sq unit **50.** 1 sq unit
51. 46.4 m^2 **52.** 732 ft^2
53. 356 cm^2 **54.** 163 km^2
55. 722.9 $in.^2$ **56.** 289.9 m^2
57. 65.94 cm^2 **58.** 84.41 m^2

44. ***Ground Distances Measured by Aerial Photography*** The distance covered by an aerial photograph is determined by both the focal length of the camera and the tilt of the camera from the perpendicular to the ground. A camera lens with a 12-in. focal length will have an angular coverage of 60°. If an aerial photograph is taken with this camera tilted $\theta = 35°$ at an altitude of 5000 ft, calculate to the nearest foot the ground distance d that will be shown in this photograph. (*Source:* Brooks, R. and D. Johannes, *Phytoarchaeology,* Dioscorides Press.)

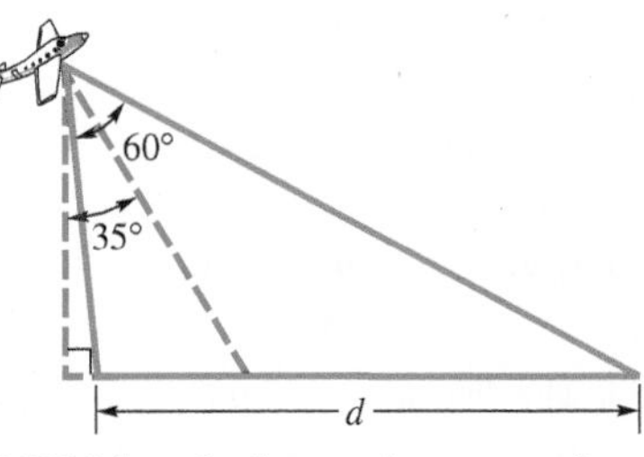

45. ***Ground Distances Measured by Aerial Photography*** Refer to **Exercise 44.** A camera lens with a 6-in. focal length has an angular coverage of 86°. Suppose an aerial photograph is taken vertically with no tilt at an altitude of 3500 ft over ground with an increasing slope of 5°, as shown in the figure. Calculate the ground distance CB, to the nearest hundred feet, that will appear in the resulting photograph. (*Source:* Moffitt, F. and E. Mikhail, *Photogrammetry,* Third Edition, Harper & Row.)

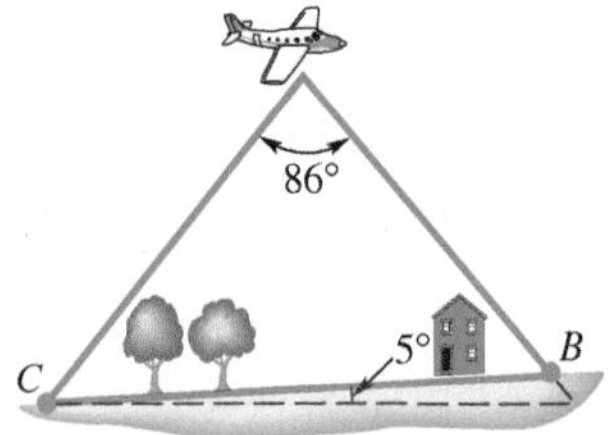

46. ***Ground Distances Measured by Aerial Photography*** Repeat **Exercise 45** if the camera lens has an 8.25-in. focal length with an angular coverage of 72°.

Find the area of each triangle using the formula $\mathcal{A} = \frac{1}{2}bh$, *and then verify that the formula* $\mathcal{A} = \frac{1}{2}ab \sin C$ *gives the same result.*

47.

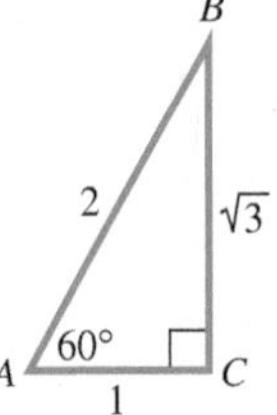

48.

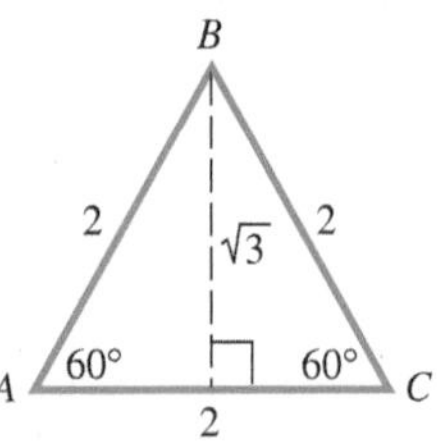

49.

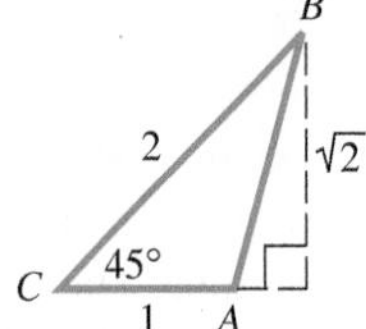

50.

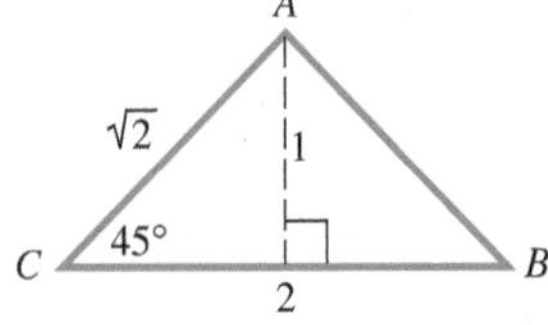

Find the area of each triangle ABC. ***See Examples 4 and 5.***

51. $A = 42.5°$, $b = 13.6$ m, $c = 10.1$ m **52.** $C = 72.2°$, $b = 43.8$ ft, $a = 35.1$ ft

53. $B = 124.5°$, $a = 30.4$ cm, $c = 28.4$ cm **54.** $C = 142.7°$, $a = 21.9$ km, $b = 24.6$ km

55. $A = 56.80°$, $b = 32.67$ in., $c = 52.89$ in. **56.** $A = 34.97°$, $b = 35.29$ m, $c = 28.67$ m

57. $A = 30.50°$, $b = 13.00$ cm, $C = 112.60°$ **58.** $A = 59.80°$, $b = 15.00$ m, $C = 53.10°$

59. 100 m^2 60. 373 m^2

61. $a = \sin A$, $b = \sin B$, $c = \sin C$

63. $x = \frac{d \sin \alpha \sin \beta}{\sin(\beta - \alpha)}$

64. (a) house: $X_H = 1131.8$ ft, $Y_H = 4390.2$ ft; fire: $X_F = 2277.5$ ft, $Y_F = -2596.2$ ft
(b) 7079.7 ft

Solve each problem.

59. ***Area of a Metal Plate*** A painter is going to apply a special coating to a triangular metal plate on a new building. Two sides measure 16.1 m and 15.2 m. She knows that the angle between these sides is 125°. What is the area of the surface she plans to cover with the coating?

60. ***Area of a Triangular Lot*** A real estate agent wants to find the area of a triangular lot. A surveyor takes measurements and finds that two sides are 52.1 m and 21.3 m, and the angle between them is 42.2°. What is the area of the triangular lot?

61. ***Triangle Inscribed in a Circle*** For a triangle inscribed in a circle of radius r, the law of sines ratios

$$\frac{a}{\sin A}, \quad \frac{b}{\sin B}, \quad \text{and} \quad \frac{c}{\sin C} \quad \text{have value } 2r.$$

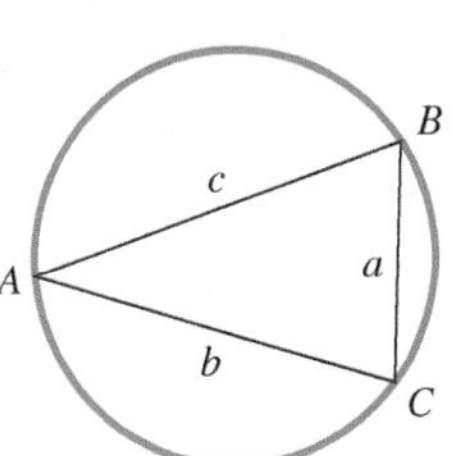

The circle in the figure has diameter 1. What are the values of a, b, and c? (*Note:* This result provides an alternative way to define the sine function for angles between 0° and 180°. It was used nearly 2000 yr ago by the mathematician Ptolemy to construct one of the earliest trigonometric tables.)

62. ***Theorem of Ptolemy*** The following theorem is also attributed to Ptolemy:

In a quadrilateral inscribed in a circle, the product of the diagonals is equal to the sum of the products of the opposite sides.

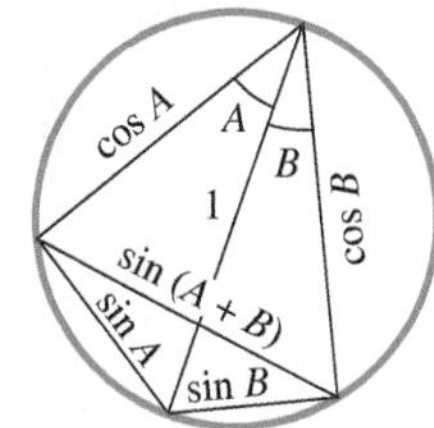

(*Source:* Eves, H., *An Introduction to the History of Mathematics,* Sixth Edition, Saunders College Publishing.) The circle in the figure has diameter 1. Use Ptolemy's theorem to derive the formula for the sine of the sum of two angles.

63. ***Law of Sines*** Several of the exercises on right triangle applications involved a figure similar to the one shown here, in which angles α and β and the length of line segment AB are known, and the length of side CD is to be determined. Use the law of sines to obtain x in terms of α, β, and d.

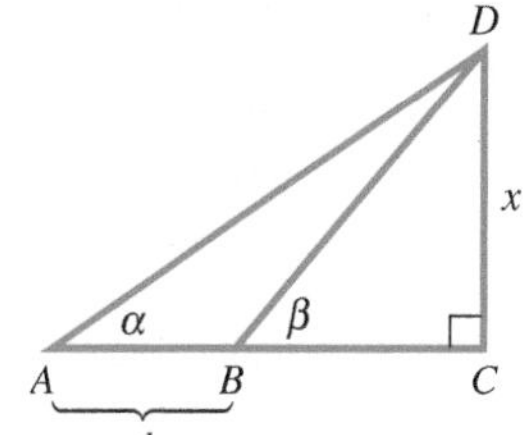

64. ***Aerial Photography*** Aerial photographs can be used to provide coordinates of ordered pairs to determine distances on the ground. Suppose we assign coordinates as shown in the figure. If an object's photographic coordinates are (x, y), then its ground coordinates (X, Y) in feet can be computed using the following formulas.

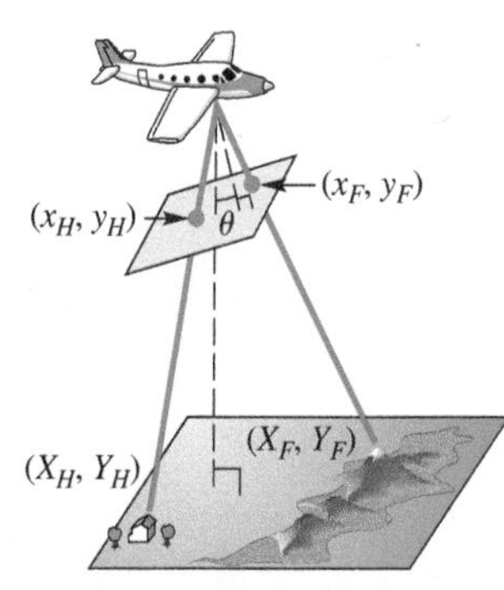

$$X = \frac{(a - h)x}{f \sec \theta - y \sin \theta}, \quad Y = \frac{(a - h)y \cos \theta}{f \sec \theta - y \sin \theta}$$

Here, f is focal length of the camera in inches, a is altitude in feet of the airplane, and h is elevation in feet of the object. Suppose that a house has photographic coordinates $(x_H, y_H) = (0.9, 3.5)$ with elevation 150 ft, and a nearby forest fire has photographic coordinates $(x_F, y_F) = (2.1, -2.4)$ and is at elevation 690 ft. Also suppose the photograph was taken at 7400 ft by a camera with focal length 6 in. and tilt angle $\theta = 4.1°$. (*Source:* Moffitt, F. and E. Mikhail, *Photogrammetry,* Third Edition, Harper & Row.)

(a) Use the formulas to find the ground coordinates of the house and the fire to the nearest tenth of a foot.

(b) Use the distance formula $d = \sqrt{(x_2 - x_1)^2 + (y_2 - y_1)^2}$ to find the distance on the ground between the house and the fire to the nearest tenth of a foot.

7.2 The Ambiguous Case of the Law of Sines

- Description of the Ambiguous Case
- Solutions of SSA Triangles (Case 2)
- Analyzing Data for Possible Number of Triangles

Description of the Ambiguous Case We have used the law of sines to solve triangles involving Case 1, given SAA or ASA. If we are given the lengths of two sides and the angle opposite one of them (Case 2, SSA), then zero, one, or two such triangles may exist. (There is no SSA congruence axiom.)

Suppose we know the measure of acute angle A of triangle ABC, the length of side a, and the length of side b, as shown in **Figure 8.** We must draw the side of length a opposite angle A. The table shows possible outcomes. This situation (SSA) is called the **ambiguous case** of the law of sines.

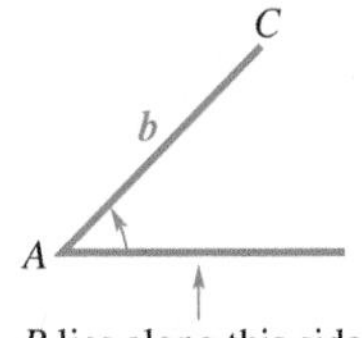

Figure 8

Possible Outcomes for Applying the Law of Sines

Angle A is	Possible Number of Triangles	Sketch	Applying Law of Sines Leads to
Acute	0		$\sin B > 1,\ a < h < b$
Acute	1		$\sin B = 1,\ a = h$ and $h < b$
Acute	1		$0 < \sin B < 1,\ a \geq b$
Acute	2		$0 < \sin B_1 < 1,\ h < a < b,$ $A + B_2 < 180°$
Obtuse	0		$\sin B \geq 1,\ a \leq b$
Obtuse	1		$0 < \sin B < 1,\ a > b$

Teaching Tip The case leading to two triangles, for acute angle A and known sides a and b, is troublesome for students. It is a good idea to draw the "triangle" below and talk through the possibilities, comparing the lengths of a, b, and h. Use a straightedge anchored at point C, and have it swing an arc. Use different lengths for the straightedge so that students can actually visualize the cases.

Emphasize making good sense of the situation. Note that $\sin A = \frac{h}{b}$, or $h = b \sin A$.

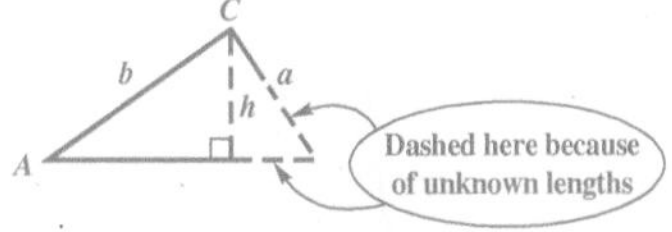

- If $a < h$, no triangle exists.
- If $a = h$, a right triangle exists.
- If $h < a < b$, two triangles exist.
- If $a \geq b$, one triangle exists.

Draw a similar sketch with obtuse angle A to show that a must be greater than b for a single triangle to exist.

The following basic facts help determine which situation applies.

Applying the Law of Sines

1. For any angle θ of a triangle, $0 < \sin \theta \leq 1$. If $\sin \theta = 1$, then $\theta = 90°$ and the triangle is a right triangle.
2. $\sin \theta = \sin(180° - \theta)$ (Supplementary angles have the same sine value.)
3. The smallest angle is opposite the shortest side, the largest angle is opposite the longest side, and the middle-valued angle is opposite the intermediate side (assuming the triangle has sides that are all of different lengths).

Solutions of SSA Triangles (Case 2)

Classroom Example 1
Solve triangle ABC if $a = 17.9$ cm, $c = 13.2$ cm, and $C = 75° 30'$.

Answer: No such triangle exists.

EXAMPLE 1 Solving the Ambiguous Case (No Such Triangle)

Solve triangle ABC if $B = 55° 40'$, $b = 8.94$ m, and $a = 25.1$ m.

SOLUTION We are given B, b, and a. We use the law of sines to find angle A.

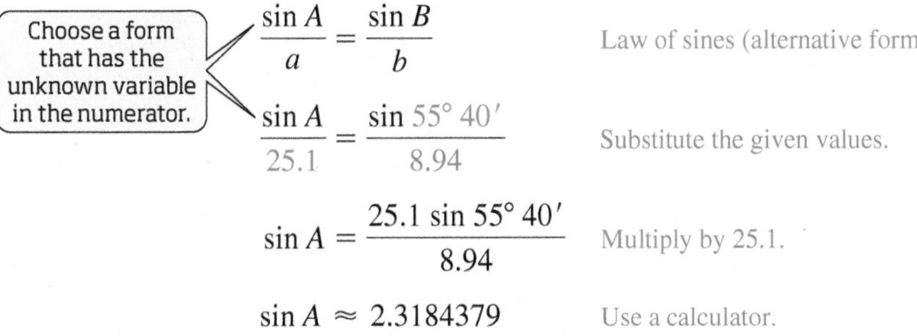

$$\frac{\sin A}{a} = \frac{\sin B}{b} \quad \text{Law of sines (alternative form)}$$

$$\frac{\sin A}{25.1} = \frac{\sin 55° 40'}{8.94} \quad \text{Substitute the given values.}$$

$$\sin A = \frac{25.1 \sin 55° 40'}{8.94} \quad \text{Multiply by 25.1.}$$

$$\sin A \approx 2.3184379 \quad \text{Use a calculator.}$$

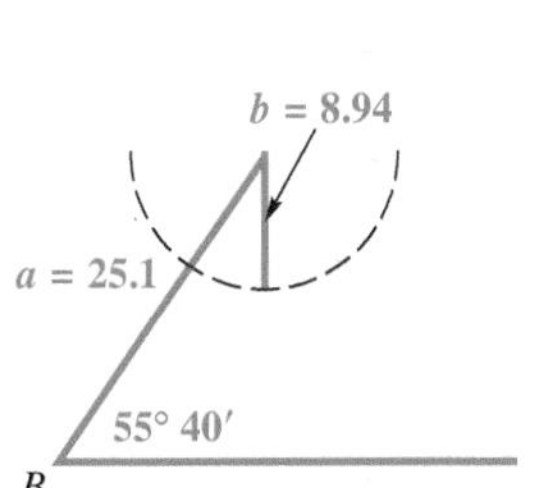

Figure 9

Because $\sin A$ cannot be greater than 1, there can be no such angle A—and thus no triangle with the given information. An attempt to sketch such a triangle leads to the situation shown in **Figure 9.**

Now Try Exercise 17.

NOTE In the ambiguous case, we are given two sides and an angle opposite one of the sides (SSA). For example, suppose b, c, and angle C are given. This situation represents the ambiguous case because angle C is opposite side c.

Classroom Example 2
Solve triangle ABC if $A = 61.4°$, $a = 35.5$ cm, and $b = 39.2$ cm.

Answer:
$B_1 = 75.8°$, $C_1 = 42.8°$, $c_1 = 27.5$ cm;
$B_2 = 104.2°$, $C_2 = 14.4°$, $c_2 = 10.1$ cm

EXAMPLE 2 Solving the Ambiguous Case (Two Triangles)

Solve triangle ABC if $A = 55.3°$, $a = 22.8$ ft, and $b = 24.9$ ft.

SOLUTION To begin, use the law of sines to find angle B.

$$\frac{\sin A}{a} = \frac{\sin B}{b} \quad \text{Solve for } \sin B.$$

$$\frac{\sin 55.3°}{22.8} = \frac{\sin B}{24.9} \quad \text{Substitute the given values.}$$

$$\sin B = \frac{24.9 \sin 55.3°}{22.8} \quad \text{Multiply by 24.9 and rewrite.}$$

$$\sin B \approx 0.8978678 \quad \text{Use a calculator.}$$

There are two angles B between 0° and 180° that satisfy this condition. Because $\sin B \approx 0.8978678$, one value of angle B, to the nearest tenth, is

$$B_1 = 63.9°. \quad \text{Use the inverse sine function.}$$

Supplementary angles have the same sine value, so another *possible* value of B is

$$B_2 = 180° - 63.9° = 116.1°.$$

To see whether $B_2 = 116.1°$ is a valid possibility, add 116.1° to the measure of A, 55.3°. Because $116.1° + 55.3° = 171.4°$, and this sum is less than 180°, it is a valid angle measure for this triangle.

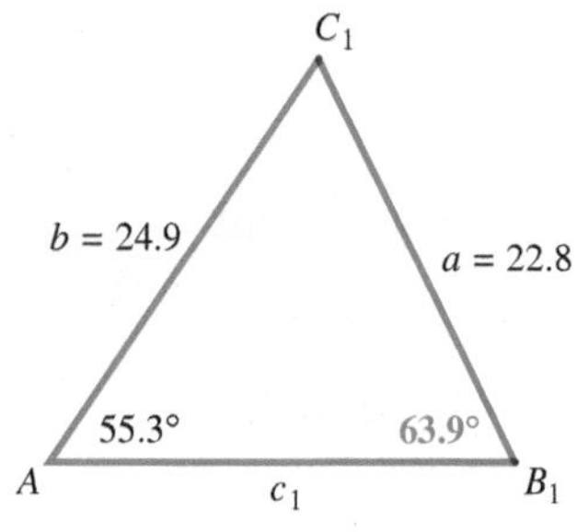

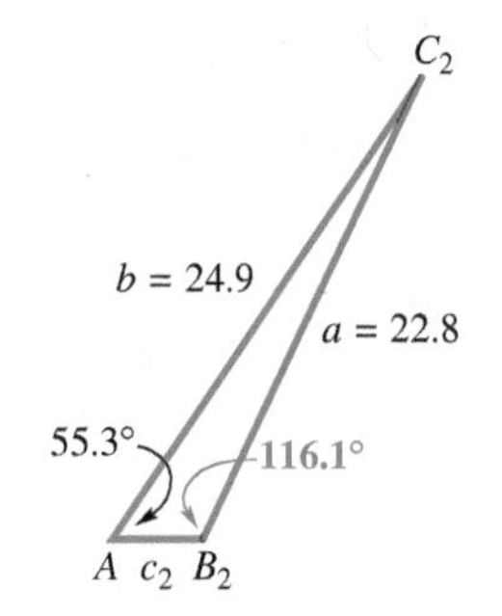

Figure 10

Now separately solve triangles AB_1C_1 and AB_2C_2 shown in **Figure 10.** Begin with AB_1C_1. Find angle C_1 first.

$$C_1 = 180° - A - B_1 \quad \text{Angle sum formula, solved for } C_1$$

$$C_1 = 180° - 55.3° - 63.9° \quad \text{Substitute.}$$

$$C_1 = 60.8° \quad \text{Subtract.}$$

Now, use the law of sines to find side c_1.

$$\frac{a}{\sin A} = \frac{c_1}{\sin C_1} \quad \text{Solve for } c_1.$$

$$\frac{22.8}{\sin 55.3°} = \frac{c_1}{\sin 60.8°} \quad \text{Substitute.}$$

$$c_1 = \frac{22.8 \sin 60.8°}{\sin 55.3°} \quad \text{Multiply by } \sin 60.8° \text{ and rewrite.}$$

$$c_1 \approx 24.2 \text{ ft} \quad \text{Use a calculator.}$$

To solve triangle AB_2C_2, first find angle C_2.

$$C_2 = 180° - A - B_2 \quad \text{Angle sum formula, solved for } C_2$$

$$C_2 = 180° - 55.3° - 116.1° \quad \text{Substitute.}$$

$$C_2 = 8.6° \quad \text{Subtract.}$$

Use the law of sines to find side c_2.

$$\frac{a}{\sin A} = \frac{c_2}{\sin C_2} \quad \text{Solve for } c_2.$$

$$\frac{22.8}{\sin 55.3°} = \frac{c_2}{\sin 8.6°} \quad \text{Substitute.}$$

$$c_2 = \frac{22.8 \sin 8.6°}{\sin 55.3°} \quad \text{Multiply by } \sin 8.6° \text{ and rewrite.}$$

$$c_2 \approx 4.15 \text{ ft} \quad \text{Use a calculator.}$$

✔ **Now Try Exercise 25.**

The ambiguous case results in zero, one, or two triangles. The following guidelines can be used to determine how many triangles there are.

Teaching Tip Ask students to outline the steps that they would take to solve each triangle. Do not actually solve.

1. $B = 85°$, $C = 40°$, $b = 26$
2. $A = 55°$, $B = 65°$, $c = 4$
3. $a = 17$, $b = 22$, $B = 32°$
4. $A = 20°$, $b = 7$, $a = 9$
5. $c = 225$, $A = 103.2°$, $B = 62.5°$

Number of Triangles Satisfying the Ambiguous Case (SSA)

Let sides a and b and angle A be given in triangle ABC. (The law of sines can be used to calculate the value of $\sin B$.)

1. If applying the law of sines results in an equation having $\sin B > 1$, then *no triangle* satisfies the given conditions.
2. If $\sin B = 1$, then *one triangle* satisfies the given conditions and $B = 90°$.
3. If $0 < \sin B < 1$, then either *one or two triangles* satisfy the given conditions.
   (a) If $\sin B = k$, then let $B_1 = \sin^{-1} k$ and use B_1 for B in the first triangle.
   (b) Let $B_2 = 180° - B_1$. If $A + B_2 < 180°$, then a second triangle exists. In this case, use B_2 for B in the second triangle.

Classroom Example 3
Solve triangle ABC, given $B = 68.7°$, $b = 25.4$ in., and $a = 19.6$ in.

Answer: $A = 46.0°$, $C = 65.3°$, $c = 24.8$ in.

EXAMPLE 3 Solving the Ambiguous Case (One Triangle)

Solve triangle ABC, given $A = 43.5°$, $a = 10.7$ in., and $c = 7.2$ in.

SOLUTION Find angle C.

$$\frac{\sin C}{c} = \frac{\sin A}{a} \quad \text{Law of sines (alternative form)}$$

$$\frac{\sin C}{7.2} = \frac{\sin 43.5°}{10.7} \quad \text{Substitute the given values.}$$

$$\sin C = \frac{7.2 \sin 43.5°}{10.7} \quad \text{Multiply by 7.2.}$$

$$\sin C \approx 0.46319186 \quad \text{Use a calculator.}$$

$$C \approx 27.6° \quad \text{Use the inverse sine function.}$$

There is another angle C that has sine value 0.46319186. It is

$$C = 180° - 27.6° = 152.4°.$$

However, notice in the given information that $c < a$, meaning that in the triangle, angle C must have measure *less than* angle A. Notice also that when we add this obtuse value to the given angle $A = 43.5°$, we obtain

$$152.4° + 43.5° = 195.9°,$$

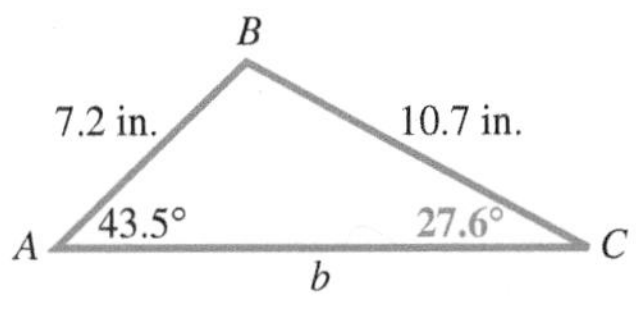

which is *greater than* 180°. Thus either of these approaches shows that there can be only one triangle. See **Figure 11.** The measure of angle B can be found next.

B, 7.2 in., 10.7 in., A, 43.5°, 27.6°, C, b

Figure 11

$$B = 180° - 27.6° - 43.5° \quad \text{Substitute.}$$

$$B = 108.9° \quad \text{Subtract.}$$

We can find side b with the law of sines.

$$\frac{b}{\sin B} = \frac{a}{\sin A} \quad \text{Law of sines}$$

$$\frac{b}{\sin 108.9°} = \frac{10.7}{\sin 43.5°} \quad \text{Substitute known values.}$$

$$b = \frac{10.7 \sin 108.9°}{\sin 43.5°} \quad \text{Multiply by } \sin 108.9°.$$

$$b \approx 14.7 \text{ in.} \quad \text{Use a calculator.}$$

✔ **Now Try Exercise 21.**

Analyzing Data for Possible Number of Triangles

Classroom Example 4
Without using the law of sines, explain why no triangle ABC exists satisfying $B = 93°$, $b = 42$ cm, and $c = 48$ cm.

Answer: Because B is an obtuse angle, it must be the largest angle of the triangle. Thus b must be the longest side of the triangle. However, we are given that $c > b$, so no such triangle exists.

EXAMPLE 4 Analyzing Data Involving an Obtuse Angle

Without using the law of sines, explain why $A = 104°$, $a = 26.8$ m, and $b = 31.3$ m cannot be valid for a triangle ABC.

SOLUTION Because A is an obtuse angle, it is the largest angle, and so the longest side of the triangle must be a. However, we are given $b > a$.

Thus, $B > A$, which is impossible if A is obtuse.

Therefore, no such triangle ABC exists. ✔ **Now Try Exercise 33.**

7.2 Exercises

1. A 2. D

3. (a) $4 < L < 5$
(b) $L = 4$ or $L > 5$
(c) $L < 4$

4. (a) none (b) $L > 5$
(c) $L \leq 5$

5. 1 6. 1
7. 2 8. 2
9. 0 10. 0

11. 45° 12. 30°

13. $B_1 = 49.1°, C_1 = 101.2°,$
$B_2 = 130.9°, C_2 = 19.4°$

14. $A_1 = 72.2°, C_1 = 59.6°,$
$A_2 = 107.8°, C_2 = 24.0°$

15. $B = 26° 30', A = 112° 10'$

16. $A = 37° 50', C = 93° 20'$

17. No such triangle exists.

18. No such triangle exists.

19. $B = 27.19°, C = 10.68°$

20. $A = 45.40°, C = 20.88°$

1. CONCEPT PREVIEW Which one of the following sets of data does *not* determine a unique triangle?

A. $A = 50°,\ b = 21,\ a = 19$ **B.** $A = 45°,\ b = 10,\ a = 12$

C. $A = 130°,\ b = 4,\ a = 7$ **D.** $A = 30°,\ b = 8,\ a = 4$

2. CONCEPT PREVIEW Which one of the following sets of data determines a unique triangle?

A. $A = 50°,\ B = 50°,\ C = 80°$ **B.** $a = 3,\ b = 5,\ c = 20$

C. $A = 40°,\ B = 20°,\ C = 30°$ **D.** $a = 7,\ b = 24,\ c = 25$

CONCEPT PREVIEW *In each figure, a line segment of length L is to be drawn from the given point to the positive x-axis in order to form a triangle. For what value(s) of L can we draw the following?*

(a) *two triangles* **(b)** *exactly one triangle* **(c)** *no triangle*

3.

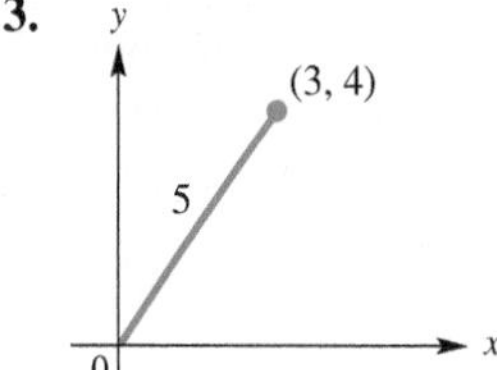

4.

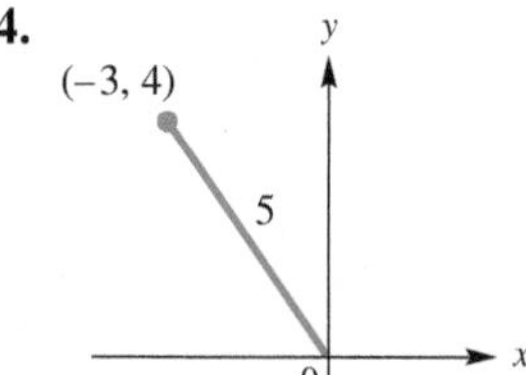

CONCEPT PREVIEW *Determine the number of triangles ABC possible with the given parts.*

5. $a = 50,\ b = 26,\ A = 95°$ **6.** $a = 35,\ b = 30,\ A = 40°$

7. $a = 31,\ b = 26,\ B = 48°$ **8.** $B = 54°,\ c = 28,\ b = 23$

9. $c = 50,\ b = 61,\ C = 58°$ **10.** $c = 60,\ a = 82,\ C = 100°$

Find each angle B. Do not use a calculator.

11.

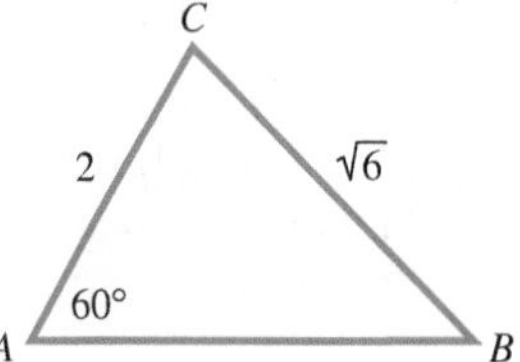

12.

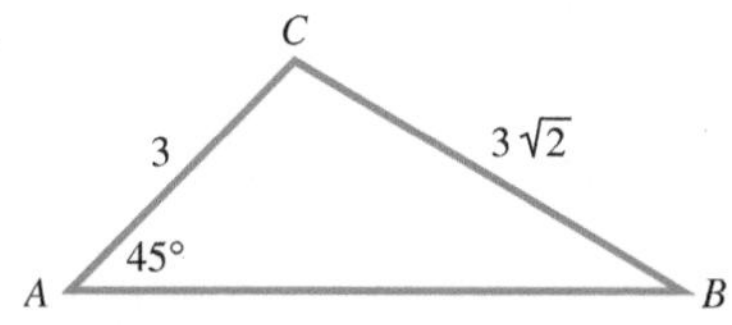

Find the unknown angles in triangle ABC for each triangle that exists. ***See Examples 1–3.***

13. $A = 29.7°,\ b = 41.5$ ft, $a = 27.2$ ft

14. $B = 48.2°,\ a = 890$ cm, $b = 697$ cm

15. $C = 41° 20',\ b = 25.9$ m, $c = 38.4$ m

16. $B = 48° 50',\ a = 3850$ in., $b = 4730$ in.

17. $B = 74.3°,\ a = 859$ m, $b = 783$ m

18. $C = 82.2°,\ a = 10.9$ km, $c = 7.62$ km

19. $A = 142.13°,\ b = 5.432$ ft, $a = 7.297$ ft

20. $B = 113.72°,\ a = 189.6$ yd, $b = 243.8$ yd

21. $B = 20.6°$, $C = 116.9°$, $c = 20.6$ ft

22. $A = 25.5°$, $B = 102.2°$, $b = 73.9$ yd

23. No such triangle exists.

24. No such triangle exists.

25. $B_1 = 49° 20'$, $C_1 = 92° 00'$, $c_1 = 15.5$ m; $B_2 = 130° 40'$, $C_2 = 10° 40'$, $c_2 = 2.88$ m

26. $A_1 = 55° 20'$, $B_1 = 94° 50'$, $b_1 = 10.4$ m; $A_2 = 124° 40'$, $B_2 = 25° 30'$, $b_2 = 4.51$ m

27. $B = 37.77°$, $C = 45.43°$, $c = 4.174$ ft

28. $B = 30.39°$, $A = 60.91°$, $a = 98.25$ m

29. $A_1 = 53.23°$, $C_1 = 87.09°$, $c_1 = 37.16$ m; $A_2 = 126.77°$, $C_2 = 13.55°$, $c_2 = 8.719$ m

30. $C_1 = 59.71°$, $B_1 = 69.09°$, $b_1 = 8640$ cm; $C_2 = 120.29°$, $B_2 = 8.51°$, $b_2 = 1369$ cm

31. 1; 90°; a right triangle

32. If angle A is acute and $\sin B > 1$, then no triangle exists. If angle A is obtuse and $\sin B \geq 1$, then no triangle exists.

33. Because A is obtuse, it is the largest angle. Thus side a should be the longest side, but it is not. Therefore, no such triangle exists.

34. Because $\sin B > 1$ for this given information, a calculator will return an error message.

35. 664 m **36.** 87.3 ft

37. 218 ft

38. 75° 00′; 274 mi

Solve each triangle ABC that exists. ***See Examples 1–3.***

21. $A = 42.5°$, $a = 15.6$ ft, $b = 8.14$ ft

22. $C = 52.3°$, $a = 32.5$ yd, $c = 59.8$ yd

23. $B = 72.2°$, $b = 78.3$ m, $c = 145$ m

24. $C = 68.5°$, $c = 258$ cm, $b = 386$ cm

25. $A = 38° 40'$, $a = 9.72$ m, $b = 11.8$ m

26. $C = 29° 50'$, $a = 8.61$ m, $c = 5.21$ m

27. $A = 96.80°$, $b = 3.589$ ft, $a = 5.818$ ft

28. $C = 88.70°$, $b = 56.87$ m, $c = 112.4$ m

29. $B = 39.68°$, $a = 29.81$ m, $b = 23.76$ m

30. $A = 51.20°$, $c = 7986$ cm, $a = 7208$ cm

Concept Check Answer each question.

31. Apply the law of sines to the following: $a = \sqrt{5}$, $c = 2\sqrt{5}$, $A = 30°$. What is the value of $\sin C$? What is the measure of C? Based on its angle measures, what kind of triangle is triangle ABC?

32. What condition must exist to determine that there is no triangle satisfying the given values of a, b, and B, once the value of $\sin A$ is found by applying the law of sines?

33. Without using the law of sines, why can no triangle ABC exist that satisfies $A = 103° 20'$, $a = 14.6$ ft, $b = 20.4$ ft?

34. If the law of sines is applied to the data given in **Example 4,** what happens when we try to find the measure of angle B using a calculator?

Use the law of sines to solve each problem.

35. ***Distance between Inaccessible Points*** To find the distance between a point X and an inaccessible point Z, a line segment XY is constructed. It is found that $XY = 960$ m, angle $XYZ = 43° 30'$, and angle $YZX = 95° 30'$. Find the distance between X and Z to the nearest meter.

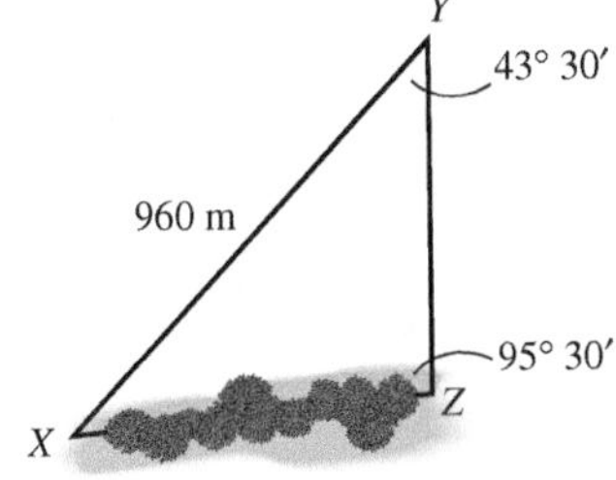

36. ***Height of an Antenna Tower*** The angle of elevation from the top of a building 45.0 ft high to the top of a nearby antenna tower is 15° 20′. From the base of the building, the angle of elevation of the tower is 29° 30′. Find the height of the tower.

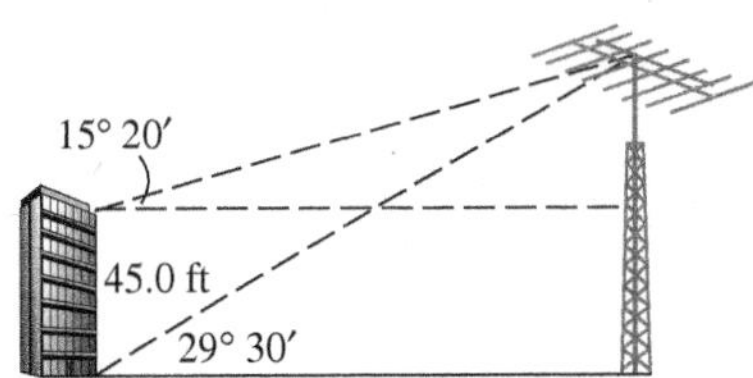

37. ***Height of a Building*** A flagpole 95.0 ft tall is on the top of a building. From a point on level ground, the angle of elevation of the top of the flagpole is 35.0°, and the angle of elevation of the bottom of the flagpole is 26.0°. Find the height of the building.

38. ***Flight Path of a Plane*** A pilot flies her plane on a bearing of 35° 00′ from point X to point Y, which is 400 mi from X. Then she turns and flies on a bearing of 145° 00′ to point Z, which is 400 mi from her starting point X. What is the bearing of Z from X, and what is the distance YZ?

Use the law of sines to prove that each statement is true for any triangle ABC, with corresponding sides a, b, and c.

39. $\dfrac{a + b}{b} = \dfrac{\sin A + \sin B}{\sin B}$

40. $\dfrac{a - b}{a + b} = \dfrac{\sin A - \sin B}{\sin A + \sin B}$

42. $\mathcal{A} = 1.12257R^2$

43. (a) 8.77 in.^{2} (b) 5.32 in.^{2}

44. red

Relating Concepts

For individual or collaborative investigation. *(Exercises 41–44)*

Colors of the U.S. Flag The flag of the United States includes the colors red, white, and blue.

Which color is predominant?

Clearly the answer is either red or white. (It can be shown that only 18.73% *of the total area is blue.)* (Source: *Banks, R.,* Slicing Pizzas, Racing Turtles, and Further Adventures in Applied Mathematics, *Princeton University Press.)*

To answer this question, work **Exercises 41–44 in order.**

41. Let R denote the radius of the circumscribing circle of a five-pointed star appearing on the American flag. The star can be decomposed into ten congruent triangles. In the figure, r is the radius of the circumscribing circle of the pentagon in the interior of the star. Show that the area of a star is

$$\mathcal{A} = \left[5\frac{\sin A \sin B}{\sin(A + B)}\right]R^2. \quad (\textit{Hint: } \sin C = \sin[180° - (A + B)] = \sin(A + B).)$$

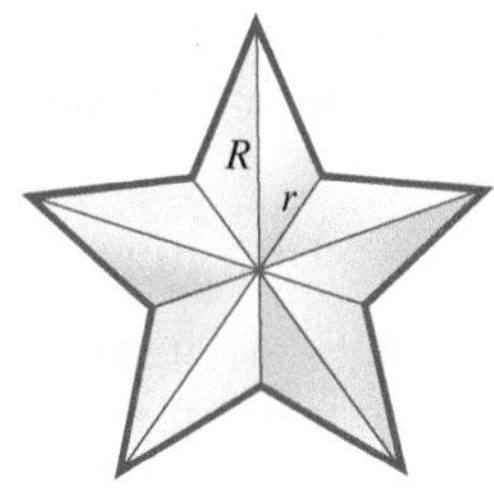

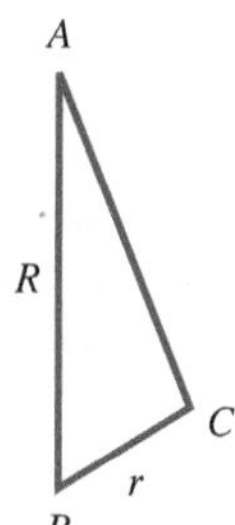

42. Angles A and B have values 18° and 36°, respectively. Express the area $\mathcal{A}$ of a star in terms of its radius, R.

43. To determine whether red or white is predominant, we must know the measurements of the flag. Consider a flag of width 10 in., length 19 in., length of each upper stripe 11.4 in., and radius R of the circumscribing circle of each star 0.308 in. The thirteen stripes consist of six matching pairs of red and white stripes and one additional red, upper stripe. Therefore, we must compare the area of a red, upper stripe with the total area of the 50 white stars.

(a) Compute the area of the red, upper stripe.

(b) Compute the total area of the 50 white stars.

44. Which color occupies the greatest area on the flag?

7.3 The Law of Cosines

- Derivation of the Law of Cosines
- Solutions of SAS and SSS Triangles (Cases 3 and 4)
- Heron's Formula for the Area of a Triangle
- Derivation of Heron's Formula

If we are given two sides and the included angle (Case 3) or three sides (Case 4) of a triangle, then a unique triangle is determined. These are the SAS and SSS cases, respectively. Both require using the *law of cosines* to solve the triangle.

The following property is important when applying the law of cosines.

Triangle Side Length Restriction

In any triangle, the sum of the lengths of any two sides must be greater than the length of the remaining side.

As an example of this property, it would be impossible to construct a triangle with sides of lengths 3, 4, and 10. See **Figure 12.**

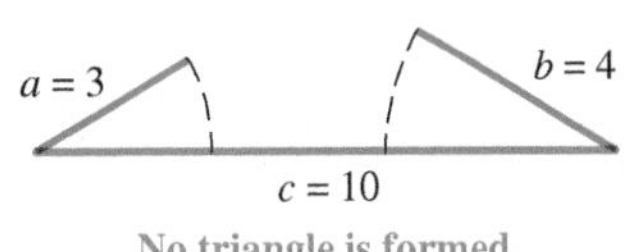

Figure 12

Derivation of the Law of Cosines To derive the law of cosines, let ABC be any oblique triangle. Choose a coordinate system so that vertex B is at the origin and side BC is along the positive x-axis. See **Figure 13.**

Let (x, y) be the coordinates of vertex A of the triangle. Then the following are true for angle B, whether obtuse or acute.

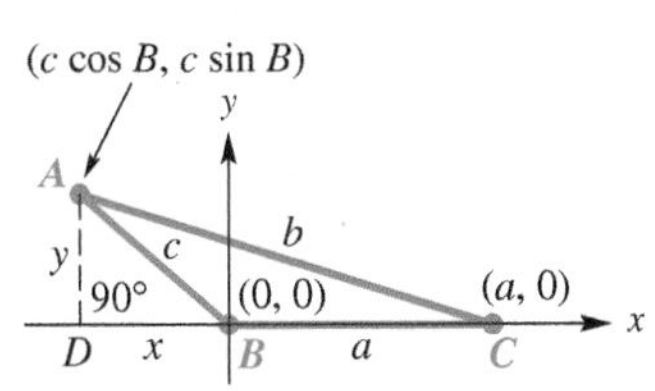

Figure 13

$$\sin B = \frac{y}{c} \quad \text{and} \quad \cos B = \frac{x}{c}$$ Definition of sine and cosine

$$y = c \sin B \quad \text{and} \quad x = c \cos B$$ Here x is negative when B is obtuse.

Thus, the coordinates of point A become $(c \cos B, c \sin B)$.

Point C in **Figure 13** has coordinates $(a, 0)$, AC has length b, and point A has coordinates $(c \cos B, c \sin B)$. We can use the distance formula to write an equation.

$$b = \sqrt{(c \cos B - a)^2 + (c \sin B - 0)^2}$$ $d = \sqrt{(x_2 - x_1)^2 + (y_2 - y_1)^2}$

$$b^2 = (c \cos B - a)^2 + (c \sin B)^2$$ Square each side.

$$b^2 = (c^2 \cos^2 B - 2ac \cos B + a^2) + c^2 \sin^2 B$$ Multiply; $(x - y)^2 = x^2 - 2xy + y^2$

$$b^2 = a^2 + c^2(\cos^2 B + \sin^2 B) - 2ac \cos B$$ Properties of real numbers

$$b^2 = a^2 + c^2(1) - 2ac \cos B$$ Fundamental identity

$$b^2 = a^2 + c^2 - 2ac \cos B$$ Law of cosines

This result is one of three possible forms of the law of cosines. In our work, we could just as easily have placed vertex A or C at the origin. This would have given the same result, but with the variables rearranged.

Law of Cosines

In any triangle ABC, with sides a, b, and c, the following hold.

$$a^2 = b^2 + c^2 - 2bc \cos A$$
$$b^2 = a^2 + c^2 - 2ac \cos B$$
$$c^2 = a^2 + b^2 - 2ab \cos C$$

That is, according to the law of cosines, the square of a side of a triangle is equal to the sum of the squares of the other two sides, minus twice the product of those two sides and the cosine of the angle included between them.

Classroom Example 1
Two boats leave a harbor at the same time, traveling on courses that make an angle of 82° 20′ between them. When the slower boat has traveled 62.5 km, the faster one has traveled 79.4 km. At that time, what is the distance between the boats?

Answer: 94.3 km

NOTE If we let $C = 90°$ in the third form of the law of cosines, then $\cos C = \cos 90° = 0$, and the formula becomes

$$c^2 = a^2 + b^2. \quad \text{Pythagorean theorem}$$

The Pythagorean theorem is a special case of the law of cosines.

Solutions of SAS and SSS Triangles (Cases 3 and 4)

EXAMPLE 1 Applying the Law of Cosines (SAS)

A surveyor wishes to find the distance between two inaccessible points A and B on opposite sides of a lake. While standing at point C, she finds that $b = 259$ m, $a = 423$ m, and angle ACB measures 132° 40′. Find the distance c. See **Figure 14.**

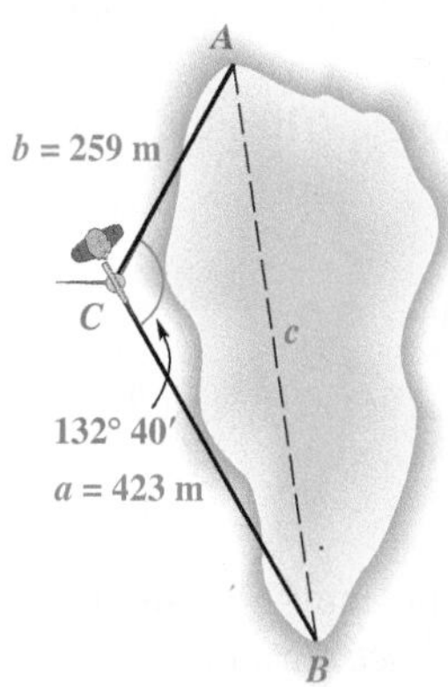

Figure 14

SOLUTION We can use the law of cosines here because we know the lengths of two sides of the triangle and the measure of the included angle.

$c^2 = a^2 + b^2 - 2ab \cos C$ Law of cosines

$c^2 = 423^2 + 259^2 - 2(423)(259) \cos 132° 40'$ Substitute.

$c^2 \approx 394{,}510.6$ Use a calculator.

$c \approx 628$ Take the square root of each side. Choose the positive root.

The distance between the points is approximately 628 m. ✔ Now Try Exercise 39.

EXAMPLE 2 Applying the Law of Cosines (SAS)

Solve triangle ABC if $A = 42.3°$, $b = 12.9$ m, and $c = 15.4$ m.

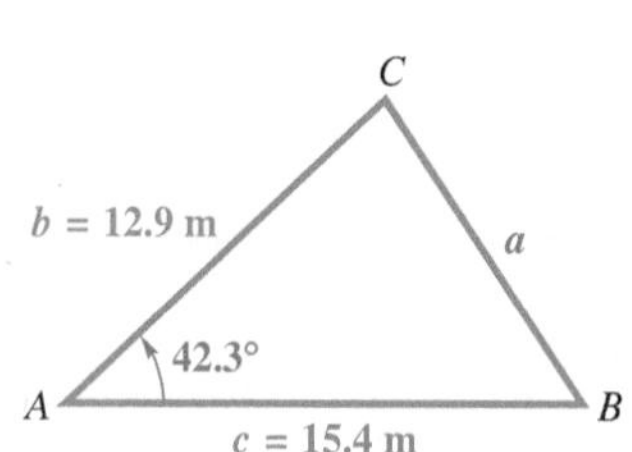

Figure 15

SOLUTION See **Figure 15.** We start by finding side a with the law of cosines.

$a^2 = b^2 + c^2 - 2bc \cos A$ Law of cosines

$a^2 = 12.9^2 + 15.4^2 - 2(12.9)(15.4) \cos 42.3°$ Substitute.

$a^2 \approx 109.7$ Use a calculator.

$a \approx 10.47$ m Take square roots and choose the positive root.

Of the two remaining angles B and C, B must be the smaller because it is opposite the shorter of the two sides b and c. Therefore, B cannot be obtuse.

$\frac{\sin A}{a} = \frac{\sin B}{b}$ Law of sines (alternative form)

$\frac{\sin 42.3°}{10.47} = \frac{\sin B}{12.9}$ Substitute.

$\sin B = \frac{12.9 \sin 42.3°}{10.47}$ Multiply by 12.9 and rewrite.

$B \approx 56.0°$ Use the inverse sine function.

Classroom Example 2
Solve triangle ABC if $B = 73.5°$, $a = 28.2$ ft, and $c = 46.7$ ft.

Answer: $b = 47.2$ ft, $A = 34.9°$, $C = 71.6°$

The easiest way to find C is to subtract the measures of A and B from 180°.

$C = 180° - A - B$ Angle sum formula, solved for C

$C \approx 180° - 42.3° - 56.0°$ Substitute.

$C \approx 81.7°$ Subtract. ✔ Now Try Exercise 19.

Teaching Tip Emphasize that once the law of cosines has been used, students do not need to be concerned about the ambiguous case of the law of sines.

CAUTION Had we used the law of sines to find C rather than B in **Example 2,** we would not have known whether C was equal to 81.7° or to its supplement, 98.3°.

Classroom Example 3
Solve triangle ABC if $a = 25.4$ cm, $b = 42.8$ cm, and $c = 59.3$ cm.

Answer: $A = 22.1°$, $B = 39.3°$, $C = 118.6°$

EXAMPLE 3 Applying the Law of Cosines (SSS)

Solve triangle ABC if $a = 9.47$ ft, $b = 15.9$ ft, and $c = 21.1$ ft.

SOLUTION We can use the law of cosines to solve for any angle of the triangle. We solve for C, the largest angle. We will know that C is obtuse if $\cos C < 0$.

$$c^2 = a^2 + b^2 - 2ab \cos C \quad \text{Law of cosines}$$

$$\cos C = \frac{a^2 + b^2 - c^2}{2ab} \quad \text{Solve for } \cos C.$$

$$\cos C = \frac{9.47^2 + 15.9^2 - 21.1^2}{2(9.47)(15.9)} \quad \text{Substitute.}$$

$$\cos C \approx -0.34109402 \quad \text{Use a calculator.}$$

$$C \approx 109.9° \quad \text{Use the inverse cosine function.}$$

Now use the law of sines to find angle B.

$$\frac{\sin B}{b} = \frac{\sin C}{c} \quad \text{Law of sines (alternative form)}$$

$$\frac{\sin B}{15.9} = \frac{\sin 109.9°}{21.1} \quad \text{Substitute.}$$

$$\sin B = \frac{15.9 \sin 109.9°}{21.1} \quad \text{Multiply by 15.9.}$$

$$B \approx 45.1° \quad \text{Use the inverse sine function.}$$

Since $A = 180° - B - C$, we have $A \approx 180° - 45.1° - 109.9° \approx 25.0°$.

✔ **Now Try Exercise 23.**

Figure 16

Trusses are frequently used to support roofs on buildings, as illustrated in **Figure 16.** The simplest type of roof truss is a triangle, as shown in **Figure 17.** (*Source:* Riley, W., L. Sturges, and D. Morris, *Statics and Mechanics of Materials,* John Wiley and Sons.)

EXAMPLE 4 Designing a Roof Truss (SSS)

Find angle B to the nearest degree for the truss shown in **Figure 17.**

Classroom Example 4
Find angle C to the nearest degree for the truss shown in **Figure 17.**

Answer: 55°

SOLUTION

$$b^2 = a^2 + c^2 - 2ac \cos B \quad \text{Law of cosines}$$

$$\cos B = \frac{a^2 + c^2 - b^2}{2ac} \quad \text{Solve for } \cos B.$$

$$\cos B = \frac{11^2 + 9^2 - 6^2}{2(11)(9)} \quad \text{Let } a = 11,\ b = 6, \text{ and } c = 9.$$

$$\cos B \approx 0.83838384 \quad \text{Use a calculator.}$$

$$B \approx 33° \quad \text{Use the inverse cosine function.}$$

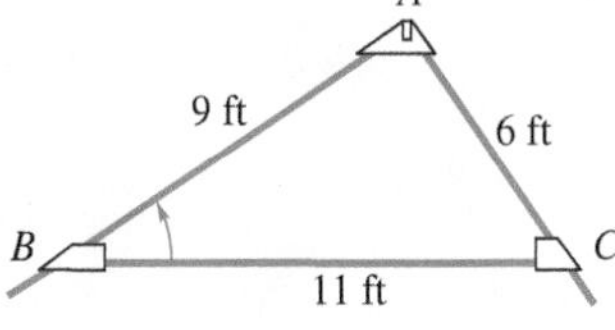

Figure 17

✔ **Now Try Exercise 49.**

Four possible cases can occur when we solve an oblique triangle. They are summarized in the following table. In all four cases, it is assumed that the given information actually produces a triangle.

Teaching Tip Ask students to outline the steps that they would take to solve each triangle. Do not actually solve.

1. $A = 32°, B = 67°, a = 20$
2. $b = 22, c = 15, A = 39°$
3. $a = 5, c = 7, C = 22°$
4. $b = 13, c = 15, B = 22°$
5. $a = 5, b = 7, c = 9$

Four Cases for Solving Oblique Triangles

Oblique Triangle	Suggested Procedure for Solving
Case 1: One side and two angles are known. **(SAA or ASA)**	***Step 1*** Find the remaining angle using the angle sum formula $(A + B + C = 180°)$. ***Step 2*** Find the remaining sides using the law of sines.
Case 2: Two sides and one angle (not included between the two sides) are known. **(SSA)**	*This is the ambiguous case. There may be no triangle, one triangle, or two triangles.* ***Step 1*** Find an angle using the law of sines. ***Step 2*** Find the remaining angle using the angle sum formula. ***Step 3*** Find the remaining side using the law of sines. *If two triangles exist, repeat Steps 2 and 3.*
Case 3: Two sides and the included angle are known. **(SAS)**	***Step 1*** Find the third side using the law of cosines. ***Step 2*** Find the smaller of the two remaining angles using the law of sines. ***Step 3*** Find the remaining angle using the angle sum formula.
Case 4: Three sides are known. **(SSS)**	***Step 1*** Find the largest angle using the law of cosines. ***Step 2*** Find either remaining angle using the law of sines. ***Step 3*** Find the remaining angle using the angle sum formula.

Teaching Tip Heron's formula, though intended for use with oblique triangles, can be easily illustrated by finding the area of a right triangle—that is, apply Heron's formula and $\mathcal{A} = \frac{1}{2}bh$.

Heron's Formula for the Area of a Triangle A formula for finding the area of a triangle given the lengths of the three sides, known as **Heron's formula,** is named after the Greek mathematician Heron of Alexandria. It is found in his work *Metrica*. Heron's formula can be used for the case SSS.

Heron of Alexandria (c. 62 CE)

Heron (also called Hero), a Greek geometer and inventor, produced writings that contain knowledge of the mathematics and engineering of Babylonia, ancient Egypt, and the Greco-Roman world.

Heron's Area Formula (SSS)

If a triangle has sides of lengths a, b, and c, with **semiperimeter**

$$s = \frac{1}{2}(a + b + c),$$

then the area $\mathcal{A}$ of the triangle is given by the following formula.

$$\mathcal{A} = \sqrt{s(s - a)(s - b)(s - c)}$$

That is, according to Heron's formula, the area of a triangle is the square root of the product of four factors: (1) the semiperimeter, (2) the semiperimeter minus the first side, (3) the semiperimeter minus the second side, and (4) the semiperimeter minus the third side.

Classroom Example 5 The distance "as the crow flies" from Chicago to St. Louis is 262 mi, from St. Louis to New Orleans is 599 mi, and from New Orleans to Chicago is 834 mi. What is the area of the triangular region having these three cities as vertices?

Answer: 40,800 mi²

EXAMPLE 5 Using Heron's Formula to Find an Area (SSS)

The distance "as the crow flies" from Los Angeles to New York is 2451 mi, from New York to Montreal is 331 mi, and from Montreal to Los Angeles is 2427 mi. What is the area of the triangular region having these three cities as vertices? (Ignore the curvature of Earth.)

SOLUTION In **Figure 18,** we let $a = 2451$, $b = 331$, and $c = 2427$.

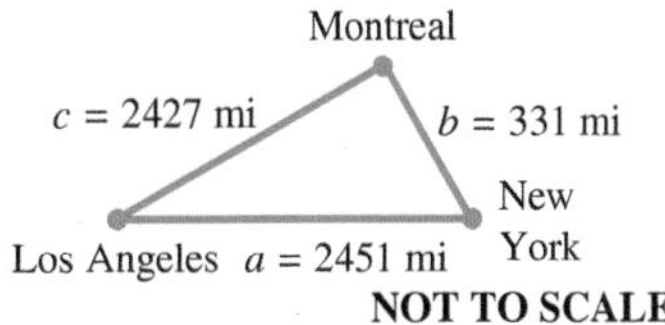

Figure 18

First, find the semiperimeter s.

$$s = \frac{1}{2}(a + b + c) \quad \text{Semiperimeter}$$

$$s = \frac{1}{2}(2451 + 331 + 2427) \quad \text{Substitute the given values.}$$

$$s = 2604.5 \quad \text{Add, and then multiply.}$$

Now use Heron's formula to find the area $\mathcal{A}$.

$$\mathcal{A} = \sqrt{s(s-a)(s-b)(s-c)}$$

Don't forget the factor s.

$$\mathcal{A} = \sqrt{2604.5(2604.5 - 2451)(2604.5 - 331)(2604.5 - 2427)}$$

$$\mathcal{A} \approx 401{,}700 \text{ mi}^2 \quad \text{Use a calculator.}$$

Now Try Exercise 73.

Derivation of Heron's Formula A trigonometric derivation of Heron's formula illustrates some ingenious manipulation.

Let triangle ABC have sides of lengths a, b, and c. Apply the law of cosines.

$$a^2 = b^2 + c^2 - 2bc \cos A \quad \text{Law of cosines}$$

$$\cos A = \frac{b^2 + c^2 - a^2}{2bc} \quad \text{Solve for } \cos A. \quad (1)$$

The perimeter of the triangle is $a + b + c$, so half of the perimeter (the semiperimeter) is given by the formula in equation (2) below.

$$s = \frac{1}{2}(a + b + c) \quad (2)$$

$$2s = a + b + c \quad \text{Multiply by 2.} \quad (3)$$

$$b + c - a = 2s - 2a \quad \text{Subtract } 2a \text{ from each side and rewrite.}$$

$$b + c - a = 2(s - a) \quad \text{Factor.} \quad (4)$$

Subtract $2b$ and $2c$ in a similar way in equation (3) to obtain the following.

$$a - b + c = 2(s - b) \quad (5)$$

$$a + b - c = 2(s - c) \quad (6)$$

Now we obtain an expression for $1 - \cos A$.

$$1 - \cos A = 1 - \underbrace{\frac{b^2 + c^2 - a^2}{2bc}}_{\cos A,\ \text{from (1)}}$$

$$= \frac{2bc + a^2 - b^2 - c^2}{2bc}$$ Find a common denominator, and distribute the − sign.

$$= \frac{a^2 - (b^2 - 2bc + c^2)}{2bc}$$ Regroup.

Pay attention to signs.

$$= \frac{a^2 - (b - c)^2}{2bc}$$ Factor the perfect square trinomial.

$$= \frac{[a - (b - c)][a + (b - c)]}{2bc}$$ Factor the difference of squares.

$$= \frac{(a - b + c)(a + b - c)}{2bc}$$ Distributive property

$$= \frac{2(s - b) \cdot 2(s - c)}{2bc}$$ Use equations (5) and (6).

$$1 - \cos A = \frac{2(s - b)(s - c)}{bc}$$ Lowest terms (7)

Similarly, it can be shown that

$$1 + \cos A = \frac{2s(s - a)}{bc}. \quad (8)$$

Recall the double-angle identities for $\cos 2\theta$.

$$\cos 2\theta = 2\cos^2\theta - 1$$

$$\cos A = 2\cos^2\left(\frac{A}{2}\right) - 1$$ Let $\theta = \frac{A}{2}$.

$$1 + \cos A = 2\cos^2\left(\frac{A}{2}\right)$$ Add 1.

$$\underbrace{\frac{2s(s - a)}{bc}}_{\text{From (8)}} = 2\cos^2\left(\frac{A}{2}\right)$$ Substitute.

$$\frac{s(s - a)}{bc} = \cos^2\left(\frac{A}{2}\right)$$ Divide by 2.

$$\cos\left(\frac{A}{2}\right) = \sqrt{\frac{s(s - a)}{bc}} \quad (9)$$

$$\cos 2\theta = 1 - 2\sin^2\theta$$

$$\cos A = 1 - 2\sin^2\left(\frac{A}{2}\right)$$ Let $\theta = \frac{A}{2}$.

$$1 - \cos A = 2\sin^2\left(\frac{A}{2}\right)$$ Subtract 1. Multiply by −1.

$$\underbrace{\frac{2(s - b)(s - c)}{bc}}_{\text{From (7)}} = 2\sin^2\left(\frac{A}{2}\right)$$ Substitute.

$$\frac{(s - b)(s - c)}{bc} = \sin^2\left(\frac{A}{2}\right)$$ Divide by 2.

$$\sin\left(\frac{A}{2}\right) = \sqrt{\frac{(s - b)(s - c)}{bc}} \quad (10)$$

The area of triangle ABC can be expressed as follows.

$$\mathcal{A} = \frac{1}{2}bc\sin A$$ Area formula

$$2\mathcal{A} = bc\sin A$$ Multiply by 2.

$$\frac{2\mathcal{A}}{bc} = \sin A$$ Divide by bc. (11)

Recall the double-angle identity for sin 2θ.

$$\sin 2\theta = 2 \sin \theta \cos \theta$$

$$\sin A = 2 \sin\left(\frac{A}{2}\right) \cos\left(\frac{A}{2}\right) \quad \text{Let } \theta = \tfrac{A}{2}.$$

$$\frac{2\mathcal{A}}{bc} = 2 \sin\left(\frac{A}{2}\right) \cos\left(\frac{A}{2}\right) \quad \text{Use equation (11).}$$

$$\frac{2\mathcal{A}}{bc} = 2\sqrt{\frac{(s-b)(s-c)}{bc}} \cdot \sqrt{\frac{s(s-a)}{bc}} \quad \text{Use equations (9) and (10).}$$

$$\frac{2\mathcal{A}}{bc} = 2\sqrt{\frac{s(s-a)(s-b)(s-c)}{b^2c^2}} \quad \text{Multiply.}$$

$$\frac{2\mathcal{A}}{bc} = \frac{2\sqrt{s(s-a)(s-b)(s-c)}}{bc} \quad \text{Simplify the denominator.}$$

Heron's formula results. →

$$\mathcal{A} = \sqrt{s(s-a)(s-b)(s-c)} \quad \text{Multiply by } bc. \text{ Divide by 2.}$$

7.3 Exercises

1. (a) SAS (b) law of cosines
2. (a) SAA (b) law of sines
3. (a) SSA (b) law of sines
4. (a) ASA (b) law of sines
5. (a) ASA (b) law of sines
6. (a) SSA (b) law of sines
7. (a) SSS (b) law of cosines
8. (a) SAS (b) law of cosines

9. 5 10. 7

11. 120° 12. 30°

13. $a = 7.0$, $B = 37.6°$, $C = 21.4°$
14. $a = 5.4$, $B = 40.7°$, $C = 78.3°$

CONCEPT PREVIEW *Assume a triangle ABC has standard labeling.*

(a) Determine whether SAA, ASA, SSA, SAS, *or* SSS *is given.*

(b) Decide whether the law of sines or the law of cosines should be used to begin solving the triangle.

1. a, b, and C **2.** A, C, and c **3.** a, b, and A **4.** a, B, and C

5. A, B, and c **6.** a, c, and A **7.** a, b, and c **8.** b, c, and A

Find the length of the remaining side of each triangle. Do not use a calculator.

9.

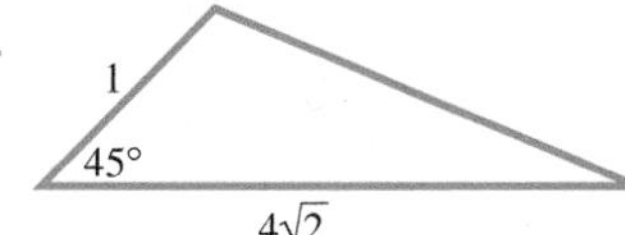

10.

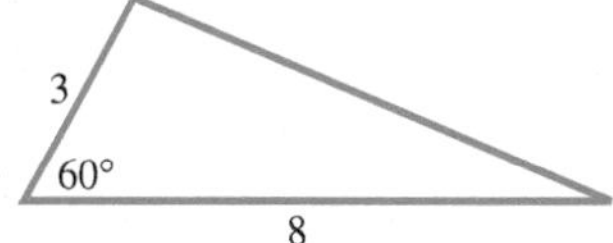

Find the measure of θ in each triangle. Do not use a calculator.

11.

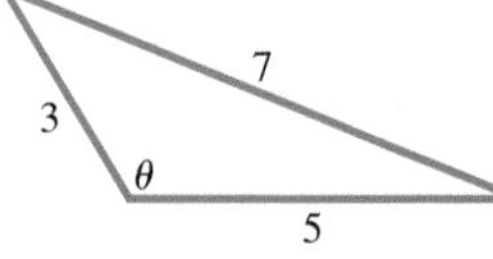

12.

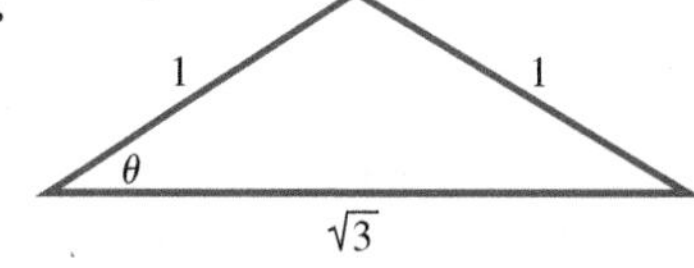

Solve each triangle. Approximate values to the nearest tenth.

13.

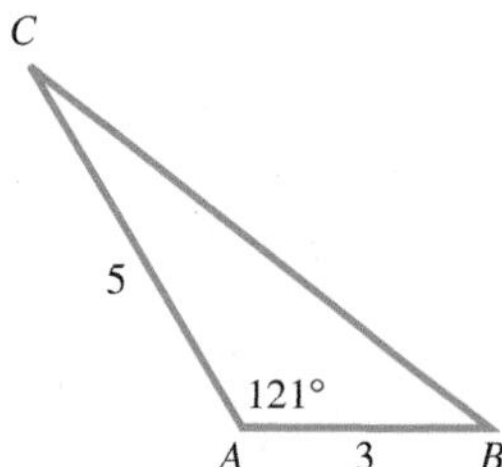

14.

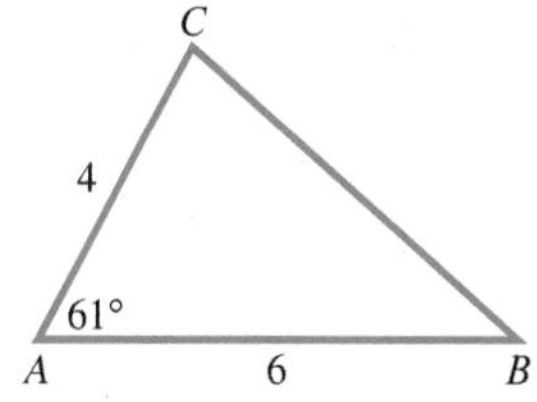

15. $A = 73.7°$, $B = 53.1°$, $C = 53.1°$ (The angles do not sum to 180° due to rounding.)

16. $A = 22.3°$, $B = 108.2°$, $C = 49.5°$

17. $b = 88.2$, $A = 56.7°$, $C = 68.3°$

18. $A = 33.6°$, $B = 50.7°$, $C = 95.7°$

19. $a = 2.60$ yd, $B = 45.1°$, $C = 93.5°$

20. $c = 2.83$ in., $A = 44.9°$, $B = 106.8°$

21. $c = 6.46$ m, $A = 53.1°$, $B = 81.3°$

22. $a = 43.7$ km, $B = 53.2°$, $C = 59.5°$

23. $A = 82°$, $B = 37°$, $C = 61°$

24. $C = 98°$, $A = 29°$, $B = 53°$

25. $C = 102° \, 10'$, $B = 35° \, 50'$, $A = 42° \, 00'$

26. $C = 107° \, 20'$, $B = 39° \, 00'$, $A = 33° \, 40'$

27. $C = 84° \, 30'$, $B = 44° \, 40'$, $A = 50° \, 50'$

28. $B = 85° \, 10'$, $C = 44° \, 50'$, $A = 50° \, 00'$

29. $a = 156$ cm, $B = 64° \, 50'$, $C = 34° \, 30'$

30. $c = 348$ ft, $A = 63° \, 50'$, $B = 43° \, 30'$

31. $b = 9.53$ in., $A = 64.6°$, $C = 40.6°$

32. $c = 4.28$ mi, $A = 48.8°$, $B = 71.5°$

33. $a = 15.7$ m, $B = 21.6°$, $C = 45.6°$

34. $b = 34.1$ cm, $A = 5.2°$, $C = 6.6°$

35. $A = 30°$, $B = 56°$, $C = 94°$

36. $A = 24°$, $B = 31°$, $C = 125°$

37. The value of $\cos \theta$ will be greater than 1. A calculator will give an error message (or a nonreal complex number) when using the inverse cosine function.

38. Any given side of a triangle represents the shortest distance between two points. If the sum of the other two sides of the triangle is not larger than the given side, then the two sides will not be able to form an angle opposite the given side.

39. 257 m

15.

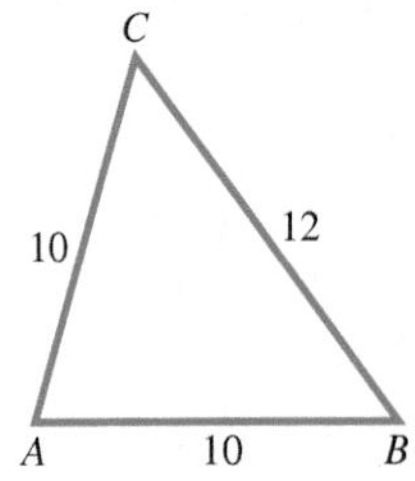

16.

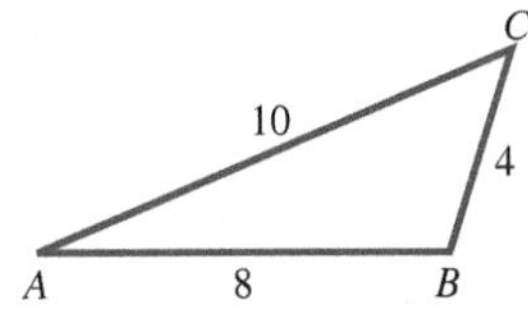

17.

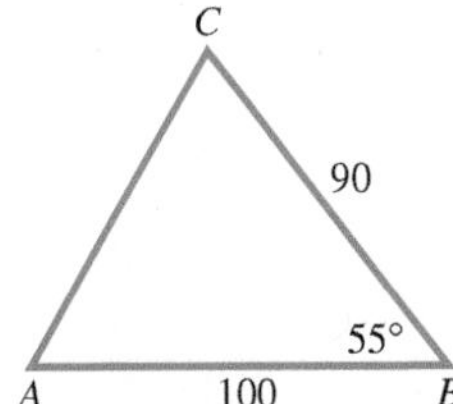

18.

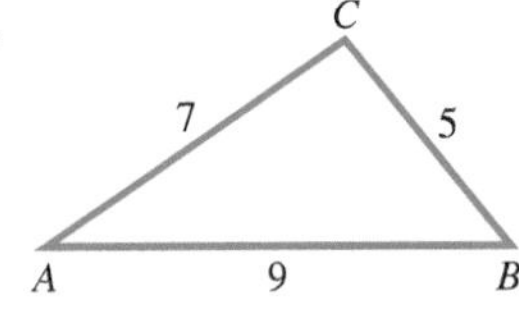

Solve each triangle. ***See Examples 2 and 3.***

19. $A = 41.4°$, $b = 2.78$ yd, $c = 3.92$ yd

20. $C = 28.3°$, $b = 5.71$ in., $a = 4.21$ in.

21. $C = 45.6°$, $b = 8.94$ m, $a = 7.23$ m

22. $A = 67.3°$, $b = 37.9$ km, $c = 40.8$ km

23. $a = 9.3$ cm, $b = 5.7$ cm, $c = 8.2$ cm

24. $a = 28$ ft, $b = 47$ ft, $c = 58$ ft

25. $a = 42.9$ m, $b = 37.6$ m, $c = 62.7$ m

26. $a = 189$ yd, $b = 214$ yd, $c = 325$ yd

27. $a = 965$ ft, $b = 876$ ft, $c = 1240$ ft

28. $a = 324$ m, $b = 421$ m, $c = 298$ m

29. $A = 80° \, 40'$, $b = 143$ cm, $c = 89.6$ cm

30. $C = 72° \, 40'$, $a = 327$ ft, $b = 251$ ft

31. $B = 74.8°$, $a = 8.92$ in., $c = 6.43$ in.

32. $C = 59.7°$, $a = 3.73$ mi, $b = 4.70$ mi

33. $A = 112.8°$, $b = 6.28$ m, $c = 12.2$ m

34. $B = 168.2°$, $a = 15.1$ cm, $c = 19.2$ cm

35. $a = 3.0$ ft, $b = 5.0$ ft, $c = 6.0$ ft

36. $a = 4.0$ ft, $b = 5.0$ ft, $c = 8.0$ ft

Concept Check Answer each question.

37. Refer to **Figure 12.** If we attempt to find any angle of a triangle with the values $a = 3$, $b = 4$, and $c = 10$ using the law of cosines, what happens?

38. "The shortest distance between two points is a straight line." How is this statement related to the geometric property that states that the sum of the lengths of any two sides of a triangle must be greater than the length of the remaining side?

Solve each problem. ***See Examples 1–4.***

39. ***Distance across a River*** Points A and B are on opposite sides of False River. From a third point, C, the angle between the lines of sight to A and B is 46.3°. If AC is 350 m long and BC is 286 m long, find AB.

40. 95.3 m
41. 163.5°
42. 5.2 cm, 8.8 cm
43. 281 km 44. 180 mi
45. 438.14 ft 46. 745 mi
47. 10.8 mi

40. *Distance across a Ravine* Points X and Y are on opposite sides of a ravine. From a third point Z, the angle between the lines of sight to X and Y is 37.7°. If XZ is 153 m long and YZ is 103 m long, find XY.

41. *Angle in a Parallelogram* A parallelogram has sides of length 25.9 cm and 32.5 cm. The longer diagonal has length 57.8 cm. Find the measure of the angle opposite the longer diagonal.

42. *Diagonals of a Parallelogram* The sides of a parallelogram are 4.0 cm and 6.0 cm. One angle is 58°, while another is 122°. Find the lengths of the diagonals of the parallelogram.

43. *Flight Distance* Airports A and B are 450 km apart, on an east-west line. Tom flies in a northeast direction from airport A to airport C. From C he flies 359 km on a bearing of 128° 40′ to B. How far is C from A?

44. *Distance Traveled by a Plane* An airplane flies 180 mi from point X at a bearing of 125°, and then turns and flies at a bearing of 230° for 100 mi. How far is the plane from point X?

45. *Distance between Ends of the Vietnam Memorial* The Vietnam Veterans Memorial in Washington, D.C., is V-shaped with equal sides of length 246.75 ft. The angle between these sides measures 125° 12′. Find the distance between the ends of the two sides. (*Source:* Pamphlet obtained at Vietnam Veterans Memorial.)

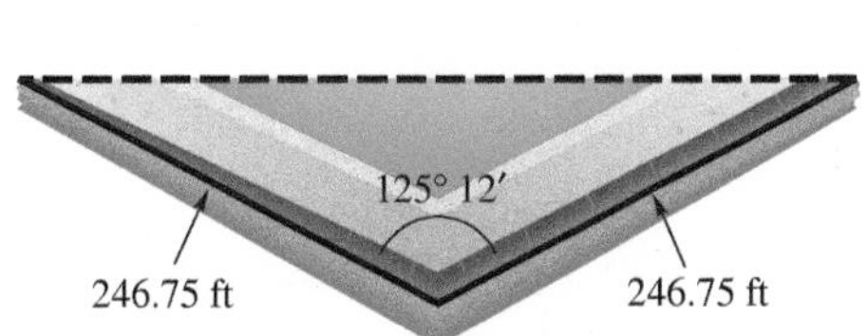

46. *Distance between Two Ships* Two ships leave a harbor together, traveling on courses that have an angle of 135° 40′ between them. If each travels 402 mi, how far apart are they?

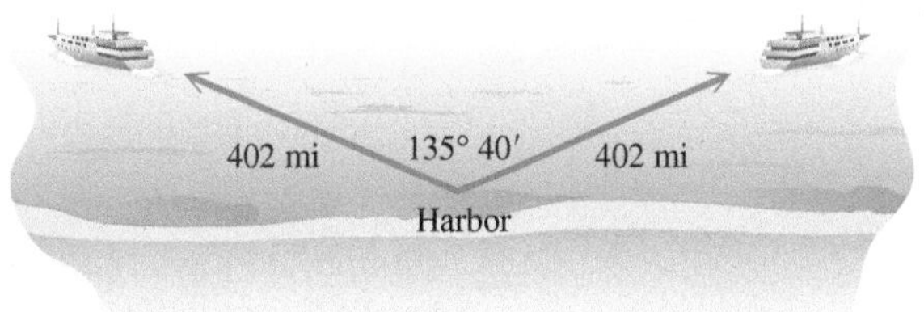

47. *Distance between a Ship and a Rock* A ship is sailing east. At one point, the bearing of a submerged rock is 45° 20′. After the ship has sailed 15.2 mi, the bearing of the rock has become 308° 40′. Find the distance of the ship from the rock at the latter point.

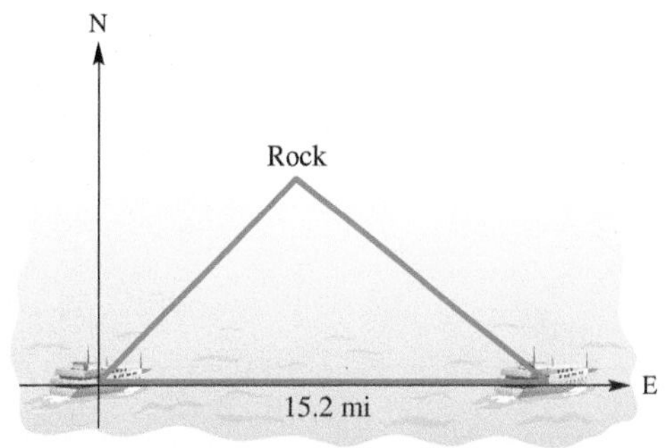

48. 1450 ft
49. 40°
50. 53°
51. 26° and 36°
52. 18 ft
53. second base: 66.8 ft; first and third bases: 63.7 ft
54. second base: 38.9 ft; first and third bases: 42.6 ft

48. *Distance between a Ship and a Submarine* From an airplane flying over the ocean, the angle of depression to a submarine lying under the surface is 24° 10′. At the same moment, the angle of depression from the airplane to a battleship is 17° 30′. See the figure. The distance from the airplane to the battleship is 5120 ft. Find the distance between the battleship and the submarine. (Assume the airplane, submarine, and battleship are in a vertical plane.)

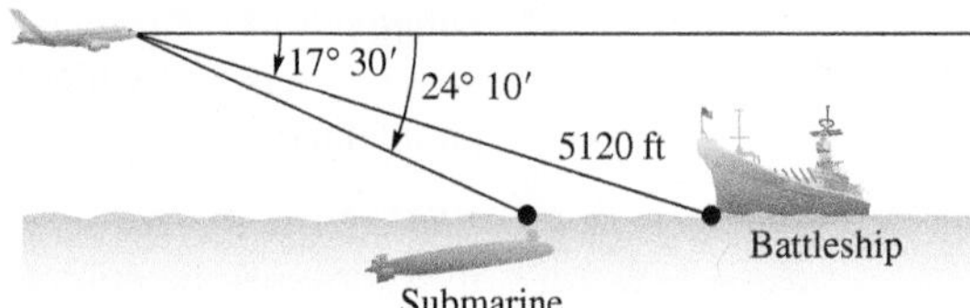

49. *Truss Construction* A triangular truss is shown in the figure. Find angle θ.

50. *Truss Construction* Find angle β in the truss shown in the figure.

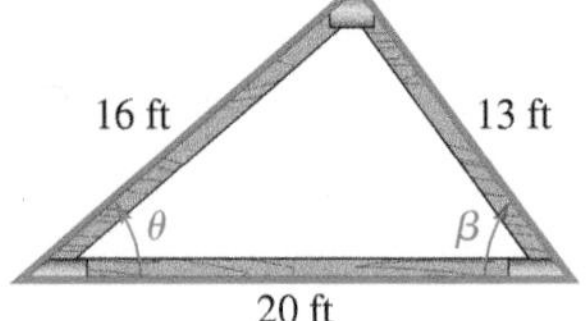

51. *Distance between a Beam and Cables* A weight is supported by cables attached to both ends of a balance beam, as shown in the figure. What angles are formed between the beam and the cables?

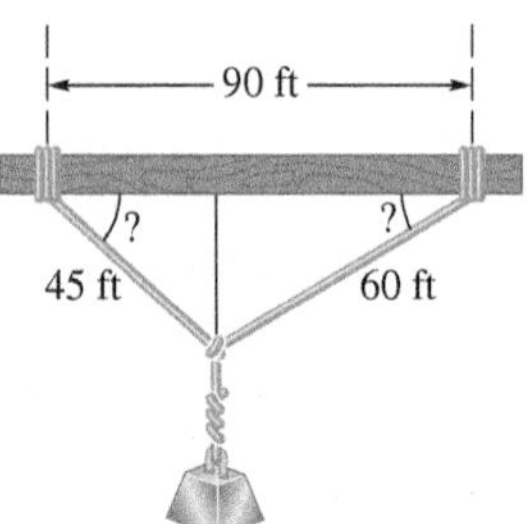

52. *Distance between Points on a Crane* A crane with a counterweight is shown in the figure. Find the horizontal distance between points A and B to the nearest foot.

53. *Distance on a Baseball Diamond* A baseball diamond is a square, 90.0 ft on a side, with home plate and the three bases as vertices. The pitcher's position is 60.5 ft from home plate. Find the distance from the pitcher's position to each of the bases.

54. *Distance on a Softball Diamond* A softball diamond is a square, 60.0 ft on a side, with home plate and the three bases as vertices. The pitcher's position is 46.0 ft from home plate. Find the distance from the pitcher's position to each of the bases.

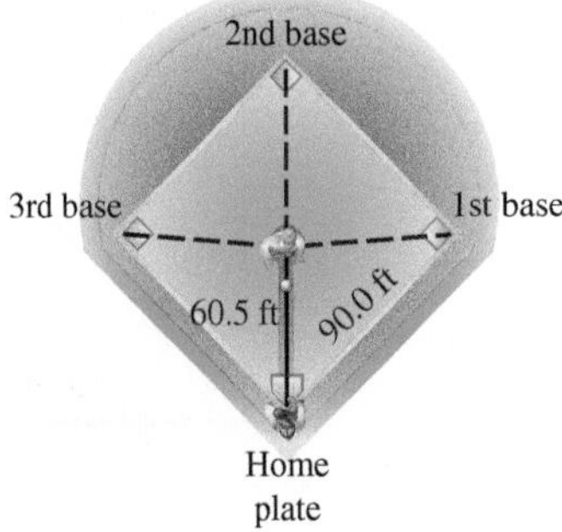

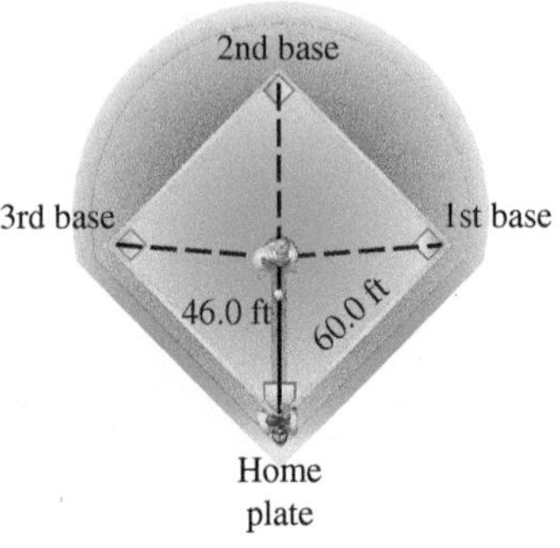

55. 39.2 km 56. 1473 m
57. 47.5 ft 58. 5.9 mi
59. 5500 m 60. 5.99 km

55. *Distance between a Ship and a Point* Starting at point A, a ship sails 18.5 km on a bearing of 189°, then turns and sails 47.8 km on a bearing of 317°. Find the distance of the ship from point A.

56. *Distance between Two Factories* Two factories blow their whistles at exactly 5:00. A man hears the two blasts at 3 sec and 6 sec after 5:00, respectively. The angle between his lines of sight to the two factories is 42.2°. If sound travels 344 m per sec, how far apart are the factories?

57. *Measurement Using Triangulation* Surveyors are often confronted with obstacles, such as trees, when measuring the boundary of a lot. One technique used to obtain an accurate measurement is the **triangulation method.** In this technique, a triangle is constructed around the obstacle and one angle and two sides of the triangle are measured. Use this technique to find the length of the property line (the straight line between the two markers) in the figure. (*Source:* Kavanagh, B., *Surveying Principles and Applications,* Sixth Edition, Prentice-Hall.)

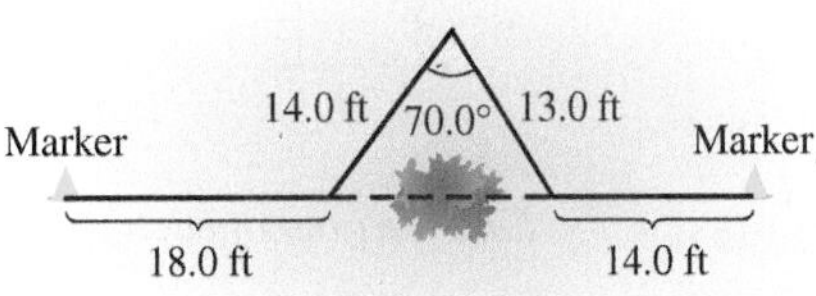

58. *Path of a Ship* A ship sailing due east in the North Atlantic has been warned to change course to avoid icebergs. The captain turns and sails on a bearing of 62°, then changes course again to a bearing of 115° until the ship reaches its original course. See the figure. How much farther did the ship have to travel to avoid the icebergs?

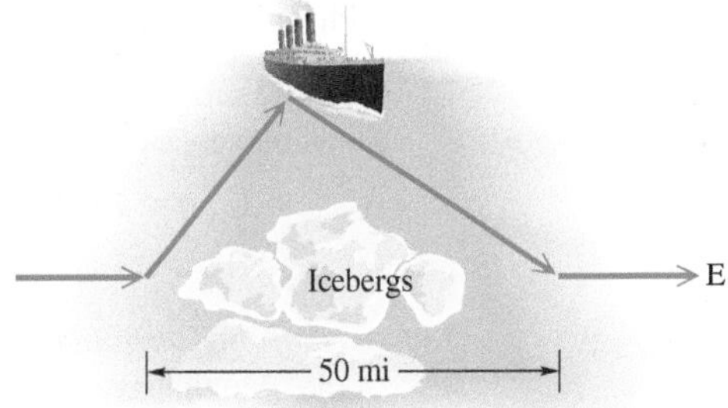

59. *Length of a Tunnel* To measure the distance through a mountain for a proposed tunnel, a point C is chosen that can be reached from each end of the tunnel. See the figure. If $AC = 3800$ m, $BC = 2900$ m, and angle $C = 110°$, find the length of the tunnel.

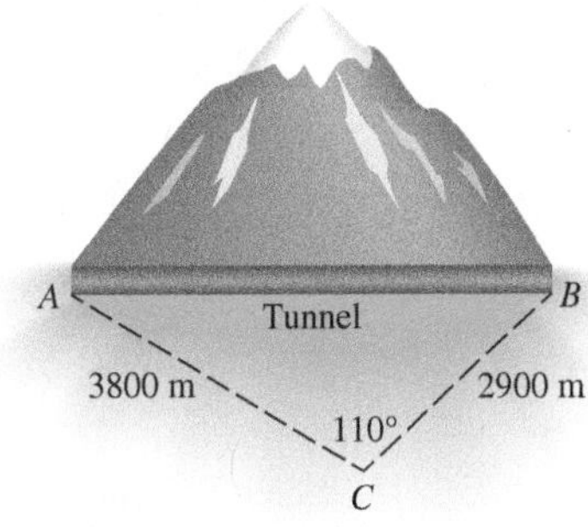

60. *Distance between an Airplane and a Mountain* A person in a plane flying straight north observes a mountain at a bearing of 24.1°. At that time, the plane is 7.92 km from the mountain. A short time later, the bearing to the mountain becomes 32.7°. How far is the airplane from the mountain when the second bearing is taken?

61. 16.26° 62. 14.25°

63. $24\sqrt{3}$ sq units

64. $15\sqrt{3}$ sq units

65. 78 m^2 66. 310 $in.^2$

67. 12,600 cm^2 68. 228 yd^2

69. 3650 ft^2 70. 82.2 m^2

71. Area and perimeter are both 36.

72. (a) Area is the integer 66.
(b) Area is the integer 84.
(c) Area is the integer 42.
(d) Area is the integer 36.

73. 390,000 mi^2

74. 33 cans

75. (a) 87.8° and 92.2° are possible angle measures.
(b) 92.2°
(c) With the law of cosines we are required to find the inverse cosine of a negative number. Therefore, we know that angle C is greater than 90°.

Find the measure of each angle θ to two decimal places.

61.

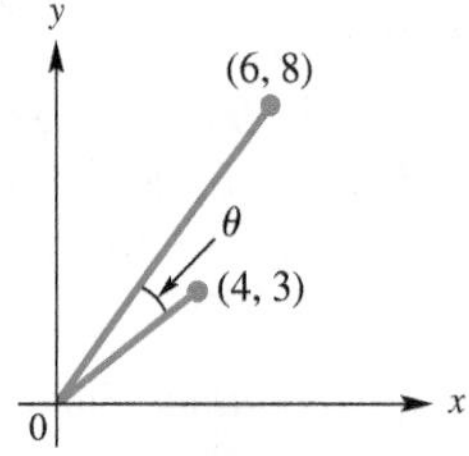

62.

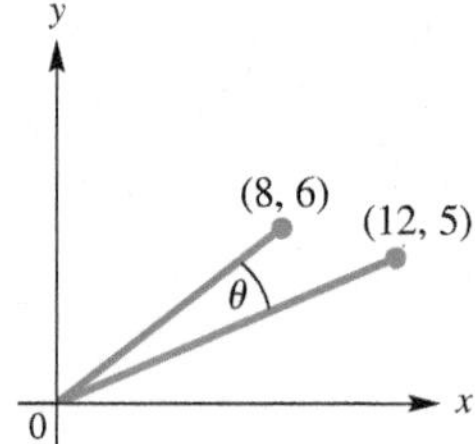

Find the exact area of each triangle using the formula $\mathcal{A} = \frac{1}{2}bh$, *and then verify that Heron's formula gives the same result.*

63.

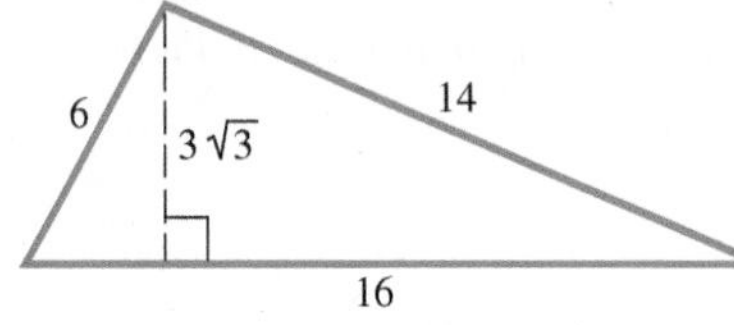

64.

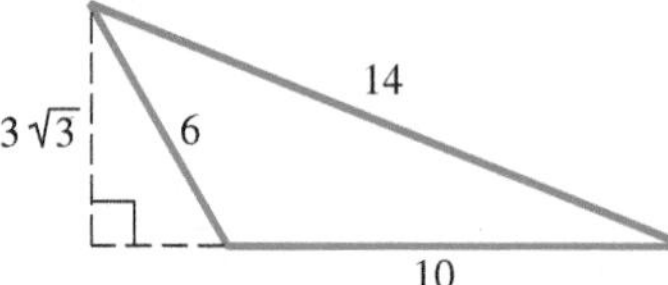

Find the area of each triangle ABC. ***See Example 5.***

65. $a = 12$ m, $b = 16$ m, $c = 25$ m

66. $a = 22$ in., $b = 45$ in., $c = 31$ in.

67. $a = 154$ cm, $b = 179$ cm, $c = 183$ cm

68. $a = 25.4$ yd, $b = 38.2$ yd, $c = 19.8$ yd

69. $a = 76.3$ ft, $b = 109$ ft, $c = 98.8$ ft

70. $a = 15.8$ m, $b = 21.7$ m, $c = 10.9$ m

Solve each problem. ***See Example 5.***

71. *Perfect Triangles* A **perfect triangle** is a triangle whose sides have whole number lengths and whose area is numerically equal to its perimeter. Show that the triangle with sides of length 9, 10, and 17 is perfect.

72. *Heron Triangles* A **Heron triangle** is a triangle having integer sides and area. Show that each of the following is a Heron triangle.

(a) $a = 11$, $b = 13$, $c = 20$

(b) $a = 13$, $b = 14$, $c = 15$

(c) $a = 7$, $b = 15$, $c = 20$

(d) $a = 9$, $b = 10$, $c = 17$

73. *Area of the Bermuda Triangle* Find the area of the Bermuda Triangle if the sides of the triangle have approximate lengths 850 mi, 925 mi, and 1300 mi.

74. *Required Amount of Paint* A painter needs to cover a triangular region 75 m by 68 m by 85 m. A can of paint covers 75 m^2 of area. How many cans (to the next higher number of cans) will be needed?

75. Consider triangle ABC shown here.

(a) Use the law of sines to find candidates for the value of angle C. Round angle measures to the nearest tenth of a degree.

(b) Rework part (a) using the law of cosines.

(c) Why is the law of cosines a better method in this case?

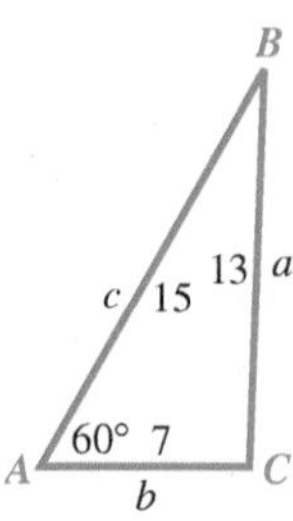

76. Show that the measure of angle A is twice the measure of angle B. (*Hint:* Use the law of cosines to find $\cos A$ and $\cos B$, and then show that $\cos A = 2\cos^2 B - 1$.)

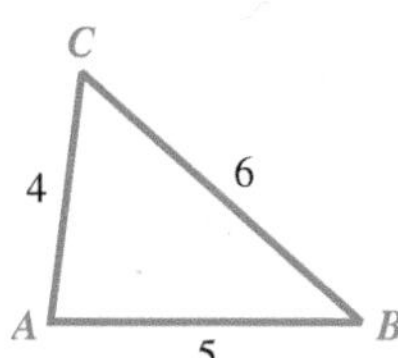

77. 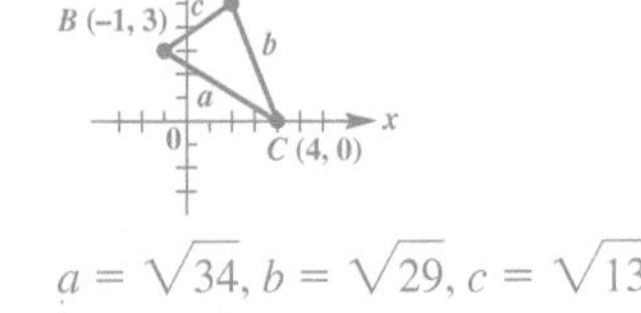

$a = \sqrt{34}, b = \sqrt{29}, c = \sqrt{13}$

78. 9.5 sq units

79. 9.5 sq units

80. 9.5 sq units

Relating Concepts

For individual or collaborative investigation *(Exercises 77–80)*

We have introduced two new formulas for the area of a triangle in this chapter. We can now find the area $\mathcal{A}$ *of a triangle using one of three formulas.*

(a) $\mathcal{A} = \frac{1}{2}bh$

(b) $\mathcal{A} = \frac{1}{2}ab \sin C \quad \left(\text{or } \mathcal{A} = \frac{1}{2}ac \sin B \text{ or } \mathcal{A} = \frac{1}{2}bc \sin A\right)$

(c) $\mathcal{A} = \sqrt{s(s-a)(s-b)(s-c)}$ (Heron's formula)

If the coordinates of the vertices of a triangle are given, then the following area formula is also valid.

(d) $\mathcal{A} = \frac{1}{2}\left|(x_1y_2 - y_1x_2 + x_2y_3 - y_2x_3 + x_3y_1 - y_3x_1)\right|$

The vertices are the ordered pairs (x_1, y_1), (x_2, y_2), and (x_3, y_3).

Work Exercises 77–80 in order, *showing that the various formulas all lead to the same area.*

77. Draw a triangle with vertices $A(2, 5)$, $B(-1, 3)$, and $C(4, 0)$, and use the distance formula to find the lengths of the sides *a*, *b*, and *c*.

78. Find the area of triangle *ABC* using formula (b). (First use the law of cosines to find the measure of an angle.)

79. Find the area of triangle *ABC* using formula (c)—that is, Heron's formula.

80. Find the area of triangle *ABC* using new formula (d).

Chapter 7 Quiz (Sections 7.1–7.3)

[7.1]
1. 131°

[7.3]
2. 201 m
3. 48.0°

[7.1]
4. 15.75 sq units

[7.3]
5. 189 km²

[7.2]
6. 41.6°, 138.4°

[7.1]
7. $a = 648$, $b = 456$, $C = 28°$
8. 3.6 mi

Find the indicated part of each triangle ABC.

1. Find *A* if $B = 30.6°$, $b = 7.42$ in., and $c = 4.54$ in.

2. Find *a* if $A = 144°$, $c = 135$ m, and $b = 75.0$ m.

3. Find *C* if $a = 28.4$ ft, $b = 16.9$ ft, and $c = 21.2$ ft.

Solve each problem.

4. Find the area of the triangle shown here.

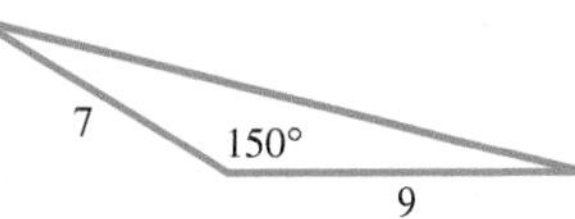

5. Find the area of triangle *ABC* if $a = 19.5$ km, $b = 21.0$ km, and $c = 22.5$ km.

6. For triangle *ABC* with $c = 345$, $a = 534$, and $C = 25.4°$, there are two possible values for angle *A*. What are they?

7. Solve triangle *ABC* if $c = 326$, $A = 111°$, and $B = 41.0°$.

8. ***Height of a Balloon*** The angles of elevation of a hot air balloon from two observation points *X* and *Y* on level ground are 42° 10′ and 23° 30′, respectively. As shown in the figure, points *X*, *Y*, and *Z* are in the same vertical plane and points *X* and *Y* are 12.2 mi apart. Approximate the height of the balloon to the nearest tenth of a mile.

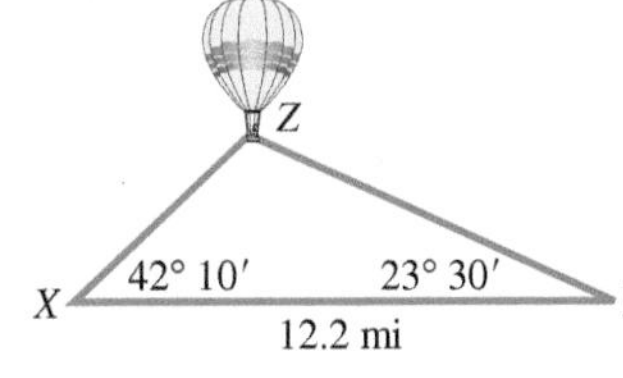

[7.3]
9. 25.24983 mi
10. 3921 m

9. *Volcano Movement* To help predict eruptions from the volcano Mauna Loa on the island of Hawaii, scientists keep track of the volcano's movement by using a "super triangle" with vertices on the three volcanoes shown on the map at the right. Find BC given that $AB = 22.47928$ mi, $AC = 28.14276$ mi, and $A = 58.56989°$.

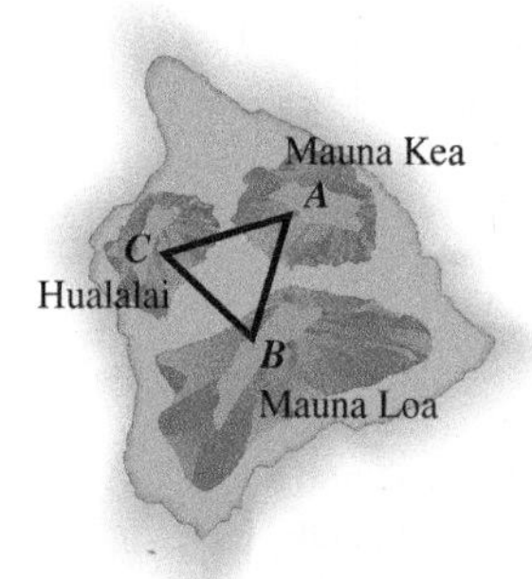

10. *Distance between Two Towns* To find the distance between two small towns, an electronic distance measuring (EDM) instrument is placed on a hill from which both towns are visible. The distance to each town from the EDM and the angle between the two lines of sight are measured. See the figure. Find the distance between the towns.

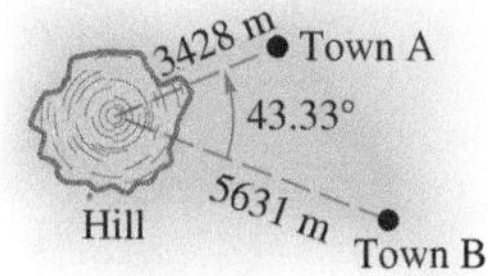

7.4 Geometrically Defined Vectors and Applications

- Basic Terminology
- The Equilibrant
- Incline Applications
- Navigation Applications

Basic Terminology Quantities that involve magnitudes, such as 45 lb or 60 mph, can be represented by real numbers called **scalars.** Other quantities, called **vector quantities,** involve both magnitude *and* direction. Typical vector quantities are velocity, acceleration, and force. For example, traveling 50 mph *east* represents a vector quantity.

A vector quantity can be represented with a directed line segment (a segment that uses an arrowhead to indicate direction) called a **vector.** The *length* of the vector represents the **magnitude** of the vector quantity. The *direction* of the vector, indicated by the arrowhead, represents the direction of the quantity. See **Figure 19.**

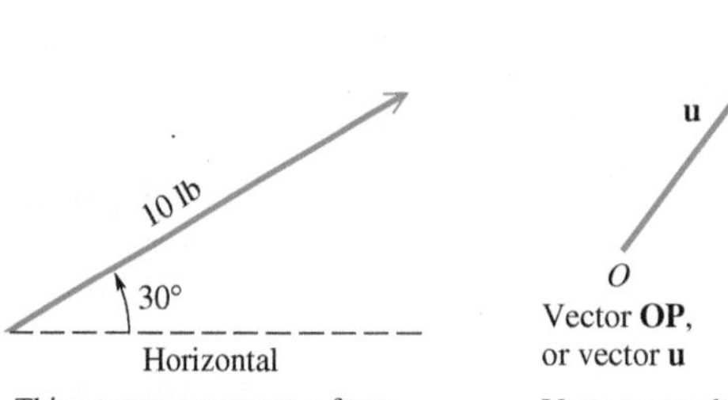

This vector represents a force of 10 lb applied at an angle 30° above the horizontal.

Figure 19

Vectors may be named with two uppercase letters or with one lowercase or uppercase letter.

Figure 20

When we indicate vectors in print, it is customary to use boldface type or an arrow over the letter or letters. Thus, **OP** and $\overrightarrow{OP}$ both represent the vector **OP**. When two letters name a vector, the first indicates the **initial point** and the second indicates the **terminal point** of the vector. Knowing these points gives the direction of the vector. For example, vectors **OP** and **PO** in **Figure 20** are not the same vector. They have the same magnitude but *opposite* directions. The magnitude of vector **OP** is written $|\mathbf{OP}|$.

Two vectors are equal if and only if they have the same direction and the same magnitude. In **Figure 21,** vectors **A** and **B** are equal, as are vectors **C** and **D**. As **Figure 21** shows, equal vectors need not coincide, but they must be parallel and in the same direction. Vectors **A** and **E** are unequal because they do not have the same direction, while $\mathbf{A} \neq \mathbf{F}$ because they have different magnitudes.

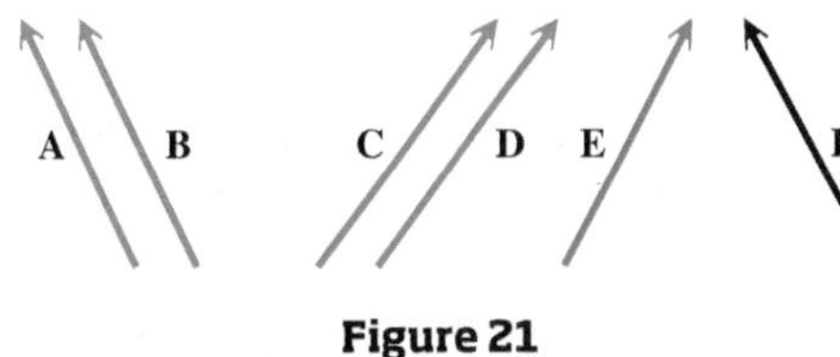

Figure 21

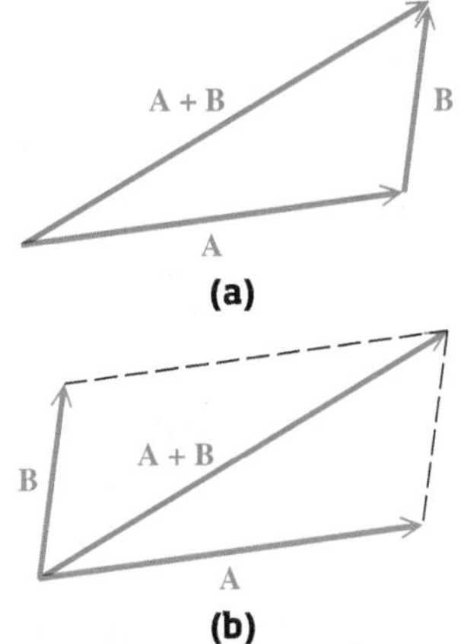

Figure 22

The sum of two vectors is also a vector. There are two ways to find the sum of two vectors **A** and **B** geometrically.

1. Place the initial point of vector **B** at the terminal point of vector **A**, as shown in **Figure 22(a).** The vector with the same initial point as **A** and the same terminal point as **B** is the sum **A** + **B**.

2. Use the **parallelogram rule.** Place vectors **A** and **B** so that their initial points coincide, as in **Figure 22(b).** Then, complete a parallelogram that has **A** and **B** as two sides. The diagonal of the parallelogram with the same initial point as **A** and **B** is the sum **A** + **B**.

Parallelograms can be used to show that vector **B** + **A** is the same as vector **A** + **B**, or that **A** + **B** = **B** + **A**, so ***vector addition is commutative.*** The vector sum **A** + **B** is the **resultant** of vectors **A** and **B**.

For every vector **v** there is a vector −**v** that has the same magnitude as **v** but opposite direction. Vector −**v** is the **opposite** of **v**. See **Figure 23.** The sum of **v** and −**v** has magnitude 0 and is the **zero vector.** As with real numbers, to subtract vector **B** from vector **A**, find the vector sum **A** + (−**B**). See **Figure 24.**

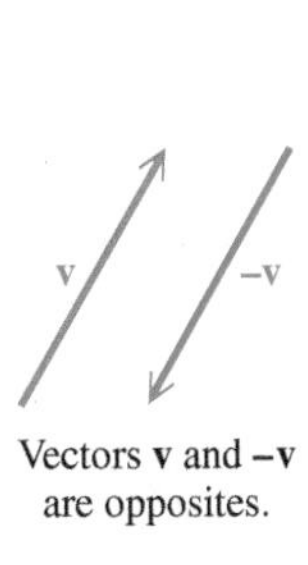

Vectors **v** and −**v** are opposites.

Figure 23

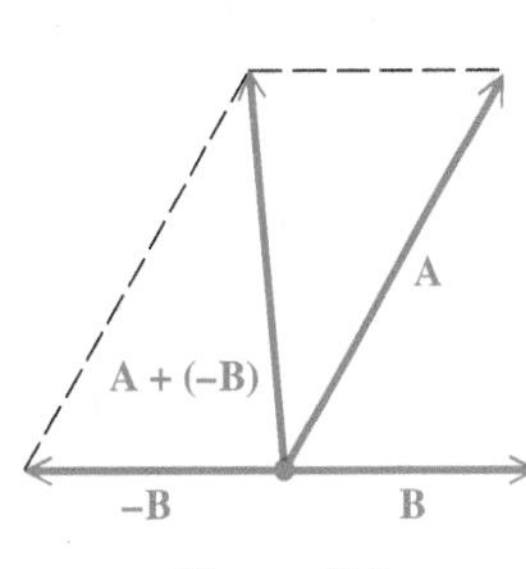

Figure 24

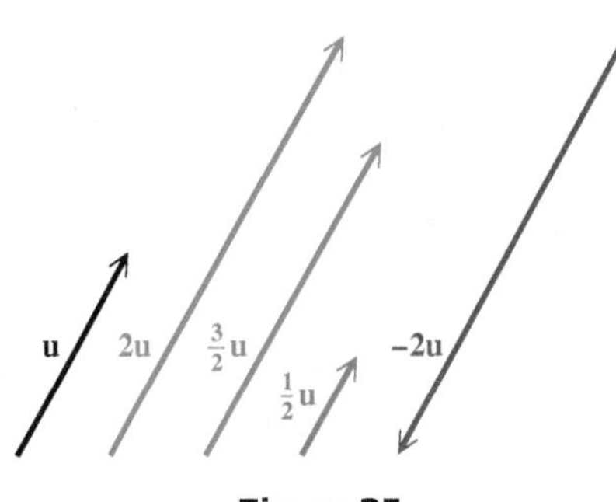

Figure 25

Teaching Tip Show by geometric analysis that vector addition is commutative. Also, mention that **A** + (−**B**) is the same as **A** − **B**. In **Figure 24,** move vector −**B** so that its initial point is the terminal point of **A** to further illustrate the meaning of **A** − **B**.

The product of a real number (or scalar) k and a vector **u** is the vector $k \cdot \mathbf{u}$, which has magnitude $|k|$ times the magnitude of **u**. The vector $k \cdot \mathbf{u}$ has the same direction as **u** if $k > 0$ and the opposite direction if $k < 0$. See **Figure 25.**

The following properties are helpful when solving vector applications.

Geometric Properties of Parallelograms

1. A parallelogram is a quadrilateral whose opposite sides are parallel.
2. The opposite sides and opposite angles of a parallelogram are equal, and adjacent angles of a parallelogram are supplementary.
3. The diagonals of a parallelogram bisect each other, but they do not necessarily bisect the angles of the parallelogram.

Classroom Example 1
Two forces of 32 and 48 newtons act on a point in the plane. If the angle between the forces is 76°, find the magnitude of the resultant vector.

Answer: 64 newtons

EXAMPLE 1 Finding the Magnitude of a Resultant

Two forces of 15 and 22 newtons act on a point in the plane. (A **newton** is a unit of force that equals 0.225 lb.) If the angle between the forces is 100°, find the magnitude of the resultant force.

SOLUTION As shown in **Figure 26,** a parallelogram that has the forces as adjacent sides can be formed. The angles of the parallelogram adjacent to angle P measure 80° because adjacent angles of a parallelogram are supplementary. Opposite sides of the parallelogram are equal in length. The resultant force divides the parallelogram into two triangles. Use the law of cosines with either triangle.

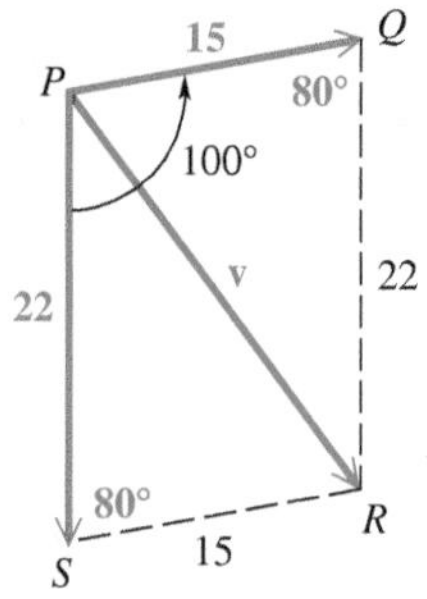

$|\mathbf{v}|^2 = 15^2 + 22^2 - 2(15)(22)\cos 80°$ Law of cosines

$|\mathbf{v}|^2 \approx 225 + 484 - 115$ Evaluate powers and cos 80°. Multiply.

$|\mathbf{v}|^2 \approx 594$ Add and subtract.

$|\mathbf{v}| \approx 24$ Take square roots and choose the positive square root.

Figure 26

To the nearest unit, the magnitude of the resultant force is 24 newtons.

Now Try Exercise 27.

The Equilibrant The previous example showed a method for finding the resultant of two vectors. Sometimes it is necessary to find a vector that will counterbalance the resultant. This opposite vector is the **equilibrant.** That is, the equilibrant of vector **u** is the vector $-\mathbf{u}$.

Classroom Example 2
Find the magnitude of the equilibrant of forces of 54 newtons and 42 newtons acting on a point A, if the angle between the forces is 98°. Then find the angle between the equilibrant and the 42-newton force.

Answer: 64 newtons; 123°

EXAMPLE 2 Finding the Magnitude and Direction of an Equilibrant

Find the magnitude of the equilibrant of forces of 48 newtons and 60 newtons acting on a point A, if the angle between the forces is 50°. Then find the angle between the equilibrant and the 48-newton force.

SOLUTION As shown in **Figure 27,** the equilibrant is $-\mathbf{v}$.

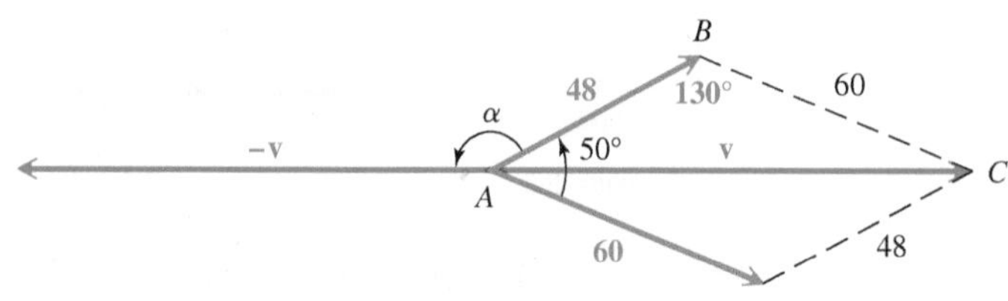

Figure 27

Teaching Tip In **Example 2,** caution students not to assume that $\overrightarrow{AC}$ bisects the angle formed by the two forces. In fact, this happens only if the forces have equal magnitudes.

The magnitude of **v**, and hence of $-\mathbf{v}$, is found using triangle ABC and the law of cosines.

$|\mathbf{v}|^2 = 48^2 + 60^2 - 2(48)(60)\cos 130°$ Law of cosines

$|\mathbf{v}|^2 \approx 9606.5$ Use a calculator.

$|\mathbf{v}| \approx 98$ Square root property; Give two significant digits.

To the nearest unit, the magnitude is 98 newtons.

The required angle, labeled α in **Figure 27,** can be found by subtracting angle CAB from 180°. Use the law of sines to find angle CAB.

$$\frac{\sin CAB}{60} = \frac{\sin 130^\circ}{98}$$ Law of sines (alternative form)

$$\sin CAB \approx 0.46900680$$ Multiply by 60 and use a calculator.

$$CAB \approx 28^\circ$$ Use the inverse sine function.

Finally, $\alpha \approx 180^\circ - 28^\circ = 152^\circ$.

✔ **Now Try Exercise 31.**

Classroom Example 3
Find the force required to keep a 2500-lb car parked on a hill that makes a 12° angle with the horizontal.

Answer: 520 lb

Incline Applications

We can use vectors to solve incline problems.

EXAMPLE 3 Finding a Required Force

Find the force required to keep a 50-lb wagon from sliding down a ramp inclined at 20° to the horizontal. (Assume there is no friction.)

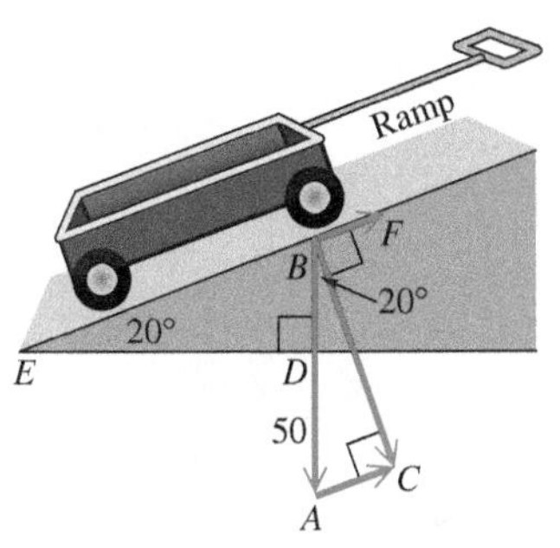

Figure 28

SOLUTION In **Figure 28,** the vertical 50-lb force **BA** represents the force of gravity. It is the sum of vectors **BC** and −**AC**. The vector **BC** represents the force with which the weight pushes against the ramp. The vector **BF** represents the force that would pull the weight up the ramp. Because vectors **BF** and **AC** are equal, $|\mathbf{AC}|$ gives the magnitude of the required force.

Vectors **BF** and **AC** are parallel, so angle *EBD* equals angle *A* by alternate interior angles. Because angle *BDE* and angle *C* are right angles, triangles *CBA* and *DEB* have two corresponding angles equal and, thus, are similar triangles. Therefore, angle *ABC* equals angle *E*, which is 20°. From right triangle *ABC*, we have the following.

$$\sin 20^\circ = \frac{|\mathbf{AC}|}{50}$$ $\sin B = \frac{\text{side opposite } B}{\text{hypotenuse}}$

$$|\mathbf{AC}| = 50 \sin 20^\circ$$ Multiply by 50 and rewrite.

$$|\mathbf{AC}| \approx 17$$ Use a calculator.

Teaching Tip Emphasize that weight is a *force* whose direction is always straight down—toward the center of Earth.

A force of approximately 17 lb will keep the wagon from sliding down the ramp.

✔ **Now Try Exercise 39.**

Classroom Example 4
A force of 18.0 lb is required to hold a 74.0-lb crate on a ramp. What angle does the ramp make with the horizontal?

Answer: 14.1°

EXAMPLE 4 Finding an Incline Angle

A force of 16.0 lb is required to hold a 40.0-lb lawn mower on an incline. What angle does the incline make with the horizontal?

SOLUTION This situation is illustrated in **Figure 29.** Consider right triangle *ABC*. Angle *B* equals angle θ, the magnitude of vector **BA** represents the weight of the mower, and vector **AC** equals vector **BE**, which represents the force required to hold the mower on the incline.

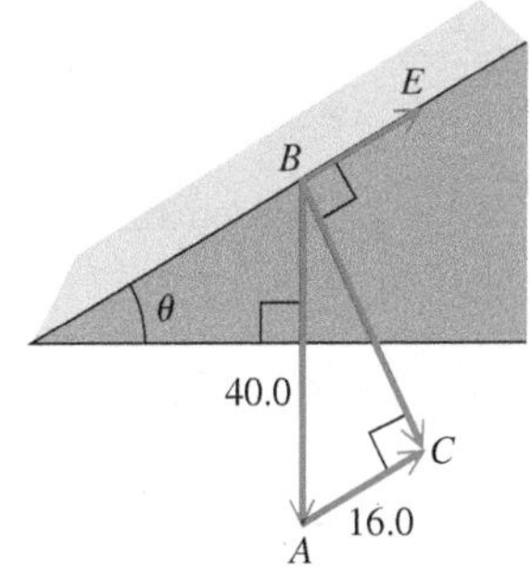

Figure 29

$$\sin B = \frac{16.0}{40.0}$$ $\sin B = \frac{\text{side opposite } B}{\text{hypotenuse}}$

$$\sin B = 0.4$$ Simplify.

$$B \approx 23.6^\circ$$ Use the inverse sine function.

The hill makes an angle of about 23.6° with the horizontal.

✔ **Now Try Exercise 41.**

Navigation Applications Problems that involve bearing can also be solved using vectors.

Classroom Example 5
A ship leaves port on a bearing of 25.0° and travels 61.4 km. The ship then turns due east and travels 84.6 km. How far is the ship from port? What is its bearing from port?

Answer: 124 km; 63.3°

EXAMPLE 5 Applying Vectors to a Navigation Problem

A ship leaves port on a bearing of 28.0° and travels 8.20 mi. The ship then turns due east and travels 4.30 mi. How far is the ship from port? What is its bearing from port?

SOLUTION In **Figure 30,** vectors **PA** and **AE** represent the ship's path. The magnitude and bearing of the resultant **PE** can be found as follows. Triangle *PNA* is a right triangle, so

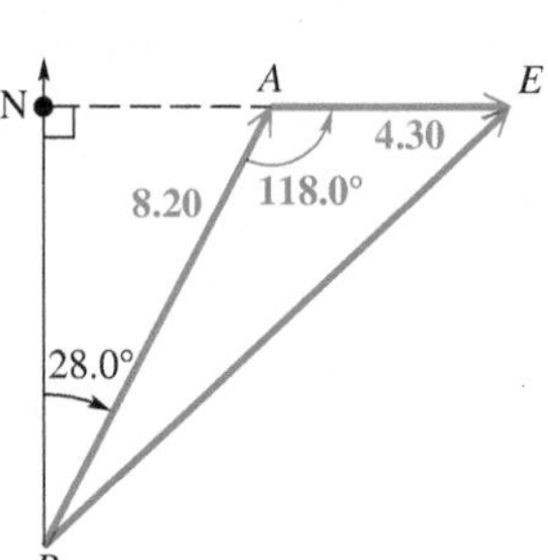

Figure 30

$$\text{angle } NAP = 90° - 28.0° = 62.0°,$$

and $$\text{angle } PAE = 180° - 62.0° = 118.0°.$$

Use the law of cosines to find $|\mathbf{PE}|$, the magnitude of vector **PE**.

$$|\mathbf{PE}|^2 = 8.20^2 + 4.30^2 - 2(8.20)(4.30)\cos 118.0° \quad \text{Law of cosines}$$

$$|\mathbf{PE}|^2 \approx 118.84 \quad \text{Use a calculator.}$$

$$|\mathbf{PE}| \approx 10.9 \quad \text{Square root property}$$

The ship is about 10.9 mi from port.

To find the bearing of the ship from port, find angle *APE*.

$$\frac{\sin APE}{4.30} = \frac{\sin 118.0°}{10.9} \quad \text{Law of sines}$$

$$\sin APE = \frac{4.30 \sin 118.0°}{10.9} \quad \text{Multiply by 4.30.}$$

$$APE \approx 20.4° \quad \text{Use the inverse sine function.}$$

Finally, 28.0° + 20.4° = 48.4°, so the bearing is 48.4°.

✔ **Now Try Exercise 45.**

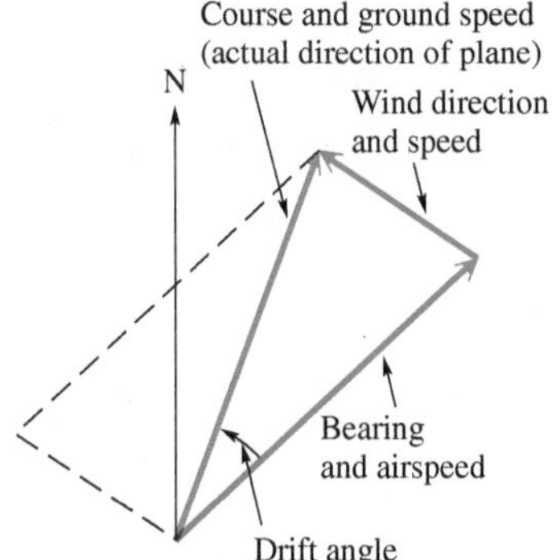

Figure 31

In air navigation, the **airspeed** of a plane is its speed relative to the air, and the **ground speed** is its speed relative to the ground. Because of wind, these two speeds are usually different. The ground speed of the plane is represented by the vector sum of the airspeed and windspeed vectors. See **Figure 31.**

Classroom Example 6
A small plane follows a bearing of 70° at an airspeed of 230 mph and encounters a wind blowing at 25.0 mph from a direction of 290°. Find the resulting bearing and ground speed of the plane.

Answer: bearing: 74°; ground speed: 250 mph

EXAMPLE 6 Applying Vectors to a Navigation Problem

An airplane that is following a bearing of 239° at an airspeed of 425 mph encounters a wind blowing at 36.0 mph from a direction of 115°. Find the resulting bearing and ground speed of the plane.

SOLUTION An accurate sketch is essential to the solution of this problem. We have included two sets of geographical axes, which enable us to determine measures of necessary angles. Analyze **Figure 32** on the next page carefully.

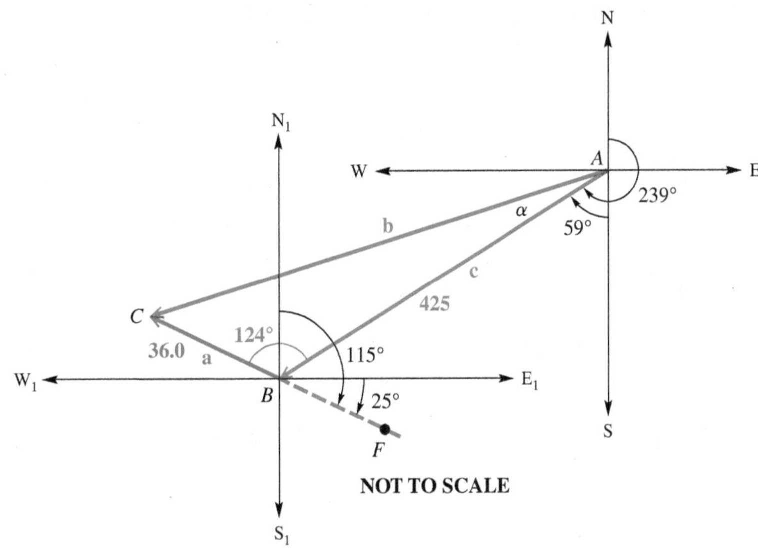

Figure 32

Vector **c** represents the airspeed and bearing of the plane, vector **a** represents the speed and direction of the wind, and vector **b** represents the resulting bearing and ground speed of the plane. Angle ABC has as its measure the sum of angle ABN_1 and angle N_1BC.

- Angle SAB measures $239° - 180° = 59°$. Because angle ABN_1 is an alternate interior angle to it, $ABN_1 = 59°$.
- Angle E_1BF measures $115° - 90° = 25°$. Thus, angle CBW_1 also measures $25°$ because it is a vertical angle. Angle N_1BC is the complement of $25°$, which is $90° - 25° = 65°$.

By these results,

$$\text{angle } ABC = 59° + 65° = 124°.$$

To find $|\mathbf{b}|$, we use the law of cosines.

$$|\mathbf{b}|^2 = |\mathbf{a}|^2 + |\mathbf{c}|^2 - 2|\mathbf{a}||\mathbf{c}| \cos ABC \qquad \text{Law of cosines}$$

$$|\mathbf{b}|^2 = 36.0^2 + 425^2 - 2(36.0)(425) \cos 124° \qquad \text{Substitute.}$$

$$|\mathbf{b}|^2 \approx 199{,}032 \qquad \text{Use a calculator.}$$

$$|\mathbf{b}| \approx 446 \qquad \text{Square root property}$$

The ground speed is approximately 446 mph.

To find the resulting bearing of **b**, we must find the measure of angle α in **Figure 32** and then add it to 239°. To find α, we use the law of sines.

$$\frac{\sin \alpha}{36.0} = \frac{\sin 124°}{446}$$

To maintain accuracy, use all the significant digits that a calculator allows.

$$\sin \alpha = \frac{36.0 \sin 124°}{446} \qquad \text{Multiply by 36.0.}$$

$$\alpha = \sin^{-1}\left(\frac{36.0 \sin 124°}{446}\right) \qquad \text{Use the inverse sine function.}$$

$$\alpha \approx 4° \qquad \text{Use a calculator.}$$

Add 4° to 239° to find the resulting bearing of 243°. ✔ **Now Try Exercise 51.**

7.4 Exercises

1. m and p; n and r
2. m and q; p and q; r and s; n and s
3. m and p equal 2t, or t equals $\frac{1}{2}$m and $\frac{1}{2}$p. Also m = 1p and n = 1r.
4. m = −1q; p = −1q; r = −1s; q = −2t; n = −1s

5. 

6. 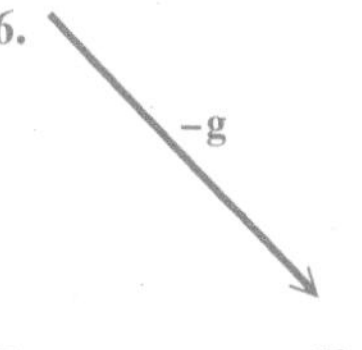

7. 

8. 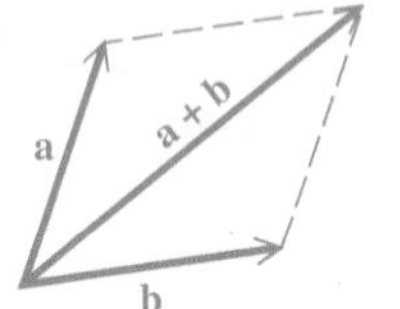

9. 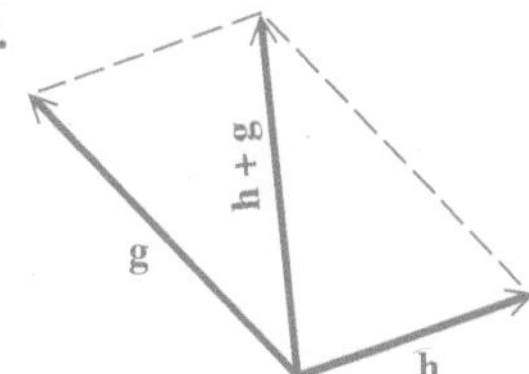

10. 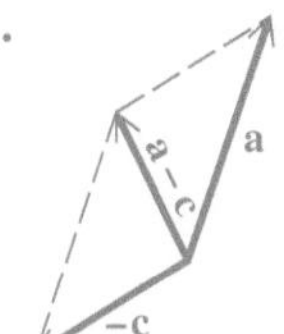

11. 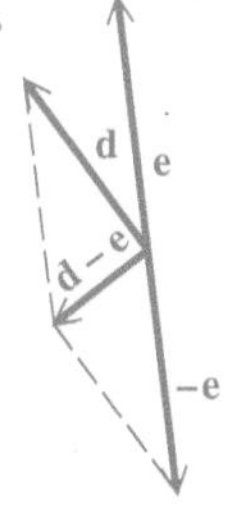

12.

CONCEPT PREVIEW *Refer to the vectors* **m** *through* **t** *below.*

1. Name all pairs of vectors that appear to be equal.

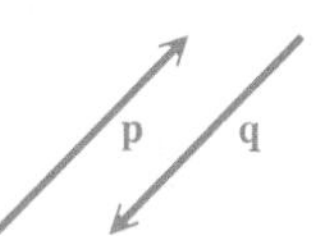

2. Name all pairs of vectors that are opposites.
3. Name all pairs of vectors where the first is a scalar multiple of the other, with the scalar positive.
4. Name all pairs of vectors where the first is a scalar multiple of the other, with the scalar negative.

CONCEPT PREVIEW *Refer to vectors* **a** *through* **h** *below. Make a copy or a sketch of each vector, and then draw a sketch to represent each of the following. For example, find* **a + e** *by placing* **a** *and* **e** *so that their initial points coincide. Then use the parallelogram rule to find the resultant, as shown in the figure on the right.*

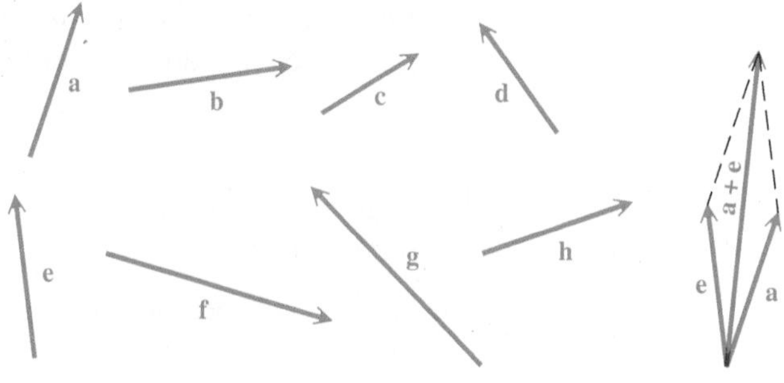

5. −**b**	6. −**g**	7. 2**c**	8. 2**h**
9. **a** + **b**	10. **h** + **g**	11. **a** − **c**	12. **d** − **e**
13. **a** + (**b** + **c**)	14. (**a** + **b**) + **c**	15. **c** + **d**	16. **d** + **c**

17. From the results of **Exercises 13 and 14,** does it appear that vector addition is associative?
18. From the results of **Exercises 15 and 16,** does it appear that vector addition is commutative?

For each pair of vectors **u** *and* **v** *with angle* θ *between them, sketch the resultant.*

19. $|\mathbf{u}| = 12$, $|\mathbf{v}| = 20$, $\theta = 27°$
20. $|\mathbf{u}| = 8$, $|\mathbf{v}| = 12$, $\theta = 20°$
21. $|\mathbf{u}| = 20$, $|\mathbf{v}| = 30$, $\theta = 30°$
22. $|\mathbf{u}| = 50$, $|\mathbf{v}| = 70$, $\theta = 40°$

Use the parallelogram rule to find the magnitude of the resultant force for the two forces shown in each figure. Round answers to the nearest tenth.

23. 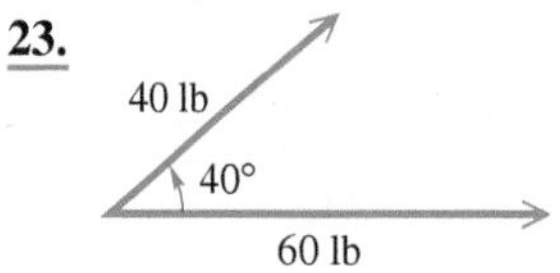

24. 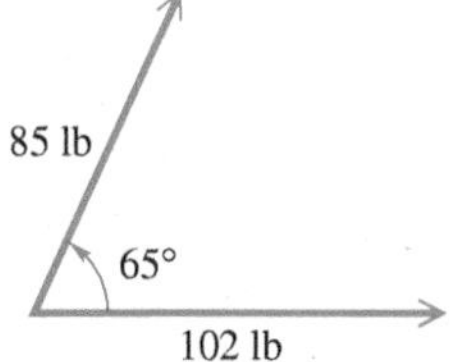

25. 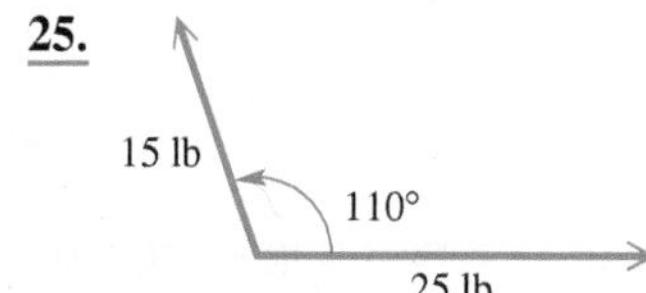

26. 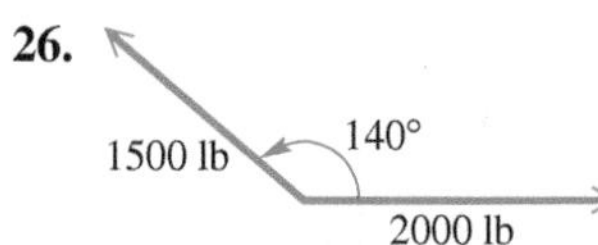

13. 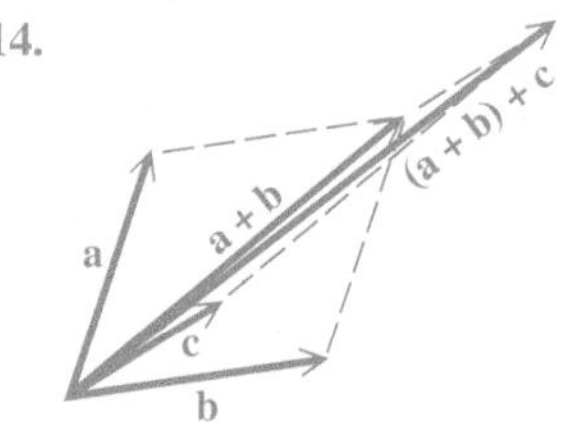

14. 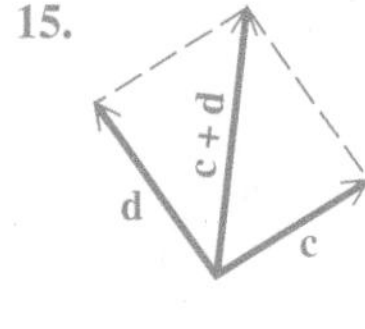

15. 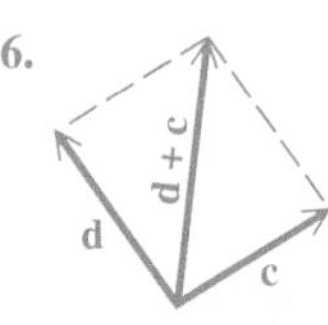

16.

17. Yes, it appears that vector addition is associative (and this is true, in general).

18. Yes, it appears that vector addition is commutative (and this is true, in general).

19. 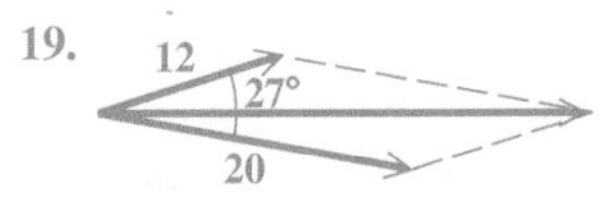

20. 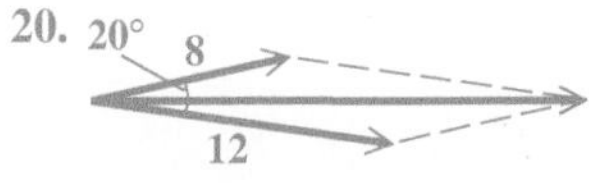

21. 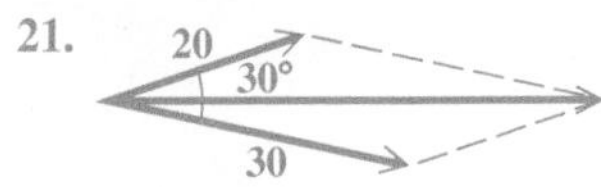

22. 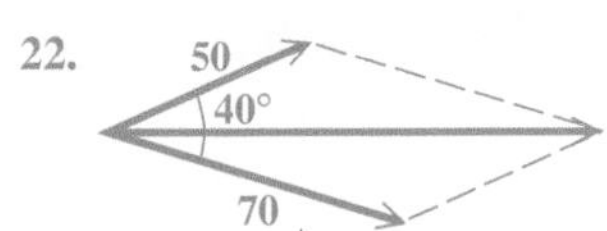

Two forces act at a point in the plane. The angle between the two forces is given. Find the magnitude of the resultant force. **See Example 1.**

27. forces of 250 and 450 newtons, forming an angle of 85°

28. forces of 19 and 32 newtons, forming an angle of 118°

29. forces of 116 and 139 lb, forming an angle of 140° 50′

30. forces of 37.8 and 53.7 lb, forming an angle of 68.5°

Solve each problem. **See Examples 1–4.**

31. *Direction and Magnitude of an Equilibrant* Two tugboats are pulling a disabled speedboat into port with forces of 1240 lb and 1480 lb. The angle between these forces is 28.2°. Find the direction and magnitude of the equilibrant.

32. *Direction and Magnitude of an Equilibrant* Two rescue vessels are pulling a broken-down motorboat toward a boathouse with forces of 840 lb and 960 lb. The angle between these forces is 24.5°. Find the direction and magnitude of the equilibrant.

33. *Angle between Forces* Two forces of 692 newtons and 423 newtons act at a point. The resultant force is 786 newtons. Find the angle between the forces.

34. *Angle between Forces* Two forces of 128 lb and 253 lb act at a point. The resultant force is 320 lb. Find the angle between the forces.

35. *Magnitudes of Forces* A force of 176 lb makes an angle of 78° 50′ with a second force. The resultant of the two forces makes an angle of 41° 10′ with the first force. Find the magnitudes of the second force and of the resultant.

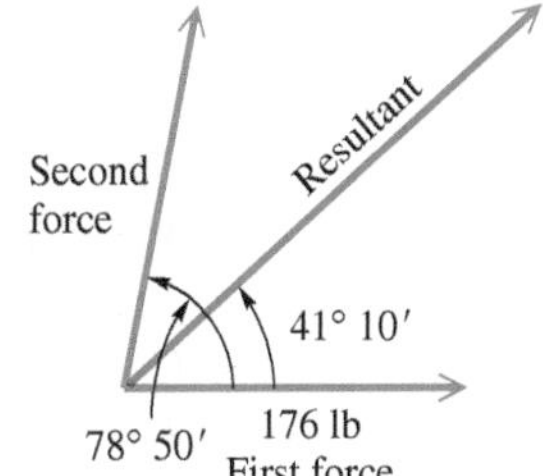

36. *Magnitudes of Forces* A force of 28.7 lb makes an angle of 42° 10′ with a second force. The resultant of the two forces makes an angle of 32° 40′ with the first force. Find the magnitudes of the second force and of the resultant.

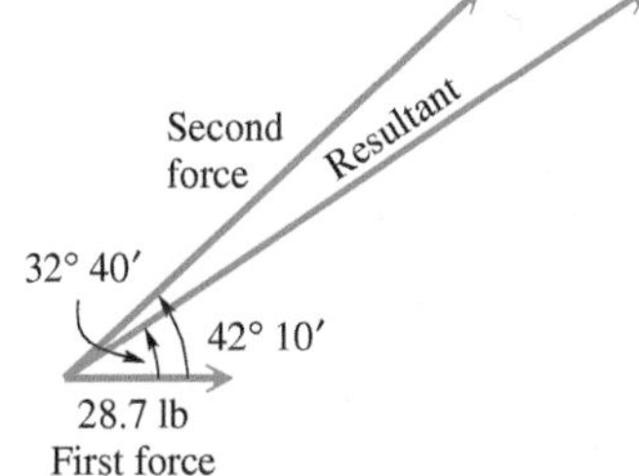

37. *Angle of a Hill Slope* A force of 25 lb is required to hold an 80-lb crate on a hill. What angle does the hill make with the horizontal?

38. *Force Needed to Keep a Car Parked* Find the force required to keep a 3000-lb car parked on a hill that makes an angle of 15° with the horizontal.

39. *Force Needed for a Monolith* To build the pyramids in Egypt, it is believed that giant causeways were constructed to transport the building materials to the site. One such causeway is said to have been 3000 ft long, with a slope of about 2.3°. How much force would be required to hold a 60-ton monolith on this causeway?

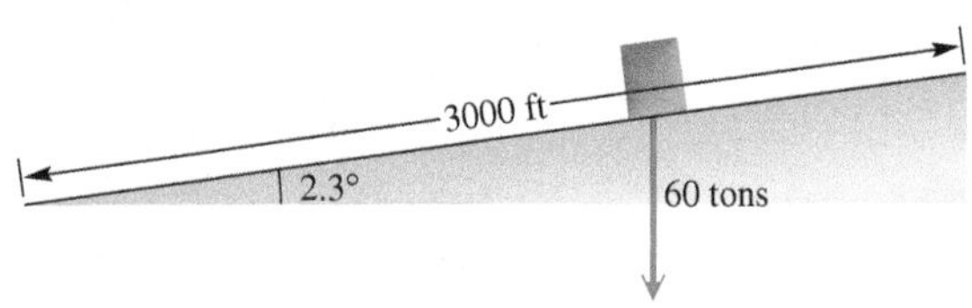

40. *Force Needed for a Monolith* If the causeway in **Exercise 39** were 500 ft longer and the monolith weighed 10 tons more, how much force would be required?

41. *Incline Angle* A force of 18.0 lb is required to hold a 60.0-lb stump grinder on an incline. What angle does the incline make with the horizontal?

42. *Incline Angle* A force of 30.0 lb is required to hold an 80.0-lb pressure washer on an incline. What angle does the incline make with the horizontal?

43. *Weight of a Box* Two people are carrying a box. One person exerts a force of 150 lb at an angle of 62.4° with the horizontal. The other person exerts a force of 114 lb at an angle of 54.9°. Find the weight of the box.

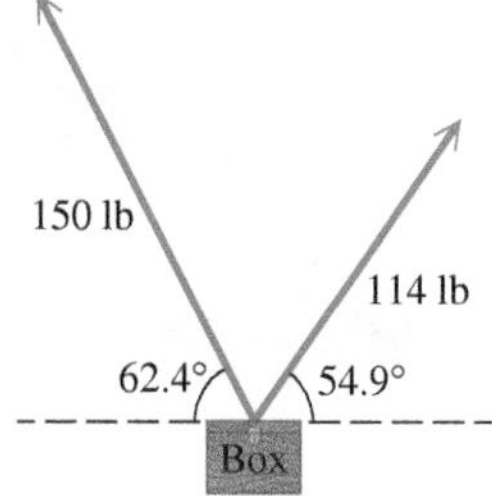

44. *Weight of a Crate and Tension of a Rope* A crate is supported by two ropes. One rope makes an angle of 46° 20′ with the horizontal and has a tension of 89.6 lb on it. The other rope is horizontal. Find the weight of the crate and the tension in the horizontal rope.

Solve each problem. ***See Examples 5 and 6.***

45. *Distance and Bearing of a Ship* A ship leaves port on a bearing of 34.0° and travels 10.4 mi. The ship then turns due east and travels 4.6 mi. How far is the ship from port, and what is its bearing from port?

46. *Distance and Bearing of a Luxury Liner* A luxury liner leaves port on a bearing of 110.0° and travels 8.8 mi. It then turns due west and travels 2.4 mi. How far is the liner from port, and what is its bearing from port?

47. *Distance of a Ship from Its Starting Point* Starting at point A, a ship sails 18.5 km on a bearing of 189°, then turns and sails 47.8 km on a bearing of 317°. Find the distance of the ship from point A.

48. *Distance of a Ship from Its Starting Point* Starting at point X, a ship sails 15.5 km on a bearing of 200°, then turns and sails 2.4 km on a bearing of 320°. Find the distance of the ship from point X.

49. *Distance and Direction of a Motorboat* A motorboat sets out in the direction N 80° 00′ E. The speed of the boat in still water is 20.0 mph. If the current is flowing directly south, and the actual direction of the motorboat is due east, find the speed of the current and the actual speed of the motorboat.

50. *Movement of a Motorboat* Suppose we would like to cross a 132-ft-wide river in a motorboat. Assume that the motorboat can travel at 7.0 mph relative to the water and that the current is flowing west at the rate of 3.0 mph. The bearing θ is chosen so that the motorboat will land at a point exactly across from the starting point.

(a) At what speed will the motorboat be traveling relative to the banks?

(b) How long will it take for the motorboat to make the crossing?

(c) What is the measure of angle θ?

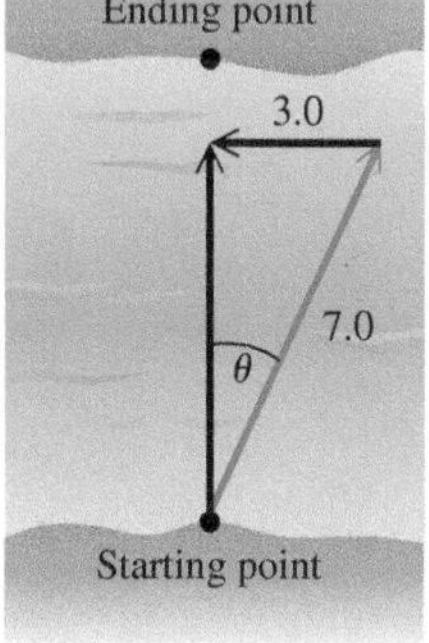

Answers

23. 94.2 lb 24. 158.0 lb
25. 24.4 lb 26. 1286.0 lb
27. 530 newtons
28. 29 newtons
29. 88.2 lb 30. 76.2 lb
31. 2640 lb at an angle of 167.2° with the 1480-lb force
32. 1800 lb at an angle of 167° with the 840-lb force
33. 93.9° 34. 70.1°
35. 190 lb and 283 lb, respectively
36. magnitude of resultant: 117 lb; force: 93.9 lb
37. 18° 38. 800 lb
39. 2.4 tons 40. 2.8 tons
41. 17.5° 42. 22.0°
43. 226 lb
44. weight: 64.8 lb; tension: 61.9 lb
45. 13.5 mi; 50.4°
46. 6.6 mi; 117.1°
47. 39.2 km 48. 14.5 km
49. current: 3.5 mph; motorboat: 19.7 mph
50. (a) 6.3 mph (b) 14 sec (c) 25°

51. bearing: 237°;
ground speed: 470 mph
52. 3:21 P.M.
53. ground speed: 161 mph;
airspeed: 156 mph
54. 173.1°
55. bearing: 74°;
ground speed: 202 mph
56. bearing: 65° 30′;
ground speed: 181 mph
57. bearing: 358°;
airspeed: 170 mph
58. ground speed: 198 mph;
bearing: 186.5°
59. ground speed: 230 km per hr;
bearing: 167°
60. (a) 56 mi per sec
(b) 87 mi per sec

51. ***Bearing and Ground Speed of a Plane*** An airline route from San Francisco to Honolulu is on a bearing of 233.0°. A jet flying at 450 mph on that bearing encounters a wind blowing at 39.0 mph from a direction of 114.0°. Find the resulting bearing and ground speed of the plane.

52. ***Path Traveled by a Plane*** The aircraft carrier *Tallahassee* is traveling at sea on a steady course with a bearing of 30° at 32 mph. Patrol planes on the carrier have enough fuel for 2.6 hr of flight when traveling at a speed of 520 mph. One of the pilots takes off on a bearing of 338° and then turns and heads in a straight line, so as to be able to catch the carrier and land on the deck at the exact instant that his fuel runs out. If the pilot left at 2 P.M., at what time did he turn to head for the carrier?

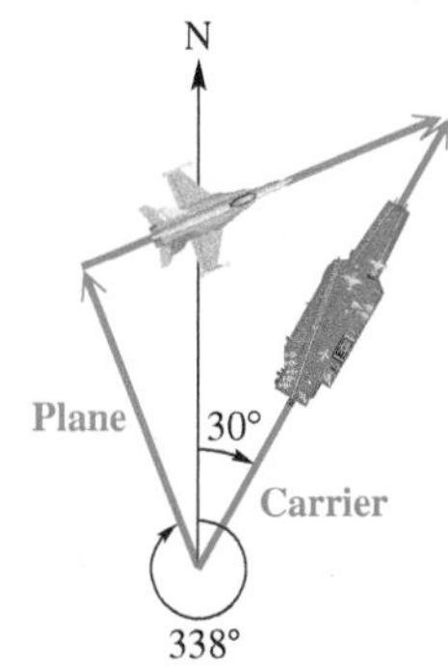

53. ***Airspeed and Ground Speed*** A pilot wants to fly on a bearing of 74.9°. By flying due east, he finds that a 42.0-mph wind, blowing from the south, puts him on course. Find the airspeed and the ground speed.

54. ***Bearing of a Plane*** A plane flies 650 mph on a bearing of 175.3°. A 25-mph wind, from a direction of 266.6°, blows against the plane. Find the resulting bearing of the plane.

55. ***Bearing and Ground Speed of a Plane*** A pilot is flying at 190.0 mph. He wants his flight path to be on a bearing of 64° 30′. A wind is blowing from the south at 35.0 mph. Find the bearing he should fly, and find the plane's ground speed.

56. ***Bearing and Ground Speed of a Plane*** A pilot is flying at 168 mph. She wants her flight path to be on a bearing of 57° 40′. A wind is blowing from the south at 27.1 mph. Find the bearing she should fly, and find the plane's ground speed.

57. ***Bearing and Airspeed of a Plane*** What bearing and airspeed are required for a plane to fly 400 mi due north in 2.5 hr if the wind is blowing from a direction of 328° at 11 mph?

58. ***Ground Speed and Bearing of a Plane*** A plane is headed due south with an airspeed of 192 mph. A wind from a direction of 78.0° is blowing at 23.0 mph. Find the ground speed and resulting bearing of the plane.

59. ***Ground Speed and Bearing of a Plane*** An airplane is headed on a bearing of 174° at an airspeed of 240 km per hr. A 30-km-per-hr wind is blowing from a direction of 245°. Find the ground speed and resulting bearing of the plane.

60. ***Velocity of a Star*** The space velocity $\mathbf{v}$ of a star relative to the sun can be expressed as the resultant vector of two perpendicular vectors—the radial velocity $\mathbf{v}_r$ and the tangential velocity $\mathbf{v}_t$, where $\mathbf{v} = \mathbf{v}_r + \mathbf{v}_t$. If a star is located near the sun and its space velocity is large, then its motion across the sky will also be large. Barnard's Star is a relatively close star with a distance of 35 trillion mi from the sun. It moves across the sky through an angle of 10.34″ per year, which is the largest motion of any known star. Its radial velocity $\mathbf{v}_r$ is 67 mi per sec toward the sun. (*Sources:* Zeilik, M., S. Gregory, and E. Smith, *Introductory Astronomy and Astrophysics,* Second Edition, Saunders College Publishing; Acker, A. and C. Jaschek, *Astronomical Methods and Calculations,* John Wiley and Sons.)

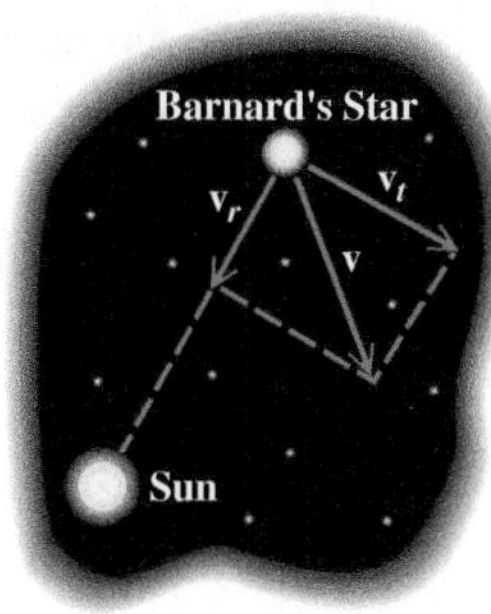

NOT TO SCALE

(a) Approximate the tangential velocity $\mathbf{v}_t$ of Barnard's Star. (*Hint*: Use the arc length formula $s = r\theta$.)

(b) Compute the magnitude of $\mathbf{v}$.

7.5 Algebraically Defined Vectors and the Dot Product

- Algebraic Interpretation of Vectors
- Operations with Vectors
- The Dot Product and the Angle between Vectors

LOOKING AHEAD TO CALCULUS

In addition to two-dimensional vectors in a plane, calculus courses introduce three-dimensional vectors in space. The magnitude of the two-dimensional vector $\langle a, b \rangle$ is given by

$$\sqrt{a^2 + b^2}.$$

If we extend this to the three-dimensional vector $\langle a, b, c \rangle$, the expression becomes

$$\sqrt{a^2 + b^2 + c^2}.$$

Similar extensions are made for other concepts.

Algebraic Interpretation of Vectors

A vector with initial point at the origin in a rectangular coordinate system is a **position vector.** A position vector **u** with endpoint at the point (a, b) is written $\langle a, b \rangle$, so

$$\mathbf{u} = \langle a, b \rangle.$$

This means that every vector in the real plane corresponds to an ordered pair of real numbers. ***Thus, geometrically a vector is a directed line segment while algebraically it is an ordered pair.*** The numbers a and b are the **horizontal component** and the **vertical component,** respectively, of vector **u**.

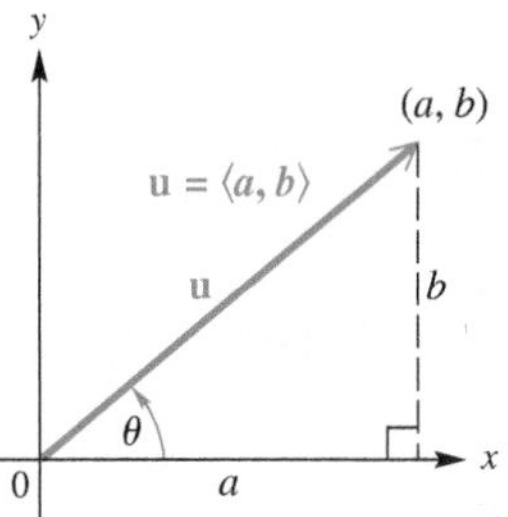

Figure 33

Figure 33 shows the vector $\mathbf{u} = \langle a, b \rangle$. The positive angle between the x-axis and a position vector is the **direction angle** for the vector. In **Figure 33,** θ is the direction angle for vector **u.** The magnitude and direction angle of a vector are related to its horizontal and vertical components.

Magnitude and Direction Angle of a Vector $\langle a, b \rangle$

The magnitude (length) of vector $\mathbf{u} = \langle a, b \rangle$ is given by the following.

$$|\mathbf{u}| = \sqrt{a^2 + b^2}$$

The direction angle θ satisfies $\tan \theta = \frac{b}{a}$, where $a \neq 0$.

Classroom Example 1
Find the magnitude and direction angle for $\mathbf{u} = \langle -5, 4 \rangle$.

Answer: $\sqrt{41}$; 141.3°

EXAMPLE 1 Finding Magnitude and Direction Angle

Find the magnitude and direction angle for $\mathbf{u} = \langle 3, -2 \rangle$.

ALGEBRAIC SOLUTION

The magnitude is $|\mathbf{u}| = \sqrt{3^2 + (-2)^2} = \sqrt{13}$. To find the direction angle θ, start with $\tan \theta = \frac{b}{a} = \frac{-2}{3} = -\frac{2}{3}$. Vector **u** has a positive horizontal component and a negative vertical component, which places the position vector in quadrant IV. A calculator then gives $\tan^{-1}\left(-\frac{2}{3}\right) \approx -33.7°$. Adding 360° yields the direction angle $\theta \approx 326.3°$. See **Figure 34.**

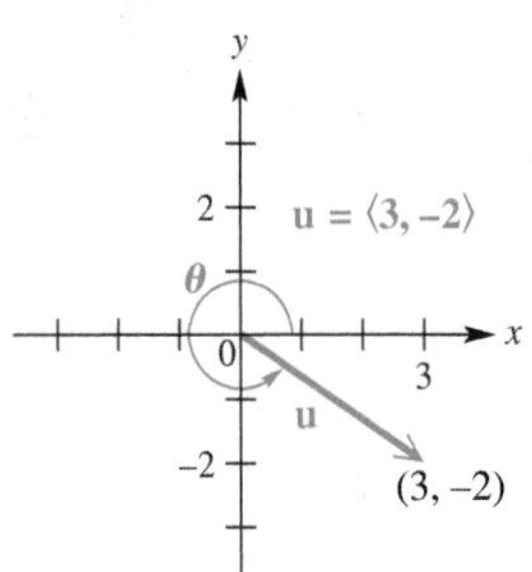

Figure 34

GRAPHING CALCULATOR SOLUTION

The TI-84 Plus calculator can find the magnitude and direction angle using rectangular to polar conversion (which is covered in detail in the next chapter). An approximation for $\sqrt{13}$ is given, and the TI-84 Plus gives the direction angle with the least possible absolute value. We must add 360° to the given value −33.7° to obtain the positive direction angle $\theta \approx 326.3°$.

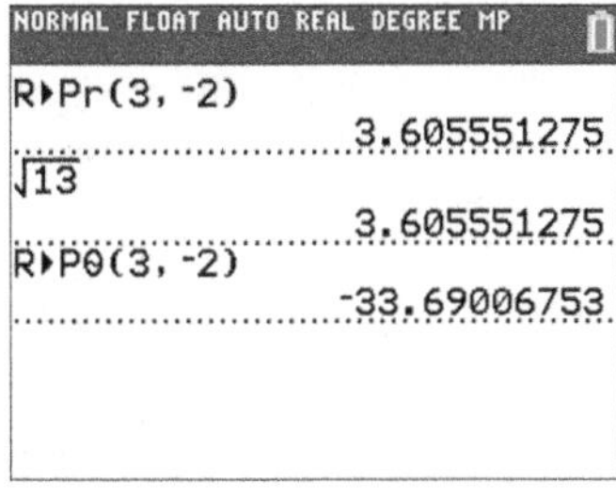

Figure 35

✔ Now Try Exercise 9.

Classroom Example 2
Vector $\mathbf{v}$ has magnitude 14.5 and direction angle 220°. Find the horizontal and vertical components.

Answer: $\langle -11.1, -9.3 \rangle$

Horizontal and Vertical Components

The horizontal and vertical components, respectively, of a vector $\mathbf{u}$ having magnitude $|\mathbf{u}|$ and direction angle θ are the following.

$$a = |\mathbf{u}| \cos \theta \quad \text{and} \quad b = |\mathbf{u}| \sin \theta$$

That is, $\mathbf{u} = \langle a, b \rangle = \langle |\mathbf{u}| \cos \theta, |\mathbf{u}| \sin \theta \rangle$.

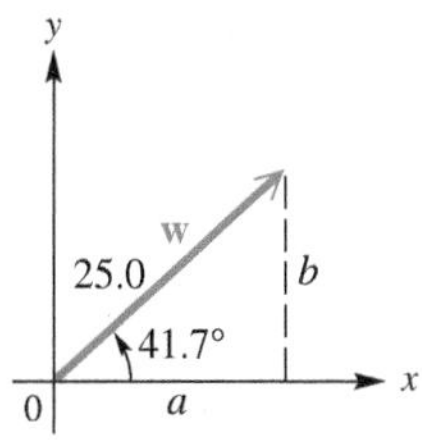

Figure 36

EXAMPLE 2 Finding Horizontal and Vertical Components

Vector $\mathbf{w}$ in **Figure 36** has magnitude 25.0 and direction angle 41.7°. Find the horizontal and vertical components.

ALGEBRAIC SOLUTION

Use the formulas below, with $|\mathbf{w}| = 25.0$ and $\theta = 41.7°$.

$a = \|\mathbf{w}\| \cos \theta$	$b = \|\mathbf{w}\| \sin \theta$
$a = 25.0 \cos 41.7°$	$b = 25.0 \sin 41.7°$
$a \approx 18.7$	$b \approx 16.6$

Therefore, $\mathbf{w} = \langle 18.7, 16.6 \rangle$. The horizontal component is 18.7, and the vertical component is 16.6 (rounded to the nearest tenth).

GRAPHING CALCULATOR SOLUTION

See **Figure 37.** The results support the algebraic solution.

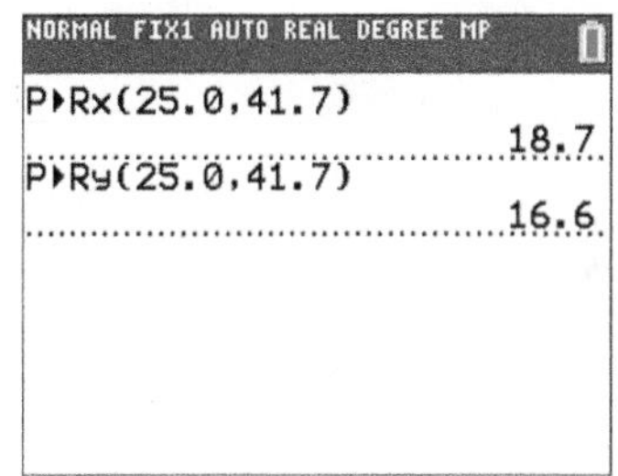

Figure 37

✔ **Now Try Exercise 13.**

EXAMPLE 3 Writing Vectors in the Form $\langle a, b \rangle$

Write each vector in **Figure 38** in the form $\langle a, b \rangle$.

Figure 38

SOLUTION

$$\mathbf{u} = \langle 5 \cos 60°, 5 \sin 60° \rangle = \left\langle 5 \cdot \frac{1}{2}, 5 \cdot \frac{\sqrt{3}}{2} \right\rangle = \left\langle \frac{5}{2}, \frac{5\sqrt{3}}{2} \right\rangle$$

$$\mathbf{v} = \langle 2 \cos 180°, 2 \sin 180° \rangle = \langle 2(-1), 2(0) \rangle = \langle -2, 0 \rangle$$

$$\mathbf{w} = \langle 6 \cos 280°, 6 \sin 280° \rangle \approx \langle 1.0419, -5.9088 \rangle \quad \text{Use a calculator.}$$

✔ **Now Try Exercises 19 and 21.**

Classroom Example 3
Write each vector in the form $\langle a, b \rangle$.
- $\mathbf{u}$: magnitude 8, direction angle 135°
- $\mathbf{v}$: magnitude 4, direction angle 270°
- $\mathbf{w}$: magnitude 10, direction angle 340°

Answer:
$\mathbf{u} = \langle -4\sqrt{2}, 4\sqrt{2} \rangle$;
$\mathbf{v} = \langle 0, -4 \rangle$;
$\mathbf{w} \approx \langle 9.3969, -3.4202 \rangle$

Operations with Vectors As shown in **Figure 39,**

$$\mathbf{m} = \langle a, b \rangle, \quad \mathbf{n} = \langle c, d \rangle, \quad \text{and} \quad \mathbf{p} = \langle a + c, b + d \rangle.$$

Using geometry, we can show that the endpoints of the three vectors and the origin form a parallelogram. A diagonal of this parallelogram gives the resultant of $\mathbf{m}$ and $\mathbf{n}$, so we have $\mathbf{p} = \mathbf{m} + \mathbf{n}$ or

$$\langle a + c, b + d \rangle = \langle a, b \rangle + \langle c, d \rangle.$$

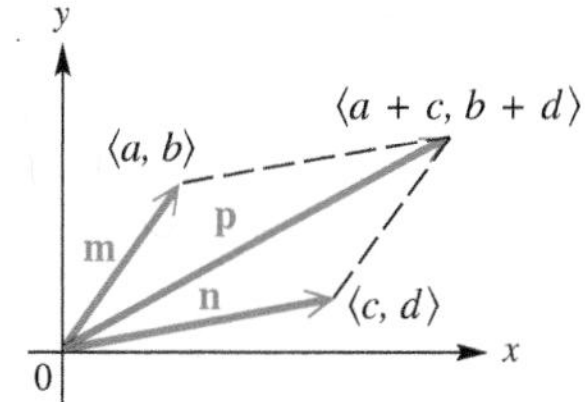

Figure 39

Similarly, we can verify the following operations.

Vector Operations

Let a, b, c, d, and k represent real numbers.

$$\langle a, b\rangle + \langle c, d\rangle = \langle a + c, b + d\rangle$$

$$k \cdot \langle a, b\rangle = \langle ka, kb\rangle$$

$$\text{If } \mathbf{u} = \langle a_1, a_2\rangle, \text{ then } -\mathbf{u} = \langle -a_1, -a_2\rangle.$$

$$\langle a, b\rangle - \langle c, d\rangle = \langle a, b\rangle + (-\langle c, d\rangle) = \langle a - c, b - d\rangle$$

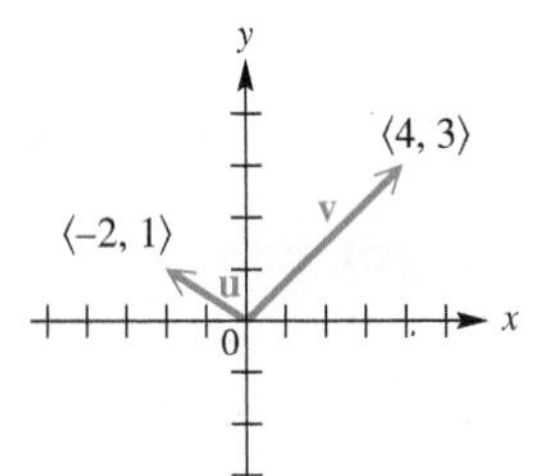

Figure 40

EXAMPLE 4 Performing Vector Operations

Let $\mathbf{u} = \langle -2, 1\rangle$ and $\mathbf{v} = \langle 4, 3\rangle$. See **Figure 40.** Find and illustrate each of the following.

(a) $\mathbf{u} + \mathbf{v}$ **(b)** $-2\mathbf{u}$ **(c)** $3\mathbf{u} - 2\mathbf{v}$

SOLUTION See **Figure 41.**

(a) $\mathbf{u} + \mathbf{v}$

$= \langle -2, 1\rangle + \langle 4, 3\rangle$

$= \langle -2 + 4, 1 + 3\rangle$

$= \langle 2, 4\rangle$

(b) $-2\mathbf{u}$

$= -2 \cdot \langle -2, 1\rangle$

$= \langle -2(-2), -2(1)\rangle$

$= \langle 4, -2\rangle$

(c) $3\mathbf{u} - 2\mathbf{v}$

$= 3 \cdot \langle -2, 1\rangle - 2 \cdot \langle 4, 3\rangle$

$= \langle -6, 3\rangle - \langle 8, 6\rangle$

$= \langle -6 - 8, 3 - 6\rangle$

$= \langle -14, -3\rangle$

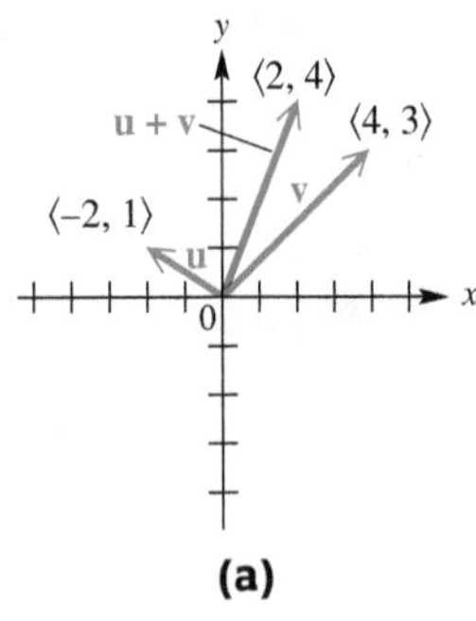

(a)

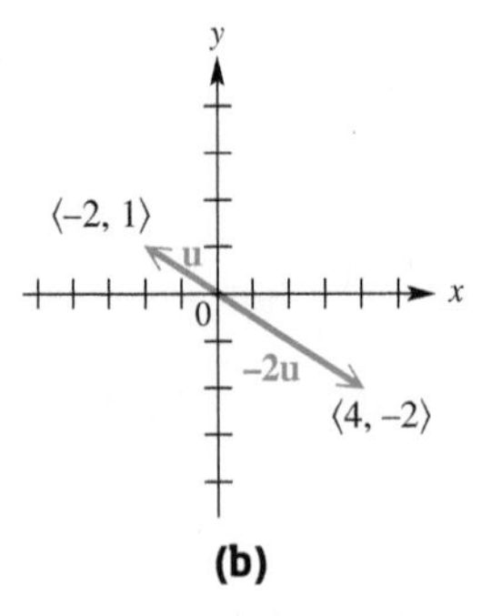

(b)

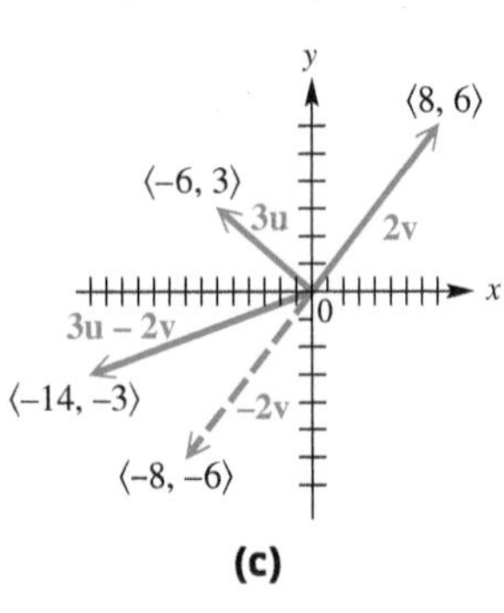

(c)

Figure 41

✔ **Now Try Exercises 35, 37, and 39.**

Classroom Example 4
Let $\mathbf{u} = \langle 6, -3\rangle$ and $\mathbf{v} = \langle -14, 8\rangle$. Find the following.

(a) $\mathbf{u} + \mathbf{v}$ (b) $-\frac{1}{2}\mathbf{v}$

(c) $5\mathbf{u} + 2\mathbf{v}$

Answers:

(a) $\langle -8, 5\rangle$ (b) $\langle 7, -4\rangle$

(c) $\langle 2, 1\rangle$

A **unit vector** is a vector that has magnitude 1. Two very important unit vectors are defined as follows and shown in **Figure 42(a).**

$$\mathbf{i} = \langle 1, 0\rangle \qquad \mathbf{j} = \langle 0, 1\rangle$$

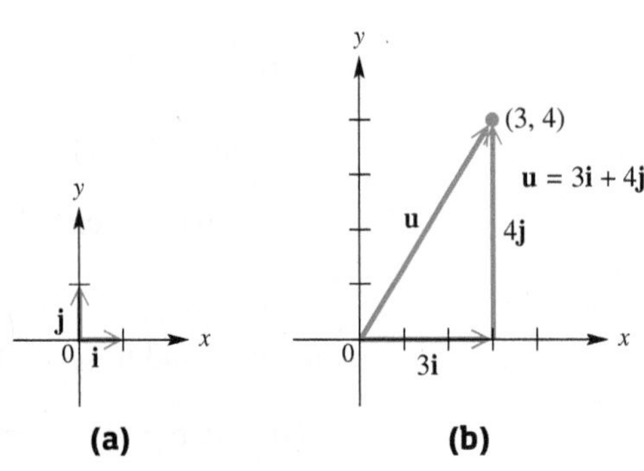

(a) **(b)**

Figure 42

Teaching Tip Point out that the boldface type for unit vectors helps distinguish the use of **i** and **j** in vector notation from the use of *i* and *j* with complex numbers.

With the unit vectors **i** and **j**, we can express any other vector $\langle a, b \rangle$ in the form $a\mathbf{i} + b\mathbf{j}$, as shown in **Figure 42(b),** where $\langle 3, 4 \rangle = 3\mathbf{i} + 4\mathbf{j}$. The vector operations previously given can be restated, using $a\mathbf{i} + b\mathbf{j}$ notation.

i, j Form for Vectors

If $\mathbf{v} = \langle a, b \rangle$, then

$$\mathbf{v} = a\mathbf{i} + b\mathbf{j}, \quad \text{where } \mathbf{i} = \langle 1, 0 \rangle \text{ and } \mathbf{j} = \langle 0, 1 \rangle.$$

The Dot Product and the Angle between Vectors ***The* dot product *of two vectors is a real number, not a vector.*** It is also known as the *inner product.* Dot products are used to determine the angle between two vectors, to derive geometric theorems, and to solve physics problems.

Dot Product

The **dot product** of the two vectors $\mathbf{u} = \langle a, b \rangle$ and $\mathbf{v} = \langle c, d \rangle$ is denoted $\mathbf{u} \cdot \mathbf{v}$, read "**u** dot **v**," and given by the following.

$$\mathbf{u} \cdot \mathbf{v} = ac + bd$$

That is, the dot product of two vectors is the sum of the product of their first components and the product of their second components.

Classroom Example 5
Find each dot product.
(a) $\langle -5, 2 \rangle \cdot \langle -2, -4 \rangle$
(b) $\langle 10, -1 \rangle \cdot \langle 1, 10 \rangle$

Answers:
(a) 2 (b) 0

EXAMPLE 5 Finding Dot Products

Find each dot product.

(a) $\langle 2, 3 \rangle \cdot \langle 4, -1 \rangle$ **(b)** $\langle 6, 4 \rangle \cdot \langle -2, 3 \rangle$

SOLUTION

(a) $\langle 2, 3 \rangle \cdot \langle 4, -1 \rangle$
$= 2(4) + 3(-1)$
$= 5$

(b) $\langle 6, 4 \rangle \cdot \langle -2, 3 \rangle$
$= 6(-2) + 4(3)$
$= 0$

✔ **Now Try Exercises 47 and 49.**

The following properties of dot products can be verified using the definitions presented so far.

Properties of the Dot Product

For all vectors **u**, **v**, and **w** and real numbers k, the following hold.

(a) $\mathbf{u} \cdot \mathbf{v} = \mathbf{v} \cdot \mathbf{u}$ **(b)** $\mathbf{u} \cdot (\mathbf{v} + \mathbf{w}) = \mathbf{u} \cdot \mathbf{v} + \mathbf{u} \cdot \mathbf{w}$

(c) $(\mathbf{u} + \mathbf{v}) \cdot \mathbf{w} = \mathbf{u} \cdot \mathbf{w} + \mathbf{v} \cdot \mathbf{w}$ **(d)** $(k\mathbf{u}) \cdot \mathbf{v} = k(\mathbf{u} \cdot \mathbf{v}) = \mathbf{u} \cdot (k\mathbf{v})$

(e) $\mathbf{0} \cdot \mathbf{u} = 0$ **(f)** $\mathbf{u} \cdot \mathbf{u} = |\mathbf{u}|^2$

For example, to prove the first part of property (d),

$$(k\mathbf{u}) \cdot \mathbf{v} = k(\mathbf{u} \cdot \mathbf{v}),$$

we let $\mathbf{u} = \langle a, b \rangle$ and $\mathbf{v} = \langle c, d \rangle$.

$$\begin{aligned}
(k\mathbf{u}) \cdot \mathbf{v} &= \left(k\langle a, b \rangle\right) \cdot \langle c, d \rangle && \text{Substitute.} \\
&= \langle ka, kb \rangle \cdot \langle c, d \rangle && \text{Multiply by scalar } k. \\
&= kac + kbd && \text{Dot product} \\
&= k(ac + bd) && \text{Distributive property} \\
&= k\left(\langle a, b \rangle \cdot \langle c, d \rangle\right) && \text{Dot product} \\
&= k(\mathbf{u} \cdot \mathbf{v}) && \text{Substitute.}
\end{aligned}$$

The proofs of the remaining properties are similar.

The dot product of two vectors can be positive, 0, or negative. A geometric interpretation of the dot product explains when each of these cases occurs. This interpretation involves the angle between the two vectors.

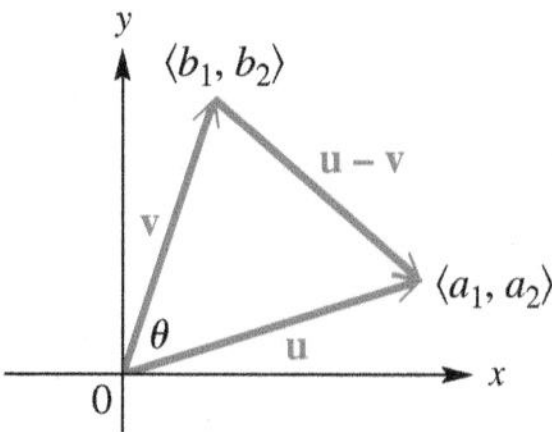

Figure 43

Consider the two vectors $\mathbf{u} = \langle a_1, a_2 \rangle$ and $\mathbf{v} = \langle b_1, b_2 \rangle$, as shown in **Figure 43.** The **angle θ between u and v** is defined to be the angle having the two vectors as its sides for which $0° \le \theta \le 180°$.

We can use the law of cosines to develop a formula to find angle θ in **Figure 43.**

$$|\mathbf{u} - \mathbf{v}|^2 = |\mathbf{u}|^2 + |\mathbf{v}|^2 - 2|\mathbf{u}||\mathbf{v}| \cos \theta$$

Law of cosines applied to **Figure 43**

$$\left(\sqrt{(a_1 - b_1)^2 + (a_2 - b_2)^2}\right)^2 = \left(\sqrt{a_1^2 + a_2^2}\right)^2 + \left(\sqrt{b_1^2 + b_2^2}\right)^2 - 2|\mathbf{u}||\mathbf{v}| \cos \theta$$

Magnitude of a vector

$$a_1^2 - 2a_1b_1 + b_1^2 + a_2^2 - 2a_2b_2 + b_2^2 = a_1^2 + a_2^2 + b_1^2 + b_2^2 - 2|\mathbf{u}||\mathbf{v}| \cos \theta \quad \text{Square.}$$

$$-2a_1b_1 - 2a_2b_2 = -2|\mathbf{u}||\mathbf{v}| \cos \theta \quad \text{Subtract like terms from each side.}$$

$$a_1b_1 + a_2b_2 = |\mathbf{u}||\mathbf{v}| \cos \theta \quad \text{Divide by } -2.$$

$$\mathbf{u} \cdot \mathbf{v} = |\mathbf{u}||\mathbf{v}| \cos \theta \quad \text{Definition of dot product}$$

$$\cos \theta = \frac{\mathbf{u} \cdot \mathbf{v}}{|\mathbf{u}||\mathbf{v}|} \quad \text{Divide by } |\mathbf{u}||\mathbf{v}| \text{ and rewrite.}$$

Geometric Interpretation of Dot Product

If θ is the angle between the two nonzero vectors **u** and **v**, where $0° \le \theta \le 180°$, then the following holds.

$$\cos \boldsymbol{\theta} = \frac{\mathbf{u} \cdot \mathbf{v}}{|\mathbf{u}||\mathbf{v}|}$$

Classroom Example 6
Find the angle θ between the two vectors.
(a) $\mathbf{u} = \langle 5, -12 \rangle$ and $\mathbf{v} = \langle 4, 3 \rangle$
(b) $\mathbf{u} = \langle 4, -3 \rangle$ and $\mathbf{v} = \langle 6, 8 \rangle$

Answers:
(a) $\theta \approx 104.25°$ (b) $\theta = 90°$

EXAMPLE 6 Finding the Angle between Two Vectors

Find the angle θ between the two vectors.

(a) $\mathbf{u} = \langle 3, 4 \rangle$ and $\mathbf{v} = \langle 2, 1 \rangle$ **(b)** $\mathbf{u} = \langle 2, -6 \rangle$ and $\mathbf{v} = \langle 6, 2 \rangle$

SOLUTION

(a)

$$\cos\theta = \frac{\mathbf{u} \cdot \mathbf{v}}{|\mathbf{u}||\mathbf{v}|}$$ Geometric interpretation of the dot product

$$\cos\theta = \frac{\langle 3, 4 \rangle \cdot \langle 2, 1 \rangle}{|\langle 3, 4 \rangle||\langle 2, 1 \rangle|}$$ Substitute values.

$$\cos\theta = \frac{3(2) + 4(1)}{\sqrt{9 + 16} \cdot \sqrt{4 + 1}}$$ Use the definitions.

$$\cos\theta = \frac{10}{5\sqrt{5}}$$ Simplify.

$$\cos\theta \approx 0.894427191$$ Use a calculator.

$$\theta \approx 26.57°$$ Use the inverse cosine function.

(b)

$$\cos\theta = \frac{\mathbf{u} \cdot \mathbf{v}}{|\mathbf{u}||\mathbf{v}|}$$ Geometric interpretation of the dot product

$$\cos\theta = \frac{\langle 2, -6 \rangle \cdot \langle 6, 2 \rangle}{|\langle 2, -6 \rangle||\langle 6, 2 \rangle|}$$ Substitute values.

$$\cos\theta = \frac{2(6) + (-6)(2)}{\sqrt{4 + 36} \cdot \sqrt{36 + 4}}$$ Use the definitions.

$$\cos\theta = 0$$ Evaluate. The numerator is equal to 0.

$$\theta = 90°$$ $\cos^{-1} 0 = 90°$

✔ **Now Try Exercises 53 and 55.**

For angles θ between 0° and 180°, cos θ is positive, 0, or negative when θ is less than, equal to, or greater than 90°, respectively. Therefore, the dot product of nonzero vectors is positive, 0, or negative according to this table.

Dot Product	Angle between Vectors
Positive	Acute
0	Right
Negative	Obtuse

Thus, in **Example 6,** the vectors in part (a) form an acute angle, and those in part (b) form a right angle. If $\mathbf{u} \cdot \mathbf{v} = 0$ for two nonzero vectors $\mathbf{u}$ and $\mathbf{v}$, then $\cos\theta = 0$ and $\theta = 90°$. Thus, $\mathbf{u}$ and $\mathbf{v}$ are perpendicular vectors, also called **orthogonal vectors.** See **Figure 44.**

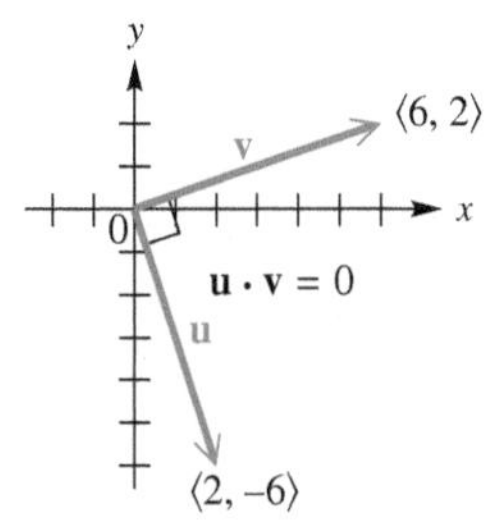

Orthogonal vectors

Figure 44

7.5 Exercises

1. 2 2. 30°
3. $\frac{\sqrt{2}}{2}$ 4. $\frac{\sqrt{2}}{2}$
5. $\langle 4, 9 \rangle$ 6. $4\mathbf{i} - 2\mathbf{j}$
7. $ac + bd$ 8. acute

9. 17; 331.9° 10. 25; 106.3°
11. 8; 120° 12. 16; 315°

13. 47, 17 14. 17, 20
15. 38.8, 28.0 16. 13.7, 7.11
17. −123, 155 18. −198, 132

19. $\left\langle \frac{5\sqrt{3}}{2}, \frac{5}{2} \right\rangle$ 20. $\langle 4, 4\sqrt{3} \rangle$
21. $\langle -3.0642, 2.5712 \rangle$

CONCEPT PREVIEW *Fill in the blank to correctly complete each sentence.*

1. The magnitude of vector **u** is ________.
2. The direction angle of vector **u** is ________.

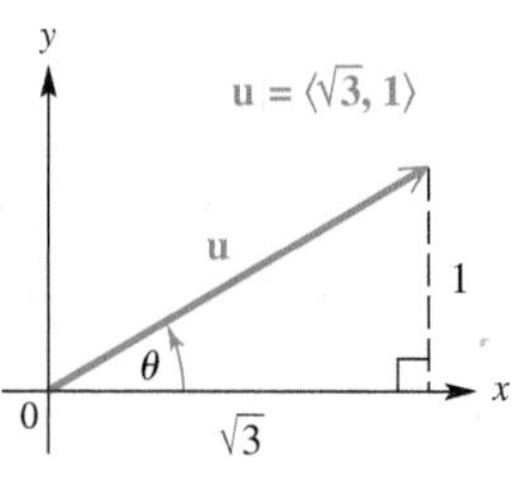

3. The horizontal component, a, of vector **v** is ________.
4. The vertical component, b, of vector **v** is ________.

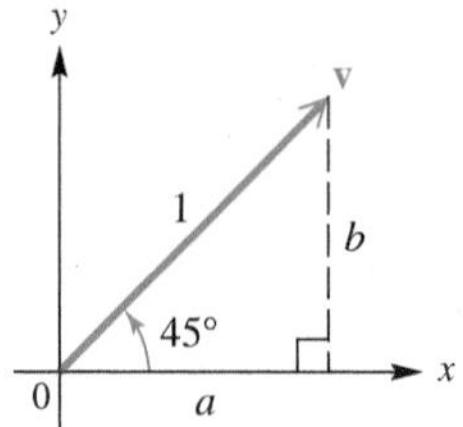

5. The sum of the vectors $\mathbf{u} = \langle -3, 5 \rangle$ and $\mathbf{v} = \langle 7, 4 \rangle$ is $\mathbf{u} + \mathbf{v} =$ ________.
6. The vector $\mathbf{u} = \langle 4, -2 \rangle$ is written in **i**, **j** form as ________.
7. The formula for the dot product of the two vectors $\mathbf{u} = \langle a, b \rangle$ and $\mathbf{v} = \langle c, d \rangle$ is

$$\mathbf{u} \cdot \mathbf{v} = \_\_\_\_\_\_\_\_.$$

8. If the dot product of two vectors is a positive number, then the angle between them is ________.
   (acute/obtuse)

Find the magnitude and direction angle for each vector. ***See Example 1.***

9. $\langle 15, -8 \rangle$ 10. $\langle -7, 24 \rangle$

11. $\langle -4, 4\sqrt{3} \rangle$ 12. $\langle 8\sqrt{2}, -8\sqrt{2} \rangle$

Vector **v** *has the given direction and magnitude. Find the horizontal and vertical components of* **v**, *if θ is the direction angle of* **v** *from the horizontal.* ***See Example 2.***

13. $\theta = 20°, \quad |\mathbf{v}| = 50$ 14. $\theta = 50°, \quad |\mathbf{v}| = 26$

15. $\theta = 35° \, 50', \quad |\mathbf{v}| = 47.8$ 16. $\theta = 27° \, 30', \quad |\mathbf{v}| = 15.4$

17. $\theta = 128.5°, \quad |\mathbf{v}| = 198$ 18. $\theta = 146.3°, \quad |\mathbf{v}| = 238$

Write each vector in the form $\langle a, b \rangle$. Round to four decimal places as applicable. ***See Example 3.***

19. 

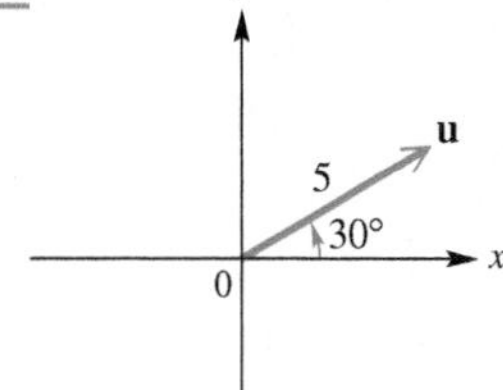

20.

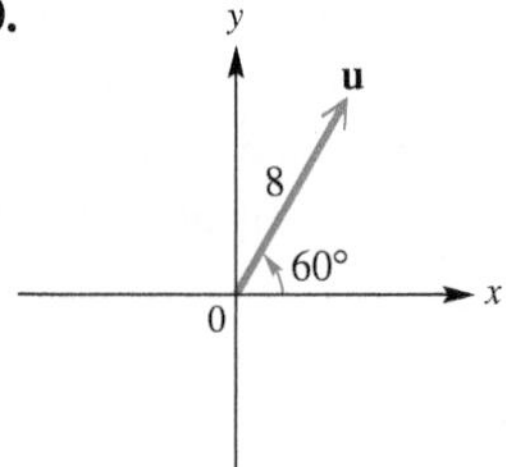

21.

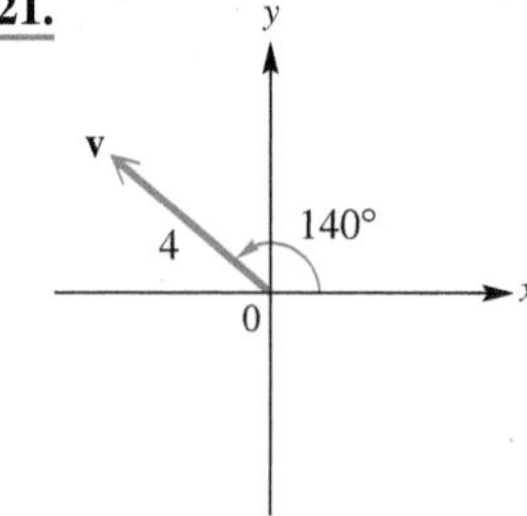

22. $\langle -1.9284, 2.2981 \rangle$
23. $\langle 4.0958, -2.8679 \rangle$
24. $\langle -1.5321, -1.2856 \rangle$

25. (a) $\langle -4, 16 \rangle$ (b) $\langle -12, 0 \rangle$ (c) $\langle 8, -8 \rangle$
26. (a) $\langle -4, -8 \rangle$ (b) $\langle 12, 0 \rangle$ (c) $\langle -4, 4 \rangle$
27. (a) $\langle 8, 0 \rangle$ (b) $\langle 0, 16 \rangle$ (c) $\langle -4, -8 \rangle$
28. (a) $\langle 4, 0 \rangle$ (b) $\langle -12, -8 \rangle$ (c) $\langle 4, 4 \rangle$
29. (a) $\langle 0, 12 \rangle$ (b) $\langle -16, -4 \rangle$ (c) $\langle 8, -4 \rangle$
30. (a) $\langle 4, 4 \rangle$ (b) $\langle 12, -12 \rangle$ (c) $\langle -8, 4 \rangle$

31. (a) $4\mathbf{i}$ (b) $7\mathbf{i} + 3\mathbf{j}$ (c) $-5\mathbf{i} + \mathbf{j}$
32. (a) $-2\mathbf{i} + 4\mathbf{j}$ (b) $\mathbf{i} + \mathbf{j}$ (c) $4\mathbf{i} - 7\mathbf{j}$
33. (a) $\langle -2, 4 \rangle$ (b) $\langle 7, 4 \rangle$ (c) $\langle 6, -6 \rangle$
34. (a) $\langle -4, -2 \rangle$ (b) $\langle -13, 4 \rangle$ (c) $\langle 3, 5 \rangle$

35. $\langle -6, 2 \rangle$ 36. $\langle 6, -2 \rangle$
37. $\langle 8, -20 \rangle$ 38. $\langle -20, -15 \rangle$
39. $\langle -30, -3 \rangle$ 40. $\langle 20, 2 \rangle$
41. $\langle 8, -7 \rangle$ 42. $\langle -24, -5 \rangle$

43. $-5\mathbf{i} + 8\mathbf{j}$ 44. $6\mathbf{i} - 3\mathbf{j}$
45. $2\mathbf{i}$, or $2\mathbf{i} + 0\mathbf{j}$
46. $-4\mathbf{j}$, or $0\mathbf{i} - 4\mathbf{j}$

47. 7 48. -61
49. 0 50. 0
51. 20 52. -4

53. 135° 54. 36.87°
55. 90° 56. 45°
57. 36.87° 58. 78.93°

59. -6 60. -6
61. -24 62. -24

22.

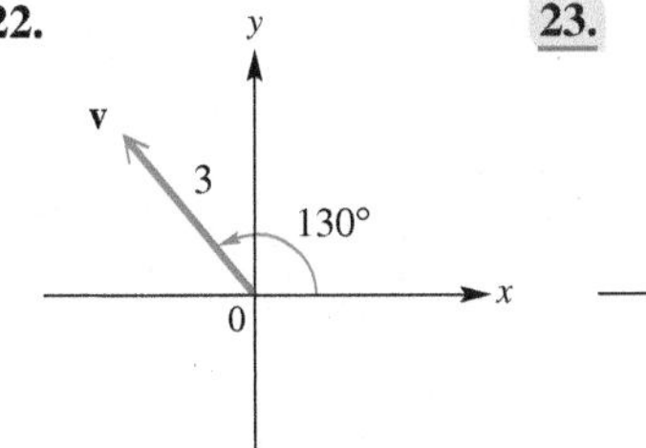

23.

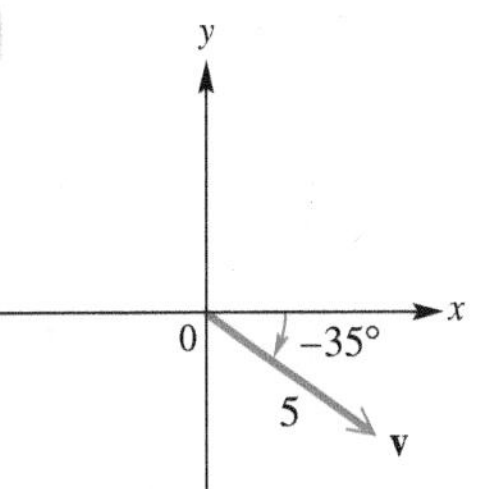

24.

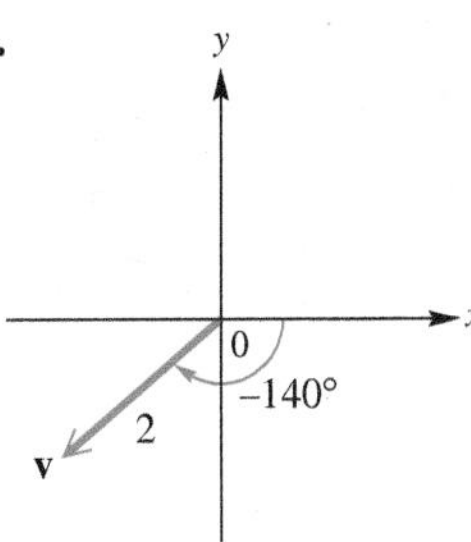

Use the figure to find each vector: ***(a)*** $\mathbf{u} + \mathbf{v}$ ***(b)*** $\mathbf{u} - \mathbf{v}$ ***(c)*** $-\mathbf{u}$. *Use vector notation as in* ***Example 4.***

25.

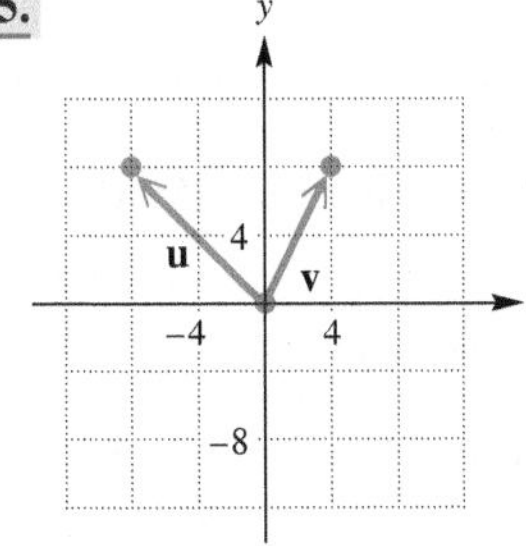

26.

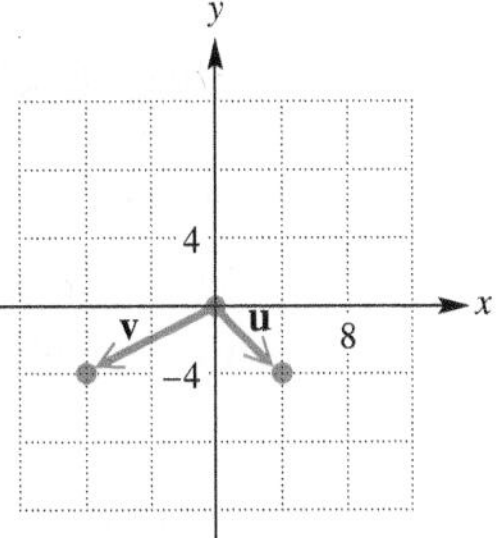

27.

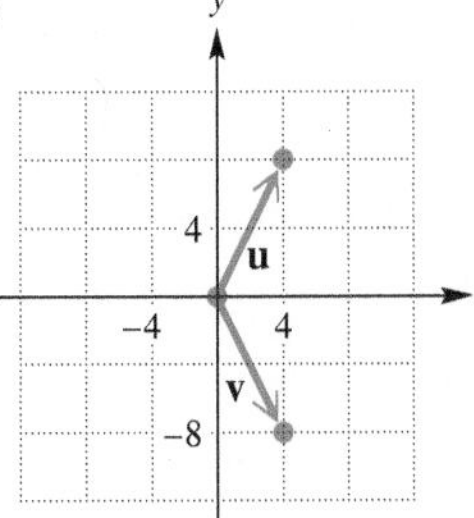

28.

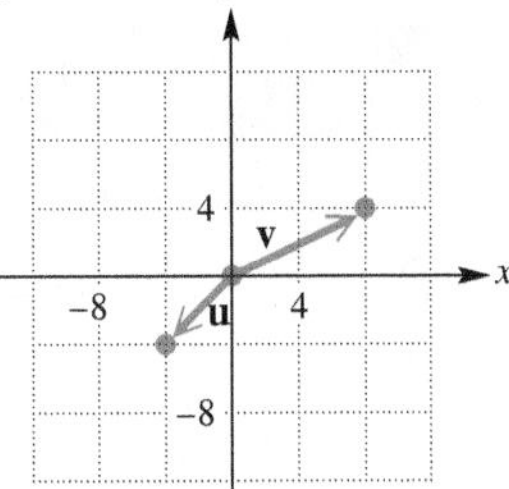

29.

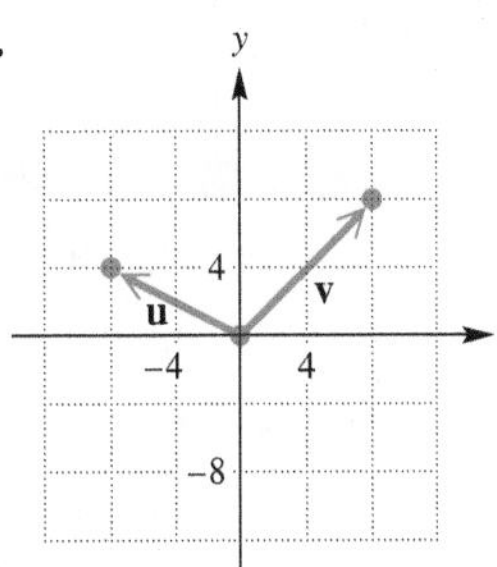

30.

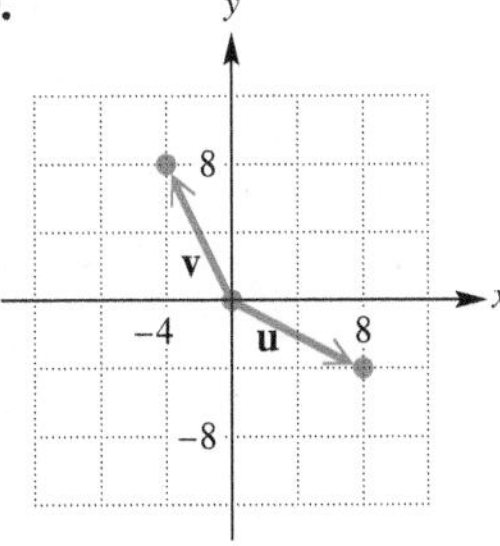

Given vectors **u** *and* **v**, *find:* ***(a)*** $2\mathbf{u}$ ***(b)*** $2\mathbf{u} + 3\mathbf{v}$ ***(c)*** $\mathbf{v} - 3\mathbf{u}$. ***See Example 4.***

31. $\mathbf{u} = 2\mathbf{i}, \quad \mathbf{v} = \mathbf{i} + \mathbf{j}$
32. $\mathbf{u} = -\mathbf{i} + 2\mathbf{j}, \quad \mathbf{v} = \mathbf{i} - \mathbf{j}$
33. $\mathbf{u} = \langle -1, 2 \rangle, \quad \mathbf{v} = \langle 3, 0 \rangle$
34. $\mathbf{u} = \langle -2, -1 \rangle, \quad \mathbf{v} = \langle -3, 2 \rangle$

Given $\mathbf{u} = \langle -2, 5 \rangle$ *and* $\mathbf{v} = \langle 4, 3 \rangle$, *find each of the following.* ***See Example 4.***

35. $\mathbf{u} - \mathbf{v}$ **36.** $\mathbf{v} - \mathbf{u}$ **37.** $-4\mathbf{u}$ **38.** $-5\mathbf{v}$
39. $3\mathbf{u} - 6\mathbf{v}$ **40.** $-2\mathbf{u} + 4\mathbf{v}$ **41.** $\mathbf{u} + \mathbf{v} - 3\mathbf{u}$ **42.** $2\mathbf{u} + \mathbf{v} - 6\mathbf{v}$

Write each vector in the form $a\mathbf{i} + b\mathbf{j}$.

43. $\langle -5, 8 \rangle$ **44.** $\langle 6, -3 \rangle$ **45.** $\langle 2, 0 \rangle$ **46.** $\langle 0, -4 \rangle$

Find the dot product for each pair of vectors. ***See Example 5.***

47. $\langle 6, -1 \rangle, \langle 2, 5 \rangle$ **48.** $\langle -3, 8 \rangle, \langle 7, -5 \rangle$ **49.** $\langle 5, 2 \rangle, \langle -4, 10 \rangle$
50. $\langle 7, -2 \rangle, \langle 4, 14 \rangle$ **51.** $4\mathbf{i}, 5\mathbf{i} - 9\mathbf{j}$ **52.** $2\mathbf{i} + 4\mathbf{j}, -\mathbf{j}$

Find the angle between each pair of vectors. ***See Example 6.***

53. $\langle 2, 1 \rangle, \langle -3, 1 \rangle$ **54.** $\langle 1, 7 \rangle, \langle 1, 1 \rangle$ **55.** $\langle 1, 2 \rangle, \langle -6, 3 \rangle$
56. $\langle 4, 0 \rangle, \langle 2, 2 \rangle$ **57.** $3\mathbf{i} + 4\mathbf{j}, \mathbf{j}$ **58.** $-5\mathbf{i} + 12\mathbf{j}, 3\mathbf{i} + 2\mathbf{j}$

Let $\mathbf{u} = \langle -2, 1 \rangle$, $\mathbf{v} = \langle 3, 4 \rangle$, *and* $\mathbf{w} = \langle -5, 12 \rangle$. *Evaluate each expression.*

59. $(3\mathbf{u}) \cdot \mathbf{v}$ **60.** $\mathbf{u} \cdot (3\mathbf{v})$ **61.** $\mathbf{u} \cdot \mathbf{v} - \mathbf{u} \cdot \mathbf{w}$ **62.** $\mathbf{u} \cdot (\mathbf{v} - \mathbf{w})$

63. orthogonal
64. orthogonal
65. not orthogonal
66. not orthogonal
67. not orthogonal
68. not orthogonal

69. (a) $|\mathbf{R}| = \sqrt{5} \approx 2.2$, $|\mathbf{A}| = \sqrt{1.25} \approx 1.1$; 2.2 in. of rain fell. The area of the opening of the rain gauge is 1.1 in.^{2}.
(b) $V = 1.5$; The volume of rain was 1.5 in.^{3}.

70. **R** and **A** should be parallel and point in opposite directions.

In Exercises 71–75, answers may vary due to rounding.

71. magnitude: 9.5208; direction angle: 119.0647°
72. $\langle -4.1042, 11.2763 \rangle$
73. $\langle -0.5209, -2.9544 \rangle$
74. $\langle -4.6252, 8.3219 \rangle$
75. magnitude: 9.5208; direction angle: 119.0647°
76. They are the same. Preference of method is an individual choice.

Determine whether each pair of vectors is orthogonal. ***See Example 6(b).***

63. $\langle 1, 2 \rangle, \langle -6, 3 \rangle$

64. $\langle 1, 1 \rangle, \langle 1, -1 \rangle$

65. $\langle 1, 0 \rangle, \langle \sqrt{2}, 0 \rangle$

66. $\langle 3, 4 \rangle, \langle 6, 8 \rangle$

67. $\sqrt{5}\,\mathbf{i} - 2\mathbf{j},\ -5\mathbf{i} + 2\sqrt{5}\,\mathbf{j}$

68. $-4\mathbf{i} + 3\mathbf{j},\ 8\mathbf{i} - 6\mathbf{j}$

69. *(Modeling) Measuring Rainfall* Suppose that vector **R** models the amount of rainfall in inches and the direction it falls, and vector **A** models the area in square inches and the orientation of the opening of a rain gauge, as illustrated in the figure. The total volume V of water collected in the rain gauge is given by

$$V = |\mathbf{R} \cdot \mathbf{A}|.$$

This formula calculates the volume of water collected even if the wind is blowing the rain in a slanted direction or the rain gauge is not exactly vertical. Let $\mathbf{R} = \mathbf{i} - 2\mathbf{j}$ and $\mathbf{A} = 0.5\mathbf{i} + \mathbf{j}$.

(a) Find $|\mathbf{R}|$ and $|\mathbf{A}|$ to the nearest tenth. Interpret the results.

(b) Calculate V to the nearest tenth, and interpret this result.

70. *Concept Check* In **Exercise 69,** for the rain gauge to collect the maximum amount of water, what should be true about vectors **R** and **A**?

Relating Concepts

For individual or collaborative investigation *(Exercises 71–76)*

Consider the two vectors **u** *and* **v** *shown. Assume all values are exact.* ***Work Exercises 71–76 in order.***

71. Use trigonometry alone (without using vector notation) to find the magnitude and direction angle of **u** + **v**. Use the law of cosines and the law of sines in your work.

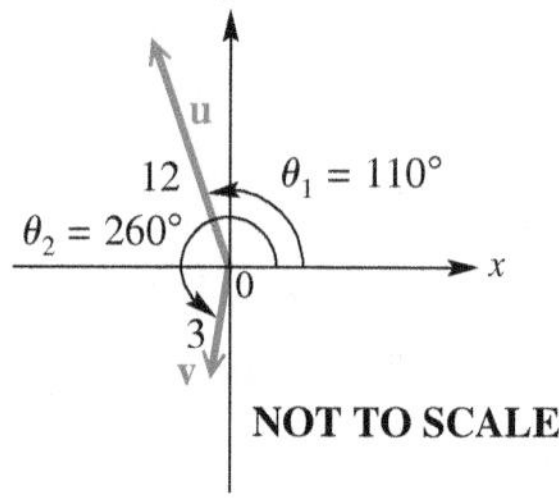

72. Find the horizontal and vertical components of **u**, using a calculator.

73. Find the horizontal and vertical components of **v**, using a calculator.

74. Find the horizontal and vertical components of **u** + **v** by adding the results obtained in **Exercises 72 and 73.**

75. Use a calculator to find the magnitude and direction angle of the vector **u** + **v**.

76. Compare the answers in **Exercises 71 and 75.** What do you notice? Which method of solution do you prefer?

Summary Exercises on Applications of Trigonometry and Vectors

1. 29 ft; 38 ft

These summary exercises provide practice with applications that involve solving triangles and using vectors.

1. *Wires Supporting a Flagpole* A flagpole stands vertically on a hillside that makes an angle of 20° with the horizontal. Two supporting wires are attached as shown in the figure. What are the lengths of the supporting wires?

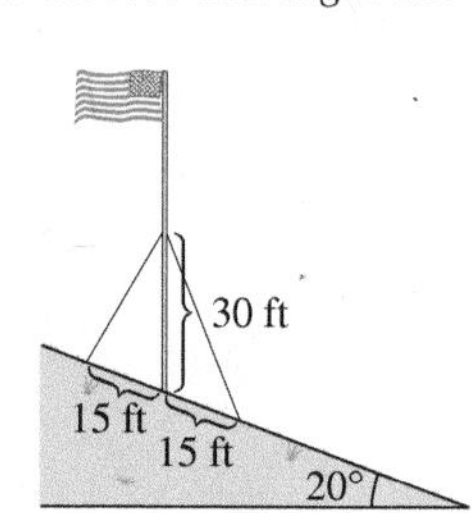

2. 38.3 cm
3. 5856 m
4. 15.8 ft per sec; 71.6°
5. 42 lb
6. 7200 ft
7. (a) 10 mph
(b) $3\mathbf{v} = 18\mathbf{i} + 24\mathbf{j}$; This represents a 30-mph wind in the direction of **v**.
(c) **u** represents a southeast wind of $\sqrt{128} \approx 11.3$ mph.
8. 380 mph; 64°
9. It cannot exist.
10. Other angles can be 36° 10′, 115° 40′, third side 40.5, or other angles can be 143° 50′, 8° 00′, third side 6.25. (Lengths are in yards.)

2. ***Distance between a Pin and a Rod*** A slider crank mechanism is shown in the figure. Find the distance between the wrist pin W and the connecting rod center C.

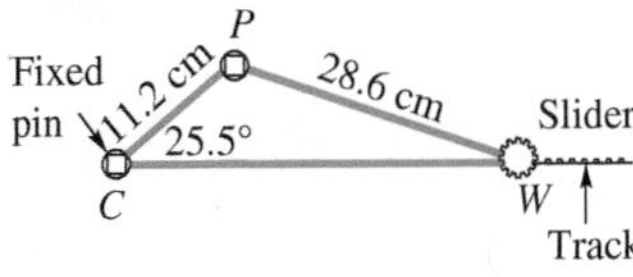

3. ***Distance between Two Lighthouses*** Two lighthouses are located on a north-south line. From lighthouse A, the bearing of a ship 3742 m away is 129° 43′. From lighthouse B, the bearing of the ship is 39° 43′. Find the distance between the lighthouses.

4. ***Hot-Air Balloon*** A hot-air balloon is rising straight up at the speed of 15.0 ft per sec. Then a wind starts blowing horizontally at 5.00 ft per sec. What will the new speed of the balloon be and what angle with the horizontal will the balloon's path make?

5. ***Playing on a Swing*** Mary is playing with her daughter Brittany on a swing. Starting from rest, Mary pulls the swing through an angle of 40° and holds it briefly before releasing the swing. If Brittany weighs 50 lb, what horizontal force, to the nearest pound, must Mary apply while holding the swing?

6. ***Height of an Airplane*** Two observation points A and B are 950 ft apart. From these points the angles of elevation of an airplane are 52° and 57°. See the figure. Find the height of the airplane.

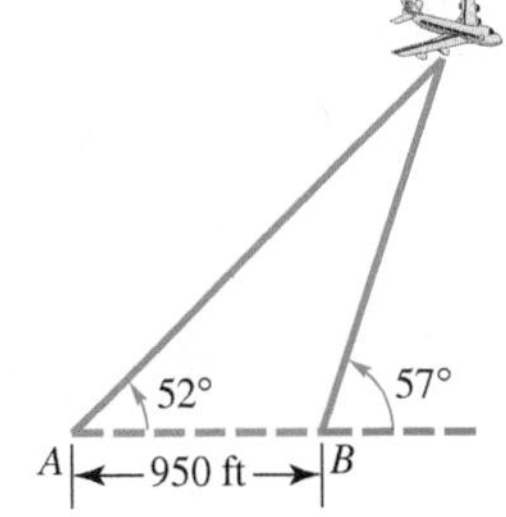

7. ***Wind and Vectors*** A wind can be described by $\mathbf{v} = 6\mathbf{i} + 8\mathbf{j}$, where vector $\mathbf{j}$ points north and represents a south wind of 1 mph.
   (a) What is the speed of the wind?
   (b) Find $3\mathbf{v}$ and interpret the result.
   (c) Interpret the direction and speed of the wind if it changes to $\mathbf{u} = -8\mathbf{i} + 8\mathbf{j}$.

8. ***Ground Speed and Bearing*** A plane with an airspeed of 355 mph is on a bearing of 62°. A wind is blowing from west to east at 28.5 mph. Find the ground speed and the actual bearing of the plane.

9. ***Property Survey*** A surveyor reported the following data about a piece of property: "The property is triangular in shape, with dimensions as shown in the figure." Use the law of sines to see whether such a piece of property could exist.

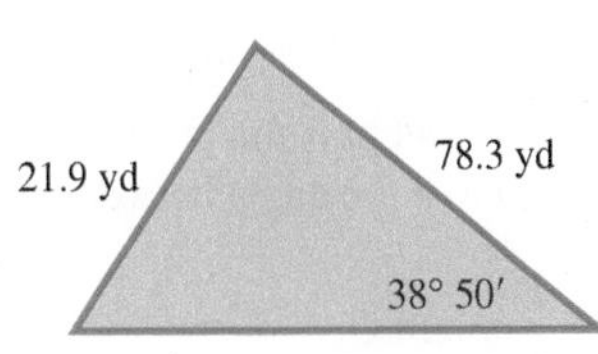

Can such a triangle exist?

10. ***Property Survey*** A triangular piece of property has the dimensions shown. It turns out that the surveyor did not consider every possible case. Use the law of sines to show why.

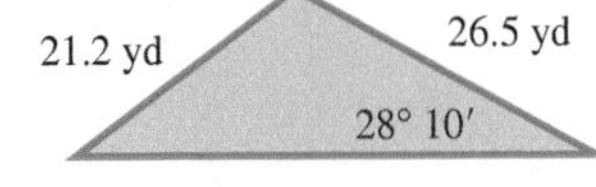

Chapter 7 Test Prep

Key Terms

7.1 Side-Angle-Side (SAS) Angle-Side-Angle (ASA) Side-Side-Side (SSS) oblique triangle Side-Angle-Angle (SAA) **7.2** ambiguous case	**7.3** semiperimeter **7.4** scalar vector quantity vector magnitude initial point terminal point parallelogram rule resultant	opposite (of a vector) zero vector equilibrant airspeed ground speed **7.5** position vector horizontal component	vertical component direction angle unit vector dot product (inner product) angle between two vectors orthogonal vectors

New Symbols

OP or $\overrightarrow{OP}$	vector **OP**	$\langle a, b \rangle$	position vector
$\lvert$**OP**$\rvert$	magnitude of vector **OP**	**i**, **j**	unit vectors

Quick Review

Concepts | **Examples**

7.1 Oblique Triangles and the Law of Sines

Law of Sines

In any triangle ABC, with sides a, b, and c, the following holds.

$$\frac{a}{\sin A} = \frac{b}{\sin B} = \frac{c}{\sin C}$$

$$\frac{\sin A}{a} = \frac{\sin B}{b} = \frac{\sin C}{c} \quad \text{Alternative form}$$

In triangle ABC, find c, to the nearest hundredth, if $A = 44°$, $C = 62°$, and $a = 12.00$ units. Then find its area.

$$\frac{a}{\sin A} = \frac{c}{\sin C} \quad \text{Law of sines}$$

$$\frac{12.00}{\sin 44°} = \frac{c}{\sin 62°} \quad \text{Substitute.}$$

$$c = \frac{12.00 \sin 62°}{\sin 44°} \quad \text{Multiply by } \sin 62° \text{ and rewrite.}$$

$$c \approx 15.25 \text{ units} \quad \text{Use a calculator.}$$

Area of a Triangle

In any triangle ABC, the area $\mathscr{A}$ is half the product of the lengths of two sides and the sine of the angle between them.

$$\mathscr{A} = \frac{1}{2}bc \sin A, \quad \mathscr{A} = \frac{1}{2}ab \sin C, \quad \mathscr{A} = \frac{1}{2}ac \sin B$$

For triangle ABC above, apply the appropriate area formula.

$$\mathscr{A} = \frac{1}{2}ac \sin B \quad \text{Area formula}$$

$$\mathscr{A} = \frac{1}{2}(12.00)(15.25) \sin 74° \quad B = 180° - 44° - 62°, \; B = 74°$$

$$\mathscr{A} \approx 87.96 \text{ sq units} \quad \text{Use a calculator.}$$

Concepts	Examples

7.2 The Ambiguous Case of the Law of Sines

Ambiguous Case

If we are given the lengths of two sides and the angle opposite one of them (for example, A, a, and b in triangle ABC), then it is possible that zero, one, or two such triangles exist. If A is acute, h is the altitude from C, and

- $a < h < b$, then there is no triangle.
- $a = h$ and $h < b$, then there is one triangle (a right triangle).
- $a \geq b$, then there is one triangle.
- $h < a < b$, then there are two triangles.

If A is obtuse and

- $a \leq b$, then there is no triangle.
- $a > b$, then there is one triangle.

See the guidelines in this section that illustrate the possible outcomes.

Solve triangle ABC, given $A = 44.5°$, $a = 11.0$ in., and $c = 7.0$ in.

Find angle C.

$$\frac{\sin C}{7.0} = \frac{\sin 44.5°}{11.0} \quad \text{Law of sines}$$

$$\sin C \approx 0.4460 \quad \text{Solve for } \sin C.$$

$$C \approx 26.5° \quad \text{Use the inverse sine function.}$$

Another angle with this sine value is

$$180° - 26.5° \approx 153.5°.$$

However, $153.5° + 44.5° > 180°$, so there is only one triangle.

$$B = 180° - 44.5° - 26.5° \quad \text{Angle sum formula}$$

$$B = 109° \quad \text{Subtract.}$$

Use the law of sines again to solve for b.

$$b \approx 14.8 \text{ in.}$$

7.3 The Law of Cosines

Law of Cosines

In any triangle ABC, with sides a, b, and c, the following hold.

$$a^2 = b^2 + c^2 - 2bc \cos A$$

$$b^2 = a^2 + c^2 - 2ac \cos B$$

$$c^2 = a^2 + b^2 - 2ab \cos C$$

In triangle ABC, find C if $a = 11$ units, $b = 13$ units, and $c = 20$ units. Then find its area.

$$c^2 = a^2 + b^2 - 2ab \cos C \quad \text{Law of cosines}$$

$$20^2 = 11^2 + 13^2 - 2(11)(13) \cos C \quad \text{Substitute.}$$

$$400 = 121 + 169 - 286 \cos C \quad \text{Square and multiply.}$$

$$\cos C = \frac{400 - 121 - 169}{-286} \quad \text{Solve for } \cos C.$$

$$\cos C \approx -0.38461538 \quad \text{Use a calculator.}$$

$$C \approx 113° \quad \text{Use the inverse cosine function.}$$

Heron's Area Formula

If a triangle has sides of lengths a, b, and c, with semiperimeter

$$s = \frac{1}{2}(a + b + c),$$

then the area $\mathcal{A}$ of the triangle is given by the following.

$$\mathcal{A} = \sqrt{s(s - a)(s - b)(s - c)}$$

The semiperimeter s of the above triangle is

$$s = \frac{1}{2}(11 + 13 + 20) = 22,$$

so the area is

$$\mathcal{A} = \sqrt{22(22 - 11)(22 - 13)(22 - 20)}$$

$$\mathcal{A} = 66 \text{ sq units.}$$

Concepts	Examples

7.4 Geometrically Defined Vectors and Applications

Vector Sum

The sum of two vectors is also a vector. There are two ways to find the sum of two vectors **A** and **B** geometrically.

1. The vector with the same initial point as **A** and the same terminal point as **B** is the sum **A** + **B**.

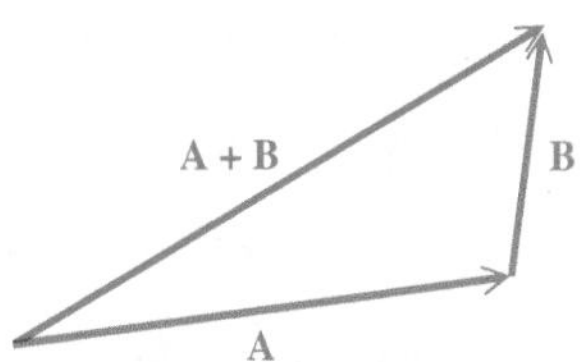

2. The diagonal of the parallelogram with the same initial point as **A** and **B** is the sum **A** + **B**. This is the **parallelogram rule.**

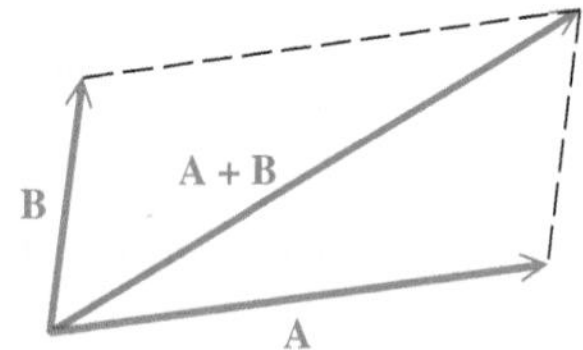

Two forces of 25 newtons and 32 newtons act on a point in a plane. If the angle between the forces is 62°, find the magnitude of the resultant force.

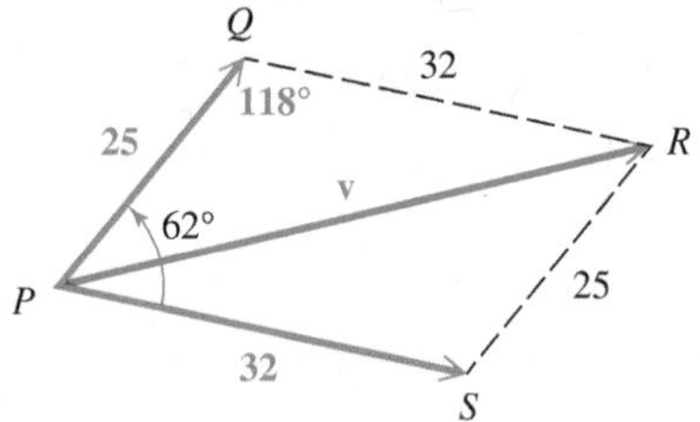

The resultant force divides a parallelogram into two triangles. The measure of angle Q in the figure is 118°. We use the law of cosines to find the desired magnitude.

$$|\mathbf{v}|^2 = 25^2 + 32^2 - 2(25)(32)\cos 118^\circ$$

$$|\mathbf{v}|^2 \approx 2400$$

$$|\mathbf{v}| \approx 49$$

The magnitude of the resultant force is 49 newtons.

7.5 Algebraically Defined Vectors and the Dot Product

Magnitude and Direction Angle of a Vector

The magnitude (length) of vector $\mathbf{u} = \langle a, b\rangle$ is given by the following.

$$|\mathbf{u}| = \sqrt{a^2 + b^2}$$

The direction angle θ satisfies $\tan\theta = \frac{b}{a}$, where $a \neq 0$.

Find the magnitude and direction angle of vector **u** in the figure.

$$|\mathbf{u}| = \sqrt{\left(2\sqrt{3}\right)^2 + 2^2} = \sqrt{16}$$

$$= 4 \leftarrow \text{Magnitude}$$

$$\tan\theta = \frac{2}{2\sqrt{3}} = \frac{1}{\sqrt{3}} \cdot \frac{\sqrt{3}}{\sqrt{3}} = \frac{\sqrt{3}}{3}, \text{ so } \theta = 30^\circ.$$

If $\mathbf{u} = \langle a, b\rangle$ has direction angle θ, then

$$\mathbf{u} = \langle |\mathbf{u}|\cos\theta, |\mathbf{u}|\sin\theta\rangle.$$

For **u** defined above,

$$\mathbf{u} = \langle 4\cos 30^\circ, 4\sin 30^\circ\rangle$$

$$= \langle 2\sqrt{3}, 2\rangle. \quad \cos 30^\circ = \frac{\sqrt{3}}{2};\ \sin 30^\circ = \frac{1}{2}$$

Vector Operations

Let a, b, c, d, and k represent real numbers.

$$\langle a, b\rangle + \langle c, d\rangle = \langle a + c, b + d\rangle$$

$$k \cdot \langle a, b\rangle = \langle ka, kb\rangle$$

If $\mathbf{u} = \langle a_1, a_2\rangle$, then $-\mathbf{u} = \langle -a_1, -a_2\rangle$.

$$\langle a, b\rangle - \langle c, d\rangle = \langle a, b\rangle + (-\langle c, d\rangle) = \langle a - c, b - d\rangle$$

Find each of the following.

$$\langle 4, 6\rangle + \langle -8, 3\rangle = \langle -4, 9\rangle$$

$$5\langle -2, 1\rangle = \langle -10, 5\rangle$$

$$-\langle -9, 6\rangle = \langle 9, -6\rangle$$

$$\langle 4, 6\rangle - \langle -8, 3\rangle = \langle 12, 3\rangle$$

i, j Form for Vectors

If $\mathbf{v} = \langle a, b\rangle$, then

$$\mathbf{v} = a\mathbf{i} + b\mathbf{j}, \quad \text{where } \mathbf{i} = \langle 1, 0\rangle \text{ and } \mathbf{j} = \langle 0, 1\rangle.$$

If $\mathbf{u} = \langle 2\sqrt{3}, 2\rangle$ as above, then

$$\mathbf{u} = 2\sqrt{3}\,\mathbf{i} + 2\mathbf{j}.$$

Concepts	Examples
Dot Product The dot product of the two vectors $\mathbf{u} = \langle a, b \rangle$ and $\mathbf{v} = \langle c, d \rangle$, denoted $\mathbf{u} \cdot \mathbf{v}$, is given by the following. $\mathbf{u} \cdot \mathbf{v} = ac + bd$	Find the dot product. $\langle 2, 1 \rangle \cdot \langle 5, -2 \rangle$ $= 2 \cdot 5 + 1(-2)$ $= 8$
Geometric Interpretation of the Dot Product If θ is the angle between the two nonzero vectors **u** and **v**, where $0° \le \theta \le 180°$, then the following holds. $\cos \theta = \dfrac{\mathbf{u} \cdot \mathbf{v}}{\lvert\mathbf{u}\rvert \lvert\mathbf{v}\rvert}$	Find the angle θ between $\mathbf{u} = \langle 3, 1 \rangle$ and $\mathbf{v} = \langle 2, -3 \rangle$. $\cos \theta = \dfrac{\mathbf{u} \cdot \mathbf{v}}{\lvert\mathbf{u}\rvert \lvert\mathbf{v}\rvert}$ Geometric interpretation of the dot product $\cos \theta = \dfrac{3(2) + 1(-3)}{\sqrt{3^2 + 1^2} \cdot \sqrt{2^2 + (-3)^2}}$ Use the definitions. $\cos \theta = \dfrac{3}{\sqrt{130}}$ Simplify. $\cos \theta \approx 0.26311741$ Use a calculator. $\theta \approx 74.7°$ Use the inverse cosine function.

Chapter 7 Review Exercises

Use the law of sines to find the indicated part of each triangle ABC.

1. Find b if $C = 74.2°$, $c = 96.3$ m, $B = 39.5°$.
2. Find B if $A = 129.7°$, $a = 127$ ft, $b = 69.8$ ft.
3. Find B if $C = 51.3°$, $c = 68.3$ m, $b = 58.2$ m.
4. Find b if $a = 165$ m, $A = 100.2°$, $B = 25.0°$.
5. Find A if $B = 39° \, 50'$, $b = 268$ m, $a = 340$ m.
6. Find A if $C = 79° \, 20'$, $c = 97.4$ mm, $a = 75.3$ mm.

Answer each question.

7. If we are given a, A, and C in a triangle ABC, does the possibility of the ambiguous case exist? If not, explain why.
8. Can triangle ABC exist if $a = 4.7$, $b = 2.3$, and $c = 7.0$? If not, explain why. Answer this question without using trigonometry.
9. Given $a = 10$ and $B = 30°$ in triangle ABC, for what values of b does A have
   (a) exactly one value **(b)** two possible values **(c)** no value?
10. Why can there be no triangle ABC satisfying $A = 140°$, $a = 5$, and $b = 7$?

Use the law of cosines to find the indicated part of each triangle ABC.

11. Find A if $a = 86.14$ in., $b = 253.2$ in., $c = 241.9$ in.
12. Find b if $B = 120.7°$, $a = 127$ ft, $c = 69.8$ ft.
13. Find a if $A = 51° \, 20'$, $c = 68.3$ m, $b = 58.2$ m.

Answers:

1. 63.7 m 2. 25.0°
3. 41.7° 4. 70.9 m
5. 54° 20′ or 125° 40′
6. 49° 30′
7. If one side and two angles are given, the third angle can be determined using the angle sum formula, and then the ASA axiom can be applied. This is not the ambiguous case.
8. Because the sum of the measures of side a and side b is not greater than the measure of side c, the triangle does not exist.
9. (a) $b = 5$, $b \ge 10$
   (b) $5 < b < 10$
   (c) $b < 5$
10. Because A is obtuse, it is the largest angle. Thus side a should be the longest side, but it is not. Therefore, no such triangle exists.
11. 19.87°, or 19° 52′
12. 173 ft
13. 55.5 m

14. 26.5°, or 26° 30′
15. 19 cm **16.** 47°
17. $B = 17.3°$, $C = 137.5°$, $c = 11.0$ yd
18. $B_1 = 74.6°$, $C_1 = 43.7°$, $c_1 = 61.9$ m; $B_2 = 105.4°$, $C_2 = 12.9°$, $c_2 = 20.0$ m
19. $c = 18.7$ cm, $A = 91° 40'$, $B = 45° 50'$
20. $A = 47.7°$, $B = 73.9°$, $C = 58.4°$
21. 153,600 m² **22.** 20.3 ft²
23. 0.234 km² **24.** 680 m²
25. 58.6 ft **26.** 11 ft
27. 13 m **28.** 15.8 ft

14. Find B if $a = 14.8$ m, $b = 19.7$ m, $c = 31.8$ m.

15. Find a if $A = 60°$, $b = 5.0$ cm, $c = 21$ cm.

16. Find A if $a = 13$ ft, $b = 17$ ft, $c = 8$ ft.

Solve each triangle ABC.

17. $A = 25.2°$, $a = 6.92$ yd, $b = 4.82$ yd **18.** $A = 61.7°$, $a = 78.9$ m, $b = 86.4$ m

19. $a = 27.6$ cm, $b = 19.8$ cm, $C = 42° 30'$ **20.** $a = 94.6$ yd, $b = 123$ yd, $c = 109$ yd

Find the area of each triangle ABC.

21. $b = 840.6$ m, $c = 715.9$ m, $A = 149.3°$ **22.** $a = 6.90$ ft, $b = 10.2$ ft, $C = 35° 10'$

23. $a = 0.913$ km, $b = 0.816$ km, $c = 0.582$ km

24. $a = 43$ m, $b = 32$ m, $c = 51$ m

Solve each problem.

25. ***Distance across a Canyon*** To measure the distance AB across a canyon for a power line, a surveyor measures angles B and C and the distance BC, as shown in the figure. What is the distance from A to B?

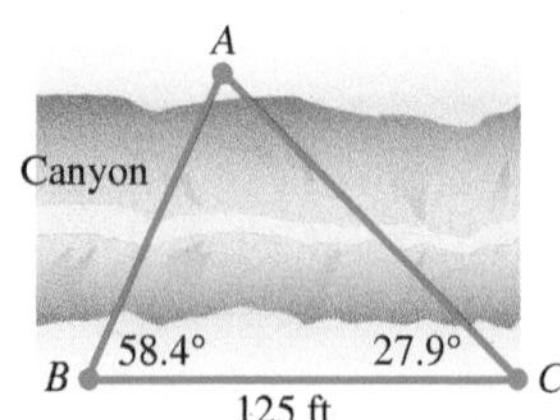

26. ***Length of a Brace*** A banner on an 8.0-ft pole is to be mounted on a building at an angle of 115°, as shown in the figure. Find the length of the brace.

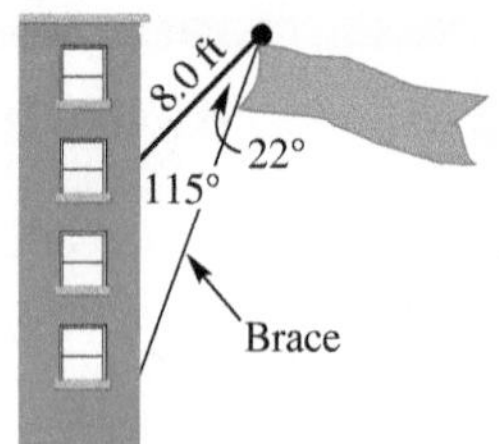

27. ***Height of a Tree*** A tree leans at an angle of 8.0° from the vertical. From a point 7.0 m from the bottom of the tree, the angle of elevation to the top of the tree is 68°. Find the slanted height x in the figure.

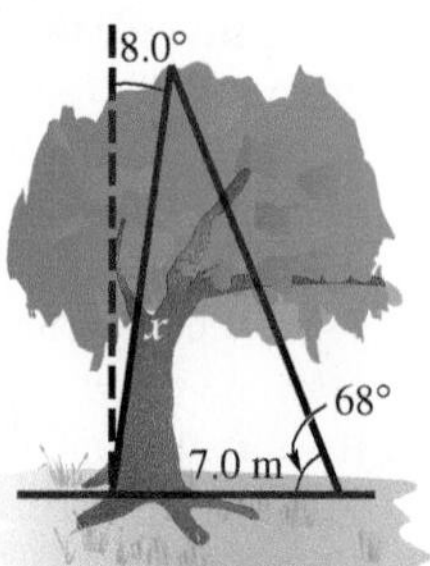

28. ***Hanging Sculpture*** A hanging sculpture is to be hung in an art gallery with two wires of lengths 15.0 ft and 12.2 ft so that the angle between them is 70.3°. How far apart should the ends of the wire be placed on the ceiling?

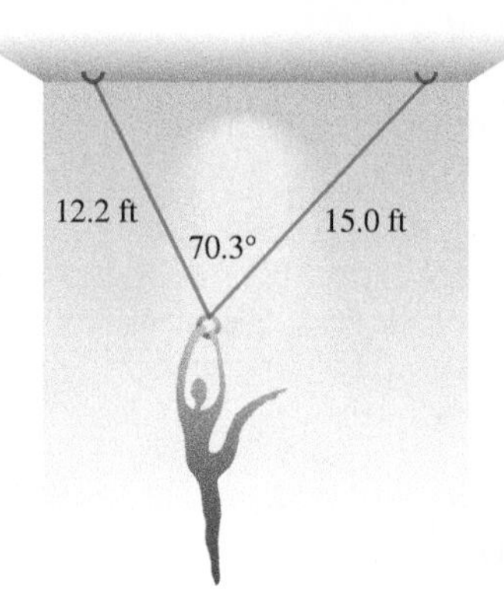

29. 53.2 ft 30. 77.1°
31. 115 km 32. 2.4 mi
33. 25 sq units
34. The sides measure 5, 10, and $5\sqrt{5}$, which satisfy the converse of the Pythagorean theorem. The area is 25 sq units.

35.

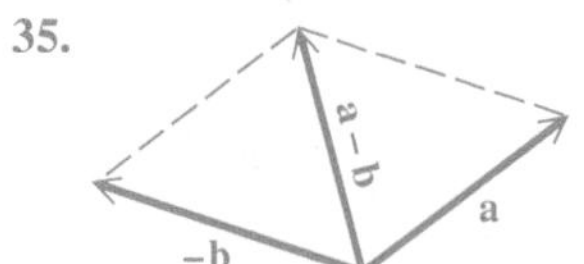

36.

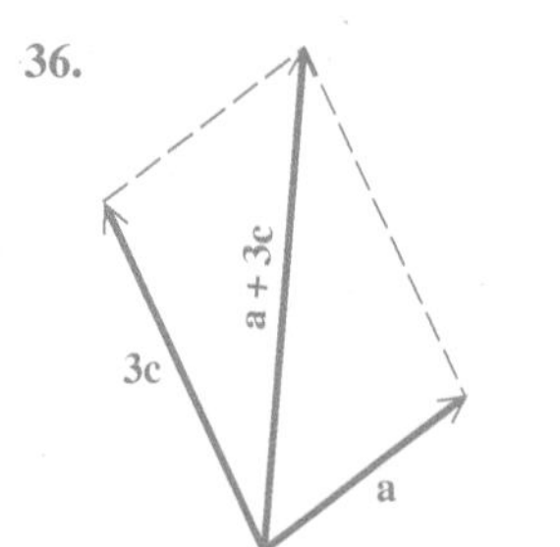

29. ***Height of a Tree*** A hill makes an angle of 14.3° with the horizontal. From the base of the hill, the angle of elevation to the top of a tree on top of the hill is 27.2°. The distance along the hill from the base to the tree is 212 ft. Find the height of the tree.

30. ***Pipeline Position*** A pipeline is to run between points A and B, which are separated by a protected wetlands area. To avoid the wetlands, the pipe will run from point A to C and then to B. The distances involved are $AB = 150$ km, $AC = 102$ km, and $BC = 135$ km. What angle should be used at point C?

31. ***Distance between Two Boats*** Two boats leave a dock together. Each travels in a straight line. The angle between their courses measures 54° 10′. One boat travels 36.2 km per hr, and the other travels 45.6 km per hr. How far apart will they be after 3 hr?

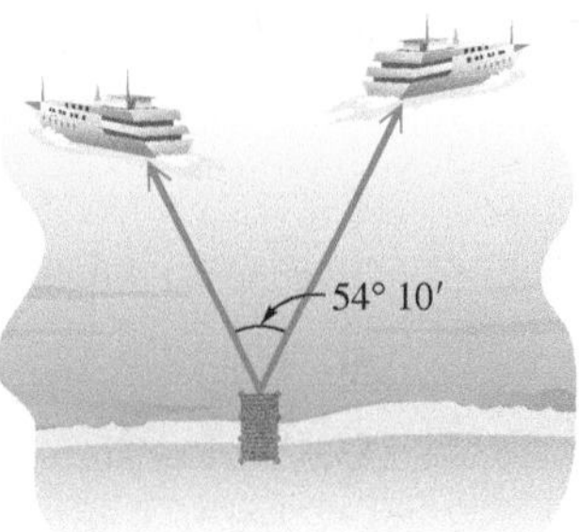

32. ***Distance from a Ship to a Lighthouse*** A ship sailing parallel to shore sights a lighthouse at an angle of 30° from its direction of travel. After the ship travels 2.0 mi farther, the angle has increased to 55°. At that time, how far is the ship from the lighthouse?

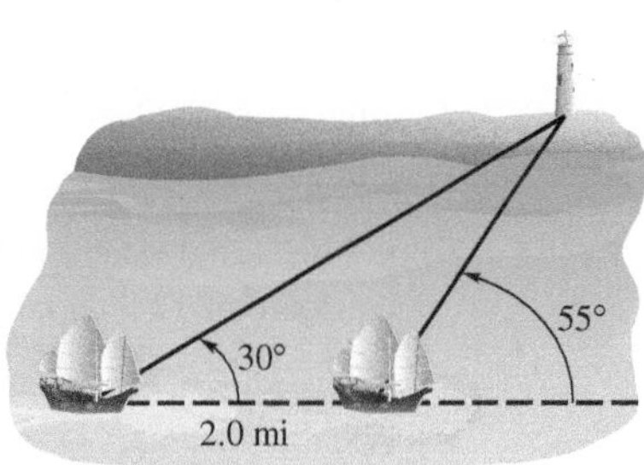

33. ***Area of a Triangle*** Find the area of the triangle shown in the figure using Heron's area formula.

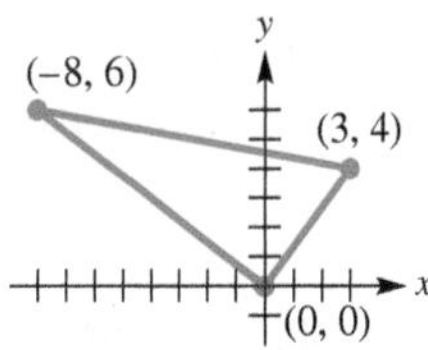

34. Show that the triangle in **Exercise 33** is a right triangle. Then use the formula $\mathcal{A} = \frac{1}{2}ac \sin B$, with $B = 90°$, to find the area.

Use the given vectors to sketch each of the following.

35. **a** − **b**

36. **a** + 3**c**

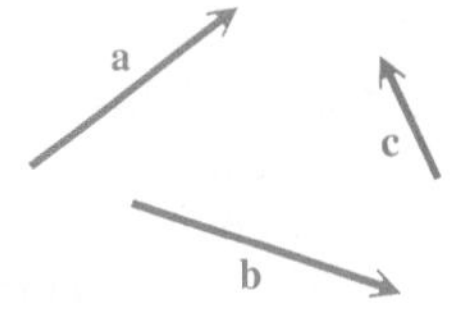

37. 207 lb
38. 209 newtons
39. −869; 418
40. $25\sqrt{2}$; $25\sqrt{2}$
41. 15; 126.9°   42. 29; 316.4°
43. (a) **i**   (b) 4**i** − 2**j**
(c) 11**i** − 7**j**
44. 142.1°
45. 90°; orthogonal
46. 45°
47. 29 lb
48. 280 newtons; 30.4°
49. bearing: 306°; ground speed: 524 mph
50. 3.8°
51. 34 lb
52. speed: 21 km per hr; bearing: 118°

Given two forces and the angle between them, find the magnitude of the resultant force.

37.

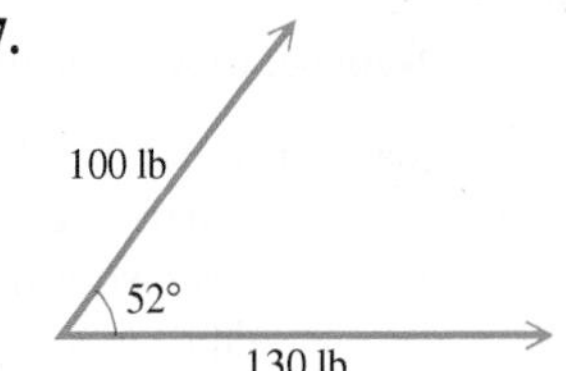

38. forces of 142 and 215 newtons, forming an angle of 112°

Vector **v** *has the given magnitude and direction angle. Find the horizontal and vertical components of* **v**.

39. $|\mathbf{v}| = 964,\ \theta = 154°\,20'$

40. $|\mathbf{v}| = 50,\ \theta = 45°$ (Give exact values.)

Find the magnitude and direction angle for **u** *rounded to the nearest tenth.*

41. $\mathbf{u} = \langle -9, 12 \rangle$

42. $\mathbf{u} = \langle 21, -20 \rangle$

43. Let $\mathbf{v} = 2\mathbf{i} - \mathbf{j}$ and $\mathbf{u} = -3\mathbf{i} + 2\mathbf{j}$. Express each in terms of **i** and **j**.

(a) $2\mathbf{v} + \mathbf{u}$ (b) $2\mathbf{v}$ (c) $\mathbf{v} - 3\mathbf{u}$

Find the angle between the vectors. Round to the nearest tenth of a degree. If the vectors are orthogonal, say so.

44. $\langle 3, -2 \rangle, \langle -1, 3 \rangle$ **45.** $\langle 5, -3 \rangle, \langle 3, 5 \rangle$ **46.** $\langle 0, 4 \rangle, \langle -4, 4 \rangle$

Solve each problem.

47. ***Weight of a Sled and Passenger*** Paula and Steve are pulling their daughter Jessie on a sled. Steve pulls with a force of 18 lb at an angle of 10°. Paula pulls with a force of 12 lb at an angle of 15°. Find the magnitude of the resultant force on Jessie and the sled.

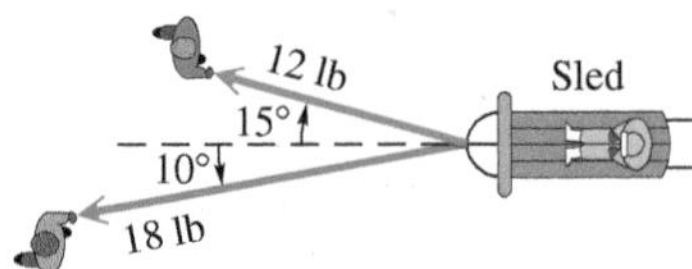

48. ***Force Placed on a Barge*** One boat pulls a barge with a force of 100 newtons. Another boat pulls the barge at an angle of 45° to the first force, with a force of 200 newtons. Find the resultant force acting on the barge, to the nearest unit, and the angle between the resultant and the first boat, to the nearest tenth.

49. ***Direction and Speed of a Plane*** A plane has an airspeed of 520 mph. The pilot wishes to fly on a bearing of 310°. A wind of 37 mph is blowing from a bearing of 212°. In what direction should the pilot fly, and what will be her ground speed?

50. ***Angle of a Hill*** A 186-lb force is required to hold a 2800-lb car on a hill. What angle does the hill make with the horizontal?

51. ***Incline Force*** Find the force required to keep a 75-lb sled from sliding down an incline that makes an angle of 27° with the horizontal. (Assume there is no friction.)

52. ***Speed and Direction of a Boat*** A boat travels 15 km per hr in still water. The boat is traveling across a large river, on a bearing of 130°. The current in the river, coming from the west, has a speed of 7 km per hr. Find the resulting speed of the boat and its resulting direction of travel.

53. Both expressions equal $\frac{1+\sqrt{3}}{2}$.

54. Both expressions equal $\frac{1-\sqrt{3}}{2}$.

55. Both expressions equal $-2+\sqrt{3}$.

Other Formulas from Trigonometry *The following identities involve all six parts of a triangle ABC and are useful for checking answers.*

$$\frac{a+b}{c}=\frac{\cos\frac{1}{2}(A-B)}{\sin\frac{1}{2}C}\qquad \text{Newton's formula}$$

$$\frac{a-b}{c}=\frac{\sin\frac{1}{2}(A-B)}{\cos\frac{1}{2}C}\qquad \text{Mollweide's formula}$$

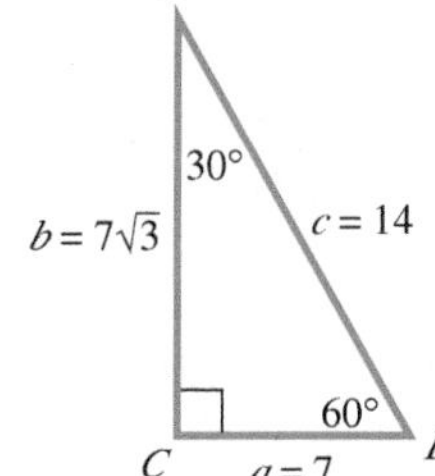

53. Apply Newton's formula to the given triangle to verify the accuracy of the information.

54. Apply Mollweide's formula to the given triangle to verify the accuracy of the information.

55. ***Law of Tangents*** In addition to the law of sines and the law of cosines, there is a **law of tangents.** In any triangle *ABC*,

$$\frac{\tan\frac{1}{2}(A-B)}{\tan\frac{1}{2}(A+B)}=\frac{a-b}{a+b}.$$

Verify this law for the triangle *ABC* with $a = 2$, $b = 2\sqrt{3}$, $A = 30°$, and $B = 60°$.

Chapter 7 Test

[7.1]
1. 137.5°

[7.3]
2. 179 km
3. 49.0°
4. 168 sq units

[7.1]
5. 18 sq units

[7.2]
6. (a) $b > 10$ (b) none (c) $b \le 10$

[7.1–7.3]
7. $a = 40$ m, $B = 41°$, $C = 79°$
8. $B_1 = 58° \, 30'$, $A_1 = 83° \, 00'$, $a_1 = 1250$ in.; $B_2 = 121° \, 30'$, $A_2 = 20° \, 00'$, $a_2 = 431$ in.

[7.5]
9. $|\mathbf{v}| = 10$; $\theta = 126.9°$

Find the indicated part of each triangle ABC.

1. Find *C* if $A = 25.2°$, $a = 6.92$ yd, and $b = 4.82$ yd.

2. Find *c* if $C = 118°$, $a = 75.0$ km, and $b = 131$ km.

3. Find *B* if $a = 17.3$ ft, $b = 22.6$ ft, $c = 29.8$ ft.

Solve each problem.

4. Find the area of triangle *ABC* if $a = 14$, $b = 30$, and $c = 40$.

5. Find the area of triangle *XYZ* shown here.

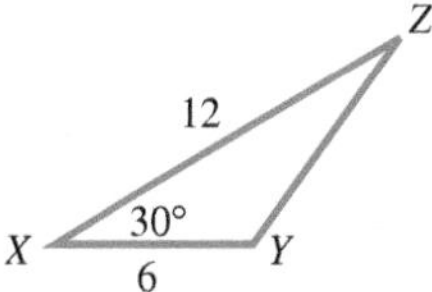

6. Given $a = 10$ and $B = 150°$ in triangle *ABC*, determine the values of *b* for which *A* has

(a) exactly one value **(b)** two possible values **(c)** no value.

Solve each triangle ABC.

7. $A = 60°$, $b = 30$ m, $c = 45$ m

8. $b = 1075$ in., $c = 785$ in., $C = 38° \, 30'$

Work each problem.

9. Find the magnitude and the direction angle, to the nearest tenth, for the vector shown in the figure.

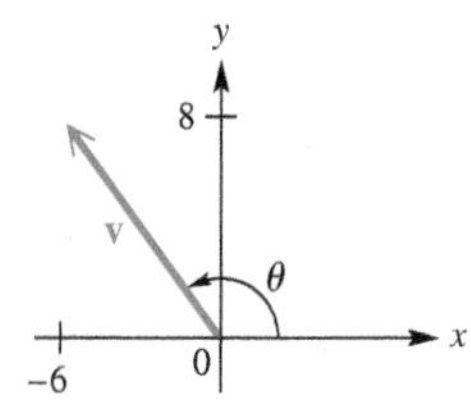

[7.4]
10.

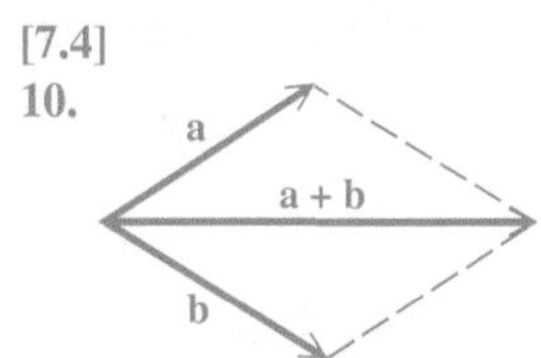

[7.5]
11. (a) $\langle 1, -3 \rangle$ (b) $\langle -6, 18 \rangle$
(c) -20 (d) $\sqrt{10}$

12. 41.8°

13. Show that $\mathbf{u} \cdot \mathbf{v} = 0$.

[7.1]
14. 2.7 mi

[7.5]
15. $\langle -346, 451 \rangle$

[7.4]
16. 1.91 mi

[7.1]
17. 14 m

[7.4]
18. 30 lb

19. bearing: 357°; airspeed: 220 mph

20. 18.7°

10. Use the given vectors to sketch $\mathbf{a} + \mathbf{b}$.

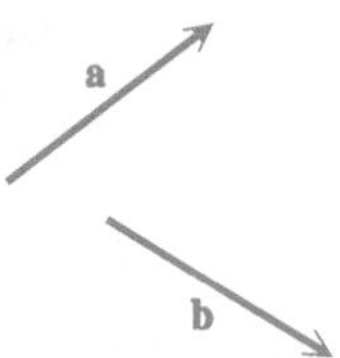

11. For the vectors $\mathbf{u} = \langle -1, 3 \rangle$ and $\mathbf{v} = \langle 2, -6 \rangle$, find each of the following.

(a) $\mathbf{u} + \mathbf{v}$ **(b)** $-3\mathbf{v}$ **(c)** $\mathbf{u} \cdot \mathbf{v}$ **(d)** $|\mathbf{u}|$

12. Find the measure of the angle θ between $\mathbf{u} = \langle 4, 3 \rangle$ and $\mathbf{v} = \langle 1, 5 \rangle$ to the nearest tenth.

13. Show that the vectors $\mathbf{u} = \langle -4, 7 \rangle$ and $\mathbf{v} = \langle -14, -8 \rangle$ are orthogonal vectors.

Solve each problem.

14. ***Height of a Balloon*** The angles of elevation of a balloon from two points A and B on level ground are 24° 50′ and 47° 20′, respectively. As shown in the figure, points A, B, and C are in the same vertical plane and points A and B are 8.4 mi apart. Approximate the height of the balloon above the ground to the nearest tenth of a mile.

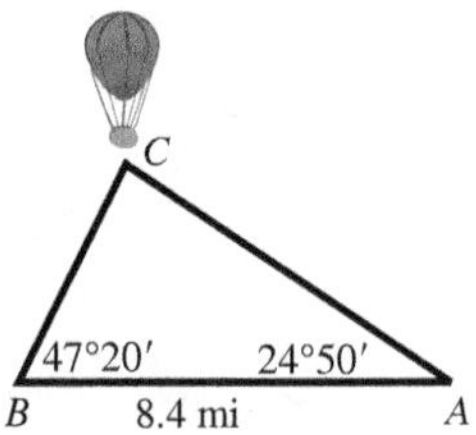

15. ***Horizontal and Vertical Components*** Find the horizontal and vertical components of the vector with magnitude 569 and direction angle 127.5° from the horizontal. Give your answer in the form $\langle a, b \rangle$ to the nearest unit.

16. ***Radio Direction Finders*** Radio direction finders are placed at points A and B, which are 3.46 mi apart on an east-west line, with A west of B. From A, the bearing of a certain illegal pirate radio transmitter is 48°, and from B the bearing is 302°. Find the distance between the transmitter and A to the nearest hundredth of a mile.

17. ***Height of a Tree*** A tree leans at an angle of 8.0° from the vertical, as shown in the figure. From a point 8.0 m from the bottom of the tree, the angle of elevation to the top of the tree is 66°. Find the slanted height x in the figure.

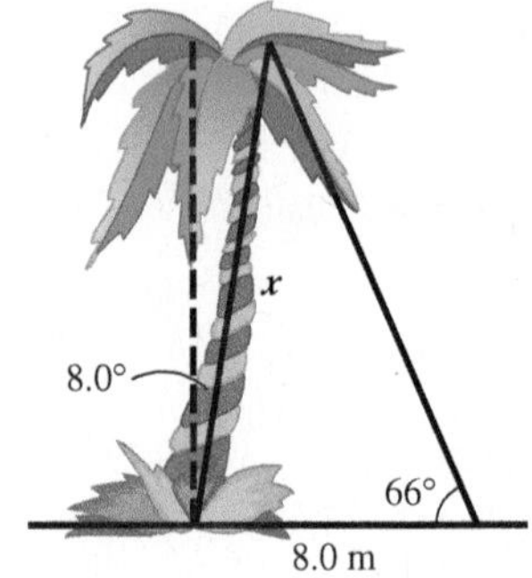

18. ***Walking Dogs on Leashes*** While Michael is walking his two dogs, Gus and Dotty, they reach a corner and must wait for a WALK sign. Michael is holding the two leashes in the same hand, and the dogs are pulling on their leashes at the angles and forces shown in the figure. Find the magnitude of the equilibrant force (to the nearest tenth of a pound) that Michael must apply to restrain the dogs.

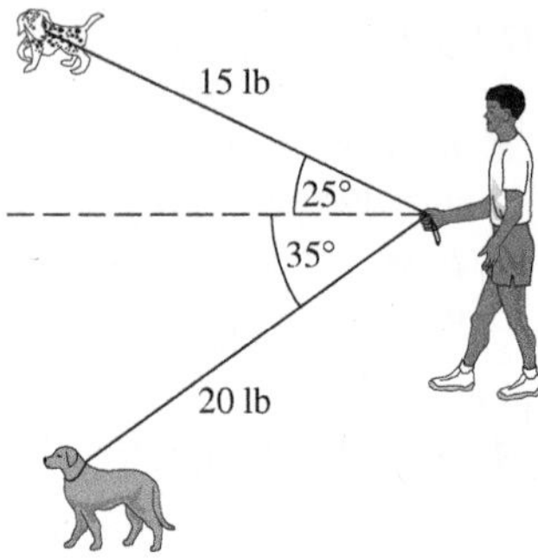

19. ***Bearing and Airspeed*** Find the bearing and airspeed required for a plane to fly 630 mi due north in 3.0 hr if the wind is blowing from a direction of 318° at 15 mph. Approximate the bearing to the nearest degree and the airspeed to the nearest 10 mph.

20. ***Incline Angle*** A force of 16.0 lb is required to hold a 50.0-lb wheelbarrow on an incline. What angle does the incline make with the horizontal?

8 Complex Numbers, Polar Equations, and Parametric Equations

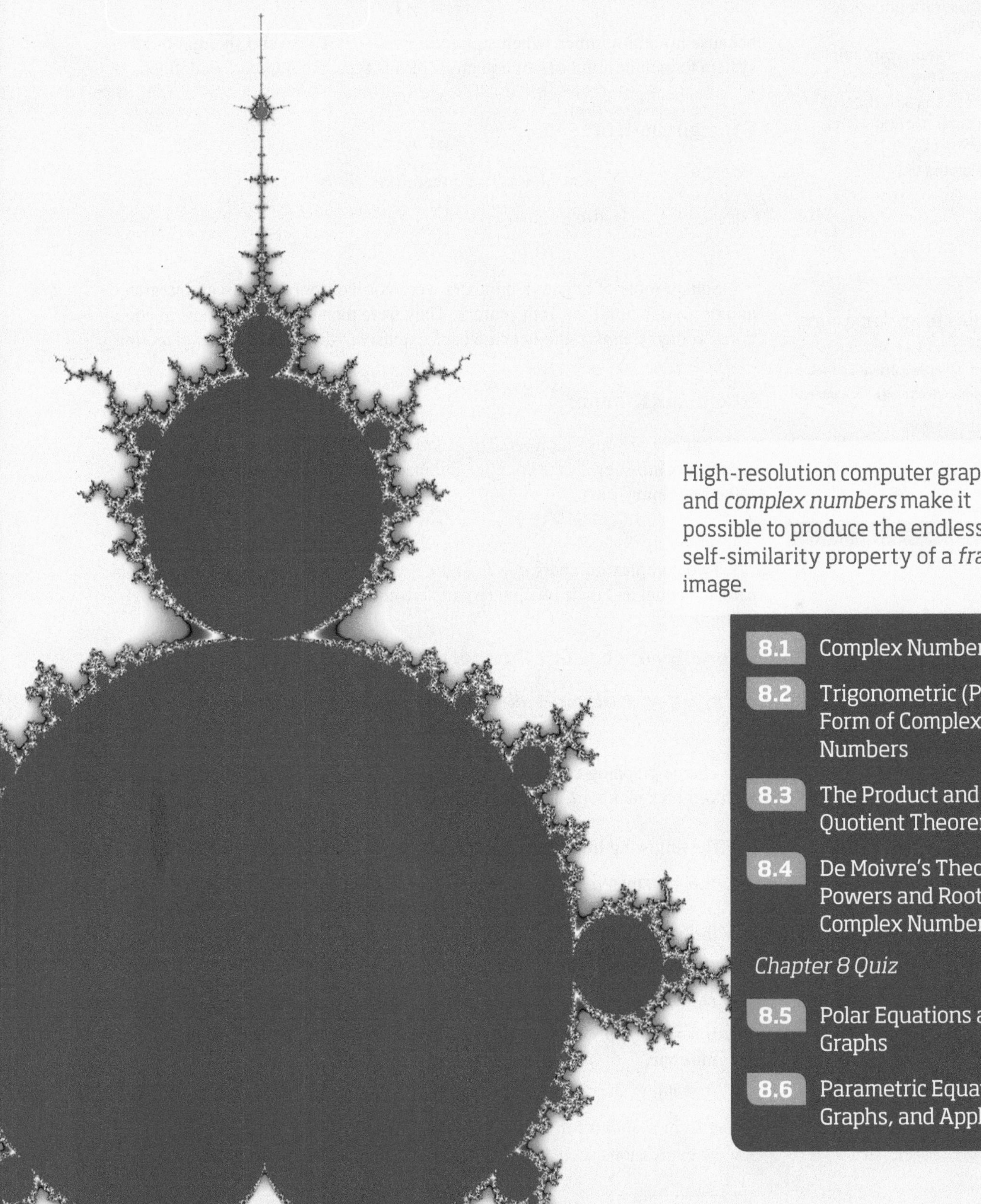

High-resolution computer graphics and *complex numbers* make it possible to produce the endless self-similarity property of a *fractal* image.

8.1 Complex Numbers

- Basic Concepts of Complex Numbers
- Complex Solutions of Quadratic Equations (Part 1)
- Operations on Complex Numbers
- Complex Solutions of Quadratic Equations (Part 2)
- Powers of i

Basic Concepts of Complex Numbers There is no real number solution of the equation

$$x^2 = -1$$

because no real number, when squared, gives -1. To extend the real number system to include solutions of equations of this type, the number i is defined.

Imaginary Unit i

$$i = \sqrt{-1}, \quad \text{and therefore} \quad i^2 = -1.$$

(Note that $-i$ is also a square root of -1.)

LOOKING AHEAD TO CALCULUS
The letters j and k are also used to represent $\sqrt{-1}$ in calculus and some applications (electronics, for example).

Square roots of negative numbers were not incorporated into an integrated number system until the 16th century. They were then used as solutions of equations. Today, *complex numbers* are used extensively in science and engineering.

Complex Number

If a and b are real numbers, then any number of the form $\boldsymbol{a + bi}$ is a **complex number.** In the complex number $a + bi$, a is the **real part** and b is the **imaginary part.***

```
NORMAL FLOAT AUTO a+bi RADIAN MP
√-1
                                 i
i²
                                -1
real(7+2i)
                                 7
imag(7+2i)
                                 2
```

The calculator is in complex number mode. This screen supports the definition of i and shows how the calculator returns the real and imaginary parts of the complex number $7 + 2i$.

Figure 1

Two complex numbers $a + bi$ and $c + di$ are equal provided that their real parts are equal and their imaginary parts are equal.

Equality of Complex Numbers

$$\boldsymbol{a + bi = c + di} \quad \text{if and only if} \quad \boldsymbol{a = c} \text{ and } \boldsymbol{b = d}.$$

Some graphing calculators, such as the TI-84 Plus, are capable of working with complex numbers, as seen in **Figure 1.** ■

The following important concepts apply to a complex number $a + bi$.

1. If $b = 0$, then $a + bi = a$, which is a real number. (This means that the set of real numbers is a subset of the set of complex numbers. See **Figure 2** on the next page.)
2. If $b \neq 0$, then $a + bi$ is a **nonreal complex number.**
   Examples: $7 + 2i$, $-1 - i$
3. If $a = 0$ and $b \neq 0$, then the nonreal complex number is a **pure imaginary number.**
   Examples: $3i$, $-16i$

The form $a + bi$ (or $a + ib$) is **standard form.** (The form $a + ib$ is used to write expressions such as $i\sqrt{5}$ because $\sqrt{5}i$ could be mistaken for $\sqrt{5i}$.)

Teaching Tip Tell students that, despite their name, imaginary (and complex) numbers have many properties that make them useful in practical applications. For example, electrical engineers use complex numbers to model and analyze electric circuits.

*In some texts, the term bi is defined to be the imaginary part.

Figure 2 shows the relationships among subsets of the complex numbers.

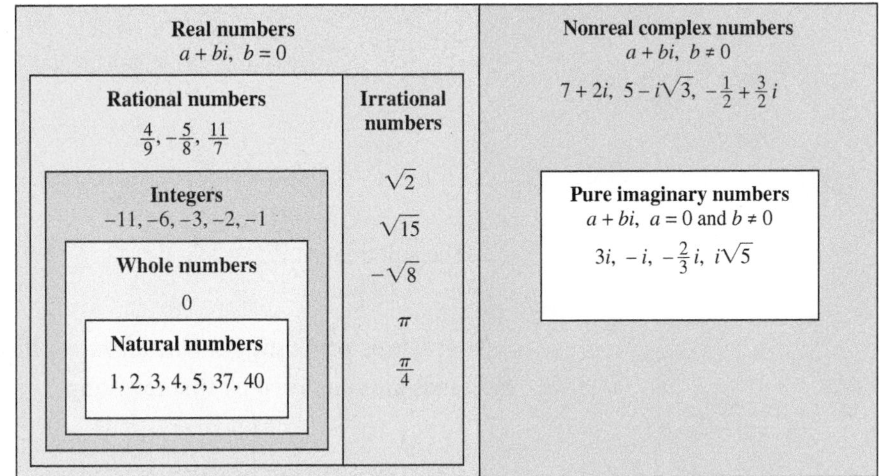

Figure 2

For a positive real number a, the expression $\sqrt{-a}$ is defined as follows.

Meaning of $\sqrt{-a}$

If $a > 0$, then $\sqrt{-a} = i\sqrt{a}$.

Classroom Example 1
Write as the product of a real number and i, using the definition of $\sqrt{-a}$.
(a) $\sqrt{-81}$ (b) $\sqrt{-55}$
(c) $\sqrt{-98}$

Answers:
(a) $9i$ (b) $i\sqrt{55}$
(c) $7i\sqrt{2}$

EXAMPLE 1 Writing $\sqrt{-a}$ as $i\sqrt{a}$

Write as the product of a real number and i, using the definition of $\sqrt{-a}$.

(a) $\sqrt{-16}$ **(b)** $\sqrt{-70}$ **(c)** $\sqrt{-48}$

SOLUTION

(a) $\sqrt{-16} = i\sqrt{16} = 4i$ **(b)** $\sqrt{-70} = i\sqrt{70}$

(c) $\sqrt{-48} = i\sqrt{48} = i\sqrt{16 \cdot 3} = 4i\sqrt{3}$ Product rule for radicals: $\sqrt[n]{ab} = \sqrt[n]{a} \cdot \sqrt[n]{b}$

✔ **Now Try Exercises 21, 23, and 25.**

Complex Solutions of Quadratic Equations (Part 1) Such solutions are expressed in the form $a + bi$ or $a + ib$.

Classroom Example 2
Solve each equation over the set of complex numbers.
(a) $x^2 = -144$
(b) $x^2 + 75 = 0$

Answers:
(a) $\{\pm 12i\}$ (b) $\{\pm 5i\sqrt{3}\}$

EXAMPLE 2 Solving Quadratic Equations (Complex Solutions)

Solve each equation over the set of complex numbers.

(a) $x^2 = -9$ **(b)** $x^2 + 24 = 0$

SOLUTION

(a)

$$x^2 = -9$$

$$x = \pm\sqrt{-9} \quad \text{Square root property}$$

Take *both* square roots, indicated by the $\pm$ symbol.

$$x = \pm i\sqrt{9} \quad \sqrt{-a} = i\sqrt{a}$$

$$x = \pm 3i \quad \sqrt{9} = 3$$

The solution set is $\{\pm 3i\}$.

(b)

$$x^2 + 24 = 0$$

$$x^2 = -24 \quad \text{Subtract 24.}$$

$$x = \pm\sqrt{-24} \quad \text{Square root property}$$

$$x = \pm i\sqrt{24} \quad \sqrt{-a} = i\sqrt{a}$$

$$x = \pm i\sqrt{4} \cdot \sqrt{6} \quad \text{Product rule for radicals}$$

$$x = \pm 2i\sqrt{6} \quad \sqrt{4} = 2$$

The solution set is $\{\pm 2i\sqrt{6}\}$.

✔ **Now Try Exercises 85 and 87.**

Operations on Complex Numbers Products or quotients with negative radicands are simplified by first rewriting

$$\sqrt{-a} \quad \text{as} \quad i\sqrt{a}, \quad \text{for a positive number } a.$$

Then the properties of real numbers and the fact that $i^2 = -1$ are applied.

CAUTION ***When working with negative radicands, use the definition $\sqrt{-a} = i\sqrt{a}$ before using any of the other rules for radicals.*** In particular, the rule $\sqrt{c} \cdot \sqrt{d} = \sqrt{cd}$ is valid only when c and d are *not* both negative. For example, consider the following.

$$\sqrt{-4} \cdot \sqrt{-9} = 2i \cdot 3i = 6i^2 = -6 \quad \text{Correct}$$

$$\sqrt{-4} \cdot \sqrt{-9} = \sqrt{(-4)(-9)} = \sqrt{36} = 6 \quad \text{Incorrect}$$

Classroom Example 3
Find each product or quotient. Simplify the answers.

(a) $\sqrt{-21} \cdot \sqrt{-21}$

(b) $\sqrt{-5} \cdot \sqrt{-30}$

(c) $\dfrac{\sqrt{-42}}{\sqrt{-3}}$ (d) $\dfrac{\sqrt{-63}}{\sqrt{21}}$

Answers:
(a) -21 (b) $-5\sqrt{6}$
(c) $\sqrt{14}$ (d) $i\sqrt{3}$

Teaching Tip Point out that the result in **Example 3(a)** follows directly from the definition of the square root of -7; that is, it is the number whose square is -7.

EXAMPLE 3 Finding Products and Quotients Involving $\sqrt{-a}$

Find each product or quotient. Simplify the answers.

(a) $\sqrt{-7} \cdot \sqrt{-7}$ **(b)** $\sqrt{-6} \cdot \sqrt{-10}$ **(c)** $\dfrac{\sqrt{-20}}{\sqrt{-2}}$ **(d)** $\dfrac{\sqrt{-48}}{\sqrt{24}}$

SOLUTION

(a) $\sqrt{-7} \cdot \sqrt{-7}$

$= i\sqrt{7} \cdot i\sqrt{7}$ (First write all square roots in terms of i.)

$= i^2 \cdot (\sqrt{7})^2$

$= -1 \cdot 7 \quad i^2 = -1;\ (\sqrt{a})^2 = a$

$= -7 \quad$ Multiply.

(b) $\sqrt{-6} \cdot \sqrt{-10}$

$= i\sqrt{6} \cdot i\sqrt{10}$

$= i^2 \cdot \sqrt{60}$

$= -1\sqrt{4 \cdot 15}$

$= -1 \cdot 2\sqrt{15}$

$= -2\sqrt{15}$

(c) $$\frac{\sqrt{-20}}{\sqrt{-2}} = \frac{i\sqrt{20}}{i\sqrt{2}} = \sqrt{\frac{20}{2}} = \sqrt{10} \quad \text{Quotient rule for radicals: } \frac{\sqrt[n]{a}}{\sqrt[n]{b}} = \sqrt[n]{\frac{a}{b}}$$

(d) $$\frac{\sqrt{-48}}{\sqrt{24}} = \frac{i\sqrt{48}}{\sqrt{24}} = i\sqrt{\frac{48}{24}} = i\sqrt{2} \quad \text{Quotient rule for radicals}$$

✔ **Now Try Exercises 29, 31, 33, and 35.**

Classroom Example 4

Write $\frac{15-\sqrt{-75}}{5}$ in standard form $a+bi$.

Answer: $3-i\sqrt{3}$

EXAMPLE 4 Simplifying a Quotient Involving $\sqrt{-a}$

Write $\frac{-8+\sqrt{-128}}{4}$ in standard form $a+bi$.

SOLUTION $\frac{-8+\sqrt{-128}}{4}$

$= \frac{-8+\sqrt{-64\cdot 2}}{4}$ Product rule for radicals

$= \frac{-8+8i\sqrt{2}}{4}$ $\sqrt{-64}=8i$

Be sure to factor before dividing.

$= \frac{4\left(-2+2i\sqrt{2}\right)}{4}$ Factor.

$= -2+2i\sqrt{2}$ Lowest terms; standard form

✔ **Now Try Exercise 41.**

With the definitions $i^2=-1$ and $\sqrt{-a}=i\sqrt{a}$ for $a>0$, all properties of real numbers are extended to complex numbers.

Teaching Tip Point out the similarities between addition of complex numbers and addition of linear polynomials.

Addition and Subtraction of Complex Numbers

For complex numbers $a+bi$ and $c+di$, the following hold.

$$(a+bi)+(c+di)=(a+c)+(b+d)i$$
$$(a+bi)-(c+di)=(a-c)+(b-d)i$$

That is, to add or subtract complex numbers, add or subtract the real parts, and add or subtract the imaginary parts.

Classroom Example 5

Find each sum or difference. Write answers in standard form.

(a) $(4-5i)+(-5+8i)$

(b) $(-10+7i)-(5-3i)$

Answers:

(a) $-1+3i$ (b) $-15+10i$

NORMAL FLOAT AUTO a+bi RADIAN MP
(3-4i)+(-2+6i)
1+2i
(-4+3i)-(6-7i)
-10+10i

This screen supports the results in **Example 5.**

EXAMPLE 5 Adding and Subtracting Complex Numbers

Find each sum or difference. Write answers in standard form.

(a) $(3-4i)+(-2+6i)$ **(b)** $(-4+3i)-(6-7i)$

SOLUTION

(a) $(3-4i)+(-2+6i)$

Add real parts. Add imaginary parts.

$= [3+(-2)]+[-4+6]i$ Commutative, associative and distributive properties

$= 1+2i$ Standard form

(b) $(-4+3i)-(6-7i)$

$= (-4-6)+[3-(-7)]i$ Subtract real parts. Subtract imaginary parts.

$= -10+10i$ Standard form

✔ **Now Try Exercises 47 and 49.**

The product of two complex numbers is found by multiplying as though the numbers were binomials and using the fact that $i^2 = -1$, as follows.

$$
\begin{aligned}
&(a + bi)(c + di) \\
&= ac + adi + bic + bidi && \text{FOIL method (Multiply First, Outer, Inner, Last terms.)} \\
&= ac + adi + bci + bdi^2 && \text{Commutative property; Multiply.} \\
&= ac + (ad + bc)i + bd(-1) && \text{Distributive property; } i^2 = -1 \\
&= (ac - bd) + (ad + bc)i && \text{Group like terms.}
\end{aligned}
$$

Multiplication of Complex Numbers

For complex numbers $a + bi$ and $c + di$, the following holds.

$$(a + bi)(c + di) = (ac - bd) + (ad + bc)i$$

To find a given product in routine calculations, it is often easier to multiply as with binomials and use the fact that $i^2 = -1$.

Classroom Example 6
Find each product. Write answers in standard form.
(a) $(5 + 3i)(2 - 7i)$
(b) $(4 - 5i)^2$
(c) $(9 - 8i)(9 + 8i)$

Answers:
(a) $31 - 29i$
(b) $-9 - 40i$
(c) 145, or $145 + 0i$

EXAMPLE 6 Multiplying Complex Numbers

Find each product. Write answers in standard form.

(a) $(2 - 3i)(3 + 4i)$ **(b)** $(4 + 3i)^2$ **(c)** $(6 + 5i)(6 - 5i)$

SOLUTION

(a) $(2 - 3i)(3 + 4i)$

$$
\begin{aligned}
&= 2(3) + 2(4i) - 3i(3) - 3i(4i) && \text{FOIL method} \\
&= 6 + 8i - 9i - 12i^2 && \text{Multiply.} \\
&= 6 - i - 12(-1) && \text{Combine like terms; } i^2 = -1. \\
&= 18 - i && \text{Standard form}
\end{aligned}
$$

(b) $(4 + 3i)^2$

$$
\begin{aligned}
&= 4^2 + 2(4)(3i) + (3i)^2 && \text{Square of a binomial: } (x + y)^2 = x^2 + 2xy + y^2 \\
&= 16 + 24i + 9i^2 && \text{Multiply; } (3i)^2 = 3^2i^2. \\
&= 16 + 24i + 9(-1) && i^2 = -1 \\
&= 7 + 24i && \text{Standard form}
\end{aligned}
$$

Remember to add twice the product of the two terms.

(c) $(6 + 5i)(6 - 5i)$

$$
\begin{aligned}
&= 6^2 - (5i)^2 && \text{Product of the sum and difference of two terms: } (x + y)(x - y) = x^2 - y^2 \\
&= 36 - 25(-1) && \text{Square; } (5i)^2 = 5^2i^2 = 25(-1). \\
&= 36 + 25 && \text{Multiply.} \\
&= 61, \text{ or } 61 + 0i && \text{Standard form}
\end{aligned}
$$

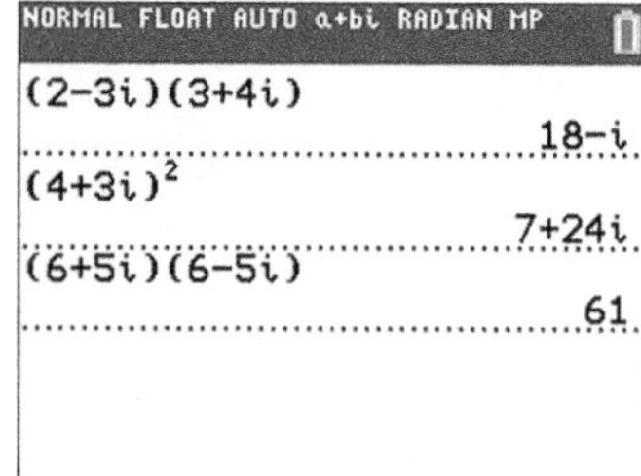

This screen supports the results found in **Example 6.**

✔ **Now Try Exercises 55, 59, and 63.**

```
NORMAL FLOAT AUTO a+bi RADIAN MP
conj(6+5i)
                    6-5i
```

The complex conjugate of $6 + 5i$ is $6 - 5i$.

Example 6(c) showed that $(6 + 5i)(6 - 5i) = 61$. The numbers $6 + 5i$ and $6 - 5i$ differ only in the sign of their imaginary parts and are called **complex conjugates.** ***The product of a complex number and its conjugate is always a real number.*** This product is the sum of the squares of the real and imaginary parts.

Property of Complex Conjugates

For real numbers a and b,

$$(a + bi)(a - bi) = a^2 + b^2.$$

To find the quotient of two complex numbers in standard form, we multiply both the numerator and the denominator by the complex conjugate of the denominator.

Classroom Example 7
Find each quotient. Write answers in standard form.

(a) $\frac{5 - 5i}{3 + i}$ (b) $\frac{15}{-i}$

Answers:
(a) $1 - 2i$
(b) $15i$, or $0 + 15i$

EXAMPLE 7 Dividing Complex Numbers

Find each quotient. Write answers in standard form.

(a) $\frac{3 + 2i}{5 - i}$ **(b)** $\frac{3}{i}$

SOLUTION

(a) $\frac{3 + 2i}{5 - i}$

$= \frac{(3 + 2i)(5 + i)}{(5 - i)(5 + i)}$ Multiply by the complex conjugate of the denominator in both the numerator and the denominator.

$= \frac{15 + 3i + 10i + 2i^2}{25 - i^2}$ Multiply.

$= \frac{13 + 13i}{26}$ Combine like terms; $i^2 = -1$.

$= \frac{13}{26} + \frac{13i}{26}$ $\frac{a + bi}{c} = \frac{a}{c} + \frac{bi}{c}$

$= \frac{1}{2} + \frac{1}{2}i$ Lowest terms; standard form

CHECK $\left(\frac{1}{2} + \frac{1}{2}i\right)(5 - i) = 3 + 2i$ ✓ Quotient × Divisor = Dividend

(b) $\frac{3}{i}$

$= \frac{3(-i)}{i(-i)}$ $-i$ is the conjugate of i.

$= \frac{-3i}{-i^2}$ Multiply.

$= \frac{-3i}{1}$ $-i^2 = -(-1) = 1$

$= -3i$, or $0 - 3i$ Standard form

✔ **Now Try Exercises 73 and 79.**

```
NORMAL FLOAT AUTO a+bi RADIAN MP
(3+2i)/(5-i)
                   .5+.5i
Ans▸Frac
                  1/2+1/2i
3/i
                     -3i
```

This screen supports the results in **Example 7.**

Complex Solutions of Quadratic Equations (Part 2)

Classroom Example 8
Solve $2x^2 - 2x = -5$ over the set of complex numbers.

Answer: $\left\{\frac{1}{2} \pm \frac{3}{2}i\right\}$

EXAMPLE 8 Solving a Quadratic Equation (Complex Solutions)

Solve $9x^2 + 5 = 6x$ over the set of complex numbers.

SOLUTION

$$9x^2 - 6x + 5 = 0 \quad \text{Standard form } ax^2 + bx + c = 0$$

$$x = \frac{-b \pm \sqrt{b^2 - 4ac}}{2a} \quad \text{Use the quadratic formula.}$$

The fraction bar extends under $-b$.

$$= \frac{-(-6) \pm \sqrt{(-6)^2 - 4(9)(5)}}{2(9)} \quad \text{Substitute } a = 9,\ b = -6,\ c = 5.$$

$$= \frac{6 \pm \sqrt{-144}}{18} \quad \text{Simplify.}$$

$$= \frac{6 \pm 12i}{18} \quad \sqrt{-144} = 12i$$

Factor first, then divide out the common factor.

$$= \frac{6(1 \pm 2i)}{6 \cdot 3} \quad \text{Factor.}$$

$$x = \frac{1 \pm 2i}{3} \quad \text{Write in lowest terms.}$$

The solution set is $\left\{\frac{1}{3} \pm \frac{2}{3}i\right\}$. $\quad \frac{a \pm bi}{c} = \frac{a}{c} \pm \frac{b}{c}i$

✔ Now Try Exercise 89.

Powers of i

Powers of i can be simplified using the facts

$$i^2 = -1 \quad \text{and} \quad i^4 = (i^2)^2 = (-1)^2 = 1.$$

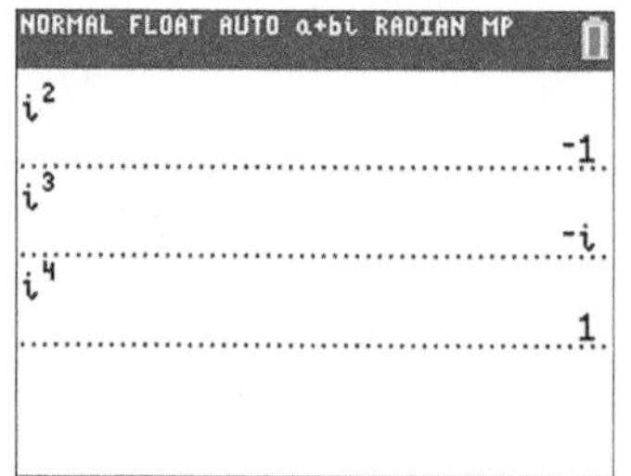

Powers of i can be found on a TI-84 Plus calculator.

Consider the following powers of i.

$i^1 = i$	$i^5 = i^4 \cdot i = 1 \cdot i = i$
$i^2 = -1$	$i^6 = i^4 \cdot i^2 = 1(-1) = -1$
$i^3 = i^2 \cdot i = (-1) \cdot i = -i$	$i^7 = i^4 \cdot i^3 = 1 \cdot (-i) = -i$
$i^4 = i^2 \cdot i^2 = (-1)(-1) = 1$	$i^8 = i^4 \cdot i^4 = 1 \cdot 1 = 1$ and so on.

Powers of i cycle through the same four outcomes (i, -1, $-i$, and 1) because i^4 has the same multiplicative property as 1. It follows that a power of i with an exponent that is a multiple of 4 has value 1.

Classroom Example 9
Simplify each power of i.
(a) i^{33} (b) i^{-14}

Answers:
(a) i (b) -1

EXAMPLE 9 Simplifying Powers of i

Simplify each power of i.

(a) i^{15} **(b)** i^{-3}

SOLUTION

(a) Because $i^4 = 1$, write the given power as a product involving i^4.

$$i^{15} = i^{12} \cdot i^3 = (i^4)^3 \cdot i^3 = 1^3(-i) = -i$$

(b) Multiply i^{-3} by 1 in the form of i^4 to create the least positive exponent for i.

$$i^{-3} = i^{-3} \cdot 1 = i^{-3} \cdot i^4 = i \quad i^4 = 1$$

✔ Now Try Exercises 97 and 105.

8.1 Exercises

1. $\sqrt{-1}$; -1 2. -4; -8
3. $10i$ 4. $6+2i$
5. true 6. true
7. true 8. true
9. false; *Every* real number is a complex number.
10. true
11. real, complex
12. real, complex
13. pure imaginary, nonreal complex, complex
14. pure imaginary, nonreal complex, complex
15. nonreal complex, complex
16. nonreal complex, complex
17. real, complex
18. real, complex
19. pure imaginary, nonreal complex, complex
20. pure imaginary, nonreal complex, complex
21. $5i$ 22. $6i$
23. $i\sqrt{10}$ 24. $i\sqrt{15}$
25. $12i\sqrt{2}$ 26. $10i\sqrt{5}$
27. $-3i\sqrt{2}$ 28. $-4i\sqrt{5}$
29. -13 30. -17
31. $-2\sqrt{6}$ 32. $-5\sqrt{3}$
33. $\sqrt{3}$ 34. $\sqrt{10}$
35. $i\sqrt{3}$ 36. $i\sqrt{2}$
37. $\frac{1}{2}$ 38. $\frac{1}{3}$
39. -2 40. -3
41. $-3-i\sqrt{6}$ 42. $-3-i\sqrt{2}$
43. $2+2i\sqrt{2}$ 44. $10+i\sqrt{2}$
45. $-\frac{1}{8}+\frac{\sqrt{2}}{8}i$ 46. $-\frac{1}{2}+\frac{\sqrt{2}}{2}i$

CONCEPT PREVIEW *Fill in the blank(s) to correctly complete each sentence.*

1. By definition, $i =$ ____, and therefore $i^2 =$ ____.
2. In $-4-8i$, the real part is ____ and the imaginary part is ____.
3. In terms of i, $\sqrt{-100} =$ ____.
4. The complex conjugate of $6-2i$ is ____.

CONCEPT PREVIEW *Determine whether each statement is* true *or* false. *If it is false, tell why.*

5. Every real number is a complex number.
6. No real number is a pure imaginary number.
7. Every pure imaginary number is a complex number.
8. A number can be both real and complex.
9. There is no real number that is a complex number.
10. A complex number might not be a pure imaginary number.

Concept Check Identify each number as real, complex, pure imaginary, *or* nonreal complex. *(More than one of these descriptions will apply.)*

11. -4 12. 0 13. $13i$ 14. $-7i$ 15. $5+i$
16. $-6-2i$ 17. π 18. $\sqrt{24}$ 19. $\sqrt{-25}$ 20. $\sqrt{-36}$

Write each number as the product of a real number and i. **See Example 1.**

21. $\sqrt{-25}$ 22. $\sqrt{-36}$ 23. $\sqrt{-10}$ 24. $\sqrt{-15}$
25. $\sqrt{-288}$ 26. $\sqrt{-500}$ 27. $-\sqrt{-18}$ 28. $-\sqrt{-80}$

Find each product or quotient. Simplify the answers. **See Example 3.**

29. $\sqrt{-13}\cdot\sqrt{-13}$ 30. $\sqrt{-17}\cdot\sqrt{-17}$ 31. $\sqrt{-3}\cdot\sqrt{-8}$
32. $\sqrt{-5}\cdot\sqrt{-15}$ 33. $\dfrac{\sqrt{-30}}{\sqrt{-10}}$ 34. $\dfrac{\sqrt{-70}}{\sqrt{-7}}$
35. $\dfrac{\sqrt{-24}}{\sqrt{8}}$ 36. $\dfrac{\sqrt{-54}}{\sqrt{27}}$ 37. $\dfrac{\sqrt{-10}}{\sqrt{-40}}$
38. $\dfrac{\sqrt{-8}}{\sqrt{-72}}$ 39. $\dfrac{\sqrt{-6}\cdot\sqrt{-2}}{\sqrt{3}}$ 40. $\dfrac{\sqrt{-12}\cdot\sqrt{-6}}{\sqrt{8}}$

Write each number in standard form $a+bi$. **See Example 4.**

41. $\dfrac{-6-\sqrt{-24}}{2}$ 42. $\dfrac{-9-\sqrt{-18}}{3}$ 43. $\dfrac{10+\sqrt{-200}}{5}$
44. $\dfrac{20+\sqrt{-8}}{2}$ 45. $\dfrac{-3+\sqrt{-18}}{24}$ 46. $\dfrac{-5+\sqrt{-50}}{10}$

47. $12 - i$ 48. $12 + 4i$
49. 2 50. 1
51. 0 52. 0
53. $-13 + 4i\sqrt{2}$
54. $-3\sqrt{7} + 2i$
55. $8 - i$ 56. $-2 + 16i$
57. $-14 + 2i$ 58. $17 + i$
59. $5 - 12i$ 60. $3 + 4i$
61. 10 62. 26
63. 13 64. 52
65. 7 66. 18
67. $25i$ 68. $53i$
69. $12 + 9i$ 70. $-120 - 35i$
71. $20 + 15i$ 72. $20 - 60i$
73. $2 - 2i$ 74. $4 - i$
75. $\frac{3}{5} - \frac{4}{5}i$ 76. $\frac{7}{25} - \frac{24}{25}i$
77. $-1 - 2i$ 78. $-2 + i$
79. $5i$ 80. $6i$
81. $8i$ 82. $12i$
83. $-\frac{2}{3}i$ 84. $-\frac{5}{9}i$
85. $\{\pm 4i\}$ 86. $\{\pm 6i\}$
87. $\{\pm 2i\sqrt{3}\}$ 88. $\{\pm 4i\sqrt{3}\}$
89. $\left\{-\frac{2}{3} \pm \frac{\sqrt{2}}{3}i\right\}$
90. $\left\{-\frac{3}{4} \pm \frac{\sqrt{7}}{4}i\right\}$
91. $\{3 \pm i\sqrt{5}\}$ 92. $\{-2 \pm i\sqrt{7}\}$
93. $\left\{\frac{1}{2} \pm \frac{\sqrt{6}}{2}i\right\}$ 94. $\left\{\frac{1}{3} \pm \frac{\sqrt{6}}{3}i\right\}$
95. $\left\{-\frac{1}{2} \pm \frac{\sqrt{3}}{2}i\right\}$ 96. $\{1 \pm i\}$
97. i 98. i
99. -1 100. -1
101. $-i$ 102. $-i$
103. 1 104. 1
105. $-i$ 106. -1
107. $-i$ 108. 1

Find each sum or difference. Write answers in standard form. ***See Example 5.***

47. $(3 + 2i) + (9 - 3i)$
48. $(4 - i) + (8 + 5i)$
49. $(-2 + 4i) - (-4 + 4i)$
50. $(-3 + 2i) - (-4 + 2i)$
51. $(2 - 5i) - (3 + 4i) - (-1 - 9i)$
52. $(-4 - i) - (2 + 3i) + (6 + 4i)$
53. $-i\sqrt{2} - 2 - \left(6 - 4i\sqrt{2}\right) - \left(5 - i\sqrt{2}\right)$
54. $3\sqrt{7} - \left(4\sqrt{7} - i\right) - 4i + \left(-2\sqrt{7} + 5i\right)$

Find each product. Write answers in standard form. ***See Example 6.***

55. $(2 + i)(3 - 2i)$
56. $(-2 + 3i)(4 - 2i)$
57. $(2 + 4i)(-1 + 3i)$
58. $(1 + 3i)(2 - 5i)$
59. $(3 - 2i)^2$
60. $(2 + i)^2$
61. $(3 + i)(3 - i)$
62. $(5 + i)(5 - i)$
63. $(-2 - 3i)(-2 + 3i)$
64. $(6 - 4i)(6 + 4i)$
65. $\left(\sqrt{6} + i\right)\left(\sqrt{6} - i\right)$
66. $\left(\sqrt{2} - 4i\right)\left(\sqrt{2} + 4i\right)$
67. $i(3 - 4i)(3 + 4i)$
68. $i(2 + 7i)(2 - 7i)$
69. $3i(2 - i)^2$
70. $-5i(4 - 3i)^2$
71. $(2 + i)(2 - i)(4 + 3i)$
72. $(3 - i)(3 + i)(2 - 6i)$

Find each quotient. Write answers in standard form. ***See Example 7.***

73. $\frac{6 + 2i}{1 + 2i}$
74. $\frac{14 + 5i}{3 + 2i}$
75. $\frac{2 - i}{2 + i}$
76. $\frac{4 - 3i}{4 + 3i}$
77. $\frac{1 - 3i}{1 + i}$
78. $\frac{-3 + 4i}{2 - i}$
79. $\frac{-5}{i}$
80. $\frac{-6}{i}$
81. $\frac{8}{-i}$
82. $\frac{12}{-i}$
83. $\frac{2}{3i}$
84. $\frac{5}{9i}$

Solve each equation over the set of complex numbers. ***See Examples 2 and 8.***

85. $x^2 = -16$
86. $x^2 = -36$
87. $x^2 + 12 = 0$
88. $x^2 + 48 = 0$
89. $3x^2 + 2 = -4x$
90. $2x^2 + 3x = -2$
91. $x^2 - 6x + 14 = 0$
92. $x^2 + 4x + 11 = 0$
93. $4(x^2 - x) = -7$
94. $3(3x^2 - 2x) = -7$
95. $x^2 + 1 = -x$
96. $x^2 + 2 = 2x$

Simplify each power of i. ***See Example 9.***

97. i^{25}
98. i^{29}
99. i^{22}
100. i^{26}
101. i^{23}
102. i^{27}
103. i^{32}
104. i^{40}
105. i^{-13}
106. i^{-14}
107. $\frac{1}{i^{-11}}$
108. $\frac{1}{i^{-12}}$

109. Every i^4 factor acts as 1, so if the remainder is R, the final product is i^R.

110. $i^{44} = (i^4)^{11} = 1^{11} = 1$, so multiplying by i^{44} is equivalent to multiplying by 1, the identity element for multiplication.

111. $110 + 32i$

112. $16.22°$

113. $E = 30 + 60i$

114. $E = 50 + 98i$

115. $Z = \frac{233}{37} + \frac{119}{37}i$

116. $I = \frac{215}{26} + \frac{95}{26}i$

Concept Check *Work each problem.*

109. Suppose a friend says that she has discovered a method of simplifying a positive power of i:

"Just divide the exponent by 4. The answer is i raised to the remainder."

Explain why her method works.

110. Why does the following method of simplifying i^{-42} work?

$$i^{-42} = i^{-42} \cdot i^{44} = i^2 = -1$$

(Modeling) Impedance ***Impedance*** *is a measure of the opposition to the flow of alternating electrical current found in common electrical outlets. It consists of two parts,* ***resistance*** *and* ***reactance.*** *Resistance occurs when a light bulb is turned on, while reactance is produced when electricity passes through a coil of wire like that found in electric motors. Impedance Z in ohms (Ω) can be expressed as a complex number, where the real part represents resistance and the imaginary part represents reactance.*

For example, if the resistive part is 3 ohms *and the reactive part is* 4 ohms, *then the impedance could be described by the complex number* $Z = 3 + 4i$. *In the series circuit shown in the figure, the total impedance will be the sum of the individual impedances. (Source:* Wilcox, G. and C. Hesselberth, *Electricity for Engineering Technology,* Allyn & Bacon.*)*

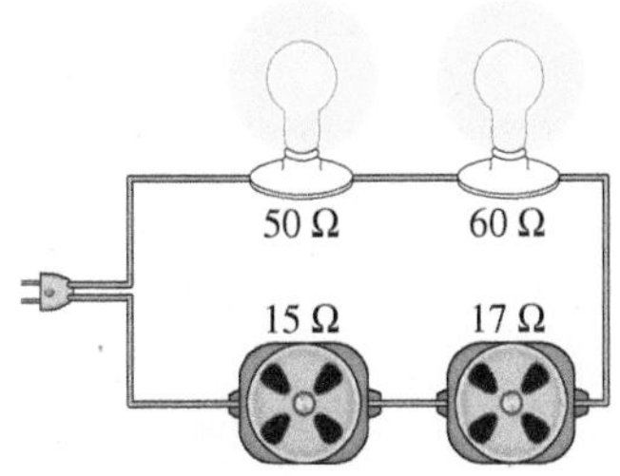

111. The circuit contains two light bulbs and two electric motors. Assuming that the light bulbs are pure resistive and the motors are pure reactive, find the total impedance in this circuit and express it in the form $Z = a + bi$.

112. The phase angle θ measures the phase difference between the voltage and the current in an electrical circuit. Angle θ (in degrees) can be determined by the equation $\tan \theta = \frac{b}{a}$. Find θ, to the nearest hundredth, for this circuit.

(Modeling) Ohm's Law Complex numbers are used to describe current I, voltage E, and impedance Z (the opposition to current). These three quantities are related by the equation

$$E = IZ, \quad \text{which is known as } \textbf{Ohm's law.}$$

Thus, if any two of these quantities are known, the third can be found. In each exercise, solve the equation $E = IZ$ *for the missing variable.*

113. $I = 8 + 6i, \quad Z = 6 + 3i$

114. $I = 10 + 6i, \quad Z = 8 + 5i$

115. $I = 7 + 5i, \quad E = 28 + 54i$

116. $E = 35 + 55i, \quad Z = 6 + 4i$

Work each problem.

117. Show that $\frac{\sqrt{2}}{2} + \frac{\sqrt{2}}{2}i$ is a square root of i.

118. Show that $\frac{\sqrt{3}}{2} + \frac{1}{2}i$ is a cube root of i.

119. Show that $-2 + i$ is a solution of the equation $x^2 + 4x + 5 = 0$.

120. Show that $-3 + 4i$ is a solution of the equation $x^2 + 6x + 25 = 0$.

8.2 Trigonometric (Polar) Form of Complex Numbers

- The Complex Plane and Vector Representation
- Trigonometric (Polar) Form
- Converting between Rectangular and Trigonometric (Polar) Forms
- An Application of Complex Numbers to Fractals

The Complex Plane and Vector Representation

Unlike real numbers, complex numbers cannot be ordered. One way to organize and illustrate them is by using a graph in a rectangular coordinate system.

To graph a complex number such as $2 - 3i$, we modify the coordinate system by calling the horizontal axis the **real axis** and the vertical axis the **imaginary axis.** Then complex numbers can be graphed in this **complex plane,** as shown in **Figure 3.** ***Each complex number*** $a + bi$ ***determines a unique position vector with initial point*** $(0, 0)$ ***and terminal point*** (a, b).

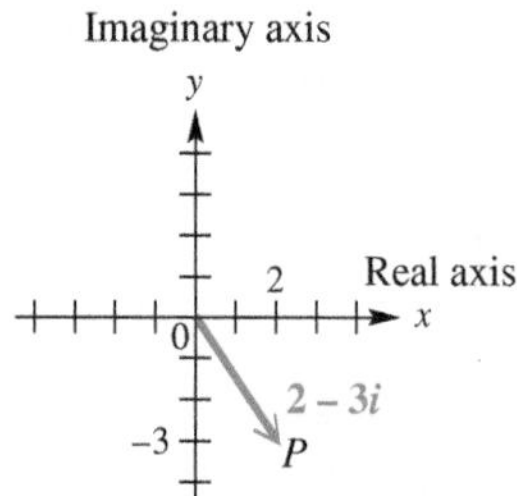

Figure 3

NOTE This geometric representation is the reason that $a + bi$ is called the **rectangular form** of a complex number. (*Rectangular form* is also known as *standard form.*)

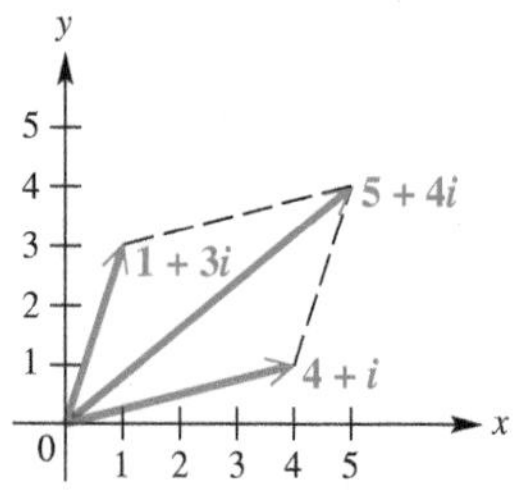

Figure 4

Recall that the sum of the two complex numbers $4 + i$ and $1 + 3i$ is

$$(4 + i) + (1 + 3i) = 5 + 4i.$$

Graphically, the sum of two complex numbers is represented by the vector that is the **resultant** of the vectors corresponding to the two numbers. See **Figure 4.**

EXAMPLE 1 Expressing the Sum of Complex Numbers Graphically

Find the sum of $6 - 2i$ and $-4 - 3i$. Graph both complex numbers and their resultant.

SOLUTION The sum is found by adding the two numbers.

$$(6 - 2i) + (-4 - 3i) = 2 - 5i \quad \text{Add real parts, and add imaginary parts.}$$

The graphs are shown in **Figure 5.**

Figure 5

✔ **Now Try Exercise 17.**

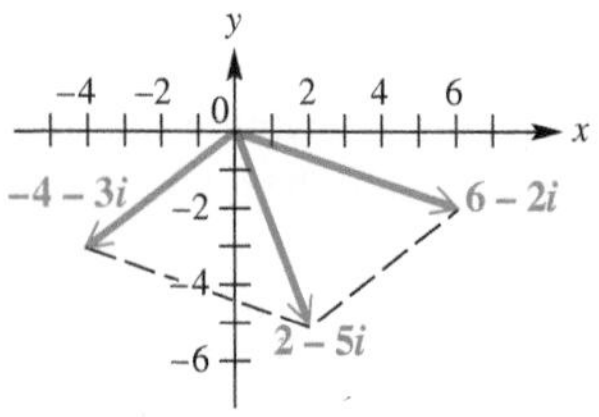

Trigonometric (Polar) Form

Figure 6 shows the complex number $x + yi$ that corresponds to a vector **OP** with direction angle θ and magnitude r. The following relationships among x, y, r, and θ can be verified from **Figure 6.**

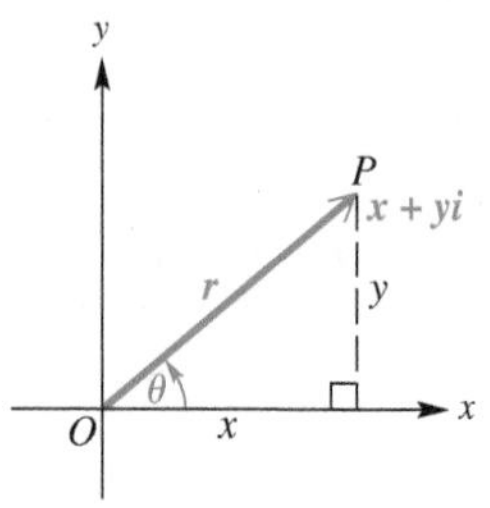

Figure 6

Relationships among x, y, r, and θ

$$x = r\cos\theta \qquad y = r\sin\theta$$

$$r = \sqrt{x^2 + y^2} \qquad \tan\theta = \frac{y}{x}, \text{ if } x \neq 0$$

Substituting $x = r\cos\theta$ and $y = r\sin\theta$ into $x + yi$ gives the following.

$$\begin{aligned} & x + yi \\ &= r\cos\theta + (r\sin\theta)i \quad && \text{Substitute.} \\ &= r(\cos\theta + i\sin\theta) \quad && \text{Factor out } r. \end{aligned}$$

Classroom Example 1
Find the sum of $2 + 3i$ and $-4 + 2i$. Graph both complex numbers and their resultant.

Answer: $-2 + 5i$

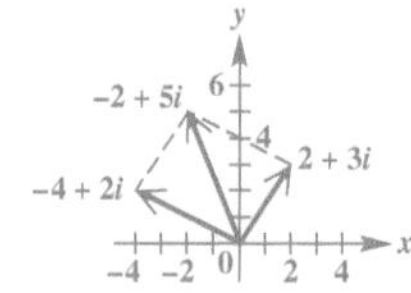

Trigonometric (Polar) Form of a Complex Number

The **trigonometric form** (or **polar form**) of the complex number $x + yi$ is

$$r(\cos \theta + i \sin \theta).$$

The expression $\cos \theta + i \sin \theta$ is sometimes abbreviated **cis θ.** Using this notation, $r(\cos \theta + i \sin \theta)$ is written r **cis** θ.

The number r is the **absolute value** (or **modulus**) of $x + yi$, and θ is the **argument** of $x + yi$. In this section, we choose the value of θ in the interval $[0°, 360°)$. Any angle coterminal with θ also could serve as the argument.

Classroom Example 2
Write $10(\cos 135° + i \sin 135°)$ in rectangular form.

Answer: $-5\sqrt{2} + 5i\sqrt{2}$

EXAMPLE 2 Converting from Trigonometric Form to Rectangular Form

Write $2(\cos 300° + i \sin 300°)$ in rectangular form.

ALGEBRAIC SOLUTION

$2(\cos 300° + i \sin 300°)$

$= 2\left(\frac{1}{2} - i\frac{\sqrt{3}}{2}\right)$ $\quad$ $\cos 300° = \frac{1}{2}$; $\sin 300° = -\frac{\sqrt{3}}{2}$

$= 1 - i\sqrt{3}$ $\quad$ Distributive property

Note that the real part is positive and the imaginary part is negative. This is consistent with 300° being a quadrant IV angle. For a 300° angle, the reference angle is 60°. Thus the function values cos 300° and sin 300° correspond *in absolute value* to those of cos 60° and sin 60°, with the first of these equal to $\frac{1}{2}$ and the second equal to $-\frac{\sqrt{3}}{2}$.

GRAPHING CALCULATOR SOLUTION

In **Figure 7,** the first result confirms the algebraic solution, where an approximation for $-\sqrt{3}$ is used for the imaginary part (from the second result). The TI-84 Plus also converts from polar to rectangular form, as seen in the third and fourth results.

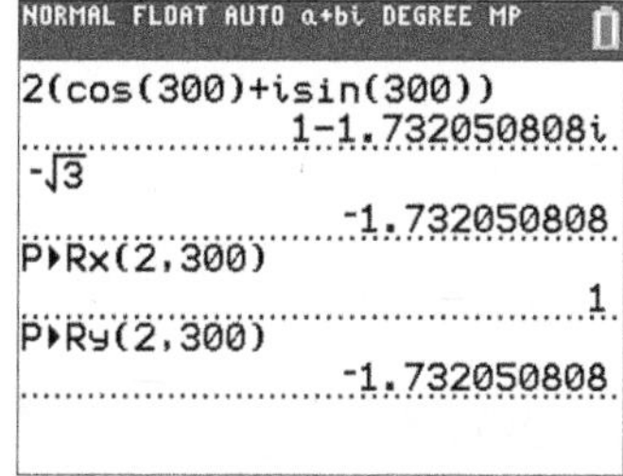

Figure 7

✔ **Now Try Exercise 33.**

Teaching Tip Relate the earlier definitions

$$\frac{x}{r} = \cos \theta, \quad \frac{y}{r} = \sin \theta,$$

and $\quad x^2 + y^2 = r^2$

to the relationships given here. ***Caution students to be especially careful about determining the correct quadrant of θ in Step 3.***

Converting between Rectangular and Trigonometric (Polar) Forms

To convert from rectangular form to trigonometric form, we use the following procedure.

Converting from Rectangular to Trigonometric Form

Step 1 Sketch a graph of the number $x + yi$ in the complex plane.

Step 2 Find r by using the equation $r = \sqrt{x^2 + y^2}$.

Step 3 Find θ by using the equation $\tan \theta = \frac{y}{x}$, where $x \neq 0$, choosing the quadrant indicated in Step 1.

CAUTION Errors often occur in Step 3. ***Be sure to choose the correct quadrant for θ by referring to the graph sketched in Step 1.***

Classroom Example 3
Write each complex number in trigonometric form.
(a) $8 - 8i$ (b) -15

Answers:
(a) $8\sqrt{2}$ cis 315°
(b) 15 cis 180°

EXAMPLE 3 Converting from Rectangular to Trigonometric Form

Write each complex number in trigonometric form.

(a) $-\sqrt{3} + i$ (Use radian measure.) **(b)** $-3i$ (Use degree measure.)

SOLUTION

(a) We start by sketching the graph of $-\sqrt{3} + i$ in the complex plane, as shown in **Figure 8.** Next, we use $x = -\sqrt{3}$ and $y = 1$ to find r and θ.

$$r = \sqrt{x^2 + y^2} = \sqrt{\left(-\sqrt{3}\right)^2 + 1^2} = \sqrt{3 + 1} = 2$$

$$\tan \theta = \frac{y}{x} = \frac{1}{-\sqrt{3}} = -\frac{1}{\sqrt{3}} \cdot \frac{\sqrt{3}}{\sqrt{3}} = -\frac{\sqrt{3}}{3}$$
Rationalize the denominator.

```
NORMAL FLOAT AUTO a+bi RADIAN MP
R▸Pr(-√3,1)
                       2
R▸Pθ(-√3,1)
             2.617993878
5π/6
             2.617993878
```

See **Example 3(a).** The TI-84 Plus converts from rectangular form to polar form. The value of θ in the second result is an approximation for $\frac{5\pi}{6}$, as shown in the third result.

Because $\tan \theta = -\frac{\sqrt{3}}{3}$, the reference angle for θ in radians is $\frac{\pi}{6}$. From the graph, we see that θ is in quadrant II, so $\theta = \pi - \frac{\pi}{6} = \frac{5\pi}{6}$.

$$-\sqrt{3} + i = 2\left(\cos \frac{5\pi}{6} + i \sin \frac{5\pi}{6}\right), \quad \text{or} \quad 2 \text{ cis } \frac{5\pi}{6}$$

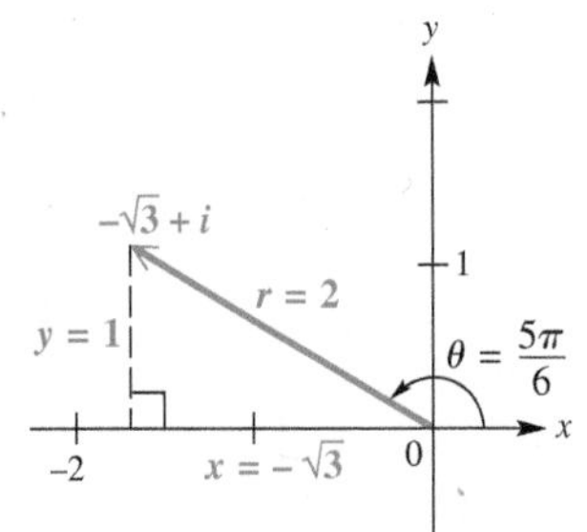

Figure 8

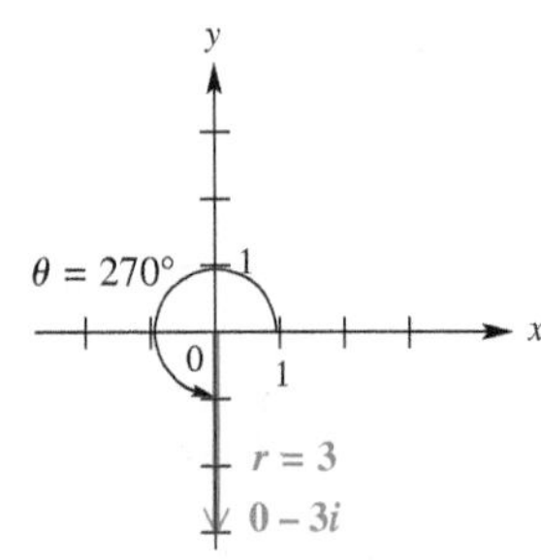

Figure 9

(b) See **Figure 9.** Because $-3i = 0 - 3i$, we have $x = 0$ and $y = -3$.

$$r = \sqrt{0^2 + (-3)^2} = \sqrt{0 + 9} = \sqrt{9} = 3 \quad \text{Substitute.}$$

```
NORMAL FLOAT AUTO a+bi DEGREE MP
R▸Pr(0,-3)
                       3
R▸Pθ(0,-3)
                     -90
```

Compare to the result in **Example 3(b).** The angle $-90°$ is coterminal with 270°. The calculator returns θ values between $-180°$ and 180°.

We cannot find θ by using $\tan \theta = \frac{y}{x}$ because $x = 0$. However, the graph shows that the least positive value for θ is 270°.

$$-3i = 3(\cos 270° + i \sin 270°), \quad \text{or} \quad 3 \text{ cis } 270° \quad \text{Trigonometric form}$$

✔ **Now Try Exercises 45 and 51.**

Classroom Example 4
Write each complex number in its alternative form, using calculator approximations as necessary.
(a) $7(\cos 205° + i \sin 205°)$
(b) $-7 + 2i$

Answers:
(a) $-6.3442 - 2.9583i$
(b) $\sqrt{53}$ cis 164.05°

EXAMPLE 4 Converting between Trigonometric and Rectangular Forms Using Calculator Approximations

Write each complex number in its alternative form, using calculator approximations as necessary.

(a) $6(\cos 125° + i \sin 125°)$ **(b)** $5 - 4i$

SOLUTION

(a) Because 125° does not have a special angle as a reference angle, we cannot find exact values for cos 125° and sin 125°.

$$6(\cos 125° + i \sin 125°)$$
$$\approx 6(-0.5735764364 + 0.8191520443i) \quad \text{Use a calculator set to degree mode.}$$
$$\approx -3.4415 + 4.9149i \quad \text{Four decimal places}$$

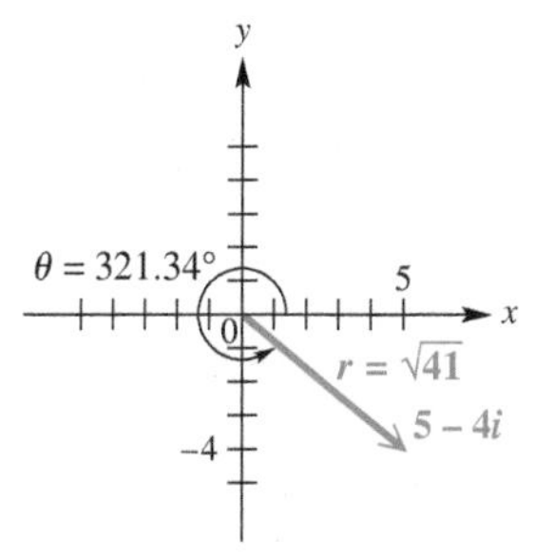

Figure 10

(b) A sketch of $5 - 4i$ shows that θ must be in quadrant IV. See **Figure 10.**

$$r = \sqrt{5^2 + (-4)^2} = \sqrt{41} \quad \text{and} \quad \tan\theta = -\frac{4}{5}$$

Use a calculator to find that one measure of θ is $-38.66°$. In order to express θ in the interval $[0, 360°)$, we find $\theta = 360° - 38.66° = 321.34°$.

$$5 - 4i = \sqrt{41} \text{ cis } 321.34°$$

✔ **Now Try Exercises 57 and 61.**

An Application of Complex Numbers to Fractals At its basic level, a **fractal** is a unique, enchanting geometric figure with an endless self-similarity property. A fractal image repeats itself infinitely with ever-decreasing dimensions. If we look at smaller and smaller portions, we will continue to see the whole—it is much like looking into two parallel mirrors that are facing each other.

Classroom Example 5
Refer to **Example 5**. Determine whether each number belongs to the Julia set.

(a) $z = 1 - 1i$

(b) $z = -1 + 0i$

Answers:

(a) no (b) yes

EXAMPLE 5 Deciding Whether a Complex Number Is in the Julia Set

The fractal called the **Julia set** is shown in **Figure 11.** To determine whether a complex number $z = a + bi$ is in this Julia set, perform the following sequence of calculations.

$$z^2 - 1, \quad (z^2 - 1)^2 - 1, \quad [(z^2 - 1)^2 - 1]^2 - 1, \quad \ldots$$

If the absolute values of any of the resulting complex numbers exceed 2, then the complex number z is not in the Julia set. Otherwise z is part of this set and the point (a, b) should be shaded in the graph.

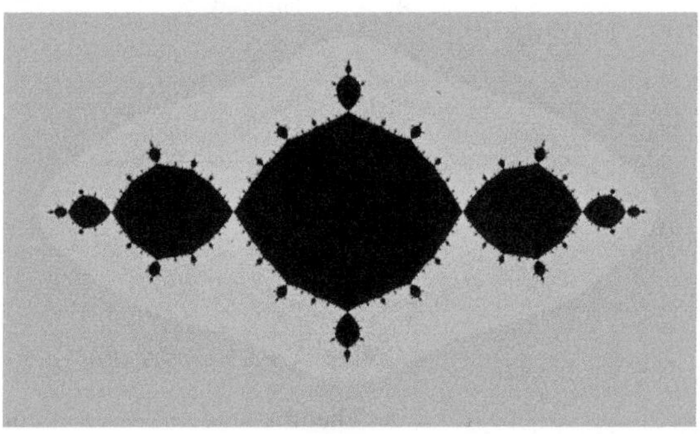

Figure 11

Determine whether each number belongs to the Julia set.

(a) $z = 0 + 0i$ **(b)** $z = 1 + 1i$

SOLUTION

(a) Here

$$z = 0 + 0i = 0,$$
$$z^2 - 1 = 0^2 - 1 = -1,$$
$$(z^2 - 1)^2 - 1 = (-1)^2 - 1 = 0,$$
$$[(z^2 - 1)^2 - 1]^2 - 1 = 0^2 - 1 = -1, \quad \text{and so on.}$$

We see that the calculations repeat as $0, -1, 0, -1$, and so on. The absolute values are either 0 or 1, which do not exceed 2, so $0 + 0i$ is in the Julia set and the point $(0, 0)$ is part of the graph.

(b) For $z = 1 + 1i$, we have the following.

$$\begin{aligned} & z^2 - 1 \\ &= (1+i)^2 - 1 && \text{Substitute for } z;\ 1 + 1i = 1 + i. \\ &= (1 + 2i + i^2) - 1 && \text{Square the binomial; } (x+y)^2 = x^2 + 2xy + y^2. \\ &= -1 + 2i && i^2 = -1 \end{aligned}$$

The absolute value is

$$\sqrt{(-1)^2 + 2^2} = \sqrt{5}.$$

Because $\sqrt{5}$ is greater than 2, the number $1 + 1i$ is not in the Julia set and $(1, 1)$ is not part of the graph.

Now Try Exercise 67.

8.2 Exercises

1. (a) 2
   (b) $2(\cos 0° + i \sin 0°)$
2. (a) -2
   (b) $2(\cos 180° + i \sin 180°)$
3. (a) $2i$
   (b) $2(\cos 90° + i \sin 90°)$
4. (a) $-2i$
   (b) $2(\cos 270° + i \sin 270°)$
5. (a) $2 + 2i$
   (b) $2\sqrt{2}(\cos 45° + i \sin 45°)$
6. (a) $-\sqrt{3} - i$
   (b) $2(\cos 210° + i \sin 210°)$
7. length (magnitude)
8. angle; x

9. $-3 + 2i$

10. $6 - 5i$

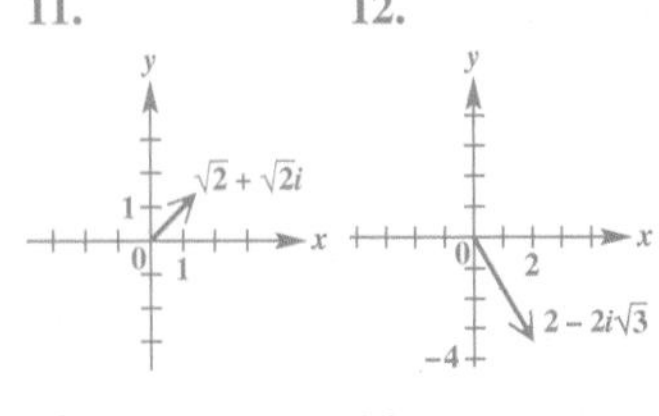

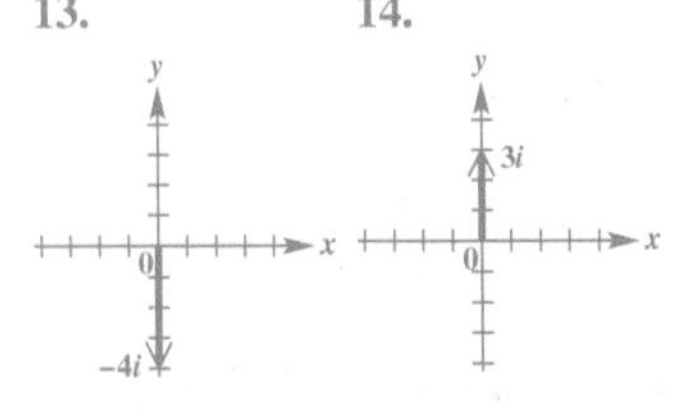

CONCEPT PREVIEW *For each complex number shown, give* **(a)** *its rectangular form and* **(b)** *its trigonometric (polar) form with* $r > 0$, $0° \le \theta < 360°$.

1.

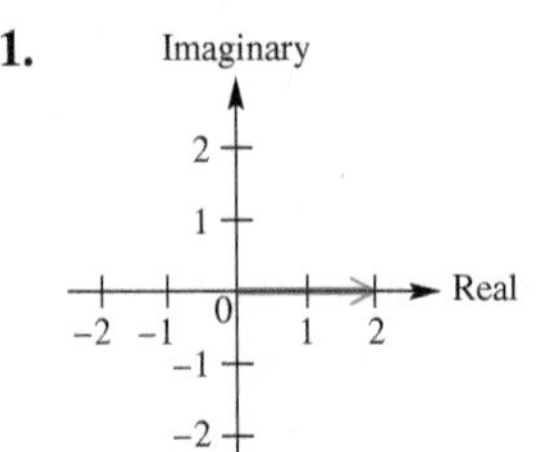

2.

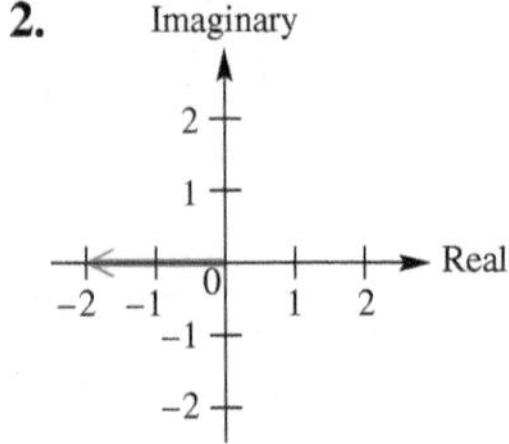

3.

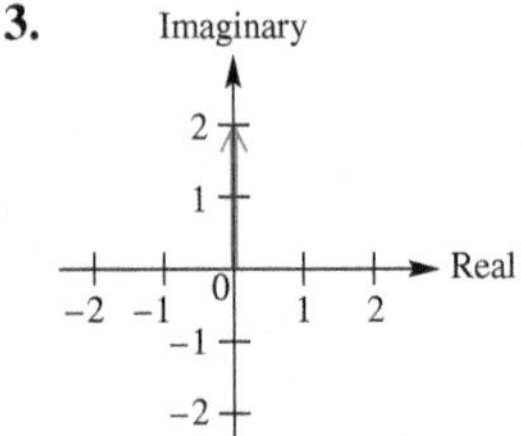

4.

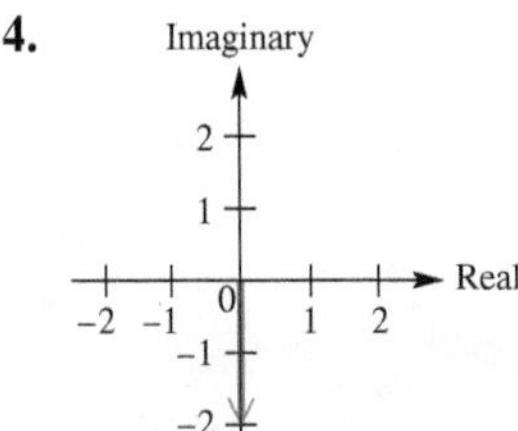

5.

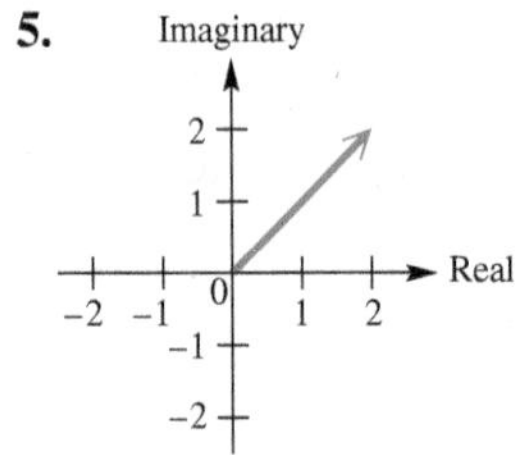

6.

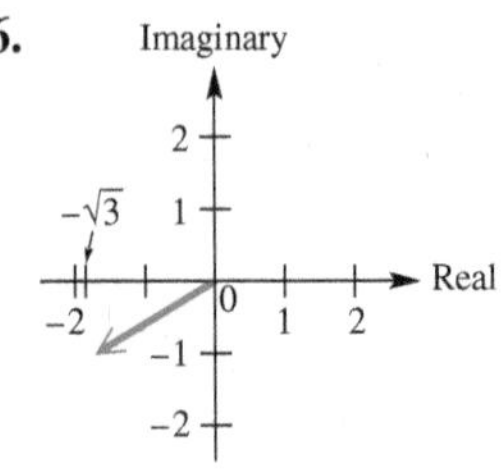

CONCEPT PREVIEW *Fill in the blank(s) to correctly complete each sentence.*

7. The absolute value (or modulus) of a complex number represents the ______ of the vector representing it in the complex plane.

8. The geometric interpretation of the argument of a complex number is the ______ formed by the vector and the positive ______-axis.

Graph each complex number. See ***Example 1.***

9. $-3 + 2i$	**10.** $6 - 5i$	**11.** $\sqrt{2} + \sqrt{2}i$	**12.** $2 - 2i\sqrt{3}$
13. $-4i$	**14.** $3i$	**15.** -8	**16.** 2

Find the sum of each pair of complex numbers. In Exercises 17–20, graph both complex numbers and their resultant. See ***Example 1.***

17. $4 - 3i,\ -1 + 2i$	**18.** $2 + 3i,\ -4 - i$	**19.** $5 - 6i,\ -5 + 3i$
20. $7 - 3i,\ -4 + 3i$	**21.** $-3,\ 3i$	**22.** $6,\ -2i$
23. $-5 - 8i,\ -1$	**24.** $4 - 2i,\ 5$	**25.** $7 + 6i,\ 3i$

15. 16.

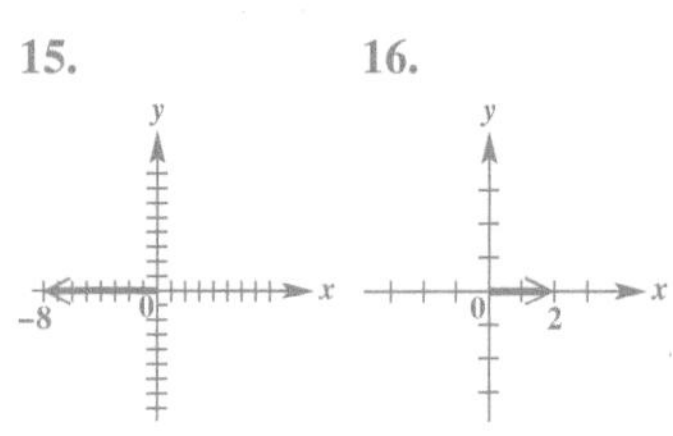

17. $3 - i$ 18. $-2 + 2i$

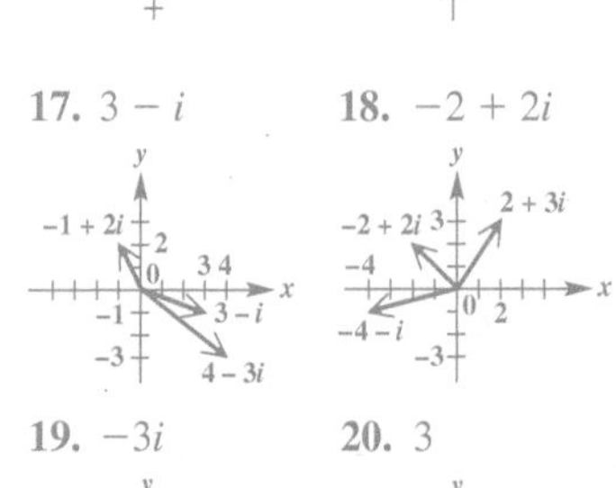

19. $-3i$ 20. 3

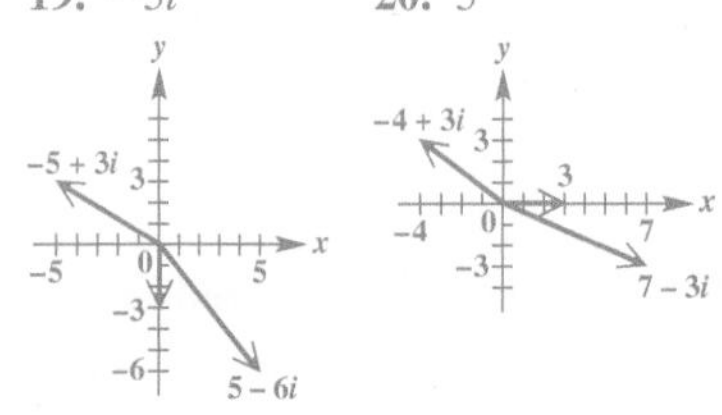

21. $-3 + 3i$ 22. $6 - 2i$
23. $-6 - 8i$ 24. $9 - 2i$
25. $7 + 9i$ 26. $2 + 4i$
27. $\frac{7}{6} + \frac{7}{6}i$ 28. $\frac{8}{35} - \frac{13}{28}i$
29. $\sqrt{2} + i\sqrt{2}$ 30. $2 + 2i\sqrt{3}$
31. $10i$ 32. $-8i$
33. $-2 - 2i\sqrt{3}$ 34. $\sqrt{3} - i$
35. $-\frac{3\sqrt{3}}{2} + \frac{3}{2}i$ 36. $\sqrt{3} + i$
37. $\frac{5}{2} - \frac{5\sqrt{3}}{2}i$
38. $-3\sqrt{2} + 3i\sqrt{2}$
39. $-1 - i$ 40. $\frac{\sqrt{6}}{2} - \frac{\sqrt{6}}{2}i$
41. $2\sqrt{3} - 2i$ 42. $\frac{\sqrt{2}}{2} - \frac{\sqrt{6}}{2}i$
43. $6(\cos 240° + i \sin 240°)$
44. $2(\cos 60° + i \sin 60°)$
45. $2(\cos 330° + i \sin 330°)$
46. $8(\cos 30° + i \sin 30°)$
47. $5\sqrt{2}(\cos 225° + i \sin 225°)$
48. $2\sqrt{2}(\cos 135° + i \sin 135°)$
49. $2\sqrt{2}(\cos 45° + i \sin 45°)$
50. $4\sqrt{2}(\cos 45° + i \sin 45°)$
51. $5(\cos 90° + i \sin 90°)$
52. $2(\cos 270° + i \sin 270°)$
53. $4(\cos 180° + i \sin 180°)$
54. $7(\cos 0° + i \sin 0°)$
55. $\sqrt{13}(\cos 56.31° + i \sin 56.31°)$
56. $0.8192 + 0.5736i$
57. $-1.0261 - 2.8191i$
58. $\sqrt{17}(\cos 165.96° + i \sin 165.96°)$
59. $12(\cos 90° + i \sin 90°)$

26. $2 + 6i, -2i$ **27.** $\frac{1}{2} + \frac{2}{3}i, \frac{2}{3} + \frac{1}{2}i$ **28.** $-\frac{1}{5} + \frac{2}{7}i, \frac{3}{7} - \frac{3}{4}i$

Write each complex number in rectangular form. ***See Example 2.***

29. $2(\cos 45° + i \sin 45°)$ **30.** $4(\cos 60° + i \sin 60°)$
31. $10(\cos 90° + i \sin 90°)$ **32.** $8(\cos 270° + i \sin 270°)$
33. $4(\cos 240° + i \sin 240°)$ **34.** $2(\cos 330° + i \sin 330°)$
35. $3 \text{ cis } 150°$ **36.** $2 \text{ cis } 30°$ **37.** $5 \text{ cis } 300°$
38. $6 \text{ cis } 135°$ **39.** $\sqrt{2} \text{ cis } 225°$ **40.** $\sqrt{3} \text{ cis } 315°$
41. $4(\cos(-30°) + i \sin(-30°))$ **42.** $\sqrt{2}(\cos(-60°) + i \sin(-60°))$

Write each complex number in trigonometric form $r(\cos \theta + i \sin \theta)$, with θ in the interval $[0°, 360°)$. ***See Example 3.***

43. $-3 - 3i\sqrt{3}$ **44.** $1 + i\sqrt{3}$ **45.** $\sqrt{3} - i$ **46.** $4\sqrt{3} + 4i$
47. $-5 - 5i$ **48.** $-2 + 2i$ **49.** $2 + 2i$ **50.** $4 + 4i$
51. $5i$ **52.** $-2i$ **53.** -4 **54.** 7

Write each complex number in its alternative form, using a calculator to approximate answers to four decimal places as necessary. ***See Example 4.***

	Rectangular Form	Trigonometric Form
55.	$2 + 3i$	________
56.	________	$\cos 35° + i \sin 35°$
57.	________	$3(\cos 250° + i \sin 250°)$
58.	$-4 + i$	________
59.	$12i$	________
60.	________	$3 \text{ cis } 180°$
61.	$3 + 5i$	________
62.	________	$\text{cis } 110.5°$

Concept Check The complex number z, where $z = x + yi$, can be graphed in the plane as (x, y). Describe the graphs of all complex numbers z satisfying the given conditions.

63. The absolute value of z is 1. **64.** The real and imaginary parts of z are equal.
65. The real part of z is 1. **66.** The imaginary part of z is 1.

Julia Set Refer to ***Example 5.***

67. Is $z = -0.2i$ in the Julia set?

68. The graph of the Julia set in **Figure 11** appears to be symmetric with respect to both the x-axis and the y-axis. Complete the following to show that this is true.

(a) Show that complex conjugates have the same absolute value.

(b) Compute $z_1^2 - 1$ and $z_2^2 - 1$, where $z_1 = a + bi$ and $z_2 = a - bi$.

(c) Discuss why if (a, b) is in the Julia set, then so is $(a, -b)$.

(d) Conclude that the graph of the Julia set must be symmetric with respect to the x-axis.

(e) Using a similar argument, show that the Julia set must also be symmetric with respect to the y-axis.

60. -3

61. $\sqrt{34}(\cos 59.04° + i \sin 59.04°)$

62. $-0.3502 + 0.9367i$

63. It is the circle of radius 1 centered at the origin.

64. It is the line $y = x$.

65. It is the vertical line $x = 1$.

66. It is the horizontal line $y = 1$.

Answers for the remaining exercises are given in the answer section at the back of the text.

Concept Check Identify the geometric condition (A, B, or C) that implies the situation.

A. *The corresponding vectors have opposite directions.*

B. *The terminal points of the vectors corresponding to* $a + bi$ *and* $c + di$ *lie on a horizontal line.*

C. *The corresponding vectors have the same direction.*

69. The difference between two nonreal complex numbers $a + bi$ and $c + di$ is a real number.

70. The absolute value of the sum of two complex numbers $a + bi$ and $c + di$ is equal to the sum of their absolute values.

71. The absolute value of the difference of two complex numbers $a + bi$ and $c + di$ is equal to the sum of their absolute values.

72. *Concept Check* Show that z and iz have the same absolute value. How are the graphs of these two numbers related?

8.3 The Product and Quotient Theorems

- Products of Complex Numbers in Trigonometric Form
- Quotients of Complex Numbers in Trigonometric Form

Products of Complex Numbers in Trigonometric Form Using the FOIL method to multiply complex numbers in rectangular form, we find the product of $1 + i\sqrt{3}$ and $-2\sqrt{3} + 2i$ as follows.

$$\left(1 + i\sqrt{3}\right)\left(-2\sqrt{3} + 2i\right)$$

$$= -2\sqrt{3} + 2i - 2i(3) + 2i^2\sqrt{3} \quad \text{FOIL method}$$

$$= -2\sqrt{3} + 2i - 6i - 2\sqrt{3} \quad i^2 = -1$$

$$= -4\sqrt{3} - 4i \quad \text{Combine like terms.}$$

```
NORMAL FLOAT AUTO a+bi DEGREE MP
angle(1+√3i)
                              60
|1+√3i|
                               2
```

With the TI-84 Plus calculator in complex and degree modes, the MATH menu can be used to find the angle and the magnitude (absolute value) of a complex number.

We can also find this same product by first converting the complex numbers $1 + i\sqrt{3}$ and $-2\sqrt{3} + 2i$ to trigonometric form.

$$1 + i\sqrt{3} = 2(\cos 60° + i \sin 60°)$$

$$-2\sqrt{3} + 2i = 4(\cos 150° + i \sin 150°)$$

If we multiply the trigonometric forms and use identities for the cosine and the sine of the sum of two angles, then the result is as follows.

$$[2(\cos 60° + i \sin 60°)][4(\cos 150° + i \sin 150°)]$$

$$= 2 \cdot 4(\cos 60° \cdot \cos 150° + i \sin 60° \cdot \cos 150° + i \cos 60° \cdot \sin 150° + i^2 \sin 60° \cdot \sin 150°) \quad \text{Multiply the absolute values and the binomials.}$$

$$= 8[(\cos 60° \cdot \cos 150° - \sin 60° \cdot \sin 150°) + i(\sin 60° \cdot \cos 150° + \cos 60° \cdot \sin 150°)] \quad i^2 = -1;\ \text{Factor out } i.$$

$$= 8[\cos(60° + 150°) + i \sin(60° + 150°)]$$

$\cos(A + B) = \cos A \cdot \cos B - \sin A \cdot \sin B$; $\sin(A + B) = \sin A \cdot \cos B + \cos A \cdot \sin B$

$$= 8(\cos 210° + i \sin 210°) \quad \text{Add.}$$

The absolute value of the product, 8, is equal to the product of the absolute values of the factors, $2 \cdot 4$. The argument of the product, 210°, is equal to the sum of the arguments of the factors, $60° + 150°$.

The product obtained when multiplying by the first method is the rectangular form of the product obtained when multiplying by the second method.

$$8(\cos 210° + i \sin 210°)$$

$$= 8\left(-\frac{\sqrt{3}}{2} - \frac{1}{2}i\right) \quad \cos 210° = -\frac{\sqrt{3}}{2};\ \sin 210° = -\frac{1}{2}$$

$$= -4\sqrt{3} - 4i \quad \text{Rectangular form}$$

Teaching Tip Use the product theorem to show how to obtain the product of two complex numbers geometrically. Whereas sums of complex numbers involve parallelograms, products involve rotations.

The product theorem often results in a value of $\theta_1 + \theta_2$ greater than 360°. In such cases, students may want to subtract a multiple of 360° to obtain an angle between 0° and 360°.

Product Theorem

If $r_1(\cos \theta_1 + i \sin \theta_1)$ and $r_2(\cos \theta_2 + i \sin \theta_2)$ are any two complex numbers, then the following holds.

$$[r_1(\cos \theta_1 + i \sin \theta_1)] \cdot [r_2(\cos \theta_2 + i \sin \theta_2)]$$
$$= r_1 r_2[\cos(\theta_1 + \theta_2) + i \sin(\theta_1 + \theta_2)]$$

In compact form, this is written

$$(r_1 \text{ cis } \theta_1)(r_2 \text{ cis } \theta_2) = r_1 r_2 \text{ cis}(\theta_1 + \theta_2).$$

That is, to multiply complex numbers in trigonometric form, multiply their absolute values and add their arguments.

Classroom Example 1 Find the product of

$$4(\cos 120° + i \sin 120°)$$

and

$$5(\cos 30° + i \sin 30°).$$

Write the answer in rectangular form.

Answer: $-10\sqrt{3} + 10i$

EXAMPLE 1 Using the Product Theorem

Find the product of $3(\cos 45° + i \sin 45°)$ and $2(\cos 135° + i \sin 135°)$. Write the answer in rectangular form.

SOLUTION

$$[3(\cos 45° + i \sin 45°)][2(\cos 135° + i \sin 135°)] \quad \text{Write as a product.}$$

$$= 3 \cdot 2[\cos(45° + 135°) + i \sin(45° + 135°)] \quad \text{Product theorem}$$

$$= 6(\cos 180° + i \sin 180°) \quad \text{Multiply and add.}$$

$$= 6(-1 + i \cdot 0) \quad \cos 180° = -1;\ \sin 180° = 0$$

$$= -6 \quad \text{Rectangular form}$$

✔ **Now Try Exercise 11.**

Quotients of Complex Numbers in Trigonometric Form The rectangular form of the quotient of $1 + i\sqrt{3}$ and $-2\sqrt{3} + 2i$ is found as follows.

$$\frac{1 + i\sqrt{3}}{-2\sqrt{3} + 2i}$$

$$= \frac{(1 + i\sqrt{3})(-2\sqrt{3} - 2i)}{(-2\sqrt{3} + 2i)(-2\sqrt{3} - 2i)} \quad \text{Multiply both numerator and denominator by the conjugate of the denominator.}$$

$$= \frac{-2\sqrt{3} - 2i - 6i - 2i^2\sqrt{3}}{12 - 4i^2} \quad \text{FOIL method; } (x + y)(x - y) = x^2 - y^2$$

$$= \frac{-8i}{16} \quad \text{Simplify.}$$

$$= -\frac{1}{2}i \quad \text{Lowest terms}$$

Writing $1 + i\sqrt{3}$, $-2\sqrt{3} + 2i$, and $-\frac{1}{2}i$ in trigonometric form gives

$$1 + i\sqrt{3} = 2(\cos 60° + i \sin 60°),$$

$$-2\sqrt{3} + 2i = 4(\cos 150° + i \sin 150°),$$ Use $r = \sqrt{x^2 + y^2}$ and $\tan \theta = \frac{y}{x}$.

and $$-\frac{1}{2}i = \frac{1}{2}[\cos(-90°) + i \sin(-90°)].$$ $-90°$ can be replaced by $270°$.

Here, the absolute value of the quotient, $\frac{1}{2}$, is the quotient of the two absolute values, $\frac{2}{4} = \frac{1}{2}$. The argument of the quotient, $-90°$, is the difference of the two arguments,

$$60° - 150° = -90°.$$

Quotient Theorem

If $r_1(\cos \theta_1 + i \sin \theta_1)$ and $r_2(\cos \theta_2 + i \sin \theta_2)$ are any two complex numbers, where $r_2(\cos \theta_2 + i \sin \theta_2) \neq 0$, then the following holds.

$$\frac{r_1(\cos \theta_1 + i \sin \theta_1)}{r_2(\cos \theta_2 + i \sin \theta_2)} = \frac{r_1}{r_2}[\cos(\theta_1 - \theta_2) + i \sin(\theta_1 - \theta_2)]$$

In compact form, this is written

$$\frac{r_1 \text{ cis } \theta_1}{r_2 \text{ cis } \theta_2} = \frac{r_1}{r_2} \text{ cis}(\theta_1 - \theta_2).$$

Teaching Tip The quotient theorem often results in a negative value for $\theta_1 - \theta_2$. In such cases, students may want to add a multiple of 360° to obtain an angle between 0° and 360°.

That is, to divide complex numbers in trigonometric form, divide their absolute values and subtract their arguments.

Classroom Example 2 Find the quotient

$$\frac{27 \text{ cis } 45°}{9 \text{ cis}(-180°)}.$$

Write the result in rectangular form.

Answer: $-\frac{3\sqrt{2}}{2} - \frac{3\sqrt{2}}{2}i$

EXAMPLE 2 Using the Quotient Theorem

Find the quotient $\frac{10 \text{ cis}(-60°)}{5 \text{ cis } 150°}$. Write the answer in rectangular form.

SOLUTION

$$\frac{10 \text{ cis}(-60°)}{5 \text{ cis } 150°}$$

$$= \frac{10}{5} \text{ cis}(-60° - 150°)$$ Quotient theorem

$$= 2 \text{ cis}(-210°)$$ Divide and subtract.

$$= 2[\cos(-210°) + i \sin(-210°)]$$ Rewrite.

$$= 2\left[-\frac{\sqrt{3}}{2} + i\left(\frac{1}{2}\right)\right]$$ $\cos(-210°) = -\frac{\sqrt{3}}{2}$; $\sin(-210°) = \frac{1}{2}$

$$= -\sqrt{3} + i$$ Distributive property

Now Try Exercise 21.

Classroom Example 3
Use a calculator to find the following. Write the answers in rectangular form.

(a) $(2.8 \text{ cis } 38.5°)(5.4 \text{ cis } 74.5°)$

(b) $\dfrac{11.37\left(\cos \frac{5\pi}{6} + i \sin \frac{5\pi}{6}\right)}{3.79\left(\cos \frac{\pi}{4} + i \sin \frac{\pi}{4}\right)}$

Answers:

(a) $-5.9079 + 13.9180i$

(b) $-0.7765 + 2.8978i$

EXAMPLE 3 Using the Product and Quotient Theorems with a Calculator

Use a calculator to find the following. Write the answers in rectangular form.

(a) $(9.3 \text{ cis } 125.2°)(2.7 \text{ cis } 49.8°)$ **(b)** $\dfrac{10.42\left(\cos \frac{3\pi}{4} + i \sin \frac{3\pi}{4}\right)}{5.21\left(\cos \frac{\pi}{5} + i \sin \frac{\pi}{5}\right)}$

SOLUTION

(a) $(9.3 \text{ cis } 125.2°)(2.7 \text{ cis } 49.8°)$

$= 9.3(2.7) \text{ cis}(125.2° + 49.8°)$ Product theorem

$= 25.11 \text{ cis } 175°$ Multiply. Add.

Multiply the absolute values and add the arguments.

$= 25.11(\cos 175° + i \sin 175°)$ Equivalent form

$\approx 25.11[-0.99619470 + i(0.08715574)]$ Use a calculator.

$\approx -25.0144 + 2.1885i$ Rectangular form

(b) $\dfrac{10.42\left(\cos \frac{3\pi}{4} + i \sin \frac{3\pi}{4}\right)}{5.21\left(\cos \frac{\pi}{5} + i \sin \frac{\pi}{5}\right)}$

$= \dfrac{10.42}{5.21}\left[\cos\left(\dfrac{3\pi}{4} - \dfrac{\pi}{5}\right) + i \sin\left(\dfrac{3\pi}{4} - \dfrac{\pi}{5}\right)\right]$ Quotient theorem

$= 2\left(\cos \dfrac{11\pi}{20} + i \sin \dfrac{11\pi}{20}\right)$ $\frac{3\pi}{4} = \frac{15\pi}{20}$; $\frac{\pi}{5} = \frac{4\pi}{20}$

Divide the absolute values and subtract the arguments.

$\approx -0.3129 + 1.9754i$ Rectangular form

✔ **Now Try Exercises 31 and 33.**

8.3 Exercises

1. multiply; add
2. divide; subtract
3. 10; 180°; 180°; −10; 0
4. 3; 90°; 90°; 0; 3
5. 0°; 1; 0
6. 0°; 5; 0

CONCEPT PREVIEW *Fill in the blanks to correctly complete each problem.*

1. When multiplying two complex numbers in trigonometric form, we ______ their absolute values and ______ their arguments.

2. When dividing two complex numbers in trigonometric form, we ______ their absolute values and ______ their arguments.

3. $[5(\cos 150° + i \sin 150°)][2(\cos 30° + i \sin 30°)]$

$=$ _____ (cos _____ $+ i$ sin _____)

$=$ _____ $+$ _____ i

4. $\dfrac{6(\cos 120° + i \sin 120°)}{2(\cos 30° + i \sin 30°)}$

$=$ _____ (cos _____ $+ i$ sin _____)

$=$ _____ $+$ _____ i

5. $\text{cis}(-1000°) \cdot \text{cis } 1000°$

$=$ cis _____

$=$ _____ $+$ _____ i

6. $\dfrac{5 \text{ cis } 50{,}000°}{\text{cis } 50{,}000°}$

$= 5$ cis _____

$=$ _____ $+$ _____ i

Find each product. Write answers in rectangular form. **See Example 1.**

7. $[3(\cos 60^\circ + i \sin 60^\circ)][2(\cos 90^\circ + i \sin 90^\circ)]$
8. $[4(\cos 30^\circ + i \sin 30^\circ)][5(\cos 120^\circ + i \sin 120^\circ)]$
9. $[4(\cos 60^\circ + i \sin 60^\circ)][6(\cos 330^\circ + i \sin 330^\circ)]$
10. $[8(\cos 300^\circ + i \sin 300^\circ)][5(\cos 120^\circ + i \sin 120^\circ)]$
11. $[2(\cos 135^\circ + i \sin 135^\circ)][2(\cos 225^\circ + i \sin 225^\circ)]$
12. $[8(\cos 210^\circ + i \sin 210^\circ)][2(\cos 330^\circ + i \sin 330^\circ)]$
13. $(\sqrt{3} \text{ cis } 45^\circ)(\sqrt{3} \text{ cis } 225^\circ)$
14. $(\sqrt{6} \text{ cis } 120^\circ)[\sqrt{6} \text{ cis}(-30^\circ)]$
15. $(5 \text{ cis } 90^\circ)(3 \text{ cis } 45^\circ)$
16. $(3 \text{ cis } 300^\circ)(7 \text{ cis } 270^\circ)$

Find each quotient. Write answers in rectangular form. In Exercises 23–28, first convert the numerator and the denominator to trigonometric form. **See Example 2.**

17. $\dfrac{4(\cos 150^\circ + i \sin 150^\circ)}{2(\cos 120^\circ + i \sin 120^\circ)}$
18. $\dfrac{24(\cos 150^\circ + i \sin 150^\circ)}{2(\cos 30^\circ + i \sin 30^\circ)}$
19. $\dfrac{10(\cos 50^\circ + i \sin 50^\circ)}{5(\cos 230^\circ + i \sin 230^\circ)}$
20. $\dfrac{12(\cos 23^\circ + i \sin 23^\circ)}{6(\cos 293^\circ + i \sin 293^\circ)}$
21. $\dfrac{3 \text{ cis } 305^\circ}{9 \text{ cis } 65^\circ}$
22. $\dfrac{16 \text{ cis } 310^\circ}{8 \text{ cis } 70^\circ}$
23. $\dfrac{8}{\sqrt{3} + i}$
24. $\dfrac{2i}{-1 - i\sqrt{3}}$
25. $\dfrac{-i}{1 + i}$
26. $\dfrac{1}{2 - 2i}$
27. $\dfrac{2\sqrt{6} - 2i\sqrt{2}}{\sqrt{2} - i\sqrt{6}}$
28. $\dfrac{-3\sqrt{2} + 3i\sqrt{6}}{\sqrt{6} + i\sqrt{2}}$

Use a calculator to perform the indicated operations. Write answers in rectangular form, expressing real and imaginary parts to four decimal places. **See Example 3.**

29. $[2.5(\cos 35^\circ + i \sin 35^\circ)][3.0(\cos 50^\circ + i \sin 50^\circ)]$
30. $[4.6(\cos 12^\circ + i \sin 12^\circ)][2.0(\cos 13^\circ + i \sin 13^\circ)]$
31. $(12 \text{ cis } 18.5^\circ)(3 \text{ cis } 12.5^\circ)$
32. $(4 \text{ cis } 19.25^\circ)(7 \text{ cis } 41.75^\circ)$
33. $\dfrac{45\left(\cos \frac{2\pi}{3} + i \sin \frac{2\pi}{3}\right)}{22.5\left(\cos \frac{3\pi}{5} + i \sin \frac{3\pi}{5}\right)}$
34. $\dfrac{30\left(\cos \frac{2\pi}{5} + i \sin \frac{2\pi}{5}\right)}{10\left(\cos \frac{\pi}{7} + i \sin \frac{\pi}{7}\right)}$
35. $\left[2 \text{ cis } \dfrac{5\pi}{9}\right]^2$
36. $\left[24.3 \text{ cis } \dfrac{7\pi}{12}\right]^2$

Work each problem.

37. Note that $(r \text{ cis } \theta)^2 = (r \text{ cis } \theta)(r \text{ cis } \theta) = r^2 \text{ cis}(\theta + \theta) = r^2 \text{ cis } 2\theta$. Explain how we can square a complex number in trigonometric form. (In the next section, we will develop this idea more fully.)

38. Without actually performing the operations, state why the following products are the same.

$$[2(\cos 45^\circ + i \sin 45^\circ)] \cdot [5(\cos 90^\circ + i \sin 90^\circ)]$$

and $\quad [2[\cos(-315^\circ) + i \sin(-315^\circ)]] \cdot [5[\cos(-270^\circ) + i \sin(-270^\circ)]]$

39. Show that $\frac{1}{z} = \frac{1}{r}(\cos \theta - i \sin \theta)$, where $z = r(\cos \theta + i \sin \theta)$.

Answers (margin)

7. $-3\sqrt{3} + 3i$
8. $-10\sqrt{3} + 10i$
9. $12\sqrt{3} + 12i$
10. $20 + 20i\sqrt{3}$
11. 4 12. -16
13. $-3i$ 14. $6i$
15. $-\frac{15\sqrt{2}}{2} + \frac{15\sqrt{2}}{2}i$
16. $-\frac{21\sqrt{3}}{2} - \frac{21}{2}i$
17. $\sqrt{3} + i$ 18. $-6 + 6i\sqrt{3}$
19. -2 20. $2i$
21. $-\frac{1}{6} - \frac{\sqrt{3}}{6}i$ 22. $-1 - i\sqrt{3}$
23. $2\sqrt{3} - 2i$ 24. $-\frac{\sqrt{3}}{2} - \frac{1}{2}i$
25. $-\frac{1}{2} - \frac{1}{2}i$ 26. $\frac{1}{4} + \frac{1}{4}i$
27. $\sqrt{3} + i$ 28. $3i$
29. $0.6537 + 7.4715i$
30. $8.3380 + 3.8881i$
31. $30.8580 + 18.5414i$
32. $13.5747 + 24.4894i$
33. $1.9563 + 0.4158i$
34. $2.0732 + 2.1684i$
35. $-3.7588 - 1.3681i$
36. $-511.3793 - 295.2450i$
37. To square a complex number in trigonometric form, square its absolute value and double its argument.
38. In both cases, the absolute value of the product is $2 \cdot 5$. The argument in the first product is 135° and that in the second product is -585°. These are coterminal and thus have the same trigonometric function values.

41. 1.18 − 0.14i
42. 1.7 + 2.8i
43. 27.43 + 11.50i
44. 22.75°
45. 2
46. $w = \sqrt{2}$ cis 135°; $z = \sqrt{2}$ cis 225°
47. 2 cis 0°
48. 2; It is the same.
49. $-i$
50. cis(−90°)
51. $-i$
52. It is the same.

40. The complex conjugate of $r(\cos\theta + i\sin\theta)$ is $r(\cos\theta - i\sin\theta)$. Use these trigonometric forms to show that the product of complex conjugates is always a real number.

(Modeling) Electrical Current Solve each problem.

41. The alternating current in an electric inductor is $I = \frac{E}{Z}$ amperes, where E is voltage and $Z = R + X_L i$ is impedance. If $E = 8(\cos 20° + i\sin 20°)$, $R = 6$, and $X_L = 3$, find the current. Give the answer in rectangular form, with real and imaginary parts to the nearest hundredth.

42. The current I in a circuit with voltage E, resistance R, capacitive reactance X_c, and inductive reactance X_L is

$$I = \frac{E}{R + (X_L - X_c)i}.$$

Find I if $E = 12(\cos 25° + i\sin 25°)$, $R = 3$, $X_L = 4$, and $X_c = 6$. Give the answer in rectangular form, with real and imaginary parts to the nearest tenth.

(Modeling) Impedance In the parallel electrical circuit shown in the figure, the impedance Z can be calculated using the equation

$$Z = \frac{1}{\frac{1}{Z_1} + \frac{1}{Z_2}},$$

where Z_1 and Z_2 are the impedances for the branches of the circuit.

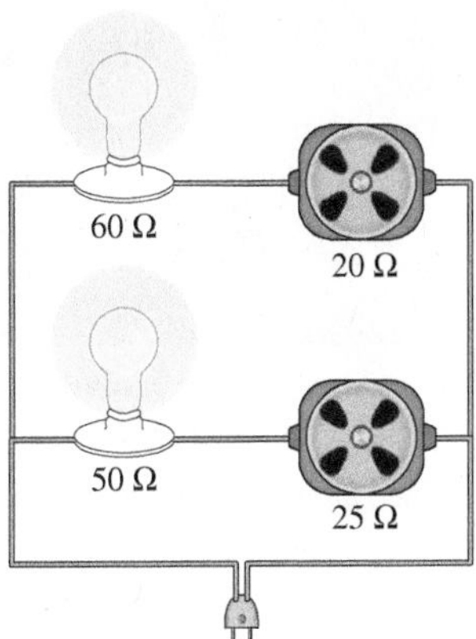

43. If $Z_1 = 50 + 25i$ and $Z_2 = 60 + 20i$, approximate Z to the nearest hundredth.

44. Determine the angle θ, to the nearest hundredth, for the value of Z found in **Exercise 43.**

Relating Concepts

For individual or collaborative investigation *(Exercises 45-52)*

Consider the following complex numbers, and ***work Exercises 45–52 in order.***

$$w = -1 + i \quad \textit{and} \quad z = -1 - i$$

45. Multiply w and z using their rectangular forms and the FOIL method. Leave the product in rectangular form.

46. Find the trigonometric forms of w and z.

47. Multiply w and z using their trigonometric forms and the method described in this section.

48. Use the result of **Exercise 47** to find the rectangular form of wz. How does this compare to the result in **Exercise 45?**

49. Find the quotient $\frac{w}{z}$ using their rectangular forms and multiplying both the numerator and the denominator by the conjugate of the denominator. Leave the quotient in rectangular form.

50. Use the trigonometric forms of w and z, found in **Exercise 46,** to divide w by z using the method described in this section.

51. Use the result in **Exercise 50** to find the rectangular form of $\frac{w}{z}$.

52. How does the result in **Exercise 51** compare to the result in **Exercise 49?**

8.4 De Moivre's Theorem; Powers and Roots of Complex Numbers

- Powers of Complex Numbers (De Moivre's Theorem)
- Roots of Complex Numbers

Abraham De Moivre (1667–1754)

Named after this French expatriate friend of Isaac Newton, De Moivre's theorem relates complex numbers and trigonometry.

Powers of Complex Numbers (De Moivre's Theorem) Because raising a number to a positive integer power is a repeated application of the product rule, it would seem likely that a theorem for finding powers of complex numbers exists. Consider the following.

$$[r(\cos\theta + i\sin\theta)]^2$$

$$= [r(\cos\theta + i\sin\theta)][r(\cos\theta + i\sin\theta)] \quad a^2 = a \cdot a$$

$$= r \cdot r[\cos(\theta + \theta) + i\sin(\theta + \theta)] \quad \text{Product theorem}$$

$$= r^2(\cos 2\theta + i\sin 2\theta) \quad \text{Multiply and add.}$$

In the same way,

$$[r(\cos\theta + i\sin\theta)]^3 \quad \text{is equivalent to} \quad r^3(\cos 3\theta + i\sin 3\theta).$$

These results suggest the following theorem for positive integer values of n. Although the theorem is stated and can be proved for all n, we use it only for positive integer values of n and their reciprocals.

De Moivre's Theorem

If $r(\cos\theta + i\sin\theta)$ is a complex number, and if n is any real number, then the following holds.

$$[r(\cos\theta + i\sin\theta)]^n = r^n(\cos n\theta + i\sin n\theta)$$

In compact form, this is written

$$[r \operatorname{cis}\theta]^n = r^n(\operatorname{cis} n\theta).$$

Classroom Example 1 Find $\left(\sqrt{2} + i\sqrt{2}\right)^5$ and write the answer in rectangular form.

Answer: $-16\sqrt{2} - 16i\sqrt{2}$

EXAMPLE 1 Finding a Power of a Complex Number

Find $\left(1 + i\sqrt{3}\right)^8$ and write the answer in rectangular form.

SOLUTION Using earlier methods, write $1 + i\sqrt{3}$ in trigonometric form.

$$2(\cos 60° + i\sin 60°) \quad \text{Trigonometric form of } 1 + i\sqrt{3}$$

Now, apply De Moivre's theorem.

$$\left(1 + i\sqrt{3}\right)^8$$

$$= [2(\cos 60° + i\sin 60°)]^8 \quad \text{Trigonometric form}$$

$$= 2^8[\cos(8 \cdot 60°) + i\sin(8 \cdot 60°)] \quad \text{De Moivre's theorem}$$

$$= 256(\cos 480° + i\sin 480°) \quad \text{Apply the exponent and multiply.}$$

$$= 256(\cos 120° + i\sin 120°) \quad 480° \text{ and } 120° \text{ are coterminal.}$$

$$= 256\left(-\frac{1}{2} + i\frac{\sqrt{3}}{2}\right) \quad \cos 120° = -\tfrac{1}{2};\ \sin 120° = \tfrac{\sqrt{3}}{2}$$

$$= -128 + 128i\sqrt{3} \quad \text{Rectangular form}$$

✔ Now Try Exercise 13.

Teaching Tip Mention that De Moivre's theorem greatly simplifies calculations involving powers and roots of complex numbers. Demonstrate this concept by evaluating $\left(1 + i\sqrt{3}\right)^3$ with and without the theorem.

Roots of Complex Numbers Every nonzero complex number has exactly n distinct complex nth roots. De Moivre's theorem can be extended to find all nth roots of a complex number.

*n*th Root

For a positive integer n, the complex number $a + bi$ is an ***n*th root** of the complex number $x + yi$ if the following holds.

$$(a + bi)^n = x + yi$$

To find the three complex cube roots of $8(\cos 135° + i \sin 135°)$, for example, look for a complex number, say $r(\cos \alpha + i \sin \alpha)$, that will satisfy

$$[r(\cos \alpha + i \sin \alpha)]^3 = 8(\cos 135° + i \sin 135°).$$

By De Moivre's theorem, this equation becomes

$$r^3(\cos 3\alpha + i \sin 3\alpha) = 8(\cos 135° + i \sin 135°).$$

Set $r^3 = 8$ and $\cos 3\alpha + i \sin 3\alpha = \cos 135° + i \sin 135°$, to satisfy this equation. The first of these conditions implies that $r = 2$, and the second implies that

$$\cos 3\alpha = \cos 135° \quad \text{and} \quad \sin 3\alpha = \sin 135°.$$

For these equations to be satisfied, 3α must represent an angle that is coterminal with 135°. Therefore, we must have

$$3\alpha = 135° + 360° \cdot k, \quad k \text{ any integer}$$

or

$$\alpha = \frac{135° + 360° \cdot k}{3}, \quad k \text{ any integer.}$$

Now, let k take on the integer values 0, 1, and 2.

$$\text{If } k = 0, \text{ then } \alpha = \frac{135° + 360° \cdot 0}{3} = 45°.$$

$$\text{If } k = 1, \text{ then } \alpha = \frac{135° + 360° \cdot 1}{3} = \frac{495°}{3} = 165°.$$

$$\text{If } k = 2, \text{ then } \alpha = \frac{135° + 360° \cdot 2}{3} = \frac{855°}{3} = 285°.$$

In the same way, $\alpha = 405°$ when $k = 3$. But note that $405° = 45° + 360°$, so $\sin 405° = \sin 45°$ and $\cos 405° = \cos 45°$. Similarly, if $k = 4$, then $\alpha = 525°$, which has the same sine and cosine values as 165°. Continuing with larger values of k would repeat solutions already found. Therefore, all of the cube roots (three of them) can be found by letting $k = 0$, 1, and 2, respectively.

When $k = 0$, the root is $2(\cos 45° + i \sin 45°)$.

When $k = 1$, the root is $2(\cos 165° + i \sin 165°)$.

When $k = 2$, the root is $2(\cos 285° + i \sin 285°)$.

In summary, we see that $2(\cos 45° + i \sin 45°)$, $2(\cos 165° + i \sin 165°)$, and $2(\cos 285° + i \sin 285°)$ are the three cube roots of $8(\cos 135° + i \sin 135°)$.

*n*th Root Theorem

If n is any positive integer, r is a positive real number, and θ is in degrees, then the nonzero complex number $r(\cos \theta + i \sin \theta)$ has exactly n distinct nth roots, given by the following.

$$\sqrt[n]{r}(\cos \alpha + i \sin \alpha) \quad \text{or} \quad \sqrt[n]{r} \text{ cis } \alpha,$$

where

$$\alpha = \frac{\theta + 360° \cdot k}{n}, \quad \text{or} \quad \alpha = \frac{\theta}{n} + \frac{360° \cdot k}{n}, \quad k = 0, 1, 2, \ldots, n - 1$$

If θ is in radians, then

$$\alpha = \frac{\theta + 2\pi k}{n}, \quad \text{or} \quad \alpha = \frac{\theta}{n} + \frac{2\pi k}{n}, \quad k = 0, 1, 2, \ldots, n - 1.$$

Classroom Example 2
Find the three cube roots of -8. Write the roots in rectangular form.

Answer: $1 + i\sqrt{3}, -2, 1 - i\sqrt{3}$

EXAMPLE 2 Finding Complex Roots

Find the two square roots of $4i$. Write the roots in rectangular form.

SOLUTION First write $4i$ in trigonometric form.

$$4\left(\cos \frac{\pi}{2} + i \sin \frac{\pi}{2}\right) \quad \text{Trigonometric form (using radian measure)}$$

Here $r = 4$ and $\theta = \frac{\pi}{2}$. The square roots have absolute value $\sqrt{4} = 2$ and arguments as follows.

$$\alpha = \frac{\frac{\pi}{2}}{2} + \frac{2\pi k}{2} = \frac{\pi}{4} + \pi k$$

Be careful simplifying here.

Because there are two square roots, let $k = 0$ and 1.

$$\text{If } k = 0, \text{ then} \quad \alpha = \frac{\pi}{4} + \pi \cdot 0 = \frac{\pi}{4}.$$

$$\text{If } k = 1, \text{ then} \quad \alpha = \frac{\pi}{4} + \pi \cdot 1 = \frac{5\pi}{4}.$$

Using these values for α, the square roots are $2 \text{ cis } \frac{\pi}{4}$ and $2 \text{ cis } \frac{5\pi}{4}$, which can be written in rectangular form as

$$\sqrt{2} + i\sqrt{2} \quad \text{and} \quad -\sqrt{2} - i\sqrt{2}.$$

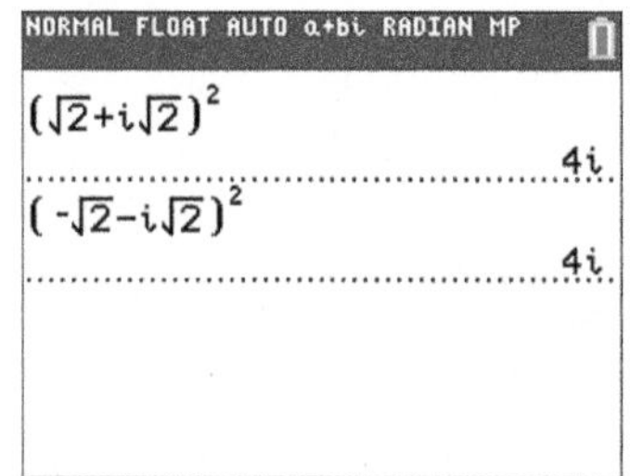

This screen confirms the results of **Example 2.**

Now Try Exercise 23(a).

EXAMPLE 3 Finding Complex Roots

Find all fourth roots of $-8 + 8i\sqrt{3}$. Write the roots in rectangular form.

SOLUTION $-8 + 8i\sqrt{3} = 16 \text{ cis } 120°$ Write in trigonometric form.

Here $r = 16$ and $\theta = 120°$. The fourth roots of this number have absolute value $\sqrt[4]{16} = 2$ and arguments as follows.

$$\alpha = \frac{120°}{4} + \frac{360° \cdot k}{4} = 30° + 90° \cdot k$$

NORMAL FLOAT AUTO a+bi DEGREE MP
R▸Pr(-8,8√3)
16
R▸Pθ(-8,8√3)
120

This screen shows how a calculator finds r and θ for the number in **Example 3.**

Classroom Example 3
Find all fourth roots of

$$-128 - 128i\sqrt{3}.$$

Write the roots in rectangular form.

Answer:
$2 + 2i\sqrt{3}$, $-2\sqrt{3} + 2i$, $-2 - 2i\sqrt{3}$, $2\sqrt{3} - 2i$

Because there are four fourth roots, let $k = 0, 1, 2,$ and 3.

$$\text{If } k = 0, \text{ then } \quad \alpha = 30^\circ + 90^\circ \cdot 0 = 30^\circ.$$
$$\text{If } k = 1, \text{ then } \quad \alpha = 30^\circ + 90^\circ \cdot 1 = 120^\circ.$$
$$\text{If } k = 2, \text{ then } \quad \alpha = 30^\circ + 90^\circ \cdot 2 = 210^\circ.$$
$$\text{If } k = 3, \text{ then } \quad \alpha = 30^\circ + 90^\circ \cdot 3 = 300^\circ.$$

Using these angles, the fourth roots are

$$2 \text{ cis } 30^\circ, \quad 2 \text{ cis } 120^\circ, \quad 2 \text{ cis } 210^\circ, \quad \text{and} \quad 2 \text{ cis } 300^\circ.$$

These four roots can be written in rectangular form as

$$\sqrt{3} + i, \quad -1 + i\sqrt{3}, \quad -\sqrt{3} - i, \quad \text{and} \quad 1 - i\sqrt{3}.$$

The graphs of these roots lie on a circle with center at the origin and radius 2. See **Figure 12.** The roots are equally spaced about the circle, 90° apart. (For convenience, we label the real axis x and the imaginary axis y.)

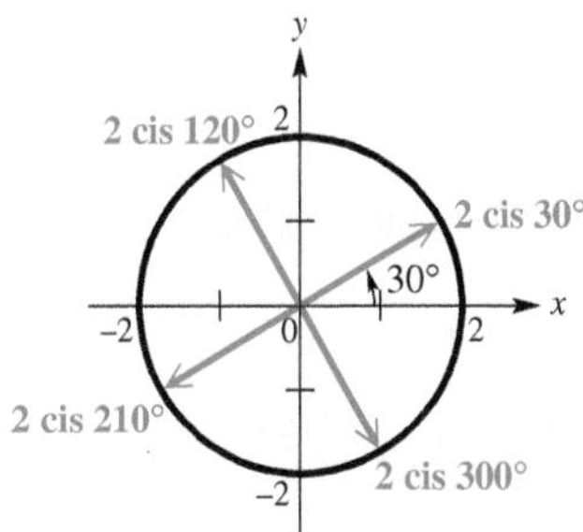

Figure 12

Now Try Exercise 29.

Classroom Example 4
Find all complex number solutions of

$$x^6 + i = 0.$$

Graph them as vectors in the complex plane.

Answer:
$\{\text{cis } 45^\circ, \text{cis } 105^\circ, \text{cis } 165^\circ, \text{cis } 225^\circ, \text{cis } 285^\circ, \text{cis } 345^\circ\}$

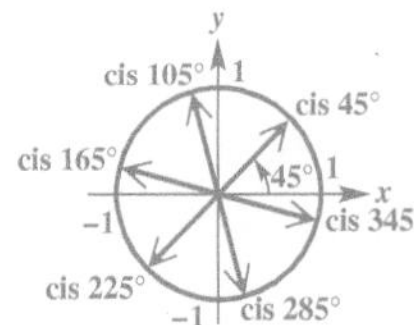

EXAMPLE 4 Solving an Equation (Complex Roots)

Find all complex number solutions of $x^5 - 1 = 0$. Graph them as vectors in the complex plane.

SOLUTION Write the equation as

$$x^5 - 1 = 0, \quad \text{or} \quad x^5 = 1.$$

Because $1^5 = 1$, there is a real number solution, 1, and it is the only one. There are a total of five complex number solutions. To find these solutions, first write 1 in trigonometric form.

$$1 = 1 + 0i = 1(\cos 0^\circ + i \sin 0^\circ) \quad \text{Trigonometric form}$$

The absolute value of the fifth roots is $\sqrt[5]{1} = 1$. The arguments are given by

$$0^\circ + 72^\circ \cdot k, \quad k = 0, 1, 2, 3, \text{ and } 4.$$

By using these arguments, we find that the fifth roots are as follows.

$$\text{Real solution} \longrightarrow 1(\cos 0^\circ + i \sin 0^\circ), \qquad k = 0$$
$$1(\cos 72^\circ + i \sin 72^\circ), \qquad k = 1$$
$$1(\cos 144^\circ + i \sin 144^\circ), \qquad k = 2$$
$$1(\cos 216^\circ + i \sin 216^\circ), \qquad k = 3$$
$$1(\cos 288^\circ + i \sin 288^\circ) \qquad k = 4$$

Teaching Tip Knowing that the roots are evenly spaced on a circle in the complex plane, students may prefer to find the first solution for α and add multiples of $\frac{360^\circ}{n}$ until all n roots are found.

The solution set of the equation can be written as

$$\{\text{cis } 0^\circ, \text{cis } 72^\circ, \text{cis } 144^\circ, \text{cis } 216^\circ, \text{cis } 288^\circ\}.$$

The first of these roots is the real number 1. The others cannot easily be expressed in rectangular form but can be approximated using a calculator.

The tips of the arrows representing the five fifth roots all lie on a unit circle and are equally spaced around it every 72°, as shown in **Figure 13.**

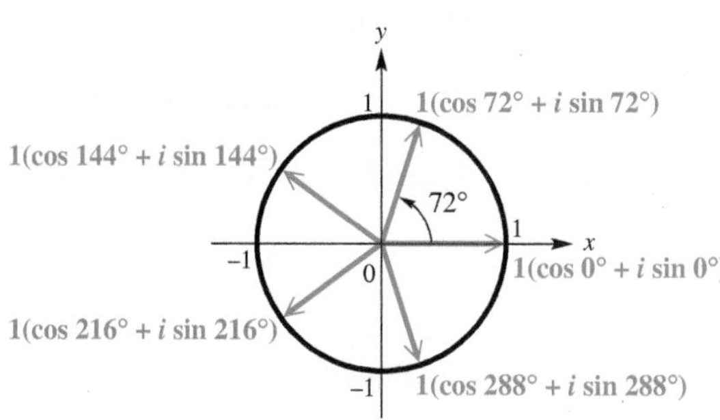

Figure 13

✔ Now Try Exercise 41.

8.4 Exercises

1. 27; 90°; 90°; 27; 0; 1; 0; 27; $27i$
2. 2; 20°
3. 180°; 180°; −1; 0
4. 30th; −1
5. two
6. eight
7. $27i$
8. −16
9. 1
10. 8
11. $\frac{27}{2} - \frac{27\sqrt{3}}{2}i$
12. $-\frac{27}{2} + \frac{27\sqrt{3}}{2}i$
13. $-16\sqrt{3} + 16i$
14. $-128 + 128i\sqrt{3}$
15. $4096i$
16. 1
17. $128 + 128i$
18. $-8 - 8i$
19. (a) $\cos 0° + i \sin 0°$, $\cos 120° + i \sin 120°$, $\cos 240° + i \sin 240°$

(b)

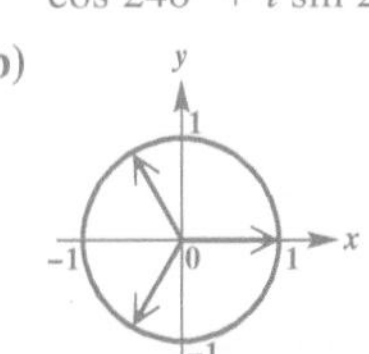

20. (a) $\cos 30° + i \sin 30°$, $\cos 150° + i \sin 150°$, $\cos 270° + i \sin 270°$

(b)

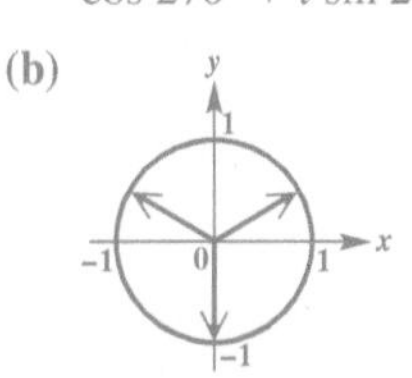

CONCEPT PREVIEW *Fill in the blanks to correctly complete each problem.*

1. If $z = 3(\cos 30° + i \sin 30°)$, it follows that

$z^3 =$ ______ (cos ______ + i sin ______)

= ______ (______ + i ______)

= ______ + ______ i, or simply ______.

2. If we are given

$$z = 16(\cos 80° + i \sin 80°),$$

then any fourth root of z has $r =$ ______, and the fourth root with least positive argument has $\theta =$ ______.

3. $[\cos 6° + i \sin 6°]^{30}$

= cos ______ + i sin ______

= ______ + ______ i

4. Based on the result of **Exercise 3,**

$$\cos 6° + i \sin 6°$$

is a(n) ______ root of ______.

CONCEPT PREVIEW *Answer each question.*

5. How many real tenth roots of 1 exist?

6. How many nonreal complex tenth roots of 1 exist?

Find each power. Write answers in rectangular form. ***See Example 1.***

7. $[3(\cos 30° + i \sin 30°)]^3$

8. $[2(\cos 135° + i \sin 135°)]^4$

9. $(\cos 45° + i \sin 45°)^8$

10. $[2(\cos 120° + i \sin 120°)]^3$

11. $[3 \text{ cis } 100°]^3$

12. $[3 \text{ cis } 40°]^3$

13. $(\sqrt{3} + i)^5$

14. $(2 - 2i\sqrt{3})^4$

15. $(2\sqrt{2} - 2i\sqrt{2})^6$

16. $\left(\frac{\sqrt{2}}{2} - \frac{\sqrt{2}}{2}i\right)^8$

17. $(-2 - 2i)^5$

18. $(-1 + i)^7$

21. **(a)** 2 cis 20°, 2 cis 140°, 2 cis 260°

(b)

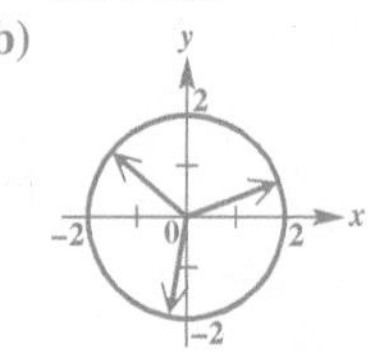

22. **(a)** 3 cis 100°, 3 cis 220°, 3 cis 340°

(b)

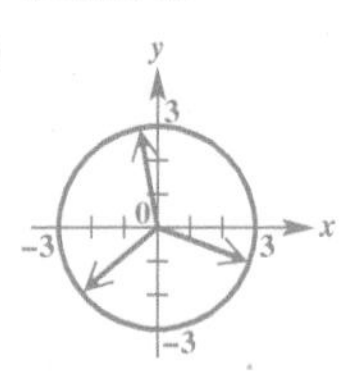

23. **(a)** $2(\cos 90° + i \sin 90°)$, $2(\cos 210° + i \sin 210°)$, $2(\cos 330° + i \sin 330°)$

(b)

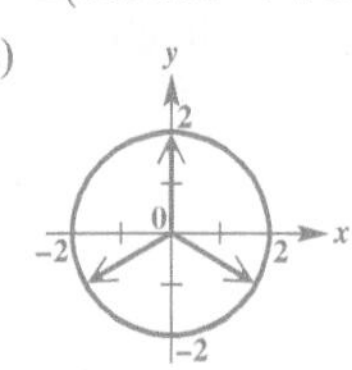

24. **(a)** $3(\cos 30° + i \sin 30°)$, $3(\cos 150° + i \sin 150°)$, $3(\cos 270° + i \sin 270°)$

(b)

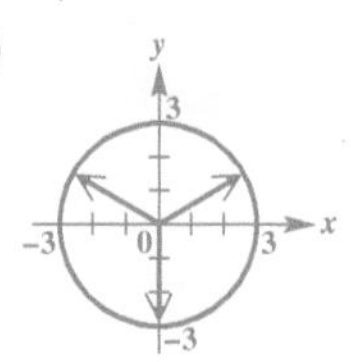

25. **(a)** $4(\cos 60° + i \sin 60°)$, $4(\cos 180° + i \sin 180°)$, $4(\cos 300° + i \sin 300°)$

(b)

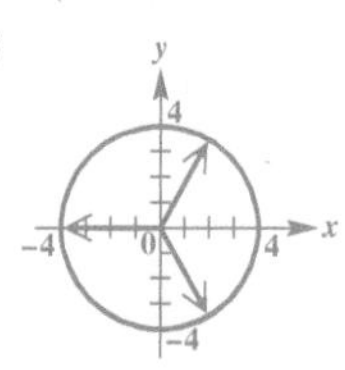

26. **(a)** $3(\cos 0° + i \sin 0°)$, $3(\cos 120° + i \sin 120°)$, $3(\cos 240° + i \sin 240°)$

(b)

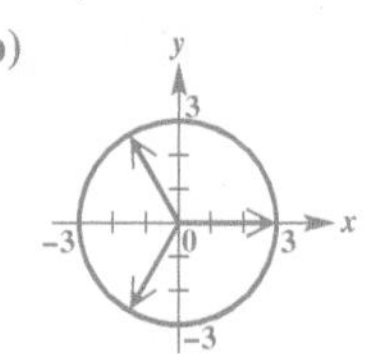

*For each of the following, **(a)** find all cube roots of each complex number. Write answers in trigonometric form. **(b)** Graph each cube root as a vector in the complex plane.* ***See Examples 2 and 3.***

19. $\cos 0° + i \sin 0°$ **20.** $\cos 90° + i \sin 90°$ **21.** 8 cis 60°

22. 27 cis 300° **23.** $-8i$ **24.** $27i$

25. -64 **26.** 27 **27.** $1 + i\sqrt{3}$

28. $2 - 2i\sqrt{3}$ **29.** $-2\sqrt{3} + 2i$ **30.** $\sqrt{3} - i$

Find and graph all specified roots of 1.

31. second (square) **32.** fourth **33.** sixth

Find and graph all specified roots of i.

34. second (square) **35.** third (cube) **36.** fourth

Find all complex number solutions of each equation. Write answers in trigonometric form. ***See Example 4.***

37. $x^3 - 1 = 0$ **38.** $x^3 + 1 = 0$ **39.** $x^3 + i = 0$

40. $x^4 + i = 0$ **41.** $x^3 - 8 = 0$ **42.** $x^3 + 27 = 0$

43. $x^4 + 1 = 0$ **44.** $x^4 + 16 = 0$ **45.** $x^4 - i = 0$

46. $x^5 - i = 0$ **47.** $x^3 - \left(4 + 4i\sqrt{3}\right) = 0$ **48.** $x^4 - \left(8 + 8i\sqrt{3}\right) = 0$

Solve each problem.

49. Solve the cubic equation

$$x^3 = 1$$

by writing it as $x^3 - 1 = 0$, factoring the left side as the difference of two cubes, and using the zero-factor property. Apply the quadratic formula as needed. Then compare the solutions to those of **Exercise 37.**

50. Solve the cubic equation

$$x^3 = -27$$

by writing it as $x^3 + 27 = 0$, factoring the left side as the sum of two cubes, and using the zero-factor property. Apply the quadratic formula as needed. Then compare the solutions to those of **Exercise 42.**

51. *Mandelbrot Set* The fractal known as the **Mandelbrot set** is shown in the figure. To determine whether a complex number $z = a + bi$ is in this set, perform the following sequence of calculations. Repeatedly compute

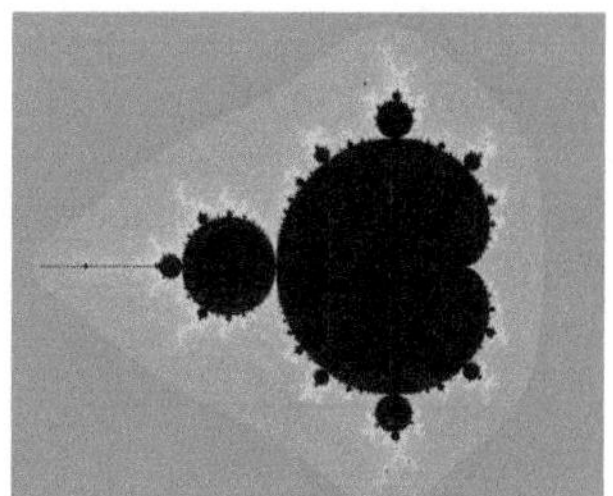

$$z, \quad z^2 + z, \quad (z^2 + z)^2 + z,$$
$$\left[(z^2 + z)^2 + z\right]^2 + z, \ldots.$$

In a manner analogous to the Julia set, the complex number z does not belong to the Mandelbrot set if any of the resulting absolute values exceeds 2. Otherwise z is in the set and the point (a, b) should be shaded in the graph. Determine whether the following numbers belong to the Mandelbrot set. (*Source:* Lauwerier, H., *Fractals,* Princeton University Press.)

(a) $z = 0 + 0i$ **(b)** $z = 1 - 1i$ **(c)** $z = -0.5i$

27. (a) $\sqrt[3]{2}(\cos 20° + i \sin 20°)$, $\sqrt[3]{2}(\cos 140° + i \sin 140°)$, $\sqrt[3]{2}(\cos 260° + i \sin 260°)$

(b)

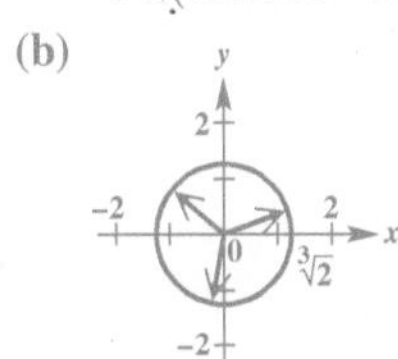

28. (a) $\sqrt[3]{4}(\cos 100° + i \sin 100°)$, $\sqrt[3]{4}(\cos 220° + i \sin 220°)$, $\sqrt[3]{4}(\cos 340° + i \sin 340°)$

(b)

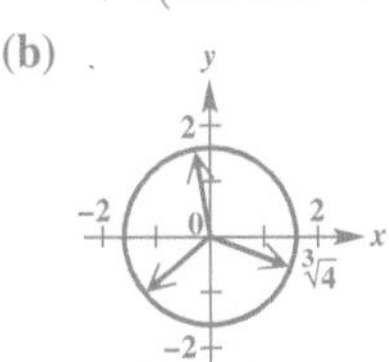

29. (a) $\sqrt[3]{4}(\cos 50° + i \sin 50°)$, $\sqrt[3]{4}(\cos 170° + i \sin 170°)$, $\sqrt[3]{4}(\cos 290° + i \sin 290°)$

(b)

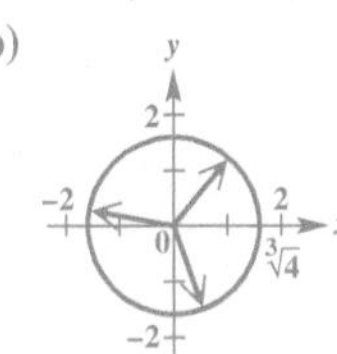

30. (a) $\sqrt[3]{2}(\cos 110° + i \sin 110°)$, $\sqrt[3]{2}(\cos 230° + i \sin 230°)$, $\sqrt[3]{2}(\cos 350° + i \sin 350°)$

(b)

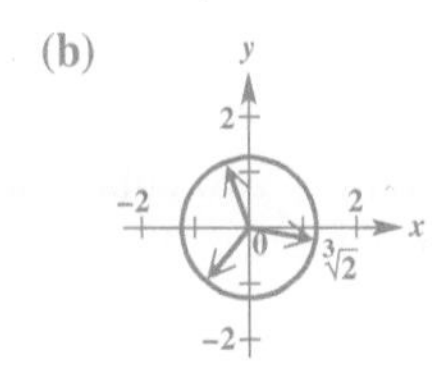

31. $\cos 0° + i \sin 0°$, $\cos 180° + i \sin 180°$

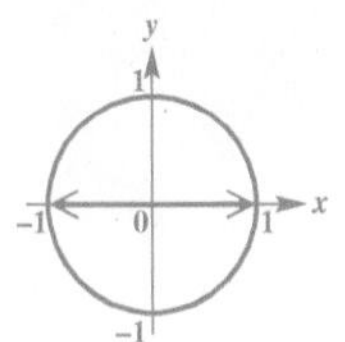

Answers for the remaining exercises are given in the answer section at the back of the text.

52. ***Basins of Attraction*** The fractal shown in the figure is the solution to Cayley's problem of determining the basins of attraction for the cube roots of unity. The three cube roots of unity are

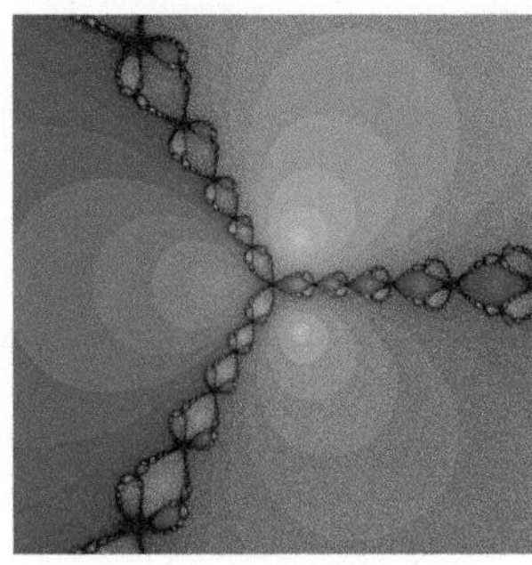

$$w_1 = 1, \quad w_2 = -\frac{1}{2} + \frac{\sqrt{3}}{2}i,$$

and $$w_3 = -\frac{1}{2} - \frac{\sqrt{3}}{2}i.$$

A fractal of this type can be generated by repeatedly evaluating the function

$$f(z) = \frac{2z^3 + 1}{3z^2},$$

where z is a complex number. We begin by picking $z_1 = a + bi$ and successively computing $z_2 = f(z_1)$, $z_3 = f(z_2)$, $z_4 = f(z_3)$, ... Suppose that if the resulting values of $f(z)$ approach w_1, we color the pixel at (a, b) red. If they approach w_2, we color it blue, and if they approach w_3, we color it yellow. If this process continues for a large number of different z_1, a fractal similar to the figure will appear. Determine the appropriate color of the pixel for each value of z_1. (*Source:* Crownover, R., *Introduction to Fractals and Chaos,* Jones and Bartlett Publishers.)

(a) $z_1 = i$ **(b)** $z_1 = 2 + i$ **(c)** $z_1 = -1 - i$

53. The screens here illustrate how a pentagon can be graphed using a graphing calculator. Note that a pentagon has five sides, and the Tstep is $\frac{360}{5} = 72$. The display at the bottom of the graph screen indicates that one fifth root of 1 is $1 + 0i = 1$. Use this technique to find all fifth roots of 1, and express the real and imaginary parts in decimal form.

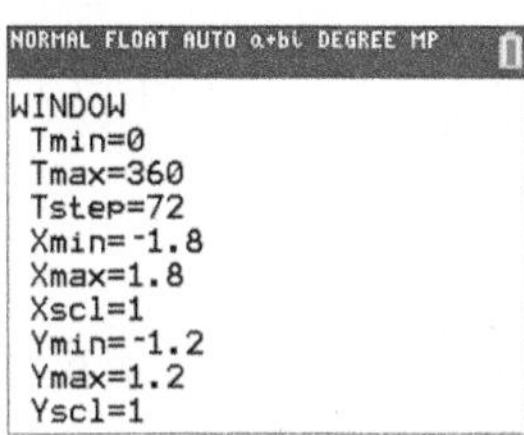

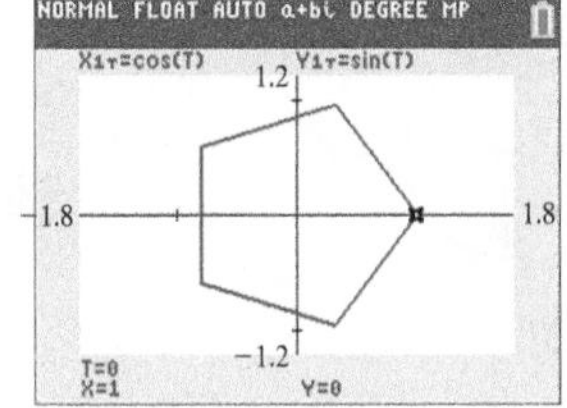

The calculator is in parametric, degree, and connected graph modes.

54. Use the method of **Exercise 53** to find the first three of the ten 10th roots of 1.

Use a calculator to find all solutions of each equation in rectangular form.

55. $x^2 - 3 + 2i = 0$

56. $x^2 + 2 - i = 0$

57. $x^5 + 2 + 3i = 0$

58. $x^3 + 4 - 5i = 0$

Relating Concepts

For individual or collaborative investigation *(Exercises 59–62)*

In earlier work we derived identities, or formulas, for $\cos 2\theta$ *and* $\sin 2\theta$*. These identities can also be derived using De Moivre's theorem.* ***Work Exercises 59–62 in order, to see how this is done.***

59. De Moivre's theorem states that $(\cos \theta + i \sin \theta)^2 =$ ________.

60. Expand the left side of the equation in **Exercise 59** as a binomial and combine like terms to write the left side in the form $a + bi$.

61. Use the result of **Exercise 60** to obtain the double-angle formula for cosine.

62. Repeat **Exercise 61,** but find the double-angle formula for sine.

Chapter 8 Quiz (Sections 8.1–8.4)

1. Find each product or quotient. Simplify the answers.

   (a) $\sqrt{-24} \cdot \sqrt{-3}$ (b) $\dfrac{\sqrt{-8}}{\sqrt{72}}$

2. Write each of the following in rectangular form for the complex numbers

$$w = 3 + 5i \quad \text{and} \quad z = -4 + i.$$

   (a) $w + z$ (and give a geometric representation)

   (b) $w - z$ (c) wz (d) $\dfrac{w}{z}$

3. Write each of the following in rectangular form.

   (a) $(1 - i)^3$ (b) i^{33}

4. Solve $3x^2 - x + 4 = 0$ over the set of complex numbers.

5. Write each complex number in trigonometric (polar) form, where $0° \le \theta < 360°$.

   (a) $-4i$ (b) $1 - i\sqrt{3}$ (c) $-3 - i$

6. Write each complex number in rectangular form.

   (a) $4(\cos 60° + i \sin 60°)$ (b) $5 \text{ cis } 130°$

   (c) $7(\cos 270° + i \sin 270°)$ (d) $2 \text{ cis } 0°$

7. Write each of the following in the form specified for the complex numbers

$$w = 12(\cos 80° + i \sin 80°) \quad \text{and} \quad z = 3(\cos 50° + i \sin 50°).$$

   (a) wz (trigonometric form)   (b) $\dfrac{w}{z}$ (rectangular form)

   (c) z^3 (rectangular form)   (d) w^3 (rectangular form)

8. Find the four complex fourth roots of -16. Write them in both trigonometric and rectangular forms.

[8.1]

1. (a) $-6\sqrt{2}$ (b) $\frac{1}{3}i$

[8.1, 8.2]

2. (a) $-1 + 6i$

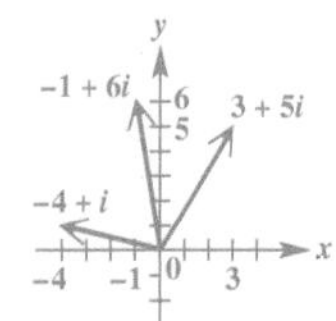

(b) $7 + 4i$ (c) $-17 - 17i$

(d) $-\frac{7}{17} - \frac{23}{17}i$

3. (a) $-2 - 2i$ (b) i, or $0 + i$

[8.1]

4. $\left\{\frac{1}{6} \pm \frac{\sqrt{47}}{6}i\right\}$

[8.2]

5. (a) $4(\cos 270° + i \sin 270°)$

   (b) $2(\cos 300° + i \sin 300°)$

   (c) $\sqrt{10}(\cos 198.4° + i \sin 198.4°)$

6. (a) $2 + 2i\sqrt{3}$

   (b) $-3.2139 + 3.8302i$

   (c) $-7i$, or $0 - 7i$

   (d) 2, or $2 + 0i$

[8.3, 8.4]

7. (a) $36(\cos 130° + i \sin 130°)$

   (b) $2\sqrt{3} + 2i$

   (c) $-\frac{27\sqrt{3}}{2} + \frac{27}{2}i$

   (d) $-864 - 864i\sqrt{3}$

[8.4]

8. $2(\cos 45° + i \sin 45°)$, $2(\cos 135° + i \sin 135°)$, $2(\cos 225° + i \sin 225°)$, $2(\cos 315° + i \sin 315°)$; $\sqrt{2} + i\sqrt{2}$, $-\sqrt{2} + i\sqrt{2}$, $-\sqrt{2} - i\sqrt{2}$, $\sqrt{2} - i\sqrt{2}$

8.5 Polar Equations and Graphs

- Polar Coordinate System
- Graphs of Polar Equations
- Conversion from Polar to Rectangular Equations
- Classification of Polar Equations

Polar Coordinate System Previously we have used the rectangular coordinate system to graph points and equations. In the rectangular coordinate system, each point in the plane is specified by giving two numbers (x, y). These represent the directed distances from a pair of perpendicular axes, the x-axis and the y-axis.

Now we consider the **polar coordinate system** which is based on a point, called the **pole,** and a ray, called the **polar axis.** The polar axis is usually drawn in the direction of the positive x-axis, as shown in **Figure 14.**

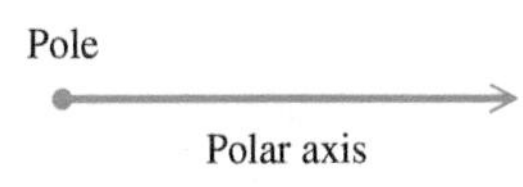

Figure 14

In **Figure 15** the pole has been placed at the origin of a rectangular coordinate system so that the polar axis coincides with the positive x-axis. Point P has rectangular coordinates (x, y). Point P can also be located by giving the directed angle θ from the positive x-axis to ray OP and the *directed* distance r from the pole to point P. The ordered pair (r, θ) gives the **polar coordinates** of point P. If $r > 0$ then point P lies on the terminal side of θ, and if $r < 0$ then point P lies on the ray pointing in the opposite direction of the terminal side of θ, a distance $|r|$ from the pole.

Figure 16 shows rectangular axes superimposed on a polar coordinate grid.

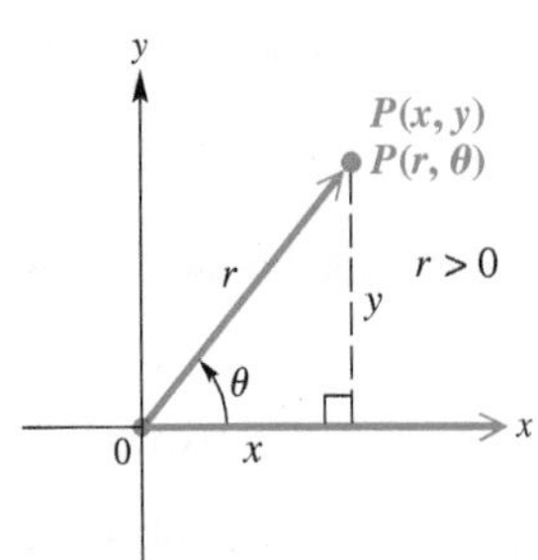

Figure 15

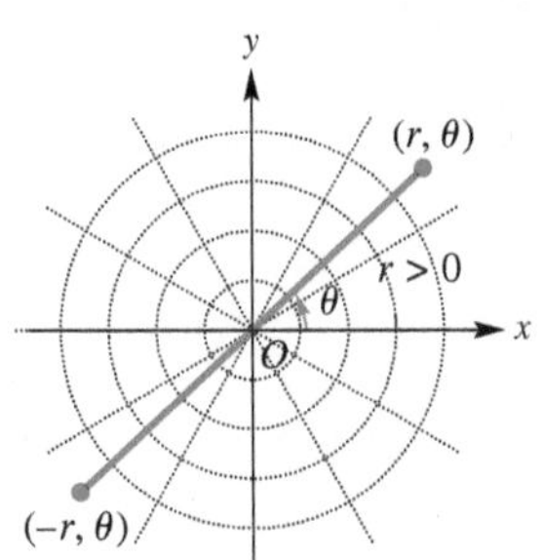

Figure 16

Classroom Example 1
Plot each point in the polar coordinate system. Then determine the rectangular coordinates of each point.

(a) $P(4, 135°)$ (b) $Q\left(-3, \frac{\pi}{6}\right)$
(c) $R\left(2, -\frac{\pi}{2}\right)$

Answers:
(a)

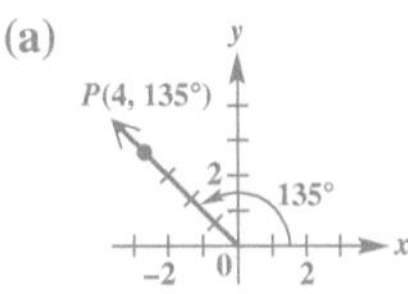

$\left(-2\sqrt{2}, 2\sqrt{2}\right)$

(b)

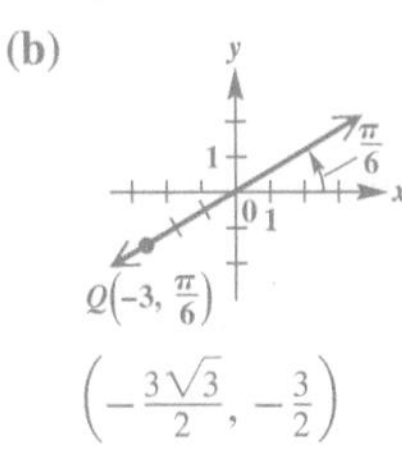

$\left(-\frac{3\sqrt{3}}{2}, -\frac{3}{2}\right)$

(c)

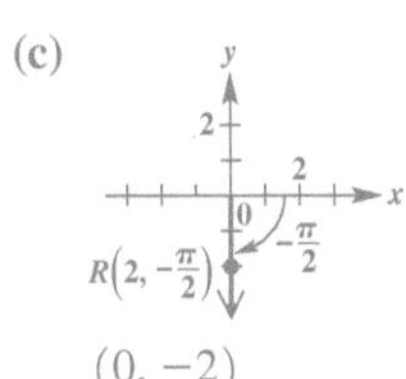

$(0, -2)$

Rectangular and Polar Coordinates

If a point has rectangular coordinates (x, y) and polar coordinates (r, θ), then these coordinates are related as follows.

$$x = r \cos \theta \qquad y = r \sin \theta$$

$$r^2 = x^2 + y^2 \qquad \tan \theta = \frac{y}{x}, \quad \text{if } x \neq 0$$

EXAMPLE 1 Plotting Points with Polar Coordinates

Plot each point in the polar coordinate system. Then determine the rectangular coordinates of each point.

(a) $P(2, 30°)$ **(b)** $Q\left(-4, \frac{2\pi}{3}\right)$ **(c)** $R\left(5, -\frac{\pi}{4}\right)$

SOLUTION

(a) In the point $P(2, 30°)$, $r = 2$ and $\theta = 30°$, so P is located 2 units from the origin in the positive direction on a ray making a 30° angle with the polar axis, as shown in **Figure 17.**

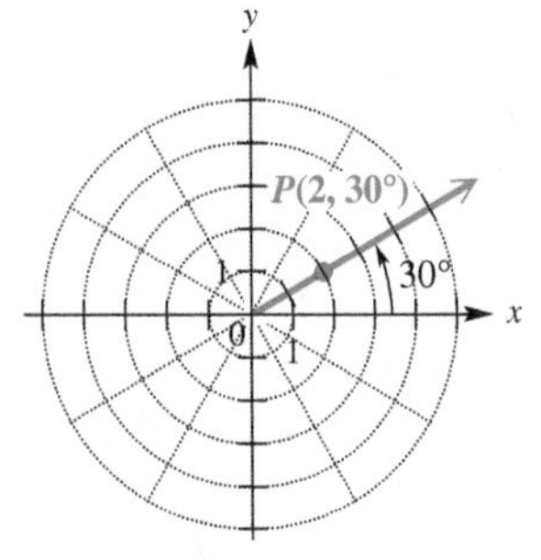

Figure 17

We find the rectangular coordinates as follows.

$x = r \cos \theta$	$y = r \sin \theta$	Conversion equations
$x = 2 \cos 30°$	$y = 2 \sin 30°$	Substitute.
$x = 2\left(\frac{\sqrt{3}}{2}\right)$	$y = 2\left(\frac{1}{2}\right)$	Evaluate the functions.
$x = \sqrt{3}$	$y = 1$	Multiply.

The rectangular coordinates are $\left(\sqrt{3}, 1\right)$.

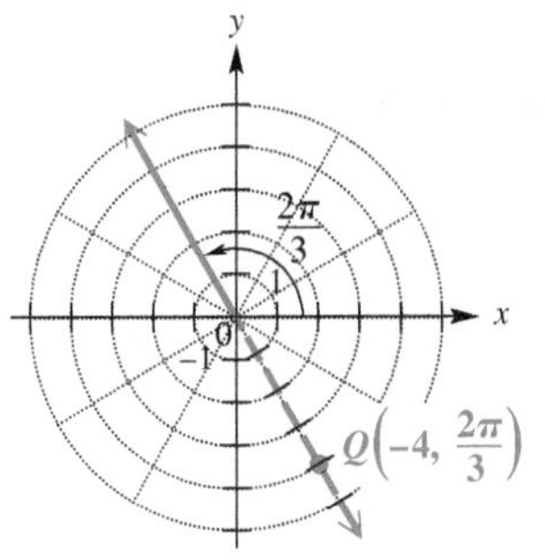

Figure 18

(b) In the point $Q\left(-4, \frac{2\pi}{3}\right)$, r is *negative*, so Q is 4 units in the *opposite* direction from the pole on an extension of the $\frac{2\pi}{3}$ ray. See **Figure 18.** The rectangular coordinates are

$$x = -4 \cos \frac{2\pi}{3} = -4\left(-\frac{1}{2}\right) = 2$$

and
$$y = -4 \sin \frac{2\pi}{3} = -4\left(\frac{\sqrt{3}}{2}\right) = -2\sqrt{3}.$$

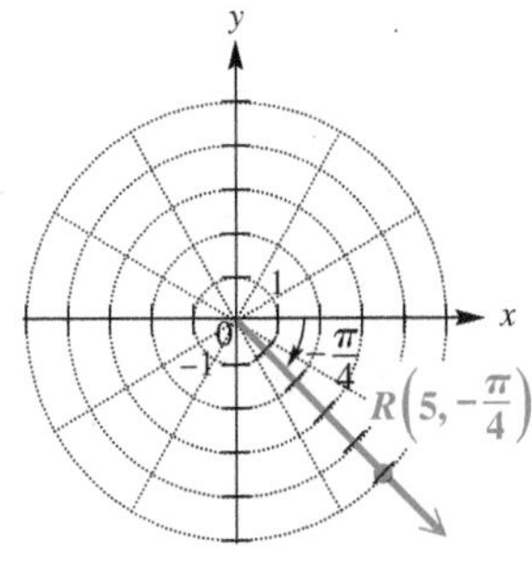

Figure 19

(c) Point $R\left(5, -\frac{\pi}{4}\right)$ is shown in **Figure 19.** Because θ is negative, the angle is measured in the clockwise direction.

$$x = 5 \cos\left(-\frac{\pi}{4}\right) = \frac{5\sqrt{2}}{2} \quad \text{and} \quad y = 5 \sin\left(-\frac{\pi}{4}\right) = -\frac{5\sqrt{2}}{2}$$

✔ **Now Try Exercises 13(a), (c), 15(a), (c), and 21(a), (c).**

While a given point in the plane can have only one pair of rectangular coordinates, this same point can have an infinite number of pairs of polar coordinates. For example, $(2, 30°)$ locates the same point as

$$(2, 390°), \quad (2, -330°), \quad \text{and} \quad (-2, 210°).$$

Classroom Example 2
Determine the following.
(a) Three other pairs of polar coordinates for the point $P(5, -110°)$
(b) Two pairs of polar coordinates for the point with rectangular coordinates $(6, -6\sqrt{3})$

Answers:
(Other answers are possible.)
(a) $(5, 250°)$, $(-5, 70°)$, $(-5, -290°)$
(b) $(12, 300°)$, $(-12, 120°)$

EXAMPLE 2 Giving Alternative Forms for Coordinates of Points

Determine the following.

(a) Three other pairs of polar coordinates for the point $P(3, 140°)$

(b) Two pairs of polar coordinates for the point with rectangular coordinates $(-1, 1)$

SOLUTION

(a) Three pairs that could be used for the point are $(3, -220°)$, $(-3, 320°)$, and $(-3, -40°)$. See **Figure 20.**

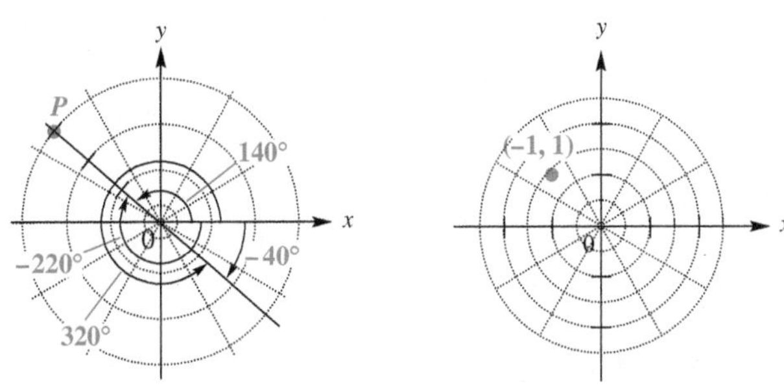

Figure 20 **Figure 21**

LOOKING AHEAD TO CALCULUS
Techniques studied in calculus associated with derivatives and integrals provide methods of finding slopes of tangent lines to polar curves, areas bounded by such curves, and lengths of their arcs.

(b) As shown in **Figure 21,** the point $(-1, 1)$ lies in the second quadrant. Because $\tan \theta = \frac{1}{-1} = -1$, one possible value for θ is 135°. Also,

$$r = \sqrt{x^2 + y^2} = \sqrt{(-1)^2 + 1^2} = \sqrt{2}.$$

Two pairs of polar coordinates are $\left(\sqrt{2}, 135°\right)$ and $\left(-\sqrt{2}, 315°\right)$.

✔ **Now Try Exercises 13(b), 15(b), 21(b), and 25.**

Classroom Example 3
For each rectangular equation, give the equivalent polar equation and sketch its graph.

(a) $y = 2x - 4$

(b) $x^2 + y^2 = 25$

Answers:

(a) $r = \frac{4}{2\cos\theta - \sin\theta}$

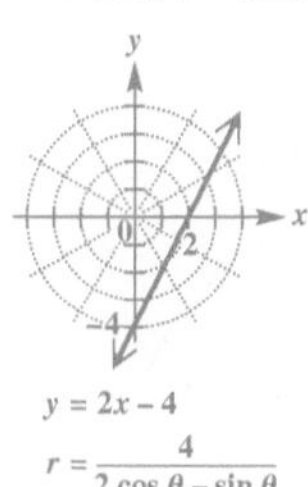

$y = 2x - 4$
$r = \frac{4}{2\cos\theta - \sin\theta}$

(b) $r = 5$ or $r = -5$

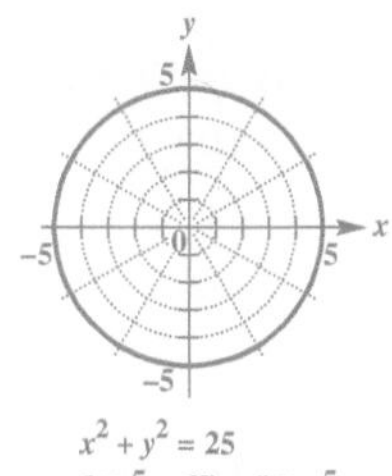

$x^2 + y^2 = 25$
$r = 5$ or $r = -5$

Graphs of Polar Equations An equation in the variables x and y is a **rectangular** (or **Cartesian**) **equation.** An equation in which r and θ are the variables instead of x and y is a **polar equation.**

$$r = 3\sin\theta, \quad r = 2 + \cos\theta, \quad r = \theta \qquad \text{Polar equations}$$

Although the rectangular forms of lines and circles are the ones most often encountered, they can also be defined in terms of polar coordinates. The polar equation of the line $ax + by = c$ can be derived as follows.

Line:

$$ax + by = c \qquad \text{Rectangular equation of a line}$$

$$a(r\cos\theta) + b(r\sin\theta) = c \qquad \text{Convert to polar coordinates.}$$

$$r(a\cos\theta + b\sin\theta) = c \qquad \text{Factor out } r.$$

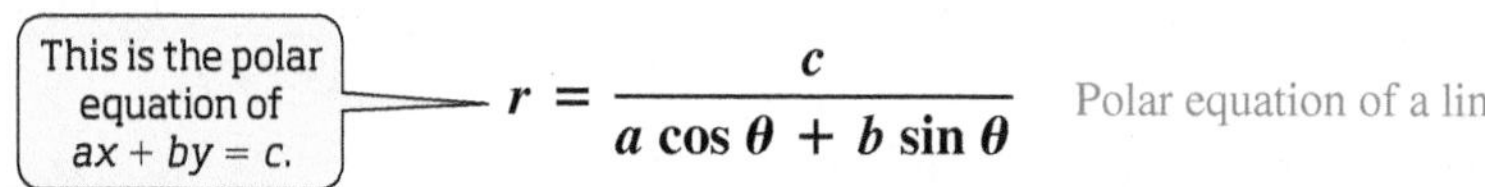

$$r = \frac{c}{a\cos\theta + b\sin\theta} \qquad \text{Polar equation of a line}$$

For the circle $x^2 + y^2 = a^2$, the polar equation can be found in a similar manner.

Circle:

$$x^2 + y^2 = a^2 \qquad \text{Rectangular equation of a circle}$$

$$r^2 = a^2 \qquad x^2 + y^2 = r^2$$

These are polar equations of $x^2 + y^2 = a^2$.

$$r = \pm a \qquad \text{Polar equation of a circle; } r \text{ can be negative in polar coordinates.}$$

We use these forms in the next example.

EXAMPLE 3 Finding Polar Equations of Lines and Circles

For each rectangular equation, give the equivalent polar equation and sketch its graph.

(a) $y = x - 3$ **(b)** $x^2 + y^2 = 4$

SOLUTION

(a) This is the equation of a line.

$$y = x - 3$$

$$x - y = 3 \qquad \text{Write in standard form } ax + by = c.$$

$$r\cos\theta - r\sin\theta = 3 \qquad \text{Substitute for } x \text{ and } y.$$

$$r(\cos\theta - \sin\theta) = 3 \qquad \text{Factor out } r.$$

$$r = \frac{3}{\cos\theta - \sin\theta} \qquad \text{Divide by } \cos\theta - \sin\theta.$$

A traditional graph is shown in **Figure 22(a),** and a calculator graph is shown in **Figure 22(b).**

$y = x - 3$ (rectangular)

$r = \frac{3}{\cos\theta - \sin\theta}$ (polar)

(a)

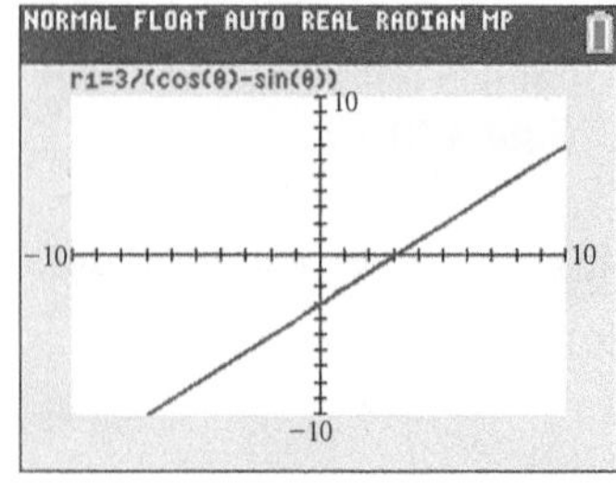

Polar graphing mode

(b)

Figure 22

(b) The graph of $x^2 + y^2 = 4$ is a circle with center at the origin and radius 2.

$$x^2 + y^2 = 4$$

$$r^2 = 4 \qquad x^2 + y^2 = r^2$$

$$r = 2 \quad \text{or} \quad r = -2$$

In polar coordinates, we may have $r < 0$.

The graphs of $r = 2$ and $r = -2$ coincide. See **Figure 23** on the next page.

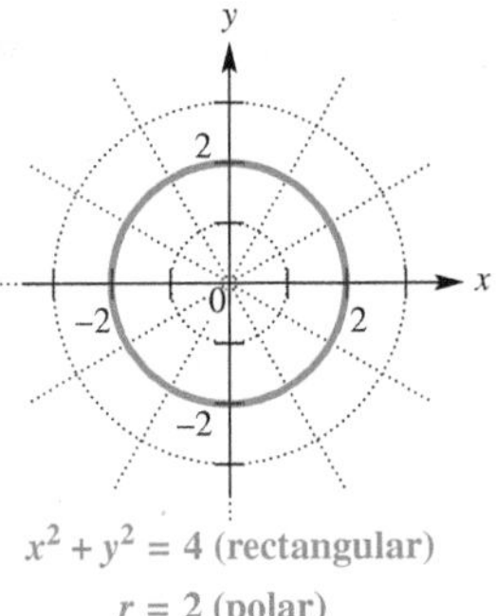

$x^2 + y^2 = 4$ (rectangular)
$r = 2$ (polar)

(a)

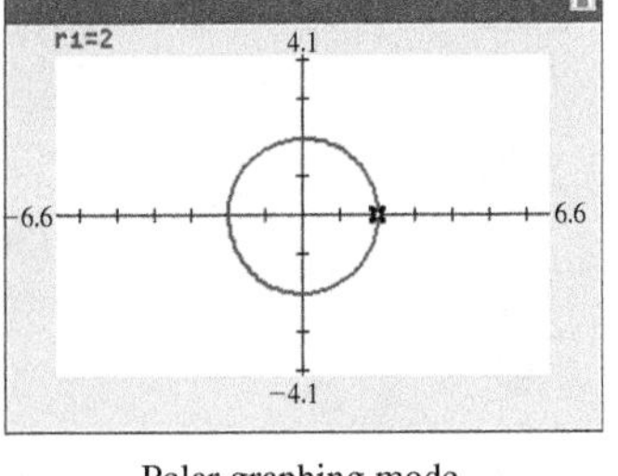

Polar graphing mode

(b)

Figure 23

Classroom Example 4
Graph $r = 1 - \sin \theta$.

Answer:

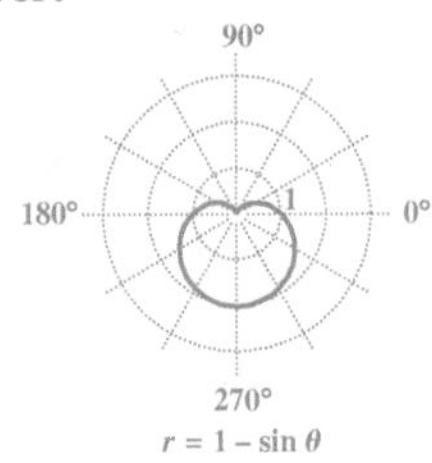

✔ **Now Try Exercises 37 and 39.**

To graph polar equations, evaluate r for various values of θ until a pattern appears, and then join the points with a smooth curve. The next four examples illustrate curves that are not usually discussed when rectangular coordinates are covered. (Using graphing calculators makes the task of graphing them quite a bit easier than using traditional point-plotting methods.)

EXAMPLE 4 Graphing a Polar Equation (Cardioid)

Graph $r = 1 + \cos \theta$.

ALGEBRAIC SOLUTION

To graph this equation, find some ordered pairs as in the table. Once the pattern of values of r becomes clear, it is not necessary to find more ordered pairs. The table includes approximate values for $\cos \theta$ and r.

θ	$\cos \theta$	$r = 1 + \cos \theta$	θ	$\cos \theta$	$r = 1 + \cos \theta$
0°	1	2	135°	−0.7	0.3
30°	0.9	1.9	150°	−0.9	0.1
45°	0.7	1.7	180°	−1	0
60°	0.5	1.5	270°	0	1
90°	0	1	315°	0.7	1.7
120°	−0.5	0.5	330°	0.9	1.9

Connect the points in order—from $(2, 0°)$ to $(1.9, 30°)$ to $(1.7, 45°)$ and so on. See **Figure 24.** This curve is called a **cardioid** because of its heart shape. The curve has been graphed on a **polar grid.**

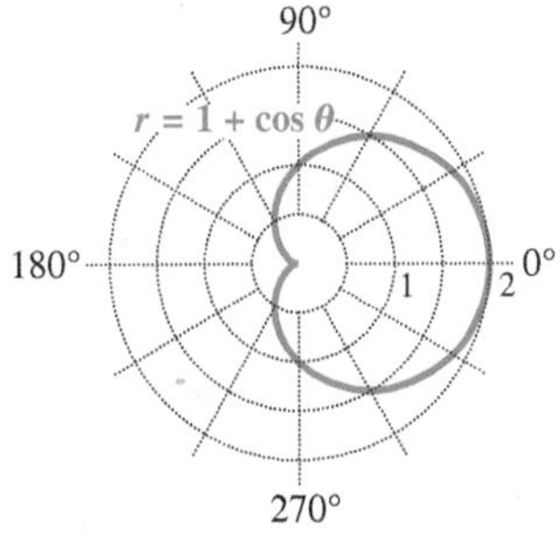

Figure 24

GRAPHING CALCULATOR SOLUTION

We choose degree mode and graph values of θ in the interval $[0°, 360°]$. The screen in **Figure 25(a)** shows the choices needed to generate the graph in **Figure 25(b).**

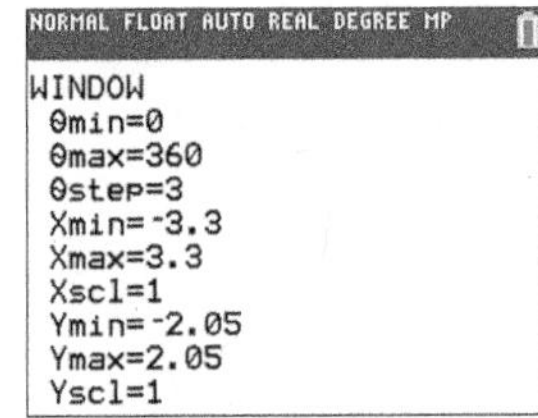

(a)

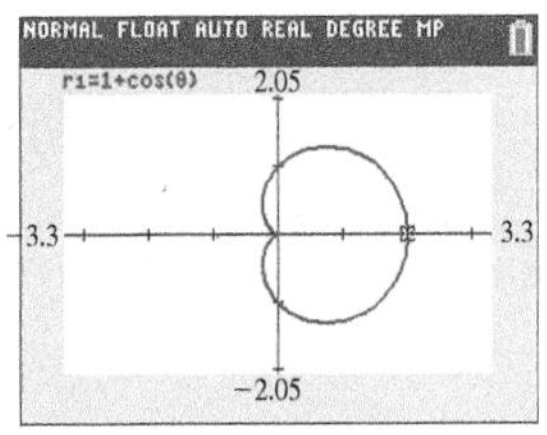

Polar graphing mode

(b)

Figure 25

✔ **Now Try Exercise 47.**

Classroom Example 5
Graph $r = 4 \cos 3\theta$.

Answer:

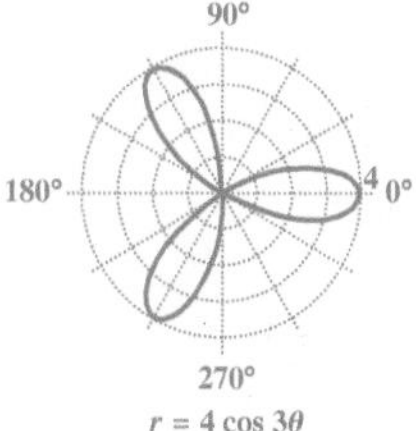

EXAMPLE 5 Graphing a Polar Equation (Rose)

Graph $r = 3 \cos 2\theta$.

SOLUTION Because the argument is 2θ, the graph requires a greater number of points than when the argument is just θ. We complete the table using selected angle measures through 360° in order to see the pattern of the graph. Approximate values in the table have been rounded to the nearest tenth.

θ	2θ	$\cos 2\theta$	$r = 3\cos 2\theta$	θ	2θ	$\cos 2\theta$	$r = 3\cos 2\theta$
0°	0°	1	3	120°	240°	−0.5	−1.5
15°	30°	0.9	2.6	135°	270°	0	0
30°	60°	0.5	1.5	180°	360°	1	3
45°	90°	0	0	225°	450°	0	0
60°	120°	−0.5	−1.5	270°	540°	−1	−3
75°	150°	−0.9	−2.6	315°	630°	0	0
90°	180°	−1	−3	360°	720°	1	3

Plotting these points in order gives the graph of a **four-leaved rose.** Note in **Figure 26(a)** how the graph is developed with a continuous curve, beginning with the upper half of the right horizontal leaf and ending with the lower half of that leaf. As the graph is traced, the curve goes through the pole four times. This can be seen as a calculator graphs the curve. See **Figure 26(b).**

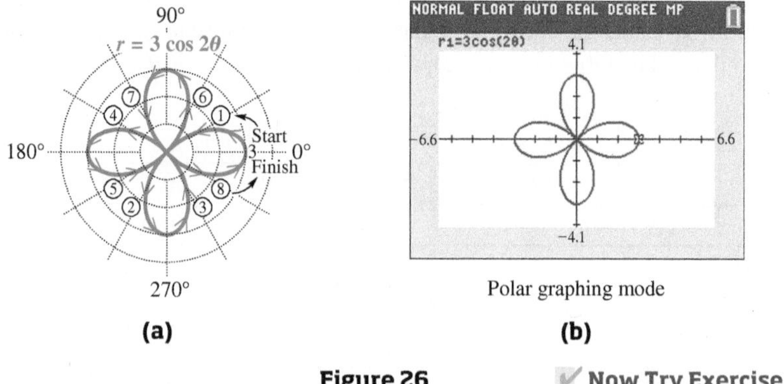

(a) (b)

Figure 26

✔ **Now Try Exercise 51.**

NOTE To sketch the graph of $r = 3 \cos 2\theta$ in polar coordinates, it may be helpful to first sketch the graph of $y = 3 \cos 2x$ in rectangular coordinates. The minimum and maximum values of this function may be used to determine the location of the tips of the leaves, and the x-intercepts of this function may be used to determine where the polar graph passes through the pole.

Classroom Example 6
Graph $r^2 = 9 \sin 2\theta$.

Answer:

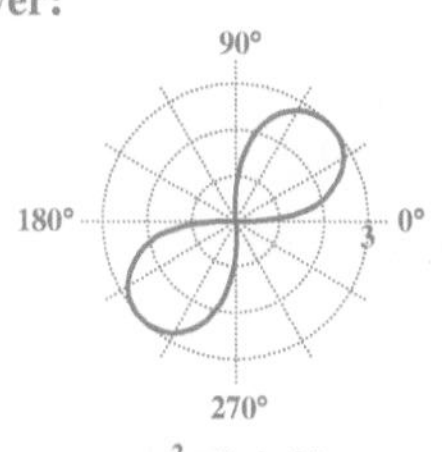

The equation $r = 3 \cos 2\theta$ in **Example 5** has a graph that belongs to a family of curves called **roses.**

$$r = a \sin n\theta \quad \text{and} \quad r = a \cos n\theta \qquad \text{Equations of roses}$$

- The graph has n leaves if n is odd, and $2n$ leaves if n is even.
- The absolute value of a determines the length of the leaves.

EXAMPLE 6 Graphing a Polar Equation (Lemniscate)

Graph $r^2 = \cos 2\theta$.

ALGEBRAIC SOLUTION

Complete a table of ordered pairs, and sketch the graph, as in **Figure 27.** The point $(-1, 0°)$, with r negative, may be plotted as $(1, 180°)$. Also, $(-0.7, 30°)$ may be plotted as $(0.7, 210°)$, and so on.

Values of θ for $45° < \theta < 135°$ are not included in the table because the corresponding values of $\cos 2\theta$ are negative (quadrants II and III) and so do not have real square roots. Values of θ greater than 180° give 2θ greater than 360° and would repeat the points already found. This curve is called a **lemniscate.**

θ	0°	30°	45°	135°	150°	180°
2θ	0°	60°	90°	270°	300°	360°
$\cos 2\theta$	1	0.5	0	0	0.5	1
$r = \pm\sqrt{\cos 2\theta}$	± 1	± 0.7	0	0	± 0.7	± 1

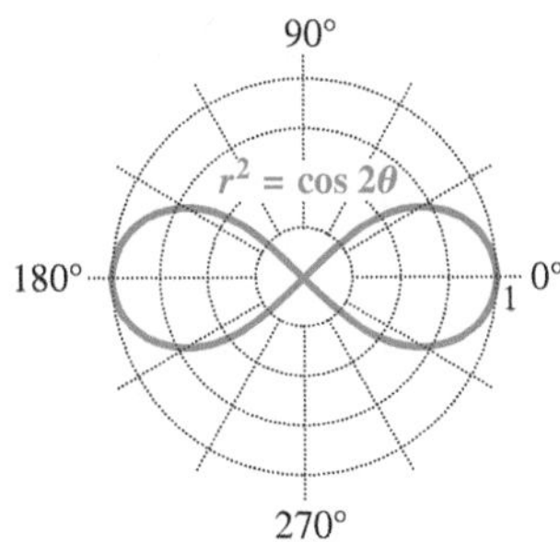

Figure 27

GRAPHING CALCULATOR SOLUTION

To graph $r^2 = \cos 2\theta$ with a graphing calculator, first solve for r by considering both square roots. Enter the two polar equations as

$$r_1 = \sqrt{\cos 2\theta} \quad \text{and} \quad r_2 = -\sqrt{\cos 2\theta}.$$

See **Figures 28(a) and (b).**

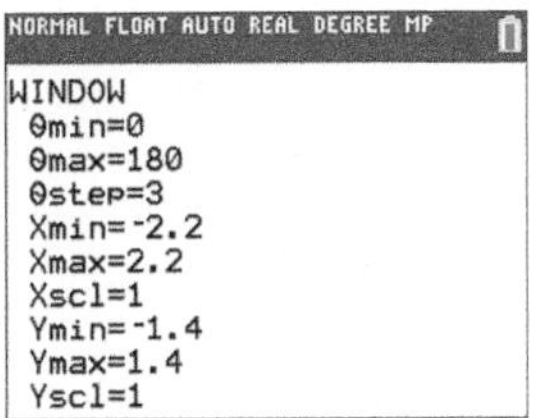

Settings for the graph below

(a)

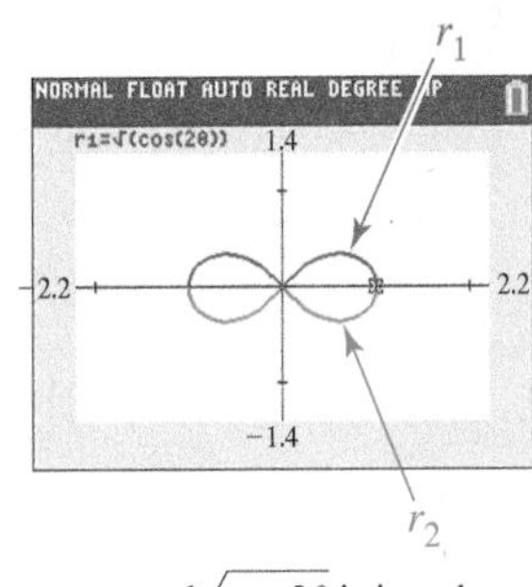

$r_2 = -\sqrt{\cos 2\theta}$ is in red.

(b)

Figure 28

✔ **Now Try Exercise 53.**

Classroom Example 7
Graph $r = -\theta$ (with θ measured in radians).

Answer:

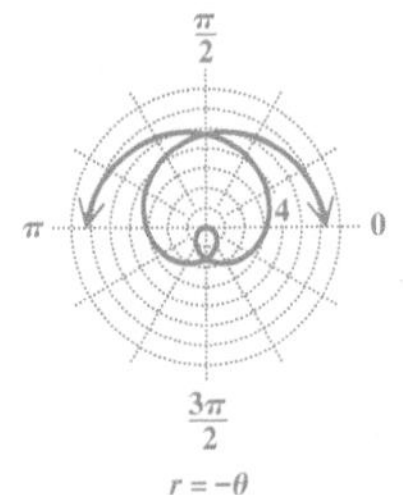

EXAMPLE 7 Graphing a Polar Equation (Spiral of Archimedes)

Graph $r = 2\theta$ (with θ measured in radians).

SOLUTION Some ordered pairs are shown in the table. Because $r = 2\theta$ does *not* involve a trigonometric function of θ, we must also consider negative values of θ. The graph in **Figure 29(a)** on the next page is a **spiral of Archimedes. Figure 29(b)** shows a calculator graph of this spiral.

θ (radians)	$r = 2\theta$	θ (radians)	$r = 2\theta$
$-\pi$	-6.3	$\frac{\pi}{3}$	2.1
$-\frac{\pi}{2}$	-3.1	$\frac{\pi}{2}$	3.1
$-\frac{\pi}{4}$	-1.6	π	6.3
0	0	$\frac{3\pi}{2}$	9.4
$\frac{\pi}{6}$	1	2π	12.6

Radian measures have been rounded.

Teaching Tip Point out that not all polar equations are easier to graph after being converted to rectangular equations.

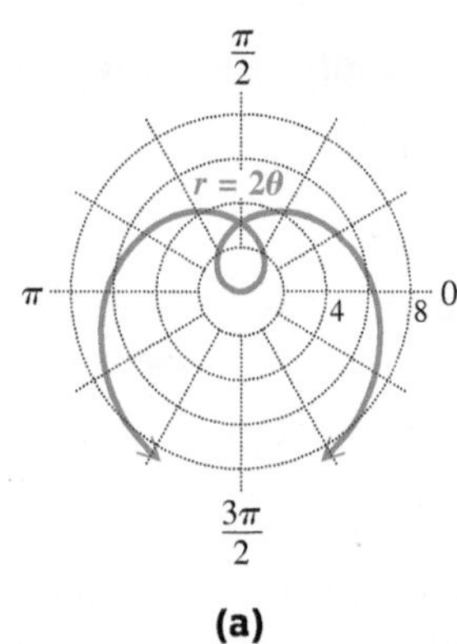

(a)

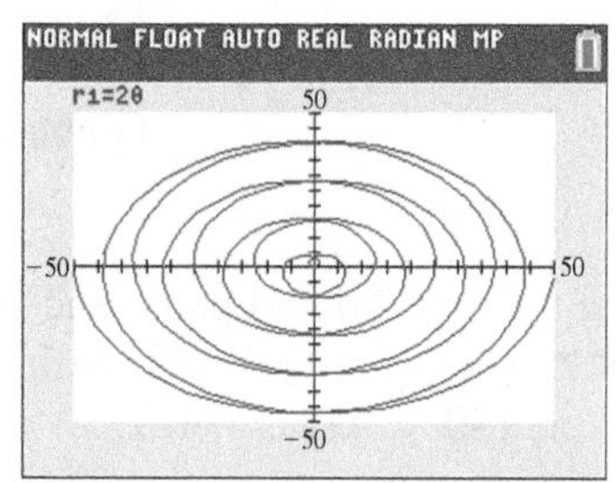

More of the spiral is shown in this calculator graph, with $-8\pi \le \theta \le 8\pi$.

(b)

Figure 29

Now Try Exercise 59.

Classroom Example 8
For the equation

$$r = \frac{1}{1 - \cos\theta},$$

write an equivalent equation in rectangular coordinates, and graph.

Answer: $y^2 = 2\left(x + \frac{1}{2}\right)$

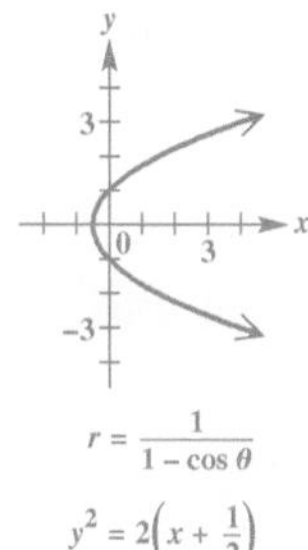

Conversion from Polar to Rectangular Equations

EXAMPLE 8 Converting a Polar Equation to a Rectangular Equation

For the equation $r = \frac{4}{1 + \sin\theta}$, write an equivalent equation in rectangular coordinates, and graph.

SOLUTION

$$r = \frac{4}{1 + \sin\theta} \qquad \text{Polar equation}$$

$$r(1 + \sin\theta) = 4 \qquad \text{Multiply by } 1 + \sin\theta.$$

$$r + r\sin\theta = 4 \qquad \text{Distributive property}$$

$$\sqrt{x^2 + y^2} + y = 4 \qquad \text{Let } r = \sqrt{x^2 + y^2} \text{ and } r\sin\theta = y.$$

$$\sqrt{x^2 + y^2} = 4 - y \qquad \text{Subtract } y.$$

$$x^2 + y^2 = (4 - y)^2 \qquad \text{Square each side.}$$

$$x^2 + y^2 = 16 - 8y + y^2 \qquad \text{Expand the right side.}$$

$$x^2 = -8y + 16 \qquad \text{Subtract } y^2.$$

$$x^2 = -8(y - 2) \qquad \text{Rectangular equation}$$

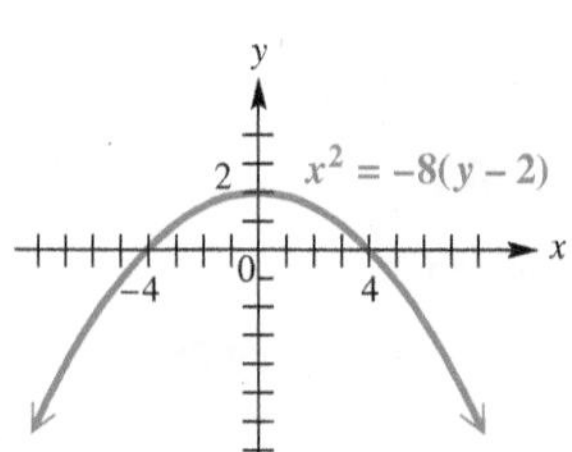

Figure 30

The final equation represents a parabola and is graphed in **Figure 30.**

Now Try Exercise 63.

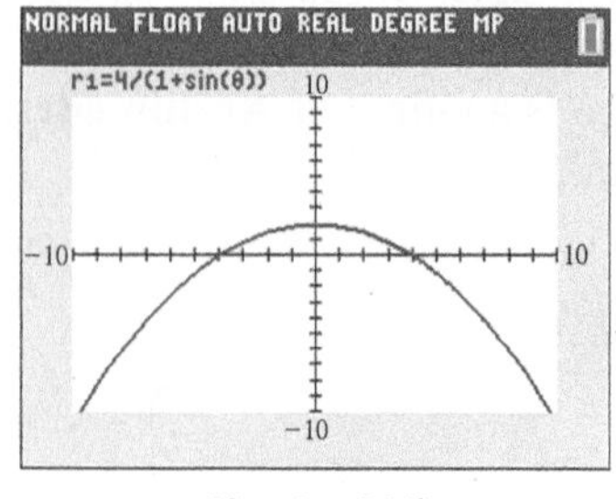

$0° \le \theta \le 360°$

Figure 31

The conversion in **Example 8** is not necessary when using a graphing calculator. **Figure 31** shows the graph of $r = \frac{4}{1 + \sin\theta}$, graphed directly with the calculator in polar mode. ■

Classification of Polar Equations The table on the next page summarizes common polar graphs and forms of their equations. In addition to circles, lemniscates, and roses, we include **limaçons.** Cardioids are a special case of limaçons, where $\left|\frac{a}{b}\right| = 1$.

NOTE Some other polar curves are the **cissoid, kappa curve, conchoid, trisectrix, cruciform, strophoid,** and **lituus.** Refer to older textbooks on analytic geometry or the Internet to investigate them.

Polar Graphs and Forms of Equations

Circles and Lemniscates

Circles

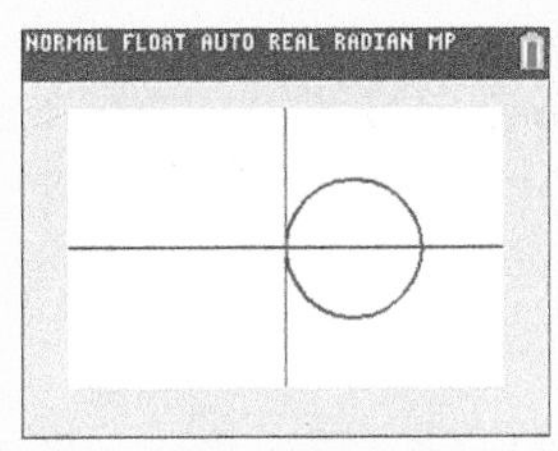

$r = a \cos \theta$

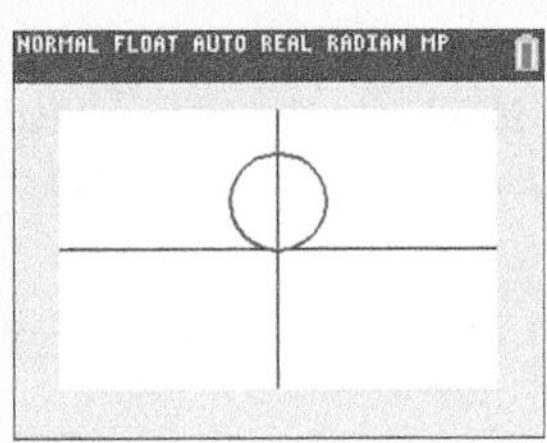

$r = a \sin \theta$

Lemniscates

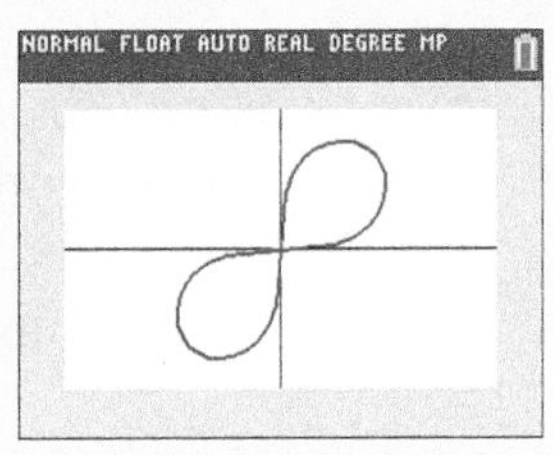

$r^2 = a^2 \sin 2\theta$

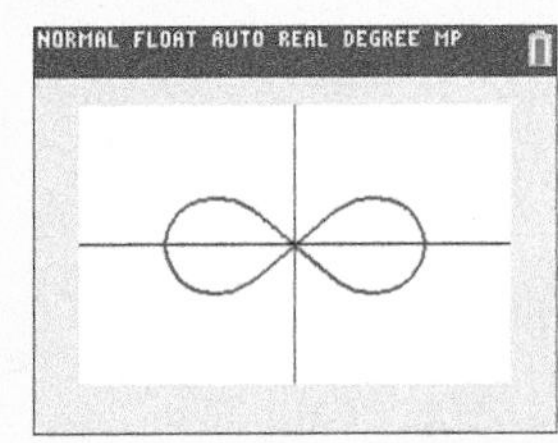

$r^2 = a^2 \cos 2\theta$

Limaçons

$r = a \pm b \sin \theta$ or $r = a \pm b \cos \theta$

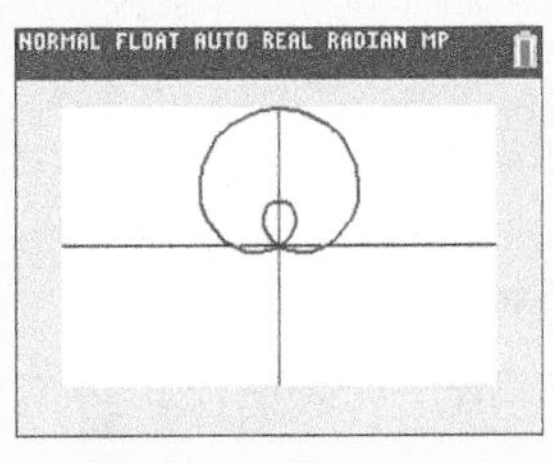

$\frac{a}{b} < 1$

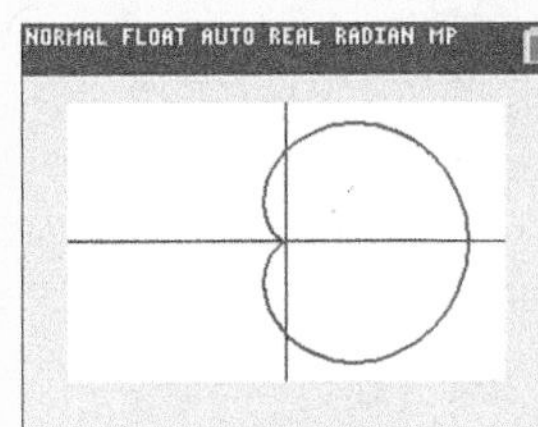

$\frac{a}{b} = 1$

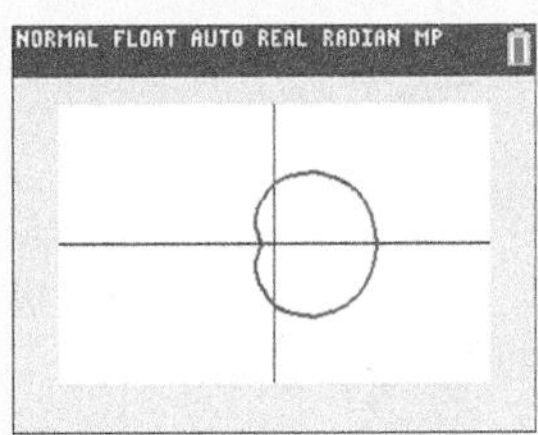

$1 < \frac{a}{b} < 2$

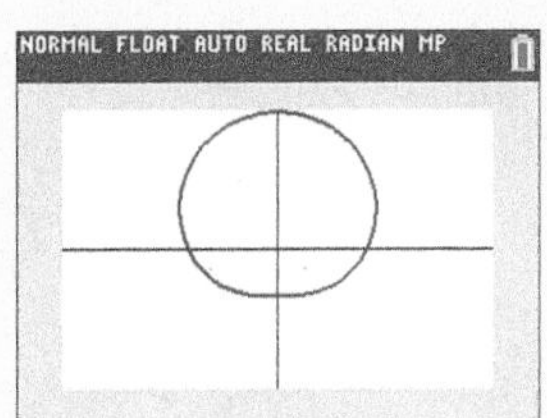

$\frac{a}{b} \geq 2$

Rose Curves

$2n$ leaves if n is even, $n \geq 2$

$n = 2$

NORMAL FLOAT AUTO REAL DEGREE MP

$r = a \sin n\theta$

$n = 4$

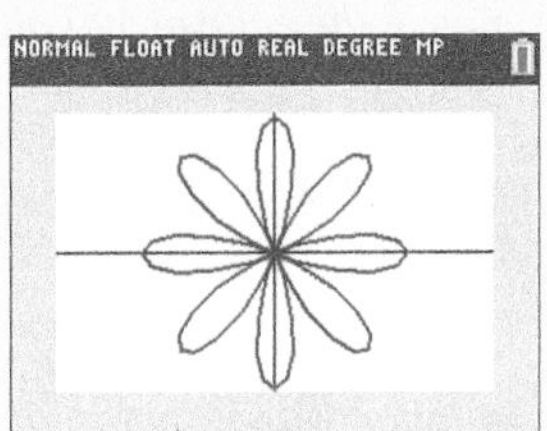

$r = a \cos n\theta$

n leaves if n is odd

$n = 3$

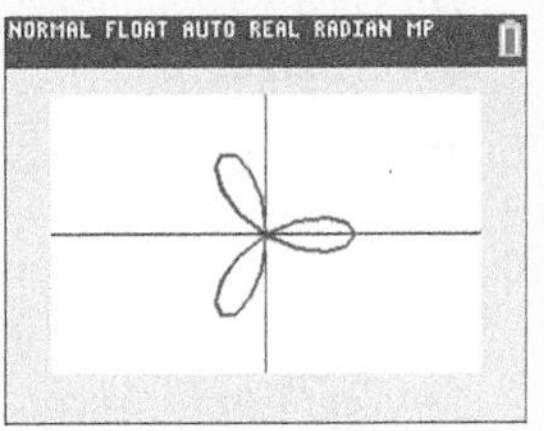

$r = a \cos n\theta$

$n = 5$

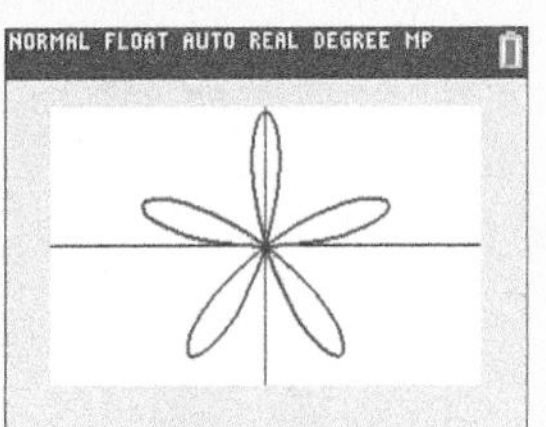

$r = a \sin n\theta$

8.5 Exercises

1. $\frac{3}{2}$ 2. 1
3. $\pm\sqrt{2}$ 4. ± 1
5. II 6. I
7. IV 8. III
9. positive x-axis
10. negative x-axis

CONCEPT PREVIEW *Fill in the blank to correctly complete each sentence.*

1. For the polar equation $r = 3 \cos \theta$, if $\theta = 60°$, then $r =$ ______.
2. For the polar equation $r = 2 \sin 2\theta$, if $\theta = 15°$, then $r =$ ______.
3. For the polar equation $r^2 = 4 \sin 2\theta$, if $\theta = 15°$, then $r =$ ______.
4. For the polar equation $r^2 = -2 \cos 2\theta$, if $\theta = 60°$, then $r =$ ______.

11. negative y-axis
12. positive y-axis

Graphs for Exercises 13(a), 15(a), 17(a), 19(a), 21(a), 23(a)

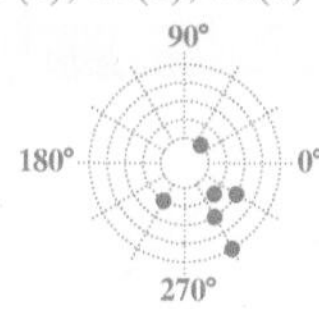

Graphs for Exercises 14(a), 16(a), 18(a), 20(a), 22(a), 24 (a)

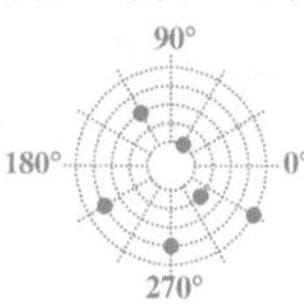

Answers may vary in Exercises 13(b)–24(b).

13. **(b)** $(1, 405^\circ), (-1, 225^\circ)$
(c) $\left(\frac{\sqrt{2}}{2}, \frac{\sqrt{2}}{2}\right)$

14. **(b)** $(3, 480^\circ), (-3, 300^\circ)$
(c) $\left(-\frac{3}{2}, \frac{3\sqrt{3}}{2}\right)$

15. **(b)** $(-2, 495^\circ), (2, 315^\circ)$
(c) $\left(\sqrt{2}, -\sqrt{2}\right)$

16. **(b)** $(-4, 390^\circ), (4, 210^\circ)$
(c) $\left(-2\sqrt{3}, -2\right)$

17. **(b)** $(5, 300^\circ), (-5, 120^\circ)$
(c) $\left(\frac{5}{2}, -\frac{5\sqrt{3}}{2}\right)$

18. **(b)** $(2, 315^\circ), (-2, 135^\circ)$
(c) $\left(\sqrt{2}, -\sqrt{2}\right)$

19. **(b)** $(-3, 150^\circ), (3, -30^\circ)$
(c) $\left(\frac{3\sqrt{3}}{2}, -\frac{3}{2}\right)$

20. **(b)** $(-1, 240^\circ), (1, 60^\circ)$
(c) $\left(\frac{1}{2}, \frac{\sqrt{3}}{2}\right)$

21. **(b)** $\left(3, \frac{11\pi}{3}\right), \left(-3, \frac{2\pi}{3}\right)$
(c) $\left(\frac{3}{2}, -\frac{3\sqrt{3}}{2}\right)$

22. **(b)** $\left(4, -\frac{\pi}{2}\right), \left(-4, \frac{\pi}{2}\right)$
(c) $(0, -4)$

23. **(b)** $\left(-2, \frac{7\pi}{3}\right), \left(2, \frac{4\pi}{3}\right)$
(c) $\left(-1, -\sqrt{3}\right)$

24. **(b)** $\left(-5, \frac{17\pi}{6}\right), \left(5, \frac{11\pi}{6}\right)$
(c) $\left(\frac{5\sqrt{3}}{2}, -\frac{5}{2}\right)$

Graphs for Exercises 25(a), 27(a), 29(a), 31(a), 33(a), 35(a)

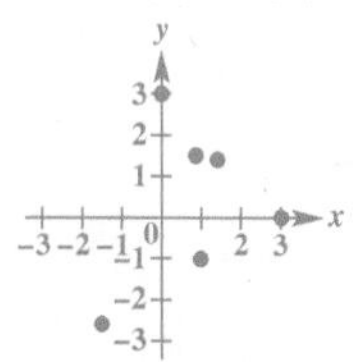

CONCEPT PREVIEW *For each point given in polar coordinates, state the quadrant in which the point lies if it is graphed in a rectangular coordinate system.*

5. $(5, 135^\circ)$ **6.** $(2, 60^\circ)$ **7.** $(6, -30^\circ)$ **8.** $(4.6, 213^\circ)$

CONCEPT PREVIEW *For each point given in polar coordinates, state the axis on which the point lies if it is graphed in a rectangular coordinate system. Also state whether it is on the positive portion or the negative portion of the axis. (For example, $(5, 0^\circ)$ lies on the positive x-axis.)*

9. $(7, 360^\circ)$ **10.** $(4, 180^\circ)$ **11.** $(2, -90^\circ)$ **12.** $(8, 450^\circ)$

*For each pair of polar coordinates, **(a)** plot the point, **(b)** give two other pairs of polar coordinates for the point, and **(c)** give the rectangular coordinates for the point.* ***See Examples 1 and 2.***

13. $(1, 45^\circ)$ **14.** $(3, 120^\circ)$ **15.** $(-2, 135^\circ)$ **16.** $(-4, 30^\circ)$

17. $(5, -60^\circ)$ **18.** $(2, -45^\circ)$ **19.** $(-3, -210^\circ)$ **20.** $(-1, -120^\circ)$

21. $\left(3, \frac{5\pi}{3}\right)$ **22.** $\left(4, \frac{3\pi}{2}\right)$ **23.** $\left(-2, \frac{\pi}{3}\right)$ **24.** $\left(-5, \frac{5\pi}{6}\right)$

*For each pair of rectangular coordinates, **(a)** plot the point and **(b)** give two pairs of polar coordinates for the point, where $0^\circ \le \theta < 360^\circ$.* ***See Example 2(b).***

25. $(1, -1)$ **26.** $(1, 1)$ **27.** $(0, 3)$

28. $(0, -3)$ **29.** $\left(\sqrt{2}, \sqrt{2}\right)$ **30.** $\left(-\sqrt{2}, \sqrt{2}\right)$

31. $\left(\frac{\sqrt{3}}{2}, \frac{3}{2}\right)$ **32.** $\left(-\frac{\sqrt{3}}{2}, -\frac{1}{2}\right)$ **33.** $(3, 0)$

34. $(-2, 0)$ **35.** $\left(-\frac{3}{2}, -\frac{3\sqrt{3}}{2}\right)$ **36.** $\left(\frac{1}{2}, -\frac{\sqrt{3}}{2}\right)$

For each rectangular equation, give the equivalent polar equation and sketch its graph. ***See Example 3.***

37. $x - y = 4$ **38.** $x + y = -7$ **39.** $x^2 + y^2 = 16$

40. $x^2 + y^2 = 9$ **41.** $2x + y = 5$ **42.** $3x - 2y = 6$

Concept Check *Match each equation with its polar graph from choices A–D.*

43. $r = 3$ **44.** $r = \cos 3\theta$ **45.** $r = \cos 2\theta$ **46.** $r = \frac{2}{\cos\theta + \sin\theta}$

A.

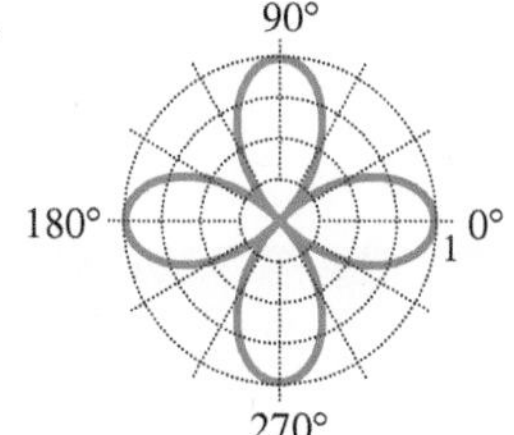

B.

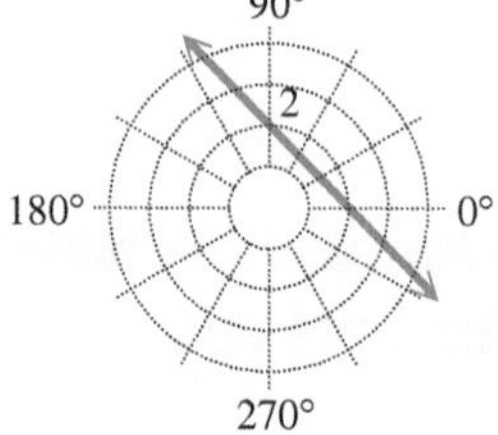

C.

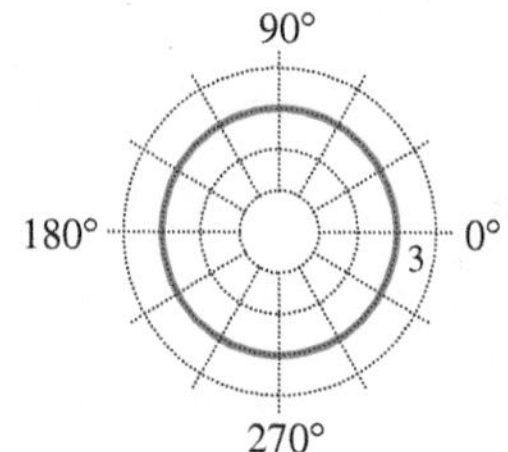

D.

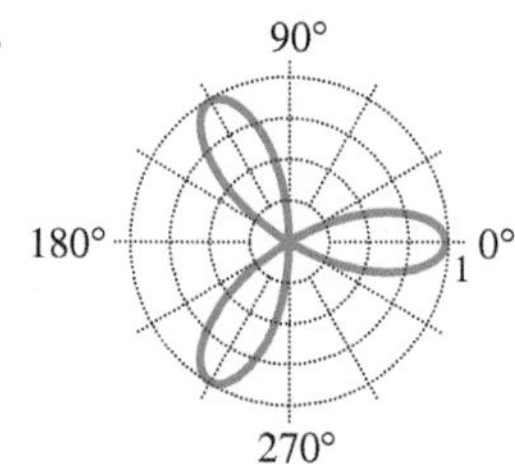

Graphs for Exercises 26(a), 28(a), 30(a), 32(a), 34(a), 36(a)

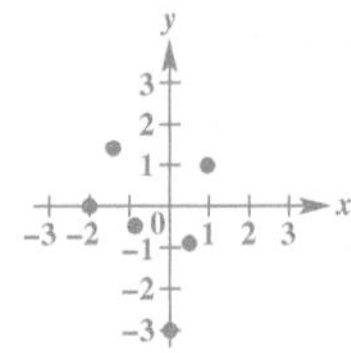

Answers may vary in Exercises 25(b)–36(b).

25. (b) $(\sqrt{2}, 315°), (-\sqrt{2}, 135°)$

26. (b) $(\sqrt{2}, 45°), (-\sqrt{2}, 225°)$

27. (b) $(3, 90°), (-3, 270°)$

28. (b) $(3, 270°), (-3, 90°)$

29. (b) $(2, 45°), (-2, 225°)$

30. (b) $(2, 135°), (-2, 315°)$

31. (b) $(\sqrt{3}, 60°), (-\sqrt{3}, 240°)$

32. (b) $(1, 210°), (-1, 30°)$

33. (b) $(3, 0°), (-3, 180°)$

34. (b) $(-2, 0°), (2, 180°)$

35. (b) $(3, 240°), (-3, 60°)$

36. (b) $(1, 300°), (-1, 120°)$

37. $r = \frac{4}{\cos\theta - \sin\theta}$

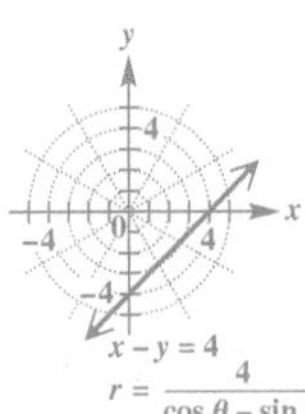

38. $r = \frac{-7}{\cos\theta + \sin\theta}$

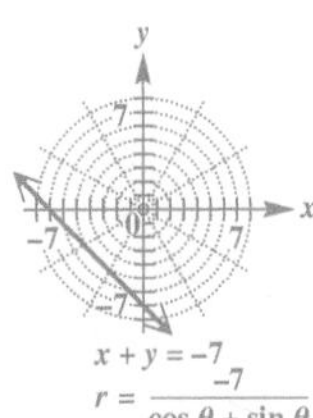

39. $r = 4$ or $r = -4$

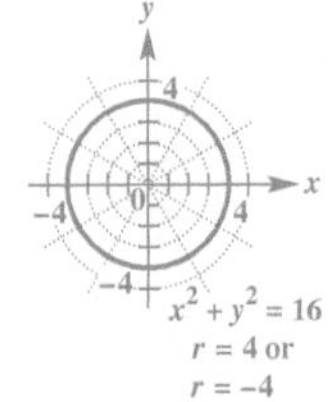

40. $r = 3$ or $r = -3$

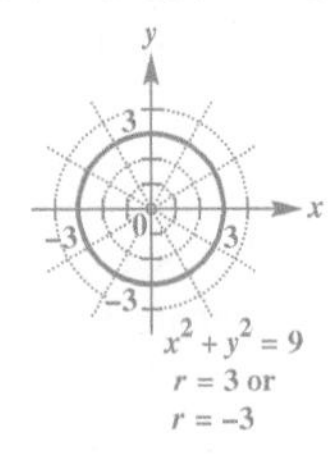

Graph each polar equation. In Exercises 47–56, also identify the type of polar graph. ***See Examples 4–6.***

47. $r = 2 + 2\cos\theta$

48. $r = 8 + 6\cos\theta$

49. $r = 3 + \cos\theta$

50. $r = 2 - \cos\theta$

51. $r = 4\cos 2\theta$

52. $r = 3\cos 5\theta$

53. $r^2 = 4\cos 2\theta$

54. $r^2 = 4\sin 2\theta$

55. $r = 4 - 4\cos\theta$

56. $r = 6 - 3\cos\theta$

57. $r = 2\sin\theta\tan\theta$ (This is a **cissoid.**)

58. $r = \frac{\cos 2\theta}{\cos\theta}$ (This is a **cissoid with a loop.**)

Graph each spiral of Archimedes. ***See Example 7.***

59. $r = \theta$ (Use both positive and nonpositive values.)

60. $r = -4\theta$ (Use a graphing calculator in a window of $[-30, 30]$ by $[-30, 30]$, in radian mode, and θ in $[-12\pi, 12\pi]$.)

For each equation, find an equivalent equation in rectangular coordinates, and graph. ***See Example 8.***

61. $r = 2\sin\theta$

62. $r = 2\cos\theta$

63. $r = \frac{2}{1 - \cos\theta}$

64. $r = \frac{3}{1 - \sin\theta}$

65. $r = -2\cos\theta - 2\sin\theta$

66. $r = \frac{3}{4\cos\theta - \sin\theta}$

67. $r = 2\sec\theta$

68. $r = -5\csc\theta$

69. $r = \frac{2}{\cos\theta + \sin\theta}$

70. $r = \frac{2}{2\cos\theta + \sin\theta}$

Solve each problem.

71. Find the polar equation of the line that passes through the points $(1, 0°)$ and $(2, 90°)$.

72. Explain how to plot a point (r, θ) in polar coordinates, if $r < 0$ and θ is in degrees.

Concept Check The polar graphs in this section exhibit symmetry. Visualize an xy-plane superimposed on the polar coordinate system, with the pole at the origin and the polar axis on the positive x-axis. Then a polar graph may be symmetric with respect to the x-axis (the polar axis), the y-axis (the line $\theta = \frac{\pi}{2}$), or the origin (the pole).

73. Complete the missing ordered pairs in the graphs below.

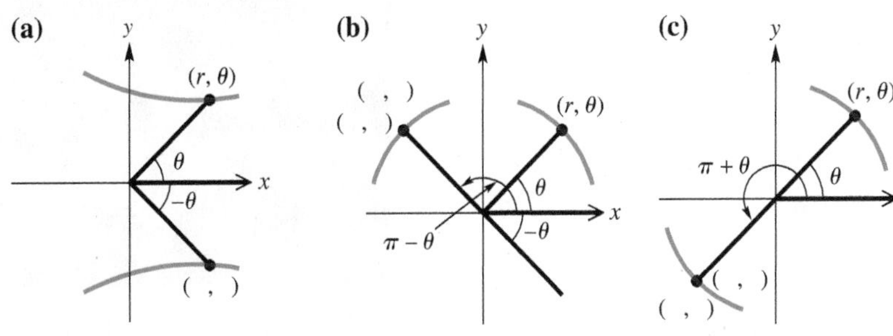

41. $r = \frac{5}{2\cos\theta + \sin\theta}$

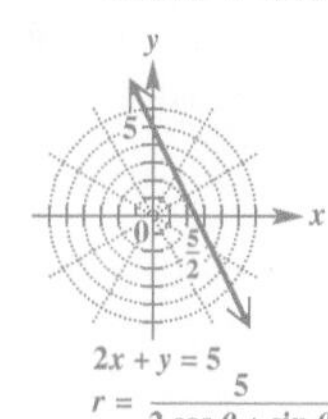

$2x + y = 5$

$r = \frac{5}{2\cos\theta + \sin\theta}$

42. $r = \frac{6}{3\cos\theta - 2\sin\theta}$

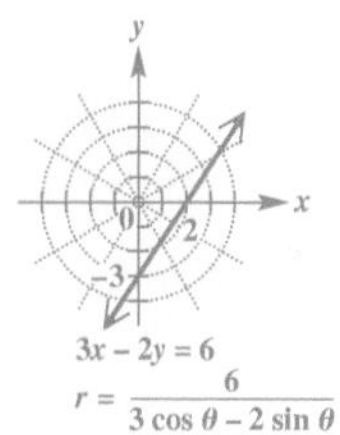

$3x - 2y = 6$

$r = \frac{6}{3\cos\theta - 2\sin\theta}$

43. C 44. D

45. A 46. B

47.

$r = 2 + 2\cos\theta$

cardioid

48.

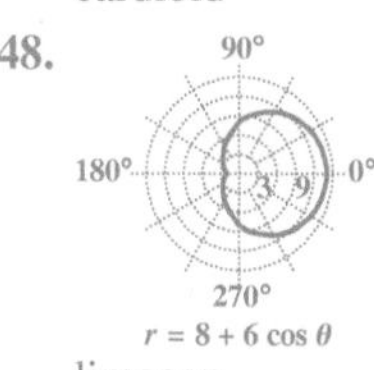

$r = 8 + 6\cos\theta$

limaçon

49.

$r = 3 + \cos\theta$

limaçon

50.

$r = 2 - \cos\theta$

limaçon

51.

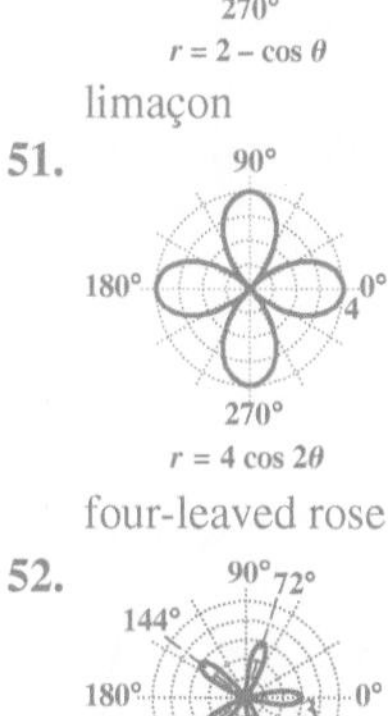

$r = 4\cos 2\theta$

four-leaved rose

52.

$r = 3\cos 5\theta$

five-leaved rose

74. Based on the results in **Exercise 73,** fill in the blank(s) to correctly complete each sentence.

(a) The graph of $r = f(\theta)$ is symmetric with respect to the polar axis if substitution of ______ for θ leads to an equivalent equation.

(b) The graph of $r = f(\theta)$ is symmetric with respect to the vertical line $\theta = \frac{\pi}{2}$ if substitution of ______ for θ leads to an equivalent equation.

(c) Alternatively, the graph of $r = f(\theta)$ is symmetric with respect to the vertical line $\theta = \frac{\pi}{2}$ if substitution of ______ for r and ______ for θ leads to an equivalent equation.

(d) The graph of $r = f(\theta)$ is symmetric with respect to the pole if substitution of ______ for r leads to an equivalent equation.

(e) Alternatively, the graph of $r = f(\theta)$ is symmetric with respect to the pole if substitution of ______ for θ leads to an equivalent equation.

(f) In general, the completed statements in parts (a)–(e) mean that the graphs of polar equations of the form $r = a \pm b\cos\theta$ (where a may be 0) are symmetric with respect to ________.

(g) In general, the completed statements in parts (a)–(e) mean that the graphs of polar equations of the form $r = a \pm b\sin\theta$ (where a may be 0) are symmetric with respect to ________.

Spirals of Archimedes *The graph of $r = a\theta$ in polar coordinates is an example of a spiral of Archimedes. With a calculator set to radian mode, use the given value of a and interval of θ to graph the spiral in the window specified.*

75. $a = 1, 0 \le \theta \le 4\pi$, $[-15, 15]$ by $[-15, 15]$

76. $a = 2, -4\pi \le \theta \le 4\pi$, $[-30, 30]$ by $[-30, 30]$

77. $a = 1.5, -4\pi \le \theta \le 4\pi$, $[-20, 20]$ by $[-20, 20]$

78. $a = -1, 0 \le \theta \le 12\pi$, $[-40, 40]$ by $[-40, 40]$

Intersection of Polar Curves *Find the polar coordinates of the points of intersection of the given curves for the specified interval of θ.*

79. $r = 4\sin\theta,\ r = 1 + 2\sin\theta;$ $0 \le \theta < 2\pi$

80. $r = 3,\ r = 2 + 2\cos\theta;$ $0^\circ \le \theta < 360^\circ$

81. $r = 2 + \sin\theta,\ r = 2 + \cos\theta;$ $0 \le \theta < 2\pi$

82. $r = \sin 2\theta,\ r = \sqrt{2}\cos\theta;$ $0 \le \theta < \pi$

(Modeling) Solve each problem.

83. ***Orbits of Satellites*** The polar equation

$$r = \frac{a(1 - \boldsymbol{e}^2)}{1 + \boldsymbol{e}\cos\theta}$$

can be used to graph the orbits of the satellites of our sun, where a is the average distance in astronomical units from the sun and $\boldsymbol{e}$ is a constant called the **eccentricity.** The sun will be located at the pole. The table lists the values of a and $\boldsymbol{e}$.

Satellite	a	e
Mercury	0.39	0.206
Venus	0.78	0.007
Earth	1.00	0.017
Mars	1.52	0.093
Jupiter	5.20	0.048
Saturn	9.54	0.056
Uranus	19.20	0.047
Neptune	30.10	0.009
Pluto	39.40	0.249

Source: Karttunen, H., P. Kröger, H. Oja, M. Putannen, and K. Donners (Editors), *Fundamental Astronomy, 4th edition,* Springer-Verlag. Zeilik, M., S. Gregory, and E. Smith, *Introductory Astronomy and Astrophysics,* Saunders College Publishers.

53. 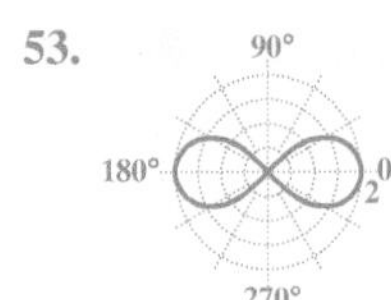

$r^2 = 4 \cos 2\theta$
lemniscate

54. 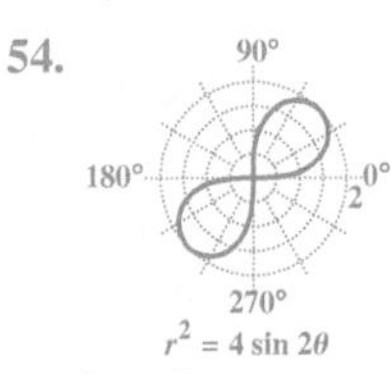

$r^2 = 4 \sin 2\theta$
lemniscate

55. 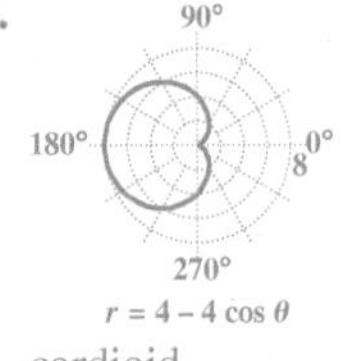

$r = 4 - 4 \cos \theta$
cardioid

56. 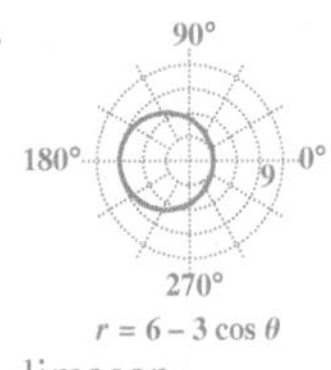

$r = 6 - 3 \cos \theta$
limaçon

57. 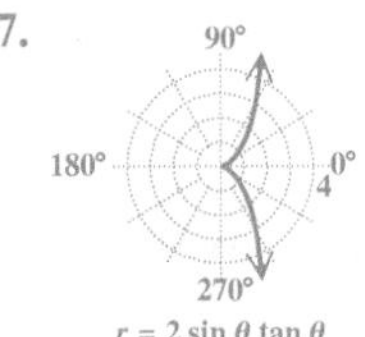

$r = 2 \sin \theta \tan \theta$

58. 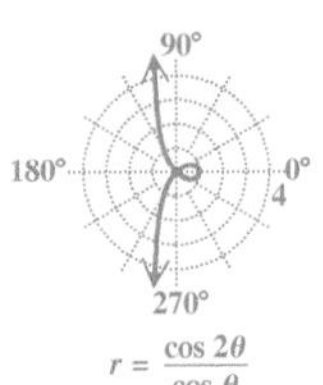

$r = \dfrac{\cos 2\theta}{\cos \theta}$

59. 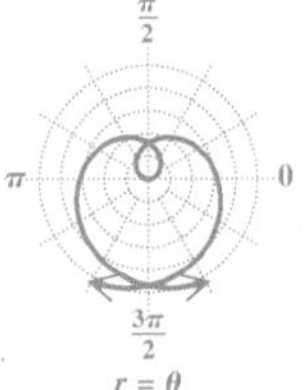

$r = \theta$

60. $r = 4\theta, -12\pi \leq \theta \leq 12\pi$

30
−30
30
−30

Answers for the remaining exercises are given in the answer section at the back of the text.

(a) Graph the orbits of the four closest satellites on the same polar grid. Choose a viewing window that results in a graph with nearly circular orbits.

(b) Plot the orbits of Earth, Jupiter, Uranus, and Pluto on the same polar grid. How does Earth's distance from the sun compare to the others' distances from the sun?

(c) Use graphing to determine whether or not Pluto is always farthest from the sun.

84. ***Radio Towers and Broadcasting Patterns*** Radio stations do not always broadcast in all directions with the same intensity. To avoid interference with an existing station to the north, a new station may be licensed to broadcast only east and west. To create an east-west signal, two radio towers are sometimes used. See the figure. Locations where the radio signal is received correspond to the interior of the curve

$$r^2 = 40{,}000 \cos 2\theta,$$

where the polar axis (or positive x-axis) points east.

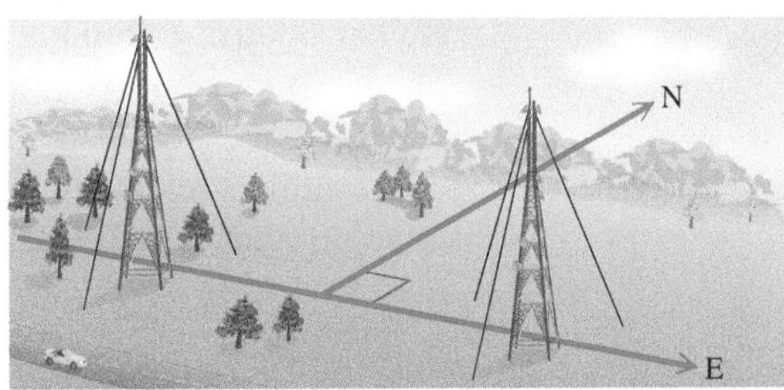

(a) Graph $r^2 = 40{,}000 \cos 2\theta$ for $0° \leq \theta \leq 360°$, where distances are in miles. Assuming the radio towers are located near the pole, use the graph to describe the regions where the signal can be received and where the signal cannot be received.

(b) Suppose a radio signal pattern is given by the following equation. Graph this pattern and interpret the results.

$$r^2 = 22{,}500 \sin 2\theta$$

Relating Concepts

For individual or collaborative investigation *(Exercises 85-92)*

In rectangular coordinates, the graph of

$$ax + by = c$$

is a horizontal line if $a = 0$ or a vertical line if $b = 0$. ***Work Exercises 85–92 in order,*** *to determine the general forms of polar equations for horizontal and vertical lines.*

85. Begin with the equation $y = k$, whose graph is a horizontal line. Make a trigonometric substitution for y using r and θ.

86. Solve the equation in **Exercise 85** for r.

87. Rewrite the equation in **Exercise 86** using the appropriate reciprocal function.

88. Sketch the graph of the equation $r = 3 \csc \theta$. What is the corresponding rectangular equation?

89. Begin with the equation $x = k$, whose graph is a vertical line. Make a trigonometric substitution for x using r and θ.

90. Solve the equation in **Exercise 89** for r.

91. Rewrite the equation in **Exercise 90** using the appropriate reciprocal function.

92. Sketch the graph of $r = 3 \sec \theta$. What is the corresponding rectangular equation?

8.6 Parametric Equations, Graphs, and Applications

- Basic Concepts
- Parametric Graphs and Their Rectangular Equivalents
- The Cycloid
- Applications of Parametric Equations

Basic Concepts We have graphed sets of ordered pairs that correspond to a function of the form $y = f(x)$ or $r = g(\theta)$. Another way to determine a set of ordered pairs involves the equations $x = f(t)$ and $y = g(t)$, where t is a real number in an interval I. Each value of t leads to a corresponding x-value and a corresponding y-value, and thus to an ordered pair (x, y).

Parametric Equations of a Plane Curve

A **plane curve** is a set of points (x, y) such that $x = f(t)$, $y = g(t)$, and f and g are both defined on an interval I. The equations $x = f(t)$ and $y = g(t)$ are **parametric equations** with **parameter** t.

Classroom Example 1
Let

$x = t^2 - 1$ and $y = t - 2$,

for t in $[-3, 3]$. Graph the set of ordered pairs (x, y).

Answer:

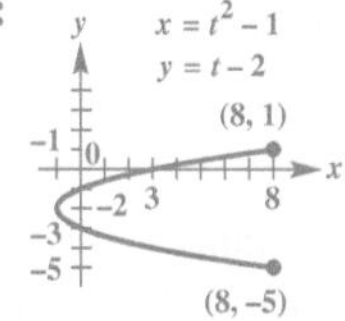

Graphing calculators are capable of graphing plane curves defined by parametric equations. The calculator must be set to parametric mode. ■

Parametric Graphs and Their Rectangular Equivalents

EXAMPLE 1 Graphing a Plane Curve Defined Parametrically

Let $x = t^2$ and $y = 2t + 3$, for t in $[-3, 3]$. Graph the set of ordered pairs (x, y).

ALGEBRAIC SOLUTION

Make a table of corresponding values of t, x, and y over the domain of t. Plot the points as shown in **Figure 32.** The graph is a portion of a parabola with horizontal axis $y = 3$. The arrowheads indicate the direction the curve traces as t increases.

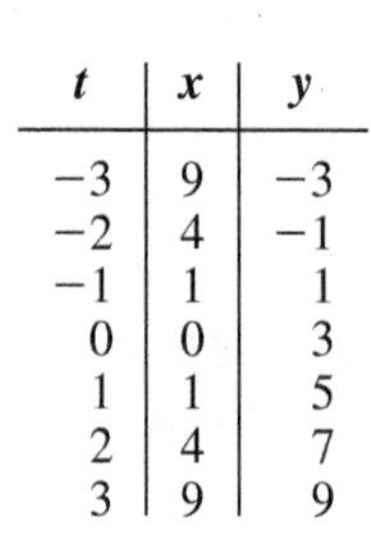

t	x	y
−3	9	−3
−2	4	−1
−1	1	1
0	0	3
1	1	5
2	4	7
3	9	9

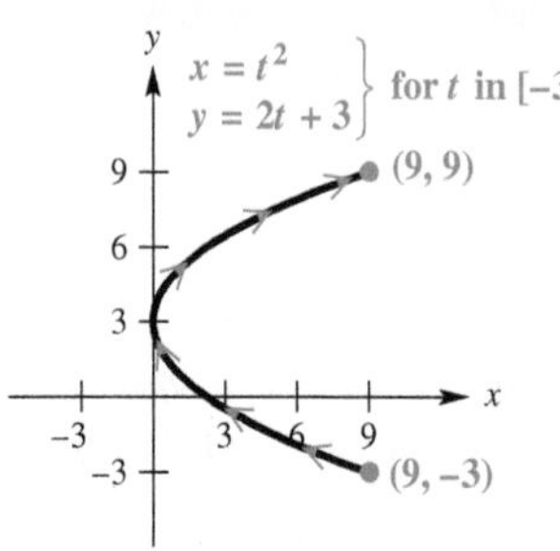

Figure 32

GRAPHING CALCULATOR SOLUTION

We set the parameters of the TI-84 Plus as shown to obtain the graph. See **Figure 33.**

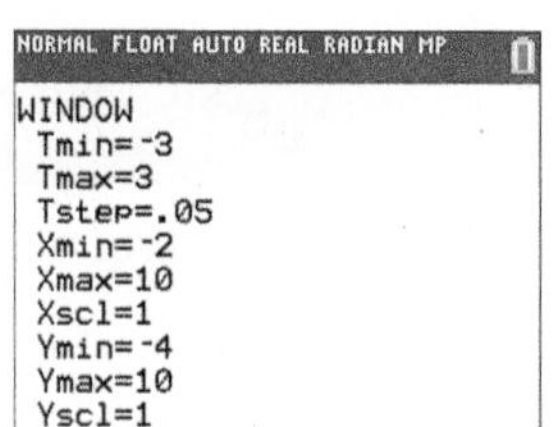

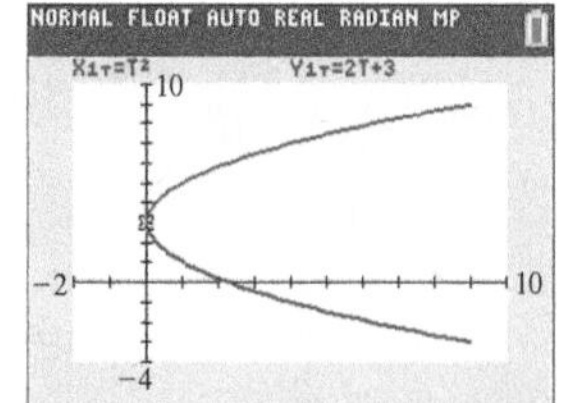

Figure 33

Duplicate this graph and observe how the curve is traced. It should match **Figure 32.**

Now Try Exercise 9(a).

Classroom Example 2
Find a rectangular equation for the plane curve of **Classroom Example 1,**

$x = t^2 - 1$, $y = t - 2$,
for t in $[-3, 3]$.

Answer: $x = (y + 2)^2 - 1$, for x in $[-1, 8]$

EXAMPLE 2 Finding an Equivalent Rectangular Equation

Find a rectangular equation for the plane curve of **Example 1,**

$$x = t^2, \quad y = 2t + 3, \qquad \text{for } t \text{ in } [-3, 3].$$

SOLUTION To eliminate the parameter t, first solve either equation for t.

This equation leads to a unique solution for t.

$y = 2t + 3$ Choose the simpler equation.

$2t = y - 3$ Subtract 3 and rewrite.

$t = \frac{y - 3}{2}$ Divide by 2.

Now substitute this result into the first equation to eliminate the parameter t.

$$x = t^2$$

$$x = \left(\frac{y-3}{2}\right)^2 \quad \text{Substitute for } t.$$

$$x = \frac{(y-3)^2}{4} \quad \left(\frac{a}{b}\right)^2 = \frac{a^2}{b^2}$$

$$4x = (y-3)^2 \quad \text{Multiply by 4.}$$

This is the equation of a horizontal parabola opening to the right, which agrees with the graph given in **Figure 32.** Because t is in $[-3, 3]$, x is in $[0, 9]$ and y is in $[-3, 9]$. The rectangular equation must be given with restricted domain as

$$4x = (y-3)^2, \qquad \text{for } x \text{ in } [0, 9].$$

Now Try Exercise 9(b).

Classroom Example 3
Graph the plane curve defined by

$$x = 4 \cos t, \quad y = 2 \sin t,$$

for t in $[0, 2\pi]$.

Answer:

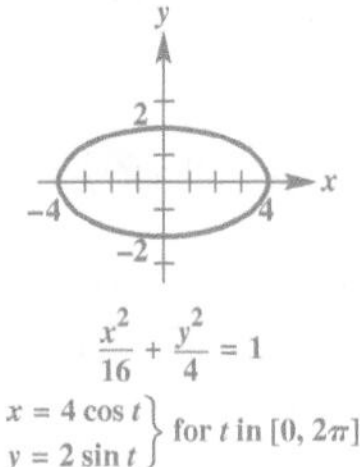

Teaching Tip If students are not sure how to proceed, have them create a table of values for t, x, and y, identify the curve represented by the parametric equations, and then use this information to determine the next step.

EXAMPLE 3 Graphing a Plane Curve Defined Parametrically

Graph the plane curve defined by $x = 2 \sin t$, $y = 3 \cos t$, for t in $[0, 2\pi]$.

SOLUTION To convert to a rectangular equation, it is not productive here to solve either equation for t. Instead, we use the fact that $\sin^2 t + \cos^2 t = 1$ to apply another approach.

$x = 2 \sin t$	$y = 3 \cos t$	Given equations
$x^2 = 4 \sin^2 t$	$y^2 = 9 \cos^2 t$	Square each side.
$\frac{x^2}{4} = \sin^2 t$	$\frac{y^2}{9} = \cos^2 t$	Solve for $\sin^2 t$ and $\cos^2 t$.

Now add corresponding sides of the two equations.

$$\frac{x^2}{4} + \frac{y^2}{9} = \sin^2 t + \cos^2 t$$

$$\frac{x^2}{4} + \frac{y^2}{9} = 1 \quad \sin^2 t + \cos^2 t = 1$$

This is an equation of an **ellipse.** See **Figure 34.**

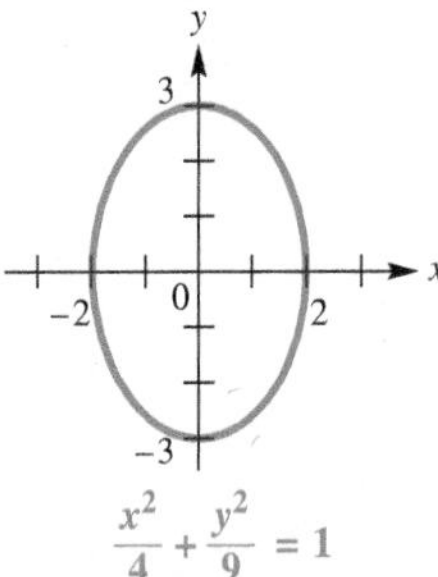

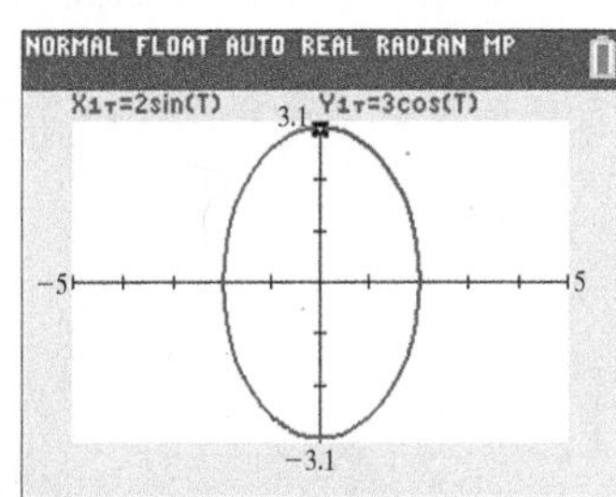

Parametric graphing mode

Figure 34

Now Try Exercise 31.

Parametric representations of a curve are not unique. In fact, there are infinitely many parametric representations of a given curve. If the curve can be described by a rectangular equation $y = f(x)$, with domain X, then one simple parametric representation is

$$x = t, \quad y = f(t), \quad \text{for } t \text{ in } X.$$

Classroom Example 4
Give two parametric representations for the equation of the parabola.

$$y = (x-3)^2 - 1$$

Answer:
$x = t,\ \ y = (t-3)^2 - 1,$
for t in $(-\infty, \infty)$;

$x = t + 3,\ \ y = t^2 - 1,$
for t in $(-\infty, \infty)$

(Other answers are possible.)

EXAMPLE 4 Finding Alternative Parametric Equation Forms

Give two parametric representations for the equation of the parabola.

$$y = (x-2)^2 + 1$$

SOLUTION The simplest choice is to let

$$x = t, \quad y = (t-2)^2 + 1, \qquad \text{for } t \text{ in } (-\infty, \infty).$$

Another choice, which leads to a simpler equation for y, is

$$x = t + 2, \quad y = t^2 + 1, \qquad \text{for } t \text{ in } (-\infty, \infty).$$

✔ **Now Try Exercise 33.**

NOTE Verify that another choice in **Example 4** is

$$x = 2 + \tan t, \quad y = \sec^2 t, \qquad \text{for } t \text{ in } \left(-\frac{\pi}{2}, \frac{\pi}{2}\right).$$

Other choices are possible.

The Cycloid The *cycloid* is a special case of the **trochoid**—a curve traced out by a point at a given distance from the center of a circle as the circle rolls along a straight line. If the given point is on the *circumference* of the circle, then the path traced as the circle rolls along a straight line is a **cycloid,** which is defined parametrically as follows.

$$\boldsymbol{x = at - a\sin t, \quad y = a - a\cos t, \qquad \text{for } t \text{ in } (-\infty, \infty)}$$

Other curves related to trochoids are **hypotrochoids** and **epitrochoids,** which are traced out by a point that is a given distance from the center of a circle that rolls not on a straight line, but on the inside or outside, respectively, of another circle. The classic Spirograph toy can be used to draw these curves.

Classroom Example 5
Graph the cycloid

$$x = 2t - 2\sin t,$$
$$y = 2 - 2\cos t,$$

for t in $[0, 2\pi]$.

Answer:

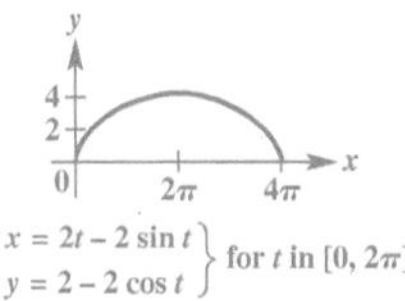

EXAMPLE 5 Graphing a Cycloid

Graph the cycloid.

$$x = t - \sin t, \quad y = 1 - \cos t, \qquad \text{for } t \text{ in } [0, 2\pi]$$

ALGEBRAIC SOLUTION

There is no simple way to find a rectangular equation for the cycloid from its parametric equations. Instead, begin with a table using selected values for t in $[0, 2\pi]$. Approximate values have been rounded as necessary.

t	0	$\frac{\pi}{4}$	$\frac{\pi}{2}$	π	$\frac{3\pi}{2}$	2π
x	0	0.08	0.6	π	5.7	2π
y	0	0.3	1	2	1	0

Figure 35

Plotting the ordered pairs (x, y) from the table of values leads to the portion of the graph in **Figure 35** from 0 to 2π.

GRAPHING CALCULATOR SOLUTION

It is easier to graph a cycloid with a graphing calculator in parametric mode than with traditional methods. See **Figure 36.**

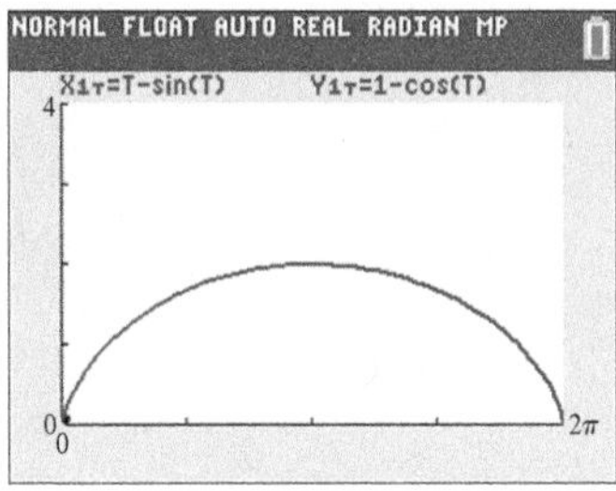

Figure 36

Using a larger interval for t would show that the cycloid repeats the pattern shown here every 2π units.

✔ **Now Try Exercise 37.**

Figure 37

The cycloid has an interesting physical property. If a flexible cord or wire goes through points P and Q as in **Figure 37,** and a bead is allowed to slide due to the force of gravity without friction along this path from P to Q, the path that requires the shortest time takes the shape of the graph of an inverted cycloid.

Applications of Parametric Equations Parametric equations are used to simulate motion. If an object is thrown with a velocity of v feet per second at an angle θ with the horizontal, then its flight can be modeled by

$$x = (v \cos \theta)t \quad \text{and} \quad y = (v \sin \theta)t - 16t^2 + h,$$

where t is in seconds and h is the object's initial height in feet above the ground. Here, x gives the horizontal position information and y gives the vertical position information. The term $-16t^2$ occurs because gravity is pulling downward. See **Figure 38.** These equations ignore air resistance.

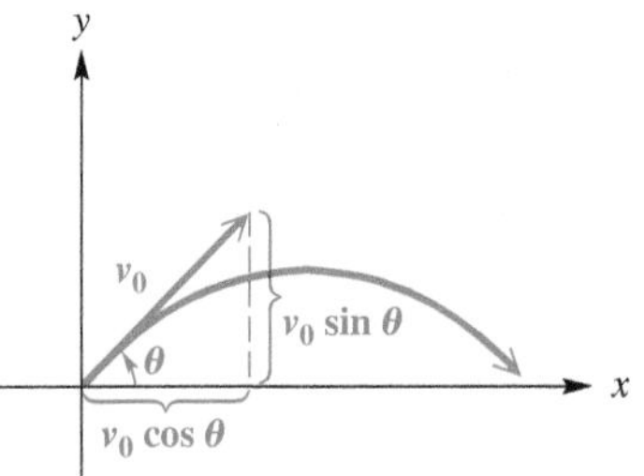

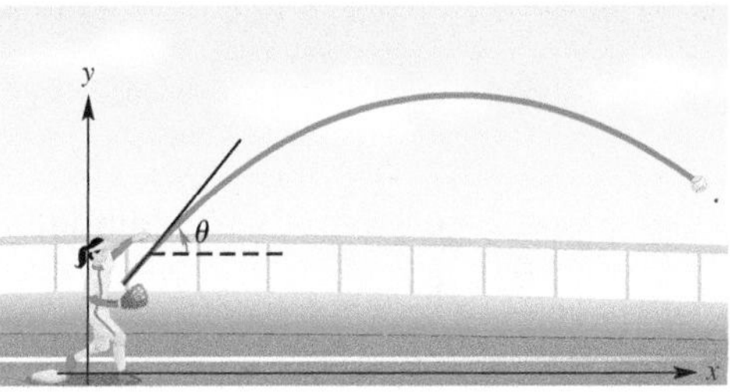

Figure 38

Classroom Example 6
Repeat **Example 6** for three golf balls hit simultaneously into the air at 120 ft per sec at angles of 25°, 45°, and 65° with the horizontal.

Answers:
(a) the ball hit at 45°; 450 ft
(b) the ball hit at 65°; 185 ft

EXAMPLE 6 Simulating Motion with Parametric Equations

Three golf balls are hit simultaneously into the air at 132 ft per sec (90 mph) at angles of 30°, 50°, and 70° with the horizontal.

(a) Assuming the ground is level, determine graphically which ball travels the greatest distance. Estimate this distance.

(b) Which ball reaches the greatest height? Estimate this height.

SOLUTION

(a) Use the following parametric equations to model the flight of the golf balls.

$$x = (v \cos \theta)t \quad \text{and} \quad y = (v \sin \theta)t - 16t^2 + h$$

Write three sets of parametric equations.

$$x_1 = (132 \cos 30°)t, \quad y_1 = (132 \sin 30°)t - 16t^2$$
$$x_2 = (132 \cos 50°)t, \quad y_2 = (132 \sin 50°)t - 16t^2$$
$$x_3 = (132 \cos 70°)t, \quad y_3 = (132 \sin 70°)t - 16t^2$$

Substitute $h = 0$, $v = 132$ ft per sec, and $\theta = 30°, 50°$, and $70°$.

The graphs of the three sets of parametric equations are shown in **Figure 39(a),** where $0 \le t \le 3$. From the graph in **Figure 39(b),** where $0 \le t \le 9$, we see that the ball hit at 50° travels the greatest distance. Using the tracing feature of the TI-84 Plus calculator, we find that this distance is about 540 ft.

(b) Again, use the tracing feature to find that the ball hit at 70° reaches the greatest height, about 240 ft.

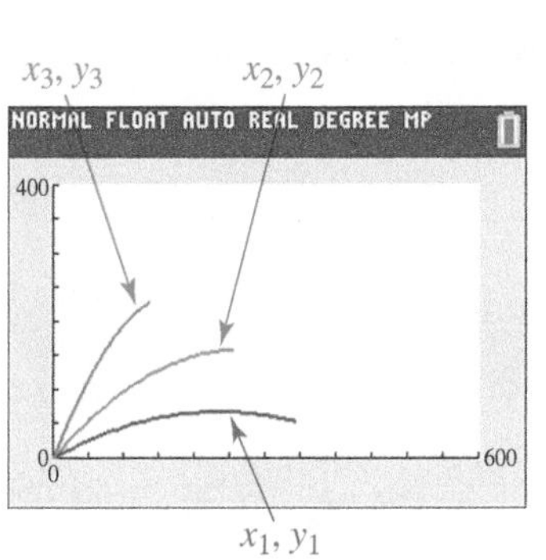

(a)

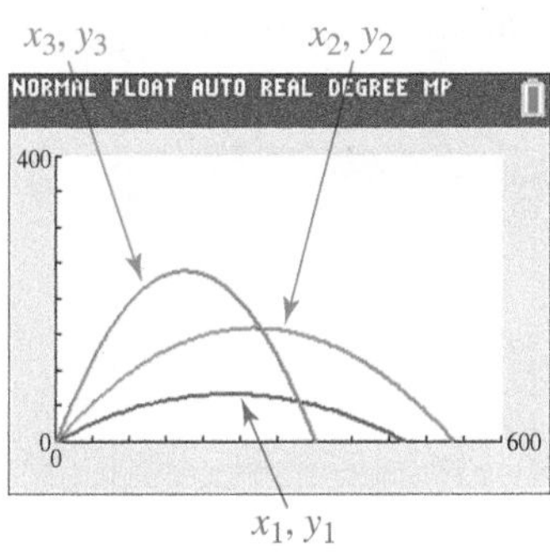

(b)

Figure 39

Now Try Exercise 43.

Teaching Tip Explain that in motion problems, the variable t acts as a control variable (specifying an amount of time appropriate to the situation), while x gives horizontal position information and y gives vertical position information.

A TI-84 Plus calculator allows us to view the graphing of more than one equation either *sequentially* or *simultaneously*. By choosing the latter, the three golf balls in **Figure 39** can be viewed in flight at the same time. ■

Classroom Example 7
A toy rocket is launched from the ground with velocity 36 ft per sec at an angle of 45° with the ground. Find the rectangular equation that models this path. What type of path does the rocket follow?

Answer:

$y = -\frac{2}{81}x^2 + x$;

It follows a parabolic path.

EXAMPLE 7 Examining Parametric Equations of Flight

Jack launches a small rocket from a table that is 3.36 ft above the ground. Its initial velocity is 64 ft per sec, and it is launched at an angle of 30° with respect to the ground. Find the rectangular equation that models its path. What type of path does the rocket follow?

SOLUTION The path of the rocket is defined by the parametric equations

$$x = (64 \cos 30°)t \quad \text{and} \quad y = (64 \sin 30°)t - 16t^2 + 3.36$$

or, equivalently, $\quad x = 32\sqrt{3}t \quad$ and $\quad y = -16t^2 + 32t + 3.36.$

From $x = 32\sqrt{3}t$, we solve for t to obtain

$$t = \frac{x}{32\sqrt{3}}. \quad \text{Divide by } 32\sqrt{3}.$$

Substituting for t in the other parametric equation yields the following.

$$y = -16t^2 + 32t + 3.36$$

$$y = -16\left(\frac{x}{32\sqrt{3}}\right)^2 + 32\left(\frac{x}{32\sqrt{3}}\right) + 3.36 \quad \text{Let } t = \frac{x}{32\sqrt{3}}.$$

$$y = -\frac{1}{192}x^2 + \frac{\sqrt{3}}{3}x + 3.36 \quad \text{Simplify.}$$

This equation defines a parabola. The rocket follows a parabolic path.

✔ **Now Try Exercise 47(a).**

Classroom Example 8
Determine the total flight time and the horizontal distance traveled by the rocket in **Classroom Example 7.**

Answer: 1.59 sec; 40.5 ft

EXAMPLE 8 Analyzing the Path of a Projectile

Determine the total flight time and the horizontal distance traveled by the rocket in **Example 7.**

ALGEBRAIC SOLUTION

The equation $y = -16t^2 + 32t + 3.36$ tells the vertical position of the rocket at time t. We need to determine the positive value of t for which $y = 0$ because this value corresponds to the rocket at ground level. This yields

$$0 = -16t^2 + 32t + 3.36.$$

Using the quadratic formula, the solutions are $t = -0.1$ or $t = 2.1$. Because t represents time, $t = -0.1$ is an unacceptable answer. Therefore, the flight time is 2.1 sec.

The rocket was in the air for 2.1 sec, so we can use $t = 2.1$ and the parametric equation that models the horizontal position, $x = 32\sqrt{3}t$, to obtain

$$x = 32\sqrt{3}(2.1) \approx 116.4 \text{ ft}.$$

GRAPHING CALCULATOR SOLUTION

Figure 40 shows that when $t = 2.1$, the horizontal distance x covered is approximately 116.4 ft, which agrees with the algebraic solution.

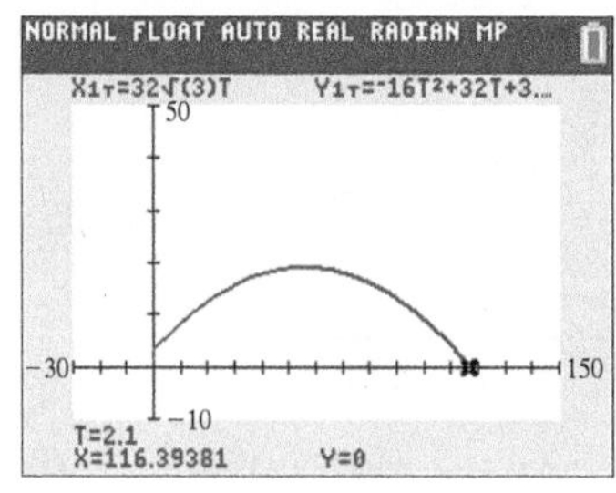

Figure 40

✔ **Now Try Exercise 47(b).**

8.6 Exercises

1. $(10, -3)$ 2. $(-6, 13)$

3. $\left(\frac{1}{2}, \sqrt{3}\right)$ 4. $(4, 259)$

5. C 6. D

7. A 8. B

9. (a)

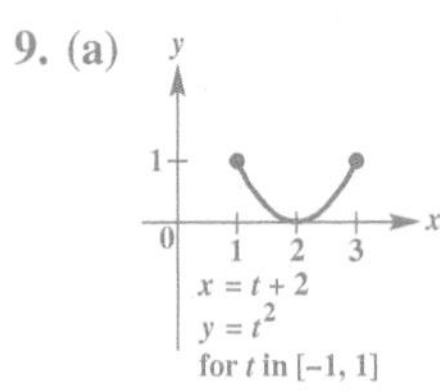

(b) $y = x^2 - 4x + 4$, for x in $[1, 3]$

10. (a)

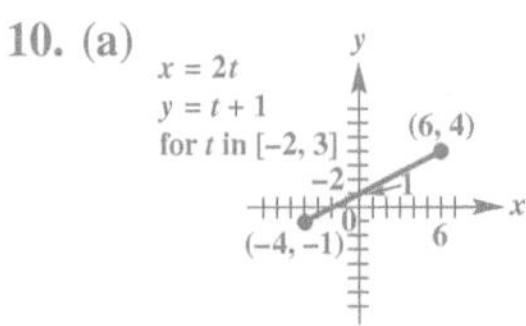

(b) $y = \frac{1}{2}x + 1$, for x in $[-4, 6]$

11. (a)

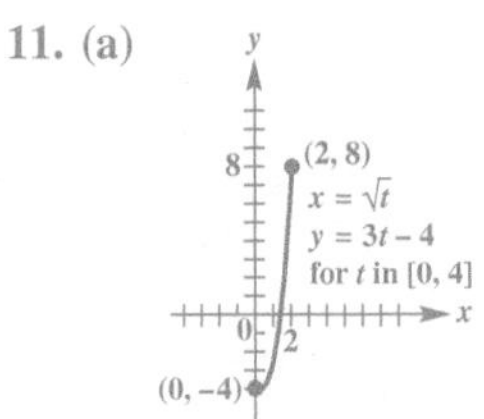

(b) $y = 3x^2 - 4$, for x in $[0, 2]$

12. (a)

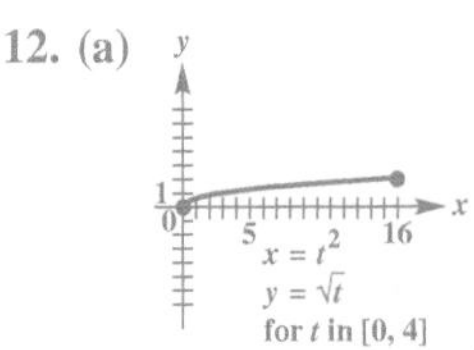

(b) $y = \sqrt[4]{x}$, for x in $[0, 16]$

13. (a)

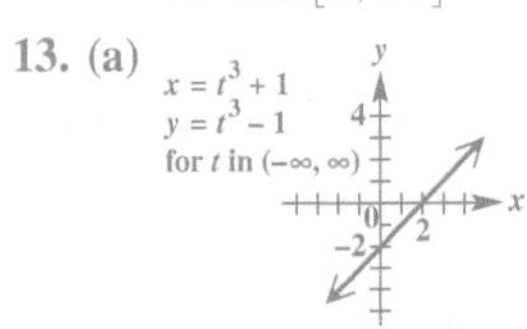

(b) $y = x - 2$, for x in $(-\infty, \infty)$

CONCEPT PREVIEW *Fill in the blank to correctly complete each sentence.*

1. For the plane curve defined by
$$x = t^2 + 1, \quad y = 2t + 3, \quad \text{for } t \text{ in } [-4, 4],$$
the ordered pair that corresponds to $t = -3$ is ________.

2. For the plane curve defined by
$$x = -3t + 6, \quad y = t^2 - 3, \quad \text{for } t \text{ in } [-5, 5],$$
the ordered pair that corresponds to $t = 4$ is ________.

3. For the plane curve defined by
$$x = \cos t, \quad y = 2 \sin t, \quad \text{for } t \text{ in } [0, 2\pi],$$
the ordered pair that corresponds to $t = \frac{\pi}{3}$ is ________.

4. For the plane curve defined by
$$x = \sqrt{t}, \quad y = t^2 + 3, \quad \text{for } t \text{ in } (0, \infty),$$
the ordered pair that corresponds to $t = 16$ is ________.

CONCEPT PREVIEW *Match the ordered pair from Column II with the pair of parametric equations in Column I on whose graph the point lies. In each case, consider the given value of t.*

I	II
5. $x = 3t + 6,\ y = -2t + 4;\quad t = 2$	A. $(5, 25)$
6. $x = \cos t,\ y = \sin t;\quad t = \frac{\pi}{4}$	B. $(7, 2)$
7. $x = t,\ y = t^2;\quad t = 5$	C. $(12, 0)$
8. $x = t^2 + 3,\ y = t^2 - 2;\quad t = 2$	D. $\left(\frac{\sqrt{2}}{2}, \frac{\sqrt{2}}{2}\right)$

For each plane curve, ***(a)*** *graph the curve, and* ***(b)*** *find a rectangular equation for the curve.* ***See Examples 1 and 2.***

9. $x = t + 2,\ y = t^2$, for t in $[-1, 1]$

10. $x = 2t,\ y = t + 1$, for t in $[-2, 3]$

11. $x = \sqrt{t},\ y = 3t - 4$, for t in $[0, 4]$

12. $x = t^2,\ y = \sqrt{t}$, for t in $[0, 4]$

13. $x = t^3 + 1,\ y = t^3 - 1$, for t in $(-\infty, \infty)$

14. $x = 2t - 1,\ y = t^2 + 2$, for t in $(-\infty, \infty)$

15. $x = 2 \sin t,\ y = 2 \cos t$, for t in $[0, 2\pi]$

16. $x = \sqrt{5} \sin t,\ y = \sqrt{3} \cos t$, for t in $[0, 2\pi]$

17. $x = 3 \tan t,\ y = 2 \sec t$, for t in $\left(-\frac{\pi}{2}, \frac{\pi}{2}\right)$

18. $x = \cot t,\ y = \csc t$, for t in $(0, \pi)$

19. $x = \sin t,\ y = \csc t$, for t in $(0, \pi)$

20. $x = \tan t,\ y = \cot t$, for t in $\left(0, \frac{\pi}{2}\right)$

21. $x = t,\ y = \sqrt{t^2 + 2}$, for t in $(-\infty, \infty)$

22. $x = \sqrt{t},\ y = t^2 - 1$, for t in $[0, \infty)$

14. (a)

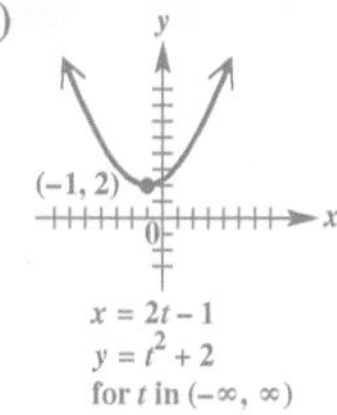

(b) $y = \frac{1}{4}(x + 1)^2 + 2$, for x in $(-\infty, \infty)$

15. (a)

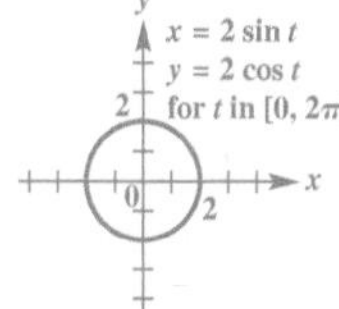

(b) $x^2 + y^2 = 4$, for x in $[-2, 2]$

16. (a)

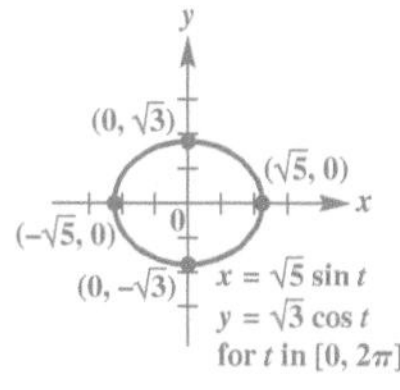

(b) $\frac{x^2}{5} + \frac{y^2}{3} = 1$, for x in $[-\sqrt{5}, \sqrt{5}]$

17. (a)

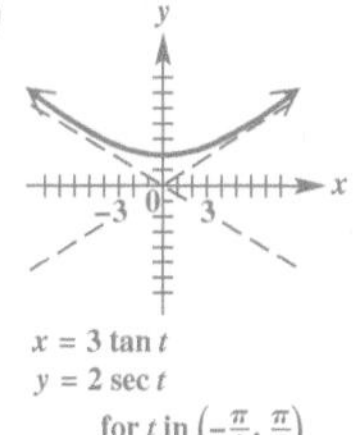

(b) $y = 2\sqrt{1 + \frac{x^2}{9}}$, for x in $(-\infty, \infty)$

18. (a)

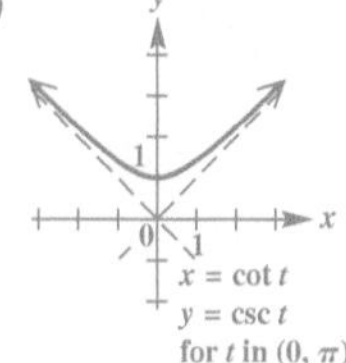

(b) $y = \sqrt{x^2 + 1}$, for x in $(-\infty, \infty)$

19. (a)

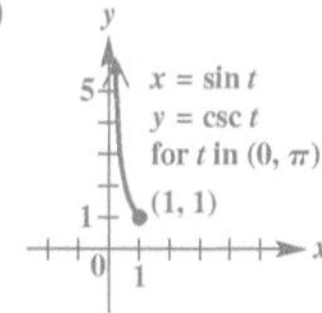

(b) $y = \frac{1}{x}$, for x in $(0, 1]$

23. $x = 2 + \sin t$, $y = 1 + \cos t$, for t in $[0, 2\pi]$

24. $x = 1 + 2\sin t$, $y = 2 + 3\cos t$, for t in $[0, 2\pi]$

25. $x = t + 2$, $y = \frac{1}{t + 2}$, for $t \neq -2$

26. $x = t - 3$, $y = \frac{2}{t - 3}$, for $t \neq 3$

27. $x = t + 2$, $y = t - 4$, for t in $(-\infty, \infty)$

28. $x = t^2 + 2$, $y = t^2 - 4$, for t in $(-\infty, \infty)$

Graph each plane curve defined by the parametric equations for t in $[0, 2\pi]$. Then find a rectangular equation for the plane curve. ***See Example 3.***

29. $x = 3\cos t$, $y = 3\sin t$

30. $x = 2\cos t$, $y = 2\sin t$

31. $x = 3\sin t$, $y = 2\cos t$

32. $x = 4\sin t$, $y = 3\cos t$

Give two parametric representations for the equation of each parabola. ***See Example 4.***

33. $y = (x + 3)^2 - 1$

34. $y = (x + 4)^2 + 2$

35. $y = x^2 - 2x + 3$

36. $y = x^2 - 4x + 6$

Graph each cycloid defined by the given equations for t in the specified interval. ***See Example 5.***

37. $x = 2t - 2\sin t$, $y = 2 - 2\cos t$, for t in $[0, 4\pi]$

38. $x = t - \sin t$, $y = 1 - \cos t$, for t in $[0, 4\pi]$

*Lissajous Figures The screen shown here is an example of a **Lissajous figure.** Such figures occur in electronics and may be used to find the frequency of an unknown voltage. Graph each Lissajous figure for t in $[0, 6.5]$ using the window $[-6, 6]$ by $[-4, 4]$.*

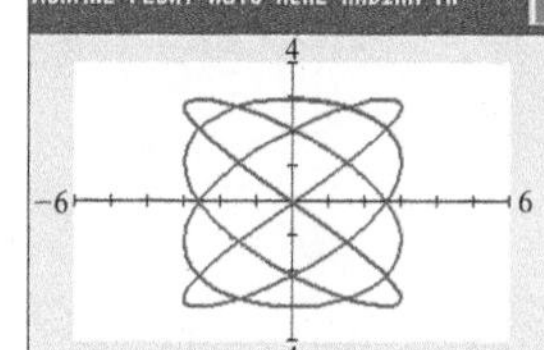

39. $x = 2\cos t$, $y = 3\sin 2t$

40. $x = 3\cos t$, $y = 2\sin 2t$

41. $x = 3\sin 4t$, $y = 3\cos 3t$

42. $x = 4\sin 4t$, $y = 3\sin 5t$

(Modeling) Do the following. ***See Examples 6–8.***

(a) Determine parametric equations that model the path of the projectile.

(b) Determine a rectangular equation that models the path of the projectile.

(c) Determine approximately how long the projectile is in flight and the horizontal distance it covers.

43. ***Flight of a Model Rocket*** A model rocket is launched from the ground with velocity 48 ft per sec at an angle of 60° with respect to the ground.

44. ***Flight of a Golf Ball*** Tyler is playing golf. He hits a golf ball from the ground at an angle of 60° with respect to the ground at velocity 150 ft per sec.

45. ***Flight of a Softball*** Sally hits a softball when it is 2 ft above the ground. The ball leaves her bat at an angle of 20° with respect to the ground at velocity 88 ft per sec.

20. (a)

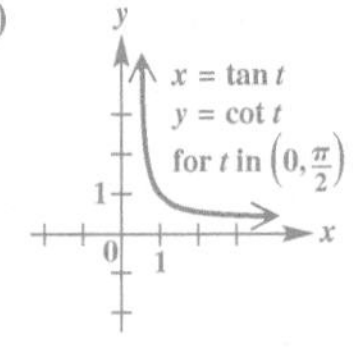

(b) $y = \frac{1}{x}$, for x in $(0, \infty)$

21. (a)

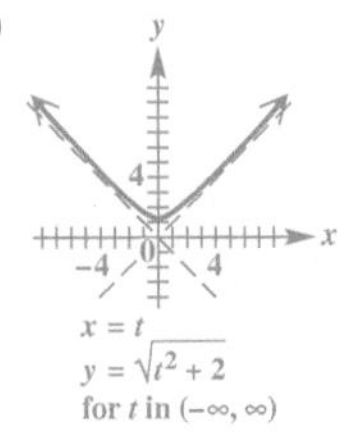

(b) $y = \sqrt{x^2 + 2}$, for x in $(-\infty, \infty)$

22. (a)

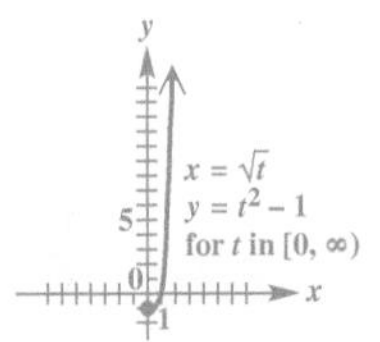

(b) $y = x^4 - 1$, for x in $[0, \infty)$

23. (a)

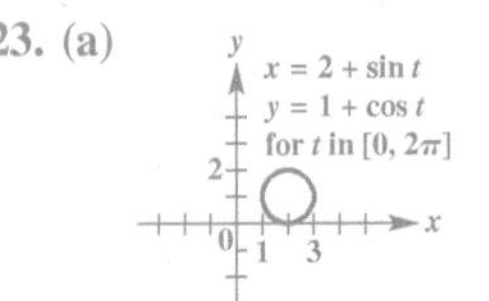

(b) $(x - 2)^2 + (y - 1)^2 = 1$, for x in $[1, 3]$

24. (a)

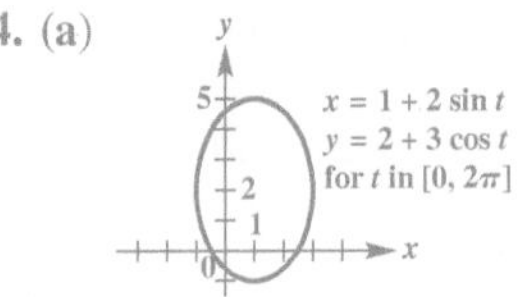

(b) $\frac{(x - 1)^2}{4} + \frac{(y - 2)^2}{9} = 1$, for x in $[-1, 3]$

25. (a)

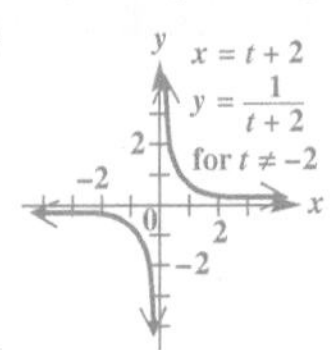

(b) $y = \frac{1}{x}$, for x in $(-\infty, 0) \cup (0, \infty)$

26. (a)

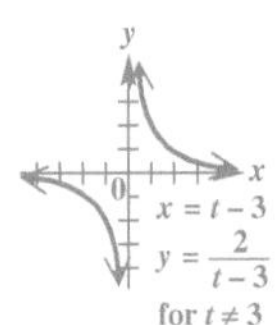

(b) $y = \frac{2}{x}$, for x in $(-\infty, 0) \cup (0, \infty)$

46. ***Flight of a Baseball*** Francisco hits a baseball when it is 2.5 ft above the ground. The ball leaves his bat at an angle of 29° from the horizontal with velocity 136 ft per sec.

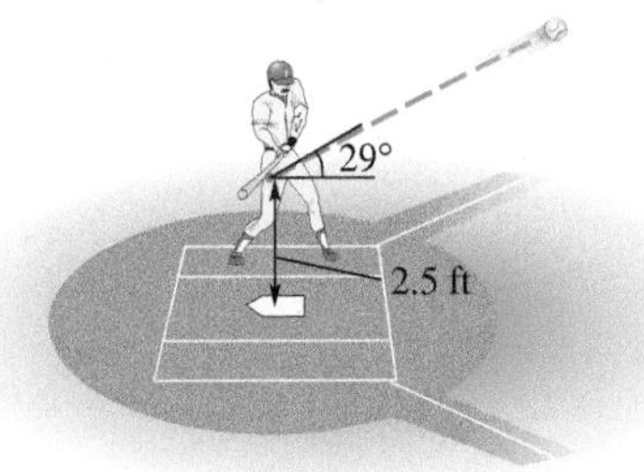

(Modeling) Solve each problem. ***See Examples 7 and 8.***

47. ***Path of a Rocket*** A rocket is launched from the top of an 8-ft platform. Its initial velocity is 128 ft per sec. It is launched at an angle of 60° with respect to the ground.

(a) Find the rectangular equation that models its path. What type of path does the rocket follow?

(b) Determine the total flight time, to the nearest second, and the horizontal distance the rocket travels, to the nearest foot.

48. ***Simulating Gravity on the Moon*** If an object is thrown on the moon, then the parametric equations of flight are

$$x = (v \cos \theta)t \quad \text{and} \quad y = (v \sin \theta)t - 2.66t^2 + h.$$

Estimate, to the nearest foot, the distance a golf ball hit at 88 ft per sec (60 mph) at an angle of 45° with the horizontal travels on the moon if the moon's surface is level.

49. ***Flight of a Baseball*** A baseball is hit from a height of 3 ft at a 60° angle above the horizontal. Its initial velocity is 64 ft per sec.

(a) Write parametric equations that model the flight of the baseball.

(b) Determine the horizontal distance, to the nearest tenth of a foot, traveled by the ball in the air. Assume that the ground is level.

(c) What is the maximum height of the baseball, to the nearest tenth of a foot? At that time, how far has the ball traveled horizontally?

(d) Would the ball clear a 5-ft-high fence that is 100 ft from the batter?

50. ***Path of a Projectile*** A projectile has been launched from the ground with initial velocity 88 ft per sec. The parametric equations

$$x = 82.7t \quad \text{and} \quad y = -16t^2 + 30.1t$$

model the path of the projectile, where t is in seconds.

(a) Approximate the angle θ that the projectile makes with the horizontal at the launch, to the nearest tenth of a degree.

(b) Write parametric equations for the path using the cosine and sine functions.

Work each problem.

51. Give two parametric representations of the parabola $y = a(x - h)^2 + k$.

52. Give a parametric representation of the rectangular equation $\frac{x^2}{a^2} - \frac{y^2}{b^2} = 1$.

53. Give a parametric representation of the rectangular equation $\frac{x^2}{a^2} + \frac{y^2}{b^2} = 1$.

54. The spiral of Archimedes has polar equation $r = a\theta$, where $r^2 = x^2 + y^2$. Show that a parametric representation of the spiral of Archimedes is

$$x = a\theta \cos \theta, \quad y = a\theta \sin \theta, \qquad \text{for } \theta \text{ in } (-\infty, \infty).$$

55. Show that the **hyperbolic spiral** $r\theta = a$, where $r^2 = x^2 + y^2$, is given parametrically by

$$x = \frac{a \cos \theta}{\theta}, \quad y = \frac{a \sin \theta}{\theta}, \qquad \text{for } \theta \text{ in } (-\infty, 0) \cup (0, \infty).$$

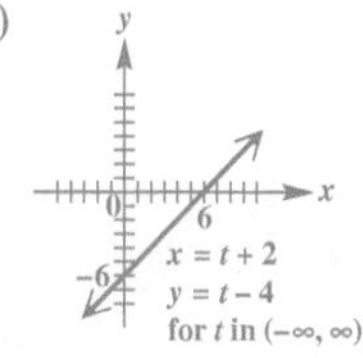

(b) $y = x - 6$, for x in $(-\infty, \infty)$

Answers for the remaining exercises are given in the answer section at the back of the text.

56. The parametric equations $x = \cos t$, $y = \sin t$, for t in $[0, 2\pi]$ and the parametric equations $x = \cos t$, $y = -\sin t$, for t in $[0, 2\pi]$ both have the unit circle as their graph. However, in one case the circle is traced out clockwise (as t moves from 0 to 2π), and in the other case the circle is traced out counterclockwise. For which pair of equations is the circle traced out in the clockwise direction?

Concept Check *Consider the parametric equations* $x = f(t)$, $y = g(t)$, *for* t *in* $[a, b]$, *with* $c > 0$, $d > 0$.

57. How is the graph affected if the equation $x = f(t)$ is replaced by $x = c + f(t)$?

58. How is the graph affected if the equation $y = g(t)$ is replaced by $y = d + g(t)$?

Chapter 8 Test Prep

Key Terms

8.1 imaginary unit complex number real part imaginary part nonreal complex number pure imaginary number standard form complex conjugates	**8.2** resultant real axis imaginary axis complex plane rectangular form of a complex number trigonometric (polar) form of a complex number absolute value (modulus) argument	**8.4** *n*th root of a complex number **8.5** polar coordinate system pole polar axis polar coordinates rectangular (Cartesian) equation polar equation cardioid polar grid	rose curve lemniscate spiral of Archimedes limaçon **8.6** plane curve parametric equations of a plane curve parameter cycloid

New Symbols

Symbol	Meaning
i	imaginary unit
$a + bi$	complex number
cis θ	$\cos \theta + i \sin \theta$

Quick Review

Concepts	Examples
8.1 Complex Numbers	
Imaginary Unit i $i = \sqrt{-1}$, and therefore $i^2 = -1$.	$3 - 4i$ (3: Real part; -4: Imaginary part)
Complex Number If a and b are real numbers, then any number of the form $a + bi$ is a complex number. In the complex number $a + bi$, a is the real part and b is the imaginary part.	

Concepts	Examples
Meaning of $\sqrt{-a}$ If $a > 0$, then $\sqrt{-a} = i\sqrt{a}$.	Simplify. $\sqrt{-4} = 2i$ $\sqrt{-12} = i\sqrt{12} = i\sqrt{4 \cdot 3} = 2i\sqrt{3}$
Adding and Subtracting Complex Numbers Add or subtract the real parts, and add or subtract the imaginary parts.	$(2 + 3i) + (3 + i) - (2 - i)$ $= (2 + 3 - 2) + (3 + 1 + 1)i$ $= 3 + 5i$
Multiplying and Dividing Complex Numbers Multiply complex numbers as with binomials, and use the fact that $i^2 = -1$.	$(6 + i)(3 - 2i)$ $= 18 - 12i + 3i - 2i^2$ FOIL method $= (18 + 2) + (-12 + 3)i$ $i^2 = -1$ $= 20 - 9i$
Divide complex numbers by multiplying the numerator and denominator by the complex conjugate of the denominator.	$\frac{3 + i}{1 + i} = \frac{(3 + i)(1 - i)}{(1 + i)(1 - i)} = \frac{3 - 3i + i - i^2}{1 - i^2}$ $= \frac{4 - 2i}{2} = \frac{2(2 - i)}{2} = 2 - i$

8.2 Trigonometric (Polar) Form of Complex Numbers

Concepts	Examples
Trigonometric (Polar) Form of Complex Numbers Let the complex number $x + yi$ correspond to the vector with direction angle θ and magnitude r. $x = r\cos\theta$ $y = r\sin\theta$ $r = \sqrt{x^2 + y^2}$ $\tan\theta = \frac{y}{x}$, if $x \neq 0$	Write $2(\cos 60° + i\sin 60°)$ in rectangular form. $2(\cos 60° + i\sin 60°)$ $= 2\left(\frac{1}{2} + i \cdot \frac{\sqrt{3}}{2}\right)$ $= 1 + i\sqrt{3}$
The trigonometric (polar) form of the expression $x + yi$ is $r(\cos\theta + i\sin\theta)$ or $r\text{ cis }\theta$.	Write $-\sqrt{2} + i\sqrt{2}$ in trigonometric form. 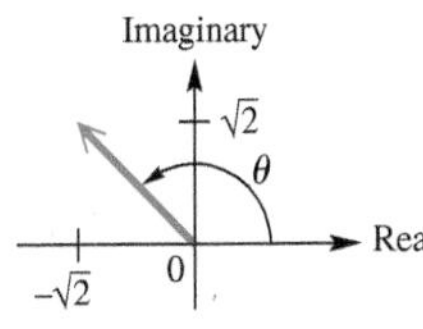 $r = \sqrt{\left(-\sqrt{2}\right)^2 + \left(\sqrt{2}\right)^2} = 2$ $\tan\theta = -1$ and θ is in quadrant II, so $\theta = 180° - 45° = 135°$. $-\sqrt{2} + i\sqrt{2} = 2\text{ cis }135°$.

8.3 The Product and Quotient Theorems

Concepts	Examples
Product and Quotient Theorems If $r_1(\cos\theta_1 + i\sin\theta_1)$ and $r_2(\cos\theta_2 + i\sin\theta_2)$ are any two complex numbers, then the following hold. $[r_1(\cos\theta_1 + i\sin\theta_1)] \cdot [r_2(\cos\theta_2 + i\sin\theta_2)]$ $= r_1r_2[\cos(\theta_1 + \theta_2) + i\sin(\theta_1 + \theta_2)]$ and $\frac{r_1(\cos\theta_1 + i\sin\theta_1)}{r_2(\cos\theta_2 + i\sin\theta_2)}$ $= \frac{r_1}{r_2}[\cos(\theta_1 - \theta_2) + i\sin(\theta_1 - \theta_2)]$, where $r_2(\cos\theta_2 + i\sin\theta_2) \neq 0$	Let $z_1 = 4(\cos 135° + i\sin 135°)$ and $z_2 = 2(\cos 45° + i\sin 45°)$. $z_1z_2 = 8(\cos 180° + i\sin 180°)$ $4 \cdot 2 = 8$; $135° + 45° = 180°$ $= 8(-1 + i \cdot 0)$ $= -8$ $\frac{z_1}{z_2} = 2(\cos 90° + i\sin 90°)$ $\frac{4}{2} = 2$; $135° - 45° = 90°$ $= 2(0 + i \cdot 1)$ $= 2i$

Concepts	Examples

8.4 De Moivre's Theorem; Powers and Roots of Complex Numbers

De Moivre's Theorem

$$[r(\cos\theta + i\sin\theta)]^n = r^n(\cos n\theta + i\sin n\theta)$$

***n*th Root Theorem**

If n is any positive integer, r is a positive real number, and θ is in degrees, then the nonzero complex number $r(\cos\theta + i\sin\theta)$ has exactly n distinct nth roots, given by the following.

$$\sqrt[n]{r}\,(\cos\alpha + i\sin\alpha), \quad \text{or} \quad \sqrt[n]{r}\text{ cis }\alpha,$$

where

$$\alpha = \frac{\theta + 360^\circ \cdot k}{n}, \quad k = 0, 1, 2, \ldots, n-1.$$

If θ is in radians, then

$$\alpha = \frac{\theta + 2\pi k}{n}, \quad k = 0, 1, 2, \ldots, n-1.$$

Let $z = 4(\cos 180^\circ + i\sin 180^\circ)$. Find z^3 and the square roots of z.

$$\begin{aligned}&[4(\cos 180^\circ + i\sin 180^\circ)]^3 && \text{Find } z^3.\\ &= 4^3(\cos 3\cdot 180^\circ + i\sin 3\cdot 180^\circ)\\ &= 64(\cos 540^\circ + i\sin 540^\circ)\\ &= 64(-1 + i\cdot 0)\\ &= -64\end{aligned}$$

For the given z, $r = 4$ and $\theta = 180^\circ$. Its square roots are

$$\begin{aligned}&\sqrt{4}\left(\cos\frac{180^\circ}{2} + i\sin\frac{180^\circ}{2}\right)\\ &= 2(0 + i\cdot 1)\\ &= 2i\end{aligned}$$

and $\sqrt{4}\left(\cos\dfrac{180^\circ + 360^\circ}{2} + i\sin\dfrac{180^\circ + 360^\circ}{2}\right)$

$$\begin{aligned}&= 2(0 + i(-1))\\ &= -2i.\end{aligned}$$

8.5 Polar Equations and Graphs

Rectangular and Polar Coordinates

If a point has rectangular coordinates (x, y) and polar coordinates (r, θ), then these coordinates are related as follows.

$$x = r\cos\theta \qquad y = r\sin\theta$$

$$r^2 = x^2 + y^2 \qquad \tan\theta = \frac{y}{x}, \quad \text{if } x \neq 0$$

Find the rectangular coordinates for the point $(5, 60^\circ)$ in polar coordinates.

$$x = 5\cos 60^\circ = 5\left(\frac{1}{2}\right) = \frac{5}{2}$$

$$y = 5\sin 60^\circ = 5\left(\frac{\sqrt{3}}{2}\right) = \frac{5\sqrt{3}}{2}$$

The rectangular coordinates are $\left(\frac{5}{2}, \frac{5\sqrt{3}}{2}\right)$.

Find polar coordinates for $(-1, -1)$ in rectangular coordinates.

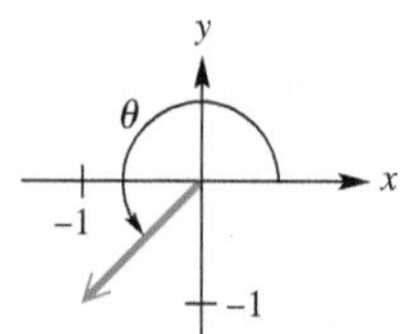

$$r = \sqrt{(-1)^2 + (-1)^2} = \sqrt{2}$$

$\tan\theta = 1$ and θ is in quadrant III, so $\theta = 225^\circ$.

One pair of polar coordinates for $(-1, -1)$ is $(\sqrt{2}, 225^\circ)$.

Polar Equations and Graphs

$r = a\cos\theta$, $r = a\sin\theta$	Circles	$r^2 = a^2\sin 2\theta$, $r^2 = a^2\cos 2\theta$	Lemniscates
$r = a \pm b\sin\theta$, $r = a \pm b\cos\theta$	Limaçons	$r = a\sin n\theta$, $r = a\cos n\theta$	Rose curves

Graph $r = 4\cos 2\theta$.

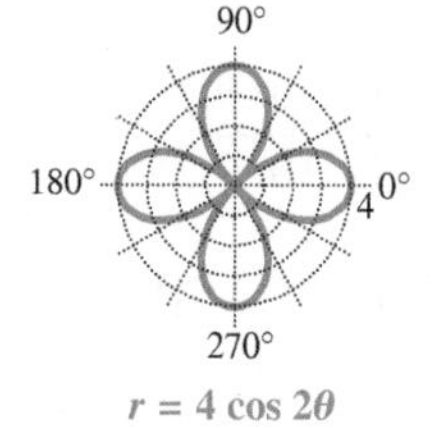

$r = 4\cos 2\theta$

Concepts	Examples

8.6 Parametric Equations, Graphs, and Applications

Parametric Equations of a Plane Curve
A **plane curve** is a set of points (x, y) such that $x = f(t)$, $y = g(t)$, and f and g are both defined on an interval I. The equations

$$x = f(t) \quad \text{and} \quad y = g(t)$$

are **parametric equations** with **parameter** $\boldsymbol{t}$.

Graph $x = 2 - \sin t$, $y = \cos t - 1$, for $0 \le t \le 2\pi$.

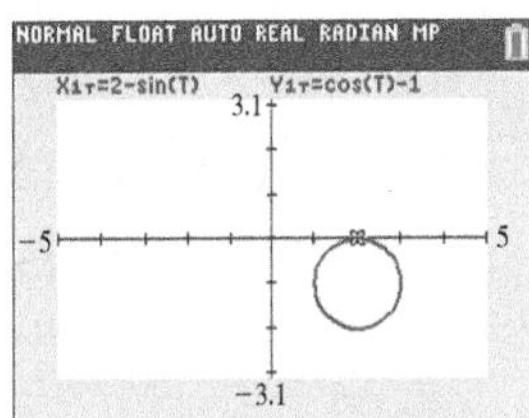

Flight of an Object
If an object has initial velocity v and initial height h, and travels such that its initial angle of elevation is θ, then its flight after t seconds can be modeled by the following parametric equations.

$$\boldsymbol{x = (v \cos \theta)t \quad \text{and} \quad y = (v \sin \theta)t - 16t^2 + h}$$

Joe kicks a football from the ground at an angle of 45° with a velocity of 48 ft per sec. Give the parametric equations that model the path of the football and the distance it travels before hitting the ground.

$$x = (48 \cos 45°)t = 24\sqrt{2}\,t$$

$$y = (48 \sin 45°)t - 16t^2 = 24\sqrt{2}\,t - 16t^2$$

When the ball hits the ground, $y = 0$.

$$24\sqrt{2}\,t - 16t^2 = 0 \qquad \text{Substitute } y = 0.$$

$$8t\left(3\sqrt{2} - 2t\right) = 0 \qquad \text{Factor.}$$

$$t = 0 \quad \text{or} \quad t = \frac{3\sqrt{2}}{2} \qquad \text{Zero-factor property}$$

(Reject)

The distance it travels is $x = 24\sqrt{2}\left(\frac{3\sqrt{2}}{2}\right) = 72$ ft.

Chapter 8 Review Exercises

1. $3i$ 2. $2i\sqrt{3}$
3. $\{\pm 9i\}$ 4. $\left\{-\frac{3}{4} \pm \frac{\sqrt{23}}{4}i\right\}$
5. $-2 - 3i$ 6. $8 - 15i$
7. $5 + 4i$ 8. $-5 - 6i$
9. $29 + 37i$ 10. $22 + 3i$
11. $-32 + 24i$ 12. $27 - 36i$
13. $-2 - 2i$ 14. $2 + 11i$
15. $2 - 5i$ 16. $-\frac{3}{2} - \frac{7}{2}i$
17. $-\frac{3}{26} + \frac{11}{26}i$ 18. $2 - 3i$
19. i 20. $-i$
21. $-30i$ 22. $-3 - 3i\sqrt{3}$
23. $-\frac{1}{8} + \frac{\sqrt{3}}{8}i$ 24. -2
25. $8i$ 26. $-128 + 128i$
27. $-\frac{1}{2} - \frac{\sqrt{3}}{2}i$

Write each number as the product of a real number and i.

1. $\sqrt{-9}$ **2.** $\sqrt{-12}$

Solve each equation over the set of complex numbers.

3. $x^2 = -81$ **4.** $x(2x + 3) = -4$

Perform each operation. Write answers in standard form.

5. $(1 - i) - (3 + 4i) + 2i$ **6.** $(2 - 5i) + (9 - 10i) - 3$

7. $(6 - 5i) + (2 + 7i) - (3 - 2i)$ **8.** $(4 - 2i) - (6 + 5i) - (3 - i)$

9. $(3 + 5i)(8 - i)$ **10.** $(4 - i)(5 + 2i)$ **11.** $(2 + 6i)^2$

12. $(6 - 3i)^2$ **13.** $(1 - i)^3$ **14.** $(2 + i)^3$

15. $\dfrac{25 - 19i}{5 + 3i}$ **16.** $\dfrac{2 - 5i}{1 + i}$ **17.** $\dfrac{2 + i}{1 - 5i}$

18. $\dfrac{3 + 2i}{i}$ **19.** i^{53} **20.** i^{-41}

28. x

29. 30.

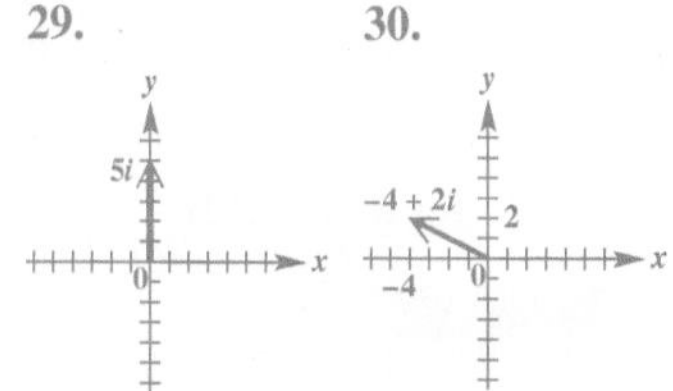

31. 32.

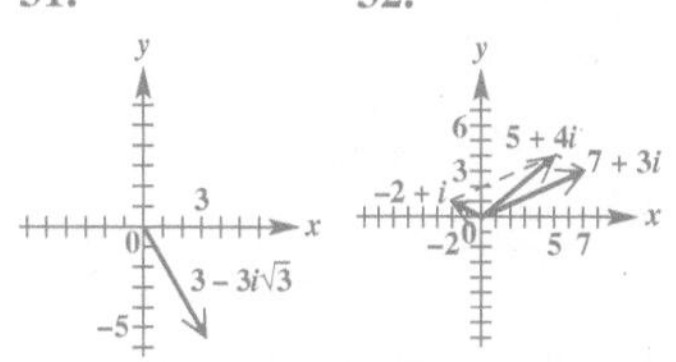

33. $2\sqrt{2}(\cos 135° + i \sin 135°)$

34. $3i$

35. $-\sqrt{2} - i\sqrt{2}$

36. $8(\cos 120° + i \sin 120°)$

37. $\sqrt{2}(\cos 315° + i \sin 315°)$

38. $-2 - 2i\sqrt{3}$

39. $4(\cos 270° + i \sin 270°)$

40. $4.4995 - 5.3623i$

41. It is the line $y = -x$.

42. It is a circle of radius 2 with the origin as center.

43. $\sqrt[6]{2}(\cos 105° + i \sin 105°)$, $\sqrt[6]{2}(\cos 225° + i \sin 225°)$, $\sqrt[6]{2}(\cos 345° + i \sin 345°)$

44. $\sqrt[10]{8}(\cos 27° + i \sin 27°)$, $\sqrt[10]{8}(\cos 99° + i \sin 99°)$, $\sqrt[10]{8}(\cos 171° + i \sin 171°)$, $\sqrt[10]{8}(\cos 243° + i \sin 243°)$, $\sqrt[10]{8}(\cos 315° + i \sin 315°)$

45. none 46. one

47. $\{2(\cos 45° + i \sin 45°)$, $2(\cos 135° + i \sin 135°)$, $2(\cos 225° + i \sin 225°)$, $2(\cos 315° + i \sin 315°)\}$

48. $\{5(\cos 60° + i \sin 60°)$, $5(\cos 180° + i \sin 180°)$, $5(\cos 300° + i \sin 300°)\}$

49. $\{\cos 135° + i \sin 135°$, $\cos 315° + i \sin 315°\}$

50. $\left(\frac{5\sqrt{2}}{2}, -\frac{5\sqrt{2}}{2}\right)$

51. $(2, 120°)$

52. It is a circle centered at the origin having radius k.

Perform each operation. Write answers in rectangular form.

21. $[5(\cos 90° + i \sin 90°)][6(\cos 180° + i \sin 180°)]$

22. $[3 \text{ cis } 135°][2 \text{ cis } 105°]$

23. $\dfrac{2(\cos 60° + i \sin 60°)}{8(\cos 300° + i \sin 300°)}$

24. $\dfrac{4 \text{ cis } 270°}{2 \text{ cis } 90°}$

25. $\left(\sqrt{3} + i\right)^3$

26. $(2 - 2i)^5$

27. $(\cos 100° + i \sin 100°)^6$

28. ***Concept Check*** The vector representing a real number will lie on the ______-axis in the complex plane.

Graph each complex number.

29. $5i$

30. $-4 + 2i$

31. $3 - 3i\sqrt{3}$

32. Find the sum of $7 + 3i$ and $-2 + i$. Graph both complex numbers and their resultant.

Write each complex number in its alternative form, using a calculator to approximate answers to four decimal places as necessary.

	Rectangular Form	**Trigonometric Form**
33.	$-2 + 2i$	______
34.	______	$3(\cos 90° + i \sin 90°)$
35.	______	$2(\cos 225° + i \sin 225°)$
36.	$-4 + 4i\sqrt{3}$	______
37.	$1 - i$	______
38.	______	$4 \text{ cis } 240°$
39.	$-4i$	______
40.	______	$7 \text{ cis } 310°$

Concept Check *The complex number z, where $z = x + yi$, can be graphed in the plane as (x, y). Describe the graph of all complex numbers z satisfying the given conditions.*

41. The imaginary part of z is the negative of the real part of z.

42. The absolute value of z is 2.

Find all roots as indicated. Write answers in trigonometric form.

43. the cube roots of $1 - i$

44. the fifth roots of $-2 + 2i$

45. ***Concept Check*** How many real sixth roots does -64 have?

46. ***Concept Check*** How many real fifth roots does -32 have?

Find all complex number solutions. Write answers in trigonometric form.

47. $x^4 + 16 = 0$

48. $x^3 + 125 = 0$

49. $x^2 + i = 0$

50. Convert $(5, 315°)$ to rectangular coordinates.

51. Convert $\left(-1, \sqrt{3}\right)$ to polar coordinates, with $0° \le \theta < 360°$ and $r > 0$.

52. ***Concept Check*** Describe the graph of $r = k$ for $k > 0$.

Identify and graph each polar equation for θ in $[0°, 360°)$.

53. $r = 4 \cos \theta$

54. $r = -1 + \cos \theta$

55. $r = 2 \sin 4\theta$

56. $r = \dfrac{2}{2 \cos \theta - \sin \theta}$

53. circle

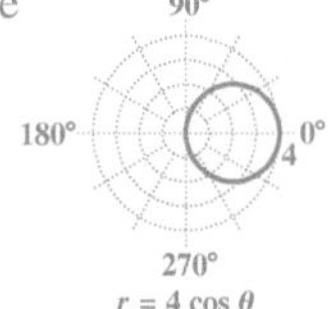

54. cardioid

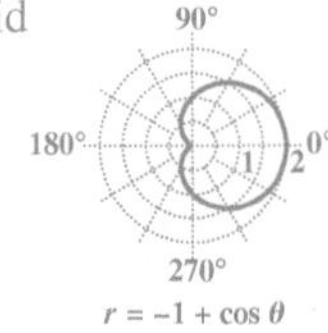

55. eight-leaved rose

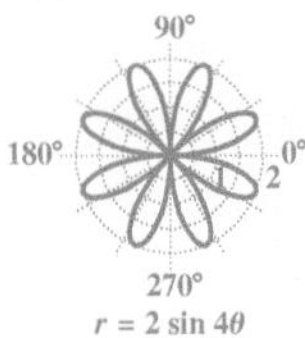

56. line

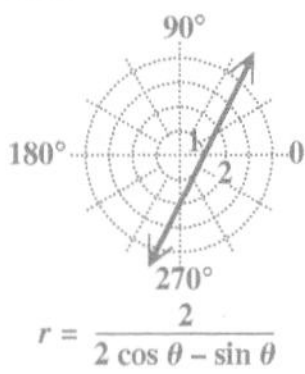

57. $y^2 = -6\left(x - \frac{3}{2}\right)$, or
$y^2 + 6x - 9 = 0$

58. $\left(x - \frac{1}{2}\right)^2 + \left(y - \frac{1}{2}\right)^2 = \frac{1}{2}$, or
$x^2 + y^2 - x - y = 0$

59. $x^2 + y^2 = 4$

60. $\sin \theta = \cos \theta$, or $\tan \theta = 1$

61. $r = \tan \theta \sec \theta$, or $r = \frac{\tan \theta}{\cos \theta}$

62. $r = 5$ or $r = -5$

63. B 64. A 65. C 66. B

67. $r = 2 \sec \theta$, or $r = \frac{2}{\cos \theta}$

68. $r = 2 \csc \theta$, or $r = \frac{2}{\sin \theta}$

69. $r = \frac{4}{\cos \theta + 2 \sin \theta}$

70. $r = 2$ or $r = -2$

71.

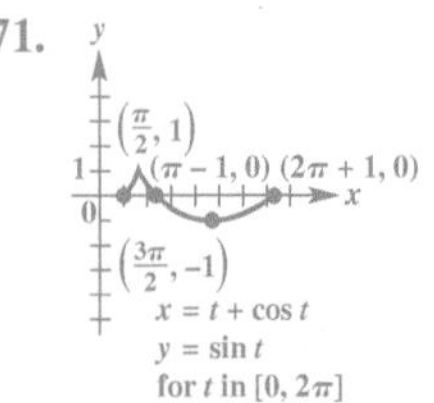

73. $y = \sqrt{x^2 + 1}$, for x in $[0, \infty)$

74. $x - 3y = 5$, for x in $[-13, 17]$

75. $y = 3\sqrt{1 + \frac{x^2}{25}}$, for x in $(-\infty, \infty)$

76. $y = \frac{1}{x - 4}$, for x in $[5, \infty)$

77. $y^2 = -\frac{1}{2}(x - 1)$, or
$2y^2 + x - 1 = 0$, for x in $[-1, 1]$

Find an equivalent equation in rectangular coordinates.

57. $r = \dfrac{3}{1 + \cos \theta}$ **58.** $r = \sin \theta + \cos \theta$ **59.** $r = 2$

Find an equivalent equation in polar coordinates.

60. $y = x$ **61.** $y = x^2$ **62.** $x^2 + y^2 = 25$

Identify the geometric symmetry (A, B, or C) that each graph will possess.

A. *symmetry with respect to the origin*

B. *symmetry with respect to the y-axis*

C. *symmetry with respect to the x-axis*

63. Whenever (r, θ) is on the graph, so is $(-r, -\theta)$.

64. Whenever (r, θ) is on the graph, so is $(-r, \theta)$.

65. Whenever (r, θ) is on the graph, so is $(r, -\theta)$.

66. Whenever (r, θ) is on the graph, so is $(r, \pi - \theta)$.

Find a polar equation having the given graph.

67.

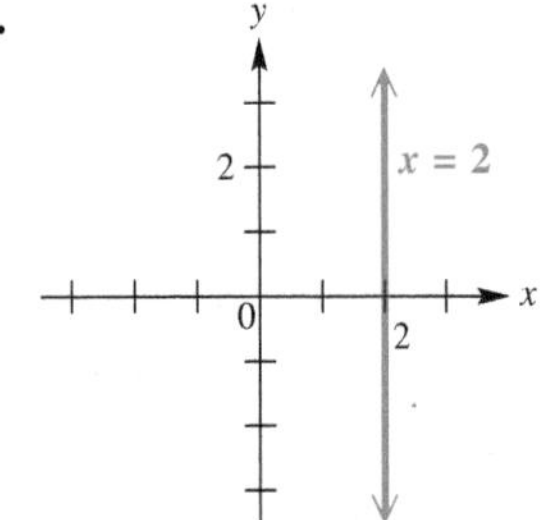

68.

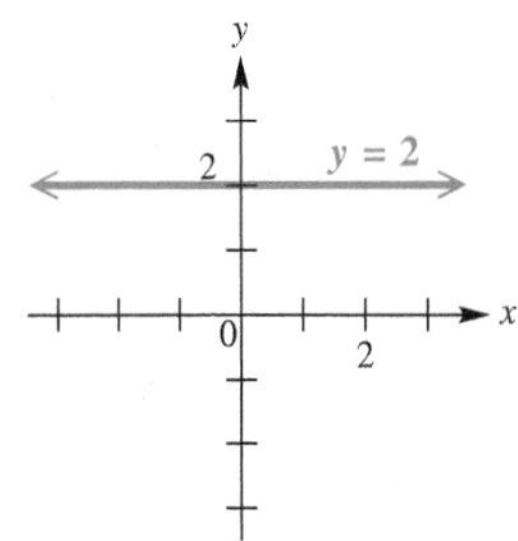

69.

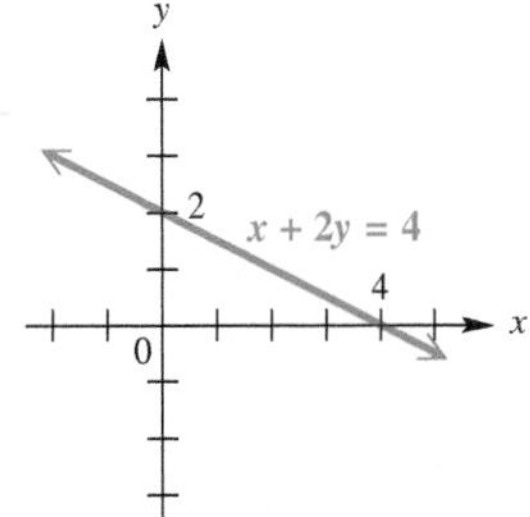

70.

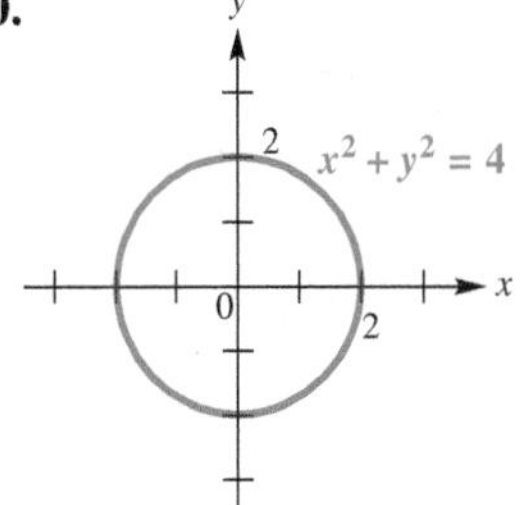

71. Graph the plane curve defined by the parametric equations $x = t + \cos t$, $y = \sin t$, for t in $[0, 2\pi]$.

72. Show that the distance between (r_1, θ_1) and (r_2, θ_2) in polar coordinates is given by

$$d = \sqrt{r_1{}^2 + r_2{}^2 - 2r_1r_2 \cos(\theta_1 - \theta_2)}.$$

Find a rectangular equation for each plane curve with the given parametric equations.

73. $x = \sqrt{t - 1}$, $y = \sqrt{t}$, for t in $[1, \infty)$

74. $x = 3t + 2$, $y = t - 1$, for t in $[-5, 5]$

75. $x = 5 \tan t$, $y = 3 \sec t$, for t in $\left(-\dfrac{\pi}{2}, \dfrac{\pi}{2}\right)$

76. $x = t^2 + 5$, $y = \dfrac{1}{t^2 + 1}$, for t in $(-\infty, \infty)$

77. $x = \cos 2t$, $y = \sin t$, for t in $(-\pi, \pi)$

78. Give a pair of parametric equations whose graph is the circle having center $(3, 4)$ and passing through the origin.

78. $x = 3 + 5 \cos t$, $y = 4 + 5 \sin t$, for t in $[0, 2\pi]$

79. (a) $x = (118 \cos 27°)t$, $y = 3.2 - 16t^2 + (118 \sin 27°)t$
(b) $y = 3.2 - \frac{4x^2}{3481 \cos^2 27°} + (\tan 27°)x$
(c) 3.4 sec; 358 ft

80. $z^2 + z = 1 + 3i$; $r = \sqrt{10}$ and $\sqrt{10} > 2$

79. ***(Modeling) Flight of a Baseball*** A batter hits a baseball when it is 3.2 ft above the ground. It leaves the bat with velocity 118 ft per sec at an angle of 27° with respect to the ground.

(a) Determine parametric equations that model the path of the baseball.

(b) Determine a rectangular equation that models the path of the baseball.

(c) Determine approximately how long the baseball is in flight and the horizontal distance it covers.

80. ***Mandelbrot Set*** Consider the complex number $z = 1 + i$. Compute the value of $z^2 + z$, and show that its absolute value exceeds 2, indicating that $1 + i$ is not in the Mandelbrot set.

Chapter 8 Test

[8.1]
1. (a) $-4\sqrt{3}$ (b) $\frac{1}{2}i$ (c) $\frac{1}{3}$

[8.1, 8.2]
2. (a) $7 - 3i$

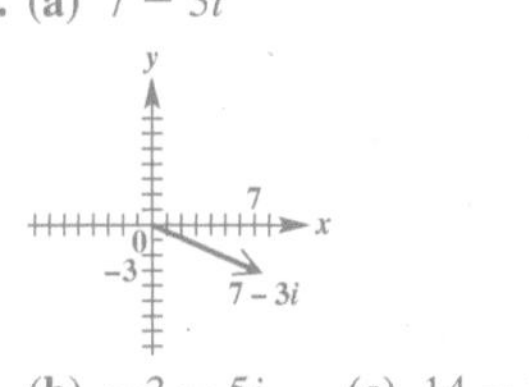

(b) $-3 - 5i$ (c) $14 - 18i$
(d) $\frac{3}{13} - \frac{11}{13}i$

3. (a) $-i$ (b) $2i$

[8.1]
4. $\left\{\frac{1}{4} \pm \frac{\sqrt{31}}{4}i\right\}$

[8.2]
5. (a) $3(\cos 90° + i \sin 90°)$
(b) $\sqrt{5}$ cis 63.43°
(c) $2(\cos 240° + i \sin 240°)$

6. (a) $\frac{3\sqrt{3}}{2} + \frac{3}{2}i$ (b) $3.06 + 2.57i$
(c) $3i$

[8.3, 8.4]
7. (a) $16(\cos 50° + i \sin 50°)$
(b) $2\sqrt{3} + 2i$ (c) $4\sqrt{3} + 4i$

[8.4]
8. 2 cis 67.5°, 2 cis 157.5°, 2 cis 247.5°, 2 cis 337.5°

[8.5]
9. Answers may vary.
(a) $(5, 90°)$, $(5, -270°)$
(b) $(2\sqrt{2}, 225°)$, $(2\sqrt{2}, -135°)$

Answers for the remaining exercises are given in the answer section at the back of the text.

1. Find each product or quotient. Simplify the answers.

(a) $\sqrt{-8} \cdot \sqrt{-6}$ **(b)** $\frac{\sqrt{-2}}{\sqrt{8}}$ **(c)** $\frac{\sqrt{-20}}{\sqrt{-180}}$

2. For the complex numbers $w = 2 - 4i$ and $z = 5 + i$, find each of the following in rectangular form.

(a) $w + z$ (and give a geometric representation) **(b)** $w - z$ **(c)** wz **(d)** $\frac{w}{z}$

3. Express each of the following in rectangular form.

(a) i^{15} **(b)** $(1 + i)^2$

4. Solve $2x^2 - x + 4 = 0$ over the set of complex numbers.

5. Write each complex number in trigonometric (polar) form, where $0° \le \theta < 360°$.

(a) $3i$ **(b)** $1 + 2i$ **(c)** $-1 - i\sqrt{3}$

6. Write each complex number in rectangular form.

(a) $3(\cos 30° + i \sin 30°)$ **(b)** 4 cis 40° **(c)** $3(\cos 90° + i \sin 90°)$

7. For the complex numbers $w = 8(\cos 40° + i \sin 40°)$ and $z = 2(\cos 10° + i \sin 10°)$, find each of the following in the form specified.

(a) wz (trigonometric form) **(b)** $\frac{w}{z}$ (rectangular form) **(c)** z^3 (rectangular form)

8. Find the four complex fourth roots of $-16i$. Write answers in trigonometric form.

9. Convert the given rectangular coordinates to polar coordinates. Give two pairs of polar coordinates for each point.

(a) $(0, 5)$ **(b)** $(-2, -2)$

10. Convert the given polar coordinates to rectangular coordinates.

(a) $(3, 315°)$ **(b)** $(-4, 90°)$

Identify and graph each polar equation for θ in $[0°, 360°)$.

11. $r = 1 - \cos \theta$ **12.** $r = 3 \cos 3\theta$

13. Convert each polar equation to a rectangular equation, and sketch its graph.

(a) $r = \frac{4}{2 \sin \theta - \cos \theta}$ **(b)** $r = 6$

Graph each pair of parametric equations.

14. $x = 4t - 3$, $y = t^2$, for t in $[-3, 4]$ **15.** $x = 2 \cos 2t$, $y = 2 \sin 2t$, for t in $[0, 2\pi]$

16. ***Julia Set*** Consider the complex number $z = -1 + i$. Compute the value of $z^2 - 1$, and show that its absolute value exceeds 2, indicating that $-1 + i$ is not in the Julia set.

Appendices

A Equations and Inequalities

Basic Terminology of Equations An **equation** is a statement that two expressions are equal.

$$x + 2 = 9, \quad 11x = 5x + 6x, \quad x^2 - 2x - 1 = 0 \quad \text{Equations}$$

To *solve* an equation means to find all numbers that make the equation a true statement. These numbers are the **solutions,** or **roots,** of the equation. A number that is a solution of an equation is said to *satisfy* the equation, and the solutions of an equation make up its **solution set.** Equations with the same solution set are **equivalent equations.** For example,

$$x = 4, \quad x + 1 = 5, \quad \text{and} \quad 6x + 3 = 27 \quad \text{are equivalent equations}$$

because they have the same solution set, $\{4\}$. However, the equations

$$x^2 = 9 \quad \text{and} \quad x = 3 \quad \text{are } not \text{ equivalent}$$

because the first has solution set $\{-3, 3\}$ while the solution set of the second is $\{3\}$.

One way to solve an equation is to rewrite it as a series of simpler equivalent equations using the **addition and multiplication properties of equality.** Let a, b, and c represent real numbers.

Teaching Tip Emphasize this important "Golden Rule" of algebra: "Do unto one entire side of the equation (on one side of the equals symbol) as you do unto the other" to keep sides equal.

If $a = b$, then $a + c = b + c$.

If $a = b$ and $c \neq 0$, then $ac = bc$.

These properties can be extended: The same number may be subtracted from each side of an equation, and each side may be divided by the same nonzero number, without changing the solution set.

Linear Equations We use the properties of equality to solve *linear equations.*

Teaching Tip To help students recognize *linear* equations, list several equations and ask which ones are linear. Here are some possible examples.
(a) $3x + 5 = 0$ [yes]
(b) $(x - 2)(2x + 3) = 0$ [no]
(c) $2(4x + 1) = 3 - (2x + 4)$ [yes]
(d) $3\sqrt{x - 1} = 6$ [no]

Linear Equation in One Variable

A **linear equation in one variable** is an equation that can be written in the form

$$ax + b = 0,$$

where a and b are real numbers and $a \neq 0$.

A linear equation is a **first-degree equation** because the greatest degree of the variable is 1.

$$3x + \sqrt{2} = 0, \quad \frac{3}{4}x = 12, \quad 0.5(x + 3) = 2x - 6 \quad \text{Linear equations}$$

$$\sqrt{x} + 2 = 5, \quad \frac{1}{x} = -8, \quad x^2 + 3x + 0.2 = 0 \quad \text{Nonlinear equations}$$

Classroom Example 1
Solve $-4(3x-5)=3-(8x+7)$.

Answer: $\{6\}$

Teaching Tip In the solution for **Example 1,** emphasize that the final line of the check does *not* give the answer, only a confirmation that the answer found is correct. Also, give an example of how a check can show that an answer is incorrect.

EXAMPLE 1 Solving a Linear Equation

Solve $3(2x-4)=7-(x+5)$.

SOLUTION

$$3(2x-4)=7-(x+5)$$

$$6x-12=7-x-5 \quad \text{Distributive property}$$

$$6x-12=2-x \quad \text{Combine like terms.}$$

$$6x-12+x=2-x+x \quad \text{Add } x \text{ to each side.}$$

$$7x-12=2 \quad \text{Combine like terms.}$$

$$7x-12+12=2+12 \quad \text{Add 12 to each side.}$$

$$7x=14 \quad \text{Combine like terms.}$$

$$\frac{7x}{7}=\frac{14}{7} \quad \text{Divide each side by 7.}$$

$$x=2$$

CHECK

A check of the solution is recommended.

$$3(2x-4)=7-(x+5) \quad \text{Original equation}$$

$$3(2\cdot 2-4)\stackrel{?}{=}7-(2+5) \quad \text{Let } x=2.$$

$$3(4-4)\stackrel{?}{=}7-(7) \quad \text{Work inside the parentheses.}$$

$$0=0 \ ✓ \quad \text{True}$$

The solution set is $\{2\}$.

✔ **Now Try Exercise 9.**

Classroom Example 2
Solve $\frac{3x+6}{10}-\frac{1}{2}x=\frac{2}{5}x+\frac{33}{5}$.

Answer: $\{-10\}$

EXAMPLE 2 Solving a Linear Equation with Fractions

Solve $\frac{2x+4}{3}+\frac{1}{2}x=\frac{1}{4}x-\frac{7}{3}$.

SOLUTION

$$\frac{2x+4}{3}+\frac{1}{2}x=\frac{1}{4}x-\frac{7}{3}$$

Distribute to *all* terms within the parentheses.

$$12\left(\frac{2x+4}{3}+\frac{1}{2}x\right)=12\left(\frac{1}{4}x-\frac{7}{3}\right) \quad \text{Multiply by 12, the LCD of the fractions.}$$

$$12\left(\frac{2x+4}{3}\right)+12\left(\frac{1}{2}x\right)=12\left(\frac{1}{4}x\right)-12\left(\frac{7}{3}\right) \quad \text{Distributive property}$$

$$4(2x+4)+6x=3x-28 \quad \text{Multiply.}$$

$$8x+16+6x=3x-28 \quad \text{Distributive property}$$

$$14x+16=3x-28 \quad \text{Combine like terms.}$$

$$11x=-44 \quad \text{Subtract } 3x. \text{ Subtract 16.}$$

$$x=-4 \quad \text{Divide each side by 11.}$$

CHECK

$$\frac{2x+4}{3}+\frac{1}{2}x=\frac{1}{4}x-\frac{7}{3} \quad \text{Original equation}$$

$$\frac{2(-4)+4}{3}+\frac{1}{2}(-4)\stackrel{?}{=}\frac{1}{4}(-4)-\frac{7}{3} \quad \text{Let } x=-4.$$

$$-\frac{10}{3}=-\frac{10}{3} \ ✓ \quad \text{True}$$

The solution set is $\{-4\}$.

✔ **Now Try Exercise 11.**

An equation satisfied by every number that is a meaningful replacement for the variable is an **identity.**

$$3(x + 1) = 3x + 3 \quad \text{Identity}$$

An equation that is satisfied by some numbers but not others is a **conditional equation.**

$$2x = 4 \quad \text{Conditional equation}$$

The equations in **Examples 1 and 2** are conditional equations. An equation that has no solution is a **contradiction.**

$$x = x + 1 \quad \text{Contradiction}$$

Classroom Example 3
Determine whether each equation is an *identity*, a *conditional equation*, or a *contradiction*. Give the solution set.

(a) $6x - 9 = 4x + 13$

(b) $10 + 14x = 7(2x - 5)$

(c) $-3(2x - 1) + 5x = 3 - x$

Answers:

(a) conditional equation; $\{11\}$

(b) contradiction; $\varnothing$

(c) identity; {all real numbers}

EXAMPLE 3 Identifying Types of Equations

Determine whether each equation is an *identity*, a *conditional equation*, or a *contradiction*. Give the solution set.

(a) $-2(x + 4) + 3x = x - 8$ **(b)** $5x - 4 = 11$ **(c)** $3(3x - 1) = 9x + 7$

SOLUTION

(a) $-2(x + 4) + 3x = x - 8$

$-2x - 8 + 3x = x - 8$ Distributive property

$x - 8 = x - 8$ Combine like terms.

$0 = 0$ Subtract x. Add 8.

When a *true* statement such as $0 = 0$ results, the equation is an identity, and the solution set is **{all real numbers}.**

(b) $5x - 4 = 11$

$5x = 15$ Add 4 to each side.

$x = 3$ Divide each side by 5.

This is a conditional equation, and its solution set is $\{3\}$.

(c) $3(3x - 1) = 9x + 7$

$9x - 3 = 9x + 7$ Distributive property

$-3 = 7$ Subtract $9x$.

When a *false* statement such as $-3 = 7$ results, the equation is a contradiction, and the solution set is the **empty set,** or **null set,** symbolized $\varnothing$.

✔ **Now Try Exercises 21, 23, and 25.**

Teaching Tip Remind students that an equation with no solution, such as the one in **Example 3(c),** has the empty (or null) set as its solution set.

Quadratic Equations

A *quadratic equation* is defined as follows.

Teaching Tip To help students see the difference between linear and quadratic equations, list several equations and have students tell whether each is linear, quadratic, or neither. Remind them that a quadratic equation has degree 2.

Quadratic Equation in One Variable

An equation that can be written in the form

$$ax^2 + bx + c = 0,$$

where a, b, and c are real numbers with $a \neq 0$, is a **quadratic equation.** The given form is called **standard form.**

A quadratic equation is a **second-degree equation**—that is, an equation with a squared variable term and no terms of greater degree.

$$x^2 = 25, \quad 4x^2 + 4x - 5 = 0, \quad 3x^2 = 4x - 8 \quad \text{Quadratic equations}$$

When the expression $ax^2 + bx + c$ in a quadratic equation is easily factorable over the real numbers, it is efficient to factor and then apply the following **zero-factor property.**

If a and b are complex numbers with $ab = 0$, then $a = 0$ or $b = 0$ or both equal zero.

Classroom Example 4
Solve $10x^2 + x - 2 = 0$ using the zero-factor property.

Answer: $\left\{-\frac{1}{2}, \frac{2}{5}\right\}$

Teaching Tip Students often make the mistake of writing the equation in **Example 4** as

$$x(6x + 7) = 3$$

and then setting $x = 3$ and $6x + 7 = 3$. Emphasize that the right side of the equation must be 0 before factoring on the left.

EXAMPLE 4 Using the Zero-Factor Property

Solve $6x^2 + 7x = 3$.

SOLUTION

$6x^2 + 7x = 3$ ← Don't factor out x here.

$6x^2 + 7x - 3 = 0$ Standard form

$(3x - 1)(2x + 3) = 0$ Factor.

$3x - 1 = 0$ or $2x + 3 = 0$ Zero-factor property

$3x = 1$ or $2x = -3$ Solve each equation.

$x = \frac{1}{3}$ or $x = -\frac{3}{2}$

CHECK $6x^2 + 7x = 3$ Original equation

$6\left(\frac{1}{3}\right)^2 + 7\left(\frac{1}{3}\right) \stackrel{?}{=} 3$ Let $x = \frac{1}{3}$.	$6\left(-\frac{3}{2}\right)^2 + 7\left(-\frac{3}{2}\right) \stackrel{?}{=} 3$ Let $x = -\frac{3}{2}$.
$\frac{6}{9} + \frac{7}{3} \stackrel{?}{=} 3$	$\frac{54}{4} - \frac{21}{2} \stackrel{?}{=} 3$
$3 = 3$ ✓ True	$3 = 3$ ✓ True

Both values check because true statements result. The solution set is $\left\{\frac{1}{3}, -\frac{3}{2}\right\}$.

✓ **Now Try Exercise 33.**

Teaching Tip Tell students that a polynomial of degree 2 can have as many as two distinct solutions. Solve the equation

$$x^2 - 16 = 0$$

by factoring as a difference of squares. This will help make the point that 4 and −4 are *both* solutions of the equation $x^2 = 16$, while at the same time making the square root property more understandable.

A quadratic equation written in the form $x^2 = k$, where k is a constant, can be solved using the **square root property.**

If $x^2 = k$, then $x = \sqrt{k}$ or $x = -\sqrt{k}$.

That is, the solution set of $x^2 = k$ is

$$\left\{\sqrt{k}, -\sqrt{k}\right\}, \text{ which may be abbreviated } \left\{\pm\sqrt{k}\right\}.$$

Classroom Example 5
Solve each quadratic equation using the square root property.
(a) $x^2 = 29$
(b) $(x - 8)^2 = 24$

Answers:
(a) $\{\pm\sqrt{29}\}$
(b) $\{8 \pm 2\sqrt{6}\}$

EXAMPLE 5 Using the Square Root Property

Solve each quadratic equation.

(a) $x^2 = 17$ **(b)** $(x - 4)^2 = 12$

SOLUTION

(a) $x^2 = 17$

$x = \pm\sqrt{17}$ Square root property

The solution set is $\left\{\pm\sqrt{17}\right\}$.

(b)

$$(x-4)^2 = 12$$

$$x - 4 = \pm\sqrt{12} \quad \text{Generalized square root property}$$

$$x = 4 \pm \sqrt{12} \quad \text{Add 4.}$$

$$x = 4 \pm 2\sqrt{3} \quad \sqrt{12} = \sqrt{4 \cdot 3} = 2\sqrt{3}$$

CHECK $(x-4)^2 = 12$ Original equation

$$\left(4 + 2\sqrt{3} - 4\right)^2 \stackrel{?}{=} 12 \quad \text{Let } x = 4 + 2\sqrt{3}.$$

$$\left(2\sqrt{3}\right)^2 \stackrel{?}{=} 12$$

$$2^2 \cdot \left(\sqrt{3}\right)^2 \stackrel{?}{=} 12$$

$$12 = 12 \quad \checkmark \text{ True}$$

$$\left(4 - 2\sqrt{3} - 4\right)^2 \stackrel{?}{=} 12 \quad \text{Let } x = 4 - 2\sqrt{3}.$$

$$\left(-2\sqrt{3}\right)^2 \stackrel{?}{=} 12$$

$$(-2)^2 \cdot \left(\sqrt{3}\right)^2 \stackrel{?}{=} 12$$

$$12 = 12 \quad \checkmark \text{ True}$$

The solution set is $\left\{4 \pm 2\sqrt{3}\right\}$.

✔ **Now Try Exercises 43 and 47.**

Any quadratic equation can be solved by the **quadratic formula,** which says that the solutions of the quadratic equation $ax^2 + bx + c = 0$, where $a \neq 0$, are given by

$$x = \frac{-b \pm \sqrt{b^2 - 4ac}}{2a}.$$

This formula is derived in algebra courses.

Classroom Example 6
Solve $x^2 + 6x = 3$ using the quadratic formula.

Answer: $\left\{-3 \pm 2\sqrt{3}\right\}$

Teaching Tip Point out that the equation in **Example 6** can be thought of as

$$1x^2 + (-4)x + 2 = 0.$$

Therefore, $a = 1$, $b = -4$, and $c = 2$.

EXAMPLE 6 Using the Quadratic Formula

Solve $x^2 - 4x = -2$.

SOLUTION

$$x^2 - 4x + 2 = 0 \quad \text{Write in standard form. Here } a = 1,\ b = -4, \text{ and } c = 2.$$

$$x = \frac{-b \pm \sqrt{b^2 - 4ac}}{2a} \quad \text{Quadratic formula}$$

$$x = \frac{-(-4) \pm \sqrt{(-4)^2 - 4(1)(2)}}{2(1)} \quad \text{Substitute } a = 1,\ b = -4, \text{ and } c = 2.$$

The fraction bar extends *under* $-b$.

$$x = \frac{4 \pm \sqrt{16 - 8}}{2} \quad \text{Simplify.}$$

$$x = \frac{4 \pm 2\sqrt{2}}{2} \quad \sqrt{16-8} = \sqrt{8} = \sqrt{4 \cdot 2} = 2\sqrt{2}$$

$$x = \frac{2\left(2 \pm \sqrt{2}\right)}{2} \quad \text{Factor out 2 in the numerator.}$$

Factor first, then divide.

$$x = 2 \pm \sqrt{2} \quad \text{Lowest terms}$$

The solution set is $\left\{2 \pm \sqrt{2}\right\}$.

✔ **Now Try Exercise 55.**

Inequalities An **inequality** says that one expression is greater than, greater than or equal to, less than, or less than or equal to another. As with equations, a value of the variable for which the inequality is true is a solution of the inequality, and the set of all solutions is the solution set of the inequality. Two inequalities with the same solution set are equivalent.

Inequalities are solved with the **properties of inequality,** which are similar to the properties of equality. Let a, b, and c represent real numbers.

1. **If $a < b$, then $a + c < b + c$.**
2. **If $a < b$ and if $c > 0$, then $ac < bc$.**
3. **If $a < b$ and if $c < 0$, then $ac > bc$.**

Replacing $<$ with $>$, $\leq$, or $\geq$ results in similar properties. (Restrictions on c remain the same.)

Multiplication may be replaced by division in Properties 2 and 3. ***Always remember to reverse the direction of the inequality symbol when multiplying or dividing by a negative number.***

Linear Inequalities and Interval Notation The definition of a *linear inequality* is similar to the definition of a linear equation.

Linear Inequality in One Variable

A **linear inequality in one variable** is an inequality that can be written in the form

$$ax + b > 0,^*$$

where a and b are real numbers and $a \neq 0$.

*The symbol $>$ can be replaced with $<$, $\leq$, or $\geq$.

Classroom Example 7
Solve $-2x + 7 < -5$.
Answer: $\{x \mid x > 6\}$

EXAMPLE 7 Solving a Linear Inequality

Solve $-3x + 5 > -7$.

SOLUTION

$$-3x + 5 > -7$$

$$-3x + 5 - 5 > -7 - 5 \quad \text{Subtract 5.}$$

$$-3x > -12 \quad \text{Combine like terms.}$$

Don't forget to reverse the inequality symbol here.

$$\frac{-3x}{-3} < \frac{-12}{-3} \quad \text{Divide by } -3. \text{ Reverse the direction of the inequality symbol when multiplying or dividing by a negative number.}$$

$$x < 4$$

The original inequality $-3x + 5 > -7$ is satisfied by any real number less than 4. The solution set can be written using **set-builder notation** as

$$\{x \mid x < 4\}, \quad \text{Set-builder notation}$$

which is read "the set of all x such that x is less than 4."

The solution set $\{x \mid x < 4\}$ is an example of an **interval.** Using **interval notation,** we write it as

$$(-\infty, 4). \quad \text{Interval notation}$$

The symbol $-\infty$ does not represent an actual number. Rather, it is used to show that the interval includes all real numbers less than 4. The interval $(-\infty, 4)$ is an example of an **open interval** because the endpoint, 4, is not part of the interval. An interval that includes both its endpoints is a **closed interval.** A square bracket indicates that a number *is* part of an interval, and a parenthesis indicates that a number *is not* part of an interval.

The solution set $(-\infty, 4)$ is graphed in **Figure 1.** ✔ **Now Try Exercise 73.**

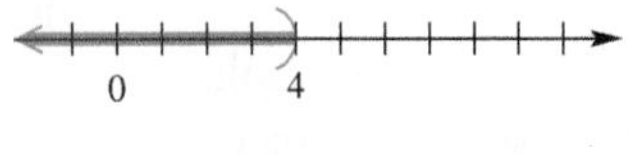

0 4

Figure 1

Teaching Tip Have students translate English phrases such as

"all values of x such that x is greater than -3 and less than or equal to 4"

into set notation and interval notation.

Summary of Types of Intervals (Assume that $a < b$.)

Type of Interval	Set	Interval Notation	Graph
Open interval	$\{x \mid x > a\}$	(a, ∞)	
	$\{x \mid a < x < b\}$	(a, b)	
	$\{x \mid x < b\}$	$(-\infty, b)$	
Other intervals	$\{x \mid x \geq a\}$	$[a, \infty)$	
	$\{x \mid a < x \leq b\}$	$(a, b]$	
	$\{x \mid a \leq x < b\}$	$[a, b)$	
	$\{x \mid x \leq b\}$	$(-\infty, b]$	
Closed interval	$\{x \mid a \leq x \leq b\}$	$[a, b]$	
Disjoint interval	$\{x \mid x < a \text{ or } x > b\}$	$(-\infty, a) \cup (b, \infty)$	
All real numbers	$\{x \mid x \text{ is a real number}\}$	$(-\infty, \infty)$	

Three-Part Inequalities The inequality $-2 < 5 + 3x < 20$ says that $5 + 3x$ is *between* -2 and 20. This inequality is solved using an extension of the properties of inequality given earlier, working with all three expressions at the same time.

Classroom Example 8
Solve $1 \leq 6x - 8 \leq 4$. Give the solution set in interval notation.

Answer: $\left[\frac{3}{2}, 2\right]$

EXAMPLE 8 Solving a Three-Part Inequality

Solve $-2 < 5 + 3x < 20$. Give the solution set in interval notation.

SOLUTION

$$-2 < 5 + 3x < 20$$

$$-2 - 5 < 5 + 3x - 5 < 20 - 5 \quad \text{Subtract 5 from each part.}$$

$$-7 < 3x < 15 \quad \text{Combine like terms in each part.}$$

$$\frac{-7}{3} < \frac{3x}{3} < \frac{15}{3} \quad \text{Divide each part by 3.}$$

$$-\frac{7}{3} < x < 5$$

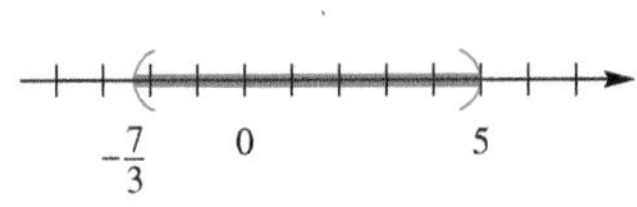

Figure 2

The solution set, graphed in **Figure 2,** is the interval $\left(-\frac{7}{3}, 5\right)$.

✔ **Now Try Exercise 83.**

Appendix A Exercises

1. equation 2. solve

Concept Check Fill in the blank to correctly complete each sentence.

1. A(n) ________ is a statement that two expressions are equal.

2. To ________ an equation means to find all numbers that make the equation a true statement.

3. first-degree equation
4. identity
5. contradiction

6. B

7. $\{-4\}$ 8. $\{-5\}$
9. $\{1\}$ 10. $\{-1\}$
11. $\left\{-\frac{2}{7}\right\}$ 12. $\left\{\frac{5}{12}\right\}$
13. $\left\{-\frac{7}{8}\right\}$ 14. $\left\{-\frac{6}{11}\right\}$
15. $\{-1\}$ 16. $\{-2\}$
17. $\{75\}$ 18. $\{3\}$
19. $\{0\}$ 20. $\{0\}$

21. identity; {all real numbers}
22. identity; {all real numbers}
23. conditional equation; $\{0\}$
24. conditional equation; $\{0\}$
25. contradiction; $\emptyset$
26. contradiction; $\emptyset$

27. D; $\left\{\frac{1}{3}, 7\right\}$
28. B; $\left\{\frac{-5 \pm \sqrt{7}}{2}\right\}$
29. A, B, C, D 30. A; $\left\{-\frac{1}{3}, 6\right\}$
31. $\{2, 3\}$ 32. $\{-4, 2\}$
33. $\left\{-\frac{2}{5}, 1\right\}$ 34. $\left\{-\frac{5}{2}, 3\right\}$
35. $\left\{-\frac{3}{4}, 1\right\}$ 36. $\left\{-\frac{5}{6}, 2\right\}$
37. $\{\pm 10\}$ 38. $\{\pm 8\}$
39. $\left\{\frac{1}{2}\right\}$ 40. $\left\{\frac{2}{3}\right\}$
41. $\left\{-\frac{3}{5}\right\}$ 42. $\left\{-\frac{5}{6}\right\}$
43. $\{\pm 4\}$ 44. $\{\pm 11\}$
45. $\{\pm 3\sqrt{3}\}$ 46. $\{\pm 4\sqrt{3}\}$
47. $\left\{\frac{1 \pm 2\sqrt{3}}{3}\right\}$ 48. $\left\{\frac{-1 \pm 2\sqrt{5}}{4}\right\}$

3. A linear equation is a(n) ________ because the greatest degree of the variable is 1.

4. A(n) ________ is an equation satisfied by every number that is a meaningful replacement for the variable.

5. A(n) ________ is an equation that has no solution.

6. *Concept Check* Which one is *not* a linear equation?

A. $5x + 7(x - 1) = -3x$ **B.** $9x^2 - 4x + 3 = 0$

C. $7x + 8x = 13x$ **D.** $0.04x - 0.08x = 0.40$

Solve each equation. ***See Examples 1 and 2.***

7. $5x + 4 = 3x - 4$ **8.** $9x + 11 = 7x + 1$

9. $6(3x - 1) = 8 - (10x - 14)$ **10.** $4(-2x + 1) = 6 - (2x - 4)$

11. $\frac{5}{6}x - 2x + \frac{4}{3} = \frac{5}{3}$ **12.** $\frac{7}{4} + \frac{1}{5}x - \frac{3}{2} = \frac{4}{5}x$

13. $3x + 5 - 5(x + 1) = 6x + 7$ **14.** $5(x + 3) + 4x - 3 = -(2x - 4) + 2$

15. $2[x - (4 + 2x) + 3] = 2x + 2$ **16.** $4[2x - (3 - x) + 5] = -6x - 28$

17. $0.2x - 0.5 = 0.1x + 7$ **18.** $0.01x + 3.1 = 2.03x - 2.96$

19. $-4(2x - 6) + 8x = 5x + 24 + x$ **20.** $-8(3x + 4) + 6x = 4(x - 8) + 4x$

Determine whether each equation is an identity, *a* conditional equation, *or a* contradiction. *Give the solution set.* ***See Example 3.***

21. $4(2x + 7) = 2x + 22 + 3(2x + 2)$ **22.** $\frac{1}{2}(6x + 20) = x + 4 + 2(x + 3)$

23. $2(x - 8) = 3x - 16$ **24.** $-8(x + 5) = -8x - 5(x + 8)$

25. $4(x + 7) = 2(x + 12) + 2(x + 1)$ **26.** $-6(2x + 1) - 3(x - 4) = -15x + 1$

Concept Check Use choices A–D to answer each question.

A. $3x^2 - 17x - 6 = 0$ **B.** $(2x + 5)^2 = 7$

C. $x^2 + x = 12$ **D.** $(3x - 1)(x - 7) = 0$

27. Which equation is set up for direct use of the zero-factor property? Solve it.

28. Which equation is set up for direct use of the square root property? Solve it.

29. Which one or more of these equations can be solved using the quadratic formula?

30. Only one of the equations is set up so that the values of a, b, and c can be determined immediately. Which one is it? Solve it.

Solve each equation using the zero-factor property. ***See Example 4.***

31. $x^2 - 5x + 6 = 0$ **32.** $x^2 + 2x - 8 = 0$ **33.** $5x^2 - 3x - 2 = 0$

34. $2x^2 - x - 15 = 0$ **35.** $-4x^2 + x = -3$ **36.** $-6x^2 + 7x = -10$

37. $x^2 - 100 = 0$ **38.** $x^2 - 64 = 0$ **39.** $4x^2 - 4x + 1 = 0$

40. $9x^2 - 12x + 4 = 0$ **41.** $25x^2 + 30x + 9 = 0$ **42.** $36x^2 + 60x + 25 = 0$

Solve each equation using the square root property. ***See Example 5.***

43. $x^2 = 16$ **44.** $x^2 = 121$ **45.** $x^2 - 27 = 0$

46. $x^2 - 48 = 0$ **47.** $(3x - 1)^2 = 12$ **48.** $(4x + 1)^2 = 20$

49. $\{1, 3\}$ 50. $\{3, 4\}$

51. $\left\{-\frac{7}{2}, 4\right\}$ 52. $\left\{-\frac{5}{4}, 2\right\}$

53. $\left\{1 \pm \sqrt{3}\right\}$ 54. $\left\{5 \pm \sqrt{7}\right\}$

55. $\left\{3 \pm \sqrt{2}\right\}$ 56. $\left\{2 \pm \sqrt{3}\right\}$

57. $\left\{\frac{1 \pm \sqrt{5}}{2}\right\}$ 58. $\left\{\frac{3 \pm \sqrt{17}}{2}\right\}$

59. $\left\{\frac{2 \pm \sqrt{10}}{2}\right\}$ 60. $\left\{\frac{3 \pm 2\sqrt{6}}{3}\right\}$

61. F 62. J
63. A 64. H
65. I 66. D
67. B 68. G
69. E 70. C

71. A square bracket is used to show that a number is part of the solution set, and a parenthesis is used to indicate that a number is not part of the solution set.
72. D

73. $[-4, \infty)$ 74. $[-5, \infty)$
75. $[-1, \infty)$ 76. $(-\infty, 1]$
77. $(-\infty, \infty)$ 78. $(-\infty, \infty)$
79. $(-\infty, 4)$ 80. $\left(-\infty, \frac{15}{7}\right)$
81. $\left[-\frac{11}{5}, \infty\right)$ 82. $\left(-\infty, \frac{1}{2}\right]$

83. $(-5, 3)$ 84. $(-3, 1)$
85. $[3, 6]$ 86. $\left[-\frac{3}{2}, 3\right]$
87. $(4, 6)$ 88. $\left(\frac{1}{6}, 1\right)$
89. $[-9, 9]$ 90. $[-12, 6]$

Solve each equation using the quadratic formula. ***See Example 6.***

49. $x^2 - 4x + 3 = 0$ **50.** $x^2 - 7x + 12 = 0$ **51.** $2x^2 - x - 28 = 0$

52. $4x^2 - 3x - 10 = 0$ **53.** $x^2 - 2x - 2 = 0$ **54.** $x^2 - 10x + 18 = 0$

55. $x^2 - 6x = -7$ **56.** $x^2 - 4x = -1$ **57.** $x^2 - x - 1 = 0$

58. $x^2 - 3x - 2 = 0$ **59.** $-2x^2 + 4x + 3 = 0$ **60.** $-3x^2 + 6x + 5 = 0$

Concept Check *Match the inequality in each exercise in Column I with its equivalent interval notation in Column II.*

I	II
61. $x < -6$	**A.** $(-2, 6]$
62. $x \leq 6$	**B.** $[-2, 6)$
63. $-2 < x \leq 6$	**C.** $(-\infty, -6]$
64. $x^2 \geq 0$	**D.** $[6, \infty)$
65. $x \geq -6$	**E.** $(-\infty, -3) \cup (3, \infty)$
66. $6 \leq x$	**F.** $(-\infty, -6)$
67. 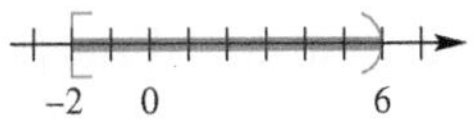	**G.** $(0, 8)$
68. 	**H.** $(-\infty, \infty)$
69. 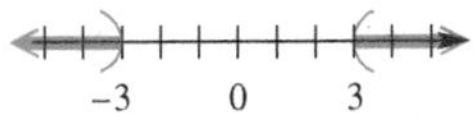	**I.** $[-6, \infty)$
70. 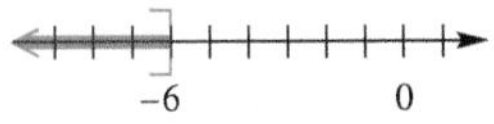	**J.** $(-\infty, 6]$

71. Explain how to determine whether to use a parenthesis or a square bracket when writing the solution set of a linear inequality in interval notation.

72. ***Concept Check*** The three-part inequality $a < x < b$ means "a is less than x and x is less than b." Which inequality is *not* satisfied by some real number x?

A. $-3 < x < 10$ **B.** $0 < x < 6$

C. $-3 < x < -1$ **D.** $-8 < x < -10$

Solve each inequality. Give the solution set in interval notation. ***See Example 7.***

73. $-2x + 8 \leq 16$ **74.** $-3x - 8 \leq 7$

75. $-2x - 2 \leq 1 + x$ **76.** $-4x + 3 \geq -2 + x$

77. $3(x + 5) + 1 \geq 5 + 3x$ **78.** $6x - (2x + 3) \geq 4x - 5$

79. $8x - 3x + 2 < 2(x + 7)$ **80.** $2 - 4x + 5(x - 1) < -6(x - 2)$

81. $\frac{4x + 7}{-3} \leq 2x + 5$ **82.** $\frac{2x - 5}{-8} \leq 1 - x$

Solve each inequality. Give the solution set in interval notation. ***See Example 8.***

83. $-5 < 5 + 2x < 11$ **84.** $-7 < 2 + 3x < 5$

85. $10 \leq 2x + 4 \leq 16$ **86.** $-6 \leq 6x + 3 \leq 21$

87. $-11 > -3x + 1 > -17$ **88.** $2 > -6x + 3 > -3$

89. $-4 \leq \frac{x + 1}{2} \leq 5$ **90.** $-5 \leq \frac{x - 3}{3} \leq 1$

B Graphs of Equations

- The Rectangular Coordinate System
- Equations in Two Variables
- Circles

The Rectangular Coordinate System Each real number corresponds to a point on a number line. This idea is extended to **ordered pairs** of real numbers by using two perpendicular number lines, one horizontal and one vertical, that intersect at their zero-points. The point of intersection is the **origin.** The horizontal line is the ***x*-axis,** and the vertical line is the ***y*-axis.** See **Figure 1.**

The x-axis and y-axis together make up a **rectangular coordinate system,** or **Cartesian coordinate system** (named for one of its coinventors, René Descartes. The other coinventor was Pierre de Fermat). The plane into which the coordinate system is introduced is the **coordinate plane,** or ***xy*-plane.** See **Figure 1.** The x-axis and y-axis divide the plane into four regions, or **quadrants,** labeled as shown. The points on the x-axis or the y-axis belong to no quadrant.

Each point P in the xy-plane corresponds to a unique ordered pair (a, b) of real numbers. The point P corresponding to the ordered pair (a, b) often is written $P(a, b)$ as in **Figure 1** and referred to as

"the point (a, b)."

The numbers a and b are the **coordinates** of point P.

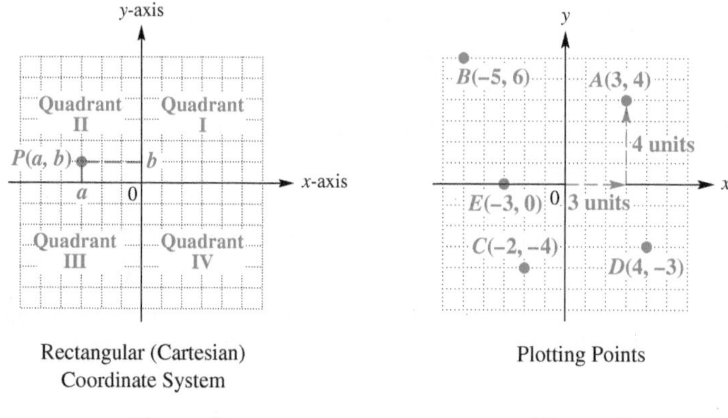

Rectangular (Cartesian) Coordinate System

Figure 1

Plotting Points

Figure 2

To locate on the xy-plane the point corresponding to the ordered pair $(3, 4)$, for example, start at the origin, move 3 units in the positive x-direction, and then move 4 units in the positive y-direction. See **Figure 2.** Point A corresponds to the ordered pair $(3, 4)$.

Equations in Two Variables Ordered pairs are used to express the solutions of equations in two variables. When an ordered pair represents the solution of an equation with the variables x and y, the x-value is written first. For example, we say that

$$(1, 2) \quad \text{is a solution of} \quad 2x - y = 0.$$

Substituting 1 for x and 2 for y in the equation gives a true statement.

$$2x - y = 0$$

$$2(1) - 2 \stackrel{?}{=} 0 \qquad \text{Let } x = 1 \text{ and } y = 2.$$

$$0 = 0 \checkmark \qquad \text{True}$$

Classroom Example 1
For each equation, find at least three ordered pairs that are solutions.

(a) $y = -2x + 5$

(b) $x = \sqrt[3]{y + 1}$

(c) $y = -x^2 + 1$

Answers:

(a) $(-1, 7), (0, 5), (3, -1)$

(b) $(1, 0), (-1, -2), (2, 7), (0, -1), (-2, -9)$

(c) $(0, 1), (-1, 0), (1, 0), (-2, -3), (2, -3)$

(Other answers are possible.)

EXAMPLE 1 Finding Ordered-Pair Solutions of Equations

For each equation, find at least three ordered pairs that are solutions.

(a) $y = 4x - 1$ **(b)** $x = \sqrt{y - 1}$ **(c)** $y = x^2 - 4$

SOLUTION

(a) Choose any real number for x or y, and substitute in the equation to obtain the corresponding value of the other variable. For example, let $x = -2$ and then let $y = 3$.

$y = 4x - 1$		$y = 4x - 1$	
$y = 4(-2) - 1$	Let $x = -2$.	$3 = 4x - 1$	Let $y = 3$.
$y = -8 - 1$	Multiply.	$4 = 4x$	Add 1.
$y = -9$	Subtract.	$1 = x$	Divide by 4.

This gives the ordered pairs $(-2, -9)$ and $(1, 3)$. Verify that the ordered pair $(0, -1)$ is also a solution.

(b)

$x = \sqrt{y - 1}$	Given equation
$1 = \sqrt{y - 1}$	Let $x = 1$.
$1 = y - 1$	Square each side.
$2 = y$	Add 1.

One ordered pair is $(1, 2)$. Verify that the ordered pairs $(0, 1)$ and $(2, 5)$ are also solutions of the equation.

(c) A table provides an organized method for determining ordered pairs. Here, we let x equal $-2, -1, 0, 1$, and 2 in

$$y = x^2 - 4$$

and determine the corresponding y-values.

x	y	
-2	0	$(-2)^2 - 4 = 4 - 4 = 0$
-1	-3	$(-1)^2 - 4 = 1 - 4 = -3$
0	-4	$0^2 - 4 = -4$
1	-3	$1^2 - 4 = -3$
2	0	$2^2 - 4 = 0$

Five ordered pairs are $(-2, 0), (-1, -3), (0, -4), (1, -3)$, and $(2, 0)$.

✔ Now Try Exercises 15(a), 19(a), and 21(a).

The **graph** of an equation is found by plotting ordered pairs that are solutions of the equation. The **intercepts** of the graph are good points to plot first. An ***x*-intercept** is a point where the graph intersects the x-axis, and a ***y*-intercept** is a point where the graph intersects the y-axis. In other words, the x-intercept is represented by an ordered pair with y-coordinate 0, and the y-intercept is represented by an ordered pair with x-coordinate 0.

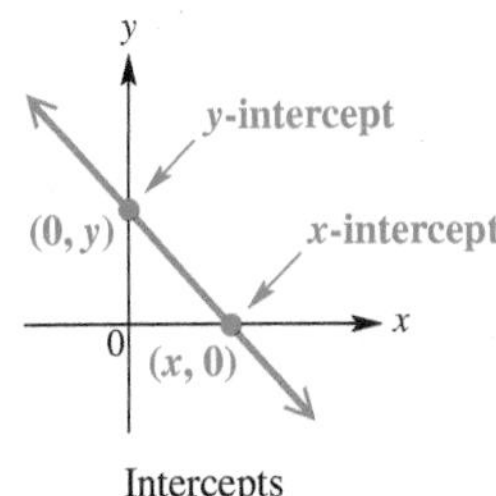

Intercepts

A general algebraic approach for graphing an equation using intercepts and point-plotting follows.

Graphing an Equation by Point Plotting

Step 1 Find the intercepts.

Step 2 Find as many additional ordered pairs as needed.

Step 3 Plot the ordered pairs from Steps 1 and 2.

Step 4 Join the points from Step 3 with a smooth line or curve.

Classroom Example 2
Graph each of the equations here, from **Classroom Example 1.**

(a) $y = -2x + 5$

(b) $x = \sqrt[3]{y + 1}$

(c) $y = -x^2 + 1$

Answers:

(a) (b) (c)

EXAMPLE 2 Graphing Equations

Graph each of the equations here, from **Example 1.**

(a) $y = 4x - 1$ **(b)** $x = \sqrt{y - 1}$ **(c)** $y = x^2 - 4$

SOLUTION

(a) *Step 1* Let $y = 0$ to find the x-intercept, and let $x = 0$ to find the y-intercept.

$y = 4x - 1$	$y = 4x - 1$
$0 = 4x - 1$ Let $y = 0$.	$y = 4(0) - 1$ Let $x = 0$.
$1 = 4x$	$y = 0 - 1$
$\frac{1}{4} = x$	$y = -1$

The intercepts are $\left(\frac{1}{4}, 0\right)$ and $(0, -1)$.*

Step 2 We use the intercepts and the other ordered pairs found in **Example 1(a):** $(-2, -9)$ and $(1, 3)$.

Step 3 Plot the four ordered pairs from Steps 1 and 2 as shown in **Figure 3.**

Step 4 Join the points plotted in Step 3 with a straight line. This line, also shown in **Figure 3,** is the graph of the equation $y = 4x - 1$.

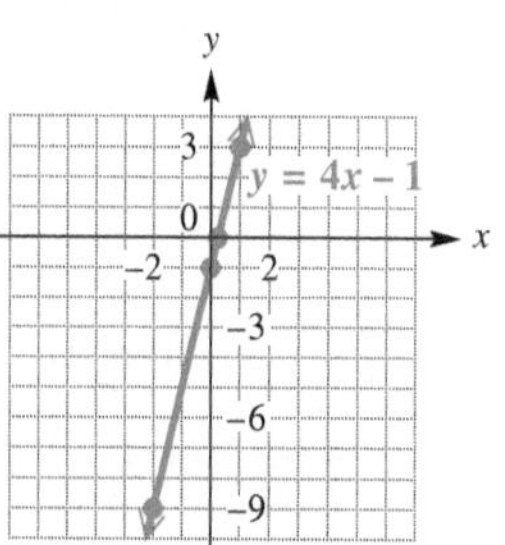

Figure 3

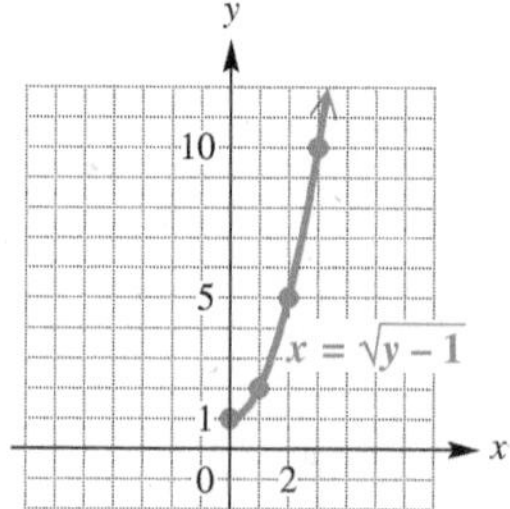

Figure 4

(b) For $x = \sqrt{y - 1}$, the y-intercept $(0, 1)$ was found in **Example 1(b).** Solve

$$x = \sqrt{0 - 1} \quad \text{Let } y = 0.$$

to find the x-intercept. When $y = 0$, the quantity under the radical symbol is negative, so there is no x-intercept. In fact, $y - 1$ must be greater than or equal to 0, so y must be greater than or equal to 1.

We plot the ordered pairs $(0, 1)$, $(1, 2)$, and $(2, 5)$ from **Example 1(b)** and join the points with a smooth curve as in **Figure 4.** To confirm the direction the curve will take as x increases, we find another solution, $(3, 10)$.

*Intercepts are sometimes defined as numbers, such as x-intercept $\frac{1}{4}$ and y-intercept -1. In this text, we define them as ordered pairs, such as $\left(\frac{1}{4}, 0\right)$ and $(0, -1)$.

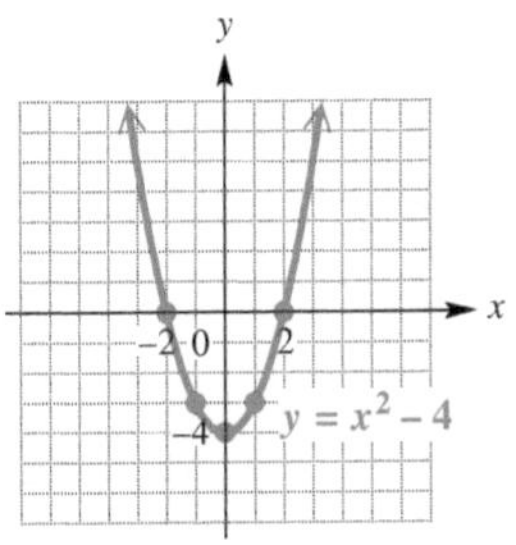

Figure 5

(c) In **Example 1(c),** we made a table of five ordered pairs that satisfy the equation $y = x^2 - 4$.

$$(-2, 0), \quad (-1, -3), \quad (0, -4), \quad (1, -3), \quad (2, 0)$$

↑ x-intercept ↑ y-intercept ↑ x-intercept

Plotting the points and joining them with a smooth curve gives the graph in **Figure 5.** This curve is called a **parabola.**

✔ **Now Try Exercises 15(b), 19(b), and 21(b).**

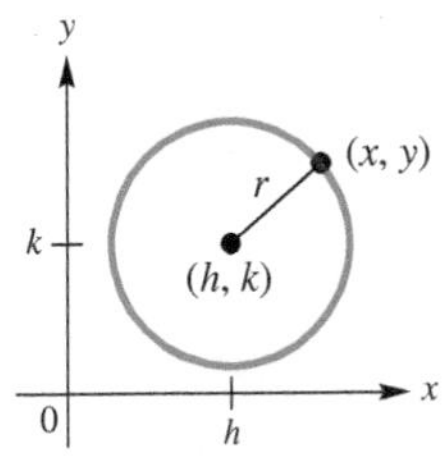

Figure 6

Circles By definition, a **circle** is the set of all points in a plane that lie a given distance from a given point. The given distance is the **radius** of the circle, and the given point is the **center.**

We can find the equation of a circle from its definition using the distance formula. Suppose that the point (h, k) is the center and the circle has radius r, where $r > 0$. Let (x, y) represent any point on the circle. See **Figure 6.**

$$\sqrt{(x_2 - x_1)^2 + (y_2 - y_1)^2} = d \qquad \text{Distance formula}$$

$$\sqrt{(x - h)^2 + (y - k)^2} = r \qquad (x_1, y_1) = (h, k), (x_2, y_2) = (x, y), \text{ and } d = r$$

$$(x - h)^2 + (y - k)^2 = r^2 \qquad \text{Square each side.}$$

Center-Radius Form of the Equation of a Circle

A circle with center (h, k) and radius r has equation

$$(x - h)^2 + (y - k)^2 = r^2,$$

which is the **center-radius form** of the equation of the circle. As a special case, a circle with center $(0, 0)$ and radius r has the following equation.

$$x^2 + y^2 = r^2$$

Classroom Example 3
Find the center-radius form of the equation of each circle described.
(a) center $(1, -2)$, radius 3
(b) center $(0, 0)$, radius 2

Answers:
(a) $(x - 1)^2 + (y + 2)^2 = 9$
(b) $x^2 + y^2 = 4$

EXAMPLE 3 Finding the Center-Radius Form

Find the center-radius form of the equation of each circle described.

(a) center $(-3, 4)$, radius 6 **(b)** center $(0, 0)$, radius 3

SOLUTION

(a)

$$(x - h)^2 + (y - k)^2 = r^2 \qquad \text{Center-radius form}$$

$$[x - (-3)]^2 + (y - 4)^2 = 6^2 \qquad \text{Substitute. Let } (h, k) = (-3, 4) \text{ and } r = 6.$$

$$(x + 3)^2 + (y - 4)^2 = 36 \qquad \text{Simplify.}$$

Be careful with signs here.

(b) The center is the origin and $r = 3$.

$$x^2 + y^2 = r^2 \qquad \text{Special case of the center-radius form}$$

$$x^2 + y^2 = 3^2 \qquad \text{Let } r = 3.$$

$$x^2 + y^2 = 9 \qquad \text{Apply the exponent.}$$

✔ **Now Try Exercises 35(a) and 41(a).**

Classroom Example 4
Graph each circle discussed in **Classroom Example 3.**

(a) $(x-1)^2+(y+2)^2=9$
(b) $x^2+y^2=4$

Answers:

(a)

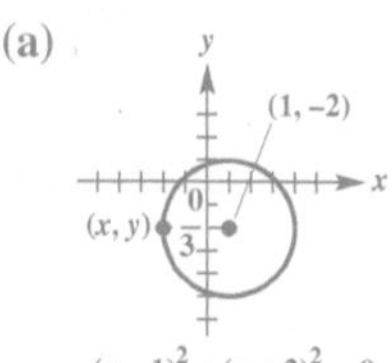

(b)

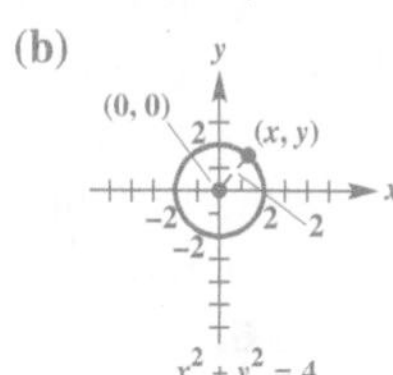

EXAMPLE 4 Graphing Circles

Graph each circle discussed in **Example 3.**

(a) $(x+3)^2+(y-4)^2=36$ **(b)** $x^2+y^2=9$

SOLUTION

(a) Writing the given equation in center-radius form

$$[x-(-3)]^2+(y-4)^2=6^2$$

gives $(-3, 4)$ as the center and 6 as the radius. See **Figure 7.**

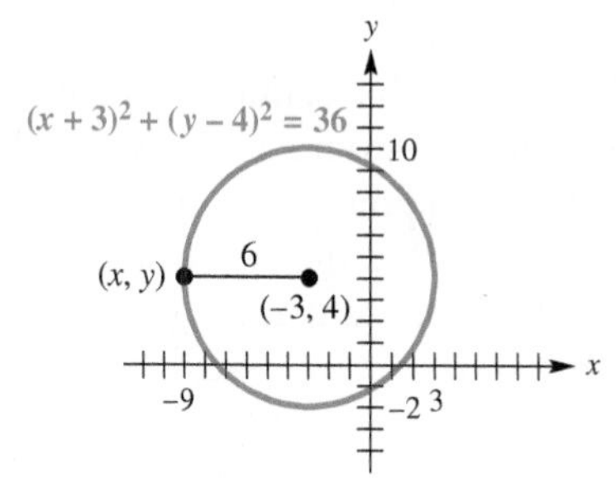

Figure 7

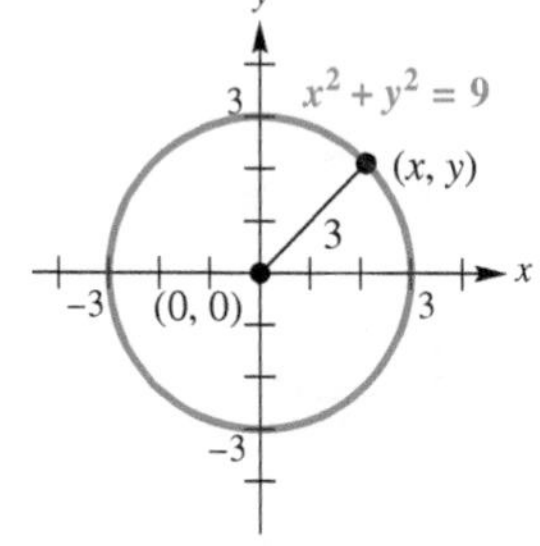

Figure 8

(b) The graph with center $(0, 0)$ and radius 3 is shown in **Figure 8.**

✔ **Now Try Exercises 35(b) and 41(b).**

Appendix B Exercises

1. II 2. 6
3. 0 4. $(0, 6)$
5. $(5, 0)$
6. any three of the following: $(2, -5)$, $(-1, 7)$, $(3, -9)$, $(5, -17)$, $(6, -21)$

7.–14.
II (−7.5, 8) y (4.5, 7) I
(−7, 6)
(0, 5) y-axis
x-axis (3, 2)
(−8, 0) 0 x
(−7, −4) (8, −5)
III IV

Other ordered pairs are possible in Exercises 15–26.

15. (a)

x	y
0	−2
4	0
2	−1

(b)

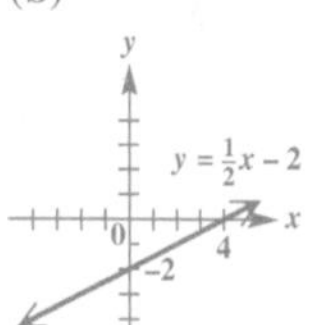

Concept Check Fill in the blank to correctly complete each sentence.

1. The point $(-1, 3)$ lies in quadrant ________ in the rectangular coordinate system.

2. The point $(4, \_\_\_)$ lies on the graph of the equation $y=3x-6$.

3. Any point that lies on the x-axis has y-coordinate equal to ________.

4. The y-intercept of the graph of $y=-2x+6$ is ________.

5. The x-intercept of the graph of $2x+5y=10$ is ________.

6. Give three ordered pairs from the table.

x	y
2	−5
−1	7
3	−9
5	−17
6	−21

Concept Check Graph the points on a coordinate system and identify the quadrant or axis for each point.

7. $(3, 2)$ **8.** $(-7, 6)$ **9.** $(-7, -4)$ **10.** $(8, -5)$

11. $(0, 5)$ **12.** $(-8, 0)$ **13.** $(4.5, 7)$ **14.** $(-7.5, 8)$

*For each equation, **(a)** give a table with at least three ordered pairs that are solutions, and **(b)** graph the equation. **See Examples 1 and 2.***

15. $y=\frac{1}{2}x-2$ **16.** $y=-\frac{1}{2}x+2$ **17.** $2x+3y=5$

18. $3x-2y=6$ **19.** $y=x^2$ **20.** $y=x^2+2$

16. (a)

x	y
0	2
4	0
−4	4

(b)

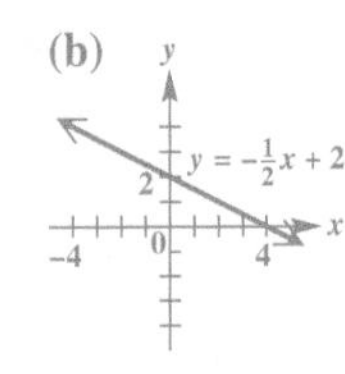

17. (a)

x	y
0	$\frac{5}{3}$
$\frac{5}{2}$	0
4	−1

(b)

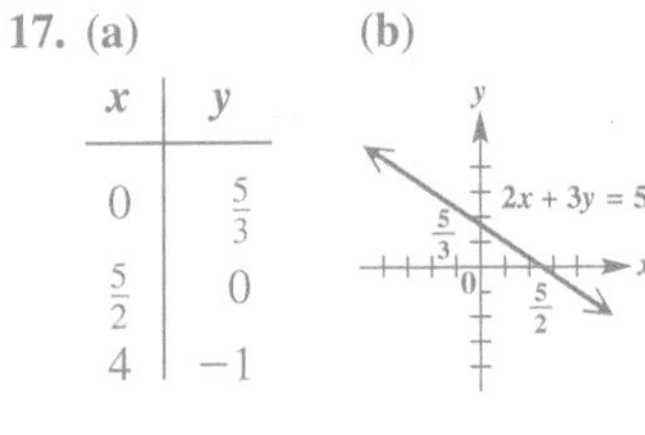

18. (a)

x	y
0	−3
2	0
4	3

(b)

19. (a)

x	y
0	0
1	1
−2	4

(b)

20. (a)

x	y
0	2
−1	3
2	6

(b)

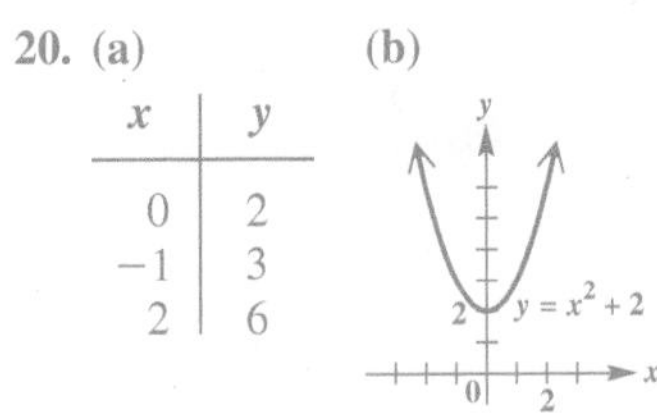

21. (a)

x	y
3	0
4	1
7	2

(b)

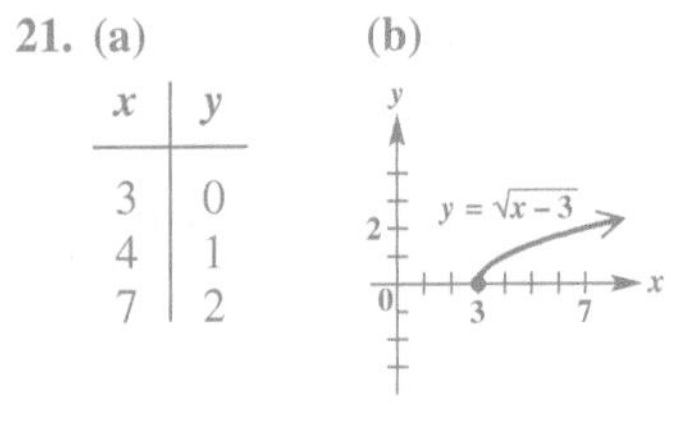

22. (a)

x	y
0	−3
4	−1
9	0

(b)

23. (a)

x	y
4	2
−2	4
0	2

(b)

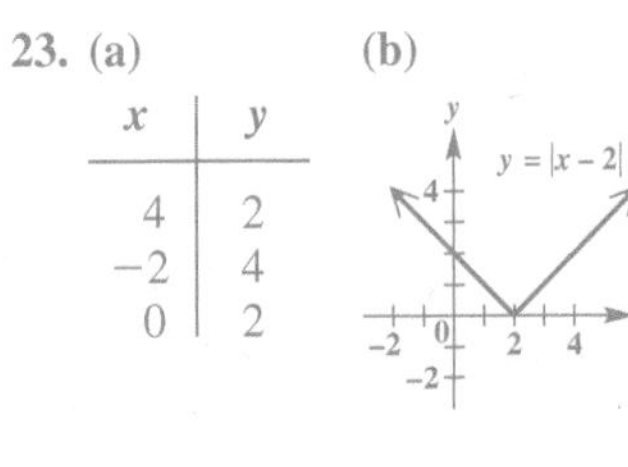

Answers to the remaining exercises are given in the answer section at the back of the text.

21. $y = \sqrt{x-3}$ **22.** $y = \sqrt{x} - 3$ **23.** $y = |x-2|$

24. $y = -|x+4|$ **25.** $y = x^3$ **26.** $y = -x^3$

Concept Check Fill in the blank(s) to correctly complete each sentence.

27. The circle with equation $x^2 + y^2 = 49$ has center with coordinates ______ and radius equal to ______.

28. The circle with center $(3, 6)$ and radius 4 has equation ______.

29. The graph of $(x-4)^2 + (y+7)^2 = 9$ has center with coordinates ______.

30. The graph of $x^2 + (y-5)^2 = 9$ has center with coordinates ______.

Concept Check Match each equation in Column I with its graph in Column II.

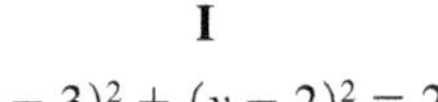
I

II

31. $(x-3)^2 + (y-2)^2 = 25$

32. $(x-3)^2 + (y+2)^2 = 25$

33. $(x+3)^2 + (y-2)^2 = 25$

34. $(x+3)^2 + (y+2)^2 = 25$

A.

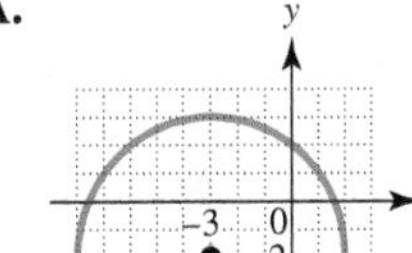

B.

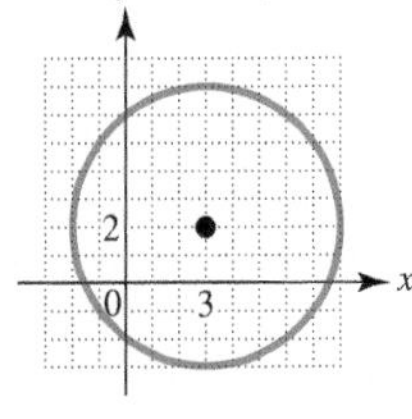

C.

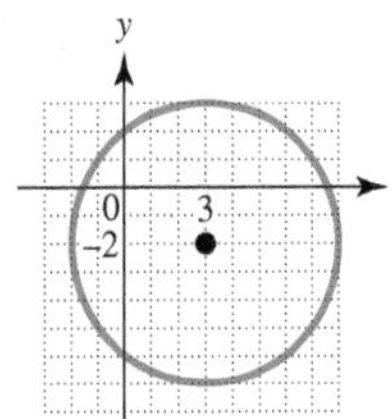

D.

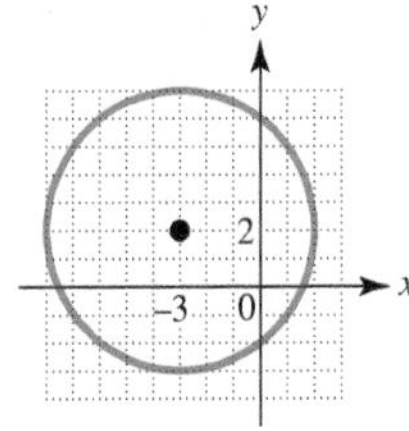

In the following exercises, ***(a)*** *find the center-radius form of the equation of each circle described, and* ***(b)*** *graph it. See* ***Examples 3 and 4.***

35. center $(0, 0)$, radius 6

36. center $(0, 0)$, radius 9

37. center $(2, 0)$, radius 6

38. center $(3, 0)$, radius 3

39. center $(0, 4)$, radius 4

40. center $(0, -3)$, radius 7

41. center $(-2, 5)$, radius 4

42. center $(4, 3)$, radius 5

43. center $(5, -4)$, radius 7

44. center $(-3, -2)$, radius 6

45. center $(\sqrt{2}, \sqrt{2})$, radius $\sqrt{2}$

46. center $(-\sqrt{3}, -\sqrt{3})$, radius $\sqrt{3}$

Connecting Graphs with Equations Use each graph to determine an equation of the circle in center-radius form.

47.

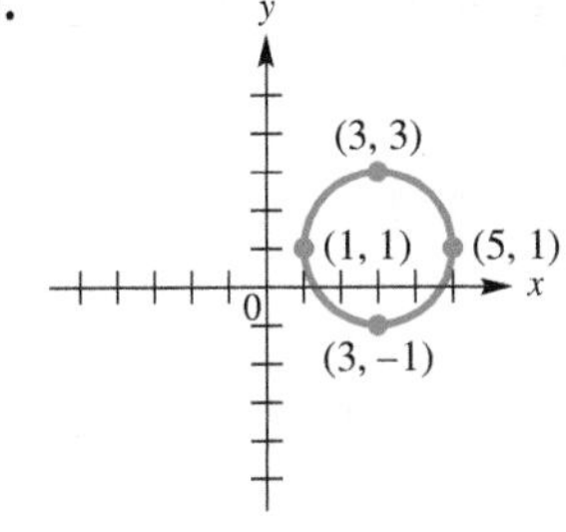

48.

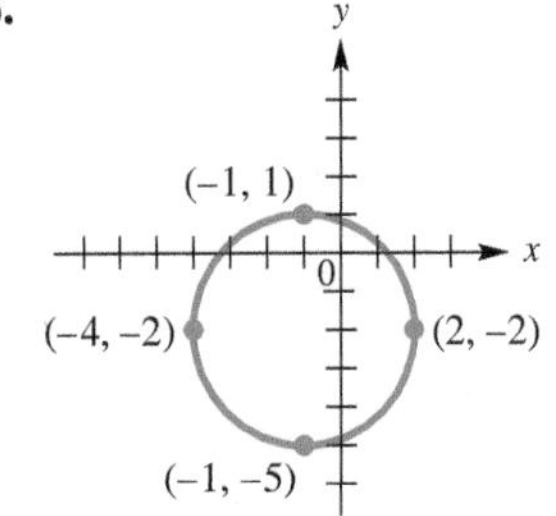

C Functions

Relations and Functions In algebra, we use ordered pairs to represent related quantities. For example, (3, \$10.50) might indicate that we pay \$10.50 for 3 gallons of gas. The amount we pay *depends* on the number of gallons pumped, so the amount (in dollars) is called the *dependent variable,* and the number of gallons pumped is called the *independent variable.*

Generalizing, if the value of the second component y depends on the value of the first component x, then y is the **dependent variable** and x is the **independent variable.**

Independent variable → x, Dependent variable → y

$$(x, y)$$

Teaching Tip An excellent classroom illustration of the *function* concept is the activity of pumping gasoline. For every number of gallons that appears on the pump display, there is one and only one price displayed. The same illustration can be used to explain *domain, range, components, independent variable,* and *dependent variable.*

A set of ordered pairs such as $\{(3, 10.50), (8, 28.00), (10, 35.00)\}$ is a *relation.* A *function* is a special kind of relation.

Relation and Function

A **relation** is a set of ordered pairs. A **function** is a relation in which, for each distinct value of the first component of the ordered pairs, there is *exactly one* value of the second component.

EXAMPLE 1 Deciding Whether Relations Define Functions

Decide whether each relation defines a function.

$$F = \{(1, 2), (-2, 4), (3, 4)\}$$
$$G = \{(1, 1), (1, 2), (1, 3), (2, 3)\}$$
$$H = \{(-4, 1), (-2, 1), (-2, 0)\}$$

Classroom Example 1
Decide whether each relation defines a function.

$M = \{(-4, 0), (-3, 1), (3, 1)\}$
$N = \{(2, 3), (3, 2), (4, 5), (5, 4)\}$
$P = \{(-4, 3), (0, 6), (2, 8), (-4, -3)\}$

Answer: M is a function; N is a function; P is not a function.

SOLUTION Relation F is a function because for each different x-value there is exactly one y-value. We can show this correspondence as follows.

$$\{1, -2, 3\} \quad x\text{-values of } F$$
$$\downarrow \quad \downarrow \quad \downarrow$$
$$\{2, 4, 4\} \quad y\text{-values of } F$$

As the correspondence below shows, relation G is not a function because one first component corresponds to *more than one* second component.

$$\{1, 2\} \quad x\text{-values of } G$$
$$\{1, 2, 3\} \quad y\text{-values of } G$$

In relation H the last two ordered pairs have the same x-value paired with two different y-values (-2 is paired with both 1 and 0), so H is a relation but not a function. ***In a function, no two ordered pairs can have the same first component and different second components.***

Different y-values

$$H = \{(-4, 1), (-2, 1), (-2, 0)\} \quad \text{Not a function}$$

Same x-value

✓ **Now Try Exercises 1 and 3.**

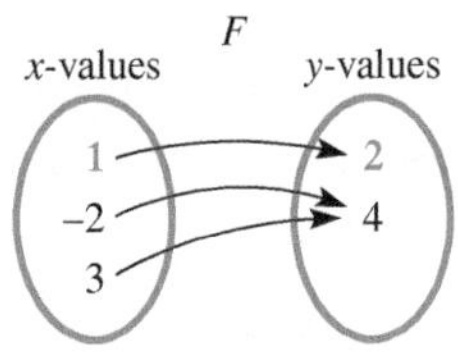

F is a function.

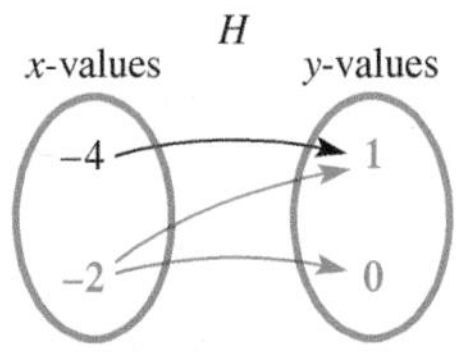

H is not a function.

Figure 1

Relations and functions can also be expressed as a correspondence or *mapping* from one set to another, as shown in **Figure 1** for function F and relation H from **Example 1.** The arrow from 1 to 2 indicates that the ordered pair $(1, 2)$ belongs to F—each first component is paired with exactly one second component. In the mapping for relation H, which is not a function, the first component -2 is paired with two different second components, 1 and 0.

Because relations and functions are sets of ordered pairs, we can represent them using tables and graphs. A table and graph for function F are shown in **Figure 2.**

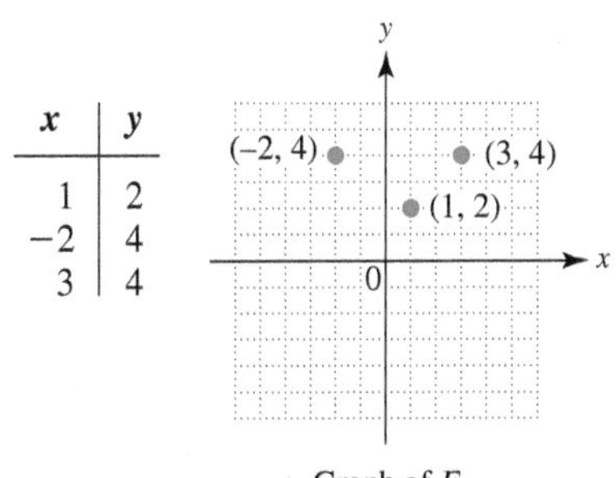

x	y
1	2
−2	4
3	4

Graph of *F*

Figure 2

Finally, we can describe a relation or function using a rule that tells how to determine the dependent variable for a specific value of the independent variable. The rule may be given in words: for instance, "the dependent variable is twice the independent variable." Usually the rule is an equation, such as the one below.

Teaching Tip In the gasoline-pumping illustration, the *price* of the amount pumped *depends* on the number of gallons pumped.

$$\text{Dependent variable} \longrightarrow y = 2x \longleftarrow \text{Independent variable}$$

In a function, there is exactly one value of the dependent variable, the second component, for each value of the independent variable, the first component.

Domain and Range We consider two important concepts concerning relations.

Domain and Range

For every relation consisting of a set of ordered pairs (x, y), there are two important sets of elements.

- The set of all values of the independent variable (x) is the **domain.**
- The set of all values of the dependent variable (y) is the **range.**

Classroom Example 2
Give the domain and range of each relation. Tell whether the relation defines a function.

(a) $\{(-4, -2), (-1, 0), (1, 2), (3, 5)\}$

(b) 1, 2, 3 → 4, 5, 6, 7

(c)

x	y
−3	5
0	5
3	5
5	5

Answers:
(a) domain: $\{-4, -1, 1, 3\}$; range: $\{-2, 0, 2, 5\}$; function
(b) domain: $\{1, 2, 3\}$; range: $\{4, 5, 6, 7\}$; not a function
(c) domain: $\{-3, 0, 3, 5\}$; range: $\{5\}$; function

EXAMPLE 2 Finding Domains and Ranges of Relations

Give the domain and range of each relation. Tell whether the relation defines a function.

(a) $\{(3, -1), (4, 2), (4, 5), (6, 8)\}$

(b)

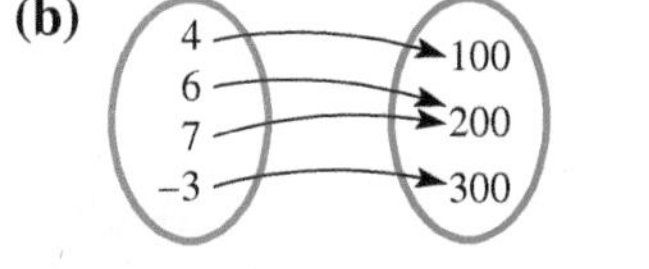

(c)

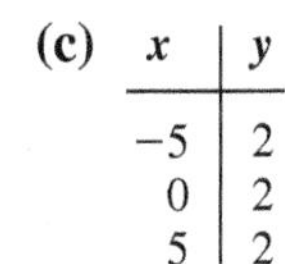

x	y
−5	2
0	2
5	2

SOLUTION

(a) The domain is the set of x-values, $\{3, 4, 6\}$. The range is the set of y-values, $\{-1, 2, 5, 8\}$. This relation is not a function because the same x-value, 4, is paired with two different y-values, 2 and 5.

(b) The domain is $\{4, 6, 7, -3\}$ and the range is $\{100, 200, 300\}$. This mapping defines a function. Each x-value corresponds to exactly one y-value.

x	y
-5	2
0	2
5	2

(c) This relation, represented by a table, is a set of ordered pairs. The domain is the set of x-values $\{-5, 0, 5\}$, and the range is the set of y-values $\{2\}$. The table defines a function because each different x-value corresponds to exactly one y-value (even though it is the same y-value).

Now Try Exercises 9, 11, and 13.

Classroom Example 3
Give the domain and range of each relation.

(a) (b)

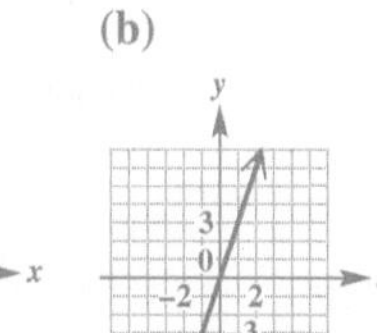

(c) (d)

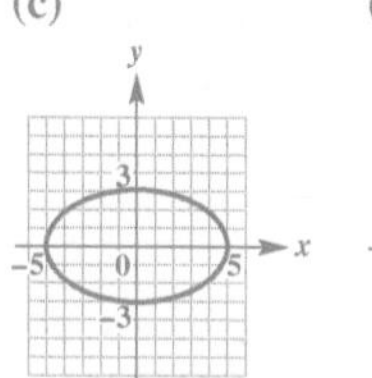

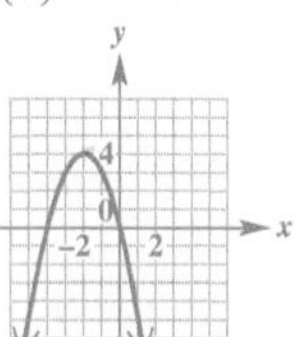

Answers:
(a) domain: $\{-2, 4\}$; range: $\{0, 3\}$
(b) domain: $(-\infty, \infty)$; range: $(-\infty, \infty)$
(c) domain: $[-5, 5]$; range: $[-3, 3]$
(d) domain: $(-\infty, \infty)$; range: $(-\infty, 4]$

Teaching Tip Use graphs of vertical and horizontal lines, parabolas, hyperbolas, and square root functions to show how to find domain and range. Give results in interval notation, paying special attention to the use of parentheses or brackets.

EXAMPLE 3 Finding Domains and Ranges from Graphs

Give the domain and range of each relation.

(a)

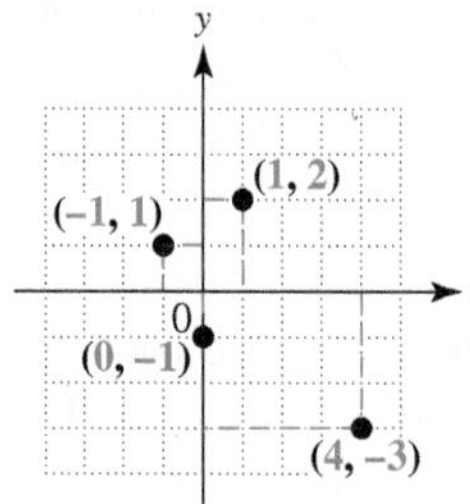

(b)

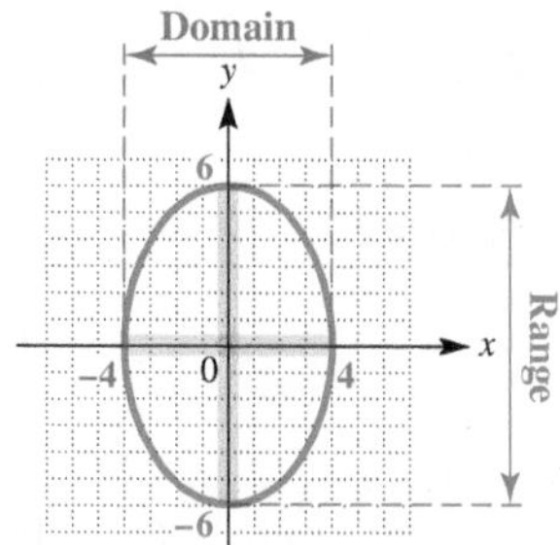

(c)

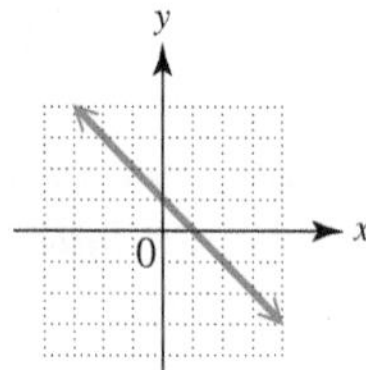

(d)

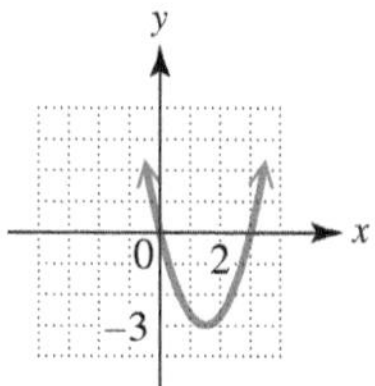

SOLUTION

(a) The domain is the set of x-values, $\{-1, 0, 1, 4\}$. The range is the set of y-values, $\{-3, -1, 1, 2\}$.

(b) The x-values of the points on the graph include all numbers between -4 and 4, inclusive. The y-values include all numbers between -6 and 6, inclusive.

The domain is $[-4, 4]$. The range is $[-6, 6]$. Use interval notation.

(c) The arrowheads indicate that the line extends indefinitely left and right, as well as up and down. Therefore, both the domain and the range include all real numbers, which is written

$$(-\infty, \infty).$$ Interval notation for the set of all real numbers

(d) The arrowheads indicate that the graph extends indefinitely left and right, as well as upward. The domain is $(-\infty, \infty)$. Because there is a least y-value, -3, the range includes all numbers greater than or equal to -3, written $[-3, \infty)$.

Now Try Exercise 19.

Determining Whether Relations Are Functions Because each value of x leads to only one value of y in a function, any vertical line must intersect the graph in at most one point. This is the **vertical line test** for a function.

Vertical Line Test

If every vertical line intersects the graph of a relation in no more than one point, then the relation is a function.

The graph in **Figure 3(a)** represents a function because each vertical line intersects the graph in no more than one point. The graph in **Figure 3(b)** is not the graph of a function because there exists a vertical line that intersects the graph in more than one point.

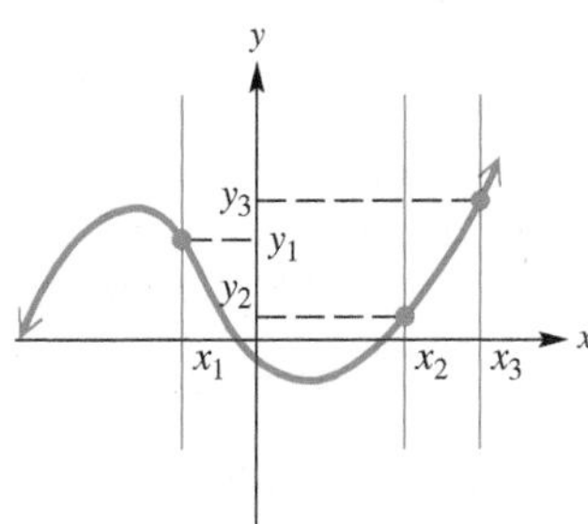

This is the graph of a function. Each x-value corresponds to only one y-value.

(a)

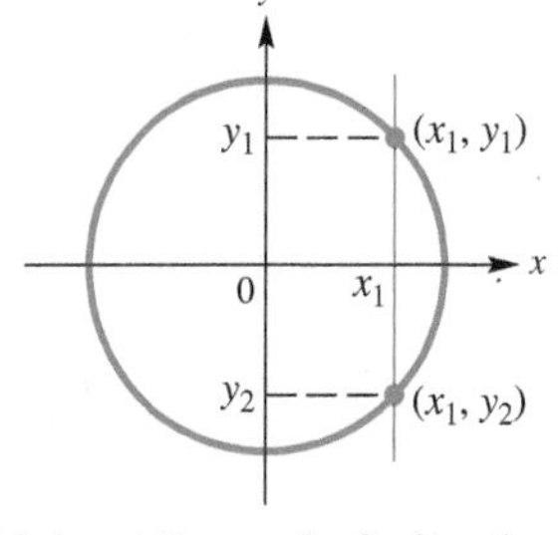

This is not the graph of a function. The same x-value corresponds to two different y-values.

(b)

Figure 3

Classroom Example 4
Use the vertical line test to determine whether each relation graphed in **Classroom Example 3** is a function.

Answers:
(a) not a function
(b) function
(c) not a function
(d) function

EXAMPLE 4 Using the Vertical Line Test

Use the vertical line test to determine whether each relation graphed in **Example 3** is a function.

SOLUTION We repeat each graph from **Example 3,** this time with vertical lines drawn through the graphs.

(a)

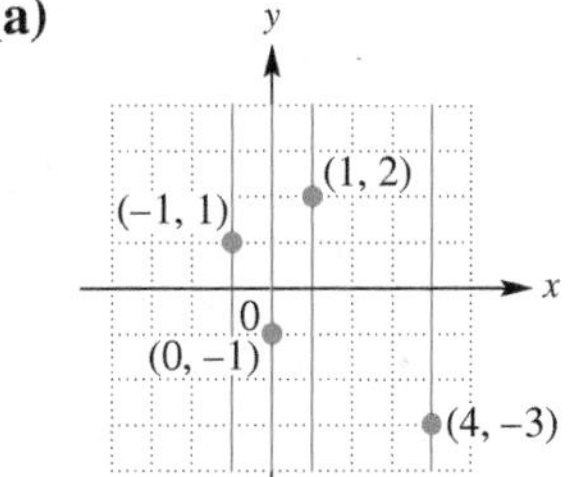

(b)

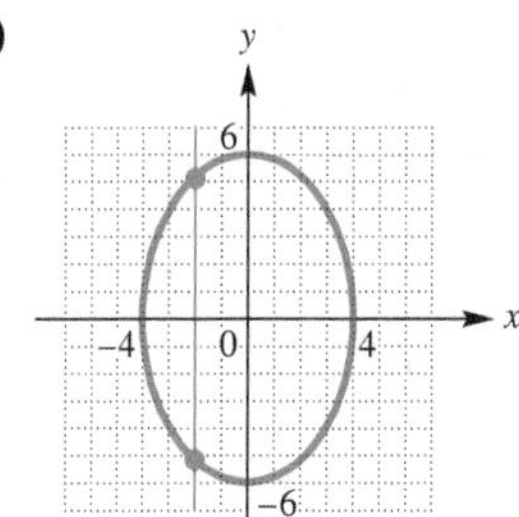

(c)

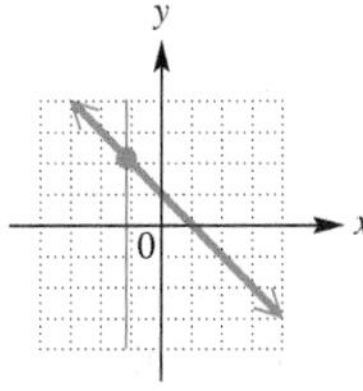

(d)

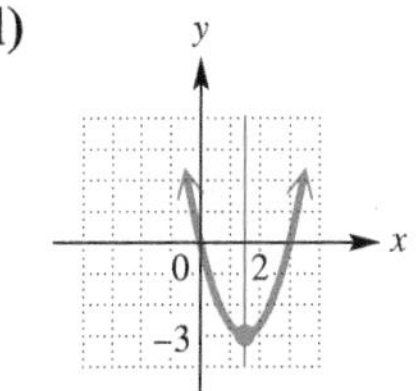

- The graphs of the relations in parts (a), (c), and (d) pass the vertical line test because every vertical line intersects each graph no more than once. Thus, these graphs represent functions.
- The graph of the relation in part (b) fails the vertical line test because the same x-value corresponds to two different y-values. Therefore, it is not the graph of a function.

Now Try Exercises 15 and 17.

The vertical line test is a simple method for identifying a function defined by a graph. Deciding whether a relation defined by an equation or an inequality is a function, as well as determining the domain and range, is more difficult. The next example gives some hints that may help.

Classroom Example 5
Decide whether each relation defines y as a function of x, and give the domain and range.

(a) $y = 2x - 5$ (b) $y = x^2 + 3$

(c) $x = |y|$ (d) $y = \frac{3}{x+2}$

Answers:

(a) function; domain: $(-\infty, \infty)$; range: $(-\infty, \infty)$

(b) function; domain: $(-\infty, \infty)$; range: $[3, \infty)$

(c) not a function; domain: $[0, \infty)$; range: $(-\infty, \infty)$

(d) function; domain: $(-\infty, -2) \cup (-2, \infty)$; range: $(-\infty, 0) \cup (0, \infty)$

EXAMPLE 5 Identifying Functions, Domains, and Ranges

Decide whether each relation defines y as a function of x, and give the domain and range.

(a) $y = x + 4$ **(b)** $y = \sqrt{2x - 1}$ **(c)** $y^2 = x$ **(d)** $y = \frac{5}{x - 1}$

SOLUTION

(a) In the defining equation (or rule), $y = x + 4$, y is always found by adding 4 to x. Thus, each value of x corresponds to just one value of y, and the relation defines a function. The variable x can represent any real number, so the domain is

$$\{x \mid x \text{ is a real number}\}, \quad \text{or} \quad (-\infty, \infty).$$

Because y is always 4 more than x, y also may be any real number, and so the range is $(-\infty, \infty)$.

(b) For any choice of x in the domain of $y = \sqrt{2x - 1}$, there is exactly one corresponding value for y (the radical is a nonnegative number), so this equation defines a function. The equation involves a square root, so the quantity under the radical sign cannot be negative.

$$2x - 1 \geq 0 \quad \text{Solve the inequality.}$$

$$2x \geq 1 \quad \text{Add 1.}$$

$$x \geq \frac{1}{2} \quad \text{Divide by 2.}$$

The domain of the function is $\left[\frac{1}{2}, \infty\right)$. Because the radical must represent a nonnegative number, as x takes values greater than or equal to $\frac{1}{2}$, the range is $\{y \mid y \geq 0\}$, or $[0, \infty)$. See **Figure 4.**

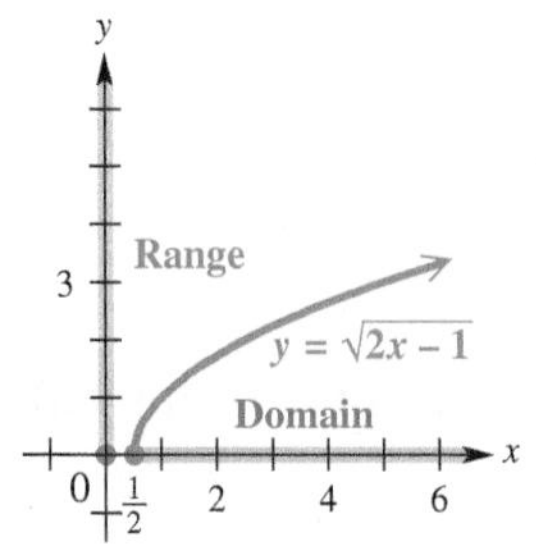

Figure 4

(c) The ordered pairs $(16, 4)$ and $(16, -4)$ both satisfy the equation $y^2 = x$. There exists at least one value of x—for example, 16—that corresponds to two values of y, 4 and -4, so this equation does not define a function.

Because x is equal to the square of y, the values of x must always be nonnegative. The domain of the relation is $[0, \infty)$. Any real number can be squared, so the range of the relation is $(-\infty, \infty)$. See **Figure 5.**

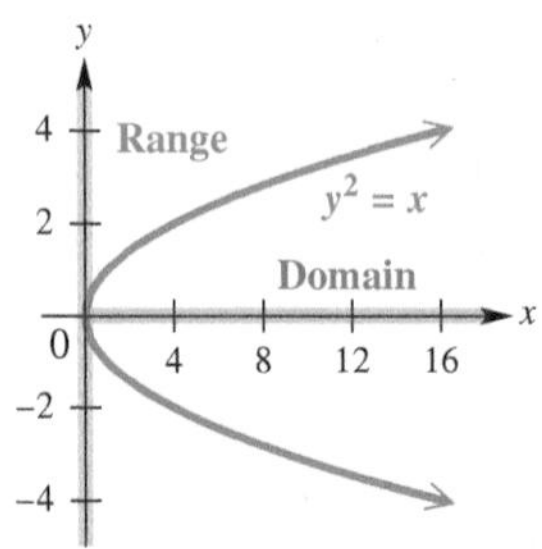

Figure 5

(d) Given any value of x in the domain of

$$y = \frac{5}{x - 1},$$

we find y by subtracting 1 from x, and then dividing the result into 5. This process produces exactly one value of y for each value in the domain, so this equation defines a function.

The domain of $y = \frac{5}{x-1}$ includes all real numbers except those that make the denominator 0. We find these numbers by setting the denominator equal to 0 and solving for x.

$$x - 1 = 0$$

$$x = 1 \quad \text{Add 1.}$$

Thus, the domain includes all real numbers except 1, written as the interval $(-\infty, 1) \cup (1, \infty)$. Values of y can be positive or negative, but never 0, because a fraction cannot equal 0 unless its numerator is 0. Therefore, the range is the interval $(-\infty, 0) \cup (0, \infty)$, as shown in **Figure 6.**

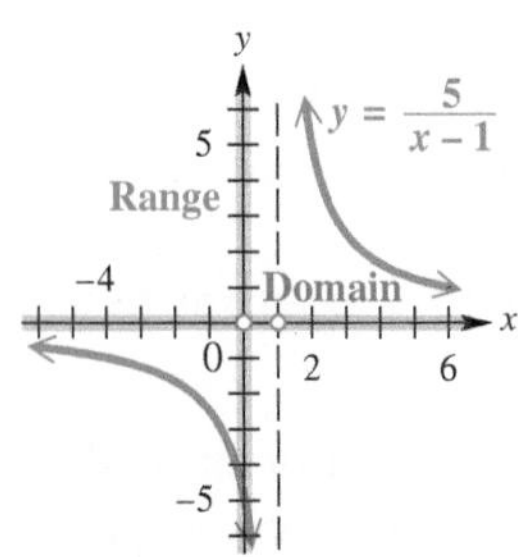

Figure 6

Now Try Exercises 23, 25, and 29.

Function Notation When a function f is defined with a rule or an equation using x and y for the independent and dependent variables, we say, "*y is a function of x*" to emphasize that *y depends on x*. We use the notation

$$y = f(x),$$

called **function notation,** to express this and read $f(x)$ as **"f of x,"** or **"f at x."** The letter f is the name given to this function.

For example, if $y = 3x - 5$, we can name the function f and write

$$f(x) = 3x - 5.$$

Note that $f(x)$ is just another name for the dependent variable y. For example, if $y = f(x) = 3x - 5$ and $x = 2$, then we find y, or $f(2)$, by replacing x with 2.

$$f(2) = 3 \cdot 2 - 5 \quad \text{Let } x = 2.$$

$$f(2) = 1 \quad \text{Multiply, and then subtract.}$$

The statement "In the function f, if $x = 2$, then $y = 1$" represents the ordered pair $(2, 1)$ and is abbreviated with function notation as follows.

$$f(2) = 1$$

The symbol $f(2)$ is read "f of 2" or "f at 2."

Function notation can be illustrated as follows.

Name of the function

Defining expression

$$y = f(x) = 3x - 5$$

Value of the function   Name of the independent variable

Classroom Example 6

Let $f(x) = -x^2 - 6x + 4$ and $g(x) = 3x + 1$. Find each of the following.

(a) $f(-3)$

(b) $f(r)$

(c) $g(r + 2)$

Answers:

(a) 13

(b) $-r^2 - 6r + 4$

(c) $3r + 7$

EXAMPLE 6 Using Function Notation

Let $f(x) = -x^2 + 5x - 3$ and $g(x) = 2x + 3$. Find each of the following.

(a) $f(2)$ **(b)** $f(q)$ **(c)** $g(a + 1)$

SOLUTION

(a) $f(x) = -x^2 + 5x - 3$

$f(2) = -2^2 + 5 \cdot 2 - 3$ Replace x with 2.

$f(2) = -4 + 10 - 3$ Apply the exponent and multiply.

$f(2) = 3$ Add and subtract.

Thus, $f(2) = 3$, and the ordered pair $(2, 3)$ belongs to f.

(b) $f(x) = -x^2 + 5x - 3$

$f(q) = -q^2 + 5q - 3$ Replace x with q.

(c) $g(x) = 2x + 3$

$g(a + 1) = 2(a + 1) + 3$ Replace x with $a + 1$.

$g(a + 1) = 2a + 2 + 3$ Distributive property

$g(a + 1) = 2a + 5$ Add.

Now Try Exercises 35, 43, and 49.

Functions can be evaluated in a variety of ways, as shown in **Example 7.**

EXAMPLE 7 Using Function Notation

For each function, find $f(3)$.

(a) $f(x) = 3x - 7$

(b) $f = \{(-3, 5), (0, 3), (3, 1), (6, -1)\}$

(c)

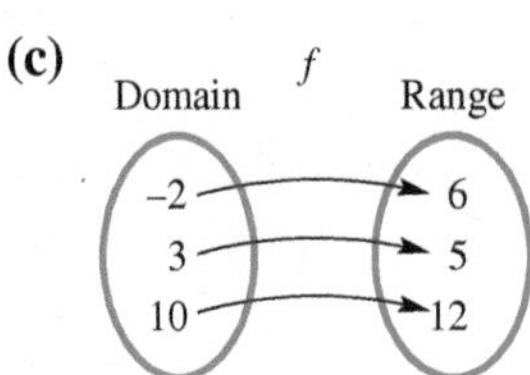

(d)

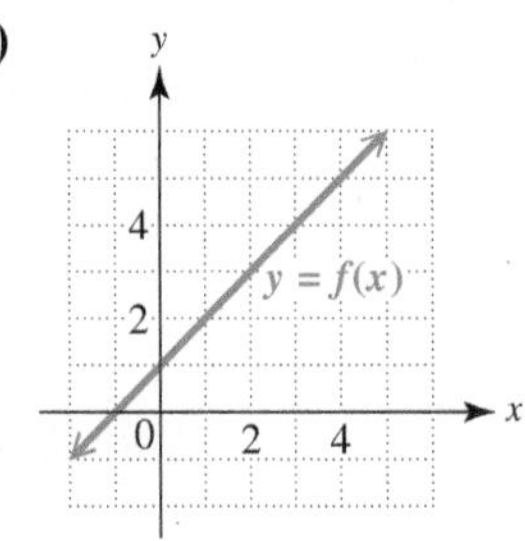

SOLUTION

(a) $f(x) = 3x - 7$

$f(3) = 3(3) - 7$ Replace x with 3.

$f(3) = 2$ Simplify.

$f(3) = 2$ indicates that the ordered pair $(3, 2)$ belongs to f.

(b) For $f = \{(-3, 5), (0, 3), (3, 1), (6, -1)\}$, we want $f(3)$, the y-value of the ordered pair where $x = 3$. As indicated by the ordered pair $(3, 1)$, when $x = 3$, $y = 1$, so $f(3) = 1$.

(c) In the mapping, repeated in **Figure 7,** the domain element 3 is paired with 5 in the range, so $f(3) = 5$.

(d) To evaluate $f(3)$ using the graph, find 3 on the x-axis. See **Figure 8.** Then move up until the graph of f is reached. Moving horizontally to the y-axis gives 4 for the corresponding y-value. Thus, $f(3) = 4$.

Now Try Exercises 51, 53, and 55.

Domain f Range

-2 → 6
3 → 5
10 → 12

Figure 7

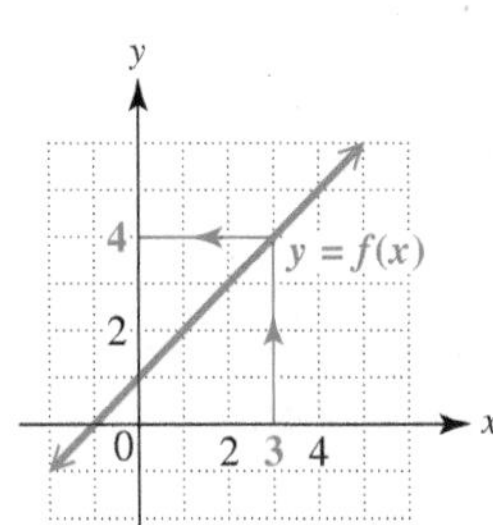

Figure 8

Classroom Example 7

For each function, find $f(-1)$.

(a) $f(x) = 2x^2 - 9$

(b) $f = \{(-4, 0), (-1, 6), (0, 8), (2, -2)\}$

(c)

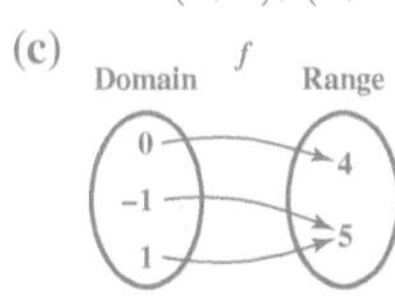

(d)

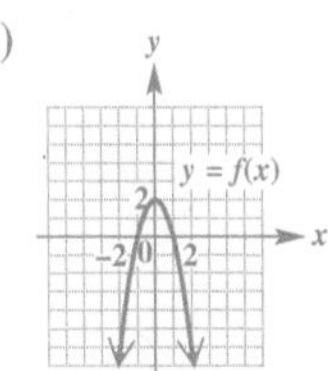

Answers:

(a) -7 (b) 6

(c) 5 (d) 0

Increasing, Decreasing, and Constant Functions Informally speaking, a function *increases* over an open interval of its domain if its graph rises from left to right on the interval. It *decreases* over an open interval of its domain if its graph falls from left to right on the interval. It is *constant* over an open interval of its domain if its graph is horizontal on the interval.

For example, consider **Figure 9.**

- The function increases over the open interval $(-2, 1)$ because the y-values continue to get larger for x-values in that interval.
- The function is constant over the open interval $(1, 4)$ because the y-values are always 5 for all x-values there.
- The function decreases over the open interval $(4, 6)$ because in that interval the y-values continuously get smaller.

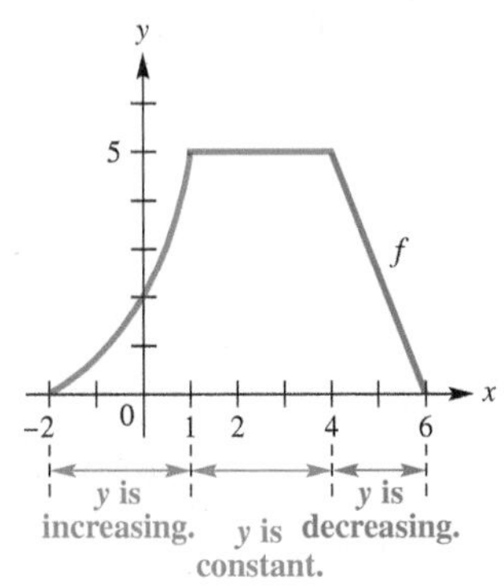

Figure 9

The intervals refer to the x-values where the y-values either increase, decrease, or are constant.

The formal definitions of these concepts follow.

Teaching Tip Point out that "whenever $x_1 < x_2$, $f(x_1) < f(x_2)$" means that for every two numbers x_1 and x_2 in the interval with x_1 to the left of x_2, the point on the graph at x_1 is lower than the point on the graph at x_2. Refer students to **Figure 10(a)** to see this.

Increasing, Decreasing, and Constant Functions

Suppose that a function f is defined over an *open* interval I and x_1 and x_2 are in I.

(a) f **increases** over I if, whenever $x_1 < x_2$, $f(x_1) < f(x_2)$.

(b) f **decreases** over I if, whenever $x_1 < x_2$, $f(x_1) > f(x_2)$.

(c) f is **constant** over I if, for every x_1 and x_2, $f(x_1) = f(x_2)$.

Figure 10 illustrates these ideas.

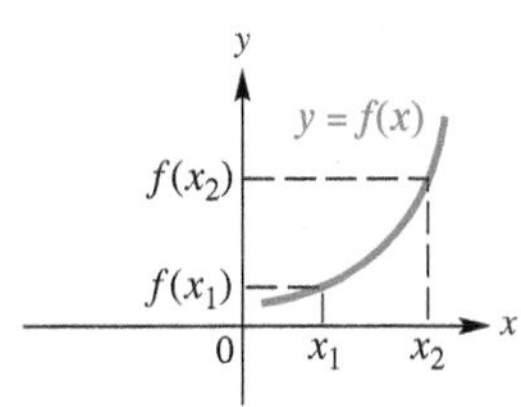

Whenever $x_1 < x_2$, and $f(x_1) < f(x_2)$, f is increasing.

(a)

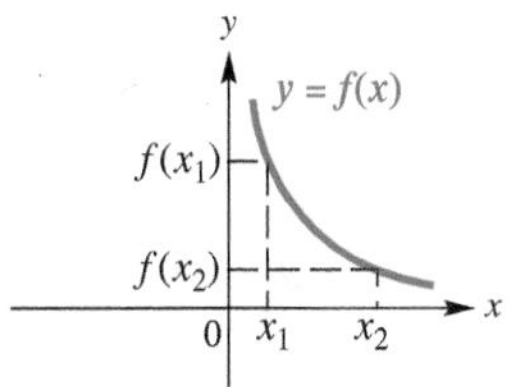

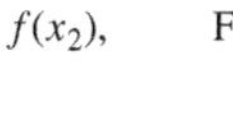

Whenever $x_1 < x_2$, and $f(x_1) > f(x_2)$, f is decreasing.

(b)

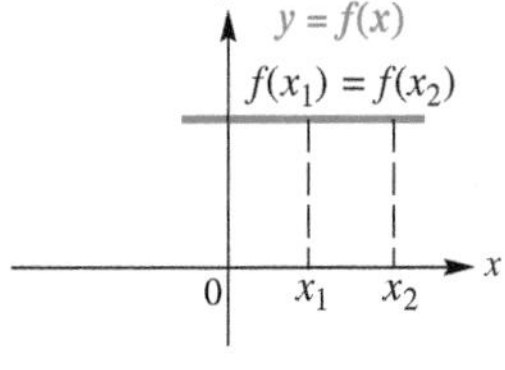

For every x_1 and x_2, if $f(x_1) = f(x_2)$, then f is constant.

(c)

Figure 10

NOTE To decide whether a function is increasing, decreasing, or constant over an interval, ask yourself, ***"What does y do as x goes from left to right?"*** Our definition of *increasing, decreasing,* and *constant* function behavior applies to open intervals of the domain, not to individual points.

Classroom Example 8
The figure shows the graph of a function. Determine the largest open intervals of the domain over which the function is increasing, decreasing, or constant.

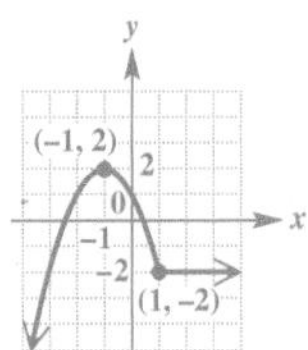

Answer:
increasing on $(-\infty, -1)$;
decreasing on $(-1, 1)$;
constant on $(1, \infty)$

EXAMPLE 8 Determining Open Intervals of a Domain

Figure 11 shows the graph of a function. Determine the largest open intervals of the domain over which the function is increasing, decreasing, or constant.

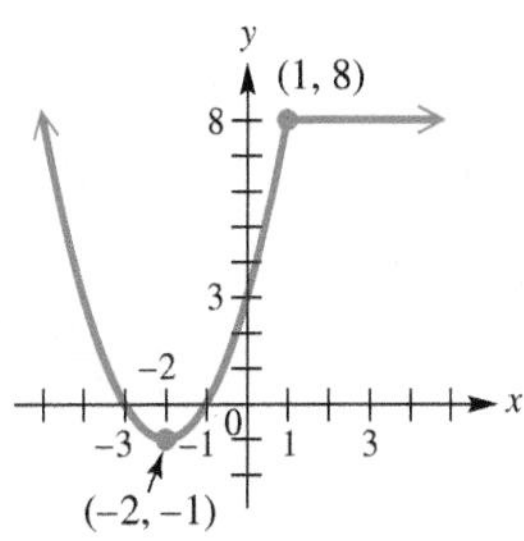

Figure 11

SOLUTION We observe the domain and ask, *"What is happening to the y-values as the x-values are getting larger?"* Moving from left to right on the graph, we see the following:

- On the open interval $(-\infty, -2)$, the y-values are *decreasing*.
- On the open interval $(-2, 1)$, the y-values are *increasing*.
- On the open interval $(1, \infty)$, the y-values are *constant* (and equal to 8).

Therefore, the function is decreasing on $(-\infty, -2)$, increasing on $(-2, 1)$, and constant on $(1, \infty)$.

Now Try Exercise 65.

Appendix C Exercises

1. function 2. function
3. not a function
4. not a function
5. function 6. function
7. function 8. function
9. not a function; domain: $\{0, 1, 2\}$; range: $\{-4, -1, 0, 1, 4\}$
10. not a function; domain: $\{2, 3, 5\}$; range: $\{5, 7, 9, 11\}$
11. function; domain: $\{2, 3, 5, 11, 17\}$; range: $\{1, 7, 20\}$
12. function; domain: $\{1, 2, 3, 5\}$; range: $\{10, 15, 19, 27\}$
13. function; domain: $\{0, -1, -2\}$; range: $\{0, 1, 2\}$
14. function; domain: $\{0, 1, 2\}$; range: $\{0, -1, -2\}$
15. function; domain: $(-\infty, \infty)$; range: $(-\infty, \infty)$
16. function; domain: $(-\infty, \infty)$; range: $(-\infty, 4]$
17. not a function; domain: $[3, \infty)$; range: $(-\infty, \infty)$
18. not a function; domain: $[-4, 4]$; range: $[-3, 3]$
19. function; domain: $(-\infty, \infty)$; range: $(-\infty, \infty)$
20. function; domain: $[-2, 2]$; range: $[0, 4]$
21. function; domain: $(-\infty, \infty)$; range: $[0, \infty)$
22. function; domain: $(-\infty, \infty)$; range: $(-\infty, \infty)$

Decide whether each relation defines a function. ***See Example 1.***

1. $\{(5, 1), (3, 2), (4, 9), (7, 8)\}$

2. $\{(8, 0), (5, 7), (9, 3), (3, 8)\}$

3. $\{(2, 4), (0, 2), (2, 6)\}$

4. $\{(9, -2), (-3, 5), (9, 1)\}$

5. $\{(-3, 1), (4, 1), (-2, 7)\}$

6. $\{(-12, 5), (-10, 3), (8, 3)\}$

7.

x	y
3	-4
7	-4
10	-4

8.

x	y
-4	$\sqrt{2}$
0	$\sqrt{2}$
4	$\sqrt{2}$

Decide whether each relation defines a function, and give the domain and range. ***See Examples 1–4.***

9. $\{(1, 1), (1, -1), (0, 0), (2, 4), (2, -4)\}$

10. $\{(2, 5), (3, 7), (3, 9), (5, 11)\}$

11. 2 → 1, 5 → 1, 11 → 7, 17 → 20, 3 → 20

12. 1 → 10, 2 → 15, 3 → 19, 5 → 27

13.

x	y
0	0
-1	1
-2	2

14.

x	y
0	0
1	-1
2	-2

15.

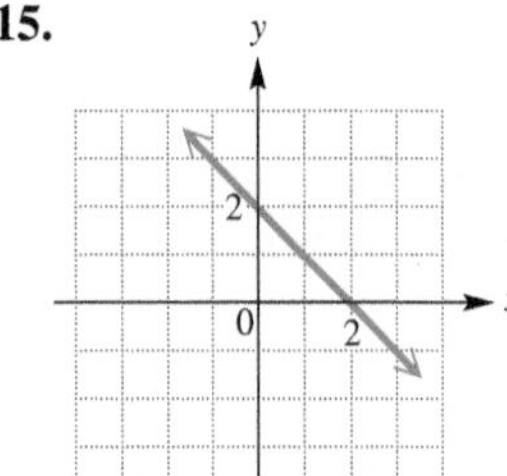

16.

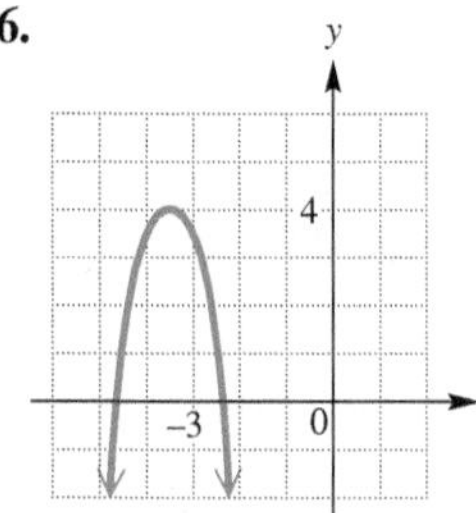

17.

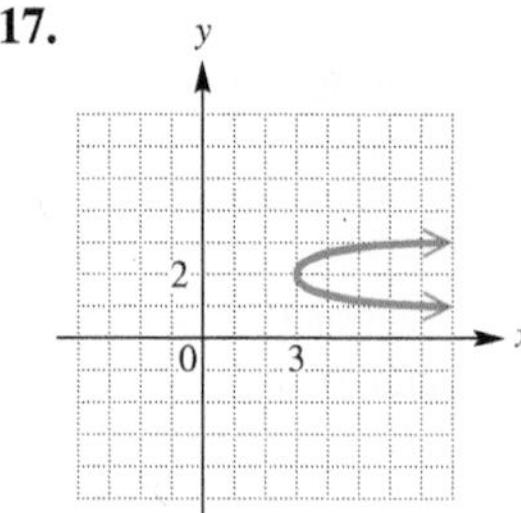

18.

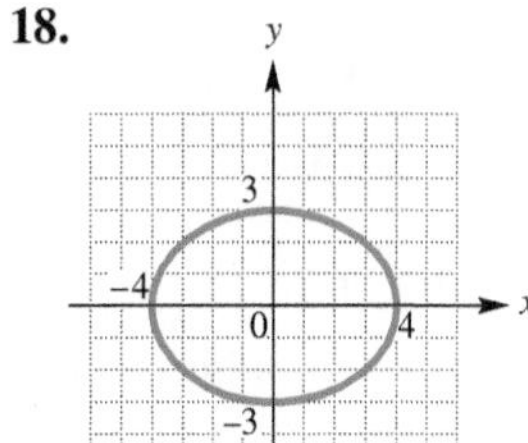

19.

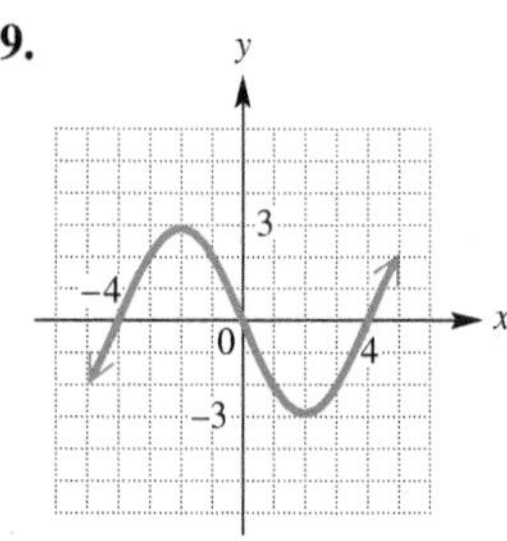

20.

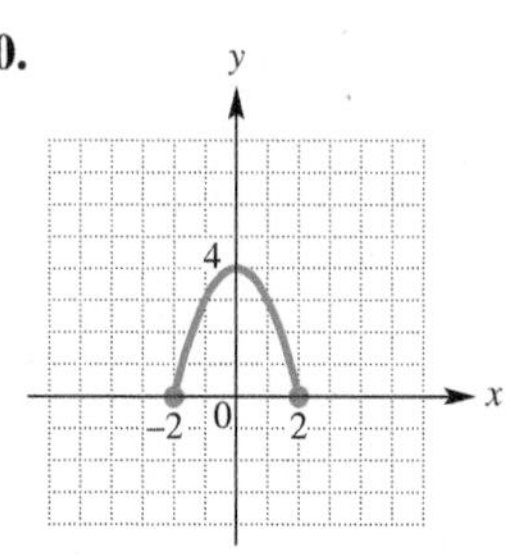

Decide whether each relation defines y as a function of x. Give the domain and range. ***See Example 5.***

21. $y = x^2$

22. $y = x^3$

23. $x = y^6$

24. $x = y^4$

25. $y = 2x - 5$

26. $y = -6x + 4$

27. $y = \sqrt{x}$

28. $y = -\sqrt{x}$

29. $y = \sqrt{4x + 1}$

30. $y = \sqrt{7 - 2x}$

31. $y = \dfrac{2}{x - 3}$

32. $y = \dfrac{-7}{x - 5}$

23. not a function;
domain: $[0, \infty)$;
range: $(-\infty, \infty)$

24. not a function;
domain: $[0, \infty)$;
range: $(-\infty, \infty)$

25. function;
domain: $(-\infty, \infty)$;
range: $(-\infty, \infty)$

26. function;
domain: $(-\infty, \infty)$;
range: $(-\infty, \infty)$

27. function;
domain: $[0, \infty)$;
range: $[0, \infty)$

28. function;
domain: $[0, \infty)$;
range: $(-\infty, 0]$

29. function;
domain: $\left[-\frac{1}{4}, \infty\right)$;
range: $[0, \infty)$

30. function;
domain: $\left(-\infty, \frac{7}{2}\right]$;
range: $[0, \infty)$

31. function;
domain: $(-\infty, 3) \cup (3, \infty)$;
range: $(-\infty, 0) \cup (0, \infty)$

32. function;
domain: $(-\infty, 5) \cup (5, \infty)$;
range: $(-\infty, 0) \cup (0, \infty)$

33. B

34. Here is one example: The cost of gasoline; number of gallons used; cost; number of gallons

35. 4 36. 13
37. -11 38. -59
39. 3 40. 11
41. $\frac{11}{4}$ 42. $-\frac{1}{16}$
43. $-3p + 4$ 44. $-k^2 + 4k + 1$
45. $3x + 4$ 46. $-x^2 - 4x + 1$
47. $-3x - 2$ 48. $-3a - 8$
49. $-6m + 13$ 50. $-9t + 10$

51. (a) 2 (b) 3
52. (a) 5 (b) 11
53. (a) 15 (b) 10
54. (a) 1 (b) 7
55. (a) 3 (b) -3
56. (a) -3 (b) 2

57. (a) 0 (b) 4 (c) 2 (d) 4
58. (a) 5 (b) 0 (c) 2 (d) 4
59. (a) -3 (b) -2 (c) 0 (d) 2
60. (a) 3 (b) 3 (c) 3 (d) 3

33. *Concept Check* Choose the correct answer: For function f, the notation $f(3)$ means

A. the variable f times 3, or $3f$.

B. the value of the dependent variable when the independent variable is 3.

C. the value of the independent variable when the dependent variable is 3.

D. f equals 3.

34. *Concept Check* Give an example of a function from everyday life. (*Hint:* Fill in the blanks: ________ depends on ________, so ________ is a function of ________.)

Let $f(x) = -3x + 4$ and $g(x) = -x^2 + 4x + 1$. Find each of the following. Simplify if necessary. ***See Example 6.***

35. $f(0)$ **36.** $f(-3)$ **37.** $g(-2)$ **38.** $g(10)$

39. $f\left(\frac{1}{3}\right)$ **40.** $f\left(-\frac{7}{3}\right)$ **41.** $g\left(\frac{1}{2}\right)$ **42.** $g\left(-\frac{1}{4}\right)$

43. $f(p)$ **44.** $g(k)$ **45.** $f(-x)$ **46.** $g(-x)$

47. $f(x + 2)$ **48.** $f(a + 4)$ **49.** $f(2m - 3)$ **50.** $f(3t - 2)$

For each function, find ***(a)*** *$f(2)$ and* ***(b)*** *$f(-1)$.* ***See Example 7.***

51. $f = \{(-1, 3), (4, 7), (0, 6), (2, 2)\}$ **52.** $f = \{(2, 5), (3, 9), (-1, 11), (5, 3)\}$

53.

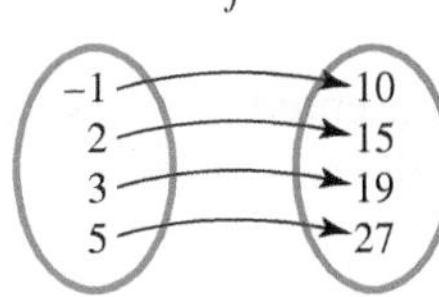

54.

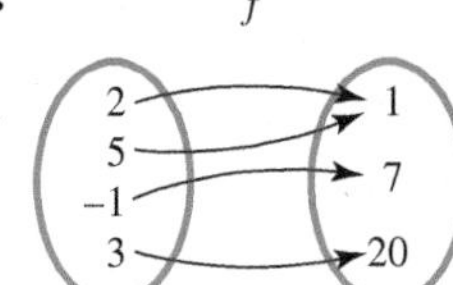

55.

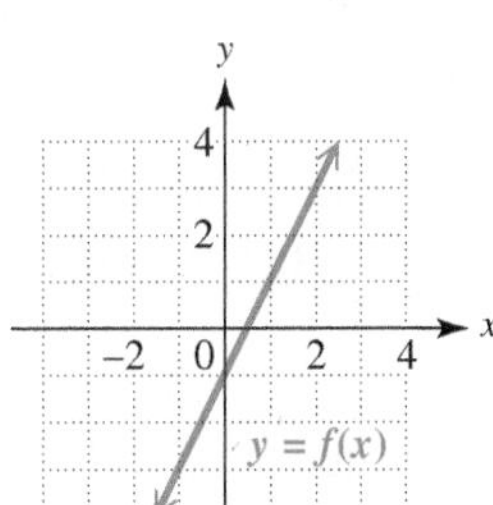

56.

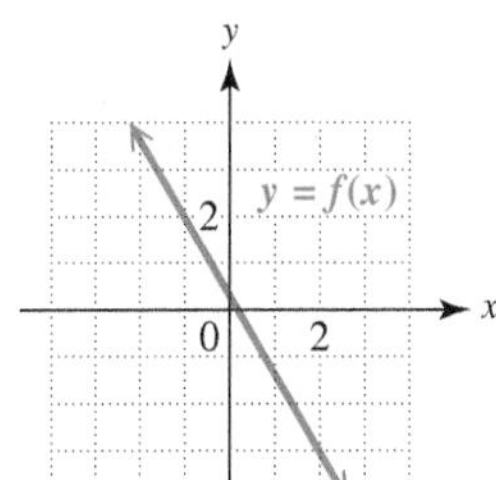

Use the graph of $y = f(x)$ to find each function value: ***(a)*** *$f(-2)$,* ***(b)*** *$f(0)$,* ***(c)*** *$f(1)$, and* ***(d)*** *$f(4)$.* ***See Example 7(d).***

57.

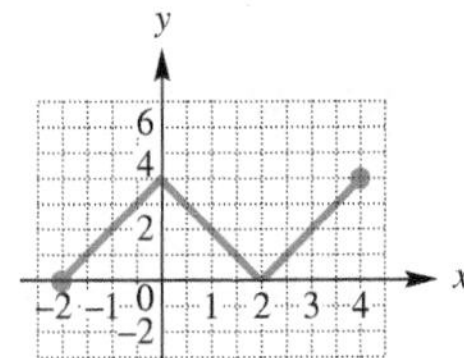

58.

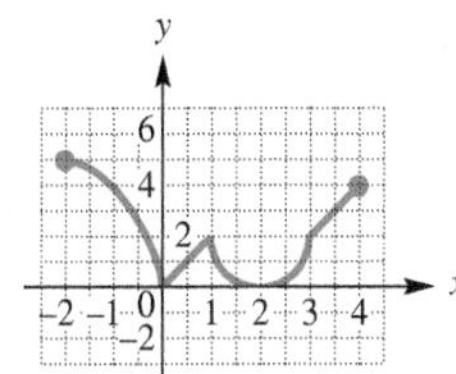

59.

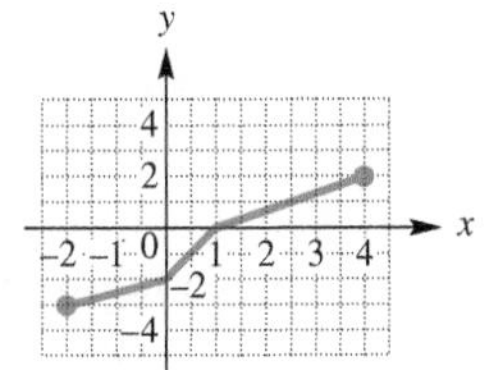

60.

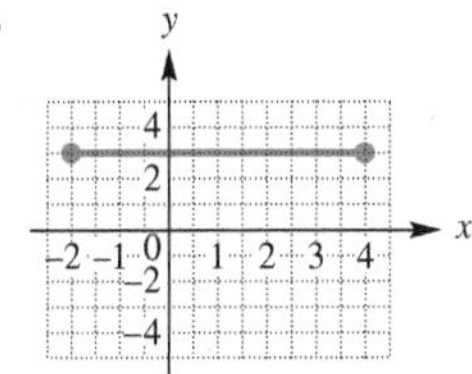

61. (a) $(-2, 0)$
(b) $(-\infty, -2)$
(c) $(0, \infty)$

62. (a) $(-3, -1)$
(b) $(-1, \infty)$
(c) $(-\infty, -3)$

63. (a) $(-\infty, -2)$; $(2, \infty)$
(b) $(-2, 2)$
(c) none

64. (a) $(-3, 3)$
(b) $(-\infty, -3)$; $(3, \infty)$
(c) none

65. (a) $(-1, 0)$; $(1, \infty)$
(b) $(-\infty, -1)$; $(0, 1)$
(c) none

66. (a) $(-\infty, -2)$; $(0, 2)$
(b) $(-2, 0)$; $(2, \infty)$
(c) none

Determine the largest open intervals of the domain over which each function is ***(a)*** *increasing,* ***(b)*** *decreasing, and* ***(c)*** *constant.* ***See Example 8.***

61.

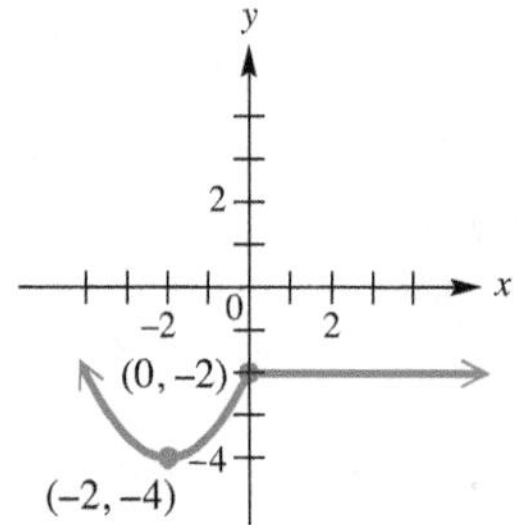

62.

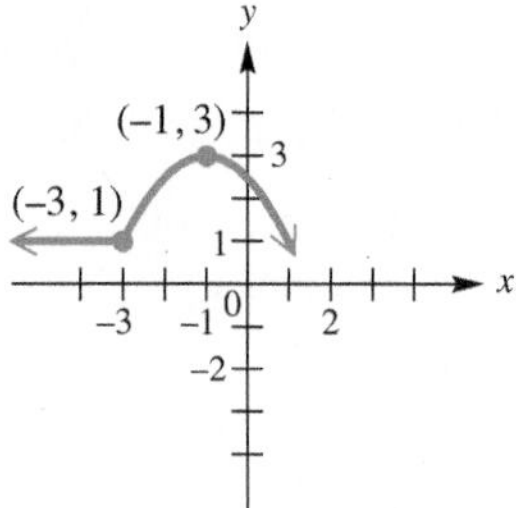

63.

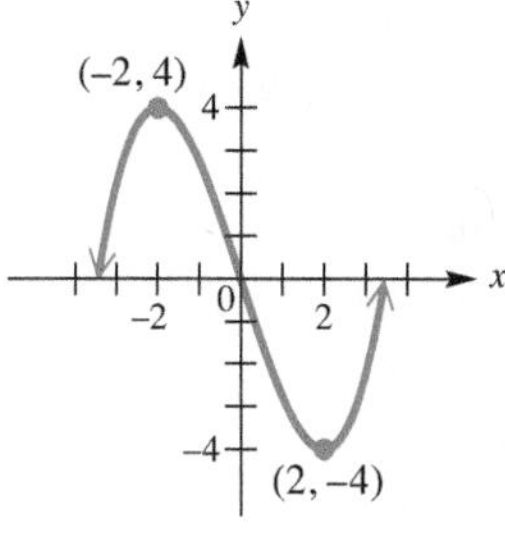

64.

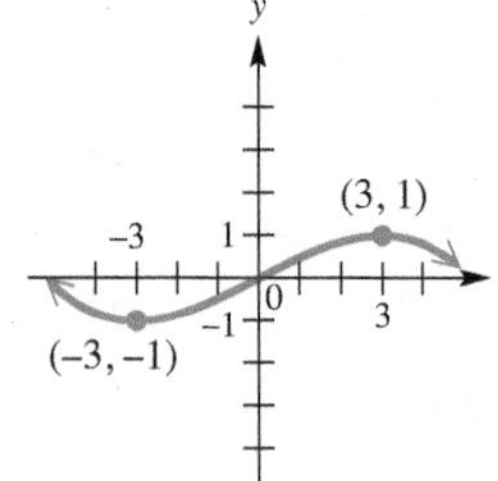

65.

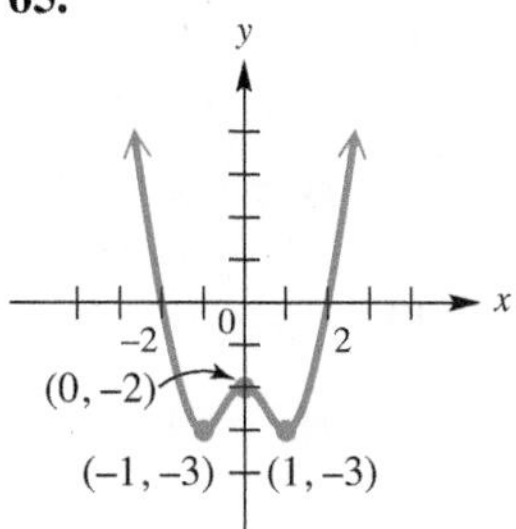

66.

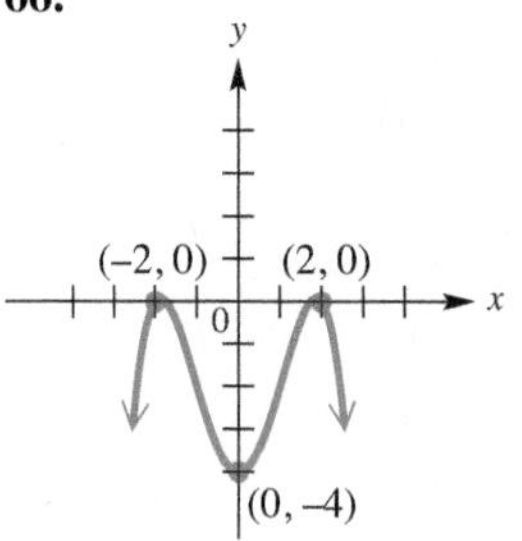

D Graphing Techniques

- **Stretching and Shrinking**
- **Reflecting**
- **Symmetry**
- **Translations**

Graphing techniques presented in this section show how to graph functions that are defined by altering the equation of a basic function.

NOTE Recall from algebra that $|a|$ is the absolute value of a number a.

$$|a| = \begin{cases} \boldsymbol{a \text{ if } a \text{ is positive or } 0} \\ \boldsymbol{-a \text{ if } a \text{ is negative}} \end{cases}$$

Thus, $|2| = |2|$ and $|-2| = |2|$.

We use absolute value functions to illustrate many of the graphing techniques in this section.

y
2
x
-2
0
$y = |x|$

Graph of the absolute value function

Stretching and Shrinking We begin by considering how the graphs of $y = af(x)$ and $y = f(ax)$ compare to the graph of $y = f(x)$, where $a > 0$.

EXAMPLE 1 Stretching or Shrinking Graphs

Graph each function.

(a) $g(x) = 2|x|$ **(b)** $h(x) = \frac{1}{2}|x|$ **(c)** $k(x) = |2x|$

SOLUTION

(a) Comparing the tables of values for $f(x) = |x|$ and $g(x) = 2|x|$ in **Figure 1** on the next page, we see that for corresponding x-values, the y-values of g are each twice those of f. The graph of $f(x) = |x|$ is *vertically stretched.* The graph of $g(x)$, shown in blue in **Figure 1,** is narrower than that of $f(x)$, shown in red for comparison.

Classroom Example 1
Graph each function.

(a) $g(x) = 2x^2$ (b) $h(x) = \frac{1}{2}x^2$

(c) $k(x) = \left(\frac{1}{2}x\right)^2$

Answers:

(a)

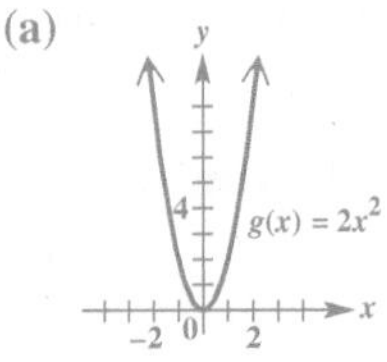

(b)

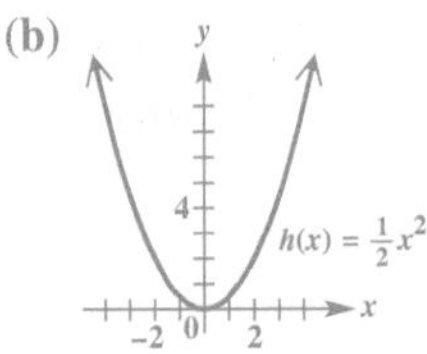

(c)

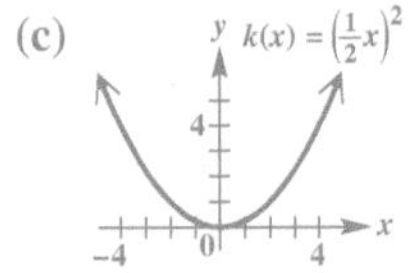

x	$f(x) = \|x\|$	$g(x) = 2\|x\|$
-2	2	4
-1	1	2
0	0	0
1	1	2
2	2	4

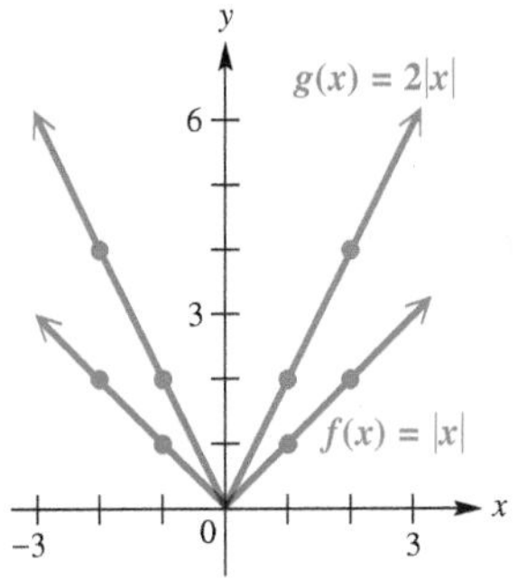

Figure 1

(b) The graph of $h(x) = \frac{1}{2}|x|$ is also the same general shape as that of $f(x)$, but here the coefficient $\frac{1}{2}$ is between 0 and 1 and causes a *vertical shrink*. The graph of $h(x)$ is wider than the graph of $f(x)$, as we see by comparing the tables of values. See **Figure 2.**

x	$f(x) = \|x\|$	$h(x) = \frac{1}{2}\|x\|$
-2	2	1
-1	1	$\frac{1}{2}$
0	0	0
1	1	$\frac{1}{2}$
2	2	1

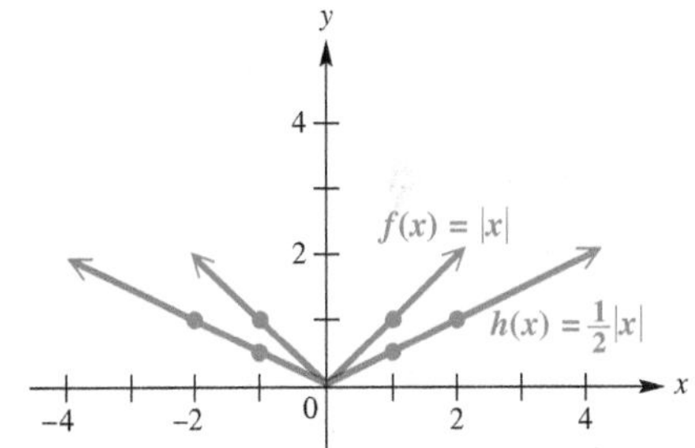

Figure 2

(c) Use the property of absolute value that states $|ab| = |a| \cdot |b|$ to rewrite $|2x|$.

$$k(x) = |2x| = |2| \cdot |x| = 2|x|$$

Therefore, the graph of $k(x) = |2x|$ is the same as the graph of $g(x) = 2|x|$ in part (a). This is a *horizontal shrink* of the graph of $f(x) = |x|$. See **Figure 1.**

✔ **Now Try Exercises 13 and 15.**

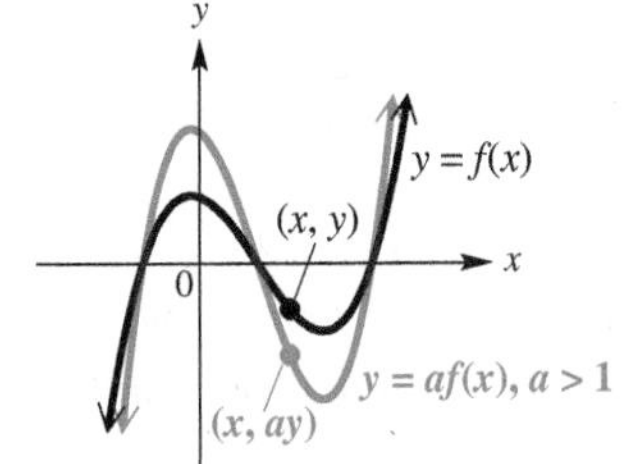

Vertical stretching
$a > 1$

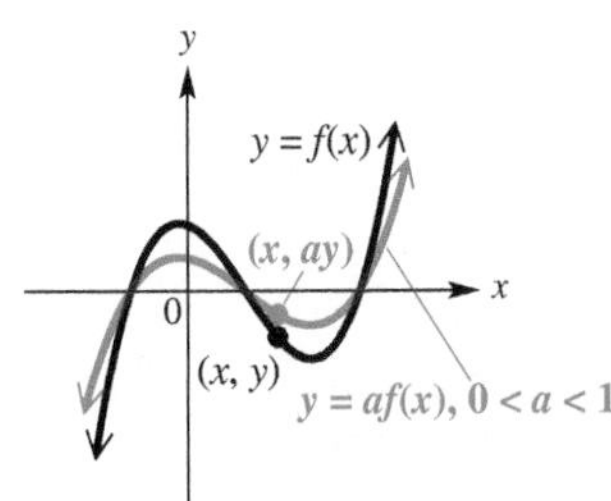

Vertical shrinking
$0 < a < 1$

Figure 3

Vertical Stretching or Shrinking of the Graph of a Function

Suppose that $a > 0$. If a point (x, y) lies on the graph of $y = f(x)$, then the point (x, ay) lies on the graph of $\boldsymbol{y = af(x)}$.

(a) If $a > 1$, then the graph of $y = af(x)$ is a **vertical stretching** of the graph of $y = f(x)$.

(b) If $0 < a < 1$, then the graph of $y = af(x)$ is a **vertical shrinking** of the graph of $y = f(x)$.

Figure 3 shows graphical interpretations of vertical stretching and shrinking. ***In both cases, the x-intercepts of the graph remain the same but the y-intercepts are affected.***

Graphs of functions can also be stretched and shrunk horizontally.

Horizontal Stretching or Shrinking of the Graph of a Function

Suppose that $a > 0$. If a point (x, y) lies on the graph of $y = f(x)$, then the point $\left(\frac{x}{a}, y\right)$ lies on the graph of $\boldsymbol{y = f(ax)}$.

(a) If $0 < a < 1$, then the graph of $y = f(ax)$ is a **horizontal stretching** of the graph of $y = f(x)$.

(b) If $a > 1$, then the graph of $y = f(ax)$ is a **horizontal shrinking** of the graph of $y = f(x)$.

See **Figure 4** for graphical interpretations of horizontal stretching and shrinking. ***In both cases, the y-intercept remains the same but the x-intercepts are affected.***

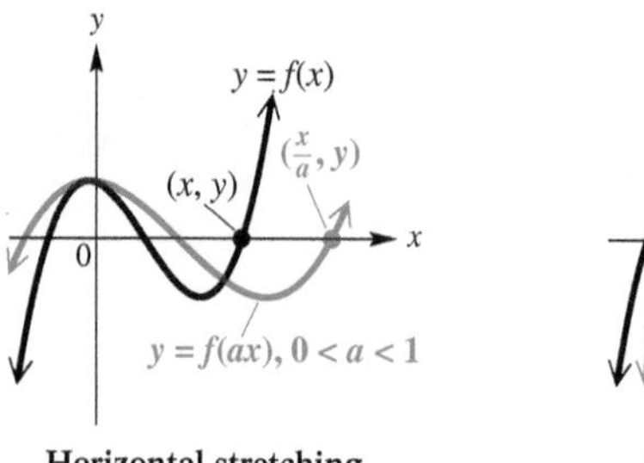

Horizontal stretching
$0 < a < 1$

Horizontal shrinking
$a > 1$

Figure 4

Reflecting Forming the mirror image of a graph across a line is called ***reflecting the graph across the line.***

Classroom Example 2
Graph each function.
(a) $g(x) = -|x|$
(b) $h(x) = |-x|$

Answers:
(a)

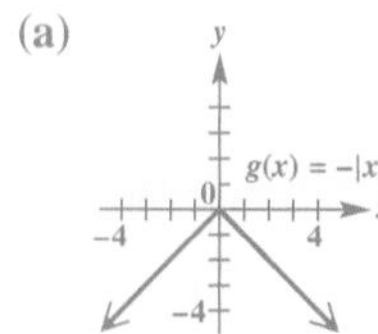

(b)

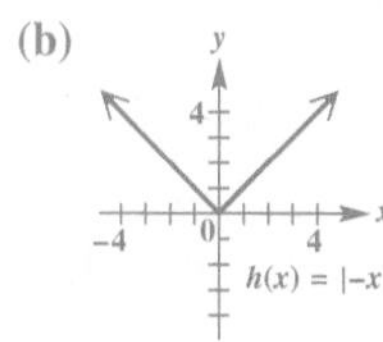

EXAMPLE 2 Reflecting Graphs across Axes

Graph each function.

(a) $g(x) = -\sqrt{x}$ **(b)** $h(x) = \sqrt{-x}$

SOLUTION

(a) The tables of values for $g(x) = -\sqrt{x}$ and $f(x) = \sqrt{x}$ are shown with their graphs in **Figure 5.** As the tables suggest, every y-value of the graph of $g(x) = -\sqrt{x}$ is the negative of the corresponding y-value of $f(x) = \sqrt{x}$. This has the effect of reflecting the graph across the x-axis.

x	$f(x) = \sqrt{x}$	$g(x) = -\sqrt{x}$
0	0	0
1	1	−1
4	2	−2

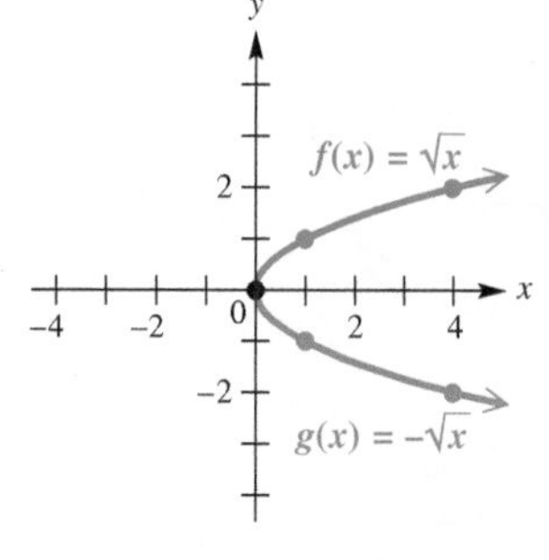

Figure 5

(b) The domain of $h(x) = \sqrt{-x}$ is $(-\infty, 0]$, while the domain of $f(x) = \sqrt{x}$ is $[0, \infty)$. Choosing x-values for $h(x)$ that are negatives of those used for $f(x)$, we see that corresponding y-values are the same. The graph of h is a reflection of the graph of f across the y-axis. See **Figure 6.**

x	$f(x) = \sqrt{x}$	$h(x) = \sqrt{-x}$
-4	undefined	2
-1	undefined	1
0	0	0
1	1	undefined
4	2	undefined

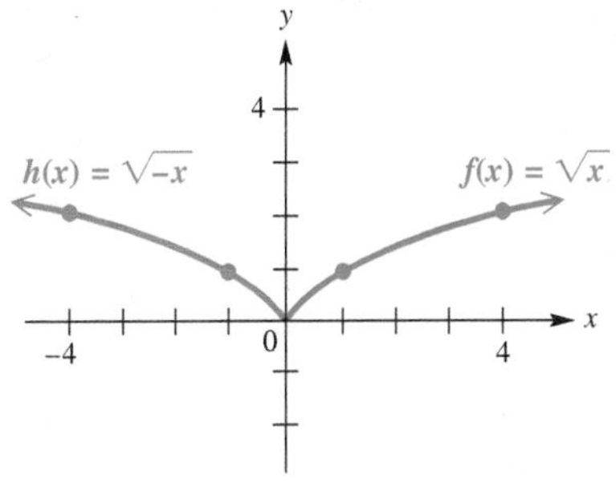

Figure 6

Now Try Exercises 23 and 29.

The graphs in **Example 2** suggest the following generalizations.

Reflecting across an Axis

The graph of $\boldsymbol{y = -f(x)}$ is the same as the graph of $y = f(x)$ reflected across the x-axis. (If a point (x, y) lies on the graph of $y = f(x)$, then $(x, -y)$ lies on this reflection.)

The graph of $\boldsymbol{y = f(-x)}$ is the same as the graph of $y = f(x)$ reflected across the y-axis. (If a point (x, y) lies on the graph of $y = f(x)$, then $(-x, y)$ lies on this reflection.)

Symmetry The graph of f shown in **Figure 7(a)** is cut in half by the y-axis, with each half the mirror image of the other half. Such a graph is *symmetric with respect to the y-axis.* ***In general, for a graph to be symmetric with respect to the y-axis, the point $(-x, y)$ must be on the graph whenever the point (x, y) is on the graph.***

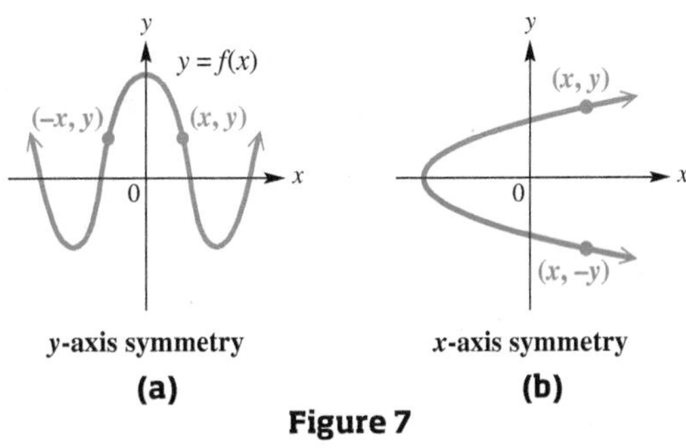

y-axis symmetry
(a)

x-axis symmetry
(b)

Figure 7

Similarly, if the graph in **Figure 7(b)** were folded in half along the x-axis, the portion at the top would exactly match the portion at the bottom. Such a graph is *symmetric with respect to the x-axis.* ***In general, for a graph to be symmetric with respect to the x-axis, the point $(x, -y)$ must be on the graph whenever the point (x, y) is on the graph.***

Symmetry with Respect to an Axis

The graph of an equation is **symmetric with respect to the y-axis** if the replacement of x with $-x$ results in an equivalent equation.

The graph of an equation is **symmetric with respect to the x-axis** if the replacement of y with $-y$ results in an equivalent equation.

Teaching Tip Use numerical coordinates to demonstrate the properties of symmetry.

Classroom Example 3
Test for symmetry with respect to the x-axis and the y-axis.
(a) $x = |y|$ (b) $y = |x| - 3$
(c) $2x - y = 6$ (d) $x^2 + y^2 = 25$

Answers:
(a) x-axis
(b) y-axis
(c) neither axis
(d) x-axis, y-axis

EXAMPLE 3 Testing for Symmetry with Respect to an Axis

Test for symmetry with respect to the x-axis and the y-axis.

(a) $y = x^2 + 4$ **(b)** $x = y^2 - 3$ **(c)** $x^2 + y^2 = 16$ **(d)** $2x + y = 4$

SOLUTION

(a) In $y = x^2 + 4$, replace x with $-x$.

$$y = x^2 + 4$$
$$y = (-x)^2 + 4 \quad \text{(Use parentheses around } -x.)$$
$$y = x^2 + 4$$

The result is equivalent to the original equation.

Thus the graph, shown in **Figure 8,** is symmetric with respect to the y-axis. The y-axis cuts the graph in half, with the halves being mirror images.

Now replace y with $-y$ to test for symmetry with respect to the x-axis.

$$y = x^2 + 4$$
$$-y = x^2 + 4$$
$$y = -x^2 - 4 \quad \text{(Multiply by } -1.)$$

The result is *not* equivalent to the original equation.

The graph is *not* symmetric with respect to the x-axis. See **Figure 8.**

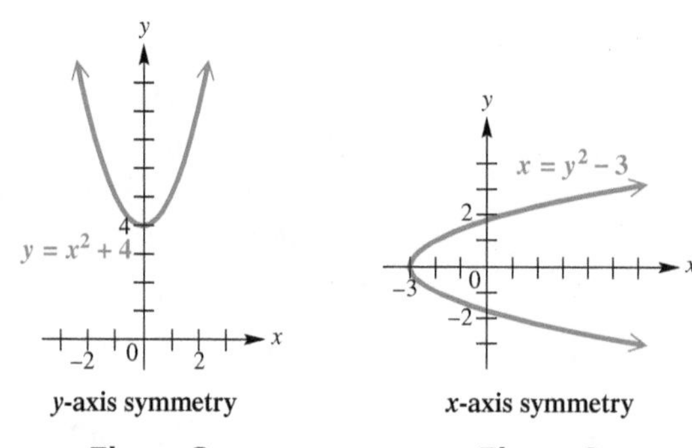

y-axis symmetry
Figure 8

x-axis symmetry
Figure 9

(b) In $x = y^2 - 3$, replace y with $-y$.

$$x = (-y)^2 - 3 = y^2 - 3 \quad \text{Same as the original equation}$$

The graph is symmetric with respect to the x-axis, as shown in **Figure 9.** It is *not* symmetric with respect to the y-axis.

(c) Substitute $-x$ for x and then $-y$ for y in $x^2 + y^2 = 16$.

$$(-x)^2 + y^2 = 16 \quad \text{and} \quad x^2 + (-y)^2 = 16$$

Both simplify to the original equation,

$$x^2 + y^2 = 16.$$

The graph, a circle of radius 4 centered at the origin, is symmetric with respect to *both* axes. See **Figure 10.**

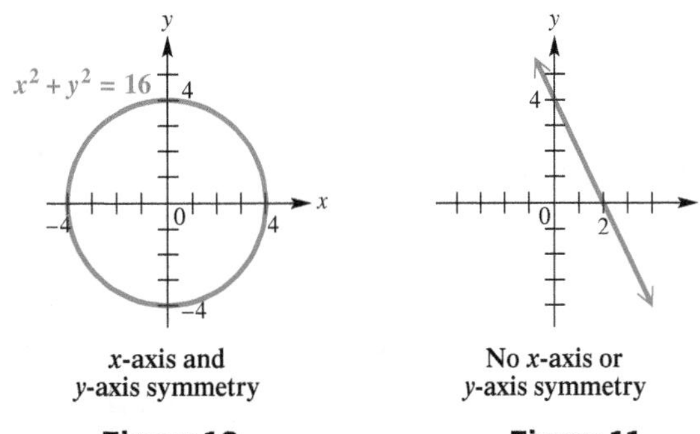

x-axis and *y*-axis symmetry

Figure 10

No *x*-axis or *y*-axis symmetry

Figure 11

(d) In $2x + y = 4$, replace x with $-x$ and then replace y with $-y$.

$2x + y = 4$		$2x + y = 4$	
$2(-x) + y = 4$	Not equivalent	$2x + (-y) = 4$	Not equivalent
$-2x + y = 4$		$2x - y = 4$	

The graph is not symmetric with respect to either axis. See **Figure 11.**

Now Try Exercise 35.

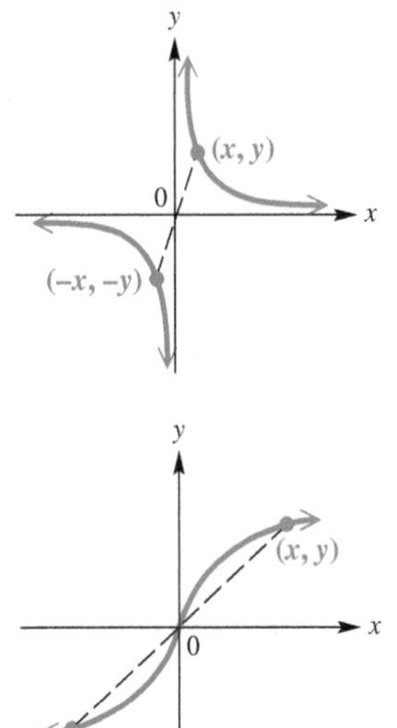

Origin symmetry

Figure 12

Another kind of symmetry occurs when a graph can be rotated 180° about the origin, with the result coinciding exactly with the original graph. Symmetry of this type is *symmetry with respect to the origin. **In general, for a graph to be symmetric with respect to the origin, the point $(-x, -y)$ is on the graph whenever the point (x, y) is on the graph.***

Figure 12 shows two such graphs.

Symmetry with Respect to the Origin

The graph of an equation is **symmetric with respect to the origin** if the replacement of both x with $-x$ and y with $-y$ at the same time results in an equivalent equation.

Classroom Example 4
Determine whether the graph of each equation is symmetric with respect to the origin.
(a) $y = -2x^3$ (b) $y = -2x^2$

Answers:
(a) yes (b) no

EXAMPLE 4 Testing for Symmetry with Respect to the Origin

Determine whether the graph of each equation is symmetric with respect to the origin.

(a) $x^2 + y^2 = 16$ **(b)** $y = x^3$

SOLUTION

(a) Replace x with $-x$ and y with $-y$.

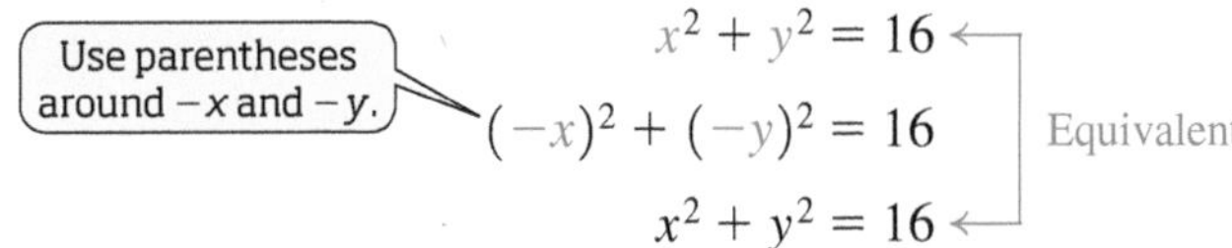

$$x^2 + y^2 = 16$$
$$(-x)^2 + (-y)^2 = 16 \quad \text{Equivalent}$$
$$x^2 + y^2 = 16$$

The graph, which is the circle shown in **Figure 10** in **Example 3(c),** is symmetric with respect to the origin.

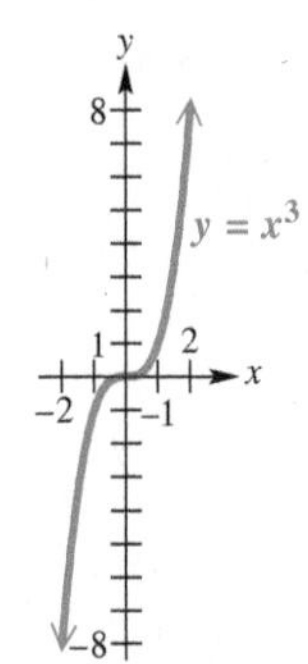

Origin symmetry

Figure 13

(b) In $y = x^3$, replace x with $-x$ and y with $-y$.

$$\left.\begin{aligned} y &= x^3 \\ -y &= (-x)^3 \\ -y &= -x^3 \\ y &= x^3 \end{aligned}\right\} \text{Equivalent}$$

The graph, which is that of the cubing function, is symmetric with respect to the origin and is shown in **Figure 13.**

✔ **Now Try Exercise 39.**

Notice the following important concepts regarding symmetry:

- A graph symmetric with respect to both the x- and y-axes is automatically symmetric with respect to the origin. (See **Figure 10.**)
- A graph symmetric with respect to the origin need *not* be symmetric with respect to either axis. (See **Figure 13.**)
- Of the three types of symmetry—with respect to the x-axis, with respect to the y-axis, and with respect to the origin—a graph possessing any two types must also exhibit the third type of symmetry.
- A graph symmetric with respect to the x-axis does not represent a function. (See **Figures 9 and 10.**)

Translations The next examples show the results of horizontal and vertical shifts, or **translations,** of the graph of $f(x) = |x|$.

Classroom Example 5
Graph $f(x) = x^2 + 2$.

Answer:

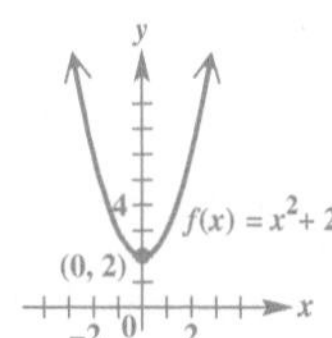

EXAMPLE 5 Translating a Graph Vertically

Graph $g(x) = |x| - 4$.

SOLUTION Comparing the table shown with **Figure 14,** we see that for corresponding x-values, the y-values of g are each 4 *less* than those for f. The graph of $g(x) = |x| - 4$ is the same as that of $f(x) = |x|$, but translated 4 units down. The lowest point is at $(0, -4)$. The graph is symmetric with respect to the y-axis and is therefore the graph of an even function.

x	$f(x) = \lvert x \rvert$	$g(x) = \lvert x \rvert - 4$
-4	4	0
-1	1	-3
0	0	-4
1	1	-3
4	4	0

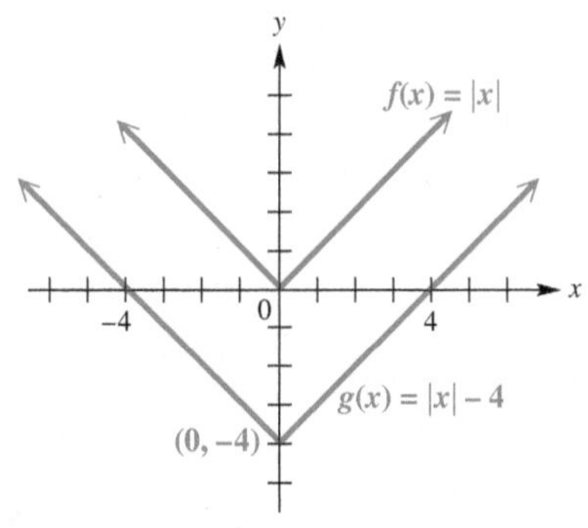

Figure 14

✔ **Now Try Exercise 51.**

The graphs in **Example 5** suggest the following generalization.

Vertical translation 2 units up
$y = f(x) + 2$
$y = f(x)$
Original graph
$y = f(x) - 3$
Vertical translation 3 units down

Figure 15

Vertical Translations

Given a function g defined by $\boldsymbol{g(x) = f(x) + c}$, where c is a real number:

- For every point (x, y) on the graph of f, there will be a corresponding point $(x, y + c)$ on the graph of g.
- The graph of g will be the same as the graph of f, but translated c units up if c is positive or $|c|$ units down if c is negative.

The graph of g is a **vertical translation** of the graph of f. See **Figure 15.**

EXAMPLE 6 Translating a Graph Horizontally

Graph $g(x) = |x - 4|$.

Classroom Example 6
Graph $f(x) = (x + 4)^2$.

Answer:

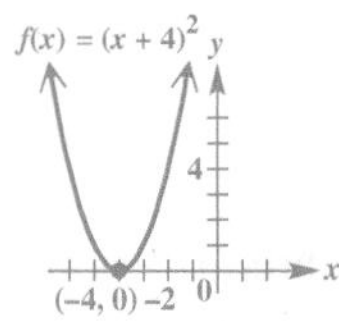

SOLUTION Comparing the tables of values given with **Figure 16** shows that for corresponding y-values, the x-values of g are each 4 *more* than those for f. The graph of $g(x) = |x - 4|$ is the same as that of $f(x) = |x|$, but translated 4 units to the right. The lowest point is at $(4, 0)$. As suggested by the graphs in **Figure 16,** this graph is symmetric with respect to the line $x = 4$.

x	$f(x) = \lvert x\rvert$	$g(x) = \lvert x - 4\rvert$
−2	2	6
0	0	4
2	2	2
4	4	0
6	6	2

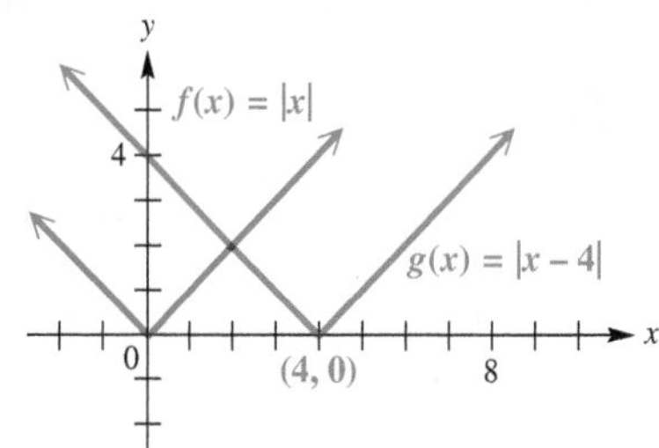

Figure 16

✔ **Now Try Exercise 49.**

The graphs in **Example 6** suggest the following generalization.

Horizontal Translations

Given a function g defined by $\boldsymbol{g(x) = f(x - c)}$, where c is a real number:

- For every point (x, y) on the graph of f, there will be a corresponding point $(x + c, y)$ on the graph of g.
- The graph of g will be the same as the graph of f, but translated c units to the right if c is positive or $|c|$ units to the left if c is negative.

The graph of g is a **horizontal translation** of the graph of f. See **Figure 17.**

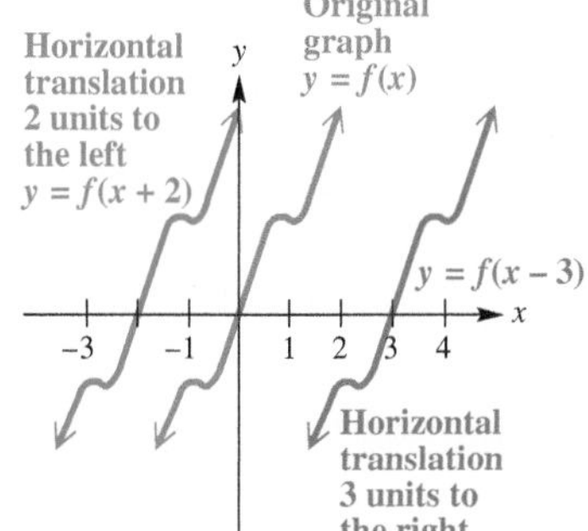

Figure 17

CAUTION ***Errors frequently occur when horizontal shifts are involved.*** Find the value that causes the expression $x - h$ to equal 0, as shown below.

$F(x) = (x - 5)^2$	$F(x) = (x + 5)^2$
Because **+5** causes $x - 5$ to equal 0, the graph of $F(x)$ illustrates a shift of **5 units to the right.**	Because **−5** causes $x + 5$ to equal 0, the graph of $F(x)$ illustrates a shift of **5 units to the left.**

EXAMPLE 7 Using More Than One Transformation

Graph each function.

(a) $f(x) = -|x + 3| + 1$ **(b)** $h(x) = |2x - 4|$ **(c)** $g(x) = -\frac{1}{2}x^2 + 4$

SOLUTION

(a) To graph $f(x) = -|x + 3| + 1$, the *lowest* point on the graph of $y = |x|$ is translated 3 units to the left and 1 unit up. The graph opens down because of the negative sign in front of the absolute value expression, making the lowest point now the highest point on the graph, as shown in **Figure 18.** The graph is symmetric with respect to the line $x = -3$.

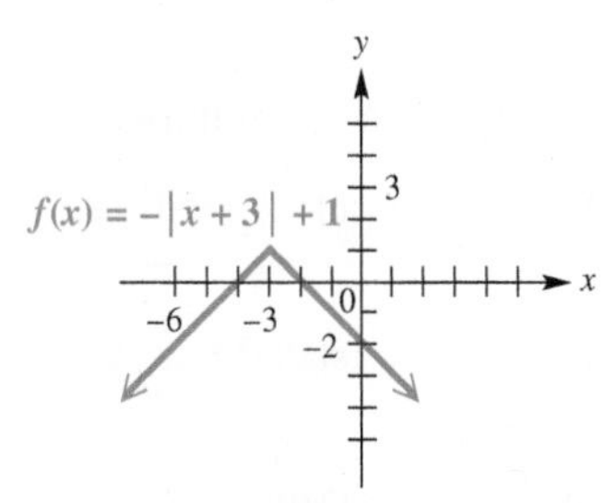

Figure 18

(b) To determine the horizontal translation, factor out 2.

$$h(x) = |2x - 4|$$
$$h(x) = |2(x - 2)| \quad \text{Factor out 2.}$$
$$h(x) = |2| \cdot |x - 2| \quad |ab| = |a| \cdot |b|$$
$$h(x) = 2|x - 2| \quad |2| = 2$$

The graph of h is the graph of $y = |x|$ translated 2 units to the right, and vertically stretched by a factor of 2. Horizontal shrinking gives the same appearance as vertical stretching for this function. See **Figure 19.**

(c) The graph of $g(x) = -\frac{1}{2}x^2 + 4$ has the same shape as that of $y = x^2$, but it is wider (that is, shrunken vertically), reflected across the x-axis because the coefficient $-\frac{1}{2}$ is negative, and then translated 4 units up. See **Figure 20.**

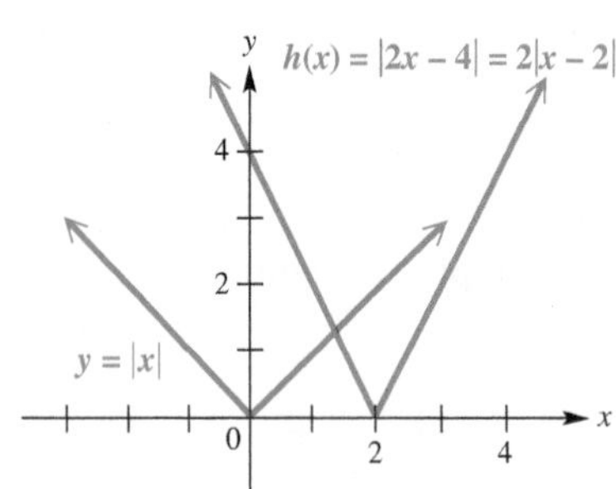

Figure 19

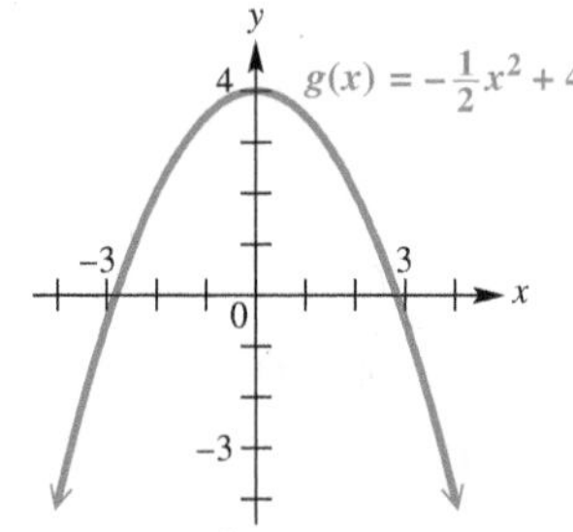

Figure 20

✔ **Now Try Exercises 55, 57, and 65.**

Classroom Example 7

Graph each function.

(a) $f(x) = -(x - 1)^2 + 4$

(b) $g(x) = -2|x + 3|$

(c) $h(x) = \frac{1}{2}\sqrt{x + 2} - 3$

Answers:

(a)

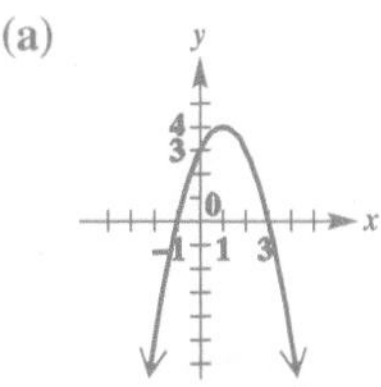

(b)

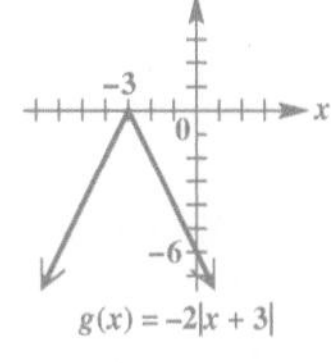

(c)

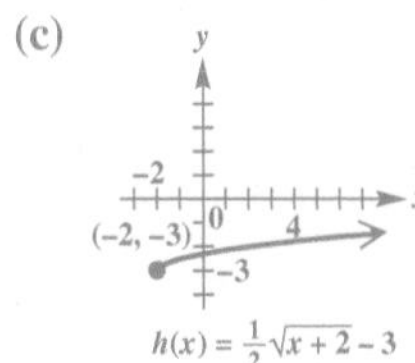

Appendix D Exercises

1. 3
2. 5
3. left
4. right
5. x

Concept Check Fill in the blank(s) to correctly complete each sentence.

1. To graph the function $f(x) = x^2 - 3$, shift the graph of $y = x^2$ down ______ units.

2. To graph the function $f(x) = x^2 + 5$, shift the graph of $y = x^2$ up ______ units.

3. The graph of $f(x) = (x + 4)^2$ is obtained by shifting the graph of $y = x^2$ to the ______ 4 units.

4. The graph of $f(x) = (x - 7)^2$ is obtained by shifting the graph of $y = x^2$ to the ______ 7 units.

5. The graph of $f(x) = -\sqrt{x}$ is a reflection of the graph of $y = \sqrt{x}$ across the ______-axis.

6. y
7. 2; 3
8. 3; 6
9. (a) B (b) D (c) E (d) A (e) C
10. (a) E (b) C (c) D (d) A (e) B
11. (a) B (b) A (c) G (d) C (e) F (f) D (g) H (h) E (i) I
12. (a) F (b) C (c) H (d) D (e) G (f) A (g) E (h) I (i) B

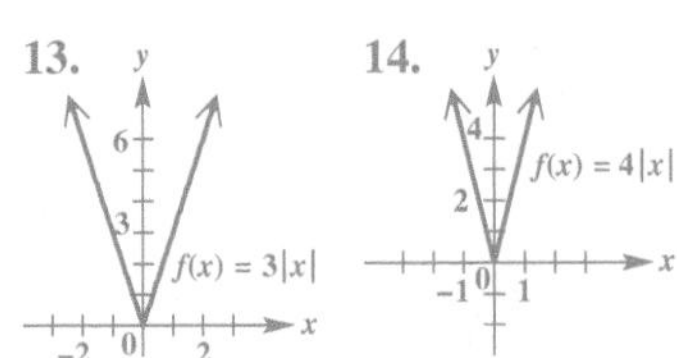

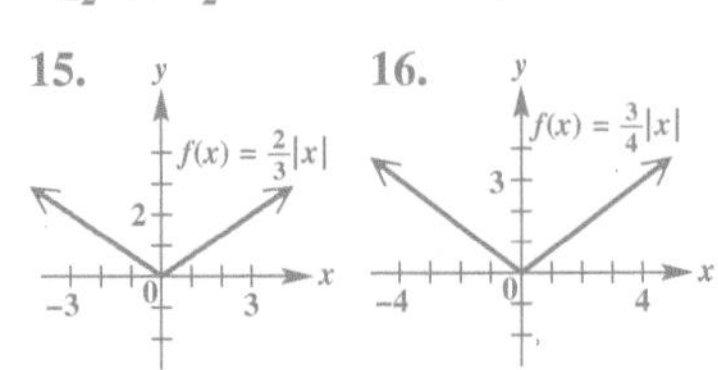

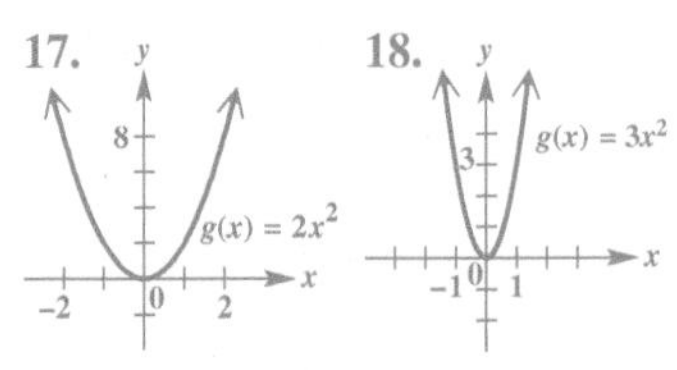

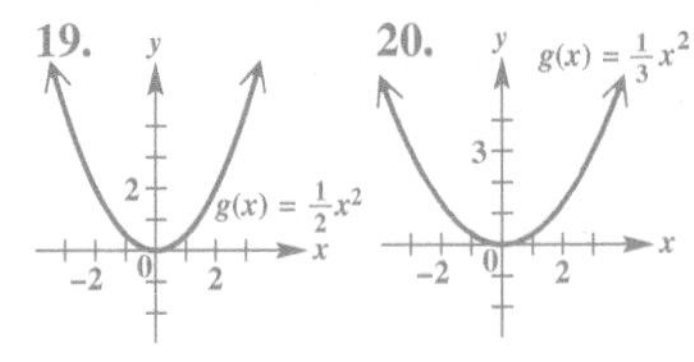

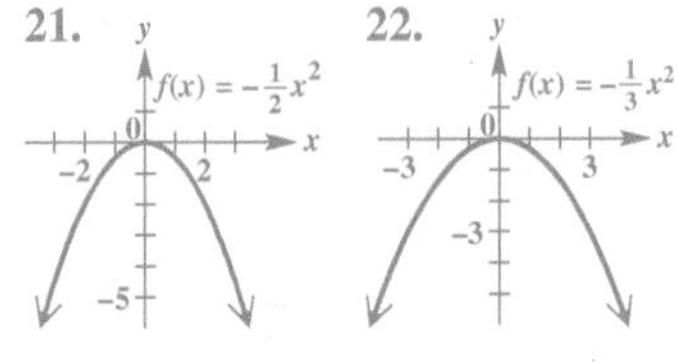

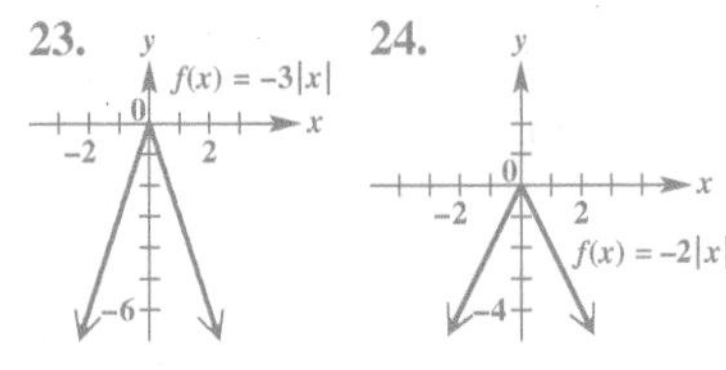

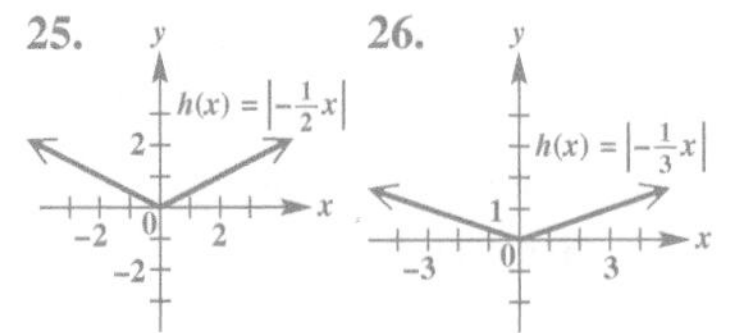

6. The graph of $f(x) = \sqrt{-x}$ is a reflection of the graph of $y = \sqrt{x}$ across the ______-axis.

7. To obtain the graph of $f(x) = (x + 2)^3 - 3$, shift the graph of $y = x^3$ to the left ______ units and down ______ units.

8. To obtain the graph of $f(x) = (x - 3)^3 + 6$, shift the graph of $y = x^3$ to the right ______ units and up ______ units.

Concept Check *Work each matching problem.*

9. Match each equation in Column I with a description of its graph from Column II as it relates to the graph of $y = x^2$.

I	II
(a) $y = (x - 7)^2$	**A.** a translation 7 units to the left
(b) $y = x^2 - 7$	**B.** a translation 7 units to the right
(c) $y = 7x^2$	**C.** a translation 7 units up
(d) $y = (x + 7)^2$	**D.** a translation 7 units down
(e) $y = x^2 + 7$	**E.** a vertical stretching by a factor of 7

10. Match each equation in Column I with a description of its graph from Column II as it relates to the graph of $y = \sqrt[3]{x}$.

I	II
(a) $y = 4\sqrt[3]{x}$	**A.** a translation 4 units to the right
(b) $y = -\sqrt[3]{x}$	**B.** a translation 4 units down
(c) $y = \sqrt[3]{-x}$	**C.** a reflection across the x-axis
(d) $y = \sqrt[3]{x - 4}$	**D.** a reflection across the y-axis
(e) $y = \sqrt[3]{x} - 4$	**E.** a vertical stretching by a factor of 4

11. Match each equation with the sketch of its graph in A–I.

(a) $y = x^2 + 2$ **(b)** $y = x^2 - 2$ **(c)** $y = (x + 2)^2$

(d) $y = (x - 2)^2$ **(e)** $y = 2x^2$ **(f)** $y = -x^2$

(g) $y = (x - 2)^2 + 1$ **(h)** $y = (x + 2)^2 + 1$ **(i)** $y = (x + 2)^2 - 1$

A. **B.**

C.

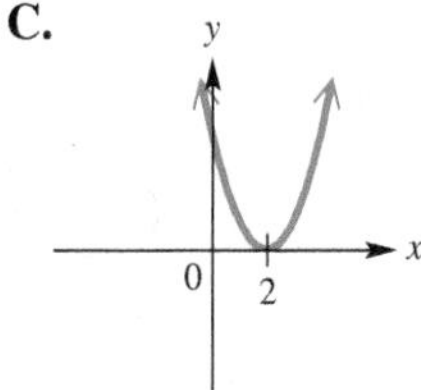

D.

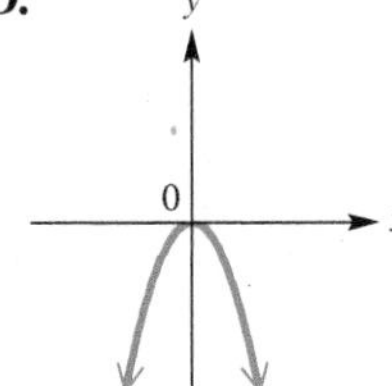

E.

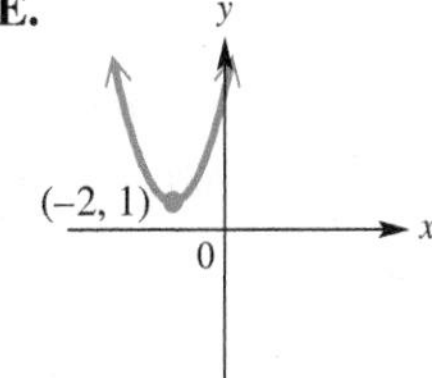

F.

G. **H.**

I.

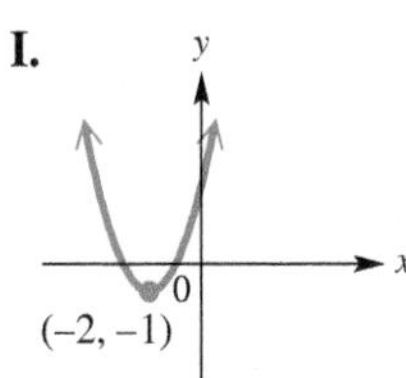

27.

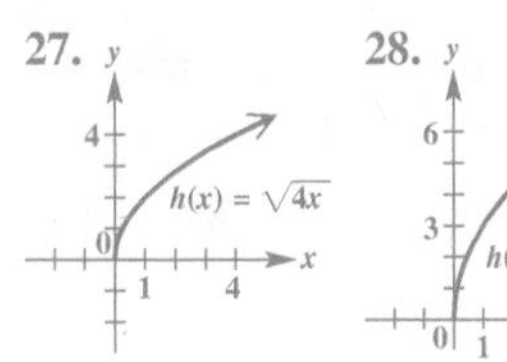

28.

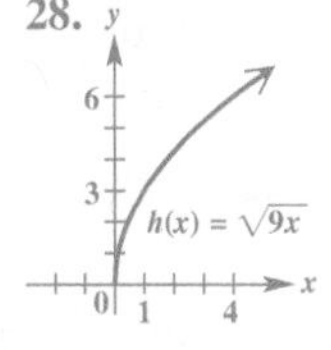

29. 30.

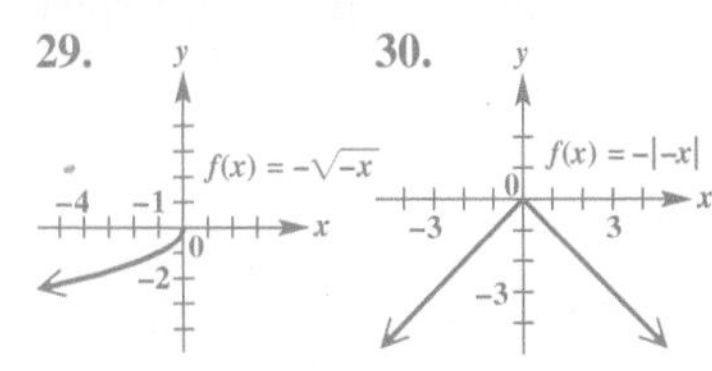

31. 32.

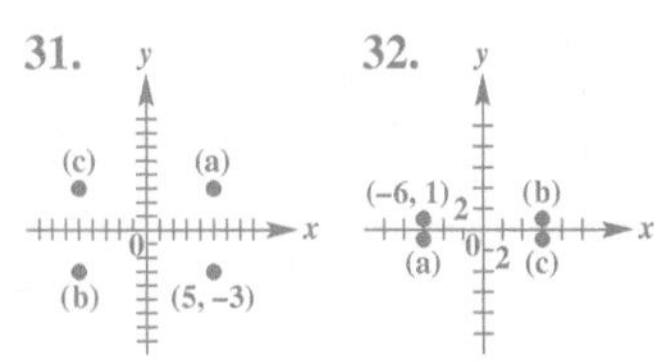

33. 34.

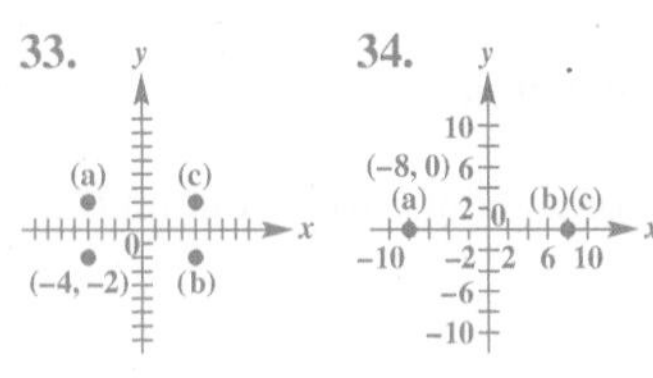

35. y-axis 36. y-axis
37. x-axis, y-axis, origin
38. x-axis, y-axis, origin
39. origin 40. origin
41. none of these
42. none of these

43. 44.

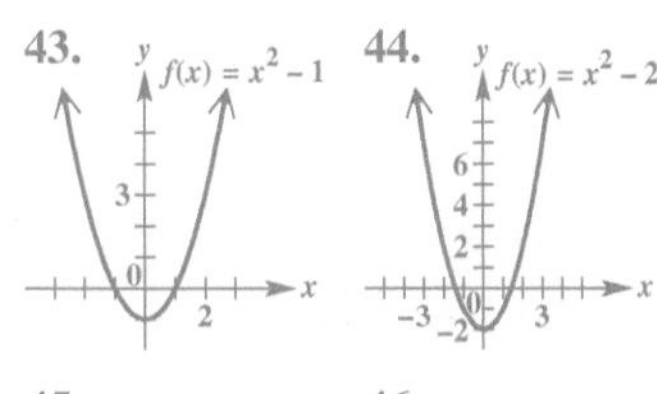

45. 46.

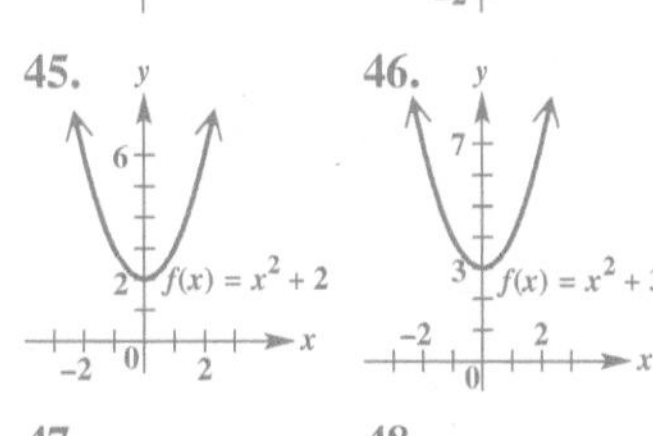

47. 48. 49. 50.

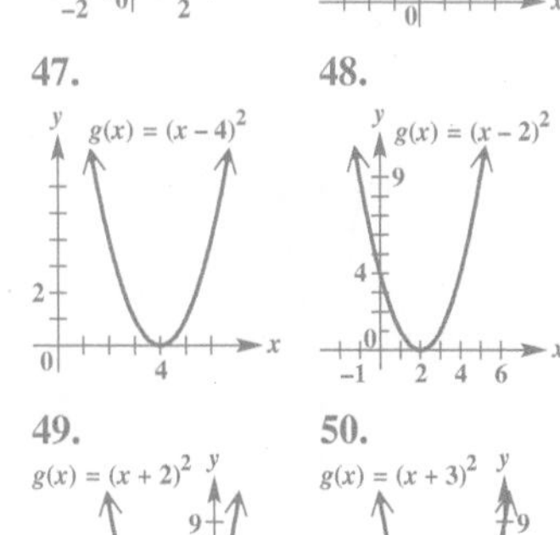

12. Match each equation with the sketch of its graph in A–I.

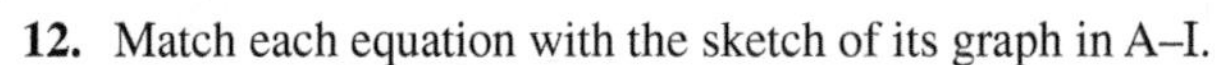

(a) $y = |x - 2|$ **(b)** $y = |x| - 2$ **(c)** $y = |x| + 2$
(d) $y = 2|x|$ **(e)** $y = -|x|$ **(f)** $y = |-x|$
(g) $y = -2|x|$ **(h)** $y = |x - 2| + 2$ **(i)** $y = |x + 2| - 2$

A.

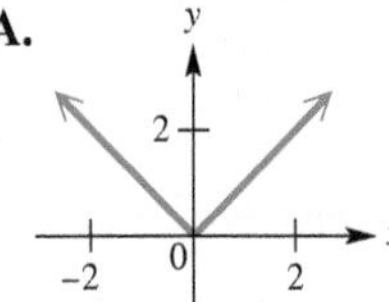

B.

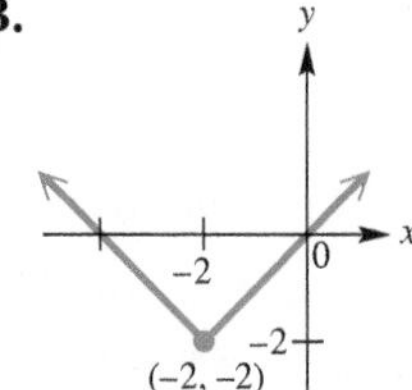

C.

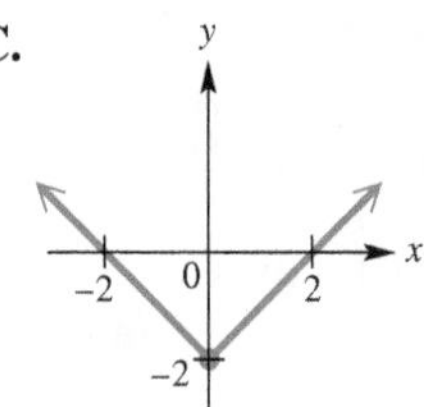

D.

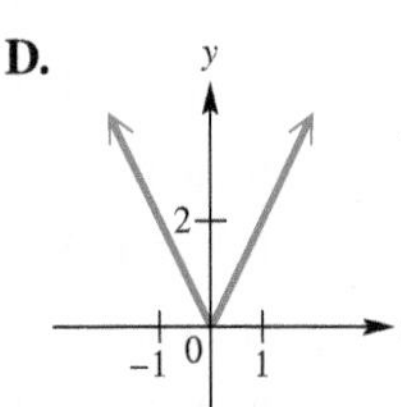

E.

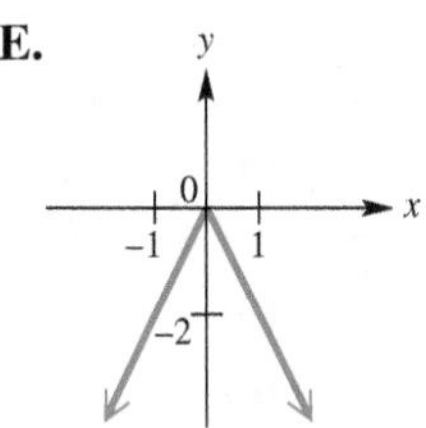

F.

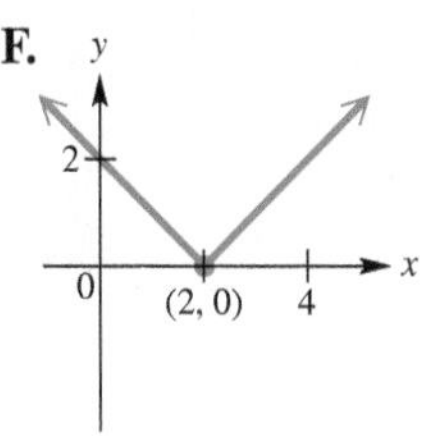

G.

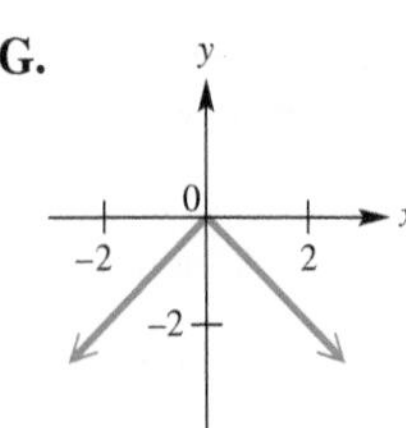

H.

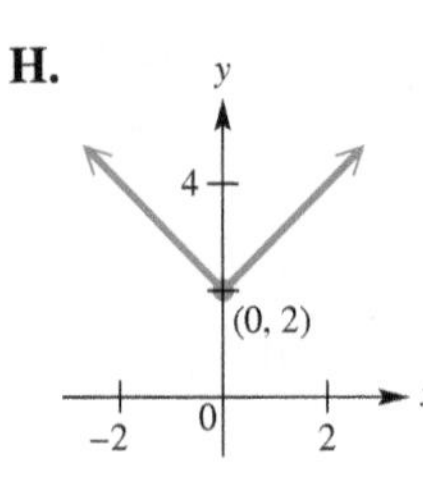

I.

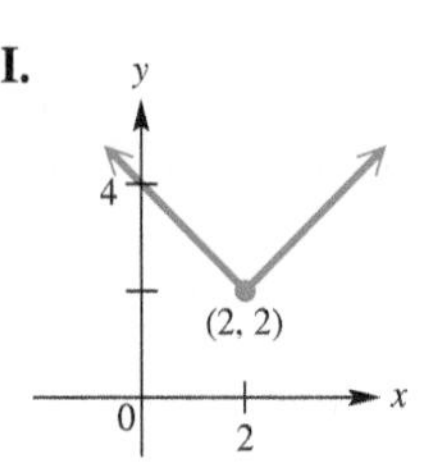

Graph each function. ***See Examples 1 and 2.***

13. $f(x) = 3|x|$ **14.** $f(x) = 4|x|$ **15.** $f(x) = \frac{2}{3}|x|$

16. $f(x) = \frac{3}{4}|x|$ **17.** $g(x) = 2x^2$ **18.** $g(x) = 3x^2$

19. $g(x) = \frac{1}{2}x^2$ **20.** $g(x) = \frac{1}{3}x^2$ **21.** $f(x) = -\frac{1}{2}x^2$

22. $f(x) = -\frac{1}{3}x^2$ **23.** $f(x) = -3|x|$ **24.** $f(x) = -2|x|$

25. $h(x) = \left|-\frac{1}{2}x\right|$ **26.** $h(x) = \left|-\frac{1}{3}x\right|$ **27.** $h(x) = \sqrt{4x}$

28. $h(x) = \sqrt{9x}$ **29.** $f(x) = -\sqrt{-x}$ **30.** $f(x) = -|-x|$

Concept Check *Plot each point, and then plot the points that are symmetric to the given point with respect to the* ***(a)*** *x-axis,* ***(b)*** *y-axis, and* ***(c)*** *origin.*

31. $(5, -3)$ **32.** $(-6, 1)$ **33.** $(-4, -2)$ **34.** $(-8, 0)$

Without graphing, determine whether each equation has a graph that is symmetric with respect to the x-axis, the y-axis, the origin, *or* none of these. ***See Examples 3 and 4.***

35. $y = x^2 + 5$ **36.** $y = 2x^4 - 3$

37. $x^2 + y^2 = 12$ **38.** $y^2 - x^2 = -6$

39. $y = -4x^3 + x$ **40.** $y = x^3 - x$

41. $y = x^2 - x + 8$ **42.** $y = x + 15$

51. $g(x) = |x| - 1$

52. $g(x) = |x + 3| + 2$

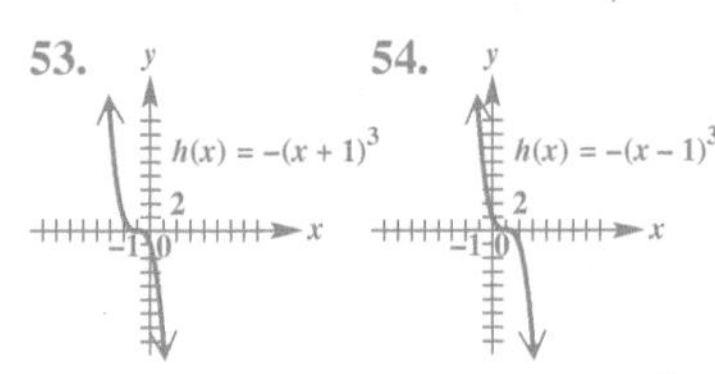

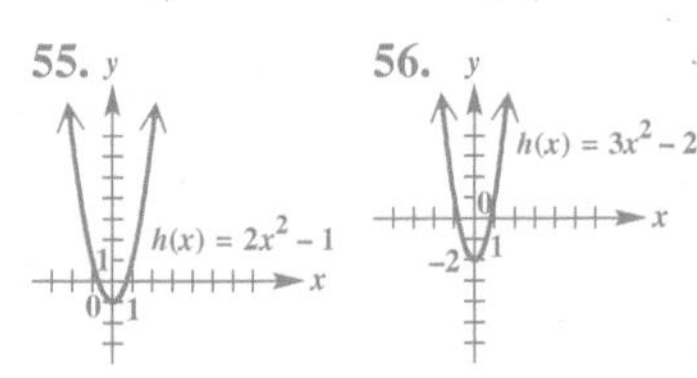

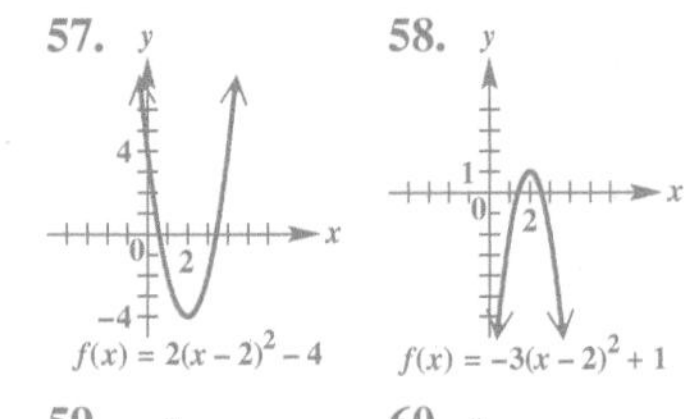

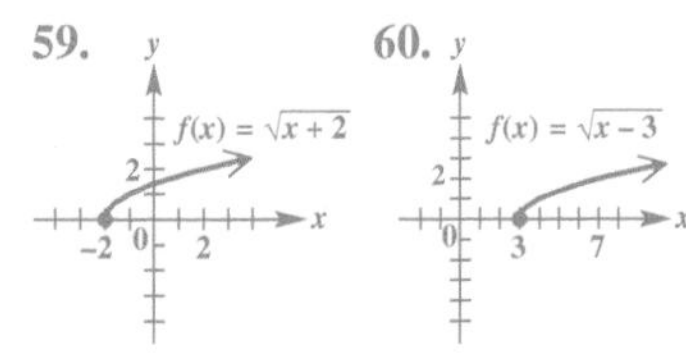

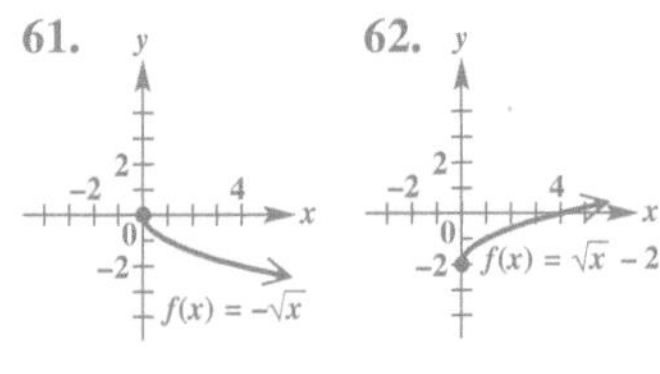

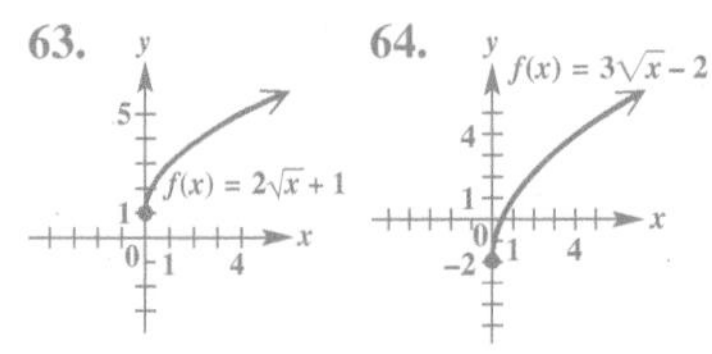

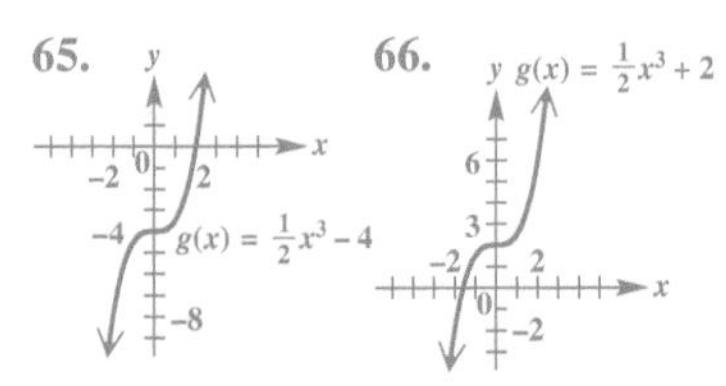

Graph each function. ***See Examples 5–7.***

43. $f(x) = x^2 - 1$ **44.** $f(x) = x^2 - 2$ **45.** $f(x) = x^2 + 2$

46. $f(x) = x^2 + 3$ **47.** $g(x) = (x - 4)^2$ **48.** $g(x) = (x - 2)^2$

49. $g(x) = (x + 2)^2$ **50.** $g(x) = (x + 3)^2$ **51.** $g(x) = |x| - 1$

52. $g(x) = |x + 3| + 2$ **53.** $h(x) = -(x + 1)^3$ **54.** $h(x) = -(x - 1)^3$

55. $h(x) = 2x^2 - 1$ **56.** $h(x) = 3x^2 - 2$ **57.** $f(x) = 2(x - 2)^2 - 4$

58. $f(x) = -3(x - 2)^2 + 1$ **59.** $f(x) = \sqrt{x + 2}$ **60.** $f(x) = \sqrt{x - 3}$

61. $f(x) = -\sqrt{x}$ **62.** $f(x) = \sqrt{x} - 2$ **63.** $f(x) = 2\sqrt{x} + 1$

64. $f(x) = 3\sqrt{x} - 2$ **65.** $g(x) = \frac{1}{2}x^3 - 4$ **66.** $g(x) = \frac{1}{2}x^3 + 2$

Connecting Graphs with Equations *Each of the following graphs is obtained from the graph of $f(x) = |x|$ or $g(x) = \sqrt{x}$ by applying several of the transformations discussed in this section. Describe the transformations and give an equation for the graph.*

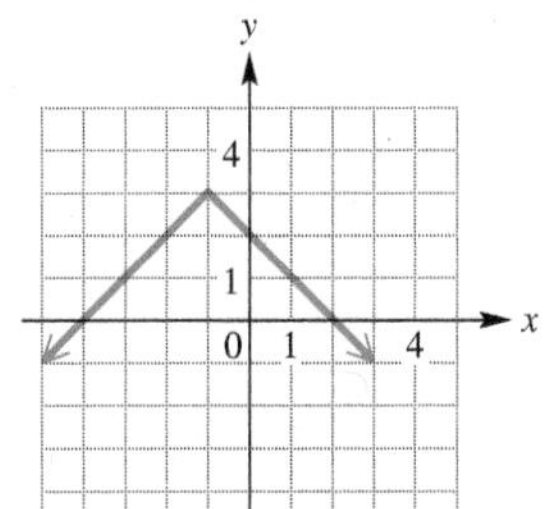

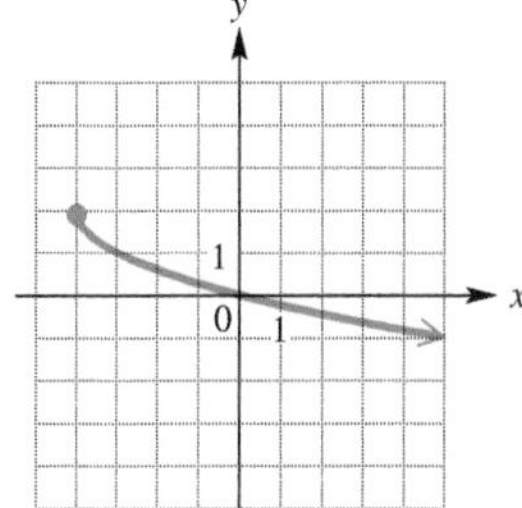

69.

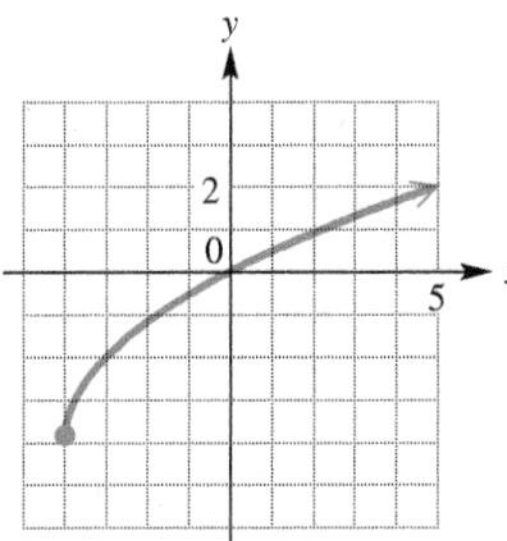

70.

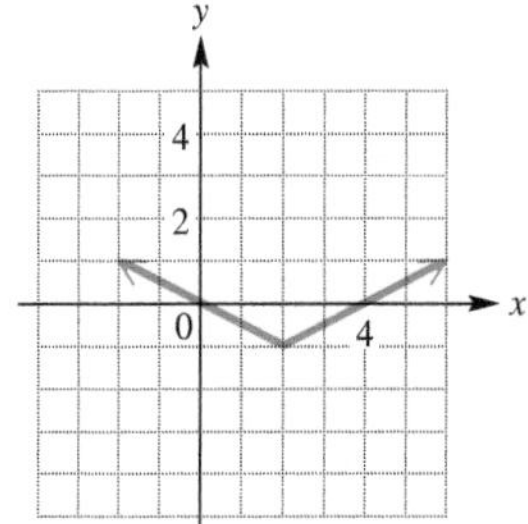

67. It is the graph of $f(x) = |x|$ translated 1 unit to the left, reflected across the x-axis, and translated 3 units up. The equation is

$$y = -|x + 1| + 3.$$

68. It is the graph of $g(x) = \sqrt{x}$ translated 4 units to the left, reflected across the x-axis, and translated 2 units up. The equation is

$$y = -\sqrt{x + 4} + 2.$$

69. It is the graph of $g(x) = \sqrt{x}$ translated 4 units to the left, stretched vertically by a factor of 2, and translated 4 units down. The equation is

$$y = 2\sqrt{x + 4} - 4.$$

70. It is the graph of $f(x) = |x|$ translated 2 units to the right, shrunken vertically by a factor of $\frac{1}{2}$, and translated 1 unit down. The equation is

$$y = \frac{1}{2}|x - 2| - 1.$$

Answers to Selected Exercises

Chapter 4

Summary Exercises on Graphing Circular Functions

1.
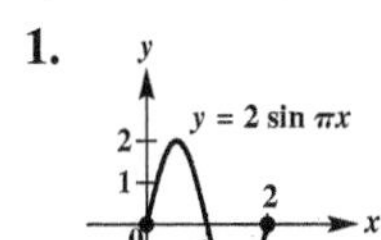

2.
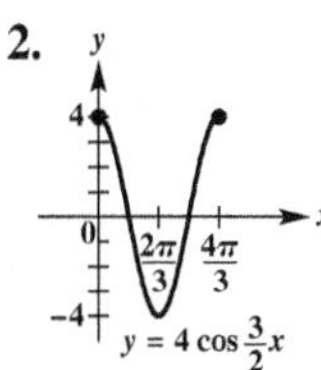

3.
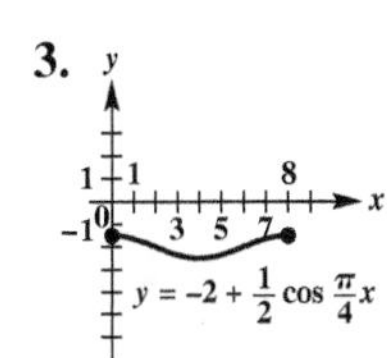

4.
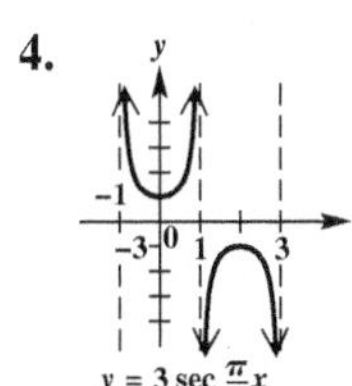

5.
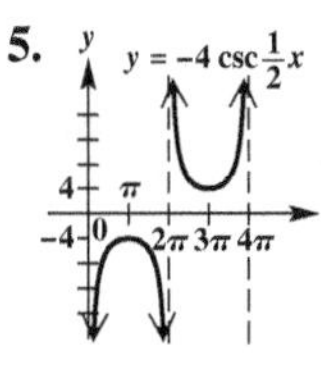

6.
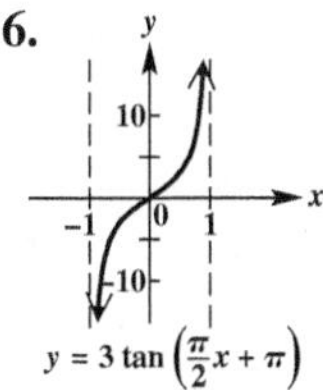

7.
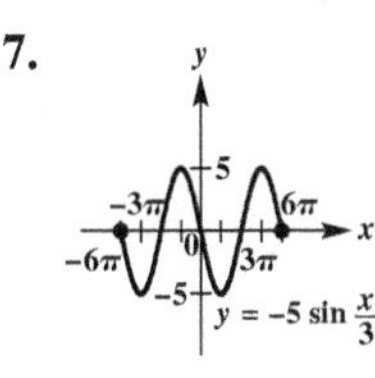

8.
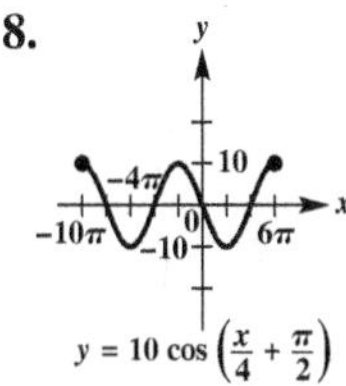

9.
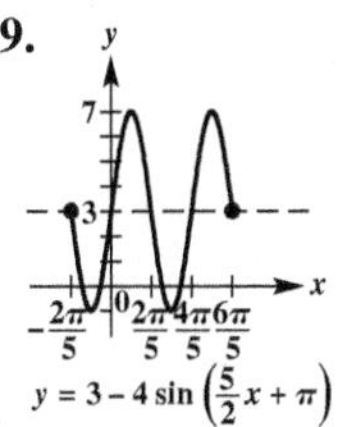

10.
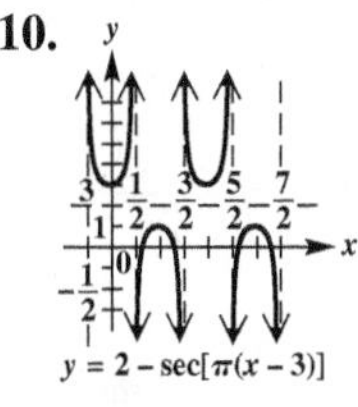

8.2 Exercises

67. yes

68. **(b)** $z_1{}^2 - 1$
$= a^2 - b^2 - 1 + 2abi;$
$z_2{}^2 - 1$
$= a^2 - b^2 - 1 - 2abi$

(c) Both have the same r value because for any a and b,

$$\sqrt{a^2 + b^2} = \sqrt{a^2 + (-b)^2}.$$

(d) The condition described in part (c) assures this.

69. B **70.** C **71.** A

8.4 Exercises

32. cos 0° + i sin 0°,
cos 90° + i sin 90°,
cos 180° + i sin 180°,
cos 270° + i sin 270°

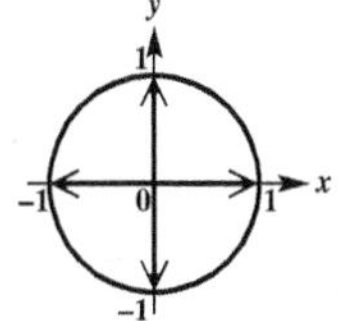

33. cos 0° + i sin 0°,
cos 60° + i sin 60°,
cos 120° + i sin 120°,
cos 180° + i sin 180°,
cos 240° + i sin 240°,
cos 300° + i sin 300°

34. cos 45° + i sin 45°,
cos 225° + i sin 225°

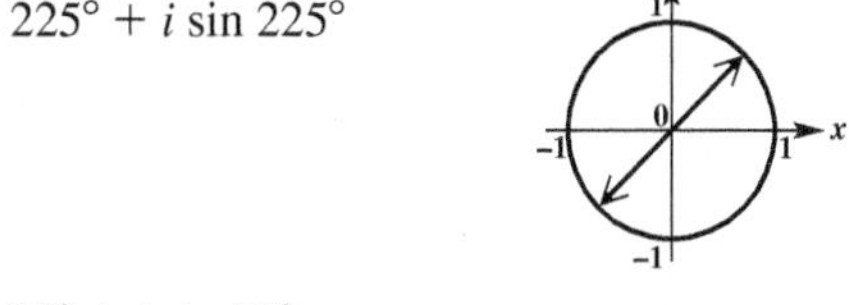

35. cos 30° + i sin 30°,
cos 150° + i sin 150°,
cos 270° + i sin 270°

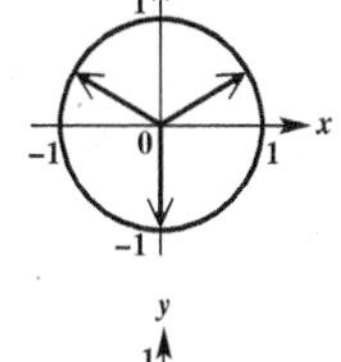

36. cos 22.5° + i sin 22.5°,
cos 112.5° + i sin 112.5°,
cos 202.5° + i sin 202.5°,
cos 292.5° + i sin 292.5°

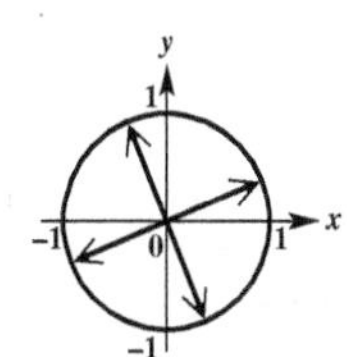

37. {cos 0° + i sin 0°, cos 120° + i sin 120°, cos 240° + i sin 240°} **38.** {cos 60° + i sin 60°, cos 180° + i sin 180°, cos 300° + i sin 300°} **39.** {cos 90° + i sin 90°, cos 210° + i sin 210°, cos 330° + i sin 330°} **40.** {cos 67.5° + i sin 67.5°, cos 157.5° + i sin 157.5°, cos 247.5° + i sin 247.5°, cos 337.5° + i sin 337.5°} **41.** {2(cos 0° + i sin 0°), 2(cos 120° + i sin 120°), 2(cos 240° + i sin 240°)} **42.** {3(cos 60° + i sin 60°), 3(cos 180° + i sin 180°), 3(cos 300° + i sin 300°)} **43.** {cos 45° + i sin 45°, cos 135° + i sin 135°, cos 225° + i sin 225°, cos 315° + i sin 315°} **44.** {2(cos 45° + i sin 45°), 2(cos 135° + i sin 135°), 2(cos 225° + i sin 225°), 2(cos 315° + i sin 315°)} **45.** {cos 22.5° + i sin 22.5°, cos 112.5° + i sin 112.5°, cos 202.5° + i sin 202.5°, cos 292.5° + i sin 292.5°} **46.** {cos 18° + i sin 18°, cos 90° + i sin 90°, cos 162° + i sin 162°, cos 234° + i sin 234°, cos 306° + i sin 306°} **47.** {2(cos 20° + i sin 20°), 2(cos 140° + i sin 140°), 2(cos 260° + i sin 260°)} **48.** {2(cos 15° + i sin 15°), 2(cos 105° + i sin 105°), 2(cos 195° + i sin 195°), 2(cos 285° + i sin 285°)} **49.** $1, -\frac{1}{2} + \frac{\sqrt{3}}{2}i, -\frac{1}{2} - \frac{\sqrt{3}}{2}i$ **50.** $-3, \frac{3}{2} + \frac{3\sqrt{3}}{2}i, \frac{3}{2} - \frac{3\sqrt{3}}{2}i$ **51.** **(a)** yes **(b)** no **(c)** yes **52.** **(a)** blue **(b)** red **(c)** yellow

53. 1, 0.30901699 + 0.95105652i, −0.809017 + 0.58778525i, −0.809017 − 0.5877853i, 0.30901699 − 0.9510565i **54.** 1, 0.80901699 + 0.58778525i, 0.30901699 + 0.95105652i **55.** $\{-1.8174 + 0.5503i, 1.8174 - 0.5503i\}$ **56.** $\{0.3436 + 1.4553i, -0.3436 - 1.4553i\}$ **57.** $\{0.8771 + 0.9492i, -0.6317 + 1.1275i, -1.2675 - 0.2524i, -0.1516 - 1.2835i, 1.1738 - 0.5408i\}$ **58.** $\{1.3606 + 1.2637i, -1.7747 + 0.5464i, 0.4141 - 1.8102i\}$ **59.** $\cos 2\theta + i \sin 2\theta$ **60.** $(\cos^2 \theta - \sin^2 \theta) + i(2 \cos \theta \sin \theta) = \cos 2\theta + i \sin 2\theta$ **61.** $\cos 2\theta = \cos^2 \theta - \sin^2 \theta$ **62.** $\sin 2\theta = 2 \sin \theta \cos \theta$

8.5 Exercises

61. $x^2 + (y - 1)^2 = 1$

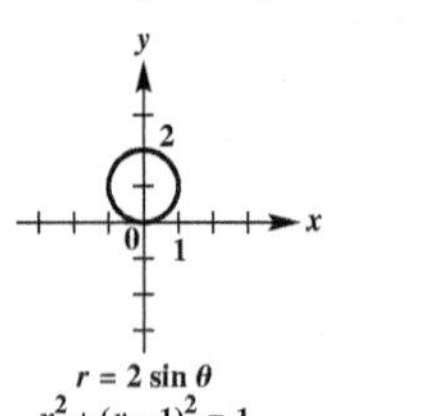

62. $(x - 1)^2 + y^2 = 1$

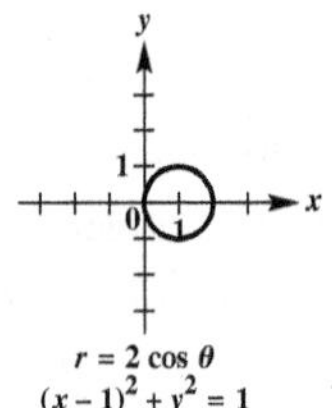

63. $y^2 = 4(x + 1)$

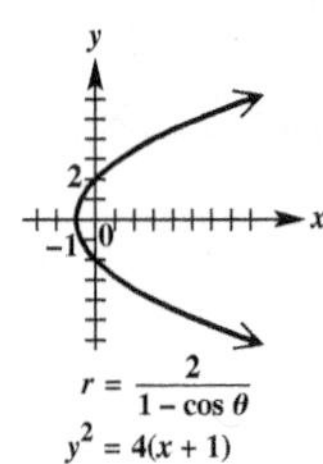

64. $x^2 = 6\left(y + \frac{3}{2}\right)$

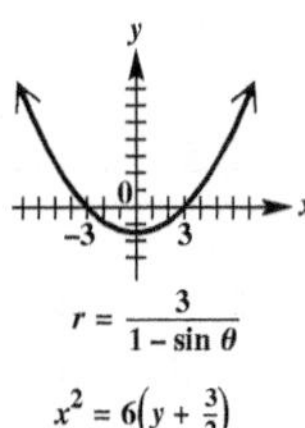

65. $(x + 1)^2 + (y + 1)^2 = 2$

66. $4x - y = 3$

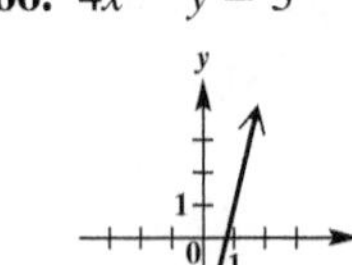

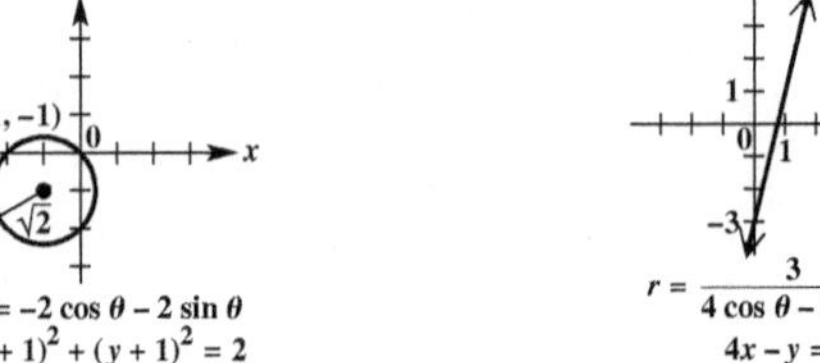

67. $x = 2$

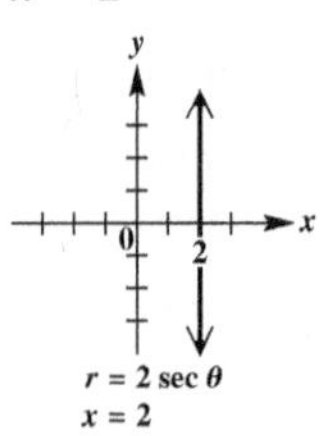

68. $y = -5$

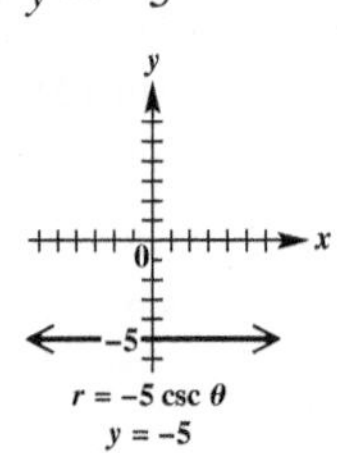

69. $x + y = 2$

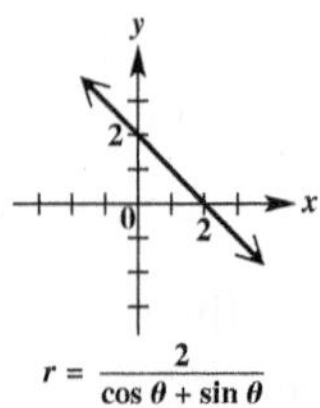

70. $2x + y = 2$

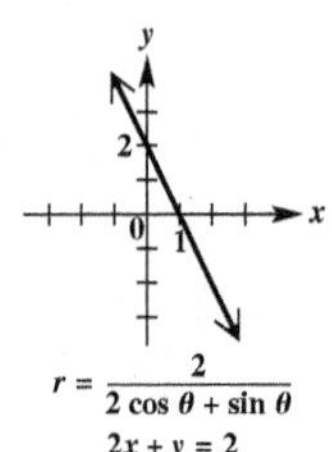

71. $r = \frac{2}{2 \cos \theta + \sin \theta}$ **72.** To plot the point (r, θ) for $r < 0$, and θ in degrees, add 180° to θ and plot $(|r|, \theta + 180°)$. (Alternatively, count $|r|$ units in the opposite direction from the pole on the extension of the ray with angle θ.) **73. (a)** $(r, -\theta)$ **(b)** $(r, \pi - \theta)$ or $(-r, -\theta)$ **(c)** $(r, \pi + \theta)$ or $(-r, \theta)$ **74. (a)** $-\theta$ **(b)** $\pi - \theta$ **(c)** $-r$; $-\theta$ **(d)** $-r$ **(e)** $\pi + \theta$ **(f)** the polar axis **(g)** the line $\theta = \frac{\pi}{2}$

75.

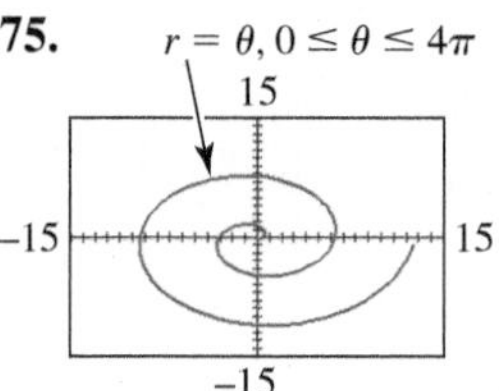

76. $r = 2\theta, -4\pi \le \theta \le 4\pi$
30
−30 30
−30

77.

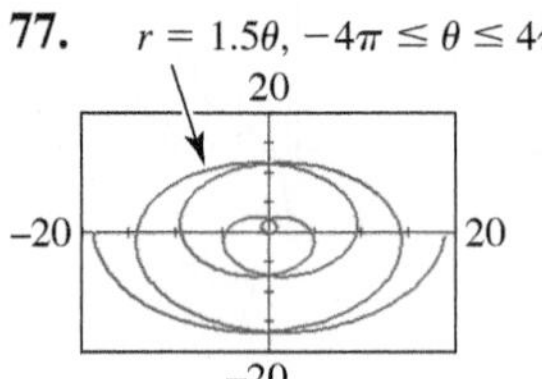

78. $r = -\theta, 0 \le \theta \le 12\pi$
40
−40 40
−40

79. $\left(2, \frac{\pi}{6}\right), \left(2, \frac{5\pi}{6}\right), (0, 0)$ **80.** (3, 60°), (3, 300°)

81. $\left(\frac{4 + \sqrt{2}}{2}, \frac{\pi}{4}\right), \left(\frac{4 - \sqrt{2}}{2}, \frac{5\pi}{4}\right)$

82. $\left(1, \frac{\pi}{4}\right), \left(0, \frac{\pi}{2}\right), \left(-1, \frac{3\pi}{4}\right)$

83. (a)

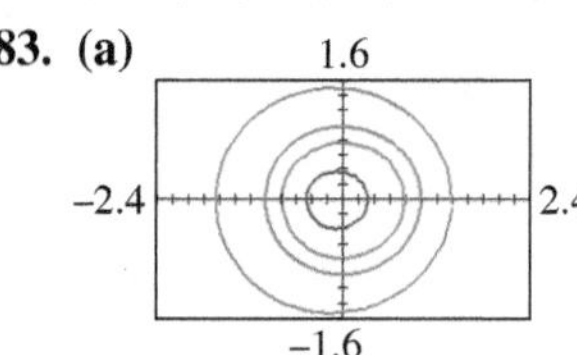

(b)

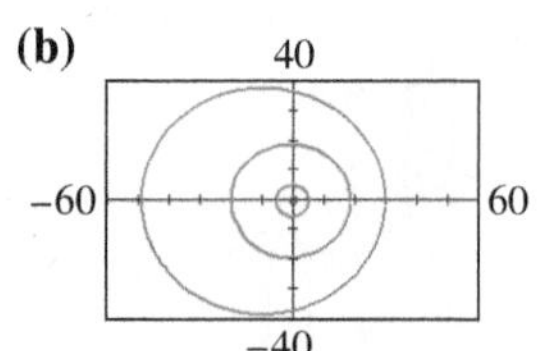

Earth is closest to the sun.

(c) no

84. (a) Inside the lemniscate the radio signal can be received. This region is generally in an east-west direction from the two radio towers with maximum distance 200 mi.

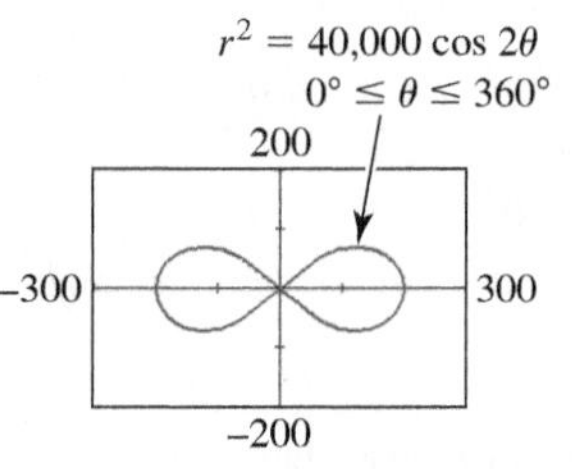

(b) The radio signal is received inside the lemniscate. This region is generally in a southwest-northeast direction from the two towers with maximum distance 150 mi.

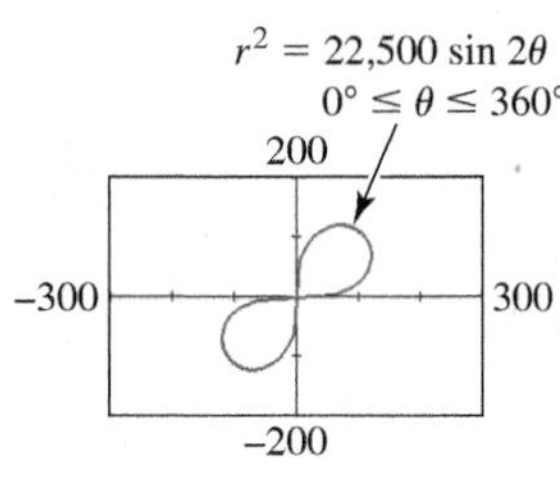

85. $r \sin \theta = k$ **86.** $r = \frac{k}{\sin \theta}$ **87.** $r = k \csc \theta$

88. $y = 3$

89. $r \cos \theta = k$

90. $r = \frac{k}{\cos \theta}$

91. $r = k \sec \theta$

92. 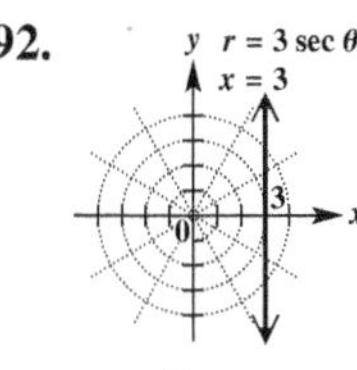

$x = 3$

8.6 Exercises

28. **(a)** 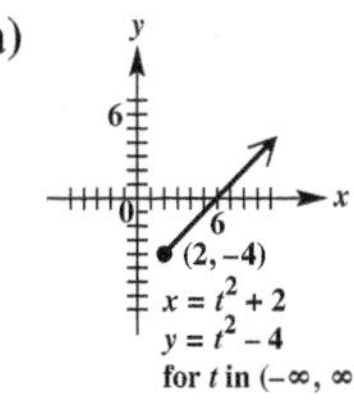

(b) $y = x - 6$, for x in $[2, \infty)$

29. 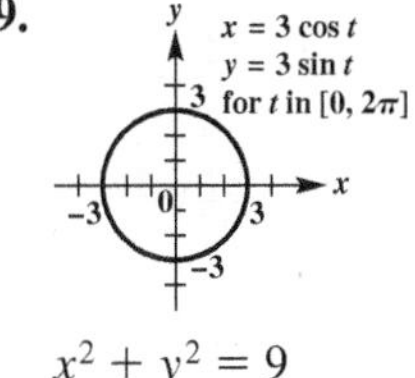

$x^2 + y^2 = 9$

30. 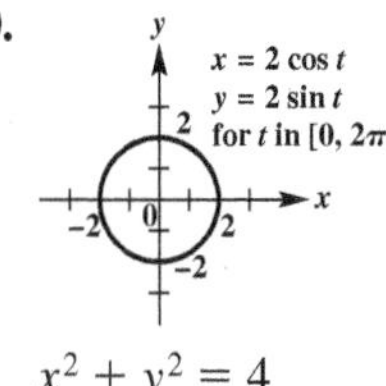

$x^2 + y^2 = 4$

31. 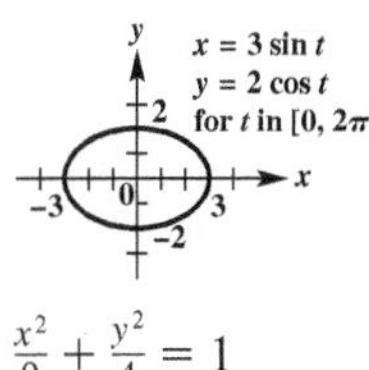

$\frac{x^2}{9} + \frac{y^2}{4} = 1$

32. 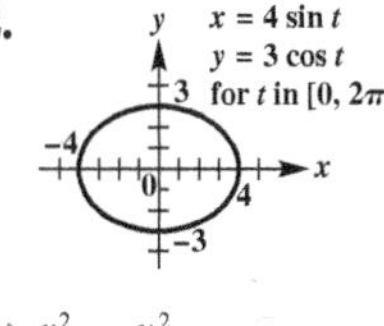

$\frac{x^2}{16} + \frac{y^2}{9} = 1$

Answers may vary for Exercises 33–36.

33. $x = t$, $y = (t + 3)^2 - 1$, for t in $(-\infty, \infty)$; $x = t - 3$, $y = t^2 - 1$, for t in $(-\infty, \infty)$ **34.** $x = t$, $y = (t + 4)^2 + 2$, for t in $(-\infty, \infty)$; $x = t - 4$, $y = t^2 + 2$, for t in $(-\infty, \infty)$ **35.** $x = t$, $y = t^2 - 2t + 3$, for t in $(-\infty, \infty)$; $x = t + 1$, $y = t^2 + 2$, for t in $(-\infty, \infty)$ **36.** $x = t$, $y = t^2 - 4t + 6$, for t in $(-\infty, \infty)$; $x = t + 2$, $y = t^2 + 2$, for t in $(-\infty, \infty)$

37. 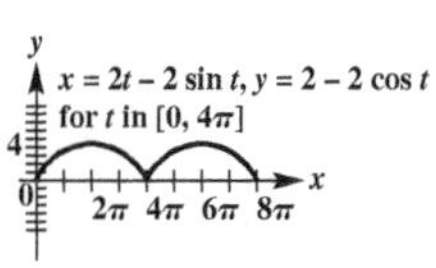

38. 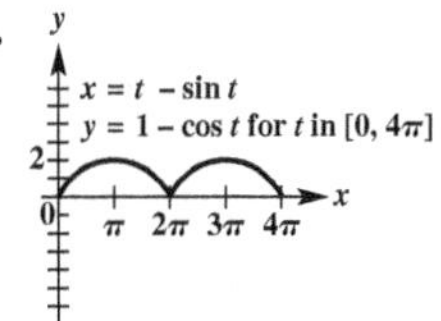

39. 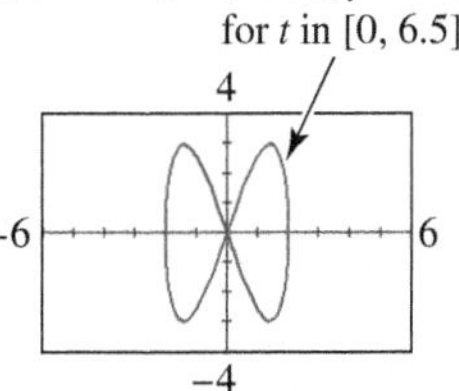

40. 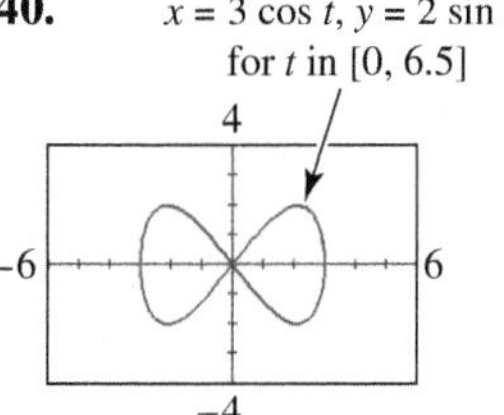

41. 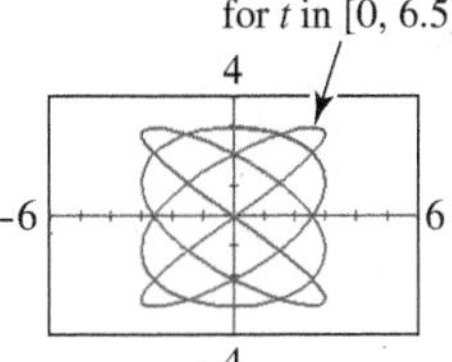

42. 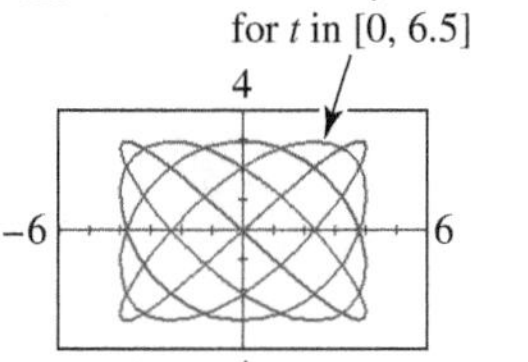

43. **(a)** $x = 24t$, $y = -16t^2 + 24\sqrt{3}t$ **(b)** $y = -\frac{1}{36}x^2 + \sqrt{3}x$ **(c)** 2.6 sec; 62 ft **44.** **(a)** $x = 75t$, $y = -16t^2 + 75\sqrt{3}t$ **(b)** $y = -\frac{16}{5625}x^2 + \sqrt{3}x$ **(c)** 8.1 sec; 609 ft **45.** **(a)** $x = (88 \cos 20°)t$, $y = 2 - 16t^2 + (88 \sin 20°)t$ **(b)** $y = 2 - \frac{x^2}{484 \cos^2 20°} + (\tan 20°)x$ **(c)** 1.9 sec; 161 ft **46.** **(a)** $x = (136 \cos 29°)t$, $y = 2.5 - 16t^2 + (136 \sin 29°)t$ **(b)** $y = 2.5 - \frac{x^2}{1156 \cos^2 29°} + (\tan 29°)x$ **(c)** 4.2 sec; 495 ft **47.** **(a)** $y = -\frac{1}{256}x^2 + \sqrt{3}x + 8$; parabolic path **(b)** 7 sec; 448 ft **48.** 1456 ft **49.** **(a)** $x = 32t$, $y = 32\sqrt{3}t - 16t^2 + 3$ **(b)** 112.6 ft **(c)** 51 ft maximum height; The ball had traveled horizontally 55.4 ft. **(d)** yes **50.** **(a)** 20.0° **(b)** $x = (88 \cos 20.0°)t$, $y = -16t^2 + (88 \sin 20.0°)t$ **51.** Many answers are possible; for example, $y = a(t - h)^2 + k$, $x = t$ and $y = at^2 + k$, $x = t + h$. **52.** Many answers are possible; for example, $x = a \sec \theta$, $y = b \tan \theta$ and $x = t$, $y^2 = \frac{b^2}{a^2}(t^2 - a^2)$. **53.** Many answers are possible; for example, $x = a \sin t$, $y = b \cos t$ and $x = t$, $y^2 = b^2\left(1 - \frac{t^2}{a^2}\right)$. **56.** the pair $x = \cos t$, $y = -\sin t$ **57.** The graph is translated c units to the right. **58.** The graph is translated d units up.

Chapter 8 Test

10. **(a)** $\left(\frac{3\sqrt{2}}{2}, -\frac{3\sqrt{2}}{2}\right)$ **(b)** $(0, -4)$

11. cardioid

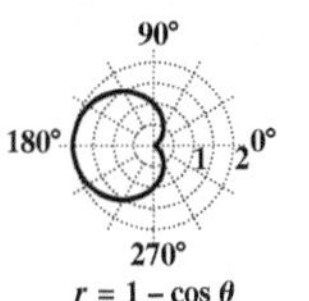

12. three-leaved rose

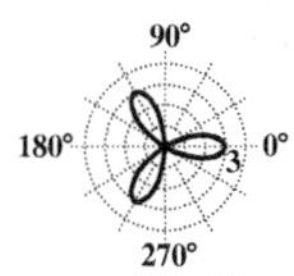

13. **(a)** $x - 2y = -4$ **(b)** $x^2 + y^2 = 36$

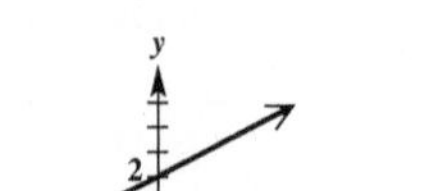
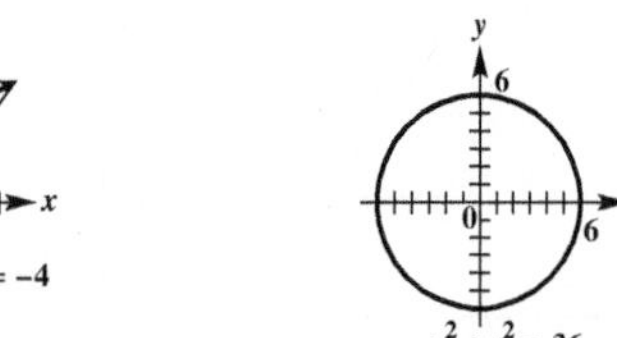

[8.6] **14.**

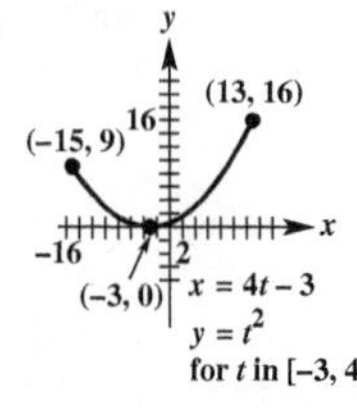

15.

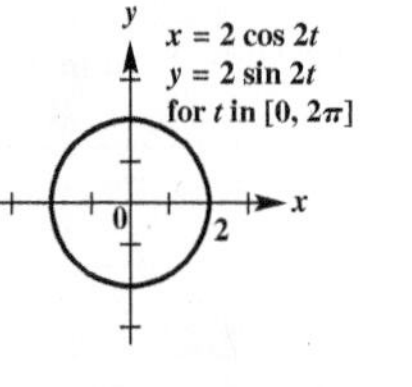

[8.2] **16.** $z^2 - 1 = -1 - 2i$; $r = \sqrt{5}$ and $\sqrt{5} > 2$

Appendix B Exercises

24. **(a)**

x	y
−2	−2
−4	0
0	−4

(b)

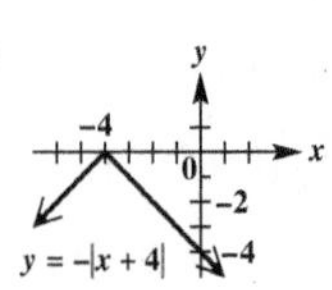

25. **(a)**

x	y
0	0
−1	−1
2	8

(b)

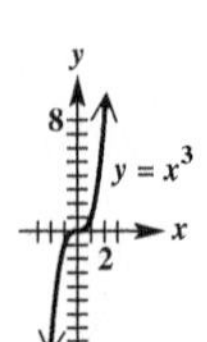

26. **(a)**

x	y
0	0
1	−1
2	−8

(b)

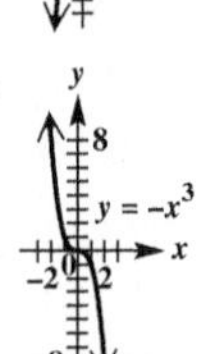

27. $(0, 0)$; 7 **28.** $(x - 3)^2 + (y - 6)^2 = 16$ **29.** $(4, -7)$

30. $(0, 5)$ **31.** B **32.** C **33.** D **34.** A

35. **(a)** $x^2 + y^2 = 36$ **(b)**

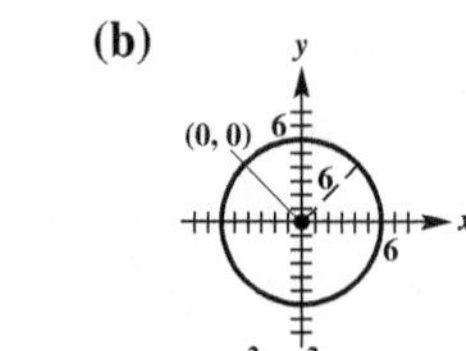

36. **(a)** $x^2 + y^2 = 81$ **(b)**

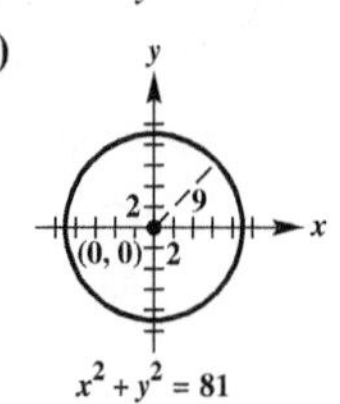

37. **(a)** $(x - 2)^2 + y^2 = 36$ **(b)**

38. **(a)** $(x - 3)^2 + y^2 = 9$ **(b)**

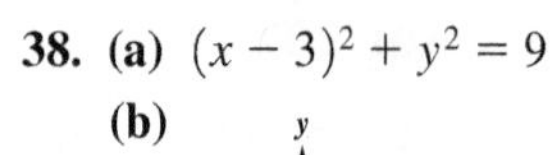

39. **(a)** $x^2 + (y - 4)^2 = 16$ **(b)**

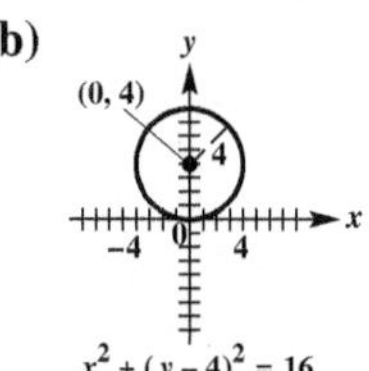

40. **(a)** $x^2 + (y + 3)^2 = 49$ **(b)**

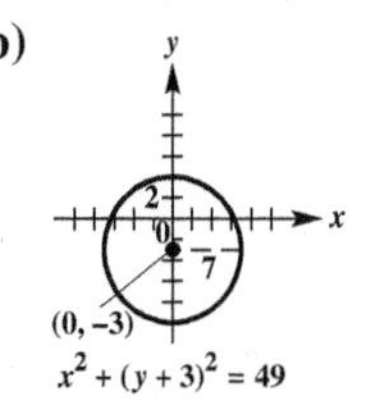

41. **(a)** $(x + 2)^2 + (y - 5)^2 = 16$

(b)

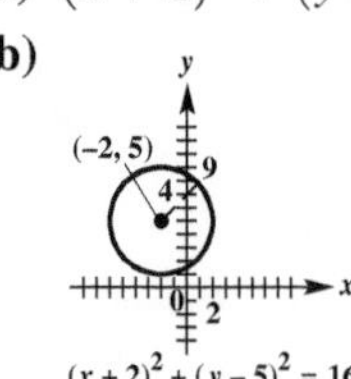

42. **(a)** $(x - 4)^2 + (y - 3)^2 = 25$

(b)

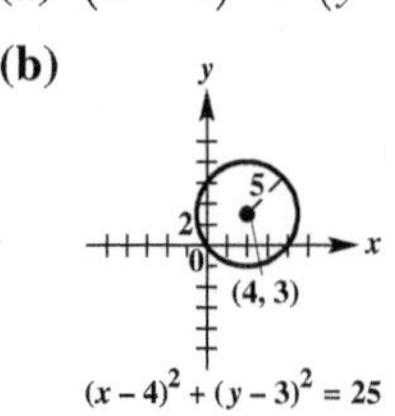

43. **(a)** $(x - 5)^2 + (y + 4)^2 = 49$

(b)

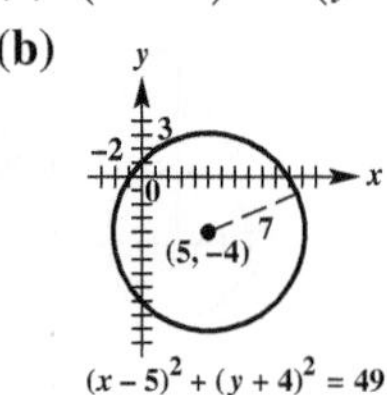

44. **(a)** $(x + 3)^2 + (y + 2)^2 = 36$

(b)

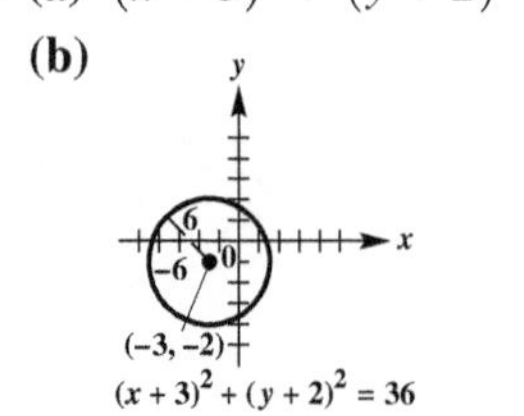

45. **(a)** $\left(x - \sqrt{2}\right)^2 + \left(y - \sqrt{2}\right)^2 = 2$

(b)

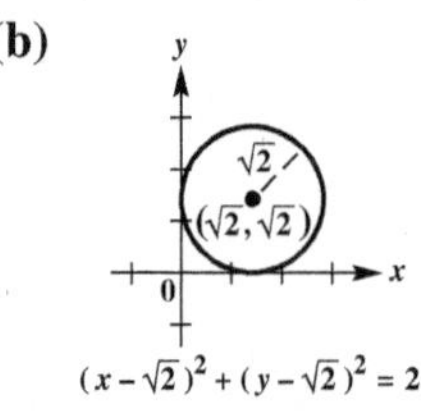

46. **(a)** $\left(x + \sqrt{3}\right)^2 + \left(y + \sqrt{3}\right)^2 = 3$

(b)

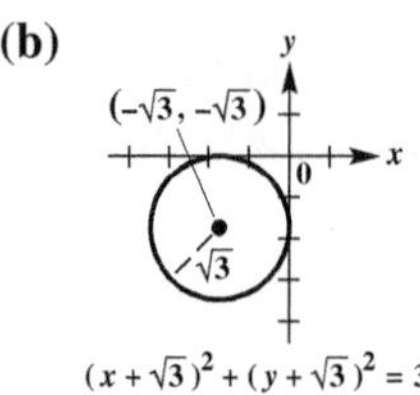

47. $(x - 3)^2 + (y - 1)^2 = 4$

48. $(x + 1)^2 + (y + 2)^2 = 9$

Photo Credits

COVER: Moaan/Getty Images

CHAPTER 1: **1** Christian Delbert/Shutterstock **13** Scott Prokop/Fotolia **19** Courtesy of Callie Daniels **20** Markus Gann/Shutterstock **21** Krzysztof Wiktor/123RF

CHAPTER 2: **47** Sergii Mostovyi/Fotolia **68** Brad Pict/Fotolia **69** Vilainecrevette/Shutterstock **70** VladKol/Shutterstock **71** Smileus/Shutterstock **72** (top left) Unknown photographer; (top right) Copyright 1899, 1900 by Elmer A. Lyman and Edwin C. Goddard; (bottom) Africa Studio/Fotolia **76** AP Images **77** Courtesy of John Hornsby **81** F9photos/Shutterstock **82** World History Archive/Alamy **83** 123RF **89** Courtesy of Callie Daniels **95** Marin Conic/Fotolia

CHAPTER 3: **99** Vadimsadovski/Fotolia **106** Plaksik13/Fotolia **109** Cecilia Lim H M/Shutterstock **127** Andres Rodriguez/Fotolia **132** J. Bell, Cornell University/M. Wolff, Space Science Center/NASA **137** Selezeni/Fotolia

CHAPTER 4: **139** Victoria Kalinina/Shutterstock **158** Gibleho/Fotolia **164** Famveldman/Fotolia **185** Alexander Sakhatovsky/Shutterstock **192** Carolina K Smith MD/Fotolia

CHAPTER 5: **195** Plampy/Shutterstock **207** Serggod/Shutterstock **220** Ysbrand Cosijn/Shutterstock **228** Mimagephotos/Fotolia **238** Lithian/123RF **243** US Navy Photo/Alamy **244** Remik44992/123RF

CHAPTER 6: **251** Stokkete/123RF **263** Robert Daly/OJO Images/Getty Images **274** Svetoslav Radkov/Fotolia **278** MindStudio/Pearson Education Ltd

CHAPTER 7: **295** Henryk Sadura/Shutterstock **304** Richard Cavalleri/Shutterstock **312** Stephanie Kennedy/Shutterstock **316** Liszt collection/Alamy **321** Jason Maehl/Shutterstock **333** Sculpies/Shutterstock **334** Gary Blakeley/Shutterstock

CHAPTER 8: **355** Florin C/Shutterstock **369** Ross Anderson **378** Pearson Education, Inc. **383** Ross Anderson **384** Ross Anderson **396** NatUlrich/Shutterstock

Index

S

T

3.1 Conversion of Angular Measure

Degree/Radian Relationship: $180° = \pi$ radians

Conversion Formulas:

From	To	Multiply by
Degrees	Radians	$\dfrac{\pi}{180}$
Radians	Degrees	$\dfrac{180°}{\pi}$

3.2 Applications of Radian Measure

Arc Length: $s = r\theta$, θ in radians

Area of Sector: $\mathcal{A} = \dfrac{1}{2}r^2\theta$, θ in radians

3.4

Angular Speed ω	Linear Speed v
$\omega = \dfrac{\theta}{t}$ (ω in radians per unit time, θ in radians)	$v = \dfrac{s}{t}$ $v = \dfrac{r\theta}{t}$ $v = r\omega$

5.1 Fundamental Identities

$$\cot\theta = \frac{1}{\tan\theta} \qquad \sec\theta = \frac{1}{\cos\theta} \qquad \csc\theta = \frac{1}{\sin\theta}$$

$$\tan\theta = \frac{\sin\theta}{\cos\theta} \qquad \cot\theta = \frac{\cos\theta}{\sin\theta}$$

$$\sin^2\theta + \cos^2\theta = 1 \qquad \tan^2\theta + 1 = \sec^2\theta \qquad 1 + \cot^2\theta = \csc^2\theta$$

$$\sin(-\theta) = -\sin\theta \qquad \cos(-\theta) = \cos\theta \qquad \tan(-\theta) = -\tan\theta$$

$$\csc(-\theta) = -\csc\theta \qquad \sec(-\theta) = \sec\theta \qquad \cot(-\theta) = -\cot\theta$$

5.3, 5.4 Sum and Difference Identities

$$\cos(A + B) = \cos A\cos B - \sin A\sin B$$

$$\cos(A - B) = \cos A\cos B + \sin A\sin B$$

$$\sin(A + B) = \sin A\cos B + \cos A\sin B$$

$$\sin(A - B) = \sin A\cos B - \cos A\sin B$$

$$\tan(A + B) = \frac{\tan A + \tan B}{1 - \tan A\tan B}$$

$$\tan(A - B) = \frac{\tan A - \tan B}{1 + \tan A\tan B}$$

5.5 Product-to-Sum and Sum-to-Product Identities

$$\cos A\cos B = \frac{1}{2}[\cos(A + B) + \cos(A - B)]$$

$$\sin A\sin B = \frac{1}{2}[\cos(A - B) - \cos(A + B)]$$

$$\sin A\cos B = \frac{1}{2}[\sin(A + B) + \sin(A - B)]$$

$$\cos A\sin B = \frac{1}{2}[\sin(A + B) - \sin(A - B)]$$

$$\sin A + \sin B = 2\sin\left(\frac{A + B}{2}\right)\cos\left(\frac{A - B}{2}\right)$$

$$\sin A - \sin B = 2\cos\left(\frac{A + B}{2}\right)\sin\left(\frac{A - B}{2}\right)$$

$$\cos A + \cos B = 2\cos\left(\frac{A + B}{2}\right)\cos\left(\frac{A - B}{2}\right)$$

$$\cos A - \cos B = -2\sin\left(\frac{A + B}{2}\right)\sin\left(\frac{A - B}{2}\right)$$

5.3 Cofunction Identities

$$\cos(90° - \theta) = \sin\theta$$

$$\sin(90° - \theta) = \cos\theta$$

$$\tan(90° - \theta) = \cot\theta$$

$$\cot(90° - \theta) = \tan\theta$$

$$\sec(90° - \theta) = \csc\theta$$

$$\csc(90° - \theta) = \sec\theta$$

5.5, 5.6 Double-Angle and Half-Angle Identities

$$\cos 2A = \cos^2 A - \sin^2 A \qquad \cos 2A = 1 - 2\sin^2 A$$

$$\cos 2A = 2\cos^2 A - 1 \qquad \sin 2A = 2\sin A\cos A$$

$$\tan 2A = \frac{2\tan A}{1 - \tan^2 A} \qquad \cos\frac{A}{2} = \pm\sqrt{\frac{1 + \cos A}{2}}$$

$$\sin\frac{A}{2} = \pm\sqrt{\frac{1 - \cos A}{2}} \qquad \tan\frac{A}{2} = \pm\sqrt{\frac{1 - \cos A}{1 + \cos A}}$$

$$\tan\frac{A}{2} = \frac{\sin A}{1 + \cos A} \qquad \tan\frac{A}{2} = \frac{1 - \cos A}{\sin A}$$

7.1 Law of Sines

In any triangle ABC, with sides a, b, and c,

$$\frac{a}{\sin A} = \frac{b}{\sin B}, \quad \frac{a}{\sin A} = \frac{c}{\sin C}, \quad \text{and} \quad \frac{b}{\sin B} = \frac{c}{\sin C}.$$

Area of a Triangle

The area $\mathcal{A}$ of a triangle is given by half the product of the lengths of two sides and the sine of the angle between the two sides.

$$\mathcal{A} = \frac{1}{2}bc\sin A, \qquad \mathcal{A} = \frac{1}{2}ab\sin C, \qquad \mathcal{A} = \frac{1}{2}ac\sin B$$

7.3 Law of Cosines

In any triangle ABC, with sides a, b, and c,

$$a^2 = b^2 + c^2 - 2bc\cos A, \qquad b^2 = a^2 + c^2 - 2ac\cos B,$$

and $$c^2 = a^2 + b^2 - 2ab\cos C.$$

Heron's Area Formula

If a triangle has sides of lengths a, b, and c, with semiperimeter $s = \frac{1}{2}(a + b + c)$, then the area $\mathcal{A}$ of the triangle is

$$\mathcal{A} = \sqrt{s(s - a)(s - b)(s - c)}.$$